JANE'S
AEROSPACE DICTIONARY

New Edition

JANE'S

AEROSPACE DICTIONARY

New Edition

Bill Gunston

JANE'S

First published in 1980
Second edition 1986

Published in the United Kingdom by
Jane's Publishing Company Limited
238 City Road, London EC1V 2PU

Distributed in the Philippines
and the USA and its dependencies by
Jane's Publishing Inc
115 Fifth Avenue
New York, NY 10003

ISBN 0 7106 0365 7

Printed in the UK by
Butler & Tanner Ltd,
Frome, Somerset

Introduction

Dictionaries are designed to help people communicate. To my knowledge there have been at least 43 aerospace dictionaries in English, if we include military, company and specialist glossaries, but I have been left in no doubt during the compilation of this volume that we need No 44. This is partly because most of the existing ones are out of date — as, with the best will in the world, this one will be next week (but it will be a boon for many years, despite that). In most of the commercially published aerospace dictionaries one can be sure of finding traditional terms such as "stagger". The nub of the problem is that my morning mail usually brings me several new entries, and I have spent so many hours collecting new entries that I have hardly had time left to write the definitions. Hence the need for this volume.

Technologists, and aerospace people in particular, are so steeped in new concepts — all of which have to be identified, named or labelled so that everyone knows exactly what is meant — that they generate a dozen new terms before breakfast. And what makes life much harder is the fact that, because these people are busy and the new ideas tend to have long names, they abbreviate them into various kinds of shorthand. By far the most common is the acronym. A few — such as Morass (modern ramjet systems synthesis) — raise a twinge of suspicion that perhaps the objective was to create a slick acronym rather than assist human communication. That abbreviations are popular in aerospace may be verified by studying the entries from MP onwards.

To some people MP means Member of Parliament. As a general rule, if a term is in an ordinary dictionary, that term or meaning will not be found here. (There are exceptions, however; several standard dictionaries on both sides of the Atlantic emphasize that to "cruise" means to travel in a carefree and seemingly random manner. Webster's wording is "To sail about touching at a series of ports as distinguished from voyaging to a set destination". Clearly this is a meaning unknown in aviation.) A few basic and generally familiar terms such as "elevator" are included, partly because the reader would expect to find them and partly because the traditional definitions no longer suffice.

Deciding what to include was a major part of the task. For every entry in the dictionary at least one failed to make it. For example, I spent hours studying the word "crater" in connection with flaws in metal parts, bombs, micrometeorites and the lunar surface before deciding not to include it at all. The conventional lexicographer may suffer from an excess of acquisitiveness, grasping new entries eagerly to make the tome more impressive. In this case my main objective was to sift and filter to try to preserve a reasonable size and price.

For this reason I have adopted a terse and hard style that at times may seem unhelpful. For example, under "crescent wing" I could have waffled on about "This is a wing that is shaped like . . . generally found on high-subsonic aircraft . . . it was adopted because . . .", and the book would have been bigger than the old-time family Bible. Instead almost every entry is in concise note form, devoid of definite or indefinite articles and wherever possible using the trade's own jargon and abbreviations, not for effect but because it saves space. This will offend some; I ask them to consider whether they would rather have a bigger and more highly priced dictionary containing no additional information.

The very jargon of modern aerospace often causes offence. One retired air marshal once berated me for using the term "hardware" and was unimpressed by the argument that there was no other word. Some English people still cannot believe that "program" can be a proper spelling, or that "analog" is a different word from "analogue". The criterion for inclusion in this volume has been whether or not the term has been used in public print without explanation. Often, like the conjuror who explains his tricks, giving the meaning is an anticlimax. Without the explanation, meaning can be elusive.

Many of the entries have more than one meaning. "Critical point" has several meanings in physics and metallurgy other than that given here. "LAP" to some means "London Air Port", but one of the harder decisions was which, if any, of the proper names to allow. Some aerospace dictionaries have entries such as "Spitfire, a British fighter plane". I am not on this wavelength at all. The book is quite large enough with no proper names other than a sprinkling where, because the product is so widely used or important, it simply would be unhelpful to leave it out. For example, "Lexan" gets in as an aerospace material and "Metalastik" is worth a line as a shock-absorbing technique.

Again, there are countless societies, organizations and military or government establishments, but few abbreviations of particular air forces or units, nor (I hope) one name of a commercial company. In general, trade names are excluded, but there is a difficult grey area in which a term or acronym invented by a company has become an accepted term. Initial capitals are reserved only for proper names, as in the case of societies or registered organizations, but there is a grey area here, too, with committees and projects.

Where there are several meanings for a single entry the alternatives are numbered, and the correct one must be found by inspection. In cross-references, which are in italics, the number is repeated in parentheses, so that "*DECM (1)*" means "refer to the first meaning of DECM". Alphabetical order is rigid; for example, one sequence reads "cross-bleed, crosscheck, cross-country, crossed controls, crossed-spring balance, cross-fall . . .". After much balancing of the pros and cons I decided usually to leave out all entries beginning with a prefix. Thus "millibar" is absent; but "milli" and "bar" are included. Everyday words used in a slightly unusual context are not included where their meaning is the normal one; but "health" is included because it appears in some definitions in connection with the fitness of hardware instead of living things. Asterisks are used to avoid repeating the entry being defined.

In some cases a suffix in parentheses indicates a national origin or other source: examples include A, Australia; arch, archaic term; BS/BSI, British Standards Institution; DoD, US Department of Defense; F, France; FAA, US Federal Aviation Administration; G, Federal Germany; ICAO, International Civil Aviation Organization; I, Italy; Int, International; J, Japan; NASA, National Aeronautics and Space Administration; obs, obsolete usage; RAF, Royal Air Force; Switz, Switzerland; UK, United Kingdom; US, United States of America; USA, US Army (and USAAF, USAF, USMC and USN for Army Air Force, Air Force, Marine Corps and Navy); USSR, Soviet Union; and WW1 and WW2 for the two world wars. Other suffixes of this kind refer to entries in the dictionary.

In a very few cases arguments persist about either basic meanings or shades of meaning. It is sad that these still extend to such fundamentals as "angle of incidence" and "track", each of which has two totally different meanings to large groups of experts in aerospace. Not relevant to an aerospace dictionary are the large number of words which aerospace — along with other groups — has tended to misuse. For example "rugged" means rough, hairy and uneven. Perhaps someone once applied the word, correctly, to the leathery face of a Hollywood hero. Suddenly the meaning was thought to be "strong and tough". So today almost every strong structure is described as "rugged," even though the adjective is usually being applied incorrectly. But we are too late here; once a few million people have latched on to the wrong meaning the clock cannot be put back. Like Sir Ernest Gowers (author of *The Complete Plain Words*, HMSO and later Pelican Books) I consider the English language a marvellous tool for humanity, and it is criminal folly to take a single clearly defined word and, by misusing it, blur its meaning.

If I have thus erred here I apologize. I greatly welcome constructive criticism and suggestions. Some of the definitions are standard, some are my own, adopted after much thought. Do not scoff, for example, at the suggestion that a combat aircraft could be a balloon. It is unlikely, certainly, but a dictionary cannot exclude possibilities and combat aircraft certainly do not stop at aeroplanes and helicopters. Suggestions for new terms will be welcome, though there are limitless borderline areas where aerospace interfaces with data processing, manufacturing, materials science, finance, the package-tour holiday and even the Universe, and one has to call a halt somewhere.

I am grateful to Mike Hardy for assistance with entries beginning with the letters D, E, F and G, and humbly indebted to the publisher for retaining interest in a much-delayed work.

Bill Gunston
June 1980

Introduction to Second Edition

The first edition of this work was generally well received. The author noted only one criticism, contained in the review published by *Aerospace*, magazine of the RAeS. The reviewer deplored the inclusion of so many terms and phrases "that have no application to aerospace" and cited as an example "barrier pattern". Ironically, this very term was included because, while the dictionary was in preparation, the author was asked by British Aerospace Manchester whether he could provide a definition! Rather belabouring his point, the same reviewer continued: "Much more room could be provided by the omission of terms that have no direct aerospace application, for example 'limacon' ...." As this entry occupies only one line we do not win a great deal, but the author continues to welcome suggestions, and this includes deletions as well as additions.

Perhaps three-quarters of the 4,000 or so additional terms in this new edition are acronyms, which continue to spread like a contagion. If the trend continues there will eventually be no plain English at all: one will simply string together the initial letters of the words. The rise of the acronym results from thinking akin to that of the ancient alchemists, who were concerned to use language incomprehensible to ordinary people in order to preserve their secrets. But surely the field of aerospace is one in which there is a need for clear, unambiguous communication?

The author admits that there were times when it seemed reasonable just to give up and leave out a few thousand of the acronyms in which A stood for "advanced" or "air force" (always the US Air Force). As before, registered trade names of materials are included, but company acronyms are generally excluded. Thus Caps (Collins adaptive processor system) is absent, but after some agonizing Finrae was included. Words found in ordinary dictionaries are normally excluded (unless they have special aerospace meanings); thus "pyrophorics" is absent despite its fostering of a new growth industry.

The introduction to the first edition omitted to explain that when acronyms are commonly spoken as a word they are written with only the first letter capitalized. Thus Nato could be offered as an alternative to the "NATO" on which the 15 nations officially insist, but in fact this, like "RAAF" for the Royal Australian Air Force, is one of hundreds of abbreviations considered obvious.

The author would like to thank many who have helped, especially Sqn Ldr Darrol Stinton MBE and John Blake, whose usage of the original volume spurred many comments and suggestions.

To end on a lighter note, the author worried for weeks about "TLC", which appeared several times in a highly technical article by Col Art Bergman USAF on how to avoid stall/stagnation with the F100 engine. The Plaps, EECs and UFCs were no problem, but TLC defeated not only the author but several certified F-15 drivers! Bergman explained how "the F100 needs a lot more TLC than the J79 .... We need to give the young pilots some understanding and TLC ...." In desperation the opinion of the USAF's European Aggressor squadron (527th TFTS) was sought. A charming female voice answered. She thought for a moment and said: "Ever think of tender loving care?"

Bill Gunston, November 1985

Addenda

If you cannot find a term in the main body of the Dictionary it may be in the Addenda at the back of the book. The Addenda permitted the incorporation of new terms even after most of the pages had gone to press.

A

A 1 General symbol for area (see *S*).
2 Aspect ratio (see A_s).
3 Ampères.
4 Atomic weight.
5 Moment of inertia about longitudinal axis, rolling mode.
6 Anode.
7 Attack-category aircraft (US DoD).
8 Degrees absolute.
9 Amber airway.
10 IFR flight plan suffix, fitted DME and 4096-code.
11 JETDS code: piloted aircraft, IR or UV radiation.
12 Airborne Forces-category aircraft (UK, 1944–46).
13 Atomic (as in A-bomb).
14 Sonobuoy standard size class, c 1 m/3 ft.
15 Air Branch (UK Admiralty).
16 Calibration (USAF role prefix 1948–62).
17 Ambulance (US category 1919–26).

Å Ångström (10^{-10}m), very small unit of length, contrary to SI.

a 1 Velocity of sound in any medium.
2 Structural cross-section area.
3 Anode.
4 (Prefix) atto, 10^{-18} (thus, ag, am).
5 (Suffix) available (thus, LD_a = landing distance available).

A0, A_0 Unmodulated (steady note) CW radio emission.

A0A1 Unmodulated (steady note) radio emission identified by Morse coding in a break period.

A0A2 Unmodulated emission identified by Morse coding heard above unbroken carrier (eg an NDB).

A1, A_1 1 Unmodulated but keyed radio emission, typically giving Morse dots and dashes.
2 Military flying instructor category; two years and 400 h as instructor.

a_1 Lift-curve slope for wing or other primary aerodynamic surface, numerically equal to $dC_L/d\alpha$.

A2 Military flying instructor category; 15 months and 250 h.

a_2 Lift-curve slope for hinged trailing-edge control surface, numerically $dC_L/d\epsilon$.

A3 AM radio transmission with double SB.

A^3 Affordable acquisition approach (USAF).

A3H AM, SSB transmission with full carrier.

A3J AM, SSB transmission with suppressed carrier.

A-25 Royal Navy form for reporting aircraft accidents.

A-battery Electric cell to heat cathode filament in valve (tube).

A-frame hook Aircraft arrester hook in form of an A; hook at vertex and hinged at base of each leg.

A-Licence Basic PPL without additions or endorsements.

A-sector Sector of radio range in which Morse A is heard.

A-station In Loran, primary transmitting station.

A-stoff Liquid oxygen (G).

A-type entry Fuselage passenger door meeting FAA emergency exit requirements; typical dimensions 42 in × 76 in.

AA 1 Anti-aircraft.
2 Airship Association (UK)
3 Acquisition Aiding, technique for matching EM waveforms (esp for ECM).
4 Air-to-air (ICAO code).

A/A Air-to-air (radar mode).

AAA 1 Airport advisory area.
2 Army Aviation Association (US).
3 Antique Airplane Association (US).
4 American Airship Association.
5 Anti-aircraft artillery (triple-A).
6 Affordable acquisition approach (usually A^3, USAF).
7 Associazione Arma Aeronautica (I).

AAAA 1 Australian Aerial Agricultural Association.
2 Army Aviation Association of America Inc.
3 Antique Aeroplane Association of Australia.

AAAD Airborne anti-armour defence.

AAAE American Association of Airport Executives.

AAAF Association Aéronautique et Astronautique de France.

AABM Air-to-air battle management.

AABNCP Advanced airborne (national) command post (DoD).

AAC 1 Army Air Corps (UK).
2 Army Aviation Centre (Middle Wallop, UK).
3 Alaskan Air Command (USAF).
4 Aviation Advisory Commission (US).
5 All-aspect capability (air-to-air missiles).

AACA Alaska Air Carriers' Association Inc.

AACC Airport Associations Co-ordinating Council (Int).

AACE Aircraft alerting communications EMP.

AACO Arab Air Carriers' Organization.

AACS Army Airways Communications Service (sometimes rendered as System) (USAAF).

AACT Air-to-air combat test (USN).

AACU Anti-Aircraft Co-operation Unit.

AAE Agrupación Astronáutica Española (Spain).

AAEE, A&AEE Aeroplane & Armament Experimental Establishment (Boscombe Down, UK).

AAEEA Association des Anciens Elèves de l'Ecole de l'Air (F).

AAF Auxiliary Air Force (UK).

AAFCE Allied Air Forces Central Europe (NATO).

AAFES Army and Air Force Exchange Service (US).

AAFRA Association of African Airlines.

AAFSS Advanced aerial fire-support system.

AAG, A/AG Air-to-air gunnery.

AAGF Advanced aerial gun, far-field.

AAHS American Aviation Historical Society.

AAI 1 Angle of approach indicator (see *VASI*), or angle of approach indication.
2 Airline Avionics Institute (US).
3 Air aid to intercept, American term for which AI (2) is preferred.

AAIRA Assistant air attaché (US).

AAL 1 (Often rendered a.a.l.) above airfield level.
2 Australian Air League.
3 Aircraft approach limitations, UK service usage specifying weather minima for particular type with particular ground aids.

AALAE See *ALAE*.

AAM 1 Air-to-air missile, fired by combat aircraft to destroy enemy aircraft.
2 Azimuth-angle measuring unit (Madge).

AAMA Association des Amis du Musée de l'Air (F).

AANCP Advanced airborne national command post (USAF).

AAO 1 Air-to-air operation.
2 Airborne area of operation.

AAP 1 Apollo Applications Program (NASA).
2 Acceptable alternative product (NATO).

AAp Angle of approach lights.

AAPP Airborne auxiliary powerplant.

AAR 1 Aircraft accident report.
2 Air-augmented rocket.
3 Air-to-air refuelling.

AARB Advanced aerial refuelling boom.

AARS 1 Automatic altitude-reporting system.
2 Attitude/altitude retention system.

AAS 1 Airport Advisory Service (FAA).
2 Army Aviation School (US).
3 American Astronautical Society.
4 Air Armament School (UK).

AASF Advanced Air Striking Force (RAF, 1939–40).

AASU Aviation Army (or Armies) of the Soviet Union.

AATA Asociación Argentina de Transportadores Aéreos.

AATC American Air-Traffic Controllers' Council.

AATF Airport and airway trust fund.

AATH Automatic approach to hovering (anti-submarine helicopters).

AATS Alternate aircraft takeoff systems.

AAU 1 Aircrew Allocation Unit (UK).
2 Aircraft Assembly Unit (UK, WW2).

AAV Autonomous aerial vehicle.

AAVS Aerospace Audio-Visual Service (USAF).

AAW Anti-air warfare.

AAWEX Anti-air warfare exercise.

AAWS Automatic Aviation Weather Service.

AB Air base (USAF).

a/b Afterburner.

ABA American Bar Association; IPC adds International Procurement Committee.

ABAC 1 Conversion nomogram, eg for plotting great-circle bearings on Mercator projection.
2 Association of British Aviation Consultants.
3 Association of British Aero Clubs and Centres.

ABB Automated beam-builder (space).

ABBCC Airborne battlefield control center.

Abbey Hill ESM for British warships, tuned to hostile air (and other) emissions.

ABC 1 Advance-booking charter.
2 Advancing-blade concept (Sikorsky).
3 Automatic boost control.
4 Airborne commander (SAC).
5 Airborne Cigar (EW, UK, WW2).

ABCA American, British, Canadian, Australian Standardization Loan Programme.

ABCCC Airborne Battlefield Command and Control Center (USAF).

ABD Airborne broadband defence (ECM).

ABDR Aircraft battle damage repair.

abeam Bearing approximately 090° or 270° relative to vehicle.

Aberporth Chief UK missile test centre, administered by RAE, on Cardigan Bay.

aberration Geometrical inaccuracy introduced by optical, IR or similar electromagnetic system in which radiation is processed by mirrors, lenses, diffraction gratings and other elements.

ABES Air-breathing engine system.

ABF 1 Annular blast fragmentation (warhead).
2 Auto beam forming (passive sonobuoys).

ABIA Associaçao Brasileira das Industrias Aeronauticas.

ABILA Airborne instrument landing approach.

ab initio trainer Aircraft intended to train pupil pilot with no previous experience.

ABITA Association Belge des Ingénieurs et Techniciens de l'Aéronautique et de l'Astronautique.

ABL Airborne laser.

ablation Erosion of outer surface of body travelling at hypersonic speed in an atmosphere. An ablative material (ablator) chars or melts and is finally lost by vaporization or separation of fragments. Char has poor thermal conductivity, chemical reactions within ablative layer may be endothermic, and generated gases may afford transpiration cooling. Main mechanism of thermal protection for spacecraft or ICBM re-entry vehicles re-entering Earth atmosphere.

ABM 1 Apogee boost motor.
2 Anti-ballistic missile, with capability of intercepting re-entry vehicle(s) of ICBM.
3 Abeam (ICAO code).

ABMA US Army Ballistic Missile Agency (defunct).

ABn Aerodrome beacon.

abnormal spin Originally defined as spin which continued for two or more turns after initiation of recovery action; today obscure.

A-bomb Colloquial term for fission bomb based upon plutonium or enriched uranium (A = atomic).

abort 1 To abandon course of action, such as take-off or mission.
2 Action thus abandoned, thus "an *".

abort drill Rehearsed and instinctive sequence of actions for coping with emergency abort situation; thus, RTO sequence would normally include throttles closed, wheel brakes, spoilers, then full reverse on all available engines consistent with ability to steer along runway.

above-wing nozzle Socket for gravity filling of fuel tanks.

ABPNL Association Belge des Pilotes et Navigateurs de Ligne.

ABR Amphibian bomber reconnaissance.

abradable seal Surface layer of material, usually non-structural, forming almost gas-tight seal with moving member and which can abrade harmlessly in event of mechanical contact. Some fan and compressor-blade ** are silicone rubber with 20% fill of fine glass beads.

Abres, ABRES Advanced ballistic re-entry systems.

ABRV Advanced ballistic re-entry vehicle.

ABS Acrylonitrile butadiene styrene, strong thermosetting plastic material.

abs Absolute scale of units.

ABSA Advanced base support aircraft.

abscissa 1 In co-ordinate geometry, X-axis.
2 X-axis location of a point.

ABSL Ambient background sound level.

absolute aerodynamic ceiling Altitude at which maximum rate of climb of aerodyne, under specified conditions, falls to zero. Usually pressure altitude amsl, atmosphere ISA, aircraft loading 1g, and weight must be specified. Except for zoom ceiling, this is greatest height attainable.

absolute alcohol Pure ethyl alcohol (ethanol) with all water removed.

absolute altimeter Altimeter that indicates absolute altitude; nearest approach to this theoretical ideal is laser altimeter, closely followed by instruments using longer EM wavelengths (radio altimeter).

absolute altitude Distance along local vertical between aircraft and point where local vertical cuts Earth's surface.

absolute angle of attack Angle of attack measured from angle for zero lift (which with cambered wing is negative with respect to chord line).

absolute ceiling Usually, absolute aerodynamic ceiling.

absolute density Theoretical density (symbol ρ) at specified height in model atmosphere.

absolute fix Fix (2) established by two or more position lines crossing at large angles near 90°.

absolute humidity Humidity of local atmosphere, expressed as g/m^{-3}.

absolute inclinometer Inclinometer reading attitude with respect to local horizontal, usually by precise spirit level or gyro.

absolute optical shaft encoder Electromechanical transducer giving coded non-ambiguous output exactly proportional to shaft angular posn.

absolute pressure Gauge pressure plus local atmospheric pressure.

absolute system Of several ** of units, or of calculating aerospace parameters, most important is reduction of aerodynamic forces to dimensionless coefficients by dividing by dynamic pressure head $\frac{1}{2}\rho V^2$.

absolute temperature Temperature related to absolute zero. Two scales in common use: absolute (°A) using same unit as Fahrenheit or Rankine scale (contrary to SI), and Kelvin (°K) using same unit as Celsius scale.

absolute zero Temperature at which all gross molecular (thermal) motion ceases, with all substances (probably except helium) in solid state. 0°K = −273·16°C.

absorbed dose Energy imparted by nuclear or ionizing radiation to unit mass of recipient matter; measured in rads.

absorption band Range of frequencies or wavelengths within which specified e.m. radiation is absorbed by specified material; narrow spread(s) of frequencies for which absorption is at clear maximum.

absorption coefficient 1 In acoustics, percentage of sound energy absorbed by supposed infinitely large area of surface or body.
2 In EM radiation, percentage of energy that fails to be reflected by opaque body or transmitted by transparent body (in case of reflection, part of radiation may be scattered). Water vapour is good absorber of EM at long wavelengths at which solar energy is reflected from Earth's surface, so ** for solar energy varies greatly with altitude.

absorption process Chemical production of petrols (gasolines) by passing natural gas through heavy hydrocarbon oils.

absorption cross-section Absorption coefficient of radar target expressed as ratio of "absorbed" energy to incident energy.

ABT About (ICAO code).

ABTA Association of British Travel Agents.

ABTJ Afterburning turbojet.

AC 1 Aligned continuous (FRP).
2 Aircraft commander.
3 Army co-operation (UK).

Ac Alto-cumulus cloud.

A/C Approach Control (FAA style).

a.c. 1 Alternating current (electricity).
2 Aerodynamic centre of wing or other surface.

a/c Aircraft (FAA = acft).

ACAAI Air Cargo Agents Association of India.

ACAB Air Cavalry Attack Brigade (USA).

ACAC Arab Civil Aviation Council (Int).

ACAN Amicale des Centres Aéronautiques Nationaux (F).

ACAP 1 Aviation Consumer Action Project (US).
2 Advanced composite aircraft (helicopter) programme.

ACARS Automatic communications and recording system.

ACAS 1 Air-cycle air-conditioning system.
2 Assistant Chief of the Air Staff.
3 Aluminium core, aluminium skin.

Acatt Army combined arms team trainer (Cobra/Apache/Scout).

ACAVS Advanced cab and visual system.

ACC 1 Area (or aerodrome) control centre.
2 Active clearance control.

accelerated flight Although aircraft that gains or loses speed is accelerating in horizontal plane, term should be used only for acceleration in plane perpendicular to flightpath, esp in vertical plane.

accelerated history Test record of specimen subjected to overstress cycling, overtemperature cycling or any other way of "ageing" at abnormally rapid rate.

accelerated stall Stall entered in accelerated flight. As common way of inducing stall is to keep pulling up nose, it might be thought all stalls must be accelerated, but in gradual entry flight path may be substantially horizontal. "High-speed stall" is possible in violent manoeuvre because acceleration in vertical plane requires wing to exceed stalling angle of attack. Stall-protection systems are generally designed to respond to rate of change of angle of attack close to stalling angle, so stick-pusher (or whatever form system takes) is fired early enough for critical value not to be reached.

accelerate-stop Simulation of RTO by accelerating from rest to V_1 or other chosen speed and immediately bringing aircraft to rest in shortest possible distance.

accelerating pump In PE (4) carburettor, pump provided to enrich mixture each time throttle is opened, to assist acceleration of engine masses.

accelerating well Originally receptacle for small supply of fuel automatically fed into choke tube by increased suction when throttle was opened. Later became small volume connected by bleed holes to mixture delivery passage. Usually absent from modern engines.

acceleration Rate of change of velocity, having dimensions LT^{-2} and in SI usually measured in ms^{-2}. As velocity is vector quantity, * can be imparted by changing trajectory without changing speed, and this is meaning most often applied in aerospace.

acceleration datum Engine N_1 corresponding to typical approach power, used in engine type testing for 2½ min rest period before each simulated overshoot acceleration (repeated 8 or 15 times).

acceleration errors Traditional direct-reading magnetic compass misreads under linear acceleration (change of speed at constant heading) and in turn (apparent vertical acceleration at constant speed); former is a maximum on E-W headings, increasing speed on W heading in N hemisphere indicating apparent turn to N; Northerly Turning Error (N hemisphere) causes simple compass to lag true reading, while Southerly Turning Error results in over-reading. Simple suction horizon misreads under all applied accelerations, most serious under linear positive acceleration (t-o or overshoot), when indication is falsely nose-up and usually right-wing down (with clockwise rotor, indication is diving left turn).

acceleration manoeuvre High-speed yo-yo.

accelerator Device, not carried on aircraft, for increasing linear acceleration on take-off; sometimes applied to catapults of carrier (1).

accelerator pump Accelerating pump.

accelerometer Device for measuring acceleration. INS contains most sensitive * possible, usually one for each axis, arranged to emit electrical signal proportional to sensed acceleration. Recording * makes continuous hard-copy record of sensed acceleration, or indicates peak. Direct-reading * generally fitted in test flying but not in regular aircraft operation.

acceptable alternative product One which may be used in place of another for extended periods without technical advice (NATO).

acceptance test Mainly historic, test of hardware witnessed by customer or his designated authority to demonstrate acceptability of product (usually military). Schedule typically covered operation within design limits, ignoring service life, fatigue, MTBF, MMH/FH and fault protection.

acceptance trials Trials of flight vehicle carried out by eventual military user or his nominated representative to determine if specified customer requirement has been met.

access door Hinged door openable to provide access to interior space or equipment.

access light Until about 1940, light placed near airfield boundary indicating favourable area over which to approach and land.

access panel Quickly removable aircraft-skin

panel, either of replaceable or interchangeable type, removed to provide access to interior.

access time 1 Time required to access any part of computer program (typically 10^{-3} to 10^{-9} s).
2 Time required to project any desired part of film or roller map in pictorial cockpit display (typically about 3 s).
3 Time necessary to open working section of tunnel and reach model installed (typically about 1,000 s, but varies greatly).

accessories System components forming functioning integral part of aircraft. Except in general aviation, term is vague; includes pumps, motors and valves, excludes such items as liferafts and furnishing. In case of fuel system (for example) would include pumps, valves, contents gauges and flowmeters, but not tanks or pipelines.

accessory drive Shaft drive, typically for group of rotary accessory units, from main engine, APU, EPU, MEPU or other power source.

accident Incident in life of aircraft which causes significant damage or personal injury (see *notifiable*).

accident-protected recorder Flight recorder meeting mandatory requirements intended to ensure accurate playback after any crash.

accident rate In military aviation most common parameter is accidents per 100,000 flying hours; other common measures are fatal accidents, crew fatalities and aircraft write-offs on same time basis, usually reckoned by calendar year. In commercial aviation preferred yardsticks are number of accidents (divided into notifiable and fatal) per 100 million passenger-miles (to be replaced by passenger-km) or per 100,000 stage flights, either per calendar year or as five-year moving average. In general aviation usual measure is fatal accidents per 100,000 take-offs.

accident recorder Device, usually self-contained and enclosed in casing proof against severe impact, crushing forces and intense fire, which records on magnetic tape, wire, or other material, flight parameters most likely to indicate cause of accident. Typical parameters are time, altitude, IAS, pitch and roll attitude, control-surface positions and normal acceleration; many other parameters can be added, and some ** on transports are linked with maintenance recording systems. Record may continuously superimpose and erase that of earlier flight, or recorder may be regularly reloaded so that record can be studied.

ACCIS Automated command and control information system (NATO).

accompanying cargo/supplies Cargo and/or supplies carried by combat units into objectives area.

ACCS Air command and control sqn (or system) (NATO and USAF).

accumulator 1 Electrical storage battery, invariably liquid-electrolyte and generally lead/acid.
2 Device for storing energy in hydraulic system, or for increasing system elasticity to avoid excessive dynamic pressure loading. Can act as emergency source of pressure or of fluid, damp out pressure fluctuations, prevent incessant shuttling of pressure regulators and act as pump back-up at peak load.
3 Device for storing limited quantity of fuel, often under pressure, for engine starting, inverted flight or other time when normal supply may be unavailable or need supplementing.
4 Portion of computer central processor or arithmetic unit used for addition.

accuracy jump Para-sport jump in which criterion is distance from target.

accuracy landing In flying training or demonstration, dead-stick landing on designated spot (= spot landing).

accuracy of fire Linear distance between point of aim and mean point of strikes.

Accu-Tune Magnetron circuit capable of being precisely tuned to different wavelengths.

ACD Automatic chart display.

ACDA Arms Control and Disarmament Agency.

ACDAC Asociación Colombiana de Aviadores Civiles.

ACDP Armament control and display panel.

ACDTR Airborne central data tape recorder (now generally called RSD).

ACE 1 Automatic check-out equipment.
2 Association of Cost Engineers (UK).
3 Air combat evaluator (CIU software).
4 Aircrew (or accelerated copilot) enrichment.
5 Allied Command Europe.
6 Association des Compagnies Aériennes de la Communauté Européenne.

ace Combat pilot with many victories over enemy aircraft. WW2 USAAF scores included strafing (air/ground) "victories". Number required to qualify has varied, but in modern world is usually five confirmed in air combat.

ACEA Action Committee for European Aerospace (international shop-floor pressure group).

ACEBP Air-conditioning engine bleed pipe.

ACEE Aircraft energy efficiency (NASA).

ACES Advanced-concept escape system.

ACESNA Agence Centrafricaine pour la Sécurité Navigation Aérienne.

ACET Air-cushion equipment transporter, for moving aircraft and other loads over soft surfaces, especially over airbase with paved areas heavily cratered (S adds "system").

acetate Compound or solution of acetic acid and alkali. * dope is traditionally based upon acetic acid and cellulose; was used for less inflammable properties (see *nitrate dope*).

acetone $CH_3.CO.CH_3$, inflammable, generally reactive chemical, often prepared by special fermentation of grain, used as solvent. Basis of many "dopes" and "thinners".

ACETS, Acets Air-cushion equipment transportation system (for post-attack airfields).

acetylene CH.CH or C_2H_2, colourless gas, explosive mixed with air or when pressurized but safe dissolved in acetone (trade name Prestolite and others). Burns with oxygen to give 3,500°C flame for gas welding; important ingredient of plastics.

Aceval Air combat evaluation.

ACF Aircraft Components Flight (RAF).

acft Aircraft (FAA).

ACGF Aluminium-coated glass-fibre (chaff).

ACGS Aerospace Cartographic and Geodetic Service (USAF MAC).

ACH Advanced Chain Home (UK, WW2).

achromatic Transmitting white light without diffraction into spectral colours; lens system so designed that sum of chromatic dispersions is zero.

ACI Air Council instruction (UK).

acid engine Rocket engine in which one propellant is an acid, usually RFNA or WFNA.

acid extraction Stage in production of lubricating oils in which sulphuric acid is used to extract impurities.

Ack-Ack Anti-aircraft (WW1, arch).

Ackeret theory First detailed treatment for supersonic flow past infinite wing (1925); suggested section must have low t/c ratio and sharp leading and trailing edges, and is normally trapezium (parallel double wedge) or biconvex.

acknowledgement Message from addressee that message is received in full and understood.

ACL 1 Anti-collision light.
2 Allowable cabin load.

ACLG Air-cushion landing gear; underside of aircraft is fitted with inflatable skirt containing ACV-type cushion.

aclinic line Isoclinic line linking points whose angle of dip is zero.

ACLS Automatic carrier landing system (Bell/US Navy).

ACLT Aircraft carrier landing trainer.

ACM 1 Air-cycle machine.
2 Air Chief Marshal (not normally abbrev).
3 Anti-armor cluster munition (US).
4 Air-combat manoeuvre (or manoeuvring).
5 Air-conditioning module.
6 Advanced cruise missile (USAF).

ACME Advanced-core military engine.

ACMI Air-combat manoeuvring instrumentation.

ACMP Alternating-current motor/pump.

ACMR Air-combat manoeuvring range.

ACMT Advanced cruise-missile technology.

ACN 1 Aircraft classification number (ICAO, proposed universal system for aircraft pavements).
2 Academia Cosmologica Nova (G).

ACO 1 Aerosat Co-ordination Office.
2 Airborne control officer.

acorn 1 Streamlined body or forebody added at intersection of two aerodynamic surfaces, such as fin and tailplane (stabilizer), especially in transonic aircraft to reduce peak suction.
2 Streamlined body introduced at intersection of crossing bracing wires to prevent chafing.
3 Streamlined fairing over external DF loop.

acorn valve Small thermionic valve (radio tube) formerly added to VHF or UHF circuit to improve efficiency.

Acoubuoy Acoustic sensor dropped by parachute into enemy land area.

acoustic Associated with sound, and hence material vibrations, at frequencies generally audible to human beings.

acoustics In ASW, sonar and other sensing systems relying on underwater sound; thus, * operators, * displays.

acoustic delay line In computer or other EDP device, subsystem for imparting known time delay to pulse of energy; typically closed circuit filled with mercury in which acoustic signals circulate (obs).

acoustic feedback Self-oscillation in radio system caused by part of acoustic output impinging upon input.

acoustic splitter Streamlined wall introduced into flow of air or gas, parallel to streamlines, for acoustic purposes. Usually inserted to reduce output of noise, for which purpose both sides are noise-absorbent. Many are radial panels and concentric long-chord rings (open-ended cylinders).

acoustic tube Miniature acoustic/electric transducer which has replaced carbon or other types of microphone in aircrew headsets.

ACP 1 Airborne command post.
2 Anti-Concorde Project.
3 Altimeter check point.
4 Armament control panel.

ACPA Adaptive-controlled phased array.

ACPL Atmospheric cloud physics laboratory.

acquisition 1 Act of visually identifying, and remembering location of, object of interest (specific ground or aerial target).
2 Detection of target by radar or other sensor (plus, usually, automatic lock-on and subsequent tracking).
3 Detection and identification of desired radio signal or other broadcast emission.
4 Act of reaching desired flight parameter, such as heading, FL or IAS, or desired point or axis in space such as ILS G/S or LOC (see *capture*).

ACR 1 Aerial combat reconnaissance.
2 Air (or airfield, or approach) control radar.
3 Advanced cargo rotorcraft.

acre Old Imperial (FPS) unit of land surface area, equal to 0·40469 ha (1 ha = 2·47105 acres). For covered area (factory buildings, etc) probable SI unit will be m^2 (= 0·000247105 acre), so 1-acre factory becomes 4,047 m^2.

acreage Major superficial area of flight vehicle,

esp spacecraft or aerospace craft, as distinct from special areas of nose and leading edge that need ablative or other protection.

acrobatics Aerobatics.

ACRR Airborne communications restoral relay.

acrylic Transparent thermosetting plastics based on polymerized esters of * acids; original trade names Perspex (ICI, UK) and Plexiglas (Rohm & Haas, US). Since 1950 improved * transparencies produced by subjecting moulded part to controlled stretching before setting.

ACS 1 Attitude control system.
2 Aeroflight control system (in Space Shuttle and other aerospace vehicles, as distinct from reaction controls used outside atmosphere).
3 Air-conditioning system.

ACSA Allied Communications Security Agency (NATO).

ACSM Advanced conventional standoff missile.

ACT 1 Actual temperature (ISA ± deviation).
2 Active-control technology.
3 Air combat tactics.
4 Anti-communications threat.
5 Atlas composing terminal.
6 Airborne crew trainer (RAF).

actinic ray EM radiation, such as short-wavelength end of visible spectrum and ultraviolet, capable of exerting marked photochemical effect.

actinometer Instrument measuring radiation intensity, esp that causing photochemical effects, eg sunlight; one form measures degree of protection afforded from direct sunlight, while another (see *pyrgeometer*) measures difference between incoming solar radiation and that reflected from Earth.

action Principal moving mechanism of automatic weapon; in gun of traditional design typically includes bolt, trigger, sear, bent, striker, extractor and ammunition feed.

action time Duration in seconds of significant thrust imparted by solid-propellant or hybrid rocket. Several definitions, most commonly the period between the point at which thrust reaches ten per cent of maximum (or averaged maximum) and that at which it decays through same level. This period is always shorter than actual duration of combustion, but longer than burn time. Symbol t_a.

action time average chamber pressure, thrust Integral of chamber pressure or thrust versus time taken over the action time interval divided by the action time; symbols P_c, F_a.

activate To translate planned organization or establishment into actual organization or establishment capable of fulfilling planned functions.

activated carbon Organically derived carbon from which all traces of hydrocarbons have been removed; highly absorbent and used to remove odours and toxic traces from atmospheres; also called activated (or active) charcoal.

active General adjective for a device emitting radiation (as distinct from passive). Also see * munition.

active aerodynamic braking Reversed propulsive thrust.

active air defence Direct action against attacking aircraft, as distinct from passive AD.

active clearance control Technique for maintaining an extremely small gap between fixed and rotating components of a machine (for example, by blowing bleed air around a turbine casing in a gas-turbine engine).

active controls Flight-control surfaces and associated operative system energized by vertical acceleration (as in gust) and automatically deflected upwards and/or downwards, usually symmetrically on both sides of aircraft, to alleviate load; thus active ailerons or tailplanes operate in unison to reduce vertical acceleration.

active countermeasures Countermeasures requiring friendly emissions. Subdivisions include microwave, IR and electro-optical.

active homing guidance Guidance towards target by sensing target reflections of radiation emitted by homing vehicle.

active jamming ECM involving attempted masking or suppression of enemy EM signals by high-power radiation on same wavelengths.

active material Many meanings, eg: 1, phosphor, such as zinc phosphate or calcium tungstate, on inner face of CRT; 2, parts of electric storage battery that participate in electrochemical reaction.

active munition One having immediate effect (as distinct from a mine, which is passive).

active noise control Noise-suppressing or countering systems triggered by noise itself and using sound energy against itself.

active runway Runway currently in use (implied that flying operations are in progress).

active satellite Satellite with on-board electrical power sufficient to broadcast or beam its own transmissions.

activity factor See *blade activity factor*.

ACTP Advanced Computer Technology Project (UK).

ACTS Advanced communications technology satellite.

ACT-TO Actual time and fuel state at takeoff.

actual ground zero Point on surface of Earth closest to centre of nuclear detonation.

actuator Device imparting mechanical motion, usually over restricted linear or rotary range and with intermittent duty or duty cycle.

ACU 1 Gas-turbine acceleration control unit.
2 Avionics control unit.

acute dose Total radioactive dose received over period so short that biological recovery is not possible.

ACV 1 Air-cushion vehicle.
2 Escort (or auxiliary) aircraft carrier (CVE

from 1943).

AD 1 Airworthiness Directive (national certifying authorities).
2 Advisory route (FAA).
3 Aligned discontinuous (FRP).
4 Aerodrome (ICAO).
5 Air defence.
6 Area-denial munition.
7 Aerial delivery (ramp-door position).
8 Autopilot disconnect.
9 Air diagram, followed by number.
10 Armament Division (AFSC).
11 Air Division (USAAF, USAF).

A/D 1 Air defence.
2 Alarm and display.

Ad Aerodrome (DTI, CAA).

ADA Advisory area.

Ada Standard common high-order language for US DoD software.

ADAC STOL (F).

ADAD Air-defence alerting device, horizon-scanning IR surveillance system.

ADAM Air deflection and modulation.

ADAP Airport Development Aid Program (US DoT).

Adaps Automatic data acquisition and processing.

adapter Interstage device to mate and then separate adjacent stages of multi-stage vehicle. Often called skirt, especially when lower stage has larger diameter.

adaptive bus Digital data highway to which (almost) any number of inputs and outputs may be connected.

adaptive control system Control system, esp of vehicle trajectory, capable of continuously monitoring response and changing control-system parameters and relationships to maintain desired result. Adapts to changing environments and vehicle performance to ensure given input demand will always produce same output.

adaptive logic Digital computer logic which can adapt to meet needs of different programs, environments or inputs.

adaptive nulling See *Adars*.

adaptive optical camouflage Active, self-variable form of camouflage which, chameleon-like, alters emitted wavelengths to suit varying background tones.

Adars Adaptive antenna receiver system; antenna (aerial) provides gain towards desired signals arriving from within a protected angle while nulling those arriving from outside that angle.

ADAS 1 Airborne data-acquisition system.
2 Auxiliary data-annotation system (for reconnaissance film, linescan or other hard-copy print-out of reconnaissance or ECM mission).
3 Airfield damage assessment system (USAF).

Adats, ADATS 1 Air-defence/anti-tank system.
2 Airborne digital avionics test system (USAF).

ADAU Air-data acquisition unit.

ADAV VTOL (F).

ADAZ Air-defence zone.

ADC 1 Air-data computer.
2 Advanced design conference.
3 Air Defense Command (former USAF command).

ADCA Advanced-design composite aircraft (USAF).

ADCC Air-defence control centre.

ADCF Aligned discontinuous carbon fibre.

Adcock aerial Early radio DF; avoids errors due to horizontal component by using two pairs of vertical conductors spaced half a wavelength or less apart and connected in phase-opposition to give figure-eight pattern.

Adcom Aerospace Defense Command (USAF).

ADCS Air-data computer system.

ADD 1 Airstream direction detector (stall-protection).
2 Long-range aviation, VVS bomber force (USSR).
3 Allowable deferred deficiency.

ADDC Air-Defence Data Centre (UK).

additive Substance added in low concentration to rocket propellant, IC fuel, lubricant, metal alloy or other product to improve performance, shelf life, smoothness of combustion or other quality.

ADDL Aerodrome (or airfield) dummy deck landings; pronounced "addle".

add-on contract Extension of existing contract to cover new work in same programme.

ADDPB Automatic diluter-demand pressure breathing.

ADDR Aeroklub der Deutschen Demokratischen Republik.

address Electronic code identifying each part of computer memory, each bit or information unit being routed to different *.

add time Time required for single (binary) addition operation in computer arithmetic unit.

ADE 1 Automated draughting (drafting) equipment.
2 Aeronautical Development Establishment (India).
3 Ada development environment.

ADEU Automatic data-entry unit (punched card input for STOL transport navigation).

ADF Automatic direction-finder. Airborne radio navaid tuned to NDB or other suitable LF/MF broadcast source. Until 1945 aerial was loop mounted in vertical plane and rotated by motor energized by amplified loop current to rest in null position, with plane of loop perpendicular to bearing of ground station. Modern receivers fed by two fixed coils, one fore-and-aft and the other transverse, suppressed flush with aircraft skin (usually on underside).

ADF sense aerial Rotatable loop null position gives two possible bearings of ground station; ** is added to give only one null in each 360° of loop rotation.

ADFC Aligned discontinuous fibre composite.

ADG 1 Auxiliary drive generator.
2 Air-driven generator.
3 Accessory drive gear.
4 Aircraft delivery group (USAF).

ADGB Air Defence of Great Britain.

ADGE, Adge Air defence ground environment.

ADI 1 Attitude director (very rarely, display) indicator. 3-D cockpit display forming development of traditional horizon and usually linked with autopilot and other elements forming flight-director system. Most can function in at least two modes, en route and ILS, and in former can provide navigational steering indications.
2 Anti-detonant injection, such as cylinder-head injection of methanol/water, for high-compression PE (4).

adiabatic Thermodynamic change in system without heat transfer across system boundary. In context of Gas Laws, possible to admit of exact * processes and visualize them happening; shockwave, though not isentropic, is not * in classical sense because thermodynamic changes are not reversible.

adiabatic flame temperature Calculated temperature of combustion products within rocket chamber, assuming no heat loss. Symbol T_c.

adiabatic lapse rate Rate at which temperature falls (lapses) as height is increased above Earth's surface up to tropopause (see *DALR*, *ELR*, *SALR*).

ADIZ Air defense identification zone; particular regions of US airspace within which all aircraft are, for reasons of national security, promptly identified (FAA).

adjacent channel Nearest frequency above or below that on which a radio system is working; can interfere with carrier or sidebands, but ** simplex operation minimizes this.

adjustable propeller Propeller with blades which can be set to chosen pitch on ground with propeller at rest.

adjustable tailplane (stabilizer) Tailplane which can be adjusted to chosen incidence when aircraft is at rest on ground.

ADL 1 Automatic drag limiter (S adds "system").
2 Arbeitsgemeinschaft Deutscher Luftfahrt-Unternehmen (G).

ADM 1 Air decoy (or defense) missile (USAF).
2 Atomic demolition munition.

ADMA Aviation Distributors & Manufacturers' Association (US).

ADMC Actuator drive and monitor computer.

administrative loading Loading of transport vehicle for maximum utilization of volume or payload, at expense of tactical need or convenience.

admittance In a.c. (1) circuit, reciprocal of impedance; "conductivity", comprises real and imaginary parts.

ADMU Air-distance measuring unit.

ADO 1 Advanced development objective.
2 Automatic delayed opening.

ADOC Air Defence Operations Centre (UK).

ADOCS, Adocs Advanced digital optical control system.

ADP 1 Acoustic data processor.
2 Automatic data processing.
3 Air-driven pump.

ADPA American Defense Preparedness Association.

ADPG Air defence planning group.

ADPS Asars deployable processing station (USAF).

ADR 1 Accident data recorder.
2 Air defence region (UK).
3 Advisory route.

ADRC Air defence radar controller.

ADRDE Air Defence Research & Development Establishment (UK).

ADR/Hum Accident data recorder and health/usage monitor, installed as single integrated package with common inputs.

ADRS Airfield Damage Repair Squadron (RAF).

ADS 1 Accessory drive system, self-contained yet integrated package.
2 Autopilot disengage switch.
3 Audio distribution system (Awacs).
4 Air-data system.

ADSEL, Adsel Address Selective. Improved SSR system in which saturation in dense traffic is avoided by interrogating each aircraft (once acquired) only once on each aerial rotation instead of about 20 times. Transponders reply only when selected by discrete address code, reducing number of replies and mutual interference and opening up space for additional information (such as rate of turn) helpful to ATC computers (see *DABS*).

ADSIA Allied Data-System Interoperability Agency.

Adsid Air-delivered seismic detection sensor.

ADSM Air defence suppression missile.

adsorption Removal of molecules of gas or liquid by adhesion to solid surface; activated carbon has very large surface area and is powerful adsorber.

ADSP Advanced digital signal processor.

ADT Approved departure time.

ADTC Armament Development Test Center (USAF, Eglin AFB).

ADU 1 Alignment display unit (INS).
2 Auxiliary display unit.
3 Avionics display unit.
4 Air data unit.

ADV 1 Arbeitsgemeinschaft Deutscher Verkehrsflughäfen eV (Federal German Airports Association).
2 Air defence variant.

advanced Generalized (overworked) adjective

meaning new, complicated and typifying latest technology.

advance/diameter ratio Ratio between distance aircraft moves forward for one revolution of propeller(s), under specified conditions, and propeller diameter.

advanced airfield, base Base or airfield, usually with minimal facilities, in or near objective area or theatre of operations.

advanced flow-control procedure Any of six theoretical or experimental techniques for ATC in high-traffic airspace.

advanced stall Stall allowed to develop fully, yet usually with some lateral control. Many definitions claim longitudinal control must remain, but nose-down rotation is invariably automatic (see *g-break*, *stall*).

advanced trainer Former military category, more powerful and complicated than ab initio/primary/basic trainer and capable of simulating or performing combat duties when fitted with armament.

advancing blade In rotary-winged aircraft in translational flight, any blade moving forward against relative wind. Each blade advances through 180° of its travel, normally from dead-astern to dead-ahead.

advection Generally, transfer by horizontal motion, particularly of heat in lower atmosphere. On gross scale, carries heat from low to high latitudes.

advection fog Fog, generally widespread, caused by horizontal movement of humid air mass over cold (below dew point) land or sea.

adversary aircraft Fighter specially purchased and configured to act role of enemy in dissimilar air-combat training.

advisory Formal recorded helpful message repeatedly broadcast from FAA AAS centre to all local aircraft.

advisory light Displayed by aircraft (esp carrier-based) to show LSO status (gear, hook, wing, speed and AOA).

advisory route Published route served by Advisory Service, but not necessarily by ATC (1) or separation monitoring and usually without radar surveillance.

Advisory Service FAA facility to provide information on request to all pilots, and advice to those who need it.

Advon Advanced echelon.

ADZ Advise (ICAO code).

A & E Airframe and engine qualified engineer.

AEA 1 Aeronautical Engineers Association (and Atomic Energy Authority) (UK).
2 Aircraft Electronics Association Inc (US).
3 Association of European Airlines.
4 Aircrew equipment assembly.

AEAF Allied Expeditionary Air Force (WW2).

AEB Avionics equipment bay.

AEC 1 Atomic Energy Commission (US).
2 Automatic exposure control.

AECC Aeromedical Evacuation Control Centre.

AECF Aéro Club de France.

AECMA Association Européenne des Constructeurs de Matériel Aérospatial (Int).

AéCS Aéro Club de Suisse.

AED Alphanumeric entry device.

AEDC Arnold Engineering Development Center (USAF, mainly air-breathing propulsion systems, at Tullahoma, Tenn).

AEEC Airlines Electronic Engineering Committee (US and int).

AEF 1 Aerospace Education Foundation (US).
2 American Expeditionary Force (WW1).
3 Air experience flight (RAF).
4 Armament engineering flight (RAF).

AEHP Atmospheric electricity hazards protection.

AE-I Aircraft Engineers International (Int).

AEL Advanced Engineering Laboratory (Australia).

AEM Air Efficiency Medal (UK).

AEO 1 Air electronics officer (aircrew trade, RAF).
2 All engines operating.

AEPT Air engineer procedures trainer.

AER Area expansion ratio in wind tunnel; ratio of cross-sections at start of diffuser to downstream end at first bend.

AERA Association pour l'Etude et la Recherche Astronautique et Cosmique (F).

Aerall Association d'Etudes et de Recherches sur Aéronefs Allégés (F).

AERE Atomic Energy Research Establishment (UKAEA, Harwell).

aerial 1 Pertaining to aircraft, aviation or atmosphere.
2 Part of radio or radar system designed to radiate or intercept energy, with size and shape determined by wavelength, directionality and other variables (US = antenna).

aerial array Assembly of aerial elements, often identical, usually excited from same source in phase and dimensioned and positioned to radiate in pencil beam or other desired pattern (not necessarily phased-array).

aerial delivery system Complete system for air transport and delivery to surface recipient (usually without aircraft landing).

aerial survey Use of aerial cameras and/or other photogrammetric instruments for the making of maps, charts and plans.

aerial work General aviation for hire or reward other than carriage of passengers or, usually, freight; includes agricultural aviation, aerial photography, mapping and survey, cable and pipeline patrol and similar duties usually not undertaken to full-time fixed schedule.

AERO Air Education and Recreation Organization (UK).

aeroballistics Science of high-speed vehicles moving through atmosphere in which both ballistics and aerodynamics must be taken into account. Often asserted aerodynamics and ballistics are applied separately to different portions of flight path, but both act as long as there is significant atmosphere present.

aerobatics Precise and largely standardized manoeuvres, unnecessary in normal flight, executed to acquire or demonstrate mastery over aircraft, for entertainment, or for competition (US = acrobatics).

aerobiology Study of distribution and effects of living matter suspended in atmosphere (small insects, spores, seeds and micro-organisms).

aerobrake Aerodynamic brake for use in extremely low-density atmospheres at Mach numbers of 5 to 25. Typically can be deployed as a saucer shape, concave side facing direction of travel.

aeroconical canopy Form of parachute canopy suitable for use at all aerospace Mach numbers.

aerodone Basic aerodyne, glider relying upon natural stability and having no moving control surfaces. Examples are paper dart and chuck glider, most simple free-flight models, and aeroplanes which continue to fly after being abandoned by their crews.

aerodonetics Science of gliding flight, with or without use of control surfaces.

aerodontalgia Toothache caused by major changes in ambient atmospheric pressure.

aerodontia, aerodontology Branch of dentistry dealing with problems of flying personnel.

aerodrome BS.185, 1940: "A definite and limited area of ground or water (including any buildings, installations and/or equipment) intended to be used, either wholly or in part, in connexion with the arrival, departure and servicing of aircraft." Now archaic, see *airfield*, *airport*, *air base*, *strip*, etc.

aerodrome traffic zone Airspace up to circuit height and within 1.5 nm, 2,000 ft/600 m of boundary (general aviation).

aerodynamic axis Imaginary line through aerodynamic centres of every longitudinal element in solid body moving through gaseous medium. In wing, runs basically from tip to tip, but in swept or slender delta can be an acutely curved, kinky line often having little practical application.

aerodynamic balance 1 Method of reducing control-surface hinge moment by providing aerodynamic surface ahead of hinge axis (see *Frise aileron*, *horn balance*).

2 Wind-tunnel balance for measurement of aerodynamic forces and moments.

aerodynamic braking 1 Use of atmospheric drag to slow re-entering spacecraft or other RV.

2 Use of airbrakes or parachute (drag chute) in passive **.

3 Use of reversed propulsive thrust (propeller or jet) in active **.

aerodynamic centre In two-dimensional wing section, point about which there is no change of moment with change in incidence; point about which resultant force appears to rotate with change of incidence. In traditional sections about one-quarter back from leading edge (25% chord) and in symmetrical section lies on chord and thus coincident with CP. Also called axis of constant moments. Abb: a.c. ac or (incorrectly) AC.

aerodynamic chord Reference axis from which angle of attack of two-dimensional aerofoil is measured. Line passing through (supposed sharp) trailing edge and parallel to free-stream flow at zero lift at Mach numbers appreciably below 1 (see *chord*, *geometric chord*, *MAC*).

aerodynamic coefficient Aerodynamic force (lift or drag, or moment) may be reduced to dimensionless coefficient by dividing by characteristic length (which must be same parameter for all similar bodies, and in a wing is invariably area) and by dynamic head (symbol q). Traditional divisor is $\frac{1}{2}\rho V^2 S$, where ρ is air density, V velocity and S area, ensuring that units are consistent throughout; if area is m^2 then V must be ms^{-1}. The $\frac{1}{2}\rho V^2$ term, difference between pitot and static pressure, is accurate only at low speeds; if M^2 (square of Mach number) is too large to ignore, a different expression must be found for dynamic head, such as H-p (pitot minus static). See *force coefficients*, *moment coefficients*, *units of measurement*.

aerodynamic damping In flight manoeuvres rotation of aircraft (about c.g. or close to it) changes direction of relative wind to provide restoring moment which opposes control demand and arrests manoeuvre when demand is removed. As altitude increases, combination of increasing TAS (for given EAS) and reduced airflow deflection angles results in ** being progressively decreased, although control demand moment and aircraft inertia do not change. Thus at high altitude pilot must apply greater opposite control movements to arrest rotation.

aerodynamic efficiency Most common yardstick is lift/drag ratio (L/D). In general ** maximized when resultant forces are as nearly as possible perpendicular to direction of motion; tend to be reduced as speed is increased, since lift-type forces may be presumed to remain substantially constant while drag-type forces may be presumed to increase in proportion to square of speed.

aerodynamic force Force on body moving through gaseous medium assumed to be proportional to density of medium (ρ), square of speed (u^2 or V^2), characteristic dimension of body (such as length L^2 or area S) and R_n (Reynolds number raised to power n). This broad relationship sometimes called Rayleigh formula. Body assumed to be wholly within homogenous gas, reasonably compact and streamlined (eg not a sheet of paper) and to have smooth surface.

aerodynamic heating As speed of body through gaseous medium is increased, surface temperature increases roughly in proportion to square of speed. Effect due variously to friction between adjacent molecules in boundary layer, to degradation of kinetic energy to heat and to local compression of gas. Maximum temperature is reached on surfaces perpendicular to local airflow where oncoming air or gas molecules are brought to rest on surface. At Mach 2 peak stagnation temperature about 120°C and at Mach 3 about 315°C; at hypersonic speeds temperature can swiftly rise to 3,000 or 5,000°C in intense shockwave around nose and other stagnation points, causing severe radiation heating, ionization and dissociation of flow.

aerodynamic mean chord, AMC, MAC, c Chord that would result in same overall force coefficients as those actually measured. Essentially, but not necessarily exactly, same as mean of aerodynamic chords at each station; very nearly same as geometric mean chord.

aerodynamic overbalance Excessive aerodynamic balancing of control surface such that deflection will immediately promote runaway to hard-over position.

aerodynamic twist Variation of angle of incidence from root to tip of aerodynamic surface, to obtain desired lift distribution or stalling characteristic (see *wash-in*, *wash-out*).

aerodynamics Science of interactions between gaseous media and solid surfaces between which there is relative motion. Classical * based upon Bernoulli's theorem, concept of boundary layer and circulation, Reynolds number, Kármán vortex street, turbulent flow and stagnation point. High-speed * (M^2 too large to be ignored) assumes gas to be compressible and introduces critical Mach number, shockwave, aerodynamic heating, and relationships and concepts of Prandtl, Glauert, Ackeret, Büsemann, Kármán-Tsien and Whitcomb. At Mach numbers above 4, and at heights above 80 miles (130 km), even high-speed * must be modified or abandoned because of extreme aerodynamic heating, violent changes in pressure and large mean free path (see *superaerodynamics*).

aerodyne Heavier-than-air craft, sustained in atmosphere by self-generated aerodynamic force, possibly including direct engine thrust, rather than natural buoyancy. Two major categories are aeroplanes (US = airplanes) and rotorplanes, latter including helicopters.

aeroelasticity Science of interaction between aerodynamic forces and elastic structures. * deflections are increased by raising aerodynamic forces, varying them rapidly (as in gusts and turbulence) and increasing aspect ratio or fineness ratio. All * effects tend to be destabilizing, wasteful of energy and degrading to structure.

aeroembolism Release of bubbles of nitrogen into blood and other body fluids as result of too-rapid reduction in ambient pressure. May be due to return to sea-level from much increased pressure ("caisson sickness", "the bends") or from sea-level to pressure corresponding to altitude greater than 30,000 ft (about 10,000 m). Potentially fatal if original, increased, pressure is not rapidly restored.

aeroflight mode Atmospheric flight, by aerospace vehicle (Space Shuttle).

aerofoil (US = airfoil) Solid body designed to move through gaseous medium and obtain useful force reaction other than drag. Examples: wing, control surface, fin, turbine blade, sail, windmill blade, Flettner rotor, circular or elliptical rotor blade with supercirculation maintained by blowing. Some authorities maintain * must be essentially "wing-shaped" in section.

aerofoil section Traditionally, outline of section through aerofoil parallel to plane of symmetry. This must be modified to "parallel to aircraft longitudinal axis" in variable-sweep and slew wings, and "perpendicular to blade major axis" in blades for rotors, turbines and propellers. None of these sections may lie even approximately along direction of relative wind, although usually assumed to.

aerofoil boat Wing-shaped surface-effect marine craft (or low-altitude aircraft).

aerogel Colloid comprising solution of gaseous phase in solid phase or coagulated sol (colloidal liquid).

aerograph Airborne meteorological recording instrument; aerometeorograph.

aero-isoclinic wing Aerofoil which, under aeroelastic distortion, maintains essentially uniform angle of incidence from tip to tip.

Aerolite Trade name, low-density bonded sandwich structure.

aerolite Stony meteorite, richer in silicates than metals.

aerology Study of atmosphere (meteorology) other than lower regions strongly influenced by Earth's surface.

aerol strut Early oleo strut relying for energy absorption and damping upon both air and oil.

aerometeograph, aerometeorograph Airborne instrument making permanent record of several meteorological parameters such as altitude, pressure, temperature and humidity.

aerometer Instrument used in determining density of gases, esp atmosphere.

aeronaut Pilot of aerostat.

aeronautical Pertaining to aeronautics.

aeronautical chart Chart prepared and issued primarily for air navigation. Chief categories include Sectional (plotting), Regional, Radio, Flight Planning, Great Circle and Magnetic. Most are to conformal projection (especially Lambert's); non-aeronautical topographic features generally excluded.

aeronautical fixed service Network of ground radio stations.

aeronautical information overprint Overprint on military or naval map or chart for specific air navigation purposes.

aeronautical light, beacon Illuminated device approved as aid to air navigation.

aeronautical mile Nautical mile (British Admiralty standard 6,080 ft = 1·85318 km); defined as length of arc of 1° of meridian at Equator.

aeronautical multicommunications See *multicommunications service*.

aeronautical satellite Satellite provided to assist aircraft by improving navigation, communications and traffic control. Abb aerosat.

aeronautical topographic chart/map Chart or map designed to assist visual or radar navigation and showing features of terrain, hydrography, land use and air navigation facilities.

aeronautics Science of study of, design, construction and operation of aircraft.

Aéronavale Air arm of French Navy.

aéronef Aircraft, any species (F).

aeroneurosis Chronic disorder of nervous origin caused by prolonged flying stress.

aeronomy Study of upper atmosphere of planets with especial reference to effects of radiation, such as dissociation and ionization.

aeropause Vague boundary between atmosphere useful to aircraft, and space where air density is too low to provide lift, or air for air-breathing engines, or aerodynamic forces for stability and control. One definition suggests boundary is layer from "12 to 120 miles"; upper limit meaningful for hypersonic aircraft only.

aeroplane (US = airplane) BS.185, 1940: "A flying machine with plane(s) fixed in flight". Modern definition might be "mechanically propelled aerodyne sustained by wings which, in any one flight regime, remain fixed". Explicitly excludes gliders and rotorplanes, but could include MPAs, VTOLs and convertiplanes that behave as * in translational flight.

aeroplane effect Error in radio DF caused by horizontal component of fixed aerial or trail angle of wire (arch).

aeropulse Air-breathing pulsejet.

aeroresonator Resonant air-breathing pulsejet.

aerosat Aeronautical satellite.

aeroshell High-drag aerodynamic-braking heat-shield for returning spacecraft or planetary lander.

aerosol Colloid of finely divided solid or liquid dispersed in gaseous (esp air) continuous phase. Natural examples: smokes, dustclouds, mist, fog. In commercial product active ingredient is expelled as aerosol by gaseous propellant.

aerospace 1 Essentially limitless continuum extending from Earth's surface outwards through atmosphere to farthest parts of observable universe, esp embracing attainable portions of solar system.
2 Pertaining to both aircraft and spacecraft, as in "aerospace technologies".
3 Activity of creating and/or operating hardware in aerospace, as in "the US leads in aerospace".

aerospace craft Vehicle designed to operate anywhere in aerospace, and especially both within and above atmosphere.

aerospace forces National combat armoury capable of flying in atmosphere or rising into space, including all satellite systems and strategic ballistic missiles.

aerospace medicine Study of physiological changes, disorders and problems caused by aerospace navigation. Among these are high accelerations, prolonged weightlessness, vertigo, anoxia, ionizing radiation, Coriolis effects, micrometeorites, temperature control, recycling of material through human body, and possibility of developing closed ecological systems to support human life away from Earth.

aerospace plane Colloquial term for space vehicle which can re-enter, manoeuvre within atmosphere and land in conventional way on Earth's surface. Generally assumed to be manned and to include some air-breathing propulsion.

aerostat Lighter-than-air craft, buoyant in atmosphere at a height at which it displaces its own mass of air. Major sub-groups are balloons and airships. In airships aerodynamic lift from hull can be significant, but not enough to invalidate classification under this heading.

aerostatics Science of gases at rest, in mechanical equilibrium.

aerostation Operation of aerostat.

aerothermodynamic border Region, at height around 100 miles (160 km), at which Earth atmosphere is so attenuated that even at re-entry velocity aerodynamic heating is close to zero and can be neglected, as also can drag.

aerothermodynamic duct Athodyd, early name for ramjet. Aerothermodynamic in this context does not necessarily conform to definition given below.

aerothermodynamics Study of aerodynamic phenomena at velocities high enough for thermodynamic properties of constituent gases to become important.

aerothermoelasticity Study of structures subject to aerodynamic forces and elevated temperatures due to aerodynamic heating.

aero-tow Tow provided for glider by powered aircraft.

Aerozine Trade name, family of liquid fuels based on MMH or UDMH.

Aeroweb Trade name, range of structural and noise-attenuating bonded honeycomb sandwich materials.

AES 1 Aeromedical evacuation system (USAF,

MAC).
2 Auger electron spectroscopy.
3 Air Electronics School.
AESC Aft equipment service centre (on aircraft).
AESF Avionics electrical systems flight (RAF).
Aesop Automated engineering and scientific optimization program (multivariable design tool).
AESU Aerospace Executive Staff Union (Singapore).
AET Airfields Environmental Trust (UK).
AETA Association des Anciens Elèves de l'Ecole d'Enseignement Technique de l'Armée de l'Air (F).
AETMS Airborne electronic terrain map system (3-D colour-coded in real time, plan or elevation).
AEW 1 Airborne early warning.
2 Air electronic warfare.
AEWC, AEW&C, AEW+C Airborne early warning and control.
AEWTF Aircrew electronic-warfare tactics facility (NATO).
AF, a.f. 1 Audio frequency, sounds audible to average human ear (20 to 16,000 Hz). In simple radio communications RF carrier is modulated so that it "carries" AF superimposed upon basic waveform.
2 Aerodynamic force.
a/f 1 airfield.
2 airframe.
AFA 1 Air Force Association (US).
2 Air Force Act (UK).
3 Aircraft Finance Association (US).
AFAA Air Force Audit Agency (US).
AFAITC Armed Forces Air Intelligence Training Center (Denver).
AFAL Air Force Avionics Laboratory (US).
AFALD Air Force Acquisition Logistics Division (US).
AFAMRL Air Force Aerospace Medical Research Laboratory (US).
AFAP Australian Federation of Air Pilots.
AFARPS see *ARPS*.
AFATL Air Force Armament Laboratory (US).
AFB Air Force Base (US).
AFBMD Air Force Ballistic Missile Division (US).
AFC 1 Air Force Cross (decoration, RAF).
2 Automatic frequency control, especially for interlocking of coherent systems.
3 Automatic feedback control (CCV).
4 After (1973) fuel crisis.
5 Aramid fibre composite.
AFCAC African Civil Aviation Commission (Int).
AFCC Air Force Communications Command (US).
AFCEA Armed Forces Communications and Electronics Association (US).
AFCMD Air Force Contract Management Division (US).
AFCOMS, Afcoms Air Force Commissary Service (US).
AFCP Advanced flow-control procedure.
AFCRL Air Force Cambridge Research Laboratories (US).
AFCS 1 Automatic flight control system.
2 Air Force Communications Service (US).
AFDC Automatic formation drone control (USN).
AFDS 1 Autopilot and flight-director system.
2 Advanced flight-deck simulator.
3 Air Fighting Development Squadron (UK).
AFEE Airborne Forces Experimental Establishment (UK, WW2).
AFESA Air Force Engineering and Services Agency (US).
AFEWC Air Force Electronic Warfare Center (US).
AFFDL Air Force Flight Dynamics Laboratory (US).
AFFF Aqueous film-forming foam.
affinity-group charter Charter flight provided for members of group having common interest; such groups are often formed solely for this purpose.
"affirmative" R/T response signifying "yes".
AFFTC Air Force Flight Test Center (US).
AFG Arbitrary function generator.
AFGE American Federation of Government Employees.
AFGL Air Force Geophysics Laboratory (US).
AFGS Automatic flight guidance system; generally synonymous with autopilot.
AFI Assistant flying instructor.
AFIS 1 Air Force Intelligence Service (US).
2 Airfield (or aerodrome) flight information service (O adds "officer").
AFISC Air Force Inspection and Safety Center (US).
AFITAE Association Française des Ingénieurs et Techniciens de l'Aéronautique et de l'Espace.
AFK, AfK Aramid fibre composite (G), also called SfK.
AFLC Air Force Logistics Command (US).
AFLO Airborne force liaison officer (stationed at departure a/f).
AFLSC Air Force Legal Services Center (US).
AFM 1 Air Force Medal (RAF).
2 Air Force Manual (US).
3 Aircraft flight-manual.
AFMA 1 Armed Forces Management Association (US).
2 Anti-fuel-misting additive.
AFMEA Air Force Management Engineering Agency (US).
AFML Air Force Materials Lab (US).
AFMPC Air Force Military Personnel Center (US).
AFMRL Air Force Medical Research Laboratory (US).
AFO Aerodrome fire officer.
A-FOD tyre Landing-wheel tyre whose tread will

not pick up and hurl objects causing FOD.

AFOLTS Automatic fire overheat logic test system.

AFOSI Air Force Office of Special Investigations (US).

AFOSR Air Force Office of Scientific Research (US).

AFP 1 Alternate (or alternative) flight plan (US, esp NASA).
2 Air Force Publication.
3 Acceleration along flight path.

AFP costs Costs relating to all flight personnel.

AFP turn Turn made after fix passage, ie after passing designated fix.

AFPM Association Française des Pilotes de Montagne.

AFPRO Air Force plant representatives office (US).

AFQ Association Française des Qualiticiens.

AFR Air/fuel ratio.

AFRAA Association of African Airlines (Int).

AfrATC African Air Traffic Conference (Int).

AFCAC African Civil Aviation Commission (Int).

AFRCU Air/fuel ratio control unit.

AFRES, Afres Air Force Reserve (US).

AFROTC Air Force Reserve Officers Training Corps (US).

AFRP Aramid-fibre reinforced plastics.

AFRPL Air Force Rocket Propulsion Laboratory (US).

AFS 1 Aeronautical fixed service (ICAO).
2 Advanced Flying School (RAF).
3 Air Force Station (US).

Afsatcom, AFSATCOM Air Force satellite communications (US).

AFSC 1 Air Force Systems Command (US).
2 Air Force Specialty code (US).

AFSD 1 Air Force Space Division (US).
2 Airframe full-scale development.

AFSS 1 Association Française des Salons Spécialises.
2 Active flutter-suppression system.

AFT Advanced flying training.

AFTA Avionics fault-tree analyser (portable malfunction tracer).

AFTEC Air Force Test and Evaluation Facility (US).

afterbody 1 Rear part of body, esp of transonic, supersonic or hypersonic atmospheric vehicle.
2 Portion of flying-boat hull or seaplane float aft of step, immersed at rest or taxiing.

afterbody angle In side elevation, acute angle between keel of *afterbody (2)* and (a) undisturbed water line or (b) longitudinal axis.

afterburn Undesired, irregular combustion of residual propellant in rocket engine after cut-off.

afterburner Jetpipe equipped for afterburning.

afterburning Injection and combustion of additional fuel in specially designed jetpipe (afterburner) of turbojet to provide augmented thrust. Fuel, usually same as in engine, burns swiftly in remaining free oxygen in hot exhaust gas. Downstream of turbine, combustion can reach temperature limited only by radiant heat flux on afterburner wall and rate at which fuel can be completely burned before leaving nozzle. Nozzle must be opened out in area, with con-di profile to give efficiently expanded supersonic jet. Can be applied esp effectively to turbofan or leaky turbojet, with greater proportion of available oxygen downstream of turbine. Very effective in supersonic flight, but less efficient at low speeds and very noisy (UK = reheat).

afterchine Rear chine along afterbody (2).

aftercooling Cooling of gas after compression, esp of air or mixture before admission to cylinders of highly blown PE designed for operation at high altitudes.

after-flight inspection Post-flight inspection.

afterglow 1 Persistence of luminosity from CRT screen, gas-discharge tube or other luminescent device after excitation removed (of importance in design of many radar displays).
2 Pale glow sometimes seen high in western sky well after sunset due to scattering of sunlight by fine dust in upper atmosphere.
3 Transient decay of plasma after switching off EM input power.

aft flight deck Rear area of flight-deck floor where this is at upper level above main floor. Not necessarily occupied by aircrew.

AFTI Advanced fighter technology integration.

aft limit of CG Rearmost position of CG permitted in flight manual, pilot's notes, certification documentation or other authority. That at which stability in yaw and/or pitch, and static and manoeuvre margins, are still sufficiently good for average pilot to handle most adverse combination of circumstances in safety. CCV concept is leading to revolution in which much reduced, or negative, natural stability is held in check by AFCS.

AFTN Aeronautical Fixed Telecommunications Network (DTI, UK).

AFTS 1 Advanced Flying Training School.
2 Air/fuel test switch.

aft wing In oblique-wing aeroplane, wing pointing rearward.

AFU Advanced Flying Unit.

AFVA Association Française de la Voltige Aérienne.

AFWAL Air Force Wright Aeronautical Laboratory (US).

AFWL Air Force Weapons Laboratory (US).

AFWR Atlantic Fleet Weapons Range.

AG 1 Air gunner.
2 Availability guarantee.
3 Assault glider (US, WW2).
4 Reconnaissance Wing (G).

A/G Air-to-ground communication(s).

Agard, AGARD Advisory Group for Aerospace

(formerly Aeronautical) Research and Development (NATO).

AGAQ Association des Gens de l'Air du Québec.

AGB Accessory gearbox.

AGC 1 Automatic gain control, property of radio receiver designed to vary gain inversely with input signal strength to hold approximately constant output.
2 Affinity-group charter.
3 Adaptive gate centroid (radar tracking algorithms).

AGCS Advanced guidance and control systems.

AGD 1 Axial-gear differential.
2 Air generator drive (ie, windmill).

AGE Aerospace ground equipment (military inventory category).

age-hardening Many metal alloys, especially high-strength aluminium alloys, need time to harden after heat treatment, usually in order that partial precipitation may take place; preferably accomplished at room temperature or chosen higher value.

ageing (US = aging) Time-dependent changes in microstructure of metal alloys after heat treatment. Some merely relieve internal stress but most improve mechanical properties.

AGEPL Association Générale des Elèves Pilotes de Ligne (F).

AGETS, Agets Automated ground engine-test system.

AGIFORS, Agifors Airline Group, International Federation of Operational Research Societies (Int).

AGIL Airborne general illumination light (USAF).

Agile Aircraft ground-induced loads excitation (simulates rough runways).

AGL 1 Above ground level.
2 Airborne gun-laying (radar).

AGM Air-to-ground guided missile (inventory category, USAF, USN).

AGMC Aerospace Guidance and Metrology Center (AFSC).

Agnis Azimuth guidance for nose-in stands; also rendered as approach guidance nose-in to stand.

agonic line Line joining all points on Earth's surface having zero magnetic variation. Two ** exist, one sweeping in curve through Europe, Asia and W Pacific and other roughly N-S through Americas.

AGOS Department of aviation, seaplanes and experimental construction (USSR).

Agpanz Agricultural Pilots' Association of New Zealand.

agravic Hypothetical environment without gravitational field. Unrelated to weightless free-fall in gravitational field, or to possible points where net gravitational field of all mass in universe is zero.

agravic illusion Apparent movement of human visual field in weightless flight due to minute displacements of structure in inner ear.

AGREE Advisory Group on Reliability of Electronic Equipment (DoD/NATO).

agricultural Colloq, of hardware, essentially primitive and crude, but not necessarily ineffective or obsolescent.

agricultural aircraft Aircraft designed or converted for agricultural aviation.

agricultural aviation Branch of general aviation concerned with agriculture, specif crop spraying, dusting, top dressing, seeding, disease inspection and, apart from transport, work with livestock.

AGS Air Gunnery (rarely, Gunners) School.

AGTA Airline Ground Transportation Association Inc (US).

AGTS 1 Automated guideway transit system (airport terminal).
2 Air (or aerial) gunnery target system.

AGTY Frequency agility.

AGU American Geophysical Union.

AGV Automated guided vehicle, part of most FMS (6); S adds "system".

AGZ Actual ground zero.

Ag-Zn, Ag/Zn Silver/zinc electrical storage battery.

AH Artificial horizon.

Ah, A-h Ampère-hour.

AHQ Air headquarters.

AHRS Attitude/heading reference system.

AHS American Helicopter Society.

AHSA The Aviation Historical Society of Australia.

AHSW Aural high-speed warning.

AHTR Auto horizontal-tail retrimming after landing.

AI 1 Airborne interception (radar).
2 Artificial intelligence.
3 Air-data/inertial.

AIA 1 Aerospace Industries Association of America Inc.
2 Associazione Industrie Aerospaziali (Italy).
3 Associazione Italiana di Aerotechnica.
4 Académie Internationale d'Astronautique.
5 Atelier Industriel de l'Air (F).

AIAA 1 American Institute of Aeronautics and Astronautics.
2 Aerospace Industry Analysts Association (US).
3 Area of intense air activity.

AIAC Air Industries Association of Canada.

AIAEA All-India Aircraft Engineers Association.

AIANZ The Aviation Industry Association of New Zealand Inc.

AIB 1 Accidents Investigation Branch (DTI, UK).
2 Aeronautical Information Bureau.

AIC 1 Aeronautical information circular.
2 Air-inlet controller (supersonic aircraft).
3 Advanced industrial countries.

AICMA Association Internationale des Constructeurs de Matériel Aérospatial.

AICS Automatic intake (inlet) control system.

AID 1 Aeronautical Inspection Directorate (UK).
2 Airport information desk (FAA).
3 Agency for International Development (US).
4 Accident/incident/deficiency.

AIDAA Associazione Italiana di Aeronautica e Astronautica.

AIDS 1 Airborne integrated data system.
2 Aircraft integrated data system.
3 Automatic integrated data system.
4 Acoustic intelligence data system.
5 Aircraft intrusion detection system (for temp parking areas).

AIEDA Association Internationale des Avocats et Experts en Droit Aérien (Int).

AIFS Advanced indirect fire system(s).

Aigasa Associazione Italiana Gestori Aeroporti e Aeroportuali.

AILA Airborne instrument landing approach system.

ailavator Control surface functioning as aileron and elevator (see *elevon*).

aileron 1 Control surface, traditionally hinged to outer wing and forming part of trailing edge, providing control in roll, ie about longitudinal axis. Seldom fitted to aircraft other than aeroplanes and gliders, and in recent years supplemented or replaced by spoilers, flaperons, elevons and tailerons, while in some high-speed aircraft conventional * are mounted inboard to counter * reversal.
2 Effectiveness of lateral-control system, as in phrase "to run out of *".

aileron drag Asymmetric drag imparted by aileron deflection, greater on down-going aileron (see *differential* *, *Frise* *).

aileron reversal As aircraft speed increases, deflection of aileron can twist wing sufficiently to reduce, neutralize and finally reverse rolling moment imparted to aircraft. Many aircraft designed for Mach numbers higher than 0·9 either have no traditional outboard ailerons or else lock these except at low speeds.

aileron roll See *slow roll*.

AIM 1 Airmen's Information Manual (FAA).
2 Air intercept missile (inventory category, USAF, USN).
3 Aerospace industrial modernization programme (US).

AIMAS Académie Internationale de Médecine Aéronautique et Spatiale.

aim-bias Error between aiming point and centre of dispersion area of statistically valid number of projectiles.

aim dot Basic command reference symbol in gunsight or HUD; can show where bullets would hit if gun fired, but usually also gives other indications.

AIMIS, Aimis Advanced integrated modular instrumentation systems (USN).

aim-off Angular allowance when firing at moving target with unguided projectile, usually because of sightline spin resulting in target changing apparent position during projectile's flight, but in air-to-air combat possibly because of lateral air drag (eg in firing at aircraft abeam at same speed and heading, when sightline spin is zero).

AIMS 1 Attitude indicator measurement system.
2 Air traffic control radar beacon, IFF, Mk 12 transponder, System.
3 Aircraft identification monitoring system (DoD, interceptors).

Aimval Air intercept missile evaluation (USAF/USN).

AINS INS with prefix "advanced", "area" or "airborne".

AIO Action information organization (mainly warships, in relation to aircraft).

AIOA Aviation Insurance Offices Association (UK).

AIP Aeronautical Information Publication.

AIPPI Association Internationale pour la Protection de la Propriété Intelectuelle (Int).

air Air near Earth's surface usually taken to be (% by volume): nitrogen 78·08; oxygen 20·95; argon 0·93; other gases (in descending order of concentration, carbon dioxide, neon, helium, methane, krypton, hydrogen, xenon and ozone) 0·04. In practice also contains up to 4% water vapour. ISA SL pressure at 16·6°C is 10·332 kg m^{-2} (= 761·848 mm Hg) and density 1·2255 kg m^{-3}.

AIR 1 French aerospace material specification code.
2 Air interceptor rocket (inventory category, USAF).
3 Air-inflatable retarder (similar to ballute).

air abort Abort after take-off.

Airac Aeronautical information regulation and control, system for disseminating air navigation information (Notams).

Airad Airmen advisory (local).

AI radar Airborne interception radar, carried by fighter for finding and tracking aerial targets.

air-augmented rocket Usual form of this propulsion system is for first stage of combustion, or primary rocket propellant or gas-generator, to yield fuel-rich range of products which then combine in second stage of combustion with atmospheric air (normally induced through ram intake). Objective is to increase specific impulse, by using oxygen from atmosphere, and also burn time and vehicle range.

air base 1 Loosely, military or general-aviation airfield (term used mainly by popular media).
2 In photogrammetry, line joining two air stations.
3 Length of (2).
4 Scale distance between adjacent perspective

centres as reconstructed in plotting instrument.

air bearing Gas bearing using air as working fluid.

air-bearing table Table supported on single spherical air bearing and thus free to tilt, without sensible friction, to any attitude within design constraints.

air-blast switch Electrical circuit-breaker in which arc formed on breaking circuit is blown away by high-velocity air jet.

air-blast transformer In this context, as in some other electric and electronic equipment dissipating large heat flux, air-blast signifies forced air cooling.

airborne Sustained by atmosphere or downward component of propulsive thrust. Implication is that vehicle is not above sensible atmosphere; term not normally used in connection with spaceflights not involving aerodynamically supported vehicles, but applicable to wingless jet-lift devices.

airborne alert Generally, long-duration mission flown by strategic bomber, in all respects ready to make real attack, to reduce reaction time and remove possibility of destruction by ICBM or SLBM attack on its base. Until World War 2 "air alert" was method of deploying interceptor (pursuit) forces, keeping them on sustained flight in likely combat area under ground vector control.

airborne early warning, AEW Use of aircraft to lift powerful search radar to greatest possible height to extend line-of-sight coverage (very approximately, LOS radius in statute miles is square root of 1·5 times observer's height in feet). Modern AEW can give a PPI covering 170,000 sq miles, throughout which two low-level aircraft in close formation can be individually distinguished against ground clutter.

airborne fog blind Translucent blind or hood admitting light to cockpit or flight deck whilst removing external visual cues.

airborne force Force constituted for airborne operations.

airborne gunlaying turret, AGLT Bomber-defence gun turret incorporating automatic provisions for aim-off and other corrections when engaging aerial targets.

airborne interception, AI Use of aircraft to find, and close with, another aircraft; specifically, use of fighter to intercept, challenge by IFF and, if dissatisfied, destroy another aircraft.

airborne operation Movement of combat forces and logistic support into combat zone by air.

airborne radio relay Use of airborne relay stations to increase range, flexibility or security of communications.

air brake Passive device extended from aircraft to increase drag. Most common form is hinged flap(s) or plate(s), mounted in locations where operation causes no significant deterioration in stability and control at any attainable airspeed. Term not normally applied to flaps, drag chute or thrust-reverse systems.

air-breathing Aspiring air, specif aircraft propulsion system which sustains combustion of fuel with atmospheric oxygen. Imposes contraints on vehicle speed and height, but invariably offers longer range than rocket system for same vehicle size or mass.

airburst Detonation of explosive device well above Earth's surface. Almost all nuclear weapons are programmed for optimized airburst height, which varies with weapon and target.

AIRC Airlines Industrial Relations Conference (US).

air carrier Organization certificated or licensed to carry passengers or goods by air for hire or reward.

air cartography Aerial survey, esp aerial photography for purposes of mapmaking.

Air Cavalry Helicopter-borne attack/reconnaissance ground troops (US Army).

AIRCON Air communications network, specif serving US air carriers. Characterized by wide geographical extent, very large information flow, "on-line, real-time, full-time" storage, and computer-compatible electronic switching.

air controller In military operations, an individual trained for and assigned to traffic control of particular air forces assigned to him within a particular sector.

air control team Team organized to direct CAS strikes in the vicinity of forward ground elements.

air-cooled Heat-generating device, esp PE (4), maintained within safe limits of temperature by air cooling. Invariably cooling is direct, in case of PE (4) by radiating heat to air flowing between fins around cylinder head and barrel, or around hot rotor casing(s) of RC engine.

air corridor 1 Defined civil airway crossing prohibited airspace.
2 Restricted air route in theatre of military operations intended to afford safe passage for friendly air traffic.

aircraft Device designed to sustain itself in atmosphere above Earth's surface, to which it may be attached by tether that offers no support. Two fundamental classes are aerodynes and aerostats. Aircraft need have no means of locomotion (balloons are borne along with gross motion of atmosphere, while kites are tethered and lifted by motion of atmosphere past them), nor any control system, nor means for aerodynamic or aerostatic lift (eg, jet VTOL aircraft need be no more than jet engine arranged to direct efflux downwards). Free-falling spacecraft qualifies as aircraft if, after re-entry, its shape endows it with sufficient L/D ratio to glide extended distance, irrespective of whether or not it can control its trajectory.

aircraft cable Specially designed tensile cable,

usually either solid wire or any of eight built-up constructions, used for operating flight control and other mechanical systems.

aircraft carrier Marine craft, traditionally large surface vessel, designed to act as mobile base for military aircraft.

aircraft categories 1 For genealogical purposes, family tree of possible classifications.
2 For certification purposes, subdivision of aeroplanes (most important family of aircraft) on basis of performance. In UK aeroplanes certificated before 1951 are categorized as No Performance Group Classification; after 1951 subdivided into Performance Group A, large multi-engined; Performance Group C, light multi-engined; and Performance Group D. Also Group X for large multi-engined aeroplanes built outside UK before specified date.

aircraft certificate In US all aeroplanes (airplanes) and most other aircraft except models are categorized and licensed according to four classes of certificate, each having status of legal document; airworthiness, production, registration and type.

aircraft dispatcher In US air transport, official charged with overseeing and expediting dispatch of each flight. Traditional post analogous to train dispatcher of US railroads. Today duties include provision of met information, flight planning, arranging unloading and loading, stocking with consumables, apron servicing and other turnaround tasks, calling for large staff.

aircraft dope See *dope*.

aircraft fabric See *fabric*.

aircraft fuel See *gasoline*, *kerosene*.

aircraft log One or more volumes recording detailed operating life of individual aircraft, listing daily and cumulative flight time, notifiable irregularities or transient unserviceability of any part, all inspections, overhauls, parts replacement, modification and repair.

Aircraftman, Aircraftwoman RAF/WRAF non-commissioned rank, with junior and senior grades, having no bearing on trade in which rank-holder is qualified.

aircraft management simulator Essentially the same as a pre-1960 simulator, equivalent to a modern FFS but without 6-axis motion or synthetic external scenes; capable of training on all cockpit instruments and systems.

aircraft missile Missile launched from aircraft.

aircraft rocket Rocket launched from aircraft.

aircrew Crew required to operate aircraft, esp crew, numbering more than one, of military aircraft. Large civil aircraft normally operated by flight crew and cabin crew; * is not used.

aircrew equipment assembly Standard modular fitting incorporating PEC and various other items carried on flying clothing, forming single "umbilical" for military flight-crew member.

air cushion Volume of air at pressure slightly above local atmospheric, trapped or constantly replenished by suitably arranged air jets (possibly issuing from base of flexible skirt) to support ACV.

air-data computer Digital computer serving as central source of information on surrounding atmosphere and flight of aircraft through it. Typical ADC senses, measures, computes or transmits (to AFCS and other aircraft systems) pressure altitude, OAT and total temperature, Mach number, EAS, angle of attack, angle of yaw and dynamic pressure. All are corrected for known errors and converted into signals of form required by supplied systems. ADC may have 60 to 90 output channels, most used throughout each flight.

air defence Defence against aerial attack, ie attack by aircraft, atmospheric missiles and RVs entering atmosphere from space.

air defence identification zone Defined airspace within which all traffic must be identified, located and controlled (see *ADIZ*).

air defence operations area Geographical area, usually large, within which air and other operations are integrated.

air defence region Geographical subdivision of an AD area.

air defence sector Geographical subdivision of an AD region.

air despatcher Person trained to supervise release or ejection of cargo from aircraft in flight.

air distance Distance flown through the air, ie with respect to atmosphere.

Air Division Largest adminstrative unit of USAF below Air Force.

air-driven horizon Artificial horizon in which gyro is driven by one or more high-velocity air jets, usually arranged to impinge on cups machined in periphery. In most, instrument case is connected to vacuum line, often generated by venturi, and jets are atmospheric air. Performance reduced at altitude and by contamination by foreign matter blocking or penetrating filter.

airdrome Incorrect corruption of aerodrome.

airdrop Delivery of personnel or cargo from aircraft in flight usually by parachute.

airdrop platform Platform designed to carry large indivisible loads for airdrop or LAE.

Airep Spoken weather report by airborne aircrew.

airfield Land area designated and used, routinely or in emergency, for take-off and landing by full-scale aerodyne(s). Definition excludes aerostats and model aircraft, but admits VTOLs and RPVs. No facilities need be provided.

airfield surface movement indicator Airfield surveillance radar.

airfield surveillance radar Radar on or near airfield with scanner well above ground level rotating continuously to give fine-definition PPI display, especially showing aircraft on ground and vehicles.

airflow Air flowing past or through body. For immersed solids moving through air, major factors are speed (IAS, EAS, CAS, TAS), angle of attack or yaw, dynamic head, OAT and total temperature. For turbine engine, * normally mass flow, ie mass per unit time passing through engine.

airfoil Aerofoil.

air force station Usually means location of an air force unit where there is no airfield.

airframe BS.185/1940: "A flying machine without the engines"; today BS.185 has added "power-driven". Better definition is: assembled structure of aircraft, together with system components forming integral part of structure and influencing strength, integrity or shape. Includes transparencies, flush aerials, radomes, fairings, doors, internal ducts, and pylons for external stores. In case of ballistic rocket vehicle would not include thrust chambers of liquid-propellant engines, nor separable solid motors, but could include payload fairings. Items where argument exists include: RVs; MAD booms; rigid refuelling booms; mission equipment carried demonstrably outside structure proper (eg, AWACS aerial); and podded engine cowlings. Airframe usually includes landing gear, but not systems, equipment, armament, furnishings and other readily removable items.

airframe-attributable Accident or notifiable incident caused by defect or malfunction in airframe.

air/fuel ratio Ratio by mass of air to fuel in air-breathing engine or other combustion system. With hydrocarbon fuels ratio usually in neighbourhood of 16:1.

air gap 1 Clearance between stationary and moving portions of electrical machine, crossed by magnetic flux.
2 Air space between poles of magnet.
3 Gap left in core of chokes and transformers used in radio or radar circuits to prevent saturation by d.c.

airglow Quasi-steady radiation, visible at night, due to chemi-luminescence in upper atmosphere energized by solar radiation. Today often taken to embrace radiation outside visible range.

air gunner Member of aircrew assigned, for whole or part of mission, to manning guns to defend aircraft [obs; today's nearest equivalent = DSO (2)].

air hardening Age hardening at room temperature.

airhead 1 Designated area in hostile or disputed territory needed to sustain air landings; normally objective of airborne assault.
2 Air supply and evacuation base in theatre of operations.
3 In undeveloped region, nearest usable airfield.
4 CTOL base for support of dispersed VTOL operation.

air hostess Stewardess.

air interception Radar or visual contact between a friendly and another aircraft.

air interdiction Air attack on enemy forces sufficiently far from friendly forces for integration with the latter not to be required; esp attack on enemy supply routes rather than theatre forces.

air/land warfare Simultaneous warfare on land and in airspace above.

air launch Release from aircraft of self-propelled or aerodynamically lifted object: missile, target or other aircraft (manned or RPV) previously attached to it (not towed).

Air League UK air-minded association; not abb.

air lever On early aeroplanes, hand throttle governing engine airflow (not fuel).

air liaison officer Tactical air force or naval aviation officer attached to surface forces.

airlift 1 Carriage by air of load, esp by means other than routine airline operation.
2 Transport operation (usually military) in which aircraft make round-trip flights to transport large load such as army division or refugee population.
3 Continuing, open-ended logistic supply operation, such as Berlin *, 1948–49.

airline 1 Certificated air carrier.
2 Public image of air carrier, created by house logo, aircraft livery and advertising, even where no such single carrier may exist.
3 Any great-circle route.
4 Ground supply pipe conveying air at typically 80 lb/in^2.

air load Aggregrate force exerted on surface by relative airflow. In case of aerofoil or control surface, force exerted on three-dimensional entity, not on just one of its surfaces.

air lock 1 Small chamber through which personnel must pass to enter or leave larger chamber maintained at atmospheric pressure significantly different from ambient. Provided with two doors in series, never more than one door being open at a time.
2 Unwanted volume of air or other gas trapped at high point in liquid system in such a way as to prevent or degrade proper operation.

air log Instrument for measuring air distance flown.

air mail Mail prepaid and sent by air where there is an alternative, cheaper surface route. European letters and postcards travel by air if this speeds delivery; no separate ** service.

airman 1 Loosely, any aviator, aeronaut or man who navigates by air.
2 Tradesman certificated by appropriate licensing authority to work on aircraft.
3 Air force rank category (but not a rank) below NCO, equivalent to Army "other ranks".
4 (Capital A) lowest uniformed rank in USAF,

with class subdivisions.

airmanship Skill in piloting aircraft. Embraces not only academic knowledge but also qualities of common sense, quick reaction, awareness and experience.

air mass Very large parcel of atmosphere which at lower levels exhibits almost uniform characteristics of temperature and humidity at any given level. According to Bergeron classification, grouped according to origin (Arctic, Polar, Tropical, Equatorial), subdivided into Continental or Maritime within each group, and then again into warm (w) or cold (k).

Airmec International Aircraft Maintenance and Engineering Exhibition and Conference.

Airmen Advisory Notice to Airmen (see *NOTAM*) normally issued locally, often verbally during pre-flight or in-flight briefing.

AIRMET In-flight weather advisory category less severe than SIGMET but potentially hazardous to simple aircraft flown by inexperienced pilot.

air mile 1 Aeronautical mile = nautical mile.
2 One mile flown through the air, following Hdg at TAS; wind must be added to give distance along Tr at G/s.

air mileage unit, AMU Mechanical calculating instrument, 1942–55, to derive continuous value for air distance flown. Output, more accurate and reliable than air log, was fed to air mileage indicator (AMI) and often other instruments.

air-minded Of general public, concerned to further aviation for prosperity, defence or sport.

air ministry In many countries, national department charged with administering military (sometimes all) aviation. In UK, replaced by MoD (RAF).

airmiss Incident reported by at least one member of aircrew who considers there was "definite risk of collision" between two airborne aircraft.

air-mobile band Band of communications frequencies assigned to air-mobile forces.

air-mobile operation Operation by ground forces carried in air vehicles.

air movement Military air transport operation involving landing and/or airdrop.

air movement table Detailed schedule of utilization of aircraft space, numbers and types of aircraft, and departure places and times.

Air National Guard, ANG Part-time voluntary auxiliary to USAF equipped with fighter, tactical strike and transport aircraft, organized as self-contained arm by each state.

air navigation Art of conducting aircraft from place of departure to predetermined destination, or along intermediate routes (eg to follow precise tracks in surveying). Originally pure pilotage (contact flying); by 1918 moved into nautical realm of dead reckoning and celestial observation (astro-nav); by 1960 nearly all ** relied upon ground and airborne aids, except in gliders and simple light aircraft.

air navigation facility Navaid; surface facility for air navigation including "landing areas, lights, any apparatus or equipment for disseminating weather information, for signalling, for radio direction-finding, or for radio or other electronic communication . . ." (FAA). Today add "for electronic position-finding".

AIRO Airborne IR observatory.

air officer, Air Officer 1 Loosely, officer commissioned in an air force.
2 Specif, officer of Air Rank in RAF.

air/oil strut Telescopic member utilizing properties of air and oil to absorb compressive shocks (rarely, tensile) with minimal or controlled rebound.

air phone HF air/ground telephony.

airplane Aeroplane (N America).

air plot 1 Continuous air navigation graphic plot constructed (usually on board aircraft) by drawing vectors of true headings for lengths equivalent to air distances flown.
2 Similar plot constructed for airborne object derived from visual or radar observation of its flight.
3 Automatic or manually constructed display showing position and movements of airborne objects (if in a ship, relative to the ship).

air pocket Colloq, sudden and pronounced gust imparting negative vertical acceleration; downdraught.

airport Airfield or marine base designated and used for public air service to meet needs of quasi-permanent community. Need be no facilities for aircraft replenishment or repair, customs facilities, nor scheduled service; but must be facilities for passengers and/or cargo. Community served can be mainly or even exclusively employees of one company (eg at oilfield).

airport advisory area Area within 5 miles of geographical centre of uncontrolled airport on which is located FSS so depicted on appropriate sectional aeronautical chart (FAA).

airport advisory service Terminal service provided by FSS located at airport where control tower is not operating.

airport code Three-letter code identifying all commercial airports (eg, LHR, JFK, LAX).

airport commission Board of management of most US airports.

airport information desk Unmanned facility at local airport provided for pilot self-service briefing, flight planning and filing of flight plans (FAA).

airport of entry Airport provided with customs facilities through which air traffic can be cleared before or after international flight.

airport surface detection/movement See *ASDE, ASMI*.

airport surveillance Airfield surveillance radar.

airport traffic area Unless otherwise designated (FAR Pt 93), airspace within 5 miles of geographical centre of airport with TWR operating, extending up to, but not including, 3,000 ft AAL (FAA).

airport traffic control service ATC service provided by airport TWR for aircraft operating on movement area and in vicinity (FAA).

airport traffic control tower Facility providing airport ATC service (FAA).

air position Geographical position airborne aircraft would occupy if entire flight was made in still air; point derived by plotting Hdg and TAS.

air-position indicator, API Instrument which continuously senses Hdg and TAS (usually not allowing for compressibility error) to indicate current air position.

air raid Aerial attack on surface target, esp against civil population.

Air Rank Senior to Group Captain; Air Officer, equivalent to naval Flag Officer.

air route Defined airspace between two geographical points, subject to navigational regulations.

air route surveillance Surface radar giving display(s) showing geographical position and height of all traffic along designated civil route (usually airway).

air route traffic control centre, ARTCC Facility providing ATC service to aircraft operating on IFR flight plan in controlled airspace and principally during en route phase (FAA).

air-run landing Final deceleration in ground effect followed by vertical landing (fixed-wing V/STOL).

air-run take-off Vertical take-off followed by horizontal acceleration in ground effect (fixed-wing V/STOL).

AIRS 1 Airborne integrated reconnaissance system (USN).
2 Advanced inertial reference sphere.

airscape Broad vista of sky, not necessarily including Earth's surface, from aerial viewpoint.

air scoop Colloq ram intake, esp projecting from exterior profile of aircraft.

airscrew BS.185, 1951: "Any type of screw designed to rotate in air". Word never common in US, but in UK used in early days of powered flight to denote rotary aerodynamic device intended to impart thrust. From about 1920–50 explicitly denoted tractor device ("propeller" being "an airscrew joined to the engine by a shaft in compression"). Today redundant. See *fan*, *propeller*, *rotor*, *windmill*.

airscrew-turbine engine Turboprop.

air/sea rescue, ASR Use of aircraft to rescue life in danger at sea, esp permanently established service for this purpose (UK, RAF; US, USCG).

airship BS.185, 1951: "A power-driven lighter-than-air aircraft". Thus need not be provided with means for controlling its path, though if * is to be of use such means must be provided. Traditional classes are: blimp, a small non-rigid; non-rigid, in which envelope is essentially devoid of rigid members and maintains shape by inflation pressure; semi-rigid, non-rigid with strong axial keel acting as beam to support load; and rigid, in which envelope is itself stiff in local bending or supported within or around rigid framework.

air snatch 1 Recovery of passive body from atmosphere by passing powered aircraft, esp recovery of space payload descending by parachute.
2 Recovery of human being from hostile territory or sea by passing aircraft unable to hover (see *Fulton*).

air sounding 1 Measurement of atmospheric parameters from sea level to specified upper level by transmitting or recording instruments lofted by rocket or aircraft (esp balloon).
2 Record thus obtained.

airspace Volume of atmosphere bounded by local verticals and Earth's surface or given flight levels. May be controlled or uncontrolled, but always an administrative unit defined by precise geographical or Earth-referred locations.

airspace denial Military mission flown by fighter to destroy all hostile aircraft entering particular airspace, usually that above friendly troops.

airspeed, air speed Relative velocity between tangible object, such as raindrop or aircraft, and surrounding air. In most aircraft measured by pitot-static system connected to airspeed indicator (ASI) to give airspeed indicator reading (ASIR). When corrected for instrument error (IE), result is indicated airspeed (IAS). When corrected for position error (PE), result is rectified airspeed (RAS). Most ASIs calibrated according to ideal incompressible flow ($\frac{1}{2}\rho V^2$), so from RAS subtract compressibility correction to give equivalent airspeed (EAS). Finally density correction, proportional to difference between ambient air density and calibration density (1,225 $g\,m^{-3}$), applied to give true airspeed (TAS). This sequence ignores errors, usually transient, due to major changes in angle of attack (eg, in manoeuvres). Some ASIs calibrated to allow for compressibility according to ISA SL, indicating calibrated airspeed (CAS). Confusion caused by fact most authorities now use "calibrated airspeed" to mean ASIR corrected for IE and PE; CAS thus defined would have to be corrected for density and compressibility. Thus since 1980 CAS must be regarded as ASIR + IE + PE; if allowing for compressibility then at ISA/SL CAS = TAS.

airspeed indicator, ASI Instrument giving continuous indication of airspeed.

airspeed transducer In flight testing, or performance measurement of unmanned vehicle in atmosphere, transducer giving electrical signal proportional to airspeed. In simple systems signal is d.c. voltage.

air spotting Correcting adjustment of friendly

surface bombardment based on air observation.

air staging unit Military unit stationed at airfield to handle all assigned air traffic calling at that airfield.

airstairs Passenger and/or crew stairway forming integral part of aircraft and, after use, folded or hinged up and stowed on board.

air start Action of starting or re-starting aircraft main propulsion or lift engines in flight, esp by windmilling.

air station Point in space occupied by camera lens at moment of exposure.

air stream 1 Moving air mass, esp that penetrates and divides more stationary mass.
2 Loosely, any localized airflow.

airstrip Prepared operating platform for aeroplanes, usually from STOL to CTOL, distinguished from airfield by either: hasty construction under battlefield conditions; lack of permanent paved surfaces; lack of permanent accommodation for personnel or hardware; or lack of facilities, other than temporary fuel supply or ATC.

air superiority Degree of airspace dominance sufficient to prevent prohibitive enemy interference with one's own operations.

air-superiority fighter Combat aircraft designed to clear airspace of hostile aircraft.

air supremacy Degree of air superiority sufficient to prevent effective enemy interference with one's own operations.

air surface zone Restricted area established to protect friendly surface vessels and aircraft and permit ASW operations unhindered by presence of friendly submarines.

air surveillance Systematic observation of airspace by visual, electronic or other means to plot and identify all traffic.

air-taxi operator Operator of light aircraft, below 12,500 lb TOGW, licensed to ply for hire for casual passenger traffic.

air taxiing Positioning helicopter or other VTOL aircraft by short translational flight at very low altitude. Standardized ** manoeuvres form part of VTOL flying instruction.

air time Elapsed time from start of take-off run to end of landing run.

air-to-air From one aerial position to another, esp between one airborne aircraft and another.

air-to-ground Between aerial position, esp airborne aircraft, and land surface.

air-to-surface Between aerial position, esp airborne aircraft, and any part of Earth's surface or target thereon.

air-to-underwater Between aerial position, esp airborne aircraft, and location below water surface, esp flight profile of ASW weapons and operating regime of ASW detection systems.

Air Track Landing Early form of ILS developed by NBS and Washington Institute of Technology.

air traffic Aircraft operating in air or on airport surface, exclusive of loading ramps and parking areas (FAA); aircraft in operation anywhere in airspace and on manoeuvring area of aerodrome (BSI). To air carriers "traffic" has entirely different meaning, but this is never qualified by "air".

air traffic clearance See *clearance (1)*.

air traffic control radar beacon system See *ATCRBS*.

air traffic control centre Unit combining functions of area control centre and flight information centre.

air traffic control service Service provided for promoting safe, orderly and expeditious flow of air traffic, including airport, approach and en route ATC service.

air train Aerial tug towing two or more gliders in line-ahead.

air transportable units Military units, other than airborne, whose equipment is all adapted for air movement.

air transport operation BS.185, 1951: "The carriage of passengers or goods for hire or reward"; this eliminates all military air transport, business flying and several other classes of ***.

air trooping Non-tactical air movement of personnel.

air-tube oil cooler Oil cooler in which air passes through tubes surrounded by oil.

air turnback Point at which mission already airborne is abandoned, for any reason.

air umbrella Massive friendly air support over surface operation or other air activity at lower level.

air vane 1 Small fin carried on pivoted arm to respond to local changes in incident airflow; arm usually drives potentiometer pick-off sending signal of angle of attack or yaw.
2 Powered surface to control trajectory of ballistic vehicle in atmosphere (see *jet vane*).

air vector In DR navigation, Hdg and TAS (air plot).

air volume In aerostat, volume of air displaced by solid body having same size and shape as envelope or outer cover. Volume used in airship aerodynamics.

airway BS.185, 1951: "An air route provided with ground organization". Most civil air routes are flown along airways, typically 10 nm wide with centreline defined by point-source radio navaids spaced sufficiently close for inherent accuracy to be less than half width of airway at midpoint. Each airway has form of corridor, of rectangular cross-section well above Earth. Airspace within is controlled, and traffic separated by being assigned different levels and from ATC having position reports and accurate forecasts of future position (typically, by ETA at next reporting point). In general, made up of a series of route segments each linking two waypoints.

airway beacon Light beacon located on or near

airway (see *NDB*).

airways-equipped Equipped with functioning statutory avionics and instruments (eg, two pressure altimeters) to satisfy ICAO requirements for flight in controlled airspace.

airways flying Constrained to dogleg along centrelines of airways instead of flying direct to destination.

airway traffic control Civil air traffic control formerly exercised on "airways" basis; today no separate system for designated airways.

Air Wheel Wheel/tyre combination introduced by Goodyear after World War 1, characterized by small wheel and fat tyre to absorb landing shocks.

air work In flying instruction, student air time as distinct from classroom time.

airworthiness Fitness for flight operations, in all possible environments and foreseeable circumstances for which aircraft or device has been designed.

airworthiness directive, AD Message from national certifying authority requiring immediate mandatory inspection and/or modification.

airworthy Complying with all regulations and requirements of national certifying authority.

AIS 1 Aeronautical Information Service (DTI, UK).
2 Advanced (or airborne) instrumentation subsystem (ACMR).
3 Avionics intermediate shop.
4 Academic instructor school.
5 Aircraft indicated (air) speed (IAS is preferred).

AI²S Advanced IR imaging seeker.

AISFS Avionics integration support facilities.

aisle Longitudinal walkway between seats of passenger aircraft.

aisle height Headroom along aisle.

AIT Alliance Internationale de Tourisme (Int).

AITAL, Aital Asociación Internacional de Transportes Aéreos Latinoamericanos (Int).

AITFA Association des Ingénieurs et Techniciens Français des Aéroglisseurs (hovercraft) (F).

AIV Accumulator isolation (or isolator) valve.

AJ Anti-jam.

A_j Nozzle throat area (occasionally AJ).

Ajax unit Device for providing artificial feel in pitching plane as function of stick displacement, altitude and airspeed.

AJJ Adaptive-jungle jammer, sophisticated ECM self-adapting to variable enemy transmissions.

AJM Anti-jam modem.

AJPAE, Ajpae Association des Journalistes Professionnels de l'Aéronautique et de l'Espace (F).

A/K Aluminium/Kevlar (armour).

AL 1 Approach and landing chart (FAA).
2 Spec prefixes for methanol or water/methanol (AL-24 PE(4), AL-28 gas turbines).

A/L Approach/land, operative mode for airborne system.

ALA 1 Alighting area (ICAO, marine aircraft).
2 Asociación de Lineas Aéreas (Spain).

ALAE Association of Licensed Aircraft Engineers (Australia, UK).

Alarm Air-launched anti-radiation missile.

ALARR Air-launched, air-recoverable rocket.

ALAT Aviation Léger de l'Armée de Terre (F).

ALAVIS Advanced low-altitude IR reconnaissance system.

albedo 1 Percentage of EM radiation falling on unpolished surface that is reflected from it, esp percentage of solar radiation (particularly in visible, or other specified, range) reflected from Moon or Earth.
2 Radiation thus reflected.

ALBM Air-launched ballistic missile.

ALC Air Logistics Center (USAF).

ALCAC Air Lines Communications Administrative Council (US).

ALCC 1 Airborne launch control centre.
2 Airlift control centre.

Alclad Trade name (Alcoa) of high-strength light alloys (usually sheet) coated with corrosion-resistant high-purity aluminium. Originally developed for marine aircraft.

ALCM Air-launched cruise missile.

alcohol Large family of hydrocarbons containing hydroxyl groups, esp methyl alcohol CH_3OH (toxic) and ethyl alcohol (ethanol, C_2H_5OH, potable), both used as fuels, anti-detonants and rocket propellants.

ALCS Active lift control system, to reduce peak wing stresses in gusts.

Alcusing Light alloy (aluminium, copper, silicon).

ALD 1 Arbitrary landing distance, standard comparison distance along runway, from touchdown to stop, using specified landing technique; used in determining field-length requirements.
2 JETDS code: piloted aircraft, countermeasures, combination of purposes.

ALDCS Active-lift distribution control system.

Aldis Patented hand signalling lamp with optical sight, trigger switch (kept on throughout use) and second trigger to tilt mirror to deflect light beam intermittently down to target.

ALE Aviazione Leggera Esercito (Italian army aviation).

ALEA Airborne Law Enforcement Association (US).

ALEK See *anchor-line extension kit*.

alert 1 Specified condition of readiness for action, esp of military unit.
2 Warning of enemy air attack.
3 ATC action taken after 30 min "uncertainty" period (5 min in case of aircraft previously cleared to land) when contact cannot be established.
4 Response by manufacturer and/or certifying authority to unacceptable incidence of service

failures by hardware item.

alert area Airspace which may contain high volume of pilot-training activities or unusual type of aerial activity (FAA).

alerting centre Centre designated by appropriate authority to perform functions of RCC where none exists (BS.185).

alerting service Service provided to notify and assist all appropriate organizations capable of aiding aircraft in need of search or rescue.

alert-level standard Agreed reliability performance below which special and urgent action must be taken (eg 0·3 IFSD per thousand engine hours).

alert(ing) unit In encoding altimeter, device triggering a warning alarm under certain potentially dangerous flight conditions.

ALF Auto/lock-follow target tracker.

AlGaAs Aluminium gallium arsenide.

algae Class of primitive plants (thallophytes) which elaborate their food by photosynthesis. Many species investigated as possible foods for life support in extended space travel.

algal corrosion Structural or system degradation caused by algae and other micro-organisms, especially those dwelling at fuel/air interface in tankage (see *fungus*).

algorithm 1 Established method of computation, numerical or algebraic.
2 Computation with relatively small number of steps in fixed preassigned order, usually involving iteration, designed for solution of particular class of problem. Fundamental in designing software.

ALH Active laser homing.

ALI Automatic line integration.

alidade Optical sight (microscope or telescope) used to read linear scale or other mensuration system.

alight To land, esp of marine aircraft on water.

alighting channel Part of water aerodrome navigable and cleared for safe alighting or taking off.

alighting gear See *landing gear*.

align 1 In INS, to rotate stable platform before start of journey until precisely aligned with local horizontal and desired azimuth.
2 In radio, radar or other equipment having resonant or tuned circuits, to adjust each circuit with signal generator to obtain optimum output at operating frequencies.
3 Normal meaning of word is relevant to erection of airframe jigging, lasers often being used when structures are large.

aligned mat Intermediate semi-prepared composite structure in which strong and/or stiff reinforcing fibres (rarely whiskers) are arranged substantially parallel in two-dimensional mat.

alignment time In INS or guidance system, minimum time required to spool-up gyros and align platform, preparatory to allowing significant movement of vehicle.

ALIMS Automatic laser inspection and measurement system.

ALJEAL, Aljeal Association of Lawyers, Jurists and Experts in Air Law (Int).

alkali metal Group of metals in First Group of Periodic Table characterized by single electron in outermost shell which they readily lose to form stable cation (thus, strongly reactive). Lithium, sodium, potassium and caesium (cesium) are important in electrical storage batteries and as working fluids in closed-circuit space power generation.

alkylation Addition of alkyl group (generally, radical derived from the aliphatic hydrocarbons); important in manufacture of gasolines (petrols) having high anti-knock (octane) rating.

ALL Airborne laser laboratory.

ALLA, Alla Allied Long Lines Agency (NATO).

all-burnt Rocket propulsion system which has consumed all propellant (where there are two, which has consumed all of either); specif, time and flight parameters when this occurs.

ALLD Airborne laser locator designator.

alleviation Reduction of structural loads (eg wing bending moment) in vertical gusts by active controls.

alleviation lag Time difference between actual and ideal response of a GAC active control system.

alleviation technique Method of reducing heat flux on atmospheric re-entry by controlling plasma sheath surrounding vehicle.

all-flying tail Term formerly used to describe variable-incidence tailplane used as primary control surface in pitch, separate elevators serving merely as additional part of surface or as a means of increasing camber.

allithium Generic name for aluminium-lithium alloys; also (capital A) trade name.

"all out" Signal signifying glider or other towrope taut, towed vehicle ready for take-off.

allowable deficiency Missing, damaged, inoperative or imperfectly functioning item which does not invalidate C of A and does not delay scheduled departure (eg, rudder bias system, fuel flowmeter and almost any item not part of structure or aircraft systems).

allowance 1 Intentional difference between dimensions (with tolerances permitted on each) of mechanically mating parts, to give desired fit.
2 Calculated quantity of fuel beyond minimum needed for flight carried to comply with established doctrine for diversion, holding and other delays or departures from ideal flight plan.
3 In sheet-metal construction, extra material needed to form bend of given inside radius and angle.

alloy Mixture of two or more metals, or metal-like elements, often as solid solution but generally with complex structure. Small traces of one element can exert large good or bad influence.

Most aircraft made principally of alloys of aluminium (about 95%, rest being copper, magnesium, manganese, tin and other metals) or titanium (commercially pure or alloyed with aluminium, vanadium, tin or other elements), with steels (alloys of iron) at concentrated loads.

all-shot Aerial target hit by every round from one gunnery pass.

all-speed aileron Lateral-control surface operable throughout flight envelope, as distinct from second aileron group on same aircraft operable at low speeds only.

"all systems go" Colloq, absence of mechanical malfunction; not authorized R/T procedure.

all-up round Munition, especially guided missile, complete and ready to fire.

all-up weight See *AUW*.

all-ways fuze Fuze triggered by acceleration in any direction exceeding specified level.

all-weather Former category of interceptor which could not, in fact, fly in **. Strictly true only for aircraft with triplex or quad AFCS and blind landing system plus ground guidance.

all-wing aircraft Aerodyne consisting of nothing but wing. Some aeroplanes of 1944–49 were devoid of fuselage, tail or other appendages, and approached this closely.

ALM Air loadmaster.

ALMS 1 Aircraft landing measurement system, typically using IR beams and geophones to produce hard-copy print-out of final approach and touch-down.
2 Air lift management system (software).

ALMV Air-launched miniature vehicle.

Alnico Permanent-magnet materials (iron alloyed with Al, Ni, Co — hence name — and often Cu) showing good properties and esp high coercive force.

ALNZ The Air League of New Zealand Inc.

ALO Air Liaison Officer.

ALOC Air line of communication, Army airlift for spare parts (US).

Alochrome Surface treatment for light-alloy structure to ensure good key for paint: chemical cleaning, light etching and final passivating.

Alodine Proprietary treatment similar to Alochrome.

Aloft Airborne light-optical-fibre technology.

along and across Configuration of track position display unit in which separate windows show continuous reading of distance to go (along) and distance off track (across), usually driven by Doppler.

ALOTS Airborne lightweight optical tracking system (precision photography of ballistic vehicles).

AlP Aluminium powder.

alpha Angle of attack of main wing (α, AOA).

alpha floor Mainly to protect against windshear, system which automatically applies full power if AOA exceeds preset value, and earlier if rate of change of TAS/GS passes preset thresholds.

alpha hinge Crossed-spring pivot (eg in tunnel balance).

alphanumeric Character representing capital letter or numeral portrayed in precisely repeatable stylized form either by electronic output (computer peripheral or display) or printed in same form for high-rate reading by OCR system.

alpha particle Nucleus of He atom: 2 protons, 2 neutrons, positive charge.

alpha speed, α-SPD Safe stall-margin speed (autothrottle mode setting).

Alply Trade name (Alcoa) of sandwich comprising polystyrene foam between two sheets of aluminium.

ALR Alerting message.

ALRAAM Advanced long-range air-to-air missile.

ALS 1 Approach light system (FAR Pt 1).
2 Alert-level standard.
3 Automatic take-off and landing system (RPV).

ALSC Advanced logistic systems center (AFLC).

ALSCC Apollo lunar-surface close-up camera.

ALSEP Apollo lunar-surface experiment package.

AL/SL Weapon capable of being air-launched or surface-launched (ie from surface vessel) (USN).

alt, ALT 1 Altitude.
2 Alternate (airfield or destination).
3 Automatic altitude hold.
4 Attack light torpedo.

ALTA 1 Association of Local Transport Airlines (US).
2 Asociación Latinoamericana de Transportadores Aéreos.

alternate Incorrectly, has come to mean "alternative" in flight operations; alternative destination, designated in flight plan as chosen if landing not possible at desired destination.

alternate hub airport Secondary civil airport at large traffic centre.

alternating light Intermittent light of two or more alternate colours.

alternator A.c. generator.

altigraph Recording altimeter; generally aneroid barograph, and thus subject to inaccuracy in pressure/height relationship assumed in calibration.

altimeter Instrument for measuring and indicating height. Pressure * is aneroid barometer or atmospheric pressure gauge calibrated to give reading in height. Sensitive * has stack of aneroid capsules, refined drive mechanism to multiply capsule movements with minimal friction or free play, and setting knob to adjust to different SL or airfield pressures (or to read zero at airfield height). In servo-assisted * mechanism is replaced with more accurate electrical one (see *engine* *, *radio* *).

altimeter errors Apart from servo-assisted altimeter, all pressure altimeters suffer significant lag, so rapid reversal of climb and descent will give a reading up to 200–300 feet in arrears; called lag or hysteresis. There are errors in drive friction and lost motion. Static pressure sensed is subject to PE(2) and compressibility errors, and to transient excursions during manoeuvres. Most significant, parameter measured depends on atmospheric pressure variation, temperature variation and variation in lapse rate between departure airfield and aircraft height (See *altimeter setting*).

altimeter fatigue Supposed tendency of aneroid system to become "set" in distorted position in long flight at high altitude; this error, not confirmed by most authorities, is called fatigue or, confusingly, hysteresis.

altimeter lag See *altimeter errors*.

altimeter setting For safe vertical separation all altimeters in controlled airspace must be set to uniform datum. Standard is 1013·25 mb (see *ISA*) throughout most en route flying. Instrument then registers vertical separation between aircraft and pressure surface 1013·25 mb, usually below local ground level and may be below local SL or MSL. Second common setting is QNH, at which reading is difference between aircraft height and MSL. Third common setting is QFE, at which reading is difference between aircraft height and appropriate airfield height AMSL; thus at that airfield instrument reads zero. Two other settings, QFF and QNL, seldom necessary.

altimetric valve Device sensitive to increasing cabin altitude (ie, falling pressure) and set to release drop-out oxygen at given level.

altitude 1 Vertical distance of level, point or object considered as point, measured from MSL (normally associated with QNH) (DTI, UK). In this context see *height*, *flight level*, *absolute* *.
2 Arc of vertical circle, or corresponding angle at centre of Earth, intercepted between heavenly body and point below it where circle cuts celestial horizon.
3 In spaceflight, distance from spacecraft to nearest point on surface of neighbouring heavenly body (in contrast to "distance", measured from body's centre).
4 In aircraft performance measurement and calculation, pressure * shown by altimeter set to 1013·25 mb.

altitude acclimatization Gradual physiological adaptation to reduced atmospheric pressure.

altitude chamber Airtight volume evacuated and temperature-controlled to simulate any atmospheric level.

altitude clearance Clearance for VFR flight above smoke, cloud or other IFR layer.

altitude datum Local horizontal level from which heights or altitudes are measured (see *true altitude*, *pressure altitude*, *height*).

altitude delay Deliberate time-lag between emission of radar pulse and start of indicator trace, to eliminate altitude hole or slot.

altitude hole Blank area at centre of radial (eg PPI) display.

altitude power factor In PE (4) ratio of power developed at specified altitude to power at same settings at ISA SL.

altitude parallax In altitude (2), angle between LOS from body to observer (assumed on Earth's surface) and LOS from body to centre of Earth.

altitude recorder Altigraph.

altitude reservation, ALTRV Airspace utilization under prescribed conditions, normally employed for mass movement or other special-user requirements which cannot otherwise be accomplished (FAA).

altitude sickness Malaise, nausea, depression, vomiting and ultimate collapse, caused by exposure to atmospheric pressure significantly lower than that to which individual is acclimatized.

altitude signal In airborne radar operating in forward search mode, unwanted return signal reflected by Earth directly below.

altitude slot Blank line at origin of SLAR display.

altitude switch Barometric instrument which makes or breaks electric circuit at preset pressure altitude; contacting altimeter.

altitude tunnel Wind tunnel whose working section can simulate altitude conditions of pressure, temperature and humidity. In view of advantages of high pressure and driest possible air, conditions chosen usually compromise.

ALTN 1 Alternate airfield.
2 Alternating (two-colour light).

altocumulus, Ac Medium cloud, about 12,000 ft in groups, lines or waves of white globules.

altostratus, As Stratiform veil 6,000–20,000 ft with ice-crystal content of variable thickness (giving mottled appearance) but usually allowing Sun/Moon to be seen.

ALTP Air Line Transport Pilot.

ALTR Approach/landing thrust reverser.

ALTRV Altitude reservation.

ALTS Altimeter setting

Alumigrip Trade name of paint used on airframe exterior.

Alumilite Trade name for sulphuric-acid anodizing process for aluminium and alloys.

alumina Aluminium oxide Al_2O_3, occurring naturally but also manufactured to close tolerance in various densities. Hard, refractory, white or transparent ceramic.

aluminium (N America, aluminum), Al Metal element, SG about 2·7, MPt 659·7°C, BPt 2,057°C, most important structural material in aerospace, commercially pure and, esp, alloyed with other metals (see *duralumin*).

aluminium-cell arrester Lightning arrester/conductor in which insulating film of aluminium plates breaks down and conducts at high applied voltage.

aluminium dip brazing Method of metallizing printed-circuit boards by closely controlled dipping in molten aluminium.

aluminum Aluminium (N America).

ALVRJ Air-launched low-volume ramjet.

ALW Air/land warfare.

ALWT Airborne (or advanced) light weight torpedo.

AM 1 Air Ministry (defunct in UK).
2 (or a.m.) ante-meridian, before noon.
3 (or a.m.) amplitude modulation, MCW or A_2 emission in which AF is impressed on carrier by varying carrier amplitude at rate depending on frequency, and degree of modulation (depth) depending on audio amplitude.
4 Asynchronous machine.

am Ambient.

AMA Air materiel area (USAF).

AMACUS Automatic microfilm aperture card updating system (Singer).

AMAD Airframe-mounted accessory (or auxiliary) drive.

amagnetic Having no magnetic properties.

AMARV Advanced manoeuvring re-entry vehicle.

AMAS Automated manoeuvring attack system.

Amatol High explosive (**AM**monium nitrate **A**nd **TOL**uene).

Ambac R-Nav system (see *Mona*).

Amber 1 Colour identifying global groups of airways aligned predominantly N-S.
2 Day/night training equipment, often called two-stage amber, in which pupil pilot is denied visual cues outside cockpit by wearing blue-lens glasses while cockpit transparency is amber; two stages together cut off 99% of light, while allowing pupil to see blue instruments and instructor to see amber outside world (obsolete).

ambient Characteristic of environment (eg that around aircraft but unaffected by its presence).

AMC 1 Aerodynamic mean chord.
2 Acceptable means of compliance.
3 Avionics Maintenance Conference (US and int).
4 Air Materiel Command (USAF).
5 Automatic modulation control.
6 Air management computer.

AMCM Airborne mine countermeasures.

AMCS 1 Adaptive microprogrammed control systems (IBM).
2 Airborne missile control system (aircraft-mounted).

AMD 1 Aerospace Medical Division (USAF).
2 Automatic map display.
3 (rare) Air mileage distance.

AMDA Airlines Medical Directors Association.

AMDP Air Member for Development and Production (UK, WW2).

AMDS Automatic manoeuvre device system.

amdt Amendment (FAA).

AME 1 Authorized medical examiner.
2 Alternate (alternative is meant) mission equipment.

AMEC Advanced multifunction embedded computer.

AMeDAS Automated met data-acquisition system (J).

amended clearance Clearance altered by ATC while flight en route, typically requesting change of altitude or hold, to avoid future conflict unforeseen when clearance filed.

American National Family of 60° screw (bolt) threads, basically divided into National Coarse (NC), National Fine (NF) and National Special (N), which have in part superseded SAE and ASME profiles.

Ames Major NASA laboratory, full title Ames Research Center, Moffett Field, Calif, mainly associated with atmospheric flight.

AMF 1 Allied [Command Europe] Mobile Force.
2 Armé Marin & Flygfilm (Sweden).

AMFI Aviation Maintenance Foundation (US).

amino Group $-NH_2$ which can replace hydrogen atom in hydrocarbon radical to yield amino acid; these play central role in metabolic pathways of living organisms; thus of interest in space exploration.

AMIR 1 Air Mission Intelligence Report, detailed and complete report on results of air mission.
2 Anti-missile infra-red.

AMIS Anti-materiel incendiary submunition.

AMK 1 FAA-approved airplane modification kit.
2 Anti-misting kerosene.

AML Adaptive manoeuvring logic.

ammeter Instrument for measuring electric current (d.c. or a.c., or both in case of "universal testers") with reading usually given in amperes.

Ammonal High explosive (**AMMON**ium nitrate + **AL**uminium, and often finely divided carbon).

ammonia NH_3, gas at ISA SL, pungent, toxic, present in atmosphere of Jupiter (ice crystals and vapour) and more distant planets (frozen solid). Ammonium chloride in "dry batteries", nitrate in many explosives, perchlorate plasticized propellants in large solid rocket motors, sulphate soldering and brazing flux and in dry cells, and several compounds in fireproofing.

ammunition Projectiles and propellants for guns; increasingly, guided weapons are logistically treated as * but term normally excludes them.

ammunition tank Compartment or container housing ammunition for airborne automatic weapon, usually in form of belt arranged in specified way; reloadable when removed from aircraft.

AMO 1 Air Ministry Orders.
2 Air mass zero, test condition for solar arrays and other space hardware.

amortization Fiscal process of writing-down value of goods and chattels over specified period.

Typically, transport aircraft * over five, seven or ten years, after which book value is zero. Rate exerts major influence on DOC.

AMP 1 Assisted maintenance period (aircraft carrier).
2 Avionics master plan (USAF).

amp Ampere.

ampere SI unit of electric current (quantity per unit time), symbol A; named for A.M. Ampère but no accent in unit. Hence: ampere-hour (1 A flowing for 1 h); ampere-turn (unit magnetizing force, 1 A flowing round 1 turn of coil).

AMPG, a.m.p.g. Air-miles per gallon, air distance flown per gallon of fuel consumed. UK gallon was Imp; distance was statute miles (not "air mile"). In US, st mi and US gal (0·83267 Imp gal). New unit must be found; SI suggests air metres per litre, unless fuel measured by mass (see *AMPP*, *NAMP*).

amphibian Aerodyne capable of routinely operating from land or water.

amplidyne D.c. generator whose output voltage governed by field excitation; formerly used as power amplifier in airborne systems.

amplification factor In thermionic valve (radio vacuum tube), ratio of change in plate voltage to change in grid voltage for constant plate current (UK, plate = anode).

amplifier Device for magnifying physical or mechanical effect, esp electronic circuit designed to produce magnified image of weak input signal whilst retaining exact waveform.

amplitude 1 Maximum value of displacement of oscillating or otherwise periodic phenomenon about neutral or reference position.
2 Angular distance along celestial horizon from prime vertical (ie due N-S) of heavenly body, generally as it rises or sets at horizon.

amplitude modulation See *AM (3)*.

AMPP, a.m.p.p. Air miles per pound (of fuel), air distance flown for each pound avoirdupois of fuel consumed, former measure of specific range (see *AMPG*, *NAMP*).

AMPSS Advanced manned precision strike system.

AMPT Air miles per tonne (of fuel).

AMPTE Active Magnetospheric Particle Tracer Explorer(s).

AMR Atlantic Missile Range, military (DoD) range run by Pan Am and RCA from Patrick AFB and also serving NASA's KSC at Cape Canaveral.

Amraam, AMRAAM Advanced medium-range AAM.

Amrics Automatic management radio and intercom system.

AMRL Aerospace Medical Research Laboratory (USAF).

AMS 1 Aeronautical Mobile Service, radiocommunication service between aircraft or between aircraft and ground stations.
2 Air maintenance squadron (USAF).
3 Aircraft management simulator (or system).
4 Academy of Military Science (US).
5 Air (or aircraft) material specification.

AMSL, a.m.s.l. Above mean sea level.

AMSO Air Member for Supply and Organization (UK, WW2).

AMSU Air-motor servo unit.

AMT Accelerated mission test(ing).

AMTC Aerospace Medicine Training Centre.

AMTE Adjusted megaton equivalent.

AMTI Airborne-moving-target indicator.

Amtorg Former organization for importing and licensing US products (USSR).

AMU 1 Air mileage unit.
2 Astronaut manoeuvring unit.
3 Aircraft maintenance unit.

amyl Family of univalent hydrocarbon radicals, all loosely C_5H_{11}, esp: amyl acetate (banana oil) solvent and major ingredient of aircraft dopes; and amyl alcohol, in lacquers.

AN 1 Air navigation.
2 Prefix to designation codes of US military hardware denoting "Army-Navy"; now rare, though code system remains.

A_n Acceleration normal to flight path, usually along OZ axis.

ANA 1 Air Navigation Act.
2 Association of Nordic Aeroclubs (Int).

anabatic Wind blowing uphill as result of insolation heating slope and adjacent air more than distant air at same level.

ANAC, Anac Automatic nav/attack control(s).

Anacna Associazione Nazionale Assistenti e Controllori della Navigazione Aerea (I).

anacoustic region Extreme upper level of atmosphere (say, 100 miles above Earth) where mean free path too great for significant propagation of sound.

analog computer 1 Computational device functioning by relating or operating upon continuous variables (in contrast to digital computers, which operate with discrete parcels of information). Simplest example, slide-rule.
2 Electronic computer in which input data are continuously variable values operated upon as corresponding electrical voltages. Actual hardware can be coupled directly in so that, for example, control response, angular movement and aeroelastic distortion of control surface can be investigated in situ and in real time.

analog/digital converter, ADC Device for converting analog output into discrete digital data according to specified code of resolution; also called digitizer and, esp for linear and rotary movement, encoder.

analog output Transducer signal in which amplitude (typically quasi-steady voltage) is continuously proportional to function of stimulus.

anaglyph Picture, generally photographic but often print-out from some other system, comprising stereoscopic pairs of images, one in one colour (eg red) and other in second colour (eg blue). Viewed through corresponding (eg blue/red) spectacles, result appears three-dimensional.

analyser In PE (4) installations, device intended to indicate mixture ratio by sampling composition of exhaust gas (hence EGA, exhaust-gas *). Some for static-test purposes depend on chemical absorption of carbon dioxide, but airborne instrument uses Wheatstone bridge to measure variation in resistance due to proportion of carbon dioxide in gas.

analysis Stress analysis.

ANAO Australian National Aerospace Organization.

anaprop Anomalous propagation.

ANB Air Navigation Bureau (ICAO).

ANC 1 Air navigation charges.
2 Air Navigation Commission.

ANCB Association Nationale Contre les Bangs (supersonic) (F).

anchor-centred Also anchor charge, anchor grain: solid rocket propellant charge in which initial combustion surface has cross-section resembling radial array of anchors, flukes outward.

anchor light Riding light.

anchor line, cable Cable running along interior of airdrop aircraft to which parachute static lines (strops) are secured.

anchor-line extension kit Assembly arranged to extend anchor line to allow airdropping through rear clamshell doors or aperture with such doors removed.

anchor nut Large family of nuts positively securable by means of screwed or bolted plate projecting from base (see *nut*, *nutplate*, *stiffnut*, *stopnut*).

ANCOA Aerial Nurse Corps of America.

AND 1 Active-nutation damping.
2 Aircraft nose-down.

ANDA Associazione Nazionale Direttori di Aeroporto (I).

ANDAG Associazione Nazionale Dipendenti Aviazione Generale (I).

Anderson shelter Small air-raid shelter assembled from sheet aluminium pressings in pit and covered with deep layer of earth (UK, WW2).

AND gate Bistable logic function triggered only when all inputs are in ON state; in computers used as addition circuit, performing Boolean function of intersection. Hence AND/OR gate, AND/NOR, AND/NOT.

AND pad Standard Army/Navy drive accessory pad.

androgynous Mating portions of docking system which are topologically identical (eg US and Soviet docking faces).

ANDS Automatic navigation differential station.

ANDVT Advanced narrow-band digital voice terminal.

anechoic Without echoes; thus, * facility, * room, in which specially constructed interior walls reduce reflections to infinite number of vanishingly small ones. Chamber can be designed to operate best at given wavelength, with sound, ultrasound, ultrasonic energy, microwaves and various other EM wavelengths. Mobile facilities used for boresighting nose radar of combat aircraft.

anemogram Record produced by anemograph.

anemograph Instrument designed to produce permanent record of wind speed (ie recording anemometer) and, usually, direction. Dines * incorporates weathercock vane carrying pitot and static tubes.

anemometer Instrument for measuring speed of wind, usually 10 m (32·8 ft) above ground level. Robinson Cup * has free-rotating rotor with three or four arms each terminating in a hemispherical or conical cup.

anemoscope Instrument for checking existence and direction of slow air currents.

ANEPVV Association Nationale d'Entraide et de Prévoyance du Vol à Voile (gliding) (F).

anergolic Not spontaneously igniting; thus, most rocket-propellant combinations comprising two or more liquids. Opposite of hypergolic.

aneroid Thin-walled airtight compartment designed to suffer precisely predictable and repeatable elastic distortion proportional to pressure difference between interior and exterior. Most are evacuated steel capsules in form of disc with two corrugated faces which can approach or recede from each other at centre. To increase displacement a stack can be used linked at adjacent centres. Common basis of pressure altimeter, ASI and Machmeter.

aneroid altimeter Pressure altimeter; aneroid barometer calibrated to read pressure altitude.

aneroid altitude Pressure altitude.

ANFCMA Associazione Nazionale Famille Caduti e Mutilati dell'Aeronautica (I).

ANG Air National Guard (US).

ANGB ANG base.

angels 1 Historic military R/T code word for altitude in thousands of feet; thus, "angels two-three" = 23,000 ft.
2 Distinct, coherent and often strong (40 dB above background) radar echoes apparently coming from clear sky. Probable cause strong pressure, temperature or humidity gradient in lower atmosphere giving even sharper gradient in refractive index.

angled deck Aircraft-carrier deck inclined obliquely from port (left) bow to starboard (right) stern to provide greater deck space, greater catapult capacity and unobstructed flight path further from island than with axial deck.

angle of . . . In general, see under operative word.

angle of attack indicator Instrument served by ***

sensing system.

angle of attack sensing system Incorporated in aircraft, esp aeroplane, to trigger stall-warning, stall-protection system or other desired output, and possibly serve an indicator. Sensing unit (SU) comprises freely pivoted vane or series of pitot tubes set at different angles of incidence and each connected to different supply pipe to give dP output. SU on wing leading edge, to sense movement of stagnation point, or on side of fuselage, repeated on opposite side to eliminate error due to sideslip. SU anti-iced and must allow for changes in aircraft configuration.

angle of depression Acute angle between axis of oblique camera and horizontal.

angle off Acute angle between own-fighter sightline and longitudinal axis of target aircraft.

angle of incidence indicator Instrument giving continuous reading of angle of foreplane, horizontal tail (esp tailplane where not primary pitch control) or wing, where incidence variable.

angle of view Angle subtended at perspective centre of camera lens by two opposite corners of format.

ANGSA Air National Guard support aircraft.

Ångström Unit Å or AU, unit of length equal to 10^{-10}m, formerly used to express wavelengths of light; nearest SI is nanometre (nm).

angular acceleration Time rate of angular velocity of body rotating about axis which need not pass through it.

angular displacement 1 Angular difference between two directions or axes, esp between reference axis of hinged or pivoted body and same axis in neutral or previous position.
2 In magneto, angular difference between neutral position of rotor pole and later position giving highest-energy spark (colloq, E-gap).

angular distance 1 Angular displacement.
2 Smaller arc of great circle joining two points expressed in angular measure.
3 In all sine-wave phenomena (radio, radar, astronomy, etc), number of waves of specified frequency between two points (numerically multiplied by 360 or 2π depending on whether unit is degree or radian).

angular measure SI unit of plane angle is radian (rad), angle subtended by arc equal in length to radius of circle on which arc centred. Thus one revolution = 2π rad, and 1 rad = 57·296°. Degree (°) defined as 1/360th part of one revolution, itself subdivided into 60 minutes (′) each subdivided into 60 seconds (″); pedantically distinguished from units of time by calling them arc-minutes and arc-seconds. Thus 1 rad = 57° 17′ 45″. For small displacements milliradian (mrad) to be used; roughly 3′ 26¼″.

angular momentum For rigid body of significant mass (not elementary particle), product of angular velocity and moment of inertia; or, if axis of rotation at some distance from it (as in axial turbine blade), mass, instantaneous linear velocity and radial distance of CG from axis. Thus, $L = I\omega = mvr$.

angular resolution Angular distance between LOS from radar, human eye or other "seeing" system to target and LOS from same system to second target which system just distinguishes as separate object; usually only a few mrad, esp if targets are pinpoints of light against dark background.

angular speed 1 Loosely, angular velocity.
2 Rate of change of target bearing, esp as seen on PPI.

angular velocity Time rate of angular displacement of body rotating about axis which need not pass through it. Preferred measure is rad s^{-1} or mrad s^{-1}; in traditional engineering most common is rpm.

anhedral 1 Negative dihedral, smaller angle between reference plane defining wing (such as lower surface or locus of AMCs) which slopes downward from root to tip, and horizontal plane through root. In early aircraft dihedral considered desirable as means to natural stability, esp in roll; in some, and many modern gliders, wing flexure converts static * into dihedral under 1g in flight. Tendency to design modern wing with * to counter excessive roll response to sideslip or side gusts, esp in high-wing or supersonic aircraft. In VG aeroplane angle may be varied with sweep.
2 Some authorities define as "absence of dihedral" (from Greek root of prefix *an* = not), and suggest "cathedral" for downward-sloping wing.

ANIAF Associazione Nazionale Imprese Aerofotogrammetriche (I).

ANIE Associazione Nazionale Industrie Elettrotechniche ed Elettroniche (I).

aniline Phenylamine, aminobenzene, $C_6H_5NH_2$, colourless, odorous amine, m.p. –6°C (thus, normally liquid), b.p. 184°C, turns gradually brown on exposure to air, reacts violently with RFNA or other strong nitric acids with which often used as rocket propellant.

anion Negative charged ion or radical, travels towards anode in electrolytic cell.

anisotropic Exhibiting different physical properties along different axes, esp different optical properties or, in structural material, different mechanical properties, esp tensile strength and stiffness.

ANK Automatic navigation kit.

ANL Auto noise limiter (communications).

ANLS Automatic navigation launch station.

ANMI Air navigation multiple indicator.

ANMS Aircraft navigation and management system.

annealing Heat treatment for pure metals and alloys to obtain desired physical properties by altering crystalline microstructure. Usually involves heating to above solid-solution or critical temperatures, followed by gentle cooling in air. General aim to make metal less brittle, tougher, more ductile and relieve interior stress.

annotation Identifying and coding reconnaissance outputs, such as visual-light photographs, IR print-outs, ECM records, etc, with digital data: date, time, place, unit, altitude, flight speed, heading and data specific to reconnaissance system.

annual Annual mandatory inspection of aircraft.

annual variation Amount by which magnetic variation at specified place on Earth varies in calendar year. In UK about −7′ (min), reducing local variation to zero in year 2140.

annular combustion chamber In gas-turbine engine, chamber in form of body of revolution, usually about major axis of engine.

annular gear Ring gear or annulus, gearwheel in which teeth project inwards from outer periphery. In annulus gear there is no centre, teeth being carried on open ring (which in turboprop/turbofan reduction gears may be resiliently mounted and torque-reacted by torque-signalling system). When shaft-mounted, teeth usually on one side only of flat or conical disc.

annular injector Rocket (or possibly other engine) injector in which liquid fuel and/or oxidant is sprayed from narrow annular orifice. In bipropellant engine numerous such orifices spaced around chamber head, alternately for fuel and oxidant.

annular radiator Cooling radiator shaped as body of revolution to fit around axis of aircraft engine, esp between propeller and PE (4) having circular cowl.

annular spring Has form of ring distorted radially under load. Sometimes called ring spring, esp when given tapered cross-section and used in multiple, in an intermeshing stack, to resist load along axis of symmetry.

annular wing Wing in form of body of revolution, designed to operate in translational flight with axis of symmetry almost horizontal.

annulus drag Most often refers to base drag of annular periphery around propulsive jet nozzle which incompletely fills base of vehicle, esp ballistic rocket rising through atmosphere.

annunciator 1 In gyrocompass (remote compass), indicator flag visible through window of cockpit instrument; with a.c. supply on and gyro synchronized with compass detector, indication should hover between dot and cross, never settle on either.

2 In aircraft system, esp on aircraft having flight deck rather than cockpit, panel of captioned warnings often distributed on schematic diagram of system.

ANO, Air Navigation Order UK statutory instrument for enactment of ICAO policy defining laws, licensing and similar fundamental issues regarding aerial navigation (see *ANR*).

anode In electrical circuit (electrolytic cell, valve, CRT), positive pole, towards which electrons flow; that from which "current" conventionally depicted as emanating.

anodizing Electrolytic (electrochemical) treatment for aluminium and alloys, magnesium and alloys and, rarely, other metals, coated with inert surface film consisting mainly of oxide(s) as protection against corrosion. Electrolyte usually weak sulphuric or chromic acid.

anomalistic period Time between successive passages of satellite through perigee.

anomalistic year Earth's orbital period round Sun, perihelion to perihelion: 365 d 6 h 13 m 53·2 s, increasing by about 0·26·s per century.

anomalous dispersion Local reversal in rule that medium transparent to EM radiation diffracts it with refractive index that falls as wavelength increases; discontinuities in absorption spectrum make index increase as wavelength increases.

anomalous propagation Of wave motions, esp EM radiation of over 30 kHz frequency or sound, by route(s) grossly different from expected, usually because of atmospheric reflection and/or refraction, sharp humidity gradients and temperature inversion.

anomaly 1 Difference between mean of measured values of meteorological parameter at one place and mean of similar values at all other points on same parallel (in practice, mean of similar values at other stations near same parallel).

2 In general, deviation of observed geodesic parameter from norm or theoretical value.

3 Specif local distortion of terrestrial magnetic field caused by local concentrations of magnetic material, used in aerial geophysical surveys and ASW (see *MAD*).

ANORAA Association Nationale des Officiers de Réserve de l'Armée de l'Air (F).

anoxaemia Hypoxaemia, deficiency in oxygen tension (loosely, concentration) of blood.

anoxia Absence of oxygen available for physiological use by the body (see *hypoxia*).

ANP 1 Aircraft nuclear propulsion; general subject and defunct DoD programme.

2 Air navigation plan (ICAO).

ANPA Aircraft Nuisance Prevention Association (J).

ANPAC Association Nazionale Piloti Aviazione Commerciale (I); other Italian associations include ANPAV (Assistenti di Volo), ANPCAT (Professionale Controllori & Assistenti Traffico Aereo), ANPI (Paracadutisti d'Italia), ANPIC (Piloti Istruttori Civili), ANPiCo (Piloti Collaudatori), and ANPSAM (Piloti Servizi Aerei Minori).

ANR Air Navigation Regulation.

ANRA Association Nationale des Résistants de

l'Air (F).

ANRS Automatic navigation relay station.

ANRT Association Nationale de la Recherche Technique (F).

ANS 1 Air Navigation School.
2 Airborne navigation system.

ANSA Advisory group, air navigation services (G).

ant Antenna (aerial).

ANTAC Association des Navigants Techniciens de l'Aviation Civile (Belg).

antenna US term for aerial (2), portions of broadcasting EM system used for radiating or receiving radiation.

anthophyllite Crystalline mineral, essentially (Mg, Fe) Si O_3.

anti-aircraft, AA Surface-based defence against aerial attack. Suggested historic word; better to introduce SA, surface-to-air, as prefix for guns as well as missiles, together with associated radars and other peripherals.

anti-aircraft artillery, AAA Guns and unguided-rocket projectors dedicated to surface-to-air use with calibre 12·7 mm (0·5 in) or greater.

anti-air warfare, AAW All operations intended to diminish or thwart hostile air power, eg air defence and interdiction against enemy airfields.

anti-balance Opposing, counteracting or reducing balance, esp in dynamic system.

anti-balance tab Tab on control surface mechanically constrained to deflect in same sense as parent surface to increase surface hinge-moment (ie to make it more difficult to move in airstream); opposite of servo tab.

anti-ballistic missile, ABM System designed to intercept and destroy hypersonic ballistic missiles, esp RVs of ICBMs. Speed of such targets, small radar cross-section, possible numbers, use of ECM and decoys, nuclear blanketing of large volumes of sky, ability to change trajectory, enormous distances, and need for 100% interception, make ABM difficult.

anti-buffet Describes measures adopted on atmospheric vehicles, esp high-speed aircraft, to reduce or eliminate aerodynamic buffet. Almost always auxiliary or locally hinged surface moved out into airflow to reduce buffet which would otherwise be caused by configuration change, eg opening weapon-bay doors, Thus * flap, panel, comb, rake, slot.

anticing See *anti-icing.*

anti-coning In most helicopters, and some other rotorcraft, ** device fitted to prevent main rotor blades from reaching excessive coning angle (being blown upwards, eg, by high wind at zero or low rotor rpm on ground), from which they could fall and exceed root design stress when suddenly arrested. Usually fixed range of angular coning permissible between ** device and droop stop.

anti-corrosion Measures taken depend on environment (see *marinizing*), and working stress and temperature of hardware; apart from choice of material and surface coating (Alclad, anodizing, painting with epoxy-based paint, etc), special agents can be introduced to fuels, lubricants, seals, hydraulic fluids and interior of stored device (inhibiting).

anti-cyclone, anticyclone Atmospheric motion contrary to Earth's rotation. Large area of high pressure, generally with quiet, fine weather, with general circulation clockwise in N hemisphere and anticlockwise in S; divided into "cold" and "warm", each group being subdivided into "permanent" and "temporary".

anti-dazzle panel 1 Rearwards extension around top of instrument panel or cockpit coaming to improve instrument visibility and at night prevent reflection of instruments from windscreen.
2 On aircraft with natural metal finish, areas of exterior painted with non-reflective black or dark blue to prevent bright reflections being visible to crew.

anti-drag wire Structural bracing filament, usually incorporated within wing, intended to resist forwards ("anti-drag") forces.

anti-freeze Most important agent ethylene glycol ($CH_2OH.CH_2OH$), usually used as aqueous solution with minor additives. Neat "glycol" (there are many) remains liquid over more than twice temperature range of water and freezes not as solid ice but as slush.

anti-friction bearing Loose term applicable to bearing suffering only rolling friction (ball, needle, roller) but esp signifying advanced geometry and high precision.

anti-frosting Measures taken to prevent frost (ice condensing from atmosphere and freezing as layer of fine crystals), esp on windscreen; typically raise temperature by hot air or fine electrical resistance grids or conductive films.

anti-g Measures counteracting adverse effects of unusual accelerations in vertical plane.

anti-glare Against optical glare, generally synonymous with anti-dazzle but esp dull non-reflective painted panels on airframe and propeller blades.

anti-gravity Yet to be invented mechanism capable of nullifying local region in gravitational field.

anti-g suit See *g-suit.*

anti-icing Measures to prevent formation of ice on aircraft; required on small vital areas where ice should not be allowed to form even momentarily (see *icing*, *de-icing*).

anti-icing correction Applied to aircraft, esp advanced aeroplane, performance with various forms of ice-protection operative. Esp necessary when engine air bleed exerts significant penalty in reduced take-off, climb-out, overshoot and en route terrain-clearance calculations. Required numerical values usually given in flight manual as percentage for each flight condition.

anti-knock rating Measure of resistance of PE (4) fuel to detonation (1) (see *octane rating*).

anti-lift In direction opposite to lift forces, eg loads experienced by wing on hard landing with lift dumpers in use. Thus ** wire (landing wire), structural bracing filament, usually within wing, to resist downloads.

anti-missile Against missiles, specif system intended to intercept and destroy hostile missiles (which may or may not include guided devices, artillery shells, bullets, mortar bombs and other flying hardware). In large-scale defence against ICBM attack, anti-ballistic missile.

anti-misting kerosene Jet fuel chemically and physically tailored so that, on sudden release to atmosphere (from ruptured tanks in a crash), it spreads in the form of droplets too large to form an explosive mixture with air.

antimony Element existing in several allotropic forms, most stable being grey metal with brittle crystalline structure. Widely used in aerospace in small quantities: with tin and other metals in bearings and applications involving sliding friction, with lead in storage batteries, as acceptor impurity in semiconductors, in type metals and electronic cathodes.

antinode 1 Points on wave motion where displacement (amplitude) is maximum.
2 Locations in aircraft structure where flexure (due to vibration or aeroelastic excitation) is maximum.
3 Either of two points in satellite orbit where line in orbit plane perpendicular to line of nodes, and passing through focus, intersects orbit (thus, essentially, points in orbit midway between nodes).

antipode Point on Earth, or other body, as far as possible from some other point or body; specif point on Earth from which line through centre of Earth would pass through centre of Moon.

antipodes Region on Earth diametrically opposite observer on Earth.

anti-radiation missile Missile designed to home on to hostile radars.

anti-rolling In rigid airship, measures intended to prevent rolling of any part relative to hull or envelope.

anti-rumble panel Small anti-buffet panel necessary on grounds of noise.

anti-snaking strip In early high-subsonic aircraft, strip of cord or metal attached to one side of rudder or elsewhere to prevent snaking (yawing oscillations).

anti-solar point Point on celestial sphere 180° from Sun; projection to infinity of line from Sun through observer.

anti-spin parachute Streamed from extremity of aeroplane or glider to assist recovery from spin; most common location is extreme tail.

antistatic Measures taken to reduce static interference with radio communications, traditionally by trailed ** wire, released from ** cartridge, which serves as pathway for dissipation of charge built up on aircraft.

anti-submarine warfare See *ASW*.

anti-surge measures 1 To prevent aerodynamic surging in axial compressor, eg redesign further from surge line, use of variable stators, blow-off valves and interstage bleeds.
2 Valves and baffles in oil cooler to maintain steady oil flow.

anti-torque drift Inherent lateral drift of helicopter due to side-thrust of tail rotor; often countered by aligning main rotor so that tip-path plane is tilted to give cancelling lateral component.

anti-torque rotor Tail rotor of helicopter, or any other rotor imparting thrust (moment) neutralizing that of main rotor.

antitrade wind Semi-permanent winds above surface trades, generally at height of at least 3,000 feet, especially in winter hemisphere, moving in opposite direction (ie westerly).

antivibration loop Closed-loop servo system designed to suppress structural or system vibration.

ANU Aircraft nose-up.

ANUA Associazione Nazionale Ufficiali Aeronautica (I).

Anvis Aviator's night-vision system.

ANVR Association of travel agents (Neth).

ANZUK, Anzuk Australia, New Zealand, UK, and SE Asia defence.

ANZUS, Anzus Australia, New Zealand and US (1951 defence pact).

AO 1 Administrative Operations, major US Federal budget heading.
2 Artillery observation.

AOA 1 Aerodrome Owners' Association (UK).
2 Angle of attack (units, thus "6 AOA").
3 Angle of arrival (ECM).
4 Air Officer i/c Administration (RAF).
5 Airborne optical adjunct (ABM).
6 Amphibious operating area (DoD).
7 At or above (FAA).

AOB 1 At or below.
2 Angle of bank.

AOC 1 Air Officer Commanding.
2 Air Operator's Certificate.
3 Autopilot omni-coupler.
4 Aerodrome obstruction chart.
5 Assumption of control message (ICAO).
6 Adaptive optical camouflage.
7 Association of Old Crows (EW, US).
8 Air operations center (US).

AOCI Airport Operators Council International, Inc (Int).

AOCM Airborne optical countermeasures.

AOCP Airborne operational computer program.

AOCS 1 Attitude and orbit control system.
2 Airline Operational Control Society (US).

AOD Aft of datum (c.g.).

AOFR Aluminium oxide fibre-reinforced.
AOG Aircraft on ground, code inserted in message (eg for spare parts) indicating aircraft unable to operate until remedial action taken.
AOI Arab Organization for Industrialization.
AOP 1 Airborne observation post.
2 All other persons (airline costings).
AOPA Aircraft Owners & Pilots Association (US).
AOPAA Aircraft Owners & Pilots Association of Australia.
AOPF Active optical proximity fuze.
AOS Acquisition of signal (telecommunications, telemetry).
AOSP Advanced on-board signal processor.
AOY Angle of yaw.
AP 1 Armour-piercing.
2 Ammonium perchlorate (solid rocket fuel).
3 Air Publication (UK).
4 Airport (ICAO).
5 Autopilot.
6 Aviation regiment (USSR).
7 Allied publication (NATO).
A/P 1 Autopilot.
2 Airplane(s).
3 Airport.
APA 1 Airline Passengers Association (US).
2 Airport (or airfield) pressure altitude.
APAC Association of Professional Aviation Consultants (UK).
Apacs Atlas prompting and checking system.
APAG Allied Policy Advisory Group (NATO).
AP/AM Anti-personnel/anti-material (last word often spelt materiel).
APC 1 Association of Parascending Clubs.
2 Approach power control (or compensator).
3 Armament practice camp (RAF).
4 Avionics planning conference.
5 Aviation Press Club (Belg).
6 Area positive control.
APCA Association Professionnelle de la Circulation Aérienne (F).
APCC Air Pollution Control Center (EPA, US).
apch, apchg Approach, approaching (FAA).
APCO Air Pollution Control Office (EPA, USA).
APCS Air Photo and Charting Service (USAF).
APD 1 JETDS code: piloted aircraft, radar, DF/reconnaissance/surveillance (usually SLAR).
2 Aerial position, digital (usually 4,096 pulses per 360°).
3 Avalanche photo-diode.
4 Amplifying photo-diode.
aperiodic Of any dynamic and potentially oscillatory system, so heavily damped as to have no period; unable to accomplish one cycle of oscillation; thus, * magnetic compass, * electric circuit.
aperture 1 Diameter of objective of optical instrument, either direct length or function of it; also angular *, minor angle subtended at principal focus by extremes of objective diameter; numerical *, n sin u, where u is refractive index between lens and object and u is objective angular radius (half angular aperture); and relative * (f-number) relating focal length to objective diameter.
2 In radio or radar aerial, either greatest dimension; or, with unidirectional aerial, greatest length across plane perpendicular to direction of maximum radiation, close to aerial, through which all radiation is intended to pass (ie all except diffuse stray radiation).
aperture card Standardized unit in microfilm filing, comprising frame of microfilm mounted in card border; stored, retrieved and projected automatically.
apex Highest point in canopy of parachute in vertical descent.
APF Association des Pilotes Françaises.
APFD Autopilot flight director.
APG JETDS code: piloted aircraft, radar, fire control.
APGC Air Proving Ground Command (USAF, defunct).
aphelion Point in solar orbit furthest from Sun.
API 1 Air-position indicator.
2 Armour-piercing incendiary.
3 American Petroleum Institute.
4 Associazione Pilote Italiane.
APIS 1 Apogee/perigee injection system.
2 Automatic priority interrupt system, for large computer systems with multi-programming.
APL 1 Acceptable performance level.
2 Applied Physics Laboratory of JHU.
APLA Asociación de Pilotos de Lineas Aéreas (Arg).
APM Aluminium powder metallurgy.
APMS Automatic performance-management system (also, in US, rendered as advanced power management system).
APN JETDS code: piloted aircraft, radar, navaid.
APNA Association des Professionnels Navigants de l'Aviation (F).
apoapsis Point in orbit furthest from primary; apocentre.
apoastron Furthest point in orbit round star.
apocynthion Point in lunar orbit furthest from Moon.
APOE Air (or aerial) port of embarkation.
apogalacticon Furthest point in orbit round galaxy.
apogee Point in geocentric (Earth) orbit furthest from centre of Earth (in near-circular polar orbit equatorial bulge could result in satellite being closer to surface at equatorial apogee than at polar perigee).
apogee motor Apogee kick motor or kick motor, small rocket designed to impart predetermined (sometimes remotely controllable) velocity change (delta-V) to satellite or spacecraft to

change orbit from an apogee position.

Apollo Applications Program, AAP Much altered and largely defunct NASA programme intended to make maximum and earliest post-Apollo use of Apollo technology. Major portion evolved into Skylab; other AAP being built in to various plans for future manned and unmanned spaceflight, include Shuttle missions.

apolune Apocynthion of spacecraft departed from Moon into lunar orbit.

apostilb Non-SI unit of luminance equal to $1/\pi$ international candle (candela) m^{-2} or 10^{-4} lambert (see *luminance*).

APP 1 Approach (DTI, UK).
2 Approach control office (ICAO).
3 Approach pattern.
4 Association des Pilotes Privés (F).
5 Association of Priest Pilots (US).

APPA 1 Associaçao de Pilotos e Proprietarios de Aeronaves (Braz).
2 Association des Pilotes Privés Avions (F).

apparent precession Apparent tilt of gyro due to rotation of Earth; vertical component = topple, horizontal = drift.

apparent solar day Length of Earth day determined by two successive meridian passages of apparent Sun; longer than sidereal day by time taken by Earth to turn additional increment to nullify distance travelled in solar orbit during this day. Basis of most human time scales, being divided into 24 h, hour being thus defined.

apparent wander Apparent precession.

APP CON Approach control (FAA).

Appleton layer F layer ($F_1$and F_2) of ionosphere, most useful for reflection of EM radiation (see *F-layer*).

approach BS.185: "To manoeuvre an aircraft into position relative to the landing area for flattening-out and alighting". Now subdivided into various categories, each of which needs pages of explanation defining circumstances, clearances and procedures. Following are brief notes. VFR * may be made with no radio at uncontrolled airport or airfield. Visual * may be made in IFR by pilot in contact with runway either not following other traffic or else in visual contact with it, with ceiling at least 500 ft above minimum vectoring altitude and visibility at least three miles. Various types of instrument * are admissible in IFR with radio TWR authorization: straight-in, circling, precision (with g/s and runway centreline guidance) and parallel (two parallel ILS runways, or, in military aviation, two parallel runways, each with PAR). In certain circumstances pilot may receive clearance for contact *, even in IFR. ILS * is most important IFR precision *. If required and available, pilot can be "talked down" in GCA or RCA, his only necessary equipment being primary instruments and operative R/T.

approach area Airspace over designated region of terminal area controlled by approach control unit (in some cases serving two or more airfields).

approach beacon 1 Historically, short-range track beacon (see *BABS*).
2 Today, beacon giving fix before or after approach gate (rare).

approach control BS.185: "A service established to provide ATC for those parts of an IFR flight when an aircraft is arriving at, or departing from, or operating in the vicinity of, an aerodrome". DTI (Air Pilot): "ATC service for arriving or departing IFR flights". FAA adds "and, on occasion, VFR aircraft".

approach control radar ACR, radar at approach control facility displaying PPI positions (and, in advanced models, height or alphanumeric data) of all aircraft within its range (which is not less than radius of furthest point in the controlled airspace).

approach coupler Electronic linkage between aircraft ILS receiver and autopilot and hence to AFCS; thus aircraft can make "hands-off" approach.

approach fix From or over which final approach (IFR) to airport is executed (FAA). On projected centreline 3–5 miles from threshold.

approach gate Point on final-approach course 1 mile beyond approach fix (ie further from airport) or 5 miles from landing threshold, whichever is greater distance from threshold (FAA).

approach indicator Ambiguous: could mean ILS or other cockpit instrument or any of several visual systems on ground indicating angle of approach.

approach lights 1 In modern large airfields, any of several systems of lights extending along projected centreline of runway in use towards approaching aircraft to provide visual indication of runway location, distances, alignment, glide path slope, and, probably, transverse horizontal.
2 In smaller or older airfields, one or more lights (often green) at, or extending from, downwind end of landing area to show favourable direction of approach.

approach noise Measured on extended runway centreline 1 nm (one nautical mile = 6,080 ft = 1,853 m) from downwind end of runway, with aircraft at height of 370 ft (112·58 m).

approach operations Flight operations within approach area, esp those of aircraft arriving or departing, designated as IFR or VFR.

approach plane Approach surface, sloping plane below which no aircraft should penetrate; in UK ** to grass airfield extends at inclination of 1:30 in all directions from periphery of landing area.

approach plate Flight-planning document relevant to specific airfield, giving details of minimum heights, safe headings and weather minimums (UK = minima), and including horizontal map and often also vertical profile for approach to each instrument runway.

approach power compensator Autothrottle, esp on combat aircraft.

approach radar See *PAR*, *GCA*, *SRE*.

approach receiver 1 ILS receiver.
2 Historically, radio receiver "capable of interpreting the special indications given by an approach beacon installation".

approach sequence Order in which aircraft are placed while awaiting landing clearance and in subsequent approach. In busy TMA traffic drawn in blocks from alternate landing stacks.

approach surveillance radar Approach control radar.

approval 1 Of manufacturer of aerospace hardware, approval by delegated national authority to design, manufacture, repair or modify such hardware, subject to specified conditions and inspection.
2 Of item of aerospace hardware, certificate issued by delegated national authority that item is correctly designed and manufactured and will thus be likely to perform within its design limits satisfactorily. In case of complete aircraft, C of A, or Type Certificate.
3 Of flight plan, signature by ATC officer or other responsible person that proposed plan does not conflict with pilot's qualifications, aircraft equipment, expected met conditions and expected air traffic, and that flight may proceed.

APQ JETDS code: piloted aircraft, radar, combination of purposes.

APR 1 JETDS code: piloted aircraft, radar, passive detection.
2 Airman performance report.
3 Auto power reserve.

APRA Air Public Relations Association (UK).

apron 1 Large paved area of airfield for such purposes as: loading and unloading of aircraft; aircraft turnaround operations; aircraft modification, maintenance or repair; any other approved purpose other than flight operations.
2 In engine cowling, any portion hinged down to act as walkway or servicing stand.
3 In ejection seat, lower forward face behind occupant's lower legs.
4 In vehicle fuelled with corrosive liquid, corrosion-resistant panel surrounding, and especially beneath, relevant supply hose coupling.

apron capacity Nominated number of transport aircraft to be accommodated on particular apron area in designated positions.

apron-drive bridge Passenger loading bridge (jetway).

apron-drive unit Self-propelled vehicular support for free end of passenger jetty (jetway), usually provided with two heavy-duty wheels steering through at least 180°.

APS 1 Aircraft prepared for service; standard weighing condition, or condition at which weight is calculated: comprises aircraft in all respects ready to take off on mission of type for which it was designed, complete with all stores, equipment (such as passenger reading material), fuel, crew and all consumable items, but with no revenue load.
2 Appearance-potential spectroscopy.
3 JETDS code: piloted aircraft, radar, search and detection.
4 Adaptive-processor sonar, or sonobuoy.
5 Armament practice station (UK).
6 Auxiliary power system.

APSE Ada programming support environment.

APSI Aircraft propulsion-system integration (USAF/USN).

apsides Plural of apsis.

apsis Extreme point of orbit, esp apocentre (furthest) or pericentre (nearest).

APT 1 Automatically programmed tool, eg in machining.
2 Automatic picture transmission; system for using vidicon cameras and data link to transmit real-time pictures from satellite to Earth with minimal degradation in picture quality.
3 Automated powerplant test.

APTS Automatic picture-transmission system.

APTT Aircrew part-task trainer.

APU Auxiliary power unit, esp airborne APU.

APV Autopiloted vehicle.

APVO See *IA-PVO*.

AQAD Aeronautical Quality Assurance Directorate (MoD-PE, UK).

AQAP Allied quality-assurance publication (NATO).

AQL Agreed quality level (material specifications, dimensional tolerances, etc).

aquaplane To run wheeled vehicle, esp landing aircraft, over shallow standing water at so high a speed that weight is supported wholly by dynamic reaction of water; tyres, out of ground contact, unable to provide steering or braking.

aqueous Pertaining to water, thus * solution.

ARA 1 Aircraft Research Association (UK).
2 Airborne-radar approach.

Araldite Trade name (Ciba) of two-component (resin + hardener) epoxy-based adhesive used in airframe structural bonding.

aramid fibre Man-made fibre of extraordinary tensile strength, so named because of its chemical and physical similarity to spider-web fibre; see *Kevlar*.

ARB 1 Air Registration Board, then Airworthiness Requirements Board; became (1972) Airworthiness Division of CAA (UK).
2 Air Research Bureau (BRA, ICAO).

arbitrary landing distance See *ALD*.

ARBS 1 Angle/rate bombing system.
2 Airline Representative Board Sweden.

ARC 1 Aeronautical Research Council (UK).
2 JETDS code: piloted aircraft, radio, communications.
3 Ames Research Center (NASA).

4 Area reprogramming capability (EW).
5 Arc, position of compass rose on EFIS.

arc Ground track of aircraft flying constant DME distance from navaid.

Arcads Armament control and delivery system.

Arcan Aeronautical Radio of Canada.

Archie Colloq, anti-aircraft gunfire, 1915–18 (allegedly from "Archie! Certainly not", music-hall song; archaic).

arc-minute, arc-second See *angular measure*.

ARCS Acquisition radar and control system.

Arctic air mass Major class of air mass most highly developed in winter over ice and snow, although surface temperature may be higher than that for Polar masses. Similar masses formed at either pole.

Arctic minimum Densest of standard model atmospheres assumed in aircraft performance calculation.

Arctic smoke Surface fog essentially caused by very cold air drifting across warmer water.

ARCTS Automated radar-controlled terminal system.

ARDC Air Research and Development Command (USAF, defunct).

ARDC model atmosphere Devised by ARDC, published 1956 (see *model atmosphere*).

ARE Airborne radar extension (surveillance C-130).

area SI unit of plane area is square metre (m^2); to convert from ft^2 multiply by 0·092903; from hectares by 10^4; from sq yd by 0·836127.

area, aerospace surfaces See *gross wing* *, *net wing* *, *disc* *, *equivalent flat-plate* *, *control-surface* *.

area bombing Bombing in which target occupies large area, such as built-up area of city, with aiming point loosely defined near centre (when expression was current, WW2, marked at night by TIs or TMs).

area defence system In general, anti-aircraft or AAW system capable of providing effective defence over large area (dispersed battlefleet, task force, ground battlefield or large tract of country containing several cities) rather than point target.

area-denial munition Explosive device, usually dispensing cluster bombs each with time-delay fuze, to deny area to enemy ground forces.

area-increasing flap Wing flap which in initial part of travel moves almost directly rearwards to increase wing chord, without significant angular movement.

area navigation, R-nav, RNAV Navaid that permits aircraft operations on any desired course within coverage of station-referenced navigation signals or within limits of self-contained system capability (FAA); thus, does not constrain aircraft to preset pathways.

area-navigation route Established R-nav route, predefined route segment, arrival or departure route (including RNAV SIDs and STARs). Route, based on existing high-altitude or low-altitude VOR/DME coverage, which has been designated by Administrator and published (FAA).

area ratio 1 In rocket thrust chamber, usually ratio of idealized cross-section area at nozzle to minimum cross-section area at throat; also called expansion ratio. In general, chambers designed to expand products of combustion into atmosphere have ** 10:1 to 25:1; those for use in upper atmosphere may exceed 50:1; SSME for Space Shuttle has ** 157:1.
2 For a wing, S/b^2; area divided by span squared, reciprocal of aspect ratio.

area rule Formulated by Richard T. Whitcomb at NACA in 1953. For minimum transonic drag at zero lift aircraft should be so shaped that nose-to-tail plot of gross cross-section areas should approximate to that of ideal body for chosen flight Mach number. Thus, addition of wing should be compensated for by reduction in section of body (which gave some early area-ruled aircraft "wasp waists", which are generally undesirable). Obviously, streamlines cannot be sharply deflected; it is not possible to have perfect area-ruling both with and without bulky external stores. In 1954–55 rules extended to Mach 2 by plotting cross-section area distributions on sloping axes approximately aligned with Mach angle.

area sterilization Seeding part of sky with chaff of such extent and density that radar operation is impossible.

Aresa Association des Radio-Electroniciens de la Sécurité Aérienne (F).

Aresti International procedures governing competitive aerobatics to set formula stipulating competing aircraft, set and free manoeuvres, judging and marking.

ARF Air Reserve Forces; suffix PDS adds personnel data systems (US).

ARFA, Arfa Allied Radio-Frequency Agency (NATO).

ARFAC Australian Royal Federation of Aero Clubs.

ARFC Aerospace Reconstruction Finance Corporation (US Government).

ARFOR Area forecast (Int Met Figure Code, ICAO).

ARGMA Army Rocket and Guided Missile Agency (USA, now Miradcom).

argon Most widespread of noble or inert gases naturally present in atmosphere; typical sea-level proportion about 0·934%. Used in some fluorescent lamps and filament bulbs and as major constituent of breathing mixtures, especially at above-normal pressures.

Argos National space-based navigation system (F).

argument Angle or arc (astronomy). Thus * of

perigee, angular arc traversed from ascending node to perigee as seen by observer at near focus, measured in orbital plane in satellite's direction of travel.

ARH 1 Active radar homing.
2 Anti-radar homing.
3 Anti-radiation homing.

ARI 1 Aileron/rudder interconnect.
2 Airborne radio installation (UK).
3 Airpower Research Institute (US).

ARIA Apollo (later Advanced) range instrumentation aircraft.

Arinc Aeronautical Radio Inc, with subsidiary Arinc Research. Non-profit research organization responsible for aeronautical radio standards and widespread ground aids, esp communications, across Pacific and in other regions.

Arinc (ARINC) 429 Standard digital data highway for civil aircraft.

ARIP Air refuelling initial point.

arithmetic unit Heart of typical digital computer; portion of central processor where arithmetical and logic functions are performed; invariably contains accumulator(s), shift and sequencing circuitry and various registers.

ARJA Association Suisse Romande des Journalistes Aéro & Astronautique (Switz).

ARJS Airborne radar jamming system.

ARL 1 Aeronautical Research Laboratories (Australia).
2 Aerospace research laboratories (US, OAR).

ARM Anti-radiation missile.

arm 1 To prepare explosive or pyrotechnic device so that it will operate when triggered.
2 Horizontal distance from aircraft or missile reference datum to c.g. of particular part of it.

armament Carried on combat aircraft specifically to cause injury, by direct action, to hostile forces. Excludes radars, laser rangers (unless they cause optical injury by chance), illumination devices, detection or tracking devices, defoliating sprays, smokescreen generators and, unless filled with napalm, drop tanks.

armature reaction In d.c. generator or alternator, distortion of main field by armature current, factor fundamental to machine design, and speed/voltage regulation.

ARM/CM Anti-radiation missile countermeasures.

armed System switched to function upon command; thus, eg, when pneumatic escape chute (slide) is * it extends immediately passenger door is opened.

armed decoy Aerodynamic vehicle launched by penetrating bomber to generate additional target for hostile radars, send out its own countermeasures and, if it finds hostile aerial target, home on it and destroy it.

armed reconnaissance Mission flown with primary purpose of finding and attacking targets of opportunity.

ARMET Area forecast, upper winds and temperatures (ICAO).

arming vanes One name for slipstream-driven windmill used in some aerial bombs, mines and other stores to unscrew safety device or in some other way arm device as it falls.

armour Typical materials used in airborne armour include thick light alloys, titanium alloys, boron carbide and several filament-reinforced composites.

armourer Trade for military ground crew specializing in armanent.

arm restraint In some types of ejection seat, automatic straps or arms energized during firing sequence to hold occupant's arms securely against aerodynamic forces until he is released from seat.

ARMTS Advanced radar maintenance training set.

Arnold Engineering Development Center, AEDC Large USAF installation in Tennessee charged with aerodynamic development, especially of air-breathing propulsion.

ARO 1 Air Traffic Services reporting office.
2 Aspheric reflective optics.
3 Airport reservation office, for arranging GA traffic slots.

AROG Auto roll-out guidance (after blind Cat 3b landing).

aromatic Hydrocarbon petrol (gasoline) fuels containing, in addition to straight-chain paraffins (kerosenes), various linked or ring-form compounds such as toluenes, benzines and xylines. Some cause rapid degradation of natural or synthetic rubbers.

ARP 1 Air report (written).
2 Aero-rifle platoon (infantry section of Air Cavalry).
3 Air raid precautions (UK, WW2).
4 Aerospace recommended practice.
5 Aerodrome reference point (ICAO).
6 Aluminium-reinforced polyimide.
7 Attack reference point (appears in practice to mean IP) (US).

ARPA Advanced Research Projects Agency (US DoD, now Darpa).

ARPC Air Reserve Personnel Center (USAF).

ARPS 1 USAF Aerospace Research Pilots' School, Edwards AFB, Calif.
2 Advanced radar processing system (AEW).

ARPTT Air-refuelling part-task trainer (USAF).

ARQ Automatic error correction (ICAO).

ARR 1 Arrival message.
2 Airborne radio relay.
3 Air-refuelling receiver.

array Transmitting or receiving aerial (antenna) system made up of two or more (often 20 or more) normally identical aerials positioned to give enormously multiplied gain in desired direction.

arrested landing Normal fixed-wing landing on

aircraft carrier, engaging arrester cable at high speed.

arrested-propeller system In aircraft with free-turbine turboprop engines, system for bringing one or more propellers to rest and holding them stationary while gas generator continues to run. Should not cause turbine overheat condition; speeds up turn-round and sustains on-board power without causing danger to passengers or others near aircraft.

arresting barrier Runway barrier.

arresting gear, arrester gear Fixed to aeroplane landing area to halt arriving aircraft within specified distance. Many systems qualified for use on aircraft carriers, rough battlefield airstrips (in this case, mainly by light STOL machines) and major military runways. In nearly all cases involves one or more transverse cables traversed by hook on arriving aircraft. Kinetic energy of aircraft dissipated by cable pulling pistons through hydraulic cylinders, driving fan through step-up gears or towing heavy free chains.

arresting hook, arrester hook Strong hook hinged to some land-based and all carrier-based combat aeroplanes for engagement of arresting gear; usually released by pilot from flight position to free-fall or be hinged under power to engage position.

arresting unit Energy-absorbing device on one, or usually both, ends of arrester wire.

ARRGp Aerospace Rescue and Recovery Group (USAF).

arrival 1 In flight planning, calculated time when destination should be reached (see *ETA*); may be determined by plotting straight line from last waypoint to overhead destination, but professional pilots refine this to take account of approach procedures.
2 Inbound unit of traffic (ie one aircraft approaching destination airfield).
3 Colloq, derogatory description of a particular landing.

arrival runway At major airfield, runway currently being used only by arrivals (2).

arrival time Time at which inbound aircraft touches down (BS, FAA); airlines sometimes use different definitions, esp time at which first door opened.

ARROW Aircraft routing right of way (not spoken as word) (US).

arrow engine PE (4) having three, or multiples of three, cylinders arranged with one (or one row) vertical and others equally inclined on either side. Also called broad-arrow or W (if inverted, M or inverted-arrow).

arrow stability Weathercock stability stemming from simple distribution of mass and side areas.

arrow wing Markedly swept wing; in his Wright Brothers lecture in 1946 von Kármán used ** exclusively, and "swept wing" did not become universal until 1948.

ARRS Aerospace Rescue and Recovery Service (USAF).

ARS 1 Special air report (written).
2 Atmosphere revitalization subsystem.
3 Auto-relight system
4 Attack radar set (USAF).

ARSR Air route surveillance radar.

ART Airborne radar (or, US, Air Reserve) technician.

Artads Army tactical data system (USA).

ARTCC Air route traffic control center (FAA).

ARTCS Advanced radar traffic control system (FAA).

Arthur An AFCS (F, colloq).

ARTI, Arti Advanced rotorcraft technology integration.

articulated blade In rotorplane, rotor blade connected to hub through one or more hinges or pivots.

articulated rod In radial PE (4), any connecting rod pivoted to piston at one end and master rod at other.

artificial ageing Ageing of alloy at other than room temperature, esp at elevated temperature.

artificial feel In aircraft control system, esp AFCS, forces generated within system and fed to cockpit controls to oppose pilot demand. In fully powered system there would otherwise be no feedback and no "feel" of how hard any surface was working. Simulates ideal response while giving true picture of surface moments insofar as response curve of each surface and their harmonization are concerned. System invariably strongly influenced by dynamic pressure q, generates force for each surface according to optimized law (not necessarily same for all axes) and prevents pilot from damaging aircraft by primary control (but rarely takes into account rapid trimmer movements).

artificial gravity Simulated gravitational effects (but not field) in space environment. Obvious method involves rotation about axis, which introduces Coriolis forces.

artificial horizon 1 Primary cockpit flight instrument which, often in addition to other functions, indicates aircraft attitude with respect to horizon ahead.
2 Simulation of Earth horizon (planet's limb) for use as uniform and accurate sensing reference in near-Earth spaceflight (Orbital Scanner is programme generating data for this ideal).

artificial satellite Man-made satellite of planetary body.

ARTS 1 Automated radar terminal system (FAA).
2 Automated remote tracking station.
3 All-round thermal surveillance.

ARU 1 Attitude retention unit (helicopter height-hold).
2 Auxiliary readout unit (ECM).

ARV Air recreational vehicle.

AS 1 Aerospace Standard.
2 Air station, headquarters or location of units other than those using aircraft.
3 Anti-submarine (role prefix).
4 Also A/S, anti-skid.

As 1 Structural aspect ratio.
2 Alto-stratus.

A_s Web average burning surface area (solid rocket).

ASA 1 Anti-static additive (ASA–3) to jet fuel.
2 American Standards Association.
3 Army Security Agency (US).

ASALM Advanced strategic air-launched missile.

ASAP Advanced survival avionics program (USAF).

ASARS Advanced synthetic-aperture radar system.

Asat, ASAT, A-Sat Anti-satellite.

Asata Asociación de Aviadores de Trabajos Aéreos (Spain).

ASB 1 Air Safety Board (UK).
2 Alert service bulletin.

asbestos Fibrous minerals widely used for their refractory qualities, esp as fire insulation; specially prepared * fibres used as structural reinforcement in aerospace composite materials, eg Durestos.

ASC 1 Ascend, ascending (ICAO).
2 Air Support Command (RAF, defunct).
3 Aviation Statistics Center (Canada).
4 American Society for Cybernetics.

Ascas Automated security-clearance approval system.

ASCB Avionics synchronized control bus, a commercial counterpart to MIL 1553 (Sperry).

ASCC Air Standards Co-ordinating Committee (US, UK, Canada, Australia, NZ).

ASCE American Society of Civil Engineers.

ascending node Point at which body crosses to north side of ecliptic.

ascent Rise of spacecraft from planetary body.

ASCII American standard code for information interchange (7-bit, plus 8th bit for parity in serial transmission).

ASCL Advanced sonobuoy communication link.

ASCM Anti-ship cruise missile.

ASCU Armament station control unit.

ASD 1 Accelerate/stop distance.
2 Aeronautical Systems Division (AFSC, USAF).
3 Average sortie duration.

ASDA, ASD$_a$ 1 Accelerate/stop distance available.
2 Association Suisse de Droit Aérien (now ASDAS) (Switz).

ASDAS Association Suisse de Droit Aérien et Spatial.

ASDE Airport surface detection equipment, airfield surveillance radar.

asdic Sonar, from "Anti-Submarine Detection Investigation Committee"; UK term originally applied to high-energy sound systems carried in surface vessels.

ASDR Avionic systems demonstrator rig.

ASDS Aircraft-sound description system.

ASE 1 Airborne support equipment.
2 Amalgamated Society of Engineers (US).
3 Autostabilization equipment.
4 Allowable steering error (HUD).
5 Auto slat extension (DC-10).
6 Aircraft survivability equipment.

Asecna, ASECNA Agence pour la Sécurité de la Navigation Aérienne (F).

ASEE American Society for Engineering Education.

ASEG All-Services Evaluation Group (NATC, USN).

ASF 1 Aeromedical staging facilities.
2 Aircraft servicing flight (RAF).

ASG 1 Airborne system, gun (US DoD hardware code).
2 JETDS code: piloted aircraft, special/combination, fire control.
3 Air Safety Group (UK).

ASH Advanced support helicopter (USA).

ASI 1 Airspeed indicator.
2 The Aeronautical Society of India.
3 Air system interrogator.
4 Augmented spark igniter.
5 Air Staff Instruction (usually plural, ASIs).
6 Aviation Safety Institute (US).

ASIA Association Suisse de l'Industrie Aéronautique.

ASID American Society for Information Display.

ASIP Aircraft structural integrity program (USAF).

ASIR Airspeed indicator reading (see *airspeed*).

ASIT Adaptable surface-interface terminal.

ASJ Automatic search jammer.

ASK Available seat-km.

ASL Above sea level.

ASLA Air Services Licensing Authority (NZ).

ASLAR, Aslar Aircraft surge launch and recovery.

ASLR Air/surface laser ranger.

ASM 1 Air-to-surface missile; invariably guided, and usually with propulsion, launched from air against land or sea-surface targets.
2 Available seat-miles (often a.s.m.).
3 Advanced systems monitor (cockpit).
4 American Society for Metals.

ASMA Aerospace Medical Association.

ASMD Anti-ship missile defence (defense) (USN).

ASME The American Society of Mechanical Engineers.

ASMI Airfield surface movement indicator (see *airfield surveillance radar*).

ASMK Advanced serpentine manoeuvrability kit.

ASMP Air/sol moyenne portée (F).

ASMR Advanced short/medium range.

ASMS Advanced strategic missile systems (previously Abres).

ASNT American Society for Non-destructive Testing.

ASO Air support operations (C adds centre).

ASP 1 Armament status panel.
2 Aircraft servicing platform.
3 Automated small-batch production.
4 Audio selector panel.
5 Airbase survivability program (US).

ASPA Asociación Sindical de Pilotos Aviadores (Mexico).

aspect change Changing appearance or signature of reflective target as seen by radar, caused by attitude changes.

aspect ratio General measure of slenderness of aerofoil in plan. For constant-section rectangular surface, numerical ratio of span divided by chord, discounting effects due to presence of body or other parts of aircraft. For most wings ** A defined as b^2/S, where b is span measured from tip to tip perpendicular to longitudinal axis (slew or VG wing as nearly as possible transverse) and S is gross area. Structural ** A_s generally defined as $b^2 \sec^2 \Lambda$ where Λ is $\frac{1}{4}$-chord sweep angle. Effective or equivalent ** increased by fitting end-plates (eg horizontal tail surface on top of fin), but there is no universally applicable formula. Generally, faster aircraft have lower **; shorter field length demands higher **. Sailplanes and man-powered aircraft need extremely high **, from 20 to 40.

ASPJ Advanced (or airborne) self-protection jamming.

ASPR Armed Services Procurement Regulations (US).

Aspro Airborne associative processor.

ASQC American Society for Quality Control.

ASR 1 Air/sea rescue.
2 Airport surveillance radar (FAA).
3 Approach surveillance radar (DTI UK).
4 Airfield surveillance radar.
5 Altimeter setting region.
6 Air Staff Requirement (UK).

ASRAAM, Asraam Advanced short-range AAM.

ASRC Alabama Space and Rocket Center, Huntsville.

ASRP Aviation Safety Reporting Program (FAA/NASA).

ASRS Aviation safety reporting system.

ASRT Air support radar team.

ASS 1 Attitude-sensing system.
2 Airlock support subsystem.
3 Air Signallers' School (RAF).
4 Anti-shelter submunition.

ASSA Aeronautical Society of South Africa, more usually in Afrikaans: LKV.

assault aircraft Aeroplanes and/or helicopters which convey assault troops to their objective and provide for their resupply.

assembly 1 Completed subsystem, portion of airframe or other part of larger whole which itself is assembled from smaller pieces.
2 Process of putting together parts of functioning item of equipment, engine, subsystem, portion of airframe or other aerospace hardware (other than complete aircraft, buildings, docks, large launch complexes and similar major structures, for which preferred term is *erection*).

assembly drawing Engineering drawing giving no information on manufacture but necessary for correct assembly; shows geometric relationships, assembly sequence, necessary tooling (jigging), fits and tolerances, and operations required during assembly.

assembly line Essentially linear arrangement of work stations in manufacturing plant for assembly or erection of finished product or major component (eg wing). Modern aircraft produced in such small numbers optimum erection-shop layout often not linear. High-rate production (eg car engines or radio sets) parts travel on belt or overhead conveyor from station to station (see *transfer machines*).

ASSET Aero-Space Structure Environmental Test; USAF research programme.

Asset Airborne sensor system for evaluation and test (Northrop).

asset Item or group of items from a nation's military inventory, especially weapons.

assisted take-off, ATO Aerodyne take-off with linear acceleration augmented by accelerator, by self-propelled trolley, by rockets attached to aircraft or by other means not forming part of normal flight propulsion. Criterion is use of external force; mere downward slope, giving component of weight in take-off direction, does not qualify.

associated (VOR, Tacan) VOR and Tacan/DME facilities either co-located or situated as closely as possible; subject to maximum aerial (antenna) separation of 100 ft in TMAs when used for approach or other purposes requiring maximum accuracy, or 2,000 ft elsewhere.

associative processor Digital computer processor operating wholly as ancillary to another, usually larger, installation. Normally has no parent computer. Future ATC computer installations may use powerful, large-memory sequential machine to resolve conflicts and exercise overall control, supplemented by ** working in parallel seeking potential conflicts.

ASST 1 Anti-ship surveillance and targeting (helicopter subsystem).
2 Advanced supersonic transport.

assured destruction Concept of measurable inevitable damage inflicted on enemy heartlands for purposes of deterrence and arms limitation.

ASSV Alternate-source select(or) valve.

AST 1 Advanced supersonic technology.

2 Advanced simulation technology.
3 Atmospheric surveillance technology.
4 Air Staff Target (UK).
5 Accelerated service test.

ASTA American Society of Travel Agents.

astable Not having a stable state.

astatic Without specific orientation or direction.

ASTE Association pour le Développement des Sciences et Techniques de l'Environnement (F).

asteroid Minor planet, esp fragments (mostly much less than 50 miles across) orbiting between Mars and Jupiter; thus, small body, such as artificial satellite, in solar orbit.

ASTF Aeropropulsion systems test facility (USAF).

ASTM American Society for Testing and Materials.

ASTO Arab Satellite Telecommunications Organization (Int).

ASTOVL Advanced Stovl.

ASTP Apollo-Soyuz Test Project (US, USSR).

Astra, ASTRA 1 Applications of Space Technology to Requirements of Civil Aviation (ICAO panel).
2 Air staff training programme.
3 Association in Scotland to Research into Astronautics.

Astrap, ASTRAP Panel on Application of Space Techniques Relating to Aviation (ICAO).

astrionics Space electronics.

Astro Air support to regular operations (police helicopters).

astrobiology Science of possible life on planets other than Earth, or elsewhere in space (note, in USSR means "space medicine").

astrocompass Non-magnetic instrument, gives direction of true North relative to celestial body which must emit light and be of known direction (see *astronavigation*).

astrodome Optically transparent dome in roof of large aircraft 1935–50 through which navigator could take astro fixes using sextant.

astro fix Fix obtained by sighting two or more stars of known direction using sextant or astrocompass.

astrogation Navigation in space, suggest colloq.

astro-inertial Navigation by means of inertial system updated or corrected by astro fixes.

astronaut One who navigates in space; ie who travels in space. Specif, one selected for space flight by NASA.

astronautics Science of study, design, construction and operation of spacecraft.

astronavigation 1 Navigation of aircraft or spacecraft by measuring declination, right ascension and/or other angular positions of stars and other celestial bodies whose location on celestial sphere is known.
2 Navigation of spacecraft by any means (usage ambiguous).

astronics Astrionics.

astronomical twilight Period between day and night when Sun's centre between 12° and 18° below sea-level horizon (see *civil twilight*, *nautical twilight*).

Astronomical Unit, AU, A.U. Unit of linear distance based on mean distance between Earth and Sun; accepted value is 149,598,500 km.

astronomy Science of celestial bodies other than Earth. Not included in this definition are celestial phenomena, such as polarization of stellar light and other measures concerned more with radiation than with "bodies". Thus, subdivisions radar *, X-ray *, IR *, UV * etc.

astrophysics Physics of observable universe, esp states of matter and energy generation and transfer.

astro-tracker Automatic sextant capable of searching celestial sphere for particular luminous body, identifying it and determining orientation in terms useful for navigation, and of repeating sequence with same and at least one other celestial body. Corrects and updates INS in long-range aircraft.

ASTS Association Suisse pour les Techniques Spatiales (Switz).

ASU 1 Aeromedical staging unit.
2 Altitude (or, confusingly, attitude) sensing unit.
3 Approval for service use.
4 Aircraft storage unit (UK).

Asupt Advanced simulator for undergraduate pilot training.

asymmetric flight Flight by aerodyne in sustained grossly asymmetric condition of lift, weight, thrust or drag, esp flight by multi-engined aircraft in which at least one engine at substantial distance from axis of symmetry is inoperative.

asymmetric loading Flight by aircraft, esp aerodyne, in which c.g. is located at substantial distance from vertical line through centre of lift with aircraft in level attitude and trimmed for normal horizontal flight (eg strike aircraft unable to release one of two heavy stores carried on outer-wing pylons).

ASV Air-to-surface-vessel. WW2 airborne search radar. Incorrectly rendered as anti-surface vessel.

ASVEH Air-surveillance vehicle.

ASW 1 Anti-submarine warfare; major category of research, weapon development and operation.
2 Aft-swept wing.
3 Confusingly, also used in US for anti-ship warfare, and in UK (WW2) for air/sea warfare.

ASWE Admiralty Surface Weapons Establishment (Portsdown, UK).

asymptote Limiting position of tangent to curve, where lines meet at infinity. Thus, asymptotic, where slope of plotted curve becomes parallel to either x or y axis.

asynchronous Not synchronized, not in frequency or phase.

asynchronous computer Electronic computer, usually digital, in which operations do not proceed according to timing clock but are signalled to start by completion of preceding operation.
ASZ Air surface zone (NATO, USAF).
AT 1 Advanced trainer (USAAF category).
2 Anti-tank.
3 Autogenic training.
4 Autothrottle.
ATA 1 Air Transport Association of America.
2 Air Transport Auxiliary (UK, WW2).
3 Actual time of arrival.
ata Atmospheres pressure; 1 ata (written atm in UK) is 101·325 kPa (kN m^{-2}); formerly most common unit of PE (4) manifold pressure or boost on European aircraft not built for export.
ATAB Air transport allocations board; joint agency in theatre of operations which assigns priorities to loads.
ATAC 1 Air Transport Association of Canada.
2 Applied-technology advanced computer (airborne EW).
3 Air-transportable acoustic communication (expendable buoy).
ATAF 1 Allied Tactical Air Force (NATO).
2 Association Internationale des Transporteurs Aériens (Int, acronym from previous title).
ATAFCS Airborne target acquisition and fire-control system.
ATAL, Atal Automatic test application language.
ATAR 1 Association des Transporteurs Aériens Régionaux.
2 Air-to-air recovery.
ATARS, Atars 1 Automatic traffic advisory and resolution service.
2 Advanced tactical air-reconnaissance system.
ATAS 1 Air Traffic Advisory Service (UK); service to provide separation between known aircraft operating under IFR in certain areas along certain routes (DTI); also Atars, * plus "Resolution".
2 Advanced target-acquisition sensor.
ATAWS, Ataws Advanced tactical air-warfare system.
ATB 1 Air transport bureau (ICAO).
2 Advanced-technology bomber.
ATBM Anti-tactical ballistic missile.
ATC 1 Air traffic control (invariably used in this sense in this book).
2 Air Training Command (USAF).
3 Air Training Corps (UK).
4 Approved type certificate (US).
5 Advanced-technology component (US DoD).
6 Aerospace Technical Council (AIAA, US).
7 Air Transport Command (USAAF).
8 Air transport conference (travel agencies).
ATCA 1 Air Traffic Conference of America.
2 Air Traffic Control Association (US).
3 Advanced tanker/cargo aircraft.
4 Allied Tactical Communications Agency (NATO).
5 Air traffic control assistant (UK).
ATCAC Air Traffic Control Advisory Committee (US Government).
ATCAP, Atcap Air Traffic Control Automation Panel (ICAO).
ATCC Air Traffic Control Centre (US, UK).
ATC clearance Authorization by ATC for purpose of preventing collision between known aircraft for aircraft to proceed under specified conditions in controlled airspace (FAA).
ATCEU Air Traffic Control Evaluation Unit (UK).
ATCO Air traffic control officer (US, UK).
ATCR Air Training Command Regulation (USAF).
ATCRBS Air Traffic Control Radar Beacon System (FAA); see *secondary radar*.
ATCRU Air traffic control radar unit (UK).
ATCS 1 Air Traffic Control Service (US, UK).
2 Active thermal control subsystem.
ATCT Air traffic control tower (FAA).
ATD 1 Actual time of departure.
2 Airline-transmitted disease.
3 Aviation technical division (USSR).
ATDMA Advanced time-division multiple access.
ATDS Air-turbine drive system.
ATE Automatic test equipment.
ATEC Automatic test equipment complex.
ATECMA Agrupación Técnica Española de Constructores de Material Aerospacial.
ATEGG, Ategg Advanced turbine engine gas generator.
Atepsa Asociación Técnicos y Empleados de Protección e Securidad a la Aeronavegación (Arg).
ATF 1 Advanced tactical fighter (programme).
2 Actual time of fall.
3 Aviation turbine fuel.
Atfero Atlantic Ferry Organization (UK, WW2).
ATGW Anti-tank guided weapon.
ATH 1 Autonomous terminal homing.
2 Automatic target handoff.
athodyd Aero-thermodynamic duct (see *ramjet*).
ATHS Airborne target handoff system.
Atigs Advanced tactical inertial guidance system.
ATILO Air technical intelligence liaison office.
ATIS Automatic terminal information service (ICAO, FAA); continuous broadcast of recorded non-control information in selected high-activity terminal areas (to improve controller effectiveness and relieve congestion by automating repetitive transmission of routine information).
ATITA Air Transport Industry Training Association (UK).
ATITB Aviation and Travel Industry Training Board (NZ).
ATK Aviation turbine kerosene.
ATk 1 Anti-tank (DoD).
2 Available tonne-kilometres.
ATLA Air Transport Licensing Authority (HK).

Atlas Abbreviated test language for avionic systems.
ATLB Air Transport Licensing Board (UK).
Atlis Automatic tracking laser illumination system.
ATM 1 Air-turbine motor.
2 Air transport movement.
atm Atmospheres pressure (see *ata*).
ATMA Association Technique Maritime et Aéronautique (F).
Atmos Ammunition, toxic material open space.
atmosphere 1 Gaseous envelope surrounding Earth, subdivided into layers (see *atmospheric regions*, *model atmosphere*).
2 Gaseous or vaporous envelope surrounding other planets and celestial bodies.
3 Theoretical model atmosphere providing standard basis for performance and other calculation.
4 Any of group of units of pressure all approximately equal to pressure of atmosphere of Earth at sea level. Most important is Standard * (abb ata on European continent, atm in UK) equal to 1013·25 mb = 1·01325 bars or hectopièze = 14·6959 lbf in^{-2} = 101·325 kN m^{-2} = 761·848 mm (29·994 in) Hg at 16·6°C. Second is Metric * (also ata) equal to 0·98642 Standard * and defined as 0·98177 bars (981·117 mb, ie acceleration due to 1g) or 14·223 lbf in^{-2}. Third is Technical * (at), usually identical with Metric. Fourth is bar (b), 1000 mb and 14·5038 lbf in^{-2} (see *pressure*).
atmospheric absorption Absorption of EM radiation due to ionization in atmosphere. Apparent loss of signal or beam power may be much greater, as result of diffraction and dispersion by vapour and particulate matter.
atmospheric braking Use of air drag, esp of upper atmosphere on re-entering spacecraft or RV, converting very high kinetic energy into heat.
atmospheric circulation Gross quasi-permanent wind system of Earth, based on bands between parallels of latitude.
atmospheric constituents See *air*.
atmospheric diffraction Of importance chiefly with sound waves, which can be substantially changed in direction and intensity distribution by changes in air velocity and density. Effect with most EM radiation is small.
atmospheric duct Almost horizontal layer or channel in troposphere apparently defined by values of refractive index within which EM radiation, esp in microwave region, is propagated with abnormal efficiency over abnormally great distances.
atmospheric electric field Intensity of electrostatic field of Earth varies enormously but on fine day may be about 100 V m^{-1} at SL falling to around 5 V m^{-1} at 10 km height. Air/Earth current continuously degrades ***; believed that thunderstorms reinforce it.
atmospheric entry Re-entry, or entry of extra-terrestrial bodies such as meteors.
atmospheric filtering Use of upper zones of ionosphere and mesosphere to filter out ICBM decoys from true warheads, it being supposed that latter will have higher density, better thermal protection and increasingly divergent trajectories, while decoys decelerate more violently, fall behind and burn up.
atmospheric pressure See *atmosphere (4)*, *atmospheric regions*.
atmospheric refraction Bending of EM radiation as it passes through different layers of atmosphere, esp obliquely. Affects radio and radar, esp when directionally beamed; visibly manifest in air over, say, hot roadway in sunshine when objects seen through this air "shimmer"; in astronavigation ** makes apparent altitude of celestial bodies falsely great.
atmospheric regions Layers of Earth's atmosphere differ in different model atmospheres; following notes are based on ISA. Lowest layer, troposphere, extends from SL to about 8 km (26,000 ft) at poles, to 11 km (36,000 ft) in temperate latitudes, and to 16 km (52,000 ft) over tropics. Throughout this region ISA characteristics of temperature, pressure and relative density are precisely plotted. Assumed lapse rate is 6·5°C km^{-1} and at tropopause, taken in ISA to be 11 km (36,090 ft), temperature is –56·5°C. From tropopause stratosphere extends at almost constant temperature but falling pressure to 30 km, stratopause, above which is mesosphere. Here there is reversed lapse rate, temperature reaching peak of about 10°C at 47·35–52·43 km, thereafter falling again to minimum of 180·65°K at mesopause (79·994–90·000 km). Above this is ionosphere, extending to at least 1,000 km, where temperature again rises through 0°C (273°K) at about 112 km and continues to over 1,000°C at 150 km and to peak of about 1,781°C at 700 km. Between 100–150 km lies E (Kennedy-Heaviside) layer; at 200–400 km is F (Appleton) layer, which at night is single band but by day divides into F1 and F2, F2 climbing to 400–500 km on summer day. E and F layers are reflective to suitable EM radiation striking at acute angle. Above ionosphere is open-topped exosphere, from which atmospheric molecules can escape to space and where mean free path varies with direction, being greatest vertically upward. Other ** are based upon composition, electrical properties and other variables.
atmospheric tides Produced by gravitational attraction of Sun and Moon. Latter exerts small influence, equal to equatorial pressure difference of 0·06 mb, but solar ** has 12 h harmonic component (apparently partly thermal) of 1·5 mb in tropics and 0·5 in mid-latitudes.
atmospheric turbulence See *gust*, *CAT*.
ATO Air task order.
ATOA Air Taxi Operators Association (UK).

Atoc(s) Allied tactical operations centre(s) (NATO).

A to F Authority to fly.

ATOL Air travel organizer's (or operator's) licence (UK).

Atoll Assembly/test oriented launch language.

atom bomb Colloq, fission bomb; very loosely, any NW (see *nuclear weapons*).

atomic number Symbol Z, number of protons in atomic nucleus or number of units of positive electronic charge it bears.

atomic weight Mass of atom of element in units each 1/12th that of atom of carbon 12 (refined to 12·01115 on 1961 table). Numerical value for each element is same as atomic mass.

atomizing Continuous conversion of solid or liquid, esp high-pressure jet of liquid, into spray of fine particles. Also called atomization.

Atops Advanced transport operating systems (NASA).

ATOS Automated technical-orders system.

ATP 1 At time, place.
2 Actual track pointer.
3 Authority to proceed.
4 Aviation technical regiment (USSR).

ATPL Airline transport pilot's licence; ALTP licence; /H adds endorsement for helicopters.

ATR 1 Air transport rating.
2 Air Transport Radio, system of standardizing dimensions of airborne radio and other electronic, AFCS and navaid boxes according to system of modules; thus ATR, ½ATR, ¼ATR etc.
3 Analog tape recorder.
4 Attained turn rate.
5 Advanced tactical radar (programme).
6 Armed turnround.

Atran Automatic terrain radar and navigator; cruise-missile guidance (obs).

ATRC Air transport (or traffic) regulation center (US).

Atrel Air-transportable reconnaissance exploitation laboratory (RAF).

Atrif Air Transportation Research International Forum (Int).

ATS 1 Air Traffic Services; thus * route (ICAO).
2 Applications technology satellite, wide research programme.
3 Suomen Avaruustutkimusseura Ry (Finnish astronautical society).
4 Automatic throttle system.
5 Armament training station (UK, WW2).
6 Aircrew training system (C-5, USAF).
7 Automatic test system (or station) for LRU check away from aircraft.

ATSC Air Technical Service Command (USAAF).

AT/SC Autothrottle/speed control.

ATSG Acoustic test signal generator.

ATSU Air traffic service unit.

ATSy Air Transport Security Section (RAF).

ATT Automatic attitude hold (AFCS).

Att Tail-on-tail aerodynamic influence coefficient.

attach To place temporarily in a military unit.

attached shockwave Caused by supersonic body having leading edge or nose sufficiently sharp or pointed not to cause shock to detach and move ahead of it. Critical values of M at which shock will just remain attached; eg for cone of 30° included angle shock will detach below 1·46; for 30° wedge M is 2·55 because wedge exerts larger obstructing effect on airflow. In most supersonic aircraft aim is to keep most shocks attached, especially at engine inlets.

attack aircraft Combat aircraft, usually aeroplane but sometimes helicopter, designed for attacking surface targets of tactical nature; missions include CAS (3) and interdiction.

attack, angle of Angle α between wing chord or other reference axis and local undisturbed airflow direction. There are several ways of measuring this crucial parameter. One is absolute angle of attack. Another is *** for infinite aspect ratio, which assumes two-dimensional flow. Effective *** varies greatly with aspect ratio; modern wings of low aspect ratio have no stall in conventional sense even at $\alpha = 40°$. Some authorities in UK cause confusion by using "angle of incidence", which already has clear meaning unconnected with angle of incident airflow.

attack avionics Navigation and weapon-aiming systems, often integrated into single "fit" for particular attack-aircraft type.

ATTC Automatic take-off thrust control.

attention-getter Prominently positioned caption in cockpit, or esp flight deck, triggered by onboard malfunction or hazardous situation (eg potential mid-air collision) immediately to flash bright amber or red. Less strident caption, warning light or other visual and/or aural circuit also triggered, enabling crew to identify cause and, if possible, take remedial action. Where CWS is fitted first task is to trigger ** while routing appropriate signal to captioned warning panel or other more detailed information.

attenuation Loss of signal strength of EM radiation, esp broadcast through atmosphere, due to geometric spread of energy through volume increasing as cube of distance, loss of energy to Earth, water vapour, air and possibly ionized E and F layers.

attenuation factor Ratio of incident dose or dose rate to that passing through radiation shield.

Attinello flap Blown flap.

ATTITB Air Transport and Travel Industry Training Board (UK).

attitude Most, if not all, aircraft * described by relating to outside reference system three major axes OX/OY/OZ (see *axes*); * of flight relates these mutually perpendicular co-ordinates to relative wind; * with respect to ground relates axes to local horizontal.

attitude control system, ACS Control system to alter or maintain desired flight attitude, esp in satellite or spacecraft to accomplish this purpose in Earth orbit or other space trajectory. Typical *** uses sensing system, referred to Earth's limb, star or other "fixed" point or line, and imparts extremely small turning moments to structure by means of gas jets or small rocket motors. In some cases passive *** used (PACS), or vehicle stabilized by spin about an axis, with portions despun if necessary.

attitude gyro Loosely, gyro instrument designed to indicate attitude of vehicle. Specif, instrument similar to artificial horizon but with 360° freedom in roll and preferably 360° freedom in pitch. Also applicable to conventional horizon with restricted indications of movement in aircraft not intended for aerobatics.

attitude jet 1 Reaction jet imparting control moments to aircraft at low airspeed (see *RCS*). 2 Sometimes applied to small thrusters or attitude motors used for same purpose on spacecraft.

attitude motor Small rocket motor used to control attitude of space vehicle (see *thruster*, *reaction control engine*).

attitude reference symbol Usually an inverted T giving heading and pitch attitude in HUD symbology.

ATTLA Air transportability test-loading agency (US).

atto Prefix, 10^{-18}, symbol a; thus, 1 am (attometre) = 0·000000000000000001 m.

attrition Wastage of hardware in operational service, esp of combat or other military aircraft.

attrition buy Additional increment of production run ordered to make good anticipated attrition over active life of system.

ATUC Air Transport Users Committee (CAA, UK).

ATV Associazione Tecnici di Volo Aviazione Civile (I).

AT-Vasi Abbreviated T-Vasi, ten light units on one side of runway in single 4-unit wing bar plus 6-unit bisecting longitudinal line.

ATVC Ascent thrust-vector control.

ATVS Advanced TV seeker.

A_{TW} Wing-on-tail aerodynamic influence coefficient.

ATWS Active tail-warning system.

ATZ Aerodrome traffic zone.

AU, a.u. Astronomical Unit.

AU Air University (Maxwell AFB, USAF).

audio box, audio control Governs voice communication and broadcast throughout aircraft and by telephone to ground crew.

audio frequency Frequency within range normally heard as sound. Limits vary widely with individuals but normally accepted 15 Hz to 20 kHz, upper limit being depressed with advancing age.

audiometer Instrument for measuring subject's ability to hear speech and tones at different frequencies.

audio oscillator Multi-valve (tube) or multi-transistor stage in superheterodyne receiver serving as local oscillator and amplifier (detector).

audio speed signal Aural indication of vehicle speed, either on board or at ground station; usually has pitch proportional to sensed velocity, indication being qualitative only (see *aural high-speed warning*).

audio warning Loud warning by horn, buzzer, bell or voice tape in headphones or cockpit loudspeaker indicating potential danger. Examples: incorrect speed (usually sensed airspeed) for regime, configuration or other flight condition; potentially dangerous excursion of angle of attack; ground proximity; incorrect configuration (gear up, wings at maximum sweep etc).

Audist Agence Universitaire de Documentation et d'Information Scientifiques et Techniques (F).

audit Comprehensive detailed examination of airframe structure of aircraft in line service, esp after fatigue problems encountered on type.

AUEW Amalgamated Union of Engineering Workers (UK).

augmentation 1 Boosting propulsive thrust by auxiliary device, esp by afterburning.
2 Percentage of thrust added by (1).
3 Increasing inadequate natural flight stability of aerodyne by on-board system driving control surfaces (rarely, ad hoc auxiliary surfaces).
4 Enhancing target signature of radar, optical, IR or other radiation by means of corner reflectors, Luneberg lenses or other * devices.

augmented turbojet, turbofan Engine equipped with afterburning (reheat) to augment thrust, esp at transonic or supersonic flight speed. Turbofan augmentation may be in hot and/or cold flows, latter offering greater density of free oxygen.

augmentor Afterburner for turbofan, usually with burning in hot and cold flows.

augmentor wing General term for STOL aeroplane wing in which engine thrust is directly applied to augment circulation and thus lift. Countless variations, but most fundamental division is into external and internal blowing. Former typically uses engine bleed air (rarely, total efflux) discharged at sonic speed through one or more narrow slits ahead of large double or triple-slotted flaps which, because of blowing, can be depressed to unusually sharp angle. Second method uses either engine bleed air or total efflux to blow through flap system itself (or propulsion engines are distributed across main flap, in some schemes there being as many as 48 small engines). Internal blowing makes flow separation impossible and gives large downwards component of thrust, but is difficult to apply and may severely compromise aircraft in cruise (see *jet* *flap*, *blown flap*, *externally blown flap*, *upper-surface blowing*).

AUM Air-to-underwater missile (US DoD weapon category).

AUR All-up round.

aural acquisition Acquisition of target by IR seeker head as confirmed by pilot's headset.

aural high-speed warning System triggered by sensed flight speed, usually presented as EAS, significantly above allowable maximum. In transport aircraft typically triggered 10 kt above V_{mo} and 0·01 Mach above M_{mo}. Not usually made to do more than warn crew.

aural null Condition of silence between large regions where sound is heard, eg in early radio DF systems, in some types of beacon passage ("cone of silence") and several ground-test procedures.

aurora Luminescence in upper atmosphere, esp in high latitudes, associated with radiation and/or particles travelling along Earth's magnetic field and at least partly coming from Sun. Exact mechanism not yet elucidated, but 12 classes identified, based on appearance and structure.

AUSA Association of the US Army.

austenitic steel Ferrous alloy with high proportions of alloying elements and with microstructure transformed by heat treatment to consist mainly of solid solution of austenite (iron carbide in iron, face-centred). Used to make highly stressed aerospace parts, such as turbine discs.

AUTEC Atlantic Undersea Test and Evaluation Center (USN with UK help, in British territory).

authority 1 Organization empowered to pronounce hardware properly designed, constructed and maintained, and to issue appropriate certificates.
2 Extent to which functioning system is permitted to control itself and associated systems. Greater * demands greater inherent or acquired reliability, typical means of acquiring reliability being to increase redundancy and provide alternative control channels or in some other way provide for failure-survival.

Autocarp, Auto-CARP Automatic computed air release point.

autoclave Pressure chamber which can be heated; oven which can be pressurized. Large * can accept major structural parts of aircraft for adhesive bonding operations.

Autodin Automatic digital network (USAF).

autodyne oscillator Multi-electrode valve or transistor stage of superheterodyne receiver serving as both local oscillator and amplifier or detector (demodulator).

autofeathering System for automatically and swiftly feathering propeller when engine fails to drive it; usually triggered by NTS.

autogenic training Psycho-physiological technique carried out by subject according to prescription by qualified therapist, to reduce stress.

Autogiro Registered name of Juan de la Cierva autogyros, 1924–45.

autogyro Rotorplane in which propulsion is effected by horizontal-thrust system, eg propeller, and lift by rotor free to spin under action of air flowing through disc from below to above (ie, autorotating). Some can achieve VTOL by driving rotor in vertical phases of flight, but true * is STOL.

auto-hover Automatic hover, usually at low altitude, by helicopter or other VTOL aircraft, using radio altimeter and AFCS.

auto-igniting propellant Hypergolic.

auto-ignition 1 Of gas turbine engine, auxiliary system which senses angle of attack or other aerodynamic parameter and switches on igniter circuits before engine is fed grossly disturbed airflow which would otherwise pose combustion-extinguishing hazard. In some aircraft flight in rough air with full flap can cause intermittent or total flame-extinction, and there are other flight conditions (eg violent manoeuvre) when turbulent flow across intake triggers *, indicated by cockpit lights.
2 Of combustible material, spontaneous combustion.

auto-ignition temperature At which auto-igniting materials spontaneously combust in air; design factor in some rocket engines and gas generators.

Autoland 1 Loosely, AFCS capable of landing aeroplane hands-off and qualified to do so in total absence of pilot visual cues.
2 Specific systems developed in UK by Smiths Industries and GEC-Marconi-Elliott.

Autolycus ASW detection system operating by sensing minute atmospheric concentrations of material likely to have come from submarine running submerged (UK).

Automap Trade name for map and/or track guide projection system, esp for single-seat combat aircraft.

automatic boost control, ABC On PE (4) servo system which senses induction pressure and so governs boost system that permissible limits of boost pressure cannot be exceeded. These were among first airborne closed-loop feedback systems.

automatic direction finder See *ADF*.

automatic extension gear Landing gear which extends by itself, typically by sensing airspeed and engine rpm, should pilot omit to select DOWN.

automatic feathering Autofeathering.

automatic flagman Electronic installation in ag-aircraft to provide precise track guidance on each run in conjunction with ground beacons.

automatic frequency control 1 Radio receiver which self-compensates for small variations in received signal or local oscillator.
2 By different method, self-governing of a time

base.

automatic gain control See *AGC*.

automatic landing Safe, precisely repeatable landing of advanced aeroplane, helicopter or other aerodyne in visibility so restricted that external visual cues are of no assistance to pilot. Basis of present systems is high-precision ILS, approach coupler, triplexed or quad AFCS, autothrottle, autoflare and ground guidance after touchdown.

automatic manoeuvre device system Automatically schedules high-lift devices, especially on variable-sweep aircraft; usually governed by AOA.

automatic manual reversion Fully powered flight-control system may be so designed that no ordinary pilot could control aircraft manually; if not, *** on one, two or three axes allows pilot to drive surfaces giving control in those axes after malfunction of powered system can no longer be accommodated by failure-survival.

automatic mixture control In PE (4), subsystem which automatically adjusts flow rate of fuel to counteract changes in air density, or which controls intake airflow by restricting carburettor air-intake duct by amount inversely proportional to altitude until wide open at height usually around 15,000 ft.

automatic parachute 1 Parachute pulled from its pack by static line (usual meaning).
2 Parachute opened by barometric device at preset pressure altitudes.

automatic pilot See *autopilot*.

automatic pitch-coarsening Facility built into propeller control system causing it to increase pitch automatically, normally from fine-pitch setting to typical cruise angle, when called for by operating regime.

automatic power reserve See *Auto power reserve*.

automatic RDF See *ADF*.

automatic reverse pitch Facility built into propeller control system causing it to reduce pitch automatically past fine-pitch stop through zero to reverse (braking) position, upon receipt of signal from microswitch triggered when main gears compress shock struts on landing. Very rare for reverse pitch to be obtainable without deliberate selection by pilot, though with *** pilot may select in air, leaving auto system to send operative signal.

automatic riveter Machine for drilling holes through parts to be joined, inserting rivet and closing (heading) it.

automatic roll-out guidance, AROG Steering guidance after automatic landing in blind conditions (ideally extended from runway turnoff to terminal parking).

automatic search jammer ECM intercept receiver and jamming transmitter which searches for and jams all signals having particular signatures or characteristics.

automatic selective feathering 1 Airborne subsystem which, in the event of engine failure in multi-engined aircraft (invariably aeroplane), decides which engine has failed and takes appropriate action to shut down and feather propeller.
2 Similar system which, when pilot presses single feathering button, routes signal automatically to failed engine and propeller.

automatic slat Leading-edge slat pulled open automatically at high angle of attack by aerodynamic load upon it. All slats were originally of this type.

automatic synchronization In multi-engined aircraft (invariably aeroplane), subsystem which electrically locks rpm governors of all engines to common speed.

automatic terminal information service See *ATIS*.

automatic touchdown release Device incorporated in sling system for external cargo carried below helicopter or other VTOL aircraft which releases load as soon as sling tension is released.

automatic tracking Although this could have meaning in system constraining aircraft to follow preset tracks over Earth's surface, universal meaning is property of directionally aimed system to follow moving target through sensing feedback signal from it. Applications found in (1) air-superiority fighter, in which essential to lock-on and track aerial target automatically, by means of radar, IR, optics or other system; and (2) ground tracking station, which may need to follow satellite, aircraft, drone target or other moving body in order to sustain command system, interception system, data-transmission system or other directionally beamed link.

automatic VHF D/F Ground D/F system in which, instead of requiring manual turning of aerial array to find null position, signal from aircraft causes aerial to rotate automatically to this position, direction being at once displayed as radial line on CRT of the equipment. Also called CRT D/F, CRT/DF.

autonomous 1 Of aircraft or other vehicle, not needing GSE.
2 Of an SSR, not co-located.
3 Of airborne equipment, not needing external sensors; not linked to other aircraft systems (though possibly under pilot control), eg * reconnaissance pod.

autonomous vehicle Vehicle, especially unmanned aircraft, which completes mission without external help.

autopilot Airborne electronic system which automatically stabilizes aircraft about its three axes (sometimes, in light aircraft, only two, rudder not being served), restores original flight path following any upset, and, in modern *, preset by pilot or remote radio control to cause aircraft to follow any desired trajectory. In advanced aircraft * is integral portion of AFCS and can be set by dial, pushbutton or other control to capture and hold any chosen airspeed,

Mach, flight level or heading. In advanced combat aircraft * receives signals from sensing and weapon-aiming systems enabling it to fly aircraft along correct trajectories to fire guns or other ordnance at aerial target or lay down unguided bombs on surface target.

autopilot-disconnect Advanced autopilots are automatically disconnected by control overloads generated within aircraft and by certain other disturbances likely to reflect wish of pilot, eg triggering of stall-protection stick-pusher.

Autoplan Portable EDP which digitizes navigation plan and combines output with other CPGS data to provide attack-aircraft pilot with complete nav/ECM/weapon-aiming information.

auto power reserve Special increased-thrust rating available on commercial turbofan engines only in emergency, and triggered automatically by loss of power in other engine in same aircraft.

autopsy Searching examination of crashed aircraft to discover cause, esp to detect fatigue failure.

autorotation 1 Loosely, condition in which airflow past aircraft causes whole aircraft or significant part of it to rotate. Propeller in this context is not significant, though windmilling propeller is autorotating.

2 In helicopter, descent with power off, air flowing in reverse direction upwards through lifting rotor(s), causing it to continue to rotate at approximately cruise rpm. Pilot preserves usual control functions through pedals, cyclic and collective, but cannot grossly alter steep "glide path". Rate of descent may exceed design ROD for landing gear, but is reduced just before ground impact by sudden increase in collective pitch; this increases lift, trading stored rotor kinetic energy for increased aerodynamic reaction by blades, and should result in gentle touchdown.

3 In aeroplane, descent in stalled condition, with general direction of airflow coming from well beyond stalling angle of attack but in grossly asymmetric condition (see *spin*).

4 In aeroplane, descent in unstalled condition under conditions apparently not greatly different from straight and level but with stabilized spiral flight path (see *spiral dive*, *spiral stability*). Distinct case of * which purist might argue is incorrect usage.

5 In helicopter flying training, range of manoeuvres designed to increase confidence and remove fear of power failure; all power-off descents, but differ in whether they are NPR (no power recovery) or terminated at height well above ground by restoring at least partial power. In latter case * terminated either by run-on landing (running *), run-on climb-out or moderate-flare climb-out.

auto-separation Automatic (often barometric) release of occupant from ejection seat.

Autosevocom Automatic secure voice communications (USAF).

Autosyn Trade name for remote-indicating system in which angular position of indicator needle precisely follows rotary sensing device moved by fluid level, mechanical displacement or other parameter which must be remotely measured. Sensor and indicator are essentially synchronous electric motors.

autosynchronization Automatic synchronization.

autothrottle Power control system for main propulsion engines linked electro-mechanically to AFCS and automatic-landing system so that thrust is varied automatically to keep aircraft on glide path and taken off at right point in autoflare; in general * will also call for reverse thrust at full power in conjunction with automatic track guidance (roll-out guidance), though this may be left to discretion of pilot.

autotrim Aircraft trim system automatically adjusted by autopilot or other stabilizing system to alter or maintain aircraft attitude according to pilot demand or changed distribution of weight or aerodynamic load. Usually governs pitch only.

Autovon Automatic voice network (USAF Communications Service).

AUVS Association of Unmanned Vehicle Systems (US).

AUW, a.u.w. All-up weight; actual aggregate weight of particular laden aircraft at moment of weighing. For generalized equivalent, more precise term should be used (MRW, MTOW, etc), but AUW has advantage of being not explicit and thus can be used to mean "total weight of aircraft, whatever that happens to be". Should never be used to mean MRW or MTOW. If latter are not known, preferred term meaning "maximum allowable weight" is gross weight.

auxiliary bus Secondary electrical bus serving one or more devices and often maintained at voltage different from that of main bus.

auxiliary fin Generally, small additional fixed fin carried, not necessarily to enhance directional stability, well outboard on tailplane

auxiliary fluid ignition In rocket engine, use of limited supply of hypergolic fluid(s) to initiate combustion of main propellants.

auxiliary parachute Pilot parachute.

auxiliary power unit, APU Airborne power-generation system other than propulsion or lift engines, carried to generate power for airborne system (electrics, hydraulics, air-conditioning, avionics, pressurization, main-engine starting, etc). In general, term restricted to plant deriving energy from on-board source and supplying constant-speed shaft power plus air bleed. This would exclude a RAT or primitive windmill-driven generator. Some *** can provide propulsive thrust in emergency (see *MEPU*).

auxiliary rigging lines Branching from main parachute rigging lines to distribute load more evenly around canopy.

auxiliary tank Fuel tank additional to main supply, esp that can readily be removed from aircraft (see *reserve tank*, *external tank*, *drop tank*).

AV 1 Air vehicle.
2 Audio-visual.

AVA, automatic voice advice In terminal ATC, computer-generated voice message broadcast to two or more aircraft warning of potential conflict. Intended primarily for VFR traffic and for all traffic not under immediate control, and intended to relieve controller of function that appears to be safely automated.

AVADS Autotrack Vulcan air-defense system (USA).

availability 1 Proportion of time aircraft is serviceable and ready for use, expressed as decimal fraction over period or as number of hours per day or days per month.
2 Period which must elapse between purchase of aircraft, usually second-hand, and handover to customer.

avalanche 1 Any of several processes involving ions or electrons in which collisions generate fresh ions or electrons which in turn go on to have their own collisions. In * tube electrons or other charged particles are accelerated in electric field to generate additional charged particles through collisions with neutral gas atoms or molecules. In semiconductor devices * effect occurs when potential in excess of critical voltage is applied across p-n junction, enormously multiplying liberation of charge carriers.
2 Aerobatic manoeuvre devised by Ranald Porteous involving rapid rotation about all axes in combined stall turn and flick roll.

AVC 1 Automatic volume control (see *AGC*).
2 Automatic variable camber.
3 Attitude, velocity and control subsystem.

Avcat Tailored kerosene military gas-turbine fuel, also call JP-5. Low freeze and flash point specified (60°C).

AVCS Advanced vidicon camera system, carried by ESSA and other Earth satellites.

AVD Atmospheric vehicle detection.

A-VDV Aviation of airborne forces (USSR).

AVG 1 Average (ICAO).
2 American Volunteer Group (China, WW2).
3 Aircraft escort vessel (later ACV, then CVE).

Avgard Additive to JP-1 and other jet fuels to produce anti-misting kerosene (ICI trade name).

Avgas Aviation gasoline, range of PE (4) petrols (see *gasoline*).

AVHRR Advanced very-high resolution radiometer.

aviation Operation of aerodynes (but only pedant would insist on "aerostation" for aerostat).

aviator Operator of an aerodyne, esp pilot. As term archaic, difficult to define; nearest modern equivalent is aircrew member.

aviatrix Female aviator.

Aviatrust Original (1923) central aviation industry management organization (USSR).

Aviavnito Aviation department of Vnito, all-union amateur scientific/technical research organization (USSR).

aviette Man-powered aircraft.

avigation Aerial navigation (suggest undesirable word). Hence also avigator.

Avim Aviation intermediate-level maintenance (USAF).

avionics Aeronautical (not aviation) electronics, not necessarily by definition restricted to aerodynes. Term implies equipment intended for use in air; purely ground-based equipment could be argued to be outside category.

AVIP Avionics integrity program (USAF).

AVLF Airborne very low frequency.

AVM Airborne vibration monitor.

AV-MF Naval aviation (USSR).

AVNDTA Aviation development test activity (USA).

AVO Avoid verbal orders.

avoid curve Plot of height/TAS below which a helicopter may not survive total engine failure; important in anti-armour operations.

AvP.970 Manual of Design Requirements for Aircraft (UK).

Avpin Aviation specification iso-propyl nitrate (monofuel).

AvPOL Aviation petrol (or petroleum), oil and lubricant.

AVR Additional validation requirements.

Avrada Aviation R&D Activity (USA).

AVT Automatic video tracker.

Avtag Aviation turbine fuel, wide-cut (see *kerosene*).

Avtur Aviation turbine fuel (kerosene).

AVUM 1 Aviation unit-level maintenance (USA).
2 Air vehicle unit maintenance.

AW 1 All-weather (often not literally true).
2 Airway (often A/W, but AWY is preferred).
3 Automatic weapon (normally gun).

Aw Total wetted area of aircraft.

AWA Aviation/Space Writers Association (esp of US).

AWACS, Awacs Airborne warning and control system.

AWADS All- (or adverse-) weather aerial delivery system.

AWCCV Advanced weapons carriage configured vehicle.

AWCLS All-weather carrier landing system.

AWD Airworthiness Division (CAA, UK).

Awdrey Atomic weapon detection, recognition and estimation of yield.

AWDS All-weather delivery (or distribution) system.

AWE All-up weight, equipped (generally, OWE).

AWESS Automatic weapons effect signature simulator.
AWG JETDS code: piloted aircraft, armament, fire control.
AWI 1 Aircraft weight indicator; airborne system often also indicating c.g. position, usually by sensing deflections of landing-gear shock struts.
2 All-weather intercept.
AWLS All-weather landing system.
AWM Average working man (in defining aircrew sleep patterns).
AWOP All-Weather Operations Panel (ICAO).
AWPA Australian Women Pilots' Association.
AWR Airborne weather radar.
AWRS Airborne weather reconnaissance system (USAF, eg as carried by WC-130B aircraft).
AWS 1 Air Weather Service (USAF MAC).
2 Audible (or advanced) warning system.
3 Automatic wing sweep.
AWSACS All-weather stand-off aircraft (or attack) control system (USN).
AWSAS All-weather stand-off attack system.
A_{WT} Tail-on-wing aerodynamic influence coefficient.
AWTSS All-weather tactical strike system.
A_{WW} Wing-on-wing aerodynamic influence coefficient.
AWY, awy Airway (ICAO, FAA).
AX Avionics, avionics control (USAF).
AXAF Advanced X-ray astrophysics facility.
axes Aircraft attitude is described in terms of three sets of *. First set are reference *, three mutually perpendicular directions originating at c.g. (point O) and defined as longitudinal (roll) axis OX, measured positive forwards from O and negative to rear; transverse (pitch) axis OY, measured positive to right and negative to left; and vertical (yaw) axis OZ, measured positive downwards and negative upwards. Position of O defined at design stage in what is considered most likely location for real c.g. in practice. Second, single, * is wind *: direction of relative wind, drawn through O, has angle determined by flight velocity, angle of attack and angle of sideslip. Third * are those known as inertia * and are imaginary lines about which aircraft would actually rotate in manoeuvres. These need not be same as reference *, although OY and OZ inertia axes are usual closely co-incident unless aircraft is asymmetrically loaded. Principal inertia *, however, may often depart substantially from geometrically drawn fore-and-aft axis OX (see *inertia coupling*).
axial cable 1 In non-rigid airship, main longitudinal member linking supporting cables, in framework carrying crew and engines.
2 In rigid airship, essentially straight cable sometimes linking extreme nose and tail of hull and central fittings of radial or diametral wires.
axial compressor Compressor for air or other gas with drum-shaped rotor carrying one or more rows of radial blades in form of small aerofoils (airfoils) arranged to rotate around central axis, with row of stationary stator blades (vanes) between each moving row. Compressed fluid moves through alternate fixed and moving blading in essentially axial direction, parallel to axis of rotation.
axial cone In rigid airship, fabric cone at front and rear of each gas cell providing flexible gas-tight connection between cell and axial cable.
axial cord In parachute, central rigging line joining apex to eyes formed at lower extremities of rigging lines.
axial deck In carrier (1) or other ship carrying or serving as operating platform for aircraft, flight deck aligned fore and aft.
axial engine Usually, PE (4) in which axes of cylinders are parallel to crankshaft and/or main output shaft. Also, loosely, axial-flow engine.
axial firing Fixed to fire directly ahead (usually on helicopter).
axial-flow engine Gas-turbine engine having predominantly axial compressor, esp one in which airflow is essentially axial throughout (ie, not reverse-flow).
axial focusing In supersonic wind tunnel, focusing of shockwaves reflected from tunnel wall on to principal axis. Usually condition to be avoided, typically by minimizing such reflections or so shaping working section that they are dispersed in different planes.
axis of rotation In rotorplane, apparent axis about which main lifting rotor rotates: line passing through centre of tip-path circle and perpendicular to tip-path plane. May be widely divergent from mechanical axis on which hub is mounted, especially in articulated rotor.
axis of symmetry Usually aeronautical *** determined by geometrical form, but in some cases dictated by mass distribution.
A_y Direct sideforce, changing heading without bank or sideslip.
Az-El, Azel, Az/El Radar presentation giving separate pictures of azimuth (PPI display, or chosen sector) and elevation (such as side view of glide path).
azication Azimuth indication.
azimuth 1 Horizontal bearing or direction; thus * angle.
2 Rotation about vertical axis (yaw is preferred term where motion is that of whole aircraft).
3 Bearing of celestial body measured clockwise from true North, often called * angle and qualified true, compass, grid, magnetic or reference depending on measure used.
azimuth aerial Ground radar aerial rotating about vertical axis, or sending out phased-array emission rotating about such axis, intended to measure target azimuth angles.
azimuth compiler Portion of SSR system, often optional or absent, which provides accurate azi-

muth information more accurately than the normal plot extractor.

azimuth control In rotorplane, cyclic pitch.

azimuth error Radar bearing error due to horizontal refraction.

azimuth marker Scale used on PPI display to indicate bearing, including electronically generated references when display is offset from central position.

AZM Azimuth.

az-ran, azran General term for target tracking or navigational fixing by means of azimuth and range; more commonly called Rθ or rho-theta.

Azusa C-band tracking system operating on short baseline and giving continuous signals of two direction cosines plus slant range (and thus giving 3-D fix and instantaneous velocity). From Azusa, Calif.

B

B 1 Pitching moment of inertia.
2 Blue (ICAO).
3 Base (of semiconductor device).
4 Bomber type of aircraft (US DoD, UK role prefix).
5 Total aircraft noise rating (Neth).
6 Boron.
7 Magnetic flux density.

b 1 Wing span.
2 Bars (unit allowed within SI).
3 Barns (unit allowed within SI).

B_1, B_2 Graduation ratings from CFS.

b_1 Control-surface hinge moment.

b_2 Rate of change of surface hinge moment $dC_H/d\epsilon$.

B-class Military and civil prototype or experimental aircraft, not certificated but flown by manufacturer under special rules and with SBAC numerical registration (UK).

B-code In flight plan, have DME and transponder with 64-code.

B-display CRT or other display in which horizontal axis is bearing and vertical axis is range.

B-licence Commercial pilot's licence (not ALTP).

B-power supply Plate circuit that generates electron current in CRT or other electron tube.

B-rating Twin-engine pilot rating.

B-station In Loran, transmitter in each pair whose signals are emitted more than half a repetition period after next succeeding signal and less than half an r.p. before next preceding signal of other (A station).

B-stoff Hydrazine hydrate (G).

BA 1 Braking action (ICAO).
2 Budget authority.

B-A gauge Bayard-Alpert ionization gauge.

b.a. Buffer amplifier.

BAA 1 British Airports Authority.
2 Bombardiers' Alumni Association (US).

BAAHS Bay Area Airline Historical Society (San Francisco region).

Babbitt (incorrectly, babbit) Family of soft tin-based alloys used to make liners for plain bearings.

babble Incoherent cross-talk in voice communications system.

Babinet point One of three points of zero polarization of diffuse sky radiation.

BABOV Bureau Aanleg Beheer en Onderhoud van Vliegvelden (airfield plans and maint) (Neth).

BABS Beam-approach beacon system. Outmoded secondary radar system which provided fixed-wing aircraft with lateral guidance and distance information during landing approach.

BACA Baltic Air Charter Association (UK).

BACE Basic automatic checkout equipment (USN).

BACEA British Airport Construction and Equipment Association.

back 1 Of drag curve, aeroplane flight below V_{IMD}, in which reduction in speed results in increased drag.
2 Of propeller or rotor blade, surface corresponding to upper surface of wing.
3 Rear cockpit of tandem two-seat aircraft, especially combat type (hence GIB).

back beam In any beam system, especially ILS localizer, reciprocal beam on other side of transmitter.

back bearing Direction observed from aircraft holding steady course of fixed object over which it has recently passed; reciprocal of track.

backboard Multilayer circuit board.

back burner To be on * = not urgent, temporarily shelved.

back contamination Contamination of Earth by organisms introduced by spacecraft and crews returning from missions.

back course Course flown along back beam, on extended centreline of runway away from airfield.

backfire Premature ignition of charge in PE (4) cylinder such that flame travels through still-open inlet valve(s) and along induction manifold.

backfit Retrofit (US usage).

background Ambient (usually supposed steady-state) level of intensity of a physical phenomenon against which particular signal is measured. If signal amplitude never exceeds that of background, it cannot be detected. Thus: * clutter (radar), * count (radiation), * luminance, * noise (this has two meanings: noise in electronic circuit, and ambient level of aural noise at airport or elsewhere).

backing 1 General effort by hardware manufacturer to support his products after they reach customer.
2 Change in direction of prevailing wind in counter-clockwise direction viewed from above; thus, from S to SE.

backing up Rearwards taxiing of freight aircraft to loading dock, using reverse thrust.

backlash In any mechanism, lost motion due to loose fitting or wear.

backlog In manufacturing programme, items sold but not yet delivered.

backout Any reversal of countdown, usually due to technical hold or fault condition.

backpack 1 Personal parachute pack worn on back (usually thin enough for wearer to sit comfortably).
2 Any life-support system worn on back of user.

backplane Standard STD or STE bus mounting board on which computer boards are attached by multiple push-in connectors. Most * accommodate 10 boards and have 20 edge connectors.

back porch In electronic display, esp TV, brief (eg 6μs) interval of suppressed video signal at end of each line scan.

back-pressure 1 Pressure in closed fluid system opposing main flow.
2 Ambient pressure on nozzle of rocket or other jet engine or any other discharge from fluid system.

backscatter 1 Backward scatter.
2 Signal received by backward scattering.

back-shop In manufacturing plant, first shop to close on rundown of programme.

backside Moon face turned away from Earth.

backstagger Backward stagger.

backswept Swept.

backtell Transfer of information from higher to lower echelon.

backtrack 1 In aircraft operation, to turn through about 180° and follow same track in reverse direction (as allowance must be made for wind, not same as turn on reciprocal).
2 Having landed on runway in use, to turn through 180° and proceed along runway in reverse direction.

back-up 1 Complete programme, hardware item or human crew funded as insurance against failure of another.
2 Type of hardware item which could, even with degraded system performance, replace new design whose technical success is in doubt.
3 System funded to augment one already in operation (thus, BUIC).
4 Information printed on reverse of map or other sheet to supplement marginal information.

backward extrusion Extrusion by die so shaped that material being worked flows through or around it in reverse direction to that of die.

backward scatter Electromagnetic energy (eg, radio, radar or laser) scattered by atmosphere back towards transmitter. In some cases whole hemisphere facing towards transmitter is of interest; in radar, attention usually confined to small amount of energy scattered at very close to 180° and detected by receiver.

backward stagger Stagger such that upper wing is mounted further back than lower.

backward tilt Tilt such that blade tips are to rear of plane of rotation through centroids of blade roots.

backward wave In a TWT, any wave whose group velocity is opposite to direction of electron travel.

backwash Slipstream.

BACM Bootstrap air-cycle machine.

Bacon cell Hydrogen/oxygen fuel cell.

BACS Bleed-air control system.

BAD Boom-avoidance distance; thus *(A) is boom-avoidance distance measured along track on arrival and *(D) is same on departure.

BADGE, Badge Base air-defence ground environment (Japan).

BAeA British Aerobatics Association.

BAFF British air forces in France (1939–40).

baffle 1 Loosely, any device intended to disturb and impede fluid flow.
2 Shaped plates fixed around and between cylinders of air-cooled PE (4) to improve cooling.
3 Surface, usually in form of a ring, plate or grating, arranged inside liquid container to minimize sloshing.
4 In two-stroke PE (4), deflector incorporated in crown of piston.
5 Partial obstruction inside pitot tube to minimize ingress of liquid or solid matter.

BAFO British Air Forces of Occupation.

bag tank Liquid container, especially fuel tank, constructed of flexible material not forming part of airframe.

BAI Battlefield air interdiction.

bail To loan aircraft or other possession, freely but under contract, to facilitate accomplishment of specific objective; in particular, loan by owner government of military hardware to industrial contractor engaged in particular development programme for that government.

bail out To abandon dangerously unserviceable aircraft, esp in midair, by parachute. Not yet used in connection with spacecraft.

bailout bottle Emergency personal oxygen supply, usually high-pressure gox, attached to aircrew harness or ejection seat.

Bairstow number Mach number.

bakcs Back-course (ILS).

Bakelite Trade name for a phenol-formaldehyde resin plastic.

bake out In high-vacuum technology, heating to promote degassing.

Baker-Nunn Large optical camera used for tracking objects in space.

balance 1 State of equilibrium attained by aircraft or spacecraft.
2 Mechanism for supporting object under test in wind-tunnel and for measuring forces and moments experienced by it due to gas flow.
3 Mass or aerodynamic surface intended to reduce hinge moment of control surface.

balance area In aerodynamically balanced control surface, projected area ahead of hinge axis.

balance circuit In a WCS, subsystem which prevents, or warns of impending, lateral asymmetry due to unbalanced weapon load.

balanced field length 1 Hypothetical length of

runway for which TODa = EMDa (and sometimes, in addition, TORa).

2 Under CAR.4b, unfactored TOD to 50 ft following failure of one engine at V_1 = EMD to and from V_1 on dry surface.

balanced modulator Modulator whose output comprises sidebands without carrier.

balanced support Logistic supply based on predicted consumption of each item.

balanced surface Control surface whose hinge moment is wholly or partially self-balanced (usually by means of mass or area ahead of hinge axis or by tabs).

balance rod Mass distributed along or within leading edge of helicopter rotor blade.

balance station zero Imaginary reference plane perpendicular to longitudinal axis of aircraft and at or ahead of nose, used in determinations of mass distribution and longitudinal balance.

balance tab Tab hinged to, and forming part of, trailing edge of control surface, and so linked to airframe that it is deflected in opposition to main surface, and thus reduces hinge moment. Action is thus similar to that of servo tab.

ball 1 Small spheroid, or other laterally symmetric shape, in lateral glass tube of ball-type slip indicator.

2 Arbitrary unit of slip, equal to one ball-width.

ball ammunition Bullets of solid metal, containing no explosive, pyrotechnic or AP core.

ballast 1 In aerodynes, mass carried to simulate payload, and permit c.g. position to be varied (usually in flight).

2 In aerostats, mass carried for discharge during flight to change vertical velocity or adjust trim.

ballast carrier In transport aircraft, holder for metal ballast weights with locking plungers mating with floor rails.

ball bearing Any shaft bearing in which inner race is supported and located by hardened spheres.

ball inclinometer Ball turn-and-slip.

ballistic camera Photographic camera which, by means of multiple exposures on same plate or frame of film, records trajectory of body moving relative to it.

ballistic capsule Capsule enclosing environment suitable for human crew or other payload and moving in * trajectory.

ballistic flight Ballistic trajectory; arguably, not flight.

ballistic galvanometer Undamped galvanometer which, when an electrostatic charge is switched through it, causes large initial swing, taken to be proportional to quantity of electricity passing.

ballistic missile Wingless rocket weapon which, after burnout or cutoff, follows * trajectory.

ballistic parachute canopy spreader Device for accelerating deployment of drag canopy in certain types of ejection seat.

ballistic range Research facility for investigation of behaviour of projectiles or of bodies moving through gaseous media at extremely high Mach numbers; usually comprises calibrated range along which test bodies can be fired, sometimes into gas travelling at high speed in opposite direction.

ballistic re-entry Non-lifting re-entry.

ballistic trajectory Trajectory of wingless body, formerly propelled but now subject only to gravitational forces and, if in atmosphere, aerodynamic drag.

ballistic tunnel High-Mach tunnel into which free projectiles are fired in opposition to gas flow (see * *range*).

ballistic vehicle Vehicle, other than missile, describing a * trajectory. Term usually not applied to spacecraft but to vehicles used in proximity of Earth and at least mainly within atmosphere.

ballistic wind Theoretical constant wind having same overall effect on a * projectile as varying winds actually encountered.

ball lightning Rare natural phenomenon, materializing during electrical storms, having appearance of luminous balls which often appear to spin, eject sparks, travel slowly and eventually disappear (sometimes with explosion).

ballonet Flexible gastight compartment inside envelope of airship (rarely, balloon) which can be inflated by air to any desired volume to compensate for variation in volume of lifting gas and so maintain superpressure and alter trim.

ballonet ceiling Maximum altitude from which pressure aerostat with empty ballonet(s) can return to sea level without loss of superpressure.

ballon sonde See *registering balloon*.

balloon Aerostat without propulsion system.

balloon barrage Protective screen of balloons moored around target likely to be attacked by enemy aircraft.

balloon bed Area of ground prepared for mooring of inoperative captive balloon.

balloon fabric Range of fabrics of mercerised cotton meeting specifications for covering lightplanes and, impregnated with rubber, aerostats.

ballooning Colloq, sudden unwanted gain in height of aeroplane on landing approach due to lowering flaps, GCA instruction or, most commonly, flare at excessive airspeed.

balloon reflector In electronic warfare, confusion reflector supported by balloon(s).

balloon tank Tank for containing liquid or gas constructed of metal so thin that it must be pressurised for stability. Such tanks have formed airframe of large ICBMs.

ball screwjack Screwjack in which friction is reduced by system of recirculating bearing balls interposed between fixed and rotating members.

ball turn-and-slip Flight instrument whose means of indicating slip is a ball free to move within liquid-filled curved tube having its centre lower than its ends.

ball turret Gun turret on certain large aircraft of

1942–45 having part-spherical shape.
ballute Balloon-parachute; any system of inflatable aerodynamic braking system used for upper-atmosphere retardation of sounding rockets or slowing of spacecraft descending into planetary atmospheres.
BALPA British Air Line Pilots' Association.
balsa Wood of extremely low density (s.g. about 0·13), originally grown in W. Indies and Central America.
BAM Bundesanstalt für Material Prüfung (G).
BAN Beacon alphanumerics (part of FAA ARCTS).
band 1 Designated portion of EM spectrum, usually bounded by frequencies used for radio communication.
2 A (usually small) portion of EM spectrum containing frequencies of absorption or emission spectra.
3 Strip of stronger material built into non-rigid aerostat envelope to distribute stress from mooring line, car or other load.
4 Group of tracks on magnetic disc or drum.
band-elimination filter Filter that eliminates one band of EM frequencies, upper and lower limits both being finite.
B&GS Bombing and gunnery school.
band of error Band of position.
band of position Band of terrestrial position, usually extending equally on each side of position line, within which, for given level of probability, true position is considered to lie.
B&P Bid and proposal.
bandpass Width, expressed in Hz, of band bounded by lower and upper frequencies giving specified fraction (usually one-half) of maximum output of amplifier.
bandpass filter EM wave filter designed to reduce or eliminate all radiation falling outside specified band of frequencies.
bandwidth 1 Number of Hz between limits of frequency band.
2 Number of Hz separating closest lower and upper frequency limits beyond which power spectrum of time-variant quantity is everywhere less than specified fraction of its value at reference frequency between limits.
3 Range of frequencies between which aerial performs to specified standard.
4 In EDP and information theory, capacity of channel.
5 Band of position, usually expressed in nm on each side of PL (1) or track.
bang Sound caused by passage of discontinuous pressure wave in atmosphere (*sonic bang*).
bang-bang Any dynamic system, especially one exercising control function, which continually oscillates between two extreme "hard-over" positions. Also called flicker control.
bang out To eject.
bang valley Colloq, land or sea area under track where supersonic flight is permitted.
banjo Structural member having form of banjo, with open ring joined to linear portion (on either or both sides) projecting radially in same plane. Typically used to link spar booms on each side of jet engine or jetpipe aperture.
bank 1 Attitude of aerodyne which, after partial roll, is flown with wings or rotor not laterally level. Held during any properly executed turn.
2 To roll aerodyne into banked position.
3 Linear group of cylinders in PE (4).
bank and turn indicator Turn and slip indicator.
banner cloud Cloud plume extending downwind of mountain peak, often present on otherwise cloudless day.
banner sleeve Tow target in form of long tube inflated by slipstream.
banner target Air-to-air or ground-to-air firing target in form of towed strip of flexible fabric like elongated flag.
BAP Bomber aviation regiment (USSR).
BAPC British Aircraft Preservation Council.
Bapta Bearing and power-transfer assemblies (comsat).
bar Unit of pressure, allowed within SI and standard in meteorology and many other sciences. Equal to 10^5 Nm^{-2}. In units contrary to SI, 10^6 dynes cm^{-2} or 750·08 mm; 29·53 in of Hg. One bar (1,000 mb) is Normal Atmospheric Pressure.
baralyme Trade name for mixture of barium and calcium hydroxides used to absorb CO_2.
barbecue manoeuvre Deliberate intermittent half-rolls performed by spacecraft to equalize solar heating on both sides.
barber chair Chair capable of gross variation in inclination.
barber-pole instrument Indicator using rotating spirals, such as PVD (2).
barbette Defensive gun position on large aircraft projecting laterally and providing field of fire to beam.
Barcap Barrier combat air patrol between naval strike force and expected aerial threat.
Barcis British airports rapid control and indication system.
bar code Geometric patterns, no two alike, printed on document (eg flight coupon) and read by light pen connected to computer storing data (eg stolen ticket numbers).
bare base Airfield comprising runways, taxiways and supply of potable water.
Barif Bureau of Airlines Representatives in Finland.
barn Unit of area for measuring nuclear cross-sections. Equal to 10^{-28} m^2 or 10^{-22} mm^2. Symbol b.
barnstormer Formerly, itinerant freelance pilot who would operate from succession of unprepared temporary airfields giving displays and joyrides.

barogram Hard-copy record made by barograph.
barograph Barometer giving continuous hard-copy record.
barometer Instrument for measuring local atmospheric pressure.
barometric altimeter Pressure altimeter.
barometric altitude Pressure height.
barometric element Transmitting barometer carried in radiosonde payload, variation in aneroid capsules causing shift in frequency of carrier wave.
barometric fuze Fuze set to trigger at preset pressure height.
barometric pressure Local atmospheric pressure.
barometric pressure control Automatic regulation of fuel flow in proportion to local atmospheric pressure.
barometric pressure gradient Change in barometric pressure over given distance along line perpendicular to isobars.
barometric tendency Change in barometric pressure within specified time, usually the preceding three hours.
barometric wave Any short-period meteorological wave in atmosphere.
barosphere Atmosphere below critical level of escape.
barostat Device for maintaining constant atmospheric pressure in enclosed volume.
barostatic relief valve Automatic regulation of fuel flow by spilling back surplus through relief valve sensitive to atmospheric pressure.
barothermograph Instrument for simultaneously recording local temperature and pressure.
barotrauma Bodily injury due to gross or sudden change in atmospheric pressure.
barotropy Bulk fluid condition in which surfaces of constant density and constant pressure are coincident.
barrage AA artillery fire aimed not at specific targets but to fill designated rectilinear box of sky.
barrage balloon Captive balloon forming part of balloon barrage.
barrage jamming High-power electronic jamming over broadest possible spread of frequencies.
barrel 1 Of PE (4) cylinder, body of cylinder without head or liner.
2 Of rocket engine, thrust chamber or nozzle, esp of engine having multiple chambers.
3 Any portion of airframe of near-circular section, even if tapering; thus Canadair makes the F/A-18 nose *.
barrel engine PE (4) having cylinders with axes disposed parallel to engine longitudinal axis and output shaft.
barrel roll Manoeuvre in which aerodyne is flown through 360° roll while trajectory follows horizontal spiral such that occupants are always under positive acceleration in vertical plane relative to aircraft.
barrel section 1 Portion of transport aircraft fuselage added in "stretching". Alternatively called plug section (but see *barrel (3)*).
2 Parallel length of control rod.
barrel wing Wing in form of duct open at both ends and with longitudinal section of aerofoil shape.
barrette Array of closely spaced ground lights that appear to form a solid bar of light.
barricade Barrier (US).
barrier Net mounted on carrier (1) deck or airfield runway to arrest with minimal damage aircraft otherwise likely to overrun. Normally lying flat, can be raised quickly when required.
barrier crash Incident involving high-speed entry to * with or without damage. Usually applied to carrier operations.
barrier pattern Geometrical pattern of sonobuoys so disposed as to bar escape of submerged submarine in particular direction.
bar stock Standard form of metal raw material: solid rolled or extruded with round, square or hexagonal (rarely, other) section.
Barstur, BARSTUR Barking Sands Tactical Underwater Range (USN, Kauai, Hawaii).
barycentre Centre of mass of sytem of masses, such as Earth/Moon system (barycentre of which is inside Earth).
barye Unit of pressure in CGS system. Equal to 1 dyne cm^{-2} or 10^{-6} bar (hence alternative name of microbar).
Basar Breathing air, search and rescue (for diver in helo crew).
BASE Cloud base height AMSL (ICAO).
base 1 Locality from which operations are projected or supported.
2 Locality containing installations to support operations.
3 In an object moving through atmosphere, any unfaired region facing rearwards (eg rear face of bullet or shell, trailing-edge area of wedge aerofoils, and projected gross nozzle area of rocket engine after cutoff).
4 Substance constructed of ions or molecules having one or more pairs of electrons in outer shells capable of forming covalent bonds.
5 Loosely, substance that neutralizes an acid.
6 In transistor, region of semiconductor material into which minority carriers are emitted (hence, usually, between emitter and collector).
7 Underside of cloud. With cloud having grossly irregular undersurface, surface parallel to local Earth surface below which not more than 50 per cent of cloud protrudes, and which can be taken as upper limit of VFR. In mist or fog * intersects Earth.
base area Aggregate area of unfaired rearward-facing surface of aerodynamic body.
base check Examination in flight of crew on completion of conversion to new type.

base drag Drag due to base area experiencing reduced pressures.

base line, baseline 1 Yardstick used as basis for comparison, specif known standard of build for functioning system, such as combat aircraft, against which developed versions can be assessed in numerical terms. Hence, * aircraft.
2 Geodesic line between two points on Earth linked by common operative system, eg between two Loran, Decca or Gee stations.
3 In many types of visual display and pen recorder, line displayed in absence of any signal.

base metal 1 Major constituent of an alloy.
2 Metal of two parts to be joined by welding (as distinct from metal forming joint itself, which is modified or added during welding process).

base pressure Local aerodynamic pressure on base area of body moving through atmosphere.

base surge Expanding toroid surrounding vertical column in shallow underwater nuclear explosion.

base/timing sequencing Automatic sharing of transponder between several interrogators or other fixed stations by use of coded timing signals.

BASF Boron-augmented solid fuel.

basic aircraft Simplest usable form of particular type of aircraft, from which more versatile aircraft can be produced by equipment additions. In case of advanced aircraft, such as combat and large transports, ** includes IFR instruments, communications, and standard equipment for design mission.

basic cloud formations Subdivision of cloud types into: A, high; B, middle: C, low; D, clouds having large vertical development (International Cloud Atlas, 1930).

basic cover Aerial reconnaissance coverage of semi-permanent installation which can be compared with subsequent coverage to reveal changes.

basic encyclopedia Inventory of one's own or hostile places or installations likely to be targets for attack.

basic flight envelope Graphical plot of possible or permissible flight boundaries of aerodyne of particular type. Cartesian plot with TAS or Mach number as horizontal and altitude or ambient pressure as vertical. Boundaries imposed by insufficient lift, thrust or structural strength (and sometimes by social and other considerations); see *basic gust* *, *basic manoeuvring* *.

basic gust envelope Specified form of graphical plot for each new aerodyne design showing permissible limits of speed for passage through vertical sharp-edged gusts of prescribed strength (traditionally ±25, 50 and 66 ft sec^{-1}). Result is V-n diagram, with EAS as horizontal and gust load factor n as vertical.

basic load 1 Load (force) transmitted by structural member in condition of static equilibrium, usually in straight and level flight (1 g rectilinear), at specified gross weight and mass distribution.
2 Aggregate quantity of non-nuclear ammunition, expressed in numbers of rounds, mass or other units, required to be in possession of military formation.

basic manoeuvring envelope V-n diagram with EAS as horizontal and manoeuvring load factor as vertical.

basic operating platform See *bare base*.

basic operating weight Operating weight empty.

basic research See *pure research*.

basic runway Runway without aids and bearing only VFR markings: centreline dashes or arrows, direction number and, if appropriate, displaced threshold.

basic supplier Nominated supplier of hardware item in absence of specific customer option.

basic T In traditional cockpit instrument panel, primary flight instruments (ASI, horizon, turn/slip and VSI) arranged in a standard T formation.

basic thermal radiation Thermal radiation from Quiet Sun.

basic trainer American military aeroplane category used for second stage in pilot training (after primary), with greater power and flight performance. Formerly also "basic combat" (BC) category, which introduced armament and closely paralleled flight characteristics of operational type; called "scout trainer" by Navy and redesignated "advanced trainer" (obs).

basic weight Superseded term formerly having loose meaning of mass of aircraft including fixed equipment and residual fluids.

basic wing Aerofoil of known section used as starting point for modified design, often with wholly or partly different section.

basket 1 Radar-defined horizontal circular area of airspace into which dispensed payloads (eg anti-armour bomblets) are delivered by bus (5).
2 Car suspended below aerostat for payload, not necessarily of wickerwork construction.
3 Drogue on a flight-refuelling tanker hose.

basket-tube Form of construction of liquid-propellant rocket thrust chamber in which throat and nozzle is formed by welded tubes, usually of nickel or copper, through which is pumped liquid oxygen or other cryogenic propellant for regenerative cooling.

BASO Brigade air support officer.

BAT 1 Boom-avoidance technique.
2 Beam-approach training.
3 Bureau of Air Transportation (Philippines).

bathtub 1 Bath-shaped structure of heavy plate or armour surrounding lower part of cockpit or other vital area in ground-attack aircraft.
2 Temporary severe recession in production in manufacturing plant or programme, or between programmes; named from appearance on graphical plot (US, colloq).

BATOA British Air Taxi Operators' Association.
batonet Tubular or rod-like toggle forming link between rigging line and band on fabric aerostat envelope.
BATR Bullets at target range (shows location in HUD).
BATS Ballistic aerial target system.
batt Ceramic filler formed from chopped-strand mat.
batten 1 Wood or metal strip used in interlinked pairs as ground control lock.
2 Wood or metal strips arranged radially from nose of non-rigid airship to stiffen fabric against dynamic pressure, or, where applicable, mooring loads.
3 Flexible strips used in lofting drawing.
battery Enclosed device for converting chemical energy to electricity. Most aerospace batteries are secondary (rechargeable), principal families being Ni/Cd (nickel, cadmium), Ag/Zn (silver, zinc) and lead/acid. Fuel cells are batteries continuously fed with reactants.
battery booster Starting coil.
battle damage In-flight damage caused directly by enemy (not, eg, by collision with friendly aircraft).
battle formation Any of several formations characterized by open spacing and flexible interpretation.
battleship model Any model or rig used for repeated development testing, usually statically, in which major elements not themselves under test are made quickly and cheaply from "boilerplate" material to withstand repeated use. Thus battleship tank.
battleship tank Tank for liquid propellant for static testing of rocket engines having same capacity and serving same feed system as in flight vehicle but made of heavy steel or other cheap and robust material for repeated outdoor use.
BAUA Business Aircraft Users' Association (UK).
BAUAG British Airports Users' Action Group.
baud Unit of telegraphic signalling speed, equivalent to shortest signalling pulse or code element. Thus speed of 7 pulses per second is 7 bauds (pronounced "boards"); abbreviation Bd.
baulk 1 Aborted landing due to occurrence in final stages of approach (typical causes would be aircraft on airfield taxiing on to runway or a runaway stick-pusher in landing aircraft).
2 To obstruct landing of an approaching aircraft and cause it to overshoot.
baulked landing See *baulk (1)*.
bay 1 In aerodyne fuselage or rigid airship hull, portion between two major transverse members such as frames or bulkheads.
2 In biplane or triplane, portion of wings between each set of interplane struts; thus, single-* aircraft has but one set of interplane struts on each side of centreline.
3 In any aircraft or spacecraft, volume set aside for enclosing something (eg, engine *, undercarriage *, bomb *, cargo *, lunar rock *).
bayonet Electronic subsystem permitting radar of strike aircraft to home on designated ground target and providing azimuth for release of weapons equipped with radiation sensors.
bayonet exhaust Formerly, form of PE (4) exhaust stack designed to reduce noise.
BAZ Bundesamt für Zivilluftfahrt Zentrale (Austria).
BB 1 Back bearing.
2 Battleship (USN).
BBAC British Balloon and Airship Club.
BBC Broad-band chaff.
BBL Billion barrels liquid.
BBOE Billion barrels oil equivalent.
BBSU British Bombing Survey Unit.
BC 1 Boron carbide armour.
2 Back course.
3 Bus (2) controller.
4 Bomber Command (RAF, USAAF).
5 Basic combat (USA, 1936–40).
BCA 1 Board of Civil Aviation (Sweden).
2 Baro-corrected altitude.
BCARs British Civil Airworthiness Requirements.
BCAS Beacon-based collision-avoidance system.
BCD 1 Binary coded decimal.
2 Bulk chaff dispenser.
BCE Battlefield control element.
BCF Bromochlorodifluoromethane (fire extinguishant).
BCIU Bus control and interface unit.
BCL Braked conventional landing (V/STOL).
BCM Basic combat manoeuvring, series of manoeuvres simulating interceptions and close dogfights.
bcn Beacon.
BCP 1 Battery command post.
2 BIT control panel.
BCR 1 European Community Bureau of Reference (co-ordinates R & D).
2 Battle casualty replacement.
B/CRS Back course.
BCS British Computer Society.
BCST, bcst Broadcast.
BCU Bird control unit (on airfield).
BCV Belly cargo volume.
B/D Bearing and distance.
Bd Baud.
bd Candle (unit of luminous intensity), abb.
BDA Bomb (or battle) damage assessment.
BDC, b.d.c. Bottom dead centre.
BDHI Bearing/distance/heading indicator.
BDLI Bundesverband der Deutschen Luft- und Raumfahrtindustrie (W German aerospace industry association).
BDOE Barrels per day oil equivalent.
BDR Battle-damage repair.
BDRY Boundary.

beaching Pulling marine aircraft up sloping beach, out of water to position above high tide.

beaching gear Wheels or complete chassis designed to be attached to marine aircraft in water to facilitate beaching and handling on land.

beacon 1 System of visual lights marking fixed feature on ground (see *aeronautical light*).
2 Radio navaid (see *fan marker*, *homing beacon*, *NDB*, *LFM*, *marker beacon*, *Z marker*).
3 Radar transceiver which automatically interrogates airborne transponders (see *radar beacon*, *ATCRBS*).
4 Portable radio transmitter, with or without radar reflector or signature enhancement, for assisting location of object on ground (see *crash locator beacon*, *personnel locator beacon*).

beacon buoy Self-contained radio beacon carried in emergency kit. Floats on water.

beacon characteristic Repeated time-variant code of some visual light beacons, esp aerodrome beacons emitting Morse letters identifying airfield.

beacon delay Time elapsed between receipt of signal by beacon of transponder type (eg, in DME) and its response.

beacon-identification light Visual light, emitting characteristic signal, placed near visual light beacon (pre-1950) to identify it.

beacon skipping Fault condition, due to technical or natural causes, in which interrogator beacon fails to receive full transponder pulse train.

beacon stealing Interference by one radar resulting in loss of tracking of aerial target by another.

beacon tracking Tracking of aerial target by radar beacon, esp with assistance from transponder carried by target.

bead 1 Corrugation or other linear discontinuity rolled or pressed into sheet to stiffen it, esp around edges.
2 Thickened edge to pneumatic tyre shaped to mate with wheel rim and usually containing steel or other filament reinforcement.
3 Unwanted blob of weld metal.

beading Rolling or pressing sheet to incorporate beads.

bead sight Ring and bead sight.

beam 1 Structural member, long in relation to height and width and supported at either or both ends, designed to carry shear loads and bending moments.
2 Quasi-unidirectional flow of EM radiation.
3 Quasi-unidirectional flow of electrons or particles, with or without focusing to point.
4 Loosely, on either side of aircraft; specif direction from 45° to 135° on either side measured from aircraft longitudinal axis and extending undefined angle above and below horizontal. Hence, * guns (firing on either side), or surface object described as "on the port *" (90° on left side).

BEAMA British Electrical and Allied Manufacturers' Association.

beam approach Early landing systems in which final approach was directed by beam (2) from ground radio aid (see *BABS*, *ILS*, *SBA*).

beam attack Interception terminating at crossing angle between 45° and 135°.

beam bracketing Flying aircraft alternately on each side of equisignal zone of radio range or similar two-lobe beam.

beam capture To fly aircraft to intercept asymptotically a beam (2), esp ILS localizer and glide path.

beam compass Drawing instrument based on beam parallel to drawing plane having centre point and carrier for pen or other marker.

beam direction In stress analysis, direction parallel to both plane of spar web, or other load-bearing member, and aircraft plane of symmetry.

beam jitter Continuous oscillation of radar beam through small conical angle due to mechanical motion and distortion of aerial.

beam rider Missile or other projectile equipped with beam-rider guidance.

beam-rider guidance Radar guidance system in which vehicle being guided continuously senses, and corrects for, deviation from centre of coded radar or laser beam which is usually locked on to target. Accuracy degrades with distance from emitter.

beam slenderness ratio Length of structural beam divided by depth (essentially, divided by transverse direction parallel to major applied load).

beam width Angle in degrees subtended at aerial between limiting directions at which power (DoD states "RF power", NATO states "emission power") of radar beam has fallen to half that on axis. Determines discrimination.

bearer Secondary structure supporting removable part such as fuel tank or engine.

bearing 1 Angular direction of distant point measured in horizontal plane relative to reference direction.
2 Angular direction of distant point measured in degrees clockwise from local meridian, or other nominated reference. Such measure must be compass, magnetic or true. True * is same as azimuth angle.
3 Mechanical arrangement for transmitting loads between parts having relative motion, with minimum frictional loss of energy or mechanical wear.

bearing compass Portable and hand-held, used for determining magnetic bearing of distant objects.

bearing plate Simple geometrical instrument for converting bearings of distant objects into GS (1) and drift.

bearing projector Powerful searchlight trained from landmark beacon or other point towards nearby airfield (obs).

bearing stress In any mechanical bearing, with or

without relative movement, load divided by projected supporting area.

bear pads Horizontal plates added to prevent helicopter skid sinking into snow.

beat 1 Vibration of lower frequency resulting from mutual interaction of two differing higher frequencies. Often very noticeable in multi-engined aircraft.
2 One complete cycle of such interference.

beat frequency Output from oscillator fed by two different input frequencies which has frequency equal to difference between applied frequencies. (Other outputs have higher frequencies, such as sum of applied frequencies.)

beat-frequency oscillator Oscillator generating signals having a frequency such that, when combined with received signal, difference frequency is audible. Such ** is heterodyne, used in CW (1) telegraphy. Another is superheterodyne, in which local ** produces intermediate frequency by mixing with received signal.

beat reception Heterodyne reception, as used in CW (1) telegraphy.

beat-up 1 Aggressive dive by aircraft to close proximity of surface object.
2 Repeated close passes in dangerous proximity to slower aircraft.

Beaufort notation System of letters proposed by Rear-Admiral Sir Francis Beaufort (1805) to signify weather phenomena.

Beaufort scale System of numbers proposed by Beaufort to signify wind strength, ranging from 0 (calm) to 12 (hurricane).

beaver tail Tail of fuselage or other body which has progressively flattened cross-section.

beavertail aerial Radar aerial emitting flattened beam having major beam width at 90° to major axis of aerial.

BEC Boron-epoxy composite.

beefing up Strengthening of structural parts, either by redesign for new production or by modification of hardware already made (colloq). Thus, beef, beefed.

beep box Station for remote radio control of activity or vehicle, such as RPV (colloq).

beeper 1 Personal radio alerting receiver.
2 RPV pilot.
3 Manual two-way command switch, eg electric trimmer or cyclic-stick thumbswitch for fuel-valve control.

before-flight inspection Pre-flight inspection (NATO).

behavioural science Study of behaviour of living organisms, especially under stress or in unusual environments.

Beier gear Infinitely variable mechanical transmission accomplished by stacks of convex discs intermeshing with stacks of concave discs, drive ratio being varied by bringing parallel shafts closer together, for extreme ratios, or further apart.

bel Unit for relative intensity of power levels, esp in relation to sound. One bel is ratio of power to be expressed divided by reference power, expressed as logarithm to base 10. Numerically equal to 10 decibels.

Belleville washer Washer, usually of thin elastic metal in form of disc, which offers calibrated resistance to linear deflection of centre, perpendicular to plane of washer.

Bellini-Tosi First directional radio stations, using triangular loop antennas.

bellows Aneroid capsule(s).

belly Underside of central portion of fuselage.

belly-in To make premeditated landing with landing gear retracted or part-extended.

bench Static platform or fixture for manual work, system testing or any other function requiring firm temporary mounting (need not resemble a workshop *).

bench check Mandatory manual strip, inspection, repair, assembly and recalibration of airborne functional parts, made at prescribed intervals and by station and staff certificated by airworthiness authority and/or hardware manufacturer.

bench test Test of complete engine or other functional system on static testbed or rig.

bend allowance Additional linear distance of sheet material required to form bend of specified radius.

bending brake Workshop power tool for pressing metal sheet without dies.

bending moment Moment tending to cause bending in structural member. At any section, algebraic sum of all moments due to all forces on member about axis in plane of section through its centroid.

bending relief Design of aircraft to alleviate aeroelastic deflection, especially of main wing (eg, by distributing mass of fuel and engines across span).

bending stress Secondary stresses (eg, in wing skins and spar booms) which resist deflection due to applied bending moments.

bends Acute and potentially dangerous or lethal discomfort caused by release of gases within a mammalian body exposed to greatly reduced ambient pressure. Thus, a hazard of high-altitude fliers and deep-sea divers (see *aero-embolism, decompression sickness*).

bent 1 Feature of many gun mechanisms, engaging in cocked position with sear.
2 Signal code indicating that facility is inoperative (DoD).
3 Transverse frame capable of offering vertical support and transmitting bending moment.

bent beam, bent course Radio or radar beams significantly diverted from desired rectilinear path by topographic effects, hostile ECM or other cause.

benzene Liquid hydrocarbon, C_6H_6 with characteristic ring structure forming basis of large number of derivatives. Used as fuel or fuel additive, as solvent in paints and varnishes, and for many other manufactures.

benzine Mixture of hydrocarbons of paraffin series, unrelated to benzene. Volatile cleaning fluid and solvent.

benzol Benzene C_6H_6.

BER Beyond economic repair.

Bernoulli's theorem Statement of conservation of energy in fluid flow. Basis for major part of classical aerodynamics, and can be expressed in several ways. One form states an incompressible, inviscid fluid in steady motion must always and at all points have uniform total energy per unit mass, this energy being made up of kinetic energy, potential energy and (in compressible fluid) pressure energy. Making assumptions regarding proportionality between pressure and density, and ignoring gravity and frictional effects, it follows that in any small parcel of fluid or along a streamline, sum of static and dynamic pressure is constant. Thus, if fluid flows through a venturi, pressure is lowest at throat; likewise, pressure is reduced in accelerated flow across a wing.

BERP, Berp British experimental rotor programme.

beryllium Hard, light, strong and corrosion-resistant white metal, m.p. about 1,284°C, density about 3·5 (single-crystal, about 1·835). Expensive, but increasingly used for aerospace structures, especially heat sinks and shields.

b.e.s. Best-endurance speed.

best-climb speed Usually VIMD.

best-endurance speed Always VIMD.

best-fit parabola Profile of practical parabolic aerial capable of being repeatedly manufactured by chosen method. In many cases *** is made up of numerous flat elements.

best-range speed VIMR; speed at which tangent to curve of speed/drag is a minimum; hence where speed/drag is a maximum and fuel consumption/speed a minimum.

BETA, Beta Battlefield exploitation and target acquisition.

beta 1 Sideslip (ß).
2 Angle of sideslip.

beta-1, ß1 Yaw pointing angle (zero sideslip).

beta-2, ß2 Lateral translation (constant yaw angle).

beta blackout Communications interference due to beta radiation.

beta control Control mode for normally automatic propeller in which pilot exercises direct command of pitch for braking and ground manoeuvring. Also called beta mode.

beta particle Elementary particle emitted from nucleus during radioactive decay, having unit electrical charge and mass 1/1,837th that of proton. With positive charge, called positron; with negative, electron. Biologically dangerous, but stopped by metal foil.

BETT, Bett Bolt extrusion thrust-terminator.

bev Billion electron volts. No longer proper term; 10^9 ev is correctly 1 Gev.

bevel gear Gearwheel having teeth whose straight-line elements lie along conical surface (pitch cone); thus, such gears transmit drive between shafts whose axes intersect.

bezel Sloping part-conical ring that retains glass of watch or instrument; esp rotatable outer ring of pilot's magnetic compass.

BF 1 block fuel.
2 Blind-flying; thus * instrument, * panel.

BFA Balloon Federation of America.

BFAANN British Federation Against Aircraft Noise Nuisance.

BFCU Barometric fuel control unit.

BFDAS Basic flight-data acquisition system.

BFDK Before dark.

BFE 1 Buyer-furnished equipment.
2 Basic flight envelope.

BFL 1 Basic field length.
2 Balanced field length.

BFM Basic flight (or fighter) manoeuvring (or manoeuvres).

BFO 1 Beat-frequency oscillation, beat-frequency oscillator.
2 Battlefield obscuration.

BFP 1 British Flying Permit (eg, for ultra-light aeroplanes).
2 Blind-flying panel.
3 Best-fit parabola.
4 Blown fuse plug (tyre).

BFRP Boron-fibre reinforced plastics.

BFS Bundesanstalt für Flugsicherung (W German ATC 1).

BG Bomb (or bombardment) group (USAAC, USAAF, USAF).

BGA British Gliding Association.

BGFOO British Guild of Flight Operations Officers.

BG lighting Blue/green.

BGS Blasting grit, soft = Carboblast.

B/H Diagram of magnetic flux density plotted against magnetizing force.

BHAB British Helicopter Advisory Board.

BHGA British Hang Gliding Association.

BHI Bureau Hydrographique International.

BHO Black-hole ocarina (tactical aircraft IR suppressor).

BHP, b.h.p. Brake horsepower.

BHRA British Hydromechanics Research Association.

BI Basse intensité (light).

bias Voltage applied between thermionic valve (vacuum tube) cathode and control grid.

biased fabric Multi-ply fabric with one or more plies so cut that warp threads lie at angle (in general, near 45°) to length.

bias error Any error having constant magnitude and sign.

bias force Output of accelerometer when true acceleration is zero.

bias temperature effect Rate of change of bias force, usually in g $°C^{-1}$.

BIATA British Independent Air Transport Association.

Bibby coupling Drive for transmitting shaft rotation without vibration, using multiple flexural cantilevers linking adjacent discs.

biconvex Presenting convex surface on both sides. Such wings usually have profile formed from two circular arcs, not always of same radius, intersecting at sharp leading and trailing edges. Inefficient in subsonic flight.

bicycle Form of landing gear having two main legs in tandem on aircraft centreline.

bidding 1 Phase in procurement process in which rival manufacturers submit detailed proposals with prices.
2 Competitive procedure within air carrier's flying staff for licence endorsement on new type of aircraft.

BIDE Blow-in door ejector (engine nozzle).

BIDS, Bids Battlefield information distribution system.

BIFAP Bourse Internationale de Fret Aérien de Paris.

BiFET Bistable field-effect transistor (gate).

bifilar Suspension of mass by two well separated filaments; mass normally swings in plane of filaments.

bi-fuel 1 Bipropellant.
2 More rarely, heat engine which can run on either of two fuels but not both together.

bifurcated Rod, tube or other object of slender form which is part-divided into halves; fork-ended.

bigraph Two-letter code for airfield name painted on local buildings, eg gasholders.

Bigsworth Chartboard, transparent overlay, Douglas protractor, parallel rules, all integrated (obs).

bilateral Agreement between two parties.

bilateration Position determination by use of AF signal beamed to vehicle, which re-radiates it to original ground station and to second at surveyed location giving known delay path. Can be used for control of multiple vehicles beyond LOS.

billet Rough raw material metal form, usually square or rectangular-section bar, made by forging or rolling ingot or bloom.

billow Inflation of each half of Rogallo wing.

BIM 1 Ballistic intercept missile.
2 Blade indicator method (helicopters).
3 British Institute of Management.

bimetallic joint Joint between dissimilar metals.

bimetallic strip Strip made of sandwich of metals, usually two metals chosen for contrasting coefficients of thermal expansion. As halves are bonded together any change in temperature will tend to curve the strip. Principle of bimetallic switch and temperature gauge.

bi-mono aircraft Monoplane having detachable second wing for operation as biplane.

bin 1 Electronic three-dimensional block of airspace. All airspace in range of SSR or other surveillance radar is subdivided into *, size of which is much greater than expected dimensions of aircraft. Thus presence of an aircraft cannot load more than one (transiently two) * at a time.
2 Upright cylindrical receptacle, esp some early gun turrets.

binary code Binary notation.

binary notation System of counting to base of 2, instead of common base of 10. Thus 43 (sum of 2^5, 2^3, 2^1 and 2^0) is written 101011. All binary numbers are expressed in terms of two digits, 0 and 1, Thus digital computer can function with bistable elements, distinction between a 0 or 1 being made by switch being on or off, or magnetic core element being magnetized or not.

binary switch Bistable switch.

binaural Listening with both ears.

bingo 1 As an instruction, radioed command to aircrew (usually military) to proceed to agreed alternate base.
2 As information, radioed call from aircrew (usually military) meaning that fuel state is below a certain critical level (usually that necessary to return to base). Thus calls "* fuel" or "Below *".

binor Binary optimum ranging. Binary code modified for range measurement and minimizing receiver acquisition time.

bioastronautics Study of effects of space travel on life forms.

biochemical engineering Technology of biochemistry.

biochemistry Chemistry of life forms.

biodegradable Of waste material, capable of being broken down and assimilated by soils and other natural environments.

biodynamics Study of effects of motion, esp accelerations, on life forms.

biological agent Micro-organism causing damage to living or inanimate material and disseminated as a weapon.

biological decay Long-term degradation of material due to biological agents.

biological warfare Warfare involving use as weapons of biological agents, toxic biological products and plant-growth regulators.

biomedical monitoring Strictly "biomedical" is tautological, but term has come to denote inflight monitoring of heartbeat, respiration and sometimes other variables, esp of astronauts in space.

bionics Study of manufactured systems, esp those involving electronics, that function in ways intended to resemble living organisms.

bio-pak Container for housing and monitoring

life forms, usually plants, insects and small animals, in space payload. May be recoverable.

biosatellite Artificial satellite carrying life forms for experimental purposes.

biosensor Sensor for measuring variables in behaviour of living systems.

biosphere Location of most terrestrial life: oceans, surface and near-surface of land, and lower atmosphere.

biowaste All waste products of living organisms, esp humans in spacecraft.

BIP Borescope inspection port.

biplane Aeroplane or glider having two sets of wings substantially superimposed (see *tandem-wing*).

biplane interference Aerodynamic interference between upper and lower wings of biplane or multiplane.

biplane propeller Propeller having pairs of blades rotating together in close proximity, resembling biplane wings.

BIPM Bureau International des Poids et Mesures.

bipropellant 1 As adjective, rocket which consumes two propellants — solid, liquid or gaseous — normally kept separate until introduction to reaction process. In most common meaning, propellants are liquid fuel and liquid oxidizer, stored in separate tanks.

2 As noun, rocket propellant comprising two components, typically fuel and oxidizer.

BIRD Banque Internationale pour la Reconstruction et le Dévéloppement.

bird Any flight vehicle, esp aeroplane, RPV, missile or ballistic rocket (colloq).

bird gun Gun, usually powered by compressed air, for firing real or simulated standard birds at aircraft test specimens to demonstrate design compliance.

BIRDiE Battery integration and radar display equipment (USA).

bird impact Birdstrike; design case for all aerodynes intended for military or passenger-carrying use (see *standard bird*).

bird ingestion Swallowing of one or more birds by gas-turbine or other air-breathing jet engine, with or without subsequent damage or malfunction (see *ingestion certification*).

bird strike, birdstrike Collision between aerodyne and natural bird resulting in significant damage to both.

BIRMO British IR Manufacturers' Organization.

BIS 1 British Interplanetary Society.

2 Board of Inspection and Survey (US Navy).

bi-signal zone Portions of radio-range beam on either side of centreline where either A or N signal can be heard against monotone on-course background.

BISS Base and installation security system.

bistable Capable of remaining indefinitely in either of two states: thus, bistable switch is stable in either on or off position (see *flip-flop*).

BIT 1 Built-in test.

2 Bureau International du Travail (ILO).

bit Unit of data or information in all digital (binary) systems, comprising single character (0 or 1) in binary number. Thus, capacity in bits of memory is $\log_2$ of number of possible states of device. From "binary digit".

BITE, Bite Built-in test equipment.

BIT/Fi Built-in test and fault isolation.

bit rate Theoretical or actual speed at which information transmission or processing system can handle data. Measured in bits, kbits or Mbits per second.

biz Business; hence, bizjet (colloq).

BJA Baseline jamming assets.

BJSD British Joint Service Designation.

BJSM British Joint Services Mission.

BKEP Boosted kinetic-energy penetrator.

BKN Broken.

BKSA British Kite Soaring Association.

BL 1 Buttock line.

2 Base-line (hyperbolic nav).

BLAC British Light Aviation Centre.

Black SAO security classification.

black ball Traditional form of artificial horizon based on black hemisphere or bi-coloured sphere.

black body Theoretical material or surface which reflects no radiation, absorbs all and emits at maximum rate per unit area at every wavelength for any given temperature. In some cases wavelength is restricted to band from near-IR through visible to far-UV.

black-body radiation EM spectrum emitted by black body, theoretical maximum of all wavelengths possible for any given body temperature.

black box Avionic equipment or electronic controller for hardware device or sytem, removable as single package. Box colour is immaterial (colloq).

blacker-than-black In TV system, portion of video signal containing blanking and synchronizing voltages, which are below black level (with positive modulation) and prevent electrons from reaching screen.

black hole IRCM design giving engine efflux (esp of helicopter) or other heat source greatest protection.

black hot TV or IR display mode giving negative picture (warm ship looks black against white sea).

blackjack Hand tool for manually forming sheet metal; has form of flexible (often leather) quasi-tubular bag filled with lead shot too small to cause local indentations.

black level In TV system, limit of video (picture) signal black peaks, below which signal voltages cannot make electrons reach screen.

blackout 1 In war, suppression of all visible light-

ing that could convey information to enemy.
2 Fadeout of radio communications, including telemetry, as result of ionospheric disturbances or to sheath of ionized plasma surrounding spacecraft re-entering atmosphere.
3 Fadeout of radio communications caused by disruption of ionosphere by nuclear explosions.
4 Dulling of senses and seemingly blackish loss of vision in humans subjected to sustained high positive acceleration.

blackout block Block of consecutive serial numbers deliberately left unused.

black picture See *black hot.*

bladder tank Fluid tank made of flexible material, esp one not forming part of airframe.

blade 1 Aerofoil designed to rotate about an axis, as in propeller, lifting rotor, axial compressor rotor or axial turbine.
2 Rigid array of solar cells, especially one having length much greater than width.
3 Operative element of windscreen wiper.

blade activity factor Non-dimensional formula for expressing ability of blade (1) to transmit power; integral between 0·1 and 0·5 of diameter of chord and cube of radius with respect to radius. Loosely, low aspect ratio means high activity factor.

blade aerial Radio aerial, eg for VHF communications, having form of vertical blade, either rectilinear, tapered or backswept.

blade angle 1 In propeller or fan blade, angle (usually acute) between chord at chosen station and plane of rotation.
2 In helicopter lifting rotor, acute angle between no-lift direction and plane perpendicular to axis of rotation.

blade angle of attack Angle between incident airflow and blade (1) tangent to mean chord at leading edge at chosen station.

blade articulation Attachment to hub of helicopter lifting rotor blades by hinges in flapping and/or drag axes.

blade back Surface of blade (1) corresponding to upper surface of wing.

blade beam Hand tool in form of beam incorporating padded aperture fitting propeller or rotor blade; for adjusting or checking blade angle.

blade centre of pressure Point through which resultant of all aerodynamic forces on blade (1) acts.

blade damper Hydraulic, spring or other device for restraining motion of helicopter lifting rotor blade about drag (lag) hinge.

blade element Infinitely thin slice (ie having no spanwise magnitude) through blade (1) in plane parallel to axis of rotation and perpendicular to line joining centroid of slice to that axis. Thus, blade is made up of infinity of such elements from root to tip, usually all having different section profile and blade angle.

blade face Surface of blade (1) corresponding to underside of a wing. With propellers, called thrust face.

blade loading Of helicopter or autogyro, gross weight divided by total area of all lifting blades (not disc area).

blade passing noise Component of internally generated noise of turbomachinery, caused by interaction between rotating blades and wakes from inlet guide vanes and stationary blades. Generates distinct tones at blade-passing frequencies, which in turn are product of number of blades per row and rotational speed.

blade root 1 Loosely, inner end of blade (1).
2 Where applicable, extreme inboard end of blade incorporating means of attachment (see *blade shank*).

blade section Shape of blade element.

blade shank Where applicable, portion of blade of non-aerofoil form extending from root to inboard end of effective aerofoil section. Unlike root, * of propeller is outside spinner.

blade span axis 1 Axis, defined by geometry of root pitch-change bearings, about which blade is feathered.
2 Axis through centroids of sections at root and tip.

blade station Radial location of blade element, expressed as decimal fraction of tip radius (rarely, as linear distance from axis of rotation, from root or from some other reference).

blade sweep Deviation of locus of centroids of all elements of blade from radial axis tangential to that locus at centre. Was marked in early aircraft propellers, usually towards trailing edge (ie, trailing sweep); leading sweep, in which tips would be azimuthally ahead of hub, is rare.

blade tilt Deviation of locus of centroids of all elements of blade from plane of rotation. Again a feature of early aircraft propellers, more common form being backward tilt, visible in side view as propeller flat at back and tapered from boss to tip in front.

blade twist 1 Unwanted variation in pitch from root to tip caused by aerodynamic loads.
2 Natural twist which reduces blade angle from root to tip.

blade width ratio Ratio of mean chord to diameter.

blank 1 Workpiece sheared, cut, routed or punched from flat sheet before further shaping.
2 Action of cutting part from flat sheet, esp by using blanking press and shaped die.
3 Round of gun ammunition without projectile.
4 All-weather cover tailored to engine inlet or other aperture, forming part of AGE for each aircraft type.

blanket 1 Layer of thermally insulating material tailored to protect particular item.
2 Layer of heating material supplied with electrical or other energy.

blanket cover Fabric cover for aircraft machine-sewn into large sheet, draped over structure, pulled to shape and sewn by hand.

blanketing 1 Suppression, distortion or other gross interference of wanted radio signal by unwanted one.
2 In long-range radio communication, prevention of reflection from F layers by ionization of E layer.

blanketing frequency Signal frequency below which radio signals are blanketed (2).

blank-gore parachute Parachute having one gore left blank, without fabric.

blanking 1 Using press and blanking die to cut blanks (1).
2 In electron tube or CRT, including TV picture tubes, suppression of picture signal on fly-back to make return trace invisible.

blanking cap Removable cap fitted to seal open ends of unused pipe connections or other apertures in fluid system.

blanking plate Removable plate fitted to seal aperture in sheet, such as unused place for instrument in panel.

blanking signal Regular pulsed signal which effects blanking (2) and combines with picture signal to form blanked picture signal. Sometimes called blanking pulse.

Blasius flow Theoretically perfect laminar flow.

blast 1 Loosely, mechanical effects caused by blast wave, high-velocity jet or other very rapidly moving fluid.
2 Rapidly expanding products from explosion and subsequent blast wave(s) transmitted through atmosphere.

blast area Region around launch pad which, before final countdown of large vehicle, is cleared of unecessary personnel and objects.

blast cooling In rotating electrical machines and other devices, removal of waste heat by airflow supplied under pressure.

blast deflector Structure on launch pad or captive test stand to turn rocket or jet engine efflux away from ground with minimal erosion and disturbance.

blast fence Large barrier constructed of multiple horizontal strips of curved section, concave side upwards, which diverts efflux behind parked jet aircraft upwards and thus removes annoyance and danger at airfields.

blast/fragmentation Warhead, common on AAMs and SAMs, whose effect combines blast of HE charge and penetration of fragments of rod(s) or casing.

blast gate See *waste gate*.

blast line Chosen radial line from ground zero along which effects of nuclear explosion (esp blast effects) are measured.

blast-off Launch of rocket or air-breathing jet vehicle; usually, from ground or other planetary surface (colloq).

blast pad Area immediately to rear of runway threshold across which jet blast is most severe. Constructed to surface standards higher than overrun or stopway beyond.

blast pen Small pen, enclosed by strong embankments on three sides, but open above, for ground running jet or rocket aircraft or firing missile engines.

blast-pipe See *blast tube*.

blast tube Refractory tube linking rocket combustion chamber or propellant charge with nozzle, where these have to be axially separated.

blast valve Valve in air-conditioning and other systems of hardened facilities which, upon sensing blast wave, swiftly shuts to protect against nuclear contamination.

blast wave Shock wave (N-wave) of large amplitude and followed by significant (4 ata; 40 kPa or more) overpressure. Travels at or above velocity of sound and causes severe mechanical damage. Centred on explosions (local ** caused by lightning); attenuation and effective radius depend on third or fourth power of released energy.

blaugas German gas used for airship lift and fuel; mixture of ethylene, methylene, propylene, butylene, ethane and hydrogen.

BLC Boundary-layer control, esp gross control of airflow around lifting wing to increase circulation and prevent flow breakaway.

BLD 1 DSB, double sideband.
2 Berufsverband Luftfahrt-Personal in Deutschland eV (G).

BLE Boundary-layer energizer.

bleed 1 To allow quantity of fluid to escape from closed system.
2 To extract proportion (usually small) of fluid from continuously flowing supply; eg compressed air from gas-turbine engine(s).
3 To allow fluid to escape from closed system until excess pressure has fallen to lower level or equalized with surroundings.
4 To remove unwanted fluid contaminating system filled with other fluid, eg * air from hydraulic brakes.
5 To allow speed or height of aircraft to decay to desired lower level, thus *-off speed before lowering flaps.
6 To allow electronic signal or electrical voltage to decay to zero (eg, * glide path during flare in automatic landing).

bleed air Compressed air bled from main engines of gas-turbine aircraft.

bleed and burn Vertical lift system in which fuel is burned in a vertical combustion chamber fed with air from main engine (extra airflow may or may not be induced by ejector effect).

bleeder resistor Resistor permanently coupled across power supply to allow filter capacitor charge to leak away after supply is disconnected (see *bleed-off relay*).

bleeder screw Small screw in tapped hole through highest point of hydraulic or other liquid system to facilitate bleeding (4) air or vapour.

bleeding edge Edge of map or chart where cartographic detail extends to edge of paper.

bleed-off relay In laser, discharges capacitors when switched off, to render accidental firing impossible.

blended Aerodynamic (arguably, also hydrodynamic) shape in which major elements merge with no evident line of demarcation. Thus, aeroplane having * wing/body (see next entry).

blended-hull seaplane Marine aeroplane, generally called in English flying boat, in which planing bottom is blended into fuselage. Involves dispensing with chine, sacrificing hydrodynamic behaviour in order to reduce aerodynamic drag.

blend point In aerodynamic shapes having rigid and flexible surfaces mutually attached (eg Raevam, variable inlet ducts, flexible Krügers), point in section profile at which flexibility is assumed to start.

BLEU Blind Landing Experimental Unit (Bedford, UK).

BLG 1 Laser-guided bomb (F).
2 Body (-mounted) landing gear.

BLI Belgische Luchtvaart Info.

blimp Non-rigid airship (from "Dirigible Type B, limp", colloq until made official USN term in 1939).

blind 1 Without direct human vision.
2 Of radar, incapable of giving clear indication of target (eg see * *speed*).

blind bombing Dropping of free-fall ordnance on surface target unseen by aircrew.

blind bombing zone Restricted area (strictly, volume) where attacking aircraft know they will encounter no friendly land, naval or air forces.

blindfire Weapon system able to operate without visual acquisition of target.

blind flying Manual flight without external visual cues.

blind-flying panel Formerly, in British aircraft, separate panel carrying six primary flight instruments: ASI, horizon, ROC (today VSI), altimeter, DG and TB (turn/slip).

blind landing 1 Landing of manned aircraft, esp aerodyne, with crew deprived of all external visual cues.
2 Landing of RPV unseen by remote pilot except on TV or other synthetic display.

blind nut Nut inserted or attached on far side of sheet or other member to which there is no access except through bolt hole.

blind rivet Rivet inserted and closed with no access to far side of joint. Apart from explosive and rare magnetic-pulse types, invariably tubular.

blind speed effect Characteristic of Doppler MTI systems used with radars having fixed PRF which makes them blind to targets whose Doppler frequencies are multiples of PRF (see *staggered PRF*).

blind toss Programmed toss without acquisition of target (eg on DR from an offset).

blink 1 Of light or other indicator, to be illuminated and extinguished, or to present black/white or other contrasting colour indication, more than 20 times per minute.
2 In aircraft at night in VFR, manually to switch off navigation lights (typically, twice in as many seconds) as acknowledgement of message.

blinker Light or indicator that blinks (1), eg to confirm oxygen feed.

blip 1 Visible indication of target on radar display. Due normally to discrete target such as aircraft or periscope of submarine; in ground mapping mode, term used only to denote strong echo from transponder.
2 Spot, spike or other indication on CRT due to signal of interest.
3 To control energy input to early aeroplane by switching ignition on and off as necessary (normally, on landing approach).
4 To operate bang/bang control manually (eg, electric trim).

blip driver Operator of synthetic trainer for SSR or other surveillance radar with rolling ball or other means of traversing system co-ordinates to give desired blip (1) position and movement (colloq).

blisk Axial turbine rotor stage (rarely, compressor stage) in which disc and blades are fabricated as single piece of material.

blister 1 Streamlined protuberance on aerodynamic body, usually of semicircular transverse section.
2 See *blister spray*.

blister aerial Aerial projecting from surface of aircraft and faired by dielectric blister.

blister hangar Prefabricated and demountable hangar having arched roof and fabric covering.

blister spray Arching sheet of water thrown up and outwards above free water surface on each side of planing hull or float. Compared with ribbon spray, has lower lateral velocity, rises higher, is clear water rather than spray, and is much more damaging.

blister spray dam Strong strip forming near-vertical wall projecting downwards along forebody chine of hull or float.

blivet Flexible bag for transporting fuel, usually as helicopter slung load.

BLN Balloon.

BLO Below clouds (ICAO).

blob Local atmospheric inhomogeneity, produced by turbulence, with temperature and humidity different from ambient. Can produce angels (2).

block 1 In quantity production, consecutive series of identical products having same * number. In World War 2 aircraft production a * might number several hundred; with large spacecraft and launch vehicles, fewer than ten. In

general, products of two * normally differ as result of incorporation of engineering changes.
2 In research, groups of experimental items subjected to different treatment for comparative purposes.
3 In EDP, group of machine words considered as a unit.
4 In aircraft (usually commercial) operation, chocks (real or figurative) whose removal or placement defines the beginning and end of each flight.

blockbuster Large thin-case conventional bomb (colloq).

block construction Arrangement of gores of parachute such that fabric warp threads are parallel to peripheral edge.

block diagram Pictorial representation of system, other than purely electrical or electronic circuit, in which lines show signal or other flows between components depicted as blocks or other conventional symbols.

blocker door In installed turbofan engine, hinged or otherwise movable door (normally one of peripheral ring) which when closed blocks fan exit duct and opens peripheral exits directing airflow diagonally forward.

block fuel Fuel burned during block time.

blockhouse Fortified building close to launch pad for potentially explosive vehicles, from which human crew manage launch operations or perform other duties (eg photography).

block in To park transport aircraft at destination. Term spread from commercial to military transport use.

blocking 1 In wind tunnel, gross obstruction to flow caused by shockwaves at Mach numbers close to 1, unless throat and working section designed to avoid it (see *choking*).
2 Use of struts and wedges to prevent movement of loose cargo or cargo inside container.
3 Use of form block.

blocking capacitor Capacitor inserted to pass AC and block DC.

blocking layer Barrier layer in photovoltaic (ie solar) cell.

blocking oscillator Any of many kinds of oscillator which quench their output after each alternate half-cycle to generate sawtooth waveform.

blocking up To use shaped masses behind sheet metal being hammered.

block shipment Rule-of-thumb logistic supply to provide balanced support to round number of troops for round number of days.

block speed Average speed reckoned as sector distance divided by block time.

block stowage Loading all cargo for each destination together, for rapid off-loading without disturbing cargo for subsequent destinations.

block template Template for making form block.

block time Elapsed period from time aircraft starts to move at beginning of mission to time it comes to rest at conclusion. Historically derived from manual removal and placement of blocks (chocks). Normally used for scheduled commercial operations, either for intermediate sectors or end-to-end.

blood-albumen glue Adhesive used in aircraft plywood, made from dry cattle blood albumen.

blood chit 1 Form signed by aircraft passenger before flight for which no fare has been paid (eg, in military aircraft) indemnifying operator against claims resulting from passenger's injury or death (colloq).
2 Plastic or cloth message, usually in several languages, promising reward if bearer is helped to safety. Often includes representation of bearer's national flag (colloq).

bloom 1 Ingot from which slag has been removed, sometimes after rough rolling to square section.
2 Of ECM chaff, to burst into large-volume cloud after being dispensed as compact payload.

blooming 1 In surface-coating technology, to coat optical glass with layer a few molecules thick which improves optical properties (by changing refractive index and/or reducing external reflection).
2 In CRT, defocusing effect caused by excessive brightness and consequent mushrooming of beam.
3 In atmospheric laser operations, defocusing or undesired focusing of pulses or beam caused by lens-like properties in atmosphere (see *blob*).

blow 1 Rupture in case of solid rocket during firing.
2 To activate blowing system.

blowback 1 Type of action of automatic gun in which bolt, or other sliding breech-closure, is never locked and is blown back by combustion pressure on its face, or by rearwards motion of empty case, inertia sufficing to keep breech closed until projectile has left muzzle.
2 Improper escape of gas through breech during firing of gun, due to ruptured case, faulty breech mechanism or other malfunction.
3 Closure, or partial closure, of spoiler or speed brake due to aerodynamic load overcoming force exerted by actuation system.

blowback angle Maximum angle to which a spoiler can be extended under given q (dynamic pressure) without blowback (3).

blowdown During captive firing of liquid rocket, expulsion of residual propellant by gas pressure (usually nitrogen) after burnout or cutoff.

blowdown period Period in cycle of reciprocating IC engine in which exhaust valve or port is open prior to BDC.

blowdown tunnel Open-circuit wind tunnel in which gas stored under pressure escapes to atmosphere, or into evacuated chamber, through working section.

blowdown turbine Turbine driven by PE (4)

exhaust gas in such a way that kinetic energy of discharge from each cylinder is utilized.

blower 1 Centrifugal compressor with output (from NTP input) at between 1 and 35 lb/sq in gauge (6·9–240 kN m^{-2}).
2 Centrifugal fan used on PE (4), esp on radial engine, to improve distribution of mixture among cylinders.

blower pipe In an airship, duct through which propeller slipstream is rammed to pressurize ballonets.

blow-in door Door free to open inwards against spring upon application of differential pressure (eg, in aircraft inlet duct).

blowing Provision for discharging high-pressure, high-velocity bleed air from narrow spanwise slit along wing leading edge, tail surface or ahead of flap or control surface. Greatly increases energy in boundary layer, increases circulation and prevents flow breakaway.

blown flap Flap to which airflow remains attached, even at sharp angles, as result of blowing sheet of high-velocity air across its upper surface (see *boundary-layer control*, *supercirculation*).

blown periphery Parachute in which part of peripheral hem becomes blown between two rigging lines in another part of canopy, and attempts to inflate inside-out.

blown primer Percussion cartridge primer which blows rearward out of its pocket, allowing primer gases to escape.

blow-off Explosive or other enforced separation of instrument pack or other payload from rocket vehicle or other carrier. Hence, ** signal.

blow-off valve Safety valve set to open at predetermined dP at chosen point in axial compressor casing to allow escape of part of air delivered by stages upstream. Common in early axial gas turbines to prevent surging and other malfunctions, especially during starting and acceleration.

blowout 1 Flameout in any fuel/air combustion system caused by excessive primary airflow velocity.
2 In particular, afterburner flameout.

blowout disc Calibrated disc of thin metal used to seal fluid system pipe, rocket combustion space or any other device subject to large dP. If design dP limit is exceeded, ** ruptures. Also called safety diaphragm.

BLR Bundesverband der Luftfahrt Zubehör und Raketenindustrie eV (G).

BLS Blue-line speed.

BLSN Blown snow (ICAO).

BLSS Base-level supply sufficiency.

BLU 1 Bande latérale unique, = SSB.
2 Bomb, live unit (US).

Blue airway Originally in US, N-S civil airway.

blue box Pre-1945 Link trainer (colloq).

Blue Flag Command post exercises emphasizing tac-air warfare management (USAF).

Blue Force In military exercise, forces used in friendly role.

blue key Blue annotation or image on original map or diagram which does not show in subsequent reproduction.

blue-line speed For multi-engined aircraft, best climb-out speed after failure of one engine (usually thus marked on ASI).

Blue Paper Notice of proposed amendment (BCARs).

blue pole S-seeking pole of magnet.

blueprint Drawing reproduced on paper by ammonium ferric citrate or oxalate and potassium ferrocyanide to give white lines on blue background. Seldom used today, but word has common loose meanings: any drawn plan; any written plan of campaign or course of action.

Blue sector That half of ILS localizer beam modulated at 150 Hz (right of centreline).

blue suit US Air Force; usually in contradistinction to white suit = contractor personnel (colloq).

bluff body Solid body immersed in fluid stream which experiences resultant force essentially along direction of relative motion and promotes rapidly increasing downstream pressure gradient. Causes flow breakaway and turbulent wake. Broadly, bluff is opposite of streamlined.

blunt A * trailing edge or rear face of body causes turbulence immediately downstream, but main airflow cannot detect that body or aerofoil has come to an end and thus continues to behave as if in passage over surface of greater length or chord.

blushing Spotty or general milkiness or opacity of doped or varnished surface, caused by improper formulation, too-rapid solvent evaporation or steamy environment.

BLW Below (ICAO).

BM 1 Bus monitor.
2 Bubble memory.

B/M Boom/mask (switch).

BMAA British Microlight Aircraft Association.

BMC Basic mean chord.

BM/C^3 Battle management/command, control and communications (SDI).

BMD 1 US Air Force Ballistic Missile Division (later SAMSO).
2 Ballistic-missile defence.

BME Bulk memory element.

BMEWS Ballistic-missile early-warning system.

BMF Ministry of Finance (G).

BMFT Ministry for research and technology (G).

BMI Bismaleimide, high-temperature-resistant resin adhesive.

BMR Bearingless main rotor.

BMS Bureau Militaire de Standardization (F).

BMTOGW Basic mission takeoff gross weight.

BMTS Ballistic-missile target system.

BN Night bomber (F, obs).

B/N Bombardier/navigator.

B/NB Bid or no bid.
BNH battery Bipolar nickel/hydrogen.
BNK Bureau of new construction (USSR).
BO Boom operator.
Bo Boundary lights.
BOA Basic ordering agreement.
boarding card Document issued at check-in which admits passenger to aircraft.
boards Speed brakes (colloq).
boat seaplane Flying boat.
boat-tail Rear portion of aerodynamic body, esp body of revolution, tapered to reduce drag. Taper angle must be gentle to avoid breakaway.
bobbing Rare fluctuation in strength of radar echoes allegedly due to alternate attenuation and reinforcement of successive pulse waves.
Bode plot Gain and phase angle against system frequency.
body 1 Any three-dimensional object in fluid flow.
2 In most aircraft, central structure: hull of marine aircraft or airship, fuselage of aeroplane or helicopter, * of missile.
3 Any observable astronomical object, esp within solar system.
body burden Aggregate radioactive material (not dose received) in living body.
body lift Lift from fuselage of supersonic aircraft or missile at AOA other than zero.
body of revolution Body (2) having circular section at any station and surface shape described by rotating side elevation about axis of symmetry. Ideal streamlined forms are generally such.
body plan Full-scale elevations and sections of aircraft body in lofting.
body sensor Biomedical sensors worn by astronauts or aircrew to measure parameters such as body temperature, pulse rate and respiration.
body stall Gross flow breakaway from core engine and afterbody in installed turbofan.
BOE Black-out exit; predicted time in manned re-entry at which communications will be resumed.
boe Barrels of oil equivalent; thus boe/d = boe per day.
boffin Research scientist, esp senior worker on secret defence project (colloq).
bog To taxi across ground so soft that landing gear sinks in and halts aircraft (see *flotation*).
bogey Air contact (5) assumed to be enemy (UK spelling often bogy).
bogie Landing gear having multi-wheel truck on each leg.
bogie beam Pivoted beam linking front and rear axles of bogie to each other and to leg.
BOH Break-off height.
boiler Gas-turbine used as a core engine in high-ratio turbofan, as source of hot gas for tip-drive rotor or fan-lift system or any other application calling for central power source (colloq).
boilerplate Non-flying form of construction where light weight is sacrificed for durability and low cost (see *battleship*).
boiloff Cryogenic propellant lost to atmosphere through safety valves as result of heat transfer through walls of container (which may be static storage or tank in launch vehicle).
BOK Bureau of special designers (USSR).
BOL Bearing-only launch.
bold-face procedures Emergency procedures, written in flight manual in bold-face type.
bollard Mooring attachment in form of short upright cylinder on marine aircraft hull or float bow.
bolometer Sensitive instrument based on temperature coefficient of resistance of metallic element (usually platinum); used to measure IR radiation or in microwave technology (see *radiometer*).
bolt 1 In advanced airframe structure, usually precision fitted major attachment device loaded in shear.
2 In firearm, approximately cylindrical body which oscillates axially behind barrel feeding fresh rounds, closing breech and extracting empty cases (see *breech-block*).
bolter 1 In carrier (1) flying, aircraft which fails to pick up any arrester wire and overshoots without engaging barrier.
2 Verb, to perform 1.
Boltzmann constant Ratio of universal gas constant to Avogadro's number; $1{\cdot}38 \times 10^{-23}$ J °K^{-2}.
Boltzmann equation Transport equation describes behaviour of minute particles subject to production, leakage and absorption; describes distribution of such particles acted upon by gravitation, magnetic or electrical fields, or inertia. Boltzmann-Vlasov equations describe high-temperature plasmas.
bomb 1 Transportable device for delivery and detonation of explosive charge, incendiary material (including napalm), smoke or other agent, esp for carriage and release from aircraft.
2 Streamlined body containing pitot tube towed by aircraft and stabilized by fins to keep pointing into relative wind in region undisturbed by aircraft.
bomb-aimer Aircrew trade in RAF (formerly) and certain other air forces.
Bomb Alarm System Automatic system throughout Conus for detecting and reporting nuclear bursts.
bombardier Aircrew trade in US Army and Air Force (formerly).
bombardment ion engine Rocket engine for use in deep space which produces ion beam by bombarding metal (usually mercury or caesium) with electrons.
bomb bay In specially designed bomber aircraft, internal bay for carriage of bombs (in fuselage, wings or streamlined nacelles).

bomb damage assessment Determination of effects on enemy targets of all forms of aerial attack.

bomb door Door which normally seals underside of bomb bay. Can slide rearwards, sideways and upwards, open to each side or rotate through 180° about longitudinal axis to release stores from its upper face.

bomber Aircraft designed primarily to carry and release bombs. Term today reserved for strategic aircraft.

bomber-transport Former category of military aircraft capable of being used for either type of mission.

bomb fall line Bright line on HUD along which free-fall bombs would fall to the ground if they were released.

bomb impact plot Graphical picture of single bombing attack by marking all impact or detonation centres on pre-strike vertical reconnaissance photograph(s).

bombing angle Angle between local vertical through aircraft at bomb release point and line from that point to target.

bombing errors 1 50% circular error: radius of circle, with centre at desired mean point of impact, which contains half missiles independently aimed to hit that point (see *CEP (1)*).
2 50% deflection error: half distance between two lines drawn parallel to track and equidistant from desired mean point of impact which contain between them half impact points of missiles independently aimed to hit that point.
3 50% range error: half distance between two lines drawn perpendicular to track and equidistant from desired mean point of impact which contain between them half missiles independently aimed to hit that point.

bombing height Vertical distance from target to altitude of bombing aircraft.

bomblet Small bomb, usually of fragmentation type, carried in large clusters and released from single streamlined container.

bomb line Forward limit of area over which air attacks must be co-ordinated with ground forces; ahead of ** air forces can attack targets without reference to friendly ground troops.

bomb rack Formerly, attachments in bomb bay or externally to which bombs were secured; provided with mechanical or EM release, fuzing and arming circuits and sometimes other services. Replaced by universal store carriers tailored to spectrum of weapons.

bomb release line Locus of all points (often a near-circle) at which aircraft following prescribed mode of attack must release particular ordnance in order to hit objective in centre.

bomb release point Particular point in space at which free-fall ordnance must be released to hit chosen target.

bomb sight Any device for enabling aircraft to be steered to bomb release point, esp one in which aimer sights target optically and releases bombs by command.

bomb trolley Low trolley for carriage of ordnance from airfield bomb stores to aircraft (and often equipped to raise bombs into position on bomb racks).

bomb winch Manual or powered winch for hoisting bombs from trollies into or beneath racks.

bonding 1 Structurally joining parts by adhesive, esp adhesives cured under elevated temperature and/or pressure.
2 Joining together all major metal parts of an aircraft, esp an aircraft not of all-metal construction, to ensure low-resistance electrical continuity throughout. Even where metal structures are squeezed together by bolts or rivets a bond of copper strip or braided wire must link them reliably. Bonding is necessary for Earth-return systems and to dissipate lightning strikes and other electrical charges safely with no tendency to arcing.

bonding noise In older aircraft, radio interference caused by relative movement between metal parts bonded (2) together.

Bondolite Low-density sandwich of balsa faced with aluminium.

bonedome Internally padded rigid protective helmet worn by combat aircrew (colloq).

boneyard Graveyard of unwanted aircraft, usually stripped of potential spares (colloq).

bonnet Valve hood in aerostat envelope.

Boolean algebra Powerful and versatile algebra compatible with binary system and with functions AND, OR and NOT.

boom 1 Any long and substantially tubular portion of structure linking major parts of an aircraft (esp linking the tail to the wing or to a short body).
2 Longitudinal structural members forming main compression and tension members of a wing spar, having large section modulus as far as possible from wing flexural axis.
3 Device used in some air-refuelling tanker aircraft in form of pivoted but rigid telescopic tube steered by aerodynamic controls until its tip can be extended into a fuel-tight receptacle on receiver aircraft.
4 Sound heard due to passage of shockwaves from distant supersonic source, such as SST flying high overhead.
5 Spanwise pipe conveying ag-liquid to spraying nozzles; hence * width, * pivots.

boom avoidance Technique of flying aircraft for minimal boom (4) disturbance on ground.

boom avoidance distance Distance along track over which ** technique is enforced. Hence, BAD (departure) and BAD (arrival).

boom carpet Strip of Earth's surface along which

observers hear sonic boom from supersonic aircraft.

boomer Operator of refuelling boom (5) in tanker aircraft (colloq).

boom microphone Voice microphone carried on cantilever boom (slender structural beam) pivoted at side of headset so that it can be moved away from the mouth for avoiding unwanted speech broadcasting.

boom receptacle Flight refuelling socket on military aircraft with which a boom (3) forms a fuel-tight connection despite motion and changes of orientation relative to tanker.

boom throw-forward Distance along track from origin of shockwave to ground impact.

boom trough See *boom well.*

boom well Recess in deck plating of marine-aircraft float or hull to take end-fitting of struts or booms (1).

boost 1 Any temporary augmentation of thrust or power in a mechanical or propulsive system.
2 Excess pressure, over and above a datum, in induction manifold of PE (4) as result of supercharging. Datum is usually one standard atmosphere.
3 Jettisonable booster rocket for unmanned vehicle (UK, colloq).
4 Fast-burning portion of a boost/sustain motor.

boost control Control system, today invariably automatic, for maintaining suitable boost pressure in aircraft PE (4) and, in particular, for avoiding excessive boost.

boost/cruise motor Rocket motor having very large but brief thrust for vehicle launch followed by lower but long-duration thrust for aerodynamic cruise.

boosted controls Aeroplane flying control surfaces in which balance is reduced and pilot input augmented by brute force, usually by hydraulic jacks. Much simpler and cruder than a powered flying-control system (see *servo*).

booster 1 Boost rocket.
2 Sensitive high-explosive element detonated by fuze or primer and powerful enough to detonate a larger main charge.

booster APU APU capable, usually in emergency only, of augmenting aircraft propulsion.

booster coil Battery-energized induction coil to provide a spark to assist PE (4) starting.

booster magneto Auxiliary magneto, often turned by hand, for supplying hot sparks during PE (4) starting.

booster pump 1 Centrifugal pump, often located at lowest point of a liquid fuel tank, to ensure positive supply and maintain above-ambient pressure in supply line.
2 Auxiliary impeller in cryogenic propellant system to maintain system pressure and prevent vaporization upstream of main pump.

booster rocket See *boost rocket.*

boost/glide vehicle Aerodyne launched under rocket thrust and accelerated to hypersonic speed in upper fringes of atmosphere, thereafter gliding according to any of various predetermined trajectories over distances of thousands of miles.

boost motor See *boost rocket.*

boost phase Initial phase of launch and rapid acceleration in missile or other short-range aerodynamic vehicle fitted with boost rockets.

boost pressure See *boost (2)*, *international* *, *override* *.

boost rocket Rocket motor, usually solid propellant and sometimes used in multiple, used to impart very large thrust during stages of launch and initial acceleration of missile or other vehicle launched from ground or another aerial vehicle. Almost all kinds of ** burn for a few seconds only, and in some cases for only a fraction of a second. Sometimes case and chamber forms part of vehicle, but most boost rockets are separate and jettisoned after burnout.

boost rocket impact area Area within which all ** should fall during launches on a given range.

boost separation Process by which boost rocket thrust decays and becomes less than drag, causing rearward motion relative to vehicle and subsequent progressive unlocking, possibly relative rotation, and detachment.

boost/sustain motor Rocket comprising fast-burning high-thrust portion followed by slow-burning low-thrust portion.

boot Flat array of flexible tubes bonded to leading edge of wings, fins and other aircraft surfaces to break up ice. Fluid pressure is alternately applied to different sets of tubes in each boot to crack ice as it forms.

bootstrap operation Dynamic system operation in which, once cycle has been started by external power, working fluid maintains a self-sustaining process. A gas turbine, once started, sustains bootstrap operation because the turbine keeps driving the compressor which feeds it. Thus, * cycle (cold-air unit), * mainstage engine pump (turbine being fed by propellants delivered by pump), etc.

bootstrap exploration Using each space mission to bring back information to help subsequent missions, esp in lunar exploration (colloq).

Boozer Code name for British ECM carried by Mosquito aircraft in 1944.

BOP 1 Balance of payments, esp with regard to national participation in multinational programme.
2 Basic operating platform (bare base airfield).

BOPS Burn-off per sector; fuel burned on each sector, or segment, or stage (all three words are used in flight-planning documents) in commercial transport operation.

Boram Block-oriented random-access memory.

boresafe fuze Projectile fuze rendered safe by

interrupter until projectile has cleared gun muzzle.

borescope Slender optical periscope, usually incorporating illumination, capable of being inserted into narrow apertures to inspect interior of machinery.

borescope port Circular ports, fitted with openable caps, through which borescopes may be inserted (esp in aircraft engines).

boresight To align gun or other device by means of optical sighting on a target.

boresight camera Optical camera precisely aligned with tracking radar and used to assist in alignment of aerial.

boresight coincidence Optical alignment of different adjacent devices, such as radar waveguide, reflector, passive interferometer and IR or optical camera.

boresight line Optical reference line used in harmonizing guns and other aircraft weapon launchers.

boresight mode Radar is locked at one chosen angle between dead ahead and -2° or -3°.

boresight test chamber Anechoic chamber containing movable near-field test targets and aerials, capable of being wheeled over nose of radar-equipped fighter aircraft.

boring Process of accurately finishing already-drilled hole to precise dimension, usually by using single-point tool.

boron Element, either greenish-brown powder or brown-yellow crystals. Used as alloying element in hard steels, as starting point for range of possible high-energy fuels, and above all as chief constituent of boron fibre.

boron/epoxy Composite plastic materials comprising fibres or whiskers of boron in matrix of epoxy resin.

boron fibre High-strength, high-modulus structural fibre made by depositing boron from vapour phase on white-hot tungsten filament. Increasingly used as reinforcement in aerospace composite materials.

Borsic Boron fibre coated with silicon carbide. Structurally important as reinforcement in matrices of aluminium and other metals.

Boscombe Aeroplane and Armament Experimental Establishment, Boscombe Down, Wiltshire, England.

BOSS Ballistic offensive suppression system (EW).

boss 1 In traditional wooden or one-piece metal propeller, the thickened, non-aerodynamic central portion.
2 In a casting, locally thickened area to provide support for shaft bearing, threaded connection or other load.

BOT Boom-operator trainer.

BOTB British Overseas Trade Board.

bottle A JATO rocket (filled or empty case) (colloq).

bottom dead centre Position in PE (4) in which centre of crankpin is precisely aligned with axis of cylinder, with piston at bottom limit of stroke; at BDC piston cannot exert a turning moment on crankshaft.

bottom rudder In aeroplane in banked turn, applying rudder towards lower side of aircraft; among other things this will lower the nose.

bounce 1 In air combat, to catch enemy aircraft unawares; to intercept without being seen.
2 In PE (4) poppet valve gear, elastic bounce of valves on their seats.

bounce table Vibration test machine.

boundary layer Layer of fluid in vicinity of a bounding surface: eg, layer of air surrounding a body moving through the atmosphere. Within the ** fluid motion is determined mainly by viscous forces, and molecular layer in contact with surface is assumed to be at rest with respect to that surface. Thickness of ** is normally least distance from surface to fluid layer having 99 per cent of free-stream velocity. ** can be laminar or, downstream of transition point, turbulent.

boundary-layer bleed Pathway for escape of ** adjacent to engine inlet mounted close beside fuselage wall. Bleed is either open or ducted, and removes stagnant ** which would otherwise reduce ram pressure recovery and propulsion system performance.

boundary-layer control Control of ** over aircraft surface to increase lift and/or reduce drag and/or improve control under extreme flight conditions. BLC can be effected by: passive devices, such as vortex generators; ejecting high-velocity air through rearward-facing slits; sucking ** away through porous surfaces; use of engine slipstream to blow wings or flaps.

boundary-layer duct Duct to carry ** from ** bleed to point at which it can advantageously be dumped overboard.

boundary-layer energizer Low sharp-edged wall normal to airflow across aerodynamic surface (eg immediately upstream of aileron).

boundary-layer fence Shallow fence fixed axially across swept wing to reduce or check spanwise drift of ** and its consequent thickening and proneness to separation.

boundary-layer scoops Forward-facing inlets designed to remove thick ** upstream of engine inlet or other object.

boundary-layer separation Gross separation of ** from boundary surface, space being filled by undirected, random turbulence.

boundary light Visible steady light defining boundary of landing area.

boundary marker Markers, often orange cones, defining boundary of landing area.

bound vortex 1 Circulation round a wing.
2 Vortex embracing any solid body or touching a surface.

Bourdon tube Flat spiral tube, either glass filled

with alcohol or metal filled with mercury, whose radius increases (rotating the centre) with increasing temperature.

Boussinesq Formula giving distance along uniform tube necessary for laminar, viscous, incompressible flow to become fully developed.

$$x = \frac{0{\cdot}26\ U_m\ r^2}{\nu} = 0{\cdot}26\ rR$$

where U_m is mean velocity, r tube radius, ν kinematic viscosity and R Reynolds number.

bow 1 Nose of airship or marine hull or float.
2 Curvature along length of turbine blade or other slender forging, or curvature due to instability in structural compression member (in this case rhymes with slow).

bow cap Structure forming front end of airship hull or envelope. Alternatively, nose cap.

Bowen-Knapp camera High-speed strip-film camera used in vehicle flight testing.

bowser 1 Airfield fuel truck, roadable and self-propelled; unpropelled, * trailer.
2 Used as adjective, specially modified to contain overload or ultra-long-range fuel, hence * wing, * fuselage.

Bow's notation Conventional system of representing structural forces and stresses by letters and/or numbers in graphical stress analysis.

bow stiffeners Longitudinal stiffeners arranged radially around nose of aerostat envelope (esp blimp or kite balloon) to prevent buckling under aerodynamic pressure. Alternatively called battens.

bow wave 1 Shockwave from nose of supersonic body, esp one not having sharply pointed nose.
2 Shockwave caused by motion of planetary body through solar wind.
3 Form of wave caused by bows of taxiing marine aircraft.
4 Form of wave caused by landplane nose wheels running through standing water.

box 1 Tight formation of four aircraft in diamond (leader, left, right, box).
2 Structural heart of a wing comprising all major spars, ribs and attached skins (often forming integral tankage), but usually excluding leading and trailing edges, secondary structure and movable surfaces.
3 Major sections of fuselage, especially where these are of rectilinear form and thus not describable as *barrel sections*.
4 Aircraft structure formed from two or more lifting planes (wing or tail), linked by struts and bracing wires. In early aviation no other form could compete for lightness and strength.

box connector Multi-circuit connector having sockets four-sided box sockets, with linear pin engagements (2 or 3 rows, up to 240 circuits).

box girder See *box spar*.

boxing Process of assembling major airframe sections in erection jig; includes fuselage *box* (3) sections.

box kite Kite in form of rectangular- (often square-) section box, open at mid-section and at ends. Structurally related to early biplanes in form of *box* (4).

box position Rear aircraft in *box* (1).

box rib Rib assembled from left and right sides separated by a peripheral member following profile of aerofoil.

box spar Spar assembled from front and rear webs separated by upper and lower booms.

box tool Tangential cutting tool incorporating its own rest, used on automatic turning machines.

Boyle's law In an ideal gas at constant temperature, pressure and volume are inversely proportional, so that PV = f(T°) and $P/\rho = RT$.

b.p. 1 Bypass (jet engine); thus, * ratio.
2 Band-pass filter.

BPA British Parachute Association.

BPC 1 Gas-turbine barometric pressure control.
2 Benchmark pricing guide.
3 British Purchasing Commission (WW2).

BPCU Bus power control unit.

BPE Bomber penetration evaluation.

BPF 1 Band-pass filter.
2 British Pacific Fleet (WW2).

BPL Band-pass limiter.

BPPA British Precision Pilots Association.

BPR Bypass ratio.

BPS 1 Balanced pressure system; buried glycol pipes give alarm if trodden on.
2 Bistable phosphor storage.

bps Bits per second (EDP).

b_r Reference length.

BRA 1 Bureau des Recherches Aériennes (ICAO).
2 Barrel-roll attack.

bracket Limits of time/distance/altitude for one pre-planned in-flight refuelling.

bracketing 1 Obsolete method of flying a radio range in order to establish correct quadrant and hence direction to next waypoint.
2 In flak or other artillery, establishing a short and an over along desired line and then successively splitting resulting "bracket" in half.

brake 1 Device for removing energy from a moving system to reduce its speed or bring it to rest. Energy withdrawn may be rejected to atmosphere (airbrake, speed brake) or absorbed in heat sinks (wheel brake, propeller or rotor brake).
2 In sheet-metalwork, a power press for edging and folding.

brake horsepower Power available at output shaft of a prime mover.

brake mean effective pressure A measure of MEP in an operating PE (4) cylinder calculated from known BHP (BMEP is to be expressed in kPa or bars and derived from power in kW).

brake-up Aeroplane nose-over caused by harsh braking (colloq).

braking coefficient Braking force coefficient.

braking ellipse Elliptical orbit described by spacecraft on entry to a planetary atmosphere. If continued, successive ellipses are smaller, due to drag; purpose of manoeuvre is to dissipate re-entry or entry energy over a much greater time and distance and thus reduce heat flux.

braking force Linear force exerted by a braked vehicle (ie, aircraft) wheel.

braking force coefficient Coefficient of friction between wheel and fixed surface (whether rolling or sliding).

braking nod Nose-down pitch of aircraft when wheel-braked, or nose-up pitch when brakes released at full power, esp maximum angular movement thus imparted.

braking parachute Parachute streamed from aircraft to increase drag, increase dive angle or reduce landing run.

braking pitch Predetermined propeller pitch to give maximum retardation, either windmilling drag or reverse thrust under power.

braking rocket Retrorocket.

branch 1 In electrical system, portion of circuit containing one or more two-terminal elements in series.
2 In computer program, point at which * instructions are used to select one from two or more possible routines.
3 In crystal containing two or more kinds of atom, either of possible modes of vibration, termed acoustic * or optical *.

branch pipe Pipe conveying exhaust gas from PE (4) cylinder to manifold or collector ring.

brass Alloys of copper with up to 40 per cent zinc and small proportions of other elements.

brassboard Functioning breadboard model of avionic system; also adjective and verb.

Bravo exercise Combat mission called off at point when aircraft ready to taxi.

Brayton cycle Thermodynamic cycle used in most gas turbines: diagram comprises compression and expansion curves joined by straight lines representing addition or rejection of heat at constant pressure. The so-called "open cycle".

brazier head rivet Light-alloy rivet having head shallower but of larger diameter than round-head.

brazing Joining metals by filling small space between them with molten non-ferrous metal having a melting point above a given arbitrary value (originally 1,000°F = 538°C).

breadboard Preliminary assembly of hardware to prove feasibility of proposed system, without regard to packaging, reliability or, often, safety. May be laboratory rig or flyable system. Often adjective or verb.

break 1 Point at which pilot senses stall of wing.
2 Breakaway (1) (colloq).
3 Chief meaning in modern air combat: to make maximum instantaneous turn to destroy hostile fighter's tracking solution. Used as noun or as verb.
4 In carrier flying, point at which aircraft turns sharply left across bows and on to downwind leg.

breakaway 1 Point at which aircraft breaks off trajectory directed against another object, such as stern attack on enemy aircraft or gun-firing run on ground target.
2 In nuclear explosion, point in space or time at which shockfront moves ahead of expanding fireball.

breakaway thrust Engine power needed to initiate movement and reach taxiing speed.

breakdown drawing Isometric or perspective drawing showing parts separated from each other by being displaced along one or more axes. Often called exploded drawing.

breakdown potential Dielectric strength.

breaker strip 1 Linear narrow de-icing element, either thermal or mechanical, arranged along leading edge (eg of wing or engine inlet strut) to split ice accretion into two parts.
2 See *stall strip*.

break-even point 1 In any commercial aircraft operation, load factor at which total revenue equals total cost.
2 In manufacture, number of sales required to recover investment.

break lock To use ECM or other countermeasure to make hostile tracking sytem (eg IR or radar) cease to track friendly or own aircraft.

breakoff phenomenon Mental state experienced by crews of high-altitude aircraft and spacecraft of being divorced from other humanity.

break-out Point at which flight crew receive first forward visual cues after an approach through cloud.

breakout force Minimum force required to move pilot's flying controls (each axis considered separately) at zero airspeed.

breakpoint 1 In system responding to high-frequency input, the corner frequency (as seen on a Bode plot) where $f = 1/2T\ \pi$.
2 In EDP, point in program or routine at which, upon manual insertion of * instruction, machine will stop and verify progress.

break-up Separation of single radar blip into discrete parts each caused by a target.

break-up circuit Electrical circuit linking airborne portions of break-up system.

break-up shot Artillery shot designed to break into small fragments upon leaving muzzle and thus travel only a short distance.

break-up system System designed to break unmanned vehicle, such as ballistic missile, space launcher or RPV, into fragments sufficiently small to cause minor damage if they should fall on inhabited area.

break X To break (2) at point of minimum range for launch of own AAM, where X symbol appears on cockpit display.

breathing 1 Flow of air and exhaust gas through PE (4), esp the way this is limited by constraints of flow path.
2 Flow of air and/or gas into and out of aerostat in course of flight.
3 Very small air and oil vapour flow through breather holes or ducts provided to equalize pressure inside and outside an engine.

breech In a gun, end of barrel opposite muzzle, through which shot is normally loaded.

breech-block Rigid metal block normally serving to insert and withdraw shell cases, resist recoil force on fired case and seal breech during firing; usually oscillates in line with barrel axis.

breeches piece Tubular assembly, often shaped like pair of shorts, serving to bifurcate jetpipe or join two jetpipes into common pipe.

Breguet formula Rule-of-thumb formula for giving flight range of classical aeroplane, and reasonably accurate for all types of aerodyne; range given by multiplying together L/D ratio, ratio of cruising speed divided by sfc, and $\log_e$ of ratio of aircraft weight at start and finish expressed as decimal fraction greater than 1. Units must be compatible throughout.

bremsstrahlung Electromagnetic radiation emitted by fast charged particle, esp electron, subjected to positive or negative acceleration by atomic nuclei. Radiation has continuous spectra (eg, X-rays).

brennschluss Cessation of operation of rocket, for whatever cause.

Brétigny Location, southwest of Paris, of Centre des Essais en Vol (CEV).

brevet Flying badge worn on uniform, especially denoting qualification as flight-crew member.

BRF Short approach is required or desired (ICAO).

BRG Bearing.

bridle 1 Towing linkage, other than expendable strop, transmitting pull of catapult to two hard points on aircraft.
2 Assembly of electric cables or fluid system pipes which, after disconnection, can be removed from supporting structure (eg, landing gear) as a unit.
3 Rigging attached to two or more points on aerostat, esp blimp, to distribute main mooring pull.

brief To issue all relevant instructions and information in advance of flying mission (not necessarily military), static test, war game or other operation involving human decision-taking.

bright display Normally, display which can be viewed clearly without a hood in brightest daylight.

brightness control Facility provided in radar, TV and other display systems for adjusting CRT bias to control average brightness.

brilliant Describes munition having both guidance (smart) and programmable software; in practice also means with ability to guide itself to target without external help.

Brinell hardness Measure of relative hardness of solids, expressed as numerical value of load (either 500 kg or 3,000 kg) and resulting area of indentation made by hard 10 mm ball.

bring-back weight Weight at which combat aircraft recovers to airbase or carrier, with remaining fuel and unexpended ordnance.

British Thermal Unit Obsolete measure of heat: quantity required to raise temperature of 1 lb of water from 63° to 64°F. 1 Btu = 1,055 J.

brittle fracture Fracture in solid, usually metal, in which plastic deformation and energy dissipated are close to zero. Contrasts with ductile fracture, and is rare except at very low temperatures.

BRLI Bild des Raum und Luftfahrtindustrie (Federal German aerospace industry union).

broach Cutting tool having linear row(s) of teeth, each larger than its predecessor.

broad-arrow engine In-line PE (4) having three banks of cylinders with adjacent banks spaced at less than 90°; W-engine.

broadband aerial Aerial capable of operating efficiently over a spread of frequencies of the order of ten per cent of centre frequency.

broadcast Radio transmission not directed at any specific station and to which no acknowledgement is expected.

broadcast control Air interception in which interceptors are given no instructions other than running commentary on battle situation.

broadside array Aerial array in which peak polarization is perpendicular to array plane.

Broficon Broadcast flight-control management of tac-air warfare (USAF).

broken clouds Sky coverage between "scattered" and "continuous", defined by ICAO as five-tenths to nine-tenths.

bronze Alloys of copper and tin and/or aluminium.

BRS Best-range speed.

BRSL Bomb-release safety lock.

Br-stoff Avgas or benzole (G).

BRTF Battery repair and test facility (artillery, guided weapons).

BRU Bomb release unit, normally complete interface between hardpoint and munition.

brush discharge See *Corona discharge*.

BRW Brake release weight, ie at start of take-off run.

BS 1 Commercial broadcasting station.
2 British Standard, thus * parts.
3 Bomb (or bombardment) squadron (USAAC, USAAF, USAF).

BSC Beam-steering computer (EW).

BSDH Bus shared-data highway.

BSI British Standards Institution.

BSL 1 British Standard family of light alloys.
2 Base second level (servicing).

BSPL Band sound pressure level; sound pressure level in bands each one-third of an octave wide from 50 to 10,000 Hz.
BSPR Boost/sustainer pressure ratio (rocket).
BSS 1 British Standard Specification.
2 British Standard family of steels.
BST British Summer Time.
bst Boresight.
BT 1 Burn time (rocket).
2 Basic trainer, US Army Air Corps category.
3 Bathythermograph.
BTB Bus tie breaker.
BTC Bus tie connector.
BTH Beyond the horizon (radar).
BTL Between cloud layers.
BTM Bromotrifluoromethane (extinguishant).
BTN Between.
BTO Bombing through overcast (WW2).
BTP Bureau Trilatéral de Programmes (Eur).
BTR Bus tie relay.
BTU British Thermal Unit (alternatively, Btu, BThU).
BU Break-up, thus a guided-weapon * unit.
bubble Continuous tulip-shape film of fuel from burner at low flow rate.
bubble horizon Bubble turn and slip.
bubble memory Computer memory whose bits are distributed among microscopic voids (bubbles) in a 3-D volume of solid.
bubble sextant Sextant in which local horizontal is established by a bubble device. Often called bubble octant, because arc is usually not greater than 45°, restricting altitude to 90°.
bubble turn and slip Primitive flight instrument in which lateral acceleration is indicated by sideways displacement of bubble in arched glass tube of liquid.
BUCD Back-up command destruct.
bucket 1 In US, a turbine rotor blade.
2 Principal member of most types of thrust reverser, two buckets normally rotating and translating to block path of efflux and divert it diagonally forwards.
3 Graphical plot having basic U-shape resembling *, notably produced by adding one plot of negative slope (eg operating and servicing cost against MTBF) to a related plot of positive slope (eg capital cost against MTBF).
bucket brigade Integrated circuit device, comprising MOS transistors connected in series, serving as shift register by transferring analog signal charge from one storage node to next.
bucket shop Retail outlet (shop) offering non-IATA passenger tickets at cut prices.
bucking bar Shaped bar held against shank in manual riveting.
buckling Lateral deflection of structural member under compressive load; state of instability or unstable equilibrium, but may be purely elastic.
BUCS Back-up control system.
buddy pack Flight-refuelling hose reel and drogue packaged in streamlined container for carriage by standard weapon rack; thus, aircraft A can refuel buddy B flying identical aircraft (formerly colloq).
BUF Back-up facility.
buffer 1 In radio, low-gain amplifier inserted to prevent interaction between two circuits.
2 Amplifier stage having several inputs, any of which may be connected to output.
3 In EDP, temporary store used to smooth out information flow between devices, esp between I/O and main processor or core store.
buffer distance In nuclear warfare, horizontal distance (expressed in multiples of delivery error) which, added to radius of safety, will give required acceptable risk to friendly forces; alternatively, vertical distance (expressed in multiples of vertical error) added to fallout safe height to ensure that no fallout will occur.
buffet Irregular oscillation of structure caused by turbulent wake. * in aeroplanes may be caused by excessive angle of attack (due to low airspeed, extreme altitude or excessive g) or, in subsonic aircraft, an attempt to fly at too high a Mach number.
buffet boundary For any given aircraft and environment, plot of limiting values of speed and altitude beyond which buffet will be experienced in unaccelerated flight.
buffet inducer Small projection, usually in form of strake, intended to induce buffet (usually as warning in advance of dangerous buffet affecting major part of aircraft).
buffet margin For any given aircraft and environment, highest vertical acceleration (g) which can be sustained without exceeding given buffet severity (in some cases severity is zero).
buffet threshold For any given aircraft and environment, point at which buffet is first perceptible, expressed in terms of speed, altitude and vertical acceleration (g).
buffing Process for polishing sheet metal by rotary tool of soft fabric impregnated with fine abrasive.
bug 1 Heading marker on navigational instrument.
2 Fiducial index, esp on flight instrument, having appearance of *.
3 Clandestine monitoring device, esp for audiosurveillance.
4 To install and conceal (3).
5 System malfunction or other fault, esp one not yet traced and rectified; hence, to debug (colloq).
bug-eye canopy Small canopy (usually two, left and right) over each projecting pilot's head in large aircraft.
bugle bag Sick bag (colloq among cabin crew).
bug out Eject (colloq).
bug speed Speed at which ASI needle passes *bug* (2).

BUIC Back-up interceptor control; add-on to SAGE system.

build Growth in received radio signal; opposite of fade.

build standard Detailed schedule of all possible variable or unresolved items in aircraft or other complex hardware in stage of development or pre-production. Original ** may list features of airframe, development state of engines, system engineering and, esp equipment fitted or absent; altered as aircraft is modified.

built-in hold Pre-planned hold during countdown to provide time for defect correction or other activity without delaying liftoff.

built-up section Structural members having section assembled from two or more parts, rather than rolled, extruded, hogged from solid or forged. Reinforced composites are not regarded as built-up; essentially parts should be assembled by joints and could be unfastened.

bulb angle, bulb flange Structural sections, usually used as booms, having circular or polyhedral form. Nearly all were rolled from strip and had eight to 12 faces after assembly, complete ** being built up from one to five segments.

bulk cargo Homogenous cargo, such as coal. Today also means cargo carried loose; incorrectly used to mean cargo or baggage not contained in standard container or pallet.

bulk erasure Erasure of complete magnetic tape by powerful field.

bulkhead Major transverse structural member in fuselage, hull or other axial structure, esp one forming complete transverse barrier. Certain *, such as pressure * in aircraft fuselage and tank * in rocket vehicle, must form pressure-tight seal.

bulk-injection In PE (4) injection of fuel into induction airflow upstream of distribution to individual cylinders.

bulkmeter Instrument, esp in refuelling of aircraft, for measuring liquid flow, typically as mass per second or as summing indication of total mass passed. Some * measure volume and require density correction for each liquid handled.

bulk modulus Elastic modulus of solid under uniform compressive stress over entire surface, as when immersed in fluid under pressure; numerically, stress times original volume divided by change in volume.

bulk out To run out of cargo space while still within allowed weight.

bulk petroleum products Liquid products carried in tankcars or other containers larger than 45 gal (55 US gal).

bullet 1 Gun-fired projectile intended to strike target, having calibre less than 20 mm (0·7874 in).
2 Streamlined fairing having form of quasi-conical nose or forepart of body of revolution. If rotating, called spinner.

bull gear Largest gear in train, esp large gear on which aerial of surveillance radar is mounted.

bull session Informal discussion on serious aviation topics, between engineers and/or aviators.

bull's eye 1 Circular thimble.
2 Ring used to guide or secure rope.
3 Cockade having concentric rings (colloq).

bump 1 See *Gust (1)*.
2 Sensation experienced in flight through gust (1).
3 To form sheet metal on bumping hammer.

bumper bag Padded or inflated bag beneath lowest point of aerostat to absorb shock of ground impacts.

bumper rocket Pre-1955, first stage of two-stage launch vehicle.

bumper screen On spacecraft, protective screen intended to arrest micrometeorites and other macroscopic solids.

bumper wheel Wheeling machine.

bumping Practice of denying a fully booked and confirmed passenger the right to board an over-booked flight; officially called IBR (involuntary boarding refusal). Bumped pax qualify for DBC.

bumping hammer Power hammer for bumping.

bump rating Increased engine thrust rating (beyond GM or average TO) cleared for short periods; fixed-wing equivalent of contingency.

bunching 1 In traffic control, tendency of vehicles (esp aircraft) to reduce linear separation, esp to dangerous degree.
2 In klystron, separation of steady electron stream into concentrated bunches to generate required very high frequency in oscillatory circuit.

bungee Elastic cord comprising multiple strands of rubber encased in braided (usually cotton) sheath.

bunt 1 Severe negative-g manoeuvre comprising first half of outside loop followed by half roll or second half of inside loop.
2 In surface-attack missile trajectory, negative-g pushover from climb to dive in terminal phase near target.

BuOrd Bureau of Ordnance (USN).

buoyancy Upthrust due to the displaced surrounding fluid just sufficient to support a mass. Thus, in aerostat, condition in which aircraft mass equals mass of displaced air. In marine aircraft at rest, mass of aircraft equals mass of water displaced (in this case, as with all aerodynes, displaced air mass is usually ignored).

buoyant spacecraft Spacecraft designed to operate as aerostat in planetary atmosphere.

burble 1 Turbulent eddy in fluid flow, esp in proximity to, or caused by, a bounding surface.
2 Breakdown of unseparated flow (not necessarily with laminar boundary layer) across aircraft surface, esp across top of wing. First region of separated flow due to excessive angle of attack or to formation of shockwaves at M_{crit}.

burble point 1 Angle of attack at which wing first

suffers sudden separation of flow.
2 Mach number at which subsonic wing first suffers sudden separation of flow due to shock-wave formation.

Bureau Veritas International organization for surveying and underwriting vessels, including aircraft.

buried engine Engine contained within airframe, esp without causing significant protuberance.

burn 1 Operation of rocket engine, esp programmed operation for scheduled time. Thus, first *, second *.
2 Operation of main flame in burner of hot-air aerostat, Thus, a 20-second *.
3 Authorized destruction of classified material, by whatever means.

burner 1 In gas turbine, device for mixing fuel with airflow to sustain combustion inside combustion chamber.
2 Afterburner (colloq, R/T).
3 Incorrectly, gas-turbine combustion chamber.
4 In liquid or hybrid rocket, device for injecting and/or mixing liquid propellants to sustain primary combustion; more usually called injector.

burn in 1 To enter data in EDP core store so that it will subsequently resist nuclear explosion effects and other hostile action.
2 To operate avionic and other electronic equipment under severe overload conditions to stabilize it before operational service and reduce incidence of faults.

burnish To smooth and polish metal surface by rubbing (usually with lubricant) with convex surface of harder metal.

burn-off In aeroplane (esp commercial transport) operation, fuel burned between take-off and critical position for establishing terrain clearance.

burnout 1 Termination of rocket operation as result of exhaustion of propellants. Also called all-burnt.
2 Mechanical failure of part subject to high temperature as result of gross overheating, esp rocket case or chamber.

burnout plug In certain rocket motors, esp storable liquid and hybrid types, combustible plug which, when ignited, releases and fires liquid propellant.

burnout velocity Vehicle speed at burnout (theoretically, highest attainable for given trajectory).

burnout weight Mass of vehicle at burnout, including unusable fuel.

burn rate In solid rocket, linear velocity of combustion measured (usually in millimetres per second) normal to burning surface. Symbol r.

burn rate constant Factor applied to ** calculations dependent upon initial grain temperature.

burn rate exponent Pressure exponent n in burn rate law $r = aP_c^n$.

burn-through Operation of radar in face of jamming and similar ECM, esp by virtue of high transmission power to overcome interference.

burn-through range Limit of range at which *-* operation can yield useful information, if necessary in emergency short-life operation at abnormally high power.

burn time Duration in seconds of rocket motor burn. Burn time starts when chamber pressure has risen to 10% of maximum (or averaged maximum during level portion of thrust curve) and ends when pressure drops to 75%. Alternative criterion is to draw tangents to level portion and descending portion of thrust curve and measure time to point at which curve is cut by bisector of angle between tangents. Symbol t_b.

burn-time average chamber pressure Integral of chamber pressure versus time taken over burn time interval divided by burn time. Symbol $\bar{P}_c$.

burn-time average thrust Integral of thrust versus time taken over burn time interval divided by burn time. Symbol $\bar{F}_b$.

burnup 1 On entry to atmosphere from space, partial or complete destruction due to kinetic heating.
2 In nuclear reactor, esp thermal fission reactor, percentage of available fissile atoms that have undergone fission.

burst 1 In colour TV, transmission of small number of cycles of chroma sub-carrier in back-porch period (in military systems this signal not always used).
2 One round (payload) of ECM dispensed from attacking vehicle; can be active jammer or flare.
3 See *turbulent* *.

burst diaphragm Diaphragm sealing fluid system and designed to rupture either upon command or at predetermined dP.

burst height Height at which nuclear weapon is programmed to detonate.

burst order Detonation of missile warhead by command.

BUS Break-up system (of unmanned vehicle upon command).

bus 1 Spacecraft carrier vehicle for various payloads.
2 In EDP, main route for power or data. Alternatively, trunk.
3 Busbar (colloq).
4 In ICBM, carrier vehicle for MIRV payloads.
5 In any delivery system, carrier vehicle for multiple warheads.

busbar In electrical system, main conductor linking all generators and/or batteries and distributing power to operative branches.

Büsemann biplane Aeroplane, so far not built, in which at supersonic speed shockwaves and flows around upper and lower wings would react favourably.

Büsemann theory First theory for two-dimensional supersonic wing to take into account second-order terms (1935).

bush Open-ended drum tailored to fit inside hole,

eg to reduce its diameter, act as shaft bearing or serve as electrical insulation (see *grommet*).

bush aircraft Aerodyne tailored to utility service in remote (eg Canadian Arctic) regions.

bush pilot Operator (often also owner) of bush aircraft; usually freelance jobbing professional.

business aircraft Aerodyne tailored to needs of business management and executives of government and other organizations.

buster R/T command "Fly at maximum continuous power", normally to effect interception.

butterfly Distorted figure-8 pattern, looking like a butterfly, flown by orbiting combat aircraft on weapon-guidance, EW or, rarely, AWACS duty.

butterfly tail Comprises two oblique fixed stabilizer surfaces each carrying a hinged surface, the latter operating in unison as elevators or in opposition as rudders; also called V-tail.

butterfly valve Fluid-flow valve in form of pivoted plate, usually having circular form to close a pipe.

butt joint Sheet joint with edge-to-edge contact without overlap, usually with jointing strip along either or both sides.

buttock lines Profiles of intersection of vertical planes with surface of solid bodies, esp aircraft fuselages and marine floats. Zero ** is that on axis of symmetry. Used in lofting, these lines do not correspond with structural members.

button Extreme downwind end of usable runway (colloq).

buttonhead rivet Rivet with approximately hemispherical head; used where tensile load may be high.

butt rib Compression rib at joint between outer and inner wing, or wing and fuselage.

BuWeps Bureau of Weapons, combined BuAer and BuOrd in 1959 (USN).

Buys Ballot's law Professor Buys Ballot postulated that an observer with back to wind in N hemisphere has lower pressure to left (in S hemisphere, to right). True for any isobar pattern.

buy the farm To be killed in a crash, not excluding military action (colloq).

buzz 1 Oscillation of skin or other structure at frequency high enough to sound as a note.
2 Oscillation of control surface at high frequency.
3 Wake-interaction noise generated by turbomachinery, esp large fans, at 900–4,000 Hz.
4 High-frequency, often violent, pulsation of airflow at supersonic air-breathing engine inlet.
5 To fly aircraft, esp one of high performance and manoeuvrability, in way designed to harass another aircraft or ground target. Transitive, thus "to * the control tower".

buzz liner Sound-absorbent liner to fan duct or other surface bounding wake-interaction noise.

buzz number Extra-large individual aircraft number (can be unit number or aircraft serial), readable from a distance.

buzz-saw noise Buzz (3) from shock system of fan with supersonic flow over blades, composed of discrete tones at multiples of N_1.

BV 1 *Bureau Veritas.*
2 Bleed valve.

BVCU Bleed-valve control unit.

BVR Beyond visual range.

BVTRU Bleed-valve transient reset unit (controls BVs during transients).

BW 1 Biological warfare.
2 Bomb (or bombardment) wing (USAAC, USAAF, USAF).

BWB MoD procurement office (G).

BWC Biological weapons convention, 1972.

BWO Backward-wave oscillator.

BWPA British Women Pilots' Association.

BWRA British Welding Research Association.

bx Biconvex aerofoil.

BY$ Base-year dollars.

bypass 1 Capacitor connected in shunt to provide low-impedance alternative path.
2 Alternative flow path for fluid system.

bypass engine Air-breathing jet engine in which air admitted at inlet may take either of two flow paths (see *bypass turbojet*).

bypass ratio In bypass turbojet or turbofan, numerical ratio of mass flow in bypass duct divided by that through core, ie cold jet divided by hot. Some have defined as total mass flow divided by core mass flow; this is incorrect, and would always yield numbers greater than 1.

bypass turbojet Turbojet in which mass flow through LP compressor stages is slightly greater than that through HP stages, excess being discharged along bypass duct. Also called leaky turbojet. In principle difference between this and turbofan is purely of degree; turbofan has much higher bypass ratio (greater than 1) and probably at least two shafts. In general subsonic engines may be considered turbofans and supersonic engines bypass turbojets.

byte 1 Group of bits normally processed as unit.
2 Sequence of consecutive bits forming an EDP word, thus an 8-bit *.

C

C 1 Celsius.
2 Yawing moment of inertia.
3 Compass heading/bearing/course.
4 Coulombs.
5 Capacitor.
6 Capacitance, capacity.
7 Thermal conductance.
8 A constant.
9 Aggregate fuel consumption.
10 Power of carrier wave (Watts).
11 Prefix, cargo aircraft (DoD and UK).
12 Prefix, ground-service connection (BSI).
13 JETDS Code: air-transportable, carrier wave, communications use.
14 Council (ICAO).
15 Viscous-damping coefficient.
16 Fighter category (F).

c 1 Chord.
2 Collector (semiconductor, eg transistor).
3 Speed of light in vacuum.
4 Prefix, centi = 10^{-2} (non-SI)
5 Prefix, approximate.

c′ Thrust specific fuel consumption.

$\bar{c}$ Geometric mean chord.

$\bar{\bar{c}}$ Aerodynamic mean chord.

(c) Astronomical unit (see *AU*).

C* Characteristic exhaust velocity of rocket.

C1, C2 Single- and two-seat fighters (F).

C^2 Command and control (called C-squared).

C^2S Command, control and status.

C^3 Command, control and communications (called C-cubed).

C^3CM Command, control and communications countermeasures.

C^3I Command, control, communications and intelligence.

C^4 Command, control, communications and computers.

C-band Electromagnetic frequencies 3·9 to 6·2 GHz; supplanted by S and X.

C-code IFR flight plan suffix: no-code transponder and approved area navigation.

C-cycle One complete flight simulated in an engine development programme.

C-display Rectangular display in which horizontal axis is target bearing and vertical is angle of elevation.

C-duct Half a fan duct forming part of an engine pod cowl, usually pivoted at the top for access to the core.

C power supply Power supply between cathode and grid, for grid bias.

C-stoff Rocket propellant (fuel and coolant), mixture of hydrazine hydrate and methyl alcohol, often plus water (G).

CA 1 Controller, Aircraft (UK).
2 Controlled airspace.
3 Cabin attendants (in airline costings).
4 Conversion angle.

CAA 1 Civil Aeronautics Administration (US).
2 Civil Aviation Authority (UK, etc).
3 Civil Aviation Administration (Israel, etc).
4 Conformal-array aerial (antenna).

CAAA Commuter Airline Association of America.

CAACU Civilian anti-aircraft co-operation unit (UK).

CAADRP Civil Aircraft Airworthiness Data Requirements (originally "Recording") Programme.

CAAFU Civil Aviation Authority Flying Unit (UK).

CAARC Commonwealth Advisory Aeronautical Research Council (Int, office in London).

CAAS Computer-assisted approach sequencing.

CAASA Commercial Aviation Association of Southern Africa.

CAB Civil Aeronautics Board (US).

cab 1 Part of large aircraft incorporating flight deck, sometimes excluding extreme nose.
2 Airport tower.

cabane Structure, usually of braced struts, to support load above fuselage or wing. Used to carry: parasol wing; upper wing of most biplanes; engine nacelle; bracing wires for wing tips or early monoplane wings.

cabin Enclosure for aircraft occupants, formerly in contradistinction to open cockpit. Today in transport aircraft excludes flight deck.

cabin altitude Pressure height corresponding to pressure inside cabin.

cabin blower In some pressurized aircraft, mechanically driven blower to supply cabin air under pressure.

cabin crew Staff who attend to passengers in flight.

cabin pressure Ambiguous; can mean absolute pressure inside aircraft or pressure differential (dP) between cabin and surrounding atmosphere.

cabin supercharger See *cabin blower*.

cable cutter One-shot guillotine, powered by cartridge, on wing leading edge to cut barrage cables.

cable-drag drop Low-level airdrop with load extracted and arrested by ground cable installation.

cable hover Design requirement for ASW helicopter autopilot while dunking.

cable-strike protection See *wire-strike protection system*.

cabotage Freedom of air transport operator to pick up or set down traffic in (usually foreign) country for hire or reward.

cab-rank patrol Close air-support technique in which, instead of striking designated targets or targets of opportunity, aircraft loiter awaiting assignments from surface forces.

CAC 1 Combat Air Command.
2 Combat-assessment capability.
3 Computer acceleration control.

Cacas Civil Aviation Council of the Arab States (Int).

CAD 1 Computer-assisted design.
2 Cushion-augmentation device (for jet V/STOL; US = LID).
3 Cartridge-activated device.

Cadam Computer- (or computer-graphics) assisted (or -augmented) design and manufacturing.

CADB Composite air defense battalion (USA).

CADC Central air-data computer.

CADEA Confederación Argentina de Entidades Aerodeportivas.

CADES, Cades Computer-aided design and evaluation system.

CADF Commutated-aerial direction-finder.

CADM Clustered airfield defeat (or dispensed) munition.

Cadmat Computer-assisted (or -augmented) design, manufacture and test.

cadmium Soft white metal used in electroplating, low-m.p. fusible alloys and CdS IR detector cells.

CADP Central annunciator display panel.

Cads Cushion-augmentation devices; improve jet V/STOL.

CAE Computer-assisted (or -aided) engineering.

CAé Commission d'Aérologie (WMO).

CAEDM Community/airport economic development model.

CAEM Cargo-airline evaluation model.

CAeM Commission for Aeronautical Meteorology (WMO).

CAEPE, Caepe Centre d'Achèvement et d'Essais de Propulseurs d'Engins (F).

CAER See *EARC*.

CAF 1 Canadian Armed Forces.
2 Citizen Air Force (RSA).
3 Confederate Air Force (US).
4 Cleared as filed.

Cafac Commission Africaine de l'Aviation Civile (Int).

CAFATC Canadian Air Transport Command.

Cafda Commandement Air des Forces de Défense Aérienne (F).

CAFH Cumulative airframe flight hours.

CAFI Commander's annual facilities inspection.

CAFMS Computer-assisted force management system.

CAFT Combined advanced field team (evaluate new captured hardware).

CAFU Civil Aviation Flying Unit (UK).

CAG 1 Carrier air group (USN).
2 Civil Air Guard (UK, 1937–39).
3 Guided-missile heavy cruiser (USN code).
4 Auto generalized control (F).

cage To orientate and lock gyro into fixed position relative to its case.

CAGS Central attention-getting system.

CAHS Canadian Aviation Historical Society.

CAI 1 Civil Aeromedical Institute (FAA).
2 Computer-aided instruction (see *CMI*).
3 Close approach indicator (STOVL carrier landing).

CAIG Cost-analysis improvement group.

Cains Carrier aircraft (since 1982, also -aligned) inertial nav system.

CAIRA See *IAARC*.

cal Calorie.

calculated altitude Celestial altitude calculated but not observed.

Cale gear Shock-absorbing system in carrier arrester wire anchors.

Calfax Patented quick-release panel fastener, latch opened or closed by 540° rotation.

calf-garters Automatic leg-restraint straps in certain ejection seats.

calibrated airspeed IAS corrected for ASI system errors; "true indicated airspeed".

calibrated altitude Not normally used, but signifies pressure altitude or radar height corrected for instrument errors.

calibrated club propeller Club propeller whose drive torque has been measured and plotted against rpm; thus, can serve as dynamometer.

calibrated focal length Equivalent focal length adjusted to equalize positive and negative distortion over view field.

calibration card Graphical or tabular plot of instrument errors (other than compass); usually displayed near instrument.

calibration test Static run of bipropellant rocket engine to check propellant mixture ratio and performance.

calibrator Device for measuring instrument errors.

calibre Bore (ID) of tube, esp diameter of largest cylinder that fits inside (thus, in rifled barrel, touches highest points of opposing lands).

caliper Instrument for measuring or checking thickness, diameter or gap; with internal or external measuring points on tips of pivoted or sliding arms.

calipher Caliper.

Callback Confidential reporting system to attempt to record civil (especially air carrier) incidents caused by human failures (FAA via NASA).

call fire Fire against specific target delivered as requested.

call for fire Request, by FAC or other observer, for fire on specific target and containing target data.

calling out Spoken data readout by crew member or ground observer to assist pilot or other crew member; thus, co-pilot's speed/altitude checks on instrument approach.

call mission CAS (3) mission at short notice by pre-briefed pilot with pre-armed aircraft, target assigned after take-off.

call number In EDP, number code identifying subroutine and containing data relevant to it.

callout notes On engineering drawing, written notification of special features (eg material, process, tolerance or equipment installation).

call sign Pronounceable word(s), sometimes with suffix number, serving to identify a communications station (such as an aircraft). Civil aircraft ** are ICAO phonetic letters and numbers derived from international registration; ground station ** are name of airport followed by type of station (tower, departure, clearance delivery etc).

calm No sensible wind.

calorie Unit of quantity of heat, contrary to SI; apart from kg-*, various * are equal to 4·184–4·190 J.

calorific value Quantity of heat released by burning unit mass of fuel (eventually likely to be expressed in J kg^{-1}).

Calpa Canadian Air Line Pilots' Association.

Caltech California Institute of Technology .

Calvert lighting Original system of crossbar approach lighting.

CAM 1 Cockpit angle measure (flight deck vision limitations expressed as angles).
2 Catapult armed merchantman (surface ships, 1941–43).
3 Chemical-agent munition (or monitor).
4 Circulation aérienne militaire (F).
5 Computer-assisted manufacture.
6 Conventional attack missile.
7 Content-addressable memory.
8 Counter-air missile.

cam Rotating or oscillating member having profiled surface to impart linear motion to second member in contact with it.

CAMA Civil Aviation Medical Association (US).

camber 1 Generally, curvature of surface in airflow.
2 Curvature of aerofoil section, measured along centreline or upper or lower surface, positive when centreline is arched in direction of lift force (see *upper* *, *lower* *, *centreline* *, *conical* *, *reflex* *, *mean* *).
3 Centreline of aerofoil.
4 Inclination of landing wheels away from vertical plane.

cambered Krüger Krüger having flexible profile to increase camber when open.

cambered wing Wing section whose centreline is not coincident with chord.

CAMBS Command active multi-beam sonobuoy.

CAMDS Chemical agent munitions disposal system.

CAMEA Canadian Aircraft Maintenance Engineers' Association.

camera axis Perpendicular to film plane through optical centre of lens system.

camera gun Camera, usually colour ciné, aimed at target with aircraft gun and operated by gunfiring circuit; used to provide combat confirmation, intelligence information and, with unloaded gun(s), as training aid.

camera obscura Dark room equipped with lens projecting image of external scene on to wall or floor (formerly used as bombing target with roof lens for recording of bomb release position).

camera recorder One or more cameras arranged to provide continuous film of instrument panel or similar data source.

camera tube TV converter of optical scan into electrical video signals (Orthicon etc).

camfax Camera facsimile.

cam follower Driven member in sliding or rolling contact with cam.

CAMI Computer-aided (or -assisted) manufacturing and inspection.

CAML Cargo-aircraft minelayer.

cam lobe Profiled projection from straight or circular baseline.

Camloc handle Patented latch for cowlings and skin access panels.

camouflage Attempt to change appearance to mislead enemy, esp by concealment with portable material or painting to reduce visual contrast with background.

camouflage detection photography Use of film whose spectral response differs from that of human eye (eg IR-sensitive).

cam ring Ring inside crankcase of radial PE (4) geared to crankshaft and having sequence of lobes to operate inlet and exhaust valves of all cylinders in that row.

CAMS Combat aviation management system (USA).

camshaft Shaft equipped with cams aligned with valve gear of cylinders of in-line PE (4).

CAN Committee on Aircraft Noise (ICAO).

can 1 Individual flame tube of can-annular combustion chamber.
2 Complete combustion chamber of multi-combustor engine.
3 Five-sided box projecting into integral wing tank to accommodate slat track.
4 Controlled-environment weapon container.

can-annular See *cannular*.

canard 1 Tail-first aerodyne, usually with auxiliary horizontal surface at front (foreplane) but vertical surface (fin, rudder) at rear.
2 Foreplane fitted to * (1).

cancel 1 To terminate complete R&D or hardware programme.
2 To countermand order.
3 To deactivate activity (eg, * reverse thrust).
candela SI unit of luminous intensity (abb cd; in US sometimes called candle).
candle Former unit of luminous intensity (based on Harcourt pentane lamp).
2 Of parachute canopy, to become so constrained by rigging lines as to fail to deploy.
candlepower Former measure of rate of emission of light by source, usually in given direction (see *illumination*, *luminous flux*).
C&M Care and maintenance.
CANES Codes addresses numériquement à nombre d'entrées sélectionnable, = PCM (F).
canned cycle Complete routine for particular computerized process, such as drilling and reaming a hole in NC machining.
cannibalize To dismantle aircraft or other hardware to provide spare parts.
cannon Generally, gun of calibre 20 mm or greater.
cannular Annular combustion chamber containing separate flame tubes.
canoe radar Aircraft radar whose radome has canoe-like shape.
canonical time unit Time required for hypothetical satellite in geocentric equatorial orbit, with centre of satellite coincident with surface of Earth, to move distance subtending 1 radian at centre: 13·447 minutes.
canopy 1 Transparent fairing over flight crew or, in lightplanes, all occupants, which does not form part of airframe and slides or pivots for entry and exit.
2 Rarely, transparent fairing over flight crew which does form part of airframe and is not used for entry/exit.
3 Main deployable body of parachute.
CANP 1 Collision-avoidance notification procedure.
2 Civil air notification procedure(s) (UK).
CANS Civil air navigation school.
canted deck Angled deck.
canted nozzle Nozzle of jet engine, usually turbojet or rocket, whose axis is fixed and not parallel to centreline of engine or motor or to line of flight, but passes close to vehicle c.g.
cantilever Structural member, such as beam, rigidly attached at one end only. Thus * wing is monoplane without external struts or bracing wires.
cantilever ratio Semi-span divided by maximum root depth.
Canukus Standards agreed by Canada, Australia, New Zealand, UK, US.
CAP 1 Civil Air Patrol (US).
2 Combat air patrol.
3 Carrier air patrol.
4 Chloroacetophenone (tear gas, also called CN).
5 Civil air publication (various countries).
6 Contractor assessment program (US).
cap 1 Extreme nose structure of aerostat (see *bow* *).
2 Various portions of parachute system (see *petal* *, *tear-off* *, *vent* *).
3 Tension boom in form of flat strip attached along top or bottom edge or spar or around rib.
capacitance In electrical system, ratio of charge to related change in potential. Basis of fuel measurement system which gives readout of fuel mass irrespective of aircraft attitude.
capacity In an EDP installation, number of bits storable in all cores, registers and other memories.
capacity payload Payload limited by volume, number of seats or other factor apart from mass.
capacity safety valve Device on carrier catapult (dial-to-aircraft weight) to prevent overloading cylinder or strop/tow bridle.
capacity ton-mile, CTM Unit of work performed by transport aircraft with capacity payload.
Cape Canaveral On E Florida coast (US), at one time called Cape Kennedy; site of KSC.
CAPP Computer-aided programme planning.
capping membrane Flexible membrane covering hastily filled bomb crater on airstrip.
capping strip See *cap (3)*.
Capri Compact all-purpose range instrument, made by RCA.
CAPS, Caps Computer antenna pointing system (Satcom).
CAP/strike Mission is primary CAP (2), secondary strike; strike ordnance jettisoned to engage in air combat.
capsule 1 Small hermetically sealed compartment (eg aneroid).
2 Sealed compartment or container for instrumentation in space or other adverse environment.
3 Small manned spacecraft.
caption Small rectangular display bearing name of airborne system or device, visible when lit from behind.
caption panel Array of 20–60 captions, usually serving as CWP or alerting device giving indication of failure on broad system basis.
captive balloon Balloon secured by cable to surface object or vehicle.
captive firing Firing of complete vehicle, normally unmanned rocket, whilst secured to test stand.
capture 1 In flying aircraft or space vehicle, to control trajectory to acquire and hold given instrument reading.
2 In flying aircraft, to control trajectory to intercept and then follow external radio beam (eg ILS, radio range).
3 In ATC or air-defence system, to acquire and lock-on to target.

4 In interplanetary (eg Earth–Moon) flight, eventual dominant gravitational pull of destination body.
5 In automatic or self-governing system not always operative (eg yaw damper), limits of aircraft attitude and angular velocity within which its authority is complete.

CAR 1 Civil Airworthiness (or Air) Regulations (FAA, US).
2 Caribbean (ICAO).
3 Conformal-array radar.

car 1 Nacelle housing crew and/or engines suspended beneath aerostat.
2 Loosely, payload container attached to or within aerostat, esp airship.

CARA 1 Computer-aided requirements analysis.
2 Combined-altitude radar altimeter.
3 Cargo and rescue aircraft.

carboblast Lignocellulose blasting abrasive (BGS); used to clean gas path of running gas turbine and made from crushed apricot stones.

carbon/carbon Composite material: pyrolized carbon fibres in pyrolized carbon matrix.

carbon fibre Range of fine fibres pyrolized from various precursor materials (eg PAN) and exhibiting outstanding specific strength and modulus. Used as reinforcement in CFRP.

carbonizing, carbonitriding See *Nitriding*.

carbon microphone Contains packed carbon granules whose resistance, and hence output signal, is modulated by variable pressure from vibration of sound diaphragm.

carbon oxides Gases produced upon combustion of carbon-containing fuel: principally carbon monoxide CO, carbon dioxide CO_2 and wide range of valence bond variations. CO_2 present in Earth atmosphere (3 parts in 10^4); percentage much higher in exhaled breath.

carbon seal Sliding seal between moving machinery (eg turbine disc) and fixed structure; alternative is labyrinth.

carbureter See *carburettor*.

carburetion Mixing of liquid fuel with air to form optimum mixture for combustion.

carburettor Device for continuously supplying engine, esp Otto-cycle PE (4), with optimum combustible mixture. Many forms exist, some with choke tube and others injecting liquid fuel direct into cylinders (in which case injection pump can assume * function). Not fitted to most steady-burning devices such as gas turbines and heaters.

carburettor air Induced, usually via ram intake, along separate duct; intake normally anti-iced.

carburettor icing Caused by depression in venturi of choke tube giving local reduction in temperature.

carburizing Prolonged heating of fully machined steel part in atmosphere rich in CO or hydrocarbon gases to give hard, tough outer layer.

carburizing flame Oxy-acetylene flame having excess acetylene.

Carcinotron Backward wave oscillator; TWT for generating microwaves in which electron beam opposes direction of travel of a wave guided by a slow-wave structure.

CARD Civil aviation research and development (NASA, DoT).

Card Aerobatic-team formation resembling 4 (* 4) or 5 (* 5) playing cards.

card compass Simple compass with magnets attached to pivoted card on which bearings are marked.

CARDE Canadian Armament Research and Development Establishment.

cardinal altitudes, cardinal FLs Altitudes or FLs forming an odd or even multiple of 1,000 ft.

cardinal point effect Increased intensity of radar returns when target surface is most nearly perpendicular to LOS, esp in case of surface features.

cardinal points Bearings 09 (E), 18 (S), 27 (W) and 36/0 (N).

cardioid Heart-shaped; profile of cam or plot of radio signal strength against bearing.

cardioid reception Obtained by combining dipole and reflector or vertical aerial with loop in correct phase.

CARDS Computer-assisted radar display system.

Cares Cratering and related effects simulation.

CARF Central altitude reservation facility; ATC facility for special users under altitude reservation concept.

cargo Useful load other than passengers or baggage, but including live animals. In military aircraft, all load other than human beings and personal kit and weapons.

cargo conversion Passenger or other non-cargo aircraft permanently converted to carry freight (see *QC* [*2*]).

CARI Civil Aeromedical Research Institute (FAA).

CARMS Civil Aviation Radio Measuring Station (DoT, UK).

carnet Document facilitating crossing frontier by air without customs dues on aircraft; credit card valid for aviation fuel and certain services; sometimes includes medical certificates.

CARNF Charges for airports and route navigation facilities (ICAO).

Carnot cycle Ideal reversible thermodynamic cycle: isothermal compression, adiabatic compression, isothermal expansion, adiabatic expansion.

Carousel 1 (With initial capital letter) pioneer family of civil INS.
2 (Without capital) circulatory conveyor system to which baggage is delivered in arrival terminal.

CARP Computed air release point.

carpet 1 Graphical plot of three variables having appearance of flexible two-dimensional surface

viewed obliquely.
2 Strip of Earth's surface subjected to sonic boom.

carpet bombing Level bombing, using one or more aircraft, to distribute bombs uniformly over target area.

Carquals Carrier qualification tests (USN).

carrier 1 Aircraft carrier.
2 EM wave, usually continuous and constant amplitude and frequency, capable of being modulated to transmit intelligence.
3 Electronic charge *, either so-called hole or mobile electron.
4 Substance chosen to carry trace element or trace of radioactive material too small to handle conveniently.
5 Operator of commercial aircraft engaged in transport of passengers and/or freight for hire or reward.

carrier air group Two or more aircraft squadrons operating from same carrier (1) under unified command.

carrier-on-board delivery Air delivery of personnel, mail and supplies to carrier (1) at sea.

carrier suppression Communications system in which intelligence is transmitted by sidebands, carrier being almost suppressed.

carrier task force One or more carriers (1) and supporting ships intended to be self-sufficient in prolonged campaign.

carry trials Programme intended to prove carriage and release of fired, dropped or jettisoned stores.

CART Combat aircraft repair team.

Cartesian co-ordinates System of three mutually perpendicular planes to describe any position in rectilinear space.

cartridge Portable container of solid fuel or propellant, with self-ignition system, for propulsion of projectile or supplying pressure to one-shot system.

cartridge starter Main-engine starting system energized by reloadable cartridges.

cartwheel Aerobatic manoeuvre involving rotation about Z (yaw) axis, at very low airspeed, with that axis approximately horizontal.

CAS 1 Chief of the Air Staff.
2 Collision avoidance system.
3 Close air support.
4 Calibrated airspeed (see *airspeed*).
5 Corrected airspeed (obs).
6 Control augmentation system (or subsystem).
7 Controlled airspace.
8 Commission for Atmospheric Sciences.
9 Crisis action system (US JCS).
10 Cockpit avionics system (FSA/CAS).
11 Control actuation section (missiles).
12 Contract administration service (or standard[s]).

CASC Combined acceleration and speed control.

Cascade Combat air surveillance correlation and display equipment.

cascade Array of numerous (eg six or more) aerofoils superimposed to handle large gas flow (eg to turn flow round corner of tunnel circuit).

cascade reverser Thrust reverser incorporating one or more cascades to direct efflux diagonally forwards.

case 1 Outer layer of carburized, nitrided or otherwise case-hardened steel part.
2 Cartridge or shell case housing propellant.
3 Envelope containing solid rocket propellant and withstanding structural and combustion loads.

case-bonded Solid propellant poured liquid into motor case (3) and cast in situ.

case chute See *case ejection.*

case ejection Method of disposing of non-consumable case (2), usually stored on board or discharged through chute under assumed positive acceleration.

casein Cold-water glue manufactured from dehydrated milk curd.

caseless ammunition Gun ammunition in which case (2) is consumed upon firing.

casevac Casualty evacuation.

CASF Composite Air Strike Force (TAC).

CASI 1 Canadian Aeronautics and Space Institute.
2 Commission Aéronautique Sportive Internationale (Int).

cask Container for transport and storage of nuclear fuel or radioactive material.

CASS 1 Command active sonobuoy system.
2 Crab-angle sensing system.

Cassegrain Optical telescope using two parabolic mirrors in series.

cassette Standard tape container, eg for recording mission data.

CAST Civil Aircraft Study Team.

cast-block engine PE (4) with each linear row of cylinders arranged in a single cast block.

castellated nut Typically, hexagon nut with six radial slots for split-pin or other lock.

Castigliano Fundamental structural theorems relating loads, deflections and deformation energy.

casting Forming material by pouring molten into shaped mould.

cast propellant Solid rocket propellant formed into grain by casting.

CASWS Close air support weapon system.

CAT 1 Clear-air turbulence.
2 Comité de l'Assistance Technique.
3 Combat aircraft technology.
4 Computer automatic testing system.
5 Crisis action team.
6 Cockpit automation technology.

CATA Canadian Air Transportation Administration.

Catac Commande Aérienne Tactique (F fighter command).
Catalin Cast phenolic thermosetting plastic.
catalyst Substance whose presence permits or accelerates chemical reaction, itself not taking part.
catalyst bed Porous structure through which fluid passes and undergoes chemical reaction (eg HTP motor).
catapult Device for externally accelerating aeroplane to safe flying speed in short distance.
catastrophic instability Irrecoverably divergent loss of stability at dynamic head sufficient to break primary structure.
CATC 1 College of Air Traffic Control (Hurn, UK).
2 Commonwealth Air Transport Council (UK/Int).
Catca Canadian Air Traffic Controllers' Association.
CATCC Carrier air traffic control centre.
catcher Small fence-like strips around leading edge of sharply swept or delta-wing naval aircraft; designed to engage barrier.
CATE Conference on Co-ordination of Air Transport in Europe.
Cat E Category E for an aircraft, a write-off.
categories 1 For flight crew, licence authorization to qualify on all aircraft within broad groups (eg light aircraft, glider, rotary-wing).
2 For certification of aircraft, grouping based on usage (eg transport, experimental, aerobatic).
3 In bad-weather landings, combinations of decision height and visibility (RVR) agreed by ICAO: 1, 200 ft/60 m and 2,600 + ft/800 + m. 2, 200–100 ft/60–30 m and 2,600–1,300 ft/800–400 m. 3a, no DH, to and along runway with RVR 700 ft/200 m. 3b, to and along runway with RVR 150 ft/50 m (visual taxi). 3c, no external visibility.
4 For aircraft damage and repairability, see *Cat E*.
5 For runway repairs, see *rough field*.
Category 2 box Rectangular box or window in HUD defining permissible Cat 2 deviation of localizer and G/S.
catenary Curve described by filament supported at two points not on same local vertical.
Caterpillar Club Private club formed by Irvin Air Chute Co and open to all who have saved their life by using parachute of any kind.
cathedral Anhedral (pronounced cat-hedral).
cathode 1 Positive terminal of source of EMF (eg battery).
2 Electrode at which "positive current" leaves solid circuit.
3 Negative terminal of electroplating cell.
4 In CRT and similar tube, source of electron stream.
cathode ray oscillograph, cathode ray oscilloscope, CRO CRT built into device including amplifier, power pack and controls for graphic examination of waveforms and other research.
cathode ray tube, CRT Vacuum tube along which electrons (cathode rays) are projected, deflected by pairs of plates creating electric field (deflected toward positive) and impact on screen coated with electroluminescent phosphor.
cathode tuning indicator Triode amplifier and miniature CRT giving visual indication, by closure of well-defined shadow area, of changes in carrier amplitude too small to detect aurally. Also called Magic Eye.
Catia Conception assistée tridimensionelle interactive d'applications (F).
Catic (CATIC) China National Aero Technology Corporation.
cation Positively charged ion, travelling in nominal direction of current (pronounced cat-ion).
CATO Civil air traffic operations.
catoptric beam Visible light concentrated into parallel beam by accurate reflector.
cat's whisker Fine pickoff, esp from gyro (colloq).
catwalk Narrow footway along keel of semi-rigid or rigid airship.
CAU 1 Cold-air unit.
2 Canard actuation unit.
caution note Note written on approach chart warning of high ground or other hazard.
caution speeds Speeds published in flight manual as limits for each configuration.
cavitation Transient formation and shedding of vapour-pressure bubbles at surface of body moving through non-degassed liquid, or of vacuum cavities at surface of body moving through other fluid. Collapse of cavities is violent, causing extreme implosive pressures.
cavity magnetron Magnetron having resonant cavities within cylindrical anode encircling central cathode.
cavity resonator Precisely shaped volume bounded by conducting surface within which EM energy is stored at resonant frequency.
CAV-OK, Cavok Ceiling and visibility OK (for VFR), or better than predicted.
CAVU Ceiling and visibility unlimited.
CAWC Combined air-warfare course.
CAWP Central annunciator warning panel.
CB 1 Circuit breaker.
2 Chloro-bromo-type fire extinguishants.
3 Citizens' band radio.
4 Centre of balance (or c.b., C-B).
5 Chemical/biological.
Cb Cumulo-nimbus.
CBAA Canadian Business Aircraft Association.
CBC Carbon/BMI composite.
CBD Central business district of city.
CBG Carrier battle group (USN).
CBH Chemical/biological hardening.
CBI 1 Computer-based instruction.
2 Component burn-in.
C-Bite Continuous built-in test equipment.

CBI theatre China, Burma, India (WW2).
CBLS Carrier, bomb, light store.
CBM 1 Chlorobromomethane.
2 Confidence-building measure.
CBN Cubic boron nitride.
CBP Contact-burst preclusion.
CBR 1 California Bearing Ratio; system for assessing ability of soft (ie unpaved) surfaces to support aircraft operations; contains terms for aircraft weight, tyre characteristics, landing gear configuration and rutting after given numbers of sorties.
2 Chemical, biological, radiological warfare.
CBTE Conventional bomb triple ejector.
CBU Cluster bomb unit.
CBW Chemical and biological (or bacteriological) warfare.
CC 1 Central computer.
2 Critical crack.
3 Coastal Command (RAF).
4 Communications (UK role prefix).
5 Composite command (USAF).
Cc 1 Equivalent centreline chord.
2 Cirro-cumulus.
CC^3 Counter-C^3.
CCA 1 Current cost accounting.
2 Carrier-controlled approach.
CCAQ Consultative Committee on Administrative Questions (ICAO).
CCAT Carrier control approach trainer.
CCB Configuration change (or control) board (software).
CCC 1 Customs Co-operation Council (ICAO).
2 Combat control centre.
3 See *C*^{3}.
CCD 1 Charge-coupled device (solid-state electronics).
2 Computerized current density.
CCDA Cockpit-control driver actuator.
CCDR Contractor cost data report (US).
CCE 1 Communications control equipment.
2 Commission des Communautés Européennes (Int).
CCF Combined Cadet Force (UK).
CCFG Compact constant-frequency generator.
CCI Chambre de Commerce Internationale (Int). See *ICC*.
CCIA Comitato Coordinazione Industria Aerospaziale (I).
CCIG Cold-cathode ion gauge.
CCIL Continuously computed impact line; HUD snapshoot presentation with fully predicted bullet flight profile with various range assessments.
CCIP Continuously computed impact point; HUD display for air-to-ground weapon delivery with impact point indicated for any manual release from laydown to steep dive. See also *Delayed* *.
CCIP/IP CCIP with preliminary designation of an initial point.
CCIR 1 Comité Consultatif International des Radiocommunications (UIT), assigns wavebands, frequencies.
2 Commission Consultative Internationale pour la Radio (CCITT = telegraphy and telephones) (Int).
CCIS Command control information system.
CCl Commission for Climatology (WMO).
CCM 1 Conventional cruise missile.
2 Counter-countermeasures.
CCMA Comite de Compradores de Material Aeronautico de America Latina (Int).
CCN Contract change notice.
CCOC Combustion-chamber outer casing.
CCP Cutter-centre path (machining).
CCPC Civil Communications Planning Conference (NATO).
CCPDS Command and control processing and display system.
CCPR Cruise compressor pressure ratio.
CCR 1 See *ACC*.
2 Circulation-controlled rotor.
CCRP Continuously computed release point; HUD display for air-to-ground weapon delivery with steering command and auto weapon release in any attitude from laydown to OTS, system controlling entire firing sequence and triggering release or firing mechanism automatically.
CCS 1 Communications control system (aircraft R/T and i/c selection and audio routeing).
2 Conformal countermeasures system.
3 Computer and computation subsystem (ACMI).
CCT 1 Combat control team.
2 Combat crew training (S adds "squadron").
CCTV Closed-circuit TV.
CCTW Combat crew training wing (USAF).
CCV 1 Control-configured vehicle; aircraft in which aerodynamic controls and AFCS partially supplant natural stability, with reduced gust response and structural stress and enhanced manoeuvrability.
2 Chamber coolant valve.
CCW Counter-clockwise.
CD 1 Certification demonstration.
2 Circular dispersion.
3 Cycle-dependent (costs directly proportional to number of flights, eg use of reverser).
4 Convergent/divergent (nozzle).
5 Controlled diffusion (aerofoil).
Cd 1 Total drag coefficient.
2 Candela.
CD-2 Common digitizer 2 (FAA).
CDAA Circularly disposed aerial (antenna) array.
CDAI Centre de Documentation Aéronautique Internationale.
CDB Cast double-base rocket propellant, allows case-bonding, varied formulation and charge configuration, low smoke emission etc.
CDBP Command Data Buffer Program (USAF).

CDC 1 Course and distance calculator.
2 Concorde Directing Committee (Comité Direction Concorde).
CDD Credible delicious decoy.
CDDT Countdown demonstration test.
CDE Chemical Defence Establishment.
CDF Clearance delivery frequency.
CDF Drag coefficient for zero lift.
CDfric Frictional drag coefficient, usually close to CDF.
CDH Constant delta height.
CDI 1 Course/deviation indicator.
2 Collector diffusion isolation.
3 Capacitor-discharge ignition.
4 Compass director indicator.
CDIP Continuously displayed impact point (HUD).
CDL Lift-dependent drag coefficient, subcritically equal to induced drag coefficient.
CDMA Code-division multiple access.
CDmin Minimum drag coefficient.
CDMLS Commutated-Doppler microwave landing system.
CDMT Central design and management team.
CDN 1 Co-ordinating message (ICAO).
2 Certificat de Navigabilité (C of A, F).
Cdo Commando (UK).
CDP 1 Central data processor.
2 Contract-definition phase.
3 Countermeasures dispenser pod.
4 Critical decision point.
CDPI Crash data position indicator.
CDR Critical design review.
CDRB Canadian Defense Research Board.
CDRL Contract data requirements list.
CDRS 1 Control and data-retrieval system.
2 Container design retrieval system (USAF).
CDS 1 Coefficient of slush drag; numerical factor derived from basic precipitation drag, spray-impingement drag and landing-wheel arrangement.
2 Container delivery system (USAF).
3 Chief of Defence Staff (UK).
4 Controls and displays subsystem.
CdS Cadmium sulphide (US, sulfide), basis of family of cryogenic IR detectors.
CDTC Controlled descent through cloud.
CD trainer Crew drill trainer.
CDU 1 Control and display unit.
2 Cockpit display unit.
C_e Specific fuel consumption (USSR).
CEA 1 Code of European Airworthiness.
2 Commissariat à l'Energie Atomique (F).
3 Circular error average.
4 Combined electronics assembly.
CEAA Commandement des Ecoles de l'Armée de l'Air (F).
CEAC 1 Commission Européenne de l'Aviation Civile.
2 Committee for European Airspace Coordination.
CEAM Centre d'Expériences Aériennes Militaires, Mont de Marsan (F).
CEAT Centre d'Essais Aéronautiques de Toulouse (F).
CEB 1 Curve of equal bearings (from NDB).
2 Combined-effects bomb(let).
CECAI Conference of European Corporate Aviation Interests (Int).
CED 1 Continued engineering development.
2 Competitive engineering definition.
Cedocar Centre de Documentation de l'Armement (F).
CEF Cost/effective factor (materials).
CEFH Cumulative engine flight hours.
CEI 1 Critical engine inoperative.
2 Council of Engineering Institutions (UK).
3 Commission Electrotechnique Internationale.
4 Combat efficiency improvement.
ceiling 1 Of aircraft, greatest pressure height that can be reached (see *absolute* *, *absolute aerodynamic* *, *service* *, *zoom* *).
2 Of cloud, height above nearest Earth's surface of lowest layer of clouds or obscuring phenomena that is reported as "broken," "overcast" or "obscuration" and not "thin" or "partial" (FAA).
ceiling balloon Small free balloon, whose rate of ascent is known, timed from release to give measure of ceiling (2).
ceiling climb Aircraft flight authorized for express purpose of measuring ceiling (1).
ceiling height indicator Device for measuring height of spot produced by ceiling projector.
ceiling light See *ceiling projector*.
ceiling projector Source of powerful light beam projected vertically to form bright spot on underside of cloud.
ceiling unlimited Sky clear or scattered cloud, or base above given agreed height (in US 9,750 ft, 2,970 m).
ceiling zero Fog.
ceiliometer Device for measuring ceiling (2), esp ceiling projector whose beam oscillates about horizontal axis like metronome, linked with photocell with readout in tower.
CEL Centre d'Essais des Landes (F).
Celar Centre d'Electronique de l'Armement (F).
celestial altitude Altitude (2) of heavenly body or point on celestial sphere.
celestial body Meaning arguable, but normally all bodies visible or supposed other than Earth and man-made objects. Diffuse bodies (eg nebulae) often called "structures".
celestial equator Great circle formed by extending Earth equatorial plane to celestial sphere.
celestial fix See *astro fix*.
celestial guidance Guidance of unmanned vehicle by automatic star tracking.
celestial horizon Great circle formed on celestial sphere by plane passing through centre of Earth normal to straight line joining zenith and nadir.

celestial mechanics Science of motion of celestial bodies.
celestial meridian Meridian on celestial sphere.
celestial navigation See *astronavigation.*
celestial pole Terrestrial pole projected upon celestial sphere.
celestial sphere Imaginary hollow sphere of infinite radius centred at centre of Earth (for practical purposes, at eyes of anyone on Earth).
celestial triangle Spherical triangle on celestial sphere, esp one used for navigation.
cell 1 Combination of electrodes and electrolyte generating EMF; basic element of battery.
2 Portion of structure having form of rigid box, not necessarily completely enclosed.
3 In biplane or other multi-wing aircraft, complete assembly of planes, struts and wires forming structural box.
4 Gasbag of aerostat, esp in airship having multiple bags in outer envelope.
5 In EDP, elementary unit of storage.
6 Self-contained air mass of violent character (eg TRS).
7 In military operations, smallest tactical aircraft element flying together (often three); another definition is a small unit of airborne military aircraft which can if necessary operate independently.
Cellophane Early cellulose-base transparent plastics film.
cellule Cell (3) or assembly of two or more wings on either side of centreline (arch).
Celluloid Early transparent plastics film from nitrocellulose treated with camphor.
cellulose Carbohydrate forming major structural constituent of plants; precursor of many aeronautical materials.
cellulose acetate Thermoplastic from cellulose and acetic acid; used in rayon, film, lacquer, etc.
cellulose dope See *dope.*
cellulose nitrate Thermoplastic from cellulose and nitric acid; used in explosives, dopes and structural plastics.
Celsius Scale of temperature, symbol °C. Unit is same as SI scale K, but numbers are lower by 273. Until 1948 called Centigrade.
CELT Combined emitter location (or locator) testbed.
CEM 1 Centre d'Essais de la Méditerranée (F).
2 Combined-effects munition (cluster dispenser).
Cementite Iron carbide; carbon form in annealed steel.
CEMS Centre d'Etudes de la Météorologie Spatiale (F).
CEMT Conférence Européenne des Ministres des Transports.
CEN Centre d'Etudes Nucléaires (F).
CENA Centre d'Etudes (formerly d'Experimentation) de la Navigation Aérienne (F).
Cenipa Centro de Investigaçao e Prevençao de Acidentes Aeronàuticos (Braz).
Centigrade See *Celsius.*
centimetre 0·01 metre; measure of length not used in SI.
centimetre Hg Used as unit of manifold pressure = 1·33 kPa.
centimetric radar Radar operating on wavelengths around 0·01 m, with frequencies 3–30 GHz.
centipoise See *viscosity.*
centistoke See *viscosity.*
CENTO, Cento Central Treaty Organization (defunct).
central altitude reservation See *CARF.*
central warning panel See *CWP (1).*
centrebody Streamlined body in centre of circular, semicircular or quasi-circular supersonic intake (inlet) to cause inclined shock.
centre controls To move primary flight controls from deflected to neutral position.
centre engine Engine on centreline of multi-engined aircraft.
centreline 1 Principal longitudinal axis; usually also axis of symmetry (eg of aircraft, missile, runway).
2 In aerofoil section, line joining leading and trailing edges and everywhere equidistant from upper and lower surfaces, all measures being normal to line itself.
centreline camber Ratio of maximum distance between chord and centreline (2) to chord.
centreline gear Main landing gear on aircraft centreline (eg DC-10-30).
centreline lighting On runway, flush lights at 50 ft (15 m) intervals terminating 75 ft (23 m) from each threshold.
centreline noise plot Plot of aircraft noise, usually EPNdB, along runway centreline extended (usually 6 st miles, 9·6 km) in each direction covering approach and climb-out.
centre of area See *centroid.*
centre of buoyancy Point through which upthrust of displaced fluid acts; c.g. of displaced fluid in case of aerostats and marine aircraft afloat.
centre of burst Mean point of impact.
centre of dynamic lift In aerostat, point on centreline through which lift force due to motion through atmosphere acts.
centre of gravity, c.g. Point through which resultant force of gravity acts, irrespective of orientation; in uniform gravitational field, centre of mass. For two-dimensional forms, centroid.
centre of gravity limits In nearly all aircraft, esp aerodynes, published fore and aft limits for safe c.g. position; in case of aeroplanes expressed as percentages of MAC.
centre of gravity margin H_n, distance along aircraft major axis from c.g. to neutral point, expressed as % SMC.
centre of gravity travel Fore and aft wander of c.g. in course of flight due to consumption of fuel,

release of loads, etc.

centre of gross lift In aerostat, usually centre of buoyancy.

centre of gyration For solid rotating about axis, point at which all mass could be concentrated without changing moment of inertia about same axis.

centre of lift Resultant of all centres of pressure on a wing.

centre of mass Point through which all mass of solid body could act without changing dynamics in translational motion; loosely, but not always correctly, called c.g. Alternative title, centre of inertia.

centre of pressure, c.p. On aerofoil, point at which line of action of resultant aerodynamic force intersects chord. Almost same as aerodynamic centre, but latter need not lie on chord. In general, c.p. is resultant of all aerodynamic forces on surface or body.

centre of pressure coefficient Ratio of distance of c.p. from leading edge to chord.

centre of pressure moment Product of resultant force on wing (or section) and distance from c.p. to leading edge (or leading edge produced at aircraft centreline). Inapplicable to very slender wings.

centre of pressure moment coefficient As above but divided by dynamic pressure; not same as coefficient of moment.

centre of pressure travel 1 Linear distance through which c.p. travels along chord over extreme negative to positive operating range of angles of attack, ignoring compressibility in subsonic flow.
2 Linear distance through which c.p. travels along chord over complete aircraft operating range of Mach numbers (supersonic aircraft only).

centre of thrust Thrust axis, for one or multiple engines.

centre punch Hand tool for making accurate conical depressions.

centre section In most winged aircraft, centre portion of wing extending symmetrically through or across fuselage and carrying left and right wings on its tips. Certain aircraft have wing in one piece, or in left and right halves joined at centreline; such have no **, though some authorities suggest it is then wing inboard of main landing gear.

centrifugal clearance Radial clearance between rotating mass and surrounding fixed structure at peak rotating speed.

centrifugal compressor Rotary compressor in form of disc carrying radial vanes to accelerate working fluid radially outward to leave periphery at very high speed, this being converted to pressure energy in fixed diffuser.

centrifugal separator See *centrifuge filter*.

centrifugal twisting moment Moment tending to rotate propeller blades towards zero pitch (opposing coarsening of pitch).

centrifuge 1 Device for whirling human and other subjects about vertical axis, mainly in aerospace medicine research.
2 Device for imparting high unidirectional acceleration to hardware under test.
3 Device for imparting high unidirectional acceleration to mixtures of fluids with or without particulate solids to separate constituent fractions.

centrifuge filter Action, inherent in turbofans and some other machines, which in rotating fluid flow causes unwanted particulate solids to be centrifuged outwards away from core.

centring, centering Radial and, usually, also axial constraint of floated gyro to run equidistant at all points from enclosure.

centring control System which, on demand, centres a variable in another system.

centripetal Acceleration toward axis around which body is rotated; usually equal and opposite to centrifugal (see *coriolis*).

centrisep Centrifugal separator (*centrifuge filter*).

centroid In geometrical figure (one, two or three-dimensional), point whose co-ordinates are mean of all co-ordinates of all points in figure. In material body of uniform composition, centre of mass.

CEO Chief executive officer of corporation or company.

CEOA Central Europe Operating Agency (NATO pipelines).

CEP 1 See *circle of equal probability*; also called circular error probable.
2 Concurrent evaluation phase.
3 Chromate-enriched pellet.

CEPA Commission d'Evaluation Pratique d'Aéronautique (F).

CEPr Centre Essais de Propulseurs (Saclay, F).

CEPS Central European pipeline system (NATO).

CEPT Conférence Européenne des Postes et Télécommunications.

CER Cost-estimating relationship.

ceramal See *cermet* (from ceramic + alloy).

ceramel See *cermet* (from ceramic metal).

CERCA Commonwealth and Empire conference on Radio for Civil Aviation.

ceria glass Amorphous semiconductor glass doped with cerium dioxide.

Cerma, CERMA Centre d'Etudes et de Recherches de Médecine Aéronautique (F).

cermet 1 Composite material attempting to combine mechanical toughness of metal with hardness and refractory qualities of ceramics. Early examples cemented carbides, eg tungsten carbide sintered with cobalt.
2 Incorrectly, part made of ceramic bonded to metal.

CERP Centre-Ecole Régional de Parachutisme.

Cerrobase US lead-bismuth alloy, MPt 125°C.
Cerrobend US bismuth-tin-lead-cadmium alloy, MPt 68°C.
Cerromatrix US lead-tin-antimony alloy, MPt 105°C.
CERS 1 Centre Européen de Recherche Spatiale (see *ESRO*).
2 Carrier evaluation and reporting system.
CERT Centre d'Etudes et de Recherches de Toulouse (ONERA).
certification For aircraft, issue of ATC (4) for US, C of A for UK or equivalent by national certifying authority, stating type meets all authority's requirements on grounds of safety. Other aircraft certificates include Production and Registration.
certification pilot Test pilot employed by national certification authority to evaluate all types of aircraft proposed for use by operator in that country, from whatever source.
certification test Test conducted by certification authority prior to issue of certificate.
CES 1 Catapult-end speed.
2 Civil Engineering Squadron (USAF).
cesium, caesium Extremely soft silver metal, highly reactive, used as working fluid as jet of charged ions in space thrusters.
CESR Centre d'Etude Spatiale des Rayonnements (space radiation).
cetane rating Numerical scale for ignition quality of diesel fuels.
CEV Centre d'Essais en Vol (Brétigny, Istres, Cazaux, F).
CEWI Combat electronic warfare and intelligence.
CF 1 Centrifugal force.
2 Concept formulation.
3 Carbon-fibre.
C_f 1 Measured thrust coefficient of rocket.
2 Skin friction coefficient (see C_{Dfric}).
cf Cubic feet.
C_f^0 Theoretical thrust coefficient of rocket.
CFA 1 Cross- (or crossed-) field amplifier.
2 Chief flight attendant.
3 Caribbean Federation of Aeroclubs (Int).
CFAC Constant-frequency alternating current.
CFAO Conception et fabrication assistée par ordinateur (F).
CFAR Constant false-alarm rate (ECM).
CFAS Commandement des Forces Aériennes Stratégiques (F).
CFB Cruise fuel burn.
CFC 1 Cycles to first crack (structures).
2 Cold fog clearing.
3 Carbon-fibre composite.
4 Central fire control.
CFD 1 Chaff/flare dispenser.
2 Centralized fault display (IU adds "indicator unit", S adds "system").
CFE 1 Contractor-furnished equipment.
2 Central Fighter Establishment (RAF).
CFF 1 Cost plus fixed fee, sometimes CPFF.
2 Critical flicker frequency (electronic displays).
CFFT Critical flicker fusion threshold.
CFI Chief flying instructor.
CFII Civilian flight instrument instructor.
CFK, CfK CFRP (G), usually KfK.
cfm Cubic feet per minute.
CFMO Command flight medical officer.
CFO Coherent-fibre optics.
CFP Cost plus fixed price.
CFR 1 Co-operative Fuel Research.
2 Crash fire rescue.
3 Code of federal regulations.
CFRP Carbon-fibre reinforced plastics.
CFS Central Flying School (RAF).
CFSO Command flight safety officer.
CFT Conformal fuel tank.
CFTR Cold-fan thrust reverser.
CF weight Contractor-furnished.
CG 1 Cargo glider (USA, WW2).
2 Lethal gas phosgene.
C_g Ratio of actual/ideal propulsive thrust.
cg, c.g. Centre of gravity.
CGAO Conference of General-Aviation Organizations (UK).
c.g. arm Arm (2) obtained by adding all individual moments and dividing sum by aircraft total mass.
CGB Central gearbox.
CGI 1 Chief ground instructor.
2 Computer-generated image.
CGIVS CGI (2) visual system.
c.g. limits Forward and aft limits, usually expressed as percentage MAC, within which aircraft c.g. must fall for safe operation.
CGM load Combined gust and manoeuvre load.
CGN Guided-missile cruiser (USN).
CGP Coal-gasification plant, eg for jet fuel.
CGS 1 Centimetre, gramme, second system of units, superseded by SI.
2 Central Gunnery School (RAF).
3 Computer-generated simulation.
CGT Consolidated ground terminal.
CGV Computer-generated voice.
CGW Combat gross weight.
CH Channel (ICAO).
C_H 1 Hinge coefficient.
2 High cloud.
chaff Radar-reflective particulate matter or dipoles sized to known or suspected enemy wavelengths (ECM).
chaff dispenser Tube, gun, projector or other system for releasing chaff either in discrete bursts or in measured stream, manually or upon command by RWS.
chafing patch Local reinforcement of aerostat envelope where likely to suffer abrasion.
chain Geographical distribution of radio navaid stations linked together and emitting synchronized signals (eg Decca, Loran).
chain gun High-speed automatic cannon with

external power supply and open rotating bolt.

chain radar beacon Beacon having fast recovery time, thus can be interrogated by many targets up to prf.

chain reaction In fissile material or other reactant, process self-perpetuating as result of formation of materials necessary for reaction to occur. Nuclear chain reaction arranged in weapons to have optimum uncontrolled (accelerating) form.

chair chute Parachute permanently incorporated into seat (not necessarily ejection seat, but removable).

chalk Loosely, one load, especially stick or squad of troops, often 6–10.

chalk number In military logistics, number assigned one complete load and carrier vehicle; hence chalk commander, chalk troops.

chamber In liquid rocket engine, enclosed space where combustion takes place, between injectors and throat. In solid rocket, enlarging volume in which combustion takes place, varying in form with design of motor.

chamber pressure See *burn time average*, *action time average*, *MEOP*.

chamber volume In liquid rocket, total volume as defined; solid motor varies during burn.

chamfered Bevelled (edge or corner, eg of sheet).

Chance light Formerly, mobile airfield floodlight illuminating landing area and apron.

chandelle Flight manoeuvre (see *stall turn*); another definition, not necessarily synonymous with stall turn, is a manoeuvre in which speed is traded for altitude whilst reversing flight direction (see also *Immelmann*).

channel 1 Band of frequencies in EM spectrum, esp at radio nav/com frequencies (thus 20 * allotted to ILS in all countries, each with published frequency for loc and g/p).
2 Single end-to-end "route" in dynamic system, esp one exerting control authority (eg collective pitch in helicopter AFCS).
3 Structural member of channel form (eg top-hat or U).
4 In EDP, several meanings: any information or data highway, one or more parallel tracks treated as unit, portion of store accessible to given reading station.
5 In semiconductor device (eg transistor), flow bypassing base.
6 Pathways for energetic ions or atoms along crystal lattices.
7 In airport terminal, single routing for departing or arriving passengers.
8 Take-off and alighting path at marine airport.

channel nut One forming integral part of channel (3).

channel patch Channel-shaped reinforcement to aerostat envelope to anchor rigid spar.

channel section See *channel (3)*.

channel wing Wing curved in front elevation to fit closely round lower half of propeller disc.

Chapi, CHAPI Carrier helicopter approach path indicator.

char Ablative material charred and eroded during re-entry (see *ablation*).

characteristic Sense (up or down) in which barometric pressure changes in preceding 3 h.

characteristic curve 1 Curve of atmospheric sounding results plotted on Rossby diagram.
2 Curve of primary characteristic of aerofoil when plotted (see *characteristics*).

characteristic exhaust velocity, C* Measure of rocket performance, numerically g_cA_t/W_p multiplied by integral of chamber pressure over action time.

characteristic length 1 In rocket, ratio of chamber volume to nozzle throat area.
2 In rocket, length of cylindrical tube of same diameter as chamber, having same volume as chamber.
3 Convenient reference length (eg chord).

characteristics 1 Of aerofoil, primary * are: coefficients of lift and drag, L/D ratio, cp position and coefficient of moment, each plotted for all operating AOAs.
2 In electronics, relationships between basic variables (eg anode current/voltage, anode current/grid voltage) for valves (tubes) and correspondingly for transistors.

characteristic velocity 1 Sum of all changes in velocity, positive and negative all treated as positive, in course of space mission.
2 Velocity required for given planetary (esp Earth) orbit.

charge 1 Total mass of propellant in solid rocket.
2 Propellant of semi-fixed or separate-loading ammunition.
3 To fill high-pressure gas or cryogenic (eg Lox) container.
4 Quantity of electricity, measured (SI) in Coulombs.

chargeable Malfunction (eg IFSD) clearly due to fault in design, workmanship, material or technique by supplier.

charging point Standard coupling through which aircraft fluid system is replenished or pressurized.

Charles' law Perfect gas at constant pressure has volume change roughly proportional to absolute temperature change.

Charlie Preplanned landing-on time in carrier operations.

Charpy Destructive test of impact resistance of notched test bar.

chart Simplified map, typically showing coasts, certain contours, woods, water, and aeronautical information (symbols vary and not yet internationally agreed).

chart board Rigid board, suitably sized for chart or topographic map; often provided with protractor on parallel arms.

charted delay In Loran, published delay.

chase Accompany other aircraft, esp one on test, to observe behaviour and warn of visible malfunction; hence * plane, * pilot.

chassis Rigid base on which electronics are mounted; for airborne equipment, mates with racking.

chatter 1 Multiple conversations or signatures, most being of no interest, all heard on same frequency.
2 High-frequency vibration energized by intermeshing gear teeth.

CHD Crutching heavy-duty (stores carrier).

check 1 To verify flight progress (eg ETA *).
2 To examine pilot for proficiency.
3 Examination conducted in (2).
4 Programmed investigation of aircraft for malfunction (before-flight *, after or post-flight *, etc).
5 Programmed routine from entering cockpit to start of flight (cockpit *).
6 To reduce in-flight trajectory variable (* rate of descent).

checking station 1 Designated and marked location used in rigging airframe. Also called checking point.
2 Reference station on propeller blade, typically at 42 in radius on lightplanes.

checklist Written list of sequenced actions taken in check (5), usually printed on plastics pages.

check-out 1 Programmed test by user immediately before operational use of missile, radar or other system.
2 Sequenced tasks to familiarize operator with new hardware.

check point, checkpoint 1 See *waypoint*.
2 Geographical point happening to offer precise fix.
3 Planned strategic review of major industrial programme for government customer, esp collaborative.
4 Predetermined geographical location serving as reference for fire adjustment or other purpose.
5 Mean point of impact.

check six! Look astern for hostile fighters.

check valve One-way valve in fluid line.

cheese aerial (antenna) Shaped roughly like D in side elevation with quasi-circular or parabolic reflector and flat parallel sides nodding in elevation; propagates more than single mode.

chemical fuel Has ambiguous connotation excepting common fuels; hence, "exotic" fuel, esp for air-breathing engines.

chemical laser Strictly, one in which chemical reaction occurs, eg perfluoromethane/hydrogen.

chemical milling Use of acid or alkaline bath to etch metal workpiece in controlled manner.

chemical munition Non-explosive ordnance operating by chemical reaction: incendiary, smoke, irritant or lethal gas, defoliant, flare or flash, dye marker etc.

chemical propulsion Propulsion by energy released by chemical reaction, eg fuel+oxygen, with or without air breathing.

chemical rocket Rocket operating by chemical reaction; not ion, photon or nuclear.

chemical warfare Use of major chemical munitions, esp irritant or lethal gases.

chemosphere Region of upper atmosphere (say 15–120 miles, 24–190 km) noted for photochemical reactivity.

cherl Change in Earth's rate (tangential velocity) of rotation with latitude.

Cherry rivet Tubular rivet inserted blind and closed by internal mandrel (shank) which then breaks and is removed.

chest-type parachute Pack separate from harness and secured by quick clips on chest.

Cheyenne Mountain Location of USAF/Norad Space Command HQ.

CHG Change, ie modifying previous message.

chilldown Pre-cooling of tanks and system hardware before loading cryogenic propellant.

chilled casting Made in mould of metal, usually ferrous, giving rapid cooling and thus surface hardness.

chin Region under aircraft nose; hence * blister, * intake, * radome, * turret.

china-clay Technique for distinguishing regions of laminar and turbulent boundary layer from changed appearance of thin coating of * suspension.

chine In traditional marine aircraft hull or float, extreme side member running approximately parallel to keel in side elevation. In supersonic aircraft, sharp edge forming lateral extremity of fuselage, shedding strong vortex and merging into wing.

chined tyre Tyre, esp for nosewheels, with chines to depress trajectory of water or slush.

chin fin Fixed destabilizing fin under nose.

chip 1 Single completed device separated from slice, wafer or other substrate of single-crystal semiconductor.
2 Metal fragment, visible to eye, broken from engine or other machinery.

chip chart Rectangles of paint showing colours available.

chip detector Device, often magnetic, for gathering every chip (2), usually from lube oil.

chip width Length of PN code bit, T_c.

chirp 1 Radar/communications pulse compression or expansion (colloq).
2 Confidential human-factors incident reporting procedure (or programme) (CAA).

chisel window Small oblique nose window for LRMTS or camera.

ch.lat. Change of latitude (pronounced sh-lat).

ch.long. Change of longitude (pronounced sh-long).

Chobert Tubular rivet inserted blind and closed by withdrawal of re-usable mandrel.
chock Portable obstruction placed in front of and/or behind landplane wheel(s) to prevent taxiing.
chock-to-chock See *block time*.
choke 1 Inductance used to offer high reactance at chosen frequency to pass d.c. or lower a.c. frequencies only.
2 In typical car (auto) engine, manual control for reducing inlet airflow to enrich mixture when cold.
choked flow Flow of compressible fluid in duct (eg tunnel) in which local Mach number has reached 1 and velocity cannot be increased significantly by increasing upstream pressure.
choked inlet Containing normal shock and suffering choked flow.
Chol, CHOL Common high-order language.
cholesteric LCD with layers each aligned in preferred, different direction.
chomp Changeover (from one waypoint VOR, NDB etc to next) at midpoint (of leg).
chop 1 Changeover point (from one navaid to another, not at midpoint of leg).
2 Change of operational control, precisely promulgated time.
3 To get the *, sudden termination (of human life, project, etc, colloq).
4 To close throttle(s) completely and suddenly.
5 Mild atmospheric turbulence (see *light* *).
chopped fibre Reinforcing fibre chopped into short lengths.
chopped random mat Chopped fibre made into mat (two-dimensional sheet) with random orientation.
chopper 1 Rotary-wing aircraft, esp helicopter (colloq).
2 Mechanical device for periodically interrupting flow, esp light beam, or switching it alternately between two sources.
3 Device for modulating signal by making and breaking contacts at frequency higher than frequencies in signal.
chord 1 Straight line joining centres of curvature of leading and trailing edges of aerofoil section.
2 Loosely, breadth of wing or other aerofoil from front to rear.
3 Boundary members of structural truss.
chord direction In stress analysis, usually parallel to chord at aircraft centreline (of wing, or wing produced to centreline).
chord length Length of chord (1), not measured round profile.
chord line See *chord (1)*. Ambiguously, sometimes line tangent at two points to lower surface (see *geometric chord*).
chord plane Plane containing chord lines of all sections forming three-dimensional aerofoil (assuming no twist).
chord position Defined by location of quarter-chord point and inclination to aircraft x-y plane, point being defined on primary centreline coordinates.
chord wire Wire tying vertices of airship frame.
chordwise Parallel to chord (normally also to longitudinal axis).
chromate enriched pellet Small source of antifungal chemical in integral-tank low point (possible water trap).
chromate primer Anti-corrosive, anti-microbiological surface treatment, esp for water traps in airframe.
chromatic aberration Rainbow effect caused by simple lens having different focal length for each wavelength.
chrome steels Steels containing chromium; often also with vanadium, molybdenum etc.
chromic acid Red crystalline solid, used in solution as cleaner and etchant, and as electrolyte (Alocrome) for Cr plating and anodizing.
chromium Hard silvery metal taking brilliant polish (abb Cr).
Chromoly Alloy steels containing chromium and molybdenum.
chromosphere Thin (under 15,000 km) layer of gas surrounding Sun's photosphere.
chronograph Device for producing hard-copy readout of variable against time, with particular events recorded, typically by pen and disc or drum chart.
chronometer Accurate portable clock with spring drive and escapement.
CHT Cylinder-head temperature.
chuck Rotating vice (US "vise") for gripping tool or workpiece.
chuffing Pulsating irregular rocket combustion.
chutai Squadron (J).
chute 1 Inflatable slide for emergency escape from high-floor aircraft on ground.
2 Parachute (colloq).
3 Axial/radial ducts around engine periphery to guide fan air into afterburner or engine airflow through silenced nozzle.
4 Duct for ventral crew escape or discharge of ECM, leaflets etc.
CI 1 Chief instructor.
2 Compression ignition.
Ci 1 Cirrus.
2 Curie.
ci Cubic inches (not recommended).
CIA 1 Contractor interface agreement.
2 Central Intelligence Agency (US).
3 Commission Internationale d'Aérostation (FAI).
CIAA 1 Consorzio Italiano di Assicurazioni Aeronautiche.
2 Centre International d'Aviation Agricole.
CIACA Commission International des Aéronefs de Construction Amateur (FAI).
CIAM 1 Computerized integrated and automated manufacturing.

2 Commission Internationale d'Aéromodélisme (FAI).

CIAP Climatic Impact Assessment Program (US DoT).

CIARA See *IAARC*.

CIAS Comitato Interministeriale Attività Spaziali (I).

CIE 1 Commission Internationale de l'Éclairage (= ICI).

2 Centro de Investigaciones Espaciales (Arg).

CIEA Commission Internationale d'Enseignement Aéronautique (FAI).

CIEM Computer-integrated engineering and manufacturing.

CIFMS Computer-integrated flexible manufacturing systems.

CIFRR Common IFR room (New York, civil/military).

CIFS Computer-interactive flight simulation.

CIG 1 Computer image-generation.

2 Control/indicator group.

3 Commission Internationale de Giraviation (FAI).

cigar Aeroplane from which wings have been removed (colloq).

cigarette-burning Solid propellant ignited at one end across entire section and burning towards other end.

CIGFTPR Controls-instruments-gas-flaps-trim-prop-run-up (US).

CIJ Close-in jamming.

CIM Computer-integrated manufacturing.

CIME Commission Intergouvernementale des Migrations Européennes (Int, arranges mass flights, eg for refugees).

Cinch Compact inertial navigation combining HUD.

CIP 1 Cold iso-pressing (beryllium).

2 Component improvement program(me).

3 Commission Internationale de Parachutisme (FAI).

CIPR Cubic inches per revolution.

CIR Constant infra-red (heat source on target).

CIRA Cospar International Reference Atmosphere.

circadian rhythm Change in physiological activity on approximate 24-hour cycle.

circle marker On unpaved airfield, white circle indicating centre of landing area.

circle of confusion Image of any distant point on lens focal plane (eg on film).

circle of equal probabilities, CEP Radius of circle within which half the strikes (eg bullet impacts on single aiming point) fall or within which probability is equal that one bullet, bomb or RV will fall inside or outside.

circle of latitude On celestial sphere through ecliptic poles, perpendicular to ecliptic.

circle of longitude On celestial sphere parallel to ecliptic.

circle of origin Normally, Equator or Prime Meridian.

circle of position Circle on Earth's surface centred on line joining centre of Earth to heavenly body, from which altitude of body is everywhere equal. Sometimes called circle of equal altitude.

circuit 1 Short flight linking take-off to landing, typically straight climbout to 300 m/1,000 ft, 90° turn to first crosswind leg, second 90° turn to downwind leg (passing airfield parallel to runway), third 90° turn to descending crosswind leg preparing to land, and fourth 90° turn on to finals.

2 Closed loop of electrical conductors.

circuit breaker Switch for opening circuit (2) while carrying large electrical load.

circuit length Length round closed-circuit wind tunnel traversed by streamline always equidistant from walls.

circuits and bumps Repeated circuits (1) with landing at end of each (colloq).

circular approach Precision training manoeuvre, chiefly associated with carrier flying, where circuit (1) is circular.

circular dispersion Diameter of smallest circle within which 75 per cent of projectiles strike.

circular error probable See *bombing errors (1)*.

circular isotropic Aerial (antenna) radiation polar equal in all directions (normally in azimuth).

circularization Refinement of satellite orbit to approach perfect circle, usually at given required height.

circular mil Area of circle of one mil (1/1,000 inch) diameter = $5{\cdot}067 \times 10^{-10}$ m^2 ($7{\cdot}85 \times 10^{-7}$ in^2).

circular mil foot Unit of resistivity; resistance of one foot of wire of one circular mil section, equal to ohm-mm $\times$ $6{\cdot}015 \times 10^5$.

circular milliradian Conical beam or spread of fire having angle (not semi-angle) of one milliradian.

circular nose Control surface whose leading-edge section is semi-circle about hinge axis.

circular velocity At given orbital height, velocity resulting in circular orbit.

circulation 1 Rotary motion of fluid about body or point; vortex.

2 Ideal flow around (not past) circular body, with streamlines concentric circles and velocity inversely proportional to radius (body needed to avoid infinite V at centre).

3 Streamline flow around body of any form, defined as integral of component of velocity along closed circuit with respect to distance travelled around it. Wing lift created by * superimposed on rectilinear flow past surface (see *bound vortex*, *Magnus*, *Joukowski*).

4 Gross motions of planetary (eg Earth) atmosphere.

circulator Non-reciprocal device in microwave

circuit to produce phase-shift as function of direction of wave flow (see *duplexer*).
circulatory flow Rectilinear flow past lifting body inducing circulation (3) (see *Joukowski*).
CIRF Consolidated intermediate repair facility.
CIRM International Radio Medical Centre.
CIRO Centre Interarmée de Recherches Opérationelles (F).
cirro-cumulus Layer of globular cloud masses at about 6,000 m/20,000 ft (abb Cicu). Also known as mackerel sky.
cirro-stratus High milky-white or grey sheet cloud, 7,000 m/23,000 ft (abb Cist).
cirrus High white cloud; detached, fibrous, silky, 7,500–12,000 m/25,000–40,000 ft (abb Ci).
CIS 1 Combat identification system.
2 Computer interface system.
cislunar Between Moon's orbit and Earth.
CISPR Comité International Spécial des Peturbations Radiophoniques.
CIT 1 Compressor inlet temperature (flight envelope limit).
2 Central integrated testing.
3 Cranfield Institute of Technology.
CITA 1 Commission Internationale de Tourisme Aérien.
2 Confederación Interamericana de Transportadores Aéreos.
CITE Computer integrated test equipment (USAF).
Citeps Central integrated test experimental parameter subsystem.
CITS Central integrated test subsystem (eg Shuttle).
city pair Pair of cities studied from viewpoint of mutual passenger/cargo traffic.
city pair ranking Lists of ** in order of current or projected traffic generation.
CIU 1 Coupler (or central) interface unit.
2 Central Interpretation Unit (UK).
CIV 1 Crossbleed isolation valve.
2 Coannular inverted-velocity (nozzle).
Civil Aeronautics Administration Since 1958 FAA (1).
Civil Aeronautics Board, CAB US Government (DoC) agency responsible for civil aviation, including CARs, licensing, routes and US mail rates.
Civil Air Patrol, CAP US para-military organization using pilot and lightplane resources of general aviation for national ends.
civil day Day of constant 24 hours (sometimes counted as two periods of 12 hours); mean solar day.
Civil Reserve Air Fleet US airline transport aircraft and flight crews predesignated as available at any time for reasons of national emergency.
civil time See *mean solar time*.
civil twilight Period at sunrise or sunset when Sun's centre is between 0° 50′ and 6° below horizon.
CIVL Commission Internationale de Vol Libre (FAI hang-gliding organization).
CIVRES Congrès International des Techniques du Vide et de la Recherche Spatiale.
CIVV Commission International de Vol à Voile (gliding).
CIWS Close-in weapon system (air-to-air).
CJATP Commonwealth Joint Air Training Plan (1939–45).
CJCS Chairman of the Joint Chiefs of Staff.
CKD Component knock-down, parts imported for assembly in importing country.
CL 1 Centreline.
2 Checklist.
3 Chemical laser.
4 Catapult-launched.
5 Charge limit, ie limit payload (RAF).
C_L 1 Coefficient of lift.
2 Low cloud.
CLA 1 Clear ice formation.
2 Centreline average (surface roughness).
CLAC Comisión Latino Americana de Civil Aviación (Int).
clack valve Fluid one-way valve having freely hinged flap seated on one side.
clag Widespread low cloud, mist and/or rain (colloq).
CL/AL Catapult-launched, arrested landing.
CLAMP Closed-loop aeronautical management programme.
clamp Weather unfit for flight (colloq).
clamping To hold either or both peaks of waveform or signal at desired reference potential (d.c. restoration). Increasingly used in processing sensor images; black-level * references all black levels to darkest point of image.
clamshell 1 Cockpit canopy hinged at front or rear.
2 Nose or tail of cargo aircraft hinged into lower and upper or left and right halves.
3 Reverser opening in upper and lower halves meeting on jet centreline behind nozzle.
clandestine aircraft Aircraft designed to overfly without detection, having minimal noise, IR and radar signatures.
clang box Jet-engine switch-in deflector for V/STOL comprising an internal valve and side nozzle with deflecting cascade.
Clansman Army tactical radio communication system (UK).
CLAP Centre Laïque d'Aviation Populaire (F).
CLASB Citizens' League Against the Supersonic Boom (US).
class action Litigation in US courts in which plaintiffs represent a class, eg airline passengers, or passengers of a particular carrier.
classical aeroplane Aeroplane having clearly defined fuselage, nacelles and aerodynamic surfaces, not necessarily with all tail surfaces at rear. Opposite of integrated aeroplane.
classical flutter Occurring because of coupling —

aerodynamic, inertial or elastic — between two degrees of freedom.

classify 1 To protect official information from unauthorized disclosure.
2 In ASW to sort sonar returns according to types of source.

claw 1 Accelerator hook.
2 Operative part of arrester hook.

CLB Crash locator beacon.

CLBR Calibration.

CLC Command launch computer.

CLD 1 Cloud (ICAO).
2 Crutching light-duty (stores carrier).

clean 1 Of aircraft, not carrying external stores (tanks, ordnance etc).
2 Of aircraft, in cruise configuration with gear and high-lift systems retracted.
3 Nuclear weapon designed for reduced, or minimal, residual radioactivity compared with normal weapon of same yield.

cleaning In prolonged glide with PE (4) aircraft, to open up engine briefly to high power to clear over-rich mixture and gummy or carbon deposits.

clear 1 To authorize hardware as fit for use.
2 To authorize person to receive classified information.
3 To rectify stoppage in automatic weapon.
4 To unload weapon and demonstrate no ammunition remains.
5 To empty core store, register or other memory device.
6 In flight operations, authorized to take off, land or make other manoeuvre under ground control.
7 En route, to pass over waypoint.
8 To destroy all hostile aircraft in given airspace.
9 Of local sky, devoid of clouds ("the *"), but may be above or between cloud layers.
10 To clean PE (4); see *cleaning*.
11 To fly out of a local area, eg a flying display.

clear air turbulence, CAT Significant turbulence in sky where no clouds present, normally at high altitude in high windshear near jetstream.

clearance 1 Authorization by ATC (1), for purpose of preventing collision between known aircraft, for aircraft to proceed under specified conditions within controlled airspace (see *abbreviated* *, *SIDs*, *STARs*, * *delivery*, * *items*, * limits).
2 Minimum gap between portions of hardware in relative motion (eg fan blade and case).
3 Transport of troops and material from beach, port or airfield using available communications.

clearance amendment Change in clearance (1) made by controller to avoid foreseeable conflict.

clearance delivery ATC service, with assigned frequency, for issuing pre-taxi, taxi and certain other pre-flight clearances.

clearance limit Fix or waypoint to which outbound flight may be cleared, there to receive clearance to destination.

clearance volume Minimum volume remaining in PE (4) cylinder at TDC.

cleared flight level FL to which flight is cleared, though possibly not yet reached.

clear ice Glossy, clear or translucent accretion from slow freezing of large supercooled water droplets.

clearing manoeuvre Change of aircraft attitude, on ground or in flight, to give better view of other traffic.

clearing procedure Clearing manoeuvre, often combined with vocal callouts (esp when pupil under instruction) before take-off or any other flight operation (eg scrutiny of airspace beneath prior to spin).

clearing turn Turn in which pilot checks local airspace, esp below, before stall or spin.

clear-vision panel See *DV panel*.

clearway 1 Rectangular area at upwind end of runway or other take-off path devoid of obstructions and prepared as suitable for initial climbout.
2 Specif, area beyond runway, extending not less than 250 ft/76 m wide on each side of centreline, no part of which (other than threshold lights away from centreline and not over 26 in/660 mm high) projects above * plane.

clearway plane Plane extending from upwind end of runway at slope positive and not exceeding 1·25 per cent.

clevis joint Fork and tongue joint (eg between solid motor cases) secured by large-diameter pin.

CLGE Cannon-launched guidance electronics.

CLGP Cannon-launched guided projectile.

climatic test Static test in simulated adverse environments (rain, ice, temperature extremes, salt, sand, dust) to demonstrate compliance with requirements.

climb 1 Any gain in height by aircraft.
2 More commonly, deliberate and prolonged gain in height by appropriate trajectory and power setting (ie not zoom).

climb corridor Positive controlled military airspace of published dimensions extending from airfield.

climb gradient Vertical height gained expressed as percentage of horizontal distance travelled.

climb indicator See *VSI*.

climbing cruise, climb cruise Compromise between speed and range, typically at 1·15 V_{md} planned from published tables for peak efficiency higher than attainable in constant-height cruise.

climbing shaft Access hatch and ladder leading from bottom to top of airship hull.

climb out 1 Loosely, flight from unstick to setting course (lightplane in VFR).
2 Specif, flight from screen height (35 ft/11 m) to 1,500 ft/460 m. Comprises six segments: 1, 35 ft to gear up (V_2); 2, gear up to FRH (V_2); 3, level

(accelerate to FUSS); 4, FRH to 5-minute power point (FUSS); 5, level (accelerate to ERCS); 6, to 1,500 ft/460 m (ERCS) (see *NFP*).

clinker-built Marine hull or float constructed from diagonal or longitudinal planks overlapping at edges.

clinodromic Holding constant lead angle.

clinometer Instrument for measuring angle of elevation, used in some ceilometers and, formerly, as lateral-level flight instrument.

clipped wing Aircraft having wing modified by removal of tips or outer portions (eg for racing).

clipper Clipping (1) circuit.

clipping 1 Limiting positive and/or negative parts of waveform to chosen level.
2 Mutilation of communications by cutting off or distorting beginnings and/or ends of words or syllables.
3 Limitation of frequency bandwidth.
4 Reduction of amplification below given frequency.

C_LM Centreline of missile.

C_{Lmax} Maximum lift coefficient.

CLNC Clearance.

CLO Logistics/training command (Neth).

clobber To knock out a ground or air target (colloq).

CLOS Command-to-line-of-sight missile guidance.

close air support, CAS Air attack on targets close to friendly surface force, integrated with latter's fire and movement.

close-controlled interception One in which interceptor is under continuous ground control until target is within visual or AI radar range.

closed-circuit tunnel Wind tunnel which recirculates given mass of working fluid.

closed-circuit TV Camera/microphone linked to TV receiver/speaker by wires.

closed competition Procurement competition in which prices, performances and design details are not disclosed to rival bidders.

closed-jet tunnel Tunnel, not necessarily closed-circuit, in which working section is enclosed by walls.

closed-loop system Dynamic system in which controlled variables are constantly measured, compared with inputs or desired values and error signals generated to reduce difference to zero.

closed thermodynamic cycle Cycle which can transfer energy but not matter across its boundary.

close flight plan To report safe arrival to appropriate ATC authority and thus terminate flight plan. (Failure to close may trigger emergency.)

close out 1 To seal spacecraft, esp manned; task performed by ad hoc ** crew who are last to leave pad area.
2 To complete manufacturing programme.

closest approach 1 Time, location or separating distance at which two planets are closest.
2 Same for fly-by spacecraft.

close support See *close air support, CAS*.

closure Relative closing velocity between two air or space vehicles.

cloud Large agglomeration of liquid droplets (water in case of Earth) or ice crystals suspended in atmosphere.

cloud absorption Absorption of EM radiation by planetary cloud depends on cloud structure, size and EM wavelength, long waves reflected from planet surface being strongly absorbed even by thin layers.

cloud amount Estimated as apparent coverage of celestial dome, as seen by observer; expressed in oktas and written in symbolic form on met chart.

cloud attenuation Reduction in strength of microwave or IR radiation by cloud, usually due to scattering rather than absorption.

cloud banner See *banner cloud*.

cloud-break approach Final approach beginning in cloud and ending in visual contact (though possibly with precipitation).

cloud chamber Sealed chamber filled with saturated gas which, when cooled by sudden expansion, gives visible track of fog droplets upon passage of ionizing radiation or particle.

cloud/collision warning See *weather radar*.

cloud cover See *cloud amount*.

cloud-cover satellite Satellite equipped to measure by spectral response cloud cover on Earth or planet below.

cloud deck Cloud layer, esp visibly dense, seen from above.

cloud 9 To be on, feeling of elation and/or haziness.

cloud point Temperature at which cooling liquid becomes cloudy.

cloud seeding Scattering finely divided particles into cloud to serve as nuclei for precipitation (rainmaking).

CLP Club der Luftfahrtpublizisten (Austria).

CLR Clearance, or cleared to (given height).

CLS Contingency landing site.

clsd Closed.

CLT Centreline tracking (ILS/ILM).

C_{Lto} Take-off lift coefficient.

club layout Pairs of seats facing each other, often with table between.

club propeller Propeller having stubby coarse-pitch blades for bench-testing engine with suitable torque but reduced personnel danger and slipstream.

cluster 1 Two or more parachutes linked to support single load.
2 Several bombs or other stores dropped as group.
3 Several stars or other pyrotechnic devices fired simultaneously from single container.
4 Several engines forming group controlled by single throttle.
5 Several rocket motors fired simultaneously to

propel single vehicle.

cluster joint Structural joint of several members not all in same plane.

cluster munition Container which, after release from aircraft, opens to dispense numerous bomblets (rarely, ECM or other payloads).

cluster weld See *cluster joint*.

clutter Unwanted indications on display, esp radar display, due to atmospheric interference, lightning, natural static, ground/sea returns or hostile ECM.

CM 1 Command module.
2 Configuration management (EDP, software).
3 Crew member, thus *1, *2, etc.
4 Cluster munition.

C_M Coefficient of pitching moment about half-chord.

C_m Coefficient of pitching moment about quarter-chord.

C_{mac} Coefficient of pitching moment about aerodynamic centre.

CMAG Cruise-missile advanced guidance.

CMB 1 Continuous monofilament, braided.
2 Concorde Management Board.

CMC Cruise-missile carrier (A adds "aircraft").

C_{mcg} Coefficient of pitching moment about c.g.

CMD 1 Command, ie total autopilot authority.
2 Countermeasures duties.

CMDR Coherent monopulse Doppler radar.

CMDS Countermeasures dispenser system.

CME ECM (1) (F).

CMG Control moment gyro.

CMI 1 Computer-managed instruction (see *CAI [2]*)
2 Cruise-missile interface.

CMIK Cruise-missile integration kit.

CMIS Command management information system.

CMM 1 Computerized modular monitoring (of health of hardware).
2 Condition-monitored maintenance.

CMMCA Cruise-missile mission control aircraft.

C_{mo} Coefficient of pitching moment (¼-chord) at zero lift.

CMOS Complementary metal-oxide silicon (or, less often, semiconductor).

CMP 1 Countermeasures precursor (aircraft penetrating hostile airspace ahead of attacking force).
2 Counter-military potential (strategic balance).
3 Central maintenance panel.

CMPL, cmpl Completion, completed.

CMRB Composite main-rotor blade.

CMRS Countermeasures receiver system.

CMS 1 Continuous monofilament, spun.
2 Commission de Météorologie Synoptique.
3 Cockpit management system.

CMSAF Chief master sergeant of the Air Force (USAF).

CMT Cadmium mercury telluride (IR detector).

C_n Directional stability, yawing moment coefficient due to sideslip.

c/n Constructor's number.

CN_{2D} Coefficient of usable lift (variable section profile).

CNAD Conference of National Armaments Directors (NATO).

CNATS Controller of National Air Traffic Services (UK).

CNC Computer numerical control (NC machining).

CNCS Central Navigation and Control School (RAF).

CND/RTOK Could not duplicate, retest OK.

CNEIA Comité Nationale d'Expansion pour l'Industrie Aéronautique (F).

CNEL Community noise equivalent level.

CNES Centre National d'Etudes Spatiales (F).

CNI 1 Communications, navigation, identification.
2 Chief navigational instructor.

CNIE Comision Nacional de Investigaciones Espaciales (Arg).

CNL Cancel.

CNO Chief of Naval Operations (US).

CNPI Communication(s), navigation and position(ing) integration.

CNR 1 Community noise rating.
2 Consiglio Nazionale Ricerche (I).

CNRA Certificat de Navigabilité Restreint (homebuilts, F).

CNRE Centre National de Recherches de l'Espace (F).

CNRS Centre National de la Recherche Scientifique (F).

CNS 1 Continuous.
2 Communications network simulator.

CNSAC Comité National de Sûreté de l'Aviation Civile (F).

CNT Certificat de Navigabilité de Type (F).

CO 1 Commanding officer.
2 Crystal oscillator.
3 Checkout.
4 Aerodynamic mean chord.
5 Corps observation (USA, obs).

CoA Circle of ambiguity.

coach Formerly, US domestic high-density configuration.

coalescing filter Works by coalescing finely divided liquid droplets (eg water in fuel) into removable masses.

coaming Edge of open-cockpit aperture, often padded.

Coanda effect Tendency of fluid jet to adhere to solid wall even if this curves away from jet axis.

Coanda flap Flap relying on Coanda effect for attachment of flow to upper surface even at extreme angles.

coannular inverted nozzle Nozzle of variable-cycle jet engine with low-velocity core and high-velocity surrounding jet.

coarse pitch Making large angle between blade

chord and plane of disc, thus giving high forward speed for given rotational speed.

coarse-pitch stop Mechanical stop to prevent inefficient over-coarse setting (removed when feathering).

coast 1 Radar memory technique tending to slave to original target trajectory and avoid lock-on to stronger target passing same LOS.
2 Unpowered phase of trajectory, esp in atmosphere (usually verb).

coastal refraction Change in direction of EM radiation in crossing coast; also called shoreline effect, land effect.

coast-boost Period of coasting followed by rocket burn.

coastline refraction See *coastal refraction*.

COAT Corrected outside air temperature (OAT minus TAS/100).

co-axial Propeller or rotor having two or more sets of blades on same axis rotating in opposite senses.

co-axial cable Comprises central conductor wire and conducting sheath separated by dielectric insulator.

COB 1 Co-located operating base.
2 Certificated operational base.

cobalt Hard, silver-white metal, important in steels and esp in high-temperature engine alloys. Co-60 is dangerous radioisotope theoretically producible in large amount by nuclear weapons.

cobblestone turbulence 1 High-frequency * due to large mass of randomly disturbed air without significant gross air movement.
2 Buffet experienced by jet V/STOL descending into ground effect.

COBY Current operating budget year.

COC 1 Common operation centre, for tactical control of all arms in theatre.
2 Catalytic ozone converter.

cockade National insignia worn by military aircraft, esp one of concentric rings.

cocked Aircraft, especially combat type, preflighted through all checklists to point of starting engines.

cocked hat Triangle formed by three position lines that do not meet at a point.

cockpit Space occupied by pilot or other occupants, esp if open at top. Preferably restricted to small aircraft in which occupants cannot move from their seats; most * contain only one seat.

cockpit alert State of immediate readiness with combat aircrew fully suited, in * and ready to start engine.

cockpit cowling Aircraft skin around cockpit aperture.

cockpit voice recorder Automatic recycling recorder storing all crew radio and intercom traffic during previous several missions.

Coco exercise Combat mission exercise called off when aircraft are lined up on runway.

Cocomo Constructive cost model (software).

Cocraly Anti-oxidation coating for hot metal, from Co, Cr, Al, Yttrium.

COD See *carrier on-board delivery*.

Codan Carrier-operated device anti-noise.

Code Bambini Literally "child's talk", multilingual tactical radio language (Switz).

code block Standardized format of data identifying each frame in visual, IR or SLAR film, with provision for high-speed computer recall.

codem Coded modulator/demodulator.

coder Part of DME transponder which codes identity into responses.

code rate Ratio of actual data bits to total information digits transmitted in radar or communications system having deliberate redundancy. Symbol R.

Codes Common digital exploitation system.

codes Numbers assigned to multiple-pulse reply signals transmitted by ATCRBS and SIF transponders.

coding Arrangement of problem-solving instructions in format and sequence to suit particular computer.

COE Certification of equivalency (USAF).

COEA Cost and operational effectiveness analysis.

coefficients Except for next four entries, see under appropriate characteristics.

Coefficient A In simple magnetic compass, deviations on cardinal and quadrantal points summed and divided by 8.

Coefficient B In simple magnetic compass, deviation E minus deviation W divided by 2.

Coefficient C In simple magnetic compass, deviation N minus deviation S divided by 2.

coefficient conversion factor Formerly, multiplier 0·00256 required to convert absolute to engineering coefficients.

COF Centrifugal oil filter.

C of A Certificate of airworthiness.

C of F Construction of facilities.

coffin 1 Missile launcher recessed into ground but not hardened.
2 Symbol which appears in place of downed aircraft (ACMI).

COGT Centre-of-gravity towing.

coherent Radiation in which, over any plane perpendicular to direction of propagation, all waves are linked by unvarying phase relationships (common simplified picture is of waves "marching in step" with all peaks in exact alignment).

coherent echo Radar return whose amplitude and phase vary only very slowly (from fixed or slowly moving object).

coherent pulse radar, coherent radar Incorporates circuitry for comparing phases of successive echo pulses (one species of MTI).

coherent transponder Transmitted pulses are in phase with those received.

coherer RF detector in which conductance of imperfect part of circuit (eg iron filings) is improved by received signal.

Cohoe Computer-originated holographic optical elements.

Coin, Co-In, CO-IN Counter-insurgency; aircraft designed for guerilla war.

Coincat Community of Interests in Civil Air Transport (G).

coincidence circuit Gives output signal only when two or more inputs all receive signals simultaneously or within agreed time.

coke To modify aircraft with Küchemann "Coke bottle" fuselage.

col In atmosphere isobar field, saddle-shaped region separating two highs on opposite sides and two lows on remaining sides.

cold air mass Colder than surrounding atmosphere.

cold-air unit Air-cycle machine, usually in an ECS, which greatly reduces temperature of working fluid by extracting mechanical energy in expansion through a turbine.

cold bucket In aft fan with double-deck blades, outer blades handling cold air.

cold cathode Highly emissive coating and operates at ambient temperature.

cold-cockpit alert Combat aircraft has no ground power supplies and is "cold" until pilot enters and initiates start sequence for engine, gyros and systems.

cold cordite charge Does not detonate but burns to give high-pressure flow of gas.

cold drawing Drawing workpiece at room temperature.

cold flow test Static test of liquid rocket propulsion system to verify propellant loading and feeding but without firing engine(s).

cold front Front of advancing cold air mass moving beneath and lifting warmer air, esp intersection of this front with Earth's surface.

cold gas Reaction-control jet or rocket using as working fluid gas released from pressure or monopropellant decomposed without combustion.

cold launch Launch of missile or other ballistic vehicle under external impulse, usually from tube (in atmosphere, in silo or on sea bed) with vehicle's propulsion fired later.

cold mission Mission or test judged non-hazardous, thus not interfering with other activities.

cold plate In high-vacuum technology, refrigerated plate used to condense out last molecules of gas in chamber.

cold plug Spark plug having short insulated electrode keeping relatively cool (because rate of carbon deposit from oil or fuel is very low).

cold rating Cold thrust; rated output of jet engine without afterburning.

cold rocket Operating on pressurized gas or monopropellant, without combustion.

cold rolling Performed on steels to harden and increase strength, at expense of ductility.

cold round Test missile launched without active propulsion.

cold shut Porosity due to premature surface freezing in casting or formation of gas bubble in weld.

cold soak 1 Test of complete aircraft by prolonged exposure to lowest terrestrial temperature available before flying a mission.
2 Test of cryogenic propulsion system by prolonged passage of propellant.

cold test Determines lowest temperature at which oil or other liquid will flow freely.

cold thrust Max without afterburner.

cold wave Sudden major fall in surface ambient temperature in winter.

cold working Forming metal workpiece at room temperature; increases hardness and often strength but reduces ductility (increases brittleness).

Coleman theory Derived by NACA's R. P. Coleman and A. M. Feingold, basic explanation of ground resonance of helicopters with articulated rotors; hence such resonance called Coleman instability.

coleopter Aircraft having annular wing with fuselage at centre; usually tail-standing VTOL.

collaborative programme Undertaken by industrial companies in two or more countries as result of legal agreements between those companies or between their national governments.

collar Impact-absorbent ring around bottom of balloon gondola (usually lightweight foamed polystyrene).

collateral damage 1 Refers esp to injury to friendly eyes from clumsy use of powerful lasers in warfare.
2 Damage caused to anything other than the intended target.

collation Selection in correct sequence and stacking in exact register of pre-cut plies to make part in composite material.

collective pitch Pilot control in rotary-wing aircraft directly affecting pitch of all blades of lifting rotor(s) simultaneously, irrespective of azimuth position. Main control for vertical velocity. Colloq = "collective".

collective stick Collective-pitch lever (colloq).

collector 1 Bell-mouth intake downstream of working section of open-jet tunnel.
2 Region of transistor between * junction and * connection carrying electrons or holes from base.

collector ring Circular manifold collecting exhaust from cylinders of radial PE (4) engine.

collimate To adjust optical equipment to give parallel beam from point source or vice versa.

collimating tower Carries visual and radio/radar target for establishing axes of aerials (antennas) with minimal interference from other electrical fields. Alternatively collimation tower.

collision-avoidance system Provides cockpit indication of all conflicting traffic, without latter carrying any helpful equipment or co-operating in any way, and increases intensity of warning as function of range and rate of closure.

collision beacon Powerful rotating visual light, normally flashing Xenon tube, carried by IFR-equipped aircraft (normally one dorsal, one ventral).

collision-course interception Aimed at point in space which target will occupy at a selected future time; interceptor may approach this point from any direction.

collision-warning radar See *weather radar*.

collision-warning system See *collision-avoidance*.

colloidal propellant Having colloidal structure, with particles never larger than 5×10^{-3} mm and apparently homogeneous to unaided eye.

COLT CO_2 laser technology.

COM Computer output on microfilm (direct recording).

com Communications (FAR = comm).

comb 1 Rake, usually linear, of pressure heads.
2 IFF aerial (antenna) with linear array of dipoles often sized to match spread of wavelengths.

combat aircraft Aircraft designed to use its own armament for destruction of enemy forces; thus includes ASW but not AEW or transport (definition controversial).

combat air patrol, CAP Maintained over designated area for purpose of destroying hostile aircraft before latter reach their targets.

combat camera Colour ciné camera aligned with fighter armament to film target.

combat control team Air force team tasked with establishing and operating navaids, communications, landing aids and ATC facilities in objective area of airborne operation.

combat gross weight See *weight*.

combat load Aggregate of warlike stores carried (includes guns/ammunition but excludes radars, lasers/receivers and drop tanks carried for propulsion).

combat mission Mission flown by balloon, airship, kite, aeroplane, helicopter or other aircraft such that it may expect to encounter enemy land, sea or air forces.

combat plug Manual control of fighter engine permitting TET limit to rise to new higher level for period of emergency (typically 30 sec to 3 min).

combat spread Variable loose formation affording best visual lookout.

combat thrust loading Thrust loading assumed for fighter in typical combat.

combat trail Combat aircraft, usually interceptors, in loose trail formation, maintaining position visually or by radar.

combat wing loading Wing loading assumed for fighter in typical combat.

combat zone 1 Geographic area, including air-space, required by combat forces for conduct of operations.
2 Territory forward of army rear boundary.

combi, Combi Transport aircraft furnished as both passenger and freight carrier (from "combination"). Proportion devoted to freight usually variable.

combination propulsion See *mixed-power aircraft*.

Combined Involving armed forces of two or more allied nations. Thus * common user item, * forces, * staff etc.

combined display Presents information from two or more sources, usually radar superimposed on moving-map display.

combined-effects munition One having anti-armour, anti-personnel and incendiary effects.

combined stresses Two or more simple stresses acting simultaneously on same body.

combiner Optical element in HUD for aligning, collimating or focusing at infinity all displayed elements on single screen.

combining gearbox Reduction gearbox driven by two or more engines and driving single or co-axial propeller or lifting rotor.

combustion Chemical combination with oxygen (burning).

combustion chamber 1 In PE (4), space above piston(s) at TDC, arguably extended over part of stroke depending on progress of flame front.
2 In gas turbine, entire volume in which combustion takes place, including that outside flame tube(s) occupied by dilution air.
3 In liquid rocket or ramjet, entire volume in which combustion takes place, bounded by injector face, walls of chamber and plane of nozzle throat (not nozzle exit).
4 In solid or hybrid rocket, inapplicable.

combustion efficiency Ratio of energy released to potential chemical energy of fuel, both usually expressed as a rate.

combustion ratio Ratio of fuels or propellants actually achieved; in case of fuel/air usually termed mixture ratio.

combustion ring Combustion chamber of annular (eg Aerospike) liquid rocket engine.

combustion space See *combustion chamber (1)*.

combustion starter Engine-start system energized by burning fuel, either fuel/air, monopropellant (eg Avpin) or solid cartridge.

combustion test vehicle Free-flight vehicle (RPV or missile) whose purpose is test or demonstration of propulsion performance.

combustor 1 See *combustion chamber (2)*.
2 Combustion chamber (2) together with fuel manifolds, injectors, flameholders and igniters.
3 Rarely, afterburner burning region, with fuel spray bars, flameholders and ignition system.

Comdac Command, display and control (USCG).

Comed Combined map and electronic display (pronounced co-med).

Comeds Conus meteorological data system.
Comest European colour-TV satellite management consortium.
comfort chart Plot of dry-bulb T° against humidity (sometimes modified to include effect of air motion).
comint Communications intelligence.
comlo Compass locator.
comm Communications (FAA).
command 1 Intentional control input by flight crew or remote pilot.
2 Electrical or radio signal used to start or stop action.
3 In EDP, portion of instruction word specifying operation to be performed.
4 Authority over precise flight trajectory exercised by ATC or military authority (hence * altitude, * height, * heading, * speed etc).
command airspeed A target airspeed displayed as a *command parameter.*
command bars Principal reference index on flight director instruments, giving attitude in pitch and roll.
command destruct System which, at range safety officer's discretion, can explode malfunctioning missile, RPV or other unmanned vehicle, or trigger BUS, thereby averting hazard to life or property.
command dot *Command marker* in form of bold dot or small disc.
command guidance Steering by remote human operator.
command marker Reference index (line, bug, arrow or other shape) indicating target value, set by pilot on tape (sometimes dial) instrument and then flown to centre reference line. (See *command reference symbol.*)
command parameter Variable subject to command (1), (2), (4) and thereafter displayed as target value on instrument or display.
command reference symbol HUD symbology in form of ring or other shape showing a point at which to aim ahead of aircraft, eg landing touchdown point or an aerial point for optimum AOA on overshoot (go-around).
comma rudder Rudder shaped like comma, with balance area ahead of hinge axis, used without a fixed fin.
commercial In military use, purchasable from civilian source (eg aircraft rivet).
commercial aircraft Aircraft flown for hire or reward.
commercial electrics Electrical systems serving passenger functions only (eg steward call circuits, PA system, cabin lighting).
commercial support Assistance to operator of civil aircraft given or sold by original manufacturer or dealer.
committal height See *decision height.*
commodity loading All cargo of one kind grouped together, without regard to destination.
commodity rate Price charged to fly specified kind of cargo, typically per kilogram over particular route.
commonality 1 Hardware quality of being similar to, and to some degree interchangeable with, hardware of different design.
2 Objective of using one basic design of aircraft, or other major system, to meet needs of more than one user service in more than one role (with economies in training, spares and other areas).
common-flow afterburner Augmented turbofan in which fan and core flows mix upstream of afterburner.
common infrastructure Financed by two or more allies, eg by all members of NATO.
common mark Marking assigned by ICAO to aircraft of international agency (eg UN) on other than national basis. Hence ** registering authority.
common module(s) Use of identical "black box" subsystems as building blocks for different major equipments, eg * IR components to build night-vision, recon, weapon guidance and other systems for different armed forces or civilian customers.
common servicing Performed by one military service for another without reimbursement.
common-user airlift In US, provided on same basis for all DoD agencies and, as authorized, other Federal Government agencies.
communication deception Interference with hostile communications (including ATC and navaids) with intent to confuse or mislead.
communication language Complete language structure for linking otherwise completely separate (and possibly dissimilar) EDP (1) systems.
communications intelligence Gained by listening to hostile communications.
communications satellite Vehicle, normally man-made, orbiting planetary body, usually Earth, for purpose of relaying intercontinental telecommunications (telephone, telex, radio, TV) (see *active* **, *passive* **, *synchronous* **).
communications security Made up of physical security of transmitter and receiver, emission security of transmitter, transmission security en route and cryptosecurity of message.
community boundary Drawn around inhabited or urban areas surrounding airport or airfield.
community noise level Flyover, sideline and approach NLs measured at designated points on or beyond community boundary (see *noise*).
commutated Doppler Form of MLS in which beam is frequency-coded and/or linearly commutated instead of scanned in azimuth and elevation.
commutation 1 Repeated reversal of current flow in winding of electrical machine, esp to change output from a.c. to d.c.
2 Transfer of current between elements of polyphase rectifier to produce unidirectional output.

commutator Typically, radially separated series of conductors forming ring round rotating generator shaft, opposite pairs of which are touched by brushes in external circuit to give d.c. output by commutation (1).

commuter aircraft See *feederliner.*

commuter airline In theory, air carrier operating between outlying regions and major hub(s). In practice, applied to anything from air-taxi operator to — in undeveloped regions — national carrier (see *third-level*).

comp Component of WV along Tr (strictly, along flight-plan track between check points).

Compacta tyre Landing wheel tyre of reduced diameter and greater than normal width (Dunlop).

companion body Hardware from launch system accompanying space vehicle or satellite on its final trajectory.

comparative cover Reconnaissance coverage of same scene at different times.

compartment marking Stencilled subdivisions of cargo aircraft interior to assist compliance with floor loading and c.g. position limits.

compass acceleration error See *acceleration errors.*

compass base Area on airfield, usually paved disc, on which aircraft can conveniently be swung.

compass compensation See *compensating magnets.*

compass course See *heading.*

compass error 1 Vector sum of variation E plus variation W.
2 Sum of deviation, variation and northerly turning error.

compass heading See *heading.*

compass points 32 named directions comprising cardinal points, quadrantal points and 24 intermediate points.

compass rose Disc divided into 360°, either on simple magnetic compass or on compass base.

compass swing See *swing.*

compass testing platform See *compass base.*

compass variation See *variation.*

compatibility Ability of materials (solids, liquids and gases) and dynamic operating systems to interface for prolonged periods without interference under prescribed environmental conditions.

compatible 1 Colour TV transmission capable of being received as monochrome by monochrome receiver.
2 Language and software capable of being used in given computer.

compensated gyro Incorporates correction for apparent wander.

compensating magnets Two pairs carried on arms rotatable about axis of magnetic compass to correct or minimize deviation.

compensation manoeuvres Aircraft manoeuvres required for accurate use of compensator (2), always involving four orthogonal headings, and sometimes circle or cloverleaf.

compensator 1 Instrument for measuring phase difference between components of elliptically polarized light (Babinet* has pair of quartz wedges with optical axes perpendicular).
2 Device, manually or computer-controlled, carried in ASW aircraft to eliminate false readings caused by permanent (airframe and equipment hardware), induced and eddy-current interference signals.

compiler EDP (1) program more powerful than assembler, for translating and expanding input instructions into correctly assembled subroutines.

complementary shear Induced in tension field (eg aircraft skin) at right angles to applied shear, in plane of field.

completion business Process of taking green airframes from manufacturer and equipping and furnishing to each customer's specification (principally in field of executive or commuter transports).

complex See *launch complex.*

compliance Demonstrated fulfilment of requirements of certificating authority.

compliance limit Time (usually GMT) by which compliance must be demonstrated.

compliant member Capable of substantial elastic or otherwise recoverable deflection.

compliant volume Trapped body of fluid, usually oil, having predetermined stiffness resulting from fluid's bulk modulus. Often sealed by diaphragm or piston having small bleed, to even out pressures over a period (see *stiffness*).

component 1 One of assemblage of structural members.
2 One of assemblage of parts used to build hardware system.
3 Major subdivision of prime mover, esp gas turbine (eg fan, compressor, combustor, turbine, afterburner, nozzle); hence * efficiency.
4 Force, velocity or other vector quantity along reference axis, such that components along two mutually perpendicular axes sum vectorially to actual vector. Thus, crosswind * on landing.

component efficiency Measure of performance of part of machine, normally on basis of energy output × 100 divided by energy input. Thus overall efficiency of gas turbine is product of ** of each part, considered on both mechanical and thermodynamic basis.

component life Authorized period of usage without attention, as stipulated by manufacturer or other authority. At expiry may be discarded or overhauled. Period may be extended from time to time.

composite aircraft Comprising two aircraft joined together at take-off but separated later in flight.

composite beam Composed of dissimilar materials bonded together.
composite cloud Combination of, or intermediate between, basic forms, eg cirro-cumulus.
composite double-base Solid rocket filling of combined double-base and composite types (eg AP (2) + AlP in matrix of NC + NG).
composite flight plan One specifying VFR for one or more portions and IFR for remainder.
composite flying Long-range navigation along great circle but modified (eg to avoid high mountains) by inserting sectors using other methods.
composite launch Single launch vehicle carrying two or more distinct payloads.
composite material Structural material made up of two or more contrasting components, normally fine fibres or whiskers in a bonding matrix. Unlike an alloy, usually anisotropic.
composite power See *mixed power*.
composite propellant Solid rocket filling comprising separate fuel and oxidizer intimately mixed.
compound aircraft Having wing(s) and lifting rotor(s).
compound curvature Sheet or surface curved in more than one plane, thus not formable by simple bending.
compound die Performs two or more sheet-forming operations on single stroke of press.
compound engine Expands working fluid two or more times in two or more places, eg in HP and LP cylinders or in PE (4) followed by gas turbine or blow-down exhaust turbine.
compound helicopter Having propulsion (usually turbofan or turbojet) in addition to thrust component of lifting rotor.
compound stress Not simple tension/compression, torque, bending or shear but combination of two or more of these.
compressed-air starter Expands HP airflow through PE (4) cylinders or ATM or turbine-blade impingement jet. In multi-engined aircraft cross-bleed can start second and subsequent engines.
compressed-air tunnel Closed-circuit tunnel filled with gas or air under pressure; can be smaller, and cheaper to run, than one at atmospheric pressure for given M and R.
compressibility In aerodynamics, phenomena manifest at speeds close to local sonic speed, when air can no longer be regarded as incompressible. Loosely, behaviour of airflow subject to pressure/density changes of 50 per cent or more of free-stream values.
compressibility correction From RAS to EAS (see *airspeed*).
compressibility effects Manifest as local speed, at peak suctions, exceeds that of sound in surrounding flow; include abnormally rapid increase in drag, rearward shift of CP (2) on lifting wings, appearance of shockwaves, tendency to boundary-layer breakaway and, in improperly designed aircraft, control buzz and other more severe losses of stability and control.
compressibility error Manifest in all instrument readings derived from simple pitot/static system at high subsonic Mach numbers; typically, progressive under-reading until pressure and static orifices have penetrated bow shock.
compression Control of signal gain, esp to increase it for small signal voltages and reduce it for large.
compression ignition Combustion of fuel/air mixture triggered by high temperature due to compression in diesel cylinder or in highly supersonic ramjet with suitable internal profile.
compression lift Lift gained at supersonic speed by favourable flow field by forcing flow to accelerate beneath wing (accentuated by down-turned wingtips).
compression pressure Gauge pressure in PE (4) cylinder at TDC (in absence of combustion).
compression ratio Ratio of entrapped volume above piston at BDC to volume at TDC.
compression rib Provided inside fabric-covered wing to withstand tension of drag bracing.
compression ring(s) Top ring(s) on piston, of plain rectangular section, serving to seal mixture into combustion space on compression stroke.
compression wave See *blast wave*.
compressor Machine for compressing working fluid (see *axial* *, *centrifugal* *, *skew* *, *Roots* *, *positive-displacement* *). In general, term used for device handling large mass flow at moderate pressure (say, up to 40 ata, 400 kPa); small flow at high pressure = pump.
compressor blade 1 Loosely, rotor blade or stator vane in axial compressor.
2 Precisely, operative aerofoil from axial compressor rotor.
compressor casing Fixed casing closely surrounding compressor rotor.
compressor diffuser Passage for working fluid immediately downstream of compressor wherein pressure is increased at expense of flow velocity.
compressor efficiency Useful work done in delivering fluid at higher pressure, in assumed adiabatic operation, expressed as percentage of power expended in driving rotor.
compressor map Fundamental graphical plot of compressor performance showing variation of pressure ratio (ordinate) against mass flow (abscissa) for each rpm band.
compressor pressure ratio Ratio of total-head pressure at delivery to that at inlet (if ratio is 24:1, conveniently written as 24, for example).
compressor rotor Main moving part in compressor of rotary form (ie, not reciprocating type).
compressor stator Stationary part of axial compressor carrying fixed vanes.
compressor vane Stationary blade attached to stator (case), one row of such vanes preceding

each row of rotor blades.

compromised 1 Classified information known or suspected to have been disclosed to unauthorized persons.
2 Of serial number or civil registration, one inadvertently applied to two aircraft.

Compu-Scene Add-on visual system for existing simulators (General Electric).

computed air release point Air position at which first paratrooper or cargo item is released to land on objective.

computer 1 Machine capable of accepting, storing and processing information and providing results in usable form; function may be direct control of one or more operating systems.
2 Simple mechanical device for solving problems (eg Dalton *).

computer acceleration control Use of airborne computer linked to AFCS to limit (close to zero) unwanted flight accelerations, esp in vertical plane, on aeroplanes and helicopters.

computer-assisted approach sequencing Use of one, or several interlinked, computers in ATC system to solve problem of feeding arrivals automatically into optimized trajectories so that each arrives at destination runway at correct spacing and with minimal delay.

computer board Component part of a computer or similar device, each being a driver, RAM, EPROM, A/D converter, video interface or similar self-contained unit which can be assembled with others on to a bus (eg, backplane) to form a purpose-designed EDP system.

computer-programmable Capable of being controlled by digital computer without additional interfacing (typical item would be microwave signal generator for radar testing).

computing gunsight Automatically compensates for most predictable or measurable variables in weapon aiming.

comsat See *communications satellite*.

Comsec Office of Communications Security (US, NSA).

comsnd Commissioned (of facilities on airfield charts).

COMSS Coastal/oceans monitoring satellite system.

CON, Con Consol beacon.

Conaero Consorzio Italiano Compagnie Lavoro Aereo (I).

Conc Concrete surfaced runway (ICAO).

concentrated force, load See *point force*, *load*.

concentration ring 1 In balloon, ring, usually rigid, attached to envelope or (if applicable) surrounding net, and from which basket is suspended.
2 In airship, ring to which several mooring lines may be secured (sometimes also helping support car, if this is suspended below hull).

concentric Having common centre or central axis.

concession 1 Allowable departure from drawing in manufacture of part (eg on material spec, surface finish or mfg tolerance).
2 Allowable non-compliance with certification or other requirement, esp in emergency (eg take-off permitted with one engine or one altimeter inoperative).

concurrence Policy adopted for reasons of national emergency in which most, or all, parts of major system programme are implemented simultaneously, even though several large portions may need to be grossly modified or updated (eg Atlas ICBM hurriedly deployed above ground, then in surface shelters and finally in silos).

concurrent forces Acting through common point.

condensation Physical change from gaseous or vapour state to liquid.

condensation level Height at which rising parcel of air reaches saturation; cools at DALR and reaches 100% RH at ** at intersection of DALR and DPL.

condensation nuclei Minute particles, solid or liquid, upon which nucleation begins in process of condensation; most effective ** are hygroscopic.

condensation shock Sudden condensation of supersaturated air in passage through normal or inclined shock, rendering shock field visible, often showing elliptic lift distribution around transonic aircraft.

condensation trail Visible trail, usually white but sometimes darker than sky background, left by winged or propelled vehicle, esp when flying above condensation level. May be due to reduced pressure (eg in tip vortices), but nearly all persistent ** due to condensation (and probable freezing) of water vapour formed by combustion of fuel.

condenser 1 Capacitor.
2 Device for changing flow of vapour to liquid by removing latent heat of evaporation. Essential feature of closed-cycle space power systems in which working fluid must be used repeatedly.

condenser-discharge light Gives very short flashes of great intensity caused by capacitor discharge through low-pressure gas tube (eg collision beacon).

con-di nozzle Jet engine nozzle having cross-section which converges to throat and then diverges; subsonic flow accelerates to throat, becomes supersonic and then accelerates in divergent portion.

conditionally unstable Unsaturated air above or through which temperature falls with height faster than SALR but less than DALR; thus if air becomes saturated it will be unstable.

condition monitoring Health inspection of operative hardware, eg engine, using intrascope, X-ray photography, oil sampling and BITE.

Condor Confidential direct occurrence reporting, system for non-attributably ensuring that nothing having a direct bearing on flight safety is kept hidden (RAF, CAB etc).

conductance 1 Real part of admittance in electric circuit.
2 In circuit having no reactance, ratio of current to potential difference, ie reciprocal of resistance. Symbol G.
3 In vacuum system, throughput Q divided by difference in p between two specified cross-sections in pumping system.
4 Several meanings in electrolytes (little aerospace relevance).

conduction band Band of electron energies corresponding to free electrons able to act as carriers of negative charges.

conductivity Measure of ability of material to transmit energy, eg heat or electricity. Thermal *, symbol k or λ, measured in Jm/m^2s °C. Electrical *, symbol δ, measured in mhos/m (per cube); reciprocal of resistivity.

conductor Material having very low electrical resistivity, esp such material fashioned in form useful for electric circuits.

cone 1 Drag and stabilizing member trailed on end of HF aerial wire (trailing *) or on end of air-refuelling hose.
2 Drag and stabilizing member incorporating pressure and/or static heads trailed beneath aircraft under test in supposed undisturbed air.

cone angle Semi-angle of right circular cone having same increase in surface area per unit length as diffuser; hence diffuser **.

cone of confusion Inverted cone of airspace with vertical axis centred on VOR or other point navaid.

cone of escape Volume in exosphere with vertex pointing directly to Earth centre through which atom or molecule could theoretically escape to space without collision. Opens out in angle to infinity at critical level of escape.

cone of silence Inverted cone of airspace with vertical axis centred on certain marker beacons, NDBs and other point navaids within which signal strength reduces close to zero.

cone passage Flight through cone (of confusion or of silence) above point navaid.

cone yawmeter Cone flying point-first, with pitot holes spaced at 90° intervals, to obtain yaw indication at supersonic speeds (avoids averaging effect of wing-type yawmeter).

confidence level Used in statistical sense, eg as percentage probability that an actual MTBF will exceed estimated or published MTBF. Value of ** increases with number of samples. Sometimes called confidence limit.

confidence manoeuvres Set pattern of ground and air tasks easily mastered by new and inexperienced pupil pilot (eg, swinging propeller, letting aircraft recover from unnatural flight attitude hands-off); devised to ease problem of apprehension and tension. Sometimes called confidence actions.

configuration 1 Gross spatial arrangement of major elements, eg in case of aircraft disposition of wings, bodies, engines and control surfaces.
2 Aerodynamic shape of aircraft where variable by pilot command, eg position of landing gear, leading/trailing-edge devices and external stores. Thus high-lift *, clean *.
3 Standard of build or equipment for task. Thus helicopter in dunking ASW *, passenger transport converted to all-cargo *.
4 Apparent positions of heavenly bodies, esp in solar system, as seen from Earth at particular time.
5 Used, incorrectly, to mean "application", eg "Chaparral is the Sidewinder missile in ground-to-air *". This would be correct if hardware was physically changed in *.

configuration bias Channel or subsystem in stall-protection or stick-pusher system allowing for changes in configuration (2).

conflict In ATC (1), two aircraft proceeding towards potentially dangerous future situation. Hence, * situation.

conflicting traffic With respect to one aircraft, other traffic at or near same FL heading towards future conflict.

conformal-array aerial Electronically scanned, fits exterior surface of vehicle.

conformal-array radar Having plurality of small or light ES aerials covered by radomes fitting vehicle shape (eg wing or rotor leading and trailing edges etc).

conformal gears Having teeth whose mating profiles conform, both sets having instantaneous centres of curvature on same side of contact. Usually applied to WN gears.

conformal projection Having all angles and distances correct at any point, but with scale changing with distance from point.

confusion reflector Designed to reflect strong echo to confuse radar, proximity fuze, etc. Form of passive ECM.

conical camber Applied to wing leading edge so that, from root or intermediate station to tip, it is progressively drooped, centreline following surface of cone with vertex at root (or at start of ** if this is some distance along semi-span).

conical flow Theory for supersonic flow over thin flat plate having corner (apex), with flow perpendicular to rear edge: constant pressure, velocity, density and temperature along any radius (to infinity) from apex.

conical scanning Common search mode for radar, esp AI radar, in which beam is mechanically or electronically scanned in cone extending ahead of aerial, often using beam-switching to give az/el data.

conical sleeve Cone-shaped flexible sleeve extending inwards into gas cell of airship from aperture for line, providing near gas-tightness with freedom for line to move axially through envelope.

conic apogee Apogee of satellite if all mass of primary were at its centre.

conic perigee Perigee of satellite if all mass of primary were at its centre.

conic sections Perpendicular to axis = circle; parallel to axis = parabola (eccentricity 1); eccentricity less than 1 = ellipse; eccentricity greater than 1 = hyperbola. All are found in trajectories of bodies moving in space.

Conie Comision Nacional de Investigacion de Espacio (Spain).

coning angle 1 Angle between longitudinal axis of blade of lifting rotor and tip-path plane (assuming no blade bending). Symbol β.
2 Incorrectly, sometimes given as average angle between blade and plane perpendicular to axis of rotation.

conjugate Many specialized meanings in theory of groups, complex numbers and geometry of curved surfaces.

conjugate beam Hypothetical beam whose bending moment assists determination of deflection of real beam.

conjugate foci In optics, interdependent distances object/lens and lens/image.

conjunction Alignment of two heavenly bodies sharing same celestial longitude or sidereal hour angle.

connecting rod Joins reciprocating piston to rotary crank in PE (4), reciprocating pump etc.

connector Standard mating end-fitting for fluid lines, multi-core cables, co-ax cables and similar transmission hardware, providing automatic coupling of all circuits. Term preferred for multi-pin electric *; with fluid systems prefer "pipe coupling".

conplan Contigency plan.

consensus Majority vote concept in logic systems, multi-channel redundant systems etc; thus, * can command landing flare against presumed failed channel.

Consol Simple long-range navaid providing PLs (within range of two * stations, a fix) over N Atlantic. LF/MF receiver is tuned to identified * station and operator counts dots and dashes in repeated "sweep" lasting about 30 seconds; PL is then obtained by reading off * chart.

Consolan Consol-type system based at Nantucket (US).

console 1 Control station for major device or system, normally arranged for seated operator.
2 Control and instrument installation for pupil navigator, esp when such * repeated along fuselage (but not used for pilot station on flight deck).
3 Single bank of controls and/or instruments on flight deck, eg roof *, left side *.
4 Station for manual input/output interface with large system, eg air defence, ATC, EDP (1).
5 Tailored box for storage of maps, cameras and other items, eg "The Cessna 210 has centre-aisle * as an option"; misleading and ambiguous.

consolute Of two or more liquids, miscible in any ratio.

constantan Alloy of copper with 10–55 per cent nickel; resistivity essentially unchanged over wide range of temperature.

constant-colour Philosophy for cockpit warning systems, usually: no caption illuminated = no fault, all buttons normal; blue = normal-temperature operation; white = button abnormal, either from mis-select or to rectify/suppress fault; red or amber = fault.

constant duty cycle Device or system whose rate of operation is unvarying despite variable demand; eg DME ground transponder beacon has *** behaving as though continuously interrogated by 100 aircraft.

constant-flow oxygen Crew-breathing system in which gox is fed at steady rate, in contrast to demand-type supply.

constant-g re-entry RV uses aerodynamic lift in skip trajectory to impose constant total acceleration down to relatively low velocity.

constant-heading square Helicopter pilot training manoeuvre: large square described at low level with helicopter constantly facing into wind (so one leg forwards, one backwards and two sideways).

constant-incidence cruise Transport aircraft flight plan calculated on basis of constant angle of attack over major portion, angle being chosen for best L/D or other optimized point between time and fuel consumption.

constant-level balloon Designed to float at constant pressure level.

constant of gravitation See *gravitational constant.*

constant-pressure chart Plot of contours showing height above MSL of selected isobaric surfaces.

constant-speed drive, CSD Infinitely-variable-ratio gear between two rotating systems, esp variable-speed aircraft engine and constant-frequency alternator.

constant-speed propeller, c/s propeller Propeller whose control system incorporates governor and feedback which automatically adjust pitch to maintain selected rpm.

constant-speed unit, CSU Engine-driven governor controlling c/s propeller.

constant wind 1 W/V assumed for navigational purposes, until updated or refined.
2 Used in contradistinction to gust (2).

constellation 1 Traditional conspicuous group of fixed stars having supposed resemblance to Earth object.
2 Arbitrary portion of celestial sphere containing a * (1) bounded by straight lines, whole sphere being thus divided for use as reference index.

constituent day Period of Earth rotation with respect to hypothetical fixed star.

constrictor 1 Obstruction in pipe or other fluid-flow constraint pierced by small hole giving precisely known mass flow per unit pressure difference.
2 Annular or distributed constriction in nozzle of air-breathing jet engine, esp ramjet or pulsejet.

consumables Materials aboard spacecraft which must undergo once-only irreversible change during mission, eg propellants, foods (in present state of art) and some other chemicals such as in SPS.

consumables update Regular housekeeping chore, reporting to Earth mission control exact quantities (usually masses) remaining.

cont Continuous, continuously.

contact 1 Visual link between pilot (rarely, other aircrew) and ground or other external body. Thus, in * = seen.
2 Unambiguous radar link (radar *).
3 Single positive mechanical hook-up between FR tanker and receiver aircraft (dry * if no fuel to be transferred).
4 Shouted by pilot of simple aircraft to person swinging propeller of PE (4), indicating ignition about to be switched on.
5 Unidentified target appearing on radar or other surveillance system (rarely, seen visually).

contact altimeter See *contacting altimeter*.

contact approach Visual approach to airfield requested by, and granted to, pilot making IFR flight.

contact-burst preclusion Nuclear-weapon fuzing system which, in the event of failure of desired air burst, prohibits unwanted surface burst.

contact flying Aircraft attitude and navigation controlled by pilot looking at Earth's surface. (Certain authorities, questionably, include clouds as source of visual cues.)

contact height That at which runway is first glimpsed during landing approach.

contacting altimeter Makes or breaks electrical circuit (eg warning or radio transmission) at chosen reading(s).

contact ion engine Space thruster stripping electrons from caesium or other supply material infiltrated in substrate (eg tungsten). Bombardment ion engine more common.

contact lights White lights on either side of runway in use, parallel to centreline (obsolescent, see *runway edge lights*).

contact lost Situation in which contact (5) can no longer be seen, though target believed still present.

contactor Electric switch having remote (usually electromagnetic) control.

contact point In CAS (3), geographical or time point at which leader establishes R/T contact with FAC or ground ATC.

contact print Photograph made from negative or diapositive in contact with sensitized material; optical, radar or IR.

container Standard rigid box for baggage or cargo: maindeck *, ISO 96 in × 96 in × 10, 20, 30 or 40 ft; SAE 10, 96 in × 96 in × 125 in; SAE 20, 96 in × 96 in × 238·5 in; underfloor *, IATA A1 (LD3) 92 in × 60·4 in × 64 in; A2 (LD1) 79 in × 60·4 in × 64 in.

container delivery Standard military airdrop supply of from one to 16 bundles of 1,000 kg (2,200 lb) each.

containment Demonstrated ability to retain every part within machine, following mechanical breakup of portion or whole of moving machinery, esp rotating portions of gas turbine engine.

contaminate Aerospace meanings include transfer of terrestrial germs and other organisms to spacecraft sterilized for mission, transfer of unwanted atoms to single-crystal (eg semiconductor) material, and deposit and/or absorption and/or adsorption of any NBC material on friendly surfaces.

contaminated runway Surface all or partly covered with water, snow, slush, blown sand or foreign objects capable of causing damage.

Contap Consol Technical Advisory Panel.

Conticell Proprietary low-density sandwich structure.

contingency air terminal Mobile air-transportable unit providing all necessary functions to handle air transport at combat airfield.

contingency plan Drawn up and implemented by military commander or civil manager in event of failure of original plan, for anticipated reasons.

contingency rating Power levels required of helicopter and VTOL engines in emergency conditions, normally requiring subsequent special inspection (see *maximum* **, *intermediate* **).

contingency retention item Surplus to requirements but authorized for retention to meet unpredicted contingency.

contingent effects Those of nuclear detonation other than primary effects.

continuation trainer Trainer aircraft for experienced pilots, esp those in desk jobs.

continuity line Portion of line system diagram in cockpit or other human interface superimposed on push-button or magnetic indicator.

continuous beam Single structural member having more than two supports.

continuous-flow system See *constant-flow oxygen*.

continuous-path machining Shaping of workpiece by cutter traversing unbroken path, esp this form of NC control and machine program.

continuous strip Film produced by ** photography, using ** camera, in which ** film passes at constant speed, related to speed of aircraft, past slit in optical focal plane.

continuous wave, CW, c.w. EM waves repeated without breaks indefinitely, usually with constant amplitude and length (frequency); ie, not pulsed.

continuum flow See *free-molecule flow.*

contour 1 On topographic map or chart, line joining all points of equal surface elevation above datum (eg MSL).
2 On * chart, line joining all points of equal elevation (height above or below datum, eg MSL, and above or below ground or sea surface) of selected pressure surfaces. Thus can plot * of 1,000 mb surface at −120 ft, MSL and +120 ft.

contour capability 1 Of mapping radar, ability to display all ground above selected height above MSL or other datum.
2 Of weather radar, ability to make contour display.

contour display Radar display in which all echoes above given strength are cancelled. Normally used in viewing storm clouds. With CONTOUR operative cloud echo has black centre showing region of greatest precipitation (and assumed greatest gust severity). With colour radar each contour has distinctive hue.

contour flying Normally denotes holding constant small height AGL, ie not following contours (1) but terrain profile (see *NOE*).

contour interval Difference in height between adjacent contours (1, 2).

contour template Hard copy of profile of 2-D or 3-D shape, eg for tunnel throat, press tool, form block.

contract definition phase Important process in procurement linking end of feasibility study and other conceptual phases with full hardware development. CDP involves collaboration with one or more industrial contractors and can involve detailed computer study and hardware test to establish what is to be bought and on what terms.

contraction Duct of diminishing cross-section through which fluid is flowing; eg front part of venturi.

contraction ratio 1 In subsonic tunnel, ratio of maximum cross-section to that at working section.
2 In supersonic tunnel, ratio of cross-sectional area just ahead of contraction to that at throat (can be variable).

contractor-furnished equipment, CFE Hardware, software or, rarely, specialist knowledge or experience, supplied by contractor to support programme; esp items normally GFE, bought-out or supplied from other source.

contractor-furnished weight, CF weight Total mass of aircraft in precise state in which ownership transferred to customer.

contra-flow engine Loosely, any engine involving fluid flow in opposite directions; specif, gas turbine having compressor and turbine back-to-back, with flows (1) axially towards each other and radially out together, (2) radially out from compressor and radially in through turbine, or (3) forwards through compressor and back through ducts to turbine.

contrail See *condensation trail* (abbn not admitted by NATO).

contra-injection Upstream injection of fuel droplets into airflow or of one liquid rocket propellant against another.

contra-orbit defence Supposed technique of defending area by launching missile along predicted trajectory or orbit of hostile weapon.

contraprop Contra-rotating propeller.

contra-rotating 1 Two or more propellers or gas-turbine spools rotating in opposite directions on common shaft axis, and sharing common drive.
2 Installation of similar tandem PE (4)/propeller combinations back-to-back on opposite ends of common nacelle. (Not to be used for propellers rotating in opposite directions but not on common axis. See *handed.*)

contrast Difference in luminous intensity between different parts of picture (photograph, radar display, synthetic display or TV).

control 1 Exercise of civil or military authority, eg over air traffic.
2 In hardware system, device governing system operation.
3 In man/machine system, device through which human command is transmitted across interface.
4 In photogrammetry, points of known position and elevation.
5 In research experiment, unmodified test subject used as yardstick.

control airport See *tower airport.*

control and reporting centre Subordinate air-control element of tactical air control centre from which radar control and warning operations are conducted within its area of responsibility (USAF, NATO etc).

control and reporting system Organization set up for (1) early warning, tracking and identification of all air and sea traffic, and (2) control of all active air defence.

control area Controlled airspace extending upwards from specified height (ICAO prefers "limit") above Earth (NATO adds "without upper limit unless specified").

control cable Physical connection between human control (3) and operating system, esp between pilot's flying controls and control surfaces.

control car Housing pilot or coxswain of airship.

control centre See *launch* **.

control column Aerodyne trajectory control (flight control input) normally exercising authority in pitch and roll. May be stick, wheel or spectacles (see *yoke*).

control-configured vehicle See *CCV.*

control feel See *feel.*

control jet See *reaction control jet.*

controllable-pitch propeller Capable of having

blade pitch manually altered in flight, either to set positions or over infinite range (but not c/s).

controllable rocket Having rate of combustion of liquid, solid or hybrid propellants capable of being varied at will during burn.

controllable twist Helicopter rotor blade capable of changing angle of incidence in predetermined manner from root to tip in course of flight.

controlled aerodrome One at which ATC service is supplied to aerodrome traffic (does not imply existence of control zone).

controlled airspace Airspace of defined dimensions within which ATC service is provided (ICAO adds "to controlled flights").

controlled environment One in which such variables as temperature, pressure, atmospheric composition, ionizing radiation and humidity are maintained at levels suitable for life or hardware.

controlled flight Provided with ATC service.

controlled interception One in which interceptors are under positive control (from ground, ship or AWACS).

controlled leakage Environment for life or hardware in which harmful products (eg carbon dioxide) are allowed to leak away and be replaced by fresh oxygen or other material.

controlled mosaic One in which distances and directions are accurate.

controlled response Chosen from range of options as being that giving best all-round result.

control line 1 Connection between operator and ** aircraft.
2 Connection between control car of airship and controlled item.

control-line aircraft Model aircraft whose trajectory is controlled by varying tensions in two or more filaments linking it with ground operator.

control lock Physical lock preventing movement of control surface, either built into aircraft or brought to it and fastened in place.

control panel Self-contained group of controls, indicators, test connections and other devices serving whole or portion of aircraft system, either accessible in flight or only during ground maintenance.

control pattern In SSR/IFF, governs reply code for each mode selected.

control point Fixed position, marked by geographic feature, electronic device, buoy, aircraft or other object, used as designated aid to navigation or traffic control (NATO, USAF).

control reversal In aircraft flight control system, dangerous state in which pilot demand causes response in opposite sense. Normally caused by either mechanical malfunction (eg crossed controls) or aeroelastic distortion of airframe.

control rocket Usually small and intermittently fired thruster for changing spacecraft attitude and refining velocity.

controls As "the *", primary flight control input devices, esp in aerodyne; typically stick and rudder pedals.

control stick Control column (colloq).

control-stick steering Control of aircraft trajectory by input to AFCS by means of primary flight controls. Not same as *-*wheel* *.

control surface Aerofoil or part thereof hinged near extremities of airframe so that, when deflected from streamwise neutral position, imparts force tending to change aircraft attitude and thus trajectory.

control surface angle Measured between reference datum on control surface and chord of fixed surface or aircraft longitudinal axis.

control system In missile, RPV or aircraft flying on AFCS, serves to maintain attitude stability and correct deflections (NATO, USAF). Also, not included in this definition, translates guidance demands into changes in trajectory.

control tower ATC organization, normally located on tower or near airfield, providing ATC service for airfield traffic and possibly within other airspace.

control vane Refractory surface, usually small, pivoted in jet of rocket or other propulsion system to control attitude, and hence trajectory, of vehicle when deflected from neutral setting.

control-wheel steering Autopilot mode giving manual control of heading while holding velocity and/or attitude.

control zone Controlled airspace extending upwards from Earth's surface (NATO, USAF). SEATO has long and involved definition including "and including one or more airdromes" (*sic*). ICAO adds "to a specified upper limit".

Conus Continental US, ie US and its territorial waters between Mexico and Canada but excluding overseas states.

convection 1 In fluid dynamics, transfer of fluid property by virtue of gross fluid motion.
2 In atmosphere, transfer of properties by vertical motion, normally thermally induced.

convection cooling Method of cooling hot hardware, esp gas turbine rotor blades, by removing heat from within bulk of material by flow of cooler air passing through system of holes or passages (see *film cooling*, *transpiration cooling*).

convective cloud Cumuliform, CuF, triggered by convection; normal vertical development fairweather cumulus; extreme form is cumulonimbus. Bottom lies at condensation level; top can be in stratosphere.

conventional Not nuclear, ie HE.

conventional take-off and landing, CTOL Aeroplanes other than STOL, VTOL and other short-field forms.

convergence 1 Condition in which, at least reckoned on surface winds, there is net inflow of air into region.
2 Of mathematical series, one having a limit.
3 Of vector field, contraction.

4 Of terrestrial meridians, angular difference between adjacent pair at particular position.

convergence factor Ratio of convergence (4) and change of latitude (zero at Equator, max at poles).

convergent Of oscillation — eg sinusoidal motion, phugoid or structural vibration — tending to die out to zero within finite (possibly small) number of cycles.

converging flight rule Aircraft approaching from right has right of way.

conversion angle That between great-circle and rhumb-line bearings.

convertible aircraft Transport aircraft designed for rapid conversion from passenger to all-cargo configuration or vice versa.

convertiplane Aerodyne capable of flight in at least two distinct modes, eg vertical flight supported by lifting rotor and forward translational flight supported by wing.

convertor Among many other meanings;
1 Rotary machine for changing alternating into direct current.
2 Self-regulating boiler for drawing on Lox storage and supplying flow of Gox.

cook-off Inadvertent firing of automatic weapon due to round being detonated by residual heat in breech.

coolant Liquid circulated through closed circuit to remove excess heat, eg from PE (4).

cooldown See chilldown.

cooling drag That due to need to dump excess heat to atmosphere (with skill can be made negative).

cooling gills Hinged flaps forming partial or complete ring around rear edge of cowling of air-cooled PE (4) to control airflow.

co-operative aircraft In ATC, one carrying transponder for SSR.

Co-operative Fuel Research Permanent committee of SAE including fuel and engine representatives with especial brief to measure and improve anti-knock ratings.

Cooper scale Scale for quantified Pilot Opinion Rating.

Co-ops CO_2 observational platform system.

co-ordinated turn One in which controls about three axes are used to avoid slip or skid.

co-ordinates Inter-related linear and/or angular measures by which the position of a point may be defined with reference to fixed axes, planes or directions.

co-ordination In a pilot, ability to control simultaneous unrelated motions by left and right hands and feet.

COP Changeover point from one navaid to next (US = chop).

COPA Canadian Owners and Pilots Association.

copal Natural resin from tropical trees used in some varnishes.

co-pilot Licensed pilot serving in any piloting capacity other than (1) PIC or (2) being on board solely to receive instruction.

COPR Cruise overall pressure ratio.

Copy "I read you" (radio voice code).

copy machining Using machine tool having means for copying shape of template or master part.

copy milling See *copy machining*.

cordite Gun propellant prepared mainly from nitrocellulose (gun-cotton) dissolved in nitroglycerine.

Cords Coherent on-receive Doppler system.

cordtex Blasting or cutting cord comprising high explosive in flexible filament form.

Cordwood Electronic technology (1948–55) designed to achieve maximum packing density of discrete components in pre-semiconductor era.

core 1 Gas-generator portion of turbofan, term esp applied when * small in relation to fan, and less relevant to bypass or "leaky turbojet" engines.
2 Central part of launch vehicle boosted by lateral or wrap-round rockets.
3 Low-density stabilizing filling inside honeycomb, foam-filled or other two-component structure.
4 High-density penetrative filling in armour-piercing projectile.
5 Magnetic circuit of transformer or inductor.
6 Central portion of nuclear reactor in which reaction occurs.
7 Solid shape(s) which make casting hollow.
8 Loosely, EDP (1) memory of magnetic type, from * (5).
9 Interior of carburized or nitrided part unaffected by surface treatment.

core deposits Solids deposited on metal surfaces of core (1).

coring Uneven flow of oil through oil cooler due to reduced viscosity of oil in hot central core.

coriolis acceleration Acceleration of particle moving in co-ordinate system which is itself accelerating, eg by rotating. In Earth-referenced motion, ** is experienced in all motion parallel to local surface except for that on Equator.

coriolis correction Applied to all celestially derived fixes to allow for coriolis acceleration.

coriolis effect 1 Physiological response (eg vertigo, nausea) felt by persons moving inside rotating container (eg space station with rotation-induced gravity) in any direction other than parallel to axis.
2 According to AGARD: "The acceleration, due to an aircraft flying in a non-linear path in space, which causes the displacement of the apparent horizon as defined by the bubble in a sextant". This definition is inadequate.

coriolis force Apparent inertial force acting on body moving with radial velocity within a rotating reference system. Such a force is necessary if Newtonian mechanics are to be applicable. On

Earth, ** acts perpendicular to direction of travel, towards right in N hemisphere and towards left in S hemisphere. Also called deflecting force, compound centrifugal force, geostrophic force.

coriolis parameter Twice component of Earth's angular velocity about local vertical, ie twice Earth rate multiplied by sin lat.

coriolis rate sensor Instrument based on beam vibrating in plane of aircraft-referenced vertical, sensing any disturbance about longitudinal axis.

corkscrew Evasive manoeuvre, esp when subjected to stern attack by fighter; interpretation variable but * axis basically horizontal.

corncob Descriptive generic name for multi-row radial or multi-bank in-line PE (4) (colloq).

corner point Instantaneous change in slope of graph; eg kink in payload/range curve, esp limiting range for max payload.

corner reflector Passive device for giving strong radar echo, based on three mutually perpendicular metal plates or screens.

corner speed Lowest airspeed at which a fighter can pull structure- or aerodynamic-limiting g.

Corogard Vinyl-modified polysulphide paint resistant to hydraulic fluid, usually silver from added aluminium powder.

corona discharge Electric discharge occurring when potential gradient around conductor is sufficient to ionize surrounding gas. Unlike point discharge, can be luminous and audible, but unlike spark discharge there are an infinity of transmission paths carrying continuous current. Also called brush discharge, St Elmo's fire (see *static wick*).

co-rotating wheels Landing-gear wheels on live axle and thus constrained to rotate together.

corpuscular cosmic rays Cosmic rays are primary particles (protons, alpha particles and heavier nuclei) which react with Earth atmosphere to yield particles and EM radiation. Term corpuscular is redundant.

corrected airspeed No defined meaning (see *airspeed*).

corrected altitude No defined meaning, other than "true height above SL" (see *altimeter errors*).

corrected gyro Normally taken to be one corrected (by latitude nut) for apparent wander due to Earth rotation.

correlation Confirmation that aircraft or other target seen visually or on radar display or plotting table is same as that on which information is being received from other source(s).

correlation criterion Statistical basis for defruiting or decoding raw IFF, typically on ** of 2/7, ie 2 valid synchronous replies detected within any 7 successive interrogations.

correlation factor In nuclear warfare, ratio of ground dose-rate reading taken at approximately same time as one at survey height over same point.

correlation protection Development by RAE with industry of method of avoiding false ILS indications caused by spurious signals reflected from large objects near runway; localizer and glidepath aerials duplicated (respectively horizontally and vertically) and emit signals which, if not received almost simultaneously at aircraft, are suppressed.

corridor 1 Geographically determinate path through atmosphere, typically curved-axis cone with apex at surface, along which space vehicle must pass after launch.
2 Path through atmosphere, geographically determinate for given entry point, along which space vehicle must pass during re-entry; has precisely defined upper and lower limits, above which vehicle will skip back into space and below which it will suffer severe deceleration and risk injuring occupants or burnup through heating.
3 Assumed safe track in low penetration of hostile territory.
4 Path through atmosphere, usually at low level, along which defences are assumed handicapped by prior seeding with chaff and decoys.
5 Region of any shape on graph within which solution to problem is possible.

corrugated skin Stabilized against local bending by uniform rolled corrugations which, when used as external skin of aircraft, are aligned fore- and aft (incorrectly assumed parallel to local airflow).

corruption Degradation of EDP (1) memory, typically from severe EM interference or, with volatile memory, from switching off power.

CoS Chief of Staff.

Cosim Variometer (colloq, obs).

Coslettizing Anti-corrosion treatment involving a wet deposition of Zn.

cosmetic RFP Issued for sake of appearance, contract award being already decided.

cosmic speeds Those sufficiently high for interstellar exploration, similar to that of light; even allowing for relativistic time effects these are wholly unattainable at present.

cosmodrome Space launching site (USSR).

cosmology Science of the Universe.

cosmonaut Member of spacecraft crew (USSR).

cosmonautics See *astronautics*.

COSPAR, Cospar Committee on Space Research (Int, office in Paris).

cost In procurement main elements may include R&D, T&E, flyaway, spares provisioning, ground equipment, base, crew and publications. Operating adds fuel and other consumables, depreciation and various indirect *.

costa Rib, translated in aviation not as wing rib but as fuselage frame.

costal Pertaining to frames or ribs; hence intercostal.
cost/economical Cruise conditions for minimum trip cost.
cost-effectiveness Measure of desirability of product, esp a weapon system, in which single quantified figure for capability (including reliability, survivability and other factors) is divided by various costs (total ownership, acquisition etc).
cost plus fixed fee Reward invariant with actual costs but fee may be renegotiated.
cost plus incentive fee Reward covers actual costs plus a fee which depends on contractor performance and possibly costs.
cost-sharing No fee, contractor merely reimbursed agreed percentage of costs.
Cotal Confederación de Organizaciones Turisticas de la America Latina (Int).
CoTAM Commandement du Transport Aérien Militaire (F).
cotter pin 1 Wedge-shaped pin used in joining parts.
2 In US, often split pin.
coulomb SI unit for quantity of electricity or electric charge, = 1 As, symbol C.
coulomb damping That due to opposing force independent of distance or velocity; also called dry friction damping.
coulomb excitation Raising of energy level as a result of charged particle passing outside range of nuclear interactions.
Coulomb's law Force between two magnetic or electric charges is proportional to product of charges and inversely to square of distance apart.
countdown Oral telling-off of time, usually at first in minutes, then in seconds, remaining before launch of vehicle or other event.
counter 1 Portion of ship hull from stern overhanging water; thus applicable to undersurface of rear fuselage above and behind jet nozzles or other lower section.
2 Electronic circuit which counts bits, impulses, waves or other repeated signals.
counter air Defensive and offensive actions against enemy air power.
counterforce Attack directed against enemy ICBMs and SLBMs or other strategic forces.
counter-insurgent, Coin Directed against supposed primitive guerrilla forces.
countermeasures All techniques intended to confuse or mislead hostile sensors such as radar, IR, visual, TV or noise.
counter-pointer Dial indication comprising rotating pointer(s) and counter readout in same instrument.
counter readout Numerical display generated by numerals on adjacent rotating drums.
counter-reflector Metal mesh or other radio reflector arranged in pattern under VOR or other ground station to nullify interference and give radiation as from perfect level-surface site.
counter-rotating See *contra-rotating*.
countersilo Counterforce attack against ICBM silos.
countersink Form or cut conical depression in workpiece to receive rivet or bolt head flush with surface.
countersurveillance All active or passive measures to prevent hostile surveillance.
countervalue Attack directed against enemy homeland society and industry.
country cover diagram Small-scale map and index showing availability of air reconnaissance information of whole country for planning purposes.
countup Oral telling-off of time, usually in seconds, elapsed since liftoff.
couple Two parallel opposing forces not acting through same point, producing rotative force equal to either force multiplied by perpendicular distance separating axes.
coupled engines Geared to same propeller(s) but not necessarily mechanically joined.
coupled flutter In which energy is transferred through distorting structure linking two fluttering masses to augment flutter of either or both.
coupling 1 Inertia *, tendency for inertia forces in manoeuvres to overcome stabilizing aerodynamic forces, esp in long, dense aircraft having large inertia in pitch and yaw; eg, rapid roll results in violent cyclic oscillation in pitch about principal inertial axis, increasingly marked with altitude owing to divergence of this axis from relative wind.
2 Connection (electrical, electronic or mechanical) between flight-control system and other onboard system such as ILS or TFR.
coupon Small extra piece formed on casting or, rarely, forging or extrusion, to provide metallurgical test specimen.
courier See *delayed repeater comsat*.
course See *heading*.
course corrections Allowances for deviation and variation.
course light Visual beacon on airway (obs).
course line Locus of points nearest to runway centreline in any horizontal plane along which DDM is zero.
course sector Horizontal sector in same plane as course line limited by loci of nearest points having DDM of 0·155.
course selector See *OBS*.
course/speed computer See *vector computer*.
cove Local concave curved region where two structures meet, eg wing/pylon or pylon/pod.
cover 1 Protection of friendly aircraft by fighters or EW platforms at higher level.
2 Ground area shown in imagery, mosaics etc.
3 To maintain continuous EM receiver watch.
4 To use fighters to shadow hostile contact from designated BVR distance.
coverage index See *covertrace*.
cover search To select best cover (2) for air recon-

naissance for particular requirement.

covertrace Map overlay listing all air reconnaissance sorties over that ground area, marking tracks and exposures.

covert search Patrol using advanced sensors from high level so that aircraft's presence is undetected from ground, esp in offshore patrol for customs, immigration or fishery protection.

Covos Comité d'Etudes sur les Conséquences des Vols dans la Stratosphère (Int).

cowl Covering over installed engine or other device, normally mainly of hinged or removable panels.

cowl flaps See *gills*.

cowling See *cowl*.

CP 1 Critical point.
2 Centre of pressure (often c.p.).
3 Controllable pitch (not constant speed).
4 Chlorinated paraffin.
5 Circularly polarized.
6 General call to several specified stations (ICAO).
7 Co-pilot.
8 Cadet pilot (or C/p).
9 Cathode protection.
10 Command post.
11 Centre-perforate (rocket grain).

C_p 1 Pressure coefficient.
2 Specific heat at constant pressure.

CPA 1 Critical-path analysis (see *critical path*).
2 Continuous patrol aircraft.

C/PA Cost/performance analysis.

CPACS Coded-pulse anti-clutter system.

CPAM Committee of Purchasers of Aviation Materials (Int).

CPCI Computer program-configuration item.

CPE Central Photographic Establishment (RAF).

CPF Complete power failure.

CPFF Cost plus fixed fee.

CPG Co-pilot/gunner.

CPGS Cassette-preparation ground station.

CPI 1 Cost plus incentive (F adds "fee").
2 Chief pilot instructor.

CPIFT Cockpit procedures and instrument flight trainer (Pacer Systems).

CPILS Correlation-protected ILS.

CPL 1 Commercial pilot's licence.
2 Current flight plan message (ICAO).

CPL/H Commercial pilot's licence, helicopter.

CPL/IR Commercial pilot's licence, instrument rating.

CPL/SEL Commercial pilot's licence, single-engine limitation.

CPM 1 Capacity passenger-miles.
2 Critical-path method.

CPP 1 Cost per passenger.
2 Critical parts plan (ECPP).

CPR 1 Coherent-pulse radar.
2 Crack-propagation rate.

CPRTM Cents per revenue ton-mile.

CPS Central processing system (or site).

cps Cycles per second (Hz is preferred).

CPT 1 Cockpit procedure trainer.
2 Central passenger terminal complex.
3 Civilian pilot training program (US).

CPU 1 Contractor payment unit.
2 Central processing unit.

CPX Command-post exercise.

CQ Carrier (ship) qualification.

CqS Constant-q stagnation trajectory.

CR 1 Aerial victory credit.
2 Cost reimbursable.
3 or C/R, counter-rotation; usually refers to handed engines with single-rotation propellers.
4 Fighter-reconnaissance (F).
5 ATC request (FAA).

C_R 1 Resultant-force coefficient.
2 Range constant, velocity × wt/fuel flow.

CRA Centro Ricerche Aerospaziali, Rome.

crab 1 To fly with wings level but significant drift due to crosswind.
2 To fly with wings level but significant yaw due to asymmetric thrust.
3 To fly with wings level but significant yaw imparted by rudder to neutralize effect of crosswind.
4 Miniature trolley driven by Link trainer and certain other simulators which reproduces aircraft track on map on instructor's desk.

crab angle 1 Drift angle.
2 In landing, angle between runway axis and aircraft heading.
3 Angle between fore/aft camera axis and track.

crab-pot Fabric non-return valve in circular duct in airship, controlled by bidirectional pull of cord attached to centre.

crack 1 Microscopic rupture in stressed metal part which under repeated loads progressively grows longer, without deformation of structure, until remaining material suddenly breaks.
2 To break down hydrocarbons by cracking.

cracking Application of heat and usually pressure, sometimes in presence of catalysts, to break down complex hydrocarbons, esp petroleums, into desired products.

crack-stopper Structural design feature, such as assembly of part from several components with joints perpendicular to expected crack directions, to prevent crack progressing right across.

CRAF Civil Reserve Air Fleet (US).

crafted Made (US usage).

crane helicopter Designed for local lifting and positioning of heavy or bulky items rather than normal transport; characterized by vestigial fuselage with payload attached externally or slung.

Cranfield Formerly College of Aeronautics, now Cranfield Institute of Technology (UK).

cranked wing Has acute anhedral inboard, dihedral outboard, usually with abrupt change at about 30 per cent semi-span.

cranking Turning engine (any type) by external power.

crash Unpremeditated termination of mission at any point after start of taxi caused by violent impact with another body, with or without pilot in control, usually causing severe damage to aircraft. Term never used in official language.

crash arch Strong structure above or behind pilot(s) head(s), esp in open cockpit or small cabin aircraft, able to bear all likely loads in overturning and sliding inverted on ground.

crash barrier See *barrier*.

crash gate Gate in airfield periphery through which crash/fire/rescue teams can most quickly reach nearby crashed aircraft.

crash landing Emergency forced landing with severe features such as rugged terrain or incapacitated pilot, resulting in more than superficial damage to aircraft.

crash locator beacon Automatic radio beacon designed to be ejected from crashing aircraft, thereafter to float and survive all predictable impacts, crushing forces or fire while broadcasting coded signal.

crash pan Secondary structure under paradropped load, esp vehicle or artillery, which absorbs landing shock by plastic deformation.

crashproof tank Euphemistic, denotes fuel or other tank designed not to rupture, leak or catch fire in all except most severe crash.

crash pylon Structure having same purpose as crash arch.

crash switch Electrical switch triggered by various crash symptoms to shut off fuel, activate fire/explosion suppression, release CLB, etc.

crashworthiness Generally unquantifiable ability of aircraft to crash without severely injuring occupants or preventing their escape.

crate Aerodyne, esp aeroplane (colloq, derogatory, archaic).

CRAW Carrier replacement air wing (USN).

CRC 1 Control & reporting centre.
2 Carbon-fibre reinforced composite.
3 Communications Research Center (Canada).

CRCO Central Route Charges Office (ICAO).

CRD Controller, Research & Development (MAP, WW2).

CRD/F Cathode-ray direction-finding; ground D/F receiver in which aerial automatically rotates to null azimuth as soon as pilot transmits, bearing being instantly shown on circular display.

creamed Shot down, destroyed (colloq).

credible Of deterrent, demonstrably capable of being used and having desired effect; depends on its ability to penetrate and on government's resolution.

credit Unit of aerial victory scores made up of air-combat plus strafing (aircraft destroyed on ground), with fractions for targets shared (US).

creep 1 Slow plastic deformation under prolonged load, greatly accelerated by high temperatures.
2 Gradual rotation of tyre around wheel; hence * marker, white index marks on wheel and tyre initially in alignment.

creep strength Stress that will produce specified elongation over given period (typically 0·1% over 1,000 h) at given temperature.

crescent wing Has progressive reduction in both t/c ratio and sweep angle from root to tip, usually in discrete stages.

Crest 1 Comprehensive radar effects simulator trainer.
2 Consolidated reporting and evaluating subsystem, tactical.
3 Crew escape technology.

crevice corrosion Initiated by presence of crevice in structure in which foreign material may collect; eliminated by modern structural coating and assembly methods.

crew Divided into flight * to fly aircraft, mission * to carry out other duties in flight, cabin * to minister to passengers and, arguably, instructors; all assigned to these duties by appropriate authority.

crew duty time Measured from reporting for duty to completion of all post-flight duties.

crewing 1 Make-up of flight crew by trade or appointment.
2 Make-up of flight crew by individual rostered names.

crew ratio Number of complete air crews authorized per line aircraft (civil) or per aircraft in unit complement (military).

crew room Room reserved for (usually military) flight crews, some on standby and others relaxing after a mission, where publications are kept and notices promulgated.

crew trainer Aircraft designed to train whole flight crew, esp of traditional military aircraft requiring several flight-crew trades: pilot/navigator/bombardier/signaller/engineer/gunner.

CRIP Coat-rod inches per passenger.

CRIT Centre de Recherches Industrielles et Techniques (F).

critical altitude See *decision height*.

critical angle 1 Angle from local vertical at which radio signals of given frequency do not escape through ionosphere but just return to Earth.
2 Incorrectly used to mean stalling angle of attack.

critical case That combination of failures (of propulsion, flight controls or systems) giving worst performance (see *critical engine*).

critical crack One of * length.

critical engine Engine the failure of which is most disadvantageous, due to asymmetric effects, loss of system power or other adverse factors; failure of ** at V_1 is basis of take-off certification in most multi-engine aircraft.

critical frequency 1 That corresponding to natu-

ral resonance of blade, control surface or other structure.
2 Helicopter main-rotor blade-passing frequency at which whole machine resonates on landing gear.
3 Frequency at which critical angle becomes zero; highest at which vertical reflection is possible.

critical-length crack Crack of length at which application of limit load causes failure.

critical line Locus of critical points (when track is not known precisely).

critical Mach number 1 M_{crit}; Mach number at which most-accelerated flow around a body first becomes locally supersonic; for thin wing might be M 0·9 while thick wing may have * below 0·75.
2 Mach number at which compressibility effects significantly influence handling.

critical mass That of fissile material in which chain reaction becomes self-sustaining.

critical path That traced through number of tasks proceeding both consecutively and concurrently (as during turn-round of aircraft) that determines minimum total elapsed time.

critical-path technique Minimization of total elapsed time by concentrating on those elements that form critical path.

critical point That from which two fixed bases, such as departure airfield and destination, are equidistant in time.

critical position That over large city or mountain range at which propulsion failure would be most serious.

critical pressure In fluid flow through nozzle, that final pressure below which no further reduction results in increase in flow from fixed initial pressure; usually rather more than 50% of initial pressure (fixed ratio for any given medium and temperature).

critical pressure coefficient C_{pc}; pressure coefficient at critical Mach number, approximately given by Prandtl-Glauert.

critical pressure ratio That at which particular axial compressor suddenly ceases to operate efficiently due to choking, stall or other flow breakdown.

critical speed 1 See *V_1*.
2 That rotational speed at which machinery (eg, engine) suffers dangerous resonance or whip of shafting.

critical static pressure That at critical Mach number; symbol P_c.

critical temperature That below which gas or vapour may be liquefied by pressure alone.

critical velocity Speed at which fluid flow becomes sonic, ie locally reaches Mach number of unity;

$$V_{cr} \text{ or } V_{crit} = a_0 \sqrt{\frac{(\gamma - 1) M_0^2 + 2}{\gamma + 1}}$$

CRO 1 Civilian repair organization.
2 Cathode-ray oscilloscope.
3 Community relations officer (RAF).

Crocco Luigi Crocco (1932) derived equation:

$$T = T_w - (T_w - T_f)\frac{u}{U_f} + \frac{u(U_f - u)}{2C_p}$$

where T is temperature within boundary layer, T_w temperature of adjacent solid surface, T_f free-stream temperature, u local velocity, U_f free-stream velocity, and C_p specific heat at constant pressure.

crocodile Control surface, usually aileron, which can split apart into upper and lower halves as airbrake; see *deceleron*.

cropped-fan engine Turbofan whose fan has been reduced in diameter to match reduced thrust requirement and permit LP turbine and other parts to be simplified.

cropped surface Wing, tail or other surface whose tip is cut off diagonally at Mach angle appropriate to particular supersonic flight condition.

cropped tip *Cropped surface.*

cross-bar System of approach lighting using straight rows of white lights perpendicular to runway centreline. Calvert and some other systems use several bars decreasing in width to threshold while US practice is single white bar followed by red undershoot zone.

cross-bleed Pneumatic pipe system connecting all engines so that bleed from one can start, or drive accessories on, any other.

crosscheck Brief message from one pilot to another, or another crew member, in same aircraft giving or confirming situation, eg "Inner marker" or "crosscheck, I have the yoke".

cross-country Flight to predetermined destination, where landing may or may not be made, esp one to gain practice in map-reading and navigation.

cross-deck Operations by two or more aircraft carriers, not necessarily of same navy, in which aircraft operate from unfamiliar decks on exchange basis; hence * ing.

crossed controls Application of flight-control movements in opposite sense to those in normal turns or manoeuvres, eg right stick and left rudder; rarely required.

crossed-spring balance Wind-tunnel balance whose pivots are made up of two or more leaf springs crossing diagonally and giving virtually frictionless flexure through defined axes.

cross-fall Transverse slope of runway surface, to ensure sufficiently quick run-off of water to avoid aquaplaning except in particular adverse crosswinds.

cross-feed Feeding items (eg, engines) on one side of aircraft from supply (eg, fuel) on opposite side; abnormal condition under pilot control.

cross-flow Having two fluids flowing past each other at 90° while separated by thin metal walls.

cross-level Lateral clinometer, instrument formerly used to indicate direction of local vertical as aircraft manoeuvres in rolling plane.

cross-modulation Unwanted modulation from one carrier being impressed on another in same receiver, usually resulting from inability to filter out certain sidebands.
cross-needle Instrument display based on two pivoted needles which pilot attempts to keep crossed at 90° in centre of display.
cross-over exhaust Gas from inboard cylinders of multi-PE (4) aircraft is piped to discharge on outboard side to reduce noise in fuselage.
cross-over struts Inclined radial gas paths in one form of coannular inverted-flow engine to convey high-V core jet to outer periphery and low-V fan flow to centre.
cross-qualification Among other meanings, qualification of pilots on a type of aircraft with characteristics and flight deck similar to that habitually flown, but (except for simulator) not actually flown; eg, A300B/A310, B737-300/B757.
cross-radial navigation Routing not on a radial constituting a promulgated airway; ie, RNav using VOR and/or DME to fly direct from A to B (see *GNav*).
cross-range Approximately at 90° to axis of missile or space launch range.
cross-section 1 Transverse section through object, eg fuselage or structural member.
2 Measure of radar reflectivity of object, usually expressed as area of perfect reflector perpendicular to incident radiation; depends on structural materials, incident angles, physical size of target, radar wavelength and possibly other factors.
3 In nuclear or atomic reactions, area (expressed in barn) giving measure of probability of process occurring.
cross-servicing Between-flights routine maintenance and replenishment of aircraft at base of different armed force or different nation; * guide is manual facilitating operational turnround at locations where relevant documents are not available.
crosstalk Unwanted signals generated in one set of circuits in communications or EDP (1) system by traffic in another.
cross-trail Distance bomb or other free-fall object falls downwind measured perpendicular to track (or track at release point projected ahead).
cross-trail angle Angle in horizontal plane measured at release point between track and line to point of bomb impact.
cross-turn Rapid 180° in which each half of formation turns towards remainder.
crosswind One blowing more or less at right angles to track, to runway direction, or to other flown direction.
crosswind axis Straight line through c.g. perpendicular to lift and drag axes.
crosswind component Velocity of wind component at 90° to runway, track or other direction; = WV sin A where A is angle between WV and direction concerned.
crosswind force Component along crosswind axis of resultant force due to relative wind.
crosswind landing gear One whose wheels can be castored or prealigned with runway while aircraft crabs on to ground with wings level.
crosswind leg In landing circuit, that made at 90° to landing direction from end of downwind leg to start of approach.
crowbar Unswept wing (c 1950 colloq).
CRP 1 Carbon-fibre reinforced plastics.
2 Control and reporting point (or post).
3 Counter-rotation propfan.
CRPAE Cercle des Relations Publiques de l'Aéronautique et de l'Espace (F).
CRPM Compressor rpm.
CRPMD Combined radar and projected-map display.
CRS Container release system.
2 Control and reporting squadron (or section).
3 Component repair squadron (US).
4 Computer reservation system.
crs Course.
CRT Cathode-ray tube.
CRTE Combat rescue training exercise.
CRTS CRT scope; term not recommended.
cruciform Having approximate form of a cross, in aerospace usually when viewed from front; thus * wing missile has four wings arranged radially (often at 90°) at same axial position round body.
cruise 1 To fly from top of climb to top of descent en route to destination, usually at altitudes, engine settings and other factors selected for economy and long life.
2 Tour of operations by naval air unit aboard carrier.
cruise configuration Describes not only aerodynamic (normally fully clean) status but also systems status and possibly location and duties of flight crew, during cruise (1).
cruise missile Long-range pilotless delivery system whose flight is wing-supported within atmosphere.
cruise motor Propulsion, of any kind, used to sustain speed of missile from boost burnout onwards.
cruising altitude That assigned to or selected by captain for flight from top of climb to top of descent; varies with type of aircraft, sector distance, take-off weight, ATC rules and other traffic, winds, and other factors.
cruising boost With PE (4), that available in weak mixture for continuous operation giving best time or lowest fuel burn.
cruising ceiling Formerly, greatest height at which 1·35 V_{imp} could be maintained at max WM cruise power.
cruising speed That selected for cruise (1).
cruising threshold 1·35 V_{imp}, considered (1935–50) practical lower limit to cruising speed.
crutches Lateral arms carrying pads which are

screwed down on upper sides of bomb, missile or other store to prevent movement relative to rack, pylon or other carrier.
CRV Centre-reading voltmeter.
cryogen See *refrigerant*.
cryogenic Operating at extremely low temperatures.
cryogenic materials Limited range of highly specialized materials suitable for sustained structural or other use at below -180°C.
cryogenic propellants Gases used in liquid state as oxidants and/or fuels in rocket engines, esp Lox, LH_2, Fl and various Fl compounds or mixtures.
cryopump High-vacuum pump operating by cooling chamber walls so that residual gas molecules are condensed on to them, leaving vapour pressure below that required.
cryostat Usually small lab rig for experiments at ultra-low temperatures, eg NMR, superconductivity etc.
crystal laser One whose lasing medium is a perfect-lattice crystal, eg ruby.
crystal lattice Three-dimensional orthogonal space lattice whose intersections locate the atoms of a perfect crystal (except on small scale, most crystals contain important imperfections).
crystal oscillator One with added subcircuit containing piezo-electric crystal (eg quartz) whose extremely rigid response gives high frequency-stability.
crystal transducer Transducer containing piezo-electric crystal which translates mechanical strain into electrical voltage.
CS 1 Constant-speed (c/s is preferred).
2 Cassegrain system.
3 Chemical harassing agent OCBM.
Cs Cirro-stratus.
c/s 1 Constant-speed.
2 Course/airspeed.
3 Cycles per second (Hz is preferred).
4 Centre-section of wing.
5 Call-sign.
CSAM Conseil Supérieur de l'Aviation Marchande (F).
CSAR Combat search and rescue.
CSAS 1 Command stability augmentation system.
2 Common-service airlift system.
CSATC Central School for Air Traffic Control (RAF).
CSAV Academy of Sciences (Czech).
CSB Closely spaced basing (ICBM).
CSBPC Control-stick boost and pitch compensator.
CSC 1 Course and speed calculator.
2 Constant symbol contrast (HUD).
3 Centreline stowage cabinet.
4 Chief sector controller.
CSCE Conference on Security and Co-operation in Europe.
CSCP Computer software change proposal (usually SCP).
C/SCSC Cost/schedule control systems criteria.
CSD Constant-speed drive.
CSDC Computer signal data converter.
CSDE Central Servicing Development Establishment (RAF).
CSDS Constant-speed drive starter.
CSE 1 Central Studies Establishment (Australia).
2 Central Signals Establishment (UK).
CSEU Control-system electronics unit.
CSG 1 Centre Spatial Guyanais (F).
2 Constant-speed generator.
CSI Combined speed indicator, displaying airspeed and Mach.
CSII Centre for Study of Industrial Innovation (UK).
CSINA Conseil Supérieur de l'Infrastructure et de la Navigation Aérienne (F).
CSIRO Commonwealth Scientific and Industrial Research Organization (Int).
CSM Command/service module.
CSO Command signals officer (RAF).
CSOC Combined space (or satellite) operations centre (pronounced C-sock).
CSP Common signal-processor.
CSPA Canadian Sport Parachuting Association.
CSR Covert strike radar.
CSRL Common strategic rotary launcher.
CSS 1 Control-stick steering.
2 Cockpit system(s) simulator.
C/SSR Cost/schedule status report.
CSSS Combat service support system.
CST 1 Combined station and tower.
2 Centre Spatial de Toulouse (F).
C_sT_e Caesium telluride, photocathode material.
CSTI Control-surface tie-in.
CSTM Centro Studi Trasporti Missilistici (I).
CSU 1 Constant-speed unit.
2 Central suppression unit, prevents mutual interference in complex avionics.
3 Command sensor unit.
4 Cabin service unit.
CSV Capacity (or catapult) safety-valve.
CSVTS Scientific and technical society (Czech).
CSW Conventional standoff weapon
CSWS Corps-support weapon system.
CT 1 Carry trials.
2 Counter-terrorist.
3 Current transformer.
c_t Equivalent tip chord.
CTA 1 Control area.
2 Companion trainer aircraft.
3 Centro Tecnico Aeroespacial (Braz).
CTAA Commandement des Transmissions de l'Armée de l'Air (F).
CTAF Comité des Transporteurs Aériens Français.
CTAM Climb to and maintain.

CTAR Comité des Transporteurs Aériens Complémentaires (F).
CTB Comprehensive test ban (T adds "treaty").
CTC 1 Canadian Transport Commission.
2 Command track counter.
ctc Contact.
CTD Compound turbo diesel.
CTDC Civil Transport Development Corporation (J).
CTK, ctk 1 Capacity tonne-kilometres.
2 Command track.
CTL 1 Control.
2 Tactical air command (Neth).
CTM 1 Capacity ton-miles (unless otherwise stated, short tons, statute miles).
2 Centrifugal twisting moment.
3 Cost per ton-mile.
CTN Caution.
CTO 1 Chief technical officer.
2 Crypto operator.
CTOL Conventional take-off and landing, ie ordinary aeroplane.
CTP 1 Chief test pilot.
2 Command track pointer.
CTPA Comité Technique des Programmes d'Armement (F).
CTPB Carboxy-terminated polybutadiene rocket propellant.
CTR 1 Control zone.
2 Controllable-twist rotor.
CTS 1 Central tactical system.
2 Common termination system.
CTT Capital-transfer tax (UK).
CTTO Central Tactics & Trials Organization (RAF).
CTU Control terminal unit.
CTV Curved trend vector.
CTVS Cockpit TV sensor.
CTZ Corps tactical zone.
CU 1 Conversion unit.
2 Cage/uncage; gyro system control.
Cu Cumulus.
cubage Total volume of rectilinear cargo that can be accommodated; typically 0·7 of pressurized above-floor cargo volume.
Cuban eight Manoeuvre in vertical plane normally comprising ¾ loop, half-roll, ¾ loop, half-roll.
cube out To run out of payload volume (either pax, cargo or both) at less than MSP (5).
cue 1 Glimpse of Earth's surface through cloud or darkness giving helpful attitude and distance information.
2 To slave homing seeker of missile to target, using information from other source.
CuF Cumuliform cloud.
cuff 1 Secondary structure added around propeller blade root, usually for aerodynamic reasons.
2 Structure added ahead of wing LE extending chord 3–5%, with sharp inboard end.
CUG Computer Utilization Group (OECD).
culture Man-made terrestrial features.
cumulo-nimbus Extremely large cumuliform clouds whose tops reach stratosphere and spread in form of fibrous ice-crystal anvil.
cumulus Dense white clouds with almost horizontal base and large vertical development, dome-shaped tops (cauliflower) showing growth in strong upcurrents.
cumulus mammatus Cumuliform clouds having pendulous protuberances on underside.
cupping Re-rigging to increase angle of incidence.
curie, Ci SI-accepted unit of radioactivity equal to $3{\cdot}7 \times 10^{10}$ disintegrations or transformations per second.
curie point Critical temperature, different for each material, above which ferromagnetic materials lose permanent or spontaneous magnetization.
curing Process by which most synthetic rubbers, plastics and solid-propellant binders are converted to compositions of higher molecular weight; may involve heating (condensation polymerization), chain reaction via free-radical or ionic mechanism (addition polymerization), or use of catalysts. In solid rocket motors semi-liquid is often cured in case, solidifying and becoming case-bonded.
curl Vector resulting from action of operator del (differential operator in vector analysis) on vector; sometimes called rotation.
curling die Used with curling punch to bend sheet edges to tubular form.
current Qualified on particular type and routinely flying it.
cursive writing Rounded, flowing writing with strokes joined; hence formed in display by actual strokes rather than TV-type raster scanning.
curtain Non-gastight partition in aerostat.
curved approach 1 Adopted by some aircraft, notably WW2 fighters, because of inadequate forward view straight ahead at low airspeeds.
2 Any of numerous possible quasi-elliptical paths followed when using MLS or other system offering such approach paths on either side of straight centreline.
curved trend Turn information imparted by three future track-lines on EHSI terminating 30 sec, 60 sec and 90 sec hence; these are straight with wings level but in banked turn show *, in extreme case linking in 360° circle (does not allow for drift).
curve of pursuit Followed by any aircraft chasing another and continuously steering toward latter's present position; with non-manoeuvring target curve soon becomes asymptotic with target straight-line course.
curvic coupling Joint between driving and driven shaft systems which transmits torque perfectly; allows for small errors in alignment or angle but does not secure one to other. In simplest form comprises two sets of meshing radial teeth of

smooth curving profile.

curvilinear flight Accelerated flight, ie not straight and level.

cushion See *ground cushion*.

CUSRPG Canada/US Regional Planning Group.

customer Usually purchaser and operator are synonymous; where purchaser is government agency and operator an air force, or purchaser is finance company or bank, * normally applies chiefly to operator.

customer base Total list of customers (term usually refers to civil air carriers) committed to purchasing or leasing new type.

customer mock-up Exact reproduction of aircraft interior, or part thereof, furnished with materials, fabrics, colours, seats and other equipment as specified by customer.

customizing 1 Finishing GA aircraft to customer's spec, eg furnishing, avionic kit, external paint.
2 Finishing avionics or instruments for particular task with chosen language, labels, IC chips and self-test.

cutback 1 Sudden partial closure of throttles at end of first climb segment for noise-abatement reasons.
2 Reduction in existing or planned procurement.
3 Reduction in manufacturing rate.

cut-back nozzle Normally, one (not of con-di type) shortened in length and terminating obliquely to give thrust slightly inclined to pipe axis.

cutback speed ASIR at top of first segment.

cutlet Cutlet-shaped flattened outer arm of hub forging of rigid rotor.

cutoff 1 Termination of rocket propulsion before burnout because desired trajectory and velocity have been reached.
2 Flying shortest track to intercept an air target.

cutoff ports Circular apertures in forward face of solid motor which can swiftly be blown open to terminate combustion.

cut-out 1 Aperture in pressurized fuselage, for door, window, hatch or other purpose.
2 Absence of rear inner part of elevator, terminating in diagonal edge, to allow full rudder to be applied.

cut-out switch One isolating or inactivating circuit or subsystem.

CV 1 Metric horsepower, = 0·7355 kW.
2 Fleet carrier (USN).

C_v Specific heat at constant volume.

CV(A), CVA Fleet carrier, attack.

CVBG Carrier battle group.

CVD Chemical vapour deposition.

CVE Escort carrier.

CVFR Controlled visual flight rules.

CVL Controlled-vortex lift.

CV(N), CVN Fleet carrier, nuclear-propelled.

CVR Cockpit voice recorder.

CV(S), CVS Carrier, escort/ASW.

CVW Carrier air wing (USN).

CW 1 Continuous wave.
2 Ambiguously, carrier wave.
3 Clockwise.
4 Chemical warfare.

CWC Crosswind component.

CWCS Common weapons control system.

CWD Chemical warfare defence.

CWFS Crashworthy fuel system.

CWG Charges Working Group (IATA).

CWP 1 Central warning panel.
2 Contractor's working party.
3 Compact when packed (antennas).

CWR Coloured weather radar.

CWS 1 Caution/warning system.
2 Central warning system.
3 Control-wheel steering.
4 Container weapon system.

CWV Crest working voltage.

CWW Cruciform-wing weapon.

CWY Clearway.

C_x Longitudinal force coefficient (= C_d cos A minus C_L sin A, where A is angle of attack).

CY Calendar year.

cyaniding Surface hardening of steel by immersion in bath of cyanide (and other) salts producing nitrogen as chief agent.

cycle One complete sequence of events making up portion of life of machine; thus for PE (4), four strokes of Otto *, while for aircraft usually start-up, taxi, take-off, climb, cruise, possibly combat, descent, landing, thrust-reverse, taxi, shutdown.

cycle efficiency Measure of performance of heat engine derived from PV or entropy diagram; usually synonymous with thermal efficiency.

cyclic pitch In most helicopters main-rotor blade pitch progressively increases from minimum (a very small) angle when head-on to airstream (momentarily occupying position of wing) to maximum 180° later (when in position of wing trailing-edge-on to airstream); this makes blade fall on advancing side of rotor and rise on retreating side, effectively decreasing and increasing angle of attack to even-out lift on both sides. This is also called feathering, and results in blade flapping (see * *control*).

cyclic-pitch control Primary helicopter flight control. Usually governed by stick, similar to aeroplane control column, which in central position causes basic cyclic variation as described above by tilting stationary and rotating stars on rotor hub. Pilot demand is passed through mixing unit and output tilts fixed star in desired direction to superimpose additional cyclic variation causing disc to tilt in desired direction to cause helicopter to rotate about pitch or roll axis.

cyclic rate Rate at which automatic gun fires, expressed in shots per minute, measured after maximum rate has been attained and not necessarily attainable except for brief periods.

cyclic stick Cyclic-pitch control stick.

cyclic testing Repeated application of supposed operating cycle, usually of exceptionally severe nature, under arduous environmental conditions to prove endurance or life of hardware.

cycling Cyclic testing.

cyclogiro Aerodyne, never successfully achieved, lifted and propelled by pivoted blades rotating about substantially horizontal transverse axes.

cyclone Tropical revolving storm.

cyclostrophic force That experienced by wind following curved isobars acting in addition to geostrophic force to give resultant wind along isobar according to Buys Ballot's law. At equator geostrophic force vanishes, leaving pure cyclostrophic wind.

cyclotron Family of magnetic-resonance particle accelerators, many extremely large.

cyclotron resonance Motion of moving charged particle in magnetic field on which is superposed alternating electric field normal to magnetic field.

cylinder One unit of PE (4), or, specif, surrounding cylinder enclosing combustion space and guiding piston.

cylinder block Single unit enclosing row of liquid-cooled in-line PE (4) cylinders.

cylinder head Usually removable top of PE (4) cylinder containing plugs, inlet/exhaust connections and (except with sleeve valve) valves.

cylinder liner Hard abrasion-resistant lining inserted into cylinder of light alloy or other soft material.

Cytac Loran-C.

C_z Normal force coefficient, $C_L \cos A + C_d \sin A$, where A is angle of attack; rarely called C_N.

cZ Fore/aft magnetic VSI component.

CZCS Coastal zone colour scanner.

D

D 1 Electric flux density.
2 Danger area (ICAO).
3 Duration of phenomenon in seconds, eg D_{60}.
4 Drift.
5 Diameter.
6 Dual (wheels, landing gear).
7 Drone director (USAF).
8 Drone (UK).

d 1 Distance.
2 Diode.
3 Deci, prefix, multiplied by 10^{-1} (not recommended).
4 Clear distance between contact areas of landing-gear wheels, including inner/outer wheels of bogie.

D-100 figures Drag at nominal speed of 100 ft/second.

D-code In flight plan, have DME.

D-gun Detonation gun, fires suspended particles of hard surface coating by detonation of oxy-acetylene.

D-layer Region of increasing electron and ion density in the ionosphere, existing only in day-time and merging with bottom of E-layer.

D-nose Strong aerofoil leading edge formed from heavy skin and spar, often forming structural basis of whole aerofoil (eg helicopter rotor blade).

DA 1 Drift angle.
2 Diplomatic authorization.
3 Double attack.
4 Long-range (bomber) aviation, predecessor of ADD (USSR).
5 Delayed-action (bomb).
6 Dual-alloy (turbine disc).
7 Direct action (fuze).

da Deca, prefix, multiplied by 10.

DAA 1 Directorate of Air Armament.
2 Digital/analog adaptor.

DAACM Direct airfield-attack cluster munition.

DAAT Digital angle-of-attack transmitter.

DABM Defence against ballistic missile(s).

Dabs Discrete-address beacon system; can address individual aircraft via transponder, pointing a narrow beam at it to transmit messages via data-link.

Dabsef Dabs Experimental Facility (Lincoln Laboratories, US).

dac Digital-to-analog converter.

Dacron A commercial polyethylene glycol terephthalate fibre and woven fabric related to Terylene and Mylar.

DACS Directorate of Aerospace Combat Systems (Canada).

DACT Dissimilar air-combat training (or tactics).

DAD Deep air defence.

DADC Digital air-data computer.

DADS 1 Deployable air-data sensor.
2 Digital air-data system.

DAeC Deutsche Aero Club (G).

DA fuze Direct-action fuze; designed to explode on impact.

DAG Deutsche Angestellten-Gewerkschaft Bundesgruppe Luft- und Raumfahrt.

DAI Direction des Affaires Internationales (F, MoD).

DAIS Digital avionics information system.

Dalmatian effect Increase in number of spots in map of V/STOL air bases compared with airfields.

DALR Dry adiabatic lapse rate.

Dalton's Law Empirical generalization that, for many so-called perfect gases, a mixture will have pressure equal to sum of partial pressures each would have as sole component within same volume and temperature, provided there is no chemical interaction.

DAM Dollars per aircraft-mile.

DAMA Demand-assigned multiple access.

damage assessment Determination of effect of attacks on targets.

damage cycle Loss of life of engine or other hardware.

damage limitation Ability to limit effects of nuclear destruction by using offensive and defensive measures to reduce weight of enemy attacks.

damage-tolerant Structure so designed as to continue to bear normal in-flight loads after failure (through fatigue, external damage or other cause) of any member (see *fail-safe*).

DAME Designated aviation medical examiner.

damped natural frequency Frequency of free vibration of damped linear system; decreases as damping increases.

damped wave Wave whose amplitude decreases with time or whose total energy decreases by transfer to either frequencies.

damper (crankshaft) Mass, fixed or free to oscillate, attached to crank web or other part to damp vibration at what would otherwise be critical frequencies.

damping factor Ratio of peak amplitudes of successive oscillations.

damping moment Tends to restore aircraft to normal flight attitude after upset.

DAMS Drum auxiliary memory subunit.

Danac Decca area-navigation airborne computer

(1984 also appeared as digital air-navigation control).

D&D 1 Distress and diversion (ATC).
2 Diesel and dye (smoke-making).

D&F Determination and findings.

danger area Airspace of defined dimensions in which activities dangerous to flight may exist at specific times.

dangle Angle between local horizontal at glider and end of tow-rope (usually air tow).

DAO Defence Attaché Office.

DAP 1 Distortion of aligned phases (LCD).
2 Director(ate) of Aircraft (originally Aeroplane) Production (UK).

DAR 1 Design assurance review.
2 Defense Acquisition Regulation.

dark cockpit All lights out, ie correct configuration and all systems normal.

darkfire Missile system operable at night in clear visibility.

dark-trace Display phosphor creating image through reflection/absorption of light instead of light emission from phosphor (see *skiatron*).

Darlington pair High-gain amplifier stage using two transistors in which base of second is fed from emitter of first.

DArmRD Directorate of Armament Research and Development (UK).

DARPA Defense Advanced Research Projects Agency (DoD).

DARS 1 Digital attitude-reference system.
2 Drogue air-refuelling system.

Dart, DART 1 Directional automatic realignment of trajectory (ejection seat).
2 Dual-axis rate transducer.
3 Deployable automatic relay terminal.

dart 1 Aerodynamic vehicle in which stabilizing and/or central surfaces are at tail.
2 Free-flight vehicle having stabilizing fins at tail but no means of trajectory control.

DAS 1 Defensive avionics system.
2 Directorate of Aerodrome Standards (UK).

DASA Defense Atomic Support Agency (US).

DASC Direct air support centre.

DASE Digital autostabilizer equipment.

DASH 1 Drone anti-submarine helicopter.
2 Differential airspeed hold.

dash Part of flight through defended hostile territory at full throttle and probably at low altitude, ignoring temporary very high fuel consumption.

DASP Discrete analog signal processing.

DAST Drones for aerodynamic and structural testing (NASA).

data link 1 Any highway or channel along which messages are sent in digital form.
2 Communications channel or circuit used to transmit data from sensor to computer, readout device or storage.

data-logger Short-term store for digital or analog information, eg for one flight or one week, periodically read back to build up service history of system, engines or other devices.

data recorder Equipment, usually electronic or electromechanical, for recording data in digital or analog form for subsequent playback and analysis (see *flight recorder*, *maintenance recorder*).

Datsa, DATSA Depot automatic test system for avionics.

DATT, DAtt Defense Attaché (US).

datum 1 Numerical, geometric or spatial reference or base for measurement of other quantities.
2 Vertical (rarely, horizontal or other) reference line from which all structural parts are measured and identified. Most * lines are exactly at, or close in front of or behind, nose; thus, frame 443 is a nominal 443 in or mm behind *; wing * is often aircraft centreline.

DAU Directly administered units (RAF).

Davis barrier Retractable crash barrier across carrier (1) deck.

Davis wing High-aspect ratio wing designed by David R. Davis; intended to cruise at low angle of attack with low drag.

day Mean solar * is defined as $8{\cdot}64 \times 10^4$ s; sidereal * is approximately $8{\cdot}616 \times 10^4$ s.

Day-Glo Family of dyes and paints with property of converting to visible light wavelengths outside normal visible spectrum, thus giving unnaturally bright hues.

day/night Equipment giving cheap and convenient IFR training, using tinted pilot goggles and complementary tinted cockpit transparency (eg blue goggles and amber canopy or red + green); pilot sees clear but tinted cockpit while outside world appears black.

DB 1 Development batch.
2 Direct broadcast.
3 Data base.
4 Data bus.
5 Double base (rocket propellant).

dB Decibel.

dBA Decibels absolute.

DBC Denied boarding compensation for bumped passengers; for pax reaching their destination within 4 h of original booked time 50% of flight-coupon value in Europe, 200% in US. No compensation for aircraft under 60 seats.

DBF Doppler beat frequency.

DBFM Defensive basic flight manoeuvres.

dBm Decibel meter, or 1 milliwatt.

DBM/C Data bus monitor/controller.

DBMS Data-base management system.

DBNS Doppler bombing/navigation system.

DBS 1 Doppler beam-sharpening.
2 Direct-broadcast satellite.

DBT Diffusion-bonded titanium.

DBTF Duct-burning turbofan.

DBUF Defence buildup plan, or programme (J).

DBV Diagonally braked vehicle, for runway friction measures.

DC, d.c. 1 Direct current.
2 Direct cycle.
3 Display controller.
4 Directionally cast.
5 Drag control.

DCA 1 Defense Communications Agency (US).
2 Directorate of Civil Aviation.
3 Defense Contre Avions (F).
4 Department of Civil Aviation (Aust, Braz).
5 Dual-capable aircraft.
6 Document content architecture (IBM).

DCAA Defense Contract Audit Agency (US).

D-carts Decoy cartridges.

DCAS 1 Deputy Chief of the Air Staff (UK).
2 Digital cōre avionics system.
3 Defense Contract Administration Service (DoD).

DCAV STOVL (F).

DCC Drone control centre.

DCD Data collector and diagnoster.

DCDI Digital course deviation indicator.

DCFS Digitally controlled frequency service.

DCH Destination change (input button).

DCIU Digital control and interface unit.

DCMAA Direction Centrale du Matériel de l'Armée de l'Air (F).

dcmsnd Decommissioned.

DCMU Digitally coloured map unit.

DCOS, DCoS Deputy Chief of Staff.

DCP Distributed communications processor.

DCPG Defense Communications Planning Group (DoD).

dcpu Display control power unit (IFF).

DCR Digital(ly) coded radar.

DCS 1 Deputy Chief of Staff (DCOS preferred).
2 Defense Communications System (US).

dcs Data-collection system.

DCSO Deputy Commander for Space Operations (USAF).

DCS/R&D Deputy Chief of Staff for Research and Development.

DCT Digital communications terminal.

DCTE Data circuit-terminating equipment.

DCTL DC to light, ie entire usable EM spectrum.

DCU Digital (engine) control unit.

DCV Demonstrated crosswind velocity.

D_{cw} Crosswind drag component.

DD Direct drive (servo valve).

DDA Design deviation authorization.

DDBS Distributed data-base system.

DDC Digital display console.

DDI Digital display indicator.

DDL 1 Down data-link.
2 Dummy deck landing.

DDM Difference in depth of modulation.

DDMS DoD Manager for Space Shuttle (Support) Operations.

DDPE Digital data-processing equipment.

DDR&E Director of Defense Research & Engineering (DoD).

DDS Display and debriefing subsystem.

DDT Direct digital targeting.

DDT&E Design, development, test and evaluation.

DDVR Displayed-data video recorder.

DE 1 Directed energy.
2 Direct-entry (RAF).

D_E Effective drag.

DEA Delegated engineering authority.

dead engine One that cannot be operated after IFSD.

deadeye Circular block pulled by surrounding cable or rope to exert tension on other cables passing through transverse holes.

deadface To cut off all system power by circuit interrupters at interface between modules, stages or spacecraft, prior to separation.

deadhead 1 To fly to maintenance base off-route.
2 Of aircrew, to ride as passenger(s).

dead reckoning Plotting aircraft position by calculations of speed, course, time, effect of wind, and previous known position.

dead-rise Difference in height from keel to chine of float or flying-boat hull.

dead-rise angle That between line joining keel and chine and transverse horizontal through keel.

dead side Side away from aircraft formation, eg left seat when in echelon to right.

dead spot In a system, region centred about neutral position where small inputs produce no response.

dead-stick landing Landing of powered aircraft will all engines inoperative.

dead vortex Remnants of vortex after breakup and decay.

dead zone 1 Surface area within maximum range of weapon, radar or observer which cannot be covered by fire or observation because of obstacles, nature of ground, or trajectory characteristics or pointing limitations of weapon.
2 Zone within range of radio transmitter in which signal is not received.
3 Region above gun or missile into which weapon cannot fire because of mechanical or electronic limitations.
4 Area(s) next to surfaces of aircraft plate for integrally machined parts which cannot be ultrasonically inspected and for which ultrasonic-inspection thickness allowances can be removed.

DEB Digital European backbone (major NATO programme).

deboost Retrograde or braking manoeuvre which lowers either perigee or apogee of orbiting spacecraft.

debrief To interrogate aircrew or astronauts after mission to obtain maximum useful information.

debug To isolate, correct or remove faults or malfunctions, especially from computer program.

deca See *da*.

decal Insignia or other mark applied by transfer,

usually to a model.

decalage Difference in angles of incidence of wings of biplane or multiplane; angle between chord of upper plane and that of lower plane in section parallel to plane of symmetry.

decalescence point Temperature, characterized by sudden evolution of heat, at which definite crystalline transformation takes place when heating steel.

decametric Having wavelengths in the order of 10 m.

decanewton 10 Newtons, of force or thrust.

decant Drain dregs of fuel from lowest point of integral or other tank, hence * hole, * assembly.

decarburizing Heating iron or carbon steel to temperature sufficiently high to burn out or oxidize carbon.

decay 1 Progressive, accelerating reduction in orbital parameters, esp apogee and perigee, of body in orbit affected by an atmosphere.
2 Progressive reduction in intensity of many natural processes, eg radioactivity, phosphorescence.

decay curve Plot of radiation intensity against time.

decay orbit Usually, final orbit terminating in re-entry.

decay product Usually, radioactive nuclides.

decay rate Rate of disintegration with respect to time.

decay time 1 Time for electronic pulse to fall to 0·1 of peak.
2 Time for charge in storage tube to fall to given fraction of initial, usually 1/e where e is e(5).
3 Estimated lifetime of satellite in low orbit.

Decca chain Single system of master and three slave Decca Navigator stations giving guidance over one geographic region.

Decca Flight Log Pictorial presentation of Decca Navigator inputs on roller-map display.

Decca lane In original Navigator, any hyperbolic region between two adjacent position lines.

Decca Navigator Pioneer hyperbolic navaid using CW.

Decca Omnitrac Airborne digital computer which eliminates Flight Log chart distortion, sets pen accurately after chart change and enables system to be coupled to autopilot.

deceleration limit That sustained value allowed for fully equipped astronauts or aircrew, normally −10 g.

deceleron Aileron which splits into upper/lower halves to serve as speed brake (originally Northrop patent).

decentralized control In air defence, normal mode whereby higher echelon merely monitors unit actions, making direct target assignments only when necessary.

deception Measure designed to mislead enemy by manipulation, distortion or falsification of evidence, eg by DECM (1).

decibel Fundamental unit of sound pressure (see *noise*).

decimetric Having wavelengths in the order of 10^{-1} m (not recommended).

decimillimetric Having wavelengths in the order of 10^{-1} mm (not recommended).

decision height Specified height AGL at which missed approach must be initiated if required visual reference to continue approach to land has not then been established; normally but not exclusively ILS, PAR or MLS approach.

decision speed Usually, V_1.

deck 1 Any ground or water surface (colloq).
2 From 1966, FL just above such surface.

deck-edge elevator Lift built into side of aircraft carrier for moving aircraft between decks.

decking Top surface of fuselage.

deck letter Identifying letter painted on flight deck of aircraft carrier.

deck park Parking area for aircraft or other vehicles on aircraft carrier flight deck.

declaration Size of force committed by government to special purpose, esp to support multinational alliance; hence declared force.

declared Numerical or factual data published or filed before flight.

declared alternate Airfield specified in flight plan to which flight may proceed when landing at original airfield becomes inadvisable.

declared destination Airfield specified in flight plan at which flight is intended to terminate.

declared thrust Generally, those ratings published by manufacturers.

declared weight Generally, that filed in flight plan.

declination 1 Angular distance to body on celestial sphere measured north or south through 90° from celestial equator along hour circle of body. Comparable to latitude on terrestrial sphere.
2 Magnetic variation.

DECM 1 Deceptive ECM.
2 Defensive ECM.

decode To translate into plain language aeronautical telecommunications and other signals from ground to air, esp in Notam and Q-Code.

Decometer Original dial-type instrument, one per lane, in Decca Navigator before 1953.

decompression chamber Capsule or chamber in which human beings or hardware can undergo process of decompression.

decompression sickness See *aeroembolism*.

decompression stress Human stress arising from decompression syndrome.

decoupler Large-amplitude elastic connection separating two systems of masses which, if rigidly linked, would be prone to dangerous flutter; hence * pylon for separating vibration of wing and heavy stores hung below it.

decoy Device or technique used to simulate attacking aircraft and their defensive systems. Usually operates at radar or IR wavelengths.

decrab To yaw crabbing aircraft landing in cross-wind to align wheels with track.

decrement Quantified decrease in value of variable.

Dectrac Decca Navigator display for GA aircraft.

DECU Digital engine (or electronic) control unit.

dedicated Available only for one declared application.

dedicated runway That permanently assigned as main instrument runway.

DEE Di-ethyl ether.

DEEC Digital electronic engine control.

DEEP Dangerous-environment electrical protection system.

deep cycling Pre-delivery test of electronics (rarely, other hardware) by subjecting circuits to slightly excessive voltages.

deepening Decreasing pressure in centre of existing low.

deep space Not in vicinity of Earth.

deep stall Condition associated with T-tail, rear-engined configuration characterized by rapid increase in angle of attack to point where effectiveness of horizontal tail is inadequate for longitudinal control; also known as locked-in stall, superstall.

DEFA Direction des Etudes et Fabrications d'Armement (F).

DEFCS Digital electronic flight-control system.

defence suppression Secondary objective of air attack on enemy territory, to reduce or eliminate anti-aircraft defences.

defensive combat spread Loose pair 1–2 miles (1·6–3·2 km) apart and slightly separated fore/aft and vertically.

defensive electronics Airborne ECM and EW equipment used to protect aircraft against hostile defences.

defensive spiral Accelerating high-g continuous-roll dive to negate attack.

defensive split Controlled separation of target element into different planes to force enemy interceptors to commit to one of them.

defensive turn Basic defensive manoeuvre to prevent an attacker from achieving a launch or firing position.

deficiency Fault condition (known or suspected), equipment shortage or other imperfection which may or may not render aircraft unairworthy.

definition 1 Clarity and sharpness of image in display.
2 In contract proposal, complete description by contractor of product offered.

deflected slipstream Horizontal slipstream from propulsion system deflected downwards by mechanical means to augment lift for STOL or V/STOL performance.

deflecting yoke Mutually perpendicular coils around neck of CRT which control position of electron beam, enabling it to scan screen.

deflection Bending or displacement of neutral axis of structural member due to external load.

deflection crash switch One triggered by any significant impact changing shape of structure.

defruiting Elimination of fruiting by rejecting all non-synchronous replies; PRFs varying by 2·5 μs can be eliminated.

DEG Dressed engine gearbox.

degarbling Elimination of garbling by trying to extract interleaved replies, differentiating between the exact leading and trailing edges of the pulses.

degraded performance Performance reduced by internal shortcomings, eg airframe tiredness, engine gas-path deposits, etc. Not normally used for external influences, eg hot-and-high conditions.

degraded surface Airfield covered with snow, ice or standing water.

degreasing Removal of grease, oil or related residue by solvent, either liquid such as naphtha or vapour such as trichlorethylene.

degree of freedom 1 Mode of motion, angular or linear, with respect to co-ordinate system; free body has six possible ***, three linear and three angular.
2 Specif, of gyro, number of orthogonal axes about which spin axis is free to rotate.
3 In unconstrained dynamic system, number of independent variables required to specify state at given moment. If system has constraints, each reduces *** by one.
4 Of mechanical system, minimum number of independent generalized co-ordinates required to define positions of all parts at any instant. Generally, *** equals number of possible independent generalized displacements.

DEI Development engineering inspection.

de-icing Removal of ice accretion by thermal, mechanical or chemical means. Note: anti-icing prevents.

DEIMS Defense economic-impact modelling system (DoD).

de-ionization time Time for gas-discharge tube to return to neutral condition after interruption of anode current.

DEL Delay, delay message; also DLA (ICAO code).

de Laval nozzle Con-di nozzle used in steam turbines, certain rockets, and some tunnels.

delay 1 Distance from point directly beneath aircraft to beginning of area visible to its radar.
2 Electronic delay at start of time base used to select particular segment of total.
3 Difference in phase between two EM waves of same frequency.

delayed automatic gain control Applied only to received signals above predetermined level, so permitting only weak signals to be fully amplified.

delayed CCIP CCIP for highly retarded bomb.

delayed drop Live parachute descent begun by prolonged free fall.

delayed repeater Comsat which stores messages and retransmits later, usually at high rate in brief period of time.

delay line Passive network, such as closed loop of mercury, capable of delaying signal without introducing distortion.

delay rate Unusual measure of airline punctuality: number of delays (usually 5min) per 100 scheduled departures.

delivery error Overall inaccuracy of weapon system resulting in dispersion of shots about aiming point.

delta (δ) Surface deflection angle, thus $e\delta$ = elevator deflection angle.

delta aircraft One fitted with delta wing.

delta hinge Helicopter flapping hinge; that about which blade is free to flap in plane parallel to hub axis.

delta wing One of basically triangular planform: low aspect ratio, sharply tapered leading edge, straight trailing edge, pointed tips.

deluge pond Facility at large rocket launch site into which cooling water is flushed after lift-off; also called skimmer basin.

demand mask Mask through which oxygen or other gas flows only on inspiration of wearer.

demand oxygen System in which oxygen flows to user during inspiration only.

demodulation Detection of received signals; process of extracting modulating signal from carrier.

demonstrate 1 To display new hardware according to detailed test schedule before certificating authority or sponsoring military customer.
2 More specifically, to show compliance with numerical performance values, reliability or maintainability.

demonstrator programme 1 Showing of new civil aircraft in visits to potential customers.
2 Agreed schedule of tests of new hardware, including complete aircraft, before military customer in advance of any decision on procurement and often to establish what is possible.

DEngRD Directorate of Engine Research and Development (UK).

denitrogenation Removal of nitrogen dissolved in blood and body tissues, usually by breathing pure oxygen for extended period, to minimize aeroembolism.

densified wood Multiple laminates bonded under high pressure.

densitometer 1 Instrument for measurement of optical density, generally of photographic image.
2 Instrument for measuring fuel density, usually part of fuel measurement system.

density 1 Mass per unit volume; often needs qualifying for temperature and pressure (see *absolute* *, *relative* *).
2 Of aircraft, MTOW divided by total aircraft volume calculated from external envelope.

density altitude Pressure altitude corrected for non-ISA temperature.

density error Correction to EAS to give TAS (see *airspeed*).

deorbit Deliberately to depart from spacecraft orbit, usually to enter descent phase or change course.

DEP 1 Department of Employment and Productivity (UK).
2 ICAO code for depart, departure, departure message.
3 Departure airfield.
4 Design eye position (usually of pilot).

departure 1 Any aircraft taking off from airport (as distinct from other airfields) under departure control.
2 In air navigation, distance made good in E/W direction, usually expressed in nm.
3 *Divergence* (1) or *disturbance* (2).

departure alternate Alternate airfield specified in flight plan filed before take-off.

departure control Function of approach control providing service for departing IFR aircraft and, on occasion, VFR aircraft in such matters as runway clearances, vectors away from congested areas and radar separations.

departure pattern That flown in 3-D by departure (1).

departure point Navigational check point, such as VOR or visual fix, used as a marker for setting course.

departure procedures ATC procedures (usually SID) flown by departing aircraft during climb-out to minimum en route altitude.

departure profile Flight profile flown by departure (1) to suit needs of vertical and horizontal separation, noise abatement, obstacle clearance, etc.

departure runway That from which departures (1) are cleared.

departure strip Flight progress strip recording callsign, ETD and route of departure.

departure track That followed by departure (1).

departure traffic Total number of departures, scheduled and non-scheduled, from one airport, usually expressed in movements per hour or per day.

DEP CON Departure control.

depigram Plot of variation of dewpoint with pressure for given sounding on tephigram.

depletion layer In semiconductor, region in which mobile carrier charge density is insufficient to neutralize net fixed charge density of donors (N-type) and acceptors (P-type). Also known as depletion barrier or zone.

deployment 1 Strategic relocation of forces to desired area of operation.
2 Extension or widening of front of military unit.

3 Change from cruising approach, or contact disposition, to formation for battle.
4 Process from pulling ripcord to fully opened parachute.
5 Extension of solar panels from spacecraft.

depreservation run Test run of machinery after storage to validate performance.

depressed trajectory Flight profile of ballistic missile, esp SLBM, fired over relatively short range with altitude kept low to reduce exposure to defending radars.

depression 1 Region of relatively low barometric pressure, also known as cyclonic area or low; secondary * is small low accompanying primary.
2 Negative altitude, angular distance below horizon.

derated engine One whose maximum power is governed at a lower than normal value.

deregulation Removal of rules regarding admission to air-transport industry of new carriers, routes and equipment.

derivatives, resistance 1 Lateral ** give variation of forces and moments caused by small changes in lateral, rolling and yawing velocities.
2 Longitudinal ** give variation of forces and moments caused by small changes in longitudinal, normal and pitching velocities.

derivatives, stability Quantities expressing variation of forces and moments on aircraft due to any disturbance to steady motion.

DES Design environmental simulator (USAF).

descent fuel Fuel burned from TOD until either hold or approach.

descent orbit insertion Start of lunar or planetary landing procedure from orbit, with retrograde thrust into descent transfer orbit.

descent propulsion That providing trajectory control for soft lunar or planetary landing.

descent stage Lower part of two-way lunar or planetary lander which, when mission is completed, acts as launch pad for ascent stage.

descent transfer orbit Highly elliptical around Moon, can be circular around planet, in which soft lander is placed before descent to surface.

design Entire process of translating hardware requirement or specification into final production drawings.

designated target One at which friendly designator (2) is pointed.

designation marking Use of laser or other designator.

designator 1 Letter/number code identifying each flight by a scheduled carrier.
2 Laser or other device pointed at target to make latter emit signals on which missile can home.

design gross weight Anticipated MTOW used in design calculations; design take-off weight.

design landing weight Anticipated MLW used in design calculations.

design leader 1 Individual leading design team.
2 Nation in collaborative project said to have political dominance.

design load Specified load below which structural member or part is designed not to fail, usually expressed as probable maximum limit load, unfactored.

design maximum weight Assumed weight used in stressing structure for flight loads.

design office That in which design takes place, and authority vested therein.

design points Specific combinations of variables upon which design process is based; together these cover every combination of air density, airspeed, Mach, dynamic pressure, structural loads (including free or accelerated take-off and normal or arrested landing) and system demands aircraft can encounter.

design verification First item built to new design to prove compliance with drawings and demonstrate correct functioning (see *DVA*).

design wing area Area enclosed by wing outline (including flaps in retracted position and ailerons, but excluding fillets or fairings) on surface containing wing chords, extended through nacelles and fuselage to plane of symmetry.

Desir Direct English statement information retrieval (EDP).

despatch 1 To supervise exit of parachutists, or to unload stores with parachutes attached, from aircraft in flight.
2 Process of supervising readiness of civil transport for next flight, with departure on schedule.

despatch deficiency Malfunction, failure, breakage, missing equipment item or other irregularity which does not prohibit on-time departure.

despatch delay Any notifiable delay, measured variously from either 5 or 15 min, in departure of scheduled flight.

despatch deviation Any reportable irregularity other than deficiency which does not prohibit on-time departure.

despatcher One who is responsible for despatching an airline flight.

despatch reliability Percentage of all scheduled flights by particular aircraft or all aircraft of that type, often over specified period or for particular operator, that departed on time (measured as within 5 or 15 min).

de-spin To rotate part or whole of satellite or other spacecraft to neutralize spin previously imparted (see *despun*).

despun antenna One mounted on satellite spun for reasons of stability which, because it must point continuously towards an Earth station, must rotate relative to satellite.

dessyn Synchro (trade name).

destruct To destroy vehicle after launch because of guidance or other failure making it dangerous.

destructive test One which destroys specimen.

destruct line Map boundaries which vehicle must not cross; any which does is immediately destructed.

destructor Device, explosive or incendiary, for intentionally destroying all or part of vehicle such as wayward missile or aircraft down in enemy territory.
DETA Di-ethylene triamine.
detachable Capable of being removed from aircraft with normal hand tools.
detached shockwave One proceeding ahead of body causing it.
detail 1 To design small part such as attachment bracket.
2 Drawing (can be inset on main design drawing) giving graphical representation of features.
3 Small military detachment for particular task.
4 To assign to special task or duty.
detail part One not normally broken down during service or storage.
detectable crack Nominal length 100 mm, 4 in.
detector Sensitive receiver for observing and measuring IR.
deterrence Prevention of aggression through fear of consequences.
detonation 1 Violent and irregular combustion in PE (4) cylinder resulting from excessive compression ratio or supercharging, or using inferior fuel; also known as knocking or pinking.
2 Correct triggering of explosive.
detonator Explosive device usually sensitive to mechanical or electrical action and employed to set off larger charge of explosive.
detotalizing counter Indicates total remaining of substance being measured, such as rounds for a gun or kg of fuel.
DEU Display electronics unit.
deuterium Isotope of hydrogen (heavy hydrogen) whose nucleus contains a neutron as well as a proton; used as projectile in nuclear processes. Forms heavy water with oxygen.
development 1 Process of converting first flight article into mature product ready for delivery.
2 Determining by mathematical calculation, computer graphics or drafting methods, size, shape and other pertinent characteristics of non-flat parts.
3 Opening of parachute canopy.
development contract Calls for development (1) of particular hardware item.
development stage Begins as soon as hardware to new design is available; main phase complete at service (production) release or certification, but continues throughout active life of aircraft.
deviation 1 Distance by which impact misses target.
2 Angular difference between magnetic and compass headings caused by magnetic fields other than that of Earth.
3 In statistics, difference between two numbers (also known as departure), difference of variable from its mean (esp standard *), or difference of observed value from theoretical.
4 In meteorology, angle between wind and pressure gradient.
5 In radio, apparent variation of frequency above and below unmodulated centre frequency.
6 In flying, sudden excursion from normal flightpath.
7 Any significant variation from plan.
deviation card Records compass courses corresponding to desired magnetic headings.
DEW 1 Distant early warning.
2 Directed-energy weapon.
dew Atmospheric moisture condensed upon cold objects, esp at night.
dewar Thermally insulated container, eg for cryogenics.
DEWIZ Distant early-warning identification zone, extends from surface north of DEW Line and around Alaska.
DEW Line Distant early-warning radar stations at about 70th parallel across North American continent.
dewpoint Temperature at which, under ordinary conditions, condensation begins in cooling mass of air.
DF, D/F 1 Direction-finding (or finder).
2 Digital filter.
3 Directed-flow (reverser).
D_F Zero-lift drag, usually of whole aircraft.
DFA 1 Deutsche Flug-Ambulanz Gemeinnützige GmbH (G).
2 Delayed-flap approach.
DFAD Digital-feature analysis data.
DFBW Digital fly-by-wire.
DFC 1 Distinguished Flying Cross.
2 Direct force control, eg on F-16.
DFCL Director(ate) of flight-crew licensing.
DFCS Digital flight-control system.
DFD Digital frequency discriminator.
DFDAU Digital flight-data acquisition unit.
DFDR Digital flight-data recorder.
DFDS Digital fire-detection system.
DFF Display failure flag.
DF/GA Day fighter/ground attack.
DFGS Digital flight-guidance system.
DFIC Duty-free import certificate.
DF loop Direction-finding aerial consisting of one or more turns of wire on vertically pivoted frame, giving maximum response in plane of frame, and thus PL through ground station (see *Adcock*, *ADF*).
DFLS Day Fighter Leaders' School (RAF).
DFM 1 Distinguished Flying Medal.
2 Digital frequency measurement.
3 Distortion-factor meter.
4 Direct-force mode.
DFMS Digital fuel-management system.
D_{form} Form drag.
DFQI Digital fuel-quantity indicator.
DFR Dynamic flap restraint.
DFRC Dryden Flight Research Center (US, NASA).

D_{fric} Total frictional drag, at low speeds almost equal to D_F.
DFS 1 Directorate of Flight Safety (UK).
2 Digital frequency synthesis.
3 Deutsche Forschungsanstalt für Segelflug.
DFT 1 Distance from threshold.
2 Discrete Fourier transform.
DFTI Distance-from-touchdown indicator.
DFV Deutsche Flugdienstberater Vereinigung (G).
DFVLR Deutsche Forschungs- und Versuchsanstalt für Luft- und Raumfahrt.
DFWD Discrete flight warning display.
DFWES Direct-fire weapons effects simulation.
DG Directional gyro, = DI (2).
DGA 1 Dispersed ground alert.
2 Délégation Générale pour l'Armement (F).
DGAC 1 Direction Général à l'Aviation Civile (F).
2 Directorate-General of Air Communications (Indonesia).
3 Direzione Generale dell'Aviazione Civile (I).
DGI Directional gyro instrument or indicator.
DGLR Deutsche Gesellschaft für Luft- und Raumfahrt.
DGLRM Deutsche Gesellschaft für Luft- und Raumfahrtmedizin.
DGM Distance-gone meter (Doppler).
DGON Deutsche Gesellschaft für Ortung und Navigation.
DGRR Deutsche Gesellschaft für Raketentechnik und Raumfahrt.
DGRST Direction Générale à la Recherche Scientifique et Technique (F).
DGS Disc-generated signal.
DGSI Drift and groundspeed indicator (Doppler).
DGU Display generator unit.
DGZ Desired ground zero; point on Earth's surface nearest to centre of planned nuclear detonation (see *actual ground zero*, *ground zero*).
DH Decision height.
DHDA Digicon header diode array.
DHS Data-handling system.
DHUD Diffraction-optics HUD.
DI 1 Daily inspection.
2 Direction indicator.
3 Director of Intelligence.
4 Duty instructor.
D_i Induced drag.
DIA 1 Documentation Internationale des Accidents (DocIntAcc).
2 Defense Intelligence Agency (US).
3 Document interchange architecture (IBM).
diabatic process Process in thermodynamic system with transfer of heat across boundaries.
DIAC Data-Interpretation Analysis Center (US).
diagnostic routing equipment Automatic or semi-automatic fault-isolating tester with ability progressively to narrow down location of fault.
diagonal-flow compressor One in which air flows diagonally to plane of rotation, centrifugal with axial component.
Dials Digital integrated automatic landing system.
dial your weight Small computer on whose keyboard is manually inserted all fuel, crew, payload and other on-board items, displaying MTOW and c.g. position (colloq).
diamagnetic Reacting negatively to magnetic field, developing magnetic moment opposed to it, with permeability less than 1; includes aluminium, non-ferrous alloys and corrosion and heat-resistant steels.
diameter 1 That of any circular arcs making up fuselage external cross-section.
2 In optics, unit of linear measurement of magnifying power.
3 Of parachute, that while fully spread out on flat surface.
diametral pitch Ratio of number of teeth on gearwheel divided by pitch diameter.
Diamond C Highest proficiency award for which sailplane pilots can qualify.
diamond landing gear Tandem centreline mainwheels, and outriggers.
diamonds See *shock diamonds*.
diaphragm Fabric partition within aerostat; may be gastight (ballonet *) or non-gastight (stabilizer *).
diathermy Generation of heat by HF power, usually at 0·5/1·5 MHz.
DIB De-icer boot.
dibber Weapon intended to penetrate concrete runway before exploding.
Dicass Directional command-activated sonobuoy system.
dice Semiconductor chips or IC after scribing and separation.
dichroic mirror One coated with molecular-thickness layer of reflector, usually metal, so as to transmit some EM wavelengths (esp visible colours) and reflect others.
diddler CRT auxiliary electrostatic plates which can collapse elongated blips to sharp spots.
die 1 Press tool, often in mating male/female halves, which cuts sheet or imparts three-dimensional shape to workpiece.
2 Shaped tools used in *-casting.
3 Shaped tools used in *-forging.
4 Shaped female mould used in explosive or magnetic forming.
5 Shaped male tool used in ultrasonic, ECM and related mechanical, chemical or electrochemical shaping.
6 Tool with shaped aperture used in extrusion.
dielectric Substance capable of supporting electric stress, sustaining electric field and undergoing electric polarization; includes all insulators and vacuum.
dielectric heating Generated in dielectric subjected to HF field, resulting from molecular friction

due to successive reversals of polarization; power dissipated is dielectric loss.

dielectric strength Measure of resistance of dielectric to electrical breakdown under intense electric field; SI unit is Vm^{-1}; also known as breakdown potential.

dielectric tape camera TV recording camera (Vidicon) giving output on tape in form of varying electric field.

DIELI Direction des Industries Electroniques et de l'Informatique (F).

diergolic Non-hypergolic, thus requiring an igniter system.

diesel IC engine utilizing heat of compression to ignite fuel oil injected in highly atomized state direct into cylinder, with piston nearly at TDC.

dieseling Any spontaneous ignition of combustible gaseous mixture due solely to temperature caused by compression.

diesel ramjet Ramjet operating at Mach number high enough for fuel to ignite by heat of air compression.

Difar Directional acoustic frequency analysis and recording (ASW).

differential ailerons Ailerons interconnected so that upgoing aileron travels through larger angle than downgoing. This increases drag of wing with upgoing aileron and minimizes extra drag of other wing.

differential ballistic wind In bombing, hypothetical wind equal to difference in velocity between ballistic and actual winds at release altitude.

differential controls Control surfaces on opposite sides of body or fuselage which move in opposition to cause or arrest roll.

differential fare Difference in airline fare levels usually reflecting time-saving and passenger appeal of new aircraft.

differential laser gyro Two lasers of opposite polarization operate in same cavity; comparison of outputs gives twice angular-measure sensitivity of normal laser gyro.

differential pressure Pressure difference between two systems or volumes (abb dP). That of fuselage or cabin is maximum design figure for pressurization system, at which point spill valves open.

differential spoilers Wing spoilers used as primary or secondary roll control.

differential tailplane See *taileron*.

differential tracker Radar that can simultaneously measure angular separation of target and friendly missile, so that guidance system can reduce this value to zero.

differentiating circuit Circuit delivering output voltage in approximate proportion to rate of change of input voltage or current.

diffraction Phenomena which occur when EM wave train, such as beam of light, is interrupted by opaque object(s). Rays passing through narrow slit, or a grating made of slits, are bent slightly as they pass edges; thus waves can "bend" around obstacle.

diffraction-optics HUD Uses a precise 3-D array of microminiature grids, or light apertures, to create a volume hologram which makes possible a wide-angle HUD suitable for all-weather low-level navigation and weapon-aiming.

diffuser Expanding profiled duct or chamber, sometimes with internal guide vanes, that decreases subsonic velocity of fluid, such as air, and increases its pressure, downstream of compressor or supercharger, upstream of afterburner, and in some wind tunnels.

diffuser area ratio Ratio of outlet to inlet cross-section area of diffuser, esp of ramjet.

diffuser, compressor Ring of fixed vanes or expansion passages in compressor delivery of gas turbine to assist in converting velocity of air into pressure.

diffuser efficiency Ratio of total energy at exit to entry or achieved/theoretical pressure rise.

diffuser tunnel Wind tunnel containing section in which velocity is converted into pressure.

diffuser vanes Guide vanes inside diffuser that assist in converting velocity into pressure.

diffusion 1 In atmosphere or gaseous system, exchange of molecules across border between two or more concentrations so that adjacent layers tend towards uniformity of composition.
2 Of stress, variation along length of structure of transverse distribution of stress due to axial loads.
3 In materials, movement of atoms of one material into crystal lattice of adjoining material.
4 In ion engines, migration of neutral atoms through porous structure prior to ionization at emitting surface.
5 Of light, scattering of rays either when reflected from rough surface or during transmission through translucent medium.
6 In electronic circuitry, method of making p-n junction in which n- or p-type semiconductor is placed in gaseous atmosphere containing donor or acceptor impurity.
7 Of uranium, repeated gaseous-phase concentration of fissile U-235.

diffusion bonding Use of diffusion (3) to join solids with high surface finish in uncontaminated intimate contact.

diffusion coefficient Absolute value of ratio of molecular flux per unit area to concentration gradient of gas diffusing through gaseous or porous medium, evaluated perpendicular to gradient.

DIFMR Digital instantaneous frequency-measurement receiver.

DIFOT Duty involving flight operations and training.

DIG 1 Display/indicator group.
2 Directional gyro (usually DG).

digibus Any digital multiplex data highway.
digicon Diode array giving a light input to electrical signals which are then amplified and analysed.
Digilin Digital plus linear functions on one chip (trade name).
Digitac Digital tactical aircraft control (USAF).
digital Operating on discrete numbers, bits or other individual parcels of data.
digital/analog converter Device which converts analog inputs (eg varying voltages) into digits. Also known as **digitizer**.
DIH Department of Information Handling (ESOC).
dihedral Acute angle between left (port) and right (starboard) mainplanes or tailplanes (measured along axis of centroids) and lateral axis.
dihedral effect Sideslip effect in variable-sweep aircraft that causes change in rolling moment; too much augments roll response while too little (adverse sideslip) opposes it.
DIL Digital integrated logic.
Dilag Differential laser gyro.
DILS Doppler ILS.
dimensional similarity Of physical quantities, made up of same selection of fundamental M (mass), L (length) and T (time) raised to same indices.
DIMM Dual-part integrated memory monitor.
dimpled tyre Landing-wheel tyre whose contact surface is covered with small recesses, mainly to provide visual index of wear.
dimpling Countersinking thin sheet metal by tool which dimples (recesses) without cutting, so that rivethead is flush with surface.
DIN 1 Deutsches Institut für Normung eV (G equivalent of BSI, NBS).
2 Digital inertial navigation.
DINAS, Dinas Digital inertial nav/attack system.
DINFIA Direccion Nacional de Fabricaciones e Investigaciones Aeronauticas (Arg).
dinghy Life-raft, usually of inflatable rubberized fabric, for use by crew and passengers after aircraft has ditched.
dinking Use of thin blade-like shaped die(s) to cut soft sheet materials such as leather, cloth, rubber or felt, and to cut lightening holes in thin sheet-metal; inexpensive die is used, and cutting action is by steady pressure or hand hammer.
DINS Digital inertial navigation system.
diode Two-electrode thermionic valve containing cathode and anode, or semiconductor device having unidirectional conductivity.
diode lamp Semiconductor diode which, when subject to applied voltage, emits visible light. Smaller than most switchable light sources. Also known as light-emitting diode (LED).
DIP 1 Digital image processing.
2 Defense industrial plant (US); EC adds "equipment center".
dip 1 Angle between magnetic compass needle perfectly poised or on horizontal axis and local horizontal plane. Also known as magnetic inclination.
2 Vertical angle at eye of observer between astronomical horizon and apparent line of sight to visible horizon.
3 Angle between local horizontal and lines of force of terrestrial magnetic field.
4 Salutation by briefly rolling aircraft towards observer; to * wing in salute.
diplexer Device permitting antenna (aerial) system to be used simultaneously or separately by two transmitters.
diplomatic authorization Authority for overflight or landing obtained at government level.
dipole 1 System composed of two separated and equal electric or magnetic charges of opposite sign.
2 Antenna (aerial) composed of two conductors in line, fed at mid-point. Total length equal to one half wavelength.
dipper See *fuel dipper*.
dipping sonobuoy One designed to be suspended but not released from helicopter and immersed in selected places in sea.
DIPS Dipole inches per second (chaff dispenser).
dipsydoodle Official term for rollercoaster manoeuvre performed by SR-71 and some other supercruise aircraft following inflight refuelling, comprising dive to supersonic speed followed by accelerating climb back to operating height.
DIQAP Defence Industries Quality-Assurance Panel (UK).
Dircen Direction des Centres d'Expérimentations Nucléaires (F).
direct-action fuze See *DA fuze*.
direct approach Unflared landing.
direct coupling Association of two circuits by having an inductor, condenser or resistor common to both.
direct-cranking starter Hand crank geared to crankshaft to start engine.
direct current Electric current constant in direction and magnitude.
direct damage assessment Examination of actual strike area by air or ground observation or air photography.
directed-energy weapon One whose effect is produced by a high-power beam, normally of EM radiation, having essentially instantaneous effect at a distance.
directed slipstream Means of achieving STOL in which slipstream created by propellers or fans is blown over entire wing. Also known as deflected slipstream.
direct flight 1 Portions of flight not flown on radials or courses of established airways.
2 Point-to-point space flight, without rendezvous, docking or other manoeuvre.
direct force control Control of aeroplane trajec-

tory by application of force normal to flightpath without prior need to rotate to different attitude; eg lateral force by combined rudder and chin fin, vertical by tailerons/flaps or vectored thrust.

direct frontal Air-combat tactic for double attack in which one interceptor closes head-on on each side of enemy force.

directional aerial, antenna Aerial which radiates or receives more efficiently in one direction than in others.

directional beacon Transmitter emitting coded signals automatically to enable aircraft to determine their bearing from the beacon with a communications receiver.

directional gyro Free-gyro instrument for indicating azimuth direction.

directional instability Tendency to depart from straight flight by a combination of sideslipping and yawing.

directional marker Ground marker indicating true north and direction and names of nearest towns.

directional solidification Casting metal alloys in such a way that all transverse grain boundaries are eliminated, leaving long columnar crystals aligned with direction of principal stress.

directional stability Tendency of an aircraft to return at once to its original direction of flight from a yawing or sideslipping condition; also known as weathercock stability.

direction-finder Automatic or manually operated airborne receiver designed to indicate bearing of continuous-wave ground radio beacon (see *ADF*).

direction indicator See *directional gyro*.

direct lift control Use of aerodynamic surfaces, esp symmetric spoilers, to provide instantaneous control of rate of descent without need to rotate aircraft in pitch.

direct operating cost Costs of operating transport aircraft, usually expressed in pence or cents per seat-mile, per US ton-mile or per mile, and including crew costs, fuel and oil, insurance, maintenance and depreciation. Excluding indirect expenses, such as station costs or advertising; usually taken as 100 per cent of direct costs.

director 1 Aircraft equipped to control RPV or missile.
2 In air traffic control, a radar controller.
3 Fire-control tower in warships.

direct shadow photo Simplest and oldest shadow photography: bright point source of light (in former days, spark) throws shadow of body and shockwaves on to photographic plate.

direct side-force control DFC (2) flight-control system in which aircraft (heavier than air) can be translated sideways without yaw or change of heading by application of direct lateral force.

direct-view storage tube CRT storage tube needing no visor in bright sunshine.

direct-vision panel Flight-deck window or part of window that can be opened.

dirigible Capable of being guided or steered; thus an airship.

DIRP Defense industrial reserve plant (US).

dirty Aircraft configuration in which aerodynamic cleanness is spoilt by extension of drag-producing parts, eg landing gear, flaps, spoilers, airbrakes.

DIS Distributed-intelligence system (MMI).

disc 1 Ring on which one stage of compressor blades is carried.
2 Hub carrying blades of fan or turbine.
3 Circular area swept by lifting rotor.

disc area Of propeller or helicopter rotor, area of circle described by tips of blades.

dischargeable weight 1 All masses which may be jettisoned overboard in emergency.
2 Of airship, total weight that can be consumed or jettisoned and still leave ship in safe condition with specified reserves of fuel, oil, ballast and provisions.

discharge correction factor Of rocket nozzle, ratio of mass flow to that of ideal nozzle which expands identical working fluid from the same initial conditions to same exit pressure.

discing Operation of propeller in ground fine pitch to cause aerodynamic drag.

disc loading Helicopter weight divided by main-rotor disc area.

Discon Defence integrated secure communications network (Australia).

discontinuity 1 Sudden break in the continuity of mathematical variable.
2 In meteorology, zone within which there is rapid change, as between two air masses.

discontinuous fibre Chopped roving as distinct from yarn or tow.

discounting Illegal selling of airline tickets, for affinity group and other promotional fares, at below agreed tariffs.

discrimination Of radar, minimum angular separation at which two targets can be seen separately.

discriminator Stage of FM receiver which converts frequency deviations of input voltage into amplitude variations.

discus Of variable-geometry wing, part-circular portion of upper surface of fixed glove on which swinging portion can slide.

dish Reflector for centimetric radar waves whose surface forms part of paraboloid or sphere.

dishing Pressing regular depressions in thin sheet to increase stability and resistance to bends.

dismounted flight training Hands-off training on ground using hand-held model aircraft, particularly for air-combat tactics.

dispatcher See *despatcher*.

dispensing 1 Release of ECM payloads in controlled manner.
2 Supply of fuel to aircraft via hydrant.

dispensing sequence Graphical or tabular plan for ECM to meet expected threats.

dispersal 1 Geographical spreading out of aircraft, material, establishments or other activities to reduce vulnerability to enemy action.
2 Dispersal area.

dispersal areas Areas usually on airfield to which aircraft and support equipment can be dispersed in wartime.

dispersion 1 Average distance from aim point of bombs dropped under identical conditions or by projectiles fired from same weapon or group of weapons with same firing data.
2 In AAA, scattering of shots about target.
3 In chemical operations, dissemination of agents in liquid or aerosol form from bombs and spray tanks.
4 In rocketry, deviation from prescribed flight path; circular dispersion.
5 Measure of scatter of data points around mean value or around regression curve, usually expressed as standard-deviation estimate or standard error.
6 Process in which EM radiation is separated into its components.
7 Measure of resolving power of spectroscope or spectrograph, usually expressed in Å/mm.

dispersion error Distance from aim point to mean point of impacts.

dispersion hardening Scattering of fine particles of different phase within metallic material, resulting in overall strengthening.

dispersion pattern Distribution of series of rounds fired from one weapon or group of weapons on fixed aim under conditions as nearly identical as possible.

dispersion warhead Discharging bomblets, FAE or other multiple or dispersed payload.

displaced threshold Threshold not at downwind end of full-strength runway pavement.

displacement 1 In air interception, separation between target and interceptor tracks to provide interceptor acquisition space.
2 Distance from standard point (usually origin) measured in given direction.
3 Of IC engine, total volume swept by pistons during 180° crankshaft rotation. Also known as swept volume.
4 Of airship or balloon, mass of air displaced by gas, expressed as weight or volume.

displacement thickness Dimension characteristic of all boundary layers and equal to thickness of completely stagnant fluid having same overall effect. Equal to distance through which each streamline is displaced from position it would have assumed had fluid been inviscid.

$$\delta^* = \int_0^\infty \left(1 - \frac{u}{V}\right) dy \simeq \sqrt{\frac{3\nu l}{V}}$$

where u is local boundary layer velocity, V freestream velocity, y distance from solid surface, ν kinematic viscosity and l characteristic length; actual boundary-layer thickness is nearly three times δ^*.

display Graphic presentation of data for human study.

disposable lift Gross lift less fixed weight of an aerostat.

disposable load Maximum ramp weight minus OEW.

disreef system Timing system for automatically releasing reef of parachute.

disrupter-type spoiler Maximizes local turbulence.

dissimilar air-combat training. Mock air combat with friendly fighters of different type(s) acting part of enemy aircraft, chosen for performance similar to that of enemy types and usually painted to resemble them.

distance Standard airline unit is nm (contrary to SI); up to 1,200 nm airline * calculated as D (great-circle distance) + (7 + 0·015D); above 1,200 nm measure is D + 0·02D.

distance bar Rigid bar linking tow vehicle with aircraft.

distance marker 1 Numbers painted on runway side to indicate thousands of feet to upwind end.
2 Reference marker on radar display; usually one of series of concentric circles. Also known as range marker.

distance-measuring equipment Airborne secondary radar sending out paired pulses (interrogation) received at ground transponder; time for round trip is translated into distance.

distillate 1 Any petroleum product.
2 Fuel oil, eg for diesels.

distortion 1 Undesired change in shape.
2 Undesired change in waveform.
3 In radio or sound reproduction, failure exactly to transmit or reproduce received waveform.
4 Variation of flow velocity or temperature across transverse plane through gas turbine.

distress frequency Internationally 121·5 kHz.

distress signal Signal transmitted by vehicle in imminent danger.

distributed data-processing Distribution of EDP (1) capability among a number of positions in a geographically large system.

distributed load One which has no single point of application but is distributed over a line or area, such as air load on a surface.

distributed mass-balance One distributed along span of control surface.

distributor Rotary switch feeding HT in sequence to spark plugs.

disturbance 1 Upset to normal flight involving uncommanded change in AOA (α), normally quantified as change in $C_L = a\Delta\alpha$.
2 Situation involving unpremeditated loss of control, eg pitch-up or stall/spin.
3 Local departure from normal wind conditions; often used to mean cyclone or depression.

disturbing moment Moment which tends to rotate aircraft about an axis.

DITC Department of Industry, Trade and Commerce (Canada).

ditching Emergency alighting of aircraft on water.

ditching characteristic Way in which aircraft behaves on being ditched, dynamically and structurally.

ditching device Causes RPV to land or crash land when control has been lost.

ditching drill Emergency procedures for aircraft crew and passengers, performed before and after ditching.

dither Signal applied to keep servo motor or valve constantly quivering and unable to stick in null position.

DITS Data information transfer set (centralized control of military aircraft communications).

dits Digital information transport standard.

DIVADS Division air-defense system (USA).

dive Steep descent with or without power.

dive bomber Aircraft designed to release bombs at end of steep dive towards objective.

dive brakes Extensible surfaces designed to enable aircraft to dive steeply at moderate airspeed.

dive-recovery flap Simple plate flap hinged at leading edge on underside of wing at about 30% chord and opened to assist recovery from dive by changing pitching moment, removing local compressibility effects and increasing drag. Common c 1942–50.

divergence 1 Disturbance which increases without oscillation.
2 Expansion or spreading out of vector field; considered to include convergence, or negative divergence.
3 Aeroelastic instability which results when rate of change of aerodynamic forces or couples exceeds rate of change of elastic restoring forces or couples.

divergence, lateral Divergence in roll, yaw or side-slip; tends to a spin or spiral descent with increasing rate of turn.

divergence, longitudinal Non-periodic divergence in plane of symmetry; leads to nose dive or stalling.

divergence speed Lowest EAS at which aeroelastic divergence occurs.

divergent oscillation One whose amplitude increases at accelerating rate.

diversion 1 Change in prescribed route or destination made because of weather or other operational reasons.
2 Traffic diverted or claimed to be diverted from one airline by another, or to non-scheduled, charter or supplemental operators on same route. Frequently called material *.

diversity receiver See *spaced diversity*.

divided landing gear Traditional fixed main gear but with no axles or horizontal member linking wheels.

divided shielding Nuclear radiation shield in two or more separated layers.

division, air 1 Air combat organization normally consisting of two or more wings of similar type units.
2 Tactical unit of naval aircraft squadron, consisting of two or more sections.

DJE Deception jamming equipment.

DK Docked.

DL Delay line.

DLA 1 Delay, delay message; also DEL (ICAO code).
2 Defense Logistics Agency (US).

DLB Data-link buffer.

DLC 1 Direct lift control.
2 Downlink communication (sonobuoy).

DLCRJ Detect, locate, classify, record and jam.

DLF Design load factor.

DLI Deck-launched intercept.

DLL Design limit load.

DLLR Deutsche Liga für Luft- und Raumfahrt.

DLME Direct lift and manoeuvre enhancement.

DLMS Digital landmass simulation (or, DoD, system).

DLP Deck landing practice.

DLS 1 DME landing system.
2 Defect location system.

DLU Data logger unit.

DLW Design landing weight.

DMA 1 Délégation Ministérielle pour l'Armement (F).
2 Direct memory access (or addressing).
3 Defense Mapping Agency (US).
4 Dimethylene amine.

DMAB Defended modular array basing.

DMC Direct maintenance costs.

DMD 1 Deployment manning documents (USAF).
2 Digital message device.

DME Distance-measuring equipment.

DMEA Defect mode and effect analysis.

DMED Digital-message entry device.

DME fix Geographical position determined by reference to navaid which provides distance and azimuth information.

DMEP Data-management and entry panel.

DMES Deployable mobility execution system.

DME separation Spacing of aircraft on airway in terms of distance determined by DME.

DMET DME with respect to time.

DMF 1 Digital matched filter.
2 Département Militaire Fédéral (Switz).
3 Database, menu, function (software).

DMFV Deutscher Modellflieger Verband eV (G).

DMG Deutsche Meteorologische Gesellschaft eV (G).

DMI Department of Manufacturing Industry (Australia).

DMMH/FH Direct maintenance man-hours per flight hour.
DMO 1 Dependent meteorological office.
2 Development Manufacturing Organisation (modifies aircraft as system development vehicles).
DMOR Digest of mandatory occurrence reports.
DMOS 1 Double-diffused MOS.
2 Diffusive mixing of organic solutions (spaceflight).
DMP Display management panel.
DMPI Desired mean point of impact.
DMPP Display and multi-purpose processor.
DMR Delayed multipath replica.
DMS 1 Defensive management subsystem.
2 Defense Mapping School.
3 Data-management system.
4 Data multiplexer subunit.
5 Domestic military sales.
DMSP Defense Meteorological Satellite Program (DoD).
DMU 1 Distance-measurement unit.
2 Data-management unit.
DMZ De-militarized zone.
DNA 1 Defense Nuclear Agency (US).
2 Do not approach area.
3 Direction de la Navigation Aérienne (SGAC, F).
DNAC Direction Nationale de l'Aviation Civile (Mali).
DNC Direct (or digital) numerical control (NC machining).
DND Department of National Defence (Canada).
DNI Director of National Intelligence.
DNIF Duty not involving flying.
DNTAC Dirección Nacional de Transporte Aéreo Civile (Arg).
DNW Deutsch-Niederländischer Windkanal (G/Neth).
DO 1 Drawing office.
2 Drop-out (mask).
D_0 Wing profile drag.
DOA 1 Delegation option authorization; FAA document authorizing company to do its own aircraft type certification.
2 Defence Operational Analysis Establishment (UK).
3 Direction of arrival (ECM).
4 Dominant obstacle allowance.
DOB Dispersed operating base.
DOC 1 Direct operating cost.
2 Delayed-opening chaff.
3 Designed operational capability.
dock Structure surrounding whole or portion of aircraft undergoing maintenance, to provide easy access for ground crew to reach all necessary parts.
docking 1 Mechanical linking of two spacecraft or payloads.
2 Process of manoeuvring airship into its shed after landing.
documentation 1 In EDP (1), formal standardized recording of detailed objectives, policies and procedures.
2 Any hard-copy media, such as aircraft servicing manual, illustrated parts catalogue, repair manual, servicing diagram manual, cross-servicing guide, aircrew or flight manual (usually three volumes), flight reference cards, equipment manuals, servicing cards, and possibly checklists.
DoD Department of Defense (US).
Dodac DoD ammunition code.
DoE 1 Department of the Environment (UK).
2 Department of Energy (US).
DOF Degree(s) of freedom.
dog 1 Bad individual example of particular aircraft type.
2 Aircraft type all examples of which exhibit bad flying qualities (both meanings colloq).
dogbone Bone-shaped tie linking rigid-rotor blade to hub.
dogfight Air-to-air combat at close visual range.
doghouse Fairing for instrumentation, esp on rocket (colloq).
dogleg 1 Track over several waypoints away from direct route.
2 Directional turn in space launch trajectory to improve orbit inclination.
dogtooth Discontinuity at inboard end of leading-edge chordwise extension, generating strong vortex.
DOI Descent (or docking) orbit insertion.
DoI Department of Industry (UK, now DTI).
DOL Dispersed operating location.
DOLE Detection of laser emitters.
doll's eye Cockpit magnetic indicator which when triggered clicks to a white warning aspect.
dolly 1 Airborne data-link equipment.
2 Metal back-up block used in hand riveting or hammering out dents in sheet.
3 Pneumatic-tyred truck tailored to elevate and grasp engine, radar or other item and transport it on ground.
dolly roll Dolly with payload carried on two roller-supported rings for rotation to any desired angle.
DOLRM Detection of laser, radar and millimetre (millimetric) waves.
domestic Involving one's own country only.
domestic reserves Fuel reserves for scheduled domestic flight.
domestic service Airline service within one country.
door bundle Para-dropped load immediately preceding stick of parachutists.
door-hinge rotor Articulated blades on flapping hinges visually similar to door hinge.
DoPAA Description of proposed actions and alternatives.

dope 1 Liquid applied to fabric to tauten it by shrinking, strengthen it and render it airtight by acting as filler. Usually compounded from nitrocellulose or cellulose acetate base, and soluble in thinners.
2 Ingredient added to fuel in small quantities to prevent premature detonation (colloq).

doping 1 Treatment of fabric with dope.
2 Addition of impurities to semiconductor to achieve desired electronic characteristics.
3 To prime PE (4) with spray of neat fuel prior to starting from cold.

doploc Doppler phase lock; active tracking system which determines satellite orbit by measuring Doppler shift in radio signals transmitted by satellite.

Doppler beam sharpening As aircraft radar aerial points anywhere other than dead-ahead, computer breaks each reading into small pieces and reassembles them as high-resolution map using Doppler correction to eliminate background clutter.

Doppler correction Numerical correction to observed frequency or wavelength to eliminate effect of relative velocity of source and observer (eg removal of sea wave velocity from Doppler G/S[2]).

Doppler effect Increase or decrease in frequency of wave motion, such as EM radiation or sound, sensed by observer or receiver having relative speed with respect to source (train whistle seems to drop in pitch as it passes stationary observer at high speed). Also known as Doppler shift.

Doppler error In Doppler radar, error in measurement of target radial velocities due to atmospheric refraction.

Doppler hover VTOL, esp helicopter, hover controlled over desired geographical spot by Doppler coupler to AFCS.

Doppler navigation Dead reckoning by airborne navaid which gives continuous indication of position by integrating along-track and across-track velocities derived from measurement of Doppler effect of radar signals sent out (usually in four diagonal directions) and reflected from ground.

Doppler radar Radar which measures Doppler shift to distinguish between fixed and moving targets, or serve as airborne navaid by outputting G/S (2) and track.

Doppler ranging (Doran) CW trajectory-measuring system which uses Doppler effect to measure velocities between transmitter, vehicle transponder and several receiving stations; obviates necessity of continuous recording by making simultaneous measurements with four different frequencies.

Doppler shift 1 See *Doppler effect*.
2 Magnitude of Doppler effect measured in Hz or (astronomical) in terms of visible-light spectrum.

Doppler velocity and position (Dovap) CW trajectory-measuring using Doppler effect; ground transmitter interrogates a frequency-doubling transponder and output is received at three or more sites for comparison with interrogation frequency, intersection of ellipsoids formed by transmitter and each receiver providing spatial position.

DOR Directorate of Research (previously of Operational Research), under Chief Scientist (CAA, UK).

Doran See *Doppler ranging*.

dorsal Pertaining to the back, interpreted as upper surface of vehicle body.

dorsal fin Vertical surface on upper centreline sloping gradually upwards to blend with main fin.

dorsal spine Ridge running along top of fuselage from cockpit to fin for aerodynamic or system-access purposes.

DoS 1 Department of Supply (Australia).
2 Department of Space (India).

Dosaaf Voluntary society for support of Army, Air Force and Navy (USSR).

Dosav Voluntary society for assisting Air Force (USSR).

DOSC Direct oil-spray cooled.

dose rate Incident rate of ionizing radiation, measured in röntgens or mrem per hour.

dose rate contour line Line joining all points at which dose rates at given time are equal.

dosimeter 1 Instrument for measuring ultraviolet in solar and sky radiation.
2 Device worn by persons which indicates dose to which they have been exposed (each Apollo astronaut wore four passive * and carried a fifth personal-radiation * in sleeve pocket).

DoT 1 Department of Transport (Canada, etc).
2 Department of Transportation (US).
3 Designating optical tracker.
4 Day of training.

dot Electronic dot displayed on CRT for cursive writing, providing steering guidance or other information.

DoTI Department of Trade and Industry (UK, DTI is preferred).

Dotram Domain-tip random-access memory.

double attack Co-ordinated air/air operation by two partners making repeated synchronized yo-yos (lo and hi-speed) and BRA (2)s as an effective ACM system.

double-base propellant Solid rocket propellant using two unstable compounds, such as nitrocellulose and nitroglycerine, which do not require a separate oxidizer.

double blank Parachute with two gores removed.

double-bubble Fuselage cross-section consisting of two intersecting arcs with floor forming common chord.

double curvature Curvature in more than one plane; also known as compound curvature.

double delta Delta wing with sharply swept leading edge inboard changing at about mid-span to less sharply swept outer section.

double designation Nomination by a national government of two of that country's airlines as national flag-carriers operating scheduled services on same international route.

double drift Method of determining wind velocity by observing drift on three true headings flown in specific pattern.

double engine Power unit containing two engines driving co-axial propellers; usually one half can be shut down for cruising flight.

double-entry compressor Centrifugal or radial-flow compressor that takes in fluid on both sides of impeller.

double-flow engine Usually, bypass turbojet or turbofan.

double-fluxe Bypass or turbofan engine (F).

double-headed Warship having SAM systems both fore and aft of central superstructure.

double-hinged rudder Rudder with additional hinge near mid-chord for maximum effectiveness at low speeds with large camber and total angle.

double modulation Carrier wave of one frequency is first modulated by signal wave and then made to modulate second carrier wave of another frequency.

double-protection honeycomb Honeycomb structure protected by chemical surface conversion process and then varnish dip.

doubler Additional layer of sheet or strip to reinforce structural joint.

double-root blade Rotating blade or vane with root fitting at each end to enable it to be fitted into disc either way.

double-row engine See *twin-row*.

double slot Passive suction system in upper surface of wing of high-subsonic aircraft in which air is continuously extracted immediately downstream of shockwave (to stabilize and weaken it) and discharged immediately upstream of it. Improves C_l and buffet boundary.

double-slotted flap Flap with vane and thus two slots, one between wing and vane, and the second between vane and flap.

doublet In fluid mechanics, source and sink of equal strength whose distance apart is zero.

doublet antenna Antenna (aerial) composed of two similar elements in line but separated and fed in centre, and having total length equal to half wavelength; half a dipole.

double taper Taper of aerofoil which incorporates change in angle at part-span.

double-wedge aerofoil Section suitable for straight supersonic wings and blades of supersonic axial compressors, characterized by sharp wedge-like taper and sharp leading and trailing edges.

doughnut 1 Common shape for plasma contained in toroidal bottle.
2 Low-pressure tyre on small wheel for soft airfields.

DOV 1 Data over voice.
2 Discreet operational vehicle (low-profile, does not appear to be military or para-military), usually non-flying.

Dovap Doppler velocity and position.

DOW Dry operating weight.

Dowgard Aviation ethylene glycol (Dow Chemicals).

down Faulty and unusable.

downburst Local but potentially dangerous high-velocity downward movement of air mass, eg when arrested by sea-breeze front; chief cause of windshear.

downcutting Milling so that teeth enter upper surface of workpiece.

downdraft Bulk downward movement of air such as commonly found on lee side of mountain or caused by descending body of cool air.

downdraft carburettor One in which air is taken in at top and travels downwards; reduced fire hazard, and less risk of foreign-object ingestion.

downed aircraft Aircraft that has made forced landing or ditching, esp through battle damage.

downgrade To reduce security classification of a document or item of classified material.

downlink data Transmissions to Earth from spacecraft giving such information as astronaut respiration or cabin temperature; computers alert flight controllers to any deviation (7).

download 1 Any load acting downwards, eg on wing at negative angle of attack.
2 To remove unexpended ordnance or camera magazines from aircraft after operational sortie or aborted mission.

down-look radar Radar capable of detecting targets close to ground when seen from above.

downrange Away from launch site in direction of target or impact area.

downstage From upper stages downwards (signals, vibration, fluid flow, pressure etc).

downtime Period during which hardware is inoperative following technical failure.

downward identification light White light on underside of aircraft for identification, with manual keying for Morse transmission.

downwash 1 Angle through which fluid stream is deflected down by aerofoil or other lifting body, measured in plane parallel to plane of symmetry close behind trailing edge; directly proportional to lift coefficient.
2 Angle through which fluid stream is deflected by rotor of rotary-wing aircraft, measured parallel to rotor disc.

downwind Direction away from source of wind.

downwind leg Leg of circuit flown downwind, 180° from landing direction.

downwing side Inner side of aeroplane in banked turn.

DP 1 Direction des Poudres (F).
2 Display processor.
3 Differential protection.

D_p Parasite drag.

dP Differential pressure.

DPA Diphenyl amine.

DPAC Direction des Programmes Aéronautiques Civiles (F).

DPC 1 Defence Production Committee (J).
2 Defence Planning Committee (NATO).

DPEM Depot-purchased equipment maintenance.

Dpews Design-to-price electronic warfare suite.

DPFG Data-processing functional group.

DPI Direct petrol injection.

DPL Dewpoint line.

DPLL Digital phase-lock loop.

DPM 1 Digital pulse modulation.
2 Development program manual (US).
3 Digital processing module.

DPMA Data Processing Management Association (US).

DPMAA Direction du Personnel Militaire de l'Armée de l'Air (F).

DPP Deferred-payment plan.

DPS 1 Differential phase shift.
2 Dynamic pressure sensor (brake pedals).
3 Descent propulsion system (of planetary lander or lunar module).

DPSK Differentially coherent pulse shift keying.

DQAB Defence Quality Assurance Board (UK).

DR 1 Dead reckoning.
2 Dispense rate (lb/s) of chaff or aerosol (ECM).

Dr Drift.

dracone Large inflatable fluid (eg fuel, water) container, towable through sea and usable on soft land (strictly, Dracone).

Drads, DRADS Degradation of radar defence systems (active jamming by MIRVs).

drag Retarding force acting upon body in relative motion through fluid, parallel to direction of motion. Sum of all retarding forces acting on body, such as induced *, profile *.

drag area Area of hypothetical surface having absolute drag coefficient of 1·0.

drag axis Straight line through centre of gravity parallel to direction of relative fluid flow.

drag bracing Internal bracing commonly used in fabric-covered wings to resist drag forces; may consist of adjustable wires, rods or tubes between front and rear spars or between compression ribs.

drag chute Parachute streamed from aircraft to reduce landing run, or steepen diving angle. Also called brake parachute, deceleration parachute or drogue parachute.

drag-chute limit Maximum EAS at which drag chute may be deployed.

drag coefficient 1 Non-dimensional coefficient equal to total drag divided by $\frac{1}{2}\rho V^2 S$ where ρ is fluid viscosity, V relative speed and S a representative area of the body, all units being compatible.
2 Coefficient representing drag on given body expressed in pounds on one square foot of area travelling at speed of one mile per hour (arch).

drag curve Curve of lift coefficient plotted against drag coefficient; also known as drag polar.

drag divergence Mach number at which shock formation becomes significant from viewpoints of drag, buffet and control; usually same as drag rise.

drag hinge Approximately vertical hinge at root of helicopter main-rotor blade, allowing limited freedom to pivot to rear in plane of rotation.

drag link Structural tie bracing a body, such as landing gear, against drag forces.

drag manoeuvre Air-combat manoeuvre in which one of a pair draws hostile aircraft into a firing position for his partner; can be used as verb.

drag member Structural component whose purpose is to react drag forces.

drag parachute 1 Braking parachute.
2 See *drogue (3)*.

drag rib See *compression rib*.

drag rise Sudden increase in wing drag on formation of shockwaves.

drag rope Thrown overboard from balloon to act as brake or variable ballast when landing; also called trail rope or guide rope.

drag strut Strut reacting drag forces, esp one incorporated in a wing.

drag-weight ratio Ratio of total drag at burnout to total weight of missile or rocket.

drag wires 1 Wires inside or outside wing(s) to react drag.
2 Wires led forward from car or other nacelle of airship to hull or envelope to react drag.

Drapo Dessin et réalisation d'avions par ordinateur (F).

DRB Defense Resources Board (US).

DRC Defence Review Committee (NATO).

DRD Dry-runway distance.

DRDB Dual-redundant data bus.

DRDF Vhf/uhf Radio D/F system (UK).

DREA Defence Research Establishment Atlantic (Canada).

DRED Ducted-rocket engine development.

Drem lighting Visual landing approach aid with red, amber and green lights in carefully inclined tubes pointing up glidepath (from RAF Drem, Scotland, 1937).

DREO/P/S Defence Research Establishments: Ottawa/Pacific/Suffield (Canada).

dressed Equipped with all externally attached accessories, piping and control systems, esp of an aircraft engine or accessory gearbox.

DRET Direction des Recherches, Etudes et Techniques (F).

DREV Defence Research Establishment, Valcartier (Canada).

DRF 1 Dual-role fighter.
2 Deutsche Rettungsflugwacht eV (G).

DRFM Digital RF memory.

DRI/DRO Dolly roll-in to dolly roll-out; engine-change elapsed time.

drift 1 Lateral component of vehicle motion due to crosswind or to gyroscopic action of spinning projectile.
2 Slow unidirectional error movement of instrument pointer or other marker.
3 Slow unidirectional change in frequency of radio transmitter.
4 Angular deviation of spin axis of gyro away from fixed reference in space.
5 In semiconductors, movement of carriers in electric field.

drift angle Angle between heading (course) and track made good.

drift correction Angular correction to track made good to obtain correct track (for navigation, bombing, survey etc).

drift-down Gradual en route descent from top of flight profile (now rare).

drift error Change in output of instrument over period of time, caused by random wander.

drift indicator See *drift meter*.

drift meter Instrument indicating drift angle; in simple optical form a hair line is rotated until objects on ground travel parallel with it.

drift sight See *drift meter*.

drill 1 Correct procedure to be followed, eg in particular phase of flight such as take-off.
2 Training flight by a formation, including formation changes.

drill card Checklist for correct procedures to be followed in particular phase of flight.

drill round Dummy missile or gun ammunition used for training.

D-ring 1 Ring in shape of capital D to which suspension ropes from balloon or other lighter-than-air craft are attached.
2 Handle for pulling parachute ripcord.

drip flap Strip of fabric secured by one edge to envelope or outer cover of balloon or other lighter-than-air craft to deflect rain from surface below it and prevent it from dripping into basket or car; also helps to keep suspension ropes dry and non-conducting. Sometimes called drip band or drip strip.

dripshield Any tray for collecting fluid under machinery.

drip strip See *drip flap*.

DRIR Direct-readout infra-red.

DRIRU Dry-rotor inertial reference unit.

DRISS Digital read-in sub-system.

Drive Documentation review into video entry.

drive surface In NC machining or GPP, real or imaginary surface that defines direction of cutter travel.

driving band Band of soft metal around projectile fired from rifled gun which deforms into barrel rifling to impart spin.

Drivmatic riveter Patented power-driven riveter which closes aircraft rivets at high speed; can drill, countersink, insert rivet, close and mill head in sequence.

drizzle In international weather code, precipitation from stratus or fog consisting of small water droplets.

DRK Flugdienst Red Cross flying service (G).

DRLMS Digital radar land-mass simulation (for TFR, GM, Nav-weaps, EW).

DRM Ducted rocket motor (USAF).

DRME Direction des Recherches et Moyens d'Essais (F).

DRO Daily routine orders.

drogue 1 Conical funnel at end of in-flight refuelling hose used to draw hose out and stabilize it, and guide probe of receiver.
2 Fabric cone used as windsock, or towed behind aircraft as target for firing practice, or as sea anchor by seaplanes.
3 Conical parachute attached to aircraft, weapon or other body to slow it in flight, to extract larger parachute or cargo from aircraft hold, or for stabilizing the towing mass such as a re-entry body or ejection seat.
4 Part of connector on a spacecraft (eg Apollo lunar module) into which a docking probe fits.

drogue recovery Recovery system for spacecraft in which one or more small drogues are deployed to reduce aerodynamic heating and stabilize vehicle so that large recovery parachutes can be safely deployed.

drone RPV or pre-programmed pilotless aircraft, usually employed as airborne target; either pilotless version of obsolete combat aircraft or smaller aircraft designed as a target.

droop 1 Downward curvature of leading edge of aerofoil to provide increased camber.
2 See *droop leading edge*.
3 Limited downward movement under gravity of door or access panel on underside, sufficient to have measurable effect on total aircraft drag.

droop balk Mechanical interlock prohibiting (a) selection of all engines at take-off power with droops (2) up, or (b) selection of droops up in flight at above a specified angle of attack.

drooping ailerons Ailerons arranged to droop about 15° when flaps are lowered to increase lift while preserving lateral control.

droop leading edge Wing leading edge hinged and rotated down to negative angle relative to wing for high-lift low-speed flight (esp take-off and landing); colloquially called droops.

droop nose Nose designed to hinge down for low-speed flight and landing of slender delta aircraft, and to provide crew forward vision at high angles of attack. Also called droop snoot.

droops Droop leading-edge sections.

droop snoot 1 Droop nose (colloq).
2 Aircraft (esp modification of familiar type)

having extended down-sloping nose.

droop stop Buffer incorporated in helicopter rotor hub to limit downward sag of blades at rest.

drop 1 Dropping of airborne troops, equipment or supplies on specified * zone.
2 Correction used by airborne artillery observer or spotter to indicate desired decrease in range along spotting line.

drop altitude 1 Altitude above MSL at which air drop is executed (DoD, NATO, CENTO).
2 Altitude of aircraft above ground at time of drop (see *drop height*).

drop forging Forcing of metal or other materials in hot and plastic state to flow under pressure of blow(s) from drop hammer into mould or die to form parts of accurate shape.

drop height Vertical distance between drop zone and aircraft (in SEATO, drop altitude).

drop interval Time interval between drops (1).

dropmaster 1 Person qualified to prepare, perform acceptance inspection, load, lash and eject items of cargo for air drop. Also called air despatcher.
2 Air crew member who, during drop, will relay required information between pilot and jumpmaster (USAF).

drop message Written message dropped from aircraft to ground or surface unit (probably arch).

drop-out 1 Discrete variation in signal level during reproduction of recorded data which results in data-reduction error.
2 Of systems, automatic off-line disconnection following fault condition.
3 Oxygen mask which drops out automatically for passengers and crew following sudden loss of cabin pressure.

dropsonde Radiosonde dropped by parachute from high-flying aircraft to transmit atmospheric and weather measures at all heights as it descends. Used over water or other areas devoid of ground stations.

drop tank Auxiliary fuel tank carried externally and designed to be jettisoned in flight.

drop test 1 Test of landing gear by dropping it from a height under various loads, with wheels at landing rpm.
2 Of models of future aircraft, free drop (usually from balloon) to check spinning or other characteristics.

drop-test rig Rig designed to drop-test (1) any structure.

drop zone Specified area upon which airborne troops, equipment and supplies are dropped by parachute (abb DZ).

DRP Design review panel.

DR plot Chart on which successive fixes, winds, positions and courses of an aircraft calculated by dead reckoning are shown.

DR position Position of aircraft calculated by dead reckoning; symbol, dot in small square.

DRS 1 Detection and ranging set (radar, US).
2 Dead-reckoning subsystem.

DRS/CS Digital range safety/command system.

DRTS Detecting, ranging and tracking system (IR + laser).

DRU 1 Data-retrieval unit.
2 Direct-reporting unit.

drum Cylinder or series of discs upon which rotating blades of axial compressor are mounted; rotor minus its blades.

drum-and-disc construction Axial compressor rotor in which some stages of blading are held in single-piece multi-stage drum and remainder in discrete discs.

DRVS Doppler radar (or radial) velocity system.

dry adiabatic lapse rate Rate of decrease of atmospheric temperature with height, approximately 1°C per 100 m (1·8°F per 328 ft). This is close to rate at which ascending body of unsaturated air (clean air) cools due to adiabatic expansion.

dry-bulb thermometer Ordinary thermometer to determine air temperatures, as distinct from wet-bulb type; has glass capillary of uniform bore and bulb partially filled with a fluid (usually mercury) and sealed with vacuum above.

dry contact AAR hook-up between tanker and receiver without transfer of fuel.

dry emplacement Rocket of missile launch emplacement without provision for water cooling during launch.

dry filter One on which filtration is effected by a dry matrix, without oil.

dry fog Haze due to dust or smoke.

dry fuel See *solid propellant*.

dry gyro Rotor has mechanical bearings and is not floated in fluid.

dry hub Rotor hub with elastomeric bearings needing no lubrication.

dry hydrogen bomb One without heavy water or other deuterium compounds.

dry lease Transport-aircraft lease by one airline to (usually) another, without flight or cabin crew or supporting services. Sometimes called bare-hull charter.

dry run Pre-firing operation and actuation of engine (esp liquid-propellant rocket motor) control circuits and mechanical systems without causing propellants to flow or combustion to take place. This enables instrumentation and control circuits to be checked.

dry squeeze Simulated operation of a system or device (from ** of gun trigger to operate combat camera only).

dry sump Engine lubrication system in which oil does not remain in crankcase but is pumped out as fast as it collects and passed to outside tank or reservoir.

dry weight 1 Weight of engine exclusive of fuel, oil and liquid coolant, and including only those

accessories essential to its running.
2 Weight of liquid rocket vehicle without fuel, sometimes including payload.
Drzewiecki theory Treatment of propeller blade as infinite number of chordwise elements; idealized, assumes each blade has no losses and meets undisturbed air.
DS 1 Data sheet.
2 Directionally solidified alloy.
3 Documented sample.
DSA 1 Distributed system architecture.
2 Defense Supply Agency (US).
DSAC Defence Scientific Advisory Council (UK).
DSARC Defense Systems Acquisition Review Council (US).
DSASO Deputy senior air staff officer (RAF).
DSB Defense Science Board (US).
DSB, d.s.b. Double sideband.
DSC 1 Defect-survival capability.
2 Distinguished Service Cross (UK).
3 Digital selective calling.
DSCS Defense Satellite Communications System (US).
DSD Defence Support Division (NATO).
DSEO Defense Systems Evaluation Office (US).
DSES Defense Systems Evaluation Squadron (simulates hostile bombers attempting to penetrate Norad/ADC).
DSF Domestic supply flight (RAF).
DSFC Direct side-force control.
DSIF Deep Space Instrumentation Facility (US).
DSIR Department of Scientific and Industrial Research (NZ, UK).
DSL 1 Digital simulation language.
2 Depressed sightline.
3 Deutsche Schätzstelle für Luftfahrtzeuge eV (G).
DSLC Dynamic-scattering liquid crystal.
DSMAC Digital scene-matching area correlation (or correlator).
DSMC Direct seat-mile cost.
DSM/YOF Distinguished Staffel Member/Ye Olde Fokker.
DSN Deep Space Network.
DSNS Division of Space Nuclear Systems (AEC, US).
DSO 1 Distinguished Service Order (UK).
2 Defense systems operator (US).
3 Defence Sales Organization (UK).
DSP Defense support program (O adds "office") (US).
dsplcd Displaced.
DSR Director of Scientific Research.
DSRTK Actual track between two waypoints.
DSS Department of Space Science (Estec).
DSSP Deep Submergence System Program.
DST Dry specific thrust.
DSTO Department of Defence Science & Technology Organization (Australia).
DSU Direct support unit (USAF).
DT 1 Day television (TADS).
2 Development test.
3 Dual tandem (landing gear).
D_T Pressure drag.
DTA 1 Direction des Transports Aériens, now Service des Transports Aériens (F).
2 Dynamic test article.
DT&E Development, test and evaluation.
DTAT Direction Technique des Armements Terrestres (F).
DTC Design to cost.
DTCA Direction Technique de Constructions Aéronautiques (F).
DTCN Direction Technique de Constructions Navales (F).
DTCS Drone tracking and control system.
DTD 1 Directorate of Technical Development (UK).
2 Data-transfer device.
3 Damage-tolerant design.
DTDMA Distributed (or distributive) time-division multiple access.
DTE Data terminal equipment.
DTED Digital terrain elevation data.
DTEn Direction Technique des Engins (F).
DTG 1 Distance to go.
2 Dynamically tuned gyro.
DTI 1 Department of Trade and Industry (UK).
2 Digital timebase interval.
DTK Desired track.
DTL Diode/transistor logic.
DTLCC Design to life-cycle cost.
DTM pt Decision-to-miss point.
DTMC Direct ton-mile cost.
DTO Detailed test objective.
DTP Design to price.
DTR Damage tolerance rating.
DTRD Two-shaft turbojet (USSR); **DTRDF**, with afterburner.
DTRI Direction des Télécommunications du Réseau International (F).
DTS Dynamic test station.
DTSC Digital Transmissions Standards Committee (UK).
DTU Data-transfer unit.
DTUC 1 Design to unit cost.
2 Data-transfer-unit cartridge.
DTUPC Design to unit production cost.
DTV Daytime (or daylight) TV.
DTWA Dual trailing-wire antenna (AABNCP).
DU Depleted uranium.
dual 1 Duplication of cockpit controls permitting either pilot or co-pilot, instructor or pupil, to fly manually.
2 Pilot instruction in dual aircraft.
dual-axis rate transducer Gyro based on electrically spun sphere of mercury, in which floats cruciform crystal sensor with 98% buoyancy; future rate sensor capable of operating at up to ± 60 g.
dual-capable 1 Equally good at fighter and

attack missions.
2 Can carry both conventional and nuclear bombs.

dual compressor Compressor split into low-pressure and high-pressure parts, each driven by separate turbine; also known as two-spool compressor.

dual ignition PE (4) ignition system employing two wholly duplicate means of igniting mixture.

dual magneto Single magneto with dual HT outputs to duplicate harness and plugs.

dual-mode EPU Emergency power unit plus bleed from main engine.

dual persistence CRT coated with phosphors giving bright display and dim previous indications.

dual-plane separation Space vehicle staging in which interstage(s) fall away separately.

dual propellant Ambiguous, can mean liquid bipropellant, double-base solid, or different propellants in two stages.

dual-purpose Designed or intended to achieve two purposes, eg gun or missile (esp on surface warship) intended to destroy surface or aerial targets.

dual-rotation Two-spool engine in which LP and HP shafts rotate in opposite directions.

dual-thrust motor Solid rocket motor giving two levels of thrust by use of two propellant grains. In single-chamber unit boost grain may be bonded to sustainer grain, with thrust regulated by mechanically changing nozzle throat area or by using different compositions or configurations of grain. In dual-chamber unit, chambers may be in tandem or disposed concentrically.

DUC 1 Dense upper cloud.
2 Distinguished unit citation (US).

duck soup Easy, a pushover (US flying colloq).

duct 1 Passage or tube that confines and conducts fluid.
2 Channel or passage in airframe through which electric cables are run.
3 Portion of atmosphere (see * *propagation*).

ducted cooling System in which cooling air is constrained to flow in duct(s) to or from power plant.

ducted-fan engine Aircraft propulsor incorporating fan or propeller enclosed in duct; more especially jet engine in which enclosed fan is used to ingest ambient air to augment propulsive jet, thus providing greater mass flow. Most forms are better known today as turbofans.

ducted propulsor Multi-bladed fan (usually with 5 to 12 variable-pitch blades) running in propulsive duct integrated with engine and airframe to give efficient and quiet low-speed propulsion. Source of power, PE (4) or gas turbine.

ducted radiator One installed in duct, esp to give propulsive thrust.

ducted rocket See *ram-rocket*.

ductility Property of material which enables it to undergo plastic elongation under stress.

duct propagation Stratum of troposphere bounded above and below by layers having different refractive index which confine and propagate abnormally high proportion of VHF and UHF radiation, giving freak long-distance communications. Can be on surface or at height up to 16,000 ft (5 km); thickness seldom greater than 330 ft (100 m).

Dufaylite Airframe sandwich material with core of light plastics honeycomb.

dumb Of munition, unguided or "iron"; not smart or brilliant.

dumb-bell Manoeuvre in vertical plane resembling *, usually by helicopter, in making repeated low-level passes over same point.

dummy deck Facsimile of aircraft carrier deck marked out on land.

dummy round Projectile, usually air-to-air or air-to-ground missile, fitted with dummy warhead for test or practice firing.

dummy run 1 Simulated firing practice, esp air-to-ground gunnery or dive-bombing approach without release of a bomb. Also called dry run.
2 Trial approach to land, to release or pick up a load, as practice before actually performing operation.

dump 1 Emergency deactivation of whole system.
2 Of computer operation, to destroy accidentally or intentionally stored information, or to transfer all or part of contents of one section into another.
3 To jettison part of aircraft's load for safety or operational reasons, eg * fuel.
4 Of spacecraft, overboard release of waste water, astronaut waste products, etc.
5 Temporary storage area for bombs, ammunition, equipment or supplies.

dump door Quick-acting hatch under hopper or tank for emergency release of chemicals, or under water tank for firefighting.

dump valve 1 Automatic valve which rapidly drains fuel manifold when fuel pressure falls below predetermined value.
2 Large-capacity valve fitted to any fluid system to empty it quickly for emergency or operational reasons.

dunk To lower into water on tether or communication cable.

duo-tone 1 Colour or camouflage scheme with two main hues.
2 Two notes emitted by variometer.

duplex Circuit or channel which permits telegraphic communication in both directions simultaneously.

duplex burner Alternative fuel entries and single exit orifice.

duplexer Device which permits single antenna to be used for both transmitting and receiving (see *diplexer*).

Düppel Chaff (G, WW2).

duralumin Wrought alloy containing 3–4·5 per cent copper, 0·4–1·0 per cent magnesium, up to 0·7 per cent manganese and the rest aluminium. Originally trade name.

Duramold Low-density structural material of thermosetting plastic-bonded wood (Clark, Fairchild, 1936).

duration 1 Maximum time aircraft can remain in air.
2 Time in seconds of operation of rocket engine.

duration model Traditional rubber-powered model aeroplane designed for flight endurance.

Durestos Composite structural material comprising asbestos fibres bonded with adhesive, usually formed by moulding under pressure.

dustbin Gun turret lowered through aircraft floor to provide ventral defence (colloq, arch).

dusting Controlled spreading of insecticide, fertilizer or other chemical by agricultural aircraft.

Dutch roll Lateral oscillation with both rolling and yawing components; fault of early swept-wing aircraft.

DUTE Digital universal test equipment (computer-guided probes check circuit boards).

duty crew Crew detailed to fly specific mission.

duty cycle Ratio of pulse duration time to pulse repetition time.

duty factor 1 In EDP (1), ratio of active time to total time.
2 In carrier composed of pulses that recur at regular intervals, product of pulse duration and PRF.
3 In electronic jamming,

duty ratio In radar, ratio of average to peak pulse power.

duty runway Runway designated for use by aircraft landing or taking off.

DUV Data under voice.

DV 1 Distinguished visitor (US).
2 Direct-vision (window).

DVA Design verification article.

DVFR Defense Visual Flight Rules; procedures for use in an ADIZ (US).

DVI Direct voice input.

DVK Deutsche Versuchsanstalt für Kraftfahrtzeug und Fahrtzeugmotoren (G).

DVL Deutsche Versuchsanstalt für Luft- und Raumfahrt.

DVO Direct-view optics (or optical or option).

DVOR Doppler VOR.

DVS 1 Doppler velocity sensor.
2 Deutscher Verband für Schweisstechnik (G).

DVST Direct-vision storage tube.

DW 1 Double wedge (aerofoil).
2 Drop weight.

D/W Depth-to-width ratio in electron-beam hole drilling.

dwarf sonobuoy Smallest size, same 124 mm diameter as A-size but length only 305 mm (12 in).

DWD Deutscher Wetterdienst (G).

dwell Brief rest period at end(s) of sinusoidal or other oscillatory motion, eg spot scanning CRT tube, bottom of drop-forging stroke, TDC and BDC in piston engine.

dwell mark Caused by skin-mill, routing or similar tool remaining too long over one spot.

DWG Digital word generator.

DWI Installation for generating giant magnetic pulses to explode mines (from cover name Directional Wireless Installation).

DWT Deutsche Gesellschaft für Wehrtechnik eV (G).

DX See *duplex*.

dyn See *dyne*.

dynamic amplifier Audio amplifier whose gain is proportional to average intensity of audio signal.

dynamic aquaplaning Landplane tyres running at high speed over shallow standing water and riding up out of contact with runway.

dynamic balance Rotating body in which all rotating masses are balanced within themselves so that no vibration is produced.

dynamic component See *dynamics*.

dynamic damper Device intended to damp out vibration by setting up forces opposing every motion.

dynamic factor Ratio of load carried by structural part in acceleration flight to corresponding basic load (see *load factor*).

dynamic head See *dynamic pressure*; also called kinetic head.

dynamic heating Heating of the surface of an aircraft by virtue of its motion through the air. Heat is generated because the air at the surface is brought to rest relative to the aircraft, either by direct impact in the stagnation region or by the action of viscosity elsewhere.

dynamic lift Aerodynamic lift due to the movement of the air relative to a lighter-than-air aircraft.

dynamic load A load imposed by dynamic action, as distinct from a static load. Specifically, with respect to aircraft, rockets or spacecraft, a load due to acceleration as imposed by manoeuvring, landing, gusts, firing rockets, etc.

dynamic meteorology The branch of meteorology that treats of the motions of the atmosphere and their relations to other meteorological phenomena.

dynamic model A model of an aircraft (or other object) in which linear dimensions, mass and inertia are so represented as to make the motion of the model correspond to that of the full-scale aircraft.

dynamic particle filter One which separates particulate solids from moving fluid by making them more relative to fluid; most common are centrifugal filter and momentum separation.

dynamic pressure 1 Pressure of a fluid resulting from its motion when brought to rest on a surface, given by $q = \frac{1}{2}\rho V^2$; in incompressible flow,

difference between total pressure and static pressure.
2 Pressure exerted on stagnation point(s) on a body by virtue of its motion through a fluid.

dynamics Main rotating parts of helicopter airframe.

dynamic sampling Concentration by flight recorder on one particular aspect of aircraft behaviour, esp a malfunctioning channel.

dynamic scale Scale of flow about a model relative to a flow about full-scale body; if two flows have same Reynolds number, both are at same **.

dynamic similarity 1 Relationship between model and full-scale body when, by virtue of similarity between dimensions, mass distributions or elastic characteristics, aeroelastic motions are similar.
2 Similarity between fluid flows about a model and full-scale body when both have same Reynolds number.

dynamic soaring Soaring by making use of kinetic energy of air movements.

dynamic stability Characteristic of a body that causes it, when disturbed from steady motion, to damp oscillations set up and gradually return to original state; used esp of helicopters.

dynamic storage In EDP (1), storage in which information is continuously changing position, as in delay line or magnetic drum.

dynamic system See *dynamics*.

dynamic thrust Work done by fan or propeller in imparting forward motion, equal to mass of air handled per second multiplied by ($V_s - V$), where V_s is slipstream velocity and V aircraft speed.

dynamic viscosity Coefficient of viscosity divided by fluid density, $\nu = \frac{\mu}{\rho}$; also called coefficient of kinematic viscosity.

dynamometer Device designed to measure shaft power; torque or transmission * measures torque or torsional deflection of shaft; electric and hydraulic * measure power by electrical or fluid resistance; fan brake * uses air friction, and prony brake * applies torque by mechanical friction.

dynamotor A d.c. generator and electric motor having common set of field poles and two or more armature windings; one armature is wound for low-voltage d.c. and serves as motor; other serves as generator for high-voltage d.c.

dyne CGS unit of force sufficient to accelerate 1 g at 1 cm s^{-2}; = 10^{-5} N or $\frac{1}{981}$g or $2{\cdot}248 \times 10^{-6}$ lb.

dynode Electrode of electron multiplier which emits secondary electrons when bombarded by electrons.

dysbarism Decompression sickness, usually caused by release of nitrogen bubbles into blood at very low pressure.

DZ Drop (or dropping) zone.

Dzus fastener Countersunk screw with slotted shank anchored in removable or hinged panel, which hooks with half-turn into wire anchor on airframe.

Dzus rail Mounting rail for LRUs.

E

E 1 Energy.
2 Electrical field strength.
3 Electromotive force.
4 Endurance (usually safe endurance).
5 Modulus of elasticity (Young's modulus).
6 Effective (performance of machine).
7 US DoD prefix, electronic (early warning) aircraft.
8 ECM training aircraft (UK).
9 Role designation (not prefix) electronic surveillance (US).

e 1 Eccentricity of orbit or ellipse.
2 Strain rate.
3 Emitter.
4 Electron charge.
5 Base of natural logarithms, about 2·71828.
6 Elevator (or slab tailplane).

e, E Expansion ratio of rocket nozzle.

E^3 Energy-efficient engine (NASA).

E-display Target az/el form x/y axes on rectangle.

E-glass Standard high-strength glass used for fibre reinforcement in advanced engineering structures (not in commercial GRP).

E-layer Ionized layer of ionosphere, usually found at 100–120 km, most pronounced in daytime. Also called Kennelly-Heaviside layer, Heaviside layer or E_1 layer; some evidence to indicate higher layer known as E_2.

EA 1 Enemy aircraft.
2 Engine-attributable.

EAA 1 Experimental Aircraft Association (US).
2 East Anglian Aircraft Preservation Society.

EAAS Empire Air Armament School (UK).

EAC 1 Expected approach clearance time.
2 Experimental apparatus container.

EACS Electronic automatic chart system.

EACSO East Africa Common Services Organization.

EADB Elevator-angle deviation bar.

EADI Electronic attitude director (or display) indicator.

EAEM European Airlines Electronics Meeting.

EAM Emergency action message.

EAMDS European Airlines Medical Directors Society.

E&ST Employment and Suitability Test Program (USAF).

EANS Empire Air Navigation School (UK).

EAPS Engine air particle separator.

EAR 1 Electronically agile radar.
2 Espace Aérien Réglementé (= danger, restricted or prohibited areas) (F).

EARB European Airlines Research Bureau.

EARC Extraordinary Administration Radio Conference (UIT).

early turn Turn made ahead of fix in airway to ensure that aircraft does not, because of speed, course change required, wind, etc, fly outside airway boundaries.

early warning Given by radar surveillance to LOS limit.

earnings Profit (US).

EAROM Electrically alterable read-only memory.

EARSEL European Association of Remote Sensing Laboratories.

earth Grounded side of electrical circuit or device.

Earth funnel Funnel-like shape of lines of force above magnetic poles.

Earth inductor compass Compass whose indication depends on current generated in coil revolving in Earth's magnetic field.

Earth/Moon system Regarded as two-planet system whose centre of rotation is on line joining the centres.

Earth pendulum See *Schuler pendulum*.

Earth-rate correction Command rate applied to gyro to compensate for apparent precession caused by Earth's rotation.

earth return Electrically bonded part of aircraft structure used in electrical circuit.

earthshine Illumination of dark part of Moon produced by sunlight reflected from Earth's surface and atmosphere. Also called earthlight.

earth system All metallic parts of aircraft interconnected to form low-resistance network for safe distribution of electric currents and charges.

EAS 1 Equivalent airspeed (see *airspeed*).
2 Espace Aérien Supérieur (upper airspace, UIR).

Eastern Test Range Test range extending from Cape Canaveral across Atlantic into Indian Ocean and Antarctic.

EAT 1 Estimated (or expected) approach time.
2 Electronic angle tracking.

EAU European accounting unit; see *International Accounting Unit*.

EB 1 Essential bus.
2 Electron beam.
3 End-bend.

EBA Engine bleed air.

EBA/H EBA plus hydrazine.

EBF Externally blown flap.

EBIC Electron-bombardment-induced conductivity.

EBM 1 Electron-beam machining (or microanalysis).

2 Electronic battle management.

ebonite Hard, brittle substance composed of sulphur and hard black rubber; possessing high inductive and insulating properties.

EBR Electron-beam recording.

EBSC European Bird Strike Committee.

EBSV Engine-bleed shutoff valve.

EBU European Broadcasting Union.

ebullism Formation of bubbles, esp in liquid rocket propellant or in biological or body fluids, caused by reduced ambient pressure.

EBW Electron-beam welding.

EC 1 Eddy current.
2 Environmental control (system).
3 Escadrille de Chasse (fighter squadron) (F).
4 Elliptic-cubic (wing profiles).
5 Electronic combat.
6 Engine-caused.

E_c Compressive (bearing) strain.

ECAC 1 European Civil Aviation Conference.
2 Electromagnetic-compatibility analysis centre.

ECAM Electronic centralized aircraft monitor (presents all information on two CRTs in FFCC).

ECB Electronic control box (CAA).

eccentricity 1 Deviation from common centre or central point of application of load.
2 Of any conic, ratio of length of radius vector through point on conic to distance of point from directrix.
3 Of ellipse, ratio of distance between centre and focus to semimajor axis. Also called numerical *.
4 Also of ellipse, distance between centre and focus. Also called linear *.

ECCM Electronic counter-countermeasures.

ECD Excusable contract delay (no penalty).

ECDES Electronic combat digital evaluation system (USAF).

ECE Economic Commission for Europe (UN).

ECFS Empire Central Flying School.

ECG Electrochemically assisted grinding.

ECGD Export Credits Guarantee Department (UK).

ECH Electrochemically assisted honing.

echelon 1 Aircraft formation in which each is above, behind, and to left or right of predecessor; such formation is said to be in * to port or starboard.
2 Subdivision of headquarters, forward or rear.
3 Level of command.
4 Servicing unit detailed to provide ground support and maintenance facilities.

echo 1 Pulse of reflected RF energy.
2 Appearance on radar display of such energy returned from target; also called blip.

ECIF Electronic Components Industry Federation (UK).

ECIM Electronics computer-integrated manufacturing, ie, CIM of electronics.

ECL 1 Emitter coupled logic.
2 Electro-generated chemiluminescence.
3 Engine-condition lever (CAA).

ecliptic Apparent path of Sun among stars because of Earth's annual revolution; intersection of plane of Earth's orbit with celestial sphere, inclined at about 23° 27′ to celestial equator.

ECLSS Environmental control and life-support subsystems.

ECM 1 Electronic countermeasures.
2 Electrochemical machining.
3 Engine-condition monitoring.

ECMJ Escadrille de chasse multiplace de jour (multi-seat day fighter squadron) (F).

ECMO ECM officer (aircrew).

ECMT European Conference of Ministers of Transport.

ECN Escadrille de chasse de nuit (night fighter squadron) (F).

ECO 1 Electron-coupled oscillator.
2 Engineering change order.

ECOM 1 Earth centre of mass.
2 Electronics Command (USA).

economical cruise mixture PE (4) mixture with which AMPG is maximum.

economizer Reservoir in continuous-flow oxygen system in which oxygen exhaled by user is collected for recirculation.

economizer valve Assists in regulating fuel flow through PE (4) carburettor, opened by increased airflow.

economy Originally a passenger fare cheaper than first class, with less luxurious standards of cabin service, meals, seat pitch etc. IATA airlines introduced * class over North Atlantic in April 1958.

ECP 1 Engineering change proposal, for introducing modification.
2 Etablissement Cinématographique et Photographique des Armées (F).
3 Effective candlepower (non-SI).

ECPNL Equivalent Continuous Perceived Noise Level (see *noise*).

ECPP Effective critical parts plan.

ECR 1 Electronic combat and reconnaissance.
2 Embedded computer resources.

ECS Environmental control system; provides aircrew or astronauts with controlled environment and life-sustaining atmosphere.

ECSVR Engine-caused shop visit rate.

ECU 1 Engine change unit; complete power plant, often with cowl.
2 Engine control unit.
3 Environmental control unit.
4 Exercise control unit (a military formation).
5 Electronic control unit (CAA).

ECWL Effective combat wing loading.

ED 1 Emergency distress signal.
2 Engineering development (part of progress schedule).
3 End of descent (Lockheed uses "EoD").

4 Explosive device.
EDA Effective disc area (helicopter).
EDAU Engine data-acquisition unit.
EDB Extruded double-base.
EDC European Defence Community.
EDD Electronic data display (ATC flight data, tabular callsigns, heights, tracks and position information).
eddy 1 Local random fluid circulation drawing energy from flow of much larger scale and brought about by pressure irregularities, eg from passage of unstreamlined body.
2 In meteorology, developed vortex constituting local irregularity in wind producing gusts and lulls.
eddy current Generated in conductor by varying magnetic field; to reduce ** cores are built up of insulated laminations, iron dust or magnetic ferrite.
eddy damping Automatic damping by eddy currents generated by moving conductor.
EDG Electrical-discharge grinding.
edge alignment Distance, parallel to chord of propeller section, from centreline of blade to leading edge at any station.
edge flare Rim of abnormal brightness around edge of video picture.
edge-type elevator Deck-edge elevator; on side of aircraft carrier.
EDIP European Defence Improvement Programme (NATO).
EDL 1 Engage/disengage logic.
2 Electrical-discharge laser.
EDM Electrical-discharge machining.
EDP 1 Electronic data-processing.
2 Experimental data-processor (Eurocontrol).
3 Engine-driven pump.
EDS European distribution system (A adds "aircraft") (USAF).
EDT Eastern Daylight Time (US).
Edwards AFB At Muroc, California; USAF Flight Test Center and NASA Dryden Flight Research Center.
E/E 1 Emergency equipment.
2 Electrical/electronic.
EEA Electronic Engineering Association (UK).
EEC 1 European Economic Community.
2 Engine electronic (or electronic engine) control.
3 Extendable exit cone.
EECS Electrical/electronics cooling system.
EED Electro-explosive device.
EEE Energy-efficient engine (also E^3).
EEI 1 Electrical engineering instruction.
2 Essential elements of information (reconnaissance request).
EEMAC Electrical & Electronic Manufacturers Association of Canada.
EEP Experimental electronics package.
EERM Etablissement d'Etudes et de Recherches Météorologiques (F).
EES Electrical engineering squadron (RAF).
EET Estimated elapsed time.
EEW Equipped empty weight.
EEZ Exclusive economic zone (coastal patrol).
EFA European Fighter Aircraft.
EFAS En route flight advisory service.
EFATO Engine failure after take-off.
EFC 1 Expected further clearance time.
2 Elevator-feel computer.
EFCS Electrical (ie FBW) flight control system.
efctv Effective.
EFCU Electronic flight control unit.
EFDAS Electronic flight data system (CAA).
EFDC Early-failure detection centre.
EFDPMA Educational Foundation of DPMA (US).
EFEO European Flight Engineers Organization (was EFFE).
EFFE European Federation of Flight Engineers.
effective angle of attack Angle at which aerofoil produces given lift coefficient in two-dimensional flow. Also called angle of attack for infinite aspect ratio.
effective angle of incidence See *effective angle of attack*.
effective aspect ratio That of aerofoil of elliptical planform that, for same lift coefficient, has same induced-drag coefficient as aerofoil, or combination of aerofoils, in question.
effective atmosphere That part of planetary atmosphere which measurably influences particular process of motion. For an Earth satellite limit is 120 miles, 193 km (see *mechanical border*, *sensible atmosphere*).
effective current Difference between impressed current and counter-current.
effective exhaust velocity Velocity of rocket jet after effects of friction, heat transfer, non-axially directed flow, etc.
effective helix angle Angle of helix described by point on propeller blade in flight through still air measured relative to Earth.
effective horsepower Power delivered to propeller.
effective pitch Distance aircraft advances along flightpath for one revolution of propeller.
effective profile drag Difference between total wing drag and induced drag of wing with same aspect ratio but elliptically loaded.
effective propeller thrust Net propulsive force; propeller thrust minus increase in drag due to slipstream.
effective range Maximum distance at which weapon may be expected to strike target.
effective span Span minus correction for tip losses; usually defined as horizontal distance between tip chords.
effective terrestrial radiation Amount by which IR radiation from Earth exceeds counter-radiation from atmosphere. Also called effective radiation or nocturnal radiation.

effective wavelength That corresponding to effective propagation velocity.
effector Any device used to manoeuvre a vehicle in flight (rare).
efficiency Ratio of output to input, usually expressed in percentage form.
efficiency of catch Proportion of total water droplets in path of aircraft which actually strike it.
effusion Flow of gas through holes sufficiently large for velocity to be approximately proportional to square root of pressure difference.
EFH 1 Earth far horizon.
2 Engine flight hours.
EFI Electronic flight instrument(ation) (S adds "system"), primary flight and nav info on 8 × 8 colour CRTs.
EFL 1 Emitter function logic.
2 External finance limits.
EFTS 1 Elementary flying training school (UK).
2 Electronic funds transfer systems.
EGA Exhaust-gas analyser.
EGADS Electronic ground automatic destruct sequencer button.
eggbeater Intermeshing-rotor helicopter (colloq).
Eglin AFB Situated in Florida, largest AFB and home of USAF Systems Command Air Proving Ground Center.
EGR Engine ground run.
egress 1 Procedure for getting out of spacecraft in orbit or after planetary or lunar landing, whether for working in space or any other reason. Begins with putting on spacesuits, and includes depressurizing and opening hatch.
2 Departure of combat aircraft from target area.
egress handle Handle which fires ejection seat.
EGS Elementary gliding school.
EGSE Electrical ground support equipment.
EGT Exhaust-gas temperature.
EGTP External ground test program.
EGW Ethylene glycol and water.
E_h Total energy at given speed and height.
EHA European Helicopter Association (Int).
EHDD Electronic head-down display.
EHF Extremely high frequency.
EHL Environmental health laboratory (USAF).
EHM Engine health monitoring (or monitor).
ehp, e.h.p. Equivalent horsepower. Usually total equivalent shaft horsepower.
EHR Engine history recorder.
EHSI Electronic horizontal situation indicator.
EHT Electrothermal hydrazine thruster.
eht Extra high tension (volts).
EI 1 Earth (atmosphere) interface.
2 Entry interface.
EIA 1 Electronic Industries Association (US).
2 Environmental impact assessment.
EICAS, Eicas Engine indication (and) crew-alerting system.
EICMS Engine in-flight condition-monitoring system.
Eiffel-type tunnel Open-jet, non-return-flow wind tunnel in which whole working section is open.
eigen values Discrete values of undetermined parameter involved in coefficient of differential equation, such that solution, with associated boundary conditions, exists only for these values; also called characteristic values or principal values.
eight Flight manoeuvre in which aircraft flying horizontally follows track like large figure eight (see *Cuban* *, *lazy* *).
eight-ball Artificial horizon or attitude indicator (colloq).
eight-point roll Roll executed in eight stages, with aircraft held momentarily after each roll increment of 45°.
eight pylon Manoeuvre used in air racing in which aircraft is flown around pylons so that wingtip appears to pivot on pylon.
EIRA Ente Italiano Rilievi Aerofotogrammetrici.
EIRP Effective isotropically radiated power.
EIS 1 Entry into service.
2 Environmental impact statement.
3 Ejection initiation subsystem.
4 Electronic instrument(ation) system.
EISW Equivalent isolated single-wheel load (LCN).
ejectable Able to be ejected from aircraft, esp capsule, crew seat, sonobuoy, dropsonde or flight recorder.
ejection Escape from aircraft by ejection seat.
ejection angle Angle at which ejection seat leaves, measured relative to aircraft.
ejection capsule 1 Detachable compartment serving as cockpit or cabin, which may be ejected as unit and parachuted to ground.
2 Box containing recording instruments or data ejected and recovered by parachute or other device.
ejection chute Parachute(s) used to decelerate ejection seat or capsule; often ballute or drogue.
ejection seat Seat capable of being ejected in emergency to carry occupant clear of aircraft.
ejector Device comprising nozzle, mixing tube and diffuser, utilizing kinetic energy of fluid stream to pump another fluid from low-pressure region.
ejector exhaust PE (4) pipe(s) disposed or shaped to produce forward thrust, not necessarily incorporating an ejector.
ejector ramjet See *ram-rocket*.
ejector seat See *ejection seat*.
EJS Enhanced JTIDS.
Ekman layer Transition between surface boundary layer and free atmosphere.
ekW Equivalent shaft power of turboprop.
EL 1 Electroluminescent.

2 Ejector (augmented) lift.
3 Emitter locator (or location).

elastance Inability to hold electrostatic charge.

elastic axis Spanwise line along cantilever wing along which load will produce bending but not torsion.

elastic centre 1 Point within wing section at which application of concentrated load will cause wing to deflect without rotation.
2 Point within wing section about which rotation will occur when wing is subjected to twist.

elastic collision Collision between two particles in which no change occurs in their internal energy or in sum of their kinetic energies.

elastic instability Condition in which compression member will fail in bending before failing compressive strength of material is reached.

elasticity Property of material which enables a body deformed by stress to regain original dimensions when stress is removed.

elasticizer Elastic substance or fuel used in solid rocket propellant to prevent cracking of grain and bind it to case.

elastic limit Maximum stress withstood by material without causing permanent set or deformation. Hooke's Law asserts that within * * ratio of stress to strain is constant.

elastic model Linear dimensions, mass distribution and stiffness are so represented that aeroelastic behaviour of model can be correlated with that of full-scale aircraft.

elastic stability Able to bear compressive yield stress of materials without buckling.

elastic stop nut Nut in which self-locking is ensured by ring of fibre in which threads are formed as nut is screwed down.

elastivity See *specific elastance.*

elastomers Rubber-like compounds used as pliable components in tyres, seals, gaskets etc.

el-az Elevation/azimuth.

ELB Emergency locator beacon.

elbow 1 Angled section of piping used where change of direction is necessary.
2 Hollow fixture used for joining two lengths of electric conduit at an angle.

ELC Engine-life computer.

ELCU Electrical control unit (CAA).

ELD Electroluminescent display.

ELDO European Launcher Development Organization.

electrical engine Rocket in which propellant is accelerated by electrical device; also called electric rocket (see *electric propulsion*).

electrical interference Undesirable and unintended effects on equipment due to electrical phenomena associated with other apparatus, cables, materials or meteorological conditions.

electrically suspended gyro Rotor suspended by electrostatic force.

electric altimeter Indicates height by variation of electrical capacitance. Also called electrostatic or capacity altimeter.

electric bonding Interconnection of metallic parts for safe distribution of electrical charges.

electric gyro One whose rotor is driven electrically.

electric propeller Pitch-change mechanism is actuated electrically.

electric propulsion General term describing all types of propulsion in which propellant consists of charged electrical particles accelerated by electric or magnetic fields or both; eg electrostatic, electromagnetic or electrothermal.

electric starter Electric motor used to crank engine for starting.

electric steel Steel made in electric furnace (induction or arc-type) which possesses uniform quality and higher strength than open-hearth steel of same carbon content.

electric tachometer See *tachogenerator.*

electric welding Welding by electric arc or passing large current through material.

electric wind Emission of negative charge from sharp corner or point of conductor carrying high potential current. Also known as electric breeze.

electrochemical treatment Process involving application of electrical energy to produce chemical change in surface of material to be treated, such as anodization of aluminium alloys.

electrode 1 Terminal at which electricity passes from one medium into another; positive is called anode and negative cathode.
2 Semiconductor element that performs one or more of the functions of emitting or collecting electrons or ions, or of controlling their movements by electric field.
3 In electron tube, conducting element that performs one or more of the functions of emitting, collecting or controlling, by electromagnetic field, movement of electrons or ions.

electrodynamics Science dealing with forces and energy transformations of electric currents, and associated magnetic fields.

electroforming Building up a metal part of complex but thin form as an electroplated layer on a substrate, eg nickel on expanded polystyrene.

electro-hydrostatic Using hydraulic power to provide output force but in localized system with all command and power provided by multi-redundant electric channels.

electrojet Current sheet or stream moving in ionized layer in upper atmosphere; move around Equator following sub-solar point and around polar regions, where they give rise to auroral phenomena.

electrokinetics Science dealing with electricity in motion, as distinguished from electrostatics.

electroluminescence Emission of light caused by electric fields; gas light is emitted when kinetic energy of electrons or ions accelerated in field is transferred to atoms or molecules of gas.

electrolysis Chemical decomposition or change

in chemical state produced by electric current.

electrolyte Liquid or paste conductor in electrolytic cell or battery; when acid, base or salt is dissolved in water dissolved material ionizes, so that solution has electric potential and, when current is passed, will have different potential from metal immersed in it; solution used for anodizing aluminium and alloys, sulphuric or chromic acids being most common.

electrolytic corrosion Corrosion resulting from electrochemical action of dissimilar metals in presence of electrolyte.

electromagnet Magnet whose flux is produced by current in coil which encircles ferromagnetic core; temporarily magnetized while current flows.

electromagnetic Pertaining to magnetic field created by current; combined magnetic and electric fields accompanying movements of electrons through conductor. Abb EM.

electromagnetic focusing Control and concentration of electrons in narrow beam by magnetic fields.

electromagnetic frequency bands For administrative purposes various EM bands allotted letters (see *frequency band*).

electromagnetic induction Establishment of current in conductor cutting flux of electromagnet; principle of rotary electrical machines and transformers.

electromagnetic intrusion Intentional insertion of EM energy into transmission paths with object of causing confusion.

electromagnetic radiation Radiation made up of oscillating electric and magnetic fields and propagated with speed of light; includes gamma radiation, X-rays, ultra-violet, visible light, infra-red radiation, radio and radar waves.

electromagnetic riveting Closing rivets by violent EM pulse.

electromagnetic rocket See *electrical engine*, *plasma rocket*.

electromagnetic spectrum EM radiation extending from gamma rays down through broadcast band and long radio waves.

electromagnetic waves Waves associated with EM field, with electric and magnetic fields perpendicular to each other. Also known as electric waves, radio waves, light, X-rays, and by other names.

electromechanical Using electricity as sole source of power and of command/control functions. Such systems are gradually displacing hydraulics and other secondary power services, partly because of rare-earth magnets.

electrometallurgy Use of electricity for smelting, refining, welding, annealing and other processes, and for electrolytic separation of metals and deposition from solutions.

electromotive force External electrical pressure (measured at source) which tends to produce flow of electrons in conducting medium; volt is ** required to maintain current of one ampere through resistance of one ohm.

electron Subatomic particle that possesses smallest negative charge, and which is so-called "fundamental particle" assumed to be building block of the universe; mass at rest $9{\cdot}11 \times 10^{-28}$ gm, and negative charge $1{\cdot}63 \times 10^{-19}$ coulombs.

electron beam Stream of electrons focused by magnetic or electrostatic field and used for neutralization of positively charged ion beam and to melt or weld materials with high melting points. Also called cathode ray.

electron charge Unit, symbol e, $= 1{\cdot}602 \times 10^{-19}$ J.

electron gun Electrode structure which produces and may control one or more electron beams to produce TV picture or weld material.

electronic combat See *electronic warfare.*

electronic counter-countermeasures Subdivision of EW; actions to ensure effective use of electromagnetic radiation despite enemy use of countermeasures.

electronic countermeasures Subdivision of EW; actions to reduce or exploit effectiveness of enemy electromagnetic radiation.

electronic data-processing System using electronic computer(s) and other devices in gathering, transmission, processing and presentation of information.

electronic deception Deliberate radiation, reradiation, alteration, absorption or reflection of electromagnetic radiation, to mislead enemy in interpretation of data or present false indications; manipulative ** is alteration or simulation of friendly electromagnetic radiations to accomplish deception; imitative ** is introduction into enemy channels of radiation which imitates his own emissions.

electronic defence evaluation Mutual evaluation of radar(s) and aircraft by means of aircraft trying to penetrate radar through ECM.

electronic interference Disturbance that causes undesirable response in electronic equipment.

electronic jamming Deliberate radiation, reradiation or reflection of electromagnetic signals with object of impairing use of electronic devices by enemy.

electronic line of sight Path traversed by electromagnetic waves not subject to reflection or refraction by atmosphere.

electronics Branch of physics that treats of emission, transmission, behaviour and effects of electrons.

electronic scanning Scanning by cathode ray tube, or sequenced emission from large planar aerial array, instead of by mechanical means.

electronic warfare (also **electronic combat**) Use of electromagnetic emissions as a weapon or a source of intelligence.

electron multiplier Electron tube which delivers more electrons at output than it receives at input, because of secondary emission.

electron tube Gas-filled tube having anode, cathode and sometimes other electrodes for controlling flow of electrons.
electro-optical guidance EO guidance makes use of visible (optical) contrast patterns of target or surrounding area to effect seeker lock-on and terminal homing. Three such systems are contrast edge tracker (Mk 84 EOGB and Walleye); contrast centroid tracker (Maverick); and optical area correlator, which scans contrast patterns in large area surrounding target.
electroplating Coating metal with deposit removed from electrode and carried by electrolyte in which object to be coated is immersed.
electrostatic capacity Ability to hold electric charge.
electrostatic deflection Bending of electron beam during passage through electric field between two parallel flat electrodes; beam is deflected towards positive electrode.
electrostatic focusing Use of electric field to focus stream of electrons to small beam.
electrostatic precipitation Use of high voltages (large potential gradients) to remove particulate matter from gas flow, smoke or other volumes.
electrostatic rocket See *ion rocket*, *ion engine*.
electrostatics Study of electricity (charges) at rest.
electrostatic storage Storage of information as electrostatic charges.
electrostatic unit, ESU Unit of electric charge, amount of charge which repels similar charge in vacuum with force of one dyne; a statcoulomb.
Elektron Magnesium alloy with 3–12% aluminium, 0·2–0·4% manganese and often 0·3–3·5% zinc.
element 1 In electron tube, constituent part that contributes to electrical operation.
2 In circuit, electrical device such as inductor, resistor, capacitor, generator, line, electrode or electron tube.
3 In semiconductor device, integral part that contributes to its operation.
4 Parameters defining orbit of body attracted by central, inverse-square force: longitude of ascending node, inclination of orbit plane, argument of perigee, eccentricity, semimajor axis, mean anomaly and epoch.
5 Flight of two or three aircraft (US) or basic fighting unit of two aircraft (UK).
6 Component parts of aircraft sufficiently distinctive and specific in type, shape or purpose as to be of major importance in design.
elementary trainer Ab initio, also known as primary, trainer.
element leader Lead aircraft or pilot of element or flight.
elephant ear 1 Thick plate on rocket or missile used to reinforce hatch or aperture.
2 Air intake consisting of twin inlets, one on each side of fuselage.
elevation 1 Side or front view as drawn in orthographic projection.
2 Vertical distance of point or level, measured from mean sea level.
3 Height of airfield above mean sea level.
4 Angle in vertical plane between local horizontal and line of sight to object.
elevation rudder Elevator (arch).
elevator 1 Movable control surface for governing aircraft in pitch.
2 Effectiveness of pitch control, as in expression "to run out of *".
3 In air intercept, code meaning "take altitude indicated (in thousands of feet), calling off each 5,000ft increment" (DoD).
elevator angle Angle between chord of elevator and that of tailplane or aircraft longitudinal axis.
elevator tab Trim (or other) tab attached to elevator.
elevons Wing control surfaces combining functions of ailerons and elevators, esp on delta-wing aircraft.
elex Electronics (colloq).
ELF 1 Extremely low frequency.
2 Electronic location-finder.
Elfin ATR racking and module for housing instrument, electronic unit or other equipment.
ELGB Emergency Loan Guarantee Board.
Elint Electronic intelligence.
ELJ External-load jettison.
elliptic loading Ideal form of spanwise loading of wing, lift vectors forming semi-ellipse seen in front elevation.
elongation 1 Change in length of hardware.
2 Angle at Earth between lines to Sun and another celestial body of the solar system.
ELR Environmental lapse rate.
ELS 1 Emitter location system.
2 Emergency landing strip.
3 (Electron) energy-loss spectroscopy.
Elsa Electronic lobe-switching antenna.
ELSSE Electronic sky screen equipment; indicates departure of rocket from predetermined trajectory.
ELT 1 Emergency locator transponder.
2 Enforcement of laws and treaties.
ELV Expendable launch vehicle.
EM 1 Electromagnetic.
2 Energy manoeuvrability.
EMA 1 Electronic missile acquisition.
2 Electron microprobe analysis.
EMAA Etat-Major de l'Armée de l'Air (F).
EMAS Electromechanical actuation system.
embedded Computer or other processor forming integral part of device or subsystem and thus unable to communicate directly with bus or highway or be used for any other purpose.
embodiment loan Loan of government property to private industry, research organization or individual, usually to enable recipient to fulfil government contract.
EMC Electromagnetic compatibility.

EMCDB Elastomer-modified cast double-base propellant.
Emcon Emission monitor control.
EMCS Energy monitoring and control system.
EMD 1 Emergency distance.
2 Eidgenossische Militär-department (Switz).
3 Energy-management display.
4 Engine model derivative.
EMDa Emergency distance available.
EMDP Engine model derivative program (US).
EMDr Emergency distance required.
emergency air Compressed air for energizing hydraulic or pneumatic circuit in event of failure of normal power supply.
emergency cartridge Combustion products energize hydraulic or pneumatic circuit in event of failure of normal power supply.
emergency ceiling Highest altitude for multi-engined aircraft at which best rate of climb is 50 ft per minute with throttle of one engine closed; also known as usable ceiling.
emergency combat capability Condition exclusive of primary alert status whereby elements essential to combat-launch an ICBM are present and can effect launch under conditions of strategic warning (USAF).
emergency descent Premature descent from operating altitude because of in-flight emergency.
emergency distance Distance sufficient for all take-off or landing emergencies, such as runway plus stopway and possibly clearway.
emergency exit Door or window designed to be opened after emergency landing or aborted take-off for passenger and crew evacuation.
emergency flotation gear Inflatables fitted to aircraft in emergency to provide water buoyancy.
emergency landing Landing made as result of in-flight emergency.
emergency locator/transmitter Radio beacon giving position of crashed aircraft; fixed ***, portable ***, and survival *** (armoured and can float).
emergency parachute Second stand-by parachute.
emergency power unit On-board source of electrical and/or hydraulic power sufficient to continue controlled gliding flight following loss of main engines; commonly self-contained package using hydrazine monofuel (hence *MEPU*).
emergency rating PE (4) rating for emergency sprint periods, with aid of high boost, water/methanol injection, etc.
emergency scramble Aircraft carrier CAP launch of all available fighter aircraft; if smaller number required, numerals and/or type may be added (DoD).
emf, e.m.f. Electromotive force.
EMI 1 Electromagnetic interference.
2 Electromagnetic impulse(s).
EMIH EMI (1, 2) hardening.
EMIO Egyptian Military Industrialization Organization.
emission 1 Process by which body emits EM radiation as consequence of temperature only.
2 Sending out of charged particles from surface for electric propulsion.
3 Loosely, any release from solid surface of electrical signal.
emissivity Ratio of radiation emitted by body (if necessary in specified band of EM wavelengths) to that of perfect black body under same conditions; only luminescent can exceed 1, value for black body.
emitter Device releasing radiation, usually in usable optical, IR or RF wavelengths.
EMP 1 Electromagnetic pulse (nuclear).
2 Electric motor pump.
EMPASS Electromagnetic performance of air and ship system (USN).
empennage Complete tail unit.
empirical Based on observation and experiment rather than on theory; used esp of mathematical formulae.
employment Tactical usage of aircraft in desired area of operation; in airlift, movement of forces into a combat zone, usually in assault phase (USAF).
empty weight Measured weight of individual aircraft less non-mandatory removable equipment and disposable load. OEW is preferred.
EMR Electromagnetic riveting.
EMRU Electromechanical (or electromagnetic) release unit.
EMS 1 Entry monitor system.
2 Emergency medical service (helicopter).
3 Equipment maintenance squadron (US).
ems Electromagnetic-pulse shielding; also known as hardening.
EMSG European Maintenance System Guide.
EMSP Enhanced modular signal processor.
EMT Equivalent megatons.
EMU 1 Extravehicular mobility unit; suit worn during exploration of lunar surface.
2 Engine maintenance (or monitoring) unit.
3 Electromagnetic unit(s).
EMUX, Emux Electrical multiplexing.
ENA Escuela Nacional de Aeronáutica (Arg).
ENAC École Nationale de l'Aviation Civile (F).
encastré Structural beam ends are not pinned but fixed.
encoder Analog-to-digital converter, eg converting linear or angular displacement, temperature or other variable to digital signals.
encoding altimeter Presents usual display but in addition incorporates digitized output to transponder for transmission to ATC.
encounter Time-continuous action between airborne friendly and hostile aircraft.
end-bend blading Gas-turbine compressor blading whose tips are progressively bent upwards at the trailing edge (a form of washout) for increased efficiency.
end-burning grain Solid-propellant charge which

burns only on transverse surface at one end, usually facing nozzle.

end effects Aerodynamic effects due to fact wing span is finite.

end-fire Linear aerial array whose direction of maximum radiation is along axis.

end instrument Converts data into electrical output for telemetry. Also called end organ or pickup.

end item End-product ready for use.

endo-atmospheric Within an atmosphere.

endplate 1 Small auxiliary fins at or near tips of tailplane.
2 *effect, aerodynamic effect of T-tail on fin, or of tanks, pods, missiles or fairings on wingtips.

end play Unwanted axial movement of shaft.

end speed Speed of aircraft relative to carrier at release from catapult.

end thrust Thrust along axis of shaft.

endurance Maximum time aircraft can continue flying under given conditions without refuelling.

endurance limit Highest structural stress that permits indefinite repetition or reversal of loading; always less than yield stress (see *fatigue limit*).

endurance on station Maximum time maritime aircraft can patrol in designated areas.

ENEC Extendable nozzle exit cone.

Enema Etablissement National pour l'Exploitation Météorologique et Aéronautique (Algeria).

energy Capacity to do work.

energy absorption test See *drop test*.

energy conversion efficiency Ratio of kinetic energy of jet leaving nozzle to that of hypothetical ideal jet leaving ideal nozzle using same fluid under same conditions.

energy density Sound energy per unit volume (usual unit is non-SI: ergs/cc).

energy level Any specific value of energy which a particle may adopt; during transitions from one level to another, quanta or radiant energy are emitted or absorbed, frequencies depending on difference between levels.

energy management Monitoring to minimize fuel expenditure for trajectory control, navigation, environmental control, etc

energy manoeuvrability Flight manoeuvres in which full use is made of kinetic energy of aircraft, normally in trading speed for altitude.

energy state Total kinetic plus potential energy possessed by aircraft, particularly a fighter; normally expressed as altitude from SL reached (without propulsion) if all such energy were converted to potential (height) energy.

engage 1 In air interception, order to attack designated contact (DoD usage).
2 To contact arrester wire or barrier.

engagement Encounter which involves hostile action by at least one participant.

engagement control Exercised over functions of air-defence unit related to detection, identification, engagement and destruction of hostile targets.

engaging speed Speed of aircraft relative to arrester wire at engagement.

engin Missile (F).

engine altimeter Indicates altitude corresponding to manifold pressure of supercharged engine.

engine-attributable Caused by fault in an engine.

engine car Airship car wholly or mainly devoted to propulsive machinery.

engine change unit Aircraft PE (4) removable as single unit with all accessories, cooling and oil systems.

engine cowling Hinged or removable covering around aircraft engine shaped to keep drag to minimum and optimize flow of cooling air.

engineering 1 Department responsible for detail design and development.
2 Hardware design and development.

engineering mock-up Full-scale replica of new aircraft or major part thereof, made in metal with high precision, partly in hard tooling, to check three-dimensional geometry of structure, systems and equipment.

engineering time Number of man-hours required to complete engineering task.

engineering units Pre-SI (suggested obsolete) system of units for expressing lift and drag of wing or component part in lb/sq ft at 1 mph at specified angle of attack.

engine mounting Structure by which engine is attached to airframe.

engine-out Condition in which one engine of multi-engined aircraft gives no propulsive thrust.

engine-plus-fuel weight A criterion of propulsive efficiency, heavy engines generally burning less fuel.

engine rating Power permitted by regulations for specified use: maximum take-off, combat, maximum continuous, weak mixture etc.

engine speed Revolutions per minute of main or other specified rotor assembly.

English bias Missile aiming error at launch, and temporary guidance commands to overcome it.

ENH Earth near horizon.

ENJJPT Euro-NATO Joint Jet Pilot Training.

ENNK Endo-atmospheric non-nuclear kill.

enrichment 1 Adjustment by PE (4) mixture control to produce richer mixture.
2 Artificial increase in percentage of isotope; thus, enriched uranium contains more than natural 0·7% of fissile U235.

en route 1 Between point of departure and destination.
2 Portion of flight on airways or desired track, excluding initial departure and approach phases.

en route base Air base between origin and destination of air force mission which has capability of supporting aircraft operating route.

en route climb Climb to designated FL or cruising altitude on desired track.

en route height See *cruise altitude.*

en route support team Selected personnel, skills, equipment and supplies necessary to service and perform limited specialized maintenance on tactical aircraft at en route base (USAF).

en route time Time en route (1), normally measured from initial cruise altitude to TOD.

en route traffic control service Provided generally by ATC centres, to aircraft on IFR flight plan operating between departure and destination terminal areas.

ENSA École Nationale Supérieure de l'Aéronautique (F).

ENSAE New designation of ENSA, with addition of "et de l'Espace".

ENSE Enemy situation correlation element (US, intelligence).

ENSIP, Ensip Engine structural integrity program (USAF).

ENTG Euro/NATO Training Group.

enthalpy Total energy (heat content) of system or substance undergoing change from one stage to another under constant pressure, expressed as H = E + PV, where E is energy, P pressure and V volume.

entomopter Flying machine based on insect aerodynamics.

entrainment Sucking-in of induced fluid flow by high-velocity jet through duct.

entrance cone Portion of Eiffel-type tunnel upstream of working section.

entropy 1 In physics and thermodynamics, measure of unavailability of energy. Thus, in irreversible process, such as occurs in any real engine, * always increases. Any system or process having constant * is said to be insentropic.
2 In communications theory, measure of information disorder.

entry 1 Penetration of planetary atmosphere by spacecraft or other body travelling from outer space.
2 Fore part of aerofoil, esp wing.

entry corridor Limits of route through atmosphere which spacecraft must follow. With too steep a trajectory spacecraft would burn up; with too shallow, spacecraft would bounce off atmosphere and be unable to return.

entry gate Point(s) of entry for incoming airways traffic to TMA.

entry interface Point during re-entry at which returning spacecraft encounters sensible atmosphere. Traditionally (non-SI) = 400,000 ft.

entry point 1 Ground position at which aircraft entering control zone crosses boundary.
2 Where supersonic track crosses coast inbound.

envelope 1 Of variable, curve which bounds values but does not consider possible simultaneous occurrences or correlations between different values.
2 Curve drawn through peaks of family of curves or through all limiting valves.
3 Glass or metal casing of electronic tube.
4 Gas container of non-rigid aerostat.
5 Outer cover of airship.
6 Volume of airspace bounded by limits of effective use of weapon.

envelope diameter 1 Diameter of circle encompassing engine or other irregular object.
2 Diameter of airship envelope.

environmental chamber Chamber in which humidity, temperature, pressure, solar radiation, noise and other variables may be controlled to simulate different environments.

environmental control system, ECS Produces environment in which human beings and equipment can work satisfactorily.

environmental lapse rate Measured rate of decrease of temperature with height; determined by vertical distribution of temperature at given time and place, and distinguished from process lapse rate of individual parcel of air.

environmental mock-up Mock-up cabin intended to assist design of ECS.

environmental system Environmental control system.

EO Electro-optical.

EOAR European Office of Aerospace Research (USAF).

EOB Electronic order of battle.

EOCM Electro-optical countermeasures.

EOD Explosive-ordnance disposal.

EOFC Electro-optical fire control.

EO guidance Electro-optical guidance.

EOIS Electro-optical imaging system.

EOL Engine-off landing.

EOL power End-of-life power.

EOM Earth observation mission.

Eonnex Aircraft fabrics; name registered by Eonair Inc.

EOP Engine oil pressure.

EOR Extend/off/retract.

Eorsat Elint ocean-reconnaissance satellite.

EOTDS Electro-optical target-detection system.

EOW Electro-optical warfare.

EPA 1 Environmental Protection Agency (US).
2 Epoxy polyamide.
3 Extended planning annex.
4 Economic price adjustment (US contracting).
5 Experimental (or European) prototype aircraft.

EPC External power contactor (or connector).

EPCA Energy Policy and Conservation Act (US).

EPD Exhaust-plume dilution.

Epera Extractor-parachute emergency-release assembly.

EPG European participating governments (or groups).

ephemeris Periodical publication tabulating future daily positions of celestial bodies and other astronomical data (plural = ephemerides).

ephemeris time Uniform time defined by laws of

dynamics, determined in principle by observed orbital motions of Earth and other planets (see *universal time*).

EPI Engine performance indicator.

EPIC, Epic 1 Epitaxial passivated integrated circuit.
2 Engineering and production information control (management team).
3 Emergency procedures information centre (BAA).

EPMS Engine performance monitoring system.

EPN European participating nations.

EPNdB Equivalent Perceived Noise Decibel; unit of EPNL (see *noise*).

EPNL Equivalent perceived noise level; measure of effect of noise on average human beings which takes into account sound pressure level (intensity), frequency, tonal value and duration.

EPO Earth parking orbit.

epoxy resin Complex organic adhesive and electrical insulating material; addition of hardeners, plasticizers and fillers tailors its properties.

EPP Emergency power package.

Eppler Family of wing sections for competition sailplanes; tailored to small R, high IAS for penetration.

EPR 1 Engine pressure ratio.
2 External power receptacle.

EPRL Engine pressure ratio limit.

Eprom An ROM alterable only by UV light.

EPS Emergency power system (or supply).

Epsilam Copper-coated flexible substrate of ceramic-filled Teflon.

EPSU European Public Service Union (Int).

EPU Emergency power unit.

EQD Electrical Quality-assurance Directorate (UK MoD).

equal deflections Principle used in analysis of statically indeterminate structure: two members rigidly attached must deflect an equal amount at point(s) of attachment under load.

equalizer 1 Filter network which compensates over specified frequency band for distortion introduced by variation of attenuation with frequency.
2 Connection between generators in parallel to equalize current and voltage.

equalizing pulses Signals sent before and after vertical synchronizing pulses to obtain correct start of lines in iconoscope, vidicon and display tubes.

equation of time Before 1965, difference between mean time and apparent time, usually labelled + or – to obtain apparent time. After 1965, correction applied to 12 hours + local mean time (LMT) to obtain local hour angle (LHA) of Sun.

equations of motion Give information regarding motion of a body or point as a function of time when initial position and velocity are known.

equator Primary great circle of sphere or spheroid, such as Earth, perpendicular to polar axis.

equatorial bulge Excess of Earth's equatorial diameter over polar diameter.

equatorial satellite One whose orbit plane coincides, or almost coincides, with Earth's equatorial plane.

equilibrium flow Fluid flow in which energy is constant along streamlines, and composition at any point is not time-dependent.

equilibrium glide Hypersonic gliding flight in which sum of vertical components of aerodynamic lift and centrifugal force is equal to weight at that height.

equilibrium height At which, under given conditions, equilibrium is established between lift and weight of free aerostat without power.

equilibrium vapour Vapour pressure of system in which two or more phases coexist in equilibrium; in meteorology reference is to water unless otherwise specified.

equinex 1 Instant that Sun occupies one equinoctial point.
2 One of two points of intersection of ecliptic and celestial equator, occupied by Sun when declination is 0°; also called equinoctial point.

equi-period transfer orbit Orbit differing from first but having same period, eg that of lunar module following separation from command module.

equipment Type or class of aircraft used or to be used on particular air-transport route(s).

equipment interchange Agreement allowing aircraft to fly long routes over sectors of two or more carriers, crew being changed so that each carrier flies its own sectors.

equipment operationally ready Weapon system is capable of safe use and all subsystems necessary for primary mission are ready (USAF).

equipped empty weight Measured weight of individual aircraft including removable and other equipment but less disposable load.

equi-signal zone Zone within which aircraft receives equal signals from left and right intersecting lobes, giving continuous on-track signal.

equivalence ratio Ratio of stoichiometric to experimental air-fuel ratios.

equivalent airspeed See *airspeed*.

equivalent brake horsepower See *equivalent horsepower*.

equivalent circuit Theoretical circuit diagram electrically equivalent to practical circuit or device.

equivalent drag area See *equivalent flat-plate area*.

equivalent flat-plate area Area of square flat plate, normal to free-stream relative airflow, which experiences same drag as the body or bodies under consideration.

equivalent horsepower In turboprop, sum of horsepower, usually measured as brake hp, available at propeller shaft and equivalent power derived from jet thrust by applying numerical factor to measure of thrust (abb ehp). See *equivalent power.*

equivalent monoplane Monoplane wing having same lift and drag properties as combination of two or more wings under consideration.

equivalent monoplane aspect ratio Wings and tip vortices of biplane mutually interfere; Prandtl showed increase in induced drag of each wing is:

$$\Delta \, Di = \frac{\sigma L_1 L_2}{\frac{1}{2}\rho V^2 \pi b_1 b_2}$$

where σ is Prandtl interference factor, L wing lifts, b spans, and $\frac{1}{2}\rho V^2$ dynamic head. Total added induced drag is twice that of single wing, so ****

$$= \frac{b_1^2}{S}\left[\frac{\mu^2(1+r)^2}{\mu+2\sigma\mu r+r^2}\right]$$

where b_1 is longer span, S total area, μ ratio $\frac{\text{shorter span}}{\text{longer span}}$, and r ratio $\frac{\text{shorter-span lift}}{\text{longer-span lift}}$.

equivalent pendulum Freely gimballed platform, usually incorporating gyros and accelerometers, which has same period of oscillation as simple pendulum of particular length.

equivalent perceived noise level L_{PNeq}, $= L_E - 10 \log T/t_0$ where L_E is aircraft exposure level, T is total period of noise and t_0 is (usually) 1s (see *noise*).

equivalent potential temperature Temperature given sample of air would have if brought adiabatically to top of atmosphere (ie to zero pressure) so that all water vapour is condensed and precipitated, remaining dry air then being compressed adiabatically to 1,000 millibars. *** is therefore determined by absolute temperature, pressure and humidity.

equivalent power See *equivalent horsepower*; in SI units power is measured in W or multiples thereof; to a first-order approximation ekW = kW + 68 F_n where F_n is residual jet thrust in kN. In Imperial units jet thrust (lb force) is typically multiplied by 0·3846.

equivalent shaft horsepower See *equivalent horsepower.*

equivalent single-wheel load Mass which, supported by single wheel of size just large enough not to sink significantly into surface, causes same peak bending moment in airfield pavement as particular truck, bogie or other multi-wheel gear of actual aircraft.

equivalent temperature Temperature particle of air would have if brought adiabatically to top of atmosphere (ie to zero pressure) so that all water vapour is condensed and precipitated, remaining dry air then being compressed adiabatically to original pressure.

equivalent wing In stress analysis, same span as actual wing, but with chord at each section reduced in proportion to ratio of average beam load at that section to average beam load at section taken as standard.

ER Extended-range.

E/R Extend/retract.

ERA European Regional Airlines (association).

eradiation See *Earth radiation.*

ERAM Extended-range anti-tank mine (or anti-armour munition).

ERAP Earth-resources aircraft program (US).

ERAPS, Eraps Expendable reliable-acoustic-path sonobuoy.

erase In EDP (1) to expunge stored information, usually without affecting storage medium.

Erat En-route absorption of (expected) terminal delay.

ERAU Embry-Riddle Aeronautical University (Daytona Beach, US).

ERBM Electronic range/bearing marker.

ERBS Earth Radiation Budget Satellite.

ERC 1 Electronics Research Center, NASA, Cambridge, Massachusetts.
2 Extended runway centreline.
3 Engine-related causes.

ERCS Emergency rocket communications system.

ERDA Energy Research and Development Administration (US).

ERDE Explosives Research and Development Establishment (UK).

ER/DL Extended-range, data link.

erection 1 Assembly and rigging of aircraft from component parts or from dismantled state; eg after crated shipment.
2 Of gyro, acceleration from rest to operating speed with axis in desired alignment. (Thus re-*, to restore proper axis alignment after being toppled.)

erector transporter Vehicle used to convey ballistic rocket, elevate it for firing and act as launcher; also known as transporter erector.

E-region Region of ionosphere in which E-layers and Sporadic E-layer tend to form.

ERFA Conference on Economics of Route Air Navigation Facilities and Airports.

erg Unit of energy in CGS (not SI) system; work done by force of one dyne acting through distance of 1 cm = 10^{-7}J.

ergometer exerciser Device for exercising astronauts on long missions and measuring muscular work.

ERGP Extended-range guided projectile.

ERHAC En-route high-altitude chart.

ERIS, Eris Exoatmospheric re-entry vehicle interceptor system.

ERJ External-combustion ramjet; one in which airflow and combustion are outside vehicle with profiled exterior surface.

erk RAF slang (WW2) for airman ground crew possessing minimal skills and lowest rank (AC2 or ACH).

ERL 1 Environmental Research Laboratories (NOAA).
2 Electronics Research Laboratory (Australia).

ERLAC En-route low-altitude chart.
Eros Earth-resources orbiting satellite.
erosion gauge Instrument for measuring erosion by dust and micrometeorites on materials exposed to space environment.
ERP Effective radiated power.
ERRB Enhanced radiation, reduced blast.
error 1 In mathematics, difference between true value and calculated or observed value.
2 In EDP (1), incorrect step, process or result, whether due to machine malfunction or human intervention.
3 In air/ground bombing or photography, various definitions mainly concerned with linear miss-distance.
error band Error value, usually expressed in per cent of full-scale, which defines maximum allowable error permitted for specified combination of transducer parameters.
error signal Voltage proportional to difference between actual and desired condition. Thus, in radar, ** obtained from selsyns and AGC circuits and used to control servo to correct error.
ERS Earth-resources satellite.
ERTS Earth-resources technology satellite.
ERU 1 Ejector release unit for external stores.
2 Emergency reaction unit (USAFSS).
ERW Enhanced-radiation (neutron) weapon.
ES Escape slide.
E_s Specific energy, $h + v^2/2g$.
ESA 1 European Space Agency.
2 Enhanced signal average.
ESAS Electronically steerable antenna system.
ESC European Space Conference.
escalation 1 Increase in scope, violence or weapons of conflict.
2 Increase in cost due to incorrect cost estimation, inflation, advances in technology, changes in specification or other factors.
escape To achieve velocity and flightpath outward from primary body sufficient neither to fall back nor orbit.
escape capsule See *ejection capsule*.
escape chute Inflatable rubber slide ejected from transport aircraft on ground to enable passengers and crew to evacuate quickly in emergency.
escape hatch Hatch in aircraft, usually jettisonable, intended for use in abandoning aircraft; ventral for use in flight, dorsal after belly landing or ditching.
escape orbit Any of several paths body escaping from central force field must follow in order to escape.
escape rocket Small rocket used to accelerate and separate payload near pad following launch-vehicle malfunction.
escape spoiler Aerodynamic baffle extended upstream of crew escape door or chute.
escape tower Connects escape rocket(s) to vehicle; separated if ascent is normal.
escape velocity Speed body must attain to escape from gravitational field. Earth 25,022 mph, 36,700 ft/s, Moon 7,800 ft/s, Mars 16,700 ft/s and Jupiter 197,000 ft/s.
Esces Experimental Satellite-Communication Earth Station (India).
ESCS Emergency satcom system.
ESD Electronic Systems Division (USAF Systems Command).
Esdac European Space Data Centre (now DIH).
ESG Electrostatically suspended gyro.
ESGM ESG monitor.
eshp Equivalent shaft horsepower, ehp.
ESI Engineering staff instruction.
ESIC Environmental Science Information Center (NOAA).
ESIP Engine structural integrity program (US).
ESL Earth-Sciences Laboratories (NOAA).
Eslab European Space Laboratory; now DSS of Estec.
ESLE Electronic survivor-location equipment.
ESLR Electronically scanned laser radar.
ESM 1 Electronic support measures (UK).
2 Electronic surveillance measures (US).
3 Electronic surveillance monitoring.
ESO Engineering standards order (FAA).
Esoc, ESOC European Space Operations Centre, Darmstadt (Int).
ESP External starting power.
Esprit European strategic programme for research into information technology.
ESR Electro-slag refined (or remelt).
Esrange Former European (now Swedish) space launch range, Kiruna.
ESRO European Space Research Organization, now part of ESA.
ESRRD E-scope radar repeater display.
ESSA Environmental Science Services Administration (now NOAA).
essential bus Electrical bus (bus-bar) on which are grouped nothing but essential electrical loads.
ESSL Emergency speed select lever.
ESSS, ES^3 External stores support system.
Esswacs Electronic solid-state wide-angle camera system.
EST 1 Eastern Standard Time (US).
2 En route support team (USAF).
3 See *E&ST*.
4 Elevation, slope, temperature.
ESTA Electronically scanned Tacan antenna.
established Aircraft confirmed as being stable at a prescribed flight condition, notably at a given FL or on a particular glidepath.
ESTAe École Spéciale de Travaux Aéronautiques (F).
ESTC European Space Tribology Centre.
Estec European Space Technical Centre.
ester Compound which reacts with water, acid or alkali to give an alcohol plus acid; important in many aerospace lubricants and other materials.
ESTL European Space Tribology Laboratory (ESRO).

ESWL Equivalent single-wheel load (of multi-wheel landing gear).
ET Emerging technology.
e_t Tensile strain.
ETA 1 Estimated time of arrival.
2 Estimated time of acquisition.
3 Ejector thrust augmentor.
4 Effective turn angle.
ETACCS European theatre air command and control study.
ETADS Enhanced transportation automated data system.
Etalon Small interferometer which reflects/refracts laser light to form interference pattern giving unique signature, rejecting all other sources.
ET&E European test and evaluation (USAF).
eta patch Fan-shaped patch of fabric and webbing secured to aerostat envelope.
ETBS Etablissement Technique de Bourges (F).
ETD Estimated time of departure.
ETE 1 Estimated time en route.
2 Environmental test and evaluation.
ETF 1 Electronic time fuze.
2 Enhanced tactical fighter.
ethyl alcohol Alcohol prepared from organic compound such as grain, starch or sugar; withstands high compression ratios but compared with conventional fuel costs more, has lower heat value and vapour pressure, and affinity for water.
ethylene glycol Principal additive in cooling systems of liquid-cooled engines, composed of saturated solution of ethylene oxide and water ($C_2H_6O_2$).
ethylene oxide Petroleum-derived gas used in FAE devices.
ETI Elapsed-time indicator.
ETNAS Electro-level theodolite naval alignment system.
Etops Extended-range twin (engine) operations.
ETP 1 Equal time point.
2 Estimated time of penetration.
ETPS Empire Test Pilots School (UK).
ETR Eastern Test Range.
ETS Experimental test site.
EU Electronic unit (fuel management).
EuIG Europium iron garnet.
Euler formula Maximum load W of strut or long column, where $W = \frac{kEI.}{l^2}$ E is modulus of elasticity, I moment of inertia of strut section, k constant, and l^2 length between supports.
Eulerian angles Systems of three angles which uniquely define with reference to one co-ordinate system (Earth axes) orientation of a second (body axes); orientation of second system is obtainable from first by rotation through each angle of turn, sequence being important.
Eulerian co-ordinates System in which properties of fluid are assigned to points in space at each time, without attempting to identify individual parcels from one time to next.
EUM, Eumed Europe — Mediterranean Air Navigation Region (ICAO).
EUMS Engine-usage monitoring system; EULMS adds "life".
Eurac European aircraft-cost formula (includes landing, navigation and interest charges).
Eureka Ground beacon responding to Rebecca radar homing and distance-measuring system.
Euricas European Research Institute for Civil Aviation Safety.
EuroCAE European Organization for Civil-Aviation Electronics.
Eurocard Standard single-sided PCB, 160 × 100 mm.
Eurocontrol European Organization for Safety of Air Navigation, comprising Belgium, Netherlands, Luxembourg, France, West Germany, Britain and Ireland.
Eurogroup Informal group of European defence ministers (NATO).
eutectic point Lowest temperature at which mixture can be maintained in liquid phase; lowest melting or freezing point of alloy.
EUV Extreme ultra-violet, just beyond FUV (ie shorter wavelength).
EV EAS (F).
eV Electron-volt, gain in energy acquired by electron gaining one volt in potential, $1{\cdot}6021 \times 10^{-19}$ J.
EVA 1 Extravehicular activity; carried on outside spacecraft or on lunar surface.
2 Equipement vocal pour l'aéronef, cockpit human voice control (F).
evaluation 1 Appraisal of information in terms of credibility, reliability, pertinency and accuracy; for US and NATO letters A–F indicate reliability and numbers 1–6 accuracy; thus B–2 indicates probably true from usually reliable source, while E–5 means improbable from unreliable source.
2 Process of assessing proposal, design or hardware, usually on comparative basis in course of commercial or military procurement.
evaporative cooling Cooling system which uses latent heat of evaporation by allowing coolant to boil, then condensing and recycling it.
evaporative ice Ice formed in engine induction system on surface cooled by evaporation; can form from water or vapour at air temperatures up to 25°C.
Evasion Ensemble de Visualisation et d'Affichage au Service de l'Instructeur à Organisation Numérique.
evasive action Flight manoeuvre performed by aircraft to evade defending forces.
EVCS Extravehicular communications system.
EVED Eidg. Verkehrs und Energiewirtschafts Departement (Switz).
event marker Time-dependent indicator in HUD.

Everling number $N_O = \frac{n}{C_D} = \frac{V_C^3}{96{,}000\sqrt{\sigma}} \cdot \frac{W_O/P}{W_O/S}$ where n is propulsion efficiency, C_D total drag coefficient, V_C max level KEAS, σ relative density, W_O/P power loading and W_O/S wing loading.

Evett's Field Airfield serving WRE (Australia).

EVG Electrostatically supported vacuum gyro.

EVM Engine-vibration monitor.

EVR Electronic video recording.

EVS Electro-optical viewing system.

EVT 1 Extravehicular transfer.
2 Educational and vocational training (for return to civilian life).

EW 1 Electronic warfare.
2 Early warning.

EWACS, Ewacs Electronic wide-angle camera system.

EWAISF Electronic warfare avionics integrated support facility.

EW&C Early warning and control.

EWAS Electronic warfare analysis system.

EWAU Electronic warfare avionics unit (RAF).

EWEDS Electronic warfare evaluation display system.

EWEP Electronic warfare evaluation program (USAF).

EWG Executive working group.

EW/GCI Early warning and ground-controlled intercept.

EWO 1 Electronic warfare officer.
2 Emergency war order.

EWPA European Women Pilots' Association.

EWSM Confusingly, this is in current use for: electronic-warfare (or early-warning) support measures (or surveillance measures).

exact orbit That Earth satellite must follow if exact sought-after data are to be obtained.

exact-point symbology That showing a point rather than a locus; eg, in HUD, point where one projectile would have hit, rather than tracer line.

Excap Expanded capability (= better).

excess power Difference between horsepower available and horsepower required; determines rate of climb. When horsepower available and required are equal, rate of climb falls to zero and absolute ceiling has been reached (see *SEP*).

exchange rates Conversion factors used in calculating influence of variables in aircraft performance, most being guesses or assumptions; thus there are ** linking engine s.f.c., engine weight, parts cost, engine price and similar factors; ** will vary with stage length, operating conditions, costing formulae etc.

exciter 1 Source of small current, such as battery or d.c. rotary generator, which supplies current for field windings of large electrical machine.
2 Oscillator which supplies carrier voltage to drive subsequent frequency-multiplying and amplifying circuits of transmitter.
3 Source of light used to stimulate photo-emissive cell.

EXCM External countermeasures.

excursion Undesired short-term variation of variable, such as instrument reading or flight path, away from correct value.

excursion fare Promotional fare offered by airlines to stimulate traffic; usually applicable only to round trips, with limits on season, days available and/or trip duration.

excursion level 1 In glidepath, maximum vertical or angular variation of centreline voltage/signal.
2 In glidepath, lowest safe angle of centreline voltage/signal.

exercise option To convert option(s) into firm order(s).

exhaust branch Short pipe from cylinder to exhaust manifold.

exhaust collector ring Circular duct into which exhaust from radial engine is discharged.

exhaust cone Assembly of outer pipe and inner cone which leads gas from turbine to jetpipe.

exhaust duct 1 Tunnel through which gas is expelled from underground missile launcher.
2 Fan duct of aft-fan engine.

exhaust flame-damper Expanding and shrouded pipes designed to prevent exhaust gas or stacks being seen at night.

exhaust gas analyser Electrical instrument for indicating proportion of carbon monoxide and so indicating efficiency of combustion and correctness of fuel/air mixture.

exhaust manifold Duct into which gas is led from number of cylinders.

exhaust plug Streamlined body in exhaust nozzle for adjusting backpressure and giving propulsive thrust.

exhaust reheater See *afterburner.*

exhaust stack Exhaust pipe.

exhaust stator blades Whole or partial ring of blades behind turbine to remove residual whirl from gas.

exhaust stroke Fourth stroke in four-stroke cycle, in which piston moves up to expel burnt gases.

exhaust stub Short pipe linking cylinder direct with atmosphere.

exhaust turbocharger See *turbocharger.*

Eximbank Export-Import Bank (US).

exit cone Portion of wind tunnel into which air flows from working section.

exo-atmospheric Beyond the atmosphere.

exosphere Outermost layer of atmosphere where collisions between molecular particles are so rare that only gravity will return escaping molecules; lower boundary is critical level of escape (region of escape) at 500–1,000 km.

exotic fuel Any unusual fuel for air-breathing engine intended to produce greater thrust.

exotic material Structural material seldom used in conventional applications; esp one with melting point above 1,800°C.

expandable structure One packaged in space vehicle and erected to full size and shape outside atmosphere.

expanded foam Low-density material, usually rigid but of low mechanical strength, produced by chemical reaction in liquid state; often formed inside hollow metal airframe part.

expanding balloon Kite balloon encircled by rubber cords or other devices to control shape when not full of gas; also known as dilatable balloon.

expanding brake One whose segments are forced radially against drum by flexible sac.

expanding reamer One with slotted flutes expanded by tapered pin.

expansion joint Pipe joint so constructed as to allow limited axial movement between sections held together.

expansion ratio Ratio of cross-sectional area of rocket nozzle exit to area of nozzle throat.

expansion stroke See *power stroke.*

expansion wave Simple wave or progressive disturbance in compressible fluid, such that pressure and density decrease on crossing wave in direction of its motion; also known as rarefaction wave.

expected approach clearance, EAC Time at which arriving aircraft should be cleared to begin approach for landing; also known as expected approach time (EAT).

expected further clearance, EFC Time at which it is expected additional clearance will be issued to aircraft.

expendable construction Rocket propellant tanks are divided into sections jettisoned in sequence.

expendables Missiles, RPVs, drones, and stores and materials consumed in action or in flight, esp in space.

experimental aircraft Aircraft whose objectives are fundamental research, or development of hardware having general application to many types of aircraft.

experimental mean pitch Distance through which propeller advances along its axis during one revolution when giving no thrust.

exploding bridgewire Metal wire which melts at high temperature, produced by large electrical impulse.

explosion turbine Turbine rotated by gas from intermittent combustion process taking place in constant-volume chamber.

explosive bolt One incorporating explosive charge so that, when detonated, whatever it secures in position is released.

explosive cladding Use of explosive welding to clad one material with another.

explosive decompression Rapid reduction of pressure caused by catastrophic leak in pressure cabin (eg loss of window).

explosive forming High-energy-rate forming of sheet metal by using controlled explosive energy to blow workpiece against die.

explosive rivet Blind rivet with partially hollow shank charged with black gunpowder which, when detonated, causes shank to bulge.

explosive welding Effecting near-perfect bond between dissimilar metals by using explosion to drive them together under such pressure that joint melts and sweeps away previous surface impurities.

exposure level L_E, = k log Σ $10L_{EPN}$ + 10; can be amplified using L_{EPNi}/k where L_{EPNi} is i'th event and k is usually 10, with addition of $10T_0/t_0$ where t_0 usually 1s and T_0 may be 10s (see *noise*).

express Property transported under air express tariffs filed with CAB; conducted on basis of agreements between Railways Express Agency and airlines.

extendable nozzle Rocket exit cone retracted or extended to alter area ratio; also called extendable exit cone.

extended overwater operation As defined by USFAR (Pt 1), an operation over water at horizontal distance more than 50 nm from nearest shore.

extended-range Dovap, Extradop Baseline extension of Dovap to provide coherent reference to ground transmitter and all Dovap receivers located beyond line of sight.

extended-root blade 1 Gas-turbine rotor blade in which aerofoil is carried on long platform in disc of reduced diameter.
2 Propeller blade with root extended in chord.

extension contract Industrial contract formed as extension of previous contract in either scope or timing.

extension flap See *area-increasing flap.*

extensometer Instrument for measuring small amounts of deformation.

external aileron Aileron mounted clear of wing surfaces but deflected conventionally.

external energizer Portable motor used to supply motive power to engine inertia starters.

external input System input from source outside system.

externally blown flap Flap over upper surface of which air bled from engine is blown to prevent separation. Not same as USB.

externals External inspection of aircraft carried out by pilot before take-off.

external storage EDP (1) storage media separate from machine but capable of retaining information in form acceptable to it.

external supercharger Impeller (manifold-pressure booster) located upstream of carburettor.

extinction 1 Attenuation of light through absorption and scattering.
2 Cessation of combustion (see *flameout*).

extinction coefficient In meteorology, space rate of diminution of transmitted light; attenuation coefficient applied to visible radiation.

extraction parachute Extracts cargo from aircraft and deploys main parachutes.

extraction zone Specified ground area upon which supplies are delivered by extraction technique from low-flying aircraft.

extractor 1 Part of firearm which engages rim or base of cartridge to pull it from chamber.
2 Computer-controlled device for automatic initiation and maintenance of all desirable radar contacts in ATC or air-defence system.

extra section Extra flight by airline to take overflow of fully booked flight.

extravehicular activity See *EVA*.

extrusion Hot or cold forming of metals, rubbers and plastics by forcing through die of appropriate cross-sectional shape.

ExW Explosive welding.

eye In centrifugal compressor that portion through which fluid enters.

eyeball 1 Passenger-controlled spherical valve outlet for fresh air, usually overhead.
2 To search visually, or keep eyes on a target (colloq).

eyeballs down Jargon for severe positive acceleration.

eyeballs in Acceleration from behind when subject is seated upright, or below when prone.

eyeballs out Deceleration when subject seated upright.

eyeballs up Negative-g, downward acceleration.

eyebrow panel Panel of instruments or controls in flight deck roof, above and behind windscreen.

eyebrow window In roof of flight deck, also called VIT window.

eyelids Jet-engine reverser or afterburner nozzle halves similar to eyelid in appearance and action.

eye relief Distance from eyeball to NVG eyepiece.

F

F 1 Fahrenheit (contrary to SI).
2 US DoD and UK mission prefix letter for fighter aircraft or operating unit.
3 Flap angle.
4 Net propulsive force.
5 Farad.
6 Sonobuoy size, 0·3 m (1 ft) long.
7 Photographic category (USAAF, 1930–47).
8 First class (seating).

f 1 Femto (10^{-15}).
2 Frictional force.
3 Fuel
4 Equivalent parasite area of aircraft.
5 Symbol meaning a function of.

F^3 Form/fit/function, called F-cubed.

F4 Airfield subgrade, standard asphalt.

F-code In flight plan, aircraft has 4096-code transponder and approved R-nav.

F-display Target centred in rectangle, blip gives az/el aiming errors.

F-layer One layer of ionosphere, at 150–300 km divided into F_1 and F_2 layers, F_2 being always present and having higher electron density.

FA 1 Frontal (tactical) aviation (USSR).
2 Free-air (tunnel).

F/A Fix/attack (display).

F_A Flaperon (rarely, flap) angle.

FAA 1 Federal Aviation Administration (US).
2 Fleet Air Arm (UK).
3 Functional analysis and allocation.

FAAM Family of AAMs.

FAAN Fight Against Aircraft Nuisance (UK).

FAAR Forward-area alerting radar; measure against low-altitude aircraft strikes (US).

fabric Cloth or linen material of two main types: biased, multi-ply with one or more plies so cut that warp threads lie at angle; and parallel, multi-ply with warp threads of all plies parallel.

fabrication 1 Process of manufacturing hardware.
2 Assembly by welding metal components.

FAC 1 Forward air control(ler).
2 Flight augmentation computer.

face 1 Open end of duct ready for joining to next; eg ring round front of inlet of gas-turbine or other air-breathing jet engine.
2 Either surface of propeller or helicopter rotor blade: pressure *, side of blade formed by lower surface of aerofoil elements; suction * formed by upper surface of elements.

face alignment Distance perpendicular to chord from propeller or rotor blade chord centreline to flat face of blade at any station.

faceblind firing Method of firing ejection seat in which occupant pulls roller blind at top of seat down over his face, thus shielding latter from airstream on leaving aircraft.

faceplate 1 Disc mounted on nose of lathe spindle for rotating work between centres or for gripping asymmetric item of short length.
2 Accessory mounting pad.
3 Transparent front of pressurized helmet.

facility 1 Physical plant, buildings and equipment (previously US usage).
2 Any part of adjunct of a physical plant or installation which is an operating entity.
3 An activity or installation which provides specific operating assistance to military or civil air operations.

facility performance category See *categories (3).*

FACP Forward air control post.

facsimile Telecommunications process in which picture or image is scanned and signals used locally or remotely, sent by telephone or TV, to reproduce * or likeness of subject image.

Factar Follow-up action on accident reports.

factored field lengths Any distance relative to CTOL operations (TOR, EMD, TOD etc) multiplied by factor to take account of engine failure at V_1, slippery surface or any other hazard.

factoring Process of selecting and applying appropriate factors (of safety) in such areas as design and stress calculations, performance estimates etc.

factor of safety 1 Factor by which limit load is multiplied to produce load used in design of aircraft or part; intended to provide margin of strength against loads greater than limit load, and against uncertainties in materials, construction, load estimation and stress analysis.
2 Ratio of ultimate strength to actual working stress or maximum permissible stress in use of material component.

factory loaded Propellant charge or explosive filling added in plant before delivery.

Facts FLIR-augmented Cobra Tow sight.

FAD 1 Fleet air defense (US).
2 Fast-action device.

FADA Federación Argentina de Aeroclubes.

FADD Fatigue and damage data.

fade Decrease in received signal strength without change of receiver controls.

Fadec Full-authority digital engine (or electronic) control.

faded Radio word meaning "air-intercept contact has disappeared from reporting station's scope, and further information is estimated" (DoD).

fade-out Fading in which received signal strength

is reduced below noise level of receiver. Also known as radio fade-out, Dellinger effect, Mogel-Dellinger effect (see *blackout*).

fading Variation of radio field strength caused by change in transmission medium.

FADS, Fads Flexible air-data system.

FAE Fuel/air explosive; large class of ordnance devices.

Faeshed FAE store, helicopter delivery.

FAF Final approach fix.

FAFT Fore/aft fuselage tankage (LH_2).

Fahrenheit Temperature scale in which ice point is 32° and boiling point of water 212°.

FAI Fédération Aéronautique Internationale.

fail-active Quality of a dynamic functional system of remaining correctly operational after any single failure.

Fail-hard 1 Describes part or component, notably primary structure or other unduplicated load path, fracture of which would be catastrophic.
2 Describes system component whose failure renders system immediately misleading, incorrect or dangerous.

failing load That which, when applied, will just cause structural member to fail.

fail link Deliberate weak link to prevent overload damage to costly structure.

fail-operational Any single system failure has no (or limited) effect on operation, though warning is given.

fail-passive Failure inactivates system, thus preventing dangerous spurious signal or hardover output.

fail-safe Normally structural design, rather than system, technique in which no crack can cause catastrophic failure of whole structure but is allowed to occur and be detected.

fail-soft System failure does not inhibit operation but authority or limits of travel are reduced.

FAIP First-assignment instructor pilot.

fairing Secondary structure whose function is to reduce drag; eg streamlined cover, or junction between two parts.

fairing wire Wire provided as point of attachment for outer cover to maintain contour of airship envelope.

fairlead Streamlined tube through which trailing aerial or other cable exits aircraft.

fair over To reduce drag of excrescence by fitting fairing over or around it.

faker Strike aircraft engaged in air-defence exercise (DoD).

FAL Facilitation of air transport (ICAO).

Falac Forward-area liaison and control.

fallaway section Part of rocket vehicle that separates; one that falls back to Earth.

fallback 1 Immediate return of malfunctioning ballistic vehicle after vertical launch.
2 Material carried into air by nuclear explosion that ultimately drops back to Earth.

fallback area Area to which personnel retire once missile is ready for firing.

fallback programme Second project undertaken as insurance against failure of first.

falling leaf Aerobatic manoeuvre in which aircraft is stalled and then forced into spin; as soon as spin develops, controls are reversed; process is repeated, resulting in oscillations from side to side with little apparent change in heading.

fall off on a wing See *stall turn*.

fallout 1 Rain of radioactive particulate matter from nuclear explosion. Local * settles on surface within 24 hr; tropospheric * is deposited in narrow bands around Earth at about latitude of injection; stratospheric * falls slowly over much of Earth's surface.
2 See *spin-off*.

fallout contours Lines joining points of equal radiation intensity.

fallout pattern Distribution as portrayed by fallout contours.

fallout prediction Estimate before and immediately after nuclear detonation of location and intensity of militarily significant fallout.

fallout safe height Altitude of detonation above which no militarily significant fallout will be produced.

fallout wind plot Wind vector diagram from surface to highest altitude affecting fallout pattern.

false cirrus Cirrus-like clouds in advance of and at summit of thunder cloud, lacking feathery texture. Also known as thunderstorm cirrus.

false-colour film 1 Has at least one emulsion layer sensitive to radiation outside visible spectrum (eg infra-red), in which representation of colours is deliberately altered.
2 Modified film whose dye layers produce assigned colours rather than natural ones.

false cone of silence Radio-range phenomenon similar to cone of silence above transmitting station and likely to occur over mountains, ore deposits or other factor that causes dead spot in reception; lacks four characteristics of true counterpart; build-up, deadspot, surge and fade.

false glidepath Loci of points in vertical plane containing runway centreline at which DDM is zero, other than those forming ILS glidepath.

false heat Emitted by flares and other IR decoys.

false lift Additional aerostat lift caused by positive superheat, temperature difference between gas and surrounding air.

false nosing Built up of nose-rib formers and D-skin, attached to front spar to form leading edge.

false ogive Rounded fairing added to nose of vehicle to improve streamlining. Also known as ballistic cap.

false ribs Auxiliary nose ribs between main ribs forward of front spar to support fabric covering and improve contour.

false spar Secondary spar not attached to fuselage, used as mounting for movable surfaces.

FALW Family of air-launched weapons.

FAM Fighter attack manoeuvring.
FAMG Field artillery missile group (USA).
familiarization Training to acquaint technical personnel with specific system.
famished Air-intercept code: "Have you any instructions for me?" (DoD).
Famos Floating-gate avalanche-injection MOS.
FAN Forward air navigator.
fan 1 Vaned rotary device for producing airflow.
2 Multi-bladed rotor, usually with single stage, serving as first stage of blading in turbofan (not relevant to aft-* engine).
3 Propeller, when function is moving air rather than providing thrust.
4 Assembly of three or more reconnaissance or mapping cameras at such angles to each other as to provide wide lateral coverage with overlapping images.
F&F Fire and forget.
fan engine 1 See *turbofan*.
2 Three-cylinder engine with one cylinder vertical and others at about 45° to it.
fan exit case Casing surrounding fan (2) carrying reverser and often accessory gearbox.
fan jet Turbofan or aircraft powered thereby (colloq).
fan lift Jet V/STOL system using large axial fans inside wings and fuselage covered by shutters above and below which are opened only in hovering mode.
fan mapping Aerial survey using fan of cameras.
fan marker Radio position-fix beacon radiating in vertical, fan-shaped pattern, keyed for identification (see *radio beacon*, *Z-marker*, *FM-marker*).
fanned-beam antenna Unidirectional antenna so designed that transverse cross-sections of major lobe are approximately elliptical.
fanning beam Radiant-energy beam (eg radar) which sweeps back and forth over a limited arc (see *scan*).
fan ramjet See *augmented turbofan*.
fan straightener Radial vanes in front of and/or behind fan in wind tunnel to introduce or remove flow rotation usually counteracting that of fan.
FAO Fabrication assistée par ordinateur (F).
FAP Fleet average performance.
FAR 1 Federal Aviation Regulations; eg FAR Pt 25 performance, Pt 36 noise.
2 False-alarm rate.
3 Fighter/attack/recon (pilot grading, USAF).
4 Field assessment (or functional area) review (US).
5 Force d'Action Rapide (F).
FARA Formula Air Racing Association.
farad SI unit of electrical capacity; capacity of condenser (capacitor), which has potential difference of 1V when charged with 1C. More commonly used: microfarad and picofarad.
fare dilution Dilution of airline revenue yield by excursion, affinity, group, seasonal or off-peak and other types of promotional fares, and by discounted or free travel to employees, or passengers on particular sectors.
fare structure Complete range of airline fares, either approved by licensing authority such as CAB for domestic use or agreed at IATA traffic conferences for international use.
far-field boom Supersonic N-wave boom after long travel has changed form, esp by reducing rate of change of pressure at front and rear.
farm Compact group of large number of aerials (antennas), especially protruding from aircraft.
farm-gate operations Operational assistance and specialized tactical training provided to friendly foreign air force by United States armed forces; includes, under specified conditions, flying of operational missions by combined US and foreign aircrew as part of training when such missions are beyond recipient's capability.
Farnborough Location of Royal Aircraft Establishment (RAE), previously Royal Aircraft Factory (UK).
Farnborough indicator Pioneer indicator for continuously recording pressure cycles in cylinder of PE (4).
Farnham roll Large powered machine for two-dimensional bending of sheet metal.
FARRP Forward-area rearming and refuelling point (or FARP, forward arming and refuelling point).
FAS 1 Frequency-agile subsystem.
2 Flight-attendant station.
3 Forward acquisition sensor.
4 Federation for Air Sport (USSR).
FASA Friendly aircraft simulating aggressors.
FASS, Fass Forward-and-aft scanner system.
Fast 1 Fan and supersonic turbine.
2 Fuel and sensor tactical (pack).
Fast-action device Thyristor-controlled switch which brings battery on-line immediately on failure of main generation or TRUs.
Fast CAP Combat air patrol for fighters or fast strike aircraft.
fast erection Provision to accelerate gyro with much more than normal input power (usually electrical).
Fast FAC Forward air controller in high-speed combat aircraft.
fast mover Jet combat aircraft, esp in FAC role.
FATAC Force Aérienne Tactique (F).
Fate, FATE 1 Fuzing, arming, test and evaluation; RVs and nuclear weapons generally.
2 Factory acceptance test equipment.
FATG Fixed air-to-ground, ie against non-moving target.
fathom Nautical unit of sea depth, 6 ft, 1·828 m.
fatigue 1 Weakening or deterioration of metal or other material under load, esp repeated cyclic load; causes cracks and ultimately failure.
2 Progressive decline in human ability to carry out appointed task apparent through lack of

enthusiasm, inaccuracy, lassitude or other symptoms.

fatigue life Minimum time, expressed in thousands of hours or specified number of load cycles, that structure is designed to operate without fatigue failure.

fatigue strength Maximum stress that can be sustained for specified number of cycles without failure. Also known as fatigue limit.

fatigue test Test in which specimen is subjected to known reversals of stress, such as alternate tension and compression, or cycle of known loads repeatedly applied and released.

FAUSST Franco-Anglo-US SST Committee (defunct).

fav Fuel available.

favourable unbalanced field A T-O with excess power available, so that surplus acc/stop distance can give screen height greater than 35 ft.

FAW Fighter, all-weather (role prefix, UK).

faying surface Overlapping of adjoining skin surfaces with edges exposed to airstream or to water.

FB Fighter-bomber (role prefix, UK).

F_B Aerodynamic loading due to buffeting pressure field.

F_b Burn time at average thrust (rocket).

FBL Fly-by-light.

FBM Fleet ballistic missile (S adds "submarine" or "system").

FBO 1 Fixed-base operator.
2 Flights between overhauls.
3 Federal budget outlays.

FBS 1 Forward-based system.
2 Fixed-base simulator.

FBVL Fédération Belge de Vol Libre.

FBW Fly-by-wire.

FC 1 Fuel cell.
2 First class.

Fc Fracto-cumulus.

FCA 1 Future cycle accumulation (engine parts).
2 Functional configuration audit (software).

FCB Frequency Co-ordinating Body.

FCBA Fédération des Clubs Belges d'Aviation.

FCC 1 Flat conductor cable.
2 Federal Communications Commission (US).
3 Flight control computer.

FC cost Flight-crew cost.

FCDC Flight-control digital computer.

FCE Flight-control electronics.

FC-ECY First-class and economy seat configuration.

FCES Flight-control electronics system.

FCF Functional check flight.

FCI 1 Fuel-consumed indicator.
2 Flight command indicator.

FCL Flight-crew licensing.

FCNP Fire-control navigation panel.

FCO Formal change order (contract).

FCOM Flight crew operating manual.

FCP Fuel cell powerplant.

FCR Fire-control radar.

FCRS Federal Contract Research Center (US).

FCS Flight control system.

FCSS Fire-control sight system.

FCST Federal Council for Science & Technology (US).

Fcst, FCST Forecast.

FCT First configuration test.

FCU 1 Fuel control unit.
2 Fighter control unit.

FD 1 Flight director system.
2 Frequency duplex.

F_d Take-off distance correction factor for slush, equal to TOD slush divided by TOD dry runway (varies with slush depth and s.g.).

FDAI Flight director attitude indicator.

FDAS Flight-data acquisition system (flight recorder).

FDAU Flight-data acquisition unit.

FDC Frequency-to-digital converter.

FDE Fire detection and extinguishing.

FDEP Flight-data entry and print-out (or FDE panel).

FDF 1 Føreningen Danske Flyvere (Danish pilots' association).
2 Fachverband Deutsche Flugdatenbearbeiten.

FDFF Føreningen af Danske Fabrikanter af Flymaterial (Danish industry assoc).

FDI Flight director indicator.

FDL 1 Full-drawn line (symbology right across display).
2 Flight Dynamics Laboratory.
3 Fast deployment logistic (ship).

FDM Frequency-division multiple (or multiplex).

FDMA Frequency-division multiple access.

FDO Flight-deck officer.

FDP Funded delivery period.

FDR 1 Flight-data recorder.
2 Flat-deck runway.

FDS 1 Flight director (or data) system.
2 Fence disturbance sensor.

FDSO Full dispersed-site operations.

FDSU Flight-data storage unit.

FDTE Force development test and experimentation.

FE Ferroelectric.

F_e Static thrust per engine at sea level.

FEA Federal Energy Administration (US).

FEAF Far East Air Force.

FEAR Failure-effect analysis report.

feasibility study Determines whether plan is within capacity of, or makes best use of, resources available.

feathering 1 Turning propeller blades to feathering angle, following engine failure or apparent malfunction, to minimize drag and prevent further damage.
2 Of helicopter rotor, cyclic pitch.

feathering angle See *feathering pitch*.

feathering button Used to feather propeller.

feathering hinge Helicopter rotor-blade pivot which allows blade angle to be varied.
feathering pitch Angular setting giving zero windmilling torque for stopped propeller, thus minimum drag.
feathering pump Provides hydraulic pressure to feather propeller.
feathers Wing movables: slats, Krügers, droops, flaps, ailerons, spoilers (colloq).
feature-line overlap Series of overlapping air photographs which follow ground feature such as river or road.
FEBA Forward edge of battle area (replaced by FLOT).
fecal canister Sealable container for human solid wastes in spaceflight.
fed Of radio, supplied with RF oscillations.
feed 1 To provide signal.
2 Point at which signal enters circuit or device.
3 Signal entering circuit or device; input.
4 Means of supplying ammunition to gun, or chaff through dispenser.
feedback 1 Return of portion of output to input; positive * adds to input, negative * subtracts from it.
2 Information on progress, results, field performance, returned to originating source.
3 Transmissions of aerodynamic forces on control surfaces or rotor blades to cockpit controls; also forces so transmitted.
feedback control loop Closed transmission path containing active transducer, forward path, feedback path, and one or more mixing points arranged to maintain prescribed relationship between input and output signals.
feedback control system One or more feedback control loops to combine functions of controlled signals with functions of commands to tend to maintain prescribed relationships between them.
feedback path Transmission path from loop output signal to loop feedback signal.
feeder 1 Transmission line which connects aerial to transmitter or receiver.
2 Air route or service that feeds traffic to major domestic or international routes (see *commuter*, *third-level*).
feederliner Transport aircraft used to operate feeder, commuter or third-level services.
feed pipes Pipe supplying any liquid.
feel Subjective pilot assessment of aircraft response to flight-control commands, stability, attitudes and other factors influencing his opinion.
feel system Mechanism in which control feel is augmented, improved or simulated artificially rather than provided only by aerodynamic forces on control surfaces (see *artificial feel*).
feet dry Code: "I am, or contact designated is, over land" (DoD).
feet wet Code: "I am, or contact designated is, over water" (DoD).
FEFA Future European Fighter Aircraft (project).
FEFI Flight engineers fault isolation (technique and handbook).
FEIA Flight Engineers' International Association (US).
FEL Free-electron laser.
FELC Field-effect liquid crystal.
felt Non-woven materials used when properties of uni-directional fibre-reinforced plastics are not required; built up from fibres or whiskers of carbon, glass, asbestos, etc.
felt strip See *moleskin*.
FEM Force effectiveness measure.
FEMA Federal Emergency Management Agency (US).
femto Prefix: multiplied by 10^{-15}.
fence 1 Line of readout or tracking stations for communication with satellite.
2 Line or network of radar stations for detecting enemy aircraft or missiles.
3 Wall-like plate mounted on upper surface of wing, often continuing around leading edge, substantially parallel to airstream and used to prevent spanwise flow, esp over swept wing at transonic speeds.
Fenda Federación Nacional de los Desportes Aéreos (Spain).
FEO Federal Energy Office (US).
Ferpic Ferroelectric photoconductive image ceramic.
ferret Aircraft, ship or other platform equipped for detection, location, recording and analysis of hostile EM radiation (Elint mission).
Ferris scheme Carefully designed paint scheme using two shades or colours to make it difficult to ascertain aircraft attitude (secondarily, aircraft type and direction of travel).
ferrite Magnetic ceramics composed of salts of iron and another divalent metal; because of low eddy-current losses, cores constructed of sintered powders of these materials are widely used for rod aerials and cores of inductors for RF and video.
ferritic Of ferrite.
ferrous Derived from iron.
Ferroxcube Proprietary non-metallic insulating magnetic materials which have extremely high resistivity and low eddy-current losses but do not become permanently magnetized.
ferrule Small metal fitting or wire wrapping used to prevent loosening of wire terminal.
ferry flight Flight whose purpose is to reposition aircraft at a different place.
ferry pilot One responsible for delivering aircraft from one place to another; eg from manufacturer to customer.
ferry range Distance unladen aircraft can be ferried; specified with or without ferry tanks.
ferry tank Extra fuel tank for ferry flight over

range greater than normal limit.

fertile material Not itself fissile by thermal neutrons, can be converted into fissile material by irradiation; two are U-238 and Th-232, partially converted into Pu-239 and U-233.

FESC Forward electrical/electronic service centre.

FET Field-effect transistor.

FETAP Fédération Européenne des Transports Aériens Privés.

FEW Fighter escort wing.

few Seven hostile aircraft or less (DoD).

FEWP Federation of European Women Pilots.

FF, f.f. Fuel flow.

F/F First flight.

FFA Flygtekniska Forsoksanstalten (Aeronautical Research Institute of Sweden).

FFAM Fédération Française d'Aéro-Modélisme (F).

FFAR 1 Folding-fin aircraft rocket.
2 Forward-firing aircraft rocket.
3 Free-flight aircraft rocket.
4 Feel forces/stick angle relation.

FFCC Forward-facing crew cockpit.

FFCS Free-fall control system (for air-dropped ICBM).

FFM True-mass fuel flowmeter.

FFNAC Fédération Française des Navigants de l'Aviation Civile.

FFO 1 Furnace fuel oil.
2 Fixed-frequency oscillator.

FFP 1 Firm fixed price.
2 FOV/focus/polarity.
3 Flight fine pitch.
4 Fédération Française de Parachutisme (F).

FFPB Free-fall practice bomb.

FFPVL Fédération Française de Parachutisme Vol Libre (hang gliding).

FFR 1 Fuel-flow regulator.
2 Full flight regime, ie operative throughout each flight.
3 Full flight release (engine certification).

FFRAT Full-flight-regime autothrottle.

FFS Full flight simulator.

FFT Fast Fourier transform.

FFTx Fuel-flowmeter transmitter.

FFVV Fédération Française de Vol à Voile (gliding).

F_g Gross thrust.

FGA Fighter, ground attack (role prefix, UK).

FGIH Federal government in-house.

FGR Fighter, ground attack, reconnaissance (role prefix, UK).

FGV Field-gradient voltage.

FH 1 Flight hour(s).
2 Frequency-hopping.

FHA Fleet hour agreement (engine support).

FHU Force helicopter unit (helicopter portion of detached autonomous force).

FI 1 Fault isolation.
2 Fluid injection (TVC).
3 Fatigue index.

FIA Fédération Internationale d'Astronautique.

FIAS Formation Internationale Aéronautique et Spatiale.

FIAT First installed article test.

FIATA Fédération Internationale des Associations de Transitaires ou Assimilés (International Federation of Forwarding Agents' Associations).

FIBDATD "Fix it but don't alter the drawings".

fibre Word used loosely of FBL and other systems employing optical fibres for all data transmission.

Fibrefrax H High-temperature ceramic-fibre insulating material, available as bulk fibre, blanket, felt or paper.

fibreglass Glass-fibre, either raw fibre (many forms) or bonded into matrix and moulded or otherwise formed (capital F in UK).

Fibrelam Plastics honeycomb sandwich panel mechanically resistant to spike heels and not affected by galley or other spillage (Ciba/Geigy).

fibre optics Branch of optics concerned with propagation of light along thin fibres each comprising core and sheath of different glasses or other transparent material; light entering one end is transmitted by successive internal reflections. In practice extremely fine fibres a few microns in diameter are made up into bundles of 100,000 or more.

fibre-optics gyro Instrument (not a gyro at all) for measuring rotations by means of coherent light passed simultaneously both ways around a loop (typically 300–500 m long) of monomode optical fibre, rotation being measured instantly and precisely by phase shift at output.

fibrescope Fibre-optic borescope.

FIC 1 Flight information centre.
2 Finance committee (ICAO).
3 Film integrated circuit.

Fick's law Basic law of gaseous diffusion: mass flux j = diffusion coefficient D times differential dC_1/dy where C_1 is concentration of gas 1 and y is distance from surface.

Fidag Federazione Italiana Dipendenti Aviazione Generale.

fidelity Accuracy with which electronic or other system reproduces at output essential characteristics of input signal.

FIDO 1 Flight dynamics officer.
2 Fog Investigation and Dispersal Operation, UK method of dispersing fog in WW2 by burning fuel along runway edges.

fiducial marks Index marks on camera which form images on negative to determine position of optical centre or principal point of imagery; collimating marks.

field 1 Airfield, as in * length.
2 Region of space within which each point has definite value; examples are gravitational, magnetic, electric, pressure, temperature, etc. If

quantity specified at each point is vector, field is said to be vector *.

3 Customer service, thus * service, * rep, * report.

4 Operation at advanced base with austere facilities (military), thus * maintenance.

field alignment error In ground DF station, error introduced by incorrect orientation of aerial elements.

field coils Two fixed coils of DF goniometer at right angles to each other and connected to two halves of aerial system.

field extension Organizational element performing operating functions that must be retained under direct control of parent staff office (USAF).

field-handling frame Portable frame attached to airship on ground to afford grasp to large handling crew.

field length Distance required for take-off and landing, accelerate/stop, RTO and other operations as specified in flight manual (see *balanced* *).

field maintenance That authorized and performed in field (4) in direct support of operational squadrons and other units; normally limited to replacement of unserviceable items.

field modification One made in field (3), usually by FMK.

field of regard Total angular coverage of sensor; with fixed installation same as FOV, but if gimballed depends on FOV plus slewing and elevation limits.

field of view 1 Angle between two rays passing through perspective centre (rear nodal point) of camera lens to two opposite sides of format. Not to be confused with angle of view.

2 Total solid angle available when looking through sight, HUD or other optical system.

field operation From forward airfield, esp with unpaved runways.

field performance That associated with take-off and landing, esp in context of certification.

field strength 1 Flux density, intensity or gradient; also called field intensity, although this term does not follow strict radiometric definition of intensity (flux per unit solid angle).

2 Electric field strength.

3 Signal strength; magnitude of electric or magnetic component in direction of polarization.

field training detachment Established to provide maintenance-orientated technical training, at operational location, on new systems and their aerospace ground equipment (USAF).

FIF Fluorescent inspection fluid.

FIFO Fail-isolated/fail-operative.

FIFOR, Fifor Flight forecast (Int).

fifth-freedom traffic Picked up by airline of country A from country B and flown to country C (see *freedoms*).

FIG Fighter interceptor group.

fighter Aircraft designed primarily to intercept and destroy other aircraft.

fighter affiliation Training exercise carried out by bombers, other heavy aircraft, ground or naval forces, in co-operation with fighters.

fighter-bomber Fighter able to carry air-to-surface weapons for ground attack and interdiction.

fighter controller Officer on staff of tactical air controller charged with co-ordination and evaluation of air warning reports and operational control of aircraft allocated to him. Also known as fighter director (see also *air controller*, *tactical air controller*, *tactical air director*).

fighter cover Patrol of fighter aircraft over specified area or force for purpose of repelling hostile aircraft.

fighter-direction aircraft Equipped and manned for directing fighter operations.

fighter escort Force of fighters detailed to protect other aircraft from attack by enemy aircraft.

fighter sweep Offensive mission by fighter aircraft to seek out and destroy enemy aircraft or targets of opportunity in allotted area.

fighting top Cockpit box for gunner(s) on upper wing of large early bombers, access by ladder from fuselage.

fighting wing Combat formation which allows wingman to provide optimum coverage and maintain manoeuvrability during max-performance manoeuvres.

FIJPAé Fédération Internationale des Journalistes Professionels de l'Aéronautique.

Filac Federazione Italiana Lavoratori Aviazione Civile.

filament winding Manufacture of pressure vessel (eg rocket-motor case) by winding continuous high-strength filament on mandrel, bonded by adhesive.

File Feature identification and location element (OSTA).

filed flightplan That filed by pilot or his designated representative, without any subsequent changes.

fill, filling Threads in fabric which run perpendicular to selvage; weft.

fillers 1 Paste or liquid used for filling pores of wood prior to applying paint or varnish.

2 Pulse pairs generated by random noise in unsaturated DME beacon to maintain 2,700/s.

fillet 1 Aerodynamic fairing giving radius at junction of two surfaces.

2 Fill which traditional weld makes at intersection of two parts.

filling Increase in pressure in centre of low (meteorological); opposite of deepening.

filling sleeve See *inflation sleeve*.

film chip One incorporating thin or thick-film technology.

film cooling Cooling of body by maintaining thin fluid (liquid, vapour or gas) layer over surface.

filter 1 See *centrifuge* *, *momentum separation, dynamic particle* *.
2 Capacitance and/or inductance and resistance designed to pass given band of RF only. High-pass, low-pass, band-pass and band-stop * pass frequencies respectively above, below, between and outside desired frequencies. Frequencies at which attenuation falls by more than 3 dB are termed cut-off frequencies.
3 To study air warning information and eliminate any not of interest.

filter centre Location in aircraft control and warning system at which information from observation posts is filtered (3) for further dissemination to air-defence control and direction centres (DoD).

filter crystal Quartz crystal resonator used to control filter characteristics.

filter element Cleansing medium in filter (1) with dry matrix or liquid (often oil) film.

filtering 1 Analysis of signal into harmonic components.
2 Separation of wanted component of time series from unwanted residue (noise).
3 Suppression or attenuation of unwanted frequencies.
4 Cleansing of fluid flow of solid particles.
5 Process of interpreting reported information on vehicle movements to determine probable true tracks and, where applicable, heights or depths.

FIM 1 Fault-isolation monitoring.
2 Field-ion microscope.

fin 1 Vertical or inclined aerofoil, usually at rear or on wingtip to increase directional stability.
2 Projecting flat plate to increase surface available to reject unwanted heat.
3 Those parts of stabilizers of kite balloon providing stability in pitch.

final approach 1 IFR, flightpath inbound, beginning at ** fix and extending to airport or point where circling for landing or missed-approach procedure is executed.
2 VFR, flightpath in direction of landing along extended runway centreline from base leg to runway; hence "on finals".

final approach altitude Height at start of final approach.

final approach fix That from or over which final IFR approach is executed.

final approach gate Position on extended runway centreline above which landing aircraft is required to pass at time assigned by approach control.

final assembly Assembly of major structural and sub-units which form completed aircraft; erection.

final-assembly drawing Undimensioned drawing calling out all major installations on aircraft; complete index to particular model or sub-type (see *callout notes*).

final controller Radar controller employed in transmission of PAR (previously GCA) talk-down instructions, and in passing monitoring information to pilots not using PAR.

final mass Mass of rocket after burnout or cutoff.

final procedure turn Links base leg to approach.

finals Final approach (colloq).

final trim Exact adjustment of ballistic missile or space launcher to desired cutoff velocity.

fin carrier Frame laced to channel patches on aerostat to distribute loads from fin.

fine data channel Channel of trajectory-measuring system delivering accurate but ambiguous data; coarse channel resolves ambiguity.

fineness ratio Ratio of length of streamlined body to maximum diameter, or some equivalent transverse dimension.

fine pitch Governed propeller-blade angle most suitable for take-off and low-speed flight, between ground fine and range of coarser cruising settings.

fine-pitch stop Sets limit of blade rotation into fine pitch.

fin flash Rectilinear marking on fin, usually comprising stripes in national colours.

finger-bar controller Pilot flight-control input in which fingers rotate cylinder forward or backward for pitch control and rock sideways for roll.

finger four See *fingertip*.

finger lift Finger-operated latch on front or rear of throttle lever to prevent inadvertent selection of afterburner or reverser.

finger patch Aerostat envelope patch having radial "fingers" to distribute load into fabric.

fingers Long corridors projecting at about 90° from airport terminal to provide sufficient length for large number of gates.

finger-tight Assembled so that item can readily be part-dismantled or stripped; usual state of prototype engine stored after cancellation.

fingertip Formation in which four aircraft occupy positions suggested by fingertips of hand held horizontally.

fin girder Main vertical fin member in rigid airship.

finish 1 External coating or covering of aircraft or part.
2 General appraisal by eye or touch of external surface quality of aircraft or part, esp of all-metal construction.

finite-amplitude wave Shockwave generated at front or rear of supersonic body of finite dimensions.

finite wing Wing having tips, thus all real wings other than annular.

Finrae Ferranti inertial-navigation rapid-alignment equipment.

FIP Flight instruction program (AFROTC).

FIR 1 Flight information region.
2 Finite impulsive response.

fire 1 To ignite rocket engine; start of main-chamber burn.

2 To launch rocket.
fire-and-forget missile One with IR seeker or other self-homing capability.
fireball 1 Luminous sphere formed a few millionths of a second after detonation of nuclear weapon.
2 Meteor with luminosity which equals or exceeds that of brightest planets.
fireblocker Furnishing fabric meeting specific requirements as barrier to fire.
fire-control radar One providing target-information inputs to a weapon fire-control system.
fire-control system System including radar(s) mounted on land, sea or air platform to provide exact data on target position and velocity before engagement with guns, missiles or other weapons.
fire deluge system Remotely controlled pipes, hoses and spray outlets, situated throughout launch-pad area of large missile or space launching site, which operate if there is a fire or explosion in the area.
fire point Temperature at which material will give off vapour that will burn continuously after ignition (see *flashpoint*).
fire pulse Signal for remote control of firing (1); for fire (2) usually called launch pulse.
Fires Firefighters' integrated response equipment system (USAF).
fire unit Basic subdivision of large SAM system (ie not infantry-operated), usually with four to 12 launchers at one location.
firewall Fire-resistant bulkhead designed to isolate engine from rest of aircraft.
firing chamber Test cell for static firing of small horizontally mounted rocket or missile.
firing console Human interface with rocket engine or vehicle launch.
firing order Sequence in which PE (4) cylinders fire, invariably 1–3–4–2 or 1–5–3–6–2–4.
firing pass Flight of combat aircraft towards air or ground target in which weapons are fired.
firing pit Encloses rocket test stand on all sides except nozzle.
firing test Static operation of rocket.
first-angle projection Side and plan show nearest faces of front elevation (US).
first buy Lockheed term for firm order.
first hop Briefly airborne during high-speed taxi test prior to first flight.
first law of thermodynamics Statement of conservation of energy for thermodynamic systems (not necessarily in equilibrium); fundamental form requires that heat absorbed serves either to raise internal energy or do external work.
first-line life 1 Operational life of hardware in first-line service.
2 Time between delivery of missile or RPV and its destruction or withdrawal. For planning purposes usually five years.
first-line service Active operation in original design role with combat or training unit, or revenue service with air carrier.
first motion First visible motion of vehicle at start of mission.
first pilot See *PIC*.
first stage Lowest of two- or multi-stage launch vehicle, first to be fired.
first-tier supplier One supplying direct to prime contractor.
firtree root Usual gas-turbine rotor-blade root of tapered form with broached serrated edges providing multiple load-bearing faces.
FIS 1 Flight information service.
2 Federal Inspection Services.
3 Flight instructors' school.
4 Fighter interceptor squadron.
FISA Fédération Internationale des Sociétés Aérophilatéliques (Int).
fiscal year Financial year; for US, 1 July to 30 June; for Britain, 1 April to 31 March; for France and Germany, calendar year.
fishbone antenna Coplanar elements in colinear pairs, coupled to balanced transmission line.
fishtailing Using coarse rudder to swing tail from one side to another; alternative manoeuvre to sideslipping to steepen approach.
fishtail nozzle Ends in triangular portion with narrow slit nozzle.
fissile Fissionable by slow neutrons.
fission Splitting of heavy nucleus into two approximately equal parts (nuclei of lighter elements), accompanied by release of large amount of energy and generally one or more neutrons.
fissionable See *fissile*.
fission yield Amount of energy released by fission in thermonuclear explosion as distinct from that released by fusion. The fission yield ratio, frequently expressed in per cent, is ratio of yield derived from fission to total.
Fist Fire support team.
FIT Floating-input transistor.
fit 1 Desired clearance, if any, between mating surfaces; eg push *, shrink *, force *.
2 Total complement of avionic equipment in aircraft.
FITAP Fédération Internationale des Transports Aériens Privés.
fitter Skilled metalworking tradesman.
fitting Assembly of parts mating with specified fit (1).
FITVC Fluid-injection thrust-vector control.
five by five Radio reception loud and clear. First figure denotes volume and second intelligibility, so "five by three" means loud but not very clear.
Fix 1 Aircraft position established by any independent means unrelated to a previous position.
2 In particular, aircraft position established by intersection of two position lines. See *absolute* *, *running* *.
fixed-area nozzle One whose cross-section cannot be adjusted.

fixed-base operator Business operation at American airport usually including flying school, charter flights, sales agency for particular light aircraft and accessories, maintenance and overhaul facilities and, sometimes, third-level or commuter airline.

fixed-base simulator One in which cab does not move.

fixed-displacement pump Fluid pump handling uniform volume on each repetitive cycle.

fixed distance marker Located 300 m (1,000 ft) from threshold to provide marker for jet aircraft on other than precision instrument runway.

fixed-geometry 1 Aircraft that does not have variable-sweep wings.
2 Engine that does not have variable inlet or nozzle.

fixed gun Gun fixed in aircraft to fire in one direction, usually forwards.

fixed-gun mode One for close-range snapshooting in which sightline is boresighted to gun line.

fixed landing gear One not designed to retract.

fixed light Constant luminous intensity when observed from fixed point.

fixed-loop aerial Not rotating relative to aircraft.

fixed munition 1 One used against a fixed target.
2 Gun ammunition in which projectiles are held in propellant cases.

fixed-pitch propeller One that has no provision for changing pitch of blades, and hence efficient at only one flight speed.

fixed point 1 Positional notation in computer operations in which corresponding places in different quantities are occupied by coefficients of same power of base (see *floating point*).
2 Notation in which base point is assumed to remain fixed with respect to one end of numeric expressions.

fixed-price contract One which either provides for firm price, or under appropriate circumstances may provide for adjustable price; several types, designed to facilitate proper pricing under varying circumstances (DoD).

fixed-price incentive contract Has provision for adjustment by formula based on relationship which final negotiated total cost bears to negotiated target cost as adjusted by approved changes (DoD).

fixed satellite See *geostationary*.

fixed slat Forward portion of aerofoil ahead of fixed slot built into structure.

fixed station Telecommunication station in aeronautical fixed service.

fixed target One that does not move relative to local Earth's surface.

fixed weight Total mass of aerostat in flying order without fuel, oil, dischargeable weight or payload.

fix end End of holding pattern flown over fix (1) at which aircraft enters pattern.

fixer network Radio or radar direction-finding stations which, operating in conjunction, plot positions of aircraft; fixer system (obs).

fixture Small jig for detail subassembly.

FL 1 Flight level, usually expressed in hundreds of feet; thus FL96 = 9,600 ft.
2 Foot-lambert (see *luminance*).
3 Fan lift.
4 Flight line.

flade Fan blade.

FLAG Floor level above ground.

Flag Flemish Aerospace Group (Belg).

flag Small brightly coloured plate, often Day-Glo red or orange, or diagonally striped yellow/black, which flicks into view in panel instrument, spacesuit instrument or any other subsystem or device, to give visual warning of fault or impending difficulty such as loss of electric power or low fuel level.

flag carrier 1 Airline designated as part of bilateral agreement to fly international route(s).
2 National state airline.

flagman (1) Person carrying chequered flag (formerly) employed at US airports to direct arrivals to signalman or airport marshaller for parking.
2 Person carrying bright flag on tall mast to guide ag-aviation pilot towards end of each run.

flag stop Special unscheduled stop by scheduled airlift mission aircraft to load or unload traffic (USAF).

flag tracking Method of tracking helicopter rotor by holding fabric flag against blade tips coated with wet paint.

flak AAA fire.

flak-suppression fire Air-to-ground fire used to suppress AA defences immediately before and during air attack on surface targets (DoD). This definition should be amended to allow use of ASMs, cluster bombs etc.

flame attenuation Attenuation of radio signal by ionization in rocket exhaust.

flame bucket See *flame deflector*.

flame chute Concrete and metal duct carrying flame and gas from bottom of silo or test pit to surface.

flame damper 1 Pre-radar shroud or extension to PE (4) exhaust pipe to prevent visual detection at night.
2 See *flame trap*.

flame deflector Deflects hot gas of vertical-launch rocket engine from ground or from launching structure.

flame float Pyrotechnic marker that burns on water surface.

flame front Boundary of burning zone progressing through combustible mixture.

flame hardening Hardening metal surface by flame.

flameholder Body mounted in high-velocity combustible flow to create local region of turbulence and low velocity in which flame is stabilized.

flameout Cessation of combustion in gas turbine

or other air-breather from cause other than fuel shutoff.

flamer Aircraft on fire in air, especially in air combat.

flame-resistant Not able to propagate flame after ignition source is removed.

flame trap Filter in PE (4) induction system to prevent passage of flame upstream after blowback or backfire.

flame tube Perforated tube designed for mixing of fuel and air, in which fuel is burnt in gas turbine; usually inserted as inner liner in combustion chamber for diluting and cooling flame.

flanging machine Metal-forming machine with high-speed plunger which bends up successive small portions of flange on moving workpiece.

flank 1 Lower side of fuselage or other aerodynamic body.
2 Sides of lower (inner radius) end of compressor or turbine blade.

flap Movable surface forming part of leading or trailing edge of aerofoil, esp of wing, able to hinge downwards, swing down and forwards, translate aft on tracks or in some other way alter wing camber, cross-section and area in order to exert powerful effect on low-speed lift and drag. See following types: *double-slotted*, *dive-recovery*, *Fowler*, *Gouge*, *Junkers*, *Krüger*, *leading-edge*, *manoeuvre*, *plain*, *slotted*, *split*, *triple-slotted*, *Youngman* and *Zapp*.

flap angle Angle between chord of flap and that of wing.

flap blowing Discharge of HP compressor bleed air over lowered flaps to prevent airflow breakaway. Normally air issues at about sonic speed through slit facing across flap upper surface, flow attaching to flap through Coanda effect. Also called Attinello flap (see *BLC*, *supercirculation*).

flaperon Surface combining roll-control function of aileron with increased lift and drag function of flap; can be differentially operated.

flap fan Experimental concept in which flaps carry small fans driven by engine bleed air (perhaps eight fans on each flap) to maintain attachment and provide powered lift.

flapping Angular oscillation of rotor blade about flapping hinge.

flapping angle Angle between tip-path plane and plane normal to axis.

flapping hinge Sensibly horizontal pivot on helicopter main-rotor hub which allows blade tip to rise and fall.

flapping plane Plane normal to plane of each flapping hinge axis.

flaps-extended speed The highest speed permissible with flaps in a prescribed extended position.

flare 1 Final nose-up pitch of landing aeroplane to reduce rate of descent approximately to zero at touchdown.
2 Distance sides of planing bottom of marine float or hull flare out from centreline.
3 Pyrotechnic aerial device for signalling or illumination; parachute * illuminates large area when released at altitude; wingtip * illuminates ground when landing.
4 Inverse taper (ie opening out) at tail of cylindrical body, as at base of rocket vehicle.
5 Eruptions from Sun's chromosphere, which may appear within minutes and fade within an hour; eject high-energy protons, cause radio fadeouts and magnetic disturbances on Earth.
6 Fixed source of ground or water illumination, of several types, usually burning kerosene or related fuel (generally obs).

flare demand Coded Autoland signal commanding flare (1).

flare dud Nuclear weapon which detonates with anticipated yield but at altitude appreciably greater than intended; a dud in its effects on target (DoD).

flare out See *flare (1)*.

flare path Line of flares (6) or lights down one side or both sides of runway to provide illumination (generally obs).

flare-path dinghy Attends flare path laid over water for marine aircraft.

flash 1 Basically rectangular pattern of vertical bars in national colours painted on military aircraft, usually covering portion of fin.
2 White semicircular or circular badge worn in headgear of aircrew cadet.

flashback Sudden upstream travel of flame in flow of combustible mixture in enclosed system.

flash burn Caused by radiation from nuclear explosion.

flasher unit Regular make/break switch in circuit of light which flashes rather than rotates.

flashing light Intermittent light in which light periods are clearly shorter than dark, with repeated cycle. Usually has published frequency.

flashing off Drying of surface of film or finish until safe to gentle touch.

flashpoint Temperature at which vapour of substance, such as fuel, will flash or ignite momentarily. Lower than fire point.

flash suppressor Attachment to muzzle of gun which reduces or eliminates visible light emitted.

flash welding Electric welding by partial resistance welding under low pressure, heating by electric arc, and application of large compressive pressure forcing surplus weld out of joint.

flatbed Transport aircraft (so far only conceptual) for conveying 150 containers on flatbed fuselage.

flat diameter Diameter of circle enclosing canopy of flat parachute when spread out on plane surface.

flat four Four-cylinder, horizontally opposed PE (4).

flat-H PE (4) with two superimposed rows of

opposed cylinders.

flathatting Flying at lowest possible safe height (US colloq).

flat-head rivet Thin-headed rivet used internally or where round or countersunk heads are for any reason not advisable.

flat pad Ship-mounted ballistic launcher isolated from ship motion (colloq).

flat parachute Parachute whose canopy consists of triangular gores forming regular polygon when laid out flat.

flat-rated 1 Engine throttled or otherwise restricted in output at low altitudes and thus able to give constant predictable power at all FLs up to given limit.
2 Engine restricted in output in cold ambient conditions and thus able to give constant predictable power in all air temperatures up to given published limit; eg * to 28·9°C.

flat riser VTOL aircraft able to take off with fuselage substantially horizontal (term not normally applied to helicopters and other rotorcraft).

flat six Six-cylinder, horizontally opposed PE (4).

flat spin Spin at large mean angle of attack but with longitudinal axis of the aircraft nearly horizontal; recovery difficult and prolonged because, with aircraft fully stalled, ailerons, elevators and rudder are ineffective; nose-up position and rotation carries slipstream away from tail. Anti-spin parachute can assist positive recovery.

flat template Representation on two-dimensional material of dimensions, areas and other characteristics of curved part; also known as layout template.

flat-top Aircraft carrier (colloq).

flat zone Zone within indicated course sector or ILS glidepath sector in which slope of sector characteristic curve is zero.

FLC Fuel and limitation(s) computer.

FLCH Flechette(s), for piercing armour.

fld Field.

FLEEP Flying lunar-excursion experimental platform. One-man rocket-powered platform intended to enable astronaut to make quick hops on lunar surface.

fleet All aircraft of one type used by same operator.

fleet ballistic missile submarine Submarine designed to launch ballistic missiles.

fleet leader Aircraft in fleet having greatest flight time.

fleet noise level Average noise level throughout fleet.

Flem Fly-by landing excursion mode.

Flettner control See *servo tab*.

Flettner rotor Cylinder spinning on axis normal to airstream, generating transverse thrust/lift.

Flexadyne Proprietary (Rocketdyne) formulation of solid propellant.

Flexar Flexible adaptive radar.

flexibility Ability of hardware, including aircraft, to operate efficiently over wide range of conditions; eg long or short sectors, high or low level.

flexibility factor Used in helicopter rotor stress calculations to make up for structure's flexibility.

flexible air-data system Versatile microprocessor-based DADC outputting MIL-1553B, Arinc 429, analog and IFF transponder.

flexible blade Helicopter rotor blade with trailing-edge or balance tabs.

flexible flight deck Post-WW2 concept of aircraft carrier whose aircraft would need no landing gear.

flexible take-off, FTO Take-off technique in which for TOW below MTOW, less than maximum engine thrust is selected. For given WAT condition, this thrust is computed by intersection of TOW and aircraft performance to comply with regulations, giving a theoretical "ambient temperature" Tf. Thrust selected is that which would be available at full power at Tf. FTO saves engine costs, reduces noise and extends engine life.

flexible tank Bag-type tank.

flexible wall Used in wind tunnels, engine air inlets and other ducts subject to large range of flow and Mach number; may be perforated to extract boundary layer.

flexural axis Locus of flexural centres, points at which applied load produces pure bending without twist (note, on swept wing pure bending results in apparent twist, ie loss of incidence).

flexural wash-out Apparent reduction in angle of incidence from root to tip as swept wing deflects upward under load.

flexure Bending under load.

flicker Subjective sensation resulting from periodic fluctuation in intensity of light at rates less than about 25/30 times a second, preventing complete continuity of images.

flicker control See *bang-bang control*.

flicker rate Refresh rate for CRT information below which flicker becomes noticeable; dependent on eye's persistence of vision and persistence of CRT phosphor.

flicker vertigo Caused by light occulting (flickering) at frequencies from four to 20 per second (eg with single-propeller aircraft headed towards Sun at low rpm), producing nausea, dizziness or unconsciousness.

flick roll See Addenda.

flight 1 Movement of object through atmosphere or space sustained by aerodynamic, aerostatic or reaction forces, or by orbital speed.
2 An instance of such movement.
3 Specified group of aircraft engaged in common mission.
4 Basic tactical unit of three or four aircraft.
5 Flight sergeant (colloq, abb).
6 Radio call sign, Flight Directory (NASA).

7 Particular scheduled air-carrier service, with three or four-figure identifying numbers, either routinely or on particular day.
8 Fighting formation comprising two elements each of two aircraft (US, 1981 onwards).

flight advisory Message giving advice or information broadcast to airborne aircraft or interested ground stations.

flight assist Provision of maximum assistance to aircraft lost or in distress by flight-service stations, towers and centres (US).

flight attitude 1 Defined by inclination of three vehicle axes to relative wind.
2 Defined relative to Earth, ie local vertical.

flight characteristic Feature of handling, feel or performance exhibited by aircraft type or individual example.

flight compartment See *flight deck (1).*

flight controls Those governing trajectory of aircraft in flight.

flight control system See *control system.*

flight coupon Actual ticket or ticket book issued by air carrier to passenger.

flight crew Personnel assigned to operate aircraft.

flight cycle Sequence of operations and conditions, different for airframe, propulsion and each system or equipment item, which together make up one flight.

flight data recorder See *flight recorder.*

flight deck 1 Compartment in large aircraft occupied by flight crew.
2 Upper deck of aircraft carrier.

flight despatcher See *despatcher.*

flight director 1 Flight instrument generally similar to attitude director giving information on pitch, roll and related parameters.
2 Panel controller for autopilot.
3 Most senior member of large wide-body cabin crew.

flight-director attitude indicator Manned-spacecraft display indicating attitude, attitude error and rate of pitch, yaw and roll.

flight duty period See *crew duty time.*

flight engineer Aircrew member responsible for powerplant, systems and fuel management, and also sometimes for supervising turnround servicing.

flight envelope 1 See *gust envelope.*
2 Curves of speed plotted against altitude or other variable defining performance limits and conditions within which equipment must work.

flight flutter kit Installation, together with instrumentation, of "bonkers" or other devices to induce flutter in flight-test aircraft.

flight-following Maintaining contact with specified aircraft to determine en route progress.

flight idle Lowest engine speed available in flight, set by ** stop, mechanical limit released to ground-idle position at touchdown.

flight indicator Instrument combining lateral inclinometer, fore-and-aft inclinometer and turn indicator (obs).

flight indicator board Display in airport terminal showing arrivals and departures of airline flights.

flight information centre Unit established to provide flight information service and alerting service.

flight information region, FIR Airspace of defined dimensions within which flight information and alerting services are provided by air traffic control centre.

flight information service Service giving advice and information useful for safe and efficient conduct of flights.

flight instruments Those used by pilot(s) to fly aircraft, esp those providing basic information on flight attitude, speed and trajectory.

flight integrity Close relationship between two friendly combat aircraft manoeuvring for mutual support.

flight level, FL Level of surface of constant atmospheric pressure related to datum of 1013·25 mb (29·92 in mercury), expressed in hundreds of feet; thus FL 255 indicates 25,500 ft (see *QFE*, *QNH*).

flight-line 1 Ramp area of airfield, where aircraft are parked and serviced.
2 In reconnaissance mission, prescribed ground path across targets.

flight Mach number Free-stream Mach number measured in flight.

flight management system Automatic computer-controlled system with autothrottle, possible Mach-hold, and complete control of navigation, including SIDs and STARs. Offers "menus" for minimum cost, minimum fuel burn or other objectives. Relieves workload, increases precision.

flight manual Book prepared by aircraft manufacturer and carried on board, setting out recommended operating techniques, speeds, power settings, etc, necessary for flying particular type of aircraft. Known to airlines as operations manual.

flight number See *flight (7).*

flight panel Accepted definition: panel grouping all instruments necessary for continued flight without external references. Preferable: panel grouping available flight instruments.

flightpath Trajectory of centre of gravity of vehicle referred to Earth or other fixed reference. In following five definitions H signifies * in horizontal plane and V in vertical plane.

flightpath angle Acute angle between flightpath (V) and horizontal, shown on FPA display by resolving G/S and V/S.

flightpath computer See *course-line computer* (H).

flightpath deviation Angular or linear difference between track and course of an aircraft (H).

flightpath recorder Instrument for recording angle of flightpath (V) to horizontal.

flightpath sight In HUD, direct aiming point showing distant point through which aircraft will pass (V, H).

flightpath vector Prediction of future flightpath which replaces traditional flight director in advanced EFIS, especially to protect against windshear.

flight plan 1 Specified information relating to whole or portion of intended flight (2); filed orally or in writing with air traffic control facility.
2 Common working document, both in spacecraft and at all ground stations during manned or unmanned space flight. Separate ** issued for lunar or planetary surface operations.

flight-plan correlation Means of identifying aircraft by association with known flight plans.

flight platform See *helipad.*

flight profile Plot of complete flight (2) in vertical plane, usually altitude plotted against track distance.

flight progress strip ATC aide-memoire: paper strip typically 25 mm × 200 mm, coloured for traffic direction, giving one flight's c/s, FL, ETA as amended; slid into FP board until passed to colleague at handover.

flight rating test One in which member of flight crew demonstrates ability to comply with requirements of particular licence or rating.

flight-readiness firing Short-duration test of in-service rocket vehicle on launcher.

flight recorder Device for automatically recording information on aircraft operation. Main type is flight data recorder (FDR), also colloquially called crash recorder or "black box". Records 50 or more parameters, including following mandatory channels: altitude, airspeed, vertical acceleration, heading, elapsed time at 1 s intervals (UK also requires pitch, and usually control-surface positions, high-lift surface positions, engine speeds and flight-crew speech are also included). Such recorders are designed to survive crash accelerations, impacts, crushing and fire, and often carry underwater transponders or beacons. Normally recording medium, eg multi-track steel tape, is recycled every 25 h. Cockpit voice recorder (CVR) stores all speech on flight deck or cockpit, including intercom and radio. Maintenance recorders, eg AIDS, are linked by serial data highways to hundreds of transducers and other inputs recording many kinds of information (temperatures, vibrations, pressures and electronic parameters) to yield advance information of impending fault conditions or failures and improve system operation and economy. Highways lead to various logic and acquisition units, some for quick-look and others long-term; separate highway to protected FDR often provided.

flight reference card Carried in cockpit to provide quick detailed list of vital actions in event of all system failures or emergencies commonly encountered, with recommendations, suggestions and prohibitions.

flight regime State of being airborne, governing many systems and modes unavailable on ground.

flight service station Facility providing flight assistance service (FAA).

flight simulator Electronic device that can simulate entire flight characteristics of particular type of aircraft, with faithful reproduction of flight deck; used to test and check out flight crews, esp in coping with emergencies, and (military) in completing combat missions according to role; or as design and engineering tool during aircraft development.

flight sister Female nursing officer trained for aeromedical duties.

flight space Space above and beyond Earth available for atmospheric or space flight.

flight station 1 Flight crew position away from flight deck.
2 Base for marine aircraft (WW1).

flight status Indication of whether a given aircraft requires special handling by air traffic services.

flight strip 1 Auxiliary airfield on private property, farmland or adjacent to highway.
2 See *flight progress strip.*

flight surgeon Physician (invariably not surgeon) trained in aeromedical practice whose primary duty is medical examination and care of aircrew on ground.

flight test 1 Test of vehicle by actual flight to achieve specific objectives.
2 Test of component mounted on or in carrier vehicle to subject it to conditions of flight.
3 Flight rating test.

flight test vehicle Special aircraft, missile or other vehicle for conduct of flight tests to explore either its own capabilities or those of equipment or component parts.

flight time 1 Elapsed time from moment aircraft first moves under its own power until moment it comes to rest at end of flight. For flying boats and seaplanes, buoy-to-buoy time (see *block time*).
2 For gliders and sailplanes, time from start of take-off until end of landing.
3 For vehicles released in flight from parent carrier, measured from moment of release.
4 Aggregate of ** of all flights made by same basic structure or other hardware item.

flight visibility Average forward horizontal distance from cockpit (assumed at typical light-aircraft FL) at which prominent unlighted object may be seen and identified by day and prominent lighted object may be identified by night.

flight weight Similar to production item; not a battleship test construction.

flightworthy Ready for flight; for aircraft, airworthy.

FLIP 1 Flight information publication.
2 Floated lightweight inertial platform.
flip-flop 1 Bistable multivibrator; device having two stable states and two input signals each corresponding with one state; remains in either state until caused to flip or flop to other.
2 Bistable device with input which allows it to act as single-stage binary counter.
flipper Elevator (US colloq).
FLIR Forward-looking IR.
Flit Fighter lead-in training (USAF, Holloman).
FLL 1 Flight line (two words) level (UK/NATO).
2 Fördergesellschaft für Luftschiffbau und Luftschiffahrt eV (G).
FLM Flightline mechanic (US).
Flo Floodlights available on landing.
float 1 Horizontal distance travelled between flare and landing or alighting (see *ground effect*).
2 Watertight body with planing bottom forming alighting gear of * seaplane.
3 Ability of control surfaces to trail freely in airstream except when commanded by input; reckoned negative when surface deflected away from relative wind and positive when (because of overbalance ahead of hinge axis) surface moves against it.
4 Buoyant capsule in carburettor.
floatation See *flotation.*
float displacement Mass of water displaced by totally submerged seaplane float.
float gear Floats (2) applied as modification to landplane.
floating ailerons Designed to float (3).
floating gudgeon pin Free to rotate in both piston and connecting rod.
floating gyro Mass supported by hydrostatic force of surrounding liquid.
floating lines In photogrammetry, lines connecting same two points of detail on each print of stereo pair; used to determine whether or not points are intervisible, and drawn directly on prints or superimposed by transparent strips.
floating mark Mark or dot seen as occupying position in three-dimensional space formed by stereoscopic pair, used as reference in stereoscopy.
floating point EDP (1) positional notation in which corresponding places in different quantities are not necessarily occupied by coefficients of same power; eg 186,000 can be represented as $1{\cdot}86 \times 10^5$. By shifting ** so number of significant digits does not exceed machine capacity, widely varying quantities can be handled.
floating reticle One whose image can be moved within FOV.
float light See *flare (3).*
floatplane See *float seaplane.*
float seaplane Aeroplane supported on water by separate floats (2).
float-type carburettor Head of fuel supplied to jet is controlled by float (4) and needle valve.
float valve Fluid valve regulated by float acting on level in container.
float volume Ratio of seaplane gross weight to mass of unit volume (traditionally 1 ft^3) of water.
FLOLS Fresnel-lens optical landing system.
flood flow 1 Unrestricted supply of hot high-pressure air to cockpit either for demist/de-icing, or in emergency, or by auto switch triggered by excessive cabin altitude.
2 Has similar (various) meanings in oxygen systems.
floodlight Light providing general illuminations over particular area.
flood valve Controls flow of fire extinguishant.
floor vents Pass used cabin air to pressurized or non-pressurized lower fuselage.
FLOT, Flot Forward line of troops (formerly called Feba, forward edge of battle area).
flotation Quality of a wheel landing gear of operating from soft ground.
flotation bags, collars Inflatables used to provide buoyancy and stability for sea-recovered spacecraft.
flotation gear 1 Inflatable bags carried inside RPV, target or missile test vehicle to provide buoyancy after ditching.
2 Emergency inflatable bags surrounding landing gear of shipboard helicopter.
Flo Trak Patented arrangement of large-area plates fitting around landing-wheel tyre for enhanced flotation, esp to permit combat aircraft to use soft surface.
flow augmenter Usually means ejector, but also applied to inducer at entry to centrifugal pump.
flowback Runback of water from wing leading edge in icing conditions.
flow chart Graphical symbolic representation of sequence of operations.
flow control Measures designed to maintain even flow of traffic into airspace or along route. Chief feature is acceptance of each aircraft into pre-booked slot at entry to controlled airspace, at agreed gate time, to provide orderly ATC service which does not become overloaded.
flow disrupter Small hinged or retractable plate intended to promote intentional stall.
flowmeter Instrument which measures fluid (gas or liquid) flow; numerous types based on venturi pressure drop, speed of free-spinning turbine, pitot pressure and many other principles; measure can be velocity at point, near-average velocity or, with density input, mass flow.
flow rake See *rake.*
flow regime Particular type of fluid flow (see *continuum* *, *free-molecule* *, *laminar* *, *slip* *, *turbulent* *).
FLR Forward-looking radar.
FltSatCom Fleet satellite communication system; also clumsily written all capitals (USN).
fluerics See *fluidics.*

fluid Liquid or gas.
fluid dynamics Study of fluid motion.
fluid element Second or supporting element in *fluid four*.
fluid four Tactical formation in which second element is loosely spread in both vertical and horizontal planes to enhance manoeuvrability, look-out and mutual support.
fluidics Branch of technology akin to electronics but using instead of electrons air or other fluid flowing at low pressure through pipes, valves and gates for control of external systems; one advantage is relaxed upper temperature limit.
fluidity Reciprocal of viscosity.
fluidized bed Container of finely divided solid particles supported in liquid-like state by upcurrent of air or other gas.
fluid mechanics Study of static or moving fluids and reactions on bodies (includes aerodynamics, aerostatics, hydrodynamics, hydrostatics).
fluid resistance See *drag*.
fluorescence Emission of photons, esp visible light, during absorption of radiation of different wavelength from other source; photoluminescence (see *luminescence*, *phosphorescence*, *scintillation*).
fluorocarbons Generally resemble hydrocarbons, but F instead of H makes them more stable; many uses.
fluoroscope Instrument with fluorescent screen supplied by processed signals from X-ray tube, used for immediate indirect viewing inside metal or composite structures.
fluoroscopy X-ray TV.
flush antenna One conforming with external shape of vehicle.
flush deck Whole ship upper deck at same level.
flush intake Not protruding, orifice in skin of vehicle.
flush on warning Take off immediately radar evidence suggests hostile missile attack so that, when airfield is hit, aircraft are just out of dangerous radius of thermonuclear warhead.
flush rivet Head is flush with surface into which it is countersunk.
flush weld Plug or butt weld which leaves no weld material on surfaces.
flutter 1 High-frequency oscillation of structure under interaction of aerodynamic and aeroelastic forces; basic mechanism is that aerodynamic load causes deflection of structure in bending and/or twist, which itself increases imposed aerodynamic load, structure overshooting neutral position on each cycle to cause load in opposite direction. Distinguished by number of degrees of freedom (bending and torsion of wing, aileron and other components are considered separately), symmetry across aircraft centreline, and other variables. When heating involved subject becomes aerothermoelasticity (see *classical* *, *hard* *, *soft* *).
2 Radio beat distortion when receiving two signals of almost same frequency.
flutter model Flexible model with mass distribution, flexure and other features designed so that flutter qualities simulate those of full-scale aircraft.
flutter speed Lowest EAS at which flutter occurs.
flux 1 Generally, quantity proportional to surface integral of normal (90°) field (eg, magnetic) intensity over given cross-section.
2 Volume, mass or number of fluid elements or particles passing in given time through unit area of cross-section; eg luminous *, measured in lumens (abb lm).
3 Magnetic * can be thought of as number of lines of force passing through particular coil or other closed figure (see *weber*).
4 Materials used in welding, brazing and soldering to clean mating surfaces, and/or form slag, which helps separate out oxides and impurities by flotation and exclude oxygen.
flux density 1 Unit of magnetic ** is Wb/m^2, or tesla (T).
2 Neutron **, and particle physics generally, is particles per unit cross-section per second multiplied by velocity, $\Delta\phi/\Delta t = nv$.
fluxgate Sensitive detector giving electrical signal proportional to intensity of external magnetic field acting along its axis, used as sensing element of most remote-indicating compasses; also called fluxvalve.
fluxgate compass Uses fluxgate to indicate, subject to corrections, direction of magnetic meridian.
fluxvalve See *fluxgate*.
fly-away cost Published retail price of GA aircraft, with specified avionics fit, ignoring spares, training or support.
fly-away disconnects Launch-vehicle umbilicals on rigid arms which swing clear under power.
fly-back period That during which CRT spot returns from end of one line to start of next when in raster-scan mode.
fly-back time Time, usually ns or μs, for each fly-back period.
fly-back vehicle Space vehicle intended to be reusable.
flybar Flying by auditory reference.
fly before buy Philosophy of flight evaluation of new aircraft type, esp by military (government) customer for combat aircraft.
fly-by Interplanetary mission in which TV and instrumented spacecraft passes close to target planet but does not impact or orbit it.
fly-by-light Flight-control system with signalling by optical fibres.
fly-by-wire Flight-control system with electric signalling.
flyco 1 Abbreviation for Wing Commander, Flying, at an RAF station.
2 Position aboard aircraft carrier from which all

aircraft launches and recoveries are controlled.

Flygtekniska Föreningen Society of Aeronautics (Sweden).

Flygtekniska Försöksanstalten Aeronautical Research Institute (Sweden).

fly-in Informal gathering of private and club aircraft at particular airfield, usually with a relaxed programme of events and competitions.

flying boat Seaplane whose main body is a hull with planing bottom.

flying cable Connects captive or kite balloon to winch.

flying controls See *control system*.

flying diameter Overall diameter of circular parachute canopy in normal operational descent.

flying machine Powered aerodyne; common pre-1914, today humorous or derogatory.

flying position Attitude of aircraft when lateral and longitudinal axes are level or in flight attitude; esp when aircraft on ground is supported in this attitude.

flying rigging Distributes loads into balloon from flying cable.

flying shears Rotary system for cutting long web of sheet metal or other material moving at high speed.

flying speed Loosely, minimum airspeed at which aeroplane can maintain level flight (preferably, positive climb) in specified configuration.

flying spot Rapidly moving spot of light, usually generated by CRT, used to scan surface containing visual information.

flying stovepipe Ramjet (colloq).

flying-tab control See *servo tab*, *Flettner*.

flying tail Use of whole horizontal tail as primary control surface.

flying testbed Aircraft or other vehicle used to carry new engine or other device for purpose of flight testing.

flying the needle Navigating along airways by VOR.

flying time See *flight time*.

flying weight See *flight weight* (engine).

flying wing Aeroplane consisting almost solely of wing, reflecting idealized concept of pure aerodynamic body providing lift but virtually devoid of drag-producing excrescences.

flying wires Diagonal cables/wires placed under tension in 1g flight and used to join lower anchor (low on fuselage or within biplane cellule) to higher anchor further outboard on wing; also known as lift wires.

fly-off Competitive in-flight demonstration of performance and other qualities between two or more rival aircraft built to same requirement to determine which will be chosen for procurement.

fly-over noise Noise made by aircraft over particular point, usually near airport on inbound/outbound track, chosen for noise measurements.

FM 1 Frequency modulation; instantaneous frequency of EM carrier wave is varied by amount proportional to instantaneous frequency of modulating (intelligence-carrying) signal, amplitude and modulated power remaining constant.
2 Fan marker.

FM-9 Fuel modifier No 9, which with a carrier fluid forms Avgard.

FMA Flight-mode annunciator.

FM/AM 1 Amplitude modulation of carrier by frequency-modulated subcarrier(s).
2 Alternate FM and AM operation.

F_{max} 1 Maximum thrust.

FMC 1 Flight management computer (S adds "system", U adds "unit").
2 Fully mission-capable (USAF).

FMCW Frequency-modulated carrier wave.
2 Peak thrust of rocket engine.

FMD Flight management display.

FMEA Failure modes and effects analysis.

FMG Flight management guidance (C adds "computer", S adds "system").

FMI Functional management inspection.

FMICW FM (1) intermittent continuous wave.

FM marker Fan marker, transmits fan-shaped pattern of coded identity signals upwards, usually across one leg of radio range station.

FMOF First manned orbital flight.

FMS 1 Flight-management system.
2 Field maintenance squadron (USAF).
3 Frequency-multiplexed subcarrier.
4 Foreign military sales (DoD).
5 Federation of Materials Societies.
6 Fuel-management system.
7 Flexible manufacturing systems.

FMTAG Foreign military training affairs group (US).

FMV Försvarets Materielverk (Defence Materiel Administration, Sweden).

F_N Nozzle drag correction.

F_n Net thrust.

FNA Fédération Nationale Aéronautique (F).

FNBA Fédération Nationale Belge d'Aviation.

FNC Favoured-nation clause.

FNL Fleet noise level.

fnp Fusion point.

FNS Strategic nuclear forces (F).

FO 1 Foreign object.
2 Fibre optics (resulting in numerous other FO acronyms).
3 Fail operational (see *FOS*, *FOOS*).

foam carpet Layer of foam put down on runway or other space by fire tenders to cushion impact of aircraft making wheels-up landing.

foamed plastics Foaming agent provides minute voids to create low-density material used for insulation (thermal, mechanical shock etc) or to increase structural rigidity; often formed in place within structure.

foaming space Free vapour volume above fuel in tank.

foam strip Foam carpet.

FOB 1 Forward operating base.

2 Fuel on board (suggest undesirable usage).

FOBS Fractional-orbit bombardment system; attack by warhead making part of satellite orbit, thus approaching from any direction.

FOC 1 Foreign-object check.
2 Full (or final) operational capability.
3 Flares/off/chaff.
4 Faint-object camera.

focal length Distance from optical centre of lens or surface of mirror to principal focus.

focal plane That parallel to plane of lens or mirror and passing through focus.

focal point 1 See *focus.*
2 Air Staff agency or individual designated as central source of information or guidance on specific programme or project requiring co-ordinated action by two or more Air Staff agencies (USAF).

FOCAS 1 Fibre-optic communications for aerospace systems (USAF).
2 Flag Officer Carriers and Amphibious Ships (RN).

focus 1 Point at which parallel rays of light meet after being refracted by lens or reflected by mirror.
2 Point having specific significance relative to geometrical figure such as ellipse, hyperbola or parabola.

FOD Foreign-object damage.

FODCS Fibre-optics digital control system.

FODT Fibre-optics data transmission.

FOFA Follow-on forces attack.

FO/FO Flame-on/flame-out.

FOG 1 Fibre-optics guidance.
2 Fibre-optics gyro.

fog Form of cloud in surface layers of atmosphere caused by suspended particles of condensed moisture or smoke, reducing visibility to less than 1 km. Advection * results from arrival of warm humid air over cold surface; radiation * from cooling of water vapour created by evaporation during day by cold ground on clear night; sea * by condensation of moisture in warm air over cold sea (essentially advection).

föhn, foehn Dry wind with strong downward component, warm for season, characteristic of mountainous regions.

FOI Follow-on interceptor (USAF).

FOL Forward operating location.

fold Joint in wing of carrier aircraft enabling it to be folded. Incorporates one or more horizontal, skewed or vertical hinges.

folded dipole Two parallel, closely spaced dipole antennas connected at ends, with one fed at centre.

folded fell seam Fabric seam which has both edges folded and located in same place between fabric layers.

folding fin 1 Aircraft or missile fin hinged axially at base to lie flat prior to launch.
2 Rocket or missile fin hinged transversely to emerge from housing slot, as in FFAR.

folding wing One hinged outboard of root to enable overall dimensions to be reduced when aircraft in hangar, esp on carrier.

follow-on Anything considered to be second or subsequent generation in development, esp within same manufacturing team.

follow-on contract One whose terms repeat those for earlier stage of same programme.

follow-on developmental tests Tests during acquisition phase after completion of formal Cat II tests; consist of updating changes or additions not available previously (USAF).

follow-on production Serial production immediately subsequent to completion of development or previous production batch.

Folta Forward operating location training area.

FOM Figure of merit (sonar).

fone Telephone (FAA).

FOOS Fail-operational, fail-operational, fail-safe.

foot Non-SI unit of length, = 0·3048 m by definition.

foot-candle See *luminance.*

foot-lambert See *luminance.*

foot motor Foot-operated hydraulic motor used to energize wheel brakes.

foot-pound Unit of work.

foot-pound-second system Non-SI system of units still used in US; also called Imperial.

footprint 1 Area around airport enclosed by selected contour for LPN, EPNL, NNI or other noise measure.
2 Possible recovery area for spacecraft plotted from re-entry point.

foot-thumper Stall-warning device that triggers oscillating plunger in rudder pedal when stall imminent.

foot-to-head acceleration Ambiguous, means accelerating force acting on body from head to feet, negative g.

FOP 1 Forward operating pad.
2 Fired outside parameters.

FOR 1 Fail-operative redundant.
2 Field of regard.

Foracs Fleet operational readiness and calibration system.

Forcap Force CAP, combat air patrol maintained overhead a task force (USN).

force SI unit is Newton (N), which gives mass of 1 kg acceleration of 1 m/s^2; dyne = 10^{-5}N; kgf (kilogram force) and kilopond = 9·807 N; lbf (pounds force) = 4·448 N; kip = 4,448 N; ounce force = 0·278 N; poundal = 0·1383 N.

force balance transducer Output from sensing member is amplified and fed back to element which causes force-summing member to return to rest position.

force coefficients Aerodynamic forces, eg lift and drag, divided by dynamic pressure $\frac{1}{2}\rho V^2 S$.

force combat air patrol Patrol of fighters maintained to protect task force against enemy aircraft.

forced convection Process by which heat is transported by mechanical movement of air (cooling systems, meteorology).

force diagram Vector presentation of force(s) acting on object, length and direction of each vector representing magnitude and direction of one force. If diagram forms closed polygon, forces are in equilibrium. If diagram fails to close, gap indicates unbalanced force. Hence, force polygon.

forced landing Made when aircraft can no longer be kept airborne, for whatever reason.

forced oscillation One in which response is imposed by excitation; if excitation is periodic and continuing, oscillation is steady-state.

force fit Mating parts in which male dimension exceeds female (see *fit*).

force rendezvous Navigational checkpoint at which formation of aircraft or ships joins main force.

force-sensing controller Pilot's primary flight-control input (stick/pedals) which senses applied force without noticeable movement.

force structure Currently effective operational inventory (US).

force vector Line in force diagram representing force magnitude, direction and point of application.

fore-and-aft level Gravity-controlled indicator of pitch attitude (arch).

forebody 1 The front portion of a body in atmospheric flight; in * strake can mean front half of fuselage.
2 Planing bottom of float or hull upstream of step.

forebody strake Low-aspect-ratio extensions of wing at root along sides of fuselage; generate powerful vortices at high AOA to improve handling in extreme positive-acceleration manoeuvres.

forecast Statement of expected meteorological conditions at given place during specified period; air-navigation * include wind velocity at selected heights, cloud, visibility, precipitation, ice formation, and barometric pressures at airfields and sea level.

foreflap Leading member of double or triple-slotted flap.

foreign air carrier One registered in foreign country, except in case of multi-national carriers (eg SAS or Air Afrique) in collaborating countries.

foreign military sales Portion of United States military assistance authorized by Foreign Assistance Act (1961 as amended); differs from Military Assistance Program Grant Aid in that it is purchased by recipient country.

foreplane Horizontal aerofoil mounted on nose or forward fuselage to improve take-off and low-speed handling, esp of delta aircraft where wing lift is lost because of upward movement of elevons; * can be fixed or retractable, fixed-incidence or rotating, and have slats, flaps or elevators.

forging Shaping metal softened by heating by slow, rapid or repeated blows, with or without a shaped female die or matching male/female dies.

forked rod PE (4) connecting rod having forked bearing on crankshaft, fitting over big end of matching blade-type rod.

formability Unquantified measure of ease with which material can be shaped through plastic deformation.

format 1 Size and shape of map, chart or photo negative or print.
2 One of several selectable types of presentation for instrument (eg ADI or HSI) or display, such as moving map, radar map, alphanumerics, attitude indication, flight planning or en route.
3 Fortran matrix abstraction technique.

formation Ordered arrangement of two or more vehicles proceeding together.

formation flight More than one aircraft which, by prior arrangement between pilots, navigate and report as single aircraft; FAA formation limits are no more than one mile laterally or longitudinally and within 100 ft vertically from leader.

formation light(s) Fitted to aircraft to enable other aircraft to formate on it at night.

form block Block or die usually made of wood, zinc, steel or aluminium, over or into which sheet metal is formed.

form die One which performs bending and sometimes light drawing operations upon flat blank.

form drag Pressure drag minus induced drag.

former Light secondary structure added to maintain or improve external shape; eg around basic box fuselage to give curved cross-section, or extra false wing ribs ahead of front spar to maintain profile where curvature is sharp.

form factor Physical overall dimensions of a body, especially one carried externally, taken into account not so much aerodynamically as to avoid hardware conflicts.

forming Forcing flat material to assume desired contours and curves, esp compound curvatures, by such means as drop hammer, flanging machine, hydraulic press, stretch press, hand forming etc.

forming roll Bends sheet metal into cylinders of various diameters and other single-curvature shapes.

Form 700 In many English-speaking air forces, document signed by captain on taking over aircraft to signify that he is satisfied that every pre-flight maintenance inspection has been completed.

Formula 1 Air-racing rules, laid down by FARA, for uniform class of racers (includes engine not over 200 cubic inches, wing not less than 66 square feet, fixed landing gear and carefully defined cockpit/windshield geometry).

formula costs Direct operating costs for transport aircraft as calculated to ATA formula, which assumes indirect costs as 100 per cent of direct and standardized values for passenger, freight and mail revenues.
Formula V Lightweight FARA class with VW-derived engines.
FORS Foreign airlines representatives (Sweden).
Fortran Formula translator; computer language.
forward aeromedical evacuation Provides airlift for patients between battlefield and initial and subsequent points of treatment within combat zone.
forward air controller Member of tactical air control party who, from forward ground or airborne position, controls aircraft engaged in close air support of ground troops (DoD).
forward air control post Mobile tactical air control radar used to extend coverage and control in forward combat area.
forward lock Prevents selection of reverser except after landing; triggered by oleo deflection or, if possible, bogie-beam tilt.
forward oblique Oblique photograph of terrain directly ahead.
forward operating base Airfield used indefinitely to support tactical operations without establishing full support facilities.
forward operating location Forward operating base.
forward scatter Scattering of radiant energy into hemisphere of space bounded by plane normal to incident radiation and lying on side toward which radiation was advancing; opposite of backward scatter.
forward slip See *sideslip*.
forward speed Component of speed in horizontal plane.
forward stagnation point Point on leading edge of body in airstream which marks demarcation for airflow on either side; boundary-layer air is stationary.
forward supply point En route or turnaround station at which selected aircraft spares are prepositioned for support of assigned mission(s).
forward sweep Opposite of sweepback, wing tips being further forward than roots.
forward tilt Of helicopter rotor, forward angular deviation of locus of centroid of blade sections from plane of rotation.
forward tilt-wing See *slew-wing aeroplane*.
FOS 1 Fail-operational, fail-safe.
2 Faint-object spectrograph.
FOSA Fondation des Oeuvres Sociales de l'Air (F).
FOSS, Foss Fibre-optic sensor system(s).
FOST Flag Officer, Sea Training.
foster parent Company contracted to provide service support, posibly extending to major modification and update, for an imported type of military aircraft.
FOT Frequency of optimum transmission.
FOT&E Follow-on test and evaluation (continues after entry to service, USAF).
fount Range of characters for electronic display.
fountain Vertical rising column of hot air, usually plus entrained debris, formed by jets of VTO jet aircraft hovering at low level; on impact with fuselage can exert undesired suckdown effect.
four-bank eight Flight training manoeuvre similar to figure of eight except that outer portions of loops are not circular but consist of two 45° turns linked by short, straight flightpath.
four-course beacon See *radio range*.
4-D 3-D plus time.
4-D R-nav Terminal guidance sufficiently accurate to put arrival's wheels on runway within guaranteed window of $\pm$ 10 s.
Fourier expansion Expansion of waveform or other oscillation in terms of fundamental and harmonics.
Fourier integral Representation of (x) for all values of x in terms of infinite integrals.
Fourier series Representation of function f(x) in interval (–L, L) by series consisting of sines and cosines with common period 2L; when f(x) is even, only cosine terms appear; when odd, only sine terms appear.
four-minute turn See *standard turn*.
four-poster Jet engine for VTOL with two front fan nozzles and two rear core-jet nozzles, which in VTOL mode provides lift from four jets in rectangular pattern; or an aircraft equipped with such an engine.
four-wing configuration See *cruciform wing*.
FOV Field of view.
FOW Family of weapons.
Fowler flap Special form of split flap that moves at first rearwards and then downwards along tracks, thus producing initial large increase in lift and at full deflection giving high lift and drag for landing.
fox away Verbal code: "AAM has been fired or released" (DoD).
Fox Code Air-combat numeral code, usually: 1, SARH missile selected; 2, IR missile selected; 3, guns; 4, fired outside parameters to distract enemy.
FP 1 Flight progress.
2 Fluoro-polyamide.
fp Freezing point.
FP 70 Anti-fire foam compound, hydrolized protein plus perfluorocarbon surfactant.
FPA 1 Fire Protection Association (UK).
2 Flightpath angle.
3 Flying Physicians' Association (US).
4 Flightpath accelerometer.
5 Focal-plane array.
FP area Food-preparation area.
FPAS 1 Focal-plane array seeker.
2 Flight-profile advisory system.
FPB Fuel preburner.

FPCS Flight-path control system.
FPE Fédération des Pilotes Européennes (federation of European women pilots).
FPF Fixed-price firm.
FPG Force per g.
FPI 1 Fluoropolyimide.
2 Fixed-price incentive.
3 Fluorescent penetrant inspection.
FPIF Fixed-price incentive firm.
FPIS 1 Fixed-price incentive contract with successive targets.
2 Forward propagation by ionospheric scatter.
FPL Flight-plan message.
FPLN Flight plan.
FPM 1 Flightpath miles.
2 Feet per minute (ft/min preferred).
FPN Fixed-pattern noise.
FPOV Fuel preburner oxidizer valve.
FPP 1 Fixed-pitch propeller.
2 Ferry Pilots' Pool.
FPPS Flight-plan processing system.
FPR 1 Fan pressure ratio.
2 Fixed price, redeterminable.
FPRM Flight phase related mode.
FPS, fps 1 Foot-pound-second system.
2 Feet per second.
3 Fine-pitch stop.
FPSR Foot-pound-second-Rankine, traditional British system of engineering units.
FPTS Forward propagation by tropospheric scatter.
FQI Fuel-quantity indicator (or indication).
FQIS FQI switch (or system).
FQPU Fuel-quantity processor unit.
FQT Formal qualification testing.
FQTI Fuel-quantity totalizer indicator.
FR 1 Flight refuelling.
2 Flight recorder.
3 Fighter reconnaissance.
F/R 1 Function reliability.
2 Final run.
F_R Aerodynamic loading (force) due to Rth mode.
fr Fuel required.
fractional distillation Heating crude petroleum to moderate temperature at atmospheric pressure and condensing different fractions separately.
fractionation Use of large numbers of small (nuclear) warheads, esp in counterforce attack; hence to fractionate.
fracto- Prefix, clouds broken up into irregular, ragged fragments.
fragmentary order Abbreviated form of operation order which eliminates need for restating information contained in basic operation order (DoD). Usually a daily supplement to SOOs governing conduct of specific mission.
fragmentation bomb Charge contained in heavy case designed to hurl optimum fragments on bursting.
frag order Fragmentary order.
FRAM, Fram Fleet rehabilitation and modernization.
F_{ram} Ram drag.
frame 1 In photography, single exposure within continuous sequence.
2 Transverse structural member of fuselage, hull, nacelle or pod, following its periphery.
3 Of engine, structural transverse diaphragm carrying main shaft bearing.
4 Picture area scanned in TV, video or CRT (US = field).
frameless canopy Single blown transparency in metal peripheral frame.
framing pulses Transmitted in TV/video, SSR, IFF, ATCRBS and similar systems to synchronize transmitted and received timebases.
frangible Designed to, or likely to, shatter on impact.
fraudulent echo Radar echo produced by DECM.
FRC 1 Flight reference card.
2 Fibre-reinforced composite (or concrete).
FRCI Flexible reusable carbon insulation.
freak Verbal code meaning frequency in MHz (DoD).
freddie Air-intercept controlling unit (DoD).
free-air anomaly Difference between observed and theoretical gravity computed for latitude and corrected for elevation above or below geoid.
free-air overpressure Unreflected pressure in excess of ambient created by blast wave from explosion.
free airport One at which people and goods may be transhipped without customs charges or examination.
free-air tunnel Aerodynamic or other test with specimen moved through atmosphere, eg on aircraft or rail flatcar or rocket sled.
free atmosphere That portion of Earth's atmosphere, above planetary boundary layer, in which effect of Earth's surface friction on air motion is negligible, and in which air is usually treated (dynamically) as ideal fluid. Base usually taken as geostrophic wind level.
free balloon One floating untethered.
free-balloon concentration ring Ring to which are attached ropes suspending basket and to which net is secured; also known as load ring.
free-balloon net Distributes basket load over upper surface of envelope.
free-body principle Stress-analysis procedure that involves isolating structure, considering it to be held in equilibrium by loads acting upon it.
free canopy Cockpit canopy that is non-jettisonable.
freedoms Five basic freedoms relating to air traffic negotiated in bilateral air agreements between governments of pairs of countries. First freedom is right to fly across territory of other country with whom agreement is made. Second is right to make technical or non-traffic stop in that coun-

try. Third is right to set down in that country traffic emplaned in home country of airline. Fourth is right to pick up traffic in that country destined for airline's home country. Fifth is right to pick up traffic in other country and carry it to any third state.

free drop Dropping packaged equipment or supplies without parachute.

free electron Not bound to an atom.

free fall 1 Fall of body without guidance, thrust or braking device.
2 Free motion along Keplerian trajectory, in which force of gravity is counterbalanced by force of inertia.
3 Parachute jump in which parachute is manually activated at discretion of parachutist, or automatically at pre-set altitude.
4 Acceleration g under standard conditions, = $9{\cdot}807\ m/s^2$.

free-fall altimeter Worn by parachutist in free-fall jump to deploy parachute at pre-set altitude.

free-fall(ing) bomb Bomb without guidance.

free-fall landing gear Designed to be lowered by force of gravity and wind load.

free-fall model Unpowered model for spinning or other tests intended to fall freely after release.

free fan Name promoted by Boeing for UDF and propfan propulsion to avoid image of "propeller".

free flight Without guidance except, possibly, simple stabilizing autopilot.

free-flight model One normally ground-launched and self-propelled, usually instrumented for aerodynamic or flutter measures.

free-flight tunnel One in which model can be observed in free flight, ie unmounted or fired from gun.

free gyro Two degrees of freedom, spin axis may be oriented in any specified attitude and not provided with erection system.

free jet Fluid jet after emission from nozzle.

freelance Verbal code: "Self-control of air-intercept aircraft is being employed" (DoD).

free-molecule flow 1 Flow regime in which mean free path is at least ten times typical dimension of flying body, as at height above 180 km.
2 Flow about body in which collisions between molecules of fluid is negligible compared with collisions between these and body.

free parachute Deployed manually by parachutist, not by static line.

free-piston engine Any of several types in which hot gas is generated between pistons oscillating freely in linear or toroidal cylinders.

free radical Atom or group of atoms broken away from stable compound by application of external energy; often highly reactive.

free-return trajectory One in which crippled spacecraft would fall back to Earth.

free rocket Unguided rocket.

free space Ideal homogeneous medium possessing dielectric constant of unity and in which there is nothing to reflect, refract or absorb energy.

free-spinning tunnel Vertical wind tunnel used to test spinning characteristics using free models.

free-standing propellant Not case-bonded.

free stream Fluid outside region affected by aircraft or other body.

free-stream capture area Cross-sectional area of column of air ingested by ramjet.

free streamline One passing well away from moving body. Streamline separating fluid in motion from fluid at rest.

free-stream Mach number Mach number of body measured in free stream, unaccelerated by body's presence.

free turbine One that drives output shaft and is not connected to compressor.

free-vortex compressor Axial compressor designed to impart tangential velocities inversely proportional to radius from axis.

free-vortex flow Persisting in fluid remote from source or solid surface, eg tornado.

freeze 1 To arrest dynamic operations, eg in simulator training.
2 Radar mode which, once commanded, permits one more scan; emissions then cease and display remains active but frozen until * button is pushed a second time.

freeze-out Method of controlling humidity by condensing water vapour, and possibly carbon dioxide, over cold surface.

freezing 1 Stage in design when all major features are irrevocably settled, thus enabling detail design to start.
2 Manually arresting input to display, leaving static (prior) situation for study.

free zone Customs-free area.

freight Cargo, including mail and unaccompanied baggage but excluding express.

freight consolidating Process of receiving shipments of less than carload/truckload size and assembling them into carload/truckload lots for onward movement to ultimate consignee or break-bulk point.

freight container See *container*.

freight doors Designed to take freight, vehicles or containers.

French landing With plenty of power on.

Freon Tradename for family of halogenated hydrocarbons containing one or more fluorine atoms widely used as refrigerant medium (eg in vapour-cycle air conditioning), as fire extinguishant and as aerosol propellant.

frequency 1 Reciprocal of primitive period of time-periodic function, measured in hertz or multiple thereof.
2 Number of services operated by airline per day or per week over particular route.

frequency-agile Of radar, operating frequency can be changed with every scan of antenna by means of tuning magnetrons.

frequency band Continuous range of frequencies extending between two limiting values. The tables below give the international designations for the various families of bands.

CCIR BAND DESIGNATIONS

Band number	Frequency	Conventional name
4	(kHz) 3–30	very low (VLF)
5	30–300	low (LF)
6	300–3,000	medium (MF)
7	3,000–30,000	high (HF)
8	(MHz) 30–300	very high (VHF)
9	300–3,000	ultra-high (UHF)
10	3,000–30,000	super high (SHF)
11	30,000–300,000	extremely high (EHF)
12	300,000–3,000,000	

ELF (extremely low frequency) has been proposed for the band from 1 to 3,000 Hz. Frequency bands used by radar were at first designated by letters for military secrecy:

Band letter	Approx frequency (GHz)
P-band	0·225–0·39
L-band	0·39–1·55
S-band	1·55–5·20
X-band	5·20–10·90
K-band	10·90–36·00
Q-band	36·00–46·00
V-band	46·00–56·00

Radar requirements then resulted in a new series of bands which tended to be centred on convenient wavelengths rather than frequencies:

Band letter	Approx frequency (GHz)	Approx wavelength (cm)
L	12	30–15
S	2–4	15–7·5
C	4–8	7·5–3·75
X	8–13	3·75–2·3
Ku (Q)	13–20	2·3–1·5
K	20–30	1·5–1·0
Ka	30–40	1·0–0·75

Today the US has led in introducing unilaterally a new system covering a wider range:

Band	Approx frequency	Approx wavelength
HF	10–30 MHz	30–10 m
VHF	30–100 MHz	10–3 m
A	100–300 MHz	3–1 m
B	300–500 MHz	100–60 cm
C	0·5–1 GHz	60–30 cm
D	1–2 GHz	30–15 cm
E	2–3 GHz	15–10 cm
F	3–4 GHz	10–7·5 cm
G	4–6 GHz	7·5–5 cm
H	6–8 GHz	5–3·75 cm
I	8–10 GHz	3·75–3 cm
J	10–20 GHz	30–15 mm
K	20–40 GHz	15–7·5 mm
L	40–60 GHz	7·5–5 mm
M	60–100 GHz	5–3 mm

frequency bias Constant frequency added to signal to prevent its frequency falling to zero.

frequency channel 1 Band of frequencies which must be handled by carrier system to transmit specific quantity of information.
2 Band of frequencies within which station must maintain modulated carrier frequency to prevent interference with stations on adjacent channels.
3 Any telecommunications circuit over which telephone, telegraph or other signals may be sent.

frequency departure Variation of carrier frequency or centre frequency from assigned value.

frequency deviation 1 Maximum difference between IF (1) of FM wave and frequency of carrier.
2 In CW or AM transmission, variation of carrier frequency from assigned value.

frequency distortion Produced by unequal amplification of signals or reproduction of sounds of different frequency; usual criterion for high-fidelity reproduction is level amplification over 20–15,000 Hz.

frequency division multiplex Telecommunications allowing two or more signals to travel on one network simultaneously by modulating separate subcarriers suitably spaced in frequency to prevent interference.

frequency drift Slow change in frequency of oscillator or transmitter with time.

frequency equation Relates phase speed to wavelength and to physical parameters of system in linear oscillation; also known as dispersion equation.

frequency-hopping Unpredictable continual and rapid changes of frequency of radar or other military electronics to defeat hostile ECM.

frequency modulation, FM Instantaneous frequency of modulated wave differs from carrier by amount proportional to instantaneous value of modulating wave. Combination of phase and frequency modulation commonly referred to as FM.

frequency-modulation radar One in which range is measured by interference beat frequencies between transmitted and received FM waves.

frequency monitor Stabilized receiver giving audible or visual indication of any departure of transmitter from assigned frequency.

frequency pairing Association of each VOR frequency with a specific DME channel to reduce workload and errors.

frequency parameter Ratio of airspeed to product of frequency of oscillation and representative length of oscillating system.

frequency response 1 Portion of EM spectrum sensed within specified limits of error.
2 Response as function of excitation frequency.

frequency-selective Response only to narrow frequency band.

frequency-shift keying System of telegraph signalling in which keyed signal imposes small frequency shift on carrier; frequency changes of received signal are converted to amplitude changes.

frequency swing Peak difference between maximum and minimum frequencies of FM signal.

frequency tolerance Extent to which carrier (or centre of emission bandwidth) is permitted to depart from authorized frequency because of instability.

frequency-wild Electric power system whose frequency is not stabilized but varies with rotational speed of generators.

Fresnel zone Any spatial surface between transmitter and receiver, or radar and target, over which increase in distance over straightline path is equal to integer multiple of half wavelength.

fretting Rubbing together of solid surfaces, esp slight movement but high contact pressure.

FRFI Fuel-related fare increase.

FRH Flap-retraction height.

friction Force generated between solids, liquids or gases opposing relative motion.

friction coefficient Friction (static, on point of relative motion, or dynamic) divided by perpendicular load pressing surfaces together.

friction horsepower Indicated horsepower minus brake horsepower.

friction layer Planetary boundary layer.

friction lock Device in which friction is used to prevent movement; eg of throttle levers.

friction wake That downstream of streamlined non-lifting body.

friction welding Welding by rotating two parts together under load until surfaces are on point of melting and then forcing together to squeeze out joint material, simultaneously arresting rotation.

friendly Not hostile, contact positively identified as such (DoD).

FRIG Floated rate-integrated (or integrating) gyro.

Frise aileron Aileron having inset hinges and bevelled along upper leading edge; when lowered forms continuation of wing upper surface but when raised nose protrudes below wing, increasing drag and equalizing aileron drag in banked turn.

FRM Failure-related mode.

FRO Failure requiring overhaul.

FROD Functionally related observable differences.

FROG Free rocket over ground.

front 1 Boundary at Earth's surface between two contrasting air masses; usually associated with belt of cloud and precipitation, and more or less sharp change in wind (see *cold* *, *occluded* *, *stationary* *, *warm* *).
2 Occupant of front cockpit of tandem-seat aircraft, especially if aircraft commander (colloq, usually US/NATO).

frontal area Projected cross-section area of body viewed from front.

frontal attack Air intercept which terminates with heading crossing angle greater than 135° (DoD).

frontal weather To be expected in front: clouds, rain, temperature variation and other phenomena.

front course sector That situated on same side of localizer as runway.

front end Various meanings, including system location (actual geometry immaterial) where EDP program is developed, and editing, compiling and peripheral access take place.

front ignition Solid rocket motor with igniter at end of filling or grain furthest from nozzle.

frontogenesis Process which produces discontinuity in atmosphere or increases intensity of existing front, generally caused by horizontal convergence of air currents possessing widely different properties.

frontolysis Process which tends to weaken or destroy a front.

front stagnation point See *forward stagnation point*.

frost Small drops of dew which freeze upon contact with object colder than 0°C such as aircraft passing from cold air into warmer humid air. Glazed * (rain ice) is layer of smooth ice formed by rain falling on object at below 0°C. Hoar * is white semi-crystalline coating.

frost point Temperature of air at which frost forms on solid surface at same temperature.

Froude number Non-dimensional ratio of inertial force to force of gravity for fluid flow; reciprocal of Reech number; $N_{fr} = v^2/lg$, where v is characteristic velocity and l characteristic length (may be given as square root of this ratio).

FRP 1 Fibre-reinforced plastics.
2 Flight-refuelling probe.

FRS 1 Fan rotation speed (N_1).
2 Fighter, reconnaissance, strike (role prefix, UK).

FRSI Flexible reusable silica insulation.

FRTV Forward repair and test vehicle.

fruiting SSR responses of aircraft to interrogation by other stations or responses of different aircraft to interrogation by other stations, in either case not synchronized with desired response by aircraft interrogated by own station; hence defruiting usually simple.

FRUSA Flexible rolled-up solar array.

FrW Friction welding.

FRZ Freeze (information or display).

FS 1 Fuselage station.
2 Frame station.
3 Fail-safe.
4 Fighter squadron.

Fs Fractostratus.

FSA 1 Final squint angle.
2 Force-structure aircraft (US).
3 Fuel-savings advisory (FSA/CAS system).

FSAA Flight Simulator for Advanced Aircraft (NASA).
FSAGA First sortie after ground alert.
FSAS Fuel-savings advisory system.
FSAT Full-scale aerial target.
FSB Fasten seat belts; hence * turbulence.
FSC Force-sensing controller.
FSCL Fire-support co-ordinating line.
FSD 1 Full-scale deflection.
2 Full-scale development phase.
FSE Fleet supportability evaluation.
FSED Full-scale engineering development.
FSEU Flap-system electronic unit.
FSF 1 Fight Safety Foundation (US).
2 Svenska Flygsportförbundet (Sweden).
FSFT Full-scale fatigue test.
FSG Fluid sphere gyro.
FSI Field service instruction.
FSII Fuel-system icing inhibitor.
FSK Frequency-shift keying.
FSL Full-stop landing (aircraft brought to halt).
FSN 1 Factory serial number.
2 French-speaking nations.
3 Field service notice.
4 Federal stock number.
FSOV Fuel shut-off valve.
FSP/UPT Flight-screening program and undergraduate pilot training (USAF).
Fspil Inlet spillage drag correction.
FSR 1 Frequency set-on receiver.
2 Field service representative.
3 Flight-safety reporting.
FSS 1 Flight service station.
2 Flight Standards Service (FAA).
3 Flying selection squadron (RAF).
FSSR Federal Safety Standard Regulations.
FSVL Fédération Suisse de Vol Libre (Switz).
FSW Forward-swept wing.
FT 1 Fault-tolerant.
2 Fast track.
F-T diagram Flight-time diagram.
FTA Fatigue test article.
FTAE First-time-around echo.
FTAJ Frequency/time ambiguity jamming.
FTC, ftc 1 Fast time constant.
2 Federal Trade Commission (US).
FTCA Future tactical combat aircraft.
FTD 1 Field training detachment (USAF).
2 Foreign Technology Division (USAF).
FTEP Full-time echo protection.
FTF Flygtekniska Föreningen (Sweden).
FTG 1 False-target generator.
2 Fitting.
FTH Full-throttle horsepower.
FTI 1 Fixed time interval.
2 Fixed target imagery (or indicator).
FTIT Fan-turbine inlet temperature.
FtL Foot-lambert (non-SI unit of luminance or brightness).
FTLB Flight-Time Limitation Board (UK).
FTLO 1 Fast-tuned local oscillator.
2 False-target lock-on.
FTM 1 Frequency/time modulation.
2 Flight-test mission objective(s).
FTO Flexible take-off.
FTP Functional test procedure.
FTR 1 Failed to return.
2 Flat-tyre radius.
3 False-target rejection.
Ftr Fighter.
FTS 1 Flying training school.
2 Fatigue-test specimen.
3 Federal Telecommunications System.
4 Flexible turret system.
FTSA Fault-tolerant system architecture.
FTSC Fault-tolerant spacecraft (or spaceborne) computer.
FTV Flight test vehicle.
FTX Field training exercise.
FU Fuel uplifted.
fu Fuel used.
fudge factor See Addenda.
fuel Substance used to produce heat by chemical or nuclear reaction; usual chemical reaction is combustion with oxygen from atmosphere or from rocket oxidant (see *propellant*).
fuel accumulator Container for storing fuel expelled during starting cycle to augment flow momentarily at predetermined fuel pressure.
fuel additive Any material or substance added to a fuel to give it some desired quality; eg tetraethyl lead added as an anti-detonation ("knocking") agent.
fuel-air explosive Selected liquid fuels which on warhead impact are scattered in fine cloud of large volume alongside target; this is then detonated (combining with atmospheric oxygen) less than 1 s after impact with blast effects generally greater than that of same mass of conventional explosives.
fuel-air mixture analyser Measures PE (4) air-to-fuel ratio. Chemical *** measures absorption of CO_2 by substance such as sodium or potassium hydroxide; physical (electro-chemical *** measures difference in electrical resistance between two sampling cells, one open to air and other to exhaust.
fuel blending facility One having authority to mix hydrocarbon fractions to produce piston or turbine fuels to specification, with correct additives.
fuel bypass Maintains fuel pressure in carburettor float chamber of supercharged PE (4) at fixed level above carburettor air pressure.
fuel capacity Unless otherwise stated means actual volume of tank(s), not of system, and has no relevance to usable capacity.
fuel cell Device which converts chemical energy directly into electricity; differs from storage battery in that reactants are supplied at rate determined by electrical load.
fuel consumption Measured on volume and mass basis; 1 mpg = 0·354 km/l (UK), 0·425 km/l

(US); 1 gal/mile = 2·825 l/km (UK), 2·353 l/km (US).

fuel control unit Governs engine fuel supply in accordance with pilot demand, ambient conditions and engine limitations.

fuel-cooled Cooled by fuel, either en route to engine or recirculated back to tank (see *regenerative cooling*).

fuel cut-off Device for cutting off the supply of metered fuel to cylinders or a combustion chamber.

fuel dipper Fine-adjustment control of fuel flow to turbine engine, usually triggered to reduce flow at specific time, eg when firing guns.

fuel grade Quality of PE (4) petrol (gasoline), esp as defined by anti-knock rating; previously called octane number. Usual grades are 80 (dyed red), 100 (green), 100L (green or blue, lower TEL content) and 115 (purple).

fuel injection Inevitable in diesel or compression-ignition engines; term normally refers to Otto-cycle PE (4) with mechanical injection of measured quantities of fuel either into carburation induction system or directly to cylinders. Unlike carburettor, ** unaffected by aerobatic or evasive manoeuvres and not prone to icing or fire hazards.

fuel jettison Rapid discharge of fuel from aircraft in emergency.

fuel-jettison time That required to reduce weight from MTO to MLW.

fuel lag Deliberate short delay in injection of one propellant into rocket thrust chamber to establish particular ignition sequence.

fuel manifold Peripheral main pipe with branch pipes distributing fuel to all burners of gas turbine.

fuel-pressure switch Ensures full current is not applied to electric starter until fuel pressure has reached predetermined level.

fuel shut-off See *cut-off*.

fuel sink Total heat capacity of fuel as receptacle for surplus onboard energy; unit should be J.

fuel sloshing Gross movements of fuel or oxidant in part-empty tank.

fuel tank See *bag tank, integral tank, rigid tank, drop tank*.

fuel vent Small pipe used to equalize pressure inside and outside tank.

FUF Favourable unbalanced field.

fufo, FUFO Full-fuzing option (NW).

fuh Fuel mass at given height (usually rocket).

Fulat Federazione Unitaria Lavoratori Trasporte Aereo (I).

full annealing Heating above critical temperature, or crystalline-transformation point, followed by slow cooling through range of transformation; results in relief of residual stress, greater ductility, increased toughness but lower strength.

full flight regime Designed to operate throughout each flight, eg FFR autothrottle.

full lateral Describes FMC or FMS (1) exercising total authority via AFCS from just after take-off until final approach.

full-pressure suit Completely enclosing wearer's body and able to sustain internal gas pressure convenient to human functions (see *partial pressure suit, water suit, spacesuit*).

full rudder, aileron or elevator Hard-over demand signal by pilot or flight-control system moving surface to limit of travel.

full-scale development Period when system or equipment and items for its support are tested and evaluated; intended output is pre-production system which closely approximates final product, documentation for production phase, and test results which demonstrate product will meet requirements (USAF).

full-wave rectifier Two elements in split circuit rectify both positive and negative halves of waveform, currents combining unidirectionally at output.

fully active See *active*.

fully developed flow Flow of viscous fluid over solid surface on which boundary layer has reached full thickness and velocity distribution remains constant downstream.

fully expanded nozzle Normally, nozzle of jet engine, esp rocket, whose supersonic divergent portion expands jet to ambient pressure at exit plane.

fully factored Multiplied by ultimate factor of safety.

fully feathering See *feathering*.

fully grown dimension Dimension of part after process that results in enlargement, esp high-temperature creep under tensile load.

fully ionized plasma All neutral particles have lost at least one electron. With hydrogen no further ionization is possible; other atoms can be further excited.

Fulton system Folding recovery system on nose of aircraft for retrieving space capsules and other parachuted loads in mid-air; used in modified form for repeated pick-up of persons or other loads from ground.

fumble factor Factor of safety.

functionally shaped controls Manual controls shaped to confirm their function, eg terminating in miniature wheel or flap.

fundamental frequency 1 Of periodic quantity, lowest component frequency of sinusoidal quantity which has same period.

2 Of oscillating system, lowest natural frequency; normal associated mode of vibration is fundamental mode.

3 Reciprocal of period of wave.

4 Lowest resonant antenna frequency without added inductance or capacity.

fundamental units Mass, length and time, from which all other units are derived. Main systems in use are SI, foot-pound-second, centimetre-gramme-second and metre-kilogramme-second.

funding Process of finding and securing political approval for sums of money needed for military programmes (US).

funicular polygon See *force diagram*.

funnel 1 Safe approach area centred on ILS glidepath and normally considered to extend 0·5° above and below and 2° left and right; aircraft inside * should be able to land.
2 Pre-1950 term for final approach area to runway, esp any lighting or other guidance system for channelling aircraft to runway.

fu_0 Fuel mass at liftoff.

furaline French rocket fuel; aniline with traces of anti-corrosive additive.

furlong Non-SI unit of length, = 201·17 m.

furnishing Non-structural cabin interior trim, seats, galley, toilet, baggage racks and associated features in commercial aircraft.

furnishing mock-up Whole or part of cabin prepared for particular customer's evaluation and refinement.

FUS Far-UV spectrometer.

fuse Deliberate weak link with wire of low melting point, to protect electric circuit against excessive current. See *fuze*.

fusee Pyrotechnic squib installed in solid-propellant case to ignite charge over whole length.

fuselage Main body of an aerodyne, absent in all-wing designs; when tail is attached to booms, called nacelle; with planing bottom called hull.

fuselage reference line Straight line used as reference from which basic dimensions are laid out and major components located; usually along plane of symmetry and at convenient height.

fusible plug Several types, most important being fitted in main landing wheels; if temperature rises to dangerous value (at which tyre could burst) after excessive braking, ** blows and releases pressure.

fusion 1 So-called thermonuclear process whereby nuclei of light elements combine to form nucleus of heavier element, releasing very large amounts of energy; can be controlled and sustained.
2 Collection and integration of outputs from all of a range of sensors and warning systems, hence *IDF*.

fusion bomb One using energy of thermonuclear fusion.

fusion power density Power generated per unit volume in controlled thermonuclear plasma; using deuterium at 10^{16} particles/ml and 60 kV, *** is about 1 kW/ml, as kinetic energy of reaction products.

fusion reactor Reactor in which thermonuclear fusion takes place.

fusion welding Fusing edges of two base metals by using a welding flame to melt edges of metals, and a welding rod, which is similar in composition, to fuse or weld two metals together.

FUSS Flaps-up safety speed.

future cycle accumulation Number of LCF or operating (mission) cycles operator plans to accumulate on given hardware.

FUV Far ultra-violet.

fuze Device or mechanism designed to start detonation of high explosive under proper conditions of heat, impact, sound, elapsed time, proximity, external command, passage of electric current or other means, and usually without danger of detonation before weapon is armed.

FVA Federación Venezolana de Aeroclubes (Venezuela).

F_{VAC} Vacuum thrust.

FW 1 Fiscal week.
2 Fixed wheel (sailplane).

f.w. Full wave.

FW&A Fraud, waste and abuse.

FWC 1 Filament-wound cylinder.
2 Flight warning computer.

fwd Forward.

FWE Foreign weapons evaluation (US).

FWHM Full wave, half modulation.

FWOC Forward wing operations centre (RAF).

FWS 1 Fighter Weapons School (USAF).
2 Flight warning system.

FWWS Food, water and waste subsystems.

FY Fiscal year.

F/Y First-class and economy or tourist.

FYDP Five-year defence plan.

F/Y ratio Fission/yield ratio.

fZ Transverse VSI (magnetic) component.

FZP Fresnel zone plate.

G

G 1 Giga, multiplied by 10^9.
2 Geostrophic force.
3 Geared (US PE [4]).
4 Universal gravitational constant = 6·673 ×

10^{-11} Nm^2kg^{-2} (see *gravitation*).
5 Stress imposed on body due to applied force causing acceleration.
6 Gun (US DoD).
7 Gunner (aircrew).
8 Role prefix, parasite carrier (USAF 1949–51).
9 Glider category (US Army 1919–26).
10 Graphite.
11 Group or Geschwader.

g 1 Acceleration due to Earth gravity, international standard value being 9·80665 m/s^2, assumed at standard sea level.
2 Gramme.
3 Grid.
4 Suffix, gauge pressure.
5 Suffix, height AGL.

g-break Sudden change of aircraft trajectory away from previous straight line in lateral or upward direction, eg following fast low-level run over airfield before landing.

G-display Rectangular, target is at centre (when radar aerial is aimed at it) and grows lateral "wings" as range is closed; aiming errors result in appropriate displacements of blip from centre.

g-force Inertial force, that needed to accelerate mass, usually expressed in multiples of gravitational acceleration.

G-layer Layer of free electrons in ionosphere occasionally observed above F_2 layer.

g-meter Indicates acceleration, usually in vertical plane.

g-suit Worn by pilot or astronaut; exerts pressure on abdomen and lower parts of body to prevent or retard collection of blood below chest under positive acceleration. Not necessarily pressure suit.

g-tolerance Tolerance of subject to acceleration of specified level and duration.

GA 1 General aviation.
2 Gas analysis.
3 Gimbal angle.
4 Group accounting.
5 See *general-arrangement drawing*.
6 Go-around.
7 Military code for lethal nerve gas developed (G, 1937) as Tabun.

GAA General Aviation Association (Australia).

GaAs Gallium arsenide.

GAC 1 Gust-alleviation control.
2 Go-around computer.

GACC Ground-attack control capability.

GACW Gust above constant wind.

gadget Code, radar equipment; type of equipment indicated by letter, followed by colour to indicate state of jamming: green *, clear; amber *, sector partially jammed; red *, sector completely jammed; blue *, completely jammed (DoD).

GADO General aviation district office (FAA).

gage Gauge (US).

gaggle Group of aircraft (say, five to 20) flying together but with no semblance of formation.

gain 1 General term for increase in signal power in transmission, usually expressed in dB.
2 Increase or amplification; antenna * (* factor) is ratio of power transmitted along beam axis to that of isotropic radiator transmitting same total power; receiver * (video *) is amplification by receiver.

Gal, gal Non-SI unit of small acceleration, gal (Galileo) = $10^{-2}m/s^2$ = 1,000 mgal.

GALAT, Galat Groupement Aviation Légère de l'Armée de Terre (F).

GALCIT Guggenheim Aeronautical Laboratory, Caltech.

galena Lead sulphide, PbS, used in IR cells.

gallery 1 Gas-turbine fuel manifold.
2 Fluid conduit formed within three-dimensional volume of material, eg drilled through body of pump.

galling Pitting or marring of finished surface, esp bearing surface, because of vibration or wear.

gallon Non-SI unit of liquid volume measure; Imperial * = 1·20095 US * = 4·54596 litres; US * = 0·83267 Imperial * = 3·78533 litres.

galvanic corrosion Electrolytic action caused by contact of dissimilar metals or formation of oxygen cell in contact with metal.

galvanizing Coating of metal, esp steel sheet, by dipping in bath of molten zinc; protects from galvanic corrosion.

galvanometer Instrument which measures electric current passed through pivoted coil in magnetic field. With shunted external resistance, used as ammeter; with series resistance, voltmeter.

GAMA General Aviation Manufacturers' Association (US).

Gambit General anti-material bomblet with improved terminal effects.

gamma 1 γ, ratio of specific heats of gas.
2 Logarithmic function; ratio of contrast of transmitted scene to that of received display.

gamma $_r$, γ_r Ratio of viscous damping to critical damping.

gamma rays High-energy, short-wavelength EM radiation; very penetrating, similar to X-rays but nuclear in origin and usually more energetic. Symbol γ.

GAMTA General Aviation Manufacturers and Traders Association (UK).

gang drill Series of drill presses in close proximity on one bed, or drills operated simultaneously through interconnected chucks.

gantry Large crane for erection and servicing of launch vehicles; * straddles vehicle and runs on tracks crossing launcher and work area.

GAO 1 General Accounting Office (US).

2 Groupe Aérien d'Observation (F).

gap Distance between chords of any two adjacent superimposed wings of same aircraft, measured perpendicular to chord of upper wing at any point on its leading edge.

GAPA Ground-to-air pilotless aircraft.

GAPAN Guild of Air Pilots and Air Navigators (UK).

gap coding Precise gaps in radio transmissions, or intervals of silence, used to represent letters, words or phrases.

gap-filler Radar used to supplement long-range surveillance radar in area where coverage is inadequate.

gaping Discontinuity of skin profile caused by distortion or deflection of landing-gear doors, access doors and other separate panels intended to lie flush.

gapless ice guard Fitted within intake mouth; used in conjunction with automatic alternative inlet.

gapped ice guard Mounted forward of air intake to provide a gap which does not ice up.

gap-squaring shears Tool used for squaring and cutting sheet metal; similar to squaring shears but able to slit long sheet.

GAR Ground abort rate.

garbage 1 Miscellaneous hardware in orbit or deep space; usually material ejected or broken away from launch vehicle or satellite.

2 EDP (1) software or program which has been degraded or in any other way rendered imperfect or unreliable, or output therefrom.

garbling Indecipherable responses from two aircraft both at exactly same slant range of about 4 km (corresponding to 20·3 μs pulse interval) to SSR interrogation; caused by fact both sets of reply pulse trains are synchronized and superimposed (see *degarbling*).

garboard strake That section of plating on a seaplane hull which extends fore and aft and is adjacent to the keel.

GARC Groupe Aérien Régionale de Chasse (F).

GARD General address reading device.

gardening Aerial minelaying (RAF code name, WW2).

GAR/I Ground acquisition receiver/interrogator.

GARP Global Atmospheric Research Programme.

GARS Gyrocompassing attitude reference system.

Garteur Group for Aeronautical Research and Technology in Europe.

gas bag Flexible gas container of rigid airship; also known as gas cell.

gas-bag alarm Indicates when predetermined gas-bag pressure has been reached.

gas-bag net Mesh of cordage or wire to retain bag in position.

gas-bag wiring Mesh of circumferential and longitudinal wiring enclosing each bag to take pressure and transmit lift.

gas bearing Bearing for rotating assembly, usually high-speed, in which all forces are reacted by dynamic forces generated in contained volume of dry filtered gas, often He or N.

gas bleedoff Bleeding-off of hot combustion gas from rocket engine for pressurizing or driving turbomachinery.

GASC Ground Air Support Command (USAAF).

gas cap Gas immediately in front of hypersonic body in atmosphere; compressed and heated, and if speed is sufficiently high becomes incandescent.

Gasco 1 General-Aviation Safety Committee (UK).

2 Ground Air Support Command (US).

gas constant Constant factor R in equation of state for perfect gas = 8,314·34 J/kmol/K for particular gas; specific ** r = R/m where m is molecular weight (see *Boltzmann*).

gas dynamics Study of gases under high velocity, temperature, ionization and other extreme conditions prohibiting compliance with aerodynamic laws.

gaseous electronics Study of conduction of electricity through gases, involving Townsend, glow and arc discharges and collision phenomena on atomic scale.

gaseous fuels Those stored in aircraft in gaseous form, GH_2 being possible example.

gaseous rocket Bipropellant or monopropellant rocket utilizing gaseous fuel and/or oxidizer.

gas film Boundary layer on inner surface of combustion chamber.

gas generator 1 Device for producing gases, hot or cold, under pressure.

2 Gas-turbine core engine, eg turbofan minus fan or turboprop minus LP turbine, gearbox and propeller.

3 Major component supplying working fluid for turbopump of liquid rocket engine.

gas hood Cowl or ports in outer cover of airship through which gas can escape from inside hull.

gasifier Machine for producing flow of hot gas, normally by combustion of fuel compressed by mechanism extracting energy from flow; free-piston and other systems but not gas turbine.

gas laser One in which lasing medium is gaseous, eg HeNe, CO_2, Ar.

gas laws Thermodynamic laws applying to perfect gases: Boyle-Mariotte, Charles, Dalton, equation of state. Also called perfect or ideal **.

gas main Fabric hose running length of airship and having branches to gas bags for inflation.

gas misalignment See *jet misalignment*.

gas motor Mechanical drive system energized by high-pressure gas from combustion of solid fuel, usually single-shot.

gas oil Residual hydrocarbon fuel left after distillation of gasolines and kerosenes from particular petroleum fraction.

gasoline Blends of hydrocarbon liquids, almost all petroleum products boiling at 32°/220°C, used as PE (4) fuel.

gas-operated Automatic weapon in which initial rearward motion of moving parts, unlocking breech, is caused by piston moved by gas bled from barrel.

gasper system Low-pressure fresh-air supply piped to individual controllable outputs above each passenger or serving each crew-member.

gas pressurization Feeding liquid propellant for rocket engine by piping high-pressure gas to dispel it from tank, with or without use of flexible liner; eliminates need for turbopump.

gas producer See *gas generator (1)* or *(3)*.

gas ring Spring ring for maintaining gastight seal between piston and cylinder.

gas rudder See *gas vane*.

GASS Garde Aérienne Suisse de Sauvetage (Switz), same initials in Italian but SRFW in German.

gassing Replenishing gas balloon with fresh gas to increase purity or make up for losses.

gassing factor Quantity of gas required by aerostat over period of one year, ordinarily expressed as percentage of gas volume.

gas starter Early (1920 onwards) method of starting multi-engined aircraft using airborne compressor set supplying compressed air or over-rich mixture to cylinders of each engine in correct sequence.

Gaston Génération d'APT (auto-programmed tool) standard pour trajectoires d'outils normalisées (F).

gas triode See *thyratron*.

gas trunk Duct between gas-bag valve and gas hood.

gas tube Electronic tube (valve) containing gas under very low pressure.

gas turbine Engine incorporating turbine rotated by expanding hot gas. In usual form consists essentially of rotary air compressor, combustion chamber(s), and turbine driving compressor.

gas-turbine numerology Stations within an engine are designated by numbers used as inferior suffixes which vary with engine configuration; in a simple turbojet the numbers are: 1, entrance to inlet duct; 2, entrance to compressor; 3, compressor delivery; 4, turbine entry; 5, turbine exit; 6, entry to nozzle; 7, in plane of nozzle. A two-spool engine with afterburner has numbers 1 to 10. A different set of suffixes identifies the shafts in a multishaft engine, 1 being LP, 2 HP or IP, and 3 HP or free-turbine in a three-shaft engine.

gas vane Aerodynamic TVC in jet of rocket (see *jet vane*).

gas volume Volume of aerostat gas at SL.

gas welding Fusion welding with hot gas flame; eg oxy-acetylene on steels and oxy-hydrogen on aluminium.

GAT General air traffic.

GATCO Guild of Air Traffic Control Officers (UK).

gate 1 Point of passenger emplanement at airport.
2 Point at which commercial flight starts.
3 Removable lock to limit maximum travel of control lever (eg throttle) under normal conditions.
4 Position(s) on extended runway centreline above which inbound aircraft are required to pass at time assigned by approach control.
5 In air intercept: "Fly at maximum possible speed for limited period" (DoD).
6 Control electrode of any device, esp input connection to FET.
7 To control passage of signal in electronic circuits.
8 Circuit having output and input so designed that output is energized only when required input conditions are met, eg AND-*, OR-*, NOT-*.
9 Circuit designed to receive signals in small fraction of principal time interval in radar or control system.
10 Range of fuel/air ratios through which combustion can be started.
11 Of trajectory, eg speed-record run, specified transverse apertures in space through which aircraft must pass to comply with regulations.
12 Of space mission, transverse aperture(s) defined in width and height at particular time or distance from liftoff or related to other body in space, through which vehicle must pass if mission is to accomplish objectives.

gated EM pulse permitted to function only under control of another pulse, usually synchronized.

gate guardian Aircraft publicly displayed at entrance to air force base.

gate position Particular gate (1) numerically assigned to flight (7).

gate time 1 Agreed time at which flight enters new sector of controlled airspace.
2 Time at which aircraft passes particular point on track.

gate valve Valve controlling fluid flow by flat plate having linear motion across flow channel.

gateway Customs airport.

gathered parasheet Parasheet whose periphery is constrained by hem cord.

gathering Process of bringing guided missile into narrow pencil beam for subsequent guidance.

gating 1 Process of selecting portions of EM wave which exist during one or more selected time intervals or which have magnitudes between selected limits.
2 Use of Q-switching or other control to permit laser to emit only during exact specified time intervals, esp when used in rangefinding mode.

3 Imposing mechanical stop on PE (4) throttle below selected pressure altitude.

Gatorizing Isothermal forging process for high-nickel turbine rotor blades.

GATT Gate-assisted turnoff thyristor.

GAU Gun, aircraft unit.

gauge (US gage) 1 Any pressure-measuring instrument.

2 Hand comparator for GO/NO GO check on an exact dimension or screwthread.

3 Standard measures of sheet and wire thickness.

gauged fuel Sum of fuel-gauge readings; fuel state, including unusable.

gauge pressure Indicator reading showing amount by which system pressure exceeds atmospheric pressure.

gauss Non-SI unit of magnetic induction (magnetic flux density) = 10^{-4}T.

GAVC Ground/air visual code (FAA).

GAVRS Gyrocompassing attitude and velocity reference system.

GAZ State Aviation Factory, followed by three-digit identifying number (USSR).

GB 1 Groupe de bombardement (F).

2 Gain - bandwidth.

3 Lethal nerve gas, first produced as Sarin (G, 1938) and later for USA.

g.b. Grid bias.

GBA Groupe de bombardement d'assaut (F).

GBL 1 Ground-based laboratory.

2 Government bill of lading.

GBMD Global ballistic-missile defence.

G/BMI Graphite bismaleimide composite.

GC 1 Groupe de chasse, ie fighter wing (F).

2 Great circle.

GCA Ground-controlled approach.

GCB Generator circuit-breaker.

GC brg Great-circle bearing.

GCC Graduated combat capability.

GCF Ground-conditioning fan.

GCI Ground-controlled interception.

GCR Generator control relay.

GCS Ground control station(s).

Gc/s Gigacycles/second (=GHz).

GCT Government competitive test.

GCU Generator control and protection unit.

GD Branch General Duties Branch, to which aircrew belong (RAF).

GDC Gyro display coupler.

GDE Gas-discharge element light source.

GDL 1 Gas dynamic laser.

2 State Gas Dynamics Laboratory, and original centre of rocket research (USSR).

GDM Generalized development model.

GDP Geometrical dilution of precision.

GE, G/E 1 Graphite epoxy.

2 Groupe d'entrainment (F).

G_e Gain of radar jammer system.

GEANS Gimballed electrostatic aircraft navigation system.

gear Landing gear (colloq).

geared engine One having reduction gear to propeller. Gearing allows higher engine speeds, so increasing power, while keeping propeller speed within efficient limits.

geared fan Fan (2) driven through reduction gear.

geared propeller See *geared engine.*

geared supercharger PE (4) supercharger driven through friction clutch and step-up (speed-increasing) gears.

geared tab Balance tab mechanically linked to control surface so that its angular movement is determined by that of main surface.

gear-type pump Two intermeshing gears in close-fitting casing which pump fluid round outside of each gear in spaces between successive teeth and outer casing.

Gebecoma Groupement Belge des Constructeurs de Matériel Aérospatiale.

GEC Graphite epoxy composite.

Gee Pioneer precision navaid, using interrelated VHF pulses transmitted from ground stations; position of aircraft determined by observing intervals between pulses from pairs of stations and plotting on hyperbolic map.

Gee-H Secondary-radar navaid which enabled aircraft to determine position with precision by simultaneously measuring distances from two beacons.

GEEIA Ground electronics engineering installation agency.

gegenschein Faint light area of sky opposite Sun and celestial sphere; believed to be reflection of sunlight from particles beyond Earth's orbit.

GEH Graphite-epoxy honeycomb.

GEI Groupement d'Economique Interêt (see *GIE*, which is more common).

Geiger counter Geiger-Müller gas-filled tube containing electrodes; ionizing radiation releases short pulse of current from negative to positive electrode, frequency of pulses indicating intensity of radiation.

GEL Graphite-epoxy laminate.

GELIS, Gelis Ground-emitter location and identification system.

GEM 1 Ground-effect machine (ACV).

2 Generic electronics module.

GEMS Grouped engine monitoring systems.

Gen 2, Gen 3 New "generations" of helicopter visionics.

general air traffic All traffic excluding OAT (3), special military (eg lo-level training) and local pleasure flights not notified for ATC purposes.

general-arrangement drawing Usually three-view (front, side, plan) outline, in some cases with addition of dimensions, ground line, and broken lines giving additional information.

general aviation All civil aviation except air transport for hire or reward; largest sectors are private (including company transport), agricultural and aerial work.
general cargo Loose items, excluding large unit loads, not containerized or palletized.
general inference General meteorological situation and future forecast.
general-purpose aircraft Military aircraft intended to fill multiplicity of roles; nearest modern equivalent is armed utility.
generation Family of all examples of particular hardware species designed at same time to meet similar requirements; after first * hard to define, and word generally used to impress audience of prior experience of particular manufacturer in field concerned.
gentle turn Primary flight manoeuvre in which bank angle does not exceed 25°.
geocentric Related to centre of Earth.
geodesic line Shortest line between two points on mathematically derived surface; ** on Earth called geodetic line.
geodesic radome Spherical enclosure for large surface or ship radar whose structural members are geodetics and whose panels are perpendicular to transmissions.
geodesy Science which deals mathematically with size and shape of Earth and its gravitational field, esp with surveys of such precision that these measures must be taken into consideration.
geodetic construction Methods of making curved space frames in which members follow geodesics along surface, each experiencing either tension or compression; resulting basketlike framework does not need stress-bearing covering.
GEODSS Ground-based electro-optical deep-space surveillance system.
geographical mile One minute of arc at Equator, defined as 6,087·08 ft.
geographical position Where line from centre of Earth to a heavenly body cuts Earth's surface (astronav).
geographic poles North or south points of intersection of Earth's surface with axis of rotation, where all meridians meet.
geoid Earth as defined by that geopotential surface which most nearly coincides with MSL.
geomagnetic cavity Volume moving through solar wind occupied by Earth and surrounding magnetic field (magnetosphere).
geomagnetic co-ordinates System of spherical co-ordinates based on best fit of centred dipole to actual terrestrial magnetic field.
geomagnetic dipole Hypothetical magnetic dipole (bar magnet) located within Earth in such position as to give rise to actual terrestrial field.
geomagnetic equator Terrestrial great circle everywhere 90° from geomagnetic poles (should not be confused with magnetic equator).
geomagnetic poles North and south antipodal points marking intersection of Earth's surface with extended axis of geomagnetic dipole; north ** is 78½° N, 69° W, and south ** is 78½° S, 111° E. Should not be confused with magnetic pole.
geomagnetism 1 Magnetic phenomena exhibited by Earth and surrounding interplanetary space. 2 Study of magnetic field of Earth.
geometric pitch Distance propeller-blade element would advance in one revolution when moving along helix to which line defining blade angle of that element is tangential; ** of fixed-pitch propeller at standard radius is pitch of that propeller, and is marked on it.
geometric twist Variation along span of aerofoil of angle between chord and a fixed datum (see *aerodynamic twist*).
geometry-limited Restriction placed on aircraft attitude or configuration by geometric considerations; eg scraping tail on take-off or (variable-geometry aircraft) underwing stores fouling tailplane at max sweep.
geophones 1 Sensitive acoustic sensors for detecting sound transmitted through Earth's crust.
2 Seismic sensors buried along edge of runway to measure point of touchdown, severity of impact and bounces on landing (see *ALMS*).
geopotential height Height above Earth in units proportional to potential energy of unit mass (geopotential) relative to sea level. For most meteorological purposes same as geometric height, but geopotential metre = 0·98 metre; ** used under WMO convention for all aerological reports.
georef World Geographic Reference System.
George Automatic pilot (colloq, arch).
geosphere Solid and liquid portions of Earth; lithosphere plus hydrosphere. Above * lies atmosphere, and at interface is found almost all biosphere zone of life.
geostationary altitude That at which body is in geostationary orbit.
geostationary orbit Orbit in which satellite remains over same point on surface of Earth.
geostationary satellite Satellite in geostationary orbit; also called synchronous satellite.
geostrophic wind Wind the direction of which is determined by deflective force due to Earth's rotation.
geostrophic wind speed Calculated from pressure gradient, air density, rotational velocity of Earth and latitude, but neglecting curvature of wind's path.
geosynchronous Revolving at same angular speed as Earth.
GEP Glassfibre/epoxy/PVC foam sandwich.
GEPA Groupement d'Etudes de Phénomènes Aériens (N adds "Non-identifiés") (F).
GEPNA Groupe Européen de Planification de la Navigation Aérienne (Int).

germanium Silver-white, hard, brittle element possessing properties of both metals and non-metals; used in transistors, LEDs and IR windows (8–14 microns).
gerotor pump Gear pump in which spur gear having n teeth rotates inside internal ring gear having n + 1 teeth.
geschwader Wing (military unit, G).
GET Ground elapsed time.
get To remove gas from vacuum system by sorption.
get-away speed Airspeed at which seaplane or flying boat becomes entirely airborne.
GETI Ground elapsed time of ignition.
getter Material or device for removing gas by sorption.
GEV Ground-effect vehicle (usually = ACV).
GEW Ground-effect wing.
GF Gold film (transparency anti-icing).
GFA Gliding Federation of Australia.
GFAE Government-furnished avionic (or aeronautical or aerospace) equipment (US).
GFE Government-furnished equipment; supplied by DoD to industrial contractor for incorporation in aircraft or other large product.
GFI Government-furnished information (US).
GFK, GfK Glass reinforced plastics (G).
GFL Gesellschaft zur Förderung der Luftschiffahrt (G).
GFM Government-furnished materiel (US).
GFP Ground fine pitch.
GFRP Glass-fibre reinforced plastics.
GFS Gesellschaft zur Förderung der Segelflugforschung (G).
GFT 1 General flight (or flying) test.
2 Generalized fast transform.
GfW Gesellschaft fur Weltraumforschung (West German national space society).
GG 1 Gravity gradient.
2 Gas generator.
3 Gyro-angling gain.
GGG Gadolinium gallium garnet.
GGS 1 Gyro gunsight.
2 Gravity-gradient satellite.
GH General handling.
GH_2 Gaseous hydrogen.
GHA Greenwich hour-angle.
ghost Extra images or blips on radar, TV or other display caused by signal reflection from hills, buildings or other objects; usually to right of primary image at distance proportional to reflected and direct path lengths.
GHW Ground-handling wheels.
GI Gum inhibitor.
GIB Guy in back, ie navigator or other backseater in tandem two-seat military aircraft.
GIBEA, Gibea Guilde Belge des Electroniciens de l'Aviation (Belg).
GIE Groupement d'Interêt Economique.
GIEL Groupement des Industries Electroniques (F).
GIF Guy in front (see GIB).
GIFAS, Gifas Groupement des Industries Françaises Aéronautiques et Spatiales; aerospace industries trade association (F).
giga Prefix, multiplied by 10^9.
GIITS General imagery intelligence training system.
gills Hinged flaps at rear of engine cowling to control cooling airflow.
gimbal 1 Mounting with two and usually three mutually perpendicular and intersecting axes of rotation.
2 Gyro support which provides spin axis with degree of freedom.
3 To pivot propulsion engine for TVC.
4 To mount on *.
gimbal freedom Maximum angular displacement about gyro output axis, expressed in degrees or in equivalent angular input.
gimballed chamber Rocket-engine thrust chamber mounted on gimbal (3) so that it can swivel about one axis (or two perpendicular axes).
gimbal lock Condition of two-degrees-of-freedom gyro wherein alignment of wheel spin axis with axis of freedom removes degree of freedom, rendering gyro useless.
GIP Ground instructor pilot.
GIRA Groupe d'Instruction des Reserves de l'Air (F).
GIRD, Gird Group for study of reaction (ie, rocket) engines (USSR).
GIRTS Generic IR training system (ASD).
GISS Goddard Institute for Space Studies (US).
GIUK Greenland/Iceland/UK.
GKAP State Committee on Aviation Industry (USSR).
GKO State Defence Committee (USSR).
GL Ground level.
GLA Gust load alleviator.
GLAADS Gun low-altitude air-defense system (USA).
GLAM Groupe de Liaisons Aériennes Ministérielles (F).
gland Short tube fitted to airship's envelope or gas bag through which rope may slide without leakage.
glareshield Overhanging lip above instrument panel to protect pilot's night vision from bright reflections on windshield.
glass aircraft One with high proportion of GRP in airframe, including skin.
glass cockpit One featuring electronics displays in place of traditional instruments (colloq).
glass-fibre Produced by melting glass and spinning on revolving drum, fibres being typically 0·025 mm diameter. Fibreglass is registered name.
glass wool Produced by forcing molten glass through orifices of approximately 1 μ diameter.
Glauert factor Increase in lift coefficient due to fluid being compressible, = $(1-M^2)^{-1/2}$.

Glavkoavia Chief Administration of Aviation (USSR).
glazed frost Rain ice, layer of smooth ice formed by fine rain falling on sub-zero surface.
glaze finish Vitreous enamel coating on metal.
glaze ice Transparent or translucent coating with glassy surface formed by contact with rain; part freezes on impact, most flowing back and freezing over surface.
GLC Generator line contactor.
GLCM Ground-launched cruise missile.
glid See *GLLD*.
glide 1 Controlled descent by aerodyne, esp aeroplane, under little or no engine thrust in which forward motion is maintained by gravity and vertical descent is controlled by lift forces.
2 Flightpath of *.
3 To descend in *.
glide bomb Missile without propulsion but with aerofoils to provide lift and guidance; released from aircraft.
glide landing No-flare landing.
glide mode Flight-control system mode in which aircraft is automatically held to centre of glideslope.
glidepath 1 Flightpath of aircraft in glide.
2 Glideslope.
glidepath angle That between local horizontal and straight line representing mean of glideslope.
glidepath beacon ILS outer, middle or inner marker.
glidepath bend Aberration in electronic glidepath.
glidepath indicator ILS panel instrument.
glidepath localizer Contradiction in terms (see *localizer*).
glidepath sector Sector in vertical plane containing glideslope and extended runway centreline, limited by loci of points at which DDM is 0·175.
glider Fixed-wing aerodyne designed to glide, ordinarily having no powerplant propulsion (see *sailplane*).
glide ratio Ratio of horizontal distance travelled to height lost; TAS ÷ Vs (in same units).
glider train Two or more gliders towed in tandem behind one tug.
glider tug Aircraft used to tow gliders.
glideslope Radio beam in ILS providing vertical guidance (see *ILS*).
gliding angle Angle between local horizontal and glidepath.
gliding range Maximum distance that can be reached from given height in normal glide; also known as gliding distance.
gliding turn Spiral flight manoeuvre consisting of sustained turn during glide; also known as spiral glide.
glim lamp Source of illumination dim and local enough for use during black-out, esp airfield lighting.
glint Pulse-to-pulse change in amplitude of reflected radar signals, caused by reflection from object whose radar cross-section is rapidly changing.
glitch Small voltage surge affecting sensitive device; later general colloq for technical problem.
GLLD Ground laser locator-designator.
Glonass Global navigation satellite system (USSR).
glove 1 Fixed leading portion of wing root, esp of variable-sweep wing.
2 Additional aerofoil profile added around normal wing, usually over limited span, for flight-test purposes.
GLOW Gross lift-off weight (not spoken as word).
glow-discharge anemometer Sensitive method of measuring gas velocity, esp at low speeds or in turbulence, using cathode discharge between two pointed electrodes about 0·1 mm apart.
glow plug Electric heating element (sometimes used in semi-diesel engine) which aids starting.
glycerin Glycerol; liquid compounded of carbon, hydrogen and oxygen, soluble in water and alcohol; constituent of antifreeze compounds.
glyptal Synthetic resin made from glycerin and phthalic acid or phthalic anhydride.
GM 1 Ground map (radar mode).
2 Guaranteed minimum.
G/M Gun/missiles selector.
GM-1 Nitrous oxide PE (4) boost system.
GMC Ground movement control.
GMFSC Ground mobile forces satellite communications.
GMS 1 Geostationary meteorological satellite.
2 Groupement des Missiles Stratégiques (F).
GMT Greenwich mean time.
GMTI Ground moving-target indicator.
GN_2 Gaseous nitrogen.
G-nav Navigation direct from A to B not on promulgated airway but crossing radials yet still using VOR/DME; name from graphic navigation, using computer-produced charts or hand-held equipment to give pilot a picture derived from VOR/DME inputs. Basic method of cross-radial navigation.
GNC Graphic numerical control.
GNCS Guidance navigation control system.
gnomonic projection Created by projecting from centre of Earth surface features on plane tangent to surface; distortion severe except near origin (point where plane touches Earth) but great circles are straight lines.
GNS Global navigation system.
GO Geared, opposed (US PE [4] designation).
GO_2 Gaseous oxygen.
go-ahead Point in government programme at which prime contractor receives written authorization to proceed with full-scale development.
go-around mode Terminates aircraft approach

and programs climb; also known as auto overshoot.

Goddard Space Flight Center Greenbelt, Maryland, centre for NASA tracking and communications network.

Goes, GOES Geostationary operational environmental satellite.

go for broke To fire all weapons in one pass of target (colloq).

go gauge Dimensional gauge which must fit close, but without being forced, on or in the part for which it is intended.

go-home mode Emergency RPV flight-control mode used following loss of navigation or command link.

goldbeaters' fabric Layer of cloth cemented to one or more layers of goldbeaters' skin; gastight.

Gold C Gliding certificate second only to Diamond C, requiring flight of 300 km and other accomplishments.

goldie Verbal code: "Aircraft automatic flight-control system and ground-control bombing system are engaged and awaiting electronic ground commands" (DoD).

goldie lock Verbal code: "Ground controller has electronic control of aircraft" (DoD).

gold plating Introduction of what (generally ignorant or partisan) politicians claim to be costly and unnecessary features in weapon systems (US).

golf ball Turbulence control structure.

gondola Car of airship.

gong Medal or decoration (RAF colloq).

goniometer 1 Instrument for measuring angles between reflecting surfaces of crystal or prism.
2 Electrical transformer used with fixed and rotating aerials for determining bearing of radio station.
3 Motor-driven instrument used with four stationary aerials to deliver rotating signal field for VOR.

go/no-go Step-by-step basis on which manned spaceflights are flown, with flight crew and mission control jointly making positive decision whether to continue into each new phase of mission.

go/no-go check list Written guide for flight crews to determine go/no-go situation on any given subsystem deficiency (USAF).

go/no-go gauge Dimensional gauge for checking whether part is within upper and lower tolerance limits.

go/no-go test equipment Provides only one of two alternative answers to any question, eg whether given signal is in or out of tolerance.

Goodman diagrams Various graphical plots used to determine parts life under repeated cyclic loads, most common having per cent alternating stress/endurance strength as ordinate and per cent mean stress/rupture strength as abscissa.

Goodrich de-icer Original patented pulsating rubber de-icer for leading edges, intermittently inflated with air to break up ice.

Goodrich rivnut See *Rivnut*.

gooseneck flare Type of runway flare mounted on slender stem designed to bend easily if struck by aircraft.

GOR 1 General operational requirement.
2 Guy on the right, in side-by-side military aircraft, normally navigator or electronic-warfare officer.

gore 1 Shaped sector of parachute canopy normally bounded by two adjacent rigging lines.
2 Shaped section of airship envelope or gas bag, or balloon envelope.

GosNII State scientific research institute (USSR).

Gosport tube Flexible speaking tube used in tandem open-cockpit trainers connecting instructor's mouthpiece with pupil's helmet or vice versa.

Gothic delta Wing whose basic triangular shape is modified to resemble Gothic window; also known as ogival delta.

Göttingen-type tunnel Wind tunnel with return-flow circuit but open working section.

Gouge flap Flap whose upper surface forms part of cylindrical surface; thus as flap rotates immediate movement is rearwards to increase area.

gox Gaseous oxygen.

GP 1 General purpose (bomb, or former RAF squadron role prefix).
2 Glove pylon.
3 *Geographical position.*

GPa Gigapascal.

GPB Ground-power breaker.

GPC Government Procurement Code (US).

GPCU Ground-power control unit.

GPDC General-purpose digital computer.

GPES Ground-proximity extraction system.

GPI Ground position indicator.

GPIIA Groupement Professionel des Industriels Importateurs de l'Aéronef (F).

GPM 1 Glass polycarbonate mix.
2 Gallons per minute.

GPMG General-purpose machine gun.

GPP 1 Graphic part-programming; technique for communicating with computer by words and diagrams, conveying pictures of shape required and machining operations necessary to produce it.
2 Generative process planning, basis for implementing FMS (7).

GPS Global positioning system (Navstar).

GPSCS General-purpose satellite communications system.

GPT Glidepath tracking.

GPU Ground-power unit.

GPWS Ground-proximity warning system.

GR, G/R 1 Green run.
2 Ground attack, reconnaissance (role prefix, UK current).
3 General reconnaissance, ie Coastal Command

(role prefix, UK, WW2).
4 Ground relay.

Gr Graphite.

grab line See *handling line.*

grade Of fuel, see *fuel grade.*

Grade-A Standard aircraft cotton fabric, long staple with 80 threads per inch across both warp and weft (US = fill).

graded fibre Standard form of reinforcing-fibre raw material supplied according to diameter, length or other variable.

gradient 1 Of net flightpath, has normal meaning, h/D%; note runway * = slope.
2 Space rate of decrease of function; if in three dimensions, vector normal to surfaces of constant value directed towards decreasing values. Ascendent is negative of *.
3 Loosely, magnitude of either * or ascendent.
4 Rate of change of quantity, or slope of curve when plotted graphically.

gradient of climb See *climb gradient.*

gradient wind Along isobars with velocity exactly balancing pressure gradient; equilibrium between force directed towards region of low pressure and centrifugal forces.

gradient wind speed Calculated as for geostrophic but taking into account curvature of trajectory.

grading curve In determining propeller performance by Drzwiecki theory, forces on infinitely small blade element are determined; curve of these forces (as the ordinate) against blade radius is **, from spinner or root and reaching maximum between 70%–90% tip radius.

Graetz number, Gz Heat-transfer measure = Cp (specific heat at constant pressure) times mass flow divided by thermal conductivity and a length characteristic of body concerned.

Grafil Registered name (Courtaulds) for carbon-fibre raw materials.

grain 1 Entire cast or extruded charge for solid rocket motor.
2 Particle of granular solid propellant, usually in gun ammunition.
3 Particle of metallic silver remaining in photographic emulsion after developing and fixing; these form dark area of image.
4 Non-SI unit of weight = 0·0648 g = $\frac{1}{7{,}000}$ lb.

grain orientation Direction of solidification.

gramme Fundamental SI unit of mass.

gramme-molecule Mass in grammes of substance numerically equal to its molecular weight.

gramophone grooving Close-pitch grooves in female part to form abradable seal round high-speed rotating member.

Grandfather rights Permanent certificates for their existing route networks awarded US domestic airlines by CAB on its formation in July 1940. Hence Grandfather routes.

Grand Slam RAF 22,000 lb deep-penetration bomb of 1944 (colloq).

grand slam Verbal code: "All hostile aircraft sighted have been shot down" (DoD).

Grand Tour Planned unmanned exploration of series of outer planets with same spacecraft using planets' gravitational fields to turn spacecraft from one to another; possible only once in each 180 years.

granular snow Precipitation from stratus clouds (frozen drizzle) of small opaque grains 1 mm or less in diameter.

graphic part programming Translation of three-dimensional co-ordinates of workpiece into computer program for NC machining, invariably using computer graphic displays as human interfaces.

graphics Visual displays of any kind, esp electronics displays forming part of EDP (1), EW or similar system (eg on CAD), and designed written matter and symbology inside and on skin of aircraft.

graphic solution Using geometric construction to solve problem; eg calculating point of no-return.

graphite Soft naturally occurring allotropic form of carbon, also produced artificially and recently in form of strong fibres with perfect hexagonal crystalline structure. Large family commonly called carbon fibres includes many which in fact are graphite.

Grashof number, Gr Heat-transfer parameter = $l^3g\,(T_2-T_0)/\nu^2T_0$ where l is a length, g is gravitational acceleration, T_1 and T_0 are temperatures, and ν is kinematic viscosity.

grass Random spikes projecting from timebase of CRT, radar or other electronic display caused by noise or deliberate jamming.

grasshopper Safety-pin type of clip used to fasten cowl and other panels to perforated stud or similar anchor.

graticule Any array of lines used as a reference for aiming, measurement or determining spatial relationships, esp one of straight lines crossing at right angles on chart, map, CRT or other display, HUD or other human interface.

GRAU State rocket and artillery directorate (USSR).

gravipause Point between two bodies where their gravity fields are equal and opposite.

gravireceptors All sensors in human body for attitude, gravity and acceleration.

gravitation Assumed universal property of all masses of attracting all other masses with force GMm/r^2 where G is universal * constant, M and m are two masses and r is mutual distance apart.

gravitation constant, G Also called Newtonian constant, = $6{\cdot}6732 \times 10^{-11}$ Nm^2kg^{-2}.

graviton Hypothetical elementary unit of gravitation.

gravity Actual attraction experienced in vicinity of Earth or other body.

gravity drop angle That measured at gun at moment of firing between gun line and particular

future position of projectile.
gravity feed Relying on fact liquids tend to flow downhill, ie unassisted by pump.
gravity tank One relying on gravity feed, hence inoperative inverted.
gray Grey (US spelling).
grazing Almost tangent to curved surface; eg Sun-limb sensor or target ranging system in low-level attack.
grazing angle That between aircraft axis or sensor LOS and local Earth's surface, esp when less than about 5°.
GRC Glass-reinforced concrete.
great circle 1 One on spherical surface whose plane passes through centre of sphere.
2 Intersection of Earth's surface and plane passing through Earth's centre.
great-circle chart One on which all GCs are straight lines.
great-circle course See *great-circle route.*
great-circle route Shorter of two great circles linking all pairs of points on Earth's surface, giving minimum distance to fly; GC course is a misnomer because except along Equator or meridians course (hdg) is constantly changing.
great-circle track See *great-circle route.*
Greatrex nozzle Pioneer noise-reducing jet nozzle having several (typically six to eight) radial petal-like segments to increase length of periphery.
green aircraft Flyable but still lacking interior furnishing and customer avionics.
green airway One running essentially E-W.
green density That of compacted powder prior to sintering.
greenfield site Site considered for new airport or other facility where no structures exist at present.
Green Flag Tac-air war exercises strongly emphasizing EW (USAF).
greenhouse Long glazed canopy over tandem cockpits (colloq).
greenhouse effect Filtering and reflective effect of Earth's atmosphere on solar and other radiation akin to that of glass panes; part of incoming spectrum penetrates to Earth, where it heats surface and causes reradiation of longer wavelengths, some of which are absorbed by atmospheric water vapour and again reradiated.
Greenie Technical air groundcrew (RN).
green run First run of new or overhauled engine or other item.
Greenwich Earth's prime (0°) meridian, hence * apparent time (GAT), * hour angle (GHA), * mean time (GMT) and * sidereal time (GST).
green zone Traditionally, intersection between green and crossing airway at which it is traffic on crossing route that has responsibility for ensuring height separation.
grey body Unknown hypothetical body absorbing constant fraction of all wavelengths of incident EM radiation.
greyout Blurred vision under high positive acceleration.
grey scale Standard series of achromatic tones linking black to white.
grey wedge Standard filter whose opacity increases in known fashion across width, usually L to R; used in determining pulse distribution and other variables on CRT and other displays.
grf, g.r.f. Group repetition frequency.
GRG Ground-roll guidance.
grid 1 Perforated electrode between cathode and anode of thermionic valve controlling flow of electrons into fine beam.
2 Metal cylinder at negative potential in CRT designed to concentrate electrons.
3 System of two sets of parallel lines crossing at 90° to form pattern of squares each identified by numbers and/or letters in margins; superimposed on maps, charts, photographs and multi-sensor outputs so that any point can be located by letter/number code. Usually also permits accurate measures of distance and direction. Often called military *, though most are civil.
grid bearing Direction of one point from another measured clockwise from grid (3) north.
grid bias Constant potential in series with input circuit between grid (1) and cathode to hold operation to one part of characteristic curve.
grid convergence Angle between true north and grid (3) north.
grid co-ordinates Rectilinear measures about two axes in flat plane of grid (3) facilitating conversion of lat/long and other Earth measures on to flat sheet by routine plane surveying.
grid heading Aircraft heading measured relative to grid (3) north.
grid leak Resistor allowing grid (1) charge to drain to cathode.
grid magnetic angle Angle between magnetic north and grid (3) north, measured E/W from latter; also known as grivation (= grid variation).
grid modulation AM achieved by applying modulating signal to grid (1).
grid north Zero datum of grid (3), close to true north.
grid ticks Small marks on neatline or along grid (3) lines showing alternative grid system(s).
grid variation See *grid magnetic angle.*
Griffith wing Subsonic wing of very deep section with powerful suction slit on upper surface at about 70% chord to induce airflow to follow discontinuity between upper surface ahead of slit and thin trailing edge.
grip range Range of thickness of material joinable by particular blind rivet or other fastener.
GRM Ground-roll monitor.
GRO Gamma-ray observatory.
grommet 1 Rigid or reinforcing eyelet closed on to flexible surface.
2 Flexible ring set into rigid surface, often by peripheral groove matched with sheet thickness,

providing bearing surface for pipe, cable or other line (1, 2) or control cable.

grooved runway One whose surface is traversed by one of four standards of shallow grooves tailored to climate, crossfall and other factors, along which water can escape even in heavy rain and strong wind to make critical aquaplaning depth extremely unusual.

groover Machine with large wheel, usually diamond-dressed saw, for cutting runway grooves.

GROS, Gros Civil Experimental Aeroplane Construction Organization (USSR).

gross altitude scale Presentation of total altimeter operating range on one fixed scale (ASCC).

gross area Area of projected surface of aerofoil, edges being assumed continuous through nacelles, fuselages, pods or other protuberances. Where tapered wing meets fuselage, edges projected in to meet at centreline, except in case where angle is extreme (eg, with glove, Lerx, strake), where end of root is taken across at 90°.

gross ceiling Altitude at which gross climb gradient (see *gross performance*) is zero.

gross dry weight Traditional measure of powerplant weight which included propeller hub (metal hub on which wooden propeller was mounted), all starters, primers, exhaust systems, fluid filters, air inlets and accessories, but excluding cooling system, fluid tanks and supply systems and instruments.

gross flightpath Gross profile in climb-out segment.

gross flight performance See *gross performance*.

gross height Height of any point on gross flightpath.

gross lift Buoyancy in ISA (1) of aerostat under standard conditions of inflation and with allowance for humidity.

gross performance That actually measured on one aircraft of type, adjusted by small factor to reflect guaranteed rating and fleet minimum performance.

gross profile Side elevation of aircraft trajectory, esp following take-off, corresponding to gross performance.

gross thrust That developed by propulsion system in ideal conditions, not allowing for inlet momentum drag, inlet shock losses, duct losses, tailpipe losses, cooling drag, propeller slipstream drag, torque effects or any other effects.

gross weight Traditional measure usually defined as maximum flying weight permitted; today MTOW.

ground 1 US = earth.
2 To declare object or person unfit for flight.

ground-adjustable propeller One whose pitch can be changed only by ground crew.

ground air vehicle One designed for ground mobility but which can fly for short periods (ASCC).

ground alert Status of aircraft fuelled and armed and crews able to take off within specified period, usually 15 minutes.

ground angle That between local horizontal and major axis of parked fuselage.

ground clearance 1 Vertical distance between airfield or deck and tips of helicopter main rotor blades in no-lift position.
2 Vertical distance between airfield or deck and specified part of aircraft or external stores.

ground clutter Unwanted returns on radar display caused by direct reflection from ground.

ground contact Glimpse of Earth sufficient to assist navigation.

ground control Control tower position or other authority assigned to control all vehicles, including taxiing aircraft, on airfield movement area.

ground-controlled approach, GCA Ground radar installation able to watch approaching aircraft and direct them to safe landing by radio (so-called talkdown) in bad visibility; any landing thus directed.

ground-controlled interception, GCI Interception (1) controlled by ground radar and radio (usually voice plain-language) advice.

ground crew 1 Personnel assigned to cleaning, replenishment, servicing or maintenance of aircraft at turnround, between missions or in other routine situation.
2 Personnel assigned to manoeuvre aerostat on ground (see *landing crew*).

ground cushion 1 Region of increased pressure beneath landing aeroplane caused by forward motion, proximity of ground and trapping of air ahead of flaps and under fuselage (can affect flow over tail and, for this and other reasons, cause pronounced pitching moment).
2 Region of increased lift under helicopter or jet V/STOL in low-altitude hovering mode caused by reflection of downwash, jets, entrained air and possibly entrained solids or liquids from ground.

grounded Legally prohibited from flying.

ground effect 1 Increased lift caused by interaction of powered lift system and ground, as with ground cushion (2), used in ACV (GEM).
2 All effects, invariably unwanted, caused by interference of ground on radars, radio navaids and other EM systems.

ground elapsed time, GET Time measured from liftoff of major space mission, beginning with countup and continuing to provide one index of elapsed time unvarying with Earth time zone.

ground environment 1 Environment experienced by ground equipment (no definition except to meet particular specifications which are variable).
2 Electronic environment created by ground stations, esp for air-defence purposes.

ground equipment 1 All non-flying portions of aerial weapon system.

2 All hardware retained on ground needed to support flight operations. Appears to be no clear definition; most authorities agree every item intimately associated with flight operations but exclude those concerned with training, design/development, marketing or other peripheral areas, and never include consumables.

ground fine pitch Special ultra-fine pitch available after landing to increase drag on non-reversing installation; use of *** known as discing.

ground-fine-pitch stop Mechanical lock on hub released by compression of landing gear or other signal.

ground fire Gunfire from ground directed against aircraft (most authorities exclude all but small-arms fire).

ground fog Shallow fog caused by radiation chilling of surface at night.

ground half-coupling That part attached to GSE affording direct connection with mating half in aircraft.

ground handling equipment Ground equipment for lifting or moving large items, such as wings, missiles, spacecraft etc.

ground hold Hold (1) for ATC purposes taken on ground before starting engines.

ground horizon 1 Theoretical distance of horizon from sea level (see *horizon*).
2 Actual horizon seen from particular location.

ground idle Governed running speed for engine with throttle fully closed; lower rpm than flight idle.

ground liaison Officer specially trained in offensive air support (DoD) and/or air reconnaissance (NATO, CENTO, IADB); organized as member of team under ground commander for liaison with air and/or navy.

ground loiter Helicopter saving fuel by resting on ground between particular military tasks, in friendly or hostile territory.

ground loop Involuntary uncontrolled turn while moving on ground, esp during take-off or landing, common on tailwheel aeroplanes with large ground angle, caused by directional instability; if at high speed, landing gear would normally collapse before turn had reached 180°.

ground movement control, GMC Military unit assigned to control of transport by land, esp of air forces.

ground nadir Point on ground vertically beneath perspective centre of camera lens when exposure was made; coincides with principal point in vertical photo.

ground observer Trained person forming part of organization providing (DoD) visual and aural information on aircraft movements over defended area, (UK) information on fallout after nuclear attack.

ground-performance aircraft One able to move itself on ground without using flight propulsion system (ASCC).

ground plane Earthed system of conductors forming horizontal layer (mesh, sheet, radial rods etc) surrounding ground navaid.

ground position Point on Earth vertically below aircraft.

ground-position indicator, GPI Device fed with data from compass, ASI etc and giving continuous readout of DR position (obs).

ground power unit, GPU Source of power, usually electric and possibly pneumatic/hydraulic/shaft, supplied to parked aircraft.

ground-proximity extraction system Standard technique for low-level airdrop of palletized cargo using shock-absorbing ground coupling which engages with hook suspended from pallet.

ground radar aerial delivery Method of airdropping cargo, usually in A-22 (US) containers, from high altitude to avoid hostile fire, mountains or other hazards, with full parachute deployment delayed to increase accuracy.

ground readiness Status of aircraft serviceable and crews standing by so that arming, briefing etc can be completed within any specified period (longer than 15 min of ground alert).

ground resonance Dangerous natural vibration of helicopter on ground caused by stiffness and frequency of landing-gear legs amplifying primary frequency of main rotor; potentially catastrophic unless designed out, and even with certificated helicopter can occur as result of severe landing shock.

ground return See *ground clutter*.

ground roll Distance travelled from point of touchdown to runway turnoff, stopping or other point marking end of landing.

ground run Distance from brake-release to unstick, not same as TOR (see *take-off)*

ground signals Bold visual symbols displayed in signal area.

ground speed, GS, G/S Aircraft speed relative to ground.

ground-speed mode Flight-system mode holding constant GS.

ground spoiler Spoiler available only after landing, usually as lift dumper.

ground start Supply of propellants to large rocket vehicle from ground during ignition and hold-down so that at liftoff main-stage tanks are still full.

ground support 1 Air power deployed for immediate assistance of friendly army, ie close air support; hence designation * aircraft.
2 Hardware needed to facilitate operation of aircraft, eg ladders, chocks, refuelling, replenishing and rearming equipment, loaders, tie-downs, blanks (4) and ground conditioning and power supplies; and use thereof.

ground support equipment, GSE Ground equipment for military aircraft, RPV or missile.

ground swing envelope Plot of ground where obstructions would foul nose or tail of longest aircraft in most extreme positions on curves of taxiways or apron.

ground test Test on ground of equipment or system normally used in air.

ground test coupling Connections enabling airborne system to be tested on ground for fluid pressure and functioning, supply voltage or any other variable.

ground trace Ground track of satellite.

ground track Path on Earth's surface vertically below aircraft or satellite.

ground upset Accident caused to light aircraft or other vehicle by jet blast or large propeller slipstream.

ground visibility Prevailing visibility along Earth's surface as reported by accredited observer or measured by RVR.

groundwash Outward flow of wake turbulence from engines or wingtips of large aircraft on ground.

ground wave Radio or other EM waves taking direct path from ground transmitter to ground receiver (in practice mix of ground, ground-reflected and surface waves); subject to refraction in ducts in troposphere.

ground wire 1 Earthing wire (US).
2 Winched cable emerging from top of mooring mast and connected to airship mooring cable; US = mast line.

ground zero, GZ Point on Earth nearest to centre of nuclear detonation (which may be below, at or above GZ).

group 1 Military air formation consisting of two or more squadrons (DoD), or two or more wings (RAF), in both cases generally obs.
2 Several sub-carrier oscillators in telemetry system.
3 Major portion of aircraft (eg, wing *) assigned to * (4).
4 Team of engineers assigned to design, stress, develop and possibly cost major portion of aircraft, often remaining intact to work on same part of successive programmes; common in US, where * titles are wing, fuselage, tail/controls, weight, electrical, hydraulic, armament and often others.

group flashing light, GpFL Ground light with regular emission of two or more flashes or Morse letter(s).

group velocity Symbol U, that of entire disturbance of waves, equal to phase speed c minus wavelength l times dc/dl.

growl Missile tone heard in pilot headset indicating IR head locked on to target.

growler Test equipment for short circuits in electrical machines (colloq).

GRP Glass-reinforced plastics.

GRS Government rubber synthetic, Buna-S type.

GRSF Ground radio servicing flight (RAF).

GRU Main intelligence directorate of General Staff (USSR).

grunt manoeuvre One involving high g (colloq).

gruppe Group (G).

Gryphon FBMS/shore communications system.

GS 1 Ground speed.
2 Glideslope.
3 Ground plus station (costs).
4 Ground supply (usually electrical).
5 General schedule.

G/S 1 Ground speed.
2 Glideslope.

GSA Gunsight, surface-to-air.

GSE 1 Ground support equipment.
2 Ground swing envelope.

GSFC Goddard Space Flight Center, Greenbelt (NASA).

GSFG Group of Soviet Forces in Germany (NATO name).

GSI 1 Grand-scale integration (microelec).
2 Government source inspection.

GSOC German Space Operations Centre, Oberpfaffenhofen.

GSP Ground service plug (= socket).

GSQA Government source quality assurance (US).

GSR Ground surveillance radar.

GSSS Gyrostabilized sight system.

GT 1 Group technology.
2 Rate, eg kg/h, of fuel consumption (USSR).
3 Gas temperature.

G_t Gain of radar aerial (dB).

GTACS Ground-target attack control system.

GTC 1 In IFF, group time cycle.
2 Gyro time constant.

GTF Ground test facility.

GTPE Gun time per engagement.

GTR Gulf Test Range.

GTS 1 Gas-turbine starter.
2 Glider training school.

GTSIO Geared, turbocharged, direct injection, opposed.

GTT Ground test time.

GTV 1 Ground-test vehicle (helicopter).
2 Guidance test vehicle (missile).

g.u. Gravity unit, standard unit for geophysical and MAD calculations, = 10^{-6} ms^{-2},

GUAP, Guap Chief Administration of Aviation Industry (USSR).

guaranteed rating Minimum power or thrust which manufacturer guarantees every engine of type will reach.

guard Emergency VHF channel usually monitored as a secondary frequency by all air and ground stations in geographical area.

guarded switch One protected against inadvertent operation by hinged cover or shroud.

guard frequency Guard.

guardship 1 Armed escort helicopter.
2 Planeguard helicopter.

gudgeon pin Links piston to connecting rod (US, wrist pin).

GUGVF Chief Administration of Civil Air Fleet,

of which Aeroflot is operating branch (USSR).

guidance Control of vehicle trajectory, esp that of unmanned, or of manned but according to external inputs (see *active homing*, *beam-rider*, *command* *, *electro-optical* *, *inertial* *, *IR* *, *laser* *, *midcourse* *, *passive homing*, *radar command* *, *semi-active homing*, *wire* *).

guidance radar One dedicated to providing pencil beam for beam-rider or radar command guidance or illumination beam for semi-active homing.

guidance system Complete system providing guidance signals to flight-control system which steers vehicle.

guided bomb Free-fall missile with guidance, esp modified bomb.

guided missile Vehicle able to deliver warhead to target; normally not including those travelling over land surface or entirely through water (torpedo) but including all with some form of aerial trajectory.

guided weapon Guided missile (UK).

guide rope See *drag rope*.

guide-surface canopy Any of several families of parachute deployed from pack but able to be steered through air with translational motion.

guide vane 1 See *stator blade*.
2 Radial aerofoil struts at gas-turbine inlet designed to add or reduce swirl to airflow.

Guidonia Large aeronautical research centre formerly (pre-1944) run by Italian defence ministry.

gull wing One having pronounced dihedral from root to c15–20% semi-span, then little dihedral or even anhedral to tip.

gull-wing door One having pronounced curvature, concave on outer face, hinged parallel to aircraft longitudinal axis.

gum General term for viscous residues formed in gasolines (petrols) and to lesser extent other hydrocarbon fuels, mainly by slow oxidation.

gum inhibitor Now called anti-oxidant additive.

Gump Gas, undercarriage, mixture, propeller(s) (US, arch).

gun 1 Good general term for airborne rifled weapons of all calibres, including recoilless installations; no clear definition at what low muzzle velocity * becomes projector.
2 PE (4) throttle; hence to cut * = to close throttle, and to * engine = to apply full power (colloq, suggest arch).

gunbore line Projected axis of bore.

gun cross HUD symbol indicating gun is armed, ready to fire.

gun gas Emitted from muzzle, mix of initially incandescent gases from propellant deficient in unburned oxygen which if ingested by engine suddenly alters operating conditions.

gun jump Angle between gunbore line at firing and projectile trajectory as it leaves muzzle.

Gunk Registered commercial solvent for oils and greases.

gunlaying radar Early AI radar with mode for assisting attack with fixed guns on target seen only on display.

Gunn oscillator Major family of GaAs diodes generating microwave outputs on application of small bias voltage.

gun pack Quickly replaceable unit comprising one or more fixed guns (sometimes with barrels remaining installed in aircraft), feed systems and ammunition tanks, either in streamlined pod or contained within aircraft.

gunship Specially designed helicopter with slim two-seat fuselage, extensive protection and wide range of armament for roles in land warfare.

gunsight line LOS to aiming point through gunsight fixed optics.

gun time per engagement Usually firing duration in seconds, aggregate of separate bursts, against one aerial target.

gun-type weapon Nuclear weapon triggered by firing together at maximum velocity two or more subcritical fissile masses.

Guppy Aircraft with grossly swollen or bulged fuselage, esp family of B.377 rebuilds for conveyance of space-launcher stages and wide-body components (colloq).

gusset Small flat member used to reinforce joints and angles.

gust 1 Sudden increase in velocity of horizontal wind (see *gustiness factor*).
2 Suddenly encountered region of rising or falling air, causing moving aerodyne to experience sudden increase or decrease in angle of attack, = gust velocity u ÷ airspeed v. Vertical gust can theoretically be sharp-edged (instantaneous change from zero to maximum u) but normal design/airworthiness based on 1-cosine (gradual) gust curve to which gust alleviation factor applied.

gust alleviation Dynamic system for reducing effect of vertical gust on aeroplane (rarely, other aircraft) (see *active ailerons*, *Softride*).

gust alleviation factor As aeroplane encounters gust it pitches (depending on wing/tail or foreplane geometry) and wing does not generate full extra lift until it has travelled several chord lengths into gust, both of which reduce sudden structure load below instantaneous encounter; BCAR assumes *** 0·61, ie assumptions are based on 61% of true sharp-edged gust.

gust curve Assumed plot of gust (invariably 2) velocity relative to surrounding air mass against horizontal distance from undisturbed air to position of peak u.

gust envelope Basic aircraft design plot, vertical axis being structural load factor (1) and horizontal axis airspeed; normal boundaries are positive-stall curve, peak positive gust (normal non-SI = 50 ft/s) to V_C, line to meet gust of half this strength at V_D, then vertical V_D to negative half-strength gust, line to –50 ft/s gust at V_C and straight line at this negative gust value to meet positive stall at point less than 1 g.

gustiness factor Measure of gust (1), = difference between maximum gust and lull expressed as percentage of mean wind.

gust loading Increased structural loads caused by gust (1, 2).

gust locks Particular *control locks* preventing movement of flight controls of parked aircraft.

gust response Aircraft encountering gust (1, 2) experiences vertical acceleration made more severe by high speed, low wing loading (esp large span, discounting flexure effect of wing) and some other factors. Normal measure of ** is number of 0·5 g vertical accelerations experienced by pilot's seat per minute under specified conditions at high (Mach 0·9) speed at low level.

Guti Rare clag in Zimbabwe.

gutter Flameholder having cross-section generally in form of V, open side to rear, to create strong turbulence sufficient to keep flame attached.

Guttman Original scaling technique used to assess community noise response assuming that any positive answer implies positive answer to all questions of lower order; final Guttman scale is normally: no action; sign petition; attend meeting; contact officials; visit officials; help organize action group.

GUVVF Chief Administration of Air Force (USSR).

GVF Civil Air Fleet (USSR).

G/VLLD Ground/vehicle laser locator designator.

GVS Ground velocity subsystem.

GW 1 Guided weapon.
2 Groundwave.

Gwen Groundwave emergency network.

GWJ Garnet water jet for high-rate cutting of hard metals.

GWS Guided weapon system (RN).

GWT Gross weight.

G_x Gain of transponder RF amplifier.

gyro Gyroscope.

gyro angling gain GG = H/c, H sense (7).

gyrocompass Compass based upon space-rigidity of gyroscope; no true long-term instrument exists but see *directional gyro* and *Gyrosyn*, and sensing element of fluxgate compass is gyro-stabilized.

gyrodyne Aerodyne having engine power transmitted to lifting rotor(s) and propeller(s) used for thrust; convertiplane has wing in addition.

gyrograph Graphical plot of gyro drift against time.

gyro gunsight Sight for fixed guns using one or more gyros (and RAE-developed Hooke's joint with two degrees of freedom) to provide automatic lead computation by measuring rates of sightline spin while remaining insensitive to rotation about sight axis itself caused by roll of carrier aircraft.

gyrohorizon See *artificial horizon.*

gyro log Form used to calculate and record gyro drift and drift rate (ASCC).

gyromagnetic compass Directional gyro whose azimuth datum is maintained aligned with magnetic meridian by precession torquing from magnetic detector.

gyropilot See *autopilot.*

gyroplane See *autogyro.*

gyrostabilized Held in fixed attitude relative to space, subject to precession and wander.

gyrostat Hughes-developed technique for satellites of great length spinning about minor axis.

Gyrosyn Registered name for gyrosynchronized compass comprising DI (2) slaved to magnetic meridian by fluxgate.

gyro time constant GTC = J/c.

gyro vertical Local vertical indicated by vertical gyro.

GZ Ground zero.

H

H 1 Henry.
2 Total pressure.
3 Enthalpy.
4 High (synoptic chart).
5 High-altitude-class Vortac/Tacan.
6 NDB 50–1,999 W.
7 Angular momentum.
8 US military aircraft category, helicopter (DoD).
9 US military aircraft designation prefix, search/rescue and aerial recovery (DoD).
10 Magnetizing force; horizontal component of Earth's field.
11 Stored in silo but raised to surface for launch (DoD, missile).
12 Hard temper (light-alloy suffix).
13 G/S home from CP.
14 Airway or map prefix, helicopter route.
15 Ambulance category (USN 1929–31, 1942–44).

h 1 Hour.
2 Prefix hecto = $\times 10^2$.
3 Hexode, heptode.
4 Hangarage available.
5 High (synoptic chart).
6 Heater (electronics).
7 Height above MSL, or height difference in flight trajectory.
8 Specific enthalpy.
9 Planck constant.
10 Height of blade CP above flapping axis.
11 Operator, 120° (electrical).

H_2 Gaseous hydrogen.

H24, H_{24} Continuous-service airfield.

H_2S Original PPI mapping radar (UK, WW2).

H_2X Development of H_2S at shorter wavelength in US (see *Mickey*, *BTO*).

H-display B-display with elevation angle indicated; target appears as bright line with slope proportional to sin elevation.

H-engine PE (4) with left and right rows of vertical opposed cylinders, two crankshafts geared to central output.

H-film Kapton hi-temp polyimide.

H-tail One having twin fins on tips of tailplane.

HA 1 Height of apogee.
2 Hour angle.
3 High altitude.
4 Housing allowance.

ha Hectare.

HAA 1 Helicopter Association of America.
2 Height above airport.
3 Historic Aircraft Association (UK).

HAARS High-altitude airdrop resupply system.

HAC 1 House Appropriations Committee (US Congress).
2 Hover/approach coupler.
3 High-acceleration cockpit.
4 Hélicoptère anti-char (F).

Hacienda Office of Aerospace Research (USAF).

hack 1 Aircraft informally used as general transport and utility vehicle by military unit (often captured from enemy or retired from combat duty).
2 To be able to accomplish (mil, colloq).
3 To penetrate private network, especially a secure LAN.

HAD Hybrid analog/digital.

HADS High-accuracy digital sensor.

HAGB Helicopter Association of GB.

Hagen-Poiseuille Law for pressure drop per unit length for pure streamline (laminar) flow in constant-section pipe. Equal to

$$u = \frac{P}{4\mu}(a^2 - r^2)$$

where u is local velocity, P pressure drop per unit length, μ coefficient of viscosity, a radius of tube, and r radius at point concerned. Equation denotes parabolic velocity distribution.

HAH Hot and high.

HAHO High-altitude high-opening.

HAHST High-altitude high-speed target.

HAI Helicopter Association International.

hail Precipitation in form of hard or soft ice pellets, varying size; max for certification test is usually 2 in (25·4 mm) sphere.

Hair High-altitude IR.

hairline crack One in which the two sides are in close contact, hence inconspicuous to eye.

HAISS High-altitude IR sensor system.

half-brightness life Used in several senses, eg time from cutoff of stimulating radiation of phosphor to luminescence falling to half peak, time of cutoff of current to LED to same reduction in brightness, and various characteristics of storage tubes and displays.

half-life Time required for decomposition of half original mass or number of atoms of radioactive material.

half-period zone See *Fresnel zone*.

half-power points, rings Points, whose locus is normally closed ring, where radiated power from antenna is half lobe maximum.

half-power width Total angle at antenna between two opposite half-power points measured in plane containing lobe peak.

half-residence time Time for quantity of delayed fallout (weapon debris) deposited in particular part of atmosphere to decrease to half original value.

half-roll Rotation of aircraft sensibly about longitudinal axis through 180°, usually from upright, to inverted attitude or vice versa; can be half of barrel roll or of slow roll.

half thickness Thickness of absorbing medium which transmits half intensity of radiation incident upon it.

halftone screen Fine opaque grating usually scribed on glass to break up photographic image into halftone dot pattern, for printing or digitizing purposes.

halftone tube CRT containing conventional gun for writing separated from screen by fine-mesh electrode and storage plate.

half-view Drawing showing half a symmetrical object.

half-wave rectification Use of single-phase rectifier which passes only half of each alternate wave from input.

Hall effect When current-carrying material is subjected to magnetic field (or when conductor is moved through magnetic field) potential difference is set up perpendicular to both current (or motion) and field; small in metals but important in semiconductor systems and in study of electricity in ionosphere.

Halo 1 High-altitude, low-opening (paradrop system).
2 High-altitude large optics.

halo 1 Any of several species of part-circular phenomena caused by ice crystals in upper atmosphere, chief of which is 22° radius around Sun or Moon.
2 Coloured ring or disc seen on cloud in direction away from Sun, ie with aircraft shadow at centre; also called pilot's *.

halo effect Ability of SST service to generate or attract additional first-class subsonic traffic to same carrier on same route(s).

halon Family of halogen-based fuel-tank inerting gases.

haltere Twin vibrating prongs used in certain timekeeping systems.

hammer Sudden violent excursions in pressure caused by reflected shockwaves in closed fluid (esp hydraulic) system.

hammer stall Extreme stall turn in which aircraft rotates within c 5° of vertical plane; depending on aircraft, power retained until point of stall at apex. Also called hammerhead stall.

Hamots High-altitude multiple-object tracking system.

handbook problem One requiring in-flight consultation of flight manual for numerical answer.

hand bumping Use of hand tools and backing dollies to shape sheet metal.

hand controller Human interface to automatic or semi-automatic system, eg to HUD, multi-mode radar or large display; usually incorporates stick, rolling ball or triggers.

hand-crafted Still used to mean IC (4) is custom-designed for particular application, even though design is entirely via computer graphics.

hand cranking Direct mechanical connection between crank and PE (4) or small gas turbine (in latter case via step-up gears) for starting.

handed Items on left of aircraft are mirror-image of those on right.

handed propellers Left and right ** rotate in opposite directions.

hand flying Piloting autopilot-equipped aircraft in fully manual mode.

hand forging Forging by hand tools.

hand forming Shaping ductile sheet with hand tools and accurate form blocks tailored to flanging, beading etc.

hand inertia starter Hand crank spins flywheel which, when at full speed, is suddenly clutched to start PE(4).

Handley Page slat, slot See *slat*.

handling 1 Subjective impressions of response to controls of particular aircraft; hence * squadron, unit assigned to test basic flight characteristics.
2 Manoeuvring of aerostat by ground crew or landing party.

handling line 1 Primary line used by ground crew for handling aerostat.
2 Single or twin cables attached above c.g. of seaplane for use when hoisting by crane into or out of water.

handling pilot Pilot actually flying the aircraft.

handoff Transfer of control or surveillance from one ground radar controller to another.

handover Handoff (military).

hands-off 1 Condition in which non-autopilot aircraft, usually aeroplane, flies by itself in perfect trim with pilot(s) not touching controls.
2 Flight on autopilot.
3 Ground party release balloon basket.

hands-on 1 With human(s) interfacing with system, which can be EDP (1) or aircraft in flight; appropriate only to complex autopilot aircraft in which * is rare in cruising flight. Hence, * training = practice in hand flying, or interfacing manually with military sensors and weapon systems.
2 Ground party hold down balloon basket.

hand starter Arrangement for starting engine by hand other than by swinging propeller.

hand-starter magneto Separate hand-controlled auxiliary magneto carried to aircraft and used to supply powerful spark when starting PE (4).

hangar Shelter for housing aircraft on ground.

hangarette 1 Hangar tailored to single aircraft, esp hard shelter.
2 Weatherproof cover over missile launcher or similar installation.

hangar flying Social chat about flying by those involved in it, esp pilots.

hangar queen Particular aircraft notoriously

prone to unserviceability requiring major maintenance.

hangfire Fault condition in which rocket missile fails to fire; vehicle or missile thus affected.

hang glider Large class of simple ultra-light aerodynes, broadly divided into those with flexible wings (most of Rogallo type) and those having wings with preformed aerofoil section (called rigids); majority have no controls and manoeuvre by translation of pilot mass to shift c.g. Can be monoplane or biplane, and may have rear tail, canard or auxiliary engine. Demarcation line with glider or ultra-light aeroplane becoming blurred.

hang-up 1 Externally or internally carried store which fails to release when thus commanded.
2 Gas-turbine engine which starts but fails to spool-up; also called hung start.

HAP Hélicoptère d'appui et de protection (F).

Hapdar Hardpoint demonstration array radar.

Happ High-altitude powered platform, lightweight, solar-powered, unmanned aircraft able to hold station for months to years.

Haps Helmet-angle position sensor.

HAR Helicopter, search and rescue (role prefix, UK).

harass 1 Air attack on any target in area of land battle not connected with CAS or interdiction, with objective of reducing enemy's combat effectiveness.
2 To interfere with progress of another aircraft by making repeated close passes against it.

Harc jet Hall-effect arc jet, using magnetic field on hydrogen plasma jet (see *Hall*).

hard copy Immediately readable output, eg printed pages, as distinct from tape, microfilm or software.

hardened 1 Protected against blast, ground shock, overpressure, EMP, radiation and possibly other effects of nuclear explosion, and (DoD only) likely to be protected against chemical, biological or radiological attack (see *hardness*).
2 Of avionics, protected against EMP and any other powerful external EM effects which in particular would normally degrade or destroy most memory cores.
3 Of metal, physically * by precipitation, quenching or cold working.

hard flutter Normally well damped but extremely violent over one narrow range of conditions.

hard-iron magnetism That induced into all magnetic parts of aircraft during manufacture, esp by hammering or riveting, orientation and polarity depending on assembly heading and terrestrial latitude.

hard lander Spacecraft designed to free-fall to surface of heavenly body.

hard landing 1 Conventional aircraft landing with excessive rate of descent, esp that results in damage or overstressing.
2 Arrival of hard lander on lunar/planetary surface.

hardness 1 Various measures of physical * such as Moh's scale for non-engineering (eg geological) materials, and Rockwell, Vickers, Brinell and many other standard tests for precise quantified readings.
2 Any of five measures of nuclear hardening; eg resistance to overpressure, which varies from c 2 lb/sq in (c 14 kPa) for aircraft parked in open to over 11,000 lb/sq in (c 75 MPa) for hardest concrete silo.

hardover runaway Sudden unwanted operation of system, esp flight-control channel, to extreme limit of travel.

hardover signal Fault condition resulting in full demand for unidirectional system operation unrestrained by normal feedback.

hardpoint Anchorage built into aircraft structure for heavy external load, usually via intermediate pylon, MER or launcher.

hard radiation High penetrating power from very short wavelength; usual definition is ability to pass through 100 mm of lead.

hard recovery Landing under difficulties, eg barrier crash into net in bad weather with tower out of action (probably colloq).

hardstanding Paved parking area (US, often hardstand).

hard temper Modification of light-alloy properties, eg by cold working, denoted by scale such as ¾H (=75% fully hard); increased tensile strength is usually accompanied by reduced ductility, which in airframes tends to be equated with shorter fatigue life.

hardtop Paved with permanent all-weather surface.

hard turn Planned turn in air combat at rate governed by angle-off and range.

hard vacuum High vacuum, pressure below 10^{-9} Nm^{-2} (10^{-7} torr).

hardwall hose Does not collapse under atmospheric pressure when used for suction; not necessarily armoured.

hardware 1 Originally introduced to distinguish mechanical parts of EDP (1) systems from software, today useful to imply manufactured items of any kind that exist, as distinct from software, system concepts, paper designs and proposals, simulations, capabilities and functions.
2 In narrow sense, small fasteners and similar small parts.

hard wing 1 One with simplified leading edge compared with particular previous wings, eg no slat.
2 Any wing with fixed-geometry leading edge.

Harm High-speed anti-radiation missile.

harmonic Component of sinusoidal or complex waveform or tone whose frequency is integer multiple of fundamental frequency.

harmonic analyser Device, typically variable-frequency filter, which can resolve waveform into harmonic constituents.

harmonization 1 Boresighting of all guns fixed to fire in same basic direction (eg ahead) so that all are correctly aligned with respect to aircraft axes, usually so that all gun axes converge at specified distance from aircraft.
2 Adjustment of flight-control system so that effect of controls about each axis matches that about others at all airspeeds; in particular so that handling about each axis in terms of rate of pitch, roll or yaw, and load (real or synthetic) experienced by pilot appear to be in harmony.

harness 1 Assembly of straps with which member of flight crew can be secured to seat.
2 Assembly of straps with which parachutist can be attached to parachute, which may or may not be permanently attached.
3 System of straps and other restraining members with which cargo pallet or container is secured to cargo floor in cases where there is no inbuilt anchorage. Not restraint net.
4 Weatherproof, screened assembly of HT leads connecting ignition source to PE (4) plugs or gas-turbine igniters.

HARP, Harp Helicopter Airworthiness Review Panel.

harpoon system Any helicopter hold-down system based on firing barbed anchor into grid or other fixed base on ship platform.

Harry Clampers Clamp (colloq).

hartley Standard unit of information content generally taken as equal to $\log_2 10 = 3{\cdot}219$ bits or 1 decimal digit.

Hartmann generator Common electrically powered source of intense noise, usually with variable spectrum.

HARS, Hars Heading/attitude reference system.

Hartridge smoke unit, HSU One of most popular measures of smoke blackness; smoke is fed through chamber of known length and light meter measures light received from calibrated source at other end. Accurate for combustion engineers but for subjective assessment against sky jet diameter must be taken into account. PSU is similar (see *smoke*).

HAS 1 Hardened aircraft shelter.
2 Hover-augmentation system (reduces pilot workload by using sensitive accelerometers in hover).
3 Hood, aircrew survival.
4 Helicopter anti-submarine (UK).
5 Heading and attitude (reference) system (or heading/attitude sensors).

HASC House Armed Services Committee (US).

Hasell check Height, airframe, security, engine(s), location, lookout; prior to spin or other harsh manoeuvre.

Hasp High-altitude sampling program.

Haste Helicopter ambulance service to emergencies (US).

Hastelloy Family of US Ni-Cr-Mo-Fe alloys combining strength at high temperatures with resistance to oxidation or corrosion.

HAT Height above touchdown (or threshold) (FAA).

hat Section commonly used for stiffeners and other structural members, formerly called top-hat and resembling latter's outline.

hatchback Transport whose rear cabin area is Combi.

HATR 1 High-temperature attenuated total reflectance.
2 Hazardous air traffic report(ing).

Have Permanent first word in code names of AFSC projects (USAF).

have numbers Radio code: "I have received and understood wind and runway information for my inbound flight".

HAWC Homing and warning computer.

Hawfcar Helicopter adverse-weather fire-control acquisition radar.

hawking the deck To fly closely past carrier on right (starboard) side to check deck state before landing on.

Hawtads Helicopter all-weather target-acquisition and destruction system.

hazard alert Broadcast to all affected operators of existing or impending failure in hardware of nature likely to imperil safety of flight, usually as result of fault discovered on inspection; fault can be structural or in system operation, but invariably affects part of aircraft.

HB Symbol, Brinell hardness.

H-bomb Hydrogen bomb; species of thermonuclear weapon (colloq).

HBPR High bypass ratio.

HBS Hot ball and socket.

HBT Heterojunction bipolar transistor.

HC 1 Hexachloroethane, ECM aerosol also usually containing ZnO and Al powder.
2 Hand controller.
3 Helicopter, cargo (UK).

HCA 1 High-cycle aircraft, specimen which has completed more flights than any other of same type.
2 Hot compressed air (airfield snow clearance).
3 Historic cost accounting.
4 Helicopter Club of America.

HCF High-cycle fatigue.

HCGB Helicopter Club of Great Britain.

HCHE High-capacity HE (ammunition).

HCMM Heat-capacity mapping mission (spacecraft).

HD 1 Height difference, usually in attack mission between IP and target.
2 Distilled mustard gas code (USA).

HDA High-density acid (usually nitric).

HDB High-density bombing, rebuild of bomber designed for nuclear warfare to carry heavier loads of conventional bombs.

HDD Head-down display, ie a display inside

cockpit, usually CRT raster plus symbology and TV overlay.
HDDR Head-down display radar.
HDFPA High-density focal-plane array.
Hdg Heading.
Hdg C Compass heading.
HDIP Hazardous duty incentive pay (US).
HDMR High-density multitrack recording.
HDOC Hourly direct operating cost.
HDPE High-density polyethylene.
HDR High data rate.
HDS Helicopter delivery service.
HDTA High-density traffic airport (FAR 93).
HDU Hose-drum unit for in-flight refuelling.
HE High explosive.
h_e Energy height, $h + V^2/2g$.
head 1 Complete hub of helicopter rotor (main or tail) including flight-control linkage and all auxiliaries (nitrogen pressure signal, anti-icing connection, lights etc).
2 Cutter and positioning system in most machine tools, esp those in which workpiece is stationary.
3 Loosely, downwind end of runway.
head-down Looking into cockpit as distinct from outside aircraft.
head-down display See *HDD*.
heading Angle between horizontal reference datum and longitudinal axis of aircraft expressed as three-figure group 000°–360°; in UK also called course. Datum can be compass north, magnetic north or true north.
heading hold Flight-control mode which maintains selected heading.
heading-orientated map One held in hand so that heading is towards top of sheet.
heading select Flight-control mode in which aircraft automatically turns to and holds any inserted *.
head moment Total turning moment transmitted through head (1) in helicopter manoeuvre.
head-on Flying directly towards other aircraft on reciprocal track.
headset Receive/transmit interface with radio communications or other system (eg aural warning or missile launch tone), either worn separately or built into helmet.
heads up Airborne intercept code: "Hostile force, whole or in part, got through defences" or "I am not in position to engage" (DoD).
head-up display See *HUD*.
health Generally everyday meaning but applied to hardware; thus * monitoring, can be equated with inspection, AIDS output etc.
HEAO High-energy astronomical observatory.
Heart Health evaluation and risk tabulation.
heart-cut distillate Particular family of kerosene fuels, in particular JP-6.
HEAT High-enthalpy ablation test.
Heat Armament panel switch selection for IR AAMs.
heat barrier Supposed barrier to increasing flight Mach number due to kinetic heating.
heat engine Prime mover in which energy is extracted from thermodynamic system in which gas passes through cycle from which closed PV or T-entropy diagram can be plotted. Perfect ** has reversible (Carnot) cycle; all practical ** have lower-efficiency irreversible cycle. Nearly all aviation ** are constant-pressure gas turbines or Otto-cycle PE (4).
heat-exchanger Radiator in which two fluids are brought into close contact (eg, cold air and hot oil) so that one can reject heat to other.
heat of ablation Measure of value of ablating material, rate of heat input divided by rate of mass loss.
heat pulse Total heat to be absorbed, dissipated, radiated or otherwise transferred from original kinetic energy of body on re-entry.
heat-seeker Sensitive IR detector and homing guidance system.
heat shield Usually non-structural layer protecting primary structure from high-temperature environment which would degrade strength, such as around aeroplane afterburners, at base of large rocket vehicle and, esp, over upstream face of body on re-entry (latter is invariably ablative).
heat sink 1 Any location to which heat may be removed from thermodynamic system.
2 Specifically, mass of metal, fuel, oil or other material which can accept unwanted heat.
heat treatment Heating metals above specific temperatures followed by slow or rapid cooling to improve mechanical properties (see *carburizing*, *case hardening*, *nitriding*, *precipitation* **, *solution* **).
heavier than air Having density much greater than that of air, thus aerostatic buoyancy ignored in calculating lift, which is assumed generated wholly by aerodynamic or propulsive forces; aerodyne.
Heaviside layer D,E,F_1 and F_2 layers in upper atmosphere (ionosphere) where UV solar radiation ionizes gas molecules, allowing conduction of electricity.
heavy alloy One tailored to have exceptionally high density; most consist chiefly of tungsten, but osmium, iridium and depleted uranium are important.
heavy bomber Aircraft designed to deliver heavy load of conventional ordnance to targets in enemy heartlands (arch).
heavy dropping Delivery by system of parachutes of exceptionally bulky or heavy load, suitably packaged and cushioned.
heavy landing See *hard landing (1)*.
heavy maintenance Maintenance taking 30 days or longer.
Heavy Wagon Designated routes for lo military flights within Conus; 300 Series for various

nav/electronic systems evaluation and 400 Series for nav training and weather evaluation at 500 ft (152 m) down to MOCA, normally not over 500 kt (USAF, USN).

hectare Non-SI unit of large areas = 100 are = 10^4 m^2.

hecto Prefix, multiplied by 10^2.

hedge-hopping Lo (2) flight by light GA aircraft (colloq).

HEDP Ammunition, high-explosive, dual-purpose.

HEDR High-energy dynamic radiography.

heeling Roll due to turn while taxiing.

HEF High-expansion foam (anti-fire) usable at 1·5% concentration.

HEH Hershey, Eberhardt, Hottel, basic analysis and charts for idealized PE (4) cylinder performance.

HEI 1 Ammunition, high-explosive incendiary. 2 Hot-end inspection.

height 1 In performance, Ht defined as true vertical clearance between lowest part of aircraft (assumed aeroplane) in unbanked attitude and relevant datum.
2 General common definition: h, vertical distance of level, point or object considered as point, measured from specified datum (normally associated with QFE).
3 Vertical distance from reference ground level to specified upper extremity (close to but not necessarily highest point, which may be small whip aerial, static wick or other flexible part) of aircraft at specified weight (usually MTOW) and tyre inflation pressure when parked on flat horizontal surface.

height above airport, HAA Height of MDA above published airfield elevation.

height above touchdown, HAT Height of DH or MDA above highest obstruction in touchdown zone.

heightfinder radar, HFR One whose primary output is precise height of distant target; has nodding aerial giving multi-lobe beam.

height lock Function of autopilot or flight system in holding (after in some cases capturing) selected height h.

height ring Visible on most displays of weather/mapping/AW/cloud-collision warning and other forward-looking radars as bright ring beyond zero-range ring formed by direct reflection from Earth's surface.

height/velocity curve Fundamental plot of IAS against altitude included in helicopter flight manual; indicates region(s) from which safe autorotative descent is possible, normally assuming zero wind, sea level, MTOW.

HEI-T Ammunition, HE incendiary, tracer.

HEL High-energy laser.

Helarm Helicopter, armed (sortie).

HELCM HEL countermeasures.

Heliarc Inert (helium) gas welding technique patented by Northrop 1940.

helical compressor Diagonal-flow compressor.

helical gear Spur gear with teeth arranged diagonally; contact begins at one end of tooth and proceeds diagonally across width, two teeth normally transmitted load at any time. Double-* (herringbone) gear eliminates axial load.

Heli-Coil Patented hard-steel thread inserted in soft metal or to repair stripped female thread.

helicopter Rotorcraft deriving both lift and control from one or more power-driven rotors rotating about substantially vertical axes. Rotors are driven as long as engines operate above certain minimum input speed, and airflow through rotor(s) is downwards except in autorotative engine-failure mode.

helicopter lane Safety air corridor reserved for helicopters (DoD, NATO).

helidrop Discharge of passengers/cargo/weapons whilst hovering.

heliocentric Centred on the Sun.

helipad Prepared area designated as take-off/landing area for helicopters; need have no facilities other than painted markings.

heliport Facility for operating, basing, housing and maintaining helicopters; if civil, includes passenger facilities and usually mail/cargo channels.

helistat Aerostat with added helicopter rotor(s).

helistop Helipad served by civil helicopter.

helium Inert gas, density 0·1785 g/l, 0·0112 lb/ft^3, BPt 3·2°, 4·2° K (two isotopes).

helix angle 1 Of gear, measured between line tangent to tooth helix at PCD and shaft axis.
2 Of propeller, see *effective* *.

helmet sight Any of several systems which attempt to interface between human aircrew looking at surface (or other) target and aiming of armament system; usually wearer's head is in some way accurately aligned with rigid flight helmet whose orientation is then automatically monitored to give output signals fed to weapon-aiming system.

Helmholtz resonator Hollow volume connecting with outside via small orifice, single-frequency output.

Helms Helicopter malfunction system.

helo Helicopter (colloq).

HELP, Help Hybrid electronics lightweight packaging.

Helraps Heliborne long-range active (acoustic path) sonar.

Helras Helicopter long-range sonar.

hemispherical engine PE (4) having hemispherical combustion spaces above piston at TDC.

Hemloc Helicopter emitter/locator countermeasures.

Hemp Name of khaki-ochre colour of large RAF patrol and tanker aircraft.

HEMT High electron mobility transition.

He-Ne Helium/neon laser.

henry, H SI unit of inductance, that of closed circuit in which 1V generated when current varies at steady 1A/s; plural Henrys.

HEOB Hostile electronic order of battle.

HEPL High-energy-pulse laser.

heptane Basic member of straight-chain (alkane) hydrocarbons (see *paraffin series*) with seven carbon atoms and thus 16 of hydrogen; zero-octane reference fuel.

Hermes Helicopter energy/rotor management system; automatically computes limiting payload.

HERO, Hero Historical evaluation and research organization.

herringbone gear Double-helical (see *helical*).

HERT Headquarters emergency relocation team.

hertz, Hz SI unit of frequency, = cycles per second; hence kHz, MHz, GHz etc.

Hesh High-explosive squash-head.

hesitation roll Aerobatic rolling manoeuvre, normally based on slow roll, in form of succession of quick aileron applications between which rate of roll is suddenly arrested; positions of arrest called points, and common ** are 4-point, 8-point or 16-point, 16 being difficult and rare.

heterodyne Mixing of two alternating currents (eg radio signals) to generate third equal to sum or difference of their frequencies; verb or adjective.

heuristic Leading towards solution by trial and error (see *algorithm*).

HEUS High(er)-energy upper stage.

Hexotonal High explosive for filling munitions, mix of RDX, TNT and powdered aluminium.

HF 1 High frequency (see *frequency*).
2 Human factors.
3 High-altitude fighter (role prefix, UK, WW2).

Hf Hafnium.

hf High frequency.

HFCS Helicopter fire-control system.

HFE Human-factors engineering.

HFP High-fragmentation projectile.

HFR Height-finder radar.

Hg Symbol, mercury.

HGAS High-gain antenna system.

HgCdTe Mercury cadmium telluride; sensitive IR detector material for 8–14 microns.

HGL Hot gas leak.

HGM Hot-gas manifold.

HGR Hypervelocity guided rocket.

HGRI Hot-gas reingestion.

HGSI Hot-gas secondary injection TVC.

HGU Horizon gyro unit.

HH NDB class, over 2 kW.

HHC Higher harmonic control.

HHTI Hand-held thermal imager.

HHTR Hand-held tactical radar.

HHX Hydrogen heat exchanger.

HI High-intensity light.

hi Flight at tropopause or above, adopted by gas-turbine combat aircraft flying for range outside hostile environments.

HIA Held in abeyance.

hibernate To remain on station in space or in orbit with as many subsystems as possible switched off until reactivated by Earth command.

Hi-camp Highly calibrated aircraft measurements program (Darpa).

Hi-CAT CAT above 55,000 ft (16·8 km).

hide Locally constructed shelter of posts, net and camouflage.

Hiduminium Family of duralumin-type alloys, including RR (Rolls-Royce) formulations, often also containing nickel and iron; developed for PE (4) pistons and similar applications, also used in supersonic airframes (in Concorde also known by French designations such as AU2GN).

Hidyne See *Hydyne*.

HIE Helicopter-installed equipment.

HIG, Hig Hermetic integrating gyroscope.

HIGE Hovering in ground effect.

high Region of high atmospheric pressure, anticyclone.

high altitude 1 Above 10 km (32,800 ft) (NATO).
2 Between 25,000 and 50,000 ft (7·6–15·2 km) (DoD).

high-altitude bombing Level bombing with release at over 15,000 ft (4·57 km) (DoD).

high-altitude burst Nuclear weapon explosion at over 100,000 ft (30·5 km).

high blower See *high supercharger*.

high boost PE (4) operated at high inlet-manifold pressure, esp one well above SL atmospheric pressure for take-off.

high BPR Bypass ratio exceeding 2.

high cloud Extremely loose classification, typical band being 6–8 km (20,000–60,000 ft) in tropics, 5–13 km (16,000–43,000 ft) in temperate, 3–8 km (10,000–26,000 ft) polar. Always mainly ice crystals, Ci, Cc, Cs.

high compressor HP compressor (US).

high-cycle fatigue That due to high-frequency vibrations, flexure or rotation of machinery, typically at rate many times per second.

high-density focal-plane array EO sensor with thousands of 2-D IR elements integral in substrate with CCD readout and processing circuitry.

high-density seating Normally, that giving greatest practical number of passengers, more even than all-economy, tourist or other classifications.

high-density tunnel One having closed circuit filled with air or other gas under pressure, power required for given Mach/Reynolds combination being inversely proportional to density.

high-energy laser No lasting definition; varies with family, whether continuous, pulse or gated, and with rapid progress, which by 1980 had made GW output not uncommon.

higher harmonic control Reduction of helicopter stress, noise and vibration by using dynamic absorption methods (Hughes).

high-flotation gear Landing gear having geometry and other characteristics suitable for operation from soft soils, sand and similar surfaces.

high-flotation tyre One of low inflation pressure and very large contact area; not necessarily used on high-flotation gear.

high-gain aerial Usually means strongly directional, designed to transmit or receive along single axis.

high gate Different meaning in different programmes (lunar landing, air speed record runs, etc) but always a specified rectangle in exact relation to surface of Earth or other body through which vehicle must pass, often at precise time, for mission to be successful (see *gate [10, 11]*).

high gear High supercharger gear.

high-incidence stall Unaccelerated symmetric stall at highest angle of attack, normally with high-lift configuration.

high-lift System, device, configuration or mode giving lift greater than in clean or cruise configuration. Normal * devices include leading-edge flaps, droops, slats, Krügers, trailing-edge flaps, flap blowing and variable-sweep; other forms include BLC, EBF, USB, Jet Flap, vectored thrust and tilt-wing.

highly blown engine PE (4) in which supercharger used to achieve high inlet manifold pressure at low altitudes. Terminology can be ambiguous; high and low-blown can mean high and low rated altitudes (critical heights).

high-Mach buffet Experienced when critical Mach number is exceeded on aircraft not designed for transonic flight; can affect control surfaces and primary structure, and cause trim changes which automatically either remove or intensify condition.

high-Mach flight Flight at high-subsonic Mach number.

high-Mach trimmer See *Mach trimmer*.

high oblique Oblique reconnaissance or other aerial photograph in which portion of apparent horizon is visible.

high-octane Fuel grade.

high pitch Coarse pitch.

high-pressure area Region of atmosphere where pressure significantly exceeds that of surroundings, esp one bounded by enclosed isobar rings; anticyclone.

high-ratio engine Engine having high BPR.

high route Area-navigation route extending from 18,000 ft AMSL to FL450 (FAR.1).

high rudder See *top rudder*.

High-Shear rivet Patented close-tolerance steel threadless bolt held by swaged ring closed around groove.

high-speed alloys Brasses, steels, light alloys and other metals which can either be machined easily at high linear speed or which retain strength at high temperature and can edge cutting tools (two almost opposite meanings for same term).

high-speed stall Accelerated stall in which stalling angle of attack is reached at relatively high airspeed; height well below aircraft ceiling is implied.

high-speed tunnel Traditionally, one in which effects of compressibility can be observed.

high-speed warning ADC-triggered subsystem giving aural and visual indication (usually duplicated) of inadvertent speed excursion beyond threshold usually set at just above MMO; can be overridden to demonstrate MDF/VDF but aircraft thus equipped is not airworthy without it.

high-subsonic In the range of Mach numbers where sonic speed is locally exceeded at peak suctions and compressibility effects, if any, are manifest; in older aircraft can be Mach 0·8, today higher values can be reached with no significant change in handling or trim (beyond 0·9 there may be major trim changes automatically countered by Mach trimmer). Hence * aircraft, general category of jet aircraft other than those designed for supersonic performance.

high supercharger gear In older high-power PE (4) with mechanically driven supercharger it was common to have two drive ratios, high gear being automatically clutched in by aneroid at preselected height, thus giving power/altitude curve with two major discontinuities; high gear unavailable at lower levels.

high-tailed aircraft Aeroplane with horizontal tail on top of fin; T-tail.

high-test peroxide, HTP Not aqueous solution but almost pure hydrogen peroxide, used in MEPUs, rocket engines and other applications as monofuel rapidly decomposed by silver-plated nickel into superheated steam and free oxygen in which kerosene or other fuel is also often burned.

high-time That particular engine, aircraft or other hardware item that has flown more hours or completed more operating cycles than any other of same design.

high turbine HP turbine (US).

high vacuum Pressure less than 10^{-9}Nm^{-2} (10^{-7} torr).

high-velocity drop Airdrop in which conventional parachute system is not used and speed of descent lies between 30 ft/s (low-velocity) and free-fall; requires retarding means plus stabilization to ensure impact on cushioned face.

highway Generalized term for channel for data in EDP (1) or other dynamic system.

high wing One attached at top of fuselage.

HIK Heading index knob.

hikodan Wing (= UK group) (J).

Hill's mirror Instrument for measuring wind speed by observing smoke from aircraft or shell-burst.

Hilo 2 Hi/lo No 2, one of two worldwide programs for microcircuit design.

Himad High to medium (altitude) air defence.

Himat, HiMAT Highly manoeuvrable aircraft technology (research programme).

hinged wing General term for wing not always rigidly fixed but able to fold, vary sweep or incidence, slew or rotate in any other way relative to fuselage.

hinge moment Force required to rotate aerodynamic surface or resist incident air load on it, C_H = total aerodynamic load on surface (varies as square of airspeed) multiplied by distance from hinge axis to cp.

HIOC Hourly indirect operating cost.

Hip Hot isostatic pressing.

Hipar 1 High-power acquisition radar.
2 High-performance precision approach radar.

HIPO, Hipo Hierarchical input process output.

HIPPAG, Hippag High-pressure pure-air generator.

Hiran High-precision shoran.

HIRL High-intensity runway light/lights/lighting.

HIRM High-incidence research model.

HIRS Helicopter infra-red system.

Hirt High-Reynolds number tunnel.

HISL High-intensity strobe light.

Hisos Helicopter integrated sonar system.

Hi-Spot High-altitude surveillance platform for over-the-horizon targeting (airship, Lockheed).

HIT, Hit 1 Homing-intercept technology.
2 Hybrid/inertial technology.

hit band Plot of initial velocity against initial path angle for vehicle to make successful impact (hard/soft) on lunar or planetary surface; usually very narrow and curved.

hit dispersion See *dispersion.*

hit the silk Abandon aircraft by parachute in emergency (colloq).

hittile Direct-hitting guided missile, ie one designed to explode inside target.

HJ, H_J Sunrise to sunset (ICAO).

HJV Hot-jet velocity.

HKAOA Hong Kong Aircrew Officers' Association.

HL High-level airway, eg in North American VOR system, other than Jet Routes.

HLA Helicopter Listing Association (US).

HLH Heavy-lift helicopter.

HLL High-level language.

HLLV Heavy-lift launch vehicle.

HLSI Hybrid LSI.

HMAFV Her Majesty's Air Force Vessel (UK).

HMC Health-monitoring computer.

HMD Helmet-mounted display.

HMDS Hexamethyldisylazane.

HMEP Helmet-mounted equipment platform.

HMG 1 Her/His Majesty's Government.
2 Hydromechanical governor.
3 Heavy machine gun.

HMH Helmet-mounted HUD.

HML Hard mobile launcher.

HMP HMG (3) pod.

HMR Homer.

HMRS Hotline mobile repair shop.

HMS 1 Health-monitoring system.
2 High-modulus sheet (CFRP).
3 Helmet-mounted sight.

HMU 1 Hydromechanical unit.
2 Helmet-mounted unit.

HMX Cyclotetramethylenetetranitramine; shock-sensitive explosive.

HN, H_N Sunset to sunrise (ICAO).

H_n C.g. margin.

HN3 Nitrogen mustard gas (code USA).

HNMO Host-nation management office.

HNS Host-nation support.

HNVS Helicopter night-vision system (Hughes).

HO 1 Service available to meet operational requirements (ICAO).
2 Hardover.

HoA Heads of agreement.

hoar frost White semi-crystalline coating deposited in clear air on surface colder than frost point (see *frost*).

HOE Homing overlay experiment, to kill RVs above atmosphere.

Hoerner tip Sailplane wingtip of curved downturned shape; popular 1950–60.

Hofin Hostile fire indicator; detects shockwaves caused by projectiles.

HOG Hermann Oberth Gesellschaft eV (G).

HOGE Hovering out of ground effect.

hogging 1 Machining from solid, esp of large part out of heavy forging.
2 Deformation of airship such that both ends droop.
3 Stressing condition of fuselage caused by sag of overhanging nose or tail in flight or on ground.
4 Stressing condition of seaplane float or boat hull when weight wholly supported by large wave at mid-length.

Hohmann Describes minimum-energy transfer orbit between two bodies, using their gravitational attraction.

Hohner Sailplane wingtip (see *Hoerner tip*).

HOJ Homing on jamming; fundamental ECM technique.

HOL Higher-order language.

hold 1 To stop countdown until fault has been cleared.
2 Period of * (1).
3 To retain data in EDP (1) store after it has been copied into another.
4 To fly standard pattern as requested until given further clearance (see *holding fix*).
5 To wait on airfield at any time after arrival and before departure under tower instructions.
6 Underfloor cargo compartment, identified as container * or bulk *.
7 Above-floor compartment in all-cargo

aircraft.

8 Manual adjustment for vertical/horizontal synchronization of raster display.

holdback link Strong link between carrier aircraft and deck, severed in accelerated take-off; hence, any such link in launch of missile or RPV.

hold-down test One carried out while vehicle, rocket motor or other device is mounted on test stand.

holding Hold (4); hence * course, * reciprocal, * side, non-* side, * fix end, * outer (or outbound) end, etc.

holding area Region of holding fix.

holding bay Airfield area where aircraft in motion (taxiing or towed) can be held (5) to facilitate ground movement.

holding fix Navaid on which holding area is based, such as VOR radial intersection, Vortac, ILS outer marker etc.

holding off Final moments of flight when pilot tries to keep aircraft from landing/alighting in order to do so either at lower speed or (tailwheel aircraft) in three-point attitude.

holding pattern Racetrack pattern, two parallel legs joined by turns at 3°/s or at 30° bank.

holding point See *holding area.*

holding room Airport lounge.

holding side That side of holding course on which pattern is flown.

hole 1 Vacant space left by electron missing from crystal lattice, hence any positive charge (fixed or mobile, in latter case flow of * cloud through p-type semiconductor).

2 Air pocket (colloq).

hollow charge See *shaped charge.*

hollow spinner Propeller spinner in form of annular ring through which passes airflow for engine, cooling or other purpose.

hollow-stem valve PE (4) exhaust valve stem in form of tube filled with sodium or other material of high specific heat for conducting heat from head.

holography Photographic record produced by interference between two sets of coherent waves. Many unique properties, one of which is 3-D nature of image.

HOM High-orbit mission.

home 1 To use matched-wavelength seeker and error-sensing flight-control system in order to fly automatically towards source of radiation.

2 To steer aircraft manually towards particular point navaid or other emitter.

homebuilt Difficult to define, but generally accepted as powered aircraft designed for construction by amateur. Majority are designed professionally, with plans approved by EAA or other authority and then duplicated and sold to homebuilders. In some cases successful * is put into commercial manufacture for sale, removing it from category. Excluded: modifications of existing types, replicas, restorations or copies of commercial designs.

home-in See *home (2).*

home plate Base airfield to which aircraft are to recover after mission.

homing 1 Procedures for bringing two radio or other EM stations, at least one airborne, together.

2 Guidance which causes a vehicle automatically to fly towards a particular source of radiation.

homing beacon One on which pilot can home (2).

homing guidance System enabling unmanned vehicle to home (1), typically based on use of IR, IIR, EO/TV, ARH, SARH or SALH.

homing receiver Radio receiver indicating aurally/visually when heading (rather than track) is not towards transmitter.

homodyne Reception of DSB in which local oscillator generates output synchronized with original carrier.

homogenous Having uniform radar reflectivity at all points on surface (function of material substrate, coating[s] and orientation).

homotron Soft failure.

honeycomb 1 Low-density structural technique and materials based on hexagon-cell honeycomb sandwiched between two sheets too thin for stability alone; can be all-metal, or any of three components can be of various other materials, joined by various adhesives. Can be supplied as standard sheet or parts can be made individually with any form. Nid d'abeilles, NdA (F).

2 Grid of thin intersecting aerofoils across tunnel or other duct intended to remove turbulence from fluid flow.

hood See *canopy.*

hook Procedure used by air controller to direct EDP (1) of semi-automatic command and control system to take specified action on particular blip, signal or portion of display.

Hooke's law Up to elastic limit, strain set up in elastic body is proportional to applied stress.

hookman Member of carrier deck party charged with unhooking aircraft after recovery.

hop 1 Very short flight; eg by aeroplane on fast taxi test or by aerodyne only marginally capable of flight.

2 Full circuit in advance of official first flight.

3 Jump from one EM frequency to another by ECCM subsystem.

hopper 1 Small receptacle in oil tank for hot or diluted oil to assist cold-weather start.

2 Container in ag-aircraft for dust (powdered chemical).

Hops Helmet optical position sensor.

horizon 1 Actual boundary where sky and planetary body appear to meet: visible *.

2 Apparent boundary as modified by atmospheric refraction, terrain, fog or other influence: apparent *.

3 Great circle on celestial sphere at all points 90°

from zenith and nadir: celestial *.
4 Line resembling apparent * but above or below: false *.
5 Locus of points at which direct rays from terrestrial radio transmitter become tangent to Earth: radio *.
6 Artificial horizon (instrument).

horizon and pitch scale HUD symbology comprising a horizon line, parallel lines above and below (±2° or 5°) for pitch information, and heading marks at 5° intervals.

horizon bar Fixed horizontal reference in most types of horizon (6), HSI and attitude displays.

horizon sensor Radiometer or other sensitive passive receiver used to align one axis of spacecraft or satellite with apparent horizon of Earth or other body; also called *-seeker.

horizontal error Error in range, deflection or miss-radius which weapon system exceeds on half of all occasions when firing on target on horizontal plane. When trajectory at arrival is near-vertical ** is CEP; where trajectory slanting, giving elliptical dispersion, ** is probable error.

horizontally opposed engine PE (4) with left and right rows of horizontal cylinders and central crankshaft.

horizontal parallax Geocentric parallax of body on observer's horizon, = angle subtended at body by Earth radius at equator.

horizontal project One in which hierarchy in design/project team is minimized and each member works full-time on single project.

horizontal scanning Scanning in azimuth only at near-0° elevation; rare (see *scanning*).

horizontal situation indicator, HSI Standard pilot panel instrument for all except small GA aircraft; includes Hdg (T, M), angular deviation from VOR, INS or other track, Tacan/DME or INS display and alphanumeric readout of G/S and possibly other data.

horn 1 Operating arm of simple manual flight-control surface to which cable is attached.
2 Microwave aerial coupling waveguide to free air to give directional pattern; three basic forms are pyramid, sectoral and biconical, first two having rectilinear funnel appearance.
3 Acoustic emitter tube whose varying cross-section and final area control acoustic impedance and directivity.
4 Small area of control surface ahead of hinge axis, usually a tip, sometimes shielded at low deflection angles by fixed surface upstream and usually housing mass-balance; when surface deflected provides aerodynamic force assisting deflection.

horn aerial See *horn (2)*.

horn balance See *horn (4)*.

horsal Horizontal ventral fin (originally on XF10F).

horsepower Non-SI unit of power; traditional hp defined as 550 ft-lb/s = 0·7457 kW; metric PS defined as 75 kg-m/s = 0·7355 kW. There is also electric * defined exactly as 0·746 kW. Brake * is measured at output shaft by applying known retarding torque; indicated * calculated from area of indicator diagram and rpm; frictional * is indicated minus brake; equivalent * is turboprop bhp plus addition factored from residual jet thrust. Aircraft engines also have various * ratings, as do some nations' pilot certificates.

horsepower loading See *power loading*.

horseshoe Revetment having this shape in plan.

horseshoe vortex Combination of finite wing plus trailing vortex from each tip, forming circulation of approximate square-cornered U-shape in plan.

HOS Human operator simulator.

hose 1 Flexible conduit for fluid (liquid or gaseous).
2 Fire long burst of gunfire at or ahead of hostile or challenged aircraft.

host computer One to which remote terminals or I/O devices are connected.

hostile Target known or assumed to be enemy, esp one seen on remote display.

hostile track One which, based upon established criteria, is determined to be enemy airborne, ballistic or orbiting threat.

host nation One in whose territory infrastructure forming part of an international system is being built.

hot 1 Fast-landing (colloq).
2 With afterburner in operation.
3 With combustion in operation (pressure-jet tip drive or HTP/hydrocarbon rocket).
4 Strongly contaminated by fallout.
5 Hazardous to vicinity (see * *mission*).
6 Including propellant ignition (see * *test*).
7 Ready for firing (launch or static test).
8 High-temperature parts of gas turbine (see * *end*).
9 Thermally anti-iced, thus * prop.
10 In ground refuelling, with engine(s) running.

Hotas Hands on throttle and stick, design of cockpit of air-combat fighter so that pilot has every control switch, button or trigger needed in any combat on these two handholds.

hot bucket Turbine rotor blade (colloq, US).

hot day Standard ISA condition for engine rating, aircraft performance certification and other temperature-dependent lawmaking.

hot dimpling Dimpling of holes pre-heated to avoid cracking.

hot end Portion of gas-turbine subjected to high temperatures from combustion, normally all parts to rear of compressor (which itself can be hot enough to require special refractory alloys in final stages but is never included in this definition); called hot section in US.

hot gas system One energized by gas bled from combustion of solid fuel, or from operating solid rocket motor.

hot gas valve One used to control hot gas pressure for TVC purposes.

hot isostatic pressing Temperature/pressure cycle for compacting sintered ceramic or encapsulated powders close to precise net shape.

hot mike Microphone continuously on transmit.

hot mission Particular test (flight or otherwise) which involves hazards precluding other activity in same area.

Hotol Horizontal take-off and landing (usually applied to spacecraft launch).

hot rock Inexperienced pilot eager to show off (colloq).

hot rocket One in which fuel is burned.

hot round Rocket vehicle equipped with operative propulsion, in test programme where many are not.

HOTS Hands on throttle and stick.

hot section Total of all parts of gas-turbine engine subjected to high temperatures from combustion of fuel, in modular engine exactly defined and normally including associated external dressing which in fact remains relatively cool.

hot spot Place much hotter than environment, showing on 3–5 micron IR.

hot start Start of gas turbine abandoned because of overtemperature indication.

hot streak Method of igniting afterburner by injecting fuel from special (normally inoperative) nozzle in main combustor, causing long flame to pass through turbine into afterburner primary zone.

hot-stream nozzle That from core engine in turbofan without mixer.

hot test Static test of rocket engine in which actual firing takes place.

hotwell Tank or portion of larger tank in which hot liquid collects.

hot winchback Aircraft pulled into HAS with engines running.

hot-wire ammeter One measuring current by I^2R heating of fine wire; hence * galvanometer, * voltmeter.

hot-wire anemometer Measures wide range of airspeeds down to 10 mm/s by heating platinum wire to about 1,000°C and either measuring current I for constant T or resistance at constant I.

hot-wire ignition Use of suddenly heated (but not exploded) resistance wire to set off rocket engine or gun ammunition.

hot-wire transducer Detects and measures sound waves by change in resistance of heated wire.

hour angle Bearing of object on celestial sphere; angle at pole between hour circles of observer and object, abb HA (see *local* **, *Greenwich* **, *sidereal* **).

hour circle Great circle of celestial sphere formed by projecting Earth meridian.

house aircraft One used for research or development and property of user, normally a manufacturer.

hover 1 To fly (usually at low altitude) stationary relative to Earth, airspeed being that of local wind.
2 Exceptionally, to fly with zero airspeed, carried along at speed of wind in horizontal plane (while holding height constant).
3 Uncommon use, to be on station in geostationary orbit.
4 To operate or travel in air-cushion vehicle; obvious contradiction in terms.

hovering ceiling Greatest altitude at which helicopter under specified conditions can hover (2), normally defined as IGE (in ground effect) or OGE (out of ground effect).

hovering point Location where V/STOL aircraft picks up or sets down load without landing.

Howe truss One having upper and lower chords joined by verticals plus diagonals inclined outwards from bottom to top on each side of midpoint.

howgozit Basic graphical plot of particular flight for planning purposes, always on basis of distance and with vertical scale flight level; includes weights, W/V and ambient T, and is amended as flight progresses (often arch).

Howls Hostile-weapons location system; versatile phased-array radar.

HP 1 High pressure; suffix number (eg HP_8) denotes stage from which bleed air is taken.
2 Horsepower (undesirable).

hp 1 Horsepower.
2 High-pass filter.

HPC HP compressor.

HPF High-pass filter.

HPI High probability of intercept.

HPL High-power laser.

HPLB High-power laser blinding.

HPOT High-pressure oxidizer turbopump.

HPRP High-power reporting point (radar).

HPS 1 Horizon and pitch scale.
2 Helmet pointing system (slaves sensors to motions of wearer's head).

HPT 1 HP turbine.
2 High-precision thermostat.

HPW High-pressure water (snow removal).

HPZ Helicopter protected zone.

hpz Hectopieze, non-SI unit of pressure = 10^5 Pa.

HR Helicopter route (airway).

HRC Historical records container.

HRE Hypersonic ramjet engine.

HRGM High-resolution ground map (radar mode).

HRIR High-resolution IR radiometer.

HRL Human Resources Laboratory (USAF).

HRN High-resolution navigation mode.

HRR High-resolution radar.

HRSCMR High-resolution surface composition mapping radiometer.

HRSI High-temperature reusable surface (or silica) insulation; normally silica-fibre tiles protected by fretted borosilicate glass plus pigment for desired absorptance/emittance ratio.

HS Scheduled hours only (ICAO).

HSD 1 Runway has hard surface, dry.
2 Horizontal situation display.

HSI 1 Horizontal situation indicator.
2 Hours since inspection.
3 Hot-section inspection.

HSLA High-strength low alloy (steel).

HSLLADS High-speed low-level aerial delivery system.

HSLV High-speed low-voltage (LSI).

HSM Hard-structure munition; conventional munition intended to have maximum effect on hardened targets.

HSP 1 Hard spark(ing) plug.
2 Heat-shrinkable polyolefin.

HSR High sink rate.

HST 1 Hypersonic transport.
2 Heat-shrinkable thermoplastics.

HSU Hartridge smoke unit.

HSVD Horizontal situation video display.

HSWT High-speed wind tunnel.

HT 1 High turbine (US for HP turbine).
2 Hard time, thus * life.
3 Helicopter, training (role prefix, UK).

H/T Hub-tip ratio.

ht Height.

h.t. High tension (electrical).

HTA High-time aircraft of type.

HTADS Helmet target-acquisition and designation system.

HTC Hard-target capability.

HTCU Hover trim control unit.

HTL High-threshold logic.

HTLC High-temperature load calibration.

HTOVL Horizontal (conventional) take-off, vertical landing.

HTP High-test peroxide.

HTPB Hydroxy-terminated polybutadiene, rocket propellant.

HTS High-tensile sheet (CFRP).

HTTB High-technology testbed (Lockheed).

hub 1 Strictly, those structural members required to hold together blades of propeller or rotor; in practice meaning has come to include all central portions, including pitch and other mechanisms, de-icing and instrumentation, but not spinner or other fairing.
2 Missile body section containing attachments for cruciform wings.
3 Chief airport of major city.

hub drive Driven by shaft input rather than tip jet reaction.

hub/tip ratio, H/T Ratio of radius at root of gas-turbine fan, compressor or turbine blade, to radius at tip (usually inside tip shroud if fitted).

HUCP Highest useful compression pressure.

HUCR Highest useful compression ratio.

HUD Head-up display.

Hudwac HUD weapon-aiming computer.

Hudwass HUD weapon-aiming subsystem.

Hufford Family of large machine tools which grip large sheet at each end in jaws on hydraulic rams mounted on pivoting platforms and stretch it over male die itself thust forward on rams.

hull 1 Fuselage of flying boat.
2 Main body of rigid airship.
3 For insurance purposes, complete aircraft as defined in policy, ignoring any load or persons on board.

hull insurance That covering capital value (assigned at operative time, neither original first cost nor replacement cost) of aircraft.

Hultec Hull to emitter correlation, EW/ESM software programs.

HUM Health and usage monitor.

humidity Wetness of gas, esp amount of water vapour present in air; absolute * is mass of vapour in unit volume g/m^3, specific * is mass of vapour in unit mass of dry air g/kg, relative * is percentage degree of saturation = 100 times actual vapour pressure divided by SVP.

humidity mixing ratio Specific humidity.

Humint Human intelligence, ie countermeasures against human decision-taking.

hump speed That of marine aircraft (seaplane, flying boat) or ACV at which total resistance on water is greatest; in case of aircraft this speed (no symbol known) is typically about half V_{US}.

hundred per cent aircraft One that exactly measures up to published performance.

hung round Rocket or other missile which fails to release from aircraft.

hung start Starting of main turbine engine which, for any reason, automatically or under manual control is arrested after ignition but before self-sustaining speed is reached.

hunter/killer 1 Aircraft or other platform able to seek out and kill submarines unaided.
2 Pair of ASW aircraft, one with sensors, other with weapons.
3 Co-ordinated task force comprising aircraft and surface forces combining in ASW mission.

hunting Continual steady oscillation about neutral point: governed speed, governed position, desired flight attitude or other target regime; manifest in cyclic phugoids in aircraft pitch or yaw (rare in roll), shuttling of hydraulic spool valve, ceaseless rising and falling of speed of rotating machine or visible oscillation of instrument needle.

hurricane One name for tropical revolving storm, usually defined as one with winds whose mean speed exceeds 64 kt, 119 km/h.

hushkit Supplied by manufacturer to operator to quieten engine, invariably turbojet or turbofan, already in service; often includes inlet acoustic liners, liners for tailpipe and new nozzle with longer periphery and faster mixing.

HV, h.v. 1 High voltage.

2 High vacuum.
H/V Height/velocity (plotted curve).
HVA Horizontal viewing arc.
HVD Helmet visor display.
HVM Hypervelocity missile.
HVTA High-value target acquisition.
HVZL Hauptverwaltung der Zivilien Luftfahrt (DDR, East Germany).
h.w. Half-wave.
HX, H_x Airfield has variable working hours.
hybrid bearing Combining bearing with rolling elements (ball, roller, needle) encased in free-rotating journal supported by fluid film.
hybrid composite Matrix reinforced by fibres of two different types.
hybrid computer One using digital techniques for large and precise arithmetic and logic beyond scope of analog machines, and analog wherever possible for highest computational speed; invariably a good compromise for large simulations and other specialized tasks.
hybrid electronics Combination in single integrated circuit of epitaxial monolithic or thin-film with one or more discrete devices.
hybrid FCS Flight-control system with digital outer loop and analog inner loop.
hybrid IC, hybrid package Hybrid electronics.
hybrid propulsion 1 Aircraft propelled by two or more dissimilar species of prime mover; eg turboprop plus jet, or turbojet plus rocket.
2 Single propulsion engine capable of operating in two distinct modes; eg turborocket or ram rocket.
hybrid rocket One using both liquid and solid propellants simultaneously; usual arrangement is solid fuel and liquid oxidant.
hybrid solar array Part folding, part flexible.
hybrid trajectory Any space trajectory intermediate between that for minimum energy or minimum time and alternatives offering greater payloads, longer launch windows or other advantages.
hybrid wave EM wave in waveguide having both magnetic and electric components in plane of propagation.
Hycatrol One of several trade names (FPT Industries) for rubber/metal bonded structures.
hydrant dispenser Installation under apron or finger/gate area stand for refuelling aircraft without need for tanker vehicles.
hydrant pit Below-ground compartment, normally covered, housing connections and controls for fuel, hydraulic, lube oil or other liquid supply.
hydraulicing Abnormal resistance to movement of machine, esp PE (4), caused by hydraulic lock in lower cylinders part-full of essentially incompressible oil.
hydraulic lock Use of essentially incompressible liquid to prevent movement of mechanical part.
hydraulic motor Source of mechanical power, usually rotary, driven hydraulically.
hydraulic power unit Source of power to energize hydraulic system, eg when main engines inoperative.
hydraulics Science of liquids either in motion (hydrodynamics) or as media for transmitting forces (hydrostatics). In aerospace generally science of nearly incompressible liquids enclosed in closed-circuit pipe systems at high pressure and used both to apply forces, with little fluid motion, and supply power, with large fluid motion. Media originally mineral oils, today also phosphate esters, chlorinated silicones, silicate esters and (supersonic and missiles) alkyl silicate esters.
hydraulic system Complete aircraft installation comprising closed circuits of piping, engine-driven pumps, accumulators, valves, heat exchangers, filters and, usually, emergency input such as RAT or MEPU; normally divided into at least two systems with maximum degree of independence. Each system is assigned task of driving selected items by linear actuators, motors or other output devices.
hydrazine Family of chemicals, mostly colourless liquids, often corrosive; basic member is *, $(NH_2)_2$; common rocket fuel is unsymmetrical dimethyl *, $NH_2N(CH_3)_2$; another is monomethyl *, $NH_2HH.CH_3$.
hydrobooster Hydraulic power unit used in boosted (not fully powered) flight-control system (colloq).
hydrocarbon Compound of hydrogen and oxygen only; some millions are known, including all derivatives of petroleum, which in product form often have other elements added for specific purposes. Many aviation fuels are alkanes (paraffins), which are open chains with carbon atoms having single-valence bonds; first six members, with one to six carbon atoms respectively, are methane, ethane, propane, butane, pentane and hexane. These are prefixed n (normal), distinguished from prefix iso of more reactive branched alkanes. Alkenes have one carbon with double bond, based on ethylene $(CH_2)_2$. Aromatic * series are based on hexagon ring of benzene C_6H_6 with three double bonds.
hydrodynamics Science of fluid motion.
hydroflap Water rudder on flying boat.
hydrogenation Causing to combine with hydrogen, esp at high pressure and in presence of catalyst such as nickel or platinum, in conversion of crude petroleum distillates into tailored fuels and other products. * of coal also important to future aviation.
hydrogen bomb Any weapon whose major energy release is derived from fusion of hydrogen nuclei to form helium; thermonuclear or fusion bomb (colloq).
hydrogen economy Hypothetical situation in which LH_2/GH_2 replace petroleum.

hydroglider Glider with marine alighting gear.
hydrokinetics Science of liquids in motion.
hydromagnetics See *MHD*.
hydromechanical logic Performed with mechanical elements wherein information exists as hydraulic flows/pressures.
hydrometer Instrument for determining density of liquids.
hydrometeor Any atmospheric water in solid or liquid form; all precipitation, fog, dew, frost etc.
hydroplane Light boat which skims water surface on planing bottom when at high speed; erroneously misused for seaplane and/or hydrofoil.
hydroport Inland airport for marine aircraft.
hydropress Diverse family of hydraulic presses widely used in aerospace with large-area platen on which are mounted large or multiple tools around which sheet is shaped by rubber pad or mating tools.
hydroskis Planing surface, usually in left/right pair and retractable, used for take-off and landing of certain marine aircraft; have little buoyancy, so ski aircraft rests on water like flying boat when at rest.
hydrosphere Water resting on or within crust of Earth or other body, excluding that in atmosphere.
hydrostatic bearing Spinning shaft, sphere or other mass is supported (usually radially and axially) by filtered gas or liquid dynamic reaction, eliminating contact between fixed and moving solid surfaces.
hydrostatic drive Transmits power, usually between rotating shafts, by pumping hydraulic fluid round closed circuit; both pump and motor usually have stroke or output infinitely variable down to zero, giving perfectly flexible dynamic link from zero to maximum output power.
hydrostatic equation Applies when secondary effects (Earth curvature, friction, coriolis etc) ignored, leaving $dp/dz = -\rho g$ where p is pressure, z geometric height and ρ density.
hydrostatic extrusion Advanced technique for extrusion of steels and other materials, usually at room temperature, by forcing through die under extreme hydraulic pressure.
hydrostatic fuze Triggered by depth-dependent water pressure.
hydrostatic test Test of container (fuselage, solid rocket case) under high pressure using water or other liquid to minimize stored energy.
Hydrus FBMS ship/shore communications system.
Hydulignum Trade name (Hordern-Richmond) for densified wood.
Hydyne Storable rocket fuel in various formulations based upon 60% UDMH and 40% diethylenetriamine.
hyetograph 1 Recording rain gauge.
2 Annual rainfall chart.
Hyfil Trade name for family of CFRP raw materials marketed in standard forms or used for in-house production (Rolls-Royce).
hygrograph Recording hygrometer.
hygrometer Determines atmospheric humidity (strictly, wet and dry bulb * is psychrometer).
Hylite British titanium alloys (+ Sn, Zr, Al); creep-resistant.
hyperabrupt Device, eg diode, tailored to most rapid possible action.
hyperacoustic zone Region above 100 km (62 miles) where mean free path approximates to wavelengths of sound; limit of sound propagation.
hyperbaric Having atmospheric pressure or oxygen concentration greater than normal sea-level.
hyperbola Conic section obtained by plane cutting both nappes (normal right circular cone and its mirror-image inverted above); locus of points whose distances from two foci have constant difference, standard equation $x^2/a^2 - y^2/b^2 = 1$. Because of constant difference in distance from two foci, possible to base families of navaids on keyed radio emissions from two fixed stations.
hyperbolic navaids Based upon synchronized emissions from fixed ground stations, often called master and slave(s), which are received by aircraft at time differences which yield lines of constant time (ie range) difference in form of hyperbolic position lines. First were Gee, Loran and Decca, later developed to give instantaneous readout of position or moving-map display.
hyperbolic error That due to assumption that waves received at all antennas of an interferometer baseline are travelling in parallel directions.
hyperbolic frequencies Measured in several tens of GHz.
hyperbolic re-entry At hyperbolic speed.
hyperbolic speed Sufficient to escape from Solar System; on Earth trajectory away from Sun about 40,597 km/h, 25,226 mph.
hypergolic Of rocket propellants, those which ignite spontaneously when mixed.
hypersonic Having Mach number exceeding 5.
hypertension High blood pressure.
hyperventilation Overbreathing; specif reduced CO_2 causing * syndrome, dizziness, fainting, convulsions.
hypobaric Having atmospheric pressure much less than normal at sea level.
hypocapnia CO_2 deficiency in blood.
hypoid Bevel or helical gears transmitting power with some tooth-sliding action between shafts neither parallel nor intersecting.
hypoventilation Underbreathing.
hypoxaemia Condition resulting from hypoxia.
hypoxia O_2 deficiency in blood, from whatever cause.
hypsometric tints Colour gradations chosen for contrast by natural or artificial illumination (eg, for contour bands on topographic map).

Hyrat Hydraulic (pump driven by) ram air turbine.

hysteresis 1 Generally, condition exhibited by system whose state results from previous history, specif one whose instantaneous values lag behind prediction.
2 In ferromagnetic and some other materials, lagging of magnetic flux density T behind magnetic field strength A/m causing it (see * *loop*).
3 Effects caused by internal friction in elastic material undergoing varying (esp rapidly oscillating) stress.
4 Lag in instrument indication; eg that of barometric altimeter in dive or climb.

hysteresis loop Plot of magnetizing field strength against magnetic flux density for ferromagnetic material; traditionally called B/H curve because flux density was called induction, symbol B, measured in gauss, and field strength (formerly in oersteds) was identified by symbol H; today's units are (field strength) A/m and (flux density) T.

HYT High year of tenure (waiver programme).

Hz, hertz SI unit of frequency, = cycles per second.

I

I 1 Electric current.
2 Moment of inertia.
3 Prefix, direct-injection engine (US).
4 Luminous intensity.
5 Total heat content.

i 1 Intensity of rainfall.
2 "Square root of minus 1".

I-band EM radiation, 37·5–30 mm, 8–10 GHz.

I-beam One of I section.

I-display When radar aerial pointed at target latter appears as circle at radius proportional to range; when aerial points away from target latter appears as segment showing magnitude/direction of error.

I-section Structural beam with vertical web and flat upper and lower booms.

I^2F Intelligent influence fuze.

IA 1 Initial approach (FAA).
2 Inspection authorization (FAA).
3 Input axis.

Ia Anode current.

IAA 1 International Academy of Astronautics.
2 International Aerospace Abstracts (AIAA).

IAAA International Airforwarder and Agents Association.

IAAC 1 International Agricultural Aviation Centre (HQ in UK).
2 International Association of Aircargo Consolidators.

IAAE Institution of Automotive & Aeronautical Engineers (Australia).

IAAFA Inter-American Air Forces Academy (US).

IAAI International Airports Authority of India.

IAASM International Academy of Aviation & Space Medicine.

IABA International Association of Aircraft Brokers and Agents.

IABCS Integrated aircraft brake control system.

IABG Industrieanlagen Betriebs GmbH (G).

IAC Intelligence Analysis Center (US).

IACA International Air Carrier Association.

IACES International Air Cushion Engineering Society.

IACS Integrated avionics control system.

IACZ Inter Airline Club Zurich (Int).

IADB Inter-American Defense Board.

IADS Integrated air-defence system (eg Singapore/Malaysia).

IAE Instituto de Atividades Espaciais (Braz).

IAEA 1 Indian Air Engineers' Association.
2 International Atomic Energy Agency.

IAEM Instituto dos Altos Estudos Militares (Port).

IAF 1 International Astronautical Federation.
2 Initial approach fix.

IAFA International Airfreight Forwarders' Association (J).

IAFU Improved assault fire unit (USA).

IAGS Inter-American Geodetic Survey.

IAIN International Association of Institutes of Navigation.

IAIS Industrial Aerodynamics Information Service.

IAL International Airtraffic League.

IALPA Irish ALPA.

IAM 1 International Association of Machinists and Aerospace Workers (US).
2 Institute of Aviation Medicine (UK).

IANC International Airlines Navigators' Council.

I&C, I&CO Installation and checkout.

I&M Improvement and modernization.

IAOPA International Council of AOPAs.

IAP 1 Imagery architecture plan.
2 Fighter aviation regiment (USSR).
3 Initial approach procedure.

IAPA 1 International Airline Passengers' Association.
2 International Aviation Photographers' Associations.

IAPC International Airport Planning Consortium (UK).

IAP chart Instrument approach procedure chart.

IAPS Ion auxiliary propulsion system.

IA-PVO Fighter aviation, air defence of the homeland (USSR).

IAS 1 Institute of the Aerospace Sciences (US).
2 Indicated airspeed (see *airspeed*).
3 Integrated acoustic structure.
4 Interplanetary automated shuttle.

IASA International Air Shipping Association.

IASB Institut d'Aéronomie Spatiale de Belgique.

IAT International Association of Touristic Managers.

IATA International Air Transport Association.

IATS Intermediate automatic test system.

IATSC International Aeronautical Telecommunications Switching Centre.

IAU 1 International Accounting Unit.
2 International Astronomical Union.
3 Interface adaptor unit.

IAW In accordance with.

IB 1 Inbound.
2 Incendiary bomb.
3 Ion beam, thus * erosion, * engine.

I_b Burning-time impulse.

IBA Inbound boom avoidance.

IBAA International Business Aircraft Association (Europe).
Iberlant Iberia-Atlantic area (NATO).
IBF Internally blown flap.
IBn Identity (or identification) beacon.
IBR 1 Integrally bladed rotor (gas turbine).
2 Intra-base radio.
IBRD 1 Inflated ballute retarding device.
2 International Bank for Reconstruction and Development.
IBU Independent back-up unit.
IC 1 Internal combustion.
2 Interceptor controller.
3 Indirect cycle (sometimes i.c.).
4 Integrated circuit.
I/C, I & C Installation and checkout.
i.c. 1 Integrated circuit (sometimes IC).
2 Intercom.
ICA 1 International Committee of Aerospace Activities.
2 Initial cruise altitude.
ICAA 1 International Civil Airports Association.
2 International Committee of Aerospace Activities.
ICAF 1 International Committee on Aeronautical Fatigue.
2 Industrial College of the Armed Forces (US).
ICAM Integrated computer-aided manufacturing.
ICAN International Commission on Air Navigation (rue Geo Bizet, Paris, pre-1939).
ICAO International Civil Aviation Organization.
ICAOPA IAOPA.
ICAOTAM ICAO technical assistance mission.
ICAS International Council of the Aeronautical Sciences.
ICB International competitive bidding.
ICBM Intercontinental ballistic missile.
ICC 1 International Control Commission.
2 International Chamber of Commerce.
ICCAIA International Co-ordinating Council of Aerospace Industries Associations.
ICDS Integrated control and display system.
ICE 1 Interference cancellation equipment.
2 Internal-combustion engine.
3 In-circuit emulator, simulates portions of external hardware during debugging of control software.
icebox rivet One (eg 2024 alloy) which must be kept below 0°C until use.
ice frost Forms on cryogenic tank; if on rocket, easily shaken off at launch.
ice ingestion Class of tests to determine ability of engine to swallow various cubes of ice, ice-water slush and sometimes hailstones at specified flow rates.
ICEM Intergovernmental Committee for European Migrations.
ice needles Growth of long ice crystals of needle appearance.
IC engine Internal combustion, usually means reciprocating: Otto, Diesel, Stirling, Rankine etc.
ice plate Strong plate on fuselage skin in plane of propellers.
ice point NTP equilibrium temperature for ice/water mixture.
ice rod Standard ingestion-test size: 1·25 in diameter × 12 in.
Icerun Ice on runway.
ICF 1 Initial contact frequency.
2 Inertial confinement fusion.
ICFP Inverse-Cassegrain flat plate.
ICH Interline Club Holland (Int).
ICHE Intercooler heat-exchanger.
ICI 1 International Commission for Illumination.
2 Initial capability inspection.
icing Accretion of ice or related material on aircraft.
icing indicator Any of four families of device quantifying presence of icing and rate of accretion.
icing limits Usually upper and lower temperature limits, corresponding to one (sometimes two) flight levels.
icing meter Instrument giving rate of accretion, buildup thickness and cloud liquid-water content.
icing tanker Aircraft equipped to cause severe icing on another following close behind.
ICLECS Integrated closed-loop ECS.
ICM Intercontinental missile.
ICNI Integrated com, nav, IFF.
ICNIA Integrated communications, navigation and identification avionics (USAF).
ICNS Integrated com/nav system.
ICO Ignition cut-off.
ICPA Indian Commercial Pilots' Association.
ICR In-commission rate.
ICS 1 Improved composite structure.
2 Inverse conical scan(ning) (ECM).
3 Improved (or integrated) communications system.
4 Intercom switch.
5 Internal countermeasures system (or set).
6 Interim contractor support.
ICSC International Communications Satellite Corporation.
ICSMA Integrated Communications System Mangement Agency.
ICSU International Council of Scientific Unions.
ICU Instrument comparator unit.
ICW 1 Independent carrier wave, or see *i.c.w.*
2 Interpersonal communications workshop.
3 Intermittent continuous wave.
i.c.w. Interrupted continuous wave, or see *ICW*.
ICWAR, Icwar Improved continuous-wave acquisition radar.

ID 1 Internal diameter.
2 Inadvertent disconnect (flight refuel).
3 Identification.

IDA 1 Istituto di Diritto Aeronautico (I).
2 Intermediate dialect of Atlas.
3 Integrated digital avionics.

Idaflieg Interessengemeinschaft Deutscher Akademischer Fliegergruppen eV (G).

IDAS 1 Integrated defensive aids system (RWR plus jammer).
2 Integrated design automation system.

IDC Imperial Defence College (UK).

IDCSP Initial Defense Com Sat Program (DoD).

IDEA Instituto de Experimentaciones Astronauticas (Arg).

ideal fluid Perfect, inviscid fluid; forces are perpendicular to small-parcel boundaries, no kinetic energy can be degraded to heat, boundary layer absent.

ideal rocket Theoretical rocket with perfect operation, eg no heat transfer, no turbulence, no friction etc.

ident Special feature in ATCRBS and I/P in SIF to distinguish one displayed select code from other codes (FAA).

identification 1 Proclamation of identity, eg by SSR or squawk.
2 Visual recognition of aircraft type.

identification cable colour Each cable or pipeline in modern aircraft is colour coded to indicate function.

identification feature Characteristic built into each selected code in SSR, ATCRBS and military radars for ident purposes.

identification friend or foe Automatic interrogation and response, by coded transmission from transponder, to proclaim friendly status or identify flight on SSR.

identification light Pilot-controlled white lights visible from above or below for broadcasting identity by keying.

identification manoeuvre In primitive radar GCA, manoeuvre commanded by controller to establish positive identity of customer on radar.

IDF Intelligent data fusion, most efficient way of using astronomic amounts of data from many sources.

IDG Integrated drive generator; electric generator made as unit with variable-ratio drive for CFAC.

idle cutoff Position of PE (4) mixture control that cuts off fuel supply, thus stopping engine.

idles Repeated cycling of engine from idling to specified higher power.

idling Running at governed low speed consistent with reliable smooth operation, in most engines well above min sustaining rpm: usually two regimes, ground * being lower N_1 than flight * and obtained only when oleos compressed.

IDP 1 Individual development programme.
2 Integrated data-processing.

IDPM Institute of Data Processing Management (UK).

IDS 1 Interdiction/strike.
2 Integrated display set (USAF).
3 Improved data set (USAF).

IE 1 Institution of Electronics (UK).
2 Instrument error.
3 Initial equipment (RAF).

IEA International Ergonomics Association.

IEB Institut für Extraterrestrische Biologie (G).

IEC International Electrotechnical Commission.

IECMS In-flight engine-condition monitoring system.

IED Improvised explosive device.

IEE The Institution of Electrical Engineers (UK).

IEEE Institute of Electrical and Electronic Engineers (US).

IEMats Improved emergency message auto transmission system.

IEPG Independent European Programme Group (NATO).

IERE Institution of Electronic and Radio Engineers (UK).

IET Initial entry training.

IEU Interface electronics unit.

IF 1 Intermediate frequency (often i.f.).
2 Instrument flight.
3 Intensive flying (UK, RN).
4 Independent Force (RAF, 1918).

I/F module Inlet and fan module.

IFA International Federation of Airworthiness.

IFAA International Flight Attendants' Association.

IFALPA International Federation of Air Line Pilots' Associations.

IFast Integrated flexibility (or facility) for avionics system test (USAF).

IFATCA International Federation of Air Traffic Controllers' Associations.

IFATE International Federation of Airworthiness Technology and Engineering.

IFB Invitation for bid.

IFBS Individual flexible barrier system.

IFC Incentive-fee contract(ing).

IFCS 1 Integrated flight-control system.
2 Integrated fire-control system.

IFDAPS Integrated flight-data processing system.

IFE 1 In-flight emergency.
2 In-flight entertainment.

IFF 1 Identification friend or foe.
2 International Flying Farmers (US).
3 Institute of Freight Forwarders (Int).

IFFAA International Federation of Forwarding Agent Associations (now IFFFA).

IF/FCS Integrated fire/flight control system.

IFFFA International Federation of Air Freight Forwarders Associations.

IFIAT See *FITAP*.

I-file Intelligence (surveillance computer).

IFIS 1 Independent flight inspection system (for

ILS, Vortac etc).
2 Independent frequency-isolation system.
IFM 1 Instantaneous frequency measurement.
2 In-flight monitor.
IFMA In-flight mission abort.
IFMIS Integrated force management info system.
IFMR IFM (1) receiver.
IFN Institut Français de Navigation.
IFOV Instantaneous field of view.
IFP Initial flightpath.
IFPC Integrated flight and propulsion control (STOL).
IFPL In-flight power loss.
IFPM In-flight performance monitor.
IFR 1 Instrument flight rules.
2 In-flight repair.
3 In-flight refuelling.
4 Initial flight release (of engine).
5 Internationaler Förderkreis für Raumfahrt (G).
IFS 1 Institut für Segelflugforschung (G).
2 Inspectorate of Flight Safety (USAF, RAF).
3 Instrument flight simulator.
IFSA In-Flight Service Association.
IFSD In-flight shutdown.
IFSS 1 International flight service station.
2 Instrumentation and flight safety system.
i.f.t. Intermediate-frequency transformer.
IFTO International Federation of Tour Operators.
IFTU Intensive flying trials unit (UK).
IFU Interface unit.
IFURTA, Ifurta Institut de Formation Universitaire et de Recherche du Transport Aérien (F).
IG 1 Imperial gallon (sometimes I.g.).
2 Inspector-general.
Ig Grid current; hence Ig2 screen-grid current.
IGA Inner gimbal angle.
IGAA Inspection Générale de l'Armée de l'Air (F).
IGAC Inspection Générale de l'Aviation Civile et de la Météorologique (F).
IGB Integral gearbox (propeller).
IGE In ground effect.
IGFET Insulated-gate field-effect transistor.
igloo 1 Air cargo container in form of rigid box sized for above-floor loading.
2 Small pressurized experiment container designed for use with Spacelab orbital laboratory.
IGN Ignition.
igniter Pyrotechnic (rocket) or high-energy discharge device for starting combustion of solid or liquid fuel.
igniter plug Electric device of various kinds for starting combustion of gas-turbine engine.
ignition delay time Several definitions for elapsed period between ignition signal and stable or other desired combustion in PE (4), solid rocket etc. For solid motors elapsed time from ignition firing voltage to chamber pressure reaching 10% max, symbol t_d.
ignition harness Complete screened assembly of HT cables serving spark plugs of piston engine.
ignition interference Disturbance to radio communication caused by faulty ignition screening.
ignition screening Surrounding layer of conductive braid to prevent signal emission from HT harness.
IGO Prefix, US piston engine, direct injection, geared, opposed.
IGOR Intercept ground optical recorder; long-focal-length tele-camera.
IGPM Imperial gallons per minute.
IGS 1 Internal gun system.
2 Instrument guidance system.
IGV Inlet guide vane.
IGY International Geophysical Year.
IH Inhibition height (GPWS).
I_h 1 Mean horizontal candlepower.
2 Heater current.
IH Aviation Historical Association (Finland).
IHADSS Integrated helmet and display sighting system.
IHAS Integrated helicopter attack system.
IHC Integrated hand control.
IHE Insensitive high explosive.
IHEC Integrated helicopter emissions control.
IHOC International Helicopter Operations Committee.
II Image intensifier.
IIAE Instituto de Investigaciones Aeronauticas y Espacial (Arg).
IIC Image isocon camera tube.
IIASA International Institute for Applied Systems Analysis.
IIDS Istituto Italiano di Diritto Spaziale.
IIF Inserted in flight (data, target, destination etc).
IIN Istituto Italiano di Navigazione.
IIR 1 Imaging infra-red.
2 Infra-red imaging radar (US).
IIRS Instrument inertial reference set.
IIS 1 Infra-red imaging system.
2 Integrated instrument system.
IISA Integrated inertial sensor assembly.
IISL International Institute of Space Law.
IISS International Institute for Strategic Studies.
IITS Intra-theatre imagery transmission system.
IJMS Interim JTIDS message standard.
IK Club of Aeronautical Engineers (Finland).
IKBS Intelligent knowledge-based systems.
IL Infantry liaison (aircraft category, USA, 1919–25).
ILA International Law Association.
ILAAS Integrated low-altitude attack subsystem.
ILAF Identical location of accelerometer and force.
ILD Injection laser diode.
ILGH Interessengemeinschaft Luftfahrtgeräte-Handel (G).

ILL Internationale Luftverkehrsliga.
illuminance Intensity of illumination, luminous flux per unit area, unit lux, lx = lm/m^2.
illumination Lighting of target by radar or other signals, esp to make it an emitter for SARH missile.
ILM Independent landing monitor.
ILS 1 Instrument landing system.
2 Integrated logistic support.
ILS integrity Trust which can be placed in correctness of information supplied by facility (ICAO).
ILSMT Integrated logistic support management team.
ILS Point A On extended runway centreline 4 nm from threshold.
ILS Point B On extended runway centreline 1,050 m (3,500 ft) from threshold.
ILS Point C Intersection of straight line representing nominal (mean) glideslope and horizontal plane 30 m above threshold.
ILS Point D 6 m above centreline, 600 m upwind (ie towards localizer) of threshold.
ILS reference datum Point at specified height vertically above intersection of centreline and threshold through which passes straight line representing nominal (mean) glideslope.
ILS reliability (1) Facility: probability its signals are within specified tolerances.
2 Signals: probability signal in space of specified characteristics is available to aircraft.
ILVSI Instantaneous-lag VSI.
IM 1 Inner marker.
2 Intermediate maintenance.
IMA 1 Institut Médical de l'Aviation (Swiss).
2 Intermediate maintenance activity.
3 Individual mobilization augmentee.
IMAAWS Infantry man-portable anti-armour/assault weapon system.
IMAC Integrated microwave amplifer converter.
IMAD Integrated multisensor airborne display (Elint).
image degradation That due to error in sensor operation, processing procedure or other fault by user.
image intensifier Any of large family of electron tubes which multiply electron flow due to signal while ignoring noise.
image-motion compensation Synchronization of target image with recording sensor in vehicle, esp low-level reconnaissance aircraft.
imagery Representation of objects reproduced by optical or electronic means.
imagery collateral Reference materials supporting imagery interpretation function (ASCC).
imagery correlation Mutual relationship between different signatures on imagery of same object from different sensors.
imagery data-recording Auto record of sensor speed, height, tilt, geographical position, time and possibly other parameters on to sensor matrix block at moment of imagery acquisition (ASCC).
imagery exploitation Entire process from acquiring imagery to final dissemination of information.
imagery interpretation Process of location, recognition, identification and description of objects visible on imagery (DoD, NATO).
imagery interpretation key Diagrams, examples, charts, tables etc which aid interpreters in rapid identification of objects.
imagery pack Assembly of all records from different sensors covering common target area (ASCC).
imagery sortie One flight by one aircraft for acquiring imagery (DoD, NATO, Cento).
IMC 1 Instrument meteorological conditions.
2 Image movement compensator (reconnaissance).
3 Institute of Measurement and Control (UK).
IMD Indian Meteorological Department.
IME Indirect manufacturing expense.
IMEP International materiél evaluation program (US).
IMEWS Integrated missile early warning system.
IMFCA Institut de Mécanique des Fluides et Constructions Aérospatiales (Romania).
IMI 1 Intermediate maintenance instruction.
2 Improved manned interceptor.
IMK Increased maneuverability kit (US).
IML International micro-gravity laboratory (Spacelab).
IMLS Interim MLS.
immediate air support That meeting specific request during battle and which cannot be planned in advance.
Immelmann Air-combat manoeuvre. Two definitions current: (a) first half of loop followed by half-roll, resulting in 180° change of heading and gain in height; (b) steep climbing turn to bring guns to bear on target again following dive attack from front or rear.
IMN Indicated Mach number.
I_{mo} Motor specific impulse (solid propellant).
IMP 1 Interactive microprogrammable.
2 Indication of microwave propagation (anaprop) to exploit gaps in hostile radars and assess friendly radars in defence.
impact area Area having designated boundaries within which all ordnance is to hit ground.
impact microphone Sensitive to small vibrations, as of micrometeoroid impact.
impact point 1 Point on drop zone where first parachutist or air-dropped cargo should land (NATO, IADB).
2 Point at which projectile, bomb or re-entry vehicle impacts or is expected to impact (DoD).
3 Reference point of accident from which all surveys are plotted, easily identifiable and either central or having highest density of contamina-

tion (USAF); this need not be actual point of impact.

impact pressure See *stagnation pressure*.

impact tests Charpy, Izod and similar tests to measure resistance of material to suddenly imposed stress.

Impatt Impact-ionization avalanche transit time (solid-state oscillator).

IMPD Interactive multipurpose display.

impedance Resistance to a.c. (1), symbol $Z = \sqrt{(R^2 + X^2)}$ where R is ohmic resistance and X is reactance.

impeller Single-stage radial-flow compressor rotor.

impervious canopy One through which ejection is prohibited.

impingement Impact of high-velocity air or gas on structure (eg in reverse-thrust mode).

impingement cooling Cooling of material by high-velocity air jets directed on to surface (usually internal surface of hollow blade or vane).

impingement injector Liquid fuel and oxidant jets impact on each other to cause swift breakup and mixing.

implosion Detonation of spherical array of inward-facing shaped charges (eg to crush fissile core of nuclear weapon).

improved climb T-O Take-off at increased weight allowed by raising V_2 where second segment is limiting factor and runway distance is available.

impulse Rocket burn time multiplied by burn-time average thrust; total energy imparted to vehicle.

impulse magneto One whose drive incorporates stops and spring-loaded coupling (bypassed by centrifugal clutch in normal running) to give series of sudden rotations and thus hot sparks during starting.

impulse/reaction Common form of gas turbine, combining impulse and reaction techniques.

impulse turbine One whose working fluid enters at lowest pressure and maximum velocity and leaves at similar pressure and low velocity.

IMR Imaging microwave radiometer.

IMRO Industrial Marketing Research Organization (UK).

Imron Range of du Pont poly enamels.

IMS 1 Information management system.
2 Integrated management system.
3 Intermediate maintenance squadron.
4 International Military Services (MoD, UK) or Staff (NATO).
5 Integrated multiplex system, ties nav, weapons, air data and FBW.
6 Inertial measurement system.

IMU Inertial measurement unit.

IN 1 Inertial navigator (or navigation).
2 Instrument navigator (arch).

inactivate Of military unit, withdraw all personnel and transfer to inactive list.

INAS Inertial navigation/attack system.

inboard aileron Aileron situated on inner wing between, or in place of, flaps.

inboard profile Side-elevation drawing showing internal systems and equipment, sometimes as true section along centreline.

inboard quadrant Inner selectable position of power (throttle) lever on left side of cockpit giving operative afterburner.

inbound Approaching destination, thus * traffic, * controller etc.

inbound bearing Normally QDM, not QDR or VOR radial.

INC Insertable nuclear component.

Incas Integrated navigation and collision-avoidance system.

incendiary bomb Bomb designed to ignite enemy infrastructures.

Incerfa, INCERFA Code: phase of uncertainty (ICAO).

inch 1 To command powered actuator in rapid succession of small cycles to achieve target condition.
2 Non-SI unit of length, = 25·4 mm exactly.

incidence 1 Angle between chord of wing at centreline and OX axis.
2 Generally, the angular setting of any aerofoil or other plate-like surface to a reference axis.
3 Widely and incorrectly used to mean angle of attack.

incidence instability Divergent aerofoil load caused by wing flexure simultaneously resulting in increased incidence.

incidence wires Diagonal bracing wires in plane of biplane interplane struts.

incident report Normal report of incident; this falls short of an accident, takes place on ground or in air during flying operations, and usually stems from human error.

inclination Angle between isobar and wind or airflow at given point.

inclined shock Shockwave generated by body in airflow at Mach number significantly greater than 1, with angle such as to turn flow parallel to surface of body; in air-breathing inlet generated by centrebody or sharp-edged plate and focused on lip.

included angle That between longitudinal axis of body and free-stream vector.

Inconel High-nickel chromium-iron refractory alloys (Int Nickel and Mond).

Incos Integrated control system.

Incospar Indian National Committee for Space Research.

incremental airbrakes Capable of being controlled to intermediate positions.

incremental sensitivity Change in received signal per unit displacement of ILS receiver from mean glideslope.

Ind Indicator (for wind or landing direction).

INDAE Instituto Nacional de Derecho Aeronáutica y Espacial (Arg).

indefinite callsign C/s assigned to individual units/facilities etc and to large groupings.

indexed wings/fins Aerodynamic surfaces are at same angular setting to body, measured in transverse plane, eg all at $\pm 45°$ to horizontal.

index error That caused by misalignment of measurement mechanism of instrument (ASCC).

indicated airspeed See *airspeed.*

indicated altitude That shown by altimeter set to latest known QNH.

indicated course Locus of points in any horizontal plane at which ILS Loc needle is centred.

indicated course sector Sector in any horizontal plane between loci of points at which ILS Loc needle is at FSD left or right.

indicated glidepath Locus of points in vertical plane through runway centreline at which G/S needle is centred.

indicated hold Autopilot mode maintaining present IAS.

indicated horsepower See *horsepower.*

indicated Mach number That shown on Machmeter.

indicator diagram Plot of PE (4) cylinder pressure against piston position, often drawn by instrument attached to cylinder.

indirect air support Given friendly land/sea forces by action other than in tactical battle area, eg by interdiction and air superiority.

indirect cycle Nuclear propulsion with primary circuit, heat exchanger and secondary circuit.

indirect damage assessment Revised target assessment based on new data such as actual weapon yield and ground zero.

indirect wave One arriving by indirect path caused by reflection/refraction.

individual controls Control surfaces not attached to fixed surface but cantilevered from body (guided weapons).

induced drag Drag due to component of wing resultant force along line of flight; drag due to lift.

induced flow Fluid flow drawn in and accelerated by a high-velocity jet.

induced velocity That due to wing vortex system and downwash, normally considered proportional to lift.

inducer 1 Booster vanes at entry to centrifugal impeller, esp rocket-engine turbopump.
2 Bleed ejector to induce cooling airflow.

inductance Property of electric circuit to resist change in current as result of opposing magnetic linkage; symbol L, unit henry.

induction compass Based on induction coil pivoted to rotate in Earth's field.

induction heating Heating electrically conductive material by HF field.

induction manifold Pipe system conveying mixture to PE (4) cylinders.

induction period Specified delay between adding catalyst or hardener and applying or spraying material (coating, adhesive or thermosetting structure).

induction phase In pulsejet, portion of cycle when air is admitted.

induction stroke In PE (4), portion of cycle when air or mixture is admitted.

induction system In PE (4), entire flow path from combustion-air inlet to cylinder.

induction tunnel Wind tunnel driven by jet engine(s) or compressed air via ejector system.

inductive coupling 1 Mechanical shaft drive relying on magnetic linkage.
2 Magnetic coupling between primary and secondary coils, eg of transformer.

inductive reactance Impedance due to inductance.

inelastic collision Theoretical impact with no deformation or energy loss.

inert gas Gas incapable of chemical reaction.

inertial anti-icer Free-spinning vanes or other device for imparting rotation to engine airflow, ice or snow being flung out away from engine inlet.

inertial coupling See *coupling (1).*

inertial flight No propulsion, controls locked central or free.

inertial guidance Guidance by INS.

inertial gyro Gyro of characteristics and quality to meet INS requirements.

inertial navigation system, INS Assembly of super-accurate gyros to stabilize a gimballed platform on which is mounted a group of super-accurate accelerometers — typically one for each of the three rectilinear axes — to measure all accelerations imparted, which with one automatic time integration gives a continuous readout of velocity, and with a second time integration gives a readout of present position related to that at the start.

inertial orbit Trajectory when coasting.

inertial platform Rigid frame stabilized by gyro(s) to carry accelerometer(s).

inertia starter Rotary-drive starter whose energy is stored in flywheel.

inertia welding Welding by rapid rotation and pressure between mating surfaces.

inert round Missile or ammunition partly or entirely dummy, and lacking propulsion.

INEWS, Inews Integrated electronic-warfare system.

INF Intermediate-range nuclear force(s).

inferior planet Planet with orbit inside that of Earth.

infiltration Manufacturing stage with FRM (fibre-reinforced metal) components in which molten metal is used to fill gaps between compressed metal-coated fibres.

infinite aspect ratio Many aerofoil calculations ignore effects at the ends (tips) and accordingly are true only for a wing of infinite aspect ratio (ie, it goes on forever).

Infis Inertial-navigation flight inspection system.

inflatable aircraft One whose airframe is of flexible fabric, stabilized by internal gas pressure. Term usually applied to aerodynes.

inflatable de-icer One whose action is the repeated inflation and deflation of a flexible surface, thus breaking off accreted ice.

inflation manifold Links several sources of gas to gas-filled aerostat.

inflation sleeve Large thin-wall tube to which inflation manifold is connected.

inflection Point at which curve reverses direction.

in-flight advisory Sigmet/airmet broadcast to en-route pilots notifying conditions not anticipated at preflight briefing.

inflow ratio Ratio of rotorcraft TAS and peripheral velocity of tips of blades.

influence line Graphical plot of shear, bending moment and other variables as point load moves along structure.

informatics Word gaining some ground in English from transliteration "informatique" (F) = EDP (1).

information pulses Those repeated parts of the transmission in SSR or IFF that convey information (eg identity, flight level).

infra-red, IR Portion of EM spectrum with wavelength longer than deep red light, thus not visible, but sensed as heat.

infrasound Sound, especially at very high power, at very low frequency and very large amplitude.

infrastructure Fixed installations needed for activity (eg airfield, hangars, control tower, communications, fuel pipelines).

infrastructure authority Those responsible for airfields, communications, radar and other ground installations.

infringement Violation of air traffic control or other rules regarding operation of aircraft.

ingestion Swallowing of foreign matter by engine (usually gas-turbine engine), including birds, ice, snow/slush/water, sand, rocket gas, catapult steam and metal parts; hence * test, * certification.

ingress To re-enter spacecraft after EVA, including expeditions on lunar or planetary surface.

ingress route Attack aircraft track from base to initial point.

inherent stability In-flight stability achieved by basic shape of aircraft.

in Hg Inches of mercury, non-SI unit of pressure = 3,386 Pa.

inhibiting 1 To spray interior of machine or other item with anti-corrosion material before storage.
2 To coat inner surfaces of rocket solid-propellant grain to prohibit burning over treated areas.
3 The worst case in flight performance situations, eg instrument flight, failed engine and other adversities.

inhibitor 1 Additive to fuel or other liquid, eg Methyl Cellosolve (often +0·4% glycerine) to protect against ice and against formation of gumming residues, corrosion or fuel oxidation.
2 Refractory inert coating to control burning of solid grain.
3 Anti-corrosion oil for long-term storage.

INI Instituto Nacional de Industria (Spain).

initial altitude Altitude(s) prescribed for IA (1) segment (FAA).

initial approach 1 Segment of standard instrument arrival or STAR between IA (1) fix and intermediate fix or point where aircraft established on intermediate approach course (FAA).
2 Portion of flight immediately before arrival over destination airfield or over reporting point from which final approach is commenced (Seato, IADB).

initial-approach area See *initial area*.

initial area Ground area of defined width between last preceding fix or DR position and either intersection of ILS, facility to be used in instrument approach or other point marking end of IA (1).

initial contact frequency That used for ATC communication as aircraft enters new sector of controlled airspace.

initial heading That at start of rating period while using astro-gyro control (ASCC).

initial mass 1 That of rocket or rocket vehicle at launch.
2 That of fissile or other nuclear material before reaction.

initial operational capability Time at which particular hardware (eg weapon system) can first be employed effectively by trained and supported troops (USAF).

initial point 1 Well defined fixed surface feature usable visually and/or electronically as starting point for attack on surface target (most existing definitions state "starting point for bomb run").
2 Similar surface point where aircraft make final correction of course to pass over drop zone or other surface target.
3 Air-control point near landing zone from which helicopters are directed to landing sites.
4 First point at which moving target is located on plotting board or display system.

initial radiation That emitted from fireball within 60 s of nuclear burst, mainly neutrons and gamma rays.

initial surface That at start of solid-rocket burn; its area.

initial surface/throat area Fundamental non-dimensional characteristic of solid motor.

initial throat area Cross-section area of unused solid-motor throat.

initial-value problem One which, from given state, determines state of dynamic system at any future time.

initiation 1 Starting sequences leading to nuclear explosion.

2 Birth of new inbound track on air-defence system.

initiation phase In autoland, 10 nm from threshold, 2,500 ft (usually 205 kt, 15° flap).

initiator Person assigned to initiate tracks in air-defence system.

injection Insertion of satellite into orbit, or spacecraft on to desired trajectory.

injection carburettor One delivering metred flow to nozzle in induction system, usually without venturi and with insensitivity to flight attitude.

injection flow Primary flow in ejector pump or induced augmenter.

injection point That at which satellite enters orbit.

injection pressure Pressure in system between turbopump and injector.

injection pump 1 Ejector.

2 Injection carburettor.

3 More usually, multi-plunger pump in direct-injection engine.

injection velocity That at injection point.

injector Point at which one or both propellants in liquid rocket enters chamber (usually many are used, distributed around chamber).

inlet Usually air intake at upstream end of air-breathing system.

inlet distortion 1 Departure from ideal airflow through inlet, esp in yawed or turbulent conditions.

2 Plot of flow velocities across plane normal to axis of air-breathing engine, eg at entry to fan, entry to combustor or entry to turbine.

inlet duct Air passage linking inlet (intake) to engine.

inlet guide vane Fixed or variable-incidence stators preceding compressor rotor or turbine.

inlet separator Dynamic (usually centrifugal) device that removes solid particles from airflow entering engine.

inlet unstart Sudden gross disruption of airflow through supersonic engine inlet causing near-total loss of thrust; hard-to-cure hazard at Mach 3 and above (see *unstart*).

inline engine 1 PE (4) with all cylinders in single linear row; loosely (probably incorrectly) applied to engines with two or more rows such as the vee or opposed configurations.

2 Turbojet and ramjet mounted in tandem, sharing same airflow (usually with variable-geometry ducting).

Inmarsat International maritime satellite.

INMG Instituto Nacional de Meteo e Geofisica (Port).

inner horizontal surface Specified portion of local horizontal plane located above airfield and immediate surroundings.

inner loop Control loop via pilot, not autopilot.

inner marker ILS marker rarely installed near threshold, 3,000 Hz dots and white panel light.

inner planets Four nearest Sun: Mercury, Venus, Earth, Mars.

inner space Within Solar System.

inoperative Deliberately taken out of use (thus not a failed engine or other device).

INPE Instituto de Pesquisas Espaciais (Braz).

in-phase Occurring at the same point in each of a series of phugoids or other SHM or repeated cycles, but caused by external stimulus.

inplant Done within factory instead of subcontracted, or relative to the factory (thus * facilities, * modification).

input 1 Point at which signal, data, energy or material enters system.

2 Signal, data, energy or material entering system.

input axis Axis normal to gyro spin axis about which rotation of base causes maximum output.

INS 1 Inertial navigation system.

2 Ion neutralization spectroscopy.

InSb Indium antimonide, IR detector 3–5 μ.

insert 1 Small D-section body fixed inside propelling nozzle to trim area.

2 To place spacecraft in desired trajectory.

3 To convey friendly force (usually small and covert) to point on ground deep in enemy territory.

inset balance Mass located within movable surface.

inset hinge On conventional control surface one whose axis is to rear of leading edge.

inset light One flush with airport pavement.

inside wing That pointing towards centre of a turn.

insolation Solar radiation received, usually at Earth or by spacecraft.

inspectability Unquantifiable measure of extent to which fault, esp structural, may escape notice of inspectors because of inaccessibility or any other reason.

inspection Search of hardware for evidence of existing or impending fault condition.

inspection and repair as necessary Not to preordained schedule.

inspin yaw Yawing moment holding or accelerating aircraft into spin.

instability 1 Structural condition in which strut, web or other member buckles under compressive load.

2 Aerodynamic condition in which slightest disturbance triggers gross disruption of flight of body.

3 Meteorological, normal meaning.

installation envelope Overall three-view dimensioned outline drawings.

instantaneous readout System with zero lag between sensor and display.

instantaneous VSI One giving instantaneous readout, with accelerometer air pumps to counteract lag.

instruction to proceed Informal document accepted as guaranteeing payment in advance of contract.

instrument approach area Volume of sky in which non-visual landing aid operates.

instrument approach runway One providing non-visual directional guidance for straight-in approach.

instrument error Difference between indicated and true value.

instrument flight Using instruments in place of external cues.

instrument flight rules, IFR Rules applied in cloud or whenever external cues are below VFR minima which prohibit non-IFR pilots/aircraft.

instrument landing Ambiguous but usually IMC landing without ground aids (thus, non-ILS).

instrument landing system, ILS Standard ground aid to landing comprising two radio guidance beams (localizer for direction in horizontal plane and glideslope for vertical plane with usual inclination 3°) and two markers for linear guidance.

instrument meteorological conditions Conditions less than minima for VFR.

instrument monitor Various meanings from traditional flight-test panel camera(s) to automatic systems for notifying faults, ensuring majority vote or isolating failed instrument (rare).

instrument rating Endorsement to pilot's licence allowing flight in IMC.

instrument runway Instrument-approach or precision-approach runway.

Int Intersection (FAA).

Inta Instituto Nacional de Tecnica Aeroespacial Esteban Terradas (Spain).

intake Air inlet to propulsion or internal system.

intake duct See *inlet duct.*

intake stroke See *induction stroke.*

integral construction Made from single slab of metal by machining and/or etching, or by forging and machining, to finished shape.

integral tank Tank formed by coating structure with sealant.

integrated acoustic structure Noise-absorbent structures forming load-bearing part of main structure.

integrated aeroplane One in which single shape (eg Gothic delta with no separate fuselage) serves all functions, with no demarcation between parts.

integrated circuit Microelectronic device fabricated by successive etching, doping etc of single-crystal (usually silicon) substrate.

integrating accelerometer Accelerometer whose output signal is first or second-order integration with respect to time (viz velocity or position).

integrating circuit Electronic circuit whose function is to integrate (mathematically) one variable with respect to another (usually time).

integration Assembly of stages, boosters and payloads of spacecraft (post-* normally means after mating of payloads).

integrator Device, usually digital, for giving numerical approximation to integration (mathematical).

integrity Validity of structure or system, functioning in design role after suffering damage; usually, but not always, a mechanical quality associated with avoiding mechanical breakup.

intelligent missile Vague popular concept normally taken to mean self-homing.

Intelsat International Telecommunications Satellite Organization.

intensive flying Purpose is to log flight hours on new equipment at maximum rate under operational conditions.

intensive student area Regions of US airspace where IFR flight is restricted.

interaction parameter Basic measure of relative dominance of fluid motion or magnetic field in MHD and plasma physics.

interactive computer One capable of progressive dialogue with human operator, via displays and lightpen or other method.

intercalation Insertion of chemical compounds in plasma layers between planes of base material such as graphite to enhance electrical conductivity.

interception 1 Flight manoeuvre to effect closure upon hostile aircraft or spacecraft.
2 To capture and hold desired flight condition (eg, VOR radial or ILS).

interceptor 1 Aircraft or spacecraft designed to intercept, and if necessary destroy, others.
2 Small hinged strip to block local airflow, esp between slat and wing or immediately to rear of slat.

intercept point Computed location in space towards which vehicle (eg interceptor aircraft or spacecraft) is vectored.

interchanger Variable gearbox in one axis of powered flight-control system.

intercom Communication system within aircraft using crew headsets or loudspeakers but without any radio emission.

intercontinental ballistic missile Range over 5,500 nm (6,325 miles) (US).

intercooler Radiator for rejecting excess heat in enclosed fluid system.

intercostal Short longitudinal structural member (stringer) joining adjacent frames or ribs, usually to support access door or equipment.

intercrystalline corrosion Originating and propagating between crystals of alloy.

interdiction Attack on targets deep in enemy-held territory, well behind FLOT but not of strategic nature.

interface Boundary between mating portions of system; can be mechanical (eg inlet duct and engine) or electronic (eg central computer and navigation display).

interference 1 Mutual interaction between solid bodies in fluid flow, eg upper and lower biplane wings (see *Prandtl* *).

2 Disruption of radio communications by unwanted emission on same wavelength, as from unscreened ignition.

interference drag Drag caused by aerodynamic interference.

interference fit Fit between parts where male dimension exceeds female (several exact definitions).

interference strut Obstructs cockpit if flight controls locked.

interferogram Display or photograph of interferometer patterns for precise measurement or aerodynamic research.

interferometer Optical measuring system using divided light beam later rejoined to give phase interference seen as light/dark fringes.

interferometer array Aerial able to emit or receive simultaneously at large number of accurately related locations.

interior ballistics Branch of ballistics concerned with bodies under propulsion.

interlacing 1 Scanning technique for raster display in which all odd-number lines are scanned and then all even.

2 IFF technique in which pulse trains of different modes are transmitted sequentially (to achieve enough transponder returns per scan it is rare to interlace more than three modes).

interline Between different air carriers, hence interlining.

intermediate approach That part of approach from arrival at first navigational facility or predetermined fix to beginning of final approach.

intermediate case Gas-turbine casing; several meanings, eg between two compressor spools or over compressor spool downstream of fan.

intermediate contingency 1 Turboshaft power rating below emergency (max contingency) level, usually allowed for 30 min.

2 Of afterburning turbofan or turbojet, usually means maximum cold thrust.

intermediate maintenance No definition; at least six meanings.

intermediate pressure Refers to compressor/turbine spool in three-shaft engine, abb IP.

intermediate range Traditional figure for ballistic missiles is 1,500 n.m., about 1,727 miles, 2,780 km (US).

intermediate rated power Intermediate contingency.

intermediate shop Flightline fault-isolating system, esp for avionics, synthesizing all forms of EM signal for HUD, radar, etc.

intermittent duct Resonant air-breathing engine; also known as pulsejet.

intermodal Capable of use by more than one form of transport, ideally by air, sea, rail and truck.

internal balance By area fixed ahead of control-surface hinge fitting in vented chamber in fixed structure.

internal combustion Originally described prime mover whose fuel was burned in the cylinder, unlike steam engine; in gas turbine and nuclear era meaning is blurred.

internal efficiency Rocket thermal efficiency.

internal fuel Contained in aircraft or spacecraft, as distinct from fuel in removable or jettisonable tanks.

internally blown flap 1 Usually, large conventional flap on which main propulsion jet(s) impinge when at maximum depression.

2 Rarely, jet flap.

internal power Generated in the aircraft (electrical, hydraulic, etc).

internals Items inside cockpit.

internal service Air route wholly within one country.

internal supercharger Downstream of carburettor.

international As applied to flight, service or route, one passing through airspace of more than one country (though departure and destination may be in same country).

International Accounting Unit Used by NATO and in other multinational infrastructure programmes, originally equal to £ sterling prior to 1967 devaluation.

international altitude That shown by ISA-calibrated pressure altimeter set to 1013·25 mb.

international boost PE (4) boost control set to ISA.

international standard atmosphere, ISA That agreed by ICAO and still used as common standard; defines pressure (1013·25 mb at MSL), temperature (15°C at MSL) and relative density up to tropopause (see *atmosphere*).

inter-ocular Between eyes.

interphone Intercom serving crew stations, service areas and ground-crew jacks (2).

interplane strut Joins superimposed wings.

interplanetary Between planets.

interpolation Process of calculating or approximating values of function between known values.

interrogate To transmit IFF, SSR or ATC signal coded to trigger transponders.

interrogation mode Any mode in which signals include code to trigger transponder; eg Modes A, B and D for ATC transponders.

interrogator Radio transceiver scanning in synchronism with primary radar or SSR requesting replies from all co-operative airborne transponders to reply; replies sent to video displays.

interrupted Bite Ground test facility initiated manually via on-board control panel to enhance detection/location capability of C-Bite.

intersection 1 Point where centrelines of runways coincide.

2 Point on Earth, and vertical line through this

point, where centrelines of airways cross.

interservices Linking turbofan core to aircraft, thus * strut, * couplings, * interface.

Intersputnik Soviet Bloc comsat series.

interstage Space launcher airframe section between stages designed as major assembly housing guidance or other systems but without propulsion.

inter-tropical front, ITF The assumed giant front where tradewinds meet N and S of Equator, also called ITCZ; not a normal front.

intervalometer Mechanical or electronic system controlling spacing of events (eg reconnaissance photographs).

in the blind Without external cues, ie no visual reference outside aircraft.

in the groove In the desired flight condition, esp correctly aligned on the approach.

in the slot Correctly lined up for landing.

intl International.

into the mission Measured from T–0 at end of countdown, thus 30 s ***.

intrascope See *borescope*.

intravehicular Between two spacecraft.

intruder Aircraft engaged on interdiction, esp against hostile aircraft and airfields (not necessarily by night).

INU 1 Inertial navigation unit.
2 Inertial nav/attack unit.

Invar Alloy formulated for near-zero coefficient of thermal expansion.

inventory Complete list of hardware, esp of items assigned to military units.

inventory service Assigned to operational unit, including training units, but excluding those in evaluation, development, research and other non-operational status.

inverse monopulse Missile or other guidance radar in which target Doppler shift is detected early in RF amplifier chain instead of at late stage.

inverse-square law States point-source radiation and most other emissions fall off in intensity as square of distance travelled.

inversion Local region of atmosphere where lapse rates are negative, ie temperature increases with height.

inversion point Height at which inversion ceases and normal lapse rate begins.

inverted engine Piston engine with crankshaft above cylinders.

inverted gull wing Seen in front elevation, slopes down from body with pronounced anhedral and then (suddenly or gradually) slopes up with dihedral to tip.

inverted normal loop Normal loop begun from inverted position.

inverted outside loop See *bunt*.

inverted spin Spin in inverted position.

inverted-vee engine Two inclined banks of cylinders below crankshaft.

inverter Electrical machine that inverts polarity of each alternate AC sine wave to give DC output.

investment prediction Assessment of required inventory of spares and GSE.

inviscid Without viscosity.

in-weather Flying in conditions that are wet and/or icy, but not poor visibility (USAF).

In-WX In weather.

IO 1 Identification officer (air-defence systems).
2 Direct-injection, opposed cylinders (US piston engines).
3 Image orthicon.
4 Information officer (US).

I/O channel Input/output channel (EDP [1]).

I_o Mean spherical candle power.

IoA Institute of Acoustics.

IOAT Indicated outside air temperature.

IOC 1 Initial operational capability.
2 Indirect operating cost.
3 International order of characters.

IOI Item of interest/importance.

ion Electrically charged atom or group of atoms; can be in solid or solution or free; charge can be positive (missing electron) or negative (usually through extra electron).

IONDS Integrated operational nuclear detection system.

ionic propulsion See *ion rocket*.

ionization Conversion of atoms to ions.

ionization potential Work measured in eV necessary to remove or add electron in ionization.

ionization screen Barrier to charged particles which would otherwise damage human tissue.

ionogram Plot of radio frequency against pulse round-trip time, ie electron density level for reflection (approx equal to altitude of reflective layers).

ionopause Ill-defined base of ionosphere; also known as D-region.

ionosphere Entire ionized region of Earth's atmosphere (Kennelly-Heaviside, Appleton, E and F layers). Not Van Allen Belts.

ion rocket Propulsor, usually small thruster, generating high-velocity jet of ions in electrostatic field.

I/OP Input/output processor.

IOSA Integrated optical spectrum analyser.

IOTE Initial operational test and evaluation.

IP 1 Initial point.
2 Intermediate pressure.
3 Instructor (rarely, instrument) pilot.
4 Identification pulse (SSR).
5 Identification position (IFF).

I/P feature Identification position (US).

IPARS International programmed airlines reservation system.

IPB Illustrated parts breakdown.

IPC 1 Intermittent positive control (ATC back-up).
2 IP compressor.

IPC-ASA IPC (1) automatic separation assurance; advises conflicts to DABS/Adsel aircraft.

IPD 1 Institudo de Pesquisas e Desenvolvimento (Braz).
2 Improved point defence.

IPE Institution of Production Engineers (UK).

Ipex Immediate-purchase excursion; low-price fare without advance booking and without guarantee of seat.

IPFA Inspection des Programmes et Fabrications de l'Armement (F).

IPG Indian Pilots' Guild.

IPI 1 Intercept pattern for identification.
2 Inertial position insertion.

IPM Interplanetary medium.

IPMS International Plastic (singular) Modelers (US spelling) Society.

IPN Iso-propyl nitrate.

IPP 1 Institut für Plasma-Physik (G).
2 Information-processing panel.
3 Industrial preparedness planning (US).

IPR Intellectual property rights, has particular relevance to software.

IPS 1 Information presentation or processing system.
2 Instrument-pointing system.
3 Inlet particle separator.
4 Intelligence and planning squadron (RAF).

IPSA Infirmières-Pilotes-Secouristes de l'Air (F).

IPW Institut für Physikalische Weltraumforschung (G).

IQA The Institute of Quality Assurance (UK).

IQSY International Quiet Sun Year.

IQT Initial qualification training.

IR 1 Infra-red.
2 Instrument rating.
3 Inspection report.
4 Incident report.
5 Initial reserve (RAF).

IRA Inertial reference assembly.

IRAD Investigate, research and define.

IRAN Inspect and repair as necessary.

IRAS 1 Interdiction/reconnaissance attack system.
2 IR astronomy satellite.

iraser IR laser (not recommended).

IR augmenter Flare or other intense IR source carried by vehicle to facilitate IR tracking.

IRAWS IR attack weapon system.

IRBM Intermediate-range ballistic missile.

IRC Industrial Reorganization Corporation (UK).

IRCC International Radio Consultative Committee.

IRCCD IR-sensitive charge-coupled device.

IRCM IR countermeasures (sometimes rendered IRC).

IR countermeasures, IRCM Measures adopted to minimize IR emissions or render them misleading (eg by ejecting flares which it is hoped hostile IR-homing weapons will prefer instead of one's own aircraft).

IRCS Intrusion-resistant communications system.

IRD 1 Interoperability requirements document.
2 Independent research and development (company-funded).

IR decoy Intense heat source intended to distract IR-homing missiles.

IR detector Device containing cell sensitive to IR which raises electrons (cryogenically cooled to reduce noise) to higher energy level.

IRDF IR direction-finding.

IRDS IR detecting set.

IRDU IR detection unit.

IRE 1 Institution of Radio Engineers.
2 IFF reply evaluator.

Ireps Integrated refractive effects prediction system (anaprop).

IREW IR electronic warfare.

IRFITS IR fault-isolation test system.

IRFNA Inhibited red fuming nitric acid.

IR guidance Use of IR seeker cell and combined optics and flight control to make vehicle home on suitable heat source.

IRIA Institut de Recherche Informatique et d'Automatique (F).

Iris 1 IR imaging system.
2 Inferential retrieval index system.

IRJ IR jammer.

IRLS IR linescan.

IRM 1 Information resources management.
2 Intelligent robotics manufacturing.

IRMP Inertial reference mode panel.

iron All magnetic parts of compass, aircraft etc, of whatever material.

iron bird Ground rig to test major aircraft system.

iron bomb Simple free-fall bomb (colloq).

ironless Constructed of non-magnetic materials.

iron mike Autopilot (colloq, obs).

iron reserve Fuel for taxiing at destination.

IRP Intermediate rated power.

IRR Integral rocket/ramjet.

irreversible control Position of output governed only by input (thus, * flight control is unaffected by air load on surface).

irreversible screwjack Thread chosen so that linear position governed solely by rotation, not by operating load.

IRROLA Inflatable radar-reflective optical location aid.

irrotational flow Individual elements or parcels do not rotate about own axes.

IRRS IR reconnaissance system.

IRS 1 Inertial reference system.
2 Improved radar simulator.

IR seeker See *IR detector*.

IR signature Complete plot of IR emissions from given source or vehicle, usually in form of intensity plotted against wavelength.

IRSM IR surveillance measures.

IRSTS IR search and track system (AFSC).

IRT Instrument rating test.
IRTH IR terminal homing.
IRU Inertial reference unit.
Irvin Trade name of Irving Air Chute (parachutes).
IRVR Instrumented runway visual range.
IRWR IR warning receiver.
IS Institut für Segelforschung (G).
ISA 1 International standard atmosphere.
2 International Stewardess, Hostess and Airliner Association, now ISAA (2).
3 Instrument Society of America.
4 Instruction set architecture.
ISAA 1 Israel Society of Aeronautics & Astronautics.
2 International Steward/ess & Airliner Association.
ISAC Institute of Space and Aeronautical Science (J).
ISADS Integrated strapdown air-data system.
ISAE International Society of Airbreathing Engineers.
isallobar Line joining all places where, over given period (typically 3 h), barometric pressure at surface changes by same amount.
ISAR Inverse synthetic-aperture radar.
ISAS Institute of Space & Aeronautical Science (India, J).
ISASI International Society of Air Safety Investigators.
ISA-SL ISA (1) at sea level.
ISAW International Society of Aviation Writers.
ISB, isb Independent sideband.
ISC Integrated semiconductor circuit.
ISCS Integrated sensor control system.
ISDN Integrated services digital network.
ISDU Inertial system display unit.
ISE Interconnected stabilizer/elevator.
Iseman rivet Hollow blind rivet set by driving drift pin from manufactured-head end.
isenergic Without change in total energy per unit mass of fluid (in bulk or along a streamline).
isentropic Without change in entropy (usually along streamline, with respect to time).
ISG Inflatable survival gear (esp liferafts, escape slides).
ISI Institute for Scientific Information (international corporation, Philadelphia).
ISIS 1 Integrated spar inspection system.
2 Integrated strike and interception system.
ISJTA Intensive student jet training area (military, US).
ISL Inactive-status list.
island 1 Enclosed area on 2-D graph or plot.
2 Superstructure above aircraft-carrier deck.
ISLS Interrogation side-lobe suppression.
ISM 1 Institut Suisse de Météorologie (Switz).
2 Internationale Segelflugzeugmesse (G).
ISMLS Interim-standard MLS.
ISO International Organization for Standardization.
iso Prefix, equal.
isobar 1 Line on map or chart joining points of equal atmospheric pressure, usually reduced to MSL by ISA law.
2 Line on tunnel model or free-flight vehicle joining points of equal aerodynamic (surface stagnation) pressure.
isobaric range Band of flight altitudes over which cabin pressure can be held constant.
isobar sweep Angle between transverse axis and local direction of isobar (2).
isocentre Intersection of interior bisector of camera tilt angle with film plane.
isochoric Without change in volume.
isochrone Line joining points of equal time difference in reception of radio signals.
isoclinic line Line joining points of equal magnetic dip.
isoclinic wing One maintaining constant angle of incidence while flexing under load.
isodynamic line Line joining points of equal horizontal magnetic field intensity.
iso-echo Radar display mode in which cloud turbulence centres appear dark or coloured, width of surround indicating rain/turbulence gradient; also called contour mode.
isogonal, isogonic line Line joining points of equal magnetic declination (variation).
isogram See *isopleth* (also specif line joining places where meteorology event has same frequency of occurrence).
isogriv Line joining points of equal angular difference between grid and magnetic north (grid magnetic angle).
isohyet Line joining places of equal rainfall.
Isolane PU/AP + al (F).
isomers Compounds having same chemical formula but different structure.
isometric Drawing projection in which verticals are vertical, other two axes are equally inclined, and distances along all three axes are correct.
isometric switch Governs lock-on to aerial radar target.
iso-octane Hydrocarbon of paraffin series used as reference index for anti-knock ratings, normally C_8H_{18}.
isopleth Line joining points of a constant value of a variable with respect to space or time.
isopod Patented multi-mode packaging system.
iso-propyl nitrate Monopropellant and fuel for starters.
isopycnic Of unchanging density, or of equal densities, with respect to space or time.
isostatic Under equal pressure from all sides.
isotach Line joining points of equal wind speed, irrespective of direction.
isotherm Line joining points of equal temperature.
isothermal At constant temperature.
isothermal atmosphere Hypothetical atmosphere in hydrostatic equilibrium with constant tempe-

rature; also called exponential.

isothermal change Change of volume/pressure humidity and possibly other variables of perfect gas or gas mixture at constant temperature.

isothermal layer See *stratosphere*.

isotope Radioactive form of (normally non-radioactive) element.

isotope inspection X-ray using isotope as energy source.

isotope power Using isotope radiation as source of heat energy in closed fluid cycle.

Isotran Isolator transition, isolates Gunn diode from load mismatches and couples to waveguide and coaxial conductor.

isotropic Having same properties in all directions.

Isp Specific impulse.

ISRC International Search and Rescue Convention.

ISRO Indian Space Research Organization.

ISS 1 Inertial sensor system.
2 Instrument subsystem.
3 Indonesian Space Society.

ISSR Independent secondary surveillance radar.

ISST ICBM silo superhardening technology.

Istres Location of chief French Government flight-test establishment, the Centre des Essais en Vol (CEV), whose other locations are at Brétigny and Cazaux.

ISV Intensified silicon target vidicon.

ISVR Institute of Sound and Vibration Research (UK).

ISWL Isolated single-wheel load.

IT Inclusive tour.

I_t Total impulse.

ITA 1 Institut du Transport Aérien (Int).
2 Instituto Tecnologico de Aeronautica (Braz).

I_{tA} Motor specific impulse.

ITAA Inspection Technique de l'Armée de l'Air (F).

ITACS Integrated tactical air control system (USAF).

ITAV Ispettorato Telecommunicazioni e Assistenza al Volo (I).

ITB Integrated testbed.

I_{tB} Burn time impulse.

ITC Inclusive-tour charter.

ITCM Integrated tactical countermeasures (USA, USAF).

ITCS 1 Integrated track and control system for RPV.
2 Integrated target control system.

ITCZ Inter-tropical convergence zone (see *inter-tropical front*).

ITE Integral throat and entrance.

ITEC Involute throat and exit cone (rocket).

ITEM Integrated test and maintenance system.

Items Integrated turbine-engine monitoring system.

ITF 1 Inter-tropical front.
2 Intelligence task force (US).
3 International Transport Workers Federation.

ITGS Integrated track guidance system.

ITO Independent test organization (software V&V).

ITP 1 Instruction to proceed (UK).
2 Initial technical proposal (US).

ITR Integrated-technology rotor (USA).

ITS Initial training squadron (rarely, school) (RAF).

ITSS Integrated tactical surveillance system (USN).

ITT 1 Inter-stage turbine temperature.
2 Invitation to tender.

ITTCC International Telegraph & Telephone Consultative Committee.

ITU International Telecommunications Union.

ITV 1 Instrument test vehicle.
2 Inert test vehicle.
3 Inclination to vertical (parachute).
4 Idling throttle valve.

ITW Initial training wing (RAF).

ITX Inclusive-tour excursion.

IU Input unit.

IUAI International Union of Aviation Insurers.

IUKAdge Improved UK Air Defence Ground Environment.

IUOTO International Union of Tourist Organizations.

IUS Inertial upper stage.

IUT Instructor under training (US).

IVA 1 Input video amplifier.
2 Istituto del Valore Aeronautico (I).

Ivadize Patented process of ion-vapour deposition of aluminium on steel.

IV&V Independent verification and validation (software).

IVKDF Institut Von Kármán de Dynamique des Fluides (Int, also called VKIFD).

IVR Instrumented visual range.

IVSI Instantaneous VSI.

IVV Instantaneous vertical velocity.

IW Individual weapon.

IWAC Integrated weapon-aiming computer.

IWC Integrated weapon complex.

IWI Interceptor weapons instructor.

IWM 1 Imperial War Museum (UK).
2 Institution of Works Managers (UK).

IWS Integrated weapons system.

IYQS International Years of the Quiet Sun.

Izod Standard test for impact strength of notched specimen.

J

J 1 Jet (US military engine designation).
2 Jet route (civil airways).
3 Joule.
4 General EM signal power.
5 RPV direction aircraft (DoD).
6 Special test, temporary (DoD aircraft designations).
7 Polar moment of inertia.
8 Transport (1926–31), utility (1931–55) (USN).

j 1 Mass flux (gaseous diffusion).
2 Square root of minus 1.

J-band EM radiation 30–15 mm, 10–20 GHz.

J-display Time base is ring near edge of display, echo blip position varies with range; main use radar altimeters.

JA Judge-advocate.

JAA Japan Aeronautic Association.

JAAA Japan Ag-Aviation Association.

JAATT Joint air attack team tactics (outcome of Jaws [1]).

JAC 1 Joint airworthiness code (AICMA).
2 Junta de Aeronautica Civil (Chile).

JACIS Japan Association of Air Cargo Information Systems.

jack 1 Powered linear actuator.
2 Jack box.

jack box Socket for plugging in communications headset.

jackpad Strong plate, usually square, distributing support of lifting jack into surrounding airframe.

jackscrew Screwthread converting rotary power to linear (jack) output, usually irreversible (see *ball* *).

jackstay Hinged strut for bracing cowl or other large panel when in open position.

JACMAS Joint Approach Control Meteorological Advisory Service.

Jacobs-Relf Original equations for calculation of M_{crit} and related C_p.

Jacola Joint Airports Committee Of Local Authorities (UK).

JAFE Joint advanced fighter engine (US).

Jafna Joint Air Force/NASA (US).

JAG Judge-Advocate-General.

JAIC Japanese Aircraft Industry Council.

jam Obliterate hostile EM transmission (esp radar) by powerful emission on same wavelength(s).

Jamac Joint aeronautical materials activity (USAF).

James Air-damped pendulum driving deceleration (g) pointer carried in braked truck to give friction measure on runway (see *JBD*, *JBI*).

jammer High-power emitter to jam specific wavelengths.

jamming 1 Noise * obliterates by sheer power.
2 Deception * attempts to mislead enemy by causing false indications on his equipment.

jam nut Thin lock-nut.

JAN-TX Joint Army-Navy technical documentation (US).

Janus system Doppler radar technique in which frequency shift is measured as difference between beams to front and rear.

JAR-E Joint Airworthiness Requirements, Europe.

JAS 1 Joint Airmiss Section (UK, civil/military).
2 Fighter, attack, reconnaissance (Sweden).

JASS Joint Anti-Submarine School (UK).

JASU Jet aircraft start unit (US).

JATCCCS Joint advanced tactical command control/com system (US tri-service).

JATCRU Joint Air Traffic Control Radar Unit (UK).

Jato Jet (ie, rocket) assisted take-off.

Jats Jamming analysis and transmission selection (ECM).

JAWG Joint Airmiss Working Group (UK NATS).

Jaws 1 Joint attack weapon systems (fixed and rotary-wing against armour hostile to US).
2 Jamming and warning system (ECM).
3 Joint airport weather studies.

Jazz Rock Joint Survival System/Regional Operational Control Center (Adcom).

JB Jet-propelled bomb category (USAAF, WW2).

J-bar Jet runway barrier (FAA).

JBD 1 James brake decelerometer.
2 Jet blast deflector.

JBI James brake index.

JCAB Japanese CAB.

JCAP Joint Committee on Aviation Pathology.

J-catch Joint countering attack helicopters (US programme against hostile battlefield helicopters).

JCMPO Joint Cruise Missiles Project Office (USAF/USN).

JCR (Not JRC) Joint Research Centre of EEC based in Ispra and co-ordinating some 50 laboratories on remote sensing and related problems.

JCS 1 Joint Chiefs of Staff (US).
2 Jet-capable ship.

JCU Joint common user.

JDA 1 Japan Defence Agency.
2 Joint Deployment Agency.

JDCU Jamming detection control unit.
JEA Joint endeavour agreement.
JEDEC Joint experimental (sometimes engineering) development of electronic components.
Jengo Junior engineering officer (RAF).
Jeppesen Widely used commercially sold volumes of airway and airport charts and information, named for Ebroy Jeppesen, whose 1926–40 notes formed basis.
jerk Rate of change of acceleration, V (esp in ECM target or input motion).
Jesus nut Holds main rotor to rotor shaft on helicopter or, especially, light autogyro.
jet 1 High-velocity gas flow discharged from nozzle (eg from * 2 or from open-* wind tunnel).
2 Turbojet engine (loosely, turbofan).
3 Aircraft powered by 2 (loosely, also by rocket).
4 Calibrated orifice(s) in carburettor.
5 To travel by 3.
Jet A Turbine kerosene similar to JP–5; freezes about 40°C; available USA only.
Jet A–1 Turbine kerosene; freezes below -50°C; standard commercial fuel.
jet advisory area Specific regions along jet routes extending 14 nm each side of route segment from FL240–410 (radar) or FL270–310 + FL370–410 (non-radar).
jet advisory service That offered to certain civil jets in jet advisory areas; IFR separation or (in radar areas) other facilities (FAA).
jet age Loosely, era in which jet (3) aircraft became dominant.
jetavator See *jetevator*.
Jet B Wide-range distillate fuel similar to JP–4.
JETDS Joint electronics type designation system (NATO).
jet-edge shear Rate of change of velocity per unit radial distance at outer boundary of jet (1), including on large scale the velocity profile at edge of atmospheric jetstream.
jet engine Any propulsion system whose reaction is generated by a jet (1), thus a turbofan, turbojet, ramjet or rocket. Generally not applied to space thrusters of ES, ion, plasma and similar types.
jetevator Small power-actuated flap, spoiler or ring on skirt or nozzle exit of rocket for TVC.
jet flap Flap through which passes high-energy gas or air flow, discharged along trailing edge.
jet-fuel starter Main-engine starter burning main-engine fuel.
Jethete Proprietary refractory sheet metal (high-nickel).
jet lag Mild temporary symptoms produced in human beings by fast travel through large meridian difference, ie through five or more time zones.
jet lift Using jet-engine thrust to support V/STOL aircraft.
jet propulsion Aircraft propulsion by jet engine(s).
jet route High-altitude route system for aircraft with high-altitude navaids, normally extending from 18,000 ft AMSL to FL450 (FAA).
jet sheet High-velocity fluid flow of essentially two-dimensional nature.
jet shoes Astronaut shoes for weightless walking.
jetstream Quasi-horizontal wind exceeding 80 kt (148 km/hr) in warm air at sharp boundary with cold, high troposphere or stratosphere, mid latitudes, predominantly westerly.
jet tab See *jetevator*.
jettison Discard fuel, external stores or other mass; hence * pipe(s), * pump(s), * handle or switch.
jet vane See *jetevator*.
jetway Powered enclosed telescopic passenger gangway; also known as apron-drive bridge.
JEWC Joint Electronic Warfare Center (US).
Jezebel Passive acoustic ASW search system (see *Julie*) (US).
JFC Jet fuel control.
JFET Junction field-effect transistor.
JFS Jet-fuel starter.
JFTO Joint flight-test organization.
JG Jagdgeschwader (fighter wing) (G).
JHU Johns Hopkins University.
JIC Jet-induced circulation.
Jifdats Joint-services in-flight data transmission system (US).
jig Hard tooling; any rigid frame in which a part (eg of airframe) is assembled.
jigsaw Multi-lobed rotor forming moving element of wheel brake.
JILL Jet-induced lift loss.
jink To take sharp avoiding action, eg against AAA fire.
Jintaccs Joint interoperability of tactical command and control systems.
JIP Joint Interface Program, between different systems or services (US).
JIT production Just in time.
jitter 1 Send out DME pulses with random spacing (to avoid aircraft locking-on to another interrogating same beacon).
2 ECCM technique in which radar PRF is made to vary in way unpredictable by enemy.
JJPTP Joint jet-pilot training programme.
JMCC Johnson Mission Control Center (NASA, JSC).
JMRC Joint mobile relay centre.
JMSNS Justification of major-system new start.
JMTSS Joint multichannel trunking and switching system.
JOAC Junior officers' air course (RN).
Jo-bolt Patented internally threaded three-part rivet.
JOC Joint operations centre.
jock, jockey Combat aircraft pilot (colloq).
joggle Small vertical offset along edge of sheet or

strip to allow it to overlap adjacent component.

joggling Local squeezing of sheet-metal parts to improve abutment of mating surfaces.

Johanssen block Super-accurate metal block for reference of various dimensions and surface finishes.

Johnson noise RF thermal noise.

Joint-Stars Joint surveillance and target-attack radar system (USAF).

joint-use restricted area Restricted area in which, when not in use by using agency (eg USAF), FR or VFR clearance can be given for FAA traffic.

Joker Fuel planning level selected to warn of imminent approach of Bingo.

JONA Joint Office of Noise Abatement (DoT/NASA).

Josephson junction Formed by weakly linking superconductors at below 4°K, offering immense possibilities 0–1,000 GHz.

Jostle Powerful RAF active jammer, 1942–45.

Joukowsky theory Original (1907) description of airflow round wing in which viscous effects transmit disturbance far from solid surface and lift is direct function of airflow.

Joule SI unit of energy, 1 Nm or 1 Ws.

Joule constant Mechanical equivalent of heat, = 4·1858 J/15°C calorie, used in definition of 15° calorie.

Joule/Kelvin effect Expansion of gas from high pressure through throttling orifice or porous plug: also called Joule/Thompson effect.

Joule/Thompson coefficient Rate of change $\left(\frac{dT}{dP}\right)_H$ where T = temp, P = fluid pressure and H = enthalpy in reversible flow through porous plug.

joystick Control column (arch).

JP 1 Jamming pulse.
2 Jet propulsion/propelled.

JP–1 Original kerosene, Avtur (Jet Propulsion –1) fuel, replaced by Jet A–1, NATO F35.

JP–2 Improved kerosene, soon obsolete.

JP–3 Original wide-cut, including gasolines, kerosene and gas oil fractions.

JP–4 Wide-range distillate, Avtag, NATO F40, commonly available but reduces pump life and increases fire hazard; equivalent is Jet B fuel.

JP–5 Avcat, NATO F44, denser high-flash kerosene.

JP–6 "Heart-cut distillate", close control, good thermal stability.

JP–7 Special fuel for Mach 3 extreme-altitude aircraft (SR–71).

JP–10 High-density fuel developed to extend range of cruise missiles.

JPL Jet Propulsion Laboratory (Caltech).

JPO Joint program(me) office.

j.p.t. Jet-pipe temperature.

JRIA Japan Rocket Industry Association.

JRC Joint Research Centre (Euratom, Belgium/Germany/Italy/Netherlands).

JRCC Joint Rescue Co-ordinating Centre (UK).

JRDOD Joint research and development objective document.

JRMB Joint Requirements and Management Board (Office of JCS, US).

JRS Japan Rocket Society.

JRSC Jam-resistant secure communications programme.

JS Job sheet.

J/S Jam-to-signal ratio.

JSA 1 Jet standard atmosphere.
2 Joint security area.

JSC Lyndon B. Johnson Spaceflight Center (NASA, Houston).

JSCMPO Joint-Service Cruise-Missile Program Office (USAF/USN).

J-Sei Joint second-echelon interdiction (USA and USAF).

JSESPO Joint Surface-Effect Ship Program Office (US).

JSLO Joint Services Liaison Organization (G).

JSME The Japan Society of Mechanical Engineers.

JSOP Joint strategic-objective plan (US).

JSP.318 Military flying regulations, especially instrument, approach and departure procedures (UK).

JSR Jammer saturation range (EW).

JSRR Jam/signal ratio required (ECM).

JSS Joint surveillance system (replaced Sage).

J-Stars *Joint-Stars*, surveillance and target attack radar system (USAF).

JSTPS Joint strategic target planning staff.

JSWDL Joint-service weapon data link (US).

JTACMS Joint tactical missile system (USA/USAF).

JT&E Joint test and evaluation.

JTDE Joint technology demonstrator engine (APSI).

JTF Joint task force.

JTIDS Joint tactical information distribution system, all services, most of NATO.

JTRU Joint Tropical Research Unit.

JTTRE Joint Tropical Trials and Research Establishment (Australia).

Judy 1 Interceptor has contact with target and is assuming control of engagement (US).
2 Target aircraft in practice interception: I have been hit (UK).

JUKL Air-traffic control association (Yugoslavia).

jumbo Boeing 747 (colloq).

jumper Cable temporarily attached to terminals to bypass part of electric power circuit.

jump jet Jet VTOL aircraft (colloq).

jumpmaster Person in command of stick of parachutists.

jump take-off 1 Autogyro take-off using stored rotor energy to achieve initial lift at zero airspeed.
2 Jet VTO.
3 Launch of anti-tank missile with initial jump

to operating height.

junction Mating surface between different types (eg p and n) of semiconductor material.

Jungly Assault-transport helicopter, especially crew-member thereof (UK).

junkhead Head of sleeve-valve cylinder.

junkhead ring Gas sealing ring between head and sleeve.

JURA Joint-use restricted area.

jury strut 1 Additional or temporary strut for particular short-term purpose.
2 Short strut joining mid-point of main wing strut to wing.

Jusmag Joint US Military Advisory Group.

JV Joint venture.

JVC Jet-vane control.

JVX Joint-services advanced vertical lift aircraft (US).

K

K 1 Kelvin, SI absolute scale of temperature (absolute zero = 0K, 0°C = 273K, units = °C).
2 Telemetry, computing (JETDS).
3 Prefix (non-SI or metric) = 1,000, thus K-bit, 32 K, K-lb or Kp (kilopounds).
4 Factor of wing planform efficiency.
5 Tanker (UK and US aircraft category, US being a role prefix).
6 Airfield subgrade 300 lb/cu ft, ie dense concrete.
7 See *Knudsen number*.
8 Various ratios, eg surface area of solid grain to nozzle throat.
9 Circulation round aerofoil.

k 1 Prefix kilo (× 1,000).
2 Cathode.
3 Cold (air mass).
4 Boltzmann constant.
5 Radius of gyration.

K-band EM radiation, 15–75 mm, 20–40 GHz.

K-chart List of multipliers for calculating setback for bends other than 90°.

K-display Horizontal timebase shows two blips from target which vary in height if aerial azimuth direction is incorrect.

K-loader Standard US military cargo aircraft loader elevating to side-door sill.

K-wing Usually combination of canard plus slender delta.

K-words Kilo-words.

KAI Kazan Aviation Institute (USSR).

Kai Kaizo (modification) suffix (J Army 1932–45).

Kalman filter Powerful software routine for combining multiple inputs (eg INS output and Doppler radar) to give most accurate single answer.

kanat Underground aqueduct with surface breather tubes (NATO).

Kapse Kernel APSE (software).

Kapton Gold-coated plastic thermal insulation (spacecraft).

Kapustin Yar Soviet ICBM/space "cosmodrome" on flat territory near Caspian.

Karldap Administration centre for Karlsruhe ATC area.

Kármán-Moore Classic (1932) theory for aerodynamics of slender body of revolution travelling nose-first in supersonic gas flow.

Kármán street Endless succession of vortices, alternate left/right and clockwise/counterclockwise rotation, behind vibrating wire or strut in airflow.

Kármán-Tsien Classic (1939 and 1941) theory for compressibility effects on aerofoils, especially variation of Cp with M; most useful form is

$$Cp_M = \frac{Cp_0}{(1-M^2)^{1/2} + \frac{1}{2}Cp_0(1-[1-M^2]^{1/2})}$$

KAS Killed on active service.

katabatic wind Cold air flowing down mountain slope at night (keeps airfield on slope fog-free).

Katie Killer alert threat identification and evasion.

KB Constructor (design) bureau (USSR).

KBO Weapon-system officer (G).

kB/s, KBPS K-bits per second.

KCAB Korean (South) Civil Aviation Bureau.

kcal Kilogramme-calorie.

KCAS Knots calibrated airspeed.

kc/s Kilocycles, ie KHz.

KDC Knock-down components.

KDR Kill/detection ratio (ASW).

KEAS Knots equivalent airspeed.

keel area Term should be confined to lateral projected area of keel below water on marine aircraft; often loosely used to mean areas below OX axis on any aircraft.

keelson I-beam forming structural backbone of marine hull.

Kelvin scale Absolute (SI) temperature scale, *K* (1). Degree symbols not used.

Kennelly-Heaviside layer Original name for part of ionosphere, E-layer.

Kentucky windage Deflection shooting by rule-of-thumb.

KEP 1 Key emitter parameters (PD, PRF, frequency, etc).
2 Kinetic-energy penetrator.

Keplerian orbit Satellite orbit linking two occasions when observer sees satellite against same point in space.

kerosene Wide range of petroleum-derived hydrocarbons forming basis of virtually all air-breathing jet fuels; also often spelled kerosine.

Kerr cell Extremely fast electro-optical shutter based on glass container of nitrobenzene in electric field.

Kevlar Fibre-reinforced composite material with fibre properties superior to those of most glasses; called aramid fibre from resemblance to spider web (trade name, Du Pont).

keyed emission CW signal interrupted to convey intelligence.

keying Interrupting current or signal by make/break switch, eg Morse key.

keying solution Applied to give strong surface bond prior to application of dope, adhesive, corrosion protection or other coating.

KFI Krüger flaps indicator.

KfR Kommission für Raumfahrttechnik (G).
kg Kilogramme.
KGB State committee for security (USSR, was NKVD).
KGC Japanese Science and Technology Agency.
kgf Kilogrammes force or thrust (non-SI).
kgp Kilogrammes force or thrust (poids, pond, puissance).
KGSK Japanese Aeronautical Council.
KhAI Kharkov Aviation Institute (USSR).
kHz Kilohertz, thousands of cycles per second.
Ki Kitai (airframe type) number (J Army 1932–45).
KIA Killed in action.
KIAS Knots indicated airspeed.
K_{Ic} Fracture toughness.
kick Final impulse given by small upper-stage motor to space payload to achieve exact trajectory. Hence * motor, * stage, apogee * motor.
kickback Bribe offered in large-scale contracting.
kick-off drift In autolanding, separate control signal inserted before touchdown to yaw aircraft parallel to runway (but too late for crosswind to move aircraft laterally from centreline).
KIFIS, Kifis Instantaneous vertical speed indicator.
Ki-Gas Piston-engine hand-priming system drawing fuel from main tanks.
kill 1 Confirmed victory in air combat.
2 Destruction of missile or RV in flight.
kill probability Measure of probability that one missile or other single attack will destroy its target.
kill ratio For a particular type of air-combat aircraft, total of all its air victories divided by total of its own losses in air combat.
kilometric Having wavelength(s) in the order of kilometres.
kilopond Kilogramme-force.
kiloton weapon Having nominal explosive power equal to 1,000 short tons of TNT.
kiloword Unit of memory storage holding 1,024 words.
kinematic ranging Aiming ahead of target correct lead angle to allow for target relative motion.
kinematic viscosity Fluid viscosity divided by density, μ/ρ, symbol ν.
kinetheodolite Tracking and recording instrument comprising high-speed camera whose frames bear az/el measures.
kinetic heating Heating of boundary layer and surface beneath due to passage of body through gas, closely proportional to square of airspeed or Mach.
kinetic pressure See *dynamic pressure*.
kingpost One or more strong vertical struts above (and sometimes below) aircraft centreline providing attachment from primary flying and landing wires or struts.
kink point Any corners on a payload/range plot.
kip Kilopound, 1,000 lbf (half a short ton).
Kips Confusingly, in view of above, thousands of impulses per second.
Kirksite Zinc alloy used for large airframe dies.
Kirkwood gaps Gaps in the asteroid belts.
Kiruna Swedish space launch and communications stations.
KIS Kick-in step.
kite 1 Aerodyne without propulsion tethered to fixed point and sustained by wind.
2 Aeroplane (WW2, colloq).
kite balloon Balloon tethered to Earth or vehicle and shaped to derive stability (sometimes lift) from relative wind.
KJT Japan Air Self-Defence Force.
Klégécel GRP/foam sandwich material (Kléber-Colombes).
klystron Velocity-modulated electron-tube UHF oscillator.
kM Rare prefix, kilomega (giga).
km Kilometre.
K Monel Ni-Cr alloy; strong and corrosion-resistant, non-magnetic.
KMR Kwajalein Missile Range, in mid-Pacific (USA).
KMW Kuratorium der Mensch und der Weltraumfahrt eV (G).
Kn 1 Solid rocket motor ratio of initial surface area to throat area.
2 Knudsen number.
3 Static margin.
kN Kilonewton; SI unit of force standard for most aerospace propulsion.
knee Upper right corner on payload/range curve, or graph of similar shape.
kneeboard Notepad, stopwatch and other items strapped as unit to test pilot's leg above knee.
kneeling Some landplanes have * landing gear to tilt nose-down to reduce space (aboard carrier), adjust cargo floor to truck bed, or alter wing angle of attack for catapult launch.
knee panel Instrument or control panel at knee level, below side console.
KNMI Royal Netherlands Meteorological Institute.
knockdown kit Complete aircraft packaged for shipment in component parts for assembly by foreign customer (usually also a licensee).
knock-down path The route followed by firefighters to a crashed aircraft, esp the final few metres in which a path is blown through flames.
knock rating Standard scale of resistance to knock (detonation) of piston-engine fuels, measured relative to iso-octane (100) (see *fuel grade*).
knot Speed of 1 nm per hour = $0{\cdot}514\ ms^{-1}$ exactly, except in UK = $0{\cdot}514773\ ms^{-1}$. Abb kt.
knuckled Original adjective for levered-suspension landing gear.
knuckle pin See *wrist pin*.
Knudsen flow Gas flow in long tube at such near-zero pressure that mean free path is greater than tube radius.

Knudsen number, Kn Mean free path divided by characteristic length of body.

KNVvL Royal Netherlands Aeronautical Assoc.

KOD Kick-off drift.

Koku Kantai Air fleet (J Navy, WW2).

Kokutai Complete air corps (J Navy, WW2).

Kollsman number Traditionally, altimeter setting, usually QFE or QFF.

kombi Release on glider suitable for winch launch or aerotow.

Komsomol League of Young Communists (USSR).

Komta Committee for heavy aviation (USSR obs).

Kops Thousands of operations per second.

Kosos Department of experimental aeroplane construction (USSR).

Kovar Fe/Ni/Co alloy whose coefficient of thermal expansion is close to that of glass.

Kp Kilopound, also called kip.

kp Kilopond, 1 kgf.

kPa Kilopascal, 1,000 N/m^2, SI unit of pressure.

Krüger Leading-edge flap forming part of wing undersurface, hinged to swing down and forward to give bluff leading edge on high-speed wing.

Ks Factor of take-off distance.

KSAK Royal Swedish Aero Club.

KSC 1 Kennedy Space Center.
2 Incorrectly, kg/cm^2.

ksi, KSI Thousands of pounds per square inch (strongly deprecated).

K_T Geometric stress-concentration factor.

kT Kiloton.

kt Knot.

KTAS Knots true airspeed.

Küchemann carrot See *shock body*.

Küchemann tip Low-drag wingtip following outward-curved streamlines, with large-radius curve from leading edge to corner at trailing edge.

Kutney bump Drag-reducing bulge on inner side of pylon/wing junction.

Kutta-Joukowski Formula giving lift per unit span as $K\rho V$ (circulation × density × velocity) for two-dimensional irrotational inviscid flow.

kW Kilowatt.

kV Kilovolt.

kVA Kilovolt-ampere, measure of AC power.

KVAR kVA-reactive, measure of reactive power.

KWS Kampfwertsteigerung (G).

Kx, Ky "Engineering" drag and lift coefficients.

kymograph 1 See *barograph*.
2 Smoked-paper chart recording aircraft angular movements (arch).

kytoon Combination kite + balloon or kite-shaped balloon.

kZ Vertical component of VSI (1).

KZB Drop tank (G).

kΩ Kilohm.

L

L 1 Characteristic length of body.
2 Total lift.
3 Sound pressure level.
4 Inductance.
5 Mission launch time.
6 Distance to applied load.
7 Low (navaid category), under 18,000 ft (FAA).
8 Low (synoptic chart).
9 Rolling moment.
10 Countermeasures (JETDS).
11 Lighted (airfield).
12 Low-altitude fighter (prefix, UK).
13 US code: IFR aircraft has DME and transponder.
14 Liaison-category aircraft (USAAF, USAF, USN).
15 Cold-weather operation (designation prefix US, former suffix USN).

l 1 Litre.
2 Aerofoil section lift.
3 Aircraft overall length.

$\bar{L}$ Average peak sound pressure level.

$L_{300/600}$ Sound pressure level, octave band 300–600 Hz.

L-band EM radiation, 7·5–5 mm, 40–60 GHz.

L-display Central diametral timebase (usually vertical) on which appear transmitter blip and target blip at position giving range and offset according to pointing error.

L^3TV, L3TV Low-light-level TV.

LA Launch azimuth.

L_A A-weighted sound-pressure level.

LA^{3}S, LAAAS Low-altitude airfield attack system.

LAANC Local Authorities Aircraft Noise Council (UK).

LAAS 1 Laboratoire d'Automatique et d'Analyse des Systèmes.
2 Low-altitude attack system.

LAAT Laser-augmented airborne TOW.

LAAV Light airborne ASW vehicle.

LABRV Large advanced ballistic re-entry vehicle.

LABS, Labs Low-altitude bombing system (gives aircraft guidance to toss nuclear weapon).

labryrinth seal Gas seal between fixed and moving parts comprising series of chambers which, though their sides do not quite touch, reduce gas escape close to zero.

LAC 1 Leading Aircraftman (RAF, obs).
2 Low-altitude en route chart.

Lace Laser airborne communications experiment.

Lacie Large-area crop inventory experiment.

lacing 1 Process of intermittent wrapping to join wire bundle into tight loom.
2 Stitching fabric to aircraft structure.
3 Threading wire through holes drilled at same location in every blade of a fan or turbine rotor to damp vibration.

Lacta Light-air-cushion triphibious aircraft.

LACW Leading Aircraftwoman (WRAF).

LAD 1 Laser acquisition device.
2 Large-area display.
3 Low-altitude dispenser.

ladar See *lidar*.

LADD Low-altitude drogue delivery.

ladder network Cascade of electrical sub-circuits each controlled by its predecessor (central to test equipment).

LADS, Lads Lightweight air-defence systems.

LAE 1 Low-altitude extraction (para-drop cargo).
2 Licensed aircraft engineer (SLAET).

lag 1 Angular crankshaft movement between a reference position (TDC, BDC) and open/closure of a valve.
2 Angular movement of electrical vector between reference position and AC or other waveform.
3 Angular movement between helicopter main-rotor hub and (temporarily slower) blade; hence * hinge, * angle.
4 In instruments, eg VSI (1), normal meaning of delay.

Lageos Laser geodynamic satellite.

lag-plane damping Damping of fundamental mode of rotor system; critical for suppression of various instabilities, esp ground and air resonance; expressed as % critical damping in non-rotating condition.

Lagrangian co-ordinates Identify fluid parcels by assigning each a series of time-invariant co-ordinates such as transient spatial position. Constant-pressure meteorological balloon observations are Lagrangian.

Lagrangian points Five positions in space where free body could maintain station with respect to satellite in existing two-body (eg Earth/Moon) system.

LAH Light attack helicopter.

Lahaws, LAHAWS Laser homing and warning system.

LAHS Low-altitude high-speed route; military VFR training (US).

LAL Launch and leave.

$L\alpha$ Lift due to angle of attack (aircraft attitude),

hence $L_{\alpha WB}$ = lift of blended wing/body due to α.

LAM Long aerial mine (hence codename Mutton).

Lamars Large-amplitude multi-mode aerospace research simulator.

lambert Unit of luminance, $\frac{1}{\pi}$ cd/cm^2; thus 1 foot-lambert = 3·426 cd/m^2 (obs).

Lambert projection Map projection of modified conical type in which cone intersects Earth round two "standard parallels" of latitude.

LAME, Lame Line (or licensed) aircraft maintenance engineer.

lamina Elementary (ie infinitely thin) slice.

laminar boundary layer Comprises successive laminar layers, that adjacent to surface having zero relative velocity and successive layers adding velocity out to the free stream.

laminar flow Fluid flow in which streamlines are invariant and maintain uniform separation, with perfect non-turbulent sliding between layers.

laminated Made of thin layers bonded together.

LAMPS, Lamps 1 Low-altitude multi-purpose system (USAF).
2 Light airborne multi-purpose system (USN).

Lams, LAMS 1 Load-alleviation and mode stabilization.
2 Light-aircraft maintenance schedule.

LAN Local area network.

Lance Line algorithm for navigation in combat environment.

land Return to Earth or planetary surface of land-based vehicle; marine, prefer "alight".

land arm Display signifying system is functioning in autoland mode.

lander Spacecraft designed for soft landing on Moon or planet.

landing angle Usually means angle between OX axis and ground at moment of touchdown.

landing area Area of unpaved airfield reserved for landing and take-off.

landing circuit Term used by Royal Navy in carrier operations.

landing compass Precision magnetic compass on bubble-levelled tripod used as master when swinging aircraft.

landing crew Large team(s) of handlers used in airship operations to hold ropes and manoeuvre ship to mast or walk it into hangar.

landing distance, LD Distance from runway threshold to aircraft stopping point.

landing distance available That declared to be available by airfield authority.

landing flare Released by aircraft over unlit airfield immediately before landing.

landing forecast Met forecast for destination at ETA, or nearest equivalent.

landing fuel allowance, LDG, ldg, LFA Fuel mass required for theoretical circuit (go-around), landing and taxi at destination.

landing gear Any portions of aircraft or spacecraft whose function is to enable a landing to be made; this includes wheels/skis/floats and attachments, and hook, but not flaps or lift-dumpers.

landing ground See *airfield*.

landing light Forward-facing aircraft headlight, usually retractable, formerly to illuminate airfield and now used mainly as anti-collision or anti-bird beacon and for illuminating surface taxied over.

landing loads Loads acting through structure of aircraft or spacecraft in design ultimate severe landing (or arrested landing).

landing mat Various definitions, including flexible mat to reduce erosion or ingested debris in jet VTOL over unpaved surface.

landing-on Recovery of naval aircraft aboard carrier (sometimes, of helicopter on ship platform).

landing party Landing crew (UK).

landing point Intended or achieved point of MLG touchdown.

landing radar Radio altimeter used by soft-landing spacecraft (rarely, by aircraft).

landing run Actual achieved distance from touchdown point to stopping point.

landing runway Runway assigned to arrivals.

landing site Target for soft-landing spacecraft.

landing speed Generally, TAS (sometimes defined as minimum TAS) at touchdown; plays no part in normal operation, which is based on V_{MCL} and V_{AT}.

landing surface Those areas declared by airfield authority to be available for landings.

landing tee Large T-shaped sign in signals area of simple lightplane field, rotated to show wind direction.

landing weight Predicted total mass of aircraft at landing.

landing wires Bracing wires used to bear landing loads (eg sag of wings); also known as anti-lift wires.

landing zone Area surrounding airfield where allowable heights of obstructions are limited (arch).

landplane Aircraft designed to operate from land, including ice or snow on skis.

Landsat Earth-resources satellite.

lane 1 One channel of a fly-by-wire or autopilot system.
2 One hyperbolic track in early Gee or Decca navigation.

Langley NASA research centre near Hampton, Va. Abb LaRC.

Langmuir-Blodgett Film one molecule thick deposited on substrate in manufacture of very-high-speed devices.

Lannion French space communications ground station.

Lans, LANS Land navigation system.

Lantirn Lo-altitude navigational targeting IR for

night (autonomous pod-mounted fire-control linked to HUD).

LAP Large-scale advanced propeller.

lap 1 To fit two mating metal surfaces by rubbing together with fine abrasive.
2 Crankshaft angular movement with inlet and exhaust valves open.
3 Overlap (air reconnaissance).

Lapads Lightweight acoustic processing and display system.

Lapam Low-altitude penetrating attack missile.

Lapan National Institute for Aeronautics and Space (Indonesia).

Lapes Low-altitude parachute extraction system.

lap joint One sheet edge overlapping its neighbour.

Laplace Name given to chief class of integral transforms, to elliptic partial differential equation and basic theorem on probabilities.

lap pack Parachute pack carried on seated wearer's lap.

lapse rate Rate of reduction of temperature with height in atmosphere (see *dry adiabatic* **, *wet adiabatic* **).

LAPSS Laser airborne photographic scanning system.

lap strap Primitive seat harness with single belt across lap.

LARA 1 Light armed reconnaissance aircraft.
2 Low-altitude radar altimeter.

LARC Low-altitude ride control (Softride); system for reducing vertical acceleration on crew compartment in high-speed flight through gusts.

LaRC Langley Research Center (NASA).

large aircraft Over 12,500 lb (5,670 kg) MTOW (US).

large bird Mass 4 lb, 1·8 kg.

large-scale integration Typically c 1,000 circuits, gates or logic functions on each chip.

Larmor precession Motion of charged particle attracted to fixed point in overall weak magnetic field.

LAS Large astronomical satellite.

Las Landing aid system (Boeing).

LASC Lead-angle steering command (air-to-air HUD mode).

Lascom Laser communication (system).

Lascr Light-activated silicon-controlled rectifier.

laser Any of many families of device for emitting coherent light (visible or outside visual spectrum); from light amplification by stimulated emission of radiation.

laser altimeter Measures time for laser pulse to return from ground directly beneath aircraft.

Lasercom Laser communications.

laser gyro Any system, usually triangular, which senses rotation by measuring frequency shift of laser light trapped in closed circuit in horizontal plane.

laser ranging Radar technique but using laser to illuminate target, range being function of time of return journey to target.

laser trimming Use of laser mounted on travelling microscope to cut metal from specific sites in microelectronic circuit or device.

laser welding Uses high-power (usually CO_2) laser for micro-precision welding; technique related to EBW.

LASL Los Alamos Scientific Laboratory, centre of nuclear weapon and nuclear rocket research (University of New Mexico).

Laspac Landing-gear avionics systems package.

LASR Low-altitude surveillance radar.

LASSIE Low-airspeed sensing and indicating equipment (helicopter & V/STOL).

last, LAST Low-altitude supersonic target.

last-chance area Parking spot near military runway for final check of armament status, seat pins, etc.

last-chance screen Final point for removal of foreign particles before oil enters engine.

Lat, lat Latitude.

Latar Laser-augmented target acquisition and recognition.

Latas Laser TAS system.

latching indicator Instrument giving visual indication of security of door locks and pressure seals.

late-arming switch Sub-circuit in aircraft weapon-control system, enabling arming of weapons or firing circuits to be left until moment of firing.

latent heat Absorbed or emitted when material changes physical state.

lateral axis Transverse OY axis, axis of pitch rotation.

lateral deviation Error in radio D/F caused by reflections or refractions.

lateral force Forces acting parallel to line joining wingtips.

lateral gain Width of fresh ground covered by each photo-reconnaissance run over area.

lateral qualities Behaviour in roll.

lateral oscillation One involving periodic roll, invariably with yaw and sideslip.

lateral-rotor helicopter Main lifting rotors side-by-side.

lateral stability Stability in roll (secondarily, yaw and sideslip), measured by studying phugoid oscillation (stick free or fixed) following roll disturbance.

lateral translation Manoeuvre possible with direct-force-control aircraft in which lateral velocity is commanded without change in heading.

lateral velocity Speed component, usually relative to surrounding air, along line parallel to OY axis.

lateral tell Communication of air surveillance and air target data sideways to other units along front.

late turn One initiated at or after fix passage at waypoint.

Latis Lightweight airborne thermal imaging system.

latitude band Between two parallels of latitude around Earth.

latitude nut Screwed in or out on directional gyro (DI) to correct drift due to Earth rotation N or S of Equator.

launch In addition to obvious, also take-off of manned combat mission.

launch bar Towing link between catapult and nose leg.

launch complex Entire ground facilities for launch of large space vehicle, probably including facilities for integration.

launch control centre Manned room in launch complex from which countdown and launch, and possibly whole mission, is monitored and controlled.

launch cost 1 Sum charged for placing customer's payload in desired orbit.
2 Nominal sum estimated, but not necessarily available, for design, development, construction and test of new major aircraft or engine; usually to certification in country of origin.

launcher 1 Interface unit between aircraft and externally or internally carried store, not necessarily with propulsion.
2 Pad or other structure for land-based missile, space vehicle, RPV or other unmanned free-flight device.

launch escape Ability of human crew to escape from slow-acceleration ballistic vehicle during countdown or in first seconds of flight, thus ** tower, ** motors, ** signal.

launch opportunity Period in which all factors, including launch window, local weather and serviceability of all participating systems, is favourable.

launch pad Platform with GSE for launch of ballistic vehicle; normally a fixed installation.

launch vehicle Vehicle providing propulsion for space payload or, rarely, atmospheric free-flight device; may be winged or ballistic but must lift off from Earth and impart nearly all impulse required.

launch window Exactly defined period during which relative positions and velocities of Earth and other bodies are such that a particular interplanetary mission can be launched; may last minutes to days, and may be a unique opportunity or repeated at intervals.

LAV Least absolute value.

LAW Light anti-tank (or anti-armour) weapon.

LAWS 1 Light aircraft warning system (UK Met Office).
2 Lightweight aerial warning system (US).

lay 1 Adjust aim of weapon in azimuth, elevation or both (obs).
2 Spread aerial smokescreen.
3 Calculate or project course (obs).

laydown Release free-fall bombs in level flight at low altitude.

layered defence System for protecting fixed-base ICBMs by providing separate sensor/weapon systems for interception of hostile RVs at different altitudes.

lay off To redraw engineering part to full scale (has other meanings concerned with aiming and off-loading of (usually temporarily) surplus employees).

layout 1 Gross spatial arrangement of parts of aircraft (see *configuration* [*1*]).
2 Arrangement of above-floor payload accommodation, eg one-class *.
3 Arrangement of drawings on sheet of paper; hence * draughtsman.
4 Geometrically correct drawing of sheet-metal part allowing for all bends, setbacks and joggles.

lay-up Basic assembly of parts for FRC structure before bonding under pressure and possibly heat.

lazy eight Flight manoeuvre in which nose describes figure 8, upper half above horizon and lower half below.

LB 1 Light bomber.
2 Glider, bomb-carrying (USN, WW2).

LB film Langmuir-Blodgett.

LBA Luftfahrt Bundesamt (G).

lbf Pounds force.

LBI Low-band interrogator.

LBL 1 Left buttock line.
2 Länder-Behörden der Luftfahrt (G).

LBPR Low (under 1·5) bypass ratio.

LBR 1 Local base rescue.
2 Low bit rate.

LBRG, lbrg Laser beam-riding guidance.

lb st Pounds (force) static thrust.

LC_{50} CBW measure, lethal concentration in atmosphere required to kill 50% of exposed population.

LC 1 Cargo aircraft, cold-weather operation (US).
2 Local call (FAA).
3 Inductance/capacitance.
4 Letter contract.

LCAC Landing craft, air-cushion.

LCAS Light close air support.

LCB 1 Line of constant bearing.
2 Liquid-cooled brake.
3 Lowest compliant bidder (NATO).

LCC 1 Load-carrying composite.
2 Launch-control centre.
3 Linear cutting cord (canopy).
4 Life-cycle cost.
5 Leadless ceramic chip-carrier, becoming standard in microelectronics.

LCCA Lateral-control central actuators.

LCCC Launch-control-centre computer.

LCD Liquid-crystal display.

LCF Low-cycle fatigue.

LCFC Low-cycle fatigue counter.
LCFM Low-cycle fatigue meter.
LCG 1 Load classification group (I-VII, corresponding to LCN 120 – ⩾ 10).
2 Liquid-cooled garment.
LCH Light combat helicopter.
LCH_4 Liquid methane.
LCI Low-cost inertial.
LCL Local (FAA).
LCM 1 Laser countermeasures.
2 Landing craft, medium.
LCN Load classification number; scale of values for paved surfaces indicating ability to support loads without cracking or permanent deformation.
LCO 1 Life-cycle-oriented.
2 Launch control officer.
LCOSS Lead-computing optical sight system.
LCP 1 Leachable chromate primer.
2 Launch control post.
LCS Life-cycle cost, ie over whole useful life.
LCSS 1 Land combat support system.
2 Laser communications spacecraft (or satellite) system.
LCT Longitudinal cyclic trim.
Lctn Location (FAA).
LCTV Linac control and transit vehicle.
LCU Laser code unit.
LCZR Localizer.
LD 1 Landing distance.
2 Standard family of underfloor containers.
3 Lunar day.
LD D (daytime) weighted sound pressure level.
L/D Lift/drag ratio.
LD_{50} CBW measure of lethal dose; that which kills 50% of exposed population.
LDA Localizer-type directional aid only.
LD_a Landing distance available.
LDC Less-developed countries (ICAO).
LDEF Long-duration exposure facility (Shuttle).
L_δ, L_{delta} Lift due to deflection (aeroelastic or surface rotation), thus $L_{\delta T}$ = lift of tail due to deflection.
LDG Landing gear.
LDM Linear delta modulation.
L/D_{max} Maximum attainable L/D.
LDMX Local digital message exchange (secure terminals).
LDNS Laser Doppler navigation system.
LDP Laser designator pod.
LD_r Landing distance required.
L/D_r L/D for maximum range.
LDS 1 Layered-defence system.
2 Lithium-doped silicon.
LD/SD Look down, shoot down.
LDT 1 Lateral dispersion at touchdown (generally 10 ft/sec).
2 Laser detector tracker.
LDV 1 Limiting descent velocity.
2 Laser Doppler velocimeter.
ldw Landing weight.
LE Leading edge; now has confusing additional meaning arising from expression "* of technology", signifying the very latest advances into unknown fields.
Le Lewis number.
LEAA Law-Enforcement Assistance Administration (US).
lead 1 Angular measurement of many variables (eg crankshaft motion between opening of exhaust valve and TDC, or AC vectors related to zero-lead reference).
2 Angular distance between sightline to moving target and direction of aim to hit it.
3 First aircraft in element, or first element in large formation.
lead aircraft 1 Aircraft with greater flight time than any other of similar type or using similar airframe.
2 Obviously, that leading a formation or group.
lead angle See *lead (2)*.
lead azide Explosive triggered by mechanical deformation, used in detonators.
lead-computing sight Gyro or other sight sensitive to flight manoeuvres and providing a direct aiming mark to be superimposed over the target.
leaded fuel Containing small percentage TEL as anti-knock additive.
leader cable Electrically conductive cable buried along centreline of runway and taxiway to provide ground guidance in zero visibility.
lead-in 1 Formerly ground facilities and features between outer marker and threshold.
2 Tube through which aerial or towed MAD bird cable enters aircraft.
leading edge 1 Front edge of wing, rotor, tail or other aerofoil. Not precisely defined and, especially when made as detachable unit, extends to rear of 0% chord.
2 Rising slope of electronic pulse, esp one on precise timebase, as in CRT, IFF, video etc.
3 Frontier of knowledge (see comment under LE).
leading-edge flap Any hinged high-lift surface attached to the leading edge but not forming the leading edge itself (ie, not a droop).
leading-edge root extension Sharp increase of wing chord at LE root, often almost flat and projecting ahead of wing profile proper, to cause strong vortex at high AOA and enhance lift, control and manoeuvrability. In extreme (long-chord form) becomes a large strake.
leading-edge sweep Angle between local (or, sometimes, mean) leading edge and OY axis.
lead pole Connects cable to tow banner.
lead-pursuit Traditional air-to-air attack using fixed guns, approach from rear and aiming ahead of crossing target.
lead ship Prominently marked aircraft on which large day bomber formations formed up before setting course.
lead time Time between (a) placing order for

bought-out item, or (b) starting fabrication of major airframe part or even (c) receiving heavy plate or other raw material, and emergence of finished aircraft. Expression also, incorrectly, used for time between ordering aircraft and its delivery.

leaf brake Power tool for making radiused straight bends in sheet.

leaky turbojet Turbofan of very low BPR (under 0·5).

lean 1 Of fuel/air mixture, below stoichiometric, lacking fuel.
2 Linear distance at tip between position of backwards-leaning rotor blade (usually of gas-turbine compressor or helicopter) and position it would occupy if truly radial; the lean is sometimes along the tip-path plane and sometimes along chord line at tip.

leapfrog To delay one ranging pulse train from radar to avoid two targets being superimposed.

learner cost Extra element of direct-labour cost when work is unfamiliar.

learning curve Fundamental curve portraying fall in manufacturing time or cost with increasing familiarity; abscissa is number of aircraft completed (often log scale) and ordinate is total direct labour cost, or total manufacturing man-hours or total manufacturing cost including raw materials and bought-out parts; usually an idealized curve not allowing for inflation.

Leasat Leased satellite, or space bus hired out for different payloads.

LED 1 Light-emitting diode.
2 Leading-edge down (surface angular movement).
3 Leading-edge device(s).

LEDDM LED (1) dot matrix.

LEED Low-energy electron diffraction.

LEF Leading-edge flap.

left-hand rotation Anti-clockwise, viewed from rear.

left/right needle Needle pivoted at top or bottom of panel instrument giving steering indication; pilot steers to keep needle vertical.

left seat That of captain of aircraft; thus, ** time.

left-seater Pilot in command, usually.

leg 1 Main strut of landing gear.
2 Part of flight at constant heading between two waypoints.
3 Beam of radio range station, identified by particular flight as inbound * or outbound *.

legend 1 Any fixed printed notice in cockpit.
2 Explanatory written matter on engineering drawing.

leg restraint Strong belt automatically tightened round occupant's legs as ejection seat fires.

LEIP Leading-edge image process (auto map displays).

LEM Lunar excursion module.

Lemac Leading edge of mean aërodynamic chord.

LEMF Leading-edge manoeuvre flap.

Lemonnier Class of resonant valveless pulsejets, named for inventor.

length Aeroplanes normally measured in flight attitude along OX axis with perpendiculars aligned with extremities of fixed airframe, normally including pitot or instrument booms; helicopters, must specify whether fuselage only or "rotors turning", the latter being distance between perpendiculars to OX axis through periphery of rotor discs. Main cause of confusion is that measure is frequently taken from *Station Zero* at a location (often an arbitrary distance) in front of nose of aircraft. NATO measure is always major body * ignoring nose probes or booms, guns, FR probes, inlet centrebodies, rudder or tailplane overhang or any other projection or rotor blade.

lenticular Having shape resembling side elevation of double-convex lens, with two arcs of large curvature meeting at pointed ends; thus * blade, a supersonic fan or compressor profile, and * cloud, found at tops of waves in lee of hills.

Lenz's law Current induced in circuit moving relative to magnetic field will generate its own field opposing motion.

LEO Low Earth orbit, parameters of which depend upon satellite mass, density, lifetime and other variables.

L_{EPN} Effective perceived noise level, with tone/duration correction.

L_{eq} Energy-average sound pressure level.

LeRC Lewis Research Center (NASA).

LERX, Lerx Leading-edge root extension, pronounced "lurks".

LES Leading-edge slat.

LESS Leading-edge subsystem (Space Shuttle).

LET Launch and escape time (strategic bomber, cruise missile).

let-down Complete procedure from TOD at end of cruise through the approach to landing; term concerned mainly with controlled adoption of successively lower flight levels rather than with the landing; thus * procedure.

lethal envelope Volume, often spherical, within which parameters can be met for successful employment of particular munition.

letter-box inlet Large semi-rectangular air inlet along part of wing (or other) leading edge.

letter-box slot Fixed slot at about 8% chord, usually ahead of aileron; further aft than slot formed by open slat.

letter of intent Formal letter serving as notice by customer of intention to purchase, before negotiation of contract.

LEU Leading edge up (surface angular movement).

level Air intercept code: "Contact is at your angels".

level landing Tail-up landing by tailwheel aeroplane; also known as a wheeler.

levelling circuit AC filter circuit used to smooth out variation in bias voltage.
level of escape Base of exosphere at which upward-moving particle has probability 1/e of colliding with another on way out of atmosphere.
level off To pull out of dive or gentle let-down and hold height constant.
levels of similarity Quantified lists of differences in aerodynamics and systems between early prototypes.
leverage Ratio between variables (eg, if △ DOC due to △ sfc is 8 times the cost of engines and spares to achieve △ sfc then * of improved sfc is 8).
leveraged lease Lease of aircraft on any of several forms of sliding scale.
levered suspension Landing gear wheel(s) carried on arm pivoted to bottom of leg such that vertical travel of wheel is greater than that of shock strut.
LEVL Leading-edge vortex lift.
LEW Large eye/wheel distance, ie pilot must allow for his height above wheels at touchdown.
Lewis NASA research centre for aeronautics, Cleveland. Abb LeRC.
Lewis aerial Othogonal radar scanning sawtooth profile generated by electromechanical means to give flapping beams.
Lewis number Le = Pr (Prandtl)/Sc (Schmidt), used in hypersonics.
LEX Leading-edge extension (US terminology).
Lexan Commercially produced polycarbonate plastic, usually transparent.
LF 1 Low frequency (often l.f.).
2 Load factor (structural).
3 Load factor (traffic).
4 Local forces.
5 Launch facility.
L + F Leather and fabric.
LFA 1 Landing fuel allowance.
2 Low-flying area.
3 Luftfahrtforschungsanstalt (G).
LFC 1 Laminar flow control (see *BLC*).
2 Longitudinal friction coefficient.
LFH Lunar far horizon.
LFICS Landing force integrated communications system.
LFM Low-powered fan marker.
LFP Loaded flank pitch (fir-tree blade root).
LFR 1 Local flight regulations.
2 Low/medium-frequency radio range.
LFRED Liquid-fuelled ramjet engine development.
LFRJ Liquid-fuelled ramjet (in solid rocket case).
LFRR Low-frequency radio range.
LFSMS Logistic force-structure management system(s).
LFV Civil aviation board (Sweden).
LG 1 Landing gear.
2 Laser gyro.
3 Landing ground.
LGB Laser-guided bomb.
LGDM Laser-guided dispenser munition.
LGE speed Speed at which landing gear may be extended.
LGM US weapon category, silo-launched missile.
LGR Laser guidance receiver.
LGS 1 Laser gunfire simulator.
2 Laser gyro strapdown.
LGT Landing-gear tread.
lgt Light.
LGWB Landing gear wheelbase.
LH Left-hand.
L/H Local horizontal.
LH_2 Liquid hydrogen.
LHA 1 US Navy ship category, large helicopter assault carrier.
2 Local hour angle.
LHC Light helicopter cycle (standard cycle for US turbine engine testing).
LHe Liquid helium.
LHM Laser-hardened materials.
LHV Fuel heating value, formerly measured in BTU/lb.
LHW Laser-homing weapon.
LHX Light helicopter, experimental.
LI 1 Lane identification (early Decca).
2 Laser interrogator (or interrogation).
L_i Max weighted noise level over series of i noise events.
libration Small long-period oscillation, esp that of Moon's aspect from Earth.
LID 1 Lift-improvement device (jet V/STOL).
2 Luftfahrt Informations Dienst (DDR).
lidar Light detection and ranging, laser counterpart of radar.
life 1 Allowable total period of operation of hardware item.
2 To assign such a period; hence, a lifed part.
LIFT Lead-in fighter training.
lift 1 Total lifting force from a wing (component of resultant force along lift axis), aerostat envelope or other source excluding engine thrust. Normally, force supporting aircraft.
2 Any element of such lift, acting through particular point.
3 Whole or part of an airborne operation, thus second * means second force to be airlifted.
4 Aircraft-carrier elevator (British terminology).
lift axis Line through c.g. perpendicular to relative wind in plane of symmetry.
lift coefficient Dimensionless measure of lift of surface C_L; actual lift divided by free-stream dynamic pressure $\frac{1}{2}\rho V^2$ and surface's area S.
lift/cruise engine Turbofan or turbojet with vectoring to give jet lift or thrust.
lift curve Plot of lift coefficient against angle of attack (C_L:α).

lift curve slope Inclination of lift curve at any point, rate of change $dC_L/d\alpha$.

lift/drag ratio, L/D Ratio of total lift to total drag, fundamental measure of efficiency of aircraft; L is normally constant and equal to weight but drag varies approx as square of airspeed; thus L/D plot is curve with peak at one particular airspeed for each aircraft, L/D_{max}.

lift dumper Flat plate, usually long span and short chord, raised by powered system (rendered operative by weight on MLG) from upper surface of wing (usually inboard and at about 60% chord) after landing to destroy lift and improve wheelbrake traction. Usually synonymous with ground spoiler.

lift fan 1 Turbofan of HBPR installed only for lift thrust.
2 Free-running fan driven by tip turbine from external gas supply installed only for lift (note: 1 and 2 may have exit vanes to give a diagonal lift/thrust component).

lift-improvement device Any aerodynamic strake, dam, flap or other fixed or movable surface to assist jet VTO by reducing hot-gas reingestion, suckdown or other undesirable effects.

lifting body Aircraft whose chief or sole lift is generated by its body; usually hypersonic aircraft or spacecraft.

lifting re-entry One in which aerodynamic lift forces play a significant role.

liftjet Ultra-lightweight turbojet or turbofan of LBPR installed only for lift thrust.

lift-lift/cruise Equipped with both lift jet(s) and vectored-thrust engine.

lift motor Engine driving vertical-axis prop/rotor on airship.

lift-off Separation of any aircraft or other flight or space vehicle from ground or (eg Space Shuttle atmospheric tests) a parent vehicle. Hence * speed, V_{LOF}.

lift strut Bears tensile (rarely compression) load due to wing lift.

lift/thrust, L/T Ratio of lift to thrust of vectored-thrust engine, usually varies from infinity to zero over range of nozzle movement.

lift vector Vector drawn through point at which force acts with angle showing direction (usually normal to chord or OX axis, irrespective of aircraft attitude) and length showing magnitude.

lift wire Bears tensile load due to lift of wing.

LIG Laser image generator.

light aircraft One having MTOW less than 12,500 lb (5,670 kg).

light alloy One whose principal constituent is aluminium; some authorities add "or magnesium" but these are usually described as magnesium alloys.

light bomber Today meaningless, and never universally defined.

light-emitting bar Vertical bar of three (rarely, more) Si LEDs.

light-emitting diode Solid-state diode emitting visible light when stimulated by electronic input, giving quick-reacting shaped light source.

light-emitting strip Horizontal rectangular strip display made up of number of light-emitting bars, often used to give analog lateral-position readout.

lightening hole Cut-out in relatively unstressed region of structural sheet part to save weight.

lighter than air Buoyant in atmosphere (see *aerostat*).

light fighter Unusually small fighter usually intended chiefly for close air-combat role.

light gun Aldis lamp or other projector of visible pencil beam.

light-microsecond Almost exactly 300 m, 984 ft.

lightning Any natural electrical discharge between clouds or between cloud and ground.

lightoff, light off Ignition followed by acceleration of gas turbine.

light pen Fibre-optic device for interfacing and accessing computer via visual display.

light pipe Single or bundle of optical fibres.

lightplane See *light aircraft*.

light valve Photoconductive layer controlling areas of liquid crystal illuminated in large display.

light water Water, as distinct from heavy water.

lightweight fighter, LWF Despite USAF competition 1972–75, never defined.

LII 1 Light image intensifier.
2 Flight research institute (USSR).

LIIPS Leningrad institute for sail and communications engineers (USSR).

like on like Liquid rocket with streams of fuel impinging on each other from some injectors and streams of oxidant on oxidant from others.

limaçon Quartic curve, $r = a \cos \theta + b$.

limb Visible edge of heavenly body, esp the Sun.

limited panel Pilot instruction with key flight instruments obliterated and external cues absent (originally meant gyro instruments obliterated, and always horizon; today depends on panel).

limited remote communications outlet Unmanned satellite air/ground com facility operated as LRCO-A, VOR voice channel plus receiver, or LRCO-B, separate facility with transmit and receive capability, extending FSS service area (FAA).

limiter One meaning is control device attached to transducer to prevent critical or threshold value being exceeded.

limiting load factor See *design load factor*.

limiting runway One whose length, altitude or temperature necessitates take-off below MTOW.

limiting speed 1 Maximum IAS permitted in particular aircraft configuration, eg landing gear down.
2 Speed in any flight condition in which longitudinal acceleration is zero.

limiting velocity Terminal velocity at specified

angle to horizontal.

limit load Greatest load anticipated, unfactored.

limit of proportionality Tensile (rarely, other) stress at which material begins to suffer plastic deformation, acquiring permanent set.

limits 1 Weather minima permitted for particular pilot or flight.
2 Boundaries of flight regimes, eg IAS or g in particular configurations.

LIMSS Logistics information management support system.

Linac Linear accelerator for X-radiography.

Linas Laser-inertial nav/attack system.

Lindberg detector Fire detector with sealed network of stainless-steel tubing containing material which emits gas above set temperature, raising pressure.

line 1 Single pipe in fluid system.
2 Single cable in electrical system.
3 Horizontal scan on raster display.
4 Cable or rope anchored to aerostat with other end free.
5 Flight-line.
6 Adjective, in revenue service with air carrier.

linear aerial array Yagi or other array of dipoles on straight axis.

linear building One in which operations take place in sequence from one end to other.

linear configuration Vehicle assembled from separable stages arranged end to end in one line.

linear hold Usually, to delay landing by intercepting extended runway centreline far from airport, advising when 1,000 m from threshold.

linear-scale instrument Vertical or horizontal straight-line display, either tape or video, giving quantified output.

linear shaped charge Explosive cord whose cross-section is that of hollow (shaped) charge, for unidirectional cutting.

line check Examination of crew qualified on type but proving ability to fly new route.

line inspection Usually vague, but generally a special check not calling for aircraft to be moved to maintenance or engineering area or enter hangar.

lineman Engineer, marshaller or other flight-line worker, esp in general aviation.

line-mounted Usually, supported entirely by a pipe or cable, thus * valve.

line of position See *position line*.

liner Sheet or sprayed-on heat insulator in some (non-case-bonded) solid motors.

linescan IR graphics using raster display to generate picture.

line search 1 To examine one strip of film from straight reconnaissance run.
2 In sea reconnaissance, to search on constant heading at max height at which target identifiable.

line service In revenue operation with air carrier.

Linesman British attempt at combined air defence and ATC system (see *Mediator*).

line speed Predicted take-off ASIR.

line squall Violent cold front characterized by sudden drop in temperature, rise in pressure, thunderstorms and, especially, severe vertical and other gusts.

line vortex One in which vorticity is concentrated in a line.

linkbelt Ammunition feed using rigid inter-round links.

link chute Discharges used ammunition links overboard.

link route Authorized sector joining airways but not itself an airway.

Link's turbidity factor See *turbidity factor*.

Link trainer Traditional primitive electropneumatic flight (pilot training) simulator, not representative of aircraft type.

LINS, Lins Laser/inertial navigation system.

lip Leading edge of air inlet (other than a body-side splitter plate).

liquid-cooled Loosely, any engine cooled by liquid, including water, but preferably restricted to cooling by water/alcohol or glycol mix.

liquid crystal Organic liquids with elongated molecules which in electric fields arrange themselves to give controllable appearances.

liquid-film technique Traditional method of coating surface with volatile oil to show demarcation between laminar/turbulent boundary layer and some details of flow direction.

liquid-fuel starter Burning one or more liquids unlike that for main engine.

liquid inertia vibration eliminator Heavy liquid, damps helicopter rotor vibration.

liquid injection TVC Use of volatile fluid pumped into one side of rocket nozzle to create shockwave and deflect jet.

liquid propellant Liquid fuel, monofuel or oxidant used in rocket.

liquid rocket Rocket burning one or more liquid propellants.

Liquid Spring Dowty shockstrut filled with liquid with deformable large molecules absorbing energy internally.

LIR Laser intercept receiver.

LIRL Low-intensity runway light(s).

LIS Localizer inertial smoothing.

listening out Ready to receive broadcast transmissions on wavelength in use (US = listening watch).

LIT 1 Lead-in training (US).
2 Light intratheatre transport.

lit Litres (SI unit).

Lital Medium- and high-strength Al-Li alloys (Alcan).

litas Low-intensity two-colour approach slope system.

Lite Laser illuminator targeting equipment.

lithium-drift detector Ionizing-radiation detector

using semiconductor doped with lithium as n-type ions.
lithometeor Finely divided solid particles suspended in atmosphere.
lithosphere Earth land mass, as distinct from atmosphere, hydrosphere.
litre SI unit of volume, $10^{-3}m^3$, 10^3cm^3.
litter Stretcher (medical).
LIVE, Live Liquid inertial vibration eliminator.
live drop Release from aircraft of operative device, eg missile with propulsion and guidance and possibly warhead, as distinct from inert equivalent.
live engine Operative engine(s) in aircraft with one or more real or simulated failures.
live flight deck Subject to motion at sea, rather than in port (aircraft carrier).
LK Product support (G).
LKR Low-kiloton range.
LKV LuchtvaartKundige Vereeniging (SA).
LL 1 Low-level.
2 Limit load.
3 Flying laboratory, ie research aircraft (USSR).
L_L Loudness level (Stevens) in phons.
L-L Line to line (AC voltage).
LLAD Low-level air defence.
Llanbedr Airfield on Welsh coast serving Aberporth with targets (RAE).
LLC Lift-lift/cruise.
LLDF Low-level discomfort factor.
LLF 1 Low-level fan of reconnaissance cameras.
2 Long-lead funding.
LLGB Launch-and-leave guided bomb.
LLLGB Low-level laser-guided bomb.
LLLTV Low-light-level television.
LLTOW Landing limiting (or limited) take-off weight.
LLTV Low-light television.
LLV Lower limit of video (HUD).
LLWAS Low-level windshear alert system.
LLWD Low-level weapons delivery.
LM 1 Lunar module.
2 Last-minute (cargo).
3 Laser machining.
L/M Ratio of direct labour to material cost.
lm Lumen.
LMAE Lunar module ascent engine.
LMD Laboratoire Météorologique Dynamique (F).
LMDE Lunar module descent engine.
LMF 1 Liquid methane (or methanol) fuel.
2 Lacking moral fibre (RAF, 1939–45).
L/MF Low/medium frequency.
LMG 1 Liquid methane gas.
2 Light machine gun.
LMLF Limit manoeuvre load factor.
LMM Compass locator at middle marker.
LMN Local Mach number.
LMP Lunar module pilot.
LMRS London Military Radar Services.
LMS Least mean square.
LMT Local mean time.
L_N N (night) weighted sound pressure level.
L-N Line to neutral (AC volts).
LN_2 Liquid nitrogen.
LNAV, L-Nav Lateral navigation.
LNG Liquefied natural gas.
LNH Lunar near horizon.
LNO Limited nuclear option.
LNP, L_{NP} Noise pollution level, equal to L_{eq} + 2·56 dB standard deviation.
LO 1 Low observables.
2 Local oscillator.
lo 1 Low level, variously interpreted as 60 m and 200 ft; minimum practical safe height for transonic attack.
2 Minimum safe height to avoid obstructions, generally proportional to speed.
L/O 1 Lift-off.
2 Light off.
l.o. Local oscillator.
LO_2 Liquid oxygen.
LOA 1 Letter of offer and acceptance.
2 Launch on assessment.
LOAD, Load Low-altitude defence (of ICBMs).
load cell 1 Fluid-filled device for generating large forces accurately, eg in weighing large aircraft.
2 Capsule containing strain gauge or other force transducer used, eg, in weighing aircraft.
loader Loads computer main memory, esp from transit tape.
load factor 1 Vertical acceleration in g.
2 Stress applied to structural part as multiple of that in 1 g flight (not necessarily same definition as 1).
3 Ratio of failing load to assumed 1 g load in component.
4 Number of passenger seats occupied as percentage of those available.
5 Revenue ton-miles (or tonne-km) performed as percentage of RTM available.
loading 1 Total aircraft mass divided by wing area (wing *) or total installed power (power *).
2 Volume fraction of composite (FRC) material occupied by strong fibres.
loading bridge See *jetway*.
loading chart Chart or diagram displaying correct locations for airborne loads in transport aircraft, esp cargo.
loading coil Inserted inductance, eg to increase electrical length of aerial system.
loading diagram 1 Standard graphical plot of forces in structural part or assembly.
2 Document with detailed plan of cargo floor and underfloor holds on which responsible official marks positions and masses of all cargo and final c.g. position.
loading loop Graphical plot of transport aircraft weights against % SMC or MAC or distance from datum with fuel and payload forming closed figures.
load manifest Detailed inventory of cargo on

commercial or military flight.

loadmaster Member of military transport flight crew in charge of loading and unloading, and para-dropping etc if undertaken (but not of paratroops).

load programmer Person in charge of structural fatigue test.

load ring Rigid hoop to which balloon net and basket suspension are attached.

LOADS, Loads Low-altitude air defence system.

load sheet See *load manifest*.

load spreader Rigid pallet for distributing dense loads over larger floor area.

load waterline Waterline of marine aircraft at MTO weight.

LOAL Lock-on after launch.

lobe 1 One of two, four or more sub-beams that form directional radar beam from aerial with reflector.
2 One of the (usually symmetrical) plan systems of regions of most intense noise from jet engine.
3 Eccentric profile of cam.

lobe switching Radio D/F by time-variant switching of aerial radiation pattern.

LOBL Lock-on before launch.

LOC 1 Line of communication.
2 Letters of credit.
3 Logistics operations centre.
4 Limited operational capability.

loc Localizer.

local area network Bus, ring, PABX (telecommunications) or other transmission links for communication and EDP within an office, factory or other establishment.

local elastic instability Small region of buckling in otherwise almost undistorted structure.

local hour angle, LHA Hour angle of observed body measured relative to observer's meridian.

localizer ILS aerial and beam giving directional guidance.

localizer course Locus of points in any horizontal plane at which DDM is zero.

localizer needle Directional steering needle on ILS display.

local magnetic effects Those peculiar to a region (eg ore deposits) which distort terrestrial field.

local mean time, LMT Angle at celestial pole between local meridian and that of mean Sun; time elapsed since mean Sun's transit of observer's anti-meridian.

local meridian That passing through a particular place.

local stress concentration Intensification caused by shape of stressed part (eg at end of a crack).

local velocity Relative speed of fluid flow over small area of body.

local vertical Line from centre of Earth through a particular place.

Locap Lo combat air patrol.

Locat Low-cost aerial target.

Locate Loran/Omega course and tracking equipment.

locator 1 LF/MF NDB used as fix for final approach.
2 Portable radio beacon 121·5/243 MHz, carried on person or in parachute harness, sometimes with voice facility.

LOCC Launch operations control centre.

LOCE 1 Limited operational capability for Europe.
2 Large optical communications experiment.

Lockclad Conventional 7×7 control cable covered with swaged aluminium envelope.

locked 1 Gun bolt or breech block mechanically locked to barrel at moment of firing.
2 Flight controls mechanically locked to prevent damage by wind when parked; must be unlocked for flight.
3 Overbalanced flight-control surface driven to limit of deflection and not (or not readily) recoverable.

locked-in condition Aircraft in dangerous flight condition (eg superstall) with airflow over controls inadequate for recovery to be possible.

locking wire Fatigue-resistant wire pulled through ad hoc holes in series of nuts and finally tightened and twisted off with inspector's stamp on soft metal seal at free ends.

lock nut One that cannot loosen once tightened.

lock-on Operating mode of many radars and other sensor systems in which pencil beam, having searched and found a target, thereafter remains pointing at that target.

lock time Time from release of gun sear to firing of primer or detonator.

lock up Lock on.

lock washer Tightened under nut, prevents nut working loose (by biting into nut or by a tab manually turned up beside a flat on the nut).

lockwiring Preventing rotation of nuts or bolts by threading fatigue-free wire through their heads to apply torque opposing movement.

locpod Low-cost powered off-boresight dispenser.

locus Path traced out by moving object, esp one rotating in complex repeated orbit, eg tip of helicopter rotor blade.

LODE Large optics demo experiment.

LOE Level of effort.

LOF 1 Lift-off.
2 Line of flight.

Lofaads Lo-altitude forward-area air defense system (USA).

Lofar Low-frequency omnidirectional acoustic frequency analysis and recording (ASW).

LOFT Line-oriented flight training.

loft bombing Low-level NW delivery also called toss bombing, etc (see *low-angle* *).

loft floor Floor on which lofting is carried out.

lofting Plotting full-size exact shapes of airframe, from which master templates, jigs, tooling, forging and stamping dies and other large parts can be constructed and NC tapes prepared.

log Large ground-burning target marker.

Logair Logistics air network.

logarithmic decrement Natural log of ratio of two successive amplitudes in damped harmonic or other oscillatory response.

logbook Master history of member of aircrew, aircraft or other important functioning system in which are recorded times, events and occurrences.

logic Electronic circuits and subcircuits constructed to obey mathematical laws.

Logmars Logistics applications of automated marking and reading symbols.

Logo Limitation of Government obligation.

Logspark Logistic(al) park, housing 7 days' fuel, weapons and maintenance supplies, serving several STOVL hides.

LOH Light observation helicopter.

Lohmannizing Metal dipped in amalgamating salt, pickled and then plated with two or more protective alloy coatings.

LOI 1 Lunar orbit insertion.
2 Letter of intent to purchase.

Lolex Low-level extraction of parachute-retarded cargo.

LOM 1 Compass locator at outer marker.
2 Localizer outer marker.

Lombard Original design of "bonedome", 1947.

LON Longitude (FAA).

long Longitude.

long-dated Long lead time, ie must be ordered months to years in advance of aircraft completion.

longeron Principal longitudinal structural members in fuselage, nacelle, airship etc.

longitudinal axis OX axis, from nose to tail; roll axis.

longitudinal bulkhead Major full-depth web in plane of OX axis (probably arch).

longitudinal dihedral Angular difference between incidence of wing and of horizontal tail (latter normally being less).

longitudinal force coefficient Component of resultant wing force resolved along chord; $C_X = C_D \cos \alpha - C_L \sin \alpha$.

longitudinal oscillation Vibration along longitudinal axis (chiefly in ballistic rocket vehicles); also known as pogo effect.

longitudinal plane Any plane of which OX axis is a part.

longitudinal stability Ability to recover automatically to level flight after sharp dive or climb command; generally, all stabilities in plane of symmetry.

longitudinal wave One devoid of lateral components (eg sound).

long lead time Must be ordered months to years before aircraft completion; sum of contractual delays for item, delays in delivery (heavy forging may be years), processing time at subcontractor or in plant, and time from when finished part joins aircraft to completion of aircraft.

long-range No valid modern definition except DoD * bomber operational radius over 2,500 nm.

Lons, LONS Local on-line network system.

look angle Angular limits of vision of EO or IR seeker.

look-down angle Limiting inclination of main beam in AWACS-type aircraft, strongly dependent on aerial geometry.

look-down shoot-down Ability to destroy low-level hostile aircraft from high altitude against land or clutter background.

loom Tightly laced bundle of electric cables, instrumentation leads and other flexible wires.

loop 1 Flight manoeuvre in which aircraft rotates nose-up through 360° whilst keeping lateral axis horizontal; many variations but normal loop restores level flight on original heading but at slightly higher altitude.
2 Conductive coil in vertical plane rotating about vertical axis to give bearing to ground radio station; a D/F loop (obs).

Loose Deuce Fighter combat tactic, 2 to 4 friendly aircraft manoeuvre to provide mutual support and increased firepower.

LOP Line of position.

lopro Low probe; military aircraft mission.

LOR Lunar orbital rendezvous.

Loraas Long-range airborne ASW system.

Lorac Long-range accuracy, also called Loran-C or Cytac; Loran derivative.

Loran Long-range navigation, early but much developed hyperbolic navaid, using various on-board systems to translate time difference of reception of pulse-type transmissions from two or more fixed ground stations.

Lord mount Large family of patented anti-vibration mounts, usually metal/rubber.

Lorentz force, F_L That on charged particle moving in magnetic field, = q(VB) where q is particle charge, V velocity and B magnetic induction (flux density).

Lorentz system Pioneer beam-approach landing system.

LORO Lobe-on receive only.

Loroc Long-range optical camera.

Lorop Long-range oblique photography.

Lorv Low-observable re-entry vehicle, characterized by reduced radar cross-section.

LOS 1 Line of sight.
2 Loss of signal.

Loschmidt number Number of molecules of ideal gas per unit volume, $= 2{\cdot}687 \times 10^{19}$ per cm^3.

Loss Large-object salvage system.

lost-wax Technique for casting intricate precision shapes, derived from Benvenuto Cellini c 1550 but modified for modern refractory alloys.

Lotaws Laser obstacle terrain-avoidance warning system.
Lotex Life-of-type extension.
lounge Waiting area at airport for departure passengers between processing and gates.
LOW Launch on warning.
low Geographical region of low atmospheric pressure.
low airburst Fallout safe height of NW burst for max effect on surface target.
low altitude US military traditional, 500/2,000 ft; today see *lo (1)*.
low-altitude bombing system Early weapon-aiming electronics for tossing nuclear weapons in low attack, the high bomb trajectory giving aircraft time to escape explosion.
low-angle loft bombing Free-fall loft bombing where release angle is within 35° of horizontal.
low bidder Manufacturer offering lowest price in industry competition.
low blower See *low supercharger gear*.
low blown Piston engine supercharged for maximum powers at low altitudes at expense of poor performance at height.
low cloud Cumulus, cumulonimbus, stratocumulus, stratus and nimbostratus; base generally below 1,800 m, 5,900 ft.
low-cycle fatigue Fatigue caused by changes in material stress resulting from changes in speed of rotating machines; from idling to take-off power and back to cruise rpm could represent single completed cycle.
lower airspace No single definition; FAA usually below 14,500 ft AMSL.
low flying Flight at minimum safe (sometimes unsafe) altitude for training or sport.
low-flying area Geographical region within which low flying is authorized for training.
low gate Mechanical stop on piston-engine throttle box beyond which further opening inadvisable below rated height or other altitude limit.
low-level discomfort factor Usually measured as number of vertical accelerations exceeding specified level (0·5 g) experienced by occupants per minute.
low-light TV Vidicon tube with multiplier tubes giving useful picture in near-darkness.
low-light-level TV Generally used for same EO devices as preceding entry.
low/mid wing Mounted about one-third way up body.
low oblique Photography from lo altitude at oblique angles to either side, not vertical or ahead.
low observables Stealth.
low orbit Nominally, period 90 min or less.
low-pass filter Designed to cut off all signals above given frequency.
low pitch See *coarse pitch*.
low route Area-navigation routes not dependent on navaid-based airways, for low-level traffic, MAA (3) 4,000 ft AMSL (FAA).
low rudder In tight turn, or other occasion with wings near-vertical, depressing pedal nearest ground to lower nose.
low silhouette Squat aircraft, esp in frontal aspect, easy to hide on battlefield (usually anti-tank helicopter).
low-speed aerodynamics Not defined; below 100 ft/s has been suggested.
low-speed aircraft Several attempted definitions; UK CAA and FAR.91 do not define.
low-speed stall Normal 1 g stall.
low supercharger gear Lower of two gears in two-speed supercharger drive.
low vacuum Pressure below 101·247 kPa (760 torr) and above some lower level usually agreed as 3·33 kPa (25 torr).
low-velocity drop Paradrop or other drop with velocity below 30 ft/s (DoD, NATO).
low wing Mounted low on body, usually so that undersurfaces coincide.
lox Liquid oxygen.
loxodrome See *rhumb line*.
loxygen Liquid oxygen (arch).
loz Liquid ozone.
LP 1 Low pressure.
2 Licensing panel.
3 Low-pass filter (often l.p.).
LPA Laboratoire de Physique de l'Atmosphère (F).
LPB Loss Prevention Bulletin; each issue lists over 20,000 stolen airline ticket numbers (IATA).
LPBA Lawyer Pilots' Bar Association (US).
LPC 1 Luftfahrt Presse-Club (G).
2 LP compressor.
3 Linear predictive coding.
LPD 1 Labelled plan display.
2 Low-prf pulse Doppler.
LPDS Landing platform docking ship (US).
LPFT Low-pressure fuel turbopump.
LPH Landing platform, helicopter ship (US).
LPI Low probability of intercept.
LPIR LPI radar, multi-beam broadband coded waveform with very small sidelobes.
LPL Linear polarized laser.
LPM 1 Looks per minute.
2 Landing path monitor.
LPN LPN maximum value.
LPN Perceived noise level.
L̄PN Average peak outdoor LPN at individual's residence.
LPO Lunar parking orbit.
LPOT Low-pressure oxidizer turbopump.
LPT LP turbine.
LQA Link quality analysis.
LQG Linear quadratic Gaussian.
LR Long range.
LR^3 Laser ranging retro-reflector.
LRAS Low-range airspeed system; ASI for V/STOL, helicopters.
LRBA Laboratoire des Recherches Balistiques et

Aérodynamiques, Vernon (F).
LRBM Long-range ballistic missile (2,500 nm, 2,880 miles).
LRC 1 Long-range cruise.
2 Logistics Readiness Center (USAF).
LRCA Long-range combat aircraft.
LRCO 1 Limited remote communications outlet.
2 Lead range control officer.
LRD 1 Labelled radar display.
2 Laser ranger/designator.
LRF Laser rangefinder.
LRGB Long-range glide bomb.
LRINF Long-range intermediate nuclear force(s).
LRL 1 Lunar Receiving Laboratory.
2 Lightweight rocket launcher.
LRM 1 Long-range air-to-air missile.
2 Launching/reeling machine (towed MAD).
LRMP Long-range maritime patrol.
LRMTR Laser ranger and marked-target receiver.
LRMTS Laser ranger and marked-target seeker.
LRPA Long-range patrol aircraft.
LRRR Laser ranging retro-reflector.
LRS Load relief system.
LRSI Low-temperature reusable surface insulation (usually 99% pure silica-fibre felted tiles).
LRSOM Long-range stand-off missile.
LRU Line-replaceable unit; any item that can be replaced on flightline.
LRV Lunar rover vehicle.
LS 1 Loudspeaker.
2 Landmarks subsystem.
LSA Logistics support analysis.
LSAT Logistic shelter, air-transportable.
LSB Lower sideband.
LSC Logistics support costs.
LSD Large-screen display.
LSFFAR Low-speed folding-fin aircraft rocket.
LSI Large-scale integration (microelectronics).
LSIC Large-scale integrated circuit.
LSJ Lifesaving jacket.
LSMU Laser-communications space measurement unit.
LSO 1 Landing safety officer (manages aircraft-carrier projector sight).
2 Landing signals officer.
3 Limited strategic option.
LSP 1 Large-screen projection.
2 Logistics support plan.
LSS Local speed of sound.
LSSAS Longitudinal static-stability augmentation system.
LSSM Local scientific survey module.
LSSS Lightweight SHF satcom system.
LST Laser-spot tracker
LT 1 Low-pressure turbine.
2 Low temperature.
L_T Tail moment arm.
lt Light, lighted.
l.t. Low tension.
LTA 1 Lighter than air.
2 Lighter-Than-Air Society (US).
3 Light transport aircraft.
LTC 1 Long-term costing.
2 Lithium thionyl chloride (electric battery).
LTD Laser target designator.
LTDP Long-term defence plan (NATO).
LTDS Laser target designator set.
Lt Ho Lighthouse.
LTM 1 Load ton-mile.
2 Laser target marker.
LTMR Laser target marker and receiver.
LTP 1 Laboratorio de Tecnologia de la Propulsión (Spain).
2 Longitudinal touchdown point.
L_{TPN} Tone-corrected perceived noise level, now judged of doubtful value, to account for intense tones, eg due to rotating fan and compressor blades.
Lt V Light vessel.
L^3TV Low-light-level TV.
LTWA Long trailing wire aerial.
LUA Launch under attack.
lubber line Reference index, eg on compass, denoting aircraft heading.
lube Lubricating oil (US, colloq).
Lucero Ground beacon keyed to 1·5 m AI or ASV radars, 1942–57.
Lucid Software package for processing image data from all forms of visual sensor.
Lucite Resin produced from methyl methacrylate, widely used for transparent plastics.
LUF Lowest usable frequency.
Lufbery circle Military ring formation in which all aircraft fly gentle turn following that in front.
luhf Lowest usable high frequency.
lumen SI unit of luminous flux; that emitted in solid angle of 1 steradian by point source of intensity 1 candela; lm = cd.sr.
luminance Brightness; intrinsic luminous intensity; illuminance on unit surface normal to radiation divided by subtended solid angle.
luminescence Non-thermal emission of light, ie not incandescence but electro-*, phosphorescence, chemi-* and photo-* (fluorescence).
luminous intensity Luminous energy per unit solid angle per second; unit candela, cd.
lunar boot Astronaut footwear tailored to surface of Moon.
lunar orbit Orbit round the Moon.
lunar orbital rendezvous Lunar exploration by descent stage detached from larger spacecraft left in lunar orbit and which is rejoined for return to Earth.
Luneberg lens Device, often spherical, designed for maximum reflectivity of radar energy back along incident path, tailored to wavelength; an enhancing corner reflector.
lusec Lumen-second, quantity of luminous energy.

lux SI unit of illuminance; lx = lm/m^2.
L/V Local vertical.
L+V Leather and vinyl.
LVA Large vertical aperture (radar).
LVDT 1 Linear voltage differential transducer.
2 Linear variable differential (or displacement) transformer.
LVFE Long variable-flap ejector (supersonic nozzle).
LW Landing weight.
l.w. Long wave.
LWABTJ Lightweight afterburning turbojet.
L_{WBT} Total lift of wing/body/tail.
LWC 1 Light-water concentrate (firefighting foam).
2 Liquid-water content.
LWD Lowered.
LWF Light-weight fighter.
LWIR Long-wavelength IR.
LWLD Lightweight laser designator.
LWR Luftwaffenring eV (G).
lx Lux.
Lyman-Alpha Radiation emitted by hydrogen at 12·16 pm (1,216 Å), penetrates Earth atmosphere to base of D-region (90 km, 55 miles).
LZ Landing zone (assault in land battle).
LZE Luminous-zone emissivity (flare, IRCM).

M

M 1 Prefix mega, 10^6.
2 Mass.
3 Magnetic heading/course/bearing.
4 Mach number.
5 Prefix minus (wind component).
6 Maxwell.
7 Pitching moment.
8 Meteorological (JETDS).
9 Mutual inductance.
10 Molecular weight.
11 Bending moment, and generalized symbol for moment.
12 Mandatory (NASA).
13 Prefix, aircraft type designation: equipped to launch guided missiles (USN, obs).
14 Telecom code: "IFR aircraft has Tacan and transponder with no code capability" (FAA and others).
15 Mean anomaly of orbit.
16 Most ambiguously, thousand (ASA).
17 Prefix maximum.

m 1 Metre.
2 Prefix milli, 10^{-3}.
3 Superplasticity.
4 Modular ratio.
5 Bypass ratio (USSR).
6 Mass, especially of electron.

m² Square metres.

M-band EM radiation, 5–3 mm, 60–100 GHz.

M-carcinotron Backward-wave oscillator in which high power is possible by electron beam travelling between slow-wave structure and negative sole plate.

M-day Day on which mobilization is to begin.

M-display Has horizontal timebase along which target blip moves; operator moves second blip to line up on target by control graduated to indicate target range.

M-marker Middle marker.

M-stoff Methanol (G).

M-wave EM millimetric wavelengths.

M-wing Wing studied for low supersonic speeds with inner portions swept forward, outers swept back.

MA 1 Mission abort.
2 Naval aviation (USSR).
3 Mobilization augmentee (US).

M/A Mach/airspeed.

mA Milliampere.

ma, Ma Mass flow (airflow); eg that passing through engine.

MAA 1 Monitoring angle of attack.
2 Missed-approach action.
3 Max authorized altitude.
4 Mission area analysis.
5 Manufacturers' Aircraft Association (US).

MAAG Military Assistance Advisory Group (US).

MAAH Minimum asymmetric approach height.

MAB Modular-array basing (ICBM).

MAC 1 Mean aerodynamic chord, $\bar{\bar{c}}$.
2 Military Airlift Command (USAF).
3 Maintenance Analysis Center (FAA).
4 Maintenance allocation chart.
5 Master air control (UK, CAA).

Mac Pitching moment of aircraft about aerodynamic centre.

MACCS Marine air command and control system (DoD).

MACE Minimum-area crutchless ejector.

Mach Mach number.

Mach angle Angle between weak (ie point-source) shockwave and freestream flow; theoretically 90° at Mach 1, thereafter $\alpha = \sin^{-1} \frac{1}{M}$.

Mach-buster Anyone who has flown faster than sound, esp in pre-Concorde era when accomplishment had slight element of exclusivity.

Mach cone Conical shock front from point source moving at supersonic speed relative to surrounding fluid; locus of Mach lines.

Mach front Mach stem.

Mach hold Autopilot or AFCS mode holding Mach number at preset value.

machine Flying machine, normally aeroplane (colloq, arch).

machine language Normally compatible with particular computer but unintelligible to humans.

machining centre Major group of (usually NC) machine tools performing large number of operations on workpiece held either stationary or under positive control throughout.

Mach intersection Junction of two or more shockwaves.

Mach-limited Boundaries of flight performance set by restriction on permissible Mach number, not by thrust or other limitation, esp * ceiling.

Mach line 1 Weak (infinitesimal amplitude) shockwave.
2 Line on surface of body (ignoring boundary layer) at which accelerating free-stream flow reaches relative Mach 1.

Mach lock See *Mach hold*.

Machmeter Instrument giving near-instantaneous readout of Mach number.

Mach NO/YES Air-intercept code: "I have reached maximum speed and am not/am closing on target".

Mach number, M Ratio of true airspeed to speed of sound in surrounding fluid (which varies as square root of absolute temperature).

Mach reflection Attenuated shockwave reflected from solid surface, eg walls of tunnel or Earth's surface. Reflection and some other effects approximate laws of optics.

Mach stem Shock front (Mach front) formed by fusion of incident and ground-reflected blast waves from explosion.

Mach trim coupler Electronic subsystem of Mach trim system, which through analog (to 1970) or digital computational chain controls aircraft pitch-trim servo as function of Mach number; also contains switching, logic, monitoring and Bite.

Mach trim system Electronic/mechanical system for relieving pilot of task of correcting progressive deficiency in aircraft pitch trim and longitudinal stability at high Mach numbers; sensitive to Mach number and vertical acceleration and automatically feeds primary pitch-trim demand to keep aircraft level or in desired attitude while leaving pilot authority to feed manual trim. Also called Mach trimmer and, in US, pitch trim compensator.

MACIMS Military Airlift Command integrated management system.

mackerel sky Large area of high cloud giving dappled or banded effect (alto-cu or cirro-cu).

Maclaurin series Special case of Taylor series where f(x) and all derivatives remain finite at x = 0.

macro level Preliminary design of software.

macroscopic Generally, large enough to be visible to naked eye.

MACS 1 Marine air control squadron (DoD).
2 Multiple-applications control system.

MAD 1 Magnetic-anomaly detection.
2 Mutual assured destruction.
3 Maintenance access door.
4 Mass air delivery (attack weapons).

madapolam Fabric woven from long-staple cotton, originally from Madapollam (single l originally erroneous).

Madar Maintenance analysis, detection and recording.

Maddls Mirror-assisted dummy deck landings.

Madge Microwave aircraft digital guidance equipment.

MAE Mean area of effectiveness (USAF).

MAEO Master air electronics officer (RAF).

Mae West Inflatable aircrew lifejacket tied round upper torso (WW2).

MAF 1 Mixed-amine fuel.
2 Marine amphibious force.
3 Missionary Aviation Fellowship (UK).

MAFFS Modular airborne firefighting system.

Maflir Modified advanced forward-looking infra-red.

MAFT Major airframe fatigue test.

MAG 1 Military assistance group.
2 Marine air group.
3 Mobile arresting gear.
4 Machine gun.

mag 1 Magnetic (FAA).
2 Magneto (eg * drop).
3 Magazine (camera) (NASA).

Magat Military assistance grant-aid training.

Magcom Guidance technique similar to Tercom but using variations in terrestrial magnetic field.

mag drop Reduction in rpm when either ignition source of dual-ignition piston engine is switched off; always checked before take-off to confirm both sources operative.

Magerd MRCA AGE Requirement Document.

Magfet Magnetic Mosfet.

Maggs Modular advanced-graphics generation system, for simulator external visuals.

Magic Microprocessor application of graphics with interactive communications (USAF).

magic eye Tuning system using miniature CRT with radial illumination around sector whose size varies with strength of received signal; important in pre-crystal tuning era; also called magic-T.

Magiic Pronounced magic, Mobile Army Ground Imagery Interpretation Center (USA).

Magis Marine (or mobile) air/ground intelligence system.

maglev Magnetic levitation.

Magnaflux Non-destructive test for magnetic material using magnetic field and fluorescent ink; trade name.

magnesium Low-density white metal, used pure and as alloy (Elektron, Dowmetal etc) for structural parts such as engine bearers, wheels and forged ribs.

Magnesyn Patented remote-indicating system using induction between permanent-magnet rotor and saturable coil.

Mag Net Mobile arresting gear using fast-erecting net to catch aircraft lacking serviceable tailhook.

magnetically anchored rate damper One in which gyro spin axis is restrained magnetically.

magnetic amplifier Various arrangements of saturable reactors such that small control current governs large output load/power/voltage.

magnetic anomaly Local irregularity in terrestrial magnetic field caused by presence of magnetic material such as submerged submarine or ore deposit.

magnetic core Doughnut-shaped ferrite ring storing either 1 or 0 in either of two stable magnetic states.

magnetic course Course (heading) indicated by simple magnetic compass after correction for deviation.

magnetic crotchet Sudden change in numerical values of Earth's field usually ascribed to alteration in conductivity of ionosphere.

magnetic damping Use of eddy currents or other induced magnetic field to oppose oscillatory or vibratory motion.

magnetic declination Horizontal angle between terrestrial field and true N, ie between magnetic and geographic meridians; also known as variation.

magnetic deviation Errors in magnetic compass indication caused by local disturbance to field, esp by "iron" in aircraft.

magnetic dip Angle between local terrestrial field and local horizontal.

magnetic disc Computer storage in magnetic-oxide coating on surface of high-speed rotating disc.

magnetic drag cup Aluminium or copper cup surrounding rotating core in simple tachometer, rotated by generated eddy current.

magnetic equator Line joining points where angle of dip is zero.

magnetic field intensity Magnetizing force exerted on unit pole; unit A/m.

magnetic flux Product of area and of field intensity perpendicular to it (in effect number of lines of force); unit is Weber (Wb).

magnetic flux density Measure of magnetic induction; unit is Tesla, Wb/m^2.

magnetic inspection Many NDT methods which attempt to detect imperfections in magnetic-metal parts by the anomalies they cause in strong fields.

magnetic meridian Direction of horizontal component of terrestrial field near surface.

magnetic particle inspection NDT for ferrous parts which when magnetized form N/S poles across cracks rendered visible by iron oxide powder.

magnetic permeability See *permeability*.

magnetic tachometer Most common speed-measuring system in aero engines measures frequency signal from magnetic and electrical interaction between toothed wheel and fixed sensor.

magnetic tape Storage system in which information is recorded on and read off long strip coated with magnetic oxide.

magnetic variation Angle between magnetic meridian and true north, varies throughout globe and with time; thus used as ± correction to °M to give °T.

magneto-aerodynamics Aerodynamics at high hypersonic speed and near-vacuum conditions, where magnetic and aerodynamic forces are similar.

magneto drop See *mag drop*.

magnetohydrodynamics, MHD Science of interaction between magnetic fields and electrically conductive fluids, esp plasmas.

magnetometer Instrument for measuring magnetic field intensity and direction.

magnetomotive force Product of flux and reluctance (resistance), work needed to move unit pole against field.

magnetopause Outer boundary of geomagnetic cavity.

magnetosphere Region around Earth, from T-layer at 350 km to about 15 Earth radii, where magnetic field and ionized gas are dominant.

magnetostriction Change in dimensions when magnetic material is magnetized; most pronounced in nickel.

magnetron Pioneer resonant-cavity generator of high-power microwaves, with spinning electron beam deflected by transverse field.

magnitude Apparent brightness of stars, planets, on scale of relative luminosity where × 100 brightness reduces magnitude by 5; hence brightest bodies have negative *.

Magnus effect Force produced perpendicular to airflow past spinning body; basis of Flettner rotor and swerving golfball.

MAHL Maximum aircraft hook load (ASCC).

MAI Moscow Aviation Institute.

Maid/Miles Magnetic anti-intrusion detector/magnetic intrusion line sensor.

main airfield Permanent peacetime military airbase offering at least potential for all facilities.

main bangs Transmitted pulses in radar (colloq).

main-beam killing ECCM technique in which, while continuing to operate side lobes, main radar beam is suddenly cut off.

main bearing Supports main rotating assembly in gas turbine, crankshaft in piston engine.

main float Central float in three-float seaplane.

mainframe Central computer in large or geographically dispersed EDP (1) system.

main gear Main landing gear.

main landing gear Each or all units of landing gear supporting nearly all weight of aircraft; other units are nosewheel, or tailwheel, or lateral outriggers.

main line 1 Mooring line (airship).
2 Data highway, excluding side branches attached at nodes.

main parachute Canopy supporting load, as distinct from drogues, extraction chute etc.

mainplane Wing, as distinct from tailplane or canard.

main rotor Helicopter rotor providing lift and propulsion.

main runway Airfield's principal runway, normally in use (usually longest).

main spar Principal spar of wing, having modulus greater than others.

mainstage 1 In multistage rocket vehicle, that stage having greatest thrust, excepting short-duration boosters (see [3]).
2 In single-stage vehicle, main propulsion as distinct from verniers, roll-control and other motors.
3 In smaller, or non-ballistic vehicles, sustainer propulsion after separation of boost motor(s).
4 In large ballistic vehicle, period during which first-stage propulsion is delivering 90% or more of maximum thrust.

main step Seaplane step below c.g.

maint Maintenance.

maintenance Work required, scheduled and otherwise, to keep aircraft serviceable, other than repair. Term normally refers to minor tasks on flightline, but can also include major operations normally called overhaul.

maintenance burden Usually total maintenance cost, calculated as 1·8 times total maintenance labour cost.

maintenance recorder Wire or tape recorder for flight time, engine operation and variable number of other parameters.

maintenance status Non-operating condition deliberately imposed.

maintenance unit Military formation at fixed airbase able to store, modify, overhaul, flight-test and scrap aircraft.

main transverse Major frame of rigid airship joining all longitudinals.

main undercarriage Main landing gear.

mainwheel Wheel of main landing gear.

major axis Principal axis through solid body, usually along largest dimension or chief moment of inertia; where possible, axis of symmetry.

majority rule Philosophy whereby two or more operative channels (eg AFCS) always "outvote" single failed channel.

majority voting system Redundant system wherein outputs of three or more active channels are summed and output is fed back to each channel. When failure of one channel occurs, feedback causes all unfailed channels to act to offset failure; hence immediate shut-off of failure is not necessary.

MAL Maximum allowable; thus MAL TOW.

Mallory cell See *mercury battery.*

MALM Master air loadmaster.

MALS Medium-intensity approach light system (FAA).

MALSR MALS with runway alignment indicator lights.

mammatus Cumuliform cloud with pendulous bulged undersurface.

Mammut Luftwaffe surveillance radar, 1944.

MAMS Mobile air movements squadron (RAF).

manacle Mounting in form of calipers surrounding and gripping circular section hardware, eg generator.

mandatory To be complied with, often as AOG modification.

M&ES Magnetic and electromagnetic silencing.

M&O Maintenance and overhaul.

M&R Maintenance and repair.

mandrel 1 Centrebody, usually circular section, around which tubular or female part hand-forged, extruded or formed.
2 (capital M) High-power radar jammer (RAF, 1943).

M&S Marred and scarred (US category).

maneton Pinch-fit coupling for female part tightened by bolt around (usually smooth-surface) pin, eg crankshaft web or big end.

manganese Silver-white metal, density 7·3, important as alloying element in tough corrosion-resistant steels and with light alloys, brasses and bronzes.

manganese bronze Golden alloy, resistant to marine corrosion, sometimes used for compressor blading (DTD.197).

manifest List of all passengers and cargo for one flight.

manifold Fluid pipe system distributing from single input to multiple outputs or vice versa (thus piston engine inlet *, exhaust *).

manifold pressure Pressure in inlet manifold of piston engine, normally local atmospheric plus boost (US traditionally inches Hg).

manoeuvrable re-entry vehicle One which is capable of performing accurate preplanned flight manoeuvres during re-entry (DoD).

manoeuvre, maneuver Any deliberate departure from straight/level flight or existing space trajectory.

manoeuvre ceiling Maximum height at which aircraft, at given weight, can sustain specified load factor after onset of buffet; highest point of each buffet-boundary line.

manoeuvre diagram Usually means V-n diagram.

manoeuvre flap Usually small forward-hinged plate depressed under power from underside of inner wing at 20–40% chord to cause powerful nose-up trim change.

manoeuvre-enhancement Capability of a direct-force-control aircraft to achieve more rapid normal acceleration (upon pilot selection), typically by use of symmetric wing flaps for immediate increase in lift to eliminate difference between g (A_N) commanded and achieved.

manoeuvre-induced error Instrument error, usually pressure, gyro or magnetic, caused by flight manoeuvre.

manoeuvre load factor Load factor (2).

manoeuvre margins Two distances (stick-fixed and free) measured as % SMC from manoeuvre points to c.g. position.

manoeuvre point Point around which aircraft rotates about all axes. Stick-fixed ** is c.g. position at which stick movement per g is zero; stick-free ** is c.g. position at which stick force per g is zero.

manoeuvring area Area of airfield used for take-offs, landings and associated manoeuvres (NATO).

manoeuvring ballistic re-entry vehicle For practical purposes synonymous with manoeuvrable re-entry vehicle.

manoeuvring factor See *load factor (1).*

manometer Linked-twin or single vertical fluid tubes giving indication of pressure piped from remote source.

manometer bank Large array of manometers side-by-side.

manometric lock System for holding constant pressure-based flight parameters, esp IAS, alt, VS, M.

manpack General term for astronaut/cosmonaut load carriers, designed according to local (eg lunar) gravity.

Manpads Man-portable air-defense system (USA).
man-rated Sufficiently reliable to form part of manned spacecraft or launcher.
MANS, Mans Missile and Nudet Surveillance.
man space Assumed to include individual equipment and currently defined as 250 lb/13·5 cu ft (DoD).
Mantech Manufacturing technology (AFSC).
manual control Hand-flying aircraft with autopilot or AFCS.
manual D/F Obtaining bearing by hand-rotation of loop and visual or aural judgement of signal.
manual feedback Force experienced at pilot from manual FCS, or (rarely) other system.
manual override Condition in which pilot physically overcomes AFCS through cable and/or linkage connections and exerts flight control in excess of AFCS authority or in opposition to AFCS command.
manual reversion Ability to switch from autopilot or AFCS to hand-flying (can be automatic in event of AFCS failure) wherein pilot's forces are transmitted to control surfaces.
manufactured head Rivet head preformed when rivet made.
many Air intercept code: more than eight (DoD).
MAO Mature aircraft objective.
MAOT Mobile air operations team (British Army).
MAP 1 Ministry of Aircraft Production.
2 Military Assistance Program (DoD).
3 Machine assembly program.
4 Missed-approach point.
5 Ministry of Aviation Industry (USSR).
6 Manufacturing automation protocol.
7 Manifold absolute (or air) pressure.
8 Multiple aim point (USAF).
map-matching Navigation (usually of RPV or missile) by auto-correlation of terrain with stored strip of film; based on appearance, unlike Tercom.
Mapp Methyl acetylene-propadiene/propane (FAE).
mapping radar Producing pictorial display showing Earth's surface in detail (usually in sector ahead, sometimes PPI 360° all round aircraft).
MAPS 1 Mobile aerial port squadron (MAC).
2 Measurement of air pollution sensor (Shuttle).
Mapse Minimal APSE (Ada programming support environment).
MAR 1 March (ICAO).
2 At sea (ICAO).
3 Multiple- (or multi-function) array radar.
4 Mission abort rate.
5 Minimally attended radar.
maraging steel High-alloy (Ni, Co, Mo) steels aged in martensitic condition for maximum tensile strength.
Mareng tank Flexible fuel cell (from Martin Engineering, US).
mare's tail sky Cirrus.
marginal performance Barely able to comply with airworthiness requirements, or to fly safely.
marginal weather Only just good enough for safe or legal flight, actual conditions depending on whether flight is IFR or VFR. DoD definition: "Sufficiently adverse to military operation so [*sic*] as to require imposition of procedural limitations".
margin of lift 1 Aerostat buoyancy (gross lift) minus mass.
2 Various meaningless definitions for aerodynes.
margin of safety Percentage or ratio by which ultimate failing load of component exceeds design limit load.
maria Flat areas on Moon once thought to be seas.
marine aircraft Designed to operate from water.
maritime aircraft Designed to operate over sea areas.
maritime PDNES PDNES having extremely short pulses to reduce sea-clutter.
maritime SAR Region within which USCG exercises SAR co-ordination function (FAA).
MARK Material accountability and robotic kitting system (DoD).
Mark Air-control agency code for commanded point of weapon release, usually preceded by word "Standby" (DoD).
mark One particular version of aircraft whose basic design has been developed for different missions or as result of improvements introduced over a period, thus * 1, * 2 etc.
marked target One illuminated by laser.
marker Distinctive visual (usually pyrotechnic) or electronic aid dropped on surface location.
marker beacon Any beacon (course-indicating, fan, outer, middle, inner) giving substantially vertical radiation which gives aural and visual signal in cockpit (see *outer* *, *middle* * etc).
marker template One having profile of common stringer or similar standard section, for marking end-cuts and cut-outs.
market 1 The world total of available potential customers.
2 To promote one's product in (1).
marking panel Sheet of material displayed by ground troops for signalling to friendly aircraft.
marking team Troops dropped into tactical area to establish navaids and possibly other electronic facilities.
marking up Putting bright paint line round each airframe crack.
Marmon clamp Patented ring joint for pipe sections, with wedge action.
MARRES Manual radar-reconnaissance exploitation system.
marrying up Offering up major airframe or other large items in erection of aircraft or space launcher, esp when likelihood of imperfect fit.
Mars 1 Mid-air recovery system (USN).
2 Military affiliate radio system (DoD).

3 Multi-access retrieval system.
4 Modular airborne recorder series.
5 Minimally attended radar station.

Marsa Military accepts responsibility for separation of aircraft (DoD/FAA).

marshal Person giving visual and/or aural signals to sporting aircraft or sailplanes to ensure line-up in correct take-off sequence.

marshaller Person on apron or flightline giving visual signals to aircraft on ground, eg directions where to park.

marshalling point Place on airfield where aircraft come under control of a marshal (sporting, outbound) or marshaller (inbound).

marshalling wand Combined black flag and illuminated wand normally used one in each hand by marshaller.

Mart Modular air/ground remote terminal.

Martlesham Martlesham Heath, pre-1939 centre of military aircraft test and evaluation in UK, hence * figures (= indisputable).

Marv Manoeuvrable re-entry vehicle.

MAS 1 Military Agency for Standardization (NATO).
2 Military area services.
3 Military assistance sales (DoD).

mascon One of the mass concentrations scattered over lunar surface which distort orbits near the Moon.

MASDC Military Aircraft Storage and Disposition Center (USAF).

maser Large family of devices which pump electrons to higher energies and build up amplification of EM signals, usually microwaves (generally related to laser).

MASF 1 Mobile aeromedical staging flights.
2 Military assistance service fund (DoD).

MASI, Masi Mach/ASI.

masking 1 Layer protecting substrate against applied process, eg metal foil or special paper when painting aircraft, or hi-resist surface pattern when doping or etching microelectronic chip.
2 Deliberate use of additional emitters to hide or conceal true purpose of particular EM radiation.

Masonite Commercial formulation of compressed fibre-board, moulded to shape with accurate glassy finish on at least one surface.

MASR Multiple antenna surveillance radar.

MASS Marine air support squadron (DoD).

mass Applied force on body divided by resulting acceleration; in practice attraction of body related to standard kilogramme.

mass axis Line joining c.g. of all elements of part, esp wing.

mass balance Mass attached to flight-control surface, typically ahead of hinge axis, to reduce or eliminate inertial coupling with airframe flutter modes.

mass flow Quantity of fluid passing through closed system per unit time.

mass fraction See *mass ratio (1).*

mass ratio 1 Mass of vehicle, normally ballistic rocket, at liftoff divided by mass at all-burnt or other later condition.
2 Rarely and ambiguously, mass of propellants as fraction of total liftoff.

mass spectrograph Instrument for converting molecules to ions and then separating these to determine exact isotopic weights.

mass taper Rate of, or graphical plot of, reduction of cross-section-percentage material density from root to tip of blade (eg helicopter rotor), latter usually being externally untapered.

MAST 1 Military anti-shock trousers; pressurized, astronaut garb.
2 Military assistance to safety and traffic.
3 Married airmen sharing together (USAF).

mast 1 Tube projecting from underside of aircraft so that liquid can drain well away from airframe.
2 Rigid pillar for avionics aerial.
3 Rigid mooring structure for airship.
4 Pillar above centre of helicopter rotor for MMS (2).

Mastacs Manoeuvrability-augmentation system for tactical air combat simulation (USN).

master caption panel Flight-deck panel giving initial warning information and/or indication of AFCS mode, nav mode etc.

master caution signal Indication that at least one caution signal has been activated.

master connecting rod See *master rod.*

master contour template Full-size flat template showing all mould-line contours, rib contours or other related shapes to common reference.

master diversion airfield Large and well-equipped airfield able to handle all kinds of aircraft whose home bases are shut by weather or enemy action, and promulgated as such.

master oscillator That defining timebase and waveform for radio or radar.

master rating In certain air forces, highest level of aircrew proficiency in each trade.

master rod Connecting rod with big end and with attachments for all other rods in same plane.

master router template Flat template, usually same size as metal sheet or slab, used to locate pilot and pin holes and set up sheet for routing or skin-milling by non-NC control.

master station Various meanings, esp that housing timebase and emitting original signals in R-Nav hyperbolic systems, other stations being slaves keyed to it.

master template Reference for any form, usually two-dimensional, in manufacture; used to make jigs and tooling.

Mastif Multiple-axis spin-test inertia facility.

Mastiff Modular automated system to IFF.

mast-mounted sight Group of sensors, typically optical, FLIR, laser, mounted on anti-vibration mast above helicopter rotor hub to allow crew to see targets without exposing helicopter.

MASU Multiple acceleration sensor unit.

MAT 1 Modification approval test.
2 Moscow aviation technical high school.

mat 1 Flexible membrane laid on unprepared ground for limited number of V/STOL operations.
2 Area paved with quick-assembly metal mesh or planking.
3 Floating support towed by ship for recovery of marine aircraft.

Matcals Marine (or Mobile) air-traffic control and all-weather landing system (USMC).

Match Manned anti-submarine troop-carrying helicopter.

matching 1 Any prolonged process of developing dynamic parts to work properly together; eg inlet/engine/propelling nozzle over wide range of Mach numbers.
2 Achieving maximum energy transfer between electric circuits or devices by equalizing impedances and resistive/reactive components.

Matcu Marine air-traffic control unit (US).

Mate 1 Modular automatic test equipment (Grumman).
2 Modular Autodin terminal equipment (Astronautics Corp).

Matelo Maritime air-radio telegraph organization.

materiel US term for hardware, logistic supplies, esp military (but specif excluding ships and naval aircraft); accent is on last syllable.

mating Offering up major portions of aircraft and joining together; the mass-production equivalent of marrying up. Thus * jig.

Mato Military air-traffic operations (UK).

matrix 1 Computer memory core of rectilinear array configuration, eg 2-D ferrite three-wire type.
2 Crystalline grain structure of metal.
3 Skeletal basis on which structure is formed, esp by moulding.
4 Any 2-D rectilinear logic network.

MATS 1 Military Air Transport Service (now MAC) (US).
2 Multi-altitude transponder system.

matt Non-reflective, external finish (usually paint) tending to diffuse EM radiation incident on it.

MATTS Multiple airborne-target trajectory system.

maturity Vague condition after impressive number of flight hours in which major problems naively thought unlikely. Supposed methods exist for quantifying degree of * for both new and derived equipment.

MATZ Military air (or airport) traffic zone.

MAU Marine amphibious unit (US).

Mavar Parametric amplifier (from mixed amplification by variable reactance).

MAW 1 Mission-adaptive wing.
2 Missile approach warning, defensive system linked to RWR giving definite warning of a locked-on missile's approach.
3 Military airlift wing (USAF).

MAWS Modular automated weather system.

MAWTS Marine Aviation Weapons and Tactics Squadron.

max Maximum.

Maxaret Pioneer (Dunlop) anti-skid system for wheel brakes.

Maxi decoy ECM payload, originally bronze model aircraft with radar cross-section identical to attack aircraft.

maximum authorized altitude Highest altitude on airway jet route or any other direct route for which MEA (1) is designated in FAR.95 at which navaid reception is assured (FAA).

maximum boom-free speed Mach limit at high altitude to avoid shock reaching ground, variable with atmosphere in region of M 1·12.

maximum chamber pressure Peak pressure in complete firing cycle of solid-propellant motor, symbol Pc max.

maximum-C_L Value at peak of curve where $dC_L/dx = 0$, hence * angle of attack.

maximum cold thrust Highest thrust without using (available) afterburner.

maximum contingency Highest output turboshaft (rarely, other type of engine) can deliver or is authorized to deliver, usually for $2\frac{1}{2}$ min following failure of other engine of multi-engine aircraft; usually requires subsequent inspection.

maximum continuous Thrust or rpm limit available for unlimited period, usually using rich mixture if PE (4) and all augmentation for gas turbine; often available for certification or emergency only.

maximum cruise rating Highest power without augmentation or rich mixture; often this is highest available for normal continuous operation.

maximum cruising speed Highest speed permitted (usually in flight manual) for sustained operation; normally expressed as EAS or CAS, occasionally only as Mach.

maximum diving speed Highest speed demonstrated in certification (see *V_{DF}*).

maximum effort Using every aircraft that can be made serviceable, though usually without addition to list of allowable deficiencies.

maximum except take-off Highest available PE (4) power other than take-off rating.

maximum expected operating pressure Peak (or 1·1 times peak) of pressure throughout burn of solid motor.

maximum mean camber Maximum camber of median line of wing sections from root to tip.

maximum payload Limit for transport as defined in certification. For cargo aircraft usually same as *maximum structural payload*; for all-pax configuration, usually established by seat configuration at much lower value.

maximum-performance take-off That in which energy transfer is at highest possible rate, usually limited by sliding of locked tyres and avoidance of striking tail on runway or exceeding gear-down EAS.
maximum power For propulsion engine, power under ISA/SL conditions with engine operating at authorised limits of rpm, pressures and temperatures.
maximum-power altitude Lowest altitude at which full throttle is permissible (some definitions add: at max rpm for level flight); for supercharged PE (4), highest altitude at which maximum boost pressure can be maintained.
maximum q Highest dynamic pressure attainable; for given aircraft, function of dive limits, atmospheric pressure and structural strength.
maximum rpm In shaft-drive engines, invariably a transient overspeed condition permitted by lag in propeller or engine-speed control system.
maximum speed Highest TAS attainable in level flight.
maximum structural payload Combined weight of pax, baggage and cargo certificated to be carried on main deck and in belly holds.
maximum turn rate That giving maximum number of deg/s rate of change of heading, ie rotation of longitudinal axis; unrelated to change in trajectory.
maximum undistorted output Peak signal strength consistent with intelligible speech in early radio.
maximum usable frequency Highest for short-wave com between fixed points using ionospheric reflection (time-variant).
maximum vertical speed, Vm Rare flight limitation demonstrated in VTOL aircraft, including civil-certificated helicopters.
maximum weight See *MTOW*, *MLW*, *MZFW* etc.
maximum weak-mixture Highest power normally available from PE (4) throughout flight; rich-mixture ratings can reduce engine life and aircraft range.
Maxwell Non-SI unit of magnetic flux (conceptualized as "single line" of flux), = 10^{-8}Wb.
Mayday International call for urgent assistance, from French "m'aidez!" Hence, to declare a *, to go *; usually sent on 121·5 MHz.
MB 1 Magnetic bearing.
2 Marker beacon.
3 Mercury bromide (laser).
mb Millibar = 100 Nm^{-2} = $9{\cdot}87 \times 10^{-4}$ atm.
MBA Main battle area.
MBAR, M-bar Multi-beam acquisition radar.
MBC Microbiological corrosion.
MBCS Manoeuvre-boost control system; senses aircraft response and adjusts flight controls for fastest completion of demand.
MBDOE Million barrels per day oil equivalent.
MBE Molecular-beam epitaxy.
MBF Multi-body freighter.
MBFR Mutual and balanced force reductions.
MBFS Maximum boom-free speed.
MBM Magnetic bubble memory.
MBO Management by objectives.
MBP Max take-off regime (USSR, Cyrillic characters actually represent MVR).
MBPS, Mb/s Megabits/second.
MBRV Manoeuvrable ballistic re-entry vehicle.
MBS Millibars (alphanumeric readout).
MBX Management by exception.
MC 1 Maximum certificated (altitude).
2 Multi-combiner (HUD).
3 Manufacturing cost.
4 Main computer.
5 Medium case (bomb).
6 Mission-capable.
M_c Design cruise Mach number.
m/c 1 Machined.
2 Machine (colloq, arch: aeroplane).
m.c. Moving-coil.
MCA 1 Ministry of Civil Aviation.
2 Military/civil action.
3 Minimum crossing altitude.
4 ISA (USSR, Cyrillic characters actually represent MSA).
MC&G Mapping, charting and geodesy (USAF).
MCAS Marine Corps air station (US).
MCC 1 Meteorological communications centre.
2 Mid-course correction (NASA).
3 Mission control center (NASA).
4 Mobile command centre.
5 Main combustion chamber.
6 Missile control console.
7 Mission crew commander.
8 Manual control centre.
MCCS Multifunction command and control system.
MCD 1 Magnetic chip detector.
2 Marine craft detachment (RAF).
MCDP Maintenance control and display panel.
MCDU Multifunction controller/display unit; on flight deck, enables line maintenance crew to identify faulty LRU and check after replacement, without documentation.
MCE Modular control element (or equipment).
mcf Thousands (not millions) of cubic feet.
MC/FD Modular chaff/flare dispenser.
MCG Millimetre-wave contrast guidance.
M_{CG} Pitching moment about c.g.
McGee tube High-speed photo image tube, samples "sausage" of electrons slice by slice.
MCGS Microwave command-guidance system.
MCID Manufacturing change in design.
McKinnon Wood bubble Shape of plot of pressure distribution on upper surface of wing upstream of shockwave at subsonic M exceeding Mcrit.
MCL Microcomputer compiler language.
MCLOS Manual command to line of sight.
MCM 1 Mine countermeasures.
2 1,000 circular mils.
MCP 1 Maximum continuous power.
2 Maximum climb power.

3 Multi-channel plate (image intensifier).
4 Mode control panel.
5 Military construction program (USAF).
MCR Maximum cruise rating.
Mcrit Critical Mach number.
Mcrit D Mcrit at which subsonic C_D rises by 0·002 at constant angle of attack.
MCS 1 Minimum control speed.
2 Miniature control system (ATC).
3 Missile control system, on fighter.
4 Mixed-class seating.
Mc/s Megahertz.
MCT 1 Mercury cadmium telluride.
2 Maximum continuous thrust.
MCTL Militarily critical technologies list.
MCU 1 Matrix character unit (display symbols).
2 Management control unit (flight-data recorder).
3 Modular concept unit (= ⅛ ATR).
4 Marine craft unit (RAF).
MCW Modulated carrier (or continuous) wave, A2 emission.
MD 1 Manoeuvre-demand (control loop).
2 Military district.
3 Managing director (or m.d.).
M/D Miscellaneous data (display).
M_d Maximum design Mach number.
m_d Mass flow of dry gas.
MDA 1 Minimum descent altitude.
2 Multiple docking adapter.
3 Master diversion airfield.
MDAP Mutual Defense Assistance Program.
MDC 1 Minimum-displacement controller, = force-sensing controller.
2 Maintenance Data Centre (RAF).
3 Main display console.
4 Miniature (also micro, which is smaller) detonating cord.
5 Motor-driven compressor.
6 Multiple drone control.
7 Main-deck cargo (V adds volume).
MDF 1 MF D/F station (ICAO).
2 Mission degradation factor (USAF).
M_{DF} Maximum demonstrated diving flight Mach number.
MDFN Molybdenum-disulphide-filled nylon.
MDH Minimum descent height.
MDI Miss-distance indicator; hence MDIS = * system.
M_{DIV} Drag-divergence Mach number, usually same as M_{DR}.
MDLC Manoeuvring direct-lift control.
MDM Multiplexer-demultiplexer.
MDP Motor-driven pump.
MDPS Metric data-processing system.
MDR 1 Mandatory defect reporting.
2 Magnetic-field dependent resistor.
M_{DR} Drag-rise Mach number.
MDRC Material Development and Readiness Command (USA).
MDS 1 Minimum discernible signal.
2 Metal-dielectric semiconductor.
3 Matériels de servitude, ie GSE (F).
4 Mission/design/series, basic system of designators of military aircraft (US).
MDU 1 Microwave distribution unit.
2 Mine distributing unit.
MDWP Mutual Defense Weapons Program.
MD/WT Marine division/wing team (DoD).
ME 1 Manoeuvre enhancement mode.
2 Multi-engine.
3 Mission-essential.
M_e Merkel number.
m_e Electron rest-mass.
MEA 1 Minimum en-route IFR altitude.
2 Maintenance engineering analysis.
meacon To mislead enemy by receiving and instantly rebroadcasting radio-beacon signals from false positions.
Meads MEA (2) data system.
mean aerodynamic chord Chord of imaginary wing of constant section having same force vectors under all conditions as those of actual wing, $\bar{c}$ (in practice usually very close to mean chord $\bar{c}$).
mean blade width ratio Ratio of mean chord (width) of propeller blade to diameter.
mean chord Gross wing area divided by span, c.
mean day Time between successive transits of mean Sun.
mean effective pressure Mean pressure acting on piston during power stroke in Otto, diesel, two-stroke, Stirling and most other reciprocating IC engines.
mean fleet performance Mean (sometimes average) flight performance (sometimes other parameters, eg engine IFSD rate) of entire airline fleet of one type.
mean free path Mean distance point could move in straight line without collision with surrounding fluid molecule; greater than mean distance between fluid molecules.
mean geometric pitch Mean of geometric pitches of all elements from root to tip of blade, Pg.
mean line Locus of points equidistant between nearest points of upper and lower surface from LE to TE of wing profile.
mean point of impact Point whose co-ordinates are arithmetic means of co-ordinates of impacts of all weapons launched against same aiming point.
mean sea level Average height of sea surface, usually calculated from hourly tide readings (in US, measured over 19-year period).
means of compliance Methods and techniques formally demonstrated by manufacturer seeking certification of product from licensing authority.
mean solar day Period between transits of mean Sun, 24 h 3 min 56·555 s of mean sidereal time (see *sidereal*).
mean square value Arithmetic mean of squares of all values; * error equals sum of squares of errors divided by their number.

mean time between failures For specified time interval, total operating time of population of materiel divided by total number of failures within population (USAF).

mean time to repair For specified time interval, summation of active repair times divided by total number of malfunctions (USAF).

MEAS Mechanical engineering aircraft squadron (RAF).

measured performance That measured on one occasion under specified conditions; usually slightly higher than gross performance.

measured thrust That measured on one occasion under specified conditions, usually by force transducers on testbed (see *installed/net/gross thrust*).

measured thrust coefficient For solid-propellant rocket, thrust-versus-time integral over action-time interval divided by product of average throat area and integral of chamber pressure versus time over action-time interval; symbol C_f.

meatball 1 Colloquial American for any guidance reference; thus, on the *, to hack the *.
2 Accurately flown final approach.
3 Carrier optical landing aid as seen by correctly aligned pilot.

Mebul, MEBUL Multiple engine build-up list.

MEC Main engine control.

MECA Missile electronics and computer assembly.

Mecaplex Registered acrylic plastics sheet similar to Plexiglas.

mechanical de-icing Using distortion (eg rubber boot), centrifugal force or other physical method to dislodge ice.

mechanical efficiency Work delivered by a machine as percentage of work put in, difference being mainly frictional losses.

Meco Main-engine cutoff (large rocket vehicle).

MEDA Military emergency diversion airfield.

Medcat Medium-altitude CAT.

Medevac Military airlift of sick or wounded.

median lethal dose That over whole body which would be fatal to 50% of subjects.

median selection Automatic choice of mean and rejection of extreme values; can also mean majority voting.

Mediator Unsuccessful grand design for UK ATC (1) system embracing all air traffic.

medium altitude Between 2,000 ft and 25,000 ft (DoD).

medium-altitude level bombing Release between 8,000 ft and 15,000 ft (DoD).

medium-angle loft bombing Release at 35° to 75° from horizontal.

medium bird For impact or ingestion certification, one weighing 1·5 lb, 0·68 kg.

medium bomber Former category defined quite differently by different air forces, either by bomb load or range, and so developed 1920–50 as to make numerical values meaningless.

medium cloud, CM Cloud types prefixed by alto-; according to BSI with average height 8,000 ft to 20,000 ft.

medium frequency EM radiation with superimposed carrier at 300 kHz–3 MHz.

medium-range ballistic missile Operational range 600 to 1,500 nm (1,112–2,780 km); see *mid-range* ** (DoD).

medium-range transport Full-payload range 1,500 to 3,500 nm (2,780–6,486 km).

medium-scale integration Normally taken to mean 50–100 circuits per chip.

medium-scale map From 1:75,000 to 1:600,000 (DoD, IADB).

medium turn Most authorities define as bank angle 25° to 45°.

MEECN Minimum essential emergency communications network.

MEF Minimum essential facilities.

MEFC Manual emergency fuel control.

mega Prefix 10^6 (a omitted in megohm).

megaline 10^6 maxwells, $= 10^2$ Wb.

megaton, MT Explosive power equivalent to nominal 1,000,000 short tons of TNT.

megger Universal electrical continuity and resistance tester (colloq, arch).

MEI Maintenance engineering inspection.

MEIS Modular engine instrument system.

MEISR Minimum essential improvement in system reliability.

MEK Methyl ethyl ketone solvent.

MEL 1 Multi-engine licence.
2 Minimum equipment list, list of ME (3) items.

melée Confused close combat by numerous aircraft, opposite of one-on-one.

member Portion of structure bearing load.

memorandum of understanding Diplomatic agreement signed at ministerial level between governments agreed on collaborative programme, usually in advance of actual engineering.

MENS Mission element need statement (DoD).

menu Range of variable inputs from human operator to EDP, fire-control system, display or other electronic system; files, codes, information, modes, formats etc.

MEO Mass in Earth orbit.

MEOP Maximum expected operating pressure (solid motors).

MEOTBF Mean engine operating time between failures.

MEP 1 Multi-engine pilot.
2 Mission equipment package.
3 Marine environmental protection.
4 Management engineering program(me).

MEPT Multi-engine pilot trainer.

Mepu, MEPU Monofuel emergency power unit.

MER 1 Multiple ejector rack.
2 Mission evaluation room.

Mera Molecular-electronics radar.

Mercator Map projection: light at Earth centre

projects map on to cylinder wrapped round Equator.

Mercier French aerodynamicist with patents for close-fitting cowlings for radial engines and low-drag ailerons hinged diagonally across wingtip.

mercury The only metal liquid at NTP, SG 14·19, used (rarely) as vapour in closed-circuit space power and as ion beam (thrusters).

mercury battery Dry cell, 1·2 V, KOH electrolyte between mercuric oxide/graphite cathode and zinc anode.

meridian Great circle through any place on Earth and the poles.

meridian altitude Altitude of celestial body when on celestial meridian of observer (000° or 180° true).

meridian passage Time at which celestial body crosses observer's celestial meridian.

Merkel number Heat transfer equation Me = bS ‴ V/md where b is mass-transfer coefficient, S ‴ transport surface per unit volume, V volume and md mass flow of dry gas.

Merritt Island Site of largest KSC launch complexes (Saturn V vehicle).

Merto Maximum-energy rejected takeoff.

MES 1 Major equipment supplier (large projects).
2 Main engine start(ing).

Mesa Modular equipment stowage area (NASA).

MESC Mid electrical (or, UK CAA, equipment) service centre.

Mesfet Metal semiconductor field-effect transistor.

MESG Micro electrostatically suspended gyro.

mesocline Lower layer of mesosphere (1) where temperature rises to about 10°C at c60 km.

mesodecline Upper layer of mesosphere (1) where temperature falls to about -90°C at mesopause.

mesometeorology 1 Meteorology on scale between macro and micro.
2 Meteorology of mesosphere (1).

mesomorphic states See *liquid crystal.*

meson Family of elementary particles with energy 135–550 MeV and zero spin (possibly nine types, and corresponding anti-particles, but some may be resonances).

mesopause Boundary between top of mesosphere (1) and thermosphere, 80–85 km.

mesopeak Temperature at mesocline/mesodecline boundary.

mesosphere 1 Thermal region of atmosphere between stratopause (c30 km) and mesopause (Chapman gives lower limit as 32 km).
2 Rarely, according to Wares, region between top of ionosphere (c400 km) and supposed bottom of exosphere (500–1,000 km).

MET 1 Mission event timer.
2 Meteorological broadcast service.

metacentre Intersection of line of buoyancy (of marine aircraft) and axis of symmetry.

metacentric height Vertical distance from metacentre to c.g.

Metalastik Patented family of metal/rubber bonded devices, mainly anti-vibration.

metal deactivator Hydrocarbon fuel additive to reduce possibility of electrochemical action in tank or system.

metallic fuel additive 1 Various blends of high-energy (so-called "zip") turbine fuel incorporate boranes and metallic compounds.
2 Most high-energy solid rocket fuels incorporate finely divided aluminium powder.

metallics See *metal matrix composites.*

metallization Vacuum deposition of metal conductive paths in planar IC or other solid-state device.

metallizing Vast range of techniques involve coating with metal, usually by plasma-spraying with finely divided metal particles; typical example is electrical bonding of modern airframe, in which most major parts are isolated by non-conductive corrosion protection.

metal matrix composite Various structural or refractory composites have a metal matrix, the most common being silica-reinforced aluminium, boron/aluminium and boron/titanium.

metamic Cermet (suggest undesirable).

Metar, METAR Meteorological aeronautical radio code.

metastable Pseudo-equilibrium needing small external disturbance to trigger violent change.

meteor Solid body in free fall through Earth's atmosphere; also called *-oid.

meteor bumper Thin shield designed to protect spacecraft from penetration by meteors.

meteorite Meteor large enough to strike Earth's surface.

meteorograph Instrument for measuring and recording two or more meteorological measures, eg T, pressure, humidity.

meteorological rocket Ballistic vehicle to loft payload to great height, there to free-fall or descend by ballute or parachute.

meteorological satellite One designed to assist understanding of atmosphere and preparation/dissemination of actuals or forecasts.

meteorological wind Forecast wind.

meteorology Science of Earth's atmosphere (abbn met).

metering orifice Calibrated nozzle or tube passing exactly known flow at given pressure difference.

methane Simplest hydrocarbon fuel CH_4; gas at NTP but cryogenic liquid.

methanol Methyl alcohol, CH_3OH, used in special blends of fuel for racing PE (4) and as additive to water as anti-knock power boost fluid.

method of least work Formula for analysis of statically indeterminate structures.

methyl alcohol See *methanol.*

methyl bromide CH_3Br, common filling for hand (and some other) fire extinguishers.

methyl cellosolve Trade name for various strip-

pers, esp methylene chloride, CH_2Cl_2.
Meto Maximum except take-off power.
metre Fundamental SI unit of length, abbn m, 1,650,763·73 wavelengths in vacuum of radiation from transition $2p_{10}$ and $5d_5$ of Kr-86 atom.
metric Usually means MKS; aerospace gradually standardising on SI units.
metro Increasingly popular word meaning urban; applied to "downtown" air services, commuter routes and aircraft.
metrology Science of precise measurement, esp of linear dimensions.
metroplex routes Selected recommended high-altitude IFR inter-city (US).
METS Mobile engine test stand.
MeV Mega-electron-volts, eV × 10^6.
MEW 1 Manufactured empty weight.
2 Microwave early warning.
MEWS 1 Modular EW simulator.
2 Microwave EW system.
3 Missile early-warning station.
MEX Microelectronics.
MEXE Military Engineering Experimental Establishment (UK).
MEZ Missile (esp AAM) engagement zone.
MF 1 Medium frequency (see *frequency*).
2 Main frame.
3 Main force.
4 Major field.
M^2F^2 Multimode fire and forget.
MFA Multi-furnace assembly.
MFAR Multi-function array radar.
MFBF Multifunction bomb fuze.
MFC 1 Main fuel controller.
2 Maximum fuel capacity.
M_{FC} Maximum Mach number for satisfactory flight control and stability.
MFCD Modular flare/chaff dispenser.
MFD Multi-function display.
mfd Microfarad.
MFG Miniature flex gyro.
MFHBF Mean flight-hours between failures.
MFK Multifunction keyboard.
MFL Minimum field length, TO distance to clear standard (35 or 50 ft) screen.
MFLI Magnetic fluid-level indicator.
MFMA Multi-function microwave aperture.
MFO Multinational force and observers.
MFP 1 Mean free path.
2 Main fuel pump.
MFR Multifunction radar.
Mfr Manufacturer.
MFSK Multiple-frequency shift-keying.
MFV Main fuel valve.
MG 1 Master gauge.
2 Machine gun (G).
Mg 1 Magnesium.
2 Megagramme, ie 1,000 kg or 1 tonne.
mg Milligramme.
MGA Middle gimbal angle.
MGCS 1 Missile guidance and control system.
2 Mobile ground control station.
MGF Metallized glass-fibre (chaff).
MgF_2 Magnesium fluoride (IR seeker domes).
MGGB Modular guided glide bomb.
MGOS Metal glass oxide silicon.
MGS 1 Mobile ground station (RPV).
2 Minimum groundspeed system.
MGT 1 Motor gas temperature (solid rocket).
2 Module ground terminal.
MGTOW See *MTOW*.
MGTP Main-gear touchdown point.
MGU Midcourse guidance unit.
MGVF Ministry of civil aviation (USSR).
MGW Maximum gross weight (MTOW and MRW are more explicit).
MH NDB, less than 50 W.
mH Millihenry.
MHA 1 Minimum holding altitude.
2 Maintenance hazard analysis.
MHD Magnetohydrodynamic(s).
MHDF Co-located MF and HF D/F (ICAO).
MHVDF Co-located MF, HF and VHF D/F (ICAO).
mho Unit of conductance, reciprocal ohm.
mhp Muzzle horsepower.
MHR Monopropellant hydrazine rocket.
MHRS Magnetic heading reference system.
MHz Megahertz.
MI 1 Model improvement.
2 Medium intensity.
3 Miles (not recommended).
mi Miles.
m.i., M/I Minimum impulse.
MIA Missing in action.
MIB Minimum-impulse bit (rocket).
MIC 1 Mineral insulated cable.
2 Microwave integrated circuit.
mic Microphone.
Micap Mission-capable.
Micarta Trade name for phenolic insulator and small-part material made from resin-impregnated cloth.
Mice Microwave integrated checkout equipment.
mice Small inserts fitted by hand to tailor cross-section area of fluid flow path, esp gas-turbine jetpipe nozzle.
Michigan height That at which an air-dropped store, especially nuclear, is activated by its radar altimeter.
Mickey H_2X (colloq).
MICNS Modular integrated communications and navigation subsystem (Sotas).
MICOM, Micom Missile Command (USA).
micrad Microwave radiometer.
micro Prefix one-millionth, 10^{-6}.
micro-adjuster Small permanent magnets arranged to correct magnetic compass for coefficients P, Q, Cz, Fz.
microballoons Microscopic hollow spheres used to increase bulk of fillers and potting compounds.

microbarograph Records very small transient changes in atmospheric pressure at ground station.
microbiological corrosion Eating away by micro-organisms living at fuel/air interfaces and on surface of virtually all aeronautical materials.
microburst Most lethal form of vertical gust, in which core up to 2.5 km (1.5 miles) diameter forms vertical jet below convective cloud with downward velocity up to 20 m/s (4,000 ft/min), an almost instantaneous velocity difference of 80 kt, down to very low levels.
microcircuit Basic element of microelectronics, with numerous fabricated on each epitaxial chip.
microfiche Common system of storing written or visual information with 24 to 288 microfilmed pages on each 100 mm × 150 mm (or 4 in × 6 in) sheet of film.
microinch One millionth of an inch.
micro level Critical final design of software.
microlight Aeroplane with empty weight not greater than 150 kg (330·7 lb); US term ultra-light, limit 115 kg (254 lb), under FAR Pt 103.
micrometeorite Microscopic solid particle, many species throughout explored space.
micrometer Instrument for precise dimensional measurement, usually mechanical but some-times based on fluid escape through small clearance.
micron One-millionth of metre, 10^{-6}m, symbol μ (incorrectly, μ m).
micronavigation Guidance of aircraft or missile in relation to nearby target (usually air/ground).
micronic Concerned with micron dimensions, hence * filter.
micronlitre Quantity of gas: 1 litre at pressure 1μHg.
microprobe Instrument for investigating regions of micron size, hence laser-pulse * analyser.
microprogram Small sub-program for defining computer instructions in terms of other basic elemental operations.
microshaving Precision-machining heads of driven countersunk rivets.
microstrip Microwave guide comprising conduc-tor supported above ground plane, generally fabricated by printed circuitry.
microswitch Miniature electric switch governing system(s) according to external movement such as compression of MLG.
microtacan Microelectronic Tacan.
Microvision All-weather landing aid (Bendix) with pulsed beacons along runway energizing pattern on HUD.
microwaves Electromagnetic radiation between RF and far-IR, normally 1–300 GHz.
MICS Manned interactive control station.
Midas 1 Missile-defense alarm system (US).
2 Manufacturing information distribution and acquisition system.
midcourse 1 Space trajectory linking departure from neighbourhood of Earth and arrival near destination, corresponding to atmospheric-flight cruise; hence * guidance.
2 From missile boost burnout to start of termi-nal homing.
midcourse guidance Applies to midcourse (1), sometimes to (2).
midcruise weight Gross weight at midpoint of mission.
middle airspace Several national definitions, eg FL 180–290, FL 145–250.
Middleman Air-intercept code: VHF or UHF radio relay (DoD).
middle marker ILS marker on extended runway centreline at ILS Point B, usually 1,067 m (3,500 ft) from threshold.
mid-flap Intermediate portion of triple-slotted flap.
mid-high wing Set about three-quarters way up fuselage.
Midnight Air-intercept code: "Change from close to broadcast control" (DoD).
MIDR Maintenance (or mandatory) incident and defect report.
mid-range ballistic missile One having range of 500–3,000 nm (927–5,560 km); see *medium-range* * (DoD).
MIDS Management information and decision support.
Mids Multifunctional (or multiuser) information distribution system (NATO).
Midst Multiple interferometer determination of trajectories.
MIDU Missile-ignition delay unit.
mid-value logic Redundant system having odd number of active channels wherein system out-put is always that of intermediate-value channel, thus eliminating wild values.
mid-wing Set about half-way up fuselage.
MIE 1 Manoeuvre-induced error.
2 Managing (or management) in inflationary economy.
MIES Multi-imagery exploitation system.
Mifass Marine integrated fire and air support system.
MIFF Anti-tank mine dispensed from MW-1 sys-tem (G).
MIFG Shallow fog (ICAO).
Mifir, MIFIR Microwave instantaneous-frequency indication receiver (ECM).
Mig, MIG Miniature integrating gyro.
MiGCAP Combat air patrol directed specifically against MiG aircraft, hence MiG screen, etc.
MiG magnet Photoflash (US, colloq).
migration Movement at molecular level of one solid into another, eg of vinyl plasticizer into polyethylene core of electrical cable.
Migrator Microwave guidance radar beacon interrogator.
mike mike Millimetres.
MIL Military (ICAO, US); hence * specifica-

tions, common to all US armed services; * rating, high-power engine ratings, usually close to max-cold or Meto.

mil 1 One-thousandth of inch, 0·0254 mm; hence circular * = area of 1 mil circle = 506·7 μ^2.
2 Angular measure, $\frac{1}{6,400}$ of circle = 3·375′.
3 Military (UK, CAA).

Milcon Military construction (USAF).

MILD Magnetic-intrusion line detector.

mile Imperial unit of length, about 1,609 m; aviation more commonly uses nautical mile, about 1,853 m, except when reporting visibility.

Miles Multiple integrated laser engagement system.

military aircraft Any operated by armed service, legal or insurrectionary, no matter what aircraft type; combat aircraft retired and privately owned is not military (see *paramilitary*).

military power Normally maximum cold (unaugmented) thrust; for PE (4) METO.

military productivity Several forms, including flying rate per squadron or wing, aircraft utilization, ordnance work rates etc.

military qualification test, MQT Final hurdle hardware must pass before entering US military service; schedule varies depending on item.

mill 1 To machine fixed workpiece with rotary cutter.
2 Machine tool, often very large, for this purpose; hence skin *, spar *.

milled-block circuit Circuit (eg fluidic) machined from 3-D block (rarely, moulded in plastics).

Mill file Single-cut tapered hand file for fine metalworking.

milli Prefix one-thousandth, 10^{-3}.

milligal Former unit of gravitational attraction, = 10^{-5} ms^{-2} = 10 g.u.

millimetric Concerned with magnitudes of the order of one millimetre; hence * waves, EM radiation of this order of wavelength.

milling machine See *mill (2)*.

MIL-1553B Standard requirements for airborne digital data bus (US, now Int).

MILS, M-ILS See *MLS*.

Milsatcom Military satellite communications.

Milstamp Military standard transportation and movement procedure (DoD).

Milstar Military strategic and tactical relay.

MIL-STD Military standard(s) (US).

Milstrip Military standard requisitioning and issue procedure (DoD).

Miltracs Military air traffic control system(s).

Milvan Military-owned ISO container.

MIMD Multiple instruction, multiple data stream.

min 1 Minimum.
2 Minute (of time; abbreviation for minute of angular measure is ′).

M_{ind} Indicated Mach number.

mine countermeasures All methods of reducing damage or danger from land or sea mines.

mineral oil Lubricating oil of mineral (normally kerosine) base, thus not a modern synthetic turbine oil nor vegetable oil such as castor oil used in rotary engines; dyed red.

mineral wool Fibrous heat insulator made by blowing steam through molten slag with additives.

minfap Minimum-facilities project; Eurocontrol, Maastricht.

miniature flex gyro DTG is mounted on flex shaft with opposing strut producing destabilizing force in proportion to rotor angular displacement. At tuned speed (= f [N]^2) spring force and torques cancel.

Minigun GE multi-barrel belt-fed weapons of rifle calibre, usually 7·62 mm or 5·56 mm.

minimal flight path That for shortest time en route (ASCC).

minimally manned Aircraft available for flight as RPV but sometimes flown with safety pilot.

minimum airplane Concept, most active in US, of smallest/lightest/simplest aeroplane. By 1985 being accepted as MTOW not exceeding 100 kg, or 120 kg for two-seater.

minimum altitude Normally undefined except in same terms as lo.

minimum-altitude bombing Horizontal or glide bombing (ie not toss) with release height below 900 ft (274 m) (DoD, IADB).

minimum-control speed V_{mc}, lowest IAS at which aeroplane can always be flown safely (eg after sudden worst-case engine failure); specified as such in flight manual.

minimum crossing altitude Lowest altitudes at certain radio fixes at which aircraft may cross en route to higher IFR MEA (FAA).

minimum decision altitude Minimum descent altitude (NESN, NFSN).

minimum descent altitude 1 MDA, lowest altitude, expressed in feet above MSL, to which descent is authorized on final approach or during circle-to-land manoeuvre in execution of standard instrument approach where no electronic glideslope is provided such as ILS, PAR (FAA).
2 Decision height (UK and others).

minimum descent height, MDH Height above touchdown, at which pilot must either see to land or initiate overshoot, based entirely on topography and characteristics (including sink) of aircraft. Not used in US.

minimum design weight Not normally used; existing definitions are generally similar to operating weight.

minimum-energy orbit See *Hohmann orbit*.

minimum en route altitude That between radio fixes which meets obstruction clearances and assures good radio reception or navaid signal coverage; MEAs apply to entire width of all airways or other direct routes (FAA).

minimum equipment item Device whose failure does not delay departure; allowable deficiency, despatch deviation.

minimum flying speed Lowest TAS at which aeroplane or autogyro can maintain height; often well below V_{mc} or angle of attack at which operative stall-warning system would trigger.

minimum fuel 1 Smallest quantity of fuel with which aircraft may be authorized to fly.
2 Smallest quantity of fuel with which c.g. can fall within permitted range, normally sufficient for 30 min at max continuous power.
3 Lowest quantity of fuel necessary to assure safe landing in sequence with other traffic without ATC priority (USAF).

minimum gliding speed Lowest TAS at which aircraft can fly without propulsion; below V_{mc}, V_{mc} or best-range gliding speed.

minimum groundspeed system Subsystem in AFCS which continuously calculates correct approach speed using TAS, G/S and W/V (entered into FMS by pilot) to protect against windshear.

minimum holding altitude Lowest altitude prescribed for holding pattern which complies with obstruction clearance and assures good radio/navaid signal reception.

minimum human force At least three sets of measures in use as basis for aircraft/spacecraft design, factored to allow for injury, fatigue, g, anoxia and other influences.

minimum line of detection Arbitrary line at which hostile aircraft must be detected if defending interceptors are to destroy them before they reach vital area; usually same at ATC (1) line.

minimum line of interception Arbitrary line at which hostile aircraft should be intercepted by aircraft if friendly AAA and SAMs are to destroy all objects not thus intercepted.

minimum military requirement Specification or description of infrastructure designed to meet immediate or obvious future need and no more, on "no frills" basis (NATO).

minimum navigation performance specification, MNPS Effective from October 1978 over ocean areas FL275–400, calling for certain standards of navigation and close adherence to flightplan, and permitting major reductions in separation.

minimum normal burst altitude Height AGL below which air-defence nuclear warheads are not normally detonated (DoD).

minimum obstruction clearance altitude Specified altitude between radio fixes on VOR/LF airways, off-airway routes or segments, which meets obstruction clearances and ensures acceptable signal coverage only within 22 nm of each VOR (FAA).

minimum reception altitude Lowest altitude required to receive adequate signals to determine VOR/Tacan/Vortac fixes (FAA).

minimum rpm Engine rotational speed normally governed; with throttle on rearmost stop the speed depends on a governor, IAS and possibly other factors, such as MLG microswitch allowing a ground condition with superfine propeller pitch.

minimum runway Not defined, other than a limiting runway for particular aircraft or mission, with regard to weather, aids and surrounding terrain.

minimum safe altitude warning, MSAW Included in ARTS, monitors all controlled aircraft and alerts controller of potentially unsafe situations, usually 100 ft below MDA.

minimum safe distance Sum of radius of safety and buffer distance.

minimum speed See *minimum flying speed.*

minimum TAS Corresponds to IAS for minimum flying speed, and below MSL or on cold day can be lower.

minimum vectoring altitude Lowest altitude, expressed in feet AMSL, to which aircraft may be vectored by radar controller.

minimum warning time Sum of personnel and system reaction times.

Minitat Family of helicopter armament systems, from minimum tactical aircraft turret.

Minitrack Satellite-tracking system comprising array of surface receiver aerials forming interferometer tracking beacon in satellite.

Mintech Ministry of Technology (UK, obs).

minute 1 One-sixtieth of hour; abb min.
2 One-sixtieth of degree of angular measure, ′.

Mio Million (G).

MIP Missile impact predictor.

MIPB Mono-isopropyl biphenyl.

MIPI Material in process inventory.

MIPR Military interdepartmental purchase request (US).

MIPS Maintenance information planning system.

Mips Million instructions per second.

MIR Multiple-target instrumentation radar.

Mira Miniature infra-red alarm.

Miradcom Missile R&D Command (USA).

Mirage Microelectronic indicator for radar ground equipment.

Miran Missile ranging system interrogating at 600 MHz with beacon reply at 580 MHz.

Mirc, MIRC Missile in-range computer.

MIRL Medium-intensity runway-edge lights.

Mirls Miniature IR linescanner.

mirror sight Optical device to assist fixed-wing recovery on aircraft carrier, giving pitch-stabilized light indications to approaching pilot.

Mirv Multiple independently targeted re-entry vehicles, hence to * an existing missile, mirved warheads, * Salt limits.

MIS 1 Management information system.
2 Missing (ICAO).

MISDS Multiple instruction, single data stream.

mismatch 1 Inability of gas-turbine (invariably supersonic) inlet system to supply engine with

correct airflow in yawed flight or other disturbed conditions.
2 Upper and lower dies not perfectly aligned in closed die forging.
miss-distance scorer Indicates minimum passing distance between munition and target but does not give any co-ordinate values (ASCC).
missed approach 1 One aborted for any reason, followed by a go-around (overshoot).
2 Standard flight procedure to be followed in (1).
missile No existing definition satisfactory; general assumption is flight wholly or mainly through atmosphere, with/without propulsion, with/without guidance, with one or more warheads.
missile free Usually voice command, authority to launch AAM unless target is identified as friendly.
missile range Ambiguous, can mean range (1) or (2).
missilry Technology of (assumed guided) missiles (colloq).
mission 1 Single military operation flown by assigned force of aircraft.
2 Sortie, ie military operation flown by one aircraft.
3 Rarely (incorrectly) special flight of non-military nature.
4 Basic function or capability of missile or rocket, as shown by MDS designator (DoD).
mission-adaptive wing One whose section profile varies automatically to suit the requirements of each flight condition.
mission-capability rate Time out of each 24 h period that aircraft is available to perform its mission, expressed as percentage.
mission degradation factor Variables affecting mission, eg abort, nav error, misidentification of target etc.
mission equipment See *role equipment*.
mission-oriented items Those required following numerical assessment of enemy capabilities or targets.
mission profile Graphical or written plot of flight level from start to finish of mission, usually a succession of lo and/or hi.
mission radius Practical radius of action for aircraft equipped and loaded for mission on given profile, with allowances varying in peace or war; invariably less than half range with same load.
mission review report Intelligence report containing information on all targets covered by one reconnaissance sortie.
mission specialist Engineer or mathematician responsible for planning spaceflight, assigning payloads and assisting integration.
mist 1 Visibility reduced by water droplets to 1–10 km.
2 Mixture of gas (invariably air) and finely divided liquid (usually an oil).
3 Definition "popular expression for drizzle" is erroneous.
misting 1 Obscuration of transparency caused by condensed water droplets.
2 Tendency of fuel to form easily ignited dispersion in crash.
Mists Modular(ized) IR transmitting set.
MIT Massachusetts Institute of Technology.
MITI Ministry of International Trade and Industry (J).
MITO Minimum-interval take-off.
MIU Missile interface unit.
mix See *mixer*.
mixed-flow compressor Axial followed by centrifugal.
mixed mode Combining two EM-wave propagation modes.
mixed-power aircraft Equipped with more than one species of propulsion, notably turbojet and rocket.
mixed propellant See *dual propellant*.
mixed traffic Contrasting types of aircraft (eg jets and lightplanes) under control or otherwise in same airspace, esp on approach.
mixer 1 In communications radio, first detector in superhet receiver, which combines received signal with locally generated oscillation to yield intermediate frequency.
2 In AFCS, pitch/roll proportioning device, either mechanical or electronic, for supplying required signals to different surface power units.
mixing box Mixer (2), occasionally purely mechanical for translating flight-control input into required surface deflections on two axes.
mixture Air/fuel vapour suitable for piston engine, other than direct-injection types; hence * control, * ratio, etc.
MJP Magnetron jamming pod.
MJS mission Mars, Jupiter, Saturn mission; hence MJSU also includes Uranus.
MK Machine cannon (G).
MKC Multiple-kill capability.
Mkr Fan marker.
MKS Metre/kilogramme/second; outmoded "engineering" system of metric units.
MKSA Metre/kilogramme/second/ampere.
ml Millilitre.
MLA Manoeuvre load alleviation.
MLB 1 Multi-layer board.
2 Main-lobe blanking (ECM).
MLBM Modern large ballistic missile.
MLC Main-lobe clutter.
MLD Maintenance logic diagram.
MLE Missile launch envelope.
MLG Main landing gear; sometimes m.l.g.
MLMS Multipurpose lightweight missile system.
MLP Multi-level pegging.
MLRS Multiple launch rocket system.
MLS Microwave landing system.
MLT Munitions lift trailer.
MLU, MLUD Mid-life update.

MLV Mobile launch vehicle.
ML/V Memory loader/verifier.
MLW Maximum permitted landing weight.
MM 1 Rare prefix mega-mega (terra), 10^{12}.
2 Middle marker (ICAO).
3 Missile prefix mer/mer (sea/sea) (F).
mm Millimetre.
MMA Multimission aircraft.
MMACS Multimission aft crew station; cockpit occupied by either crew member or avionics.
MM&T Manufacturing methods and technology.
M-marker Beacon with Morse-coded emission (obs).
MMC 1 Metallic-matrix composite.
2 Monitor Mach computer.
MMD Master monitor display.
MMEL Master minimum equipment list.
MMF, mmf Magnetomotive force.
mmf Micro-microfarad = picofarad, pf.
MMFF Multimode fire and forget.
MMH 1 Monomethyl hydrazine.
2 Maintenance man-hour.
MMI 1 Mandatory modification and inspection.
2 Man/machine interface.
MMIC 1 Millimetre-wave integrated circuit.
2 Monolithic microwave integrated circuit.
MMMMM Sudden-change special weather report (ICAO).
MMO Main meteorological office.
M_{MO} Maximum operating Mach number.
MMP Metallic materials processor (company category).
MMPM MEECN message processing mode.
MMR 1 Machmeter reading.
2 Minimum military requirement.
MMRPV Multi-mission RPV.
MMS 1 Missile-management system.
2 Mast-mounted sight (or sensor).
3 Multi-mission modular spacecraft.
MMT Multiple-mirror telescope.
MMU Manned manoeuvring unit (gas-powered flight control system for Shuttle astronauts).
MMVX Medium-speed support aircraft (project).
MMW Millimetre (or millimetric) wave (seeker).
M_n Mach number.
m_n Neutron rest-mass.
MNC Major NATO command.
M_{NE} Mach number never to be exceeded, and not even approached in normal flying, but demonstrated in civil certification.
mnemonic Easily remembered sequence to assist aircrew with their pre-flight and other checks, eg HTMPFFG = hood/harness, trim, mixture, pitch, fuel, flaps, gills/gyro (largely arch except in general aviation).
MNFP Multinational fighter program (US).
Mnm Minimum (ICAO).
MNO Ministry of national defence (Czech).
M_{NO} Normal operating Mach number, now generally replaced by M_{MO}.
MNOS MOS with silicon nitride on oxide.
MNPA Minimum navigation performance airspace, allowing 60 nm lateral separation at FL275–400.
MNPS Minimum navigation performance specification.
m.o. Master oscillator.
MOA 1 Memorandum of agreement.
2 Military operations area.
MoA Ministry of Aviation (UK obs).
MoAS Ministry of Aviation Supply (UK obs).
MOB Main operational base (USAF).
MOBA Mobility operations for built-up areas (USA).
Mobidic Modular bird with dispensing container (MBB/Aérospatiale).
mobile air movements team Air force team trained for deployment on air-movement traffic duties.
mobile lounge Lounge which also conveys passengers between terminal and aircraft and vice versa (eg Washington Dulles Airport).
mobile quarantine facility Set up on vehicle, usually helicopter-capable ship, for receiving special cargo, eg lunar rock.
MOBSS Mobility support squadron.
MOC 1 Maintenance operational check.
2 Minimum operational characteristics.
MOCA Minimum obstruction-clearance altitude.
mock-up Quickly built replica of aircraft or other product, usually full scale, to solve various problems (see *customer* *, *engineering* *, *furnishing* *, *hard* *).
MOCM Multispectral ocean colour monitor.
MOCN NC machine tool (F).
MOCVD Metal organic chemical vapour deposition.
mod 1 Modification.
2 Modulation.
MoD Ministry of Defence.
Modas Modular data-acquisition system.
mode 1 Any of selectable methods of operation of device or system.
2 Number or letter referring to specific pulse spacing of signal transmitted by interrogator (IFF, SIF, SSR).
3 Each possible configuration of spatial variable, eg EM wave, flutter or aerodynamic phenomena.
model Ambiguously used in aviation:
1 Small-scale replica for testing characteristics of full-size aircraft.
2 In general aviation, improved version marketed at (often annual) intervals.
3 To reproduce a functioning system synthetically, eg in EDP program.
model atmosphere Mathematically exact numerical values closely approximating to an idealized real atmosphere.
model basin See *towing tank*.

model cart Often large and elaborate wheeled truck on which are mounted a balance or sting and aircraft model, the whole then installed in tunnel working section with large number of electrical and manometric connections.

model qualification test US military clearance of new item, esp engine, for production.

model tank 1 See *towing tank.*
2 Tank filled with fluid for flow exploration round model by electrical potential analogy.

model tunnel Ambiguous, term best reserved for any small-scale model of a future large wind tunnel.

modem Telecommunications or EDP (1) device: modulator + demodulator.

moderator In nuclear reactor, substance specifically present to slow down neutrons by collisions with nuclei.

modex Three-numeral designator for quick identification of individual aircraft within CAG or other unit (USN).

modification Temporary or permanent change to either single aircraft (for particular purpose) or all aircraft of type (to rectify fault or shortcoming or offer improved capability); can be unique, to owner's requirement, or one of planned and controlled series throughout life of aircraft.

modified close control Interceptor is told only target-position information (USAF).

modified mission MDS letter added to left of basic mission letter to indicate a permanent alteration to basic mission (DoD).

modified PAR Precision approach guidance for high-performance aircraft to flare point instead of to touchdown point (USAF).

Modils Modular ILS.

Modmor Commercially produced (Morgan Crucible) family of carbon/graphite fibres.

MoD(PE) MoD Procurement Executive (UK).

modular Designed in discrete series of major components for ease of inspection, overhaul or, esp, repair by module replacement; engine may comprise fan, LP compressor, HP compressor, combustor, turbine and accessories modules, while major avionic system may be series of * boxes.

modular lounge Airport lounge built as unit capable of (1) being moved from place to place to add capacity where required or (2) being driven to/from aircraft (mobile lounge).

modulated augmentation Afterburner (reheat) with fuel flow continuously variable to give smooth increase in thrust from max cold to max augmented.

modulated waves Electromagnetic waves on which is impressed information in form of variation in amplitude (AM) or frequency (FM).

module 1 One of assemblies of modular system, easily replaced but seldom itself torn down except at major base or by manufacturer; eg spacecraft command *, engine fan *.
2 Single box of electronic equipment replaceable as plug-in unit.
3 Any standard dimensions or standard size of container.
4 Standard-capacity building block of computer memory.

modulus of elasticity Ratio of unit stress to unit deformation of structural material stressed below elastic limit (limit of proportionality).

modulus of rigidity Shearing modulus of elasticity, proportional to angle of distortion but measured in stress per unit area.

modulus of rupture In beam loaded in bending to failing point $S_M = Mc/I$; in shaft or tube failing in torsion $S_S = T_c/J$.

Moffett Moffett Field, Calif, home of NACA lab, now known as NASA Ames Research Center.

Mogas Motor gasoline, ie ordinary automotive petrol. Usually 91 or 93 octane.

MOH Major overhaul.

Mohol Mix of Mogas and alcohol.

mold Mould (US spelling).

moleskin strip Soft strip around bonedome to eliminate scratching canopy.

Mollier diagram Plot of enthalpy against entropy.

Molochite Precision casting mould material, a Ca-Mg-Al silicate.

moly Colloquial for molybdenum disulphide lubricant.

molybdenum Hard white metal, important alloying element in steels and in "moly" (disulphide) lubricants.

MOM 1 Metal-oxide-metal.
2 Methanol/oxygen mix.

moment Turning effect about an axis: force multiplied by perpendicular distance from axis to force.

moment arm Distance from axis to force, eg from c.g. to aerodynamic centre of tailplane.

moment coefficient 1 Any moment reduced to non-dimensional form, usually by dividing by dynamic pressure.
2 Particular values such as cm (section *), C_M (pitching *), C_{Mac} (about aerodynamic centre) and C_{MCG} (about c.g.).

moment distribution Analytical method for frameworks based on bending moments at rigidly attached joints between members.

moment index Divisor, typically 10^4, to reduce numerical values in balance calculations of heavy or large aircraft.

moment of area Not normally used, though term "second ***" synonymous with moment of inertia.

moment of inertia Sum of all values of mr^2 where m is elementary particle and r its distance from axis about which I (***) is measured, or Mk^2 where M is total mass and k is radius of gyration.

moment of momentum See *angular momentum.*

momentum drag Drag due to change in momentum of air entering lateral or other non-ram inlet;

major factor in design of ACVs and small high-power jet-lift V/STOLs.
momentum separation Flow separation from convex duct wall or other surface where Coanda effect is inadequate.
momentum theory Idealized Froude treatment for propellers and other driven rotors by calculating momentum ahead of and behind disc exerting uniform pressure on uniform flow.
momentum thickness Measure of reduction in momentum of flow due to viscous forces in boundary layer.
momentum-transfer method Drag measurement by pitot traverse of wake.
Moms, MOMS Modular opto-electronic multi-spectral scanner.
MON 1 Above mountains (ICAO).
2 Mixed oxides of nitrogen.
Mon Monitor (NASA).
Mona Advanced hyperbolic/computer/display airborne navigator system (from modular navigation).
Monab Mobile naval airbase, on shore (USN).
Monel Range of corrosion-resistant Ni-Cu alloys.
monergol Hydrazine (F), often capital M.
Monica Large series of active radars used by RAF 1943–46 for rear warning of interception.
monitored system One continuously subjected to surveillance and if necessary corrective action (eg isolating failed channel) by separate avionic subsystem; feature of autoland and other highly reliable systems.
monoball mount Primary structural joint, eg pylon to wing, where loads are transmitted radially across spherical surface.
monobloc 1 Shaped from single block, usually by forging or casting.
2 Control surface in form of single aerofoil, not hinged downstream of fixed fin or tailplane; also called slab type or all-moving.
monocoque Three-dimensional form, such as fuselage, having all strength in skin and immediate underlying frames and stringers, with no interior structure or bracing. Some purists argue there must be skin only.
monofuel Fuel used alone without need for air or other oxidant; eg HTP, MMH.
monohud HUD for one eye only.
monolithic rocket Having solid grain cast or extruded in one piece.
monomer Substance consisting of single molecules, esp one which can be polymerized.
monomethyl hydrazine Rocket propellant; hydrazine in which one hydrogen atom is replaced by methyl group $N_2H_3.CH_3$.
monoplane Aerodyne having single wing.
monopropellant Monofuel used in liquid rocket, eg HTP.
monopulse Radar technique using four overlapping pencil beams, two for azimuth, two for elevation, with circuitry so arranged that, when target is at centre, output voltage vanishes.
monorail Installed along ceiling of paratroop transport, to which static lines are hooked.
monsoon 1 Winds which reverse direction seasonally.
2 In India, seasonal rains.
mooring band Reinforces upper part of envelope in line with mooring lines.
mooring drag See *trail rope*.
mooring guy Rope or cable for securing aerostat.
mooring line See *mooring guy*.
mooring mast Rigid or cable-braced high mast to which airship can be fastened and up/down which all supplies and people can pass.
Moose Manned orbital-operations safety equipment.
MOP Ministry of public works (Spanish-speaking countries).
MOP costs Maintenance and overhaul personnel costs.
Mops, MOPS 1 Minimum operational performance standards.
2 Million operations per second.
Moptar Multiple-object phase tracking and ranging; short-baseline CW phase-comparison.
MOR 1 Mandatory occurrence reporting (or report).
2 MF omni-range (365–415 kHz).
3 Military operational requirement(s).
Morag MOR (1) advisory group.
Morass Modern ramjet systems synthesis.
Morrison shelter Mild-steel "table shelter," able to support collapsed house; issued to British civilians 1942–44.
mortar Large-calibre low-velocity gun for projecting payloads, eg Shuttle drag chute.
MOS Metal oxide silicon, family of semiconductor devices.
MoS Ministry of Supply (UK obs).
mosaic Large assemblage of mating aerial photographs, from optical camera, vidicon, IR or other source, with adjacent boundaries accurately aligned, to cover whole area under surveillance (military reconnaissance or photogrammetric mapping and surveying).
Mosdap MOS dynamic analysis program.
MOSFET, Mosfet Metal oxide silicon, field-effect transistor.
Most, MOST Metal oxide silicon (or semiconductor) transistor.
MOS-VAO All-union association for experimental marine aircraft (USSR).
MOT Maximum overhaul time(s).
MoT Ministry of Transport (many countries).
Mote, MOT&E Multinational operational test and evaluation.
moteur canon Patented (Hispano Suiza) installation of cannon between banks of V-12 engine firing through geared propeller hub.
mother Used in parental capacity, such as RPV

control aircraft (* ship) or * printed-circuit board.

Motne Meteorological operational telecommunications network in Europe.

motor Best restricted to solid rockets and hydraulic/electrical/air machines, though several non-English languages have to use it for most or all main propulsion. Acceptable as adjective = powered, eg * glider.

motorboating Oscillations at sub-audio frequency in any device.

motor glider Ultra-light aeroplane derived from glider; best not used for powered or assisted sailplanes which are competition sailplanes with auxiliary propulsion (retractable, if propeller-type) used intermittently.

motoring Rotating main engine by means of starter for purpose other than starting (usually gas turbine).

motoriste Engine manufacturer (F).

motor specific impulse For solid-propellant rocket, total impulse divided by total motor loaded weight, symbol I_{Ta}, unit kNs or MNs (in US, lbf-s).

MOTR Multi-object tracking radar.

MOTU Maritime operational training unit.

MOTV Manned orbital transfer vehicle.

MOU Memorandum of understanding.

mould 1 3-D template.
2 3-D reference tool for fabrication of composite parts or thermosetting sheet (eg canopy).
3 Enclosed cavity for cast parts.

mould dimension Any dimension to or along a mould line.

mould line Line formed by intersection of two surfaces, eg sheet-metal parts.

mounting Vague term acceptable for most objects hung on vehicle but not for landing gear or strut attachments. Includes all linkage of structural nature and any ad hoc parts of airframe such as pylon, but not control or accessory systems.

mounting pad Circular attachment on engine or ADG for shaft-driven accessory.

mouse See *mice*.

moustaches Small canard foreplanes, usually retractable and used only at high-alpha regimes by supersonic delta.

mouth Inlet of circular parachute, especially where this has diameter less than canopy maximum.

mouth lock Reef.

MOV 1 Metal-oxide varistor.
2 Main oxidizer valve.

movables All flight-control, high-lift or high-drag surfaces on wing.

movement 1 One aircraft flight as recorded by particular ground location; for airport, transits overhead are not counted and a * is either an arrival or a departure.
2 Single military airlift operation by one or more aircraft.

movement area Region where aircraft may be found on ground proceeding under their own power.

move off blocks Start of an aeroplane flight from parked position, esp commercial transport; time entered in ATC and flight logs. Where parked nose-on, time is from start of backward push.

moving-armature speaker Audio loudspeaker driven by oscillation of part of soft-iron (ferromagnetic) circuit.

moving-base simulator Flight deck or cockpit can move linearly as well as rotate, usually in any direction.

moving-coil speaker Dynamic loudspeaker, driven by forces developed by interaction of currents in conductor and surrounding field.

moving-iron instrument Coil carrying current to be measured moves soft-iron armature connected to indicating pointer.

moving-map display One having topographical, radar, IR, target or other map projected optically on screen with aircraft position fixed (usually at centre).

moving-target indication Radar which incorporates circuits automatically eliminating fixed echoes so that display shows only those moving with respect to Earth. For airborne radar, eg AWACS, aircraft platform motion automatically subtracted.

moving wings Wings whose incidence is varied in flight as primary means of trajectory control.

Movlas Manually operated visual landing-aid system (carrier LSO).

MOVTAS, Movtas Modified visual target-acquisition system.

MP 1 Midpoint between fixes or waypoints.
2 Manoeuvre programmer (RPV).
3 Manifold pressure.
4 Multiphase (or pulse).
5 Manual proportional (flight-control system).
6 Manpower and Personnel (USAF).

m_p Proton rest-mass.

MPA 1 Man-powered aircraft.
2 Management problems analysis.
3 Maritime patrol aircraft.

MPa Megapascals, SI unit for high pressures.

MPAA Multi-beam phased-array antenna (or aerial).

MPAG Man-Powered Aircraft Group.

MPB 1 Multiple pencil beam.
2 Mobility procedures branch.

MPBW Ministry of Public Buildings and Works (UK, obs).

MPC 1 Multi-purpose console.
2 Military personnel centre.
3 Multilayer printed circuit.
4 Message-processing centre.

MPCD Multipurpose colour display.

MPCF Million parts/particles per cubic foot.

MPD 1 Medium-prf pulse Doppler (radar).
2 Maximum permitted dose (radiation).

3 Maintenance planning document.
4 Multipurpose display.
MPDR Monopulse Doppler radar.
MPDS Missile-penetrating discarding sabot.
MP flt Mission-planning flight (RAF).
MPG 1 Moulded propellant grain.
2 Miles per gallon (rare in aviation).
mph Statute miles per hour.
MPHT Missile potential-hazard team.
MPI 1 Mission and payload-integration office (NASA).
2 Magnetic-particle inspection.
3 Moving plan indicator.
4 Mean point of impact.
mpig Miles (statute) per Imp gallon (not recommended).
MPL Mid-Pacific landing (NASA).
MPM Multipurpose missile (USAF).
MPMS 1 Missile performance monitoring system (inertial guidance subsystem).
2 Multi-purpose sub-munition.
MPNL Master parts-number list.
MPO Mission-planning officer (RAF).
MPP Massively parallel processor.
MPPS Multi-purpose pylon system.
MPR Medium-power radar.
MPS 1 Mixed-propellant system.
2 Mid-platform ship (pad amidships).
3 Materials processing in space.
4 Main propulsion system (or stage).
5 Metres per second (ICAO, contrary to SI, which is m/s).
6 Multiple protective structure (ICBM).
7 Military postal system (US).
8 Mission payload subsystem.
MPSS Man-hours per ship set.
MPT Memory point track.
MPt Melting point.
MPTO Maximum-performance take-off (helicopter).
MPU 1 Missile power unit.
2 Master processor unit.
MPVO Local air defence (USSR).
MPW Maximum pavement width required for 180° turn.
MQAD Materials Quality Assurance Directorate (UK, MoD).
MQF Mobile quarantine facility.
MQP Moulded quality phenolic.
MQT Model qualification test.
MR 1 Maritime reconnaissance.
2 Medium range.
3 Marker ranger (laser).
4 Military regions.
mR Milliradian, should be mrad.
MRA 1 Major replaceable assembly.
2 Maximum relight altitude.
3 Minimum reception altitude.
MRAALS, Mraals Marine remote-area approach landing system.
MRAAM Medium-range air-to-air missile.
mRad, mrad Milliradian.
MRAG Metallics Research Advisory Group (MRCC).
MR&R Mobile reclamation and repair.
MRASM Medium-range air-to-surface missile.
MRB 1 Materials Review Board.
2 Maintenance review board.
3 Main rotor blade.
MRBF Mean rounds (fired) before failure.
MRBM Ambiguous, can be medium-range or mid-range ballistic missile.
MRCC Materials Research Consultative Committee (UK).
MRCS Multiple-RPV control system.
MRD Main-rotor diameter.
MRE Mean radial error (weapon delivery).
mrem Millröntgen, thus mrem/h.
MRF 1 Meteorological research flight.
2 Multi-role proximity fuze.
3 Medical red flag.
MRG 1 Master reference gyro.
2 Medium-range (ICAO).
MRGB Main-rotor gearbox.
MRI Magnetic/rubber inspection.
MRIR Medium-resolution infra-red radiometer.
MRL Modular rocket launcher.
MRM 1 Medium-range missile.
2 Map rectification machine.
MRP 1 Medium-angle rocket projectile (air/ground sight setting).
2 Multi-annual research programme (EEC).
3 Materials requirements planning.
4 Manufacturing resources planning.
5 Mobile repair party.
MRPC Mercury Rankine power conversion.
MRR 1 Mechanical reliability report.
2 Maritime radar reconnaissance.
MRRV Manoeuvring re-entry research vehicle.
MRS Master (or military) radar station.
MRS Generalized mass, esp in model coordinates of dynamic model.
MRSA Mandatory radar service area.
MRSP Multifunction radar signal processor.
MRT 1 Maximum reheat thrust.
2 Military rated thrust.
3 Miniature receive terminal.
MRTF Mean rounds to failure (gun).
MRTFB Major range and test-facility base (USAF).
MRU 1 Mountain rescue unit.
2 Mobile radio (or radar) unit.
3 Mach repeater unit.
MRUASTAS Medium-range unmanned aerial surveillance and target acquisition system.
MRV Multiple re-entry vehicles.
MRW Maximum ramp weight.
MRWS Manned remote work-station (for space-station construction).
MS 1 Minus (ICAO).
2 Military standard (US).
3 Mild steel.

m/s Metres per second.
MSA 1 Mission system(s) avionics.
2 Monolithic stretched acrylic.
3 Minimum safe (or separation) altitude.
4 Maritime Safety Agency (J).
MSAFQ Minimum speed for acceptable flying qualities.
MSAW Minimum safe altitude warning.
MSB Most significant bit (EDP [1]).
MSBS Mer/sol balistique stratégique; submarine-based strategic missile (F).
MSC 1 Manned Spacecraft Center (NASA, now Johnson Space Center).
2 Major subordinate command (NATO).
3 Manpower Services Commission (UK).
4 Military Sealift Command (US).
MSCP Mean spherical candlepower (non-SI).
MSD 1 Mass storage device.
2 Minimum separation distance.
3 Multisensor display.
MSE 1 Minimum single-engine speed.
2 Manned spacecraft engineer.
MSER Multiple stores ejector rack.
MSF Militarily significant fallout.
MSFC Marshall Space Flight Center (NASA).
MSFN Manned Space Flight Network (NASA).
MSFP Mosaic-staring focal plane (early-warning sensors).
MSFSG Manned spaceflight support group (USAF).
MSFU Merchant ship fighter unit (WW2).
MSG 1 Maintenance steering group (inter-airline).
2 Message (ICAO).
MSH Metastable helium.
MSI 1 Medium-scale integration.
2 Moon's sphere of influence (NASA).
MSIP Multinational staged improvement program.
MSL 1 Mean sea level.
2 Mapping Sciences Laboratory (NASA).
msl Missile.
MSLC Maintenance stock level case.
MSO Manager, shop operations.
MSOGS Molecular-sieve oxygen generation system.
MSP 1 Mach sweep programmer.
2 Maintenance service plan.
3 Magnetic speed probe.
4 Mosaic sensor program(me).
5 Maximum structural payload.
MSPFE Multi-sensor programmable feature extraction.
MSPT Marine silent power transmission (ASW).
MSR 1 Missile-site radar.
2 Multi-sensor reconnaissance.
3 Maximum-speed range.
MSRF Microwave Space Research Facility (USN).
MSRS Miniature sonobuoy receiver system.
MSS 1 Multi-spectral scanner.
2 Mission support systems.
MSSC Maritime surface surveillance capability.
MSSR Mars surface-sample return.
MSSS Man/seat separation system.
MST Missile surveillance technology.
MST&E Multi-service test and evaluation (US).
M-stoff Methanol (G).
MSU Mode selection unit.
MSW Maximum short take-off and landing weight.
MT 1 Megaton, megatonne.
2 Metric ton (highly undesirable usage).
3 Mechanical (ground) transport.
MTACS, Mtacs Marine tactical air control system.
MTAD Multi-trace analysis display (EW).
MTAE Multiple-time-around echoes (radar and EW).
MTAT Mean turn-around time.
MTBF Mean time between failures.
MTBF$_c$ Mean time between component failure (also MTBCF).
MTBFRO, MTBFro Mean time between failures requiring overhaul.
MTBI Mean time between incidents.
MTBMA 1 Mean time between maintenance action.
2 Mean time between mission aborts.
MTBO Mean time between overhauls.
MTBR Mean time between repair (or removals or replacement).
MTBSF Mean time between significant failures.
MTBUER Mean time between unscheduled engine removal.
MTBUM Mean time between unscheduled maintenance.
MTC 1 Mach trim computer.
2 Mach trim compensator.
MTCA Ministry of Transport and Civil Aviation.
MTDS Marine Tactical Data System.
MTE 1 Megatons equivalent, unit of area destruction.
2 Modular threat emitter.
MTEL Methyl-triethyl lead.
MTF 1 Modulation transfer function.
2 Mississippi Test Facility (NASA).
MTGW Maximum taxi gross weight.
MTI Moving target indication, or indicator.
MTIRA Machine Tool Industry Research Association (UK).
MTL Mean (rarely max) transmitter level.
MTLM Major throttle-lever movement.
MTM 1 Mission and traffic model; co-ordinates Space Shuttle payloads and customer requirements.
2 Maximum take-off mass.
MTMA Military terminal manoeuvring area.
MTMC Military Traffic Management Command.
MTO See *MTOW*.
MTOD Mean (or maximum) time on deck.

MTOE Mid-term operations estimate.
MTOW Maximum take-off weight.
MTP 1 Mandatory technical publications.
2 Maintenance test panel.
MTR 1 Missile tracking radar.
2 Marked-target receiver.
3 Military training route.
MTRE Missile test and readiness equipment.
MTS 1 Marked-target seeker.
2 Mobile training set.
3 Mobile test set.
4 Motion/time survey.
5 Maintenance training simulator.
MTT Multiple-target tracking.
MTTA Machine Tool Trade Association (UK).
MTTFSF Mean time to first system failure.
MTTR 1 Mean time to repair.
2 Multiple-target tracking radar.
MTTS Multi-task training system, versatile simulators which can be configured for particular aircraft types.
MTU Use metric units (ICAO).
MTUR Mean time to unscheduled replacement.
MTVC 1 Motor thrust-vector control.
2 Manual thrust-vector control.
MTWA Maximum take-off weight authorized.
M-type marker See *M-marker*.
MTZT Multiple time-zone travel.
MU Maintenance unit.
mud-mover Low-level close-support or attack aircraft (colloq).
MUF, muf Maximum usable frequency.
muff Exhaust heat exchanger, usually for cabin heating.
muffler Silencer (US).
mule 1 Refuelling bowser (US term).
2 Hydraulic test rig (colloq).
3 Modular universal laser equipment.
Mullite Commerically produced ablation material.
multibank engine PE (4) with several linear banks.
multi-body freighter Hypothetical aircraft with payload in detachable bodies all carried on one wing.
multi-bogey Air-combat situation in which there are many enemy aircraft.
multi-burn See *multi-impulse*.
multi-cell ACM engagement in which two or more DA cells participate.
multicellular foam Material of low density with air or gas-filled closed cells such as expanded polystyrene; term not used for honeycombs.
multichannel selector Manual controller for preselected communications channels.
multicombiner Optical system for projecting several sets of displays all focused at infinity on HUD screen.
multicommunications service Mobile private communications service on 122·9 MHz for such activities as ag aviation and forest firefighting.
multicoupler Device for making single aerial serve several receivers.
multi-impulse Capable of repeated cutoff and restart to meet propulsion demands of mission (rocket).
multi-lane airway For reasons of traffic, control, noise, disturbance or safety, airway divided into two (rarely three) parallel lanes, denoted by North/South or East/West, with Centre where three; lateral separation up to 20 km.
multilateral Involving several sovereign states.
multilayer board Any printed-circuit, hybrid or related electronic assembly fabricated as stack of subcircuits in single board.
multi-media classroom Equipped with audiovisual aids, overhead projection, tape/slide systems, blackboard and student dialogue buttons.
multimission Able to perform different military roles, especially air-combat/interception and ground attack/interdiction.
multi-mode radar Designed to function in several operating modes with quick switching between each, eg (for MMR for attack aircraft) ground map, ground map spoiled, etc.
multipath effect Anomalies in radar target range and position, and many other optical/radar situations, caused by receipt of reflected radiation from land or sea surface and other reflectors as well as direct signal from target.
multiplane 1 Aeroplane with several sets of main wings, normally superimposed.
2 Adjective: tail unit or other assembly with several superimposed horizontal surfaces.
multiple courses Narrow false courses heard on radio range, esp among mountains.
multiple ejection rack, MER Interface allowing several stores to be carried on single external-stores station, with positive ejector rams to ensure clean separation.
multiple independently targeted RV Delivery system containing several missile re-entry vehicles each having own guidance system to separate targets.
multiple independent RV Broader concept than above; system delivers several warheads which may free-fall or be independently targeted.
multiple kill capability Usual meaning is that platform has several types of armament, eg gun plus two types of missile.
multiple options Force employment alternatives depending on flexibility in tactical/strategic situations, retargeting and availability of conventional or nuclear weapons.
multiple RV Re-entry vehicle containing number of separate warheads which scatter slightly but fall on same area target.
multiple-time-around echoes Those from targets at exact multiples of the radar unambiguous range.
multiplexing 1 Act of combining signals from many sources into common channel, requiring frequency, phase and time-division.

2 In electrical wiring of whole vehicle, providing redundant pathways or alternative routes with auto switching triggered by battle damage or other discontinuity.

multiplexer, multiplexor Device, often stored-program computer, which handles I/O (input/output) functions of on-line EDP (1) system with multiple communications channels.

multi-ply Material, eg wood, fabric, metal or composite, assembled from several laminar layers.

multiplying valve, multiplier Device whose output is exact product of several inputs.

multiprocessor Multiprogrammed processor.

multiprogramming Technique allowing single computer to run several programs simultaneously.

multi-row engine PE (4) having more than two banks or radial rows of cylinders.

multi-sensor Using more than one type of signal to gather information; eg optical camera, radar, IR linescan, passive elint etc.

multiservice connector Personal coupling for suit pressure, electric heating, oxygen, radio and possibly other services.

multispectral Capable of responding usefully to wide range of EM wavelengths, including all visible hues, IR and possibly UV.

multi-stage rocket Vehicle with several stages each fired and staged in succession.

multi-step See *multi-stage rocket*.

multitube launcher External store filled with parallel tubes for air-launched rockets.

multivibrator Oscillator having two cross-coupled valves or transistors operating alternately, each input coming from other stage's output; can be bistable (flip/flop), monostable or astable.

mu-metal High-permeability alloy used to screen equipment (eg CRT) from stray magnetic influences.

mu-meter Rolling truck with pair of toed-out tyres, measures side force and hence coefficient of friction. From μ (mu), coefficient of sliding friction.

Munk factor Formula for performance of biplanes based on ratios of spans, lifts, gap and interference.

Muroc Dry California lake, site of USAF flight-test centre of same name (now Edwards AFB).

MUSA, Musa 1 Multiple-element steerable array.
2 Frag submunition for semi-hard targets, dispensed by MW-1 system (G).

mush To increase angle of attack suddenly but without immediate corresponding vertical acceleration, resulting from momentum along original path.

mushroom rivet One having thin convex head with sharp edge, for thin sheet.

music EW emissions, esp jamming (colloq).

MUSPA Area-denial mine, dispensed by MW-1 system (G).

Must Multimission UHF satcom terminal.

Mutes Multiple-threat emitter system; simulates many hostile emissions simultaneously.

mute switch Disconnects headset, microphone (term also often used for PA disconnect).

mutual interception By two friendly interceptors on each other.

mux Multiplex.

muzzle brake Any of variety of gun-muzzle gas deflector units to reduce recoil or blast on surrounding structure.

muzzle cap Frangible closure on gun muzzle to reduce drag and ingestion of precipitation.

muzzle energy Kinetic energy of each projectile as it leaves gun, measured relative to gun.

muzzle horsepower Standard non-SI measure of gun power, esp automatic weapons; muzzle energy multiplied by rate of fire (units being compatible).

MV 1 Muzzle velocity.
2 Miniature vehicle.

MVA 1 Minimum vectoring altitude.
2 Multivariate analysis.
3 Megavolt-amperes, basic unit of large AC powers.

MVAR Magnetic variation.

MVDF MF and VHF D/F facilities co-located (ICAO).

MVEE Military Vehicles Experimental Establishment.

MVL, mvl Mid-value logic.

MVM Muzzle-velocity measure(ment).

MVPS Multiple vertical protective shelter (ICBM).

MVS Minimum vector speed (light pen).

MVSRF Man/vehicle systems research facility (NASA).

MVTU Moscow higher technical school.

MVW Maximum VTOL weight.

MW 1 Medium wave.
2 Methanol/water.
3 Megawatts.
4 Microwave.

M_w Bending moment at wing root.

mW Milliwatt.

MW50 Methanol 50%, water 50%.

MWA Multiple weapons adaptor.

MWCS Multiple weapons carrier system.

MWDP 1 Mutual weapons development program.
2 Master warning display panel.

MWE 1 Manufactured (or manufacturer's) weight empty.
2 Maximum weight empty.

MW(E) Megawatts electrical.

MWL Minimum take-off distance on water to clear standard (35 or 50 ft) obstacle.

MWO Meteorological Watch Office (ICAO).

mwr, MWR Millimetre-wave radar.

MWS Master warning system.
MW(T) Megawatts (thermal).
MX Mixed types of icing, esp white and clear (ICAO).
M_x Bending moment at any station x.
MY Multi-year.
Mylar Tough transparent film, of terephthalate polyester family, usable down to extreme thinness (hence light weight per unit area).

MYP Multi-year procurement.

myriametre Non-SI term for 10^4 metres, hence myriametric.

MZFW Maximum zero-fuel weight.

N

N 1 Newton.
2 Shaft rotation speed, with suffix number (N_1, N_2 etc) for each shaft (see *gas-turbine numerology*).
3 Yawing moment.
4 Code: operates by sound in air (JETDS).
5 Code: navaid (ICAO).
6 Prefix, amount of cloud.
7 Number of load cycles, usually per second.
8 North (ICAO).
9 Synoptic chart code: air mass has had characteristics changed.
10 Telecom code: aircraft.
11 Avogadro's number.
12 Nitrogen.
13 STOL aircraft (FAI).
14 Prefix, night (USSR).
15 Prefix, nuclear (USSR).

n 1 Prefix, nano (10^{-9}).
2 Generalized symbol for an aircraft equipped with Tacan only and 64-code transponder.
3 Normal acceleration in g.
4 Number in a sample.
5 Frequency, especially of rotation.
6 Refractive index.
7 Negative, hence n-type semiconductor.

N_1 Fan or LP compressor speed.

N_2 1 Nitrogen.
2 IP compressor speed (CAA).
3 HP compressor speed (2-shaft engine).

N_3 HP compressor speed (CAA).

N′, N° Rate of change of N (2) on spool-up or rundown.

N-code ICAO code, amount of cloud.

N-display Target forms two blips on horizontal timebase (as in K-display), lateral positions giving range and relative amplitudes bearing.

N-sector Sector of radio range in which Morse N is heard.

N-strut Arrangement of interplane struts or cabane struts resembling N.

n-type semiconductor One in which charge carriers are nearly all electrons.

N-wave Shockwave remote from source; far-field boom signature, when profile of pressure/time or pressure/linear distance has settled down to N-like profile.

N/A Not applicable, not available, not authorized.

N_A Avogadro constant.

NAA 1 National Aeronautic Association of the US.
2 National Aviation Academy (US).

NAAA National Agricultural Aviation Association (US), or National Aerial Applicators Association (US).

NAAFI Pronounced "naffy", Navy, Army and Air Force Institutes (UK).

NAAP Netherlands Agency for Aerospace Promotion (NIVR).

NAAS National Association of Aerospace Subcontractors (US).

NAATS National Association of Air Traffic Specialists (US).

Nabs NATO air base Satcom.

NAC 1 North Atlantic Council.
2 Noise Advisory Council (UK).
3 National Air Communications (UK, 1939–40).
4 National Aviation Club (US).
5 Naval Air Command (UK).

NACA 1 National Advisory Committee for Aeronautics (US, renamed NASA 1958).
2 National Air Carrier Association (US).

NACA cowling Drag-reducing annular cowl for radial engines with aerofoil-profile leading edge and cylindrical main section.

NACA section Any of numerous aerofoil sections designed by NACA.

NACA standard atmosphere Original idealized atmosphere, published in 1925; later superseded by ICAO, ARDC and others.

nacelle Streamlined body sized according to what it contains, which may be an engine, landing gear, human occupants etc; when tail carried on separate booms * takes places of fuselage.

NACP Noise-abatement climb procedure.

nacreous cloud High layer cloud with iridescent appearance; also called mother-of-pearl cloud.

NADB Netherlands Aircraft Development Board.

NADC 1 US Naval Air Development Center.
2 Nuclear Affairs Defence Council (NATO).

NADEEC NATO Air Defence Electronic Environment Committee (NATO).

Nadge NATO air-defence ground environment; multinational programme for unified air-defence system of radars, computers, displays and communications from North Cape to Eastern Turkey.

Nadgeco Multinational company formed to implement Nadge plan.

Nadgemo Nadge Management Office; formed within NATO to act as unified customer.

nadir Point on celestial sphere vertically below observer, ie 180° from zenith.

NADL US Navy Avionics Development Laboratory.

NAE 1 National Aeronautical Establishment (Canada).
2 National Academy of Engineering (US).

NAEC 1 Naval Air Engineering Center (US).
2 National Aerospace Education Council (US).

Naegis NATO AEW ground-environment integration segment.

NAES Naval Aviation Engineering Services (US); plus U = unit.

NAEW NATO airborne early warning; F adds "force".

NAF 1 Norsk Astronautisk Førening (Nor).
2 Non-appropriated fund(s).

Nafag NATO Air Force Armaments Group.

Nafdu Naval Air Flying Development Unit.

NAFEC, Nafec National Aviation Facilities Experimental Center; Atlantic City, NJ (US).

NAFI National Association of Flight Instructors (US).

NAFP National aeronautical facilities program(me).

NAGr Short-range reconnaissance group (G).

NAI 1 Negro Airmen International (US).
2 Netherlands Aerospace Industries (member AECMA, title in English).

NaK Generalized term for all mixtures of sodium and potassium; used eg as liquid-metal heat-transfer media.

naked See *clean (1, 2)*.

nam, n.a.m. Nautical air miles; hence */lb of fuel.

NAMFI, Namfi NATO Missile Firing Installation.

NAMI Scientific auto-motor institute (Sweden).

NAML Naval Aircraft Materials Laboratory (UK).

NAMMA NATO MRCA Management Agency.

NAMRL Naval Aerospace Medical Research Laboratory (US).

NAMSA NATO Maintenance and Supply Agency (Luxembourg).

NAMTD Naval Air Maintenance Training Detachment (US).

NAND NOT + AND logic device, retains output until voltage at all inputs, then goes to 0.

nano Prefix, $\times 10^{-9}$.

NAP 1 Noise-abatement procedure.
2 Normal acceleration point (SST).

nap 1 Local profile of land surface; hence * of Earth, * of the Earth flying (as close to ground as possible).
2 Short fibre ends along edge of fabric.

napalm 1 Mixture of naphtha and palm oil (hence name), usually with additives, used as incendiary material.
2 Air-dropped ordnance filled with * designed to burst and distribute flame over large area.

NAPCA, Napca National Air-Pollution Control Association (US, now APCC).

nape Use napalm against (colloq).

naphtha Generalized name for inflammable oils distilled from coal tar and other sources.

NAPL National Air Photo Library (C).

NAPMA NATO AEW Programme Management Agency.

NaPO NASA Pasadena Office.

NAPP National Association of Priest Pilots (US).

NAPR NATO armaments planning review.

NAPTC Naval Air Propulsion Test Center (US).

NARF 1 Naval Air Reserve Force (US).
2 Naval Air Rework Facility; Pensacola (US).

narrow-body Commercial transport with fuselage width of approximately 10 ft (3 m) with single aisle between passenger seats.

narrow gate AAM operating mode permitting homing on target only within narrow limits of rate of closure.

NARTEL National air radio telecommunications (UK).

NARUC National Association of Regulatory Utility Commissioners (US).

NAS 1 Naval air station (US).
2 National airspace system (FAA).
3 National Academy of Sciences (US).
4 Nozzle actuation system.
5 National aerospace standard (CAA, UK).

NASA National Aeronautics and Space Administration (US).

Nasad National Association of Sport Aircraft Designers (US).

NASAO National Association of State Aviation Officials (US).

NASC 1 Naval Air Systems Command (US).
2 National Aeronautics and Space Council (US).
3 National Association of Spotters' Clubs (UK, WW2).
4 National Aerospace Standards Committee.

Nascom NASA Communications Network.

NASDA National Space Development Agency (J).

NASF Navigation and attack systems flight (RAF).

NASM National Air and Space Museum (US).

NASN National Air-Sampling Network (EPA).

NASO Naval Aviation Supply Office (US).

NASPG North Atlantic Systems Planning Group.

NASPO, Naspo Naval Air System Program Office (US).

NASS Naval Anti-Submarine School.

NASSA National AeroSpace Services Association (US).

Nastran NASA structural analysis (GP finite-element program).

NASWDU Naval Air/Sea Warfare Development Unit.

NATA 1 National Air Taxi Association (UK).
2 National Air Transportation Association (US).
3 National Aviation Trades Association (US).

NATC 1 Naval Air Test Center (Patuxent River, US).

2 National Air Transportation Conferences (US).

Natcapit North Atlantic Capacity And Inclusive Tours Panel.

NATCC National Air Transport Co-ordinating Committee (US).

NATCS National Air Traffic Control Services (UK, now NATS).

Natinad NATO Integrated Air Defence.

national airline Designated flag carrier of sovereign state.

national responsibility Part of collaborative programme wholly assigned to one country.

National Search and Rescue Plan American inter-agency agreement to facilitate application of full national resources in emergencies.

Natlas National laboratory assessment scheme (NPL, UK).

Natops Naval Air Training and Operating Procedures Standardization Program (US).

NATS 1 National Air Traffic Services (UK).
2 Naval Air Transport Service (US).

natural buffet Buffet occurring automatically near stall as result of airflow turbulence, thus giving warning to pilot.

natural finish Unpainted.

natural frequency That at which a system oscillates if given one sudden perturbation and thereafter left to itself.

natural language That spoken by human beings (usually means English).

naturally aspirated Not supercharged but left to draw in air at local atmospheric pressure.

natural wavelength That corresponding to natural frequency of tuned electronic circuit; that at which open aerial will oscillate.

nautical mile Standard unit of distance in air navigation, totally at variance with SI; aviation uses International *, 1,852 m; UK * is 1,853·18 m. A common aviation approximation is 6,000 ft (1,828·80 m).

nautical twilight Period when Sun's upper limb is below visible horizon and Sun's centre is not more than 12° below celestial horizon.

nav Navigation.

navaid Navigation aid, esp one of electronic nature located at fixed ground station.

naval aircraft 1 Loosely, one used by a navy.
2 One specially equipped for operation from aircraft carrier or other warship.

nav/attack system One offering either pilot guidance or direct command of aircraft to ensure accurate navigation and weapon delivery against surface target.

navcom Loosely, navigation and communications, or a single radio transceiver used for both functions.

navex Navigation exercise.

Navhars Navigation, heading and attitude reference system.

Navier-Stokes Basic set of equations for motion of body or flow parcel in viscous fluid.

navigation flare Bright-burning pyrotechnic dropped over open country at night to provide fixed object for measurement of drift (obs).

navigation float Navigation aid in form of clearly visible float, with or without pyrotechnics, for drift measurement over sea; hence navigation flame-float, navigation smoke-float (obs).

navigation lights Regulation wingtip lights (red on left, blue-green on right) visible from ahead through 110° to side, and white light at tail visible each side of rear centreline.

navigation stars Those used in astronavigation.

Navspasur Naval space surveillance system (US).

Navwass Navigation and weapon-aiming subsystem.

NAW Night/adverse weather.

Nawacs NATO Awacs.

Nawas National warning system (US).

Nawtol Night/all-weather take-off and landing.

Naxos German WW2 family of passive electronic systems tuned to home on H_2S.

NB Enhanced-radiation weapon (colloq, from "neutron bomb").

Nb_3Sn Niobium/tin superconductive alloy.

NBAA National Business Aircraft Association (US).

NBC Nuclear, biological, chemical (warfare).

n.b.c. Noise-balancing control (radio).

n.b.f.m. Narrow-band frequency modulation.

NBR Nitrile-based rubber.

NBS National Bureau of Standards (US).

NBSV Narrow-band secure voice.

NC 1 Numerical control (machine tool).
2 Nitrocellulose (or Nc).
3 No change (ICAO).

N_c Compressor rpm.

NCA National Command Authorities (US).

NCAA National Council of Aircraft Appraisers (US).

NCAP Night combat air patrol; also known as night cap.

NCAR National Center for Atmospheric Research (NSF).

NCAS Nomex core, aluminium skin.

NCDU Navigational control and display unit.

NCGS Nomex core, GRP skin.

NCI Navigation control indicator.

NCMA National Contract Management Association (US, now Int).

NCMC Norad Cheyenne Mountain Complex.

NCO Non-commissioned officer.

NCQR National Council for Quality and Reliability (UK).

NCS Numerical Control Society (US).

NCT 1 Non-co-operative (or co-operating) target, hence * recognition.
2 National Centre for Tribology (UK, now ESTB).

NCTR Non-cooperative target recognition.

NCUR Navigation computer unit readout.

ND 1 Nose-down (trim control).
2 Unable to deliver message, notify originator (ICAO).
3 Navigation display.

Nd Neodymium.

NDA National defense area (US, not DoD property but military or security interest).

NDB Non-directional beacon.

NDCL Nozzles-down cue line.

NDE Non-destructive evaluation (NDT research).

NDI Non-destructive inspection.

NDIC National Defence Industries Council (UK).

ND point Nominal deceleration point (SST).

NDRC National Defense Research Committee (US, WW2).

NDS Nuclear detection system.

NDT Non-destructive testing.

NDTA National Defense Transportation Association (US).

NDTS Non-Destructive Testing Society (UK).

NDV Nuclear delivery vehicle.

N/E 1 Neon/ethane mixture (for flushing payload compartments in space).
2 Northings and eastings.

Ne Neon.

N_e 1 Number of engines.
2 Shaft power (USSR).

NEA 1 National Electronics Association (US).
2 Nuclear Energy Agency (OECD).

NEAC Noise and Emissions Advisory Committee (IATA).

NEACP National-emergency airborne command post (US).

near encounter Close fly-by of planet or other body.

near field 1 Shockwave region close to source (hence ** signature).
2 Sonic boom area closest to SST track.

near miss DoD term for airmiss.

neatlines Parallels and meridians surrounding body of map (NATO); also called sheetlines.

NEB 1 National Enterprise Board (UK).
2 National Energy Board (C).
3 Nuclear exo-atmospheric burst (DoD).

NECI Noise exposure computer/integrator.

neck Lower tube-like portion of gas-balloon envelope.

necking Local reduction in cross-section of member due to plastic flow under tension.

NEEC Noise-excluding ear capsule.

needle Rotary "hand" of traditional dial-type instrument; but see plural.

needle and ball See *turn and slip*.

needles Generalized term for instrument (especially flight instrument) readings; thus "the * all dropped to zero".

needle split Divergence between helicopter engine and rotor speed indications, with normally superimposed needles.

needle valve One offering fine adjustment of fluid flow by linear translation of tapering pointed rod centred in orifice.

NEF Noise exposure forecast.

NEFC NATO electronic-warfare fusion cell.

NEFD Noise equivalent flux density.

"negative" Voice communications word meaning "no".

negative altitude Angular distance below horizon; depression (ASCC).

negative area 1 Area on tephigram enclosed between path of rising particle at all times colder than environment and surrounding air.
2 Generally, vague area (volume) surrounding colder air that happens to be rising.

negative camber Usually interpreted as concave on upper surface.

negative feedback 1 Signal either reversed in sense or otherwise out of phase and thus tending to increase departure from original condition.
2 Transfer of part of amplifier output in reverse phase to input.

negative g Subject to acceleration in the vertical plane in the opposite-to-normal sense, eg aircraft in sustained inverted flight or in pushover from steep climb to steep dive; wings are bent "downwards" (relative to aircraft attitude) and pilot can experience "red-out". This condition is intermittently inevitable in severe turbulence but is normally prohibited for non-aerobatic aircraft.

negative image Apart from photographic meaning, transposition of blacks and whites in TV, EO or IR picture.

negative pressure relief valve Prevents dP in pressurized aircraft becoming negative.

negative rolling moment Tending to rotate aircraft anticlockwise, seen from rear.

negative stability See *instability*.

negative stagger Backwards stagger; lower plane in advance of upper.

negative stall Stall under negative g; this regime provides lower left boundary of basic manoeuvre envelope.

negative sweep See *forward sweep*.

negative terminal That from which electrons flow; thus towards which "current" flows.

negative-torque signal Indication of fault characterized by driven member (eg propeller) tending to drive driving input (eg turboprop).

negative yaw Rotating aircraft anticlockwise about z-axis seen from above.

negotiation Commercial discussion preceding contract.

negotiation threshold Point in escalating conflict at which either participant is likely to draw back and initiate negotiation.

NEH Code: "I am connecting you to a station accepting traffic for station you request" (ICAO).

NEL National Engineering Laboratory (UK).

nematic See *liquid crystal.*
NEMP Nuclear electromagnetic pulse.
NEO National energy outlook (US).
neon Inert gas.
Neoprene Family of synthetic rubbers (polymerized chloroprenes) resistant to hydrocarbon fluids.
NEP Noise equivalent power.
neper Unit (Napier) expressing scalar ratio of two currents, N = nat log (I_1/I_2) nepers, = 8·686 dB. Applicable to all scalar ratios of like quantities.
nephelometer Measures light scattering by fine particles in suspension.
nepheloscope Measures temperature changes in gases rapidly compressed or expanded.
nephometer Convex mirror divided into one central and five radial parts for estimating cloud cover.
nephoscope Optical instrument for measuring direction of cloud motion.
NERC, Nerc 1 National Environmental Research Center (EPA).
2 National Environmental Research Council (UK).
NERO 1 National Energy Resources Organization (US).
2 Nederlanse Vereniging voor Raketonderzoek.
Nerva Nuclear energy for rocket vehicle applications.
NESC 1 Naval Electronics Systems Command (US).
2 National Environmental Satellite Center (NOAA).
NESN NATO English-speaking nations.
net 1 Tailored mesh forming structural link between traditional gas balloon envelope and useful load.
2 Electronic, optical or other telecommunications system(s) forming single service covering designated area accessible at any point.
net area 1 Traditionally gross area (normally of wing) minus projected horizontal area of fuselages, booms, nacelles, pods etc.
net dry weight Basic weight of engine or other device calculated according to various rules but always excluding fluids (fuel, lubricant, coolant etc) and usually all accessories, protective systems, instrumentation systems, etc, not essential for device to function.
net flightpath That followed by aircraft, esp aeroplane, after application of factors (particularly for average-aircraft performance and average-pilot skill); ie gross flightpath fully factored.
net height That at any point on net flightpath, esp during take-off and climb-out.
net performance Gross performance factored to take account of temporary variations (from whatever cause) and pilot handling skill.
net propulsive force See *net thrust*.
net radiation factor Percentage of radiant energy emitted by one surface or volume that is absorbed by another surface or volume.
nettage decrement Percentage difference between gross and net performance.
net thrust F_n, effective thrust of jet engine numerically equal to change in momentum of fluids (air/gas and fuel) passing through engine (plus, in engine operating with choked nozzle, extra aerodynamic thrust generated in nozzle).
net weight Loosely empty weight, but usually excluded from aerospace usage.
neutral axis Locus of points within structural member at which bending imposes neither tension nor compression.
neutral engine Main propulsion engine devoid of dressing peculiar to either particular aircraft type or to left or right installation in multi-engined aircraft, and thus available for quick completion for desired installation.
neutral equilibrium Normally means that system will tend to stay in most recently commanded attitude or condition, without oscillation, unstable divergence or recovery of previous condition.
neutral flame Neither oxidizing nor reducing.
neutral hole Aperture cut from sheet, esp in wall of internally pressurized container (usually taken to be cylindrical), shaped so that peak stress around periphery is minimized. Normally approximates to ellipse.
neutralized controls Usually taken to mean centralized.
neutralize track Air intercept code: target is ineffective or unusable.
neutral point 1 Location of aircraft, especially aeroplane, c.g. at which stability would be neutral; rear extremity from which static margin is measured. More strictly, stick-fixed ** is c.g. position at which stick movement to trim a change in speed is zero; stick-free ** is c.g. position at which stick force to trim a change in speed is zero.
2 Lagrangian point.
3 Any sky direction where polarization of diffuse (ie not from specific source) radiation is zero.
neutral stability See *neutral equilibrium*.
neutrino Elusive small particle, rest mass 0, spin $\frac{1}{2}$.
neutron Particle of atomic nucleus having no charge and mass $1{\cdot}675 \times 10^{-24}$ g (proton is $1{\cdot}672 \times 10^{-24}$ g).
neutron bomb Enhanced-radiation weapon.
neutron radiography NDT method similar to X-ray inspection and "photography" but using beam of neutrons.
Newton SI unit of force, = 1 kg m s^{-2} = 10^5 dyn = 7·233 pdls = 0·224809 lbf.
Newtonian flow That in extremely rarified gas where mean free path is in order of metres and hypersonic body is surrounded by incident arriving molecules passing between those bouncing

off surface; also called free-molecule flow.

Newtonian mechanics Those based on Newton's laws of motion, in which mass and energy are unrelated.

Newtonian speed of sound Relation $a = \sqrt{p/\rho}$ where p is pressure and ρ density.

Newtonian stress Fundamental shear stress in fluid, given by law $\tau = \mu \frac{du}{dy}$ where μ is fluid viscosity and u is fluid velocity at distance y from fixed surface.

Newton's laws Briefly: (1) body at rest remains at rest unless acted upon by outside force, (2) change in motion (momentum) is proportional to applied force, and (3) to every action (force or change in momentum) there is equal and opposite reaction.

Newts Naval electronic-warfare training system (US).

Nex, NEX Next-generation (prefix, DoD).

Nexrad Next-generation radar.

NF 1 French material specification prefix.
2 Night fighter.

N_f, N_F 1 Fan rpm.
2 Relationship between power turbine speed and gearbox governor (UK, CAA).

nF Nanofarad.

n.f. 1 Noise factor (radio).
2 Negative feedback.

NFCS Nuclear forces communications satellite.

NFF No fault found.

NFIP National foreign intelligence program (US).

NFKK Women's aero association (J).

n.f.m., NFM Narrow-band FM.

NFO Naval flight officer (US).

NFOV Narrow field of view.

NFP Net flightpath.

NFPA National Fire Protection Association (US).

NFRL Naval Facilities Research Laboratories (US).

NFSN NATO French-speaking nations.

NG Natural gas, hence LNG = liquid natural gas.

NG, Ng Nitroglycerine.

N_g Gas generator rpm.

NGB National Guard Bureau (US).

NGDC National Geographic Data Center (NOAA).

NGE Non-ground effect.

NGL Natural-gas liquids.

NGPS Navstar global positioning system.

NGR Night-goggle readable (ie at IR level).

NGS Naval gunfire support.

NGSP National geodetic-satellite program (NASA).

NGT New-generation trainer.

NGTE National Gas Turbine Establishment (UK).

NGV Nozzle guide vane (turbine stator).

NGW Nuclear gravity weapon.

N_H HP rpm, N_2.

NHC Navigator's hand controller.

NHCR National Flying League (J).

NHGA National Hang-Gliding Association (UK).

NHR National Flying Association (J).

NI Noisiness Index (South Africa, Van Niekerk/Muller, 1969; see *noise*).

Ni Nickel.

NIAG NATO Industrial Advisory Group.

Nial Nickel/aluminide diffusion coating.

NIAST National Institute for Aeronautics and Systems Technology (South Africa).

NIAT MAP-sponsored factory (USSR).

nib 1 Any substantially axial fore or aft-pointing fairing, usually with concave surfaces.
2 Aft-pointing fairing between, and projecting behind, two closely spaced jetpipes.
3 Forward-pointing extension at inner end of fixed glove on VG aircraft.

nibbler Machine tool for eating away edge of sheet by repeated local vertical shearing, in some cases with ability to impart lateral compression or joggling.

nibbles Stall testing of aircraft in which AOA is increased in small increments at 1 g, culminating in fully developed stalls.

NIC Newly industrializing country.

Nicalloy Nickel-iron alloy, low initial but high max permeability for transformers etc.

Nicap National Investigation Committee on Aerial Phenomena (US).

Ni/Cd, Nicad Nickel/cadmium electric battery.

Nicerol Widely used protein foam compound for firefighting.

Nichrome American heat-resisting alloys (c 85% Ni, 15% Cr), eg for resistance wire.

nickel Silver-white magnetic metal important in corrosion-resistant alloys.

nickel/cadmium battery Cell having KOH (potassium hydroxide) electrolyte, positive plates of nickel hydroxide and negative plates of cadmium hydroxide.

NICS, Nics NATO integrated communications system (O adds "Organization", MA adds "Management Agency").

NIFA National Intercollegiate Flying Association (US).

Ni/Fe, Nife Nickel/iron electric battery.

night airglow See *airglow*.

night and all-weather Strictly, interceptor can be used at night or in any weather, seldom true at time this term was in use; more accurately meant night and rain or snow but with acceptable landing minima.

Night cap, NCAP Night combat air patrol (DoD).

night effect Phenomenon most noticeable near sunrise and sunset when directional radio signals (D/F, radio range, VOR) give false readings thought to be due to variations in ionosphere.

night fighter Aircraft intended to intercept other

aircraft at night; today implicit in term "fighter" or "interceptor".

night-flying chart Special editions of regional charts, usually 1:1,000,000, eliminating all detail unseen at night but emphasizing lights, navaids and fields with night facilities (US).

Night Owl Night ground attack mission.

night vision Human seeing after eyes have had time fully to adapt to near-absence of light, with irises fully open.

NII Scientific test institute (USSR, many, each covering one subject).

NIL Code: "I have no message for you" (ICAO).

nimbostratus, Ns Thick dark blanket cloud in low/middle band (2,000–6,000 m), mainly ice crystals and supercooled water, large horizontal extent, usually rain.

nimbus Not normally used as cloud type but as adjective meaning rain-producing, as in Cb and Ns.

Nimocast British casting alloys, composition akin to Nimonics.

Nimonic Family of refractory and anticorrosive alloys based chiefly on nickel, originally Mond patents, important where creep-resistance essential; Inconel, Hastelloy, Udimet and Waspalloys related.

90-minute rule Certification requirement that twin-engined passenger aircraft may fly trans-oceanic sectors provided they are never more than 90 minutes from an emergency alternate.

nip 1 Local compression between adjacent components, esp that used to secure a third part, eg compressor rotor between discs.
2 Local compression caused by deflection under operating conditions, eg axial movement at periphery of conical compressor or turbine disc.

NIR Near infra-red.

NIS 1 NATO identification system.
2 Nose-in stand (airport ramp).

nit Name, not normally used, of SI unit of luminance, cd/m^2.

Nitralloy British steel for nitrided parts with small amounts of Cr, Al, Mn, C, Si and possibly Ni and Mo.

nitrate dope Aircraft fabric dope comprising cellulose fibres dissolved in nitric acid, plus pigment, thinner etc.

nitriding Surface hardening of steels by prolonged heating in nitrogen-rich atmosphere.

nitrogen Generally unreactive gas forming 78·03% by mass of sea-level air, symbol N; dry gas important as inert purging medium, liquid LN_2 used as cryogenic heat-transfer fluid.

nitrogen desaturation Human condition caused by nitrogen deficiency.

nitrogen narcosis Human condition caused by apparent reaction between tissue fats and nitrogen under pressure.

nitrogen tetroxide N_2O_4, most common storable liquid oxidant, often called NTO, BPt 21°C, Isp 285 with UDMH, 290 with hydrazine.

nitroguanidine Major constituent of many gun propellants, commercial name Picrite.

nitromethane CH_3NO_2, oily liquid, monopropellant.

nitrous oxide N_2O "laughing gas", used as source of oxygen in power-boosting piston engines in WW2.

NIVO Night bomber paint, later RDM2 (RAF).

NIVR Netherlands Agency for Aerospace Promotion.

NIW Night and in-weather.

NJG Night-fighter wing (G, WW2).

NJSK Private Pilots' Association (J).

NKAP State commissariat for aviation industry (USSR).

NKK Aeronautical Association (J).

NKO State commissariat for defence (USSR).

NKSK Aero Engineers' Association (J).

NKTP State commissariat for heavy industry (USSR).

NKUGK Society of Aeronautical and Space Sciences (J).

NKVD State commissariat for internal affairs (now KGB).

NL 1 Natural language.
2 Normenstelle Luftfahrt (G).

N_L LP rpm, N_1.

NLC Noctilucent cloud.

NLCM Non-lethal countermeasures.

NLG 1 Nose landing gear.
2 Noise Liaison Group (UK).

NLOS Non-line of sight.

NLR National Lucht en Ruimtevaartlaboratorium (Neth).

NM, nm Nautical mile; nm preferred except by ICAO. Note confusion with nanometre.

Nm SI unit of torque or moment, Newton-metre.

nm Nanometre (10^{-9}m).

NMC 1 Naval Missile Center (Pt Mugu, US).
2 Not mission-capable.
3 Naval Material Command (US).

NMCC National Military Command Center (US).

NMCS National military command system (US).

NMG Numerical master geometry.

NMKB Model Aeronautics Association (J).

NML Normal.

NMOS Negative (n-type) metal-oxide semiconductor.

NMP Navigation microfilm projector (arch).

NMR Nuclear magnetic resonance.

Nms SI unit of angular momentum, Newton-metre-second.

NNA Neutral and non-aligned.

NNE Noise and number exposure.

NNI Noise and number index (see *noise*).

NNK Non-nuclear kill.

NNMSB Non-nuclear munitions safety board.

NNR Hot-air Balloon League (J).

NO Nitric oxide, colourless gas.

N_0 Characteristic frequency, esp centroid of power spectral density distribution.

NO_2 Nitrogen dioxide (peroxide), pungent brown gas.

N_2O_4 *Nitrogen tetroxide.*

NOA 1 Non-operational aircraft.
2 Not organizationally assigned (aircraft stored for future use).
3 New obligation(al) authority (US).

NOAA National Oceanic and Atmospheric Administration (US).

NOACT National Overseas Air-Cargo Terminal (USN).

no-break supply One whose emergency standby system comes on line instantaneously, in theory without losing one waveform or pulse in coded train.

noctilucent cloud Literally appearing self-luminous at twilight.

NOD, Nod Night observation device.

nodding Deflection under vertical acceleration of masses cantilevered ahead of or behind main structure, eg forward fuselage (in flight only) or engine on wing pylon well ahead of leading edge.

nodding aerial One oscillating only, or principally, in vertical plane; eg HFR.

noddy cap Protective cover for delicate (eg IR-homing) missile nose (RAF, colloq).

node 1 In structures, location of point where load variation causes only rotation but no linear deflection.
2 Point, line or surface in wave system where some major variable has zero amplitude.
3 In any network, terminal point or point where two channels branch.
4 Intersection of orbit of satellite with plane of orbit of primary.
5 Location in mobility system where movement is originated, processed or terminated (DoD).

NODLR Night observation device, long-range.

no-draft forging One forged essentially to finish dimensions, thus needing little if any machining.

NODS Night observation and detection system.

nodular cast iron See *SG cast iron*.

NOE Nap of Earth, ie flight as low as possible over undulating terrain.

NOF International Notam Office (ICAO).

no-feathering axis Axis of swashplate, that about which there is no feathering movement or first-harmonic variation of cyclic pitch.

no-flare landing Aeroplane landing (rarely, other aircraft) in which approach trajectory is continued in essentially straight line until landing gear hits ground.

no-go gauge One whose linear dimension (between faces, threads or diameter) is just below smallest permitted limit for part.

no-go item One whose failure or absence from aircraft prohibits take-off according to operating rules (though not necessarily rendering it unairworthy).

NOGS, Nogs Night observation gunship system.

NOISE National Organization to Insure [*sic*] a Sound-controlled Environment (US).

noise 1 Noise in air. Basic unit is decibel, dB, 0·1 bel, measure of sound pressure above local atmosphere on logarithmic scale, usually related to starting reference pressure of 2×10^{-5} Nm^{-2}. Sound pressure level $L = 10 \log p^2/p_0^2 = 20 \log (p/p_0)$ where p_0 is reference pressure and p actual measured pressure. Alternative is to use source power level $L_w = 10 \log (W/W_0)$dB where W_0 is reference power commonly taken to be 10^{-12} W (Watts). Pressure levels are more common, and log scale allows for million-fold increase in human perceived pressures, each 6 dB increment representing doubling of pressure level. Study of aircraft noise from 1952 led to many new measures in attempts to quantify noise nuisance. In 1953 CNR (Community Noise Rating) gave single-number scale based on public response to six generally quantifiable factors, and in 1957 NC (Noise Criteria) curves attempted to portray equal-loudness contours taking into account discrete tones, impulsive nature of some sounds and other variables. By this time many workers had tried to quantify human aural response to different frequencies and tones with mixed frequencies, and curves drawn in 1959 were labelled L_{PN} (Perceived Noise Level) in units of dB(PN), sometimes written PNdB. Despite its complexity this gained major foothold, and virtually eliminated traditional measures (phon, relating sound pressure level to standard 1 kHz tone, and sone, loudness corresponding to 40 phons). Various weighted dB measures were introduced for the measures taken by meters with scales adjusted to equal-loudness contours for different overall pressure levels, these being called by various letters (thus, A-weighted = dBA = L_A). In 1961 a series of surveys measured annoyance according to new measures, $\bar{L}_{PN}$ (outdoor average peak L_{PN}), L_{PN50} or L_{PN90} (L_{PN} exceeded by 50 or 90 per cent of aircraft), D_{85} or D_{95} (10 log time in seconds when sound pressure exceeded 85 or 95 dB), and N (number of aircraft "passing over", latter criterion not being defined); result was single value for location called NNI (Noise and Number Index). Another 1961 unit was derived by splitting noise into one-third-octave bands and assigning each Noy rating by comparing with subjective noisiness of random noise centred on 1 kHz; individual Noy figures then added by method allowing for masking of one band by others and presented in PNdB. Further work allowing for particular features — such as intense pure tones, as from compressor blading, in otherwise broadband jet sound — led to use of L_{EPN} (Effective Perceived Noise Level) measured in EPNdB (Effective Perceived Noise dB) in first-draft legislation in 1966, which led to FAR Pt 36 and subsequently closely similar ICAO Annex 16. By this time at least 20 national or local authorities had published research, including Australia's AI (Annoyance Index) = $\bar{L}_{PN}$ + 10 log N; German Störindex $\bar{Q}$ based on dbA; French R-index = $\bar{L}_{PN}$ + 10 log N – 30; Dutch Total Noise Rating B based on log of summation of A-weighted pressure levels; American CNR (Community Noise Rating) based on many variables; California's CNEL (Community Noise Equivalent Level) using WECPNLs (Weighted Equivalent Continuous Perceived Noise Levels) varying with time of day and season; American NEF (Noise Exposure Forecast) = $\bar{L}_{EPN}$ + 10 log N – K where K is 88 by day and 76 by night; South Africa's NI (Noisiness Index) = $\bar{L}$ + 10 log N + 10 log T_a/T where T_a and T are times; British TNI (Traffic Noise Index) and resulting L_{eq} (Equivalent Average Sound Pressure Level), which led to L_{NP} (Noise Pollution Level) = L_{eq} + 2·56 σ where σ is standard deviation of dB fluctuations. Further measures include Senel (Single-Event Noise Exposure Level), SIL (Speech Interference Level), L_{TPN} (Tone-Corrected Perceived Noise Level) and various octave-band measures such as $L_{300\ 600}$ (sound pressure level of band 300–600 Hz). Also see *Approach* *, *sideline* *, *take-off* *.

2 Background noise, that present in electronic amplification, communication or recording system in absence of signal.

3 Thermal or Johnson noise caused by thermal agitation of charge carriers.

4 White (Gaussian) noise, constant energy per unit bandwidth.

5 Shot noise, fluctuation in charge-carrier current.

6 Random noise (eg white, shot), uniform energy versus frequency distribution.

noise (electronic) 1 Effects of unwanted signals, including those generated within the system.

2 Unwanted signals themselves.

noise abatement Deliberate procedures whose objective is minimization of noise perceived at ground.

noise-abatement climb procedure Modification of first segment to reach maximum attainable height AGL at point where ground track crosses boundary of built-up area, there cutting back power to predetermined value just sufficient to maintain positive rate of climb until either built-up area is passed (some operators add 1 nm margin) or height AGL exceeds FL50.

Noise and Number Index See *noise (1)*.

noise attenuation Design and/or constructional features whose purpose is to minimize externally perceived noise. Techniques include addition of sound energy-absorbing linings, structural and aerodynamic features to change frequencies (eg of blades passing), mechanical design to reduce noise of bearings and gear teeth, and maximization of jet-nozzle periphery to increase rapidity of jet/atmosphere mixing.

noise carpet Area along aircraft track subjected to significant noise nuisance.

noise certification Certifying authorities in all but a few states require compliance with noise (and emissions in most cases) legislation for all new civil aircraft. Older aircraft have to comply at specified future dates.

noise contour Locus of points on ground at which specified air traffic results in particular perceived noise level, NNI or other noise nuisance. Traffic may be an average of arrivals or departures, a weighted average, a "noisiest" aircraft type or an NNI figure taking frequency of flights into account.

noise exposure Not defined but related to noise levels, number of events (though Senel [see *noise (1)*] is one-event) and time of day.

noise floor Hypothetical minimum background noise level.

noise footprint Outline — generally footprint-shaped — of enclosed region around runway bounded by particular noise contour (often 90 EPNL) resulting from one landing and one take-off by particular aircraft type operating at MTOW in ISA with measures taken at standard reference points and other locations.

noise reference points In civil-aircraft certification, three locations at which noise measures are taken. See *approach noise*, *sideline noise* and *take-off noise*.

noise-shield aircraft One in which basic configuration, by design or as fortuitous bonus, places major portions of structure between main noise sources and ground.

noise shielding Portions of aircraft which, by design or fortuitously, are interposed between noise sources and distant observers.

noise suppression See *noise attenuation*.

noise suppressor Jet nozzle configured to reduce noise by increasing periphery of nozzle(s) and speeding mixing.

no joy Air intercept code: "I have been unsuccessful, or have no information".

NOL Naval Ordnance Laboratory (US).

no lift Stencilled instruction to ground personnel prohibiting application of lifting forces in local area.

no-lift angle That between no-lift direction and chord.

no-lift direction Angle of attack of two-dimensional aerofoil section at which lift is zero at low airspeeds. In practical wing ** varies from root to tip.

no-lift wire One bracing aerofoil from above; also called anti-lift wire.

NOLO, nolo No live operator, ie RPV is preprogrammed.

Nomex Family of nylon/phenolic honeycomb structures.

nominal acceleration point That geographical

location at which SST is to begin supersonic acceleration.

nominal deceleration point Location, varying with flight level and pitch attitude, at which SST is to begin deceleration to subsonic regime.

nominal dimension Various interpretations, typically that indicated on drawing before allowances, fits and tolerances.

nominal gas capacity That of gas cells of aerostat under defined conditions of inflation, ambient pressure and flight attitude.

nominal performance Published, or according to brochure.

nominal pitch See *standard pitch*.

nominal weapon Nuclear weapon having yield of approximately 20 kt.

NOMSS National operational meteorological satellite system (NOAA).

non-co-operative scorer One whose ammunition is not modified, or does not need modification, for scoring purposes (ASCC).

non-co-operative target One without emissions, transponder or enhancement device.

non-destructive testing Methods of testing structures for integrity, esp absence of manufacturing flaws or cracks, that do not impair serviceability or future life.

non-differential spoilers Main feature is that in airbrake mode all spoilers remain extended even in demand for roll (see *spoilers*).

non-effective sortie Aircraft which for whatever reason fails to accomplish mission (DoD).

non-ferrous Metals and alloys not based on iron; term usually also excludes aluminium alloys and generally means coppers and brasses.

non-fluff Lint-free.

non-flying prototype Essentially mock-up built to full flight standard but, for whatever reason, not cleared or intended for flight.

non-frangible wheel Various techniques applied to design and fabrication of turbine disc to preclude possibility of rupture in overspeed or asymmetric condition.

non-galvanic corrosion That due to causes other than formation of electric cells; two important examples are fretting and microbiological.

non-holding side That on left side of holding course inbound towards holding fix.

non-interchangeable socket Otherwise standard multi-pin or other sockets on device which ensure correct attachment of several connectors.

non-landing section That length of runway from original threshold to displaced threshold.

non-operating active aircraft Allowance, usually 10%, above UE level to make up for IRAN, mods and heavy maintenance (USAF).

non-precision approach Without electronic glideslope.

non-precision instrument runway Without ILS.

non-program aircraft Those in inventory other than active or reserve, eg experimental or withdrawn (DoD).

non-rigid airship Without rigid skeleton or stress-bearing covering around lifting cells.

non-sked Non-scheduled, ie not operating a timetable (colloq).

non-structural Other than primary or secondary structure, physical breakage of which would not imperil continued flight.

non-traffic stop Stop by transport aircraft in public service planned in advance for reasons other than to pick up or set down.

non-volatile Permanent memory.

NOPR Notice of proposed rulemaking (FAA).

NOPT No procedure turn required (FAA).

Norad North American Air Defense Command.

Norden sight Complex but highly accurate optical bombsight for high-altitude level bombing (US, 1941–49).

Norden gear Patented carrier-landing energy-absorption system.

NOR gate Logic circuit usable as either AND or OR, depending on logic levels chosen to represent 0 or 1.

NORM, Norm Not operationally ready, maintenance.

normal Perpendicular to.

normal acceleration Acceleration in vertical plane relative to aircraft, along OX axis (eg as result of rotation about OY axis).

normal axis Vertical axis (note: may not be vertical but must be at 90° to longitudinal axis in plane of symmetry); also called OZ axis. Positive direction is downward.

normal force That measured on body in fluid flow at 90° to free-stream direction, symbol Z (rarely N).

normal force coefficient Dimensionless coefficient C_Z derived from Z; also written $C_L \cos \alpha + C_D \sin \alpha$ where C_L and C_D are lift/drag coefficients and α is angle of attack.

normal glide That at which glide ratio is maximum.

normal gross weight Usually same as MTOW; excluding all overload, emergency or alternate gross weights.

normal horsepower Not defined but generally same as rated hp.

normalizing Stress-relieving heat treatment usually comprising heating to above critical temperature followed by cooling in atmosphere.

normal landing For tailwheel aeroplane, three-point landing.

normal loop Loop, as distinct from inverted loop, starting and finishing in straight and level flight in upright attitude.

normal mode Free vibration of undamped system.

normal outsize cargo That having cross-section greater than 9 ft × 10 ft, ie C-130 or C-141 size (DoD).

normal pressure drag C_{Dp}, downwind resultant

force coefficient.

normal propeller state Usual condition for helicopter under power, with rotor thrust in opposition to flow direction through disc.

normal shock Shockwave at 90° to fluid flow direction.

normal spin Intentional spin entered from upright attitude and recoverable by centralizing controls or applying opposite rudder (US usually adds "within two turns").

normal turn Procedure turn through 360° in two minutes.

Normand theorem On tephigram a dry-adiabatic line drawn through dry-bulb temp, saturated-adiabatic through wet-bulb temp, and dewpoint line through dewpoint temp all meet at point which represents condensation level.

NORS, Nors Not operationally ready, spare parts (or supply, as an order).

northerly turning error Transient errors in magnetic-compass reading caused by vertical component of magnetic field, at maximum when turning off northerly or southerly course. In N hemisphere compass is sluggish and lags behind when turning to L or R through northerly heading and races ahead when turning L or R through southerly.

north mode Display, eg Automap or moving-map display, has N at top.

NOS Night observation sight (or surveillance).

NOSC NATO operations support cell.

nose 1 Foremost part of vehicle, measured relative to direction of travel, excluding secondary structures or probes.
2 Leading portion of aerofoil, hence D-*, * rib.

nose art Decorative painting on aircraft forward fuselage or nose.

nose battens Radial stiffeners around nose of airship.

nose-cap 1 Removable nose, usually body of revolution, forming extremity of larger forebody.
2 Small spinner not extending further aft than front of blade roots.
3 Bow cap.

nosecone Essentially conical nose of high-speed vehicle, esp fairing over re-entry vehicle.

nose-dive Dive at very steep (near-vertical) angle.

nose down 1 To push over from level flight into glide or dive.
2 To fly with fuselage in ** attitude, though not necessarily losing height.

nose drive Shaft drive to auxiliaries taken off front of gas-turbine engine along axis of symmetry, eg ** generator.

nose entry Shape of aircraft nose evaluated from aerodynamic and aesthetic viewpoints.

nose gear Forward unit of tricycle landing gear, no matter how far location is from nose.

nose gearbox Gearbox mounted on front (usually on centreline) of gas-turbine engine to drive auxiliaries, helicopter shafting or propeller (ie turboprop); not mounted remotely on struts.

nose-heavy Tending to rotate nose-down when controls released.

nose in 1 To taxi and park facing terminal building or finger.
2 Aircraft thus parked (see *Agnis, Safeway, sidemarker*).

nose leg Main leg of nose gear.

nose over To overturn (eg after landing tailwheel-type aircraft on soft ground) by rotating tail-up to inverted position.

noseplane Canard foreplane mounted at or ahead of nose; not applicable to conventional modern canards.

nose radar Radar whose aerials (antennas) are in nose of aircraft pointing ahead, especially for use against targets ahead of aircraft.

nose ribs Ribs along leading edge extending chordwise only as far as front spar.

nose slots Apertures in low-pressure region of high velocity around nose for discharge of fluid flow, eg cooling air.

nose tow Standard US Navy method of catapult link for accelerated carrier take-off by pulling on nose leg.

nose up To rotate in pitch from level flight to climb.

nosewheel Wheel(s) of nose landing gear.

no-show Airline passenger who has booked ticket but fails to check in for flight.

Nosig No significant meteorological change (ICAO).

no step Stencilled warning on aircraft: do not put weight on this area.

NOT Naczelna Organizacja Techniczna (Polish federation of engineering associations).

Notal Not to all.

NOTAM, Notam Notice to Airmen, identified as notice or as Airmen Advisory, disseminated by all means to give information on establishment, condition or change in any aeronautical facility, service, procedure or hazard (ICAO).

Notam code Standard code for transmitting Notams; eg QAUED 3 MC 5813 142359 is interpreted as "Met com operating frequency of 3 MHz will be changed to 5,813 kHz on 14th of this month at 23·59".

Notar No tail rotor (Hughes).

notch Essentially chordwise or streamwise saw-cut or groove over nose of aerofoil.

notch aerial Formed by cutout in skin of vehicle, leaving aperture matched to wavelength (usually in HF com band) and covered with dielectric skin to original profile.

notifiable accident One which cannot legally go unreported, where any person suffers injury, third-party property is damaged or public are in any way put at risk; variable rules governing scale of damage to aircraft.

no-transgression zone Airspace where aircraft

under positive electronic IFR control must not penetrate, esp region 900+ m/2,000+ ft wide between aircraft making ILS approaches on to parallel runways.
NOTS Naval Ordnance Test Station (US, China Lake, Inyokern etc).
NOV-AB Non-persistent Toxic-B gas bomb (USSR).
Novcam Non-volatile charge-addressed memory.
Novoview Range of CGI (2) visual systems, some textured (Rediffusion).
no-wind position Geographical position aircraft would have occupied had wind velocity been zero.
NOx, NOX Nitrogen/oxygen breathing mixture (normally means supplied in absence of atmosphere).
noy, Noy Subjective measure of noisiness in bandwidths of one-third octave (see *noise*).
nozpos Nozzle position, ie angle.
nozzle 1 In jet-propulsion or reaction-jet control, aperture through which fluid escapes from system to atmosphere and in which as much energy as possible is converted to kinetic energy (see *choked*, *con/di*).
2 Wind-tunnel section immediately upstream of working section.
3 Primary aperture through which fuel is injected into gas-turbine engine combustion chamber.
4 Section of fluid flow duct upstream of axial turbine, in which flow is controlled to enter turbine at favourable direction, pressure and velocity. Form depends on whether turbine is impulse/reaction or pure impulse.
5 Incorrectly, nozzle guide vane.
nozzle blade See *nozzle guide vane*.
nozzle block See *nozzle (2)*.
nozzle box Assembly of nozzle guide vanes (all, or a group filling portion of periphery) and surrounding walls of gas duct.
nozzle bucket See *nozzle guide vane*.
nozzle contraction ratio Ratio of flow cross-section area at inlet or start of nozzle (esp con/di or rocket) to area at throat.
nozzle diaphragm Nozzle (4) as complete assembly.
nozzle efficiency Usually ratio of actual change in kinetic energy across nozzle to ideal value for given inlet conditions.
nozzle exit area Area of cross-section of flow at exit.
nozzle expansion ratio In supersonic nozzle, ratio of exit area to throat area.
nozzle guide vane Radial aerofoils upstream of axial gas turbine, convergent passages between which form nozzle (4) through which gas is directed on to turbine rotor blades.
nozzle insert Small blocking body fixed inside nozzle (1) to trim exit area; colloq = mice.
nozzle throat Region of con/di nozzle having smallest cross-section area.
nozzle thrust coefficient Usually defined as actual achieved thrust divided by product of chamber pressure and throat area (rocket).
NP Noise-preferential.
N_p Number of passengers.
NPA 1 Notice of proposed amendment (FAA, JARs).
2 National Packaging Authority (UK).
NPB 1 Nuclear-powered (or propelled) bomber.
2 Nadge Policy Board.
NPE Navy preliminary evaluation (US).
NPG 1 No performance group type aircraft (CAA).
2 Nuclear planning group.
NPL National Physical Laboratory (UK).
NPLOs NATO production and logistics organizations.
NPR No power recovery.
NPRM Notice of proposed rulemaking (FAA).
NPS Non-prior service.
NPT Non-proliferation treaty.
NPTR National parachute test range.
N_r 1 Helicopter main-rotor rpm.
2 Number (FAA).
NR, N_r Helicopter main-rotor rpm.
NRAG Non-metallics research advisory group (MRCC).
NRC 1 National Research Council (C).
2 Nuclear reporting cell.
NRCS Normalized radar cross-section.
NRDC National Research Development Corporation (UK).
NRDS Nuclear rocket development station (Jackass Flats, US).
NRL 1 Naval Research Laboratory (US).
NRO National Reconnaissance Office (US).
NROSS, Nross Navy remote ocean sensing system (spacecraft, USN).
NRP 1 Normal rated power.
2 Narrow programmable receiver.
NRSC National Remote Sensing Centre (in civil enclave at RAE Farnborough, UK).
NRT Near real time.
NRTS Not repairable this station (USAF).
nrv Non-return valve.
NS Non-skid.
Ns Nimbostratus.
ns Nanosecond (10^{-9}s).
NSA National Security Agency (US).
NSC National Security Council (US).
NSCA National Safety Council of America.
NSDA National Space Development Agency (J).
NSE Near-synchronous equatorial.
NSF National Science Foundation (US).
NSGr Night close-support group (G, WW2).
NSIA National Security Industrial Association (US).
NSNF Non-strategic nuclear forces.
NSO Nav/systems operator.
NSR No scheduled removal.

NSRP National Search and Rescue Plan (US).
NSS 1 National seismic station (US).
2 Near-source simulation.
NSSC Naval Ship Systems Command (US).
NST Noise, spikes and transients.
NSTL National Space Technology Laboratories (NASA, previously MTF).
NSTP National space technology programme (DTI, UK).
N-strut Interplane or other bracing system having general shape of N.
NSW Nominal specification weight.
NSWP Non-Soviet Warsaw Pact.
nT Nanotesla; terrestrial field varies from c 25,000 (magnetic equator) to c 70,000 nT (poles), vertical component usually being measured in geophysical prospecting and ASW.
nt Nit (name not usually used).
NTAOCH Notice to AOC holders (CAA).
NTC National training centre.
NTDS Naval tactical data system.
NTE Not to exceed.
NTG Nachrichtentechnische Gesellschaft im VDE (G).
Nth country Next power to possess nuclear weapons (DoD).
NTIS National Technical Information Service (US).
NTK Scientific and technical committee (many in USSR).
NTMV National technical means of verification.
NTNF Royal Norwegian council for scientific and industrial research.
NTO Nitrogen tetroxide (actually N_2O_4).
NTOS No time on station.
NTP Normal (ie standard) temperature and pressure.
NTPD NTP dry (gas bottle capacities).
NTS 1 Negative-torque signal.
2 Navigation technology satellite (USN).
NTSB National Transportation Safety Board (US).
NTU 1 State technical administration (USSR).
2 Navigation training unit.
NTWS New threat-warning system.
NTZ No-transgressor zone.
NU Nose-up.
Nu Nusselt number.
nuclear airburst Explosion at height AGL greater than maximum radius of fireball.
nuclear bomber Ambiguous but generally taken to mean aircraft able to deliver nuclear weapons.
nuclear cloud All-inclusive term for volume of hot gas, dust, smoke, and other particulate matter from nuclear bomb and environment carried aloft with fireball (DoD, NATO).
nuclear column Hollow cylinder of water and spray thrown up from underwater nuclear explosion through which hot high-pressure gases escape to atmosphere (DoD, NATO).
nuclear defence Defence against attack by nuclear or radiological weapons.
nuclear dud NW which after being triggered fails to provide any explosion of that portion designed to produce nuclear yield (DoD).
nuclear emulsion Thick layer used on photo-type plates for recording tracks of energetic particles.
nuclear energy That liberated by fission; more rarely, that liberated by fusion reaction, and some definitions include radioactive decay.
nuclear explosive Material designed to achieve greatest uncontrolled fission or fission+fusion reaction.
nuclear/heater propulsion Using nuclear reactor to heat working fluid for rocket propulsion.
nuclear incident Unexpected event short of NWA but resulting in damage to NW or associated facilities or increase in possibility of explosion or radioactive contamination (DoD).
nuclear radiation All EM and particulate radiations from nuclear processes.
nuclear reactor Device for containing controlled nuclear fission (rarely, fusion) reaction.
nuclear rocket Usually one whose working fluid is heated in nuclear reactor.
nuclear safety line Line drawn (if possible through prominent topographic features) on map to serve as reference in describing levels of protective measures, degrees of damage and limits allowed for effects of friendly NW.
nuclear surface burst One in which centre of fireball is below that height equal to maximum radius.
nuclear underground burst One in which centre of detonation lies below original ground level.
nuclear weapon One in which almost all released energy results from fission, fusion or both; abb NW.
nuclear-weapon(s) accident Unplanned occurrence resulting in loss of, or serious damage to, nuclear weapons or components resulting in actual or potential hazard to life or property (DoD, NATO).
nuclear-weapon degradation Degeneration of NW to such extent that anticipated yield is reduced.
nuclear-weapon employment time Reaction time of NW.
nuclear-weapon exercise Exercises involving real NW but excluding launching or flying operations.
nuclear yield Energy released in detonation of NW measured in terms of mass of TNT required to liberate same amount: categorized as very low (less than 1 kt), low (1–10 kt), medium (10–50 kt), high (50–500 kt) and very high (over 500 kt).
nucleon Particle of atomic nucleus, eg neutron, proton.
nucleonics Science of applications of nucleons and atomic emissions.
nucleus 1 Particle upon which atmospheric water vapour can condense and/or freeze.

2 Positively charged core of atom.

nuclide Member of family of atoms distinguished by having same nucleus (immediately decaying nuclear states are excluded).

Nudets Nuclear detonation detection and reporting system; system for surveillance of friendly target areas and immediately providing place, ground zero, burst height and yield of nuclear explosions (DoD, NATO).

nugget Obturator (US, colloq).

nuisance malfunction One caused by failure of safety system, eg auto-disconnect of AFCS, main system being serviceable throughout.

null 1 Orientation of receiver aerial (eg ADF) in which no signal is heard.

2 In DLC, angular setting (usually about 7°) about which spoilers oscillate.

3 Location in space where, at any moment, gravitational forces cancel out (eg five centres of libration near Earth/Moon system).

null position See *null (2)*.

numerical control Control of system, esp machine tool, by series of commands expressed in numeric (digital) terms giving locations, directions and speeds, and secondary information (eg cutter speed) in correct sequence. Software usually punched paper or magnetic tape.

NURK Astronautics Society (J).

nurse balloon Fabric gas container used as reservoir or to maintain constant inflation pressure in aerostat on ground.

Nusselt number Non-dimensional parameter Nu $= -qD/\lambda\delta T$ where q is quantity of heat, D is typical length, λ is thermal conductivity and δT is temperature difference.

Nut Near-unity probability.

nutating feed Microwave feed to tracking radar in which beam oscillates in one plane while plane of polarization (and usually aerial reflector) remains fixed.

nutation 1 Oscillation of axis of rotating body (eg gyro).

2 Irregularities in precession of equinoxes and other effects and of rotary precession of Earth's axis in period of 18·6 years (* period) with max displacement of 9·21 s (constant of *).

nutator Drive mechanism causing dipole or aerial feed horn to gyrate about focus of paraboloidal aerial without changing plane of polarization.

nutcracker V/STOL aeroplane whose fuselage hinges near mid-length to vector main-engine thrust.

nut plate, nutplate Nut, esp elastic stop nut, provided with flat-plate base attached permanently to airframe to provide anchor into which attachment bolt can be screwed for securing access panel or other removable item.

nut runner Large nut threaded on screwjack so that when either ** or screw is rotated a linear motion results (eg to drive trailplane).

NVG Night-vision goggles.

NVIS Near-vertical incident skywave.

NVL Netherlands air transport association.

NVLP Netherlands Aerospace Writers Association.

NVM Non-volatile memory.

NVR Netherlands astronautical society.

NVS Night-vision system.

nvt Neutron volts, measure of ionizing radiation.

NV thrust Nominal vacuum thrust.

NVvL Nederlandse Vereniging voor Luchtvaarttechniek.

NW 1 Nuclear weapon.

2 Nosewheel.

NWA Nuclear-weapon(s) accident.

NWC 1 Naval Weapons Center (China Lake, US).

2 National War College (US).

NWDS Navigation and weapons-delivery system.

NWEF Naval Weapons Evaluation Facility (Albuquerque, US).

NWS 1 National Weather Service (NOAA, US).

2 Nosewheel steering.

3 North warning system (former Dew Line).

NWSSG Nuclear weapon system safety group.

NXTK Next track.

Nycote Nylon lacquer protective coating.

nylon Long-chain synthetic polymer amide with recurrent amide groups distributed along chain (some definitions add "capable of being drawn into filament whose structure is oriented along fibre"). Large family, now often used for 3-D mouldings and many structural purposes (formerly TM for Du Pont de Nemours).

N_Z Normal load factor, ie along OZ axis.

NZAPA New Zealand Airline Pilots' Association.

NZCA NZ College of Aviation.

NZGA NZ Gliding Association.

NZMAA NZ Model Aeronautical Association.

NZMS NZ Meteorological service.

NZRA NZ Rotorcraft Association.

O

O 1 Opposed configuration (US piston engine symbol).
2 Operating cost over given period.
3 Ground speed "on" (onwards from critical point).
4 Observation aircraft category (DoD).
5 Hang-glider category (FAI).
O_2 Oxygen.
O_3 Ozone.
O-ring Flexible fluid-sealing ring having O-section in free state.
OA Output axis.
OAA Orient Airlines Association.
OAC Oceanic area control centre.
OACI ICAO (F).
OAE Optimized after erosion (helo blade profile).
OAFU Observers advanced flying unit.
OAM Office of Aviation Medicine (FAA).
OAMS Orbital attitude and manoeuvre system.
O&M Operations & maintenance.
O&S Operational and support (costs).
OANS Observers air navigation school.
OAO Orbiting astronomical observatory.
OAP 1 Organizzazione dell'Aviazione Privata e d'Affari (I).
2 Offset aiming point.
OAPEC Organization of Arab Petroleum Exporting Countries.
OAPS Oblique air photograph strip.
OAR Office of Aerospace Research (USAF).
OARN Off-airways R-nav.
OART Office of Advanced Research & Technology (NASA; now AST).
OAS 1 Offensive avionics system.
2 Offensive air support (air support directly linked to land operations, US).
3 Omnidirectional airspeed system (helo).
OASC Officer and Aircrew Selection Centre (RAF, Biggin Hill).
OASD Office of the Assistant Sec Def (US).
OASF Orbiting astronomical support facility.
Oasis 1 Oceanic and atmospheric scientific information system (NOAA).
2 Operational application of special intelligence systems (AAFCE).
3 Omnidirectional approach slope indicator system.
OAST Office of Aeronautics and Space Technology (NASA), more usually AST.
OAT 1 Outside air temperature.
2 Operational-acceptance test.
3 Operational air traffic (ie military).
4 Oxide aligned transistor.
OAVUK Society for aviation and gliding of the Ukraine and Crimea.
OB Outbound.
O/B 1 Outbound.
2 Outboard.
OBA Outbound boom avoidance (SST).
OBCO(S) On-board cargo operations (system).
OBDMS On-board data-monitoring systems.
OBE 1 Off-board expendables.
2 Overtaken by events.
OBFM Offensive basic flight manoeuvres.
OBI Omni-bearing indicator.
Obiggs On-board inert gas generating system.
objective 1 Normally, in optical, EO or IR system, first lens or lens group to receive incoming radiation.
2 Military target for capture or other action by surface forces; not used for target for aerial attack.
oblique Oblique photograph.
oblique camera One mounted in aircraft with axis between vertical and horizontal.
oblique photograph One taken by oblique camera; subdivided into high * in which apparent horizon appears and low * in which it does not.
oblique projection Map projection with axis inclined at oblique angle (say, 20° to 60°) to plane of Equator.
oblique-shock inlet Inlet designed for use in supersonic vehicle and provided with centrebody, wedge or other projecting portion intended to focus oblique shockwave on opposite lip.
oblique shockwave Inclined shock formed whenever supersonic flow has to turn through finite angle in compressive direction.
oblique wing Wing arranged to pivot at midpoint as single unit so that, as one half is swept back, other half is swept forwards. Also called slew wing.
Oboe WW2 precision navaid of SSR type with Cat station near Dover and Mouse station in Norfolk sending synchronized pulses which formed continuous note along correct flightpath over distant target; Mouse operators also sent signals to tell aircrew when to release bombs or TIs.
Obogs On-board oxygen generation (or generating) system.
OBP On-board processor.
OBRC Operating budget review committee.
OBS 1 Omni-bearing selector.
2 Observe/observed/observing (ICAO).
3 On-board simulation.
obs Obstruction lights.

OBSC Obscure/obscured/obscuring (ICAO).

obscuration methods Instrument techniques for measuring visible smoke (eg from jet engine) such as HSU and PSU where cutoff of calibrated light beam is measured over known distance.

observation 1 Many military definitions, most agreeing that * platform is for gathering all possible tactical information about an enemy.
2 Complete set of meteorological readings at one place and time.

observed value Measured, not calculated.

observer Common title for second crew member in two-seat military (especially combat) aircraft whose functions may include navigation, systems management, electronic warfare, command guidance of weapons and other tasks; title is traditional and may bear no relation to actual duties.

obstacle Obstruction.

obstn Obstruction (FAA).

obstruction Real or notional solid body forming hazard to aircraft on or near runway; obstruction. For certification purposes has height of 10·7 m (35 ft) or 15 m (50 ft).

obstruction angle Angle between horizontal and line joining highest point of object in flightway to nearest point of appropriate runway.

obstruction clearance surface Surface in form of plane or flat cone sloping at obstruction angle.

obstruction light Red light visible 360° on top of object dangerous to moving aircraft in air or on ground.

obstruction marker Object of approved shape or colour marking obstruction or boundary of hazardous surface on airfield.

obturator Rigid or flexible body tailored to preventing escape of gas under pressure from particular orifice or other leakage path.

obturator ring Piston ring of L-section intended to be gastight.

OBW On-board wheelchair (for disabled pax).

OC, O/C 1 O'clock.
2 Officer commanding.
3 Over-current.
4 Obstacle clearance (ICAO panel).
5 Optical communications.

OCA 1 Offensive counter-air.
2 Oceanic control area.
3 Obstacle clearance allowance.

OCAMS, Ocams On-board checkout and monitoring system.

OCC Operational control centre.

occluded front Warm air mass forced aloft by overtaking cold front.

occlusion 1 Atmospheric region in which cold front has overtaken slow or stationary warm front and forced warm air mass upwards.
2 Trapping of gas bubbles in solidifying material or by molecular adhesion on surface, esp removal of gas in high-vacuum technology by a getter.

occultation Complete disappearance of body, esp source of illumination such as star or aircraft lights, behind another object of much larger apparent size.

occulting Flashing, but with illuminated periods clearly predominant.

oceanic airspace Controlled airspace over ocean areas.

oceanic clearance Clearance delivery to enter oceanic airspace.

OCI Outside of clearance indicator (MLS).

OCL 1 Obstacle (or obstruction) clearance limit.
2 Optimum cruising level.

OCM 1 On-condition maintenance.
2 Optical countermeasures.

OCP Omnidirectional control pattern.

OCR 1 Optical character recognition.
2 Oceanic control region.
3 On-condition replacement.

OCS 1 Optimum-cost speed.
2 Officer candidate school (US).

oct Octane (FAA).

octa Unit of visible sky area representing one-eighth of total area visible to celestial horizon.

octal base Standard base for electronic device having eight connector pins.

octane number Standard system for expressing resistance of hydrocarbon or other fuel to detonation in piston engine, ranging from 0 (equivalent to n-heptane) through 100 (iso-octane) upwards to about 150. Also called knock rating (strictly, anti-knock) or performance number, esp when above 100.

octane rating See *octane number*.

octane test Standard test in which sample of fuel is used to run special variable-compression single-cylinder engine and compared with mixtures of n-heptane and iso-octane, or (above 100 octane number) with other reference fuels.

octant Bubble sextant able to measure angles to 90°.

octave Interval between any two frequencies having ratio 1:2.

octet Complete set of four pairs of electrons forming layers 2 and 3 (in all noble gases except He, outer shell).

Octol HE warhead filling, HMX/TNT.

OCU 1 Operational conversion unit.
2 Optical coupler unit.

OCV Organisme du Contrôle en Vol (F).

OD 1 Outside diameter.
2 Ordnance delivery.
3 Olive drab.

ODA Overseas Development Administration (UK).

Odads Omnidirectional air-data system.

ODF Operational degradation factor.

ODL Optical data-link.

ODM 1 Ministry of Overseas Development (UK).
2 Operating data manual.

odometer Digital readout of numerical quantity, esp distance, as in DME.
ODR Overland downlook radar.
ODS Oxide-dispersion strengthened.
ODT 1 Overland downlook technology.
2 Omnidirectional transmitter.
ODVF Society of friends of the air fleet (USSR).
ODW Optimum-drag windmilling.
OE 1 Opto-electronic (not necessarily synonymous with EO).
2 Over-excitation.
OECD Organization for Economic Cooperation and Development (19 member countries).
OED Operational evaluation demonstration.
OEI One engine inoperative.
OEIM Organisme Egyptien pour l'Industrie Militaire.
OEM Original equipment manufacturer.
OEO Operational equipment objective.
OEP Office of Emergency Preparedness (US).
OER 1 Operational effectiveness rate.
2 Officer effectiveness report.
oersted Non-SI unit of magnetic field strength, = 79·577 A/m, exact conversion is 1,000/4π.
OEU Operational evaluation unit (RAF).
OF Over-frequency.
O/F Oxidizer (oxidant)-to-fuel ratio.
OFA Office Fédéral de l'Air (Switz).
OFAB HE fragmentation (bomb, USSR).
OFAM Office Fédéral des Aérodromes Militaires (Switz).
OFEMA, Ofema Office Français d'Exportation des Matériels Aeronautiques.
off-design Any operating condition other than that or those for which equipment was intended.
offensive avionics Those carried in attacking aircraft to assist fulfilment of mission, eg by providing navigation, target sensing and weapon guidance.
office Cockpit or flight deck (colloq).
off-line 1 Connected to computer but not forming part of dedicated controlled system.
2 Computer peripheral not directly communicating with central processor.
3 Not on airline's route network (eg Western Airlines has an * office in Washington DC).
4 Computer or EDP (4) installation which stores input data for processing when commanded.
off-mounted SSR aerial not mounted on primary radar but on own turning gear, and can thus be either synchronized to primary radar or rotated at predetermined data rate.
offset 1 In major international purchase, eg of quantity of combat aircraft, agreement in reverse direction in which purchasing country is awarded one or more contracts for products which may or may not be connected with original hardware. This kind of * may be (1) purely window dressing to render costly import less unpalatable, (2) important commercial deal to benefit original importing country, (3) designed to bolster ailing home industry, or (4) important vehicle for transfer of advanced technology to importer.
2 Linear distances, usually small, measured from baseline to joggled or tapered edge, bevel, flange at angle other than 90°, or similar part dimension.
3 Precisely defined point on ground in surface attack (see * *bombing*) or in space in air interception (see * *point*).
4 Linear or angular difference between major axis (eg of aircraft or engine) and axis of drive shaft from gearbox.
5 In fir-tree root, distance between teeth measured parallel to major axis of blade.
6 Length of normal common to landing-gear castor axis and wheel axle.
7 Lateral and vertical distances from desired G/S and runway centreline of (a) an actual landing aircraft, or (b) ILS beam (locus of peak signal strength).
offset attack See *offset bombing.*
offset bombing Any surface-attack procedure which employs aiming or reference point other than target.
offset distance 1 See *offset (2).*
2 Distance from desired ground zero or actual ground zero to point target or centre of area target (DoD, NATO).
offset frontal Meeting between two DA (3) aircraft and two hostiles (7–10 miles ahead) on reciprocal, such that no aircraft will pass between opposing pair.
offset hinge Helicopter main-rotor hinge at substantial radial distance from axis of rotation.
offshore procurement Procurement by direct obligation of MAP funds of materiel outside US territory; may include common items financed by other appropriations (DoD).
offst Offset track.
off the shelf Already fully developed to military or commercial standards and available from industry for procurement without change to meet military requirement.
OFP Operational flight programme.
OFPP Office of Procurement and Policy (DoD).
OFS Operational flying school (RN).
OFT 1 Operational flight trainer.
2 Operational flight test.
3 Operational flying training.
4 Orbital flight test.
OFV Outflow valve.
OFZ Obstacle-free zone.
OG Observation group (USAAC, USAAF).
OGA 1 Outer gimbal angle.
2 Office Général de l'Air (F, export distribution).
OGE Out of ground effect; supported by lifting rotor(s) in free air with no land surface in proximity.
ogee See *ogive.*

ogive 1 Loosely, any shape formed by planar curve whose radius increases until it becomes straight line.
2 Wing plan having approximate form of curve becoming straight line parallel to longitudinal axis on each side of centreline; so-called "Gothic window" shape.
3 Body of revolution formed by rotation of curve as in (1) about axis parallel to line on same side of line as curve.
4 Any of various other shapes such as conical * (body formed by rotation of two straight lines forming cone/cylinder), tangent * (circular arc tangent to line) or secant * (circular arc meeting line at angle).
5 Common definition is "surface of revolution generated when circular arc and line segment are rotated about axis parallel to line"; this is incorrect since curve need not be circular arc and shape can be planar and not body of revolution.
OGO Orbiting geophysical observatory.
OGV Outlet guide vane, or straightener vane.
OH 1 Overhaul.
2 Overhead.
O/H Overheat.
OHA Operating hazard analysis.
OHC Operating hours counter.
ohm 1 SI unit of electric resistance, defined as that of conductor across which potential difference of 1 V produces current of 1 A.
2 By analogy with (1), mechanical measure of resistance derived by dividing applied force by velocity; has dimensions of g/s.
ohmmeter Usually, compact instrument for giving quick approximate indication of resistance of circuit to direct current.
OID Optical incremental digitizer.
Oil Burner routes Published routes within continental US along which USAF, USN and USMC conduct high-speed training missions at lo level in VFR and IFR.
oil-can Portion of metal skin where compressive stress imposed in manufacture has resulted in slight local bulge between rows of rivets or other attachment which, when subjected to pressure difference or perpendicular force at centre, can suddenly spring inwards noisily; potential fatigue hazard with thin skin.
oil control ring(s) Piston rings whose main purpose is to prevent loss of lubricating oil from cylinder wall up into combustion chamber; usually one or two, below main compression rings.
oil cooler Heat exchanger whose purpose is to remove heat from lubricating-oil circuit; usually cooled by air or fuel flow.
oil coring See *coring*.
oil dilution Mixing petrol with lubricating oil to reduce viscosity for starting large PE (4) at very low temperature.
oil drive Hydraulic drive, usually signifying infinitely variable ratio.
oil radiator See *oil cooler*.
oil ring Scraper ring.
OIML Organisation Internationale de la Métrologie Légale (Int).
OIN Organisation Internationale de Normalisation (Int).
OIP Optimum implementation plan.
OIPC Organisation Internationale de Protection Civile (Int).
OIRI French for CCIR.
OITS Organización Internacional de Telecomunicaciones por Satelites (Int).
OJCS Organization (or Office) of the Joint Chiefs of Staff (US).
OJT On the job training (DoD).
OKB Experimental construction bureau, where aircraft are designed and developed (USSR).
OKO Experimental construction section (various, USSR).
okta See *octa*.
OL Operating location.
OLBR Operational laser-beam recorder.
OLDP On-line data processing (or processor).
OLEEA On-line exhaust emissions analysis.
oleo Telescopic strut which absorbs energy of compressive loads by allowing hydraulic oil or other fluid to escape under pressure through small restrictor orifice, usually controlled by one-way valve to reduce rebound.
oleo-pneumatic Absorbing shock by combination of air compression and forcing liquid through an orifice.
olive Small ellipsoid resembling an olive threaded on cable either to act as bearing surface in tubular guideway or, in continuous row separated by small spheres located in recesses in olive ends, to provide two-way push/pull mechanical interconnection (eg in Bloctube system).
Olive Branch route Lo-level routes for B–52 training in continental US.
Ologs Open-loop oxygen generating system.
OLOS, Olos Out of line of sight.
OLRT On-line, real time.
OM 1 Outer marker.
2 Organizational maintenance.
3 Operation & maintenance (USAF).
4 Other management.
5 Over maximum (often O/M).
6 Occupational medicine.
7 Operations module.
OMAG Independent naval aviation group (USSR).
OMB Office of Management & Budget (US, White House).
OMCFP Optimized MAC computer flight plan.
OMCM Operational and maintenance configuration management (software).
OME 1 Operating mass empty, usually same as OWE.
2 Operational mission environment.
Omega Accurate long-range radio navaid of

VLF hyperbolic type, covering entire Earth from eight ground stations and usable down to SL.

omega (ω) 1 Angular velocity (SI unit is rad/s).
2 Angular frequency.
3 Any generalized frequency, thus ωr is undamped natural frequency of rth mode.

OMG Operational manoeuvre group (USSR).

OMI Omnibearing magnetic indicator.

OMM Organisation Météorologique Mondiale = WMO (Int).

Omni VOR (colloq).

omni-axial nozzle Rocket motor nozzle capable of being vectored to any angle within prescribed limits about pivot point defined by intersection of axes of symmetry of propelling system and nozzle.

omni-bearing selector, OBS Knob on most VOR/ILS indicators which is turned to each required VOR bearing, which appears in a three-digit window display, left/right needle thereafter showing deviation from required heading.

omnidirectional aerial One emitting to all points of compass (assumed equal signal strength throughout 360°).

omnidirectional beacon Fixed ground radio station giving non-directional signals for airborne DF receivers (obs).

omniflash beacon Airborne flashing (strobe) light visible equally through 360° in azimuth.

omni-range See *VOR*.

omnivision Uninterrupted view through 360°.

OMOS Department of experimental marine aircraft construction (USSR, defunct).

OMS 1 Crew chief (USAF).
2 Orbital manoeuvring subsystem.
3 Organisation Mondiale de la Santé (Int).

ON Omega navigation.

ONA Office of Noise Abatement (FAA).

ONAC Office of Noise Abatement and Control (EPA, US).

on-board cargo operations system Microprocessor-based system which automatically positions each item for correct c.g. and best structural integrity.

on-board oxygen generation Usually removes nitrogen from bleed air by Zeolite molecular sieve.

on call Ready for prearranged mission to be requested.

on-call target Planned nuclear target to be attacked not at specific time but on request.

on-condition maintenance Performed only when condition of item demands, instead of at scheduled intervals.

on-course signal Electronic signal indicating receiver is on desired course, usually in form of steady note created by two superimposed coded signals.

one and one-half stage Vehicle configuration, esp ballistic rocket, in which single set of propellant tanks feed two or more thrust chambers, one or more of which is jettisoned in flight.

one-dimensional flow Flow through duct in which all parameters are, or are assumed to be, constant throughout any section normal to the direction of flow.

one-dot error One scale division from centre on ILS/VOR needle or similar-type instrument, usually equal to 0·5°.

one-hundred-and-eighty-degree approach Any of several standard approach procedures involving downwind leg and 180° turn; variations include 180° accuracy landing, 180° spot landing, 180° U-turn approach and 180° semicircular approach.

one-hundred-and-eighty-degree error Misreading compass by flying reciprocal of desired course, avoided by rule "red on red".

one-minute turn Procedure turn through 360° (rarely, 180°) taking one minute to complete; bank angle is about 30°.

one-off Aircraft of which only a single example is constructed.

one-on-one Classical air combat by two opposing aircraft.

ONERA Office National d'Etudes et de Recherches Aérospatiales (F).

ONFA Opera Nazionale per i Figli degli Aviatori (I).

onglet Leading-edge root fillet.

on-line 1 Of computer or EDP (1) system, automatically receiving and instantly processing information and sending output data to required destination.
2 Of EDP (1) peripherals, operating under direct control of central processor.

on-mounted Indicates that SSR aerial, ECM aerial or similar ancillary device is mounted on, and oscillates or rotates with, main radar.

ONS Omega navigation system.

on the beam Correctly aligned on ILS glidepath or other guidance beam. Today has come to mean "on the right track" in most general sense.

on the deck At minimum safe altitude (colloq).

on top 1 Above unbroken cloud, with CAVU environment.
2 Directly above operating sonobuoy or indicated submarine position.

Ontrac II American VLF navaid of rho-rho type.

OOF Out of flatness (machined plate).

OOK Department of special construction (USSR).

OOS Department of special aircraft construction (USSR).

OP Observation post.

OPACI, Opaci Opera Pionieri e Anziani dell' Aviazione Civile Italiana.

opacity In optical systems or photographic film, reciprocal of transmittance (log O = density d); term absorbance is now recommended.

OPAL, Opal Order processing automated line.

opaque plasma One through which EM signals

cannot pass; generally plasma is opaque for EM frequencies below plasma frequency.

opaque rime White icing of porous form caused by rapid freezing of small droplets.

OPB Oxygen preburner.

OPC Operational control (ICAO).

OPEC, Opec Organization of Petroleum Exporting Countries.

open With normal control path disconnected; such failure interrupts or seriously distorts signal passing along that channel.

open angle Angle of mating part slightly greater than 90° or other angle of edge of metal structure such as angle section; fault condition in sheet metalwork.

open-centre system Hydraulic system without an accumulator.

open circuit 1 Electrical, circuit interrupted.
2 Wind tunnel, one having no return path.

open class FAI/CIVV categories for competition sailplanes, in one case with Standard Class span of 15 m and alternatively with span unrestricted but usually 17 to 20 m. In each case all refinements such as flaps are permitted.

open-coil armature Ends of each coil connected to different bars of commutator.

open competition Industrial competition in which all proposals, promises or offers are communicated to all participants.

open delta Three-phase transformer comprising two single-phase transformers linking three lines.

open-ended Spaceflight continued until either all possible information has been gained, and mission objectives met, or spacecraft systems run low or become faulty.

open-gore Parachute in which fabric is absent from one gore to assist trajectory control.

open-hearth Principal method of high-quality steelmaking in many countries, using shallow regenerative furnace with or without oxygen.

open improved site Open-air site for military storage whose surface has been graded and surfaced with topping or hard paving.

open-jaw ticket Generally, airline booking out by one route and return by another or over return route to different destination.

open-jet tunnel Wind tunnel in which working section is open and has no tunnel walls.

open-link ammunition Rounds are held in clips that do not encircle each case, which can be pushed out diagonally instead of removed only to rear.

open loop Servomechanism or other system comprising control path only, with no measurement of result or feedback to give self-correcting action.

open production line Inactive, but in a state ready to resume production should further aircraft be required.

open rate Situation in which fares on particular air-carrier route are fixed by each operator.

open rotor Propfan or UDF.

open section Structural member devoid of closed and thus uninspectable surface.

open system Using stored food/oxygen and discarding all body wastes.

oper Operate (FAA).

operand EDP (1) quantity entering or arising in instruction; can assume many forms, from argument to address code; generally a word on which operation is to be performed.

operating active aircraft Those for which funds are provided, as distinct from non-operating active aircraft (DoD).

operating envelope Plot of extreme boundaries of conditions to which hardware is subject, eg acceleration/temperature, or (in case of aircraft) altitude/Mach or V/n gust envelope.

operating speed Traditionally 87·5 per cent of rated rpm for light piston engine (US, obs).

operating weight Total mass of aircraft ready for flight, including oil, water, food and bar stocks, passenger consumable stores, flight and cabin crews and their baggage and, according to most definitions, empty baggage containers where appropriate, but excluding fuel and payload. In case of military aircraft, ADI fluid and drop tanks are excluded but EW pods are included.

operation A military action, or carrying out of mission.

operational Ready to accomplish mission.

operational aircraft In British usage, one intended and ready for combat missions as distinct from transport or training.

operational air traffic Generally, that which cannot conform to requirements of flights within airways or controlled airspace.

operational characteristics Those numerically specified parameters describing system performance; system can be aircraft, radar etc.

operational command Normally synonymous with operational control, and covers all functions involving composition of forces, assignment of tasks, designation of objectives and direction of mission; does not include such matters as administration, discipline or training.

operational control 1 Authority granted a commander to direct forces to accomplish missions. (NATO etc).
2 Exercise of authority over initiation, continuation, diversion and termination of a flight (ICAO).

operational control communications Communications required for operational control (2) between aircraft and operating agencies.

operational effect rate Usually, flying rate.

operational evaluation Test and analysis of system under operational conditions, with consideration of capability offered, manning and cost, potential enemy accomplishments and alternatives, to form basis for decision on quantity production. In practice ** is often not accomplished until long after production decision.

operational factors Those exercising constraints on flightplan, notably ATC (1), pilot workload, available aids, specified refuelling locations etc.

operational flightplan Operator's plan for safe conduct of particular flight.

operational interchangeability Ability to substitute one item for another of different composition or origin without loss of effectiveness or performance.

operational loss or damage Loss of or damage to military item caused by reason other than combat.

operational manoeuvrability ceiling Maximum height at which, for any given weight, aeroplane can sustain specified load factor (vertical acceleration) at onset of buffet.

operational mode Any of selectable alternative methods of operation for functioning system, eg computer-controlled, automatic with feedback, automatic open-loop, semi-automatic and manual.

operational performance categories Categories I to IIIC defining ILS and blind (automatic) landing installation performance.

operational phase Period from acceptance by first user to elimination from inventory.

operational readiness Capability of unit or hardware to perform assigned missions or functions. Hence ** inspection.

operational readiness platform Ramp where aircraft at various ready states (including alert states) are parked, with all required connections (eg telebrief) in place.

operational readiness training Consolidated instructional period wherein qualified personnel for operational units are given integrated training in operational mode.

operational research Generally, analytical study of problems to provide numerical (some definitions say "scientific") basis for decision-taking; in US often called operations research.

operational test and evaluation Test and analysis of system under operational conditions, promulgation of associated doctrines and procedures, and continuing evaluation against new threats or changed environment or circumstances.

operations manual Usually, definitive handbook for user of system, as distinct from engineering, repair or design manuals; for aircraft called flight manual.

operative 1 Able to operate; in true fail-* system there is no loss in performance (though there may be in integrity) after failure.
2 Employee engaged in routine operation of machine on production work.

operator 1 Person, organization or enterprise engaged in or offering to engage in aircraft operation (ICAO).
2 Licensed air carrier.
3 American term for employee other than foreman or manager.

operator's local representative Agent located to obtain meteorological information for operational purposes and to provide operational information to local met office.

Opeval Operation evaluation.

OPF Orbiter processing facility.

OPLE Omega position-locating equipment; synchronous satellite providing control beyond LOS.

OPN Optimized-profile navigation.

opn Operation.

Opos Open purchase order summary.

OPOV Oxygen preburner oxidizer valve.

opportunity servicing Servicing carried out at any convenient time, within specified flight or operating-time intervals.

opportunity target See *target of opportunity*.

opposed engine PE (4) having left and right rows of (usually horizontal) cylinders either exactly or approximately opposite to each other, with crankshaft on centreline.

opposed-piston engine PE (4) usually of two-cycle compression-ignition type, in which each cylinder has central combustion space and two pistons working in opposition and driving crankshafts at opposite extremities.

OPR Office of primary responsibility (DoD).

opsec Operational security.

optical communications Systems using coherent (laser) light, usually channelled along optical fibres; owing to extremely high frequencies such systems have enormous information-rate capacity.

optical countermeasures ECM in visible range of EM spectrum.

optical fibre Filament drawn from two kinds of glass having different refractive indices, in form of core and sheath; light entering at one end is reflected each time it encounters core/sheath boundary until it emerges at other end.

optical guidance Usually, passive guidance based on sensitive photocells which sense presence of target and cause vehicle to home on it.

optical gyro Fibre-optic gyro.

optical instrumentation Use of optical systems (telescope, theodolite, precision graticules and markings on vehicle) in conjunction with cameras and TV to provide stored information of location in three dimensions, velocities and other parameters of hardware or phenomena.

optical landing system Refined form of mirror sight in which gyrostabilized light beams indicate to approaching pilot deviation from glidepath (USN, USMC).

optical line of sight LOS through atmosphere, which in precision measures is not to be regarded as straight.

optical linescan Term used for various line-by-line scanning at optical wavelengths, including

RBV (video).

optically flat panel Window used in visual sighting systems through which light passes with near-zero distortion.

optical maser See *laser*.

optical pyrometer Instrument giving numerical readout of temperature of hot surface by measuring incandescent brightness.

optical reference point Index mark on canopy or elsewhere providing pilot with sightline guidance in vertical or near-vertical landing.

optical turbulence Fluctuating distortion of light rays caused by time-varying gradients of refractive index.

optimized profile Flight trajectory calculated to minimize energy consumption throughout flight (but for operational reasons seldom attainable).

optimum angle of attack That giving maximum L/D and thus highest cruise efficiency.

optimum coupling Matched radio impedance in which load equals output of amplifier or transmission line, for maximum power transfer.

optimum height Height of air burst or conventional explosion producing maximum effect against given target.

option 1 Acquisition of particular place in aircraft or other major equipment item production programme, with or without deposit and without agreement to purchase.
2 Variation in standard of product offering customer choice, eg avionic fit, long-range tankage, de-icers etc.
3 Alternative decisions open to commander engaged in battle.

opto-electronic Combining optical and electronic subsystems in single device; term appears to be used synonymously with EO.

optrotheodolite Doubtful designation for precision position-correlating instrument combining laser and video camera.

OR 1 Operational research.
2 Operational requirement (UK).
3 Operational readiness; also (1985) operationally ready (DoD).
4 See *OR gate*.

O/R On request.

Oracle Operational research and critical-link evaluation.

Orads Optical ranging and detecting systems (USA).

Orange Force Simulated hostile force during exercise (NATO, IADB).

Oranges Sour Air-intercept code: weather unsuitable for mission (DoD).

Oranges Sweet Air-intercept code: weather suitable for mission (DoD).

ORB, Orb 1 Operations record book.
2 Omni radio beacon.

orbit 1 Closed elliptical or quasi-circular path of body revolving around source of gravitational attraction balanced by its own centrifugal force.
2 Closed pattern, usually circular or racetrack, followed repeatedly by military aircraft, eg in air-intercept loiter.
3 To follow an * as in (1) or (2).

orbital elements See *orbital parameters*.

orbital parameters Basic numerical values describing orbit, such as apogee, perigee, inclination and period.

orbital period Time to complete one revolution (relative to space, ignoring such irrelevancies as rotation of primary body); symbol P, whose square is always proportional to cube of semi-major axis of orbit (distance from centre of primary to apogee).

orbital rendezvous See *rendezvous*.

orbit determination Calculation of orbital parameters of unknown satellite.

orbiter Portion of spaceflight system designed to orbit, as distinct from booster stages, tanks etc.

orbit point Geographically or electronically defined location above Earth's surface used as reference in stationing orbiting aircraft.

ORC Oversight and review committee.

order book Real or symbolic book containing list of customers for major equipment item, esp commercial transport, broken down into firm orders, options, leases etc and fading from importance with programme's maturity and second-hand customers.

ordinate 1 Vertical axis on graph or other geometric plot.
2 Vertical distances measured from datum, eg from chord line to upper/lower surfaces of aerofoil.

ordnance Explosives, chemicals and pyrotechnics for use against an enemy, including nuclear, together with directly associated hardware such as guns, rocket launchers, etc.

ordnance devices Explosive or pyrotechnic parts of spacecraft, RPV or other system.

ordnance work rate Various measures intended to offer single numerical value of rate of application of ordnance in particular theatre or against one target.

ORGALIME, Orgalime Organization for liaison between European electrical and mechanical engineering industries.

organic 1 Originally, living material; today broadened to cover all compounds of carbon and hydrogen and often with other elements.
2 Assigned to, and forming essential part of, military formation.
3 In many applications involving heat exchange (eg * reactor, * Rankine cycle), use of organic (1) fluid, often mixture of polyphenyls.

organometallic Compound having metal atom attached to organic radical.

OR gate Logic device which, whenever a voltage (eg a 1) appears at any input, responds with voltage at output.

ORHE On-ramp handling equipment.

ORI Operational readiness inspection.
orientation 1 Determining approximate position or attitude by external visual reference.
2 Turning of instrument or map until datum point or meridian is aligned with datum point or true meridian on Earth (ASCC).
3 Direction of fibres in lay-up, prepreg or finished FRC part.
orifice meter Fluid-flow measuring technique where pressure is measured upstream and downstream of transverse plate with calibrated hole.
origin 1 Point 0,0 on cartesian graph.
2 Base from which map projection drawn.
originator Command by whose authority a message is sent.
ORLA Optimum repair-level analysis.
ornithopter Aerodyne intended to fly by means of wings flapping or oscillating about any axis, but not rotating.
orographic Large-scale uplift and cooling of air mass caused by mountains; hence * rain or precipitation.
OROS, Oros Optical read-only storage.
ORP 1 Optical reference point.
2 On-request reporting point.
3 Operational readiness platform.
ORS 1 Off-range site.
2 Offensive radar system.
3 Occurrence reporting system.
ORT 1 Optical relay tube (TADS).
2 Optical resonance transfer.
orthicon Camera tube with secondary emission reduced or eliminated by using low-velocity electrons to scan two-sided sensitive mosaic, one side scanned and reverse face illuminated, resulting in stored charges on mosaic being removed by beam and generating output signal.
orthogonal Originally meant perpendicular; today blurred, as shown below.
orthogonal aerial Pair of transmitting and receiving aerials, or single T/R aerial, designed to measure difference in polarization between radiated signal and echo from target.
orthogonal biplane One without stagger.
orthogonal scanning Use of combined axial and lateral magnetic fields to control low-velocity electron beam in camera tubes.
orthographic Rectilinear projection of solid objects on two-dimensional sheet by use of parallel rays in contrast to conical perspective rays; result is six possible views (left, right, front, rear, top, bottom), of which four (front, top, left, right) are normally used (in different relative positions in Europe and N America, see *third-angle projection*), each representing appearance of object as seen from infinitely large distance.
orthormorphic Map projection in which meridians and parallels cross at right angles, all angles are correct and distortion of scale is equal in all directions from any point; several variations are in use, chief being Lambert's conformal and Mercator's.
ORTL Optical resonance transfer laser.
OS 1 Observation squadron.
2 Operating system(s).
O/S Overspeed.
OSA Operational support aircraft.
OSAD Outer-Space Affairs Division (UN, Int).
OSAF Office of the Secretary of the Air Force (US).
OSBV One-shot bonding verfahren (G).
oscillation Repeated fluctuation of quantity on each side of mean value.
oscillator Device for converting direct current into alternating current to give control or carrier frequency, using not rotating mechanism but transistor or thermionic valves.
oscillogram Hard-copy (eg photographic) record of oscilloscope output.
oscillograph Sensitive electromechanical device translating varying electrical quantities into visible trace on paper or film (generally superseded by oscilloscope).
Oscillogyro Patented (Ferranti) gyro having unique property of measuring displacements about two axes.
oscilloscope Device translating varying electric quantities into visible traces on screen of CRT or similar display.
OSD Office of the Secretary of Defense (US).
OSE 1 Optical shaft encoder.
2 Office of Systems Engineering (DoT, US).
3 Operations support equipment.
Oseen flow See *viscous flow*.
OSEM Office of Systems Engineering and Management (FAA, US).
OSHA Occupational Safety and Health Act (US).
OSI Open systems interconnection (in-plant manufacturing).
OSK Department of special construction (USSR).
Osoaviakhim Society of friends of the aviation and chemical industries (USSR).
OSR Optical solar reflector.
OSS 1 Office of Strategic Services (US).
2 Department of experimental landplane construction (USSR, defunct).
OSSAI Organisation Scientifique et Sportive d'Aérostation Internationale.
OST 1 Outline staff target (UK).
2 Operational suitability test (DoD).
3 Office of Strategic Trade (US).
OSTA Office of Space and Terrestrial Applications (NASA).
OSTD Office of SST Development (DoT, US).
OSTIV Organisation Scientifique et Technique Internationale du Vol à Voile (Int).
OSV Ocean station vessel.
Oswatitsch inlet Supersonic inlet/diffuser in form of body of revolution having pointed coni-

cal centrebody of progressively increasing semi-angle, thus generating succession of inclined shocks at increasingly coarse angles all focused on peripheral lip.

OTA 1 Office of Technology Assessment (Congress, US).
2 Orbital transfer assembly.

OTAES Optical technology Apollo extension system.

OTAN NATO (F).

OTC 1 Operational training course (USAF).
2 One-stop tour charter.
3 Overseas Telecommunications Commission of Australia.
4 Operational test centre.

OTD 1 Origin to destination.
2 Optimal terminal descent.

OTDA Office of Tracking and Data-Acquisition (NASA).

OTE, OT&E Operational test and evaluation.

Otevfor Operational test and evaluation for operational requirements (USN).

OTH Over the horizon; hence * radar.

OTH-B Over the horizon, backscatter.

OTHT Over-the-horizon targeting.

otolith Organ of inner ear sensitive to attitude and acceleration.

OTP 1 Office of Telecommunications Policy (White House, US).
2 On-top position (ie, above submerged submarine).

OTPI On-top position indicator.

OTR Operational turn-round.

OTS 1 Over the shoulder.
2 Orbital test satellite.
3 Out of service.

OTSI Operating time since inspection.

Otto cycle General name for four-stroke spark-ignition cycle for PE (4).

OTU Operational training unit.

OTV Orbital transfer vehicle (reusable tug shuttling between Orbiter and higher orbits).

OTW On the wing; hence * replacement of engine module.

OTWD Out-the-window display (simulators).

OU Operations unit.

OUE Operational utility evaluation.

outage Loosely, failure to function, especially of communications.

outboard 1 In direction from centreline to wingtip.
2 Further from centreline (eg * engine, in relation to other[s] on same side).
3 Carried outside main structure.

outbound 1 Towards destination; hence VOR * radial.
2 Away from Conus (US military usage).
3 In holding pattern, side opposite to inbound (which heads toward holding fix).

outbound bearing That of outbound leg of holding pattern or of VOR radial en route to destination or next waypoint.

outbound radial That linking fix or beacon to next waypoint.

outburst Lateral spread across surface of air arriving in downburst.

outer cover External skin of rigid airship.

outer fix Fix in destination terminal area, other than approach fix, to which aircraft are normally cleared by ATC and from which they are cleared to approach for final approach fix (FAA).

outer loop FCS in autopilot mode, human pilot being inner loop.

outer marker ILS marker normally on approach centreline approximately 4½ nm (8·3 km) from threshold; usually identified by 400 Hz aural dashes and synchronized blue panel light.

outer panel Left or right wing outboard of centre section.

outfall valve Usually synonymous with outflow valve.

outflow pattern Isobar pattern above wing at positive angles of attack.

outflow valve That incorporated in isobaric cabin-pressure regulator through which air is allowed to escape to atmosphere during climb to isobaric altitude.

outgassing Emission of gas from metals and other materials in high-vacuum conditions.

out of alignment In case of propeller or rotor blade, having incorrect sweep (bent forward or back in plane of rotation).

out-of-control lights Two superimposed red lights 6 ft (2 m) apart hoisted at night on mast of marine aircraft whose engines have failed and which is not anchored.

out of phase Not synchronized; two or more cycles or wavetrains which have same frequencies but pass through maxima and minima at different times.

out of pitch In case of propeller or rotor blade, having blade angle different from that of remaining blades.

out of step Not synchronized; two or more series of digital pulse trains or other discrete phenomena having essentially same frequency but occurring at different times.

out of track In case of propeller or rotor blade, having incorrect tilt so that tip is ahead of or behind tip-path plane.

output 1 Total product or yield from system, eg tonne-km for airline.
2 Delivery from computer, as hard copy or graphics display.
3 One of five basic functional sections of most EDP (1) systems.
4 Signal from transducer or other instrumentation.
5 Power developed by engine or other prime mover.
6 Shaft at which power is extracted from shaft-drive engine.

7 Verb, to deliver an * (2), hence outputting.

outriggers 1 Primary structural members carrying tail of aeroplane with short fuselage, nacelle or boat hull, also called booms.
2 Ancillary landing gear near wingtip or at other location well outboard from centreline on aircraft with bicycle or other type of centreline landing gear.

outside air temperature OAT, free-air static temperature.

outside loop Inverted loop, performed with aircraft upper surface outwards and thus under negative g.

outside roll One started from negative g, esp from inverted attitude.

outside wing That on outer side of twin, having greater airspeed than inside wing.

outspin yaw Yaw inevitably present in normal spin, usually function of pitch attitude.

OUV Oskar-Ursinus Vereinigung, association of amateur aircraft constructors (G).

OV 1 Over-voltage.
2 Aircraft mission code, observation V/STOL (DoD).
3 Outside vendor.

ovalization Structural distortion such that circular parts become ovals, esp on pylon-mounted turbofans.

ovenize To bake electronic device to simulate overload test condition.

over Radio procedure word: "My transmission is ended and I expect a response from you."

overall efficiency In case of aircraft propulsion system, usually thermal efficiency multiplied by propulsive (Froude) efficiency.

overall length Most definitions for aircraft stipulate in flying attitude; distance between local perpendiculars aligned with extremities of basic airframe (but usually excluding pitot or instrument probes, static wicks etc). See *length*.

overbalanced Provided with excessive balancing mass, aerodynamic area or other source of control-surface moment assisting pilot, so that control surface has little feel (and in extreme case can move by itself to hard-over deflection).

overbooking Practice of selling more tickets for a particular flight than there are seats, to compensate for (1) no-shows and (2) individual pax or agents making multiple reservations for one journey.

overbuilt Made stronger than necessary, usually to allow for future development.

overburn Operation of rocket or other space propulsion system for too long a period (0·1 s can be significant), resulting in incorrect cutoff velocity and/or trajectory.

overcontrol To apply more control deflection than necessary, commonly resulting in succession of excursions on each side of desired flight condition.

overdesigned 1 Stronger than necessary to meet requirements.
2 More refined than necessary, usually resulting in lighter structure at expense of high cost and complexity.

overexpanded nozzle One in which jet leaves at below ambient pressure.

overflight Flight over particular route, esp across unfriendly territory.

overfly To fly over.

overhang 1 Spanwise distance from junction of wing and bracing struts to tip; length of cantilever mainplane.
2 Spanwise distance to tip of upper wing of biplane from point vertically above tip of lower wing.

overhead approach Landing procedure in which pilot flies downwind over point of touchdown and then makes descending circular pattern centred on extended landing centreline; also called 360° landing.

overhead stowage Provision in passenger transport for hand baggage and other rigid or heavy items to be stowed in latched bins above seats.

overhead suspension Docked rigid airship hung from roof of shed.

overhung Cantilevered beyond last point of support, esp applied to fan, compressor or turbine stages mounted at extremity of shaft system beyond nearest bearing.

overlap 1 Percentage of photograph duplicated on next frame on same film (see *lap*).
2 See *overlap zone*.

overlap tell Transfer of information to adjacent air-defence facility of (normally hostile or unidentified) tracks in latter's region.

overlap zone Designated area on each side of boundary between adjacent tactical air control or defence regions wherein co-ordination and interaction are required.

overload weight 1 Alternate MTO weight authorized only in unusual circumstances.
2 Any take-off weight exceeding permitted maximum.

overpressure 1 Limiting values reached above and below normal atmospheric pressure during passage of blast wave of explosion (nuclear or conventional).
2 Peak positive pressure reached during passage of boom from supersonic aircraft.

override Facility for bypassing or overcoming normal limit on command action to obtain exceptional response, usually in emergency (eg * boost).

override boost Emergency higher boost pressure available in exceptional circumstances such as overload take-off or in combat.

overrun 1 To fail to stop within available landing distance.
2 Control facility causing camera to continue operating for preset number of frames or seconds after cutoff of control.

overrun area Paved area beyond end of runway which for any reason is not in use; usually marked with herringbone pattern.

overrun barrier Layer of material beyond end of runway provided to arrest overrunning aircraft with minimal damage; PFA is preferred to gravel.

overrunning clutch One allowing driven member (eg helicopter rotor) to run ahead of drive system.

overseer Common term for software used in managerial role in large system.

overshoe Externally attached de-icer of pulsating-rubber mechanical type; also called boot.

overshoot 1 To abandon landing and make fresh approach; in US called missed approach or go-around.
2 To land too far across available landing area and make uncontrolled excursion into region beyond.
3 To exceed desired IAS, alt or other flight condition.
4 To pass through enemy aircraft's future flight path in plane of symmetry.

overshoot area Designated area of semi-prepared ground available to overshooting (2) aircraft (usually GA only) from which they may taxi relatively undamaged.

overspeed condition Normally applied to rotating machinery, and permissible for brief specified period and within specified limit. An exceptional and usually undesired condition, and in no way synonymous (as claimed in one source) with take-off rpm.

overstow position Inoperative position of target-type reverser with geometry which precludes inadvertent operation in flight.

overswing Undesired azimuth excursion by aircraft with long, heavy body during fast turn while taxiing, esp on slippery surface.

overtake velocity Excess speed over that of aerial target ahead.

over the fence Immediately before touchdown (colloq, normally GA).

over the horizon Beyond the visible or LOS horizon; region accessible only to particular species of radar.

over-the-shoulder delivery See *toss bombing*.

over the weather At pressurized cruising levels; basic meaning is free from turbulence except CAT, but cloud may still be present and cu-nims must be avoided.

overturn boundary Limiting sea state for safe operation of float or pontoon-mounted helicopter.

overturn pylon Strong structure to protect occupants of small aircraft in overturn on ground.

overwater Particular provisions apply to commercial transports for use on sea crossings where power-off glide to land is not possible. Adjective also applied to extended-range versions, irrespective of actual intended routes.

OVHT Overheat.

OVI Department for war inventions (of RKKA, USSR).

OVTR Optical video tape recorder.

OW Oblique (slew) wing.

OWE Operating weight empty.

OWF Optimum working frequency.

OWL On-wing life (main engine).

OWRM Other war reserve materiel (US).

OWS 1 Orbital workshop.
2 Obstacle warning system (helo).

OWSF Oblique-wing single fuselage.

OWTF Oblique-wing twin fuselage.

OX, Ox Longitudinal axis about which aircraft rolls.

oxidant Chemical carried as rocket propellant for combination with fuel; examples are lox, nitrogen tetroxide, nitric acid and mixtures of fluorine and oxygen (flox).

oxidation 1 Removal of electron from atom or atomic group.
2 Combination with oxygen, as in burning or rusting.

oxidizer Oxidant (N American usage).

oxidizing flame In gas welding or cutting, one having excess oxygen.

oximeter Instrument giving numerical readout of blood oxygen concentration.

oxygen bottle Container for gaseous oxygen under pressure.

oxygen converter Liquid oxygen converter (ie boiler) supplying gox to breathing system.

oxygen mask Face mask for supplying gaseous oxygen to wearer, usually on demand, and separating exhaled breath.

oxygen microphone Microphone fitted to oxygen mask (probably arch).

OY, Oy Lateral axis about which aircraft pitches.

OZ, Oz Vertical axis about which aircraft yaws.

OZB Civil aviation directory, under Ministry of Works (Austria).

ozone O_3, allotropic form of oxygen present in minute quantities in air, faintly bluish, very reactive.

ozone layer See *ozonosphere*.

ozonosphere Layer of upper atmosphere, roughly at 20–30 km altitude, where ozone concentration is much higher than at SL and ozone plays major role in establishing radiation balance.

P

P 1 Generalized symbol for force.
2 Power, esp bhp of shaft engine.
3 Pursuit-type aircraft (USA, obs).
4 Prohibited area.
5 Poise.
6 Period, eg orbital.
7 Aircraft designation prefix: patrol (DoD).
8 Pressure; normally p but gas-turbine pressures normally written P_1, P_2 etc.
9 Pulsed, or pulse (eg with suffix numeral P1, P2 etc).
10 Pilot, with numerical suffix 1, 2, 3 etc to show hierarchical ranking of pilots in same crew.
11 Prefix, plus (wind component).
12 Production cost.
13 Ratio of full-scale length to model length, esp in radio aerials and related fields.
14 Generalized symbol for probability.
15 Telecom code: aircraft has Tacan only and 4096-code transponder.
16 Radar (JETDS).
17 Missile launch environment: soft pad (DoD).
18 Phosphorus.
19 Provisional.
20 Aerospace craft (FAI category).

p 1 Pico, 10^{-12}.
2 Generalized symbol for pressure.
3 Structural pitch.
4 Structural stress.
5 Pound mass (ambiguous, eg psi).
6 Per (ambiguous, eg lb psi).
7 Parts (eg ppm, parts per million).
8 Port (ie, left).
9 Pentode.
10 Pulse-modulated radio emission.
11 Plate (elec).
12 Generalized symbol for positive or positive values, eg *-type semiconductor.

$\bar{p}$ 1 Average pressure difference across upper/lower surfaces of aerofoil.
2 Specific weight (air-breathing jet engine).
3 Average rocket-motor chamber pressure.
4 Generalized symbol for averaged pressures.

P_{12}, $P_{1/2}$ Fan inlet pressure.

P_3 Compressor delivery pressure.

P.4, P.6 Standard magnetic compasses 1930–50 (UK).

P^3I Pre-planned product improvement.

P-band Frequency band, 225–390 MHz (obs).

P-charge Projectile-charge warhead (SAM).

P-display Radar display of PPI type.

P-lead Primary (switch) cable of magneto.

p-type semiconductor One in which majority of charge-carriers are holes (electron absences).

PA 1 Public address.
2 Performance appraisal.
3 Public affairs.
4 Proposal architecture.
5 Product assurance (especially software).

P/A Payload/altitude curve (sounding rocket).

Pa Pascal, SI unit of pressure.

P_a Action-time average chamber pressure; sometimes abb P_c.

pa, p.a. Power amplifier.

PAA 1 Partial air alert.
2 Primary aircraft authorization.

PABST Primary adhesively bonded structure technology.

PAC 1 Parachute and cable (rocket-launched "SAM", WW2).
2 Public Accounts Committee (UK, House of Commons).

PACAF, Pacaf Pacific Air Forces (USAF).

PACCS Post-attack command and control system (SAC).

Pacer Portable aircraft condition evaluation recorder.

pacing item Task whose time for accomplishment determines overall time for programme; any major item on critical path.

pack 1 Bag in which parachute is packed.
2 Complete system in demountable package which may be attached inside or outside aircraft, eg for multi-sensor reconnaissance, rocket thrust-boost, gun and ammunition, or air-conditioning.

package 1 See *pack (1)*.
2 Complete offer of international contract covering either collaborative development or manufacture of aircraft complete with associated offsets, loans, training and other attractions, and with agreed share to each partner of overall effort and/or cost.
3 Complete offer of international contract covering sale of aircraft with associated offers of offsets, training, service support, construction of support facilities etc.

package aircraft Subject of package (2, 3) deal.

packaged 1 Of petroleum products — eg POL, jet fuel — contained in drums having capacity not greater than 55 US gal (208·2 l).
2 Of electric generating plant, nuclear reactor, etc, mounted on skids and transportable thereon as operative unit.

packaged propellant Rocket motor fed by liquid propellant(s) sealed in container forming single unit with thrust chamber.

package gun One mounted in fairing outside aircraft fuselage, usually fixed.

package programme Timetable for package (2, 3).

packet 1 Single pulse of radar, digitized signal or other coded EM emission.
2 Prepacked quantity of chaff, either pre-cut or as roving, for loading into dispenser.

pack hardening Nitriding by heating inside loose pack of carbonaceous material.

PACS 1 Passive attitude control system, eg gravity gradient.
2 Pitch-augmentation control system.

PACT 1 Portable automatic calibration tracker.
2 Precision aircraft control technology (CCV).

PAD 1 Picture assembly device.
2 Point air defence.

pad 1 Platform for launch of vehicle, usually of ballistic or unmanned nature (colloq and becoming arch).
2 Platform for operation of helicopter, esp on surface vessel.
3 Attachment face on engine or other energy source for accessory, usually with central shaft drive.
4 Network of resistances or impedances to couple to transmission lines effectively.

pad abort Abort before launch from pad (1).

padder See *padding capacitor*.

padding capacitor In series with superhet local-oscillator tuning capacitor to match exactly to frequency.

paddle In a rate gyro, damping drag surfaces which generate angular difference proportional to first-order rate term.

paddle blade Propeller blade having unusually wide chord maintained to tip.

paddleplane See *cyclogiro*.

paddle switch Ambiguously used to denote electrical switches operated by dynamic air pressure, dynamic water pressure in ditching and various other activation sources.

PADIS, Padis Procedures for analysing the design of interactive solutions (Tornado F.2 fire-control software).

Padlocked "I have my vision fixed on bogies/bandits and will not look away".

pad output Power rating of shaft at each pad (3) location on engine, APU or MEPU.

PADR Product-assurance design review.

PADS, Pads 1 Passive advanced sonobuoy.
2 Position and azimuth determining system.

PAF 1 Parabolic aircraft flight.
2 Police de l'Air et des Frontières (F).

PAFAM, Pafam Performance and failure assessment monitor (independent flight guidance with flight-deck displays during autolanding).

PAFU Pilots advanced flying unit.

PAH 1 Anti-tank helicopter (G).
2 Polycyclic aromatic hydrocarbon.

Paid, PAID Parked-aircraft intrusion detector.

paint To create blip on radar display, esp one giving position of aircraft or other object.

paint-on bag Flexible vacuum bag formed in place to cover layup of composite material fabrication and seal tool in prototype moulding of large parts.

paint-stripe loading Vertical stripes inside cargo aircraft to assist loaders in achieving correct load distribution, esp with regard to c.g.

PAIR, pair Precision-approach interferometer radar.

PAL 1 Passive augmenting (or augmentation) lens, eg Luneberg.
2 Standard German TV/video system; from phase alternation line.
3 Permissive action link.

PALC Point Arguello Launch Complex.

Palea Philippine Airlines Employees Association.

pallet 1 Standard platform for air freight, eg (Imperial measures are still used) 88 in × 125 in, 88 in × 108 in and 96 in × 125 in.
2 Flat base, not necessarily of above dimensions, for combining stores or carrying single item to form unit load, in some cases retained as temporary or permanent base in operation.
3 Standard-dimension base on which can be assembled experiments, ancillaries and other payloads, eg for Spacelab.
4 Standard-dimension base carrying workpieces in mechanized and automated manufacturing.
5 Configured (shape-matching) payload carrier for attachment close against airframe exterior, eg fuel * (FAST Pack).

palletized bladder Flexible container for liquid mounted on pallet (1) for airfreighting.

palletrolley Pallet (2) with wheels for ground transport, eg for Sea Skua missile.

Palmer scan Radar scanning technique in which conical scan is superimposed on some other, eg Palmer-sector (beam oscillates to and fro over azimuth sector while continuously making small-angle conical scan), Palmer-raster, Palmer-circular etc.

Palnut Thin nut, usually pressed steel, screwed tightly above ordinary nut to prevent loosening of latter.

PALS, Pals 1 Precision approach and landing system (Boeing/Bendix).
2 Portable airfield light set (USAF).
3 Positioning and locating system.

PAM, p.a.m. 1 Pulse ampltitude modulation.
2 Payload assist module.
3 Picture assembly multiplexer.
4 Plasma-arc machining.

PAMA Professional Aviation Maintenance (formerly Mechanics) Association (US).

PAMC Provisional acceptable means of compliance (ICAO).

PAN, Pan 1 Radio code indicating uncertainty or alert, general broadcast to widest area but not yet at level of Mayday.
2 Polyacrilonitrile, precursor of carbon fibre in

most processes.
3 Originating station has urgent message to transmit concerning safety of vehicle or occupant(s) (DoD).

pan 1 Base of seat tailored to pilot with seat-type pack.
2 Paved dispersal point.
3 Carrier for air-drop parachuted load.

pancake 1 Vague term supposedly indicating landing at abnormally low forward speed and high sink rate (arch).
2 Air-intercept code: "I wish to land", usually followed by word giving reason, eg * fuel (DoD).
3 Verb, to land (arch).

P&ES Personnel and equipment supply.

P&P 1 Plans and programs (DoD).
2 Payments and progress (NATO).

P&SS Provost and Security Service (RAF).

panel 1 Single piece of aircraft skin, eg fuselage *.
2 Major section of wing as finished component, eg outer *.
3 Essentially planar base carrying instruments.
4 Complete portion of stressed-skin structure, sandwich or, rarely, other constructional form, used for fatigue or static test purposes; not necessarily portion of actual aircraft.
5 Subdivision of parachute canopy, either complete gore or portion thereof.
6 Body of experts, eg examining *.
7 Marking panel.

panel code Standard code for visual ground/air communication (IADB).

panel elements Items mounted on panel (3) such as instruments, switches and displays.

panelled Equipped with instruments and avionics, thus well * (colloq).

pan-handle Ejection-seat firing handle on seat pan.

pan-head screw One having thin, large head with slightly convex upper surface.

pannier Small 1-, 2- or 4-tube rocket launcher added to CBLS or other stores carrier.

panoramic camera One which by means of system of oscillating or rotating optics or mirrors scans wide strip of terrain usually from horizon to horizon. Mounted vertically or obliquely to scan across or along line of flight.

PANS, Pans Procedures for Air-Navigation Services (UK); * RAC adds radio com procedures.

PANS-ABS PANS abbreviations and codes.

panting In/out springy movement of thin skin under compressive stress; generally similar to oil-can effect.

pants Fixed fairing over landing gear leg and, esp, wheels; generally similar to spats but chiefly US usage.

pants duct Large-diameter Y-piece forming junction or bifurcation in duct system.

Papa Parallax aircraft-parking aid.

PAPI, Papi Precision approach path indicator.

PAPM, p.a.p.m. Pulse amplitude and phase modification; narrows signal bandwidth to quadruple carrying capacity per channel.

PAPS, Paps Periodic armaments planning system (NATO).

PAR 1 Precision approach radar, provides accurate display of approaching aircraft for GCA talkdown.
2 Progressive aircraft rework (USAF, USN).
3 Phased-array radar, ie electronically steered instead of mechanically scanned.
4 Perimeter-acquisition radar (ABM).
5 Pulse-acquisition radar.
6 Program(me) appraisal and review.

Par Precision aircraft reference (Lear-Siegler twin-gyro platform).

parabola Conic section made by cutting right circular cone parallel to any of its elements; locus of point which moves so that distance from line (directrix) equals distance from point (focus), so eccentricity = 1.

parabolic aerial One whose reflecting surface forms portion of parabola, thus converting plane waves into spherical waves or vice versa, and either emitting pencil beam from point feeder or focusing incoming radiation to single-point receiver.

parabolic trajectory As eccentricity is unity it represents least eccentricity for escape from attracting body.

paraboloid Shape in 3-D formed by rotating parabola about major axis.

parabrake Braking parachute.

parachute Any device comprising flexible drag or drag+lift surface from which load is suspended by shroud lines. Originally canopy was umbrella-shaped but today may be of Rogallo or other semi-winged types offering precision control of landing within large area. Distinction between * and hang glider is blurred, but chief features distinguishing * are instant deployment from packed condition and suspension of load well below canopy.

parachute deployment height Height AGL at which canopy fully deployed.

parachute flare Illuminating flare equipped with parachute to prolong descent.

parachute gore See *gore*.

parachute harness That to which personal parachute is attached.

parachute pack Bag containing packed parachute.

parachute tower 1 Tower or mast from which parachute descents are made for sport or instruction.
2 High-ceiling part of parachute section of airbase wherein parachutes are hung to dry after use.

parachute tray Rigid base to pack of parachute used in heavy dropping.

parachute vent Aperture in top of canopy or left by blank gore(s) to ensure stable descent.

para-circular Near-circular orbit (intended to be circular).

paradrag drop Ultra-low airdrop of cargo using drag of arrester parachute to extract and halt payload (NATO).

paradrop Delivery by parachute (from height significantly greater than paradrag drop) of personnel or cargo from aircraft (NATO).

paraffin Homologous hydrocarbon series having general formula C_nH_{2n+2}; first four (methane, ethane, propane, butane) are gases at STP, but next 11 are liquids and form main constituent of jet fuels.

parafoil Parachute able to fly as aerodyne (L/D about 3) and stall, but with canopy instantly deployed from packed condition and attached by normal harness.

parafoveal vision Used at night to see extremely dim sources; eye is oriented so that what light there is falls on area of retina populated mainly with rods, not in central (foveal) region but surrounding it.

paraglider 1 Inflatable hypersonic spacecraft having form of metallized-fabric kite of paper-dart shape.
2 Also formerly applied to various flexible-wing gliders, towed kites and parafoils but today no longer used; each device must be either a kite, hang glider, parachute or parafoil.

Paralkatone Protective sealant film sprayed on aluminium alloy aircraft, usually after completion or during erection.

parallel 1 Great circles parallel to Equator (* of latitude).
2 Circle on celestial sphere parallel to celestial equator (* of declination).
3 Connected so that current flow, signal, etc divides and passes through all components simultaneously, thereafter recombining.
4 Software connection so that each signal has its own wire; hence a 16-bit wire has 16 conductor paths.

parallel actuators Two or more in parallel to drive single load; usually physically separated and tied by load in force or torque-summing fashion (thus providing rip-stop design).

parallel aerofoil Having constant section from tip to tip; two-dimensional.

parallel burning Solid grain which ignites at centre over whole length and burns purely radially outwards to case.

parallel double-wedge Aerofoil whose section is that of flat hexagon with sharp wedges at leading and trailing edges and parallel upper and lower surfaces; poor at subsonic speeds but acceptable supersonic shape, esp for missiles.

parallel-heading square Training manoeuvre in which helicopter is flown around a square, each of the four legs being flown with helicopter aligned with that side of square.

parallel ILS Serving parallel runways, with minimum lateral separation of 1,500 m, 5,000 ft.

parallel of origin Parallel of latitude used as an origin (2).

parallel redundancy Addition of channels in parallel (3) mode purely to increase redundancy (1).

parallel resonant circuit Tuned circuit with inductive and capacitive elements in parallel (3).

parallel runways Runways at same airport on same alignment and with sufficient lateral separation to permit simultaneous use, including parallel ILS approaches.

parallel servo One located in control system so that servo output drives in parallel with major input; usually arranged to drive both cockpit controls and flight-control system and thus performing as alternative to pilot.

parallel yaw damper One connected direct into pilot/rudder control loop and, in some cases, resulting in movement of cockpit pedals. Normally superseded by series damper.

paramagnetic Possessing magnetic permeability above 1 and permanent magnetic moment; atoms tend to align in direction of external field giving resultant magnetic moment and tend to move to strongest part of field.

Paramatta Code for RAF WW2 target-marking by PFF using precision-aimed TIs of sophisticated types repeated at intervals; usual form with electronic aids called Musical *.

parameter Basic definable characteristic or quality, esp one that can be expressed numerically; specialized meanings in maths, EDP (1) and statistics.

parametric amplifier Reactance amplifier dependent on time-varying reactance (fed by signal and a pump source) forming part of tuned circuit; variable-capacitance usually Varactor or crystal diode. Also called Mavar (mixed amplification by variable reactance).

parametric study Study based on numerical values of system variables.

parametric take-off number Product of (landing wing loading $\times$ 1·11) $\times$ (MTOW $\div$ [total installed thrust $\times$ C_{L2}]).

paramp See *parametric amplifier*.

pararescue team Trained personnel who reach site of incident by parachute and render aid.

parasheet Parachute in form of regular polygon with rigging lines attached to apices; subdivided into gathered * (periphery constrained by hem cord) and ungathered *.

parasite Aircraft, invariably aeroplane, which for portion of flight relies on another for propulsion and possibly also for lift. Can ride on back of parent, or be attached elsewhere externally or internally, or be towed.

parasite drag Sum of all drag components from all non-lifting parts of body, usually defined as total drag minus induced drag. Not normally used (see *profile drag*).

parasitic element Resonant element of directional aerial excited to produce directional pattern; eg reflector added to Yagi aerial.

parasitic oscillation Generated by stray inductances and capacitances and eliminated by parasitic stopper.

parasiting Technique of using parasite.

paraski Sport combining downhill ski with lifting surface such as parachute, parafoil or hang glider; competitions test landing precision.

parasol wing One from which rest of aircraft is suspended by ties.

paravane 1 Hydrodynamic body towed on end of minesweeping cable.
2 Aerodynamic body towed on end of cable from aircraft to keep line taut and steady, eg in mid-air recovery.

paravisual director Attention-getting non-quantitative flight instrument giving bold linear indication of excursions in pitch or azimuth, generated by rotation of striped "barber pole" display; usually mounted on flight-deck coaming (Smiths Industries).

parawing Variation of basic Rogallo flexible wing developed at NASA Langley for payload recovery and marketed by Northrop; not current term in hang gliding.

Parcs Perimeter acquisition radar characterization system.

Pardop Passive ranging Doppler.

parent 1 Main aircraft in composite or carrier of parasite.
2 Aircraft carrying one or more RPVs over which it exercises control following release.

parent body Primary around which satellite orbits.

parity 1 Symmetric property of wave function or other phenomenon; value is 1 if function unchanged by inversion of its co-ordinate system.
2 Precise keying of two or more sets of data (eg EDP (1) software, tapes, recon pictures or wire recordings) so that they run to common time-base. System parities are assigned individual tracks in such devices as reconnaissance signal recorders.

Parkerizing Anti-corrosion treatment in which metal part is boiled in solution of phosphoric acid and manganese dioxide and then dipped in oil.

Parker-Kalon, PK Patented family of self-tapping screws for sheet.

parking brake That applied after aircraft is parked, to prevent subsequent rolling; also (US) called park brake.

parking catch Fitted to normal handbrake (rarely, toe brake) to convert to parking brake.

parking orbit Temporary spacecraft orbit for such purposes as waiting for correct timing or for delivery of components, spacecraft or station assembly or rectification of fault.

Parmod Progressive aircraft rework modification (USAF, USN).

parrot Code for IFF transponder (DoD).

PARS, Pars Programmed airline reservation system.

parsec Unit of length on cosmological scale equal to distance at which object has helicocentric parallax of 1″ (one second of arc) = 206,265 times semi-major axis of Earth's orbit = 3·26 light-years = $3{\cdot}0857 \times 10^{16}$ m, abb pc. Multiples are kpc, kilo-* and Mpc, mega-*.

Parsecs Program for astronomical research and scientific effects concerning space.

partial-admission turbine One in which working fluid enters around only part of periphery.

partial pressure That exerted by each gas in gaseous mixture; Dalton's law states value is same as if that gas occupied whole volume occupied by mixture at same temperature.

partial-pressure suit Skintight suit covering body except head, hands and feet and inflatable (usually by stitched-in tubing) to press on wearer and oppose internal pressures.

partial priority Unusual status accorded special traffic (SST was intended) by ATC services in special circumstances short of emergency.

partial shielding 1 Protection against micrometeorites incident from one direction only.
2 Radiation shielding of crew of nuclear-propelled aircraft whilst leaving radiation free to escape in other directions.

particle separator Device from cleaning solid (and possibly liquid) particles from air entering engine or other device, typically by centrifuging effect in vortex.

particle static Radio-communication static thought to be due to accumulation or shedding of particulate matter in flight (not defined and suggested arch).

part surface Surface being cut in GPP.

part-throttle reheat Augmentation of engine by afterburner brought in at less than max cold thrust, giving smoothly increased thrust and avoiding sudden augmentation by selecting afterburner at max rpm.

Parylene Conformal (vapour-deposited) barrier coating used to protect precision parts, especially electronics, from hostile environments (Union Carbide).

PAS 1 Performance advisory system (Simmonds subsystem offering min fuel and max propulsive efficiency).
2 Public-address system.
3 Power available, shaft (UK, CAA = spindle).

PASC Pacific Area Standards Congress.

pascal SI unit of pressure, Pa, = N/m^2.

Pascal's law In fluid at rest, all pressures acting on given point are equal in magnitude in each direction (neglects acceleration due to gravity).

Pass, PASS 1 Parked-aircraft security system.
2 Passive and active sensor subsystem.

3 Passive aircraft surveillance system.

pass 1 Single run by aircraft past point on ground or other object including aircraft in flight on same heading at markedly lower speed.
2 Short tactical run or dive by aircraft at target; single sweep through or within firing range of enemy air formation (DoD, IADB).
3 Single orbit by satellite, starting at point Equator is crossed northbound.
4 Single passage by satellite overhead.
5 Single period of time during which satellite is within radio contact of control or data-acquisition station.

passenger In addition to basic meaning of humans carried in vehicle other than vehicle's crew, normal meaning in air-carrier terminology is FPP (fare-paying *), normally excluding company employees on cheap rate. Latter, however, are normally included in traffic statistics.

passenger boarding bridge Passenger loading bridge.

passenger loading bridge Covered walkway linking terminal and parked aircraft with hinged connection to terminal and far (aircraft) end carried on powered chassis able to vary height, direction and length; also called jetty, jetway, apron-drive bridge and airside bridge.

passenger profile Distribution of pax at one airport through 24 h period.

passenger service charge Fee levied by airport authority on each departure passenger to help cover its costs.

passivating Coating, esp metal surface, with inert film to prevent electrochemical corrosion.

passive 1 Receiving but not emitting.
2 System, eg PFCS, which can only fail open.

passive air defence Includes such activities as cover and concealment, camouflage, deception, dispersion and shelters.

passive countermeasures Detection and analysis of hostile emissions (see *ESM*).

passive guidance Use of received signals from whatever source in order to navigate.

passive homing Tending to fly towards source of emission from target, usually IR.

passive jamming Use of chaff and confusion reflectors.

passive mode Use of airborne radar in receive-only mode, eg AWACS.

passive munition 1 Delivered safe and then triggered either by timing system or radio signal.
2 Inert until disturbed, eg mine.

passive paralleling Simplest and most common type of redundancy wherein two parallel functional devices are utilized such that, if one fails, second is still available. Limited to simple elements of control system which can only fail passive, such as springs and linkages. When failure of one element occurs, there is an acceptable change in performance or capability.

passive radar Misnomer; normally interpreted as passive mode.

passive ranging Trajectory-measuring systems that do not require a transmitter or transponder in vehicle, eg Pardop.

passive sonar Listens for emissions from submarine.

passive thermal control Changing attitude of spacecraft to even-out incident solar flux, esp by roll through 180°; so-called barbecue manoeuvre.

pass off To test and clear production item.

PATC Professional, administrative, technical and clerical.

patch 1 Strong fabric attachment cemented to aerostat envelope (see *eta* *, *channel* *, *split* *, *rigging* *).
2 Embroidered or printed badge worn on working (especially flying) clothing.

PATCO, Patco Professional Air Traffic Controllers Organization (US).

Patec Portable automatic test equipment calibrator.

path 1 Track of aircraft under control of TMA.
2 Projection on Earth's surface of satellite orbital plane; same as track.
3 Loosely, aircraft trajectory in 3-D.

pathfinder aircraft One with special crew plus drop-zone/landing-zone marking teams, markers and/or electronic navaids to prepare DZ/LZ for main force (NATO).

pathfinder drop-zone control Communication and operation centre from which pathfinders exercise guidance (DoD, LADB).

Pathfinder Force Elite sub-force within RAF Bomber Command 1943–45 charged with marking targets for main-force attack, abb PFF.

pathfinders Four meanings, all defined as plural:
1 Experienced aircrew who lead formation to DZ, RP or target.
2 Teams dropped or air-landed at objective to operate navaids.
3 Radar or other navaid used to facilitate homing on objective, esp in bad visibility.
4 Teams air-delivered into enemy territory to determine best approach/withdrawal lanes, LZs and sites for heliborne forces (all DoD, LADB).

path-stretching Deliberately routeing incoming traffic over longer path (1) to achieve correct spacing on approach.

Patrick AFB USAF base at Cape Canaveral supporting AMR.

patrol aircraft 1 Generally accepted term for aircraft engaged in offshore duties such as search/rescue, customs/immigration and enforcement of marine laws.
2 US designation for large ocean combat aircraft engaged in ASW, maritime reconnaissance and mining.

Pats Precision automated tracking station.

pattern 1 Replica of part used in constructing casting mould.

2 Radiation of transmitting aerial as plotted on diagram of field strength for each bearing.
3 Shape traced out on ground by track of aircraft, esp in circuit, making procedure turns, in holding stack or other circumstance demanding accurate geometry.
4 Authorized flightpath (1) to point of touchdown.

pattern aircraft One supplied to participant in production programme, eg licensee or co-producer, to serve as master to instruct engineers and solve arguments, and possibly facilitate local change orders.

pattern bombing Systematic covering of target area with carpet of bombs uniformly distributed according to plan.

pattern generator Signal generator whose output provides TV-type pattern for testing TV, video, EO or visual-display systems.

PATTS Programmed auto trim/test system.

Patuxent River US Naval Air Test Center.

PATWAS, Patwas Pilots' automatic telephone weather answering service (FSS, US).

PAV Power assurance valve (CAA).

PAW Plasma-arc welding.

Paws Phased-array warning system.

pax Passengers.

Pax River See *Patuxent River*.

payload 1 That part of useful load from which revenue is derived (BSI). See *maximum* *.
2 Load expressed in short tons, US gal or number of passengers vehicle is designed to transport under specified conditions (DoD etc).
3 In missile, warhead section (DoD), plus container and activating devices (NATO); both definitions need modifying in light of MIRV and similar techniques.
4 For spaceflight, not yet defined; * for Saturn V launch vehicle was complete Apollo spacecraft, whose LM ascent stage had its own * in terms of two crew plus lunar rock.
5 For ECM, all discrete devices released or ejected, eg chaff, flares and jammers in prepackaged * form.

payload of opportunity One put aboard space launcher because mass/volume is available.

payload/range Fundamental measure of capability of transport aircraft usually presented as graphical plot of payload (1) against range with specified operating conditions and reserves.

payload specialist Engineer skilled in designing and integrating payloads (4), esp those of academic-research or commercial nature.

PB 1 Payload bay.
2 Passenger bridge.
3 Aircraft designation prefix, patrol bomber (USN, obs).
4 Pre-brief.
5 Particle beam.

P_B Roll-rate induced angle of attack.

Pb Lead; PbO = lead oxide.

$\bar{P}_b$ Burning-time average chamber pressure (solid rocket motor).

PBA Pitch bias actuator.

PBAA Polybutadiene acrylic acid (solid rocket propellant).

PBAN Polybutadiene acrylonitrile (solid rocket propellant).

PBB Passenger boarding bridge.

PBC Practice bomb carrier.

PBI Pushbutton indicator.

PBP&E Professional books, papers and equipment.

PBPS Post-boost propulsion system.

PBSI Pushbutton selector/indicator.

PBTH Power by the hour.

PBV Post-boost vehicle.

PBW Particle-beam weapon.

PBX Atmospheric pressure (USSR).

PC 1 Production control.
2 Printed circuit.
3 Physical conditioning.

P_c 1 Chamber pressure (any rocket).
2 Polar continental air mass.

$\bar{P}_c$ Burn-time or action-time average chamber pressure (same symbol).

pc Parsec.

PC1, PC2 etc Powered flight-control systems.

PCA 1 Polar-cap absorption.
2 Positive-controlled airspace.
3 Physical configuration audit (software).
4 Photovoltaic concentrator array.

PCAS Pitch-control augmentation system.

PCB 1 Printed-circuit board.
2 Plenum-chamber burning.
3 Polychlorinated biphenyl(s).

PCBW Provisional combat bomb wing (USAAF).

PCC Parts-consumption cost.

PCCB Program configuration control board.

PCDN Process change design notice.

PCF 1 Pulse compression filter.
2 Pounds per cubic foot (not recommended).
3 Protein crystallization facility.
4 Passenger-cum-freight (CAA).

PCI 1 Pattern of cockpit indications (failure method).
2 Pounds per cubic inch (not recommended).

PCID Preliminary change in design.

PCM Pulse-code modulation.

$P_{c\,max}$ Maximum chamber pressure (solid rocket motor).

PCN Pavement classification number; part of proposed ICAO standard ACN-PCN system for relating aircraft footprints to pavement strengths.

PCO 1 Photochemical oxidant, = smog.
2 Procuring (or procurement) contracting officer (DoD).

PCRT Projection CRT.

PCS 1 Performance command system.
2 Portable control station (RPV).

3 Permanent change of station.

4 Pitch control system.

PCT Power control test.

PCU 1 Powered-control unit.

2 Pilot's control unit.

3 Pod conditioning unit.

PD, p.d. 1 Potential difference.

2 Pulse Doppler.

3 Powered descent.

4 Power-doubler (radio).

5 Preliminary design.

6 Project definition (abb results in confusion with [5]).

7 Period (full stop) in message.

8 Pre-digital.

9 Presidential decision (US).

10 Procurement document.

PDA, p.d.a. 1 Post-deflection modulation.

2 Photon detector assembly (space telescope).

3 Problem-detection audit.

PDADS Passenger digital-activated display system.

PDC 1 Performance data computer.

2 Public dividend capital.

3 Personnel despatch centre.

PDCS Performance-data computer system (Lear-Siegler/Boeing).

PDES Pulse-Doppler elevation scan; PDNES plus electronic scanning in vertical plane to give target height.

PDFN Phase-distortion at first null.

PDG 1 Precision-drop glider.

2 Président directeur-général (F).

3 Programmable display generator.

PDI 1 Powered descent insertion.

2 Primary direction indicator.

PDM 1 Pulse-duration modulation.

2 Primary development model.

3 Programmed depot maintenance.

4 Presidential decision memorandum.

PDMF Programmable digital matched filter(s).

PDMM Pulse-Doppler map matching.

PDNES Pulse-Doppler non-elevation scan; surveillance down to surface, without indication of target height.

PDO Pendulum dynamic observer.

PDP Program(me) decision package.

PDQ Photo-data quantizer (or quantifier).

PDR 1 Preliminary design review.

2 Pulse-Doppler radar.

3 Pilot's display recorder (Ferranti).

4 Predetermined routeing.

PDRC Pressure-drop ratio control.

PDRJ Pulse-Doppler radar jammer.

PDS 1 Passive detection system.

2 Prestocked dispersal site.

3 Pulse-Doppler search mode.

4 Passenger distribution system.

5 Post-design service(s).

PDSTT Pulse-Doppler single-target track mode.

PDU 1 Pilot display unit.

2 Power drive unit (CAA).

PDV Pressurizing and dump valve.

PDVOR Precision DVOR.

PDZC Pathfinder drop-zone control.

PE 1 Procurement Executive (UK, MoD).

2 Position error.

3 Pilot error.

4 Piston engine.

5 Program element (DoD).

6 Professional engineer (US).

P_E, P_e Pressure in core-engine jetpipe.

Pe See *Péclet number.*

peak In production programme, time or rate of maximum output.

peaking circuit Extends frequency response of video amplifier at highest frequencies.

peak suction Lowest pressure on upper surface of wing or on 2-D aerofoil profile; rarely, lowest pressure on other convex surface of aerodynamic body.

peaky Aerofoil section (profile) of traditional type causing large acceleration of flow over leading edge and hence very low pressure (large peak suction) over narrow strip at 8–15% chord; opposite to supercritical or roof-top.

PEC 1 Personal equipment connector.

2 Pressure error correction.

pecked line Broken or dashed line in graphics or artwork.

Péclet number $Pe = Vl/\lambda$ where V is fluid flow velocity, l a length and λ thermal conductivity; applies in heat transfer at low airspeeds.

PECM Passive electronic countermeasures.

pedestal 1 Raised box between pilots on flight deck carrying numerous control and system interfaces.

2 Pillar supporting aircraft on ski (and, it is suggested, float).

PEE Photoelectron emission.

PEEK Polyetheretherketone.

peel off To roll away from straight-and-level flight, esp from a formation, and dive away; normally manoeuvre performed in sequence by all aircraft of formation.

peen 1 To cold-work metal surface by repeated light blows of ball-pane hammer or bombardment with hard balls (shot-*).

2 To deform metal part by series of hammer blows, eg to set rivet or burr end of bolt to prevent loss of nut.

PELSS Precision emitter location strike system.

Peltier effect Generation of current by making circuit containing two different metals and keeping two junctions at different temperatures; some definitions refer to generation or absorption of heat at junctions upon passage of current.

PEM 1 Parametric estimation model.

2 Program element monitor (DoD).

penaids See *penetration aids.*

penalty Deficiency in one aspect of aircraft design or performance in return from improvement in another; eg adding thermocouple in jetpipe to indicate temperature assists pilot but creates drag in jet reflected in * in either cruising speed, range or both.

pencil beam Narrow, strongly directional beam with minimal sidelobes.

pendant 1 Arrester wire (deck cable).
2 Flag indicating centre of arrester wire.

pendulum stability That inherent in large vertical distance between high centre of wing lift and low c.g.

pendulum valves Gravity-operated flaps covering ports in housing of air-operated gyro which sense tilt and control airflow to cause restoring (erecting) precessive force.

penetration 1 Flight into hostile, esp defended, airspace.
2 Ability of sailplane to keep going by trading kinetic energy for distance with minimal loss of height or speed while heading for next thermal.
3 Flight deep into cloud, esp one with large vertical extent and severe turbulence.
4 Flight into eye of tropical revolving storm.
5 Success in previously unattacked market.
6 Progress into flight manoeuvre, esp one posing potential problem or danger, hence stall-*.
7 Portion of high-altitude instrument approach which prescribes descent to start of approach (DoD).

penetration aids Devices and systems assisting vehicle to accomplish penetration (1), eg jammers, chaff, flares, decoys and warning systems, low (lo) flight level, reduced RCS and improved vehicle hardness.

penetration area That within which enemy defences are to be neutralized to degree assisting succeeding aircraft to reach targets.

penetration factor Reciprocal of amplification factor of valve.

penetration fighter One intended to fly deep into hostile territory (obs).

penetrator 1 Tool, usually powered, for quickly cutting apertures in side of aircraft for speedy rescue of occupants.
2 Aircraft designed for penetration (1).
3 High-density projectile designed to pierce armour by kinetic energy.

pennant Mooring and haul-down wire for captive balloon.

penthouse roof Top of cylinder combustion space has sloping sides.

pentode Thermionic valve having three grids.

PEP, p.e.p. 1 Peak envelope power.
2 Pre-engine production.
3 Productivity enhancement programme.

PEPE Parallel-element processing ensemble (or element).

PEPP Planetary-entry parachute program (Martin Marietta/NASA).

PEPS, Peps Positive-expulsion propellant system.

PERA Production Engineering Research Association (Melton Mowbray, UK).

perceived noise level See *noise*.

percussion cap Detonator/igniter for ammunition activated by sudden deformation.

percussion gun Device for cheaply making loud bangs to scare birds.

perfect fluid Usually implies inviscid and incompressible, as well as homogeneous.

perfect gas One exactly obeying Joule and Boyle laws so that PV = nRT; also obeys Charles/Gay-Lussac and several other laws.

perforated Pierced by holes, in case of aerodynamic surface typically removing 25–30% of projected area; perforation of flap or airbrake can reduce actuation power requirement and increase drag of surface, esp when at large angle to airflow, but destroys value of flap as lifting surface and thus confined to divebrakes and airbrakes.

performance 1 From operational viewpoint, ability of system to perform, esp expressed in numerical values.
2 From flight-safety viewpoint, ability of aircraft to perform required functions and manoeuvres, esp in degraded condition and under adverse circumstances, again expressed numerically. Subdivided into three categories: 1, measured *, that actually recorded for particular aircraft at particular time; 2, gross *, factored from measured to allow for poorest aircraft in fleet, guaranteed-minimum instead of average thrust propulsion and possibly other degrading factors; and 3, net *, factored from gross to allow for further possible temporary variations (airframe damage or icing or certain non-critical fault conditions) and minimum standard of pilot skill and experience.
3 Narrow interpretation of numerical values of aircraft flight limits such as speeds, altitudes and payload/ranges.

performance factors Those deriving from aircraft performance which affect airline traffic achieved, as distinct from commercial and operational factors.

performance groups UK aeroplane categories: those whose first C of A was issued before 1951 are NPG (no performance group); those built since 1951 are Group A, large multi-engined; B, spare; C, light multi-engined; D, single-engined; X, foreign multi-engined imported before particular dates.

performance limiting conditions Those demanding flight to near boundaries of flight performance or point at which auto-ignition, stick-shaker or other subsystem is triggered.

performance management system See *flight management system*.

performance number Knock rating; below 100 called octane number.

performance-type glider Sailplane, esp competition sailplane.

periapsis Pericentre of orbit.

pericentre Point on orbit closest to primary.

pericynthian Point in spacecraft trajectory closest to Moon.

perifocus See *pericentre*.

perigee Pericentre of Earth orbit.

perihelion Pericentre of solar orbit.

perilune Pericentre of lunar orbit.

perimeter 1 Periphery of airfield.
2 Boundary of defended area.

perimeter track Taxi track linking ends of runways, revetments or dispersals and main hangar/apron area.

period 1 Time interval between successive passages through particular point in same direction of SHM or other wave motion.
2 Time interval between successive passages through particular bounding plane in same direction of satellite, usually time between successive northbound transits of Equator (orbital *) (see *sidereal* *, *synodic* *).

periodic inspection Inspection, with or without teardown, according to published schedule of time intervals irrespective of performance (1) of device.

peripheral hem Leading edge of parachute canopy.

Permalloy Family of American Fe/Ni alloys which are magnetically soft, with high permeability at low magnetizing forces.

permanent magnet One which retains its magnetism in absence of strong demagnetizing field.

Permaswage Patented fluid-tight method of connecting pipes.

permatron Gas tube similar to thyratron with magnetic-field control.

PERME, Perme Propellants, Explosives and Rocket Motor Establishment (Westcott, Waltham Abbey, UK).

permeability 1 In magnet, ratio B/H where B is magnetic flux induction and H is magnetizing force, symbol ρ; this is divided into absolute and relative or specific * except in emu system, where * of free space is defined as unity.
2 In aerostat, measure of rate at which gas at STP can pass through fabric, usually expressed in litres per 24 h.
3 In amphibian, ratio of volume of landing-gear compartments that can be occupied by water to whole buoyant volume.

permeability tuning Radio tuning by varying permeability of inductor core, usually with translating bar of ferrite.

Perminvar Family of low-hysteresis alloys of Fe/Ni/Co.

permittivity Dielectric constant, ratio of electric flux density in medium to that which same force would produce in vacuum, symbol F (see *Coulomb law*).

permly Permanently (FAA).

perpendicular-heading square Training manoeuvre in which helicopter is flown around square, crabbing at 90° sideways along each of the four sides.

personal equipment connector Quick-make/break multichannel coupling for aircrew oxygen, R/T, intercom, g-suit and EC. Not necessarily synonymous with AEA.

personnel locator beacon Miniature transmitter sending coded signal and worn on flying clothing.

personnel reaction time Time from nuclear warning to all defensive measures taken.

Perspex Family of methyl methacrylate plastics important as aircraft transparent material (ICI).

PERT, Pert Programme Evaluation Review Technique; critical path method.

PES 1 Passenger entertainment system, audio plus film.
2 Polyethersulphone.

PESO Product engineering services office.

PET Piston-engine time.

petal cowling One divided into large hinged segments opening like petals of flower.

PETN Pentaerythritol tetranitrate (explosive).

Pett, PETT Project engineers & technologists for tomorrow.

petticoat In airship gas duct, pleated sleeve able to seal duct yet leaving clear passage when released.

PEU 1 Power electronics unit.
2 Processing electronics unit.

PF 1 Pilot flying (others being PNF).
2 Plastics factor.

PF, P/F Powder forging (or forged).

P_f Pressure in fan duct of turbofan.

PFA 1 Popular Flying Association (UK).
2 Pulverized fuel ash; excellent for overrun barriers.

PFB Preliminary flying badge.

PFC 1 Primary flight control.
2 Powered flight (or flying) control(s).
3 Pre-flight console.

PFCES Primary flight-control electronic system.

PFCS Primary flight-control system.

PFCU Powered flying-control unit, surface power unit.

PFD Primary flight display.

PFE Purchaser-furnished equipment.

PFF Pathfinder Force.

PFFT Parallel fast-Fourier transform.

PFHE Prefragmented high explosive.

PFI Post-flight inspection (US = after-flight).

PFIAB President's Foreign Intelligence Advisory Board (US).

PFL Practice forced landing (A adds "area").

PFM Pulse-frequency modulation.

PFMGO Pre-flight message-generating officer (RAF).

PFQT Preliminary flight qualification test.

PFR Permitted flying route (CAA).

PFRT Preliminary flight rating test.
PFS Primary flying squadron.
PFTS 1 Production flight-test schedule.
2 Permanent field training site.
PG, p.g. 1 Processing gain.
2 Plastic/gas.
3 Peelable graphics, esp airline livery or logo.
4 Program management assistance group (USAF).
5 Photographic group (USAAF).
PGA Pressure-garment assembly.
PG bearing Plastic/gas.
PGM 1 Precision-guided munition.
2 Guided missile (ICBM, obs) launched from soft pad (DoD).
3 Precision ground mapping.
PGN Passenger-generated noise.
PGRV 1 Post-boost guided re-entry vehicle (has IMU and TPU).
2 Precision-guided re-entry vehicle (DoD).
PGSC Personnel guide surface canopy.
PHA Preliminary hazard analysis.
phantom contract See *phantom order*.
phantom drawing One using phantom lines.
phantom lines Geometrically accurate but incomplete lines merely giving location of item or alternative positions thereof, eg to show avionic equipment in structural airframe drawing.
phantom member Non-existent member added in pre-computer era to assist solution of structural analysis.
phantom order Draft contract with manufacturer with provision for preplanning immediate production in time of crisis or conflict (DoD).
phase 1 In any periodic cycle, fraction of period (1) measured from any defined reference.
2 In reactive circuit, relationship between current and voltage.
3 In physical chemistry, distinct homogeneous physical states separated by sharp boundaries, eg liquid/solid, solid/solid, or immiscible liquids.
4 Periodic variation in solar illumination of Moon as seen from Earth.
5 Normal meaning often applied to programme planning, amphibious assault, establishment of military government etc.
phase-advance Subsystem which senses aeroplane rate of change of pitch and triggers stick-pusher progressively earlier as rate increases, so that pitching momentum never takes AOA beyond prescribed limiting value.
phase angle 1 Phase (1) difference between two sets of periodic phenomena expressed in angular measure.
2 Angle between current and voltage in rotating-vector plot of AC.
3 Angle between sightlines to Sun and Earth measured at remote location, eg other celestial body.
phased-array aerial One with electronic scanning.
phase difference Measure of phase angle (1) from any VOR radial related to that on bearing 000°.
phase discriminator Detector of phase modulation.
phase inverter Radio or other signal-processing stage with unity gain whose output is reciprocal of input (not synonymous with half-wave rectifier).
phase modulation Carrier phase angle (1) varies from carrier angle by amount proportional to instantaneous amplitude of modulating signal and a rate proportional to modulation frequency, PhM.
phase out, phaseout Progressive withdrawal from production or active service.
phase shift Phase difference (not necessarily VOR); change of phase angle (1).
phase velocity That of equiphase surface of travelling plane wave along wave normal; also called wave speed/velocity.
PHEI Penetrator HE incendiary.
phenolic/epoxy Family of resins and adhesives much used in composites derived from phenol (carbolic acid) and characterized by oxygen bridges linking hydrocarbon radicals.
phenolics Large family of synthetic polymers (plastics/resins/adhesives) dating from 1907 and mainly unmodified phenol-formaldehydes or (esp in case of adhesives) resorcinol-formaldehydes.
PHI 1 Position and homing indicator.
2 Pitot-head inoperative.
Phillips entry Shape of leading edge of typical modern wing (Horatio Phillips, 1886).
phon See *noise*; not used in modern work.
phosphate esters Fire-resistant hydraulic fluids based on esters of P (18) acids.
phosphor Substance which is luminescent; those in radars/CRT/TV etc are commonly zinc sulphide/zinc selenide/copper compounds but cadmium and rare earths are common. Those for printing on opaque substrates are unrelated.
phosphorescence Luminescence which continues more than 10^{-8} s after cutoff of excitation, usually being visible to eye for days thereafter.
phot Non-SI unit of illuminance = $lm/cm^2 = 10^4$ lx.
photoactivated Activated by light.
photocathode Electrode for photoelectric emission.
photochemical Involving chemical change and emission/absorption of radiation.
photochromic Having colour, transmittance or other optical property changed by incident light.
photochromy Colour photography.
photoconductive Having electrical resistance varied by illumination.
photodiode Diode converting light into electricity; hence * array yields signals which when processed analyse incident light pattern.
photodrafting See *photographic lofting*.
photoelectric Involving light and electricity, usu-

ally by absorbing photons and emitting electrons.

photoelectric cell Transducer converting EM radiation in visible, IR or UV wavelengths into electricity; abb photocell.

photoelectron Electron ejected, eg from metal surface, by impact of energetic (short-wavelength) photon.

photoelectronics Involving electrons and photons (many devices).

photoemissive Emitting electrons (ie electric current) when illuminated.

photoflash Pyrotechnic cartridge producing brief but intense illumination, esp for lo-level night reconnaissance.

photogrammetry Making accurate measurements and drawings, esp surveying and mapmaking, by photographic means.

photographic layout drawing Photographic lofting.

photographic lofting Lofting entirely with photographs.

photographic transmission density Log of opacity (base 10), thus perfectly transparent film has *** of zero, while one transmitting only 10% has *** = 1.

photometer Instrument for measuring luminous intensity, luminance or illuminance.

photometry Science or technology of measuring luminous flux, luminous intensity, luminance and illuminance.

photomultiplier Tube containing photocathode, several intermediate electrodes (dynodes) and output electrode; also called multiplier phototube.

photon Elementary parcel of EM energy emitted by transition of single electron, with energy $h\nu$ (h = Planck constant, ν frequency) and momentum $h\nu/c$ where c is velocity of light.

photon rocket Theoretically achievable rocket whose working fluid is light, ie stream of photons; small thrust but in deep space very high I_{sp}, and vehicle velocity could be significant fraction of that of light.

photopic vision Using retinal cones, hence colours distinguishable.

photosensitivity Degree to which substance changes chemical or electrical state when light falls on it.

photosensor Device operating by photoconductivity, eg light valve.

photosmoke method One of two techniques for measuring smokiness of jet by direct determination of optical density (other is Hartridge); gives output in PSU.

photosphere Intensely hot, bright outer layer of Sun's atmosphere.

phototheodolite Instrument comprising camera whose azimuth and elevation are precisely recorded (usually on its own film).

phototransistor Solid-state device, originally Ge wafer, generating holes by light absorption and multiplying this photocurrent by transistor action at collector.

phototube Electron tube (vacuum tube) containing photoemissive cathode and collecting anode (usually plus other sub-devices).

photovoltaic cell Transducer which, like photoelectric cell, converts EM radiation in visible or near-visible wavelengths into electricity; unlike photocell its purpose is to generate usable current instead of merely giving signal or serving other purpose calling for very low power; example is solar cell.

phraseology Accepted forms of speech, codes, phonetic alphabet, etc, used to facilitate telecommunications, usually by voice.

PHS Precision hover sensor.

phugoid Long-period oscillation of aeroplane pitch axis, perpetually hunting about level attitude and trimmed speed; noun and also adj, eg * oscillation.

P_{hyd} Fluid system pressure, esp hydraulic test pressure of rocket motor case.

PI 1 Point of interception (nav plot).
2 Photographic interpreter (or interpretation).
3 Process (or program) instruction.
4 Practice interception.
5 Program introduction (D adds "document").

P_i Input power, esp of jammer.

PIA 1 Pilots' International Association (US).
2 Pilot-interpreted approach.

PIAG, Piag Propulsion Installation Advisory Group (Int).

piano hinge One continuous along edge of hinged item.

piano keys Black/white runway end markings (colloq).

piano wire Finest steel wire normally produced; 0·8–0·95% C, very high uts, accurate dimensionally.

PIC Pilot in command.

PICAO Provisional ICAO.

pick-a-back 1 See *composite aircraft.*
2 Superimposed printed-circuit boards.

picket 1 AEW or AWACS aircraft.
2 Instrumented oceangoing ship on missile range.

picketing Securing aircraft against movement when parked in open, normally by attachment to heavy masses or spiral rods screwed into ground.

picketing anchor Spiral rod with eye at upper end.

pickle Tactical air code: moment of manual triggering of system, esp release of ordnance on surface target.

pickle button That commanding release of air-dropped stores.

pickling Soaking in dilute acid solutions to remove oxides or other surface films or intercrystalline carbides and surface scale. Principal acids are HCl, H_2SO_4, HNO_3 and HFl.

pick-off Sensor of angular motion or position;

many types, eg electric potentiometer, angular digitizer, photocell, magnetic coil moving-iron reluctance bridge, or fluidic valve or gate.

pickoff excitation Normally a frequency.

pickoff sensitivity Usually signal voltage per unit angular travel.

pico Prefix 10^{-12}; hence one picosecond (1 ps) is one millionth of one millionth of a second.

Pics, PICS Photogrammetric integrated control system.

PID Program introduction document (DoD).

PIDP Programmable indicator data-processor (USAF).

PIDS Prime-item development specification.

pier Long corridor, usually two-level, connecting airport terminal with gates.

pierce To cut part from sheet; hence large family of presswork dies such as * and cut off, * and form, * and trim.

pierced-steel planking Standard (mainly WW2 to 1950) unit of prefab airfield surface; mild steel plates measuring 119·75 in × 16 in and weighing 65 lb (29·5 kg) with interlocking edges.

pièze Non-SI unit of pressure used in French legal system = 1 sn/m² = 1 kN/m² = 1kPa.

piezoelectric Relationship exhibited by certain crystalline substances, esp single crystals, between electric potential difference and mechanical stress; eg applying voltage (DC or AC) across opposite faces results in expansion/contraction or vibration, while applying stress or vibration results in potential difference.

Pifet Piezoelectric field-effect transistor.

PIG, Pig Pendulous integrating gyro.

Piga PIG accelerometer.

Pigeon Air-intercept code: "Your base bears X° and is Y miles away".

piggyback Composite aircraft, or aircraft carrying large vehicle superimposed.

Pigma Pressurized-inert-gas metal arc.

pigtail 1 Projecting rigid pipe, usually with 90° bend and threaded connection for attachment to fluid system.
2 Short length of any other kind of cable or transmission line projecting from device for attachment to system.

pillow tank Dracone or similar flexible fluid storage.

Pilot Piloted low-speed test (ambiguous).

pilot Person designated as *. Previous definitions involved operation of particular controls (in one case "mechanisms") or guidance of aircraft in 3-D flight, none of which need be done in advanced aircraft, though * required to monitor. In case of RPV * may be in other aircraft or on ground. For command-guided missiles preferred term is operator.

pilotage Contact flying, navigating by visible surface landmarks.

pilot balloon Met balloon; alternatively, small free balloon devoid of instrumentation, observation of which from ground enables wind at different heights to be calculated.

pilot canopy Small auxiliary canopy, ejection of which pulls out main canopy (personal, cargo and braking parachutes).

pilot certificate In many countries, title of document licensing pilot according to five to 11 categories. In UK and many other countries called licence.

piloted Supervised by human beings, usually on board, playing active and direct role in control of vehicle.

pilot hole Small but precisely located hole serving as guide to subsequent larger drilling.

pilot in command Person responsible for aircraft in flight.

pilot-interpreted system One, eg early AI radar, requiring skill and judgement on part of operator, in contrast to modern digital readout and unambiguous indications. Note: early systems were often interpreted by other members of crew but no term exists.

pilotless aircraft Ambiguous: aircraft whose pilot has departed or aircraft designed to fly unmanned (arch).

pilot opinion rating Subjective assessment of aircraft stability and handling, measured according to Cooper scale.

pilot parachute 1 Small auxiliary parachute (see *pilot canopy*).
2 Parachute worn by pilot, eg with seat-type pack or forming part of ejection seat.

pilot plane Auxiliary surface mounted ahead of main surface (some definitions add "and free to take up position in line with wind").

pilot's notes Handbook providing operating instructions, helpful advice and all significant numerical data for pilot of particular type or sub-type of aircraft.

pilot's preference kit Small bag of allowed personal effects (NASA).

pilot's reference eye position Assumed position of eyes of normal pilot looking ahead (as for landing) in particular type of aircraft, esp in designing flight deck.

pilot's trace Rough overlay to map made by pilot of reconnaissance aircraft immediately after sortie showing locations, directions, number and order of sensing runs together with sensors used on each.

PIM Previous intended movement, of aircraft carrier.

PIN, p-i-n Semiconductor p-n junction diode with interleaved layer of intrinsic semiconductor (from "positive-intrinsic-negative").

PINE, Pine Passive infra-red night equipment.

pinger 1 Acoustic transducer array or other source of ASW underwater signals.
2 Operating sonobuoy.
3 Crew member in charge of ASW sonics.

pingly ASW helicopter, or a crew-member thereof (UK, colloq).
ping-pong snow Loose aggregations similar in size to table-tennis ball.
ping-pong test Pressure test in which air is used and volume is filled with ping-pong balls to reduce stored energy.
pin joint Joint between structural members where link is pivot, thus no bending moment can be transmitted and members of structure entirely pin-jointed must all be in pure tension or compression.
pinked Cut with zig-zag edge (with pinking shears); almost universal with fabric coverings.
pinking See *knocking*, *detonation*.
Pinlite Miniature light source of discrete-device type.
pinpoint 1 Precise fix.
2 Small positively identified ground feature providing fix.
Pins Palletized INS.
pin stowage Authorized and clearly visible attachments for safety pins removed from ejection seat.
pint Non-SI measure of capacity, pt = (UK) 0·5683 l, (US) 0·4481 l.
pintle Word used in normal sense, as cantilever pivot-pin, in types of gun mount, esp on underside or in doorway of helicopters.
PIO Pilot-induced oscillation(s), usually phugoid.
PIP 1 Product-improvement programme.
2 Predicted impact point (NASA).
3 Pulse-interval processor.
pip Small blip on CRT, esp one used (usually as one of series) as timing mark.
PIPA, Pipa Pulse integrating pendulous accelerometer.
Piperack Advanced jammer for AI radars (RAF 192, 214 Sqns 1944).
pipper Aiming mark, typically 2-mil-diameter dot on HUD or other sight system.
Pip pin Patented family of connecting pins having one end headed (often with knurled drum) and other chamfered and provided with two spring-loaded round-head plungers 180° apart which keep pin in position; usually steel and not normally used as permanent fixture.
PIR 1 Precision instrument runway.
2 Pilot incident report.
3 Property irregularity report, dealing with loss/damage to baggage and other pax possessions.
piracy Used in traditional sense for unauthorized appropriation of aircraft, ie hijacking.
Pirani gauge Measures vacuum by Wheatstone bridge and resistance wire.
Piraz Positive-identification and radar advisory zone.
Pirep Automatic pilot report programme; pilot reports actual weather on discrete frequency to chosen VOR nearby where message is taped and rebroadcast by VOR until an amending Pirep is received (FAA).
piston engine One in which working fluid yields energy by expanding and driving piston along cylinder, specif IC engine of Otto (by far most common in aviation), diesel or Stirling type.
piston ring Precision-ground abrasion-resistant ring fitted in groove around piston to make spring-tight fit against cylinder (see *gas ring*, *junk ring*, *obturator ring*, *oil-scraper ring*).
piston-ring seal One of hard metal pressed against cylinder wall by its own elastic stress.
PIT Pilot instructor training.
pitch 1 Angular displacement (rotation) about lateral (OY) axis.
2 Arguably, angular displacement about that axis which, at any moment, is perpendicular to both vehicle longitudinal (OX) axis and local vertical; thus in vertical climb, according to this widespread definition, * = yaw.
3 Distance propeller would advance in one revolution if slip were zero.
4 In case of ballistic vehicle, rotation about axis perpendicular to vertical plane containing vehicle's longitudinal axis.
5 Angular setting, measured at defined station, of helicopter main or tail-rotor blade relative to axis of rotation.
6 Uniform distance between evenly spaced objects in row, eg rivets, bolt threads or passenger seats (in each case measured from same reference point on each object).
7 Rotation of camera about axis parallel to vehicle lateral axis; also known as tip.
pitch attitude Angle between vehicle longitudinal (OX) axis and defined reference plane, eg local horizontal.
pitch axis See *lateral axis*.
pitch circle See *pitch curves*.
pitch cones Contacting cones of bevel gears on which normal pressure angles are equal.
pitch control 1 That giving manual control of propeller pitch (3), eg by moving datum of CSU.
2 That, normally effected by stick, giving control in pitch (1) of VTOL aircraft at zero or low airspeed.
3 That giving control of pitch attitude of spacecraft.
4 Combined cyclic/collective systems of helicopter.
pitch curves Intersection of tooth surfaces in pitch cones.
pitch diameter That of pitch line in circular wheel.
pitching See *pitch (1)*.
pitching moment One causing pitch (1), measured as positive when nose-up or tail-heavy.
pitch jet RCJ providing low-airspeed control in pitch (1).
pitch line Locus of points at which centres, contact points or pitch (6) of gearwheel teeth or bolt threads are measured.

pitchover Pronounced departure from upwards-pointing attitude, eg at point where ballistic vehicle is programmed to pitch (4) away from vertical, aircraft in stall-turn pitches at highest point and aircraft attempting absolute-altitude record runs out of kinetic energy.

pitch plane That common to both pitch cylinders of helical or spur gears, pitch cones of mating bevel gears, or both pitch cylinders of worm-wheel (on which axial and transverse pitches are equal) and pitch line of mating gear.

pitch point Point of contact of two pitch circles.

pitch pointing Advanced FCS mode giving ability to vary pitch (1) and thus AOA at constant flightpath angle (eg feature of F-16).

pitch range Angular range of travel of blade.

pitch ratio Ratio of pitch (3) to diameter.

pitch setting 1 Act of setting up propeller or helicopter rotor so that all blades have correct pitch according to that commanded.
2 Actual pitch of blade(s) at particular time, in case of propeller usually at 0·75 radius.

pitch speed Product of mean geometric pitch and number of revolutions made in unit time (latter is s or h depending on unit used to express answer).

pitch trim compensator See *Mach trimmer.*

pitch-up Uncommanded positive pitch (1) experienced by some aeroplanes at high subsonic Mach numbers or (tip stall on swept wing or tailplane in wing downwash) at large AOA.

pitot bomb Pitot head carried on free-weathercocking mass towed on cable.

pitot comb Row of pitot tubes, eg in vertical row behind wing.

pitot head Sensing head for pitot/static system, or in case where static pressure is taken from skin vent, pitot pressure only.

pitot pressure That sensed by pitot head, intended to be close to stagnation pressure.

pitot rake See *pitot comb.*

pitot/static system Instrumentation system fed by combination or pitot pressure and local static pressure, difference giving ASIR.

pitot traverse Taking successive measures of pitot pressure under same conditions but at different places, esp along vertical (less commonly horizontal) line in wake of wing or other body; result indicates fluid momentum transfer and thus drag.

pitot tube Open-ended tube facing forwards into fluid flow, thus generating internal pressure equal to stagnation pressure (in case of supersonic flow, that downstream of normal shock).

PIU 1 Pilot-induced undulation (PIO is better).
2 Processor interface unit.

PIV, p.i.v. 1 Pressure isolating valve.
2 Peak inverse voltage.

pixel Picture element, from which electronically transmitted picture is assembled.

PJ Parajumper, in helo rescue crew.

Pj 1 Radar received-power from jammer.
2 Jetpipe pressure.

PJF Partially jet-borne flight.

PJH PLRS/JTIDS hybrid.

PK, P_k Kill probability.

PKD 1 Path of known delay.
2 Parts knock-down, aircraft or other product supplied unassembled.

PKO Cosmic (space) defence forces (USSR).

PKP Passenger-kilometres performed.

PL 1 Position line.
2 Plain language (often P/L).
3 Pulse length.
4 Pilote de ligne (F).
5 Parts list.

P/L 1 Payload.
2 Plain language.

PLA 1 Programmable logic array.
2 Power-lever angle.

PLAB Napalm (USSR).

placard value Published numerical values of aircraft performance, esp those concerned with safety or limiting speeds and often displayed on placard (small plate) fixed in cockpit.

place Seat; hence "4-* ship" = four-seat aircraft (US usage).

PLACO, Placo Planning committee (ISO).

plain bearing One in which rotating shaft is simply run in surrounding fixed support, usually lined with bearing metal, without needles, rollers, balls or dynamic pressure from air or gas, but with interposed oil film.

plain flap Simple flap in which trailing edge of wing is hinged.

plan One of three basic orthogonal views, that showing object from above; hence *-form.

planar Essentially lying in one plane, 2-D; hence * technology or * electronics include solid-state devices constructed as various deposited layers, with etching, metallization and other layer-modification.

planar-array radar One whose aerial comprises numerous (normally identical) elements in flat array; probably electronically scanned and probably synonymous with phased-array.

Planck law Fundamental law of quantum theory: $E = h\nu$ where E is value of quantum in units of energy, h is Planck constant ($6{\cdot}626196 \times 10^{-34}$ Js) and ν is frequency.

plane 1 Aeroplane (colloq, rarely used in aerospace).
2 A wing, either left or right or complete tip-to-tip).
3 To move over water at speed sufficiently high for hydrodynamic and aerodynamic lift to predominate over buoyancy.

plane-change engine Small rocket, usually MHR, whose thrust alters orbital plane of satellite.

plane flying Navigation without electronic aids over short distances such that curvature of Earth is neglectable (doubtful technique).

plane-guard Routine duty of aircraft (today helicopter) stationed off port (left) quarter (towards stern) of carrier while flying operations are in progress; rescues ditched aircrew and performs other tasks.

planemaker Aircraft manufacturer.

plane of rotation That in which tips of blades of rotating object travel; in case of helicopter main rotor synonymous with tip-path plane and thus seldom perpendicular to shaft axis.

plane of symmetry That containing OX and OZ axes, dividing aircraft into (usually mirror-image) left/right halves.

plane-polarized EM radiation, eg light, in which electric force and direction of propagation remain in one plane.

planer Machine tool, often large, whose work-piece is cut by linear motion past fixed tool. Skin mill has revolving cutter(s).

planetary boundary layer From planet surface to geostrophic wind level, including Ekman layer.

planetary gear Reduction gear in which driven sun-wheel turns planet-wheels engaging with fixed outer annulus; any gearwheel whose centre describes circular path around another.

planetary lander Spacecraft designed to (usually soft-) land on planet.

planetocentric Related to planet's centre, eg * orbit.

plane wave One whose front is normal to propagation direction.

planform Geometric shape in plan, esp of wings and other aerofoils.

planimeter Instrument for mechanically measuring area on plane surface.

planing bottom Faired smooth surface on underside of float or hull (BSI); this omits to note need for deadrise, chine, step etc, needed for planing (3).

plank wing Traditional wing as distinct from swept or other modern planform (colloq).

plan-label display Radar display on which SSR alphanumeric information can be written in association with positional echo or symbol; usually fast synthetic, or mixed-phosphor (hard for high-refresh alphanumerics and soft for raw position symbol).

planned flight One for which flightplan is filed and which has specific purpose, ie not air experience or joyride.

planned load One made up in advance and tailored to cargo-aircraft type and mission.

planning-programming-budgeting Integrated system for management of DoD budget and Five-Year Defense Program (USAF).

planometer Surface plate.

plan-position indicator P-type display in which scene appears in plan with observer, radar or other sensor at centre; objects at radial distance giving range (usually linear scale, often selectable to several values) and with correct bearing (000° usually at top or 12 o'clock position); offset PPI moves sensor to position away from centre, typically to 6 o'clock margin. Expanded-centre PPI has zero range at ring surrounding centre.

plan range In air reconnaissance, horizontal distance from sub-aircraft point (that where local vertical through aircraft intersects surface) and ground object.

PLAP, Plap Power-lever angle prime, throttle in UFC (1) to which engine responds irrespective of pilot demand.

PLASI, Plasi Pulse light approach slope indicator (white/red, pulse frequency proportional to aircraft deviation from glidepath).

plasma Assembly of neutral atoms, ions, electrons and possibly molecules in which particle motion determined by EM interactions; electrically conductive, hence responds to magnetic field. Study called MHD or hydromagnetics.

plasma engine See *plasma rocket.*

plasma jet Jet of plasma produced by MHD.

plasma panel Electronic display of gas-discharge type, usually AC, usually orange (Ne), but many other colours with different gases.

plasma plating Deposition of refractory, abrasion-resistant or anti-corrosive coating by means of intensely hot (c16,600°C) plasma jet moving at supersonic speed into which coat material is introduced as powder.

plasma rocket One whose working fluid is a plasma, accelerated by intense EM field (ie, plasma jet).

plasma sheath That surrounding re-entry vehicle or spacecraft, serving as barrier to radio communications.

plastic Not elastic, tending to remain in deformed position (see *plastics*).

plastic effect Electronic display shows relief but little tonal value.

plastic flow That caused by stress beyond elastic limit and remaining when stress is removed.

plastic gyro Wheel assembled from moulded plastics components.

plastic instability Column failure due to plastic flow in compression rather than to bending.

plasticity Ability to be deformed to new permanent shape.

plasticizer Substance added to polymer to change properties to improve mouldability or other useful properties; usually liquid of high boiling point. In solid propellants used chiefly to increase flexibility, strengthen bonding and eliminate cracking.

plastics General terms for vast range of synthetic materials made by mixing constituents of which prime members are polymers, for distribution as liquid, fibre, granules or sheet subsequently moulded (see *thermoplastic*, *thermosetting*) or used as reinforcement with adhesive bonding. Properties range from rubbers to highly crystalline fibres. Singular "plastic" is adjective; in describing part or finished product preferred usage is "plastics" or, if possible, name material, eg PTFE, PVC, GRP or CFRP.

plastics factor Additional factor, typically 1·2–1·5, applied in designing primary structure in fibre-reinforced composite; this "factor of ignorance" is being relaxed as experience is gained.

plate 1 Sheet thicker than 0·25 in (6·35 mm); in airframes invariably machined or chem-milled. Not to be confused with sheet.
2 Principal anode of vacuum tube.
3 Pocket-size sheet of paper, plastics or aluminium on which are printed details of facilities, aids and approach data for one airport.

plate brake Mechanical brake for rotating shaft, eg landing wheel or helicopter rotor, where retarded moving member is ring fabricated from heavy plate (steel, titanium or beryllium, or CFRP-based) often stacked in parallel; essential difference from disc brake is ring is gripped from both sides.

plated-wire memory Advanced and highly compact memory woven on loom from coated wire giving non-volatile storage and low-nanosec speeds.

platelet Small plate, esp one which is perforated by, or whose surface contains precision channels usually produced by, photo-etching, for fluidic or rocket-injector system.

platelet injector Rocket injector assembled from large stack of platelets to give optimum multiple paths for (usually two) liquid propellants.

platform 1 Vehicle carrying sensors and/or weapons, eg aircraft.
2 Extended root of turbine (rarely, other) blade linking root attachment to outer aerofoil.
3 Raised operating area for helicopter or V/STOL, esp on surface vessel, also called pad.
4 See *airdrop platform*.

platform drop Drop of loaded platform (4) from rear-loading aircraft with roller conveyor.

platform dynamics Those resulting from motion of platform (1), esp as they affect ECM/ESM, eg range, range-rate (velocity), acceleration and acceleration-rate (jerk); can cause receiver to lose lock or synchronization.

platform face That forming inner end of aerofoil portion of turbine rotor blade and part of inner wall of gas duct.

Plato Program logic for automatic teaching operations.

playing area Area of operations possible with digital ATC (1) simulator, typically 256, 512 or 1,024 miles square.

PLB Personnel (or personal) locator beacon.

plc 1 Public limited company.
2 Programmable logic controller.

PLD 1 Pulse-length discrimination (in MTI circuits eliminates fast-moving clouds and other moving objects whose size precludes their being targets).
2 Precision laser designator.

plenum chamber Airtight chamber, esp one containing fluid-flow sink such as operative air-breathing engine; essential for gas turbine having double-entry or reverse-flow compressor with ingestion all round periphery.

plenum-chamber door Blow-in door to increase airflow into chamber when internal depression falls below selected level, eg on take-off.

Plesetsk Soviet ICBM base and launch establishment for Cosmos and many other large ballistic systems; in Leningrad military district at 62·9°N 40·1°E.

Plexiglas Registered name (Rohm & Haas) of family of acrylic-acid resin plastics, esp transparent, widely used for blown mouldings; essentially US counterpart of Perspex, though different material.

PLF 1 Precise local fix.
2 Parachute landfall.

PLH Propeller load horsepower.

PLI Pre-load indicator (projects from head of structural bolt).

pliss See *PLSS (2)* (colloq).

PLM 1 Pulse-length modulation.
2 Pulse-length monitor.

PLN Flight plan.

plot 1 Graphical representation of two or more variables on 2-D surface.
2 Graphical construction for solving nav problems, eg triangle of velocities.
3 Map, chart or graph representing data of any sort (DoD).
4 Visual display, eg on radar, of aerial object at particular time; hence * extraction.
5 Portion of map or overlay showing outlines of areas covered by reconnaissance or survey photographs.

plot extraction Translating radar plot (4) into quantified target position information, formerly done manually.

plot extractor Electronic system which detects replies from primary or secondary radar and, after making validity check, digitizes information ready for transmission over narrow-band link equipment or high-grade telephone line; where input is SSR basic range/bearing information can be supplemented by identity and height. "Extraction" derives from fact system eliminates information not needed.

plotting board Large horizontal (rarely vertical) surface upon which positions of moving objects are shown with respect to co-ordinates or fixed reference points.

plotting chart Chart designed for graphical methods of navigation.

ploughing, plowing Taxiing marine aircraft at below planing speed.

PLP Pipeline patrol.

PLRO Plain-language readout.

PLRS Precision (or position) location (and) reporting system.

PLSS 1 Precision location strike system.
2 Portable life-support system.

PLU Position (or preservation of) location uncertainty.

plug 1 Extra section, usually of constant cross-section, added in front of or behind wing when stretching fuselage.
2 Air-refuelling contact, hence wet *, dry *.

plug door One so designed, eg with inward/upward travel or with retractable upper and lower portions, that it is larger than doorway, two mating with thick tapered edges to increase security of pressurized fuselage. Pressurization load merely forces door more tightly against frame.

plug gauge Male-type gauge, not always of circular cross-section and often tapered or threaded, for checking dimensions of holes and internal threads.

plug inlet One form of inlet for air-breathing propulsion at Mach 3·5–6 in which axisymmetric duct tapers from sharp lip to rear, and contains large plug (spike) which when translated fully aft can seal flow; in most forms rearward spike travel renders shock-on-lip operation impossible.

plug nozzle Proposed for rocket engines: combustion chamber is annular toroidal form discharging around central cone with curved profile which converts initial radial component into pure axial flow. Also called spike nozzle.

plug ring Translating ring surrounding rear of PE (4) cowl to control cooling airflow; not necessarily provided with hinged flaps or shutters (ie gills).

plug tap Final non-bevelled or bottoming tap for completing threaded hole.

plug weld One made by drilling through part of structure, eg boom splice or skin/stringer, and welding through hole to increase strength of joint.

plug window Overlarge window with tapered periphery (as plug door) used as emergency exit in pressurized fuselage.

Plumbicon Photoconductive camera tube similar to vidicon but using semiconductor PbO target doped to behave as reverse-biased PIN.

plumbing Pipework for liquid systems, eg hydraulics, lube oil, Lox etc (colloq).

plume 1 Wake from jet engine at high power on ground, variable with wind speed and direction.
2 Jet from rocket at low or near-zero (ie in space) atmospheric pressure.

plus count Forward count begun at liftoff of spaceflight, continued throughout mission to provide GET reference.

Plymetal Plywood/aluminium sandwiches, invariably non-structural.

PLZT Lead lanthanum zirconate titanate.

PM 1 Pulse modulation.
2 Phase modulation.
3 Permanent magnet.
4 Program(me) manager (management).
5 Powder metallurgy.

P_m Polar maritime.

PMA/A Probable missed approach per arrival.

P_{max} 1 Maximum power (electronic).
2 Maximum pressure (usually MEOP).

PMC 1 Plastic/metal composite sandwich, usually two metal skins bonded to low-density core.
2 President of mess committee (RAF).
3 Max continuous power (F).
4 Polymer/matrix composite.
5 Power management control.
6 Performance management computer (S adds "system").

PMD 1 Projected map display.
2 Max take-off power (F).
3 Panel-mounted display.
4 Programme management directive.

PMEL Precision-measurement equipment laboratory (USAF).

PMG Permanent-magnet generator.

PMH Patrol missile hydrofoil (US Navy).

PMI Payload margin indicator.

PMMA Poly-methylmethacrylate.

PMO 1 Program(me) management office.
2 Principal medical officer.

PMOS Positive (p-type) metal-oxide semiconductor.

PMQC Purchase-material quality control (UK, quality assurance).

PMR 1 Pacific Missile Range, from Vandenberg (WTR) and Pt Mugu (US).
2 Proton magnetic resonance.

PMRAFNS Princess Mary's RAF Nursing Service (UK).

PMRT Program management responsibility transfer (DoD).

PMS 1 Performance (or power) management system.
2 Projected map system (or subsystem).
3 Personnel management squadron (RAF).

PMU Max contingency power (F).

PN 1 See *pseudonoise*.
2 Performance number.

P/N Part number.

PNA Point of no alternate.

PNB Pilot/navigator/bomb aimer (former RAF aircrew grade, higher than SEG).

PNC Pneumatic nozzle control.

PNCS Performance and navigation computer system.

PND Pilot numerical display, including dist to go, G/S.

PNdB Perceived noise decibels (see *noise*).

pneud, pneudraulic Combined hydraulic and pneumatic operation.

pneumatic Air-operated; term usually reserved for services taking very small flow at high pressure energized by shaft-driven compressor or one-shot bottle. Ambiguously also often used for services taking very large flow at low pressure to drive turbines, air motors and cabin environmental controllers. Terminology at present inadequate.

pneumatic bearing Externally pressurized gas bearing.

pneumatic logic That used in fluidics.

PNF Pilot not flying.

PNG 1 Pseudonoise generator.
2 Passive night (-vision) goggles.

PNI Pictorial navigation indicator.

PNJ Pulse(d) noise jamming.

PNL Perceived noise level (see *noise*).

PNLT Tone-corrected PNL.

PNP Programmed numerical path.

PNR 1 Point of no return.
2 Prior notice required.

PNVS Pilot's night vision system (attack helicopters).

PO Purchase order.

P_0 Total static pressure head; stagnation pressure.

POA Position of advantage (air combat).

POB Persons on board.

POC Point of contact.

pocket Short spanwise length of light trailing edge attached to rear of helicopter rotor-blade spar, many * being required for each blade.

POCU Pre-OCU.

POD Preliminary orbit determination.

pod Streamlined container carried on pylon, strut or other attachment entirely outside airframe and housing propulsion system, reconnaissance sensors, flight-refuelling hose-reel or similar devices.

Podas Portable data-acquisition system.

podded Accommodated in pod.

podding Philosophy and technique of using pods, esp for accommodation of main engines.

pod formation Formation of friendly aircraft disposed so that ECM assets give maximum mutual protection.

PODS, Pods 1 Portable data store.
2 Portable digitizer subsystem.

POE Probability of error.

Poet 1 Portable opto-electronic tracker.
2 Primed oscillator expendable transponder.

Pogo 1 See *pogo effect*.
2 Tail-standing VTOL (colloq).
3 Local below-airways ATC linking Paris airports.
4 Air-intercept code: "Switch to preceding channel or, if unable to establish communications, to next channel after *" (DoD).

pogo effect Longitudinal oscillation or vibration, esp of vehicle having high ratio of thrust to mass and whose propulsive thrust may suffer short-term variations; characterized by significant and uncomfortable axial accelerations and, in large ballistic rocket, severe propellant sloshing.

Pohwaro Pulsated overheated water rocket (FFA, Switz).

Point Arguello At south Vandenberg, Western Test Range (US).

point defence 1 Defence of specified geographical areas, cities and vital installations; distinguishing feature is that missile guidance radars are near launch sites (USAF).
2 Defence of single surface ship, esp against incoming missile.

point-designation grid Grid drawn on map or photograph whose sole purpose is to assist location of small features.

point discharge Gaseous electrical discharge from surface of small radius (point, or tips of static wick) at markedly different potential from surrounding; unlike corona discharge, ** is silent and non-luminous.

point light Luminous signal without perceptible length (ICAO).

Point Mugu Location of Naval Missile Center and Naval Missile Range (US).

point of attachment Where balloon rigging cable joins flying cable.

point of entry Where aircraft enters control zone.

point of inversion Height at which lapse rate at last passes through zero; where temperature begins to fall.

point of no alternate Geographical position on track or time at which fuel remaining becomes insufficient to reach declared alternate.

point of no return Geographical position on track or time at which fuel remaining becomes insufficient for aircraft to return to starting point (DoD, NATO wording "to its own or some other associated base").

point target 1 One requiring accurate placement of conventional ordnance in order to neutralize or destroy it (DoD).
2 With NW, one in which target radius is not greater than one-fifth radius of damage.

point to point Linear motion of tool between NC-instructed commands.

point vortex Section of straight-line vortex in 2-D motion.

Poise Pointing and stabilization platform element (USA, RPVs).

poise Non-SI unit of dynamic viscosity, defined as 1 dyn-s/m^2 = 0·1 Ns/m^2.

Poiseuille equation Relates flow through tube (defined as "long and thin"; elsewhere as "capillary") to variables: $Q = \pi Pr^4/8l\mu$ where Q is volume per unit time (seconds), P is pressure difference across ends of tube, r is radius, l is length and μ is viscosity.

Poiseuille flow 1 Viscous laminar flow in circular-section pipe.
2 Viscous laminar flow between close bounding planes.

Poisson's equation In stressed material $\sigma = (E/2n)-1$ where σ is Poisson's ratio, E is Young's modulus and n is modulus of rigidity.

Poisson's ratio Ratio of lateral contraction (in absence of local waisting) per unit breadth to longitudinal extension per unit length for material stretched within elastic limit, symbol σ.

Poits Payload orientation and instrumented tracking system.

poke Propulsive thrust or power (colloq), hence "pokier", etc.

POL Petrol, oil and lubricant.

Pol, pol Polarity (but P in FFP).

polar 1 Air mass supposedly originating near pole, hence cold and usually dry; thus * Atlantic, * continental, * front, * maritime.
2 Parameter plotted on polar co-ordinates.
3 Basic performance curve of sailplane in which sink speed is plotted against EAS (units of two scales differ, traditionally ft/s against kt but today m/s: km/h).

polar-cap absorption Radio blackout by HF absorption in ionospheric storms.

polar continental Typically extremely cold, dry and stable air mass, Pc.

polar control Twist-and-steer flight control.

polar co-ordinates Those defining locations by means of angle of a radius vector, measured relative to agreed direction, and vector length; similar to rho-theta navigation.

polar distance Angular distance from celestial pole.

polar front That separating polar air mass from contrasting mass.

polarimeter Instrument for determining degree of polarization of EM radiation, esp light.

polariscope Instrument for detecting polarized radiation.

polarity 1 Of line segment, having both ends distinguishable.
2 Hence, of physical system, having two contrasting points, specif oppositely marked terminals of electric cell or plus/minus characteristics of ions.

polarization Many meanings but chiefly associated with EM radiation in which * can be plane, elliptical or circular. Plane * rotates all wave motion so that all E (electric) vibrations take place in one plane (plane of vibration) and all H (magnetic) vibration takes place at 90° to this (plane of *).

polarization diversity Having ability to switch from plane-polarized to circular-polarized (radar).

polar maritime Air mass that is cold and, though absolute humidity and dewpoint temperature low, relative humidity is high; abb P_m.

polar moment of inertia Moment of inertia of area about axis perpendicular to its plane. Traditional symbol not I but J.

polar navigation Navigation at high latitudes, distinguished in bygone days on account of unreliability of magnetic compass, possible electrical/radio interference and other problems such as rapid change of meridians and map-projection difficulties.

polar orbit Orbit passing over, or close to, poles of primary body.

polar Pacific Air mass originating over N Pacific or N America, seasonal characteristics; abb P_p.

polar plot Locating point, eg target, by polar co-ordinates.

Polar stereographic Map projection, that of high-latitude region projected on flat sheet touching Earth at Pole by light source at opposite pole. Parallels expand at $\sec^2$ co-lat/2.

polar triangle One formed by three intersecting great circles.

pole 1 Origin of polar co-ordinate system.
2 Point of concentration of magnetic charge (magnetic *).
3 Point of concentration of electric charge (dipole and sought-after monopole).
4 Intersection of Earth's surface and axis of rotation.
5 For any circle on spherical surface, intersection of surface and normal line through centre of circle.
6 Parts of surface of magnet through which magnetic flux emanates or enters (theoretical but necessary concept).
7 Terminal of battery.

pole piece That part of core of electromagnet which terminates at air gap.

poll 1 To ask specific questions of number of sensors; questions are normally asked sequentially, and answers constitute an update of information in system.
2 Technique used in data transmission whereby several terminals share communication channels, particular channel chosen for given terminal being determined by testing each to find one free or to locate channel on which incoming data are present. Also used to call for transmissions from remote terminals by signal from central terminal; method used for avoiding contention.

poll the room To obtain consensus in solving problem in manned space mission.

polybutadiene acrylic acid, PBAA Important solid rocket propellant and monofuel.

polybutadiene acrylonitrile, PBAN Important solid rocket propellant and monofuel.

polyconic Mapmaking by projecting latitude bands on to succession of cones, each centred on Earth's axis and each touching surface along parallel passing through centre of map, subsequent strips on single-curvature conical surface being unrolled.

Polydol Alcohol-resistant foam liquid for fire-fighting made from protein hydrolysate and an organometallic complex; mixes with fresh or seawater.

polygon warhead One having several (typically 8–16) radial faces from which are projected tailored fragments, steels balls, shaped-charge jets or other damage mechanisms. Common type of warhead for SAMs and AAMs, radial blasts being so timed that their plane passes through target.

polymerization Basic processes for making large (high-polymer) molecules from small ones, normally without chemical change; can be by addition, condensation, rearrangement or other methods.

POM 1 Printer output microfilm.
2 Program objective(s) memorandum (DoD).

Pomcus Prepositioned overseas materiel configured in unit sets (ie, grouped by user units).

Pomros Power-off minimum rate of sink.

ponding Settlement of runway subsoil leading to formation of standing-water pond in rain.

PONO Project office nominated official.

pontoon Inflatable buoyancy bag used as permanent alighting gear for helicopter.

POO 1 Payload of opportunity.
2 Pronounced poo, to clear manned-spacecraft computer display prior to solving fresh problem, from P-zero-zero (colloq).

pool fire Burning pool of fuel surrounding crashed aircraft.

poopy suit Flight-crew overwater survival suit (colloq).

POP 1 Probability of precipitation.
2 Product optimization programme.

Popeye Air-intercept code: "I am in cloud or reduced visibility."

poppet valve Common mushroom-type valve of PE (4).

Pop rivet Pioneer form of tubular rivet closed by withdrawing central mandrel which forms integral part of each rivet and is gripped by jaws of tension tool.

POPS, Pops Position and orientation propulsion system.

pop-up Manoeuvre made by attack aircraft in transition from lo penetration to altitude from which target can be identified and attacked; eliminated by BFPA (blind first-pass attack).

pop-up missile One ejected, often by gas generator/piston or other device not forming part of missile propulsion, from launch system in vertical direction, subsequently making fast pitch-over on target heading. Launcher need not move in azimuth or elevation, thus minimizing reaction time.

pop-up test Test of pop-up missile launch system.

POR Pilot opinion rating.

Poroloy Porous metal produced by incomplete sintering, mainly used as filter.

porpoise 1 Undulatory (near-phugoid) motion of marine aircraft and some landplanes with bicycle landing gear characterized by pitch oscillation and limited-amplitude vertical travel; normally problem only at particular speed(s).
2 In absence of radio contact, deliberate roller-coaster flight to indicate to pilot of friendly interceptor (more rarely, tower) that aircraft is in distress.

port 1 Left side or direction.
2 Aperture in solid rocket motor case opened for thrust termination.
3 Periodically opened apertures in PE (4) or reciprocating-compressor cylinder closed by valves.

portable data store Computerized flight-planning data carried by member of flight crew and plugged in before take-off.

POS Peacetime operating stocks.

posigrade In direction of travel, hence * rocket increases speed of satellite in order to reach higher orbit. Opposite of retrograde.

position 1 For celestial body, its bearing and altitude.
2 Location of crew member, esp in military aircraft.
3 Location of manually aimed defensive gun(s) in military aircraft (probably arch); gun *.
4 Location of AAA defending land target; gun *.
5 To fly aircraft to where it is needed; hence * or positioning flight.

position error That induced in ASI system by fact that stagnation pressure sensed is seldom that of true free stream (see *airspeed*).

position light See *navigation light*.

position line Line along which vehicle is known to be at particular moment, eg by taking VOR bearing. Two PLs are needed for fix.

position-stabilized Held to linear trajectory, eg LOS, in absence of commanding signal (eg anti-tank missiles).

positive acceleration Acceleration upwards along OZ axis, to right along OY and forwards (ie to increase speed) along OX. Axes are always vehicle-related.

positive area That enclosed on tephigram between path of rising particle and surrounding air when particle is throughout warmer than surroundings.

positive control 1 Operation of air traffic in radar/non-radar ground control environment in which positive identification, tracking and direction of all aircraft in airspace is conducted by authorized agency (DoD).
2 Command/control and release procedures, and operational procedures that provide acceptable level of assurance against misuse of nuclear warheads, when these warheads are part of a weapon system (USAF).

positive coupling Mutually inductive coupling such that increase in one coil induces rising voltage in other similar in sense to that caused by increasing current in other.

positive-displacement pump One delivering fluid in discrete parcels in irreversible way, eg pumps using pistons, vanes or gearwheels in contrast to centrifugal blower.

positive-driven supercharger Mechanical drive as distinct from turbo.

positive-expulsion propellant system One in which liquid propellants are forced from container by gas pressure, esp acting on flexible bag containing propellant(s).

positive feedback Feedback that results in increasing amplification; also called regenerative.

positive g *See positive acceleration.*

positive-identification and radar advisory zone Specified area established for identification and flight-following of aircraft in vicinity of fleet defended area.

positive ion One deficient in one or more electrons.

positive pitch 1 Nose-up.
2 Propeller set for forward flight as distinct from braking.

positive pressure cabin Arch, see *pressure cabin.*

positive rolling moment Tending to roll right-wing-down.

positive stability Aeroplane tends of own accord to resume original condition, esp level flight, after disturbance (upset or gust) in pitch, without pilot action. Term applies only to vertical plane, along OZ axis.

positive stagger Leading edge of upper wing is ahead of that of lower.

positive stall Stall under 1 g flight or positive acceleration, ie not from inverted attitude or negative g.

positive yaw Tending to rotate anticlockwise about OZ axis seen from above, ie nose to left.

positron Short-lived particle equal in mass to electron but positive in sign.

POSS Power-off stall speed.

POST 1 Portable optical sensor testbed (hardened sites).
2 Passive optical seeker technique (or technology).
3 Production-oriented scheduling technique.

post 1 Vertical primary structure, eg fin *, king-*.
2 Main landing leg, thus MLG of YC-14 described as of four-* type.

post-attack period Between termination of nuclear exchange and cessation of hostilities.

post-boost After cutoff of mainstage propulsion.

post-integration Occurring after assembly of complete space launch system but before liftoff.

post-pass After (usually soon after) satellite has passed overhead.

post-stall gyration Uncontrolled motions about one or more axes, usually involving large excursions in AOA, following departure (3).

post-strike Immediately after attack on surface target, hence * reconnaissance provides information for damage assessment.

posture Military strength, disposition and readiness as it affects capabilities (DoD).

pot PE (4) cylinder (colloq).

potential 1 In electric field, work done in bringing unit positive charge to that point from infinite distance.
2 Value atmospheric thermodynamic variable would have if expanded or compressed adiabatically to 1,000 mb (100 kPa).
3 Specialized meanings in thermodynamics, geodesy and celestial mechanics.

potential density That which parcel of air would have if adjusted adiabatically to 1,000 mb (100 kPa), given by $\rho' = p/R\theta$ where ρ' is **, p is pressure at 100 kPa, R is gas constant and θ is potential temperature.

potential energy That possessed by mass by virtue of position in gravitational field; can yield ** by "falling", ie moving towards region of lower **.

potential flow Fluid motion in which vorticity is everywhere zero.

potential gradient Local space (linear) rate of change of potential.

potential refractive index That so formulated that variation with height in adiabatic atmosphere is zero.

potential temperature That which parcel of dry air would have if adiabatically brought to pressure of 100 kPa.

potentiometer 1 Variable resistance (rheostat), esp one of precise type giving, eg, accurate radar pointing information.
2 Instrument for measuring EMF (potential difference, esp DC), usually without drawing current from circuit measured.

potted device Electronic component encapsulated in resin, mainly to give mechanical protection.

Pounce Air-intercept code: "I am in position to engage target".

pound Non-SI unit of mass, abb lb, = 0·45359237 kg. Note, plural also lb.

poundal Non-SI unit of force, abb pdl, = 0·138255 N.

pound force See pound weight.

pounds per square inch Non-SI unit of pressure, abb lbf/sq in (psi or lb psi not recommended), = 6,894·76 Pa = 6·89476 kPa.

pound weight Attraction of standard Earth gravity for mass of 1 lb, abb lbf (lb force), = 4·44822 N.

pour point Temperature established by standard pour test as lowest at which liquid, esp fuel or lube oil, will flow.

powdered Tactical code: enemy aircraft broke up in the air.

powder metallurgy Production of finished or near-finished parts by fusion under heat and pressure of metal in form of finely divided powder. Usually synonymous with sintering.

power Rate of doing work; SI unit is watt (W) = J/s.

power amplification Ratio of AC power at output to AC power at input circuit.

power amplifier Amplifier designed to deliver large output current into low impedance to obtain power gain as distinct from voltage gain.

power-assisted flight control FCS in which power inputs assist pilot by overcoming major part of hinge moments while leaving pilot to move surfaces directly and experience direct feedback, and with difficulty control aircraft in event of system failure.

power brake US term for sheet-metal press, esp one using mating male/female dies or rubber.

power bumping Sheet-metal forming on bumping hammer.

power centroid Point which will be selected by seeker (radar or IR) as centre of target; in case where both real target and decoy (eg flare) are visible, ** likely to be in space between them.

power coefficient In calculating propeller performance by the Drzewiecki method, a grading curve of torque component plotted against blade radius may be drawn for each pitch angle to arrive at function K_Q, constant for each value of advance ratio J; ** equals $K_Q J^2$, and has symbol C_P. Total power absorbed is $C_P \rho n^3 D^5$, where ρ is air density, n rotational speed and D tip diameter.

power density Power per unit volume (electronic device, nuclear reactor).

powered approach One in which aircraft propulsion is giving significant thrust.

powered ascent That from lunar or other non-Earth surface.

powered controls Powered flight-control system.

powered descent That of Lunar Module or other soft-landing device on to surface other than Earth.

powered flight Flight of vehicle while self-propelled.

powered flight-control system Vehicle flight-control system which uses irreversible actuation such that no manual reversion is possible, though pilot may have manual override over AFCS at input.

powered high-lift system High-lift system in which energy drawn from main propulsion plays direct role, either by flow augmentation, flap-blowing, USB or other technique other than jet lift or use of rotors.

powered-lift regime Flight regime in which sustained flight at below POSS is possible because part or all of lift and control moments is derived from installed powerplants.

powered slat One positively driven to extended position instead of being moved by its own aerodynamic lift.

power egg Complete ECU, usually without propeller (normally piston engine).

power factor 1 That by which product of AC current and voltage must be multiplied to obtain true load, $= \cos \phi$ where ϕ is phase angle between current and voltage.
2 Measure of dielectric loss of capacitor.
3 In wind tunnel, ratio of driving power to kinetic energy multiplied by mass flow in working section.

power feedback Feedback in which significant amount of power (electrical, electronic or acoustic) is transferred.

power folding Folding wings by power actuation, usually controlled from cockpit and sometimes by external protected control.

power gain 1 Ratio of power, usually expressed in dB, delivered by amplifier or other transducer to power absorbed by input.
2 In any direction, $4\pi \times$ ratio of radiation intensity sent out by aerial to RF power input; with strongly directional (eg pencil beam) aerial close to zero except on axis of main lobe.

power head On large machine tool, cutter and associated drive.

power in, power out Aircraft parks at airport and departs without use of towing vehicle.

power in, push out Tug is used to push aircraft from nose-in stand to taxilane.

power jet In traditional carburettor, fuel jet which comes into operation only when throttle advances beyond max-cruise position.

power lever Throttle, esp for gas turbine of any kind.

power loading For propeller-driven aeroplane, total mass (usually taken as MTOW) divided by total installed horsepower, ehp in case of turboprop. Jet equivalent = thrust loading.

power-on spin One entered from power-on stall.

power-on stall Normal stall in which propulsive power is maintained at significant (eg normal cruising) level throughout; nose is raised higher to lose speed at acceptable rate and slipstream usually adds to lift, generally resulting in more extreme attitude and more violent pitchover when stall finally occurs.

power overlap Overlap of firing strokes in multi-cylinder PE (4); for four-stroke engine more than four cylinders are needed to make this concept of continuous power effective.

power plant, powerplant 1 Those permanently installed prime movers responsible for propulsion, including their number; thus * of L-1011 TriStar is three RB.211 turbofans (some authorities would add "plus ST6 APU").
2 One prime mover, of any type, in complete form plus accessories, silenced nozzle, propeller and associated subsystem and, in some cases, surrounding cowling.

power press Any press whose actuation force does not derive from human muscle; need not have vertical motion.

power shear Shear for cutting heavy sheet or plate, with hydraulic ram(s) moving blade.

power spectrum Plot of $S_a(f)$, signal amplitude against frequency, normally symmetrical about peak at fc and with form varying with modulation.

power/speed coefficient 1 Function in propeller performance calculation, introduced by F.E. Weick;

$$C_s = \frac{J}{C_p^{0.2}} = \frac{\rho^{0.2} V}{P^{0.2} n^{0.4}}$$

where J is advance ratio, C_p power coefficient, ρ density, V slipstream velocity, P power consumption and n rotational speed.
2 Variation of installed power or thrust with airspeed, esp for jet engine; normally plotted at constant air density.

power takeoff Shaft from which power is available to drive accessory or other item.

power turbine Mechanically independent turbine (ie connected to engine only by shaft bearings and gas path) driving propeller gearbox of turboprop or output shaft of turboshaft.

power unit 1 Power plant (2) (not recommended).
2 Device which either generates electrical or EM radiative power or produces required currents or signals from raw AC or DC input.
3 Source of energy, other than propulsion, for missile, RPV or spacecraft where there is no shaft-driven alternator or other supply.
4 Prime mover of APU or EPU/MEPU.

power venturi Venturi used to operate air-driven instruments or other devices relying on suction.

PP Descent through cloud, procedure code (ICAO).

PP, p.p. 1 Peak-to-peak.
2 Push/pull.
3 Present position.
4 Pilot production.

P/P Pupil pilot.

PPA 1 Pre-planned attack.
2 Passengers per annum.

PPAC Product performance agreement center (US).

PPB Program/planning/budgeting; PPBS adds "system", PPBES adds "evaluation system" (US).

PPC Production possibilities curve.

PPE Passengers (per year) per employee.

PPF Production-phase financing.

PPFRT Prototype preliminary flight rating test.

P/PFRT Pupil-pilot flight rating test.

PPH Pounds per hour.

PPI Plan-position indicator.

PPIF Precision photographic interpretation facility (pre-1980, photo processing [or production] and interpretation facility).

PPL 1 Private pilot's licence (/H, helicopter; /IR, instrument rating).
2 Polypropylene.

PPM 1 Pulse-position modulation.
2 Pounds per minute.

ppm Parts per million.

ppn Precipitation.

PPO 1 Prior permission only.
2 President's Pilot Office.

PPP Synoptic chart: pressure referred to normal mean sea level.

PPR 1 Prior permission required.
2 Periodic performance report.
3 Prospective price redetermination.

PPS 1 Passenger processing system.
2 Photovoltaic power system.
3 Pilot's performance system.

pps Pulses per second.

PPSI Probe-powered speed indicator.

PPT Perspective-pole track (HUD).

PQAR Product quality action request.

PQE Product-quality engineer.

PR 1 Photo-reconnaissance.
2 Pitch rate.
3 Pressure ratio.
4 Polysulphide rubber.
5 Ply rating (tyre).
6 Purchase request.

Pr 1 Prandtl number.
2 Radiation pressure.

PRA Popular Rotorcraft Association (US).

PRADS, Prads Parachute/retrorocket air drop system.

PRAM Productivity, reliability, availability and maintainability.

Prandtl-Glauert equation States that lift, drag or pressure coefficients at high subsonic speeds, where compressibility must be taken into account, are equal to those at lower speeds factored by Glauert factor $(1 - M^2)^{-1/2}$ where M is Mach number.

Prandtl-Glauert law That describing effect of compressibility on lift of 2-D wing.

Prandtl interference factor Dimensionless factor σ used in determining interference between wings of biplane; dependent upon ratios $\frac{\text{gap}}{\text{mean span}}$ and $\frac{\text{upper span}}{\text{lower span}}$; where both wings have same span, *** $\simeq$ 0·5 for gap/span ratio of 0·2 and about 0·375 for gap/span ratio of 0·3.

Prandtl-Meyer expansion Original treatment of supersonic expansion round corner, in which entropy remains constant but pressure, density, temperature and refractive index all fall. Prandtl solved 2-D case 1907.

Prandtl number Ratio of momentum diffusivity to thermal diffusivity, $Pr = \mu C_p/\lambda = \nu/\alpha$, where μ is viscosity, C_p is specific heat at const p, λ is thermal conductivity, ν is kinematic viscosity and α is angle of attack.

Prandtl-Schlichting Original treatment of transition flow between laminar and turbulent boundary layers forming link curve on plot of skin-friction coefficient against Reynolds number.

prang To have accident, esp to crash aircraft; also derived noun (RAF, WW2, colloq). If written on aircraft (PRANG), Puerto Rico Air National Guard.

PRAT Production reliability acceptance test.

Pratt truss Basic braced monoplane.

PRAWS Pitch/roll attitude warning system.

prayer mat Portable anti-erosion mat for jet V/STOL operation.

PRC Program(me) review committee.

PRCA Pitch/roll control assembly.

PRCR Preliminary request for customer response.

PRD 1 Program review data.
2 Primary radar data.

PRDA Program research/development announcement (US).

pre-balanced assembly Gas-turbine module held in store until needed.

precautionary flight zone That beyond left boundary of helicopter h/V (height/velocity) curve, where safe autorotative descent is unlikely or impossible.

precautionary landing Practice forced landing; various techniques but invariably objective is to arrive at correct point with correct speed for minimum ground run, then overshoot without touching.

precautionary launch Launch of aircraft loaded with NW from airbase or carrier under imminent threat of nuclear attack, not necessarily on mission to enemy target but to preclude friendly-aircraft (ie, its own) destruction.

precession Rotation of axis of spinning body, eg gyro, when acted upon by external torque, such that spinning body tries to rotate in same plane and in same direction as impressed torque; ie axis tends to rotate about a line (axis of *) normal to both original axis and that of impressed torque. Horizontal component called drift; vertical called topple.

precession cone That described by longitudinal axis of slender body, eg ballistic rocket, devoid of attitude or roll stabilization; two cones each having apex at vehicle c.g.

precipitation Moisture released from atmosphere, esp that in large enough particles to fall sensibly, ie excluding fog and mist; examples are rain, snow, hail, sleet and drizzle.

precipitation attenuation Loss of RF signal due to presence of precipitation.

precipitation drag That caused by slush or standing water on aircraft wheels; snow not normally included in this term.

precipitation hardening Usually synonymous with ageing (aluminium alloys).

precipitation heat treatment Artificial ageing, usually preceded by solution heat treatment.

precipitation interference Static discharge, normally considered to be caused by precipitation and atmospheric dust, which increases ambient noise level, making some reception (eg NDB) difficult.

precise local fix An offset (3) near inconspicuous ground target.

precision approach 1 One in which pilot is provided by ground PAR controller with accurate guidance to extended centreline, informed of glidepath interception and thereafter provided with precise az/el guidance together with distance from touchdown at intervals not greater than 1 mile (rarely, 1 km).
2 One in which ILS is used.

precision-approach interferometer radar Low-power ground interrogator near runway triggers replies from airborne SSR transponder received at three azimuth dishes (readout within 0·02°) and four elevation dishes (within 0·03°).

precision approach radar, PAR Primary radar designed specifically to provide accurate azimuth, elevation and range information on aircraft from range of at least 10 miles (16 km) on extended centreline to threshold.

precision bombing Level bombing directed at specific point target.

precision casting One in which main surface areas are to finish dimension, needing only drilling or trimming and surface treatment.

precision-drop glider Glider, usually of Rogallo type, for delivery of cargo to point target after release from nearby aircraft.

precision instrument runway One served by operative non-visual precision approach aid, esp ILS; specially marked with non-precision instrument runway indications plus TDZ markers, fixed-distance markers and side stripes.

precision spin Accurately performed spin completed as training manoeuvre: entry must be positive, spin must be fully developed within quarter-turn and spin recovery must take place after prescribed number (or number and fraction) of turns.

precision tool One not held in hand whose positions and attitudes in relation to workpiece are established to limits closer than engineering tolerances.

precision turn Training manoeuvre often conducted above straight road or similar reference in which smoothness of entry, quick roll to correct bank attitude, absence of skid, correct rate of turn and proper exit on desired heading are mandatory.

precombustion chamber Rare ancillary chamber used in start cycle of rocket or ramjet where burning fuel is produced to act as ignition source for main chamber. Also feature of certain diesel or oil engines.

precursor 1 Material such as PAN yarn from which by carbonization process carbon or graphite fibre is made.
2 Air pressure wave ahead of main blast wave in nuclear explosion of appropriate yield over heat-

absorbing or dusty surface.

predatory pricing Price structure established by airline on particular routes alleged by rivals to be aimed at elimination of competition; such prices are often held to be below direct costs.

prediction Predicted numerical value.

prediction angle That between LOS to target and gun line when properly pointed for hit on future position.

predictive gate Radar range-gate sensitive to hostile frequency-hopping and ECCM.

predictor Mechanical computer used (1926–45) to predict future positions of target for AAA.

predominant height Height of 51% or more of structures within area of similar surface material subject to air reconnaissance.

pre-emptive attack Attack initiated on basis of incontrovertible evidence that enemy attack is imminent (DoD). Word "nuclear" absent.

preferential runway One suggested or offered for departure to minimize local noise nuisance and afford route away from urban areas even though not optimum from viewpoint of departing traffic.

preferred IFR route In effect aerial motorways listed to guide long-distance pilots in US and facilitate ATC; divided into high- and low-altitude routes, some involving SIDs or STARs and one list being unidirectional (FAA).

preflight 1 Pre-flight inspection, and verb to accomplish same.
2 After completion of aircraft when systems are checked and other work done prior to first flight.

pre-flight inspection That carried out on aircraft before each flight by flight crew, especially PIC, to broadly standard procedure which varies from type to type. Obvious tasks are to remove removables, check for fluid leaks or superficial damage, inspect tyres and landing gear and have cleared objects that could be blown by slipstream.

pre-flight line Parking line for newly completed aircraft where work is carried out prior to first flight; hence pre-flight hangar.

preformed cable One whose individual wires are preformed to final spiral, or spiral-spiral, shape; thus they do not splay out should cable be cut.

prefragmented warhead One containing large number of high-density steel balls or similar penetrative objects. Term not to be used for continuous rod or heavy casing scored by grooves.

pre-ignition Fault of IC piston (esp Otto) engine in which, due to incandescent carbon or other cause, mixture ignites before spark, and may run after ignition has been switched off.

pre-launch period Time between receipt of alarm and take-off for manned aircraft, esp NW carrier.

preliminary flight rating test Pupil's first assessment of aptitude.

premate preparation Tasks accomplished before vehicle integration, eg before Shuttle has SRBs and tank added.

PREP, Prep 1 Pilot's reference eye position.
2 Part-reliability enhancement programme.

preplanned air support Air support in accordance with a program(me) planned in advance of operations.

preplanned mission request Request for an air strike or reconnaissance mission on target which can be anticipated sufficiently in advance to permit detailed mission co-ordination and planning.

prepositioned In place prior to hostilities to ensure timely support of initial operations.

prepreg CFRP raw material in form of oriented fibres, tows or other reinforcement impregnated with resin and supplied as sheet, strip or other form ready for final cutting and moulding, usually as number of plies with different orientation, to make product. Not relevant to product made by filament winding. Note: terminology bound to spread into boron/epoxy and many other forms of FRC construction.

pre-rotation gear Landing gear whose wheels are spun-up before touchdown.

pre-rotation vanes Upstream of fan in wind tunnel to impart swirl which is eliminated by fan.

presentation Geometric form, character and style of display, esp one of electronic nature; eg PPI plus alphanumerics, line plot plus cursive writing or TV picture plus alphanumerics.

preset guidance Pre-programmed autopilot determining mission in advance.

pre-simulator training Training on Technamation and similar animated displays, with mainly dummy controls, to ensure familiarity with systems and procedures.

Press 1 Pacific Range electromagnetic signature study (USAF, DARPA).
2 Project review, evaluation and scheduling system.

Presselswitch Pressure selector with manual input.

press fit Loose term which usually means interference fit (arch).

press tooling Tooling for use in presses, such as rubber or mating male/female dies for making finished or near-finished parts in sheet; not usually applied to tools for applying curvature to plate.

pressure Force per unit area, SI unit is pascal (Pa), = N/m^2, hence kPa, MPa, GPa.

pressure accumulator Device for storing energy in form of compressed fluid, typically by gas (nitrogen or air) trapped above liquid (eg hydraulic system).

pressure altimeter Conventional altimeter driven by aneroid capsule(s) and measuring not height but local atmospheric pressure.

pressure altitude 1 Height in atmosphere measured as vertical distance above standard sea-level reference plane defined in pressure terms, invariably 1013·2 mb; thus height indicated by pressure altimeter set to QNE and corrected for IE (2) and PE (2).

2 Altitude in standard atmosphere corresponding to atmospheric pressure in real atmosphere.

3 That simulated in pressure (vacuum) environmental chamber.

4 That at which gas cell(s) of aerostat become full.

pressure breathing Respiration of gas (eg O_2 or mixture) at pressure greater than ambient surrounding wearer of mask; hence * mask.

pressure bulkhead One sealed to serve as boundary of pressure cabin or pressurized section of fuselage.

pressure cabin Volume in aircraft occupied by human beings in which pressure is always maintained at or above selected level (eg equivalent to atmospheric height of 2,500 m or 8,000 ft) for comfort of occupants no matter how high aircraft may ascend. Term derives from early (c1938) usage when ** was entity installed in fuselage. Today inappropriate for civil transports where entire fuselage is pressurized except for nose tip, extreme tail and cut-out for wing; preferable simply to use adjective pressurized.

pressure chamber Strictly, one in which environmental pressure is raised above atmospheric; decompression chamber is preferred.

pressure coefficient Local pressure (eg measured on surface of body such as wing) divided by dynamic pressure, thus $C_p = P/\frac{1}{2}\rho V^2$.

pressure cooling Cooling of heat-generating device (eg engine) by liquid maintained under pressure to raise boiling point.

pressure-demand system Demand oxygen system supplying at above wearer's local pressure.

pressure differential Difference in pressure between two volumes, eg pressurized fuselage and surrounding atmosphere, dP.

pressure drag Drag due to integral (summation) of all forces normal to surface resolved along free-stream direction; most wings in cruising flight have small ** because adverse pressures on and immediately below leading edge are largely countered by helpful pressures over rear portion.

pressure error Instruments using a pitot/static system suffer from compressibility error and position error.

pressure face Side of propeller or helicopter-rotor blade formed by lower surfaces of aerofoil elements, over which pressure usually greater than atmospheric.

pressure fatigue Structural fatigue induced in pressurized fuselage by repeated reversals of pressurization stress.

pressure flap Large fabric inwards- (rarely, outwards-) relief valve(s) in skin of airship to allow air to flow in during descent so that internal pressure shall never be significantly below that of surrounding atmosphere.

pressure garment assembly NASA terminology for types of space suit, without PLSS.

pressure gradient Rate of change of pressure along any line normal to local isobar direction, eg in atmosphere or on surface of lifting wing.

pressure head Combined pitot/static tubes.

pressure height 1 See *pressure altitude (4)*.

2 That, related to standard atmosphere, at which gas cells reach predetermined superpressure (BSI).

pressure instruments Flight instruments operated by air pressure difference, eg ASI, VSI and simple altimeter.

pressure jet Helicopter tip-drive unit comprising propulsive jet on or near tip of main-rotor blade fed with compressed air supplied along interior of blade. May or may not have provision for combustion of fuel, which if used results in combustion- or tip-burning.

pressure lubrication Lube oil supplied under pressure, usually from gear pump, and ducted through drillings and pipes to main bearings and other parts (see *pressure-relief valve system*).

pressure manometer Liquid-filled U-tube at foot of aerostat gas cell.

pressure-pattern flying Technique of long-range navigation (1944–60) in which isobar patterns provided basis for flight-planning to make maximum use of following winds.

pressure pump That supplying fluid under pressure to closed system, eg lube oil (less often, hydraulics).

pressure ratio Ratio of fluid pressures between two points in flow, notably between stages of axial compressor (PRPS, per stage) or between inlet and delivery of compressor spool or entire compressor system (eg fan, IP compressor, HP compressor). By 1960 overall ** was often 15:1, with PRPS of 1·2; today ** can reach 28:1 with PRPS as high as 1·4. (For piston engines, term is compression ratio.)

pressure recovery Also called recovered pressure, that generated in flow through duct by conversion of kinetic energy, esp after addition of heat as in cooling radiator. When heat added can exceed stagnation pressure.

pressure-relief valve system Oil supply, eg for gas-turbine engine, in which delivery oil pressure is opposed by relief valve backed by spring plus atmospheric pressure via oil tank; as engine rpm increases, delivery pressure eventually forces relief valve open, spilling excess back to tank, thus holding steady pressure and flow to bearings at all flight engine speeds.

pressure rigid airship One combining features of rigid and non-rigid design to maintain shape and skin tension.

pressure seal Inflatable seal between rigid members which is pressure-tight when inflated but when deflected offers no resistance to relative motion of parts; universal in pressurized doors and canopies.

pressure-stabilized Having structural rigidity wholly dependent on maintenance of positive internal pressure, eg Atlas missile.

pressure structure See *inflatables*.

pressure suit Suit which, when rigid-facepiece helmet is attached, completely encloses wearer's body and within which gas pressure is maintained at level suitable for maintenance of bodily function (apart from addition of comment on necessity of adding helmet, this is DoD, NATO). Also called full-pressure suit. Some US definitions call this a pressurized suit, to be distinguished from ** because, according to these authorities, ** merely "provides pressure upon the body . . . a pressure suit is distinguished from a pressurized suit, which inflates, although it may be fitted with inflating parts that tighten the garment as ambient pressure decreases". This is partial-**; distinction must also be drawn between **, g-suit and water suit.

pressure surface Curved and irregular surface corresponding to particular atmospheric pressure at any time; usually plotted for 1,000 (often part below pressure), 900, 800, 700, 500, 300 and 100 mb (see *contour chart*).

pressure switch One actuated by reaching preset pressure, esp that in starter circuit triggered by fuel pressure.

pressure test Invariably means response of operative device or, esp, inflated structure, eg fuselage; objective not to measure pressure.

pressure thrust Product of pressure difference between rocket exhaust pressure and ambient pressure multiplied by area of nozzle; value may range from negative at SL to positive at high altitude. Rocket thrust = ** plus momentum thrust.

pressure transducer Device, based on at least eight principles, for generating output current or signal proportional to fluid pressure or pressure difference.

pressure tube Tube fitted to aerostat envelope or gas cell to which pressure gauge may be coupled.

pressure vessel Container for gas under (usually high) pressure.

pressurization Inflation; increasing pressure in closed system. Term usually means pressurized fuselage; see *pressure cabin*.

pressurized blade Patented (Sikorsky) technique for early warning of crack in helicopter main-rotor blade primary structure by inflating with dry nitrogen or other gas and sealing; loss of pressure signals cockpit lamp or other alarm.

pressurized feed Supply of liquid rocket propellants under gas pressure.

pressurized ignition Piston-engine ignition system sealed and held close to sea-level pressure to reduce arcing and electrode wear at high altitude.

pressurized tank Tank for liquid expelled by pressurizing gas.

Pressurs Pre-strike surveillance/reconnaissance system.

pre-stage Intermediate stage of operation in start-up of large liquid-propellant rocket in which propellants are fed to main chamber by main turbopumps and ignited, but with propellant flow less than 25% of maximum; once proper combustion has been confirmed, main-stage operation is signalled (ie allowing full flow).

Prestal Specially formulated aluminium alloy with very exceptional elongation and malleability for large-deformation presswork.

pre-stall buffet Aerodynamic buffet induced by turbulence over wing and/or control surfaces or fixed tail giving warning of imminent stall.

pre-stocked site Storage of POL/ordnance at remote site which might in emergency (crisis or war) be operating base for V/STOL aircraft.

Prestone Trade name of family of coloured and colourless ethylene-glycol coolants.

prestretched cable Control cable loaded to 60% uts for 3 min immediately before installation.

prestrike recon Mission undertaken to obtain complete information on previously known targets.

prevailing visibility Horizontal distance at which targets of known distance are visible over at least half horizon; determined by viewing dark objects against horizon sky by day and moderate-intensity unfocused lights by night. Not necessarily RVR.

preventive maintenance Systematic inspection, detection and correction of incipient failures either before they occur or before they develop into major defects (DoD).

preventive perimeter Outer perimeter formed during emergency security operations by stationing aerospace security forces at key vantage points and avenues of approach to vital areas of base (USAF).

PRF 1 Pulse-repetition frequency.
2 Pulse-recurrence frequency; suggest synonymous with (1).
3 Primary reference fuel.

PRFD PRF (1) distribution.

PRFJ PRF (1) jitter (2).

PRFS PRF (1) stagger.

PRG Program review group (US).

PRI Pulse-repetition interval, ie gap between pulses.

PriFly Primary flying control (officer, authority or location, USN).

primary air/airflow 1 That mixed with fuel in primary combustion zone.
2 Airflow through engine core.

primary aircraft authorization Total number of

particular aircraft type procured in current FY (DoD).

primary alerting system Leased-circuit voice communications network linking SAC HQ to operating squadrons of missiles or bombers.

primary body That around which satellite orbits or to/from which spacecraft escapes or is attracted.

primary circulation Atmospheric circulation on gross planetary scale.

primary combustion That immediately surrounding burning fuel, using only minor fraction of total airflow; in gas turbine most air is used in secondary region for dilution and cooling.

primary configuration That in which weapon system is delivered or in which its primary mission capability is contained (USAF).

primary control See *primary flight controls.*

primary cosmic rays Those reaching Earth from outer space.

primary effects Of NW: blast, radiation, heat, EM pulse.

primary flight controls Those providing control of trajectory, as distinct from trimmers, drag-increasers and high-lift devices. Conventionally ailerons/spoilers (where latter are used for roll), elevators and/or tailplane (or foreplane) and rudder. DLC usually excluded.

primary frequency R/T frequency assigned to aircraft as first choice for air/ground communications in an R/T network.

primary glider Strong and simple training glider in which no attempt is made to achieve soaring capability.

primary heat-exchanger That which removes heat from source and rejects it to atmosphere or to a secondary circuit.

primary holes Those through which primary air enters gas-turbine combustor.

primary inspection Minimum scheduled periodic lubricating and servicing check applied to aircraft and its removable equipment, including examination for defects and simple functional testing of systems such as flight controls, radio and electrics (ASCC, qualified by "UK usage").

primary instability Failure of strut or other compression member through buckling (transverse movement near centre).

primary instruments Those giving basic information on flight trajectory and airspeed, eg traditionally horizon, turn/slip, ASI, VSI and altimeter. Heading information not included.

primary member Any part of primary structure, though not usually applied to sheet or machined skin.

primary modulus of elasticity That for sandwich or other multi-component material below yield point (if applicable) of weaker component.

primary nozzle 1 That through which main flow of fuel passes in fuel burner or injector.
2 That for primary air to combustor or airflow through pipe of vaporizing burner.

primary radar One using plain reflection of transmitted radiation from target, as distinct from retransmission on same or different wavelength.

primary runway Used in preference to others whenever conditions permit.

primary space launch One starting from Earth.

primary stress Basic applied tension or compression, as distinct from stresses induced by deflection of structure such as buckling.

primary structure That whose failure at any point would imperil safety.

primary target Main target of surface-attack mission.

primary trainer That on which pupil pilot begins flight instruction, also called ab initio, elementary (but not basic).

PRIME, Prime Precision recovery including manoeuvrable re-entry.

prime See *prime contractor.*

prime airlift Number of aircraft of force that can be continuously maintained in cycle: home base, on-load base, off-load base, recycle.

Prime Beef Worldwide Base Engineer Emergency Force for direct combat support or to assist recovery from natural disasters (USAF).

prime contract That for entire responsibility for design, development and (usually) assembly and test of complete functioning system, eg aircraft, missile etc. Manufacture, esp in production, may be assigned by prime contractor to other companies or shared by consortium. An essential is control of programme management. (Strict DoD definition: any contract entered into directly by DoD procuring activity.)

prime contractor That awarded prime contract.

prime crew 1 Crew of strategic bomber/tanker/ICBM flight with long experience of combat duty.
2 That assigned to fly space mission (see *back-up crew*) (NASA).

prime meridian Longitude 0°.

prime mover 1 Source of mechanical energy, eg engine.
2 In military sense, surface vehicle designed chiefly for towing and normally accommodating crew, ammunition etc.

primer Subsystem, usually energized by hand-pump in cockpit, for spraying fuel into piston-engine inlet manifold (rarely, elsewhere) to facilitate starting from cold.

priming 1 Use of primer.
2 Sensitive high explosive for detonating main charge.

principal axes 1 Reference axes OX, OY, OZ.
2 Axis of relative wind.
3 Principal inertia axis.
4 Rectilinear axes in plane of cross-section of structural member about which moment of inertia is maximum and minimum.

principal inertia axis That line passing through

c.g. and usually in plane of symmetry about which long, slender body tends to rotate when rolling; in case of aircraft actual axis of rotation normally lies between this and wind axis.

principal parallel On oblique photograph, line through principal point parallel to true horizon.

principal plane On oblique photograph, vertical plane containing principal point, perspective centre of lens and ground nadir.

principal point On oblique photograph, foot of perpendicular to photo plane through perspective centre, usually determined by joining opposite collimating or fiducial marks.

principal site concept All prototypes of new combat aircraft do entire flight-test programme at one customer central site under close customer control (USN).

principal tensile stress For cutout in pressurized cylinder, that along axis where stress is maximum; hence *** factor, = maximum ***/hoop stress. For neutral hole (ellipse in proportions $2/\sqrt{2}$) *** = hoop stress.

principal vertical On oblique photograph, line through principal point perpendicular to true horizon.

printed circuit Electric or electronic circuit formed by deposition of conductive and/or semiconductive material on insulating base; many techniques and many substrates, eg thin/thick-film, foil etc.

printed-circuit board Printed circuit on rigid substrate provided with multiple plug-in or other contact terminals and on which are mounted discrete devices.

printed communications Telecommunications network providing printout at each terminal of all messages.

print reference Identifying reference of each air-reconnaissance photograph.

print-through Unwanted transfer of strong signals from one part of magnetic tape to one pressed against it on spool.

priority designator Two-digit number from 01 to 20 resulting from combination of assigned force/activity designator and local urgency-of-need designator (USAF).

priority induction Immediate transfer to USAF of Air Force Reserve members who fail to participate in training.

prior permission 1 That granted by appropriate national authority before start of flight(s) landing in or flying over territory of nation concerned.
2 Specific permission to land at particular airfield within particular times.

Prism Programmed real-time information system for management (Northrop/USA).

private aircraft One owned by private pilot.

private pilot One licensed by national authority to fly particular type, class or group of aircraft without payment, and precluded from carrying fare-paying passengers.

private venture Major product designed, developed and tested at company risk, esp in case of military aircraft not requested by government.

PRM Presidential review memorandum.

PRMD Pilot's repeater map display.

PRN Pseudo-random noise.

PRNSA Pseudo-random noise signal assembly.

PRO Anti-rocket (ie anti-ICBM/SLBM) forces (USSR).

PROAR Area forecast, height indicated in pressure units (ICAO).

probable 1 See *probably destroyed.*
2 Qualifying term in photo interpretation where facts point to object's identity without much doubt (ASCC).

probable errors In range, deflection, height of burst etc, those which are exceeded as often as not (DoD).

probably destroyed Assessment on enemy aircraft seen to break off combat in circumstances which lead to conclusion it must be a loss, though not actually seen to crash (DoD, NATO).

probe 1 Instrument boom; see * *errors*, * *parameters*.
2 Rigid receiving tube for fuel passed by flight-refuelling drogue.
3 Any device used to obtain information (esp quantified) about environment, esp unmanned instrument-carrying spacecraft.
4 Loop or straight wire for coupling to waveguide and extracting energy from electric or magnetic component of radiation.
5 Transducer which converts shaft speed, sensed as sinusoidal magnetic field from toothed wheel on shaft, into alternating signal current; also called MSP.

probe and drogue British (Flight Refuelling) method of refuelling aircraft in flight.

probe errors Those originating in probe (1), as far as possible corrected by ADC.

probe parameters Subject to particular installation can include AOA, yaw/sideslip, total pressure, total temperature and static pressure.

procedure alpha Ceremonial manning of flight deck of carrier, eg when entering harbour, often with crew arranged to spell vessel's name or a slogan.

procedure manoeuvre Accurately flown and/or timed flight manoeuvre for identifying or ATC purpose.

procedure turn Flight manoeuvre in which turn is made at constant rate away from track followed by constant-rate turn in opposite direction to enable aircraft to capture and hold reciprocal; precision manoeuvre used in radio range, holding over point fix prior to joining stack, joining ILS without radar vectoring, or in simulator training of traditional (Link) type. Designated left or right depending on first turn direction.

process annealing Heating ferrous alloys to below

critical temperature followed by cooling in air or other medium.

Procod Production cost by drawing number.

Procru, ProCru Procedure oriented crew-station model.

procurement Process of obtaining personnel, services, supplies and equipment (DoD).

procurement lead time Interval in months between initiating procurement action and receipt into supply system of production article; composed of sum of admin lead time and production lead time (DoD).

prod Make probe/drogue inflight-refuelling contact (colloq).

product improvement Significant change in design of hardware to improve desirable features, and marketed as such; usually initiated to meet market need, either because of competition or to rectify deficiency.

production base Total national production capacity available for manufacture of items to meet material needs (DoD).

production lead time Time between placing contract and receipt of hardware; subdivided according to whether contract is initial or re-order.

production phase Period between production approval until last item is delivered and accepted.

productive potential Payload multiplied by range.

productivity In airline performance, traffic units (eg LTM) generated per hour per aircraft of particular type or per employee.

product support Assistance provided by manufacturer to all customers throughout period of product's use in form of training, publications, spare parts, modification kits, product-improvement and immediate response to difficulties.

proficiency student One on refresher course.

proficiency training Flight training for desk-bound flight personnel.

profile Outline of body in side elevation; many sub-meanings. Common use is to describe shape of cross-section of wing or other aerofoil section. Outboard * is external shape of body, esp aircraft fuselage. Inboard * is longitudinal cross-section showing how interior is utilized. Flight * is orthogonal projection of flightpath on vertical surface containing nominal track showing variation in height (either AGL or AMSL) along straight bottom axis representing track.

Profiledata Library of NC-machining software (Ferranti).

profile drag Total drag minus induced drag; sum of form drag and surface-friction drag. One interpretation suggests "drag of wing with camber and twist removed".

profile-drag power loss That expended in overcoming total profile drag of propeller blades.

profile line Profile of terrain, eg near airport drawn as map inset.

profile milling Milling variable profile (surface levels) in plate, eg in machined or integrally stiffened skin panel.

profile template Template for hand-guided machine tool, eg router.

profile thickness Maximum distance between upper and lower contours of aerofoil profile, each measured normal to mean line; essentially wing thickness.

profilometer Instrument for measuring surface roughness.

prognostic chart Forecast of met elements for specified time and location depicted graphically.

prograde Direct, or progressive orbit; satellite launched into Earth orbit with inclination from 001° to 179°.

program Vast topic, covered in EDP (1) dictionaries; simplest definition is group of related instructions which when followed by computer will solve a given problem.

Program Aircraft Total of active and reserve aircraft (USAF).

programmable Capable of being controlled by different programs by change of software.

programmable display generator Generator of 3-colour raster formats plus calligraphic symbology, all under variable software control.

programme Life history of major project, typified by such * milestones as definition of requirement, feasibility study, project definition, engineering design, hardware manufacture/development/flight test, flight development, service clearance (eg qualification or certification), production, modification and product-improvement, fault-rectification, phaseout.

programmer Human being engaged in programming.

programmer comparator Versatile automatic testing station providing serial evaluation on all kinds of analog and digital signals, eg in checking avionic systems in aircraft, missiles or spacecraft.

programming Art of preparing set of terms and instructions which EDP (1) machine can understand and obey and which when followed by that machine will result in solution to problem for which program was written.

progressive burning See *progressive propellant*.

progressive die One which performs series of operations (usually on sheet metal) at successive strokes of press.

progressive feel Artificial feel proportional to dynamic pressure.

progressive orbit See *prograde*.

progressive powder Solid fuel, usually not gunpowder, which burns increasingly fast as combustion pressure increases.

progressive propellant Solid rocket-motor grain so shaped and ignited that area of combustion, and hence speed of burning, rate of consumption and thrust, all increase throughout period of burn. Any radial burning technique tends to be * unless original grain has deep star centre such that area is constant (neutral burning).

progressive servicing Servicing performed at military airbase where major tasks are subdivided into sections performed at times fitting in best with operational readiness requirements.

progressive stall Ideal stall quality where breakdown of wing flow occurs gradually, with well signalled symptoms; to obtain it a breaker strip or fence may be needed.

progress strip See *Flight progress strip.*

prohibited area Airspace of defined dimensions identified by area on surface within which flight by aircraft is prohibited, usually for reason of national security or to safeguard wildlife. Height ranges from surface to published value such as 4,000 ft or 18,000 ft (FAA).

project Planned undertaking of something to be accomplished (DoD).

Project Blue Book Official dossier on UFOs (USAF).

project design Programme phase in which design is refined by evaluating alternative choices, making performance/capability/cost tradeoffs and ultimately arriving at optimized configuration on which engineering design can begin. Work possible in this stage includes tunnel testing, cockpit mock-up improvement and basic systems design, but excludes stressing (detail engineering design).

projected area Area projected from 3-D surface to plane, or from one plane to another.

projected blade area Area of propeller or other blade projected on to plane normal to axis of rotation; solidity is not based on this but on total area.

projected flightpath That which aircraft, esp aeroplane, will follow in immediate future in absence of further disturbance.

project engineer Engineer assigned to oversee design and technical management of specific project, reporting to chief engineer or v-p engineering.

projectile velocity Resultant of muzzle and aircraft velocities.

projection In cartography, any systematic arrangement of parallels and meridians portraying quasi-spherical planetary surface on plane of map.

project officer Military or civilian individual responsible for accomplishment of project; usually limited-duration appointment and not one already established within organizational and supervisory channels (USAF).

projector 1 Illuminating source sending out pencil beam of visible light, eg vertically up at cloudbase.
2 Long-dash broken line to show projection of line or surface from one plane to another in engineering drawing (drafting).

projector sight Mirror sight or similar landing guidance optical system on carrier.

proliferation Spread of NWs to additional nations.

PROM Programmable read-only memory.

prominent target One which predominates over chaff and other decoys.

Promis Procurement management information system.

promulgated Published openly; eg * in ACIs, or VOR beacon * range, in latter case * in Air Pilot and various flight guides.

prone Lying down; invariably * pilot positions are supported mainly on front of torso and thighs at angle of about 20° with head up to look ahead and toes supported in rear pedals.

Pronto 1 Code: as quickly as possible (DoD).
2 Program for NC tool operation, acronym.

prontour Chart used to forecast future pressure-surface contours.

prony brake Simple mechanical peripheral-band brake whose torque, multiplied by shaft speed, gives brake horsepower.

proof factor Factor of safety, proof load divided by design limit load; for combat aircraft possibly + 12 g/– 6 g, and for commercial transport typically + 2½ g/– 1 g, reduced with flaps extended to + 2 g.

proof load That load which primary structure is required to withstand while remaining serviceable; this definition, still surviving in official publications, makes no mention of cyclic application and fatigue life and is numerically equal to design limit load multiplied by proof factor of safety.

proof of concept article Prototype.

proof positive/negative Two sets of proof loads established for particular aircraft type and demonstrated in static test.

proof strength That required to survive proof loads (pos/neg).

propaganda balloon Free balloon carrying propaganda leaflets scattered at timed intervals when prevailing wind is expected to have carried it over enemy cities.

propagation constant Complex quantity of plane wave; real part is attenuation constant (nepers/unit length) and imaginary part is phase constant (radians/unit length).

propagation error In ranging system, algebraic sum of propagation velocity error and (important at long ranges and low angles) curved-path error.

propagation rate Linear velocity of structural crack.

propagation ratio Between two points in path of plane wave, ratio of complex electric field strength.

propagation velocity For EM wave (light, radio) in vacuum taken to be $2{\cdot}997925 \times 10^8$ m/s.

propagation-velocity error Difference between

assumed and effective velocities over ray path.

propane Hydrocarbon of paraffin series, $CH_3CH_2CH_3$, BPt -45°C.

prop banners Sleeves, usually bright Day-Glo colour, announcing (eg) FOR SALE, FOR RENT, slipped over blades.

propellant Medium used for propulsion, as in * charge of gun ammunition or material burned to form jet of rocket. Rocket * can be solid, liquid or gas, or combination. Where two are mixed rocket is bi-*, common mixture being fuel plus oxidant (oxidizer). Where catalyst is consumed and adds to jet this also is *. In uncommon case where single liquid is used rocket is mono-*. In rockets and thrusters where no chemical combustion takes place preferable to use term "working fluid".

propellant mass fraction See *mass ratio.*

propellant specific impulse See *specific impulse.*

propellant volume Total volume occupied by propellant, esp solid grain, V_P.

propeller Rotating hub with helical radial blades converting shaft power into aerodynamic thrust. Shaft power provided by human or prime mover. Tip-drive * possible (would require modified definition). Most existing definitions state "power-driven", implying use of engine. Left and right-hand rotation respectively mean anticlockwise and clockwise seen from behind. Can be pusher or tractor, latter at one time often being called airscrew (now arch). Types of propellers (co-axial, reverse-pitch etc) are covered separately (see also *propulsor*). Similar screw for converting energy of slipstream into shaft power is windmill or RAT (ram-air turbine).

propeller area Usually means total area of blades obtained by integrating total of areas of elementary chordwise slices, taking blades as having no thickness; ie each slice is projected in plane of its local chord. Essentially same as outline area of blades with twist removed.

propeller balance stand Trestle having two horizontal and parallel steel knife-edges (about 1–3 mm radius) on which propeller can be balanced on short slave-shaft.

propeller bar Handtool used to loosen or tighten main retaining nut on many lightplane propellers.

propeller blade Thrust-generating aerofoil of propeller.

propeller blade angle Except in feathered position (when close to 90°) acute angle between chord line of blade and plane of rotation, latter being normal to axis of rotation measured at standard radius.

propeller blade-width ratio Ratio of widest chord to propeller diameter.

propeller brake Brake to stop rotation of propeller after engine shut-down, either to speed passenger disembarkation or, in case of free-turbine engine, to prevent prolonged windmilling with aircraft parked.

propeller camber ratio Ratio of blade maximum thickness to chord at any station.

propeller cavitation Generation of near-vacuum on suction face near tip at high Mach numbers (not necessarily at high flight speed).

propeller characteristic Fundamental curve of V/nD (velocity of advance divided by rpm × diam) plotted against C_S (speed/power coefficient) (see *Weick*).

propeller disc Circular area swept out by propeller.

propeller efficiency Useful work expressed as thrust imparted divided by power input; thrust hp/brake hp = thrust × slipstream velocity divided by $2\pi nQ$ where n is rpm and Q is drive torque. Some authorities cite two values of **, one called net (net thrust hp/torque hp) and other propulsive (propulsive thrust hp/torque hp).

propeller governor Usually simple centrifugal governor which keeps shuttle valve oscillating about null position feeding oil to increase or decrease rpm and thus hold speed constant irrespective of aircraft forward speed.

propeller hub Central portion of propeller carried on drive-shaft, usually made as separate unit into which blades are inserted.

propeller interference Aerodynamic effects, mainly drag, of bluff bodies immediately downstream, eg radiator or cylinders.

propeller rake See *rake (3, 4).*

propeller root See *root.*

propeller shaft That on which propeller is mounted.

propeller solidity See *solidity.*

propeller state Normal condition of helicopter main rotor in which thrust is in opposite direction (upwards) to flow both through and outside rotor disc.

propeller tipping Metal skin on tip and outer leading edge of soft (eg wood) blade.

propeller torque Torque imparted by propeller drive shaft and reacted by aerodynamic rolling moment of aircraft (usually to some degree inbuilt) or asymmetric load on left/right main gears on ground; symbol Q.

propeller-turbine See *turboprop.*

propeller wash Slipstream, esp on ground; also called prop blast; see *propwash.*

propeller width ratio Product of blade-width ratio and number of blades; akin to solidity.

propelling nozzle That at exit from fluid jet system used for propulsion, esp turbojet or ramjet; for supersonic flight variable in shape (con-di) and area.

propfan Advanced propeller for use at high Mach numbers, characterized by having six to 12 blades each with thin, sharp-edged lenticular profile and curved scimitar shape, overall solidity exceeding unity and loading being high. Can be tractor or pusher, and for highest efficiency at Mach 0·8 has two contra-rotating units.

propjet See *turboprop* (colloq).

proportional control Effect at output, eg surface movement, is proportional to input, eg stick movement; opposite of flicker or bang-bang.

proportional navigation Control of trajectory in order to home on target by changing course by several times (typically 3·5) rate of change of sightline to target; thus angular rate of velocity vector is proportional to angular rate of line of sight to target.

proposal Formal and comprehensive document in which manufacturer sets out before government procurement officials complete technical specification and performance of proposed item, including timing and prices of development and, possibly, production programme.

proprioceptive Pertaining to stimuli produced within body, esp human body, by proprioceptors.

proprioceptor Internal receptor for stimuli, such as tendon tension, originating in somatic organ. Body balance maintained by * plus eyes and ear labyrinths.

propulsion efficiency Term not recommended (see *propeller efficiency*, *propulsive efficiency*).

propulsion system Sum of all components which are required to propel vehicle, eg engine, accessories and engine-control system, fuel system, protection devices, inlets and cooling systems.

propulsive duct Not recommended; usually means pulsejet or subsonic ramjet.

propulsive efficiency Broadly, energy imparted to vehicle as percentage of energy imparted to jet (from propeller or other prime-moving device) or expended in burning fuel. Basic equation is Froude efficiency $\eta_F = 1/(1 + \delta V/2V)$ where V is velocity of vehicle and δV is total increase in velocity of air in jet measured as difference between air well ahead of vehicle and that at fastest-moving part of jet (with propeller this is well behind plane of blades but with turbojet probably in plane of nozzle exit). Theoretically but not practically δV could be zero and jet would remain stationary with respect to free-stream; η_F then is 100%. Another expression for same relationship is $\eta_F = 2/(R + 1)$ where R is ratio of jet velocity relative to vehicle velocity, which in theoretical perfect case becomes unity.

propulsor Multi-bladed fan, usually with variable pitch or constant-speed control, used as superior alternative to propeller; in most cases surrounded by profiled duct.

propwash Airflow caused by propeller alone, usually helical but velocities are measured in axial direction only.

propylene oxide Stable liquid used in rapidly dispersed form as fuel/air explosive.

proration Actual yield to carrier of fares.

PRORO, Proro AFS code: route forecast with height indication in pressure units.

PROSAB Parachute flare (USSR).

protected angle Selected (usually solid) angular limits of DOA within which received signals are accepted and amplified.

protected system One incorporating max security to allow electrical transmission of classified plain language.

protection ratio Ratio of wanted to unwanted received signal strength, eg in VOR, ADF, NDB etc.

protoflight model Qualification model of spacecraft that is later actually flown in space.

proton Positively charged elementary particle of mass number 1 forming part of every complete atomic nucleus; charge magnitude equal but opposite to that of electron e.

proton-magnetometer Family of devices usually used in space instrumentation in which field strength measured by investigating effect upon atomic nuclei (eg NMR and Larmor precession).

prototype First example(s) built of item intended for production; as far as possible representative of definitive article but usually inevitably deficient in many respects. In case of modern aircraft traditional * now rare; with civil programmes production initiated in parallel with engineering design/development and even first examples are likely to be sold. With military aircraft first batch may be termed development aircraft. Aerodynamic * merely has correct shape. Breadboard or brassboard * used to develop avionics. Purpose of * is to assist development; secondary role is to permit customer evaluation.

PROV Provisional (ICAO).

proving flight Unscheduled flight by new type of commercial transport over intended routes by crew at least partly provided by customer to establish compatibility with sectors, aids, airfields and, esp, terminals and airline's ground equipment and staff.

proving ground Military area dedicated to testing of ordnance, esp of new types.

provisioning Precisely calculated schedule of necessary spare parts, types, numbers, prices, dates and locations to support operation of functioning system, eg commercial transport, fighter, radar or computer.

proximate splitter Term coined (Pratt & Whitney) for aerodynamic flow splitter added to eliminate afterburner light-up pulsations in afterburning turbofan from reaching compressor.

proximity fuze Fuze which initiates itself by remotely sensing presence, distance and/or direction of target or associated environment by means of signal generated by fuze or emitted by target or by detecting disturbance in natural field surrounding target (DoD). Can be radio (radar), IR, visual (EO), acoustic or magnetic.

proximity scorer Hit/miss device triggered by entry of munition into spherical volume with scorer at centre; indicates only that munition entered this volume, without giving miss-distance.

proximity switch Switch today used in place of microswitch in exposed locations (MLG, trim tab etc) to signal mechanical position; usually variable-reactance sensor feeding microelectronic module. Normal principle is proximity of target plate of high-permeability metal to sensor, giving varying voltage across bridge.

PRP 1 Premature-removal period.
2 Pulse recurrence (or repetition) period (= 1/PRF).
3 Parent rule point.
4 Power-deployed reserve parachute.
5 Personnel reliability program (DoD).

PRPA Professional Racing Pilots' Association (affiliate of NAA).

PRPS Pressure rise per stage.

PRR 1 Premature-removal rate.
2 Power ready relay.
3 Production readiness review.
4 Pulse-repetition rate.

PRRC Pitch/roll rate changer assembly.

PRRFC Planar randomly reinforced fibre composite.

PRS Pressure-ratio sensor(s).

PRSD Power reactant storage and distribution subsystem, LO_2 and LH_2 for fuel cells generating electric power (Space Shuttle).

PRSG Pulse-rebalanced strapdown gyro; INS gyro whose dynamic error reduced by compensating rebalance loop electronically.

PRSOV Pressure regulating and shut-off valve.

PRT 1 Pulse recurrence (or repetition) time.
2 Pulse rise time.

PRU Photo-reconnaissance unit (RAF, WW2).

prudent limit of endurance Time during which aircraft can remain airborne and retain given safety margin of fuel (NATO).

prudent limit of patrol Time at which aircraft must depart from its operational area in order to return to base and arrive there with given margin (usually 20%) of fuel.

PRV Pressure regulating (or relief) valve.

PRW Passive radar warning.

PS 1 Metric horsepower.
2 Pitot/static.
3 Passenger-service costs.
4 Procurement specification.
5 Photoemission scintillation.
6 Photo squadron.

P_s 1 Static pressure.
2 Radar power received from target.
3 Time rate-of-change of specific energy.
4 Standard pitch of propeller.
5 Specific excess power.

ps Picosecond (10^{-12}s).

P-S Pitot/static.

PSA 1 Prefab semi-permanent airfield.
2 Provisional site acceptance.
3 Pressure-swing absorption.

PSAC Presidential Science Advisory Committee (US).

PSAI Public Safety Aviation Institute (US).

PSAS Pitch stability augmentation system.

PSC 1 Principal site concept (USN).
2 Product-support committee.

PSD 1 Power spectral density.
2 Physiological Support Division (SR-71 ops).

PSDP Programmable-signal data-processor.

PSE 1 Passenger service equipment (reading lights, call button etc).
2 Passive seismic experiment.

PSEU Proximity slat electronics unit.

pseudo-adiabatic Process by which saturated air parcel undergoes adiabatic transformation, water being assumed to fall out as condensed.

pseudo-analog Electronic display which simulates traditional instrument, eg by using fixed LED matrix to form "dial" plus computer-driven LEDs to generate "pointer" and alphanumerics.

pseudo fly-by-wire Flight control system in which at least one axis of control is normally, at one point at least, electrical; * systems have capability for manual reversion or override.

pseudonoise Technique in which PN code generated sends out wideband noise-like signal which is then viewed as carrier on which message is imposed; usually direct-sequence modulation approach used in which PN code directly balance-modulates carrier.

pseudopursuit navigation Homing method in which missile is directed towards target instantaneous position in azimuth while pursuit navigation in elevation is delayed until more favourable attack angle (note: this is not the same as AOA) on target is achieved (DoD).

pseudo-random noise See *pseudonoise*.

pseudo stereo 1 False impression of stereoscopic relief (ASCC).
2 Common audio meaning, simulating stereophonic by using two speakers, one via brief delay, from single channel.

pseudo-3-D Large family of techniques for generating subjective impression of three dimensions on (usually) planar display. One is pseudo stereo (1); various methods used in PPI displays and several techniques in computer-driven OTW displays for simulators, some coming under heading of CGI.

PSF, psf 1 Phosphosilicate fibre (optical fibre).
2 Polystyrene foam.
3 Personnel services flight (RAF).

psf Pounds per square foot (non-SI and in any case strongly discouraged).

PSFP Pre-simulator familiarization panel.

PSG Post-stall gyration.

PSGR Passenger.

PSI, psi Pounds per square inch (not recommended).

PSIA Pounds per square inch, absolute (not

recommended).

PSID Pounds per square inch, differential (not recommended).

PSIG Pounds per square inch, gauge (not recommended).

PSK Pulse shift keying; differentially used in satcom terminals.

PSLO Product-support logistic operation.

PSM 1 Passenger statute-mile.
2 Post-stall manoeuvring.

PSN Potassium/sodium niobate.

PSO 1 Pilot systems officer, GIB in tandem-seat combat aircraft (USAF).
2 Protective service operations (mission).

psophometer Instrument which attempts to measure perceived noise level.

PSP 1 See *pierced steel planking*.
2 Programmable signal processor.
3 Personal (or personnel) survival pack.
4 Product support programme.
5 Primary special pay (DoD).

PSR 1 Primary surveillance radar.
2 Precision secondary radar.
3 Pulsar.
4 Post-strike reconnaissance.

PSS 1 Precision slab synchro.
2 Product-support services.

P_{SSK} SSKP.

PST 1 Pacific standard time.
2 Propeller STOL transport.

PSU 1 Photosmoke unit (not same as Hartridge).
2 Power supply unit (electronic).

psychological warfare Planned use of propaganda and other psychological actions having primary purpose of influencing opinions, emotions, attitudes and behaviour of hostile foreign groups (DoD); (NATO substitutes for "hostile foreign" "enemy, neutral or friendly").

psychrometer Instrument for measuring atmospheric humidity, usually comprising dry and wet-bulb thermometers.

psyops Psychological operations, psywar plus political, military, economic and ideological ops.

psywar See *psychological warfare*.

PT 1 Power turbine.
2 Primary trainer (US 1923–46).

P_t 1 Rocket chamber pressure at termination (solid propellant).
2 Total pressure.
3 Radar power transmitted.
4 Pennant number.

p_t Tensile stress.

P_{t2} etc Gas-turbine total pressures at usual numbered locations, ending with P_{t7} for nozzle exit.

PTA 1 Prepaid ticket advice in foreign currency.
2 Polskie Towarzystwo Astronautyczne (Poland).
3 Propfan test assessment.
4 Pilotless target aircraft.

PTAB Dispensed bomblet, usually followed by figure giving weight in kg (USSR).

PTAG Portable tactical aircraft guidance.

PTAS Pilotless target aircraft squadron.

PTC 1 Programming & test centre.
2 Part-through crack.
3 Pitch trim compensator.
4 Passive thermal control.

PTCV Primary temperature control valve.

PTEH Per thousand engine hours.

PTFCE, ptfce Polytrifluorochlorethylene.

ptfe Polytetrafluoroethylene.

PTIT Power-turbine inlet temperature.

PTK New hollow-charge bomblet (USSR).

PTL Primary technical leaflet.

PTM Pulse-time modulation.

PTN Procedure turn.

PTO 1 Power take-off (shaft output).
2 Permeability tuned oscillator.
3 Participating test organization.

P-tots Portable transparency optical test system (for rainbowing).

PTP Paper tape punch.

PTR 1 Part-throttle reheat.
2 Paper tape reader.
3 Power-turbine rotor.

PTRP Propfan technology readiness programme.

PTS Photogrammetric target system.

PTSA Prior to sample-approval.

PTSN Public telephone switching network.

PTT 1 Part-task trainer (simulator).
2 Public telecommunications system (many European nations).
3 Press to transmit.

PTU Power transfer unit; transmits power but not fluid between hydraulic systems.

PTV Propulsion technology validation.

PU, p.u. 1 Pick-up.
2 Propellant-utilization system.

public-address system, PA Interphone voice circuit used by captain or cabin crew to address passengers.

public aircraft Not public, ie used exclusively in government service (US).

public dividend capital Strange term for direct gift of taxpayer's money to national airline (UK).

puck Replacement pad for plate or disc brake.

pucker Local buckling of sheet metal in compression, eg on flange around inside of bend.

PUD Power unit de-icing (CAA).

Pugs Propellant utilization and gaging system (NASA).

pull 1 To operate, eg to * spoilers (colloq).
2 To engage arrester wire.
3 To engage arrester wire, causing damage or breakage to arrester system.

pulled Wire whose anchorage or shock-absorbing system has been damaged.

pull lead To pull nose of aircraft further round to aim correct distance ahead of target.

pull-off Practice parachute jump in which slipstream pulls wearer of opened parachute off wing or other suitable part of aircraft.

pull out 1 Recover from dive to level flight or zoom.
2 To extend arrester wire or airfield barrier near limits.

pull-out area Carrier deckspace clear for decelerating arrivals.

pull-out distance Distance travelled by hook between engaging wire and coming to stop; also called run-out.

pull-ring Parachute operating handle or D-ring.

pull up Short sudden climb from level flight, normally trading speed for height (usually general aviation or tactical attack).

pull-up point Geographical point at which aircraft must pull up from lo approach to gain sufficient height to make attack or execute retirement (DoD).

pulsating rubber De-icer boot.

pulsator Engine instrument showing both engine speed and oil circulation by pulsations of oil in glass dome (rare after 1917).

pulse 1 Transient phenomenon, esp in radio or other EM signal, characterized by rise, brief finite duration and decay.
2 Single solid-propellant grain, two or more of which are contained in rocket motor casing, hence two-* rocket can give two impulses.

pulse-amplitude modulation Signal is broken down into bits, amplitude of each being measured to give series of discrete values.

pulse code Sequences of pulses conveying information.

pulse-code modulation Modulation involving pulse codes, esp that which translates continuously modulating signal into stream of digital pulses all of uniform height (amplitude), information being conveyed by spacing/duration. PCM output is compatible with all digital EDP (1) and virtually eliminates transmission errors.

pulse decay time Time pulse takes to fall from high to low value, normally from 90% peak to 10%.

pulse Doppler, PD Radar mode using pulse trains and Doppler processing in which received signals are examined by mixers and band-pass filters which eliminate everything except genuine targets or objects of interest. Doppler technique injects information on relative range-rates as well as eliminating non-targets.

pulse duration Time that single pulse exceeds stated value, usually 10%, of peak; ICAO selects time over 50%.

pulse-duration modulation Also called pulse-time modulation, pulse-width modulation, translates CW signal into succession of constant-amplitude pulses of varying width (time duration).

pulse-frequency modulation More precisely called PRF modulation; CW signal is translated into succession of constant-amplitude pulses transmitted at frequency proportional to amplitude of original signal.

pulse interval Time between consecutive pulses both measured at same point.

pulsejet Air-breathing jet engine in which air is intermittently induced or allowed to enter, mixed with fuel, ignited (by electric discharge, residual combustion products or other method) and expelled as single expanded charge of hot gas giving pulse of thrust. Cyclic operation (typical frequency 30–60 Hz) may be inherent in aerothermodynamics of duct or imposed by sprung flapvalves or other oscillating one-way valving at inlet.

pulse limiting rate Highest PRF allowing time for echo to reach receiver in gap.

pulse modulation Variety of methods of translating CW signals into digital pulses to reduce bandwidth required, eliminate errors and, where possible, improve signal/noise ratio (see separate entries).

pulse packet Concept of radar signal pulse as physical entity occupying particular 3-D volume, esp particular length.

pulse-phase modulation See *pulse-position modulation*.

pulse-position modulation CW signal is translated into succession of constant-amplitude pulses whose position (ie time from start of each frame or timebase period) is proportional to amplitude of original wave at corresponding point.

pulse radar Most common type, in which signals are in form of pulses; also called pulsed radar.

pulse rise time Time required for EM pulse to rise from a low to a high value, normally from 10 to 90% of peak but occasionally from 5 to 95 or from 1 to 99.

pulse separator Receiver circuit which removes imposed (regular) pulse train.

pulse sorter ECM device which selects one pulse from many for detailed measurement.

pulse spike Erroneous sharp super-peak superimposed on pulse.

pulse-time modulation CW carrier is modulated by pulse train of lower frequency which in turn is modulated with variable characteristic, which may be amplitude, duration, PRF or position.

pulse train Succession of pulses.

pulse width Unlike duration, this measure of CW signal is normally defined as width (time) between half-power points.

pulse-width modulation CW signal is translated into train of constant-amplitude pulses whose widths are each proportional to corresponding amplitude of original signal. (Width in this context has no relevance to half-power points.)

pultrude Fabrication process combining pulling through die and extrusion under back-pressure (eg Grafil CFC).

pultrusion Raw material or finished section made by a pultrude-type process.

pump-up time Time taken to inflate gas storage in

blow-down tunnel.

PUN ICAO code: prepare new perforated tape for message.

punch, punch out To eject.

pundit 1 Aerodrome beacon with Morse identifying sequence (arch).
2 Portable ultrasonic non-destructive digital indicating tester.

punkah, punkah louvre Fresh-air jets in passenger cabin, of whatever geometry and location.

PUP 1 Pitch-up point in pop-up delivery.
2 Performance update programme.

purchase Single grip on yoke, spectacles or other aileron input; thus full aileron may be two-* task.

purchase cable That connecting pendant to arresting gear under flight deck.

Pure-clad Trade name (Reynolds) for Alclad-type material.

pure jet Not defined, but usually means turbojet as distinct from turboprop (arch).

pure pursuit-course lead Course in which velocity vector of attacking aircraft is always directed towards instantaneous position of target (ASCC).

pure research In case of aircraft, one whose purpose is to obtain knowledge of general application; type may be specially designed or modified version of familiar type.

Purex Plutonium/uranium extraction.

purge 1 To clean and flush device, eg liquid-propellant rocket, by high-rate pumping of inert gas, eg dry nitrogen. This removes potentially dangerous propellants and helps preserve hardware; secondary function may be to trigger various valves and leave inert gas occupying internal chambers and piping. Term also sometimes used to mean inhibiting.
2 To prevent admission of air to space above fuel in aircraft tank by continuously pumping in inert gas (usually dry nitrogen).

purity Volume percentage of lifting gas in aerostat gas cell.

Purple Air-intercept code: unit is suspected of carrying nuclear weapons (DoD).

purple airway Special temporary airway established and promulgated for Royal flight(s) in UK and certain other areas.

pursuit aircraft Interceptor (US, arch).

pursuit course Course in which attacker must maintain a lead angle over velocity vector of target to predict point in space at which gun- or rocket-fire would intercept target (ASCC) (see *pure pursuit*).

pursuit missile One which can be fired only from astern of air target, eg because IR homing head unreliable from any other aspect.

pushbutton indicator Pushbutton which when depressed illuminates.

pusher See *stick-pusher*.

pusher aircraft One with pusher propeller(s) only.

pusher propeller One mounted behind engine so that drive shaft is in compression.

push fit Fit just requiring light force to assemble (arch).

push-off drift See *kick-off drift*.

pushover Nose-down manoeuvre commanded by stick-pusher; in effect same as pitchover.

push/pull 1 Throttle, eg on lightplanes, having linear motion sliding through panel.
2 Amplifier having two similar valves or transistors connected in anti-phase and with I/O circuits combined about earthed centre.

push-push actuator Linear actuator having unidirectional output interleaved by weak or slow return stroke.

pushrod Rod transmitting cam motion of valve gear to rocker or other drive to poppet valve(s).

PUT Pop-up test of ICBM, SLBM, ABM or other launch system of externally energized cold-launch type.

PV 1 Pressure/volume or pressure × volume.
2 Private-venture aircraft.
3 Product verification (formerly MQT).
4 Prevailing visibility.

PVA Polyvinyl acetate.

PVB Polyvinyl butyral.

PVC Polyvinyl chloride.

PVD 1 Para-visual director.
2 Plan-view display.

PVDF Polyvinylidene fluoride.

PVF Polyvinyl fluoride.

PVO Air defence of the homeland, made up of IA (manned interceptors) and ZR (zenith rockets, ie SAMs) (USSR).

PVO-SV Troops of air defence of ground forces (USSR).

PVS Pilot's vision system.

PVT Product verification test.

PVV Proof vertical velocity (MLG demo case).

PW 1 Pulse width.
2 Plated wire (memory).

PWI 1 Pilot warning indicator.
2 Preliminary warning instruction.

PWIN Prototype WWMCCS intercomputer network.

PWM Pulse-width modulation; hence PWMI = * inverter.

PWP Pylon weight plug.

PWR 1 Power.
2 Passive warning radar.

PWRS Prepositioned war reserve stocks.

PWS 1 Proximity warning system (generally helicopter applications).
2 Performance work statement.

PWSDE Effective power delivered at shaft of turboprop = ehp.

PWT Propulsion wind tunnel.

PY Program(me) year.

pylon 1 Rigid pillar-like structure projecting upwards to carry load (eg engine) or protect occupants in overturn (crash *).

2 Streamline-section structure transmitting stress from external load to airframe, eg engine pod, ordnance, drop tank etc. Can extend above or below wing or horizontally or at other angle from fuselage. For engine pod often extended to *strut.

3 Object on surface used as landmark for race turning point.

4 Object on surface used as reference for pilots performing flight manoeuvres.

pylon select I/O device linking crew to WCS (weapon control system) or Navwass, enabling specific pylons, ie particular portions of load, to be selected for attack on particular target.

pylon spar Principal structural member of engine pylon normally (in case of wing engine) linking engine mounts direct to wing box.

pylon strut Pylon (2) linking engine pod to wing or rear fuselage.

pyranometer Actinometer which measures combined solar and diffuse sky radiation, sensor viewing entire visible sky. Also called solarimeter.

pyrgeometer Actinometer which measures terrestrial radiation.

pyrheliometer Actinometer which measures only direct solar radiation.

pyroelectric detector Sensitive detector of IR; radiation enters via precision window of Ge and light pipe directs it to microscopic flake of TGS which rises in temperature, changing polarization and generating surface charge amplified in low-noise electronics.

pyrogen Pyrotechnic generator; ignition squib for solid-propellant motors comprising electric resistance or bridgewire which ignites hot-burning powder charge.

pyrolitic carbon Allotrope of carbon derived from controlled pyrolysis of char-yielding resin (eg phenolic resoles or Novolaks) to form matrix used in carbon/carbon composite, with reinforcement by carbon or graphite fibre. Used in rocket nozzle liners, high-temperature wheel brakes, etc.

pyrolysis Chemical decomposition by heating.

pyromechanical actuator One energized by solid fuel or other combustible charge.

pyrometer Instrument for measuring high temperatures; optical, electrical resistance, thermocouple, radiation, etc.

pyron Non-SI unit of EM radiant intensity (no abb), = calories (Int)/cm^2/min.

pyrophoric Igniting spontaneously on contact with air.

pyrotechnic Today includes not only visual "firework" devices but also precision igniters for large solid motors, single-shot actuators, hot-gas generators and IR flares giving accurately controlled decoy wavelength.

pz See *pièze*.

PZD Petrolatum/zinc dust.

Q

Q 1 Quantity of electricity, esp electric charge.
2 Applied shear force.
3 JETDS code: sonar.
4 JETDS code: special, or combination of purposes.
5 Probability of failure.
6 Quantity of light or other EM radiation.
7 Static moment of area about any axis.
8 Volume of fluid (gas, liquid).
9 Modified-mission prefix: drone or RPV (US DoD).
10 Modified-mission suffix: electronic countermeasures (USN obs).
11 Generalized symbol for torque.
12 Common prefix "quiet".

q 1 Dynamic pressure.
2 Shear stress.
3 Tetrode.
4 Heat flux (rate of flow).
5 Angular velocity (rate of change) in pitch.
6 Generalized symbol for rate of flow, eg volume, energy etc.

Q Störindex; German measure of annoyance-weighted sound pressure level (see *noise*).

Q-aerial Combination of dipole plus quarter-wavelength of twin-wire line to match feeder impedance to dipole.

Q-alpha, QA Free-stream dynamic pressure.

Q-ball, q-ball Spherical or hemispherical-nosed instrument package on nose of spacecraft or aircraft sensing q, AOA, AOY, total temp and other parameters.

Q-band Obsolete EM radiation band with limits 33–50 GHz (US), 26·5–40 GHz (UK), now covered by K, L bands.

Q-code Basic telecommunications code of three-letter groups in three sections: QAA–QNZ are limits of Aeronautical Code; QOA–QQZ is Maritime; QRA–QUZ is for all services. Many of entries in this part of dictionary are from this code.

Q-correction That applied to observed altitudes of star Polaris (because not quite at N celestial pole).

Q-factor 1 Figure of merit of inductance, ratio of reactance to resistance.
2 Ratio of energy stored to energy dissipated per radian in electrical or mechanical system.
3 Generally, sharpness of resonance or frequency selectivity of vibratory system having one degree of freedom, mechanical or electronic.

Q-fan Quiet fan.

q-feel Flight-control feel synthetically made to resemble natural feedback from aerodynamic loads by making it approximately proportional to dynamic pressure.

Q-meter Instrument for measuring Q (1).

Q-series propellants Patented (Thiokol) slow-burning solid fuels for gas generation with LL-521 coolant keeping flame below 1,093°C.

q, Q-spring Mechanical connection with stiffness proportional to dynamic pressure.

q-stops Mechanical limits on flight-control system response, and thus surface movement, commanded by q-system.

Q-switching Extremely rapid switching of laser by means of Kerr cell or similar opto-electronic device; essential for shortest-duration high-energy bursts lasting only a few ns.

q-system That sensing dynamic pressure, eg drawing processed signal from ADC, and feeding it to flight-control and other q-sensitive systems or devices.

QA Quality assurance.

QAB 1 Quality Assurance Board (MoD-PE, UK).
2 Code: "May I have clearance for— from— to— at FL—?"

QAC 1 Quality action case.
2 Civil aircraft qualification (F).

QAD Quick attach/detach.

QAE Quality assurance evaluator.

QAF Code: "Will you advise me when you are/were at—?"

QAG Code: "Arrange your flight to arrive at/over at—?"

QAH Code: "What is your height above—?"

QAI Code: "What is essential traffic regarding my aircraft?"

QAK Code: "Is there risk of collision?"

QAL Code: "Are you going to land at—?"

QAM 1 Code: "What is latest met?"
2 Quadrature amplitude modulation.

QAN Code: "What is surface wind?"

QAO 1 Code: "What is wind at your location at different FLs?"
2 Quality assurance office (DoD).

QAR 1 Quick-access recorder.
2 Quality-assurance representative.

QA/RM Quality assurance and risk management.

QAT Qualified for all three.

QAVC Quiescent automatic volume control.

QBA Code: "What is horizontal vis at—?"

QBB Code: "What is amount, type and height above field of cloudbase?"

QBI Code: "Is flight under IFR compulsory?" Hence, QBI conditions = bad weather.

QC 1 Quality control.
2 Quick change, ie from pax to cargo configuration.
3 Quiet, clean (as prefix).
4 Quality circle (usually plural).
QC card Quadrantal correction.
QCDP Quality-control development programme.
QCE Quality-control engineer.
QCGAT Quiet, clean general-aviation turbofan.
QCIP Quality-control inspection procedure.
QCS Query control station (ECM).
QCSEE, QC-see Quiet, clean short-haul experimental engine (NASA).
QD 1 Quantity distance (explosives).
2 Quick dump.
QDG Quick-draw graphics.
QDL Code: "I intend to ask for series of bearings."
QDM 1 Code: "Will you indicate magnetic heading for me to steer towards you, with no wind?"
2 Quick-donning mask.
QDR 1 Code: "What is my magnetic bearing from you?"
2 Quality-control deficiency report.
QE Quality engineering.
QEC Quick engine change unit, ie ready-to-install powerplant.
QEP Quality-enhancement program.
QET Quick engine test.
QFE 1 Quiet, fuel-efficient.
2 Code: "To what should I set altimeter to obtain height above your location?" Usually requests airfield pressure.
QFF Code: "What is present atmospheric press converted to MSL at your location?"
QFI Qualified flying instructor.
QFU Code: "What is direction/designation of runway to be used?"
QGH Code: "May I land using— procedure?" Requests letdown procedure using radio aids.
QHI Qualified helicopter instructor.
QI Quality improvement.
QIS Quality information system.
QMAC Quarter-orbit magnetic altitude control.
QMW Code: "What is/are freezing levels?"
QN Quiet nacelle.
QNE Code: "What will my altimeter read on landing at— if set to 1013·2 mb?" Note: answer is pressure height of airfield.
QNH Code: "To what should I set my altimeter to read your airfield height on arrival?" Assuming ISA throughout, answer is equivalent MSL pressure as calculated by destination ATC. Regional *, or lowest-forecast *, is value below which actual * is predicted not to fall in given period and location; gives safe terrain clearance.
QNY Code: "What is present weather at your location?"
QOC Quality officer in charge (UK).
QOT&E Qualification operational test and evaluation.
QPD Quality procedural document.
QPL Qualified products list.
qpp Quiescent push/pull.
QPR 1 Quality problem report.
2 Quality procedural requirement.
QR Quiet radar.
q_r Generalized co-ordinate.
QRA 1 Code: "What is name of your station?"
2 Quick-reaction alert (RAF).
Q_{rb} Generalized force of rth mode due to buffeting pressure field.
QRC Quick-reaction capability.
QRF Quick-release fitting.
Q_{rf} Generalized forcing function.
QRGA Quadrupole residual gas analyser, instrument for measuring ultrahigh vacuum.
QRH Quick-reference handbook.
QRI Quick-reaction interceptor.
QRMC Quadrennial review of military compensation.
QRP Quick-reaction package.
QRS Quality requirements systems.
Q_{rs} Generalized force in rth mode due to sth-mode aerodynamic excitation.
QRT Code: "Shall I stop (or please stop) sending?"
QRTOL Quiet RTOL.
QS Quick scan.
QSEG Qualification systems-engineering group.
QSH Quiet short-haul.
QSR Quick strike reconnaissance.
QSRA Quiet short-haul research aircraft.
QSTOL Quiet STOL.
QSY Code: "Change radio frequency now to—".
QT&E Qualification test and evaluation.
QTE Code: "What is my true bearing from you?"
Q-tech Quality technology.
QTF 1 Code: "Will you give me position of my station according to bearings taken by D/F stations you control?" (Note: many countries, including UK, can no longer provide D/F fix.)
2 Quarter-turn fastener.
QTM Quality technical memorandum.
QTR 1 Quiet tail rotor.
2 Quality technical requirement.
quad 1 Group of four attitude-control thrusters indexed at 000°/090°/180°/270° relative to vehicle major axis.
2 Quadrant (ICAO).
3 Loosely, any group of four, especially of air-launched missiles fired from one container or pylon.
quad actuator Among other things, central summing/proportioning/switching unit between triply redundant systems to ensure that maximum redundancy and maximum authority are preserved.
quaded cable Telecom cable with two twisted pairs (four channels).

quadrant 1 Radio range area between equisignal zones (actually much more than 90°).
2 Circular-arc operating lever on control surface of airship or large, slow aeroplane (probably obs).

quadrantal error That caused by presence of metal structure (eg receiving aircraft) distorting radio signal.

quadrantal heights Specified flight levels assigned to traffic on headings in each of four 90° quadrants; intention is that traffic on conflicting headings shall be separated in height.

quadrantal points Intercardinal headings: 045°, 135°, 225° and 315°.

quadrature Quarter-phase difference between two wave trains of same frequency, ie displacement of 90°.

quadrex Quadruply redundant.

quadricycle 1 Landing gear with four wheels or wheel groups disposed at corners of rectangle in plan.
2 Loosely (suggest incorrectly) any aircraft with four landing gears even if three are in line.

quadruplane Aeroplane with four superimposed wings.

quadruple register Instrument recording wind speed/direction, sunshine and rain (not snow).

quadruplex Any fourfold system, ie four parallel channels.

quadruplex system Dynamic system that is quadruply redundant and thus provides multiple-failure capability, eg SFO/FS, DFO and DFO/FS.

quadrupole Idealized noise source made up of four equidistant sources, each diametrically opposite pair emitting positive pressure peaks while the other pair are at peak negative (all having same frequency).

qualification Clearance for service; thus also for production.

qualitative limitation In arms control, eg SALT II, concerned only with weapon-system capability.

quality control That management function by which conformance to established standards is assured, performance is measured, and, in event of defects, corrective action is initiated (USAF).

quantified Expressed in numerical terms.

quantitative limitation One concerned solely with numbers, eg of weapon systems in SALT II.

quantized Restricted to particular set of discrete values.

quart Non-SI measure of capacity, qt = (UK) 1·137 l, (US) 0·946 l.

quarter-chord Locus of all points lying at 25% chord (of wing or other aerofoil), each measurement being in plane parallel to longitudinal axis of aircraft. Normal interpretation is that measures are for wing in clean condition, ignoring ancillary items such as stall-breaker strips and anything else attached to wing but including dogtooth and similar basic modifications to outline.

quarterlights Windows at oblique angle between front windshield (UK windscreen) and side.

quarter-phase Electrically 90° (see *quadrature*).

quarter-turn fastener Cowl or panel fastener released by 90° anticlockwise turn.

quarter-wave aerial One quarter-wavelength long and resonating at slightly less than 4l where l is length; hence quarter-wave line is same length of co-ax or twin-wire transmission line.

quartz Natural and synthesized mineral, SiO_2, with many electrical, electronic and refractory structural uses but esp noted for piezoelectric properties in perfect-crystal form.

quenching 1 Sudden cooling of hot metal in water, oil or other medium to obtain desired crystalline properties.
2 Blanketing solid grain of rocket in mid-operation, by various related techniques, to obtain variable-pulse operation.

Questol Quiet experimental STOL.

quick cam Flap-track profile giving long rearwards travel to increase area/lift and with sharp downwards travel at end which, when full flap selected, rotates flap to landing angle giving high drag.

quick-change aircraft One whose interior is configured for passenger or freight operation and whose seat units, passenger-service unit pantries, toilets and often trim can be removed in minutes and replaced by freight restraints (attached to original floor rails) and protective wall panels.

quick-connect parachute Chest-type pack which user can clip to harness almost instantly.

quick-disconnect couplings Mating fluid-pipe couplings incorporating self-sealing shut-off valves to allow disconnection under pressure without loss of fluid.

quick-donning mask Simple oxygen mask, usually of drop-down type but sometimes carried in separate pack, which in theory can be put on with one hand in a few seconds.

quick dump Switch for getting rid of entire load of rockets, normally by jettisoning launchers.

quick-engine-change unit See *engine-change unit*.

quickie GCA GCA or PAR approach conducted in circumstances calling for priority (possibly notified emergency) and eliminating procedure or identifying turns.

quick-release box Parachute-harness latch released by 90° rotation and blow from hand.

quick-search procedure One in which double normal number of aircraft are used and entire area is searched on outbound leg (NATO).

Quicktrans Long-term contract airlift within Conus in support of USN, USMC and DoD agencies (US).

quiescent current Flow cathode/anode in absence of input signal.

quiescent flow Small flow of air maintained

throughout no-load period to keep bleed-air turbine running; eliminated in modern systems.

quiescent modulation Amplitude modulation in which carrier is radiated only during modulation.

quiet automatic gain control Automatic variation of bias of one or more amplifying stages preceding detector, in which output is suppressed for all signals too weak to trigger control.

quiet radar Scans scene continuously with several thousand very narrow beams in rapid random FH sequence.

quiet Sun Free from significant sunspots or unusual radiation.

quill shaft Slim drive shaft or driven shaft projecting as cantilever and terminating in splined coupling inserted into mating female portion. Requires no key, tolerates small misalignment, and absorbs torsional vibration. Can be keyed at both ends.

quilted blanket Thermal insulating blanket with insulator, eg rock-wool, sealed in stainless-steel foil layers joined along criss-cross bonds; tailored to particular application.

QUJ Code: "Will you indicate true track to reach you?"; ie zero-wind heading.

QWI Qualified weapons instructor.

QW mechanism Quick-wind (reconnaissance camera).

R

R 1 Range.
2 Generalized term for radius, from sheet-metal work to aircraft mission.
3 Resistance (electrical, fluid flow, marine aircraft on water).
4 Reynolds number (also Re, RN).
5 Resultant.
6 Gas constant.
7 US piston-engine code: radial.
8 US aircraft code: modified for reconnaissance (DoD).
9 US Navy modified-mission suffix pre-1962: transport conversion.
10 Generalized term for reliability, and probability thereof.
11 Moment of resistance (also M).
12 Radiance.
13 JETDS code: radio.
14 JETDS code: receiving only, ie passive detection.
15 Resistor.
16 Restricted area (ICAO).
17 Received (ICAO).
18 Repair facilities available (FAA).
19 Right (runway designator).
20 Suffix: radial (thus 234°R = VOR bearing).
21 Generalized term for rate (eg code rate, data rate).
22 Total rainfall.
23 Modulus of rupture.
24 US missile code for vehicle type: unguided rocket.
25 US missile code for launch environment: ship.
26 Rustsätz, field conversion kit (G, WW2).
27 Microlight aircraft category (FAI).

r 1 Radius of rotation.
2 Suffix: required.
3 Revolutions (eg rpm).
4 Common for ratio.
6 Rocket burn rate.

R* Universal gas constant, 8·31432 J k mol^{-1} K^{-1}.

°R 1 Rankine.
2 Réaumur.

R-12 Refrigerant 12.

R-stoff Mixture of xylidine and triethylamine.

RA 1 Rain (ICAO).
2 Receiver attenuation.
3 Research association(s) (UK).
4 Right ascension.
5 Reliability analysis.
6 Radio altimeter.
7 Restricted article.

RAA Regional Airline Association (US).

RAAWS Radar altimeter and altitude warning system.

rabbit Video display of beacon response to two unsynchronized interrogating radars.

Rabdart Rapid aircraft battle-damage augmentation repair team.

rabfac Radar beacon, forward air control.

RAC 1 Rules of the air and traffic control services (UK).
2 Radiometric area correlation (guidance).

RACE Real Aero Club de España.

racecourse Race pattern.

race pattern Closed flightpath comprising two semicircles joined by two parallel straight legs; one FL of stack.

race-pattern hold Normally selected for hold of longer than 2 min, latter being time for 360° turn; straight legs adjusted to give required hold unless pattern is established and published.

race rotation Gross rotation of spiral nature imparted by propeller to slipstream.

racetrack See *race pattern*.

RACH Remotely actuated cargo hook.

racing the count Inaccuracy when using Consol or related navaid when flying across position lines close to station.

rack 1 Attachment for air-dropped store (see *hardpoint*, *pylon*, *ejector rack*, *station*).
2 Framework inside aircraft for accommodation of avionic equipments.

rack control Powered flight control with surface actuation by linear rack/pinion drive (suggested obs).

racking Arrangement of installation of avionics according to boxes and attachment racks of standard dimensions published by ATR, Elfin etc.

RACO Roll attitude cut-out.

Racon Radar beacon transponder carried in aircraft for interrogation by ground station, eg SSR.

RAD 1 Radar approach aid (FAA).
2 Radiation accumulated dose.

rad 1 VOR radial.
2 Radian.
3 Radiation dose absorbed, unit of radiant energy received; non-SI, = 0·01 J/kg.

RA&D Requirements analysis and design.

RADA Random-access discrete-address.

Radag Radar aimpoint and guidance.

Radan Radiation direction and nature (ECM).

radar Use of reflected EM radiation, normally with wavelength in RF spectrum between 30 m and 3 mm, to give information on distant target.

Information may include range, range rate, bearing, height and relative velocity, or may be pictorial for reconnaissance purposes. (Originally US acronym, from radio detection and range.)

radar-absorbent material Range of surface coatings, those in use security-classified, which greatly reduce strength of RF energy reflected back along incident path. It is not known if RAMs extend to substrates and structures.

radar advisory Indicates that provision of advice and information is based on radar observation.

radar aerial Portion of radar system used to radiate or intercept signals; US, = antenna.

radar altimeter See *radio altimeter*.

radar altimetry area Large and comparatively level terrain with defined elevation which can be used in determining altitude of airborne equipment by means of radar (DoD, NATO).

radar altitude Automatic FCS mode in which height AGL is maintained constant by autopilot slaved to radio altimeter.

radar approach Approach executed under direc tion of radar controller.

radar approach control Facility providing approach control service by means of ASR and PAR (USAF).

radar beacon Transponder carried by aircraft, missile or spacecraft which, when it receives correct pulse code from ground radar, immediately retransmits identifying code on same or different wavelength.

radar bombing Level bombing using radar bombsight.

radar bombsight Sight for level bombing in which, irrespective of whether target can be viewed optically, numerical data on target relative position and velocity are fed by radar carried in aircraft.

radar boresight line That along axis of aerial.

radar calibration 1 Use of radar direction and distance information to check another system. 2 More often, use of accurately postioned aerial target to check accuracy of ground radar.

radar camouflage Use of radar-absorbent or reflective materials to change radar-echoing properties of object's surface; does not include dispensed countermeasures, jamming or other active technique.

radar clutter See *clutter*.

radar command Command guidance in which target and missile positions and velocities are continuously determined by radar.

radar contact Identification of echo on radar display, esp as that sought. When thus advised, in civil ATC, aircraft ceases normal reporting.

radar control Control of air traffic based upon position/height information supplied by radars.

radar-controlled gun AAA gun, aircraft turret or other gun system whose aim is controlled by radar and computer which feeds all information necessary (sightline spin, lead etc).

radar controller Air traffic controller whose positional information is provided by radar displays, and holding radar rating appropriate to assigned functions.

radar countermeasures See *countermeasures*, *ECM*.

radar coverage Limits within which objects can be effectively detected by radar(s) at given site or installation; may be angular, polar diagram or in other terms.

radar cross-section, RCS Apparent size of target as judged by its displayed echo, determined (in absence of ECM activity) by true size, range, aspect, geometric shape, materials, surface texture and treatment and other factors including intervening dust and precipitation. Normally defined by ratio P_r/P_s where P_r is radar power received at target and P_s is power reflected, plotted as polar (1) in horizontal plane.

radar display Visual electronic display of radar information.

radar echo Signal indication of object which has reflected energy back to radar.

radar element Radar as portion of larger system, or PAR as part of overall GCA.

radar fire Gunfire guided by by radar, or against radar-tracked target.

radar flight-following General observation of progress of identified aircraft targets sufficiently to retain their identity or observation of specific radar targets (FAA).

radar gunlaying Aiming using radar target position/velocity and, usually, radar input of range; can be manual, automatic with manual override or fully auto.

radar horizon At any location, line along which direct radar rays are tangential to Earth's surface; in practice, usually same as radio horizon.

radar identification Use of transponder or procedure manoeuvre to establish positive identification of object seen on radar.

radar imagery Imagery produced by recording radar waves reflected from target surface in air/surface reconnaissance.

radar intelligence item Feature which is radar-significant but which cannot be identified exactly at moment of its appearance.

radar map Cartographical information superimposed on radar display.

radar mapping Cartography based upon radar, esp SLAR.

radar mile Unit of time equal to 10·75 μs; time for EM radiation to reach target 1 statute mile distant and return.

radar monitoring One of three types of radar service afforded by civil controllers: radar flight-following of aircraft navigated by own pilots or crews, and advice on deviations from track, possible conflicting traffic or progress on instrument approach from final approach fix to runway.

radar navigation guidance One of three types of

radar service, ground vectoring of aircraft to provide course guidance.

radar netting Linking of several radars (surveillance, HFR or similar air-defence radars) to single centre to provide integrated target information.

radar netting station Centre which can receive data from radars and exchange these among other radar stations, thus forming netting system.

radar netting unit Optional electronic equipments converting air-defence fire-distribution system command centre to radar netting station.

radar picket Ship, aircraft or vehicle stationed at distance from surface force for purpose of increasing radar detection range. Hence, ** combat air patrol.

radar picket escort Surface vessel dedicated to ESM, ECM and electronic search facilities. USN designation DER.

radar prediction Graphic portrayal of estimated radar intensity, persistence and shape of cultural and natural features of specific area (USAF).

radar ranging Use of radar to obtain continuous input of target range.

radar receiver Ambiguous term when used alone; can mean receiver of radar that also transmits, or passive equipment used in ESM, ECM or radar warning.

radar reconnaissance Reconnaissance using radar(s) to determine nature of terrain and enemy activity.

radar reflectivity As in optics, fraction of incident radiation reflected by target, normally measured on unit area perpendicular to radiation. Modified by radar camouflage and RAM (2).

radar response Visual indication on radar display of signal transmitted by target following radar interrogation.

radar return See *radar echo.*

radar scan See *scan.*

radar scanner See *radar aerial.*

radarscope Radar display in which information is presented visually for human assessment.

radarscope overlay Transparent overlay for comparison and identification of radar returns.

radarscope photography Film record of radar display (today usually on cassette).

radar separation Radar service in which air traffic is spaced in accord with established minima.

radar signal film Film on which are recorded all signals acquired by a coherent radar and viewed or processed through optical correlator to permit interpretation.

radar signature See *signature.*

radar silence Imposed discipline prohibiting transmission by radar on some or all frequencies.

radar-sonde Met facility in which balloon carries instrumentation recording temperature, pressure and other atmospheric data which are transmitted when triggered by secondary radar whose observation of balloon position when related to time and known rate of ascent gives wind velocities at different heights.

radar surveillance Radar observation of given geographical area or airspace for purpose of performing radar function, eg traffic control or defence.

radar target designator control, RTD Moves acquisition symbology to bracket target prior to lock-on.

radar tracking station Radar facility with capability of tracking moving targets.

radar track position Extrapolation of aircraft position by computer based upon radar information and used by computer for tracking.

radar vector Heading (course to steer) issued as part of radar navigation guidance.

RAD/BAR Radio/pressure altimeter, or selector switch between both.

RADC Rome Air Development Center (USAF).

Radex Rapid-deployment exercise.

radiac Adjective meaning radioactivity detection, indication and computation; applied to radiological instruments and equipment.

radiac dosimeter Measures aggregate ionizing radiation received by that instrument.

radial 1 PE (4) whose cylinders are arranged radially, like spokes of wheel.
2 Magnetic bearing extending from point-source navaid, eg VOR, Tacan, Vortac; usually QDR.

radial compressor Centrifugal compressor.

radial displacement Distortion of tall buildings in low-level reconnaissance photos.

radial drill Machine tool having drill chuck carried on pivoted radial arm of variable length.

radial engine See *radial (1).*

radial error Distance between desired point of impact of munitions and actual point, both points projected and measured on imaginary plane perpendicular to munition flightpath.

radial error probable That circle drawn on radial error plane through which 50% of actual munitions pass, with centre at projected target position.

radial flow Fluid flow inwards or outwards along substantially radial path, usually outwards in supercharger or centrifugal compressor and inwards in (rare) * turbine.

radial flyability Unquantifiable measure of ease with which pilot can accurately hold radial (2), esp when near station; varies with terrain-induced errors and other factors which distort or otherwise influence signals.

radial GSI, RGSI Replaced DME in many aircraft; essentially DME panel instrument with inbuilt wind correction facilitating choice of FL giving best ground speed.

radial struts Those connecting inner and outer ridge main joints of airship transverse frames.

radial velocity Velocity of approach or recession

between two bodies, ie component of relative velocity along line connecting them.

radial-wing configuration Use of several, eg four, wings mounted radially to permit flight manoeuvre instantaneously along any plane containing longitudinal axis without prior need for roll.

radial wires Join vertices of airship's main transverse frames to central fitting or to each other diametrically opposite.

radian SI unit of plane angle; angle subtended at centre of circle by arc equal in length to diameter, rad = 57·2958° = 57° 17′ 44·8″.

radiant energy EM radiation, eg heat (IR), light, radio and radar. Arguably, also occasionally used for other energy, esp acoustic.

radiant-energy density Instantaneous value for amount of energy in unit volume of propagating medium, symbol u. With pulse radars depends on pulse length and position.

radiant-energy thermometer Instrument which determines black-body temperature; emitter need be "black" only over range of wavelengths studied.

radiant flux Time rate of flow of radiant energy, ϕ.

radiant-flux density Radiant flux per unit area; when applied to source, called radiant emittance, radiancy, symbol W; when applied to receiver, called irradiance or (not recommended) irradiancy, symbol H.

radiant intensity Radiant flux per unit solid angle, measured in given direction, SI unit W/sr.

radiant temperature That recorded by total-radiation pyrometer; when sighted on non-black body is less than true temperature.

radiating element Any portion of radar aerial, esp one of electronically scanned type, which emits or reflects transmitted energy.

radiation 1 Process by which EM waves are propagated.
2 Process by which other forms of energy are propagated, eg heat from solid body, kinetic by ocean waves and sound waves through atmosphere or other medium.
3 Other forms of energy propagation such as nuclear *, high-energy ionizing *, radioactivity.

radiation area Place where human being could receive 5 mrem/h or total of 150 mrem in 5 consecutive days.

radiation belts Belts of charged particles trapped within planetary magnetic field, esp Van Allen belts.

radiation burn Damage to skin caused by ionizing radiation.

radiation constants Two physical constants: First ** = $2\pi hc^2 = 3{\cdot}7418 \times 10^{-16}$ Wm²; Second ** = $hc/k = 1{\cdot}4388 \times 10^{-2}$ mK.

radiation cooling Cooling by direct radiation from surface, normally implying high temperature (eg rocket thrust-chamber skirt).

radiation dose Amount of ionizing radiation absorbed by substance, esp over short period of time (see *rad* [*3*], *Röntgen*, *rem*).

radiation dose rate Dose per unit time.

radiation efficiency Ratio of power radiated to power supplied to transmitting aerial at given frequency.

radiation field Volume occupied by radiation (1), esp that around conductor carrying AC or RF, comprising electrical and magnetic (inductive) components.

radiation fog Usually shallow fog caused by radiative cooling (often at night) of ground to below dewpoint, combined with gentle mixing, saturation and condensation.

radiation hardening Gradual hardening and embrittlement of most metals exposed to intense nuclear radiation, resulting in limited choice of metals for high-integrity structures in such environments.

radiation illness Disorders, some fatal, caused by excessive exposure to ionizing radiation.

radiation intelligence That derived from collection and analysis of non-information-bearing radiation unintentionally emitted by foreign devices, excluding that generated by detonation of NW.

radiation intensity Radiation dose rate, normally that measured in air. RI-3, eg, is figure 3 h after NW burst.

radiation laws Those describing black-body radiation: Stefan-Boltzmann, Planck, Kirchoff and Wien.

radiation lobe See *lobe*.

radiation medicine Branch dealing with radiation and human beings.

radiation pattern Graphical representation of radiation (1) of device, esp radio, radar or similar emitting aerial, plotted as field strength as function of direction. Normally plotted as plan-view polar or as vertical cross-section, but can be in plane of magnetic or electrical polarization of waves. Free-space ** depends on wavelength, feed system and reflector. Field ** takes account of real situation in which waves are reflected from ground or other objects so that direct and reflected waves interfere with each other. Also called coverage diagram, aerial (antenna) pattern, lobe pattern.

radiation pressure That exerted on solid body by incident radiation (1), symbol P_r.

radiation pyrometer Pyrometer measuring light wavelengths and giving readout in terms of temperature.

radiation scattering Diversion of radiation (EM, thermal or nuclear) caused by collision or interaction with atoms, molecules or large particles between source (esp NW explosion) and remote site; thus radiation is received from many directions.

radiation sickness See *radiation illness*.

radiation situation map One showing actual and/or predicted radiation situation, usually intensity, in particular ground area.

radiation source Generally a man-made, portable, sealed source of radioactivity.

radiator 1 Source of radiant energy, esp EM or RF, eg hostile operating radar.
2 Heat exchanger, esp for rejecting unwanted heat to a sink. Common usage restricts term to devices that dump heat overboard, eg to atmosphere, and to call those that use heat sink on board (eg fuel) heat-exchangers.

radiator header Tank in which liquid coolant is received from heat source, eg engine, and distributed to cooling elements, ie radiator(s).

radioactive ionization gauge One in which ions produced by radiation (usually alpha particles) from radioactive source.

radioactivity 1 Spontaneous disintegration of nuclei of unstable isotope yielding alpha and/or beta particles, often accompanied by gamma radiation.
2 Number of spontaneous disintegrations per unit mass per unit time, usually measured in curies.

radioactivity concentration guide Published values (DoD, NATO etc) of quantities of listed radioisotopes permissible per unit volume of air and water for continuous consumption.

radio altimeter Instrument, invariably of CW FM type, giving readout of height AGL by time-varying frequency and measuring difference in frequency of received waves, this being proportional to time and hence to height. Sometimes called radar altimeter.

radio approach aids Those which assist landing in bad visibility, notably ILS, MLS; also called radio or electronic landing aids.

radio astronomy Study of radio emissions received by Earth, esp those which can be associated with source of EM emissions on visible, X-ray or other wavelengths.

radio beacon Fixed ground station emitting RF signals, esp those containing identifying information, which enable mobile stations to determine their position relative to it.

radio bearing Usually, angle between apparent direction of fixed station and a reference direction, eg, true or magnetic N. Hence true **, magnetic **.

radio biology Often written as one word, study of effects of radiation on life.

radio channel One band of frequencies sufficient for practical radio communication; sum of emission bandwidth, sideband spread (interference guard bands) and tolerance for frequency variation.

radiochemistry Chemistry of radioactive materials.

radio command Command guidance using a radio link.

radio compass Originally, fixed-loop receiver with which aircraft could home on to any selected fixed station. Later superseded by ADF and other navaids.

radio control Vague, but generally means control of vehicle trajectory with commands transmitted over radio link.

radio countermeasures Those branches or activities of ECM concerned with telecommunications.

radio coupling box Inputs ADF, VOR, ILS, etc, to autopilot.

radio deception Use of radio to deceive enemy; includes sending false despatches, using deceptive headings and employing enemy callsigns (DoD, IADB).

radio detection Detection of object's presence by radio, without information on position.

radio direction-finding See *direction finding*.

radio direction-finding database Aggregate of information, provided by air and surface means, necessary to support radio D/F operations to produce fixes on target transmitters/emitters.

radio duct Shallow quasi-horizontal layer(s) in atmosphere wherein temperature and moisture gradients result in abnormally high refraction lapse rate, causing RF signals to become trapped within layer (see *anomalous propagation*, *skip effect*).

radio energy That which propagates at radio frequencies.

radio facility chart Original series of US airway maps based on radio range; name still common for modern air maps showing all facilities, airways, control zones, TMAs etc.

radio fix 1 Fix of mobile station, eg aircraft, obtained by use of radio navaid, esp by traditional crossing of position lines.
2 Geographical location of friendly or, esp, enemy emitter obtained by various ESM and D/F techniques.

radio frequencies, RF All those EM frequencies used for radio or related purposes. Common-use subdivisions are: VLF, below 30 kHz; LF, 30–300 kHz; MF, 300 kHz–3 MHz; HF, 3–30 MHz; VHF, 30–300 MHz; UHF, 300 MHz–3 GHz; SHF, 3–30 GHz; EHF, 30–300 GHz; unnamed, 300 GHz–3 THz.

radio goniometer See *direction-finder*.

radiography Photography using X-rays, gamma rays or other ionizing radiation; important NDT method, often using radiation source inside test object and film outside.

radio guard Radio station, eg aircraft, which listens out on assigned frequencies and handles traffic and records transmissions.

radio guidance Guidance or navigation system using radio waves, eg point-source aids, area coverage (R-Nav), global (Omega) and command methods.

radio hole Direction of propagation suffering

abnormal attenuation or fading, usually caused by local refraction.

radio horizon At any location, line along which direct rays from RF transmitter become tangential to Earth's surface; extends beyond visual horizon because of atmospheric refraction and varies according to whether propagation is sub-, normal or super-standard.

radio-inertial guidance Various systems combining inertial and radio tracking and/or command (probably obs).

radio interferometer Interferometer operating at RF.

radioisotope Unstable isotope that decays spontaneously, emitting radiation.

radioisotope thermoelectric generator Self-contained power system in which a radioisotope is used to heat one junction in a circuit containing dissimilar metals and thus generate sustained electricity.

radiolocation Original UK name for radar.

radiological defence Measures taken against radiation hazard resulting from use of NW and RW.

radiological survey Directed effort to determine distribution and dose rates of radiation in an area. Hence ** flight altitude.

radiological weapons Established forms of radioactive materials or radiation-producing devices, including intentional employment of NW fallout, to cause casualties or restrict use of terrain; abb RW.

radiology Science of ionizing radiation in treatment of disease.

radio loop D/F loop aerial.

radioluminescence Emission in visible EM range, typically with characteristic hues, in radiation from radioactive materials.

radio-magnetic indicator Magnetic and radio panel instrument in which card (dial) rotates so that heading appears opposite index mark at top of case while needle rotates to show magnetic bearing (QDM) of tuned beacon.

radio marker beacon See *marker*.

radio mast Mast projecting above or below aircraft, esp older aircraft, usually as one anchor for MF/HF wire aerial. Modern VHF blade seldom thus termed.

radiometeorograph Meteorograph transmitting readings by radio (see *radio-sonde*).

radiometer Instrument for detecting, and usually also measuring, IR and closely related radiation (see *bolometer*, *actinometer*, *photometer*).

radio navigation All uses of radio to determine location, obtain heading information and warn of obstructions or hazards.

radionuclide Nuclide spontaneously emitting radiation.

radiophotoluminescence Photoluminescence exhibited by substance after irradiation by radionuclide or other source of beta or gamma rays.

radio proximity fuze See *proximity fuze*.

radio range Original radio navaid (US), a land (rarely, ship) fixed station of aeronautical radio service broadcasting continuous coded signals which on one side of an airway are heard as a Morse N (–·) and on other as A (·–); in a less common system signals are I (··) and A. In the equisignal zone along centre of airway both signals combine, A and N forming continuous note and I and A cancelling to send no signal to pair of reeds which, as soon as aircraft strays from centreline, vibrate visually and aurally. Now replaced by VOR and later navaids.

radio rangefinding Use of radio (essentially in this case not radar) to determine range.

radio-range orientation Technique needed when flying radio range of finding and identifying station and then accurately flying correct leg to next station or destination.

radio recognition Determination by radio means of another station's identity; DoD wording is "friendly or enemy character, or individuality, of another".

radioresistance/radiosensitivity Measures of resistance or sensitivity of living cells to injurious radiation.

radio silence Period during which some or all RF emitters (eg of military force) are kept inoperative.

radio-sonde Instrumentation for measure of temperature, pressure and humidity carried aloft by balloon together with electronics for converting answers into code for RF transmission to ground station at intervals (see *radar-sonde*).

radio sonobuoy Sonobuoy which transmits information to friendly receivers.

radio telegraphy Transmission of telegraphic codes by radio; often one word.

radio telephony Transmission of speech by radio, R/T, often one word.

radio-telephony network Integrated group of aeronautical fixed stations which use and guard frequencies from same family and co-operate to maximize reliability of air/ground communications.

radio telescope Radio receiver capable of greatly amplifying, recording and determining source-direction of radio waves received on Earth from outer space.

radio waves EM radiation with wavelengths from tens of kilometres (VLF) down to tens of microns (above EHF, so-called decimillimetric waves).

radius block Steel block with edge of accurate radius for sheet-metalwork.

radius dimpling Pressing dimples in thin sheet with cone-shaped mating tools.

radius gauge Hand instrument for measuring inside or outside bend radii.

radius of action Maximum distance aircraft can travel away from its base along given course with normal combat load and return without refuelling, allowing for all safety and operating factors. Today, esp with gas-turbine propulsion, *** varies greatly with mission profile, fuel consumption being several times higher in lo regime and multiplied again with afterburner lit.

radius of damage Radius from ground zero within which there is 50% probability of achieving desired damage.

radius of gyration Distance from body's centre of gyration to selected axis; square root of ratio of moment of inertia, about selected axis, divided by mass.

radius of integration Radius from ground zero within which effects of NW and conventional weapons are to be integrated.

radius of safety Horizontal distance from ground zero within which NW effects on friendly troops are unacceptable.

radius rod Major bracing strut in retracting landing gear; normally pivoted near lower end of main leg and upper end pinned to backlink, actuator or sliding block, but many arrangements all with same name.

radius vector Vector connecting body with object which may have relative motion; specif, vector joining primary body to satellite, as in Earth/Moon system.

radix point In any number system, index separating numbers having negative powers from those having positive; eg decimal point, binary point.

radome Protective covering (and in aircraft aerodynamic fairing) over radar or other aerial, esp one with mechanical scanning; made of dielectric material selected according to operating wavelength and other factors.

Radop Radar/optical tracking methods.

Radot Recording automatic digital optical tracker.

Radpac Radar package (Panavia); versatile software to extend radar capability, esp in air-to-air missions.

rad/s Radians per second.

RADU Range and azimuth display unit.

RAE Royal Aircraft Establishment (MoD PE).

RAeC Royal Aero Club (UK).

RAeS Royal Aeronautical Society (UK).

Raevam RAE variable-aerofoil mechanism; infinitely adjustable flexible leading edge.

RAF 1 Royal Aircraft Factory (UK, 1912–18); also Royal Air Force, but military services as such are excluded from this dictionary, as are airlines.
2 Recursive adaptive filter.

RAFA Royal Air Forces Association.

RAG 1 Replacement air group, land-based unit tasked with supplying trained aircrew for carrier-based squadrons (USN).
2 Runway arrester gear.

rag-wing Fabric-covered aircraft (US, colloq).

RAHE Ram-air heat-exchanger.

RAI 1 Registro Aeronautico Italiano; national licensing authority.
2 Runway alignment indicator (ICAO).
3 Remote attitude indicator.

RAIL Runway alignment indicator light system (FAA).

rail Launcher on ground or aircraft mount for missile requiring support and/or guidance while accelerating.

rain Precipitation in form of water droplets making noticeable individual impact, diameter roughly 1 to 5·5 mm.

rain band Spiral cloud area of tropical revolving storm where there is heavy rain.

rainbow(ing) Appearance in laminated transparency of multicoloured (spectral) stress patterns.

rainfall Term for radioactive precipitation from base-surge clouds after underwater NW burst (DoD).

rain gauge Various forms of precipitation gauge designed primarily for measuring fall of water droplets, usually by funnelling fall over known area into calibrated container.

rain ice Most dangerous airframe icing; similar to glaze ice and caused by supercooled raindrops which take sufficiently long to freeze for flow-back to be extensive.

rain loop See *drip flap*.

rainmaking See *seeding (2)*.

rain-out 1 Removal of solid particulate pollution by rain.
2 Radioactive material in atmosphere brought down by precipitation and present on surface (DoD).

rain static Radio interference believed to be caused by rain bearing electrostatic charges.

RAJ Ring-around jammer (ECM).

rake 1 Angle between local vertical and swivel or castor axis of swivel-mounted wheel, eg tailwheel.
2 Angle between quasi-straight edge of wingtip and aircraft longitudinal axis, called positive * when leading edge is shorter than trailing edge and negative * when trailing edge is shorter.
3 Distance, measured parallel to aircraft longitudinal axis, between front of propeller blade at tip and plane of rotation through axial mid-point of hub (simple fixed-pitch propeller).
4 Acute angle between line joining centroids of propeller blade from root to tip and plane of rotation (often zero).
5 Comb, linear or other array of pitot heads.

raking shot Gunfire almost aligned with hostile aircraft longitudinal axis (suggested obs).

Ralacs Radio-altimeter low-altitude control system (RPV).

RALS, Rals Remote augmented lift system.

RAM 1 Random-access memory.
2 Radar-absorbing material(s).
3 Research and applications module (Space Shuttle).
4 Rapid area maintenance (AFLC).

5 Raid-assessment mode (radar).
6 Reliable/available/maintainable.
7 Rolling-airframe missile.

ram 1 Increase in pressure in forward-facing tube, duct, inlet etc as result of vehicle speed through atmosphere; if fluid flow were brought to rest in duct pressure would be q, dynamic pressure. Hence * inlet, * pressure, *-jet, * air, * effect.
2 Main movable portion of hydropress; term not encouraged for most hydraulic devices having linear force output, for which actuator is preferred.

ram air Air rammed in at forward-facing inlet.

ramark Fixed radio beacon continuously transmitting (sometimes coded signal) to cause radial line on PPI radars.

ram compression See *ram (1)*.

RAMDI Radioactive miss-distance indicator.

ram drag See *momentum drag*.

ram inlet Forward-facing inlet designed to achieve maximum ram recovery.

ramjet Air-breathing jet engine similar to turbojet but without mechanical compressor or turbine; compression is accomplished entirely by ram (1) and is thus sensitive to vehicle forward speed and non-existent at rest (hence * cannot start from rest). Inefficient at Mach numbers below 3 but extremely important for unmanned vehicles, esp in conjunction with rocket (eg ramrocket). Also called athodyd, Lorin duct; not to be confused with pulsejet or resonant ducts.

RAM net Camouflage net made of RAM (2).

ramp 1 Main aircraft parking area at airport, airfield, airbase.
2 Sharp-edged wedge with sloping wall forming inner wall of supersonic inlet duct to create oblique shock(s) and improve pressure recovery, esp at supersonic speeds; usually has variable geometry.
3 Inclined track for launch of target, RPV or missile under moderate acceleration.

ramp capacity Number of aircraft, of specified general size class, for which ramp (1) is designed, including nose-in and off-terminal parking.

ramp extension Increase in size of ramp (1) to augment capacity or allow for larger aircraft with wide turning circles.

Ramps Resource allocation and multipath scheduling.

ramp services All services needed by transport on transit stop or turnround, normally excluding mechanized freight handling and supplies (eg pantry/bar stocks) brought from terminal.

ramp status Accorded a new prototype after it has gone "out the door" and is being readied for taxi tests.

ramp-to-ramp See *block time*.

ramp weight, MRW Maximum weight permissible for aircraft, equal to MTOW plus fuel allowance for main engines and APU for start, run-up and taxi.

ram recovery Pressure actually achieved in ram inlet, expressed as absolute value or as percentage of total available dynamic pressure.

ramrocket Important species of propulsion system for unmanned vehicles, comprising rocket (solid, liquid or hybrid) and integral ramjet propulsion. Vehicle is launched by rocket, at conclusion of whose burn at supersonic speed nozzle is jettisoned, leaving larger con-di nozzle, air inlet is extended or opened and ramjet operation takes place with combustion of liquid or hybrid type in original solid motor case and chamber (see *ducted rocket*).

RAMS 1 Remote automatic multipurpose station.
2 Racal avionics management system.

ram's horn 1 Pilot flight-control yoke generally resembling ram's horn.
2 Microwave aerial of spiral (usually exponential) form.

ram void pressure Achieved pressure (pressure recovery) inside inlet duct, esp at supersonic speed; ratio **/freestream total pressure is function controlling dump doors on SSTs.

RANC Radar attenuation, noise and clutter (US, DNA).

R&A Report and accounts.

R&C Rolled and coined.

R&D Research and development.

R&M 1 Reports and memoranda (UK).
2 Reliability and maintainability (US).

random access Ability of computer memory to remember contents and addresses of all memory stores immediately; access time is independent of location of preceding record.

random-demand planning Planning for supplies necessitated by in-service failures.

random energy That of fluid particles in disordered motion, rapidly degrading to heat, eg downstream of shock.

random error Unpredictable, caused by short-period disturbances in system or in measuring method, normally having Gaussian distribution over period; excludes major failures, human errors and errors of systematic nature.

random flight Unplanned local flight by lightplane, esp one without radio.

random scatter See *scatter (1)*.

R&R 1 Repair and return.
2 Rest and recuperation.

range 1 Distance aircraft can travel, under given conditions, without refuelling in flight. By itself has little meaning, except for very small, simple aircraft. Maximum-fuel * normally taken to mean IFR reserves for multi-engine aircraft, VFR for single.
2 Limiting distance, over intercontinental * measured as great circle, missile, RPV or other unmanned vehicle can travel with specified load

and following specified flight profiles.
3 Distance between observer or weapon launch point and target.
4 Land and/or water area equipped and designated for vehicle testing, esp missile or RPV, or testing ordnance or practice shooting at targets.
5 Difference between upper and lower limits for variable, eg frequency or wavelength coverage of receiver, pitch of propeller blades or many other variables.
6 To organize aircraft on carrier flight deck into desired sequence with closest packing.

range ambiguity In several early radio navaids, eg radio range, possibility of flying reciprocal or obtaining misleading distance indication.

range analog bar Horizontal bar appearing in some optical, HUD or radar sight displays once full lock-on has been achieved, its length showing range.

range attenuation Inverse-square decrease in radar power density with range.

range/azimuth display unit Various instruments, esp HSI giving VOR/DME information.

range bar Bold, usually horizontal, failure-warning flag for panel instruments.

range bin Store location in SSR plot extractor; each range increment on given azimuth has store location from range 0 to range limit in which detected targets are stored until end of scan, when each is extracted and, together with scan azimuth, passed out as a plot.

range error Distance measured from imaginary line drawn through desired impact point and one drawn through actual impact point, both parallel and perpendicular to axis of attack (USAF); distance means perpendicular distance.

range error probable Distance between two parallel lines drawn perpendicular to axis of attack and equidistant from desired mean point of impact between which fall 50% of impact points of independently aimed weapons.

range gating Limiting radar or laser to detect targets only within upper and lower (often narrow) range limits.

range lights Row of green lights marking each end of usable runway at simple airfield.

range markers 1 Parallel lines, concentric circles or other fixed graticule on display giving indication of range.
2 Single synthetic echo(es) injected into radar display timebase to give sharp blip, circle or other clear indication at selected range.
3 Two upright markers, illuminated at night, placed so that, when aligned, they assist piloting or in beaching amphibious craft.

range mean pairs Continuous analysis of peaks and troughs of variable function.

range octagon Computer-generated octagon around HUDsight target which unwinds at rate of one side lost for each 100 m closure.

range-only tracking System for accurately measuring vehicle range by phase-comparison technique; vehicle transponder is interrogated by transmitter and replies are recorded by three or more widely separated receivers. In US operated on 387 and 417 MHz.

range resolution Ability of radar to discriminate (separate) two targets on same bearing but with small range separation; determined mainly by pulse length.

range ring Any circle or arc on PPI display indicating range.

range safety officer Person charged with supervising safety to personnel on range (4) and ensuring that no object travels beyond range boundary.

range strobe See *range markers (2).*

range tracking element Radar subcircuit which monitors received-echo times and operates range gate immediately before each return is received.

ranging Process of establishing target distance, by radar, optics/lasers, echo, intermittent, manual, gunfire, explosive-echo or navigational means.

Rankine Absolute Fahrenheit scale of temperature, °R = °F + 459·67, hence K = 5/9R.

Rankine cycle Thermodynamic cycle forming basis of modern steam plant, including many vapour-cycle machines for aerospace, all having closed circuit: boiler-prime mover-condenser-pump-boiler. In space power systems working fluid is metallic vapour, eg Hg, Cs.

Rankine-Hugoniot Relationship between pressure and density on each side of plane shockwave (Rankine 1870), from which p, V and γ for inclined shocks can be deduced.

Ranrap Random-range program (ECM).

Rant Re-entry antenna test (ABRES).

Raob Radio-sonde observation.

RAOC Regional air operations centre.

RAOD Ram-air overboard dump.

RAP 1 Reliable acoustic path (sonar).
2 Resource allocation plan.
3 Reliability analysis program(me).
4 Rocket-assisted projectile (ECM).
5 Radar-absorbent (or absorbing) paint.
6 Radio access point.
7 Recognized air picture.

RAPCON, Rapcon Radar approach control (FAA).

RAPD Recognized air picture display.

Rapec Rocket-assisted personnel ejection catapult (Martin-Baker).

Rapid 1 Real-time acquisition program of inflight data (Sikorsky).
2 Retrorocket-assisted parachute inflight delivery.

rapid area supply support Ability to deliver supply/transport/packaging teams where needed (USAF AFLC).

rapid-bloom Chaff or other ECM dispensed payload which quickly (within 0·1 s) assumes dimensions or emission properties resembling those of actual aircraft.

rapid deployment force Military force, usually trained in amphibious, urban and peacekeeping operations, ready for near-immediate deployment to distant trouble spot. American RDF includes armour, air and naval power.

rapid engineer deployment Quick-reaction civil-engineer squadrons that provide heavy construction and repair capability for theatre commander (USAF).

rapid-extraction parachute One designed to deploy in less than 0·5 s from initial mechanical or electrical signal, eg for lo-alt delivery or crew escape at min altitude.

rapid-fracture surface That left by high-rate crack propagation, showing as dark (often arrowhead) zones separated by bright zones of slow fatigue failure.

rapid prototyping Standard life-cycle method of software development.

rapid pucker-factor take-off One on a favourable unbalanced field.

rapid-reload capability Ability of single ICBM silo or SLBM tube system to fire at rapid rate; not defined in SALT II but generally taken as more than one round per 24 h. Note: in West, no reload missiles exist.

Rapids Radar passive identification system.

RAPS, Raps 1 RPV advanced payload system(s) (USAF RPV ECM).
2 Radar prediction system.

RAR Request radar blip identification (RBI) message.

RARDE Royal Armament Research & Development Establishment, Fort Halstead, Sevenoaks (UK).

rare-earth General term for electric machines using ** magnetic materials, notably SmCo.

rarefaction Supersonic flow through converging duct, in which pressure decreases while velocity increases greatly.

Rareps Radar reports.

RARF Radar, antenna and RF, integrated system in which multiple radar functions are performed with single electrically scanned aerial on time-shared non-interference basis (Emerson).

RAS 1 Rectified airspeed, see *airspeed*.
2 Rough air speed, for flight through gust (2).
3 Replenishment at sea.

RASE Rapid automatic sweep equipment (ECM).

RASH, Rash Rain showers.

RASN Rain and snow (ICAO).

RASS Rapid area supply support (USAF).

RAST, Rast 1 Recovery assist, secure and traverse (helicopter).
2 Radar-augmented sub-target.

Rastas Radiating site and target acquisition system (airborne ECM).

raster Generation of large-area display, eg TV screen, by close-spaced horizontal lines scanned either alternately or in sequence.

raster scan See *scan*.

Rasur Radio surveillance for intelligence purposes.

RASV Reusable aerospace vehicle.

RAT Ram-air turbine; emergency source of shaft power extended into slipstream.

Rat Hostile intruder aircraft flying alone.

RATCC Radar air traffic control centre.

rate Rate of change, first derivative of variable with respect to time; thus angle *, * gyro. Symbol is dot placed centrally above value, thus if a is acceleration along flightpath $\dot{a}$ is rate of change of acceleration or acceleration *.

rated altitude That at which PE (4) gives maximum power, for supercharged engine usually at height considerably above S/L; lowest altitude at which full throttle is permissible (or, usually, obtainable) or max boost pressure can be maintained at max rpm in level flight.

rated coverage Area within which strength of NDB vertical field of ground wave exceeds min value specified for geog area where situated.

rated power Any of several specified limits to gas-turbine power, eg take-off, 2½ min contingency, max continuous.

rate gyro Gyro whose indication gives a rate term, rate of change of attitude; single-degree-of-freedom gyro with primarily elastic restraint of spin axis about output axis.

rate integrating gyro Rig (Mig is "miniature") is single-degree-of-freedom gyro whose output axis is linked to spin axis by viscous restraint so that angular displacement is proportional to integral of angular rate of change of attitude.

rate of climb Rate of gain of height, vertical component of airspeed of aircraft (normally aerodyne) established in climbing flight at quasi-constant airspeed (ie not in zoom trading speed for height). For helicopter, two values: max *** and max vertical ***.

rate-of-climb indicator See *VSI*.

rate of temperature-rise indicator Fire-warning system; does not respond to temperature but to positive rate of change.

rate of turn Rate of change of heading, proportional to bank angle θ; R (turn radius), for constant speed, is exactly proportional to tan θ.

rates of exchange Tradeoff multipliers, eg numerical value linking take-off field length or MTOW for unchanged T-O field length per °C change in ambient temperature.

rate structure Comprehensive and in general internationally agreed prices for air carriage of unit mass of all commodities.

RATG Ram-air turbine generator.

rating 1 Authorized operating regimes, with limiting numerical values, for engine or other functional device or system.
2 Anti-knock value of piston-engine fuel.
3 Endorsements, additional qualifications, pri-

vileges and limitations added to airmen certificate (US) or pilot's licence (UK), eg night *, IMC *, instrument * and flying instructor *.

rating spring Spring, usually with linear output, whose change in length is accurately proportional to applied force.

ratio of specific heats For a gas, ratio of specific heat at constant pressure divided by specific heat at constant volume; for air $\gamma = C_p/C_v = 1{\cdot}401$.

Ratog Rocket-assisted take-off gear; hence RATO.

Ratrace Radar transmitter waveguide shape facilitating use of common aerial for transmitting and receiving.

RATS 1 Rapid area transportation support (USAF).
2 Rescue advanced-tactics school.

Ratscat Radar target scatter; more fully, radar target test scatter facility.

RATT Radio teletype.

raven Electronic-warfare officer, specialist flight-crew member (US).

raw Versatile word meaning ready for next stage of processing; thus * AC is ready for precise frequency control before being supplied to avionics; * data can be of many forms (instrument readings, reconnaissance photos, tabular statistics); and * material is in fact anything but * yet still unmanufactured into product.

rawin Wind velocity at different heights computed by radar tracking of balloon (usually with transponder).

rawinsonde See *radar-sonde*.

RAWS 1 Radar altitude warning system.
2 Radar attack (and) warning system.
3 Remote-area weather station.

Rayleigh atmosphere Ideal atmosphere devoid of all particles larger than about 0·1 wavelength of incident radiation.

Rayleigh flow First (1876) theory since Newton for lift of inclined-plane wing: streamlines travel direct to surface, where they are brought to rest, losing all relative energy.

Rayleigh formula Another Rayleigh equation gives loss of pitot pressure caused by presence of normal shock, in terms only of Mach numbers, pressures and γ (ratio of specific heats of gas).

Rayleigh number Non-dimensional ratio $N_{ra} = \dfrac{gd^3\ \alpha(\theta_1-\theta_2)}{\nu k}$ where g is acceleration due to gravity, d is vertical separation of two horizontal planes in fluid system (eg atmosphere), $\theta_2-\theta_1$ is temperature difference across planes, α is coefficient of thermal expansion, ν is coefficient of kinetic viscosity and k is thermal conductivity.

Rayleigh scattering Normal scattering of radiation by particles whose ruling size is 0·1 or less that of radiant wavelength; explains why sky is blue above and red/orange at sunset.

Rayleigh wave 2-D barotropic fluid disturbance or wave propagated along free solid surface.

RB 1 Rescue boat (ICAO).
2 Rapid-bloom (ECM).
3 Radar-blip identification message.
4 Reduced blast (NW).
5 See * *switch*.

R_b, r_b Solid rocket motor burn rate; also r.

RBC Rapid-bloom countermeasures (or chaff).

RB/ER Reduced blast, enhanced radiation (neutron bomb).

RBF Remove before flight.

RBI 1 Relative-bearing indicator.
2 Radar-blip identification (ICAO).

RBL Range and bearing launch.

RBN Radio beacon.

RBOC Rapid-bloom offboard countermeasures.

RBS 1 Radar beacon system (FAA).
2 Rutherford back-scattering.
3 Radar bomb scoring.

RBSJ Recirculating-ball screwjack.

RB switch Selects radio or barometric altitude, eg on HUD.

RBT Remote batch terminal.

RBV Return-beam vidicon.

RC 1 Rotating-combustion, engine of Wankel type.
2 Reaction control.
3 Resistance/capacitance.
4 Rotating components of gas-turbine engine.
5 Rate of climb at MTOW.
6 Radio-controlled.

RC-1 Rate of climb, one engine inoperative.

RC-2 Rate of climb, all engines operating.

RCA Reach cruising altitude (ICAO).

RCAG Remote center air/ground (FAA).

RCB Remelted cast bar.

RCC 1 Reinforced carbon/carbon.
2 Rescue co-ordination centre.
3 Resistance/capacitance coupling (often r.c.c.).
4 Research Consultative Committee (UK).

RCCB Reverse-current circuit-breaker.

RCDI VSI (rate of climb/descent).

RCDS Royal College of Defence Studies (UK).

RCE Rotating-combustion engine.

RCF Radio communications failure message (ICAO).

RCFCA Royal Canadian Flying Clubs Association.

RCG Reaction-cured glass, covers thermal-protection tiles (Space Shuttle).

RCJ Reaction control jet.

RCL Runway centreline.

RCLM Runway centreline marking (FAA).

RCLS Runway centreline light system (FAA).

RCM 1 Radio countermeasures.
2 Reliability-centred maintenance.
3 Rollercoaster manoeuvre.
4 Restricted corrosive material.

RCMAT Radio-controlled miniature aerial target.

RCMG Rifle-calibre machine gun.

RCMS Resonator-controlled microwave source.
RCO 1 Remote communications outlet.
2 Range control officer.
3 Remote control officer (or operator).
RC/OI Reaction control/orbital insertion subsystem.
RCP Role-change package.
RCS 1 Reaction-control system.
2 Radar cross-section.
3 Ride-control system.
RCT 1 Reply code train (IFF, SSR).
2 Royal Corps of Transport (UK).
3 Reverse-conducting thyristor.
RCU 1 Range converter unit.
2 Remote control unit.
3 Receiver control unit.
RCV Reaction-control valve.
rcv Receive.
rcvr Receiver.
RCW Roll-control wheel.
RD 1 Ramp door, thus * uplock.
2 Random discontinuous fibre.
3 Rigged dry, for dusting rather than spraying (ag-aviation).
RD&A Research, development and acquisition.
RDAU Remote data acquisition unit.
RDB Requirements data bank.
RDBM Relational database management (more often RDM).
RDC 1 Rate-of-descent computer.
2 Regional dissemination center (NASA, technology transfer).
3 Radar data converter.
4 Ram door control.
RDD Routine dynamic display (maritime/ASW navigation as distinct from tactical situation).
RDDBS Redundant distributed data-base system.
RDE, R&DE Research and development engineering (or evaluation).
RDEC Rapid data-entry cassette.
RDF 1 Radio direction-finding (security cover for radar, 1936–42).
2 Rapid Deployment Force (US).
RDJTF Rapid Deployment Joint Task Force (US).
RDM Random-deflection monitor.
RDMI Radio directional (or distance) magnetic indicator.
RDMS Relational database management systems.
RDO Redistribution order.
rdo Radio (FAA).
RDP Radar data processor.
RDPS Radar data-processing system.
RDS Rudder-disconnect speed (Autoland).
RDSS Rapidly deployable surveillance system.
RDT&E Research, development, test and engineering (or evaluation) (DoD).
RDX Powerful explosive $(CH_2N.NO_2)_3$; also called Hexogen, Cyclonite. Now family of explosives still used as bomb, mine and other fillings, though under other names and designations.
RE 1 Recent, ie qualifying as "actual" met data.
2 Rain erosion, hence * resistance.
3 Re-engined.
4 Role equipment.
R/E Receiver/exciter.
REA Radar echoing area (suggest = RCS [2]).
REAs Responsible engineering activities.
reach Length of threaded portion of sparking plug.
React Reliability evaluation and control technique.
reactance Component of impedance in AC circuits due to inductance or capacitance; value is $2\pi fL$ for inductance and $1/2\pi fC$ for capacitance where f is frequency, L is henrys and C is farads.
reactant Any substance consumed in reaction, esp in rocket motor combustion or other generation of working fluid.
reactant ratio Ratio of mass flow of oxidant to fuel.
reacted pressure See *stalled pressure*.
reaction See *positive feedback*, *regeneration (2)*.
reaction chamber One in which chemical reaction takes place, specif that in which reactants produce gas to drive turbopump or to power MEPU.
reaction-control system Primary attitude/trajectory control system of spacecraft, or of V/STOL aircraft at speeds too low for conventional controls to be effective. Control moments are imparted by *thrusters* or *reaction-control valves.*
reaction-control valve Small nozzles supplied with hot, high-pressure main-engine bleed air at extremities of V/STOL aircraft; at low airspeed pilot operates * to control attitude and trajectory of aircraft.
reaction engine Vague; all aerospace propulsion is by Newtonian reaction.
reaction sphere Spacecraft orientation control system comprising free-running dense sphere surrounded by three magnetic coils, one for each attitude axis. Also called free sphere.
reaction time Elapsed time between stimulus and response, eg between command to launch missile and actual launch, between first receipt of warning and dispatch of weapon system, or between pilot first seeing conflicting aircraft and initiation of manoeuvre to avert collision.
reaction turbine One in which main tangential driving force comes from acceleration of working fluid through converging passages between rotor blades; rare in gas turbines, which are usually impulse/reaction or impulse.
reactor 1 Choke or high-inductance coil.
2 Core in which nuclear fission (thermal or fast) or fusion reaction takes place.
read 1 To obtain information from EDP (1) storage and transfer it to another device or address.
2 Of humans, to receive and understand telecommunications message.

read across To have direct relevance to quite different programme or problem; eg use of LH_2 in space propulsion solved problems which ** to LH_2 in future commercial airline operation.

readback Procedure whereby receiving station repeats all or part of message to originator to verify accuracy.

readout 1 Data played back from tape or other store, eg from flight recorder.
2 To broadcast data either from storage or as received.
3 In EDP (1), to transfer word(s) from memory to another location.

ready Weapon system is available for use with no delay save reaction time.

ready CAP Combat air patrol aircraft on standby status; ready for take-off.

ready light Indicates particular munition is ready for use.

ready position Designated place where stick (4) waits for helicopter or for order to emplane.

ready room Where hardware is prepared for use, esp missiles, RPVs and other unmanned vehicles and esp aboard warship.

real fluid One exhibiting viscosity.

real precession That induced in gyro by applied force, eg friction or imbalance, and not by Earth's rotation.

real time 1 In EDP (1), operation in which event times are controlled by portions of system other than computer and cannot be modified for convenience in processing; not necessarily simultaneous with time in everyday sense.
2 Absence of delay, except for time required for transmission by EM energy, between occurrence of event or transmission of data, and knowledge of event or reception of data at some other location (DoD). Neither definition has anything to do with time computer is processing as distinct from warm-up time or rest periods, which incorrectly figure in some popular definitions.

real-time clock One indicating actual time.

real wander See *real precession*.

rear area That behind combat and forward areas in battle.

rear bearing In single-shaft gas turbine, that supporting turbine.

rear-box wing Usually means three-spar.

rear echelon Elements of air-transport force not required in objective area.

rear-loading Provided with door(s) and ramp extending across full cross-section of fuselage interior for loading of vehicles and other bulky cargo, and for air dropping.

rearming Replenishment of consumed ordnance between missions.

rearplane Rear wing of tandem-wing aircraft. Not justified when foreplane is clearly for control rather than lift (eg Viggen) but correct term where front and rear wings are comparable in size.

rear port Aperture, usually circular, in solid rocket motor case at end or on face opposite to nozzle which can be opened to reduce internal pressure for thrust cutoff.

rear stagnation point That at which fluid is at rest on surface of body on downstream side and from which a streamline emanates.

rear step That at trailing edge of seaplane (flying boat) afterbody, where planing bottom terminates; in some designs a point, but usually either vee-step or a sharp-edged discontinuity sufficient to break hydrodynamic attachment to skin.

Réaumur Non-SI temperature scale on which ice point (1 atmosphere) 0° and water boiling point 80°.

rebated blade Solid light-alloy (rarely, wood laminate or GRP) propeller blade whose leading edge is cut away to accommodate encapsulated electrothermal anticing element.

Rebecca Pioneer DME used by airborne interrogation of Eureka beacon (1942). Still in production in 1970 as * 12 for airborne forces and used in conjunction with MR 343 air-dropped beacon.

rebreather Closed-circuit oxygen system from which CO_2 and water are continuously removed, pressure being maintained by adding fresh oxygen (see * *bag*).

rebreather bag Gastight sac in oxygen line near mask so that incoming gas is diluted with exhaled breath.

REC 1 Received, receiving (ICAO).
2 Radio-electronic combat.

recalescence point Temperature, exhibited only by iron and some other ferromagnetic materials, at which on cooling from white heat exothermic change in crystalline structure halts cooling and can cause momentary rise in temperature.

recce Reconnaissance (UK, colloq, pronounced recky).

reception All arrangements for receipt of air drop.

recession Linear rate at which ablating material is eroded, normally measured normal to local surface.

Rechlin Location near Neubrandenburg of largest and oldest German aircraft/air armament test establishment before 1945.

reciprocal Heading 180° from that intended.

reciprocal bearing Bearing of observer from remote station.

reciprocal latching Use of phase-shifters to generate computer-controlled sequence and distribution pattern for electronically scanned radar transmission.

reciprocating engine One in which piston(s) oscillate to and fro in cylinder, eg Otto, diesel, Stirling (by common usage diesel without capital letter).

recirculating ball system Mechanical friction-reducing technique in which contact between spiral thread of screwjack and surrounding nut or runner is transmitted via no-gap stream of bearing balls with rolling contact only, which after reaching end of nut are returned via external tube to start.

Reclama Please reconsider proposed action or decision (DoD).

RECMF Radio and Electronic Components Manufacturers' Federation (UK).

recognition In imagery interpretation, determination of type or class of object without positive identification (ASCC).

recoil For automatic weapons in aircraft usual measures are average, peak and counter-*; all are reduced by fitting muzzle brake.

recon Abb, reconnaissance.

reconnaissance by fire Firing on suspected enemy to draw retaliatory fire.

recon pallet Multi-sensor pallet (1) installed in or under tactical aircraft in lieu of other load.

recon pod Pod housing reconnaissance sensors carried externally.

recon reference point Conspicuous geographic location from which reconnaissance objectives can be found.

recon slipper Fairing housing reconnaissance sensors carried flush with exterior skin.

record as target Code for listing target for reference or future engagement (DoD).

recorder Device translating data into hard-copy record on magnetic tape, wire, film, punched-paper tape or other medium.

recorder data package Integrated airborne systems providing record of subsystem performance in flight, eg arming and fuzing of Mirved warheads, for recovery and ground analysis.

recording accelerometer Counts and stores vertical accelerations.

recording storage tube Electronic tube which accepts CRT or other display picture, stores for period (eg 12 h) and reads out as often as required for analysis, monitor activation or various conversion processes.

recovered pressure That measured inside duct downstream of ram inlet.

recovery 1 Retrieval, in air or on surface, of part or whole of used RPV, target, test missile, spacecraft, instrument capsule or other inert body.
2 Retrieval of glider (sailplane) that has landed away from own airfield.
3 Return to base and safe landing (on land or aircraft carrier) of combat aircraft.
4 Return of combat aircraft from post-strike base to home base or designated recycle airfield.
5 Retrieval, normally by crane helicopter, of crashed or shot-down aircraft, in friendly or enemy territory.
6 Completion of flight manoeuvre and resumption of straight/level flight.
7 Conversion of kinetic to pressure (potential) energy in fluid flow (see *ram* *).

recovery airfield One at which aircraft might land post-H-hour but from which combat missions are not expected to be conducted (DoD).

recovery base Rear-area airfield used for maintenance and servicing to eliminate need for such services at airfields in combat zone (USAF).

recovery capsule Capsule containing reconnaissance pictures, instrumentation records or other data designed to separate from satellite, ICBM or other carrier and survive re-entry at preplanned location.

recovery footprint Area within which recovery capsule, returning manned spacecraft or other object is expected to fall.

recovery package Contains devices to assist recovery and retrieval of re-entry body, eg brightly coloured buoyant balloons, radio beacons and coloured pyrotechnics.

recovery temperature See *adiabatic recovery temperature.*

recovery time Time for gas-discharge tube to return to neutral under grid control.

rectangular co-ordinates See *cartesian.*

rectenna Rectifying aerial (antenna), especially of directionally beamed microwave power.

rectification 1 Process of projecting tilted or oblique photograph on to horizontal reference plane.
2 See *rectifying*.

rectified airspeed See *airspeed.*

rectified altitude Sextant altitude corrected only for inaccuracies in reference level (dip, coriolis) and reading (index, instrument and personal).

rectifier Static device exhibiting strongly asymmetric conduction properties such that it can convert AC to DC.

rectifying Elimination of errors which convert compass heading into track, eg deviation, variation and wind.

rectifying valve Thermionic valve which rectifies by virtue of unilateral conductivity of cathode/anode path.

rectilinear flight Straight and level, with 0 g imposed and thus in aeronautical sense called + 1 g.

rectilinear propagation Sent out in straight lines (ignoring relativistic effects).

rectilinear scanning TV raster (see *scan*).

recycle 1 To stop countdown and re-enter count at earlier point.
2 To return to start of EDP (1) program without entering fresh data.
3 To give completely new checkout to missile or other device (USAF).
4 To remove parts or complete module from engine, avionic device or other hardware and put through remanufacturing or inspection sequence to clear it for reuse with undiminished life.

recycle airfield One from which combat aircraft can be prepared for reuse away from home base.

RED Reconnaissance Engineering Directorate (US AFSC).

Red airway One running more or less east/west.

Redcap Real-time electromagnetic digitally controlled analyser processor.
redeployment airfield One not occupied in peacetime but available upon outbreak of war for use by units redeployed from peacetime locations; substantially same facilities as main airfield (DoD, NATO).
Red Flag Major tac-air exercises based on Nellis AFB, taking place several times per year over instrumented air/air and air/ground ranges under various "real war" scenarios.
Red Horse Rapid engineering deployment and heavy operations repair squadron, engineering (USAF).
red-line value One never to be exceeded.
red-on-red Traditional rule for setting up simple magnetic compass and avoiding mistakenly flying reciprocal.
red-out Loss of vision in powerful and sustained negative acceleration, ie where subject is restrained in seat only by harness over shoulders.
red pole North-seeking.
redrive Propeller drive incorporating speed-reducing gears or belt (homebuilts).
reduced frequency Ratio of product of frequency of oscillation and representative length of oscillating system to airspeed (nl/V); dimensionless parameter determining flutter amplitude.
reduced modulus of elasticity Theoretical value expressing relationship between modulus of elasticity and tangent modulus beyond limit of proportionality.
reducing mask Metal-sheet stamping enabling small instrument to be mounted in space on panel for larger one.
reducing valve Valve which reduces pressure in fluid system to precise lower value at * output.
reduction 1 Removal of oxygen or other electronegative atom or group; hence reducing flame.
2 Conversion of raw measured values of flight performance or met observations into standard forms, limiting values and, esp, graphical plots and derived values for comparative purposes.
reduction factor Large number, usually power of 10, used as divisor of all values in calculations to reduce size of whole numbers involved.
reduction gear Speed-reducing gear, ie output turns slower than input.
redundancy 1 Provision of two or more means of accomplishing task where one alone would suffice in absence of failure. In parallel * several usually similar systems all operate together so that failure of one either leaves remainder operative or, in majority-rule *, can always be overpowered by remainder. Standby * has primary system(s) operative and secondary or standby system(s) automatically switched in by malfunction-detection system.
2 Design of primary structure so that even after failure of any component there will remain enough load paths to carry all expected loads with adequate margin of safety.
3 In EDP (1) or information handling, amount by which logarithm of number of symbols available at source exceeds average information content per symbol.
redundant structure Basically, one possessing too many members; in practice one not amenable to simple stress analysis, eg because it has joints that are fixed instead of pinned or more than one member sharing load in way calling for elegant analytical solution. Not synonymous with fail-safe structure.
Redux Family of adhesives for most structural materials, including metals, normally applied in form of sheet, powder or liquid and cured (bonded) under heat and pressure (CIBA).
Redwood Together with * Admiralty, a traditional instrument for measuring kinematic viscosity; rare in aerospace.
Red zone Zone of intersection on Red airway (US) in which Red traffic maintains height and conflicting traffic procedure is published.
Reed & Prince screw Recess-head crosspoint screw driven by tool with single taper.
reed valve Leaf valve, usually of thin spring steel, giving unidirectional flow of fluid, eg in air-conditioning system or pulsejet.
reefed parachute One in which canopy is (usually temporarily) restrained against full deployment by encircling cord.
reefing sequence Systematic timed deployment of parachute, first in reefed condition and later to full deployment (eg Apollo CM recovery).
re-entrant angle Sudden change in direction of external surface (of aircraft skin, structural member etc) such that angle measured externally is less than 180°; in case of ** on surface of body, one causing local acceleration of airflow and, in supersonic flight, local attached shockwave.
re-entry Process of travelling from outer space into and through planetary (esp Earth's) atmosphere, in case of Earth proceeding down to planetary surface. Misnomer; should be entry.
re-entry body Body designed to survive extreme aerodynamic heating, high temperatures and large sustained deceleration of re-entry, eg ICBM RV, manned spacecraft, instrument or reconnaissance capsule.
re-entry plasma Plasma inevitably formed around all bodies arriving in Earth's atmosphere from outer space due to kinetic heating and ionization, forming barrier to radio signals.
re-entry system That portion of ballistic missile designed to place one or more RVs on terminal trajectories so as to arrive at selected targets. Includes penaids, spacers, deployment modules and associated programming, sensing and control devices (USAF).
re-entry vehicle That part of a space vehicle designed to re-enter Earth's atmosphere in terminal portion of its trajectory; can be manoeuvrable

**, or one of several multiple ** or multiple independently targeted ** (DoD).

refan To replace original fan or LP compressor of turbojet, bypass turbojet or turbofan with fan having larger diameter but fewer stages, for greater economy and less noise.

reference area Area used in components or coefficients of aerodynamic forces acting on a body.

reference axes Those in relation to which a body's attitude and motion are described; in case of aircraft they normally include body axes, all normal to each other and passing through c.g., called O (OX fore/aft, longitudinal; OY laterally, lateral or axis of pitch; OZ vertically, vertical or axis of yaw); wind axis (direction of free-stream relative wind); and principal inertia axis passing through c.g. in plane of symmetry and at small angle to OX.

reference datum 1 Arbitrary location at or beyond extremities of structure, eg aircraft, or on centreline or other major axis, from which all distances are measured and station numbers derived. Normally ** remains throughout life of aircraft even if stretching of fuselage or alteration of span of wing changes actual distances to established stations.
2 Imaginary vertical plane at or near nose of aircraft from which all horizontal distances are measured for balance purposes; called balance station zero (NATO).

reference eye position Typical position of pilot's eyes used in design of cockpit or flight deck.

reference fix Known position inserted into INS at start of flight.

reference fuel Piston-engine (Otto) fuel of known anti-knock rating used as reference to fuel whose octane or performance number is to be established.

reference humidity That specified for mandatory performance information; whichever is lesser of 70% RH to 33°C or 35 mb vapour pressure.

reference landing distance, RLD Resultant comparison landing distance (esp for civil transport aircraft) if different recommended landing technique is used.

reference line 1 Convenient line on surface used by observer, eg FAC, as line to which spots are related.
2 Single horizontal level showing correct operation by group(s) of systems or measured parameters indicated on vertical-tape instruments; thus a malfunction immediately stands out as a discontinuity.

reference meridian That selected to establish grid north or local time (ASCC).

reference phase Non-directional signal emitted by VOR having constant phase through 360° azimuth.

reference plane See *datum plane*.

reference plate Minimum-size plate cut from cheap material or scrap containing witness holes or slots cut from every tool in NC program and incorporating part of every sub-routine in program.

reference point Fixed datum near centre of airfield landing area (obs except general aviation).

reference pressure $\frac{1}{2}\rho V^2$ where ρ is fluid density and V is relative velocity.

reference section Traditional definition: a section of structure, displacements of which are taken as co-ordinates in a semi-rigid representation.

reference signal That against which telemetry data signals are compared to check differences in time, phase etc.

reference sound Commonly a random noise of one octave bandwidth centred on 1,000 Hz presented frontally.

reference speed V_{ref}, a reference speed used for comparative purposes in take-off and landing modes; commonly stall speed V_s or target threshold speed V_{AT} but needs to be defined whenever used.

reference zero Datum point from which horizontal and vertical distances are measured to each point on take-off net flightpath.

refire time Time required after initial launch to fire second missile (ICBM is assumed) from same silo or other launcher (see *rapid-reload*).

reflectance Ratio of reflected to incident radiant flux, symbol ρ.

reflected interrogation A misnomer, as reply is fresh signal (SSR).

reflected memory bus High-speed parallel data bus connecting all nodes or subsystems in a large EDP system.

reflected shockwave That reflected when a shockwave strikes a boundary between its original medium and one of greater density; part of energy generates shock which continues through denser medium but remainder is reflected in original medium, important case being in boundary between tunnel wall and tunnel working fluid.

reflection coefficient Measure of mismatch between two impedances; ** $a = (Z_1 - Z_2)/(Z_1 + Z_2)$ where Z_1, Z_2 are impedances.

reflection-interference waves Intermittent peaks in sea state caused by reinforcement of incident and reflected waves near cliff or sea wall.

reflection interval Time for radar pulse to reach target and return.

reflection-plane model Tunnel model comprising one half of model aircraft sliced down axis of symmetry which rests on floor of tunnel.

reflection suppression False echoes (ghost aircraft) in radars, esp SSR, are usually suppressed by a suppression transmitter feeding a separate Yagi fixed towards airspace which, at moment main interrogator scans reflecting surface, sends two pulses, first larger than second; this suppression pair take direct path to aircraft and suppress its transponder before arrival of interrogation signal via reflector; thus no spurious reply is received.

reflectometer Instrument for measuring transmission-line reflection coefficient.

reflector 1 Reflecting surface, usually copper gauze, so sited and shaped as to reflect radiation from primary radiator (eg of radar) in correct phase relationship to reinforce forward and reduce backward radiation; usually paraboloidal but flat in modern fighter radars.
2 Parasitic element located near primary radiator to reduce emission in all directions other than main lobe.
3 Repeller electrode of reflex klystron and similar tubes.
4 Material of high scattering cross-section surrounding core of nuclear reactor.

reflector sight Gunsight (rarely, for anti-tank missiles) in which reticle aiming mark(s) are projected as bright points on glass screen through which pilot or other aimer views target. Lead angle (aim-off) is assessed by gyro-electronics that measure rate of sightline spin and, in conjunction with range set by pilot, adjust aiming mark so that rounds should hit if reticle is superimposed on target.

reflex camber Shape of aerofoil in which mean line curves upwards towards trailing edge; eg to provide download well aft of c.g., increasing with airspeed, in tailless aircraft.

reflex ratio Measure of structural flexibility of propeller blade tested as cantilever beam (loaded in direction parallel to axis of rotation; US standard is that all blades of same propeller must exhibit tip movement uniform within 0·4 in [10 mm]). Test not applicable to certain blade types, eg hollow steel.

reflex sight ASCC term for reflector sight.

reflex trailing edge See *reflex camber*.

refraction Change in direction of travel of supposed linear radiation, eg EM radiation, sound, due to variation in properties (eg refractive index, air temperature) of transmitting medium. Can be gradual over a distance or instantaneous at boundary between two media; for radio/radar important forms are atmospheric * (low-altitude temperature, pressure, humidity; also responsible for errors in apparent altitude of celestial bodies), coastal * (change in propagation path at land/sea boundary) and ionospheric * (change of direction in passage through ionized layer).

refractive index Ratio of phase velocity of EM radiation in free space (vacuum) to phase velocity in medium considered; normally related to air, though in fact ** for air is not unity, common S/L value being 1·003. For radio normally refined to modified ** (n+h/a) where n is ** at height h and a is radius of Earth.

refractory Resistant to high temperatures; normally implies able to retain precise structural shape and bear appreciable stress at temperatures up to 2,200°C (4,000°F) without significant long-term change.

refresh rate Rate at which data on electronic display, eg radar, are resupplied in order to maintain bright picture free from flicker; early radar had no refresh and picture was generated only once on each scan; today data are updated at each revolution but are refreshed 100 to 200 times between updates.

refrigerant Working fluid pumped round closed circuit to extract heat, usually by repeated evaporation/condensation.

Refrigerant 12 Difluorodichloromethane (DDM).

refrigerant injection Water injection into gas turbine.

refrigeration capacity Common unit is "tons", normally a heat-extraction rate sufficient to convert one short ton of water at 0°C to ice every 24 h, equivalent to 3·517 kW = 4·715 hp.

refrigeration icing Caused by sudden drop in temperature of airflow, notably by depression in choke tube and/or evaporation of fuel.

refrigeration tunnel Wind tunnel used for icing tests.

regard Total solid angle of "vision" of a sensor trainable to point in different directions; FOV may be much less.

regeneration 1 Introduction of closed-circuit operation in gas-turbine plant in which by various methods part of exhaust heat is extracted and used to increase temperature of incoming airflow; hence regenerative gas turbine.
2 Increase in radio detector sensitivity by positive feedback.
3 Favourable heat transfer between rocket thrust chamber nozzle and propellant (see *regenerative cooling*).
4 In EDP (1) rewriting to prevent memory deterioration.

regenerative cooling Use of a cool incoming liquid, eg rocket engine propellant, to remove heat from hot hardware, eg rocket nozzle skirt and exit cone. Essential feature is that heat transfer is beneficial to both cooled item and coolant.

regime One defined mode of operation, clearly distinguished from other types of operation of same device; eg F-111 flight-suit pressure regulator has lo * maintaining 3 lb/sq in and hi * maintaining 6·5.

region See *flight information region*.

regional airport One serving a number of (usually modest-sized) communities.

regional carrier Civil operator whose route network covers only a minor part of a large country; often same as third-level.

regional QNH See *QNH*.

register Small array of bistable circuits storing one EDP (1) word.

registering balloon Small free balloon carrying recording met instruments; ballonsonde.

registration Entry of civil aircraft into records of national certification authority, with allocation of letter/number code displayed on aircraft and * certificate which in most countries must be displayed inside aircraft.

regression Precession of nodes; eg Moon completes revolution in 18·6 years.

regression rate Linear rate at which solid-propellant grain burns, measured normal to local surface; in 1970 typically 0·25 mm/s, in 1980 up to 5 mm/s.

regressive burning Solid-motor combustion in which burn surface area and hence chamber pressure and thrust all fall throughout period of burn.

regressive orbit See *retrograde*.

regroup airfield Military or civil airfield at which, post-H-hour, aircraft would reassemble for rearming, refuelling and resumption of armed alert, overseas deployment or further combat missions (DoD, IADB).

regular airfield One which may be listed in flight-plan as intended destination.

regular airport That at which scheduled service calls (UK usage).

regulated take-off weight See *WAT-limited*.

reheat See *afterburning*.

REI 1 Repair engineering instruction.
2 Reusable external insulation.

REIL Runway-end identification lights (FAA).

reinforced carbon/carbon Composite material comprising high-strength carbon fibres bonded in matrix of pyrolitic carbon; or can be thought of as pyrolitic carbon reinforced with CF.

reingestion stall Gas-turbine compressor stall induced by reingestion of hot gas during reverse-thrust mode.

Reins Radar-equipped inertial navigation system (Autonetics 1956).

Rejac Receiver/jammer capability.

reject To dump heat out of system into supposed sink, eg atmosphere.

rejected take-off One aborted after it has begun, ie between brakes-release and decision point.

rejector Inductance/capacitance in parallel to reject one resonant frequency.

relateral tell Relay of air-defence information between facilities via third; appropriate between automated centres in degraded communications environment.

relative altitude See *vertical separation*.

relative bearing Normally means bearing of surface feature or other aircraft relative to current heading.

relative-bearing indicator, RBI Shows bearing of tuned fixed station related to aircraft longitudinal axis; unlike RMI, does not show heading.

relative density Density at height in atmosphere related to that at S/L, ρ/ρ_0, symbol σ.

relative efficiency of biplane Ratio of wing loadings of upper and lower wings.

relative humidity Water content of unit volume of atmosphere expressed as percentage of saturation water content at same temperature.

relative inclinometer Flight instrument indicating attitude with respect to apparent gravity (resultant of gravity and applied acceleration).

relative scatter intensity Ratio of radiant intensity scattered in given direction to that in direction of incident beam; symbol $f(\phi)$, relative scattering function.

relative target altitude Vertical difference between altitudes of interceptor and target.

relative wind Velocity of free-stream air measured with respect to body in flight, in case of aircraft normally same as true airspeed but seldom exactly aligned with longitudinal axis.

relaxation time 1 Elapsed time between removal of disturbance and restoration of equilibrium conditions among molecules of fluid (eg air), operative parts of system or other dynamic components of system.
2 Time for exponentially decaying quantity to decrease in amplitude by $1/e = 0{\cdot}36788$.

relaxed static stability CCV-derived manoeuvre enhancement for air-combat fighter, involving rear c.g., longer nose, smaller fin etc, with artificial stability imparted by AFCS.

relay Device in which small control signal, usually electrical, is made not only to operate at distance but also to control large and possibly high-power devices; most serve switching functions or to protect devices against supply faults.

relay time Elapsed time between instant message is completely received and that when it is completely transmitted.

release Clearance of aircraft for line service, eg after overhaul.

release altitude Altitude of aircraft AGL at actual time of release of ordnance, tow target etc (DoD).

released 1 Of aircraft to unit, clearance for inventory service.
2 Of drawing, approval for transmission to manufacturing or production department.
3 Of air-defence unit, crews and/or weapons no longer needed at readiness; when * they will be informed when state of readiness will be resumed (NESN).

release point Point on ground directly above which first paratroop or cargo item is air-dropped (NATO).

reliability Probability that hardware will operate without failure for specified time; in practice also a result of operations already accomplished (see *despatch* *, *MTBF*).

reliability coefficient Percentage probability that aircraft can fly route on particular day with x% load factor and y% likelihood of diversion. Note: not a measure of hardware reliability.

relief hole Drilled in metal sheet to allow intersecting bends to be made to that point without buckling.

relief tube Personal urinal pipe normally discharging overboard.

relief valve Fluid system valve which releases pressure at preset value.

relight To restart combustion in gas turbine after mid-air flameout.

relight envelope Published diagram of permissible limits of TAS and height outside which engine relight should be attempted only in emergency. ** forms part of flight manual of civil gas-turbine aircraft, and for military aircraft is given for main combustion and afterburner.

relighting altitude That up to which safe and reliable restarting of power unit (gas turbine is implied) is possible (CAA).

reluctance Opposition to magnetic flux, property of magnetic circuit akin to resistance in electric circuit which limits value of flux for given MMF; numerically = l/μ A where l is length of magnetic path, μ is permeability and A cross-sectional area.

REM Rocket engine module.

rem Quantity of ionizing radiation which, when absorbed by body, produces same physiological effect as 1 roentgen of X-ray or gamma radiation; from roentgen-equivalent man.

remanence Flux density remaining (residual) in material after removal of magnetizing force. Also called retentivity.

remanufactured Aircraft completely stripped and inspected, usually by original builder, and with new structure added where necessary before reassembly and delivery with clearance for specified long life free from fatigue. Usually opportunity is taken to update systems and equipment also.

Rembass Remotely monitored battlefield-area surveillance (or sensor) system.

Remco Reference Materials Committee (ISO).

Remdeg Reliability military data exchange guide.

remote augmented lift system Arrangement for providing enhanced jet lift for V/STOL on demand in which large airflow bled from engine(s) is piped to auxiliary combustion chamber, usually near nose of aircraft, to provide additional high-energy lift jet. Latter may have means for modulation and limited vectoring, and RALS always requires large pilot-controlled diverter valve(s).

remote communications outlet Remotely controlled unmanned satellite air/ground com station providing UHF and VHF transmit-and-receive capability to extend range of FSS (FAA).

remote fan Mechanically independent fan driven by tip-drive turbine blades fed by hot gas from main engines via switch-in deflectors; used to provide jet lift for V/STOL or, rarely, to increase thrust at low speeds.

remote indicating compass Magnetic compass whose sensing element is installed in extremity of aircraft where deviation is minimal, with transmitter system serving repeater dial(s) facing crew.

remotely piloted vehicle, RPV Aerodyne usually of aeroplane type whose pilot does not fly with it but controls it from another aircraft or from station on surface. Authority of remote pilot is usually absolute, though to ease pilot workload some RPVs have choice of preselected autopilot/computer programs for at least part of each mission.

remoting Transmission of ATC, SSR or air-defence radar display by landline, microwave link or other means to distant centre.

removables All items flight crew must remove from outside of aircraft before take-off.

rendezvous Meeting of two aircraft or spacecraft at preplanned place and time. For aircraft can include air-refuelling hook-up but for spacecraft physical connection is docking.

rendezvous orbit insertion Establishment by one spacecraft of orbit almost identical to that of another before actual rendezvous.

rendezvous radar Esp in spacecraft, small ranging and range-rate radar carried to facilitate rendezvous and subsequent docking.

René alloys Family of American high-temperature alloys with Ni base plus Cr, Mo, Co, Ti, Fe, Al, and small amounts of C, Bo.

reneg To go back on previous agreement (US, colloq).

renegotiation Procedure not uncommon in US where manufacturer is required (usually by government) to renegotiate an existing active contract; usual cause is allegation of excessive profit.

renewal Procedure, usually annual, for inspection of civil aircraft for certification * and of civil pilot for * of certificate or licence. Hence * rate, total of fees payable, and * inspection, by authority-approved organization.

R/EO Radar/electro-optical.

REP 1 Reporting point (ICAO).
2 Reference eye position (or point).

repair-cycle aircraft Those in active inventory in or awaiting depot maintenance (DoD).

repairman certificate Issued by FAA to skilled tradesman engaged in repair or maintenance of aircraft or parts.

Repairnet Reconstitution post-attack interoperable radio network.

repeater jammer ECM receiver and transmitter which receives hostile signals and amplifies, multiplies and retransmits them for purposes of deception or jamming. Basic deception mode is to give false indication of range, azimuth or number of targets.

repetition rate 1 In radar, number of pulses per second.
2 In automatic gun, cyclic rate.

replaceable panel Aircraft maintenance access panel which has to be replaced; not an interchangeable panel, which can be opened in about one-tenth the time.

replacement factor Estimated percentage of hardware items that over given period will need replacing from all causes except accidents.

replenishing Refilling of aircraft with consumables such as fuel, oil, liquid oxygen and compressed gases to authorized pressure; rearming is excluded (NATO).

replenishing phase Part of operating cycle of pulsejet in which depression in duct induces fresh charge.

reply code That repeated series of pulses transmitted by SSR transponder in aircraft when interrogated; typically up to 12 information pulses between two framing pulses 20·3 μs apart.

reply-code evaluator Automatic avionic subsystem which reads reply code and determines if valid, what identity and in case of military IFF whether friendly or hostile.

reporting point Geographical point in relation to which aircraft position is reported.

repressurant Material, eg compressed air, oxygen and water vapour, stored in spacecraft outside pressurized volume and fed in to refill interior after depressurized operations.

req On request.

requalification To requalify approved system to meet more severe demand or environment.

request for proposals Document sent by central government to one or more industrial contractors outlining future requirement stated by armed force(s) and inviting suggestions on how this should best be met; RFP calls for analysis of problem by manufacturer and for submission of general scheme for hardware which in case of aircraft and most other equipment includes three-view drawing, basic description, estimated weights and performance, timescale and costs.

required flightpath That necessary to satisfy immediate task of pilot; not a recognized performance parameter but general objective weighed against projected flightpath.

required track Path aircraft commander wishes to follow; refers always to future intention. Often a ruled line joining two waypoints.

requirement Predicted future need spelt out by armed force, usually after long process of refinement and assessment of alternatives.

RES Radio emission surveillance.

RESA Runway-end safety area.

Rescap Rescue combat air patrol.

rescue co-ordination centre Initiates, manages and terminates rescue efforts within particular area by all branches offering help.

research and development Generalized term covering process of development of specific items of hardware; research involved is usually minimal and always applied, main effort being directed at solving engineering problems which are invariably unpredicted and occasionally require new fundamental knowledge; abb R&D, but US favours RDT&E, which is unnecessarily clumsy.

research coupling Disseminating research results to maximize early and widespread use.

research rocket Usually unguided ballistic rocket whose purpose is to lift scientific payload to high altitude for free-fall or parachute descent. Very occasionally purpose is to advance technology of rockets.

research vehicle Atmospheric or space vehicle, manned or otherwise, whose purpose is to provide answers to research problems; not normally prototype of production article but often associated with specific programme.

reseau Group of met stations operating under common direction; hence international *.

reserve aircraft Those accumulated in excess of immediate needs for active aircraft and retained in inventory against possible future needs (DoD).

reserve buoyancy Additional mass or applied force needed to immerse completely floats or hull of seaplane or flying boat already at specified (usually MTO) weight.

reserve factor Ratio of actual strength of structure to minimum required for a specified condition.

reserve fuel See *reserves*.

reserve parachute Second, standby parachute usually worn by professional parachutists and many others who make frequent deliberate descents; 7·3 m/24 ft ** mandatory for sport parachuting.

reserve power See *specific excess power*.

reserves Quantities of consumables, esp fuel, planned to be unconsumed when aircraft arrives at destination and available for holding (stacking), go-arounds, diversions and other contingencies.

reservoir Storage (not header) tank in fluid system, eg hydraulics.

reset To restore device, eg bistable gate, memory address or fire-warning system, to original untriggered state.

residence time 1 Time fuel droplet or gas particle remains in either gas-turbine combustor or afterburner.
2 Time, usually expressed as a halftime (not to be confused with half-life), radioactive material, eg fallout, remains in atmosphere after NW detonation.

residual magnetism See *remanence*.

residual propulsive force Net propulsive force minus total aircraft drag, F-D.

residuals 1 Any fluid left in spacecraft tanks after use.
2 Difference (plus/minus) between intended δV

(velocity increment) for a burn and that achieved (NASA).

residual stress That in structural component in absence of applied load, due to heat treatment, fabrication or other internal source.

residual thrust 1 That produced by jet engine (turbojet or turbofan) at flight or ground-idle setting.

2 That produced by any other propulsion engine after deliberate shutdown or cutoff.

resilience Measure of energy which must be expended in distorting material to elastic limit.

resin Vast profusion of natural and, increasingly, synthetic materials used as adhesives, fillers, binders and for insulation. Various types used in nearly all fibre-reinforced composites.

resistance derivatives Quantities, generally dimensionless, which express variation in forces and moments acting on aircraft after upset. In general case there are 18 translatory and 18 rotary.

resistance welding Using internal resistance of metal workpiece to very large electric current to produce heating which, under pressure, forms weld (see *spot weld*, *seam weld*).

resistivity Specific resistance; electrical resistance of unit length of material of unit cross-section. Convenient unit ohm/cm^3, though not strictly SI.

Resistojet Various patented (eg Avco, Marquardt) space thrusters using liquid ammonia or other working fluid, including biowastes, accelerated by solar-powered electrothermal chamber with de Laval nozzle.

resistor Electrical device offering accurately specified resistance.

resojet Usually means resonant pulsejet.

resolution 1 Measure of ability of optical system, radar, video/TV or other EO system, photographic film or other scene-reproducing method to reveal two closely spaced objects as separate bodies; normally defined in terms of angle at receiver subtended by two objects which can just be distinguished as separate.

2 Ability of device, as in (1), to render barely distinguishable pattern of black/white lines; expressed in number of lines per mm which can just be distinguished from flat grey tone.

Both (1) and (2) also called resolving power.

3 Measure of response of gyro to small change at input; max value of min input change that will cause detectable change in output for inputs greater than threshold, expressed as % of half input range.

4 Separation of vector quantity into vertical/horizontal components.

resolver Subsystem in spacecraft INS which measures changes of attitude, esp rotation about longitudinal axis, and informs guidance computers. In general a rotary digitizer converting small angular movements to digital signals.

resolving power 1 See *resolution (1, 2)*.

2 Ability of radar set or camera to form distinguishable images (ASCC); term unhelpful and indistinguishable in practice from resolution (1, 2).

3 Reciprocal of unidirectional aerial beamwidth measured in degrees (not synonymous with resolution [1], which is affected by other factors).

resonance 1 Condition in which oscillating system such as free wave or aircraft structure oscillates under forcing input at natural frequency, such that any change in frequency of impressed excitation causes decrease in response.

2 Condition of AC circuit when, at given frequency, inductive and capacitive reactances are equal.

3 In specif case of rotary-wing aircraft, particularly helicopter, condition in which natural frequency of landing gear corresponds with main-rotor rpm.

resonance test Structural exploration of natural frequencies by excitation over slowly varying wide range of frequencies.

resonant duct See *resonant pulsejet*.

resonant pulsejet Pulsejet in which intermittent operation occurs at natural frequency of operating air/fuel-burning duct system, without need for one-way flap valves.

resonating cavity Closed hollow space of precise geometry having electrically conductive walls in which microwaves are generated when excited by EM field or electron beam; examples are magnetron, klystron, rhumbatron; also called resonant cavity.

resonator 1 See *resonating cavity*.

2 Magnetostrictive ferromagnetic rod excited to respond at several distinct frequencies.

3 Lecher wire.

4 Piezoelectric crystal.

5 Acoustic enclosure having single-frequency response.

responder Receiving unit in transponder.

responsor Electronic device used to receive electronic challenge and display a reply thereto (DoD).

REST, Rest 1 Radar electronic scan technique; spherical/planar lens array aerial, RF angle-error sampling circuits, auto search/detection/confirmation, and display/recording.

2 Re-entry environmental and systems technology.

restart Start of burn after previous cutoff.

restart time Time between completion of adjustment of tunnel model and taking first readings.

rest-EVA period Scheduled period in manned space mission when in absence of assigned duties person may rest or conduct EVA.

restitution Determination of true planimetric position from reconnaissance photographs (NATO).

rest mass Mass of body when at rest (absolute in

cosmological terms); other masses $m = m_0 \sqrt{1-(v^2/c^2)}$ where m_0 is **, v is velocity and c is speed of light.

restoring couple Couple producing restoring moment.

restoring moment Moment generated by upset, ie rotary excursion from original or desired condition, which tends to restore original condition.

rest period Time on duty on ground during which flight crew is relieved of all duties.

restrained aircraft One undergoing dynamic structural test or analysis, eg flutter, with one or more parts anchored.

restraint 1 Standard series of tiedowns, webbing and nylon-cord nets to prevent movement of bulk cargo.

2 Process of binding, lashing and wedging items into one unit on to or into transport to ensure immobility during transit (DoD, NATO).

restricted airspace See *restricted area*.

restricted area 1 Airspace above surface area of published dimensions within which flight of aircraft is subject to restrictions caused, eg, by "unusual and often invisible hazards" such as AAA or SAM activity; in US published in FAR 73, in UK by CAA and in all Notams, charts and commercial publications.

2 Area where restrictions are in force to minimize interference between friendly forces.

restricted burning See *restricted propellant*.

restricted data Those pertaining to design, manufacture or use of NW, or special nuclear material.

restricted fire plan Safety measure for friendly aircraft which establishes airspace "reasonably safe from friendly surface-delivered non-nuclear fires" (DoD).

restricted propellant Solid-propellant grain whose surface is only partly available for ignition, remainder being protected by restrictor.

restrictor Layer of solid-propellant fuel containing no oxidant (oxidizer) or of non-combustible material bonded firmly to inner surface of grain to prevent that part being ignited except by flame travelling within propellant under *.

resultant Sum of two or more vectors. Hence * action, * force, * lift, * velocity.

ret NASA code for time between routine events.

retard 1 To cause PE (4) ignition to occur later in each cycle (normally well before TDC).

2 AT/SC mode in which throttles bleed off at programmed rate during landing flare.

retardation probe Tapering rod thrust into hydraulic aperture to offer increasingly great resistance to carrier arrester-wire pull-out.

retarded bomb Free-fall bomb with airbrakes, drogue, parachute or other high-drag device deployed automatically on release.

retarder parachute Small auxiliary parachute to pull main-parachute rigging lines out in advance of canopy.

retention area Highly loaded parts of turbine or compressor disc around blade roots.

reticle 1 Any kind of mark, such as black ring, illuminated cross or ring of bright diamonds (to give three examples) used to assist any form of optical aiming, eg aerial gunnery, spacecraft docking or airdrop on marker.

2 In photogrammetry, cross or system of lines in image plane of viewing apparatus used singly as reference mark in monocular instruments or in pair to form floating mark in certain stereoscopes (ASCC).

reticulated Having form of fine network; hence * plastic or * foam are 3-D volumes of low-density fire-resistant foam which can be foamed in place inside or outside fuel tanks and other items to prevent build-up of fuel/air mix and, even in presence of severe combat damage or post-crash rupture, prohibit explosion or swift spread of fire.

Retimet Patented (Dunlop) reticulated metal, low-density 3-D mesh of various metal strips or filaments.

retirement life Aggregate of running time of engine or other device at which decision is taken, usually by operator but sometimes suggested by manufacturer, that further overhauls are uneconomic; main reasons are obsolescence of design or onset of fatigue problems.

retrace American term for flyback.

retractable Capable of being withdrawn into aircraft so that it no longer protrudes, or protrudes only partially; applies to many devices which are * into all parts of aircraft, such as landing gear, inlet spikes, hooks, MAD gear, Fowler flaps, spoilers, sensor pods, radars and, formerly, gun turrets.

retraction lock Mechanical device to prevent inadvertent retraction of landing gear; today is removable but backed up by second lock actuated by compression of undercarriage oleos.

retread crew Military flight crew returned to OCU after tour of combat duty.

retreating blade That on side of a lifting rotor, eg on helicopter, moving relative to aircraft in same sense as slipstream; thus its airspeed is difference between its own speed and true airspeed, which is normally positive at tip, zero at a particular part-span radius and negative near root. Inboard of zero-airspeed radius, airflow is from trailing to leading edge.

retreating-blade stall Stall of retreating blade at high helicopter forward speeds, when angle of attack of retreating blade is excessive, especially towards tip; exceeding flight manual forward speed causes stall over near-rectangular area of disc whose centre is behind c.g. and thus effect is to cause nose-up roll towards retreating side.

retrieval Mid-air snatch of parachuted load, eg spacecraft.

retrieve Task of following sailplane on cross-country to goal, dismantling it and bringing it back in trailer.

retro Usually means retrofire or retrorocket (colloq, abb).

retrofire To fire retrorocket.

retrofit Modification, esp involving addition of new or improved equipment, to item already in service; hence * action, * mod.

retrograde 1 Orbit in direction different from normal; eg of a planet apparently moving westward against fixed stars.
2 Orbit in direction opposite to rotation of primary body, eg Triton around Neptune or Earth satellite launched at inclination from 180° through 270° to 360/000°.
3 In traffic direction opposite to normal, eg cargo of military logistic type moving towards United States or Soviet Union, or military command away from enemy.

retroreflection Reflection, eg of EM radiation, parallel to incident rays; hence retroreflector, device for accomplishing same, eg corner reflector, Luneberg lens.

retrorocket Rocket fitted to vehicle to oppose forward motion, eg to bring satellite out of orbit and back to Earth. Loosely used to mean what are more precisely called separation or staging motors or, on aircraft, braking rocket.

retrosequence Event sequence before, during and after retrofire.

retrothrust Thrust opposing motion; can be used for aircraft reverse thrust.

return Echo (radar), esp that due to clutter sources; often plural.

return-flow Combustor in which incoming air and issuing gas travel parallel in opposite directions; also called reverse-flow.

return grab Returns shuttle of catapult (deck accelerator) to start position.

return line 1 That traced on CRT by flyback.
2 Fluid pipe bringing fluid back from device to pump.

return load 1 That transmitted to aircraft by stopping forward motion of action in gun, in direction opposite to recoil.
2 Personnel and/or cargo to be transported by returning carrier (DoD); "carrier" means any vehicle, not aircraft carrier.

returns See *return*.

return to base Code: proceed to point indicated by displayed information which is being used as point from which aircraft can return to place where they can land; command heading, speed and altitude may be used (DoD).

revenue Air-carrier income from traffic sales.

revenue yield Rate per unit of traffic, eg cents per ton-mile.

reverberation Persistence of sound in enclosed space as result of continued multiple reflections, with or without continued emission by source.

reverberation time Time between cut-off of source and diminution of sound, measured as time-average of acoustic energy density, to fall to 10^{-6} of original.

reversal 1 Half a cycle of oscillating applied load, ie from max load in one direction to max load in opposite direction.
2 Control reversal (see *reversed controls*).

reversal speed Lowest EAS at which control reversal is manifest.

reversal temperature That at which characteristic spectral lines of incandescent gas disappear against black-body spectrum.

reversal zone Zone within ILS glideslope or course sector in which slope of sector characteristic curve is negative.

reverse bias That which reduces current.

reverse blindness Obscuration of flight-deck vision by snow or other material in reverse-thrust mode.

reverse breakdown voltage That at which reverse current across p/n junction increases rapidly with little increase in reverse voltage.

reversed controls Flight-control axis about which, in particular severe conditions, application of pilot input demand causes aircraft response in opposite sense, normally due to aeroelastic distortion of structure. Usual axis is roll, where under very high EAS (ideally not within limits of flight manual) large aileron deflection causes opposite twist of wing which more than neutralizes rolling moment due to aileron; in effect aileron acts as tab and wing as aileron.

reversed rolling moment That due to reversed control in rolling plane; also called roll reversal.

reverse-flow combustor One in which air enters at front, travels to rear-mounted fuel burners and then returns as hot gas within flame tube to leave radially inward from front; also called folded combustor, return-flow.

reverse-flow engine Gas turbine incorporating axial compressor which draws in air around rear end and compresses it in forwards direction, before turning flow radially outwards (often by added centrifugal stage) to flow back to rear through combustor(s).

reverse-flow region 1 Quasi-circular region near hub of helicopter main rotor disc, on retreating side, within which relative airflow is from trailing to leading edge, ie helicopter airspeed is greater than blade speed due to rotation.
2 Any region in turbulent boundary layer in which there is a majority-flow reversal.

reverse idle Power-lever setting at which engine is at idle (usually flight-idle because prior to touchdown on committed landing) with reverser buckets in reverse-thrust mode.

reversement See *reverse turn*.

reverse pitch Special ground-only setting available on some propellers and ducted propulsors, including several variable-pitch turbofans, in which blades accelerate air forwards, creating retrothrust proportional to engine power without change in direction of rotation.

reverser Device for deflecting some or all of efflux from jet engine to give reverse thrust (retrothrust); can take form of pivoting clamshell buckets, blocker doors and peripheral cascades, or other forms, and may be on turbofan core and/or on fan exit. Angle through which jet is turned seldom exceeds 135°.

reverse thrust Operating mode for jet engine equipped with reverser, obtainable only by overcoming gate or detent which may be locked until weight is on oleos for specified period, eg 1·5 s; normally obtained by moving power levers past idle down to ** mode, further movement in this direction opening throttle to full power to give max retrothrust.

reverse turn Opposite of Immelmann: half-roll followed by half loop. Also called reversement.

reversible propeller One in which reverse pitch may be selected.

reversing layer Thin lower part of Sun's atmosphere; cooler than photosphere and source of Fraunhofer lines.

reversion Change of operating mode (eg, but not exclusively, of flight-control system from normal powered to a degraded or manual mode).

reversionary facility Facility for changing operating mode, either automatically or upon human command, esp one following failure or degradation of existing channel or subsystem.

reversionary lane Back-up or standby channel.

reversionary mode Normally means advanced integrated flight system is available for pilot input of selected navigation mode from choice of several unrelated systems, eg INS, local R-Nav, Doppler, VOR/DME or Omega.

reverted rubber Rubber heated beyond critical point at which it loses basic mechanical properties, esp elasticity, and becomes sticky and permanently deformable. In one of the three aquaplaning modes, lack of anti-skid system causes locked wheel(s), reverted rubber in contact with runway covered in standing water, rapid steam generation and aquaplaning on steam layer above water.

revetment Area protected on three sides by blast-resistant wall of concrete, sandbags, compacted earth or other material, either to protect occupants and parked combat aircraft or other stores against external attack or to protect occupants, eg launch crew, against hazardous rocket or similar tests.

revival Restart of spacecraft systems after period of rest or shutdown, eg after long midcourse en route to planet when * commanded by characteristic telecom signal.

revolution 1 In engineering terms, one rotation of shaft or rotary system.
2 In spaceflight, motion of body about axis remote from itself, eg of planet around Sun.
3 One complete orbit starting and finishing at same point. In practice this is unattainable concept because of rotation of Earth, Earth's revolution (2) around Sun, Sun's motion through local galaxy, etc. Time for Earth-satellite revolution, as distinct from period (see *sidereal* and *synodic orbit*), is of meaning only in specif case where inclination is 090° or 270° along Equator when, because of Earth rotation, time is shortened or extended by 6 min over orbital period. At all other inclinations satellite follows fresh track on each orbit and never makes * in this sense.

revolver cannon One whose ammunition is fed to chamber via rotary cylinder driven by main action and in whose several chambers successive rounds pass through complete firing sequence, in one position being fired down a single fixed barrel. So-called "Gatling"-type guns do not come under this description but lack a proper descriptive terminology, such as "multiple rotating barrels" or MRB.

REW Radio-electronic warfare.

Reward Reporting working and reliability data.

Reynolds number Most important dimensionless coefficient used as indication of scale of fluid flow, and fundamental to all viscous fluids; $R = \rho Vl/\mu$ where ρ is density, V velocity, l a characteristic length (eg chord of wing) and μ viscosity $= Vl/\nu$ where ν is kinematic viscosity. Expression is ratio of inertia to viscous forces. It shows, eg, that for dimensional similarity model tests in tunnels should be run at pressures greater than atmospheric.

Reynolds stress 1 Shear stress in laminar boundary layer in viscous fluid (see *skin friction*).
2 Term(s) representing momentum transfer due to turbulence.

RF 1 Radio frequency.
2 Aircraft designation prefix, reconnaissance/fighter (DoD).
3 Regional forces.

RFA 1 Request for alteration.
2 Royal Fleet Auxiliary; ship usually with helicopter pad.

RFACA Royal Federation of Aero Clubs of Australia.

RF-ATE RF (1) auto test equipment.

RFC 1 Reinforced fibre composite.
2 Radio-frequency choke.
3 Royal Flying Corps (UK, 1912–18).
4 Request for change.
5 Retirement for cause (USAF).
6 Radio facilities chart.

RFDS Royal Flying Doctor Service of Australia.

RFF Research Flight Facility (NOAA).

RFFE RF (1) front end.

RFFP Rescue and Firefighting Panel (ICAO).

RFI, Rfi 1 RF (1) interference.
2 Request for information.

RFIS Receiver fire-control computer interface software.

RFL Restricted flammable liquid.
RFNA Red fuming nitric acid.
RFP 1 Request for proposals.
2 Request for procurement.
RFQ Request for quote (quotation).
RFR Request for revision.
RFS 1 Reserve flying school.
2 Restricted flammable solids.
RFS/ECM RF surveillance and ECM.
RF surveillance Maintaining continuous monitor on all hostile RF frequencies to record, analyse and interpret signals; also known as ESM.
RFT Ready for training (US).
RG 1 Retractable-gear.
2 Range (lights, ICAO).
3 Reconnaissance group.
RGB 1 Reduction-gearbox module.
2 Red/green/blue (systems ident and colour TV).
RGCS Review of the general concepts of separation (ATC).
RGF Range-gate filter.
RGO Royal Greenwich Observatory.
RGPI/PGPO (Ranrap) Random combination of RGPO and range-gate pull-in; ECM technique calling for predictive gate but effective against manual operator.
RGPO Range gate pull-off; basic ECM jamming technique usually used to pull hostile radar off target and thus provide infinite JSR for angle jamming; JSRR depends on many variables and is seldom effective technique when radar is manually controlled.
RGSI Radial groundspeed indicator (DME).
RGT Remote ground terminal.
rgt Right.
RGV Rotating guide vane.
RGWS Radar-guided weapon system.
RH, rh, r.h. 1 Right-hand.
2 Relative humidity.
3 Reheat.
RHAG Rotary hydraulic arresting gear.
RHAWS Radar homing and warning system.
rheostat Infinitely variable resistance for control of current.
rho Generalized term for radio-derived distance, usually a radial distance from fixed station.
rhombic aerial Short-wave directional aerial comprising two dipoles forming horizontal rhombus emitting travelling wave, thus having reflector at one end and non-inductive resistance at other.
rhombus wing One whose section is symmetrical double wedge, ie for supersonic flight only.
rho-rho Radio navaid giving distances from two fixed stations, thus a radio fix.
rho-rho-rho Radio navaid, eg Omega, giving simultaneous distances from three fixed stations.
rho-theta Radio navaid providing fix by one distance from fixed station and also bearing from that station; not normally written $\rho\theta$.
RHS Right-hand seat.
rhumbatron Common type of resonant cavity.
rhumb line Line drawn on Earth cutting all meridians at same angle.
rhumb-line course One flown at constant heading.
RHWR Radar homing and warning receiver.
RHWS Radar homing and warning system; basic part of ECM kit of penetrating aircraft; also written RHAWS.
RI 1 Radar interrogator.
2 Remote indicator.
3 Radio/inertial (often R/I).
RIA 1 Range instrumentation aircraft.
2 Rapid inertial alignment.
RIAP Revised instrument approach procedure (usually plural).
RIB Rigid inflatable boat (RAF).
rib 1 Primary structural member running across wing or other aerofoil essentially in chordwise direction; in highly swept wing axis may occasionally be aligned more with aircraft longitudinal axis but essential feature is that * joins leading and trailing edges and maintains correct section profile.
2 Light peripheral member not part of primary structure whose purpose is to maintain profile of aerofoil and support fabric or thin wood covering (see *compression* *, *nose* *).
ribbon microphone Comprises thin corrugated strip of aluminium alloy suspended between poles of permanent magnet; output is signals generated by strip vibrating perpendicular to field.
ribbon parachute One whose canopy is formed from rings (rarely, spiral) of ribbon, giving high porosity but reduced opening shock and good stability.
ribbon spray Water flung sideways by planing bottom at high speed; caused by first contact of hull or float with water and leaves at high speed at shallow angle; also called velocity spray.
riblet 1 Portion of rib, eg extending only from front spar to LE.
2 Carefully profiled microgroove, no larger than fine scratch, which, repeated millions of times to cover entire non-laminar part of aircraft skin, can reduce drag up to c 10 per cent.
RIC Reconnaissance interpretation centre.
rich Having excess of fuel (well above stoichiometric) for given flow of air or other oxidant. Hence * mixture.
Richardson effect See *thermionic emission*.
rich cut Sudden loss of PE (4) power caused by over-rich mixture, notably caused by flooding of float-chamber carburettor under negative g.
rich extinction Failure of combustion caused by excessively rich mixture.
RICS Rubber-impregnated chopped strands.
RIDE, Ride Radio communications intercept and D/F equipment.
ride control Automatically commanded aerody-

namic control system which reduces, and attempts to eliminate, vertical accelerations caused by flight through gusts, esp by penetrating aircraft at high (possibly transonic) speed at lo level. Typically includes sensitive g-sensors, computer and foreplanes (possibly augmented by forerudder or section of main rudder) to minimize vertical acceleration of crew compartment. In B-1 called LARC, later SMCS.

ridge Narrow extended portion of anticyclone or other high.

ridge girder Structural member forming part of stiff-jointed main transverse frame of airship, usually qualified as inner or outer and separated by main radial struts. Each ** links two longitudinals.

riding lights Those displayed by marine aircraft moored or at anchor.

RIDS, Rids Radio information distribution system (digital airborne CNI systems).

RIF Reduction in force (military).

RIG, Rig Rate integrating gyro.

rig 1 To adjust wing angular setting, wash-in/wash-out, dihedral, control-surface neutral positions and other aerodynamic shape determinants to obtain desired flight characteristics; normally applied only to light GA aircraft, in which it is possible to * by adjusting tensions of bracing wires and even alter shape of fuselage.
2 To prepare a load for airdrop (NATO).
3 Purpose-designed test installation for development of jet-lift V/STOL aircraft (in which case * may fly) or complete aircraft system, eg fuel, hydraulics, environmental, landing gear or propulsion. Usually full-scale and non-flying and often incorporating flight-quality hardware; eg fuel-system * can test entire aircraft fuel system in extreme attitudes and under abnormal environmental conditions.

rigging 1 See *rig (1)*; esp adjustment of flight-control system, even in modern powered system, so that all surfaces have exactly correct rest angles and system responses.
2 Complete system of wires, cables and cords by which aerostat (esp kite balloon or other moored type) is secured to main cable(s) or handling guys, and by which crew operate valves etc.
3 Equipment for dusting (dry *) or spraying (wet *) on ag-aircraft.

rigging angle of incidence See *angle of incidence*.

rigging band Strong tape band around kite balloon or other moored aerostat envelope to which all rigging (2) and payload are attached.

rigging lines Those connecting parachute canopy to load.

rigging patch Patch connecting rigging (2) to aerostat, in place of rigging band.

rigging position Aircraft attitude in which lateral axis and an arbitrary longitudinal axis (possibly actual longitudinal axis) are both horizontal. Esp relevant to tailwheel aeroplanes.

rigging tab Ground-adjustable tab.

right ascension Angular distance measured E from vernal equinox; arc of celestial equator, or angle at celestial pole, of given celestial body measured in hours (h) or degrees (°).

right-hand rotation Clockwise; in case of engine/propeller, seen from behind. No ruling exists for pusher engines, one school claiming that ** engine is unchanged when installed as pusher and another claiming that ** means seen from direction in which propeller is beyond engine.

right wind One blowing on aircraft from right, causing drift to left (suggest arch).

rigid airship One having rigid framework or envelope to maintain desired shape at all times.

rigid frame One having fixed joints, and thus statically indeterminate.

rigidity Property of gyro of maintaining axis pointing in fixed direction; also called gyroscopic inertia.

rigid rotor Rotor of helicopter (in theory also autogyro) whose blades cannot pivot with respect to hub except to change angle of pitch.

rill Deep, narrow depression across lunar surface.

Rilsa Resident integrated logistic support activity.

RIM US guided-missile designation prefix, ship-launched for aerial interception.

rime Icing type; rough, milky and opaque, formed by instantaneous freezing of supercooled water droplets. Definition of BS, under heading "frost": "ice of feathery nature on windward side of exposed objects when frost and fog occur together".

Rimpatt Read impatt diode.

RIN Royal Institute of Navigation (UK).

ring 1 Common structural part of most gas turbines.
2 Frame of fuselage of circular or oval section.

ring and bead sight Rudimentary gunsight in which aimer aligns target with bead foresight and correct part of ring backsight for required aim-off.

ring around Self-interrogation of beacon due to insufficient isolation between transmitter and receiver, or to triggering of airborne transponders at close range by unwanted sidelobes; in either case result is disastrous ring around display origin. Suppressed by ISLS.

ring counter Three or more bistable devices and gates so interconnected that only one is ON at one time; input pulses advance ON around ring.

ring cowling Simple ring, usually with concave inner profile (in longitudinal plane) and convex outer, surrounding radial engine (see *Townend ring*, from which stemmed long-chord NACA cowl).

Ringelman Basic method of assessing jet smoke by subjective opinion, in use since 1896 for matching smoke plumes against grey (gray) tones; hence * number, numbered scale from white to black. Said to be free from subjective variation but affected by angle, eg if four exhaust plumes are seen from below or side.

ring frame Main transverse structural frame(s) of rigid airship.

ring gauge Hand gauge for checking outside diameter of circular work, in some cases threaded; can be go/not-go type.

ring rolling Forming rings, eg for engine, by rolling on specially set-up machine which produces required radius.

ring-sail parachute One having construction based not on gores but on rings, usually arranged in form of upper disc and large skirt ring with separating gap. Unlike ribbon parachute, there is usually only one gap (sometimes two).

ring-slot parachute Family of parachutes with basic ring construction with concentric slots, merging into ribbon form.

ring spring Spring assembled from stack of steel rings with mating chamfered edges whose friction on compression reduces rebound.

Ringstone Ceramic or jewel pivot for instrument arbor (shaft).

Rint Radiation intelligence.

RIO Radar intercept officer.

riometer Relative ionospheric opacity meter.

RIP 1 Resin impregnation, or resin-impregnated plastics.
2 Remote instrument package.

ripcord 1 Cord which deploys parachute, pulled by wearer, static line or automatic (eg barostat control) system.
2 Cord, usually manually pulled, for tearing open aerostat rip panel.

rip panel Panel or patch on aerostat envelope, usually near top, which for emergency deflation can be ripped open: parachute ** right round, sealable in flight; velcro ** $\frac{3}{4}$ way round, not resealable in flight.

ripple 1 To fire large battery of rockets in timed sequence, interval typically 0·01 s.
2 Residual small alternating component superimposed on DC output.

ripster Carrier offering competitive fare (colloq).

rip-stop Structural design technique prohibiting unchecked growth of crack, eg by making part in parallel sections. Extends throughout airframe and also systems, eg to prevent single crack from affecting two hydraulic systems.

RIS 1 Range information system (ECM).
2 Radar information service.
3 Range instrumentation system.

riser 1 Quasi-vertical channel in casting mould either for admitting poured metal or for escape of air.
2 VTOL aircraft, esp aerodyne (colloq).

rise time 1 Time taken for a pulse, waveform or other repeated phenomenon to increase from a minimum to a maximum (usually measured from 10 to 90 per cent of peak value).
2 Time taken for flare to reach 90 per cent peak radiant power, usually measured from start of emission.

rising-sun magnetron One in which resonators for two frequencies are arranged alternately for mode separation.

Rita RF thruster assembly.

RITE Right, direction of turn (ICAO).

RIV Rapid-intervention vehicle.

rivet Essential feature is deformation of shank or surrounding collar to make permanent joint, but even this no longer true of some complex types, a few of which can be reused. Definition needed.

rivet gun Simple hand tool, usually pneumatic, with cup-ended striker which closes rivet by repeated blows.

rivet hammer Plain hand hammer with small flat steel face.

riveting machine Numerous forms of machine tool, most of which are fixed and through which work is fed; most close rivet between powered tool and anvil and some can drill, dimple/countersink and insert rivet, close it and then mill head.

rivet mandrel Rod passing through tubular rivet closed on withdrawal.

rivet set Hand tool having female shape which forms correct head as rivet is closed.

rivet snap See *rivet set*.

rivet squeeze(r) Hand tool which heads solid rivets by single quiet deformation, usually by mechanical leverage from operator.

Rivnut Patented (B.F. Goodrich) tubular blind rivet with threaded shank which, after closure, acts as nut (eg for panel fasteners).

RIW 1 Repairable in works.
2 Reliability interim (or improvement) warranty.

RJ Ramjet.

RJ-5 Conventional-type jet fuel for US expendable engines; high energy per unit volume.

RJT 1 Technical rejection message (ICAO).
2 Ground Self-Defence Force (J).

RKKA Ministry of heavy industry (USSR).

RL 1 Rhumb line.
2 Radioluminescence.

RLA Repair level analysis.

RLD 1 Rijksluchtvaartdienst (Netherlands certification authority).
2 Reference landing distance.

RLG 1 Ring laser gyro.
2 Relief landing ground.

RLM Reichsluftfahrtministerium (German air ministry to 1945).

RLS 1 Reservoir level sensor.
2 Raster-line structure.
3 Radius of landing site (lunar).

RM 1 Reference materials in calibration of R&D measures.
2 Radio maintenance.

3 Reflected memory.

RMA 1 Reliability, maintainability and availability.

2 Royal Mail Aircraft (followed by individual name of UK airliner, now obs).

R_{max} Maximum range, especially of weapon system.

RMCB Remote-control circuit-breaker.

RMDI Radio (or remote) magnetic direction indicator.

RME Rocket Motor Executive (UK, PERME).

RMetSoc Royal Meteorological Society; not RMS.

RMF Reconfigurable modular family.

RMI Remote (or radio) magnetic indicator.

RMM 1 Read-mostly memory.

2 Remote maintenance monitor (Wilcox).

RMOS Refractory MOS.

RMP 1 Root mean power.

2 Range mean pairs.

3 Reprogrammable microprocessor.

4 Radio management panel.

RMR Remote map-reader.

RMS 1 Root mean square.

2 Route management system.

3 Range measurement system.

4 Remote manipulator system (Shuttle).

5 Roof-mounted sight.

6 Reusable multipurpose spacecraft.

7 Rocket management system.

RMSE Root-mean-square error.

RMV Remotely manned (or managed, or manipulated) vehicle.

RMVP Reliability maintenance and validation program(me).

RN Recovery net (RPV).

RNARY Royal Navy Aircraft Repair Yard.

RNAS Royal Naval Air Station; shore airbase, also known by ship name (prefaced HMS), usually name of seabird. Previously (1912–18) Royal Naval Air Service.

RNAV, R-nav Area navigation.

RNDZ Rendezvous.

RNEFTS Royal Navy Elementary Flying Training Squadron.

RNF Radio navigation facilities.

RNG Radio range (ICAO, FAA).

RNGSA Royal Naval Gliding and Soaring Association.

RNII Reaction-engine scientific research institute (USSR).

rnwy Runway.

RNZAC Royal New Zealand Aero Club.

RO 1 Range-only-type sonobuoy.

2 Radio (or radar) operator.

3 Royal Ordnance (UK).

R/O Radio officer.

roach Wake of marine aircraft when deep in water; characterized by dense, almost vertical up-flow immediately behind float or hull.

ROAD Retired on active duty.

roadable aircraft One which, usually after some alteration, can be driven on public highway as a car.

Robinson anemometer Standard three or four-cup, with worm drive.

robot In addition to usual meaning, guided missile (Sweden).

robotics Use of intelligent machines to replace human beings in routine mechanical tasks.

ROC 1 Rate of climb.

2 Required operational capability.

3 Royal Observer Corps (UK, civilian NW/ fallout monitoring organization).

ROCC Region(al) (or range) Operations Control Center (Norad).

Roche's limit Critical radius within which planetary satellite cannot form; equal to 2·44 times radius of primary.

rock Rotary movement of wing(s) about fuselage, usually in plane perpendicular to longitudinal axis, caused by worn root anchor pins and various other faults (usually old GA aeroplanes).

rockair Small high-altitude sounding-rocket technique involving launch from aircraft, eg jet fighter, in vertical attitude at high altitude (probably arch).

rocker arm See-saw arm pivoted to PE (4) cylinder head transmitting push-rod actuation to valve(s). Hence rocker box, enclosing valve gear.

rocket 1 Propulsion system containing all ingredients needed to form its jet, or vehicle thus propelled. Most involve combustion, but examples which do not are monopropellants (eg HTP) and pressurized boiling water (eg Pohwaro).

2 See *rocket ammunition.*

rocket ammunition Rocket-powered projectiles of relatively small size fired from aircraft or other platform, mobile or stationary (USAF). Unguided, normally finned and/or spin-stabilized.

rocket artillery Artillery in which projectiles are propelled by rocket power but given guidance only during launch by amount of thrust and direction of take-off (USAF). This definition is meant to cover tube-type airborne launchers.

rocket-assisted take-off Take-off in which horizontal acceleration, and usually vertical component of thrust, are augmented by one or more rockets which may form part of aircraft or be jettisoned when spent. Ambiguously given US term JATO.

rocket/athodyd See *ram rocket.*

rocket cluster Large group of parallel rockets normally all fired together.

rocket engine Term best reserved for rocket using liquid propellants, esp with pump feed and control system.

rocket fuel Ambiguous; can mean fuel as distinct from oxidant or a solid propellant.

rocket grain See *grain.*

rocket igniter Device for starting combustion in any form of rocket; usually an electrically fired pyrotechnic, but many contrasting varieties.

rocket launcher Anything for launching rockets, specif aircraft pod, pylon or rail from which rocket ammunition is fired.

rocket loop Flight manoeuvre in which loop is made with excess initial speed, traded in loop for height.

rocket motor Term preferred for all solid-propellant rockets, of any size, together with hybrids of modest complexity.

rocket power See *rocket propulsion.*

rocket propellant See *propellant.*

rocket propulsion Propulsion of vehicle by rocket, excluding rocket drive of gyro wheels, turbo-pumps and other shaft-output devices.

rocketry Technology of rockets (colloq).

rocket sled Sled guided by straight track and propelled by rocket(s) to high (often supersonic) speed for use as test platform.

rocket thrust Measured either for sea-level or vacuum conditions, latter being about 26% greater.

rocking Uncommanded lateral and directional oscillation, chiefly in roll.

rockoon Sounding rocket fired from helium balloon.

rockover Nose-down rotation on landing, normally by tailwheel aircraft, in extreme case by 180°.

Rockwell number Measure of hardness, esp of metals, derived by impressing tool (for soft metals Type B, 1/16-in/1·5875 mm steel ball, result being R_B number; for very hard Type C, 120° diamond cone, result R_C number) first with 10 kg load and then with 90 or 140 kg, and measuring change in indent area.

Rococa Rate of change of cabin altitude.

ROD Rate of descent.

rod Photoreceptor in retina for scotopic (night) vision and detection of movement.

Rodar Helicopter rotor-blade radar (Ferranti).

ROE Rules of engagement.

ROEA Read-only, electrically alterable memory.

roentgen Unit of exposure to ionizing radiation, quantity producing 1 e.s.u. in 1 cm^3 dry air at 0°C; non-SI, $= 2{\cdot}58 \times 10^{-4}$ C/kg (C = coulombs).

roentgen equivalent man See *rem.*

ROF Royal Ordnance Factories (UK, various locations).

ROFOR Route forecast (ICAO etc).

Rogallo Originator of patented family of flexible-wing aircraft characterized by delta wing plan with three rigid members in form of arrowhead joined by flexible fabric which inflates upwards to arch under flight loads; originally paragliders with poor L/D, later developed to include powered aircraft.

Roger Voice code: "I have received and understood all of your last transmission".

rogue Aircraft which displays flight characteristics and/or performance markedly inferior to others of same type or, in particularly dangerous specimens, which is unpredictable.

Rohacell Low-density cellular plastics material of thermosetting type used to stabilize sandwich structures, usually with CFRP skins.

Rohrbond Patented metal adhesive bonding, especially for noise-absorbent sandwich.

ROI Return on investment.

role-change package Removable equipment fit (eg flight-refuelling probe installation) stored available for use when required; does not necessarily change role.

role equipment Equipment attached to or carried in aircraft for particular duties or mission(s) which can be removed subsequently; eg helicopter can offload ASW gear or anti-tank weapon/sight system and take on winch, buoyancy bags and furnishing for rescuees.

roll Rotation about longitudinal (OX) axis.

roll bar, roll cage See *roll pylon.*

roll cloud See *rotor cloud.*

roll coupling See *inertia coupling.*

roll damper Flight-control damper operating on ailerons or differential spoilers, either to preclude Dutch roll or because aircraft is difficult to hold wings-level in turbulence.

roll electrodes Freely rotating mating discs used for resistance-welding continuous seams or evenly spaced spots.

roller Landing by tailwheel aircraft with fuselage substantially level, eg because three-pointer is difficult or risks severe bounce, or because of gusty conditions.

roller-blind instrument Panel instrument whose display makes use, usually as backdrop, of roller blind (eg giving black/white sky/Earth indication of horizon). Not synonymous with tape instrument, which is analog linear scale.

roller cloud See *rotor cloud.*

roller drive Patented (TRW) reduction or step-up gear in which gear teeth and most bearings are eliminated, smooth preloaded rollers (Sun and two-step rings of planets) transmitting torque without lubrication.

roller-map display Nav display based on moving map (usually air topographic printed on film at reduced scale) projected by optics on circular screen on which heading appears as vertical central line (usually with present position as small ring near 6 o'clock position). In advanced forms, eg Ferranti Comed, combined with electronic displays and information readouts.

rolleron Flight control serving as primary for pitch or pitch/yaw and roll, esp in radial-winged missiles, which have four * to handle all manoeuvres.

rollgang Any of many makes of roller system for facilitating movement of cargo, especially in removable floor panels of standard size.

rolling balance Tunnel balance measuring forces and moments while model rolls about axis parallel to airflow.

rolling ball Compact two-axis human input to dynamic system, eg ATC or SSR radar, air-defence plot or computer display, in which partially recessed sphere drives two potentiometers giving rotary output proportional to rotation of ball about each of two perpendicular horizontal axes; output can be analog voltage or digital pulses. Smallest types fit on end of cockpit levers to command panel displays or HUD. In US often called track ball.

rolling engagement Air-combat training in which one aircraft (eg Aggressor role) takes over when predecessor is at bingo.

rolling instability Lateral instability; depending whether neutral or positive instability, an upset about OX will fail to be restored or become divergent.

rolling moment Component about longitudinal (OX) axis of couple due to relative airflow; measured positive if rotates right-wing down, negative if vice versa.

rolling plane Vertical plane normal to OX axis, ie containing line through both wingtips.

rolling radius Effective radius of landing wheels allowing for deflection (function of aircraft weight and tyre inflation pressure).

rolling tailplane Taileron.

rolling take-off Any take-off by helicopter, tilt-rotor or other VTO aircraft which, for any reason, accelerates forwards before leaving ground.

rolling vertical take-off Take-off with vectored nozzles aft for short distance (USAF 50 ft, in practice more helps), at which point nozzles are vectored to 60° or thereabouts for near-vertical ascent. Often shortened to rolling take-off.

roll-in point Point in space where aircraft enters diving attack.

roll jet Reaction-control jet for roll.

roll off the top See *Immelmann*.

roll on top Half loop followed by 360° roll followed by remaining half of loop.

rollout 1 Ground roll after landing, esp when continued to later turnoff to ease brake wear.
2 Termination of flight manoeuvre designed to place (normally combat-type) aircraft in optimal position for completion of intended activity (USAF).

rollout RVR Readout values from RVR equipment located nearest runway end (FAA).

rollover Apart from generalized meaning in business investment, replacement of airline line equipment by larger aircraft, in response to or anticipation of increased traffic.

Rollpin Press-fit pin made of roll of spring steel, forming self-retaining bearing pivot, axle or hinge-pin.

roll pylon Strong structure on ag-aircraft to give protection to pilot's head in event of aircraft becoming inverted on ground.

roll reversal Causes, apart from basic aeroelastic reversal of control, include jack stall, spoiler blowback, adverse yaw and loss of lift of accelerated wing due to increased compressibility.

Rollspring Patented (Lockheed Georgia) mechanical drive using belt of thin spring steel.

roll-stabilized Prevented, eg by autopilot or vertical gyro, from rolling; common characteristic of cruciform-winged missiles as an alternative to deliberate roll at known rate.

roll whiskers Short diagonal lines showing bank-angle limits (HUD symbology).

rolometer Extra-sensitive bolometer (ASCC).

ROM Read-only memory.

Romag Remote map generator.

RON 1 Receive only (ICAO).
2 Remain overnight (FAA).

Ronchi Modification of Schlieren technique to give quantitative results, with second slit removed and grid of parallel wires added to give light/dark shadow planes.

röntgen See *roentgen*.

roofline Generalized term for that level in passenger cabin at which lights, fresh-air (punkah) louvres, call buttons and drop-down oxygen are located. Hence * locker = overhead stowage bin.

rooftop Aerofoil profile which at typical cruise AOA generates lift more or less uniformly across large middle part of chord (chordwise plot has semi-elliptic form). Not precisely defined and generally synonymous with supercritical.

root Junction of aerofoil with supporting structure, including unspecified small innermost portion of aerofoil itself.

root mean square, RMS Square root of arithmetic mean of squares of all possible values of given function.

root-mean-square error Square root of arithmetic mean of squares of deviations from arithmetic mean of whole; standard deviation σ.

Roots One of many types of positive-displacement fluid compressors (eg superchargers, cabin blowers) in which two mating rotors are driven by external gears to revolve together inside casing; rotors in this case are identical and resemble fat figure-eights.

root section Profile of aerofoil at root; in both wings and propellers often significantly different from remainder of surface.

ROP Rotor overspeed protection.

ROPA Reserve Officers Personnel Act (ROPMA adds "Management") (US).

Ropar Regional operators' program for airframe reliability; co-operative effort by US carriers.

rope Element of chaff consisting of long roll of metallic foil or wire designed for broad low-frequency response.

rope chaff Chaff which contains one or more rope elements (DoD).

ROPW Read-only, permanently woven memory.

ROR 1 Range-only radar.

2 Rocket on rotor, helicopter tip-drive method.
3 Red on red.

Ro/Ro Roll-on roll-off, ship configured for transport of vehicles driven on or off under own power. See *T-AKF*.

Rorsat Radar ocean-reconnaissance satellite.

ROS 1 Rotor outside stator (electrical machines).
2 Read-only storage.

rose Polar plot, normally subdivided from 000° to 360° clockwise but often including only quadrantal, cardinal and possibly other points and occasionally only radial lines with variable angular intervals. Term applies to any such diagram, eg in panel instrument (ADF, RMI, HSI etc), compass or wind *.

rosette 1 Standard series of patterns, generally resembling petals of flower, for applying strain-gauges to signal stress in all directions, usually in sheet.
2 Similar-shaped hand weld made in hole through sheet or tube to secure second member inside.

Rossby chart Basic diagram assisting in establishing character of air masses and in met forecasting generally; energy diagram in which specific humidity is plotted against equivalent potential temperature, result being named characteristic curve.

Rotachute Patented (Hafner/ML Aviation) free-running personal rotor used WW2 instead of parachute for accurate paratroop descent.

rotaglider Generalized term for gliders with free-running rotary wing in place of fixed wing.

rotaplane Flying machine whose support in flight is derived from reaction of air on one or more rotors which normally rotate freely on substantially vertical axes (BSI); today called autogyro, making this term redundant.

rotary bombdoor Door to aircraft internal weapon bay made as single stress-bearing beam pivoted at extremities and itself loaded with ordnance or, alternatively, fuel tank or reconnaissance sensors; to drop ordnance door is rotated 180°, thereafter normally reversing direction to close.

rotary derivatives Stability derivatives expressing variation in forces and moments resulting from changes in aircraft rate of rotation about all axes.

rotary engine 1 PE (4) (usually Otto but occasionally two-stroke) with radial cylinders which drive themselves round a fixed crankshaft; propeller is thus attached to crankcase.
2 Today, an RC (1) engine.

rotary scan See *scan*.

rotary shears Hand or power tool used for slitting sheet along straight lines or curves of variable radius.

rotary-wing aircraft See *rotorcraft*.

rotate Voice command, usually P2 to P1, at V_R.

rotating beacon 1 Bright aircraft-mounted light steady or flashing while rotating continuously in azimuth, usually synonymous with strobe; today tends to be illuminated whenever engine run on ground.
2 Ground transmitter having directional radiation pattern rotated continuously at predetermined rate (BSI); suggested arch term overtaken by more precise names, eg VOR.

rotating-combustion engine IC engine of positive-displacement type in which main moving parts are not reciprocating but rotary; usually based on Otto cycle. According to best-known worker in this field, Felix Wankel, 864 possible configurations divided into single-rotation machines (SIM), planetary-rotation machines (PLM), SIM-type rotating-piston machines (SROM) and PLM-type rotating-piston machines (PROM). Unfortunately acronyms confuse with equally important EDP (1) meaning of PROM etc.

rotating-cylinder flap High-lift wing flap in which circulation is assisted by high-speed rotary cylinder mounted along flap leading edge (Coanda and Flettner relevant), eg in NASA YOV-10A.

rotating guide vane Curved inner leading edge of centrifugal (rarely, other types) of compressor or impeller rotor, main purpose of which is to smooth change of direction of flow from axial to radial.

rotating-wing aircraft See *rotorcraft*.

rotation Positive, ie nose-up, rotation of aeroplane about lateral (pitch) axis immediately before becoming airborne; in transports commanded at V_R.

ROTC Reserve Officers' Training Corps (USAF).

ROTHR Relocatable over-the-horizon radar.

rotochute Free-rotor airbrake fitted to certain air-dropped sonobuoys.

rotodome Slowly rotating radome of discus shape used to fair in antenna of AWACS-type aircraft, usually housing main and IFF/link antennas across diameter, back-to-back.

rotor 1 System of rotating aerofoils (ASCC); this includes propeller, so should be qualified by adding "whose primary purpose is lift".
2 Main rotating part of machine, eg gas turbine engine, turbopump or alternator.
3 Local air mass rotating about substantially horizontal axis.

rotor cloud Unusual cloud usually of Ac type and often dangerously turbulent; found in rotor flow, normally in lee of mountain range or ridge. Also called roll cloud.

rotorcraft Aerodyne deriving lift from rotor(s).

rotor damping Damping of blades about any pivot axis in helicopter main rotor (see *lag-plane*, *soft in plane*).

rotor disc 1 Circular area swept by blades of helicopter or autogyro rotor.
2 Structural disc holding compressor or turbine rotor blades.

rotor flow Large-scale rotary flow of atmosphere about substantially horizontal axis in lee of mountain or sharp ridge, which when wind very strong is turbulent to point of being dangerous, with vertical gusts of 30 m (100 ft)/s (see *rotor streaming*).
rotor force Resultant imparted by lifting rotor, analysed into lift and drag, or thrust (perpendicular to disc) and in-plane (H-) force.
rotor governing mode Control mode in which rotor (1) speed is held constant.
rotor head Complete assembly of rotating components at centre of lifting rotor of helicopter or autogyro, including hub, blade attachments and pivot bearings, if any, and complete control mechanism, as well as such adjuncts as anti-icing, electrics and pressurization instrument leads rotating with hub.
rotor hub Primary structure connecting blades of helicopter or autogyro rotor. Some authorities define hub to mean same as head, which is unhelpful.
rotor incidence Angle between plane normal to axis of rotation and relative wind (BSI).
rotor kite Towed engineless autogyro.
rotor mast Pylon carrying rotor in small rotorcraft.
rotorplane See *rotorcraft*.
rotor streaming Shedding of turbulent rotors (3) downstream, often near ground and very dangerous to aircraft.
rotor torque That imparted to airframe by rotor, esp main lifting rotor of hub-driven type, which has to be countered by tail rotor or use of two main rotors.
ROTR Rate of temperature rise.
Rotte Fighter aircraft in loose pair (G).
rough-air speed Recommended speed for flight in turbulence, V_{RA} (hence rough-air Mach, M_{RA}); lies between V_A and V_C and usually near V_B.
rough field Defined by standard categories of surface profile, including post-attack damage repair: A, single blister up to 1·5 in H (height) over aircraft-travel distance of 80 ft; B, H up to 3 in; C/D, two 3 in blisters within 80 ft; E, two up to 4·5 in; 2E, two up to 9 in.
roughness Criterion affecting surface finish: irregularities that are closely spaced (see next entry).
roughness factor Rayleigh criterion $\Delta_h = \lambda/8 \sin \psi$ where λ wavelength and ψ angle of incidence.
roughness width cutoff Max width of surface irregularities to be included in measure of surface height; in Imp measure usually 0·03 in, but occasionally 0·003, 0·010 or 0·10.
rouleur Ground trainer with wing too small for flight (F).
round Single munition, missile or device to be loaded on or in delivery platform (USAF).
round-down Rear terminator of aircraft carrier flight deck, usually normal to landing direction (term derives from older ships where deck fell away over stern in large-radius curve).
roundel National marking for military aircraft in form of concentric rings.
round-head rivet Usual rivet for thin sheet where countersinking not required; head OD larger than button-head.
route 1 Defined path, consisting of one or more courses, which aircraft traverses in horizontal plane over surface of Earth (FAA).
2 Published route linking traffic points of air carrier and used in traffic and rights negotiations.
3 Path followed by channel of AFTN network.
route flight 1 One from A to B.
2 Military mission, usually transport, over established route.
route package Geographical division of enemy landmass for purposes of air-strike targeting.
route sector Route (1) between two traffic points.
route segment Route (1) between two waypoints.
route speed En route speed, usually synonymous with block speed.
router Machine tool having high-speed cutter rotating about vertical axis capable of being positioned anywhere over horizontal work; cutter can be side or end-cutting and in large machines may be NC. (Rhymes with doubter.)
routine 1 Series of step-by-step EDP (1) instructions forming part of program; hence portions of same are subroutines.
2 Complete sequence of aerobatic manoeuvres planned by competitor.
routing Itinerary followed by message in AFTN.
routing list That showing AFTN centre which outgoing circuit to use for each addressee.
roving Traditional meaning was slightly twisted hank of textile fibre; modern meaning in aerospace is continuous fibre (eg carbon, graphite, boron, glass) for filament winding, or continuous raw material for ECM chaff to be cut to required response length in dispenser.
row Group of cylinders of radial engine in one plane, all driving one crankpin (strictly, not a row but planar array); hence two-* engine has two planes one behind other.
Roydazide Hot-isostatic-pressed sinide.
RP 1 Reporting post.
2 Rocket projectile (WW2).
3 Rocket propellant (fuel designation prefix).
4 Route package.
5 Rapid prototyping.
R/P 1 Rocket projectile (WW2 usage).
2 Receiver processor.
RP-1 1 Common kerosene fuel for rockets and ramjets.
2 One-minute racetrack pattern for holding.
RPA Remotely piloted aircraft.
RPB Retarded practice bomb.
RPCC Rotating-parts cycle count.
RPD Rapid inertial alignment.
RPDL Receiver pilot director light.

RPE Rocket Propulsion Establishment (Westcott, UK).
RPF Reticulated-plastics foam.
RPFT Rapid pucker factor take-off.
RPG Rounds per gun.
RPGT Radar procedure ground trainer.
RPH Remotely piloted helicopter.
RPK 1 Revenue passenger-kilometres.
2 Fragmentation bomblet (as well as a series of Kalashnikov LMGs) (USSR).
RPL 1 Radiophotoluminescence.
2 Rated power level (electronic).
3 Runway planning length.
RPM 1 Revolutions per minute (often rpm).
2 Revenue passenger-miles.
3 Rounds per minute (often rds/min or spm).
4 Reliability prediction manual.
5 Research and program(me) management.
6 Remotely piloted munition.
RPMB Remotely piloted mini-blimp.
RPMD Repeater projected map display.
RPO 1 Rotorcraft program office.
2 Resident project officer.
RPOADS Remotely piloted observation aircraft designator system (USA).
RPOW Relative power, one-way (SSR and other radars).
RPRV Remotely piloted research vehicle.
RPS Radar position symbol.
RPSTL Repair parts and special tool list.
RPT, Rpt 1 Repeat, or I repeat.
2 Revenue passenger transport.
R/Pt Reporting point.
RPV Remotely piloted vehicle, term normally confined to fixed-wing aerodynes.
R/P/Y Roll, pitch and yaw.
RQ Preface: request (ICAO).
RQS Request, eg supplementary flightplan.
RR 1 LF or MF radio range (FAA).
2 Rendezvous radar.
R/R Remove and replace.
RRAB Cluster-dispensed incendiary (USSR).
RRC Regional radar centre.
RRCC Rotorcraft Requirements Co-ordinating Committee (CAA).
RRCM Rudder-ratio control module.
RRE Royal Radar Establishment, now RSRE.
RRF Ready Reserve Force (US).
RRP 1 Runway reference point.
2 Rapid-reinforcement plan.
RRPS Ready reinforcement personnel section.
RRR, R^3 1 Rapid runway repair(s).
2 Reduced residual radiation.
RRTC Roll-response time constant; subjective assessment from 0·1–10 s of apparent lag in response to large roll demand suddenly applied.
RS 1 Role prefix, reconnaissance strike (US).
2 Reserve squadron (RFC).
3 Re-entry system.
RSA Réseau du Sport de l'Air (F).
RSAF Royal Small Arms Factories (UK).
RSB Recovery speed brake.
RSC 1 Rescue sub-centre (ICAO).
2 Radar-scattering camouflage.
RSCU Ramp spill control unit.
RSI 1 Reusable surface insulation.
2 Rationalization, standardization, interoperability (NATO).
RSL Range safety launch (RAF).
RSM Runway surface and markings.
RSO Reconnaissance systems officer (or operator).
RSP 1 Radar start point.
2 Responder beacon (ICAO).
3 Radar signal processor.
RSPD Rapid-solidification plasma deposition.
RSR 1 En route surveillance radar.
2 Rapid solidification rate (alloys).
RSRA Rotor-systems research aircraft.
RSRE Royal Signals and Radar Establishment (Malvern, UK).
RSRS Radio and Space Research Station (Slough, UK, now Appleton Lab).
RSS 1 Relaxed static stability (longitudinal plane, air-combat fighters).
2 Reflection suppressor system (SSR).
RSSK Rigid seat survival kit, opened during descent.
RSSN Reaction-sintered silicon nitride.
RST 1 Recording storage tube.
2 Reheat specific thrust.
RSTA Reconnaissance, surveillance and target acquisition.
RSTR Restricted.
RSU Runway supervisory unit.
RSV 1 Reserve fuel.
2 Reparto Sperimentale di Volo (I).
RSVP 1 Restartable solid variable-pulse (rocket).
2 Rotating surveillance vehicle platform.
RT 1 Real time; hence numerous other acronyms, third words: control, display, environment, interface, language, management, processor, system etc.
2 Replaceable tile(s), ie RSI (1).
3 Reaction time.
4 Remote terminal.
5 Resistance training.
R/T Radio telephone.
RTA 1 Rotation target altitude.
2 Receiver/transmitter antenna.
RTB Returned (or return) to base.
RTCA Radio Technical Commission for Aeronautics (US).
RTD Radar target designator control.
RTDT Real-time data transfer.
RTF Radiotelephone (ICAO).
RTG 1 Radiotelegraph (ICAO).
2 Radioisotope thermoelectric generator.
R-theta Radio navaid family giving distance (R) and bearing (theta, θ) from fixed stations.
RTIRL Real-time IR linescan.

RTK Revenue tonne-kilometres.
RTLS Return to launch site.
RTO 1 Rejected take-off.
2 Resident technical officer, of government in plant (UK).
3 Responsible test organization.
4 Range training officer (ACMI).
RTOG Regulated take-off graph, plot on which actual weight is assessed against T°, wind etc.
RTOL Reduced (or, sometimes, restricted) take-off and landing, conventional transport with field length from 900 m (3,000 ft) to 1,500 m (5,000 ft).
RTOR Right turn on red.
RTOW Regulated take-off weight.
RTP 1 Reporting and turn point.
2 Routine technical publication.
RTPA Room-temperature parametric amplifier.
RTR 1 Real-time reconnaissance.
2 Reserve thrust rating.
RTRCDS Real-time reconnaissance cockpit display system.
RTS 1 Radar target simulator.
2 Replacement training squadron.
3 Radar test subsystem.
4 Remote tracking station.
5 Returned to service (ICAO).
RTT 1 Reserve take-off thrust.
2 Radioteletypewriter (ICAO).
3 Radio telephone terminal.
RTTL Range target-towing launch (RAF).
RTTS Robotic target training system.
RTTY, rtty Radioteletype.
RTU Replacement training unit (USAF).
RTY Raw total yield (megatons).
rubber aircraft Aircraft in early project stage when gross variation in design is still possible (colloq).
rubber-base propellant Solid rocket propellant in which fuel is related to synthetic-rubber latex (wide range of resins, plasticizers and other additives), mixed with oxidant often of AP or other perchlorate type, finally cured into rubber-like grain. Examples include PBAA, PBAN.
rubber boot See *de-icer boot*.
rubberdraulics Rubber presswork technology, especially when flexible medium flows like liquid.
rubber press Press for forming sheet-metal parts by forcing thick pad of rubber down under high pressure on to cut-out flat parts (or, if blanking strips added, uncut sheet) arranged above male dies against which parts are forced with minimal wrinkling, rubber acting sideways as well as downwards.
rubber slick Stripe of rubber melted off tyre at touchdown and adhering to runway. Resulting blackened region shows touchdown zone preferred by actual traffic.
rubbing seal Gastight seal between fixed and moving members actually in contact; usually at least one mating face is carbon.
rubbing strip Numerous non-structural parts whose purpose is to accept impacts from doors, ground equipment (eg steps and vehicles) and abrasion by inlet blanking plates or rescue winch cable.
rub rail Mount and launch rail for missile shipped and launched from canister; sometimes four, each locating a wingtip.
rudder Primary control surface in yaw; when nose-mounted, prefaced by nose-. Term also includes fixed fin of kite balloon, usually ventral, providing weathercock stability.
rudder bar Centre-pivoted bar providing pilot rudder input in simplest ultralights and some other aircraft, Traditional term for rudder input even when pedals fitted.
rudder-bias strut Simple engine-out device comprising piston in rudder circuit with engine-bleed air piped to each side; failure of either engine causes immediate application of rudder.
rudder pedal Left/right pedals for pilot's feet acting as manual input to rudder.
rudder post Traditional (suggested arch) term for leading-edge member of rudder carrying hinges; with modern inset hinges place taken by internal spar.
rudder reversal Roll reversal using rudder only, usually in maximum-performance high-alpha air-combat manoeuvres.
rudder roll Unwanted roll produced solely by coarse use of rudder.
rudder torque Twisting moment exerted by rudder on rear fuselage.
ruddervator Movable flight-control surfaces of butterfly tail, combining duties of rudder and elevator. Sometimes "ruddevator".
ruling material That used for most of airframe structure.
rumble Rocket combustion instability audibly obvious from low growl or rumble.
rumble seat Occasional seat, eg for flight-deck observer or for stewardess on landing/take-off.
run 1 Ground or distance traversed by wheels, or water by floats/hull, on take-off and landing.
2 Number of production articles built to common type, or elapsed time to produce same.
3 Single flight over target (also called pass) for assessment, release of ordnance or operation of reconnaissance sensors.
4 Single flight past designated point(s), eg in attempt on speed record.
runaway Undesired operation of device, eg PFCU, when not commanded; in dangerous extreme case continuing to limit of travel, giving hard-over condition.
rundown 1 Fall off to zero rpm after flameout or engine failure (also called spool-down).
2 Decay in production rate due to falling demand or imminent termination.
run-flat tyre Various aircraft tyres, pioneered by

Goodrich, designed for high-speed landings after deflation caused by take-off blowout, fire or combat damage.

running fit Slight clearance between mating parts allowing rotation or other relative motion.

running fix Approximate fix obtained by taking a bearing of fixed station, or in any other way obtaining PL, then obtaining second PL and adjusting to common time.

running in Act of running newly built or completely overhauled engine or other machine to ensure parts run together under controlled gentle operating conditions.

running landing 1 Helicopter landing made into wind with groundspeed and/or translational lift at touchdown; with skid gear demands careful collective after touchdown to avoid abrupt stop.
2 Jet V/STOL landing with significant forward speed.

running mate New transport aircraft type, usually smaller, to accompany trunk-route type already in service and offering same advanced-technology appeal.

running order Traditional (poss arch) condition for measuring mass of engine: including radiator, coolant, internal oil, external pipes and controls, but excluding fuel, oil, tanks, reserve coolant, exhaust tailpipes and instruments.

running rigging Rigging for kite balloon or other aerial object which by system of vee-lines and pulleys automatically adjusts to direction of pull.

running take-off Started without lineup and hold, speed never slackening on arrival at runway in use.

run-out 1 Distance travelled by carrier aircraft between engaging hook and coming to stop, also called pull-out.
2 Distance travelled by gun or barrel, on recoil stroke.

run-up 1 To accelerate engine under own power.
2 Portion of flight immediately preceding target run.
3 To test piston engine, briefly at high power and to check dual ignition, before take-off.

run-up area Portion of airfield near taxiway designated for run-up (3).

run-up drag 1 Drag caused by windmilling propeller in air-start.
2 Drag caused by need to accelerate landing wheels on touchdown.

runway Paved surface, usually rectangular and of defined extent, available and suitable for aeroplane take-off and landing. Equipped * includes stopway, clearway, surface markings and designators. All-weather * includes lighting. Instrument * adds electronic aid, eg ILS, MLS. Unpaved * = airstrip.

runway alignment Direction of runway centreline, published as first two digits, in both directions, eg 13/31; also called direction number.

runway alignment factor Maximum angular departure from alignment admissible for straight-in approach, normally 30°.

runway alignment indicator Group of flush lights offering directional guidance on take-off (to some degree on landing) in bad visibility.

runway basic length That length selected for aerodrome planning purposes required for TO or landing in ISA for zero wind, elevation and slope.

runway capacity Frequency of landings and/or take-offs, or mixture, possible or permissible with minimal approach spacing, published for IFR and VFR. In complete airport varies greatly with number and arrangement of runways, and conflicts of crossing traffic.

runway controller ATC controller stationed, usually in caravan with checkerboard markings, close beside downwind end of runway in use; armed with signal pistol, Aldis and telephone to tower (rare since WW2).

runway designator Numerical alignment plus qualifier left/right if necessary; thus New York JFK offers 4R/22L, 4L/22R, 13R/31L and 13L/31R.

runway direction numbers Numerical values of alignment.

runway-edge lights White lights grouped according to intensity: fixed-intensity LIRL (low-intensity runway lights) and variable MIRL and HIRL. Designed to withstand flying stones.

runway end End of runway in use, identified by markings; beyond may be similar-surfaced blast pad, overrun (RESA) or stopway and (unpaved) clearway.

runway-end lights, REIL Often pair of flashing white lights, at major airport continuous transverse row of bidirectional lights showing green towards approach and red towards runway.

runway-end safety area UK term for overrun area; area adjoining runway in use, symmetrical about extended centreline, intended to minimize damage to undershoot or overrun aircraft.

runway floodlight Appears incorrectly in existing dictionaries; does not exist.

runway gradient Not used; correct word is slope.

runway localizer See *localizer*.

runway occupancy Elapsed time particular aircraft is on runway, on arrival or departure.

runway separation Time and distance intervals between arrivals and/or departures on one runway. Distance between parallel runways is spacing.

runway spacing Perpendicular distance between centrelines of parallel runways at same airport.

runway strip Defined area including runway and stopway intended to reduce risk of damage to aircraft that run off runway in any direction and to protect aircraft flying over it during take-off or landing.

runway visibility value, RVV Determined for particular runway by transmissometer with readout in tower; generally being replaced by RVR.

runway visual range, RVR Value representing horizontal distance pilot will see centreline or edge lights or runway markings down runway from approach end. Once recorded by observer 76 m from centreline, now by RVR system.

RUSA Roll-up solar array.

RUSI Royal United Services Institute (UK).

ruslick Anti-corrosive coating for bright metal in saline (ocean) environment.

RUT Standard regional route transmitting frequencies (ICAO).

Ruticon Family of photoconductive/liquid-crystal devices used in large-screen projection, typically with potential across photoconductor and elastomer with mirror surface for readout, scanned on opposite face by CRT input.

RV 1 Radar vector.
2 Re-entry vehicle.
3 Residual value.
4 Rescue vessel (RAF, ICAO).

RVA 1 Radar vectoring area.
2 Régie des Voies Aériennes (Belgium).

RVDP Radar video data processor.

RVDT Rotating variable-differential transformer (servocontrol position sensor).

RVL Rolling vertical landing.

RVO Runway visibility by observer.

RVP 1 Reid vapour pressure.
2 Ramp void pressure.

RVR 1 Runway visual range.
2 Rear-view radar.

RVR system Various electro-optical instruments for measuring RVR without subjective interpretation, usually by calibrated light source and transmissometer receiver separated by distance great enough to avoid too much error from local smoke.

RVSN Strategic rocket forces (USSR).

RVV Runway visibility value.

RV/WH Re-entry vehicle/warhead.

RW 1 Radiological weapon(s).
2 Retractable wheel (sailplanes).
3 Rigged wet, ie for spraying (ag-aircraft).
4 Reconnaissance wing.

R/W 1 Marine aircraft on water total resistance divided by weight, equivalent to an L/D ratio.
2 Runway.

RWA Reaction wheel assembly.

RW/D Rigged wet and dry, ie for both spraying and dusting (ag-aircraft).

RWM 1 Code for SAR helicopter (ICAO).
2 Read-only wire (or composite-wire) memory.

RWR 1 Rear-warning radar.
2 Radar warning receiver.

RWS Radar warning system.

RW+S Rigged wet plus spreader (ag-aircraft).

RWT Radar warning trainer.

RWY Runway (ICAO).

Rx Receiver, or reception only.

R(X) Correlation function.

RZ Recovery zone.

RZI Real-zero interpolation.

S

S 1 Generalized symbol for area, eg gross wing area.
2 Entropy (not UK and some other countries).
3 US piston engines, supercharged (hence, TS = turbocharged).
4 US military-aircraft designation category, surveillance.
5 US military-aircraft designation prefix, ASW.
6 UK military-aircraft designation category, strike.
7 JETDS code, special.
8 JETDS code, detection/range-bearing/search.
9 South, southern latitudes (ICAO).
10 Section modulus (alternatively Z).
11 Common missile code for both launch location and target, surface.
12 Single (landing gear).
13 Distance between contact areas of dual wheels.
14 Serviceable (CAA).

s 1 Second (time).
2 Stress.
3 Generalized symbol for linear measure, length, distance; esp used for aerofoil semi-span (b/2).
4 Starboard.
5 Square (eg in rms).
6 Spherical (eg in scp).

S^3 S-cubed, stick-shaker speed.

S-band Former common-use radar band originally 19·35–5·77 cm, 1·55–5·2 GHz, later rationalized to 15–7·5 cm, 2–4 GHz; now occupied by E and F.

S-code IFR flightplan code, aircraft has 64-code transponder and approved R-nav.

S-duct Curved duct supplying air to centre engine in trijets.

S-ing Series of S-turns.

S-manoeuvre To weave in horizontal plane.

S-pattern Wavy track resulting from S-ing.

S-stoff Nitric acid (G).

S-turn To describe S in horizontal plane.

SA 1 Safety altitude.
2 Sand or dust storm (ICAO).
3 Standby altimeter.
4 Shaft angle.
5 Surface-to-air.
6 Submerged-arc (weld).
7 Spin axis.
8 Single-aisle (passenger transport aircraft internal layout).
9 Structural audit.
10 Structured analysis (software).
11 Stand-alone.

SAA 1 Society of Airline Analysts (US).
2 Safety and arming.
3 Swiss Aerobatic Association.

SAAA Sport Aircraft Association of Australia.

SAAC Simulator for air-to-air combat.

SAAFA South African Air Force Association.

SAAFI South African Association of Flying Instructors.

SAAHS Stability-augmentation attitude-hold system.

SAAM Special-assignment airlift mission.

SAAPA South African Airways Pilots' Association.

SAB Scientific Advisory Board (USAF).

SABAR Satellites, balloons and rockets.

Sabatier 1 Reversal phenomenon in photo processing occurring when developed image is exposed to diffuse light and redeveloped (not encountered with X-rays).
2 Fully developed reaction process for recovery of pure oxygen from human exhalation/excretion, with byproducts such as methane and carbon dioxide.

Sabin Unit of acoustic absorption equal to 1 ft^2 of surface absorbing all sound energy falling on it. Appears to be no SI unit available.

Sabmis Surface-to-air ballistic-missile interception system.

sabot Annular driving mechanism to enable sub-calibre projectile to be fired at very high muzzle velocity from gun; usually has form of drum open at one end to receive projectile and divided radially to separate into sections beyond muzzle.

SAC 1 Strategic Air Command (USAF).
2 Space Applications Centre (India).
3 Space Activities Commission (Japan).
4 Seldom used for Soaring Assoc of Canada.
5 Supplemental air carrier.
6 Subsecretaria de Aviación Civil (Spain).
7 Standing advisory committee.
8 Standard arbitration clause.

SACCA Scottish Advisory Council for Civil Aviation.

SACCS SAC (1) automated command and control system.

Sacdin SAC digital information network (USAF).

Saceur Supreme Allied Commander, Europe.

SACL Strobe anti-collision light.

Saclant Supreme Allied Commander, Atlantic.

Saclos Semi-automatic command to line of sight.

SaCo Samarium cobalt (magnet material).

SACP Surface/air courte portée; short-range SAM (F).
sacrificial corrosion Metal is protected against corrosion by being coated with metal less noble than itself, which is attacked preferentially.
SACS 1 Speed/attitude control system (take-off).
2 Secondary attitude and compass system.
SACU Stand-alone communications unit.
SAD 1 Submarine anomaly detector.
2 Spares advanced data.
3 Self-adhesive decal.
4 Solar-array drive.
Sadarm Sense (or search) and destroy armour (USA/Aerojet).
saddle Shaped wooden former on which keel of marine aircraft rests during manufacture.
SAdO Senior administration officer (RAF).
Sadram Seek and destroy radar-assisted mission; locks on for long period after hostile emitter silent.
SADRG Semi-active Doppler radar guidance (SADRH substitutes "homing").
SADT Structured analysis and design techniques.
SAE 1 Society of Automotive Engineers (US).
2 Servicing Appraisal Executive (RAF).
3 Semi-actuator ejector.
SAE ratings Lube-oil ratings (10 to 70) based on Saybolt viscosity.
SAF Specified approach funnel.
SAFE 1 Survival & Flight Equipment Association (US).
2 Soil airfield fighter environment (DoD).
safe burst height That above which damage/fall-out is locally acceptable.
Safe Flight kit AOA-sensitive stall-warning system normally displaying green arrows on glare-shield of GA aircraft.
safe life Basic design and certification philosophy for primary structure; whether or not there are redundant load paths or fail-safe provisions, a total acceptable life (flight time) is published in flight manual and relevant structure must then be replaced, even if no crack is visible.
Safer System for aircrew flight extension and return; lifting/propelled ejection capsule.
safety altitude Loosely, one at which collision with surface is unlikely at approximate location; not accepted term (see *DH*, *MDA*, etc).
safety barrier Emergency arresting barrier across carrier flight deck (US often barricade net).
safety belt For passenger, normally called seat belt or lap strap; for flight (occasional cabin) crew, seat harness.
safety disc Disc of accurately known strength sealing fluid system, eg cartridge-operated, serving as safety valve.
safety factor See *factor of safety*.
safety height Not accepted term (see *DH*, *MDA*, etc).
safety imagination Ability to extrapolate from a real near-accident to a hypothetical accident.
safetying 1 Rendering explosive or pyrotechnic device safe by positive means.
2 Installing locking wire or other device which prevents an attachment from becoming loose.
safety net One designed to catch flying object with minimal damage, eg ejection seat on test, model in tunnel (esp spinning tunnel) or flying rotor blade in rotor test rig.
safety pilot One present in cockpit to prevent accident, eg to radio-controlled aircraft or RPV, or with pupil on instrument practice.
safety pins Those inserted to disarm ejection seat except in flight.
safety plug Blow-out plug serving as safety valve in case of excess pressure (eg JATO bottle) or excess temperature (eg tyre after prolonged braking).
safety speed That above which aircraft is safe to fly with given load and configuration; formerly several, eg FUSS, but most important today is V_2.
safety thread That connecting D-ring (ripcord) to parachute pack.
safety wire Locking wire passed through holes in nuts, turnbuckle barrels and other fasteners in such a way that they cannot loosen subsequently.
safety zone Area reserved for non-combat ops by friendly forces (DoD).
Safeway Pneutronic Guidance installation for nose-in parking; pavement pressure pads sense nosewheel(s) and illuminate progress lights facing flight deck while arrows, lights and other displays give guidance.
safing Process of rendering potentially dangerous device or system inoperative; eg complex procedure in Space Shuttle Orbiter post-landing.
Safoc Semi-automatic flight operations centre.
SAF/OI Secretary of the Air Force/Office of Information (US).
SAF/PA As above, Public Affairs.
SAFR Schweizerische Arbeitsgemeinschaft für Raumfahrt (previously Raketentechnik) (Switz).
SAFU Safety, arming and fuzing (sometimes functioning) unit.
SAG 1 Support air group (Royal Navy).
2 Scientific advisory group (US).
3 Surface action group (USN).
4 Survivability analysis group (DoD).
5 Semi-active guidance.
sag See *sagging*.
SAGA Studies Analysis and Gaming Agency (DoD).
SAGE, Sage 1 Semi-Automatic Ground Environment; pioneer (US) computer-controlled radar and communications system for defence of large airspace and management of interceptors and SAMs.
2 Stratospheric aerosol and gas experiment.
sagging 1 Distortion of airship (any kind) caused by upward loads near ends or lack of lift at

centre.
2 Bending stress on seaplane float or flying-boat hull caused by water support concentrated near ends due to swell or waves.
Sags Semi-active gravity-gradient stabilization.
SAGW Surface-to-air guided weapon (UK usage).
SAH 1 Semi-active homing.
2 Sample and hold (EDP [1]).
SAHR Semi-active homing radar (SAHRG adds "guidance").
SAHRS Standby (or secondary) attitude/heading reference system.
SAI 1 Standby airspeed indicator.
2 Spherical attitude indicator.
SAIF Standard avionics integrate fuzing (sets dispenser payload fuzes milliseconds before release).
SAIG Single-axis integrated gyro.
SAIL Shuttle-Avionics Integration Laboratories.
sail 1 Flat surface pointed towards Sun or other celestial object and attached to spacecraft, eg carrying solar cells.
2 Very large lightweight reflective surface proposed for space propulsion by pressure of sunlight.
3 To navigate seaplane (see *sailing* [*2*]).
4 Projecting structure above hull of submarine; also called fin, bridgefin or, formerly, conning tower.
5 (Usually plural) Small winglets arranged at different angles around wingtip, typically four disposed from front to rear.
sailing 1 Undesired rotation of helicopter rotors or aeroplane propeller in high wind.
2 Navigation of seaplane on water, esp in conditions of wind and current.
sailplane Glider designed for soaring.
sails See *sail (5)*.
sailwing Aerodyne whose wing assumes lifting (near-aerofoil) profile only in presence of suitable relative wind; class includes parawings but normally has flexible surface(s) restrained by rigid periphery.
Saint Satellite inspection technique.
SAL Strategic arms limitation.
Salbei See *SV-stoff*.
salmon Streamlined bulge forming tip to wing of aircraft having centreline gear and no outrigger (eg sailplane); designed to withstand rubbing on ground (obs).
Salomon damper Dynamic damper for crankshaft balance weights to remove oscillatory loads.
SALR Saturated adiabatic lapse-rate.
SALS 1 Short approach-light system (FAA).
2 Separate-access landing system.
SALT Strategic arms limitation talks.
salted weapon NW which has, in addition to normal components, extra elements or isotopes which capture neutrons at time of explosion and produce additional radioactive products; generally, opposite to clean NW.
Salto di Quirra NATO air firing range, including AAA and SAMs, in Sardinia.
salvage 1 To scrap complete aircraft or other equipment but recover parts or material.
2 To retrieve aircraft after landing away from any airfield.
3 To retrieve potentially dangerous flight situation, eg high rate of sink.
salvo 1 Simultaneous group of ECM bursts, esp from dispensed payloads; not necessarily all of same species, eg could be two chaff, one IR, one jammer.
2 In close air support/interdiction, method of delivery in which all weapons of specific type are released or fired simultaneously (DoD).
3 See *salvos*.
salvos Air-intercept code: "I am about to open fire"; can add further word specifying weapon(s), eg * mushroom = "I am about to fire special weapon", ie Genie (DoD).
SAM 1 Surface-to-air missile.
2 School of Aerospace Medicine (USAF).
3 Sound-absorbing material(s).
4 Standard (or standardized) assembly module.
5 Société Aérostatique et Météorologique, 1865 (F).
6 Special air missions (sqn) (USAF).
SAMA Sistema aeronautico modulare anticendio (I).
samarium Sm, rare-earth metal, SG 7·8, MPt 1,330°C, alloying element in magnets, etc.
SAMI System acquisition management inspection.
SAMOS 1 Stacked-gate avalanche-injection MOS.
2 Satellite and missile observation station.
SAMPE Society of Aerospace Material and Process Engineers (US).
sample length Length of specimen studied in examination of variable, eg surface finish.
Sams, SAMS 1 Six-axis motion system (simulator).
2 Software automated management support.
SAMSO Space and Missile Systems Organization, Air Force Systems Command (USAF).
Samson Strategic automatic message-switching operational network (Burroughs).
SAMT Simulated aircraft maintenance trainer.
SAMTEC Space and Missile Test Center (DoD).
sand and spinach Green/brown camouflage (colloq).
sandbag bumping American term for hand-shaping sheet metal by hammering against tough sandbag.
sandbag line Rope joining sandbag loops to prevent wear.
sandbag loops Cord loops over aerostat (esp kite balloon) envelope carrying sandbag at each end;

also called sandbag bridle.

sandbag ring Cable or rope round balloon basket from which sandbags are (suggest were) hung, forming easily jettisoned load.

sandblasting Scouring material surface with high-velocity jet of sand or other abrasive for various purposes; not common in modern aerospace, replaced by tailored steel shot or glass beads.

sand casting Casting metal parts in sand mould.

S&E Scientists and engineers.

S&I Safety and initiating.

sanding coat Heavy-bodied paint or dope coat which fills irregularities in surface, leaving smooth base for top coats.

S&M Supply and movements (RAF).

Sandow cord See *bungee*.

SANDT School of Applied Non-Destructive Testing.

sandwich aircraft Positioned horizontally or vertically between two enemy aircraft.

sandwich construction Large family of constructional methods, most of them patented, in which two load-bearing skins are joined to stabilizing low-density core. In most types both skins are locally flat and parallel, but most can be shaped to single or compound curvature before or after bonding three components together, and a few are tailored so that * thickness varies to meet requirement of part. Cores can be metal or paper/plastics honeycombs, foamed plastics, balsa or many other choices, including corrugated sheet.

sandwich manoeuvre One fighter, usually DA partner, turns away and accelerates while other falls in behind enemy.

sandwich moulding Polymer A injected inside polymer B in mould.

sandwich plate Plate, part of primary structure, located between two other members, eg on centreline between left/right keel girders.

Sandy Search and rescue aircraft (colloq).

San Marco Italian-operated offshore space launch platform.

SAO Special access only (high security classification).

SAOCS Submarine/aircraft optical communication system.

SAOWP Structural analysis and optimization working party.

SAP 1 Seaborne air platform, eg for jet V/STOL.
2 Semi-armour-piercing.
3 Silicon avalanche photodiode.
4 Simulated attack profile.

SAPC South African Parachute Club.

SAR 1 Search and rescue.
2 Synthetic-aperture radar.
3 Selected acquisition report (US one each FY).
4 Stand-alone radar.
5 Semi-active radar.

Sarah Search and rescue and homing; personal radio beacon (Ultra).

Sarbe Search and rescue beacon equipment; military/civil (Burndept).

Sarcap, SarCAP Combat air patrol which covers a rescue operation.

SARH Semi-active radar homing (SARG substitutes "guidance").

Sarie Semi-automatic radar identification equipment; part of Abbey Hill (EMI).

Saris Synthetic-aperture radar interpretation system.

SARO Supply aero-engine record office (UK).

SARP, Sarp 1 Semi-automatic radar plotting.
2 Signaal auto radar-processing system.

SARPS, Sarps Standards and recommended practices (ICAO).

SARS 1 Support and restraint system.
2 Static automatic reporting system.

Sarsat SAR (1) satellite-aided tracking.

SART Semi-active radar target.

Sart SAR (1) transponder.

Sartaf SAR (1) task force.

SAS 1 Stability-augmentation system.
2 Satellite Applications Section (NOAA).
3 Small-angle scattering.
4 Stall-avoidance subsystem.
5 Staring-array seeker.
6 Single audio system.
7 Special Air Service (UK).
8 Survival avionics system.

SASC Senate Armed Services Committee.

SAS/CSS SAS (1) plus control-stick steering.

SASE Semi-automatic support equipment.

SASI The Society of Air Safety Investigators (Canada).

SASO Pronounced sasso, Senior Air Staff Officer (RAF).

SASP Single advanced signal processor.

SASRS Satellite-aided search and rescue system.

SAT 1 Software audit team.
2 Static air temperature.

Sataf Site-activation task force.

Satair Sea acceptance trials, air.

SATCC Southern African Transportation Coordinating Commission.

SATCO Senior air traffic control officer.

satcom Generalized term for satellite communications.

Satcoma Satellite Communications Agency (US).

Satcra Stress in ATC Research Association (Neth).

satellite 1 Body revolving in equilibrium orbit around primary, natural or man-made.
2 Man-made device intended to become * (1).
3 Military airfield auxiliary to nearby main airfield and relying on latter for admin and most services.
4 Sub-terminal at airport to disperse processing and bring pax nearer relevant gates.

satellite/pier layout Airport terminal is connected to some satellites (4) and some piers.

satelloid Satellite whose orbit is within planetary atmosphere and thus requires continuous or intermittent thrust.

SATF 1 Strike and terrain-following (radar).
2 Shuttle activation task force (also rendered Sataf).

Satin SAC automated total-information network (USAF).

Satka Surveillance, acquisition, tracking and kill assessment.

Satnav Satellite navigation, ie satellite-assisted.

Satnet Satellite network.

SATO Scheduled airlines ticket office.

SATS 1 Small airfield for tactical support (USMC).
2 Small-arms target system.
3 Shuttle avionics test set.

Satsim Saturation countermeasures simulator (USAF).

saturable reactor Soft-core inductor control for pulsed radars and magnetic amplifiers.

saturated adiabatic lapse rate Rate of decrease of temperature with height for parcel of saturated air.

saturated air Air containing greatest possible density of water vapour, such that in any given period number of molecular break-ups equals number of recombinations; RH (2) = 100%.

saturated beacon Ground transponder beacon, eg DME, interrogated by so many aircraft (usually 100 simultaneously is limit) that its AGC cuts out replies except to 100 strongest interrogators.

saturation diving Undersea submergence at depths in order of 300 m for prolonged period when blood is saturated with chosen breathing mixture.

saturation vapour pressure Vapour pressure of particular substance, variable with temperature, which at given temperature is in equilibrium with plane surface of same substance in liquid or solid phase.

SAU Safety/arming unit.

saunter Air-intercept code: "Fly for best endurance".

sausage Kite balloon (colloq, arch).

SAvA Society of Aviation Artists (UK).

SAVAC Simulates, analyses, visualizes activated circuitry (Chrysler/USAF).

Savasi Simplified abbreviated Vasi.

save Rescue of a downed pilot, esp behind hostile lines.

save-list item Item of equipment to be salvaged from aircraft at "boneyard" site.

SAVR Strapdown attitude/velocity reference.

SAW 1 Surface acoustic wave.
2 Society of Aviation Writers (several countries).
3 Submerged-arc welding.
4 Special air warfare (USAF).

sawcut Chordwise slot in leading edge of aerofoil to promote chordwise flow, energize boundary layer or serve other aerodynamic function.

SAWE Society of Aeronautical Weight Engineers (US).

SAWRS Supplementary aviation weather-reporting station (NOAA).

SAWS Satellite attack warning system.

SAWSS Shipboard aircraft weight subsystem.

sawtooth 1 Voltage or timebase which when plotted has appearance of saw edge with series of linear-rate climbs and near-vertical descents.
2 See *dogtooth*.

Saybolt Standard test for viscosity of liquid, esp lube oil, in which sample heated to known temperature is poured through calibrated orifice and time in seconds recorded for 60 ml (cm^3).

SB 1 Service bulletin.
2 Sideband.
3 Scrieve board.
4 Sonic boom.

sb Stilb.

SBA 1 Standard beam approach, pioneer electronic landing aid which provided azimuth (lateral guidance) and marker (distance) information; superseded by ILS from 1945.
2 Spot-beam aerial (antenna).
3 Smaller Businesses Association (and Small Business Administration) (US).

SBAC Society of British Aerospace Companies (previously Society of British Aircraft Constructors).

SBC 1 Sonic-Boom Committee (ICAO).
2 Small bayonet cap.

SBG Standby gyro.

SBL/BMD Space-based laser/ballistic-missile defense (US).

SBN Strontium barium niobate.

SBO 1 Specific behavioural objective.
2 Sideband only.

SBR 1 Signal-to-background ratio.
2 Space-based radar.

SBS Satellite business systems.

SBSS 1 Standard base-supply system (USAF).
2 Space-based surveillance system.

SBTC Sino-British Trade Council.

SBW Search bubble window, ie bulged window.

SC 1 Single-crystal.
2 Speed-control (system).
3 Short circuit (also s.c.).
4 Subgrade code (ICAO).
5 Service ceiling.
6 Stratified charge.
7 Statement of capability.

Sc 1 Stratocumulus.
2 Schmidt number.

S/c Spacecraft.

S_c Unit compressive stress.

SC-1 One engine inoperative.

SC-2 All engines operating.

SCA 1 Single-crystal alloy.

2 Services de la Circulation Aérienne = ATC (F).
3 Simulation control area.

scab External payload carried flush against pylon or aircraft skin; hence to *-on, scabbed.

SCAD, Scad Subsonic-cruise armed decoy.

SCADC Standard central air-data computer.

Scads Shipboard containerized air-defence system (BAe).

SCAF Supply, control and accounts flight (RAF).

scalar Quantity having magnitude only, as distinct from vector.

scale To reproduce on different scale of size, esp to produce gas-turbine engine larger or smaller than original but broadly similar aerodynamically.

scale altitude effect Compressibility error, so called because at high speeds ASIR over-reads as if it were at lower altitude.

scaled Basically unchanged in design but smaller or larger than original.

scale effect Sum of all effects of change in size of body in fluid flow, keeping shape same; more specif, effect of alteration in Reynolds number.

scale factor Output for given input; eg for accelerometer ** is output current in mA per unit of applied acceleration g.

scale model Model forming exact miniature of original.

scaler Electronic counter producing one output pulse for given number of inputs; thus binary * has scaling factor of 2 and decade * has SF of 10.

scale strength Relationship between actual structural strength of model and its size and same relationship for full-scale aircraft or other item; hence scale stiffness.

scaling factor Number of input pulses for one scaler output pulse.

scaling law Mathematical equations permitting effects of NW explosion of given yield to be determined as function of distance from GZ provided that corresponding effect is known for reference explosion, eg 1 kt.

scalloping VOR bearing error due to distortion of propagation over uneven terrain; also called bends.

scalping Rough-machining surface layers off ingot.

SCAM, Scam Strike camera.

SCAMA, Scama Switching, conferencing and monitoring arrangement (NASA).

scan 1 Motion of electronic beam through space searching for target (see * *types*).
2 In EM or acoustics search, one rotation of sensor; this may determine a timebase (NATO).
3 Air intercept code: "Search sector indicated and report any contacts".
4 In TV and other video systems, process of continuously translating scene into picture elements and thus varying electrical signal(s).
5 To make one complete cycle or sweep of eyes across either selected flight instruments or external scene ahead.

scanner Device which scans (usually 1 or 4), including electron beams in image tubes and CRTs, TV cameras and many electronic display systems, but esp including radar transmitter/receiver aerials (antennas) which are scanned mechanically or electronically.

scanner column Vertical member carrying two or more superimposed radar aerials, eg in nose of C-5A.

scanning 1 Process by which radar aerial scans, either by physical rotary movement (usually driven hydraulically) or by electronic scanning.
2 Process by which electron beam scans, accomplished by electrostatic or electromagnetic plates or coils.
3 Action of keeping eyes sweeping over external scene and/or flight instruments and other internals.

scanning-beam MLS See *MLS*.

scanning field Area, usually rectangular, scanned by electron beam in TV camera, image tube, CRT or electronic display.

scanning generator Timebase controlling scanning (2).

scanning sonobuoy One whose acoustic sensors (either passive or also active) scan to give directional information and/or to filter out noise from sources other than target.

scanning spot Point of light where electron beam strikes face of CRT or other scanning field.

scan period Time period of basic scan types other than conical and lobe-switching, or period of lowest repetitive cycle in more complex combinations; basic units in US are °/s, mils/s or s/cycle.

scan stealing Appropriating a main scan for writing symbols and alphanumerics when interscan period is too brief; term not recommended.

scan transfer Sudden switching of scanning (3) from external to internal scene or vice versa, esp immediately before bad-visibility touchdown.

scan types There are many varieties of radar scan (1) patterns. What follows is descriptive of motions of centreline axis (boresight line) of main lobe only. Each * is associated with particular type(s) of display, to which aerial (antenna) az/el information is supplied by potentiometer or other servo system. Fixed-scan radar points in one direction only. Manually controlled, points where directed. Conical, traces out circular path forming small-angle cone with radar at apex; a variation, spiral, traces spiral path beginning and ending at centre of circle. Sector, scans through limited az angle (unidirectional, from L to R or R to L only, snapping back to start after each scan; bidirectional, L-R-L-R). Circular, rotates continuously in horizontal plane (common AWACS mode). Helical, scans continuously in az while winding up and down elevation from 0° to 90° and back. Palmer is conical scan superimposed on another, eg Palmer-circular or Palmer-sector covers 360° periphery or arc respectively with conical scans. Raster is TV method of horizontal lines, usually interlaced; eg an air-intercept 6-bar raster might scan to R along line 1 (top), to L along 4, to R along 3, to L along 6, to R along 2, to L along 5 and back to start. Palmer-raster is another air-intercept scan with conical scan along two or more horizontal bars; thus 3-bar PR makes quick rings along top bar, next along middle and back along bottom, then back via middle to top. Track while scan (TWS) uses an az radar and an el radar simultaneously scanning in both planes.

SCAR, Scar Strike control and reconnaissance, co-in mission (USAF).

scarf cloud Thin cirrus draping summit or anvil of Cb.

scarfed Cut off at an oblique angle; hence * joint, inlet, nozzle etc.

Scarff ring Standard British cockpit mount for hand-aimed machine gun 1917–40 with ring-mounted elevating U-frame.

scarf joint Structural joint, invariably in wood, in which mating members are given flat taper to give large glued/pinned area.

SCART Syndicat des Constructeurs d'Appareils Radio-récepteurs et Téléviseurs (F).

SCAS Stability and control augmentation system.

SCAT 1 Supersonic civil air transport.
2 Space communications and tracking.
3 Scout/attack helicopter.

Scatana Security control of air traffic and air navigation aids; special provisions and instructions in time of defence emergency (FAA).

Scatha, SCATHA Spacecraft charging at high altitude.

scatter 1 Distribution, either ordered or, more usually, random of measured values about mean point (eg of 1,000 measures of wing span of similar type aircraft).
2 Distribution of impact points of projectiles aimed at same target.
3 See *scattering*.

scattered cloud Not used; cloud amount reported only in oktas.

scattering 1 Diffusion of radiation in all directions caused by small particulate matter in atmosphere; effect varies according to ratio between wavelength and particle size; when this exceeds about 10 Rayleigh scattering occurs (see *back-**, *tropospheric* *).
2 Trajectory changes of sub-atomic particles caused by collisions or various interactions; can be elastic or (if there is energy transfer) inelastic.

scattering loss That part of transmission loss due to scattering (1) or to target's rough surface.

scattering power Ratio of total radar power scattered by target to total power received at target; also called scattering cross-section.

scatter point Geographical point where race competitors cease to be constrained to narrow take-off corridor.

scatter propagation See *back-scatter*, *tropospheric scatter*.

scatter tolerance That allowed on dimensions of die forging, often measured at random locations.

scatter weapon One releasing or dispensing many bomblets or mines.

scavenge pipe Carries lube oil from machine, eg engine, to tank. In US, often scavenger.

scavenge pump Pumps lube oil out of machine, in case of engine fitted with wet sump, from base of sump. Gas turbines generally have several scavenge (return) gears on same shaft as pressure gears. In US, often scavenger.

scavenge(r) system Exit ducting from wind tunnel for removal of contaminants, eg from smoke apparatus, combustion products from burning tests or exhaust from combustion devices.

SCC 1 Standing Consultative Commission (arms control).
2 Security consultative committee.

SCCS Source-code control system (software).

SCD Speed computing display; airspeed (often TAS) needle plus Mach counter.

SCDA Software cost-driver attribute.

SCDU Selective control decode unit (IFF).

SCE Signal conditioning equipment.

scenaric computer One able to assemble visual scenes, eg in Tepigen.

SCEPS Stored chemical energy propulsion system.

Sceptr Suitcase emergency procedures trainer (cheap erasable ROM).

SCF Satellite control facility.

SCH 1 Sonobuoy cable hold (autopilot selector).
2 Simplified combined harness.

schedule 1 Precisely controlled mechanical movements to meet system demands, carried out automatically by system usually provided with feedback; eg variable inlets and nozzles in supersonic airbreathing engine installation.
2 Preplanned sequence of timed events, eg aircraft inspections and overhauls or timetabled civil flights.

scheduled service Air-carrier service for any kind of payload run to timetable.

scheduled speed Any of type-specific speeds published in flight manual, eg V_S, V_{AT}; in no way connected with speed in commercial use, which is defined as block speed.

scheduling Numerical, analog or graphical description of sequence of scheduled (1) movements, eg of turbojet inlet spike.

schedule inventory List of all safety equipment and other removable items carried on board.

schematic diagram Drawing which explains functions and general spatial relationships but which uses standard symbols and makes no attempt to portray visual appearance. Often includes only one subsystem, rest of machine, aircraft or other device being in outline or phantom line only. Alternatives are exploded drawing, block diagram.

Schlichting Original (1936) theory treating of flat plate in supersonic flow.

schlieren German word (nearest English equivalent is "striations") for various shadowgraph-like techniques for optical investigations based on 1859 method of Foucault. Basic feature is small light source, parallel rays of light through region under investigation and opaque cut-off at focus of second lens projecting image on screen or photographic film (either point source and pinhole cut-off or line source and line cut-off). Variations in density, eg in flow through shock-waves, Prandtl-Meyer expansions and supersonic flow generally, are sharply visible as tonal gradations.

Schmidt camera Elegant astronomical camera/telescope with objective in form of thin plate of glass and rear concave spherical mirror focusing on curved film. Objective plate has one surface figured (thicker and convex at centre, thin and concave around periphery) to correct mirror's aberration, its own chromatic aberration being slight because of small thickness.

Schmidt duct Pioneer flap-valve pulsejet.

Schmidt number $Sc = \mu/\rho D_{12}$ where μ is viscosity, ρ is density and D_{12} diffusion coefficient; ratio of viscous and mass diffusivity, or kinematic viscosity divided by mass diffusivity.

Schottky defect Atom missing from crystal lattice.

Schottky diode Barrier-layer device based on rectification properties of contact between metal and semiconductor due to formation of barrier layer at point of contact.

Schottky effect Small variation in electron current of thermionic valve caused by variation in anode voltage affecting work done by electrons in escaping.

Schuler pendulum One whose length equals radius of Earth, and thus when carried in vehicle moving near Earth's surface always indicates local vertical. In practice any pendulum having same period of approx 84 min, achieved at particular relationship between c.g. and pivot, such that centre of rotation of pendulum is always at centre of Earth. Used in stable platform of INS.

Schuler tuning Adjusting period of Schuler pendulum so that its centre of rotation exactly corresponds with centre of Earth.

Schultz-Grunow Standard treatment for turbulent flow in viscid fluid at R from 10^6 to 10^{10}.

Schwarm Two Rottes, fighters in two loose pairs (G).

SCI 1 Smoke curtain installation, for laying smokescreen.
2 Switched collector impedance.

science pilot Experienced researcher in a scientific discipline subsequently qualified as pilot, eg of Space Shuttle (NASA).

scimitar wing One whose planform is curved; usually means same as crescent wing but has been applied to early wings curved across whole span with "sweep" at tips and zero sweep at root.

scintillation 1 See *glint* (radar).
2 Rapid and random variation in appearance of small light source viewed through atmosphere, esp variation in luminance.
3 Brief light emission by single event (eg impingement on phosphor) (see * *counter*).

scintillation counter Instrument for measuring alpha, beta or gamma radiation by counting scintillations (3); also called scintillator or scintillating counter.

scintillation spectrometer Scintillation counter plus pulse-height analyser for radiation energy distribution.

scintillometer Photoelectric photometer for measurement of wind speed near tropopause by various Schlieren-type measures of stellar scintillation. Also called scintillation meter.

SCIRP Semiconductor IR photography.

scissors Flight manoeuvre performed by two or more pairs of aircraft crossing at angle like two halves of *; basically a series of turn reversals intended to make following enemy overshoot.

scissor wing Wing made as single plane from tip to tip pivoted to fuselage at mid-point; usually synonymous with slew wing.

SCL Space-charge limited (C adds "current").

sclerometer Hand instrument measuring hardness by load needed to make scratch of standard depth with diamond point rotated along arc.

scleroscope Hand instrument measuring hardness by rebound height of hard-tipped (steel or diamond depending on pattern) rod-hammer dropped inside tube from standard height on to surface. Common commercial model is Shore *.

SCM 1 Single-crystal material (or metal).
2 Single-chip microprocessor.
3 Software configuration management.
4 Spoiler control module.
5 Surface contamination module.

SCMR Surface-composition mapping radiometer.

SCN 1 Specification change notice.
2 Satellite control network.

SCO Sub-carrier oscillator.

Scoff Society for Conquest of the Fear of Flying (US).

scooter bogie One with two wheels only, in tandem.

SCOPE Simple checkout program (language).

Scope Spacecraft operational-performance evaluation.

scope 1 Electronic display, esp one supplying

output of radar (colloq).
2 Generalized term for optical viewing instrument, eg microscope, CRT, etc (colloq, vague).
3 According to Webster, distance within which missile carries; unknown in aerospace.

scopodromic Headed in direction of target.

Score Signal communications by orbital relay experiment.

scoring Apart from normal use (eg in military exercises), the process by which a customer numerically evaluates bidders' proposals.

Scot Satellite-communications on-board (ship) terminal.

scotopic Vision with retinal rods associated with extremely low light intensity and detection of gross movement.

scout Single-seat aircraft, armed for air combat, operating in patrol or reconnaissance role (WW1).

SCP, scp 1 Spherical candlepower.
2 System-concept paper.

SCPL Senior commercial pilot's licence.

SCR 1 Silicon-controlled rectifier.
2 Signal/clutter ratio.
3 Stratified-charge rotary (engine).

SCRA Single-channel radio access.

scramble 1 Take off as quickly as possible (usually followed by course and altitude instructions) (DoD). Any urgent call for military (usually combat) aircraft to take off and leave vicinity of base, either as training manoeuvre or for sudden operational reason.
2 To attempt to provide telecom security by rendering transmission unintelligible to third parties (ie by scrambling), eg by speech inversion.

scramjet Supersonic-combustion ramjet; one in which flow through combustor itself is still supersonic.

scrap 1 Workpiece containing defect, even trivial, sufficient to cause it to fail an inspection.
2 To break up old unwanted aircraft or one beyond economic repair.
3 Residue of (2) other than produce or salvaged items.

scraper ring Spring piston-ring with sharp-angled lower periphery for removing oil from cylinder wall.

scrap view Small inset showing detail (eg item from different viewpoint or in different configuration) added where room permits on main drawing.

SC/RC Stratified-charge rotating-combustion engine.

SCRE Syndicat des Constructeurs de Relais Electriques (F).

screaming Undefined term descriptive of high-frequency combustion instability in rocket characterized by high-pitched noise (see *screeching*).

screeching Undefined term describing high-frequency combustion instability in rocket or afterburner characterized by harsh, shrill noise more irregular than screaming.

screeding Levelling and smoothing resin adhesive in bonding operation; hence * tool.

screen 1 Imaginary obstruction having form of level-top wall normal to flightpath which aeroplane would just clear on take-off or landing with landing gear extended and wings level.
2 Arrangement of ships, submarines and aircraft for protection of ship(s) against attack (DoD).
3 Wire-mesh gauze or sieve (ASCC).
4 Electrically conducting or magnetically permeable enclosure which shields either contents or exterior against unwanted magnetic/electrical fields.
5 Face of electronic display, eg CRT, TV, projection system, etc.

screen burn See *ion burn*.

screen captain Senior training or supervisory pilot in role of examiner. May be employee of certification authority.

screened Provided with screen (4), thus * ignition has enveloping earthed conductive covering to prevent escape of R/F interference.

screened pair Twin electric cable incorporating earthed screen (4).

screen filter Fluid filter whose element is a fine metal mesh screen.

screen grid Screen (3) between anode and control grid in thermionic valve to reduce electrostatic influence of anode.

screen height Height above ground of top of screen (1), normally 35 ft (10·67 m), occasionally 10 m and rarely 50 ft, on take-off; on landing usually 30 ft (often interpreted as 10 m). Depends upon aircraft performance group and particular case considered.

screen speed Speed at moment aeroplane passes over screen (1), either assumed or as target value, on take-off or landing. Does not have V-suffix abbreviation.

screw Generalized term for threaded connector rotated into workpiece and not held by nut; many quick-fasteners have *-thread and no sharp dividing line is possible.

screw gauge Hand instrument for measuring major diameter of screw, typically by graduated line scales forming small-angle notch.

screwjack Actuator having rotary input and linear output obtained by screwthread, often with interposed recirculating balls.

screw pitch gauge Hand instrument for checking thread on metalworking screw or bolt, usually with selection of blades each having one threaded edge.

scrieve board Portable board on which lofting lines are recorded either by scribing or some other undeformable method. Often used with locating blocks for actually assembling flight hardware, eg frames, thus becoming a jig. Becoming arch.

SCRJ Supersonic combustion ramjet.

scrub To abandon project, esp a planned flight or military mission.
scrubbing 1 Lateral sliding of landing-gear tyre on hard pavement, eg of inner bogie wheels in sharp turn.
2 Significant contact between tips of rotating members, esp blades, in gas turbine and casing; also called rubbing.
3 Rapid wear in PE (4) caused by detonation.
scrubbing torque limit Maximum permitted torque imposed on landing-gear leg by scrubbing (1), which usually determines limit of steering angle of nose gear, and hence min turn radius.
SCS 1 Stabilization control system.
2 Speed control system.
3 Sea control ship.
4 Survivable control system.
5 Single crystal sapphire (IR domes).
SCSC Strategic conventional standoff capability.
SCS–51 Signal Corps Set 51, original form of ILS.
SCT 1 Scattered (ICAO).
2 Scanning telescope (NASA).
3 Surface-charge transistor.
4 Single-channel transponder.
5 Staff continuation training (CAA).
SCTI Service Centrale des Télécommunications et de l'Informatique (F).
SCTV/GDHS Spacecraft TV ground data-handling system.
SCU Signal conditioning unit.
Scud Subsonic cruise unarmed decoy.
scud Shredded or fragmentary cloud, typically Fs, moving with apparent greater speed below solid layer of higher cloud.
scupper Fuel-tight recess around gravity filler, usually with its own drain.
scuttle 1 Hatch in top of fuselage (US usage).
2 According to Webster: "An airport. *Brit.*"; unknown in aerospace.
SD 1 Shaft delivery, ie rotary output.
2 Shipping document.
3 Structured design (software).
4 Service dress (RAF).
S_D Distance between centres of contact areas of diagonally opposite wheels of bogie.
SDA Strategic defence (defense) architecture.
SDAC System data analog converter.
SDAS Source-data automation system.
SDAT Silicon diode array target.
SDAU Safety data and analysis unit.
SDBY Standby (ICAO).
SDC 1 Synchro-to-digital converter.
2 Space Defense Center (Adcom, USAF).
3 Signal data convertor.
SDD Synthetic dynamic display.
SDDM SecDef decision memorandum.
SDF 1 Simplified directional facility (FAA).
2 Self-destruct fuze.
3 Single degree of freedom.
SDG Speed-decreasing gearbox.
SDI 1 Strategic (sometimes rendered as Space) Defense Initiative (US).
2 System discharge indicator.
SDIO SDI Organization, or Office.
SDIP SDI Program.
SDLV Shuttle-derivation launch vehicle.
SDM 1 Scatter-drop mine (or munition); helicopter weapon.
2 Site-defense of Minuteman; anti-intruder system.
SDMA Space-division multiple-access.
SDN System descriptive note.
SDO Serial digital output.
SDP System-definition phase.
SDR 1 System design report (or review, or responsibility).
2 Service difficulty report.
3 Signal data recorder.
4 Special drawing rights; assist airlines in interline fare transactions.
5 System development requirement.
SDRS Splash-detection radar system.
SDS 1 Software development system.
2 Satellite data system.
SDT System development tool.
SDU 1 Selective decode unit (SSR).
2 Safety data unit (CAA).
SDVP System demonstration validation program(me).
SDW Symmetrical double-wedge aerofoil.
SE 1 Support equipment.
2 Systems engineering.
3 Servicing echelon.
SEA 1 Single-engine asymmetry.
2 Stored-energy actuator.
sea/air/land team Group trained and equipped for unconventional and paramilitary ops including surveillance/reconnaissance in and from restricted waters, rivers and coastal areas (DoD).
sea anchor Anchor for marine aircraft in deep water, typically canvas or rubberized-fabric bucket or drogue trailed from stern.
sea bias See *water bias*.
sea breeze Onshore wind during day.
SEAC 1 Support-Equipment Advisory Committee (inter-airline).
2 South-East Asia Command (WW2).
SEAD Suppression of enemy air defence.
sea disturbance See *sea state*.
Seafac Systems engineering avionics facility.
sea fog Fog over sea usually caused by moist air over cold water.
SEAGA Selective employment of ground and air alert (SAC).
Seagull EDP for flight simulators using distributed computers linked by reflective-memory bus (Rediffusion/Gould).
SEAL Sea/air/land team.
sealant Material tailored to particular duty of ensuring fluid-tight (liquid or gas) seal between

mating materials; often applied as continuous layer, eg inside integral tank.

Sealdrum Portable, collapsible rubber container for POL, water and other liquids (US Royal).

sealed Officially closed, eg by lock wire with lead seal, until completion of particular flight or test programme.

sealed-balance control Flight-control surface whose gap is made airtight by flexible strip.

sea level 1 Actual height of sea surface, used as local height reference.
2 More often, height corresponding to pressure of 1,013·25 mb.

sea-level corrections See *STP (1)*.

seam Joint in fabric, eg overlap or three fell-types.

sea marker Anything dropped on sea to provide visible indication of drift.

sea-motion corrector See *sea-surface correction*.

seam weld Continuous, hopefully pressure-tight, weld made by roll electrodes; in theory can be accomplished by close repeated spot welding or by hand process, but unusual.

Seapac Sea-activated parachute automatic crew release.

seaplane 1 Float seaplane, marine aircraft having one, two or (rarely) more separate floats and conventional fuselage (UK).
2 Marine aircraft of any fixed-wing type (US).

seaplane basin 1 See *towing tank*.
2 Some dictionaries define this as a place having sheltered water for seaplanes.

seaplane marker Buoyant or bottom-fixed marker at marine aerodrome, also called taxi-channel marker.

seaplane tank See *pitching tank*, *towing tank*.

seaplane trim Angle between mean water surface and aircraft longitudinal axis or other reference when parked on water.

sear Generalized term for catch, pawl or other latching device which holds breechblock or bolt of gun at open position.

sea rating See *sea state*.

search Systematic reconnaissance of defined area such that all parts pass within visibility.

search and rescue facility One responsible for maintaining search and rescue service for persons and property in distress.

search jammer Automatic jammer.

searchlight Any powerful directional surface light for illumination of cloudbase or other aerial objects.

searchlight scanning Sector scanning (see *scan types*).

search mission Air reconnaissance to search for specific surface object(s).

search radar Vague; generally means surveillance radar operating in search mode, and often applied to weather radar.

search rate Reciprocal of average time to search one PN chip-width, ie chips/s (ECM).

sea return See *clutter*.

Sears-Haack Profile of body having min supersonic drag, normally body of revolution having pointed ends and fineness ratio appropriate to Mach number; complicated by addition of aerofoils.

seasat Generalized name for oceanographic satellite.

sea smoke See *steaming fog*.

sea state Condition of sea surface as related to standard list of reference conditions, invariably Beaufort scale.

sea-surface correction Electronically applied correction to Doppler output to remove false velocities due to motions of waves.

SEAT 1 Site-equipment acceptance test.
2 Status evaluation and test.

SEATO Pronounced Seato, SE Asia Treaty Organization.

seat pack Parachute pack worn in such a way that it forms seat cushion.

seat selection Offering customer for particular commercial flight right to select seat from those remaining, subsequently shown on boarding pass.

seat-type parachute One with seat pack.

sec Seconds (contrary to SI but still ICAO).

SEC 1 Secondary emission cathode.
2 Secondary-electron conduction.
3 Securities and Exchange Commission (US).
4 Secondary (NASA).

Secant Separation and control of aircraft, non-synchronous techniques.

secant modulus E_s, slope of stress/strain curve to any point; thus equal to modulus of elasticity up to limit of proportionality, thereafter varying with applied stress to point of rupture.

SecDef Secretary of State for Defense (US).

SECL Symmetrical emitter-coupled logic.

Seco, SECO S-IVB engine cut-off (Apollo).

second 1 SI unit of time, symbol s.
2 Unit of angular measure, symbol ″.

secondary Small area of low pressure on periphery of large cyclone.

secondary airflow That used to dilute and cool flame in gas-turbine combustor.

secondary airport Smaller airport at city possessing hub airport.

secondary battery Electric battery rechargeable by reverse DC current.

secondary bending In a beam or column whose chief applied stress is transverse, that bending moment due to axial load.

secondary controls See *secondary flight controls*.

secondary electron emission Flow of electrons from metal surface under bombardment by high-energy electrons or protons.

secondary emission Ejection of subatomic particles and/or photons from atoms or particles subjected to primary radiation, eg cosmic rays.

secondary explosion Explosion at surface target

caused by air attack but additional to explosions of air-dropped ordnance.

secondary fan airflow Airflow through fan which does not pass through engine core.

secondary flight controls Those used intermittently to change lift or speed but not trajectory; eg flaps, slats, Krügers, airbrakes, droops and, except in DLC, spoilers.

secondary front One formed within an air mass.

secondary glider Not defined, but training glider more advanced than primary and with enclosed cockpit. Generally any glider intermediate between primary and Standard Class sailplane.

secondary great circle See *meridian.*

secondary heat-exchanger That which rejects heat to atmosphere from secondary circuit heated by primary heat-exchanger.

secondary holes Those admitting secondary air to combustor.

secondary instruments 1 Those giving information unconnected with gross flight trajectory, ie not a primary flight instrument.

2 Those whose calibration is determined by comparison with an absolute instrument.

secondary members See *secondary structure.*

secondary modulus Modulus of elasticity for composite or other two-component material (esp two-metal components) beyond point at which weaker material yields.

secondary nozzle Annular nozzle surrounding primary nozzle of jet engine through which may pass cooling airflow, inlet excess and various other flows around engine.

secondary power system Mechanical power system on board aircraft other than main engines, eg shaft-driven accessories, gearboxes, gas-turbine starter, APU, EPU or MEPU, inter-engine cross-shafting and major bleed ducting with air-turbine drives if fitted.

secondary radar Radar in which interrogatory pulse is sent to distant transponder which is triggered to send back a different pulse code to originator. Examples are airborne DME triggering ground DME facility and ground SSR triggering airborne transponder.

secondary radiation Usually synonymous with secondary emission.

secondary stress That resulting from deflection under load, eg of end-loaded column in bending.

secondary structure Structural parts of airframe whose failure does not immediately imperil continued safe flight.

secondary surveillance radar Ground radar which interrogates air traffic with identifiable codes of pulses, triggers distinctive response from each target, extracts plots and assigns identity to each, normally presented as flight number, altitude and other information beside target on radar display. Aerial normally slaved in azimuth to main surveillance (primary) radar and may be on-mounted.

secondary winding That of transformer, magneto or other electrical device from which output is supplied.

second buy See *option (1)*; originally Lockheed term for option with paid deposit.

second-line servicing That carried out over planned period when aircraft is out of line service, sometimes at special off-base facility.

second moment of area Moment of inertia of a structural section whose mass is unity.

second pilot Person designated as second pilot in flight crew to assist PIC; in commercial crew usually has rank of First Officer. In ASCC definition: "not necessarily qualified on type".

second segment Second segment of normal take-off for large or advanced aeroplanes beginning at gear retraction at V_2 and maintaining this speed in climb to top of climb at end of segment when aircraft is levelled out, or climbed less steeply, to accelerate to FUSS or for power cutback in noise-abatement procedure.

second source Manufacturer assigned by government to augment output of major hardware item, eg aircraft or missile, with assistance of original design company, which remains in production as first source; no royalty is normally payable, and ** has no commercial rights to design, nor permission to sell to third party.

second strike Strategic attack with NW mounted after enemy's nuclear attack has taken place; objective of hardening is to confer a ** capability. DoD: "The first counterblow of a war; generally associated with nuclear operations".

second throat That downstream of working section and upstream of exhaust diffuser in simple blow-down supersonic tunnel. In operation traversed by weak normal shock.

Secop Single-engine climb-out procedure.

Secor Sequential collation of range. Usually */DME, which collates range with distance derived from DME. A basic technique in global mapping with geodesic satellites.

SECS Sequential-events control system.

section 1 Cross-section.

2 Major portion of airframe, eg nose *, tail *, but becomes dangerously ambiguous with wing *, body * when meaning normally (1).

3 Small subdivision of military air unit, normally (DoD) two combat aircraft.

4 Raw material rolled or extruded to standard (often complex) cross-section, as distinct from sheet, billets or strip.

5 Major subdivision of missile or rocket vehicle, eg guidance *.

section modulus Moment of inertia of structural member cross-section divided by perpendicular distance from neutral axis to outermost surface of section, ie most highly stressed fibre.

sector 1 Subdivision of air-defence frontier.

2 Limited range of azimuth, eg through which radar may scan (see * under *scan types*).

3 Subdivision of airspace by radio range characterized by letter A or N, also called quadrant.
4 Portion of commercial route between two traffic points.

sector controller Air-defence controller in charge of sector (1).

sector distance Length of air route sector (4).

sector fuel That allowed for in flightplan for one sector (4).

sector scan See *scan types*.

sector temperature/wind Those met values assumed in flight-planning one sector (4).

sector time In commercial operation, scheduled or actual time for sector (4).

secular Having a very long time period, eg a century or more.

secure Proof against interference by enemy and esp against information content of signals being deciphered by enemy. Various shades of meaning, eg * air refuelling is one beyond enemy radar range in which all communication is by lamp.

SED 1 Safe escape distance; min radius from airbase at which aircraft tail-on is assumed safe against hostile NW with GZ at airbase.
2 Scanning electron diffraction.
3 Sensor evolutionary development.

SEDIS, Sedis Surface-emitter detection identification system (ESM).

SEE 1 Society of Environmental Engineers (UK).
2 Secondary electron emission.

Seebeck effect Generation of EMF or current by dissimilar metals in circuit with junctions at different temperatures (see *Peltier*).

seeding 1 Aerial dispensing, at controlled rate per unit time or unit distance, of ECM payload such as chaff, flare pellets or other dispersed medium including aerosols. Hence * rate, * distance.
2 Aerial dispensing, at controlled rate per unit time, of condensation nuclei such as crystals of silver iodide or dry ice (solid CO_2) to trigger precipitation in rainmaking.

seeker Device able to sense radiation from target, lock-on and steer towards it, using radar (active or semi-active), optics (usually passive), laser or IR; rarely other methods. Normal sensor for terminal guidance of missiles and other guided ordnance.

seenot Air/sea rescue, prefix to longer words (G).

see-saw rotor See *teetering*.

SEFT Section d'Etudes et Fabrications des Télécommunications (F).

SEG Signaller, engineer, gunner (RAF aircrew grade, WW2).

segment 1 One of six standard subdivisions of take-off flightpath: 1, screen to gear-up at V_2; 2, gear-up to top of initial climb at V_2; 3, level acceleration to FUSS; 4, flaps-up to 5-min power point at FUSS; 5, level acceleration to en route climb speed; 6, climb to 1,500 ft at ERCS; rest is en route climb.
2 Portion of en route flight between two waypoints.
3 Portion of missile flight, eg boost, cruise (or midcourse) and terminal.

SEGS Selective (inclination) glideslope.

SEI 1 Small-engine instruction.
2 Standby engine indicator (or instrument) (CAA).

SEL 1 Selcal (ICAO).
2 Space Environment Laboratory (NOAA).
3 Single-engine licence.
4 Select.

Selcal Selective calling (see *Adsel*, *Dabs*).

Selecta-vision Registered name for holographic video recording.

select code That code displayed when ground interrogator (SSR, or in US ATCRBS) and airborne transponder are operating on same mode and code simultaneously.

selected track In traditional US-style airways structure, link route off-airways authorized to particular flight equipped with R-nav. Where R-nav systems (eg Decca) in use for decades term has little meaning because commercial and IFR pilots are not confined to airways and VOR radials.

selective address See *selective interrogation*.

selective fading Distortion of signal caused by variation in ionospheric density which causes fading that varies with frequency.

selective feathering Manual feathering using selector to pick out correct propeller.

selective fit Non-standard and rare engineering fit selected by hand to achieve desired mating of parts.

selective identification facility (or feature), SIF Airborne pulse-type transponder which provides automatic selective identification of aircraft in which installed (NATO); early form of selective interrogation used with Mk X ATCRBS.

selective interrogation With automated ATC systems, once aircraft acquired by computer, interrogation necessary only as radar scans exact sector containing that aircraft; if each transponder has its own (discrete) address code, computer can order interrogation on that code in particular sector, so transponders reply only when selected, number of replies cut by perhaps 99% and interference negligible (saturation avoided); two current systems are Adsel (UK) and Dabs (US).

selective jamming Jamming on single frequencies; often synonymous with spot jamming but can cover several specif frequencies.

selective loading Arranging cargo load to facilitate unloading in desired sequence.

selective pitch Early term for variable pitch.

selectivity Measure of ability of radio receiver to discriminate between wanted signal and interference signals on adjacent frequencies.

selector Manual demand input offering choice, eg * valve may have four or more fluid-flow positions, flap * may offer four or more settings.

selenium Non-metallic crystalline element (Se) with high electrical resistance except when illuminated, thus used in photocells, resistance bridges and, with semiconductor properties, in rectifiers.

seleno- Prefix, of the Moon; hence *-centric orbit, *-id (lunar satellite), *-logy (study of Moon).

self-aligning bearing Ball (occasionally other forms) bearing with outer race having part-spheroidal track allowing variation in shaft attitude.

self-bias See *automatic grid bias*.

self-contained night attack Capability of single aircraft navigating by night to acquire and strike designated point target and return to base (USAF omits word "point").

self-destroying fuze One designed to explode projectile before end of its flight.

self-destruct Ability of missile, RPV or similar air vehicle to explode into harmless fragments at particular time or geographical location.

self-erecting Of gyro, capable of erecting automatically after being toppled or started from rest.

self-FAC To carry out entire surface-attack mission without help of FAC or other aircraft giving aiming directions.

self-forging fragment Warhead which punches through thin top armour of AFVs: transverse disc of explosive converts disc of dense metal ahead of it into streamlined globule moving at over 9,000 ft (2,750 m)/s, its impact energy exceeding shear strength of armour.

self-generated multipath effect Multipath radio/radar problems caused by change in aircraft configuration, eg carriage of large stores.

self-guided missile One guided by means other than command.

self-induced vibration Caused by internal conversion of non-oscillatory to oscillatory excitation; also called self-excited vibration.

self-inductance Ratio of magnetic flux linking circuit to flux-producing current, ie EMF induced per unit rate of change of current; analogous to inertia in mechanical system.

self-manoeuvring stand Airport apron arranged so that gate parking and departure are accomplished by aircraft taxiing under own power. Normally nose-in parking avoided, and gate parking is accomplished with fuselage approximately parallel to adjacent wall of finger or terminal.

self-noise Internally generated noise within system, esp within sonar.

Selfoc Sheet electric-light focusing; patented (Nippon) two-component optical-fibre system.

self-repairing Having inbuilt ability automatically to adjust itself to best possible condition (eg operative channels, gain, feedback, power management) following any malfunction or damage.

self-rescue system Patented (Bell Aerospace) jet-propelled foldable parawing to enable combat aircrew to eject and fly to friendly area.

self-sealing tank Fluid, esp hydrocarbon fuel, tank constructed with layer of soft synthetic rubber sandwiched between main structural layers so that combat damage causes leak, bringing fuel into contact with rubber, which swells swiftly to block hole.

self stall Self-induced stall brought about by inherent tendency when near stalling AOA to pitch-up, eg by stall well outboard on swept wing with root still lifting strongly, thus progressively increasing nose-up tendency.

self-test Sequential test program performed by equipment (usually electronic) upon itself to determine whether it is operating correctly and which subsystems or circuits are faulty. In advanced form pinpoints each fault to particular device or PCB and stops.

selint Selective interrogation; not a particular system but technique.

selscan Selective-call scanning.

selsyn Patented (US General Electric) synchro system for transmitting remote compass indication to cockpit; fluxvalve current fed to three 120° coils driving central coil carrying indicator needle. From self-synchronous.

selvedge Woven edge of fabric from mill, does not unravel (US = selvage).

SEM 1 Scanning electron microscope.
2 Space environment monitoring.
3 Service engineered modification.
4 Standard electronic module.
5 Superconducting electrical machine (or machinery).

SEMA Special electronic mission aircraft.

semaphore TVC Thrust-vector control using oscillating blades pivoted to move across propulsive jet transversely and by partial blocking of flow path cause angular deflection.

semi-active homing Homing on to radiation reflected or scattered by target illuminated by radar forming part of system but mounted on fighter or surface platform.

semi-angled deck Carrier deck whose axis is at maximum diagonal angle permitted by original hull and deck without addition of large overhanging portion.

semi-annular wing Wing whose front elevation forms lower half of circle centred on fuselage or on each of two engine nacelles.

semi-armour-piercing Important category of gun-launched projectiles with AP properties conferred by high muzzle velocity and choice of materials but not relying entirely upon KE for penetration and with internal explosive charge. Some have added incendiary or tracer.

semi-automatic gun One which ejects spent case and reloads but fires only upon command; same as repetition or single-shot.

semi-automatic machine tool Usually, one that must be commanded to begin cycle of operations but thereafter runs through that cycle to completion, thereupon awaiting next start order.

semi-cantilever Not defined; loose term often meaning a braced cantilever, as in many lightplane wings.

semicircular separation ATC separation at over FL270 when magnetic tracks 000°–180° have one set of assigned FLs and 180°–360° have remainder.

semiconductor Electronic conductor whose room-temperature resistivity lies between that of metals and insulators (say, in range 10^{-2}–10^{9} ohm/cm) and which compared with metals has very few charge carriers, energy bands being full or empty (except for a few electrons or holes excited by heat in intrinsic * or provided by impurity doped in extrinsic *); n-type * has free-electron (negative) charge carriers and p-type has free-hole (positive) charge carriers.

semi-controlled mosaic One made up of photographs on approximately same scale so arranged that major ground features match geographical co-ordinates.

semi-hardened Given some protection against nuclear attack, eg buried in ground but without concrete, or enclosed in concrete but not buried; aircraft shelters in NATO are an example at bottom end of hardening scale.

semi-monocoque Structure in which loads are carried part by frame/stringer combination and part by skin. Almost all modern fuselages are of this type.

semi-permanent runway One paved with prefabricated metal by any of several standard methods.

semi-rigid airship Non-rigid with a single longitudinal member to assist in maintaining shape of envelope and distribute loads.

semi-rigid rotor Main lifting rotor of helicopter in which there are no hinges but flexible root behaves as leaf spring in vertical and horizontal planes to allow some freedom in flapping and lag modes.

semi-rigid theory Approximate treatment for elastic structure in which only finite number of degrees of freedom are allowed, each with one fixed mode.

semi-sonic blading Rotor blading (normally of axial compressor or fan) in which tips are just supersonic.

semi-span Half wingspan, theoretically measured from centreline but occasionally measured from side of fuselage, esp in calculating wing bending modes. Symbol b/2.

semi-stalled Stalled over portion of total aerofoil while remainder continues with attached lifting flow.

semi-transparent photocathode Radiation, eg light, on one side produces photoelectric emission on other.

SEMMS Solar electric multimission spacecraft.

Senel Single-Event Noise Exposure Level (see *noise*).

SENGO Senior engineering officer (RAF).

senior pilot Second-in-command of RN air squadron (UK).

sense aerial Fixed vertical receiver aerial with output same as max obtained from D/F or ADF loop and in phase with loop only over 180° to resolve 180° ambiguity of basic radio D/F method.

sense indicator Direct-reading to/from readout on VOR/ILS and similar panel instruments.

sensible atmosphere That part offering measurable aerodynamic resistance.

sensible horizon Circle of celestial sphere formed by its intersection with plane through eye of observer and perpendicular to local vertical.

sensitive altimeter Pressure (aneroid) instrument more accurate than simple altimeter with aneroid stack, corrective adjustments and three-needle or counter-pointer readout; not as accurate as servo-assisted.

sensitivity Generalized term for output divided by input, eg instrument or system response per unit stimulus.

sensitivity time control, STC Automatically reduces gain of weather radar as aircraft approaches cloud to avoid near clouds appearing brilliant and distant clouds faint; normally operates within radius of 45 km/25 nm.

Senso Sensor operator aboard ASW aircraft.

sensor According to DoD: "A technical means to extend man's senses; equipment which detects and indicates terrain configuration, presence of military targets, and other man-made and natural objects and activities by means of energy emitted or reflected by such objects; energy may be nuclear, EM (visible and invisible portions of spectrum), chemical, biological, thermal or mechanical, including sound, blast and vibration". Definition could be even broader: any transducer converting an input stimulus into a usable output.

Sentai 1 Division or flotilla comprising aircraft of one or (usually) two carriers (Japanese Navy, WW2).
2 Basic combat unit equivalent to UK wing or US group (Japan).

SEP 1 Specific excess power.
2 Single-engine performance.
3 Separate, separation.

Sepak Suspension of expendable penaids by kite.

separated flow Flow no longer attached to surface of immersed body.

separation 1 Breakdown of attached fluid flow round body into gross turbulence, occurring at particular time (stall) or place (* point); possible

to have sustained equilibrium with attached (laminar or turbulent) flow upstream and complete * downstream.
2 Authorized lateral, longitudinal and vertical clearances (distances) between aircraft under positive control.
3 Severing of links between rocket stages or other fall-away sections, also called staging.
4 Time when (3) occurs.
5 Discharge, release from active duty (USAF).
6 Distance, along any axis or direction, between interceptor and target.

separation manoeuvre Energy-gaining manoeuvre at low-alpha, high thrust, to close (reduce) or extend (increase) separation in air combat.

separation minima Minimum longitudinal, lateral or vertical distances by which aircraft are spaced through application of ATC (1) procedures (FAA).

separation motor Thruster to assist separation (3).

separation point In 2-D flow, point at which velocity of boundary layer relative to body becomes zero and flow separates from surface.

separation standards ICAO term for separation (2) minima.

separation test vehicle, STV Air vehicle for assisting development of separation (3), esp of tandem or wrap-round boost motors.

Sepla, SEPLA Sindicato Español Pilotos Lineas Aéreas (Spain).

SEPS Solar electric propulsion system (or stage).

SEQ Sequence.

sequenced ejection 1 Automatic small delays are built in between events, eg canopy, stick, calf garters, seat, drogue, harness release etc.
2 Ejection from multiseat aircraft in which crew members are fired in close-spaced series, captain or aircraft commander last.

sequence valve Fluid-flow controller scheduled to perform series of actions in sequence, each completion starting that following. Common US term: sequencer.

sequencing Assignation by ATC or radar controllers of strict order in which aircraft under control are to proceed, eg by selecting arrivals from holding points and, with path-stretching if necessary, achieving correct time/distance separation as they join localizer.

sequential collation of range, Secor Long-baseline system for determining vehicle trajectory by phase-comparison of responses of vehicle transponder to interrogation by three ground stations.

sequential computer Connected in series with other equipment, eg SSR or air-defence radars, eg to predict conflicts and advise on courses of action.

SERC Science and Engineering Research Council (UK).

SERD Support-equipment recommended data.

SEREB, Sereb Société pour l'Etude et la Réalisation d'Engins Balistiques (F).

Serf Studies of the economics of route facilities (ICAO).

serial 1 Element or group of elements within series given numerical or alphabetical designation; also that designation (DoD, NATO).
2 Numerical identity of particular hardware item, eg aircraft. Usually displayed on item concerned and recorded in all events concerning item (eg entered in pilot's log book, used as radio callsign in US and many other air forces).
3 In sequence as distinct from parallel; hence * data, * wiring.

serial data Successive signals passed over single wire or channel.

serial number See *serial (2)*.

serial rudders Rudder made in front and rear portions, latter hinged to former and deflecting through greater angle (eg on Dash 7). Also called serially hinged.

series 1 General term for subdivision or group within a larger related group, eg aircraft type Halifax II Series 1A, in this case corresponding to block number, modification state and other national terms.
2 In routine sequence as in manufacture of successive identical articles, eg * production, * aircraft; in this context often redundant word.
3 Mathematical expression with sequence of terms having form $a_1 + a_2 + \ldots a_n$.
4 Connected in succession on same line, wire or channel and thus all carrying same signal, current or flow.

series burn Consecutive burns of single or multiple rocket motors, eg on Space Shuttle Orbiter.

series loading Addition of inductance in series to increase electrical length of aerial and reduce natural frequency of system.

series modulation Connection of modulator in series with amplifier.

series/parallel redundancy Connection of fluid pipes or other lines to give particular item (eg control valves of LMAE) choice of series or parallel redundancy, either on command or automatic and switched by sensed failure.

series production Manufacture of successive identical (or near-identical) articles.

series redundancy Connection of two or more similar items, eg control valves, in series so that failure of one does not imperil functioning system.

series resonant circuit One in which inductances and capacitances are connected in series.

series servo Servo located in control system so that its output adds to that of a major input. Commonly used with SAS actuators to superimpose control on primary commands without motion at major input.

series yaw damper One connected into rudder circuit at PFCU, driving surface only but having no effect upstream and thus not felt at pedals; may be operative at all times, including take-off and landing.

SERL Services Electronics Research Laboratory (UK, MoD PE, Baldock).

Serrate Family of similar passive receivers carried by RAF WW2 night intruders and giving bearing of hostile night-fighter radars.

SERT Space electric (or electrostatic) rocket test.

service 1 Use, employment for design function; thus squadron *, line * (civil airline) etc.
2 Major branch of national armed forces.
3 To carry out routine maintenance and replenishment.
4 Facility offered to aviators, eg radar *, ATC *.

service area 1 Geographical extent of coverage of radio navaid or other surface-based electronic system.
2 Part of airfield assigned to routine servicing.
3 Part of airfield dedicated to support services, eg crash/fire/rescue, transport vehicles, trolley and stairway parking, etc.

service ceiling Basic performance parameter for (usually military) aircraft: height at which maximum rate of climb has fallen to lowest value practical for military operations, in UK and US traditionally equal to 100 ft/min.

service engineering Function of determining integrity of materiel and services to measure and maintain operational reliability, approve design changes and assure conformance with approved specs and standards (USAF).

service loads Structural loads actually met in service.

service module Major element of spacecraft supplying secondary power and consumables.

service stand Place assigned to a particular flight (7); can be gate position or marshalled location on distant apron. Some servicing is normally performed here before departure.

service tank Fuel tank located near engine to which fuel from other tanks is pumped and from which fuel is supplied to engine (arch).

Service Technique See *STAé*.

service test Test of hardware or technique under simulated or actual operational conditions to confirm satisfaction of military requirements.

service-test model Model (full-scale item is implied) used to determine characteristics, capabilities and limitations under simulated or actual service operational conditions (ASCC).

service tower Tower used to afford access to whole length of tall (eg ballistic or Shuttle) vehicle before liftoff; generally synonymous with gantry.

servicing To carry out service (3).

servicing appraisal exercise Formal study of servicing of particular hardware item (eg combat aircraft) in simulated combat conditions, to yield job times, difficulties, conflicts and shortcomings and make recommendations (UK usage).

servicing instruction Issued to remedy or prevent defect in military hardware item when action required may be urgent and recurrent (UK usage). May be issued when defect is suspected but not confirmed; roughly equivalent to Airworthiness Directive.

serving cord Usually seven-strand machine cord, used for wrapping control-cable splices.

servo Servomechanism, but now word in own right.

servo-assisted altimeter Pressure altimeter in which capsule movement is measured by sensitive EM pickoff whose output is amplified and used to drive motor geared to display.

servo-assisted controls See *servo controls.*

servo controls Not defined, and not recommended. BSI definition: a control devised to reinforce pilot's effort by a relay. A US definition: a ** is practically identical to a trimming tab. Appears to be general vague term for primary flight controls (not mentioned in BSI) where surface deflection is produced by force other than, or additional to, that of muscles. Such added force may come from PFCU or various types of tab (see *servo tab*).

Servodyne Registered (Automotive Products/Lockheed UK) family of pioneer PFCUs with mechanical signalling and hydraulic output.

servoed Operated by servo.

servo link See *servo loop.*

servo loop Control system in which human input is amplified by servomechanism provided with feedback so that, as desired output is attained, demand is cancelled.

servomechanism Force-amplifying mechanism such that output accurately follows input, even when rapidly varying, but has much greater power. Motions can be rotary but usually linear, and can be controlled by input only (open-loop) or by follow-up feedback (closed-loop) forming servo loop. Essential feature is that * constantly compares demand with output, any difference generating an error signal which drives output in required direction to reduce error to zero.

servomotor Rotary-output machine providing power locally; not necessarily part of servomechanism.

servo rudder Auxiliary rudder driven directly by pilot's pedals and moving main rudder by twin cantilever beams attaching ** to trailing edge of main surface. Common 1922–38. Precursor of servo tab but not a servomechanism.

servo system Servomechanism with feedback.

servo tab Tab in primary flight-control surface moved directly by pilot to generate aerodynamic force moving main surface.

SES Surface-effect ship.

SESA Society for Experimental Stress Analysis (US).

SESC Special environmental sample container.

SESP Space Experiment Support Program (USAF).
Sespo Support Equipment Systems Project Office.
sesquiplane Biplane whose lower wing has less than half area of upper.
SESS Space environmental support system.
Sessia Société d'Etudes de constructions de Souffleries, Simulateurs et Instruments Aérodynamiques (F).
SET Split engine transportation (fan and core travel as two items).
set 1 Drift (US, suggest arch).
2 To place storage device, binary cell or other bistable switch or gate in particular state; condition thus obtained.
3 Complete kit of special tools and/or parts for particular job, eg field modification; also called shop *.
4 Permanent deflection imparted by straining beyond elastic limit, esp lateral bending of saw teeth.
5 Hand tool/hammer for closing rivets (rare in aerospace).
SE/TA Systems engineering technical assistance.
Setac German augmented sector-Tacan system used as precision approach aid.
setback Distance from mould line (edge) and bend tangent line to allow for radius of bend in sheet metalwork.
SETE Supersonic expendable turbine engine.
SETOLS Surface-effect take-off and landing system.
set-on Assignment of offensive electronics, especially ECM jammer, to counter a particular threat, specifying frequency, signal modulation and, if possible, direction.
SETP Society of Experimental Test Pilots (US).
SETS Seeker evaluation test system.
setting Angle of incidence of wing, flap, tailplane (or trimmer) or other surface.
setting hammer See *peening*.
setting the hook Preventing roll reversal in air combat by centralizing stick and using rudder.
settling Sink of helicopter on take-off (if pilot does not increase collective) caused by lift ceasing to exceed weight as rotor rises out of ground effect.
settling chamber Section of wind tunnel upstream of working section in which large increase in cross-section results in great reduction in flow velocity and allows turbulence to die out before acceleration into working-section throat.
SEU Sensor electronics unit.
SEV Surface-effect vehicle, usually synonymous with ACV.
seven-bar format Standard format for presenting alphanumeric numerals in electronic displays, all numerals being created by illuminating some of four vertical and three horizontal bars (eg LEDs).
7 by 19 Standard high-strength steel cable made up of seven strands each of 19 twisted wires.
7 by 7 Standard steel cable made up of seven strands each of seven twisted wires.
seven flyings Seven types of flight by dead reckoning, two plane (plane and traverse) and five spherical (composite, great circle, Mercator, middle-altitude and parallel) (US usage, arch).
720° precision turn Standard US training manoeuvre; two complete circles flown at full power at as near as possible constant height at bank angle of 60°.
SEWS Satellite early-warning system.
SEWT Simulator for electronic-warfare training.
sextant Optical instrument for measuring altitude of celestial bodies.
Seybolt Hand tester of aircraft fabric which measures force required to punch a controlled hole.
SEZ Selector engagement zone.
SF 1 Signal frequency (often s.f.).
2 Scheduled freight.
3 Stick force.
S/F Ratio of system operating time divided by flight hours.
SFA 1 Société Française d'Astronautique (F).
2 Sous-direction de la Formation Aéronautique (F).
SFACT Service de la Formation Aéronautique et du Contrôle Technique; responsible for civil aircraft and aircrew licensing (F).
SFAR Special Federal Aviation Regulation(s) (US).
SFC 1 Specific fuel consumption, also sfc.
2 Surface (ICAO).
3 Side-force control.
SFCA Service des Fabrications du Commissariat de l'Air (F).
SFCC 1 Side-facing crew cockpit, ie has flight engineer panel.
2 Slat/flap control computer.
SFCS 1 Survivable flight-control system.
2 Safety flight-control system; protects aircraft (eg Concorde) against excessive AOA or jammed control column.
3 Secondary flight-control system.
SFDAS Société Française de Droit Aérien et Spatial.
SFDB Superplastic forming and diffusion bonding; important manufacturing technique in which structure is welded from sheet into gas-tight envelope and then inflated in heated mould until diffusion-bonded into desired shape.
sferics Study of atmospheric radio interference, esp from met point of view; sometimes spelt spherics.
SFF Self-forging fragment.
SFFL Standard foreign fare level (IATA).
SF/g Stick force per g.
SFI Special flying instruction (CAA).
SFIRR Solid-fuel integrated rocket/ramjet.
SFK Aramid (spider-fibre) reinforced plastics

composite (G).

SFL 1 Safe fatigue life.
2 Sequenced flashing lights.

SFM 1 Sensor-fuzed munition.
2 Self-forging munition.

SFOV Sensor field of view.

SFPA Staring focal-plane array; hence SFPAS adds seeker.

SFPCA Society for the Preservation of Commercial Aircraft (US).

SFPM Surface feet per minute; linear speed measure for machining or grinding.

SFPMAC Société Française de Physiologie et de Médécine Aéronautique et Cosmonautique.

SFRJ Solid-fuel ramjet.

SFSO Station flight-safety officer (RAF).

SFTS 1 Service flying training school.
2 Spaceflight telecommunications system.
3 Synthetic flight training system.

SFW Sensor-fuzed weapon(s).

SG 1 Specific gravity.
2 Spheroidal-graphite (cast iron).
3 Screen grid.
4 Shell gun, = cannon.
5 Sortie generation, or sorties generated.
6 Schlachtgeschwader, close-support attack wing (G, WW2).

SGAC Secrétariat Général à l'Aviation Civile (C adds "Commerciale") (F).

SGC Symbol-generator computer.

SGCI SG cast iron.

SGDN Secrétariat Général de la Défense Nationale (F).

SGEMP System-generated electromagnetic pulse.

SGL 1 Signal (ICAO).
2 Static ground line.

SGLS Space ground link subsystem (USAF).

SGME Self-generated multipath effect.

SGT Satellite ground terminal.

SGU Symbol-generator unit.

SH Showers (ICAO).

S/H Sample and hold; maintains present analog velocity until next sampling.

SHA System hazard analysis.

shadow factor Multiplication factor derived from Sun's declination, target latitude, and time in determining object heights from reconnaissance picture shadows; also called tan alt.

shadowgraph Technique, or photograph made by it, in which point-source light is focused parallel through tunnel working section and on to film; density gradients are visible as changed tonal values, proportional to second derivative of refractive index (Schlieren = first derivative).

shadow-mask tube Three-colour TV tube with three guns projecting red/green/blue beams through mask with about 500,000 holes.

shadow region Region where EM signals, eg radio or radar, are poorly received, usually because of LOS difficulties.

shadow shading Aircraft camouflage (UK, 1936–39).

shaft horsepower Horsepower measured at an engine output shaft.

shaft power Power available from rotating shaft.

shaft turbine Turboshaft engine, gas turbine providing power at an output shaft, in some cases driven by free turbine.

shaker See *vibration generator*.

shake-table test Any of various standard test schedules for delicate items conducted on vibration generator to stimulate vibration in service (according to USAF "during launch of missile or other vehicle").

shale fuel Aviation jet fuel, notably JP–4, derived from oil-bearing shales.

Shanicle Radio guidance system of hyperbolic type used on TM–61B cruise missile 1954; name from short-range navigation vehicle.

shank Inner portion of some propeller blades where section is not aerofoil but circular.

Shape Supreme HQ Allied Powers Europe (Int).

shape Appearance of electronic (eg radar) signal pulse when plotted in form of amplitude against time. This shape is not related to electronic shaping.

shaped-beam aerial One emitting beam whose main lobe is shaped by electronic phasing.

shaped-charge Hollow charge; warhead whose target-facing surface has form of re-entrant cone to generate armour-piercing jet.

shaped-charge accelerator Propulsion system using a shaped-charge propellant, a near-explosive, to achieve highest possible speed of man-made object in atmosphere after prior acceleration by rockets.

shaper Machine tool having single unidirectional cutter with horizontal (occasionally vertical) reciprocating action.

shaping Particular electronic meaning is tailoring shape of pulses or signal waveform, eg to assist manual command guidance of RPV or missile.

SHAR, Shar Shriharikota Range (India).

shared aerial One used by several receivers.

Shares Shared airline reservations system (LA-based).

sharp-edged gust Gust characterized by high rate of change of vertical air velocity per unit horizontal distance at particular place, thus aircraft encounters full gust vertical velocity almost instantaneously.

shaving Finishing machine-tool cut removing very small amount of metal; hence shave die, female die with cutting edges for finish-cutting cams etc.

SHC Synthesized hydrocarbons.

shear Stress in which parallel planes of loaded material tend to slide past each other; hence also deliberate cutting action by cutting edges which slide past each other.

shear centre See *flexural centre*.

shearing Cutting workpiece or material by shears, without formation of chips.

shear lag Structural stress diffusion in which lag of longitudinal displacement of one part of longitudinal section relative to another is result of shear applied parallel to length of structure.

shear load Load (force) applied in shear, eg of engine pod on wing spar.

shear modulus E_s, modulus of rigidity, approx half modulus of elasticity.

shear nut Thin nut used on bolt where load is entirely in shear, merely to retain bolt.

shear plan Lofting plan of body, eg fuselage, flying-boat hull, showing half-sections at numerous transverse planes.

shear slide Free-sliding piston moved by pressurant along length of propellant tank to force liquid propellant into (usually rocket) engine.

shear spinning Method of forming solid rocket case from preformed thick tubular billet by rolling against rotating mandrel, using two rollers 180° apart, normally performed at room temperature.

shear strength Stress required to produce fracture in plane of cross-section by two opposed forces with small offset.

shear stress Component of any stress lying in plane of area where stress is measured; for fluid, equal to τ, $\mu \frac{du}{dy}$ (see *Newton's laws*). Existence of ** in fluid is evidence of viscosity.

shear wave Wave in elastic medium causing any element of medium to change shape but not volume; in isotropic medium a transverse wave, mathematically one whose velocity field has zero divergence.

sheath 1 Metal tip to soft-blade propeller; also called tipping.
2 Envelope of plasma surrounding re-entry body.

sheathing See *sheath (1)*.

shed Traditional term for shelter (hangar) for aerostats, esp airship.

shedding Action for removal of ice from aircraft in flight, rain from windshield (windscreen) and non-vaporized material separated from ablating surface on re-entry.

sheep dipping Process whereby CIA pilots were given fully documented false professional backgrounds.

sheet Standard form of raw material; in case of metal, uniform sheet not over $\frac{1}{8}$ in (0·125 in, 3·175 mm) thick; thicker metal = plate.

Shelf Super-hard, extremely low frequency; military communications system.

shelf Figurative location where items are stored before use; thus * life, published maximum period during which item will not deteriorate in suitable storage; off-the-*, standard commercial product already available.

shell 1 Bare monocoque structure, eg fuselage or nacelle; shade of meaning includes thin-skinned and deformable, thus engine carcase excluded. Some definitions state "curved".
2 Ordnance projectile launched from gun or other tube, with or without own propulsion, and containing explosive, incendiary or other active filling; calibre normally greater than 20 mm. Word also applicable to AP projectiles of such calibres.
3 Supposed hollow spheres at different radii from atomic nucleus occupied by electrons, 2 in innermost *, 8 in next, 18 in next etc, all electrons in each * sharing similar energy level.

shellac Naturally derived resinous varnish.

Shelldyne Family of related synthetic fuels developed by Shell for USAF expendable turbojets, mainly characterized by high energy per unit volume.

shelter 1 Unhardened (generally recessed or underground) accommodation for civilians faced with air attack, in some cases attempting to offer some protection against nuclear attack (fallout *).
2 Unhardened reinforced-concrete structure accommodating (usually single) combat aircraft at dispersal and offering protection against conventional attack, eg Tab-Vee.

Sherardizing Anti-corrosion treatment similar to case hardening but employing Zn dust.

Sheridan tool Family of large stretch-presses often able to apply double curvature to thick plate.

SHF Super-high frequency, usually taken as 3–30 GHz (see *frequency*, *radio frequency*).

SHI Standby horizon indicator.

Shi Experimental number, with numerical prefix for year of Emperor's reign; thus 16-* = 1941 (Japanese Navy, 1931–45).

shield See *shielding (1)*.

shielded bearing Ball/roller/needle race with metal ring on each side to reduce ingress of dirt.

shielded cable See *screened*.

shielded configuration Aircraft deliberately designed so that parts of major structure, eg wing, are often interposed between sources (of noise or IR radiation) and ground observers or defences.

shielding 1 Material of suitable thickness and physical characteristics used to protect personnel from radiation during manufacture, handling and transport of radioactive and fissionable materials (DoD, NATO).
2 Obstructions which tend to protect personnel or materials from effects of NW (DoD).
3 Design philosophy of installing crucial parts of primary structure, whether damage-tolerant or otherwise, as far as possible behind others or in some other way geometrically protected from in-service damage.
4 See *screen (4)* (US usage).

shift Ability to move origin of radar P-type display away from centre of display; limit of * usually to periphery.

shim Thin spacer, from piece of paper (usually unacceptable) to large precision part tailored to specific application, to fill gap or adjust separation between parts; examples, to obtain exact rig (1) neutral and full-deflection settings of powered flight control surface, and to adjust separation of two ends of recirculating ball screwjack so that when bolted together balls have no play and no friction.

shimmy Rapid lateral angular oscillation of a trailing castoring wheel running over surface where coefficient of friction exceeds critical value, cured by twin-tread tyre or proper design of landing gear; usually affects unsteered nosewheel or tailwheel.

shimmy damper Add-on damper, usually with pneumatic/hydraulic dashpot, resistant to rapid variation of castor angle.

ship Aeroplane (US, colloq).

shipboard aircraft Aircraft designed to operate from surface vessel or submarine, including marine aircraft and rotorcraft (eg rotor-kite). Some definitions equate term with land aeroplane based on carrier.

shipment Complete consignment of hardware (probably not conveyed by sea), eg from manufacturer to operator or logistic base to user unit.

ship-plane Imprecise; most definitions restrict term to land aeroplane for operation from deck of carrier but USN includes catapulted seaplanes formerly used from surface warships.

ship-set Complete inventory of particular items for one aircraft.

shirtsleeve environment Popular and literally true description of desired environment in high-flying aircraft and spacecraft in which human performance is improved if special clothing does not have to be worn.

SHM Simple harmonic motion.

SHNKUK Roman initials of Society of Japanese Aerospace Companies.

SHNMO Shape host-nation management office (NATO).

shock 1 Shockwave.
2 Single large-energy pressure wave (see *shock front* [2]).
3 Often used to mean impact, single large externally applied impulse causing acceleration.

shock-absorber Device for dissipating energy by resisting vertical movements between landing gear (wheels or floats) and aircraft when running across surface, usually with unidirectional quality to reduce rebound and bounce (see *oleo*); other methods include simple steel or composite leaf springs, steel ring springs, rubber blocks in compression and bungee in tension.

shock body Streamlined volume added (eg on rear of wing) to improve area rule distribution; also called Whitcomb body, Küchemann carrot, speed bump, etc.

shock cloud Localized cloud caused by violent changes in flow conditions in close proximity to supersonic aircraft, notably in Prandtl-Meyer expansions, eg over wing and canopy.

shock compression Fluid flow compression occurring virtually instantaneously in passage through shockwave; for normal shock, ratio of pressures $p_1/p_2 = 7M^2/6–1/6$ where M is initial Mach.

shock cord See *bungee*.

shock diamonds Approx diamond-shaped reflections, brilliantly luminous in hot jet (eg from rocket or afterburner), caused by reflection of internal inclined shocks from edge of jet at boundary with atmosphere.

shock drag That drag associated with a shockwave (which always causes loss in total or static pressure), normally varying as fourth power of velocity or pressure amplitude.

shock excitation Generation of oscillations in circuit at natural frequency by external pulses, eg for sawtooth generation.

shock expulsion Faulty operating condition of inlet to supersonic airbreathing engine in which, for various reasons, gross flow breakdown occurs and inlet shock system is expelled forwards; accompanying large and possibly dangerous increase in drag. Often used synonymously with inlet unstart.

shock front 1 See *shockwave*.
2 Boundary between pressure disturbance created by explosion (in air, water or earth) and ambient surrounding medium.

shock isolator Device, usually mechanical and assembled from solid parts such as deformable rubbers/plastics or metal deflecting well within elastic limit, which absorbs input movements (eg to accommodate input vibration while keeping output still) or cushions large impacts (by permitting output to travel over a distance which absorbs energy within permitted limits of acceleration). Thus, some absorb vibration, usually of small amplitude, while others absorb shock, which in case of ICBM suspended in silo may require travel in order of 1 m.

shock mount Shock isolator on which delicate object is mounted.

shock softening Reduction of linear rate of pressure rise through shockwave, esp in proximity to subsonic boundary layer; ie increase in thickness of shock (1) from about 10^{-3} mm by several orders of magnitude.

shock spectrum Plot of peak amplitude of response of single-degree-of-freedom system to various single applied shocks (3).

shock stall Gross breakdown of flow behind shockwave on wing (esp one of large t/c ratio or for any other reason causing large airflow acceleration) at about critical Mach number, causing symptoms of loss of lift and turbulent wake resembling stall, but at normal AOA.

shock tube Wind tunnel for hypersonic studies in

which fluid at high pressures, usually involving rapid combustion to increase energy, is released by rupturing diaphragm and accelerates through evacuated working section containing model. Many varieties, most having stoichiometric gas mixture as driver and large-expansion-ratio (over 200) supersonic nozzle upstream of working section, giving M up to 30 and T around 18,000° K.

shockwave 1 Surface of discontinuity between free-stream fluid and that affected by body moving at relative velocity greater than speed of sound in surrounding fluid. As fluid accelerates round body, if it eventually reaches local Mach 1 a weak shock forms perpendicular to flow, called a normal shock. Pressure difference $(p_1-p_0)/p_0$ is zero and flow downstream is subsonic. As M increases, shock leans back, becoming an inclined shock, at angle $\alpha = \sin^{-1} 1/M$, while pressure ratio and velocity of propagation V_0 increase according to M and angle of deflection, a property of geometry of body; for 15° deflection (ie wedge or cone of 15° semi-angle) at Mach 3 static $(p_1-p_0)/p_0 = 5$ and $V_0/a_0 = 2{\cdot}1$, ie * moves at twice speed of sound.
2 Continuously propagated pressure pulse formed by blast from explosion in air by air blast, underwater by water blast and underground by earth blast (DoD, NATO).

Shodop Short-range Doppler.

shoe Detachable interface between pylon or hardpoint and store, often specific to latter.

shoot bolt Linear bolt type of panel latch.

shooting the breeze Engaging in casual shop-talk (US).

shop head End of rivet upset when rivet is used.

shop visit Removal of item from aircraft for repair or other attention in specially equipped workshop, usually of customer.

Shorads Short-range air-defense system (USA).

Shoran From short-range navigation, precision radio navaid based on timing pulsed transmissions from two or more fixed stations; in conjunction with suitable computer used for blind bombing.

shore Strut supporting airship during manufacture.

shoreline Is drawn straight across all inlets less than 55·6 km (30 nm) wide (ICAO).

short-distance navaid One usable within 320 km/200 miles (NATO).

short field Limiting field or runway demanding special take-off procedure.

short finals Last part of approach, usually defined as that commencing at inner marker.

short-haul Several definitions, eg max-payload range (knee of graph) 1,609 km (1,000 statute miles) or less; see also *short-range transport*.

short-life engine One designed for single flight or any other purpose not requiring prolonged use, and normally qualified for running time of 50 h.

short-range attack missile ASM launched at range not exposing launch aircraft to terminal defences (USAF).

short-range ballistic missile Up to about 600 nm (1,112 km, 691 miles) (DoD).

short-range Doppler Trajectory measurement using Dovap plus Elsse.

short-range transport Range at normal cruising conditions not to exceed 1,200 nm (2,224 km, 1,382 miles).

short round 1 Round of ammunition deficient in length (DoD "in which projectile has been seated too deeply"), causing stoppage.
2 Ordnance delivered on friendly troops.

short stacks Briefest form of PE (4) exhaust for cowled engine.

short take-off and landing, STOL Usually defined as able to take off or land over 50 ft screen (note, not 35 ft) within total distance of 1,500 ft (457 m).

short trail Towing position for sleeve (or presumably other forms) of target in which target is immediately astern of towing aircraft.

short wave Not defined and rare in aerospace: traditional radio meaning is decametric (10–100 m) corresponding to 30–3 MHz; FAA meaning is frequencies 7·7–2·8 MHz; scientific is 0·4–1 μ wavelengths.

shot 1 Commercial lead * for shotguns, normally used as cheap variable mass.
2 Tailored hardened steel balls of graded sizes.
3 Solid-projectile ammunition, eg for air-firing practice or AP type.
4 Report indicating a gun has been fired (DoD).
5 Single flight of unguided ballistic rocket, eg probe.

shot-peening Bombarding metal surface with air-propelled shot (2), usually to harden and relieve internal stress.

shoulder Area immediately beyond edge of pavement so prepared as to provide transition between pavement and adjacent surface.

shoulder bolt Thread of smaller diameter than shank; for attaching plastics parts where over-tightening must be avoided.

shoulder cowl Usually means cowling panel(s) hinged upwards near top of sides.

shoulder harness Seat harness including straps passing over shoulders to prevent body jack-knifing forward.

shoulder season Intermediate demand between low and peak (pax traffic).

shoulder wing Wing attached between mid and high positions. Original German *schulterdecker* implied wing depth more than half that of fuselage, with blended wing/body junction.

shower(s) Precipitation from convective cloud characterized by sudden onset, rapid variation in intensity and sudden stop, with intervening periods of part-clear sky.

showerhead Liquid-propellant rocket injector in

which numerous fuel and oxidant sprays are distributed (with various forms of impingement) over flat or curved surface.

show finish Glass-like finish (on homebuilts, usually) achieved by repeated doping and rubbing down.

SHP, shp Shaft horsepower.

shrimpboat Small marker of clear plastic on which controller writes flight identity, FL and other information, subsequently moved by hand to remain adjacent to blip on display.

shrinkage Natural reduction in dimensions of most castings on cooling (see *shrink rule*).

shrink fit Force or interference fit between two metal parts obtained by heating outer, cooling inner, or both.

shrink rule Casting mould made 0·010 (linear) oversize to allow for shrinkage.

shroud 1 Plate formed integrally with gas-turbine fan, compressor or turbine blade usually in plane perpendicular to blade major axis. In fan and upstream compressor blades usually as part span, * being formed in halves on each side of blade and mating with those adjacent to damp vibration. In turbine rotor invariably on tip, serving as ring minimizing gas leakage around periphery.
2 Covering plate on face of centrifugal impeller enclosing flow passages and preventing leakage.
3 Heat-resistant aerodynamic fairing over space payload or any forward-facing projection on space launch vehicle, ICBM or other hypersonic vehicle.
4 Extensions of fixed surface of aerofoil (eg wing, tailplane) projecting behind hinge line of movable surface (eg flap, aileron) to reduce drag or improve flight control.

shroud coolant Refrigerant cooling volume in which cryogenic cooling takes place, eg LH_2 serves as ** surrounding liquefaction of helium.

shrouded blade Blade fitted with shroud (1).

shrouded impeller Centrifugal impeller, eg of supercharger, enclosed by shroud (2).

shrouded insulator Radio aerial insulator (eg HF wire) fitted with overlying shroud (normal meaning of word) to prevent ice connection to metal structure.

shroud lines Main suspension cords of parachute connecting load to canopy.

SHSS Short-haul system simulation.

shunt connection Bypass circuit, usually taking most of current.

shunt excitation Feeding mast radiator (aerial) about 0·25 of way up, earthed at base.

Shup Silo-hardness upgrade program (USAF).

shutdown 1 For rocket engine, cutoff.
2 For conventional aircraft-propulsion engine, reducing power to zero and rendering inactive, eg by turning off HP cocks. If in flight becomes IFSD. Normally follows obvious or signalled failure.
3 Event in which (2) occurs.

shuttered fuze Inadvertent initiation of detonator will not initiate booster or burst charge.

shuttle 1 Generalized term for reusable space launch vehicle recovered by aeroplane-type flight.
2 High-frequency trunk-route service characterized by no-reservations, payment on board and, in some cases, aircraft always boarding, and departing if full before announced time.
3 Sliding drive member of flight-deck accelerator (catapult).

shuttle bombing Use of two bases, aircraft making one-way bombing missions between them.

shuttle valve Fluid-flow valve of bistable type which passes flow from one line and isolates other or vice versa.

SI 1 Système International d'Unités; standardized system of units adopted (but not yet fully implemented) by all industrialized nations.
2 Servicing instruction.
3 Straight-in (approach).

Si Silicon.

SIA 1 Structural-integrity audit.
2 Service de l'Information Aéronautique (F).

SIAG Salons Internationaux de l'Aviation Générale (F).

Sialon Silicon-aluminium-oxinitride.

Siam Self-initiated anti-aircraft missile; carries out IFF interrogation and handles subsequent interception automatically.

siamese To join two similar items into single paired unit or to bifurcate duct into two equal parts; hence siamesed, adj.

SIAR Service de Surveillance Industrielle de l'Armement (F).

SIAT Service instructor aircrew training.

sibilant filter One removing hissing frequencies from speech on R/T.

SIC 1 Steady initial climb, ie V_4.
2 Standards Information Center (NBS).
3 Service instruction circular.

SICBM Small ICBM.

SICM Small intercontinental missile.

SID 1 Standard instrument departure.
2 System-integration demonstration.
3 Spray-impingement drag (marine aircraft).
4 Supplemental inspection document.
5 Switch-in deflector.

SIDE Suprathermal ion detector experiment.

side-arm controller Primary flight-control input in form of miniature control column at side of cockpit (of combat aircraft) on console incorporating armrest to facilitate accurate flight under conditions of large applied acceleration.

sideband Band of frequencies produced above and below carrier frequency by modulation; sum and difference frequency products are called upper * and lower *.

side-by-side 1 Two-seat aircraft in which seats are in same transverse plane at same level.

2 PE (4) in which connecting rods are both same, with big ends side-by-side on crankpin (thus, cylinders of opposed banks are not quite in line); constructional form of most modern light-aircraft engines.

side car Airship car suspended away from centre-line plane.

side direction Normal to plane of symmetry.

side elevation Portrayed as seen from side, in case of drawing as seen from infinite distance, ie orthographically.

side float Usually means sponson.

side force Force acting normal to plane of symmetry.

side-force control Aircraft flight-control system capable of exerting side force, normally by vertical surfaces in front of as well as behind c.g. Aircraft with ** can almost instantly change track by flying diagonally, without need to roll or change fuselage axis; make immediate lateral corrections to line of fire of fixed gun; move laterally out of hostile gunfire without prior roll.

side frequencies Carrier plus or minus audio frequency.

sideline noise Measured beside TO run at distance from centreline of 450 m (ICAO Annex 16) or 0·35 nm (FAR 36 CAN 5, 4-engined aircraft) or 0·25 nm (2-, 3-engined). See *noise*.

sidelobe Lobes of aerial radiation propagated at angle to main lobe, normally unwanted and cause of clutter or false returns.

sidelobe suppression Various techniques for eliminating not presence of sidelobes but their effects.

side-looking Scanning to either side of aircraft track; hence SLAR, radar whose output is detailed picture of terrain near track, either all on one side or equally on both sides, depending on aerial arrangement.

side marker board Display beside airport gate arranged for nose-in parking giving indicator marks, usually vertical white bars on black board; when aligned with captain's left shoulder, airbridge is aligned with passenger door.

side oblique Photograph taken with camera oblique and perpendicular to longitudinal axis of aircraft.

sidereal Pertaining to stars, but see following entries. (Rhymes with material.)

sidereal day Time for one rotation of Earth as defined by period between successive transits of vernal equinox (in ASCC wording, alternative name, First Point of Aries) over upper branch of any chosen meridian, equal to 24 h of mean sidereal time or 23 h 56 min 4·09054 s of mean solar time.

sidereal hour angle Angular distance west of vernal equinox; arc of celestial equator or angle at celestial pole between vernal equinox and hour circle of observer (see *right ascension*).

sidereal month Average period of revolution of Moon with respect to stars: 27 days 7 h 43 min 11·5 s.

sidereal period 1 Time taken by planet to complete revolution around primary as seen from primary and referred to fixed stars.
2 Interval between two successive returns of Earth satellite to same geocentric right ascension.

sidereal time Time measured from rotation of Earth related to vernal equinox, called local time or Greenwich time depending on choice of meridian; when adjusted for nutation inaccuracy called mean time.

sidereal year Period of Earth's rotation around Sun, related to stars; in 1980 equal to 365 days 6 h 9 min 9·5525 s and increasing at about 0·000095 s per year (see *tropical year*).

sideslip Flight manoeuvre in which controls are deliberately crossed, eg to * to left aeroplane is banked to left while right rudder is applied; result is not much change in track but flight path inclined downwards, ie steady loss of height without significant change in airspeed and with longitudinal axis markedly displaced from flightpath. Angle of * is angle between plane of symmetry and direction of motion (flightpath, or relative wind). Rate of * is component of velocity along lateral axis.

sidewall treatment Addition of sound-absorbent material along sides of passenger cabin.

sideways translational tendency Characteristic of single-rotor helicopter to drift to L or R under thrust of anti-torque tail rotor, unless main rotor tip-path plane is tilted in opposition to neutralize this thrust.

SIDs Standard instrument departures; SIDS, standard instrument departure system.

SIE Self-initiated elimination.

SIERE Syndicat des Industries Electroniques de Reproduction et d'Enregistrement (F).

Sierra Prefix, supersonic (esp SST) airway (ICAO).

Sierracote Family of patented glass and/or acrylic transparencies which may include anti-icing heating by transparent film.

SIF Selective identification facility (or feature).

Sifet Società Italiana Fotogrammetria e Topografia.

SIFL Standard industry fare level (US).

SIFTA Sistema Interamericano de Telecommunicaciones para las Fuerzas Aereas (Int).

SIFV Sensor instantaneous field of view.

SIG 1 Signature (DoD, ICAO).
2 Significant (ICAO).
3 Stellar/inertial guidance.
4 Simplified inertial guidance.

sight 1 Optical device for measuring (eg drift *) or aiming (gun-*), often incorporating magnification or combined with HUD (Hud-*).
2 To take observation with sextant.

sight gauge Simple fluid-level gauge.

sighting Visual contact; does not include other forms of contact, eg radar, sonar.

sighting angle Angle between LOS to aiming point and local vertical; at time of bomb release, same as dropping angle (ASCC).

sighting out of wind Rigging by eye (arch).

sighting pendant Vertical plumb wire ahead of airship control car to aid in steering and drift estimation.

sighting station Crew station from which sight (2) taken or other optical measures made.

sight tracking line LOS from a computing gunsight reticle image to target.

Sigint See *signal intelligence*.

Sigmet Weather advisory service to warn of potentially hazardous (significant) extreme conditions dangerous to most aircraft, eg extreme turbulence, severe icing, squall lines, dense fog.

signal 1 Anything conveying information, eg by visible, audible or tactile means.
2 Pulse(s) of EM radiation conveying information over communications system.
3 Information conveyed by (1, 2).
4 Any electronic carrier of information, as distinct from noise.
5 Standard visual symbols displayed on simple airfield.

signal area Plot of ground at simple airfield, usually adjacent to tower, set aside and equipped for display of ground-to-air signals (5) for informing pilots of aircraft not equipped with radio, eg of circuit direction, local hazards, prohibitions, etc.

signal data General term for elint output or for any recorded (if possible quantified) information on received signals (2).

signal flare Flare pyrotechnic whose use conveys known meaning, eg two-star red.

signal frequency Frequency of transmitted carrier as distinct from component parts, or received carrier as distinct from intermediate frequency to which it is converted.

signal generator Versatile oscillator capable of outputting any desired RF or AF at any selected amplitude and with output modulated to simulate actual transmission.

signal intelligence General term for communications intelligence plus electronic intelligence; art of detecting, recording, analysing and interpreting all unknown or hostile signals.

signalman Authorized person using hands, wands and/or lights to marshal aircraft on apron or elsewhere on airport manoeuvring area.

signal pistol Projector of pyrotechnics, eg Very pistol.

signal star Pyrotechnic of distinctive character emitted from cartridge or signal pistol.

signal-to-noise ratio Ratio of amplitude of desired signal to amplitude of noise signals at a given point in time (DoD). Ratio, at selected point in circuit, of signal power to total circuit noise power (ASCC). Number of dB by which level of fully modulated signal at max output exceeds noise level, all values being rms.

signature Characteristic pattern of target displayed by detection and identification equipment (DoD, NATO etc). Like thumb-print, * analysed with sufficient accuracy can identify source as to type and even to specific example of emitter. Can be EM (radar, IR, optical), acoustic (eg SST boom *) or velocity (Doppler *).

significant tracks Tracks of aircraft or missiles which behave in an unusual manner which warrants attention and could pose a threat to a defended area (DoD, NATO).

significant turn Change of heading large enough for explicit account to be taken operationally of reduction of climb gradient.

Sigsec Signals security (USA).

s(i,k) Slope of sound pressure level; change between ⅓-octave SPLs at i-th band at k-th moment in time. Hence Δs (i,k) = change in SPL slope, s′ (i,k) is adjusted slope between adjacent adjusted bands, and $\bar{s}$ (i,k) is average slope (see *noise*).

SIL 1 Suomen Ilmailu Liitto (Finnish Aeronautical Society).
2 Site d'intégration lanceurs (F); launcher assembly site.
3 Systems integration laboratory.

Silastic Proprietary range of silicone rubbers and sealants including many resistant to hydrocarbon fuels.

silent target Non-emitting, ie no radio, radar, IR, laser, TV or other detectable radiation.

silica Silicon dioxide, quartz, SiO_2.

silica gel Numerous stable sols of colloidal particles, some important as drying agents.

silicon Si, abundant versatile solid with many forms, eg dark crystalline form or brown powder; important in alloys, in vast range of compounds and in pure crystalline form as semiconductor; basis of major part of microelectronics.

silicon chip Popular name for microcircuit fabricated in chip scribed from slice of single-crystal epitaxial silicon. Formerly each chip contained devices and subcircuits, today can be a complete equipment requiring only packaging and human interfaces.

silicone Generalized term for polymeric organosiloxanes of form $(R_2\ SiO)_n$, including elastomers, rubbers, plastics, water-repellants and finishes, eg resins, lacquers and paints.

silicone cork Relatively cheap ablator and heat insulator, eg on Shuttle external tank.

silicon nitride Refractory ceramic, Si_3N_4; increases strength slightly to 1,200°C, good thermal shock resistance because of small thermal expansion, and resists many forms of chemical attack.

silo Missile shelter that consists of hardened vertical hole in ground with facilities either for lifting missile to a launch position or for direct launch from shelter. Normally closed by hardened lid and provided with shock-isolating missile supports.

Silver C Intermediate certificate and badge for gliding requiring flight of at least five hours, gain in height of at least 1,000 m (3,281 ft) and straight flight of at least 50 km (31 miles).

silver cell Silver/zinc.

silver solder Jointing alloy of silver, copper and nickel.

silver tux Astronaut suit (US, colloq).

silver/zinc Ag/Zn electric dry battery common in applications calling for single-shot high-power electric supply.

SIM SAM (1) intercept missile; bomber defence.

Sim Simulation, simulator.

SIMA Scientific Instruments Manufacturers Association of GB (UK).

SIMD Single instruction, multiple data stream.

Simmonds nut Pioneer stop nut incorporating tightly held fibre washer.

simple flap Hinged wing trailing edge, with or without shroud (4) but without intervening slot.

simple harmonic motion Regular oscillation as exemplified by alternating current or drag-free swinging pendulum; projection on any axis in same plane of point moving round circle with constant angular velocity; expressed by $y = a \cos (2\pi nt + b)$ where y is distance from origin (at centre) at time t, n frequency and b a phase constant such that at $t = 0$, $y = a \cos b$.

simple stress Either pure tension, pure compression or pure shear.

simplex With no provision for redundancy.

simplex burner Simple gas-turbine fuel burner fed by single pipe leading to nozzle surrounded by air swirl vanes and with flow proportional to square root of supply pressure.

simplex com Communications technique in which signals pass in one direction only at any one time; can be single or double channel and switched by press-to-speak, manual T/R switch or voice-operated.

simplified directional facility ILS localizer (108·1–111·9 MHz) with aerial offset from runway and emitting beam usually not exactly aligned with it nor providing G/S information.

simply supported beam Pin-jointed at both ends.

Sims Secondary-ion mass spectrometry.

SIMUL Simultaneously (ICAO).

simulated forced landing Includes all actions except landing.

simulated-operations test Operational test needed to support statements of new requirements and support positions and programmes (USAF).

simulator Dynamic device which attempts to reproduce behaviour of another dynamic device, eg aircraft or missile, under static and controlled conditions for research, engineering design, detail development or personnel training. Those used for research are similar in appearance to other large electronic items; those for training often incorporate a complete cockpit or flight deck carried on hydraulic rams giving motion about all three linear axes and three rotary axes (eg exactly reproducing asymmetric swing on take-off engine failure or buffet at approach to stall with correct fuselage pitch attitudes), and model-form or electronically generated external scene for flight in neighbourhood of particular selected airport(s).

simultaneous engagement Concurrent engagement of target by interceptors and SAMs (DoD).

simultaneous pitch control See *collective*.

simultaneous range Radio range which simultaneously broadcasts voice messages.

SIN Significant-item number.

Sinaga Sindacato Nazionale Gente dell'Aria (I).

SINCGARS, Sincgars Single-channel ground and airborne radio subsystem.

sine curve Obtained by plotting sine on linear axis, graphically identical to plot of SHM.

sine wave Wave of SHM form, eg EM radiation.

S-ing Performing succession of S-turns.

single A single-engine aircraft, normally used for GA aeroplanes.

single-acting Actuator pushing in one direction only, with spring return; push-push.

single-aisle aircraft Narrow-body, having twin or triple seats on each side of one axial aisle.

single-axis autopilot One offering stability or control about one aircraft axis only, eg pitch, roll or yaw.

single-axis head Homing head whose sensor scans only in one plane.

single-base propellant Traditional term originating in so-called smokeless powders based on either Nc or Ng alone (see *double-base*).

single-bay Having only one set of interplane struts joining wings of biplane on each side of centreline.

single-channel simplex Same frequency in both directions.

single-crystal alloy Complete workpiece formed in piece of metal grown as single crystal, possibly containing occasional atomic imperfections but devoid of gross intercrystalline joints and as far as possible with lattice orientation selected to increase strength in direction of greatest applied stress.

single curvature Curved only in one plane, as surface of regular cylinder.

single-entry compressor Radial or centrifugal impeller with vanes on one side only.

single-face repair Repair to sandwich structure involving core and one face only.

single-flare joint End of rigid pipe flared out but not turned back on itself.

single-float seaplane Large central float and small stabilizing floats outboard.

single ignition One coil or magneto, feeding one plug per PE (4) cylinder.

single-pass heat-exchanger Each fluid passes once through without turning.
single-regime engine One designed to operate always under same conditions, eg jet VTOL.
single-rotation Composed of one unit (propeller or propfan), as distinct from two equal units rotating in opposite directions.
single-row engine Radial engine with all cylinders in same plane driving one crankpin.
single-shaft engine Gas turbine in which all compressor stages are connected to same turbine. There may be a concentric shaft linking free turbine to shaft output.
single-sideband, SSB Reduction of bandwidth by transmitting only one sideband and suppressing other (and usually carrier also); receiver heterodynes at original carrier frequency.
single-sink flow Fluid flow capable of exact representation using a single sink.
single-spar wing Wing in which primary flight loads are borne by one spar (as distinct from a box), possibly made with two booms on at least one edge or having U or circular section. Does not preclude secondary spanwise member(s) for trailing-edge surface loads.
single-spring flexure Tunnel balance in which model is supported by single (usually vertical transverse) member locally thinned at one transverse point to serve as pivot sensitive to forces or couples about that axis only.
single-stage compressor Compressor achieving total overall pressure ratio in one operation, eg by centrifugal impeller or single row of axial blades.
single-stage turbine Turbine having only one set of axial rotor blades or inward radial vanes.
single-stage vehicle Air vehicle (aircraft, RPV, missile, rocket) with only one propulsion system; term not common for vehicles with sustained cruise propulsion, as in manned aircraft.
single tipping Sheath (1) made in one piece.
single-wedge aerofoil Cross-section has sharp leading edge, flat sides and blunt or square trailing edge, eg vertical tail of X–15.
sinide Silicon nitride.
sink 1 Theoretical point in fluid flow at which fluid is consumed at constant rate.
2 Large mass to which waste heat can be rejected.
sinking Rapid uncontrolled increase in rate of descent with little change in horizontal attitude, usually caused by increasing AOA at approach to stall.
sinking speed Vertical component of velocity of aircraft without propulsive or sustaining power in still air; for glider or engine-out aeroplane, = TAS × sin gliding angle = TAS ÷ glide ratio.
sink rate Rate of descent of free-fall unpowered lifting-body, especially glider at best L/D. Usually same as sinking speed.
SINS, Sins Ship's inertial navigation system.
sintering Bonding powder or granules under heat and pressure; no melting takes place and product may be ceramic, cermet, metal or many other types of material, and compact or having any desired degree of controlled porosity. Related terms: diffusion bonding, hot isostatic pressing, powder metallurgy.
sinusoidal Having form of sine waves; characteristic of SHM.
SIOP Single integrated operational plan (US for use of NW).
SIP 1 Surface impact (or impulsion) propulsion.
2 Self-improver pilot.
Sipac Sindicato Italiano Piloti Aviazione Civile.
Sipaer Serviço de Investigação e Prevenção de Acidentes Aeronauticos (Brazil).
Sipri Stockholm International Peace Research Institute.
SIPU Super-integrated power unit, engine designed to run either on conventional fuels or, in emergency, stored-energy supplies, such as MOM or MEPU fuel.
SIR 1 Strip, inspect(ion) and rebuild.
2 Search and interrogation radar(s).
3 Shuttle imaging radar.
SIRE Satellite infrared experiment.
Sirpa Service d'Information et de Relations Publiques des Armées (F).
SIS 1 Semiconductor/insulator/semiconductor.
2 Stall-identification (or, rarely, inhibition) system.
SISD Single instruction, single data stream.
SIT 1 Spontaneous ignition temperature.
2 Surplus inventory tag.
3 Silicon-intensified target.
SITA Société Internationale de Télécommunications Aéronautiques (Int).
site error Radio navaid inaccuracy caused by radiation reaching destination by indirect, ie reflected, routes; eg NDB or VOR radiation reflected from building near beacon.
Sitelesc Syndicat des Industries de Tubes Electroniques et Semi-Conducteurs (F).
SIU Sensor (USAF, systems) interface unit.
SIWL Single isolated-wheel load.
six Six-o'clock position, ie directly behind one's own aircraft; hence "check *!"
6 by 7 cable Flexible aircraft cable with cotton core surrounded by six cables each twisted from seven wires.
six-component balance Wind-tunnel balance that simultaneously measures forces and moments (couples) about all three axes.
sized fibre Virgin carbon fibre sized with resin binder.
SJ Ski-jump.
SJAC Society of Japanese Aircraft Constructors; Japanese-language acronym = SHNKUK.
SJB Semi-jetborne (jet VTOL).
SJI Stores jettison indicator.

skate Platform(s) with castoring wheels or air-cushion pads for moving large aircraft on ground.
SKB Student construction (design) bureau (USSR).
SKC Sky clear (ICAO).
SKE Station-keeping equipment.
sked Scheduled (ICAO).
skeg 1 Small fixed fin at rear of afterbody step of marine aircraft to improve stability when taxiing.
2 ACV sidewall.
3 Ventral strip along seaplane afterbody serving as support on land.
skew aileron One whose hinge axis is markedly not parallel to transverse axis but diagonally across tip.
skew compressor Fluid compressor intermediate between centrifugal (radial) and axial.
SKF Superkritischer flügel; supercritical wing (G).
Skiatron Dark-trace CRT or related display.
skidding 1 Sliding outward in turn because of insufficient bank or excess rudder, opposite of slip.
2 Incorrect operation of ball or roller bearing in which sliding friction occurs instead of pure rolling, for various reasons.
Skiddometer Towed runway-friction measurer with 17%-slip braked centre wheel.
skid fin Fixed fin mounted high above c.g., eg above upper wing of biplane, to reduce skidding (1).
skid landing gear 1 In early aeroplanes, rigid ski-shaped member projecting ahead of landing gear to prevent nosing-over.
2 In helicopters, fixed tubular landing gear often provided with small auxiliary wheels (eg winched down by hand) to confer ground mobility.
skid-out See *skidding (1).*
skid transducer Input sensor of anti-skid wheel brake system, able to sense any sudden variation in wheel rotational speed.
skiing glider Skier also wearing lifting aerofoil or parafoil.
ski-jump Take-off over a * ramp.
ski-jump ramp Curved ramp terminating at (ideally) about 12° to horizontal providing large benefits to rolling take-off by vectored-thrust STOVL aircraft, including shorter run and/or greater weight of fuel/weapons, and increased safety (particularly off ship) in event of failure of engine or nozzles.
ski landing gear Designed for ice or compacted snow, often with heating to prevent adhesion.
skimmer 1 Missile, eg anti-ship category, programmed to fly just above crests of waves.
2 ACV (colloq).
skin 1 Outer covering of air vehicle, ACV or spacecraft, except that in case of vehicle covered with ablative layer or thermal-insulating tiles * is underlying structural layer. Can be made of any material including fabric or Mylar.
2 Outer component of sandwich.
skin Doppler Determination of air-vehicle velocity by radar.
skin drag See *surface-friction drag.*
skin effect Concentration of AC (electron flow) towards surface of conductor.
skin friction See *surface-friction drag.*
skin-friction coefficient Non-dimensional form of skin-friction drag on body immersed in a laminar, viscous, incompressible flow,

$$\gamma = \frac{\tau_0}{\frac{1}{2}\,\rho\, U_m^2}$$

where τ_0 is shearing stress at solid surface, ρ density and Um mean flow velocity $\left(= \frac{16}{R} \text{ where R is Reynolds no}\right)$.
skin mill Large machine tool with revolving cutter(s) under which passes workpiece with linear motion; cutter axis can vary from horizontal to vertical. Often NC machine and able to sculpt complete wing skin.
skin paint 1 Radar indication caused by reflected radar signal from object (DoD); ie blip.
2 Fix obtained by ground radar on aerial target.
skin temperature Temperature of outer surface of body, esp in sustained supersonic flight.
skin tracking Tracking of object by means of a skin paint (1) (DoD).
skip See *skip re-entry.*
skip bombing Method of aerial bombing in which bomb is released from such a low altitude that it slides or glances along surface of water or ground and strikes target at or above water or ground level (DoD).
skip distance Distance from transmitter at which first reflected sky wave can be received, increasing with frequency.
skip/glide See *boost/glide.*
skip it Air intercept code: "Do not attack, cease interception".
skip re-entry Atmospheric entry by lifting-body spacecraft in which energy is lost in penetrating atmosphere in curving trajectory reminiscent of stone skipping on pond, possible only with very accurate trajectory control if first skip is not to result in permanent departure from Earth (see *lifting re-entry*).
skirt 1 Lowest part of body of large ballistic vehicle surrounding rocket engine(s).
2 Lowest part of parachute canopy.
3 Flexible structure surrounding and containing cushion of amphibious and many other types of ACV enabling vehicle to run over waves or rough ground.
skirt fog Steam cloud during launch from wet pad.
ski-toe locus Imaginary object shaped like front of ski which TFR runs along terrain by electronic means; size and form are often variable to give

soft to hard ride.

SKRVT State Commission for Development and Co-ordination of Science and Technology (Czech).

SKT Specialty knowledge tests (USAF).

sky compass Instrument for determining azimuth of Sun from polarization of sunlight.

sky conditions Amount of cloud in oktas.

sky diver Sport parachutist.

Skydrol Synthetic non-flammable ester-based hydraulic fluids.

Skyhook Plastic balloon for met observation and rocket launch (see *rockoon*).

skyjack Aerial hijack (colloq).

skyline To get low enemy aircraft visibly above horizon with sky background, or to expose own aircraft in same way.

sky screen Simple optical (camera obscura) device showing range safety officer if vehicle departs from safe trajectory; often one for track and another for vertical profile.

sky wave That portion of a radiated wave that travels in space and is returned to Earth by refraction in ionosphere (ASCC). Several other authorities use word "reflection".

sky-wave correction, SWC Factor to be applied to some hyperbolic navaids, eg Loran, if sky waves used instead of ground waves; varies with relative distances to master and slave(s).

sky writing Writing, if possible against blue-sky background, using oil added to exhaust or other system and forming characters by accurately flying along their outlines. Invariably associated with PE (4), and becoming a lost art; see *smoke generation.*

SL 1 Sea level.
2 Space-limited.
3 Service letter.
4 Schätzstellen für Luftfahrtzeuge (G).

S/L Shoot/lock.

slab-sided Having essentially flat (usually near-vertical) sides.

slab tailplane Horizontal tail formed as single pivoted surface and used as primary flight control; no fixed tailplane or hinged elevators (US: horizontal stabilizer). Called taileron when installed in left/right halves capable of being driven in opposite directions.

slab trailing edge Blunt trailing edge, esp squared off normal to line of flight.

SLAEA The Society of Licensed Aircraft Engineers Australia.

SLAET The Society of Licensed Aircraft Engineers & Technologists (UK).

slag refining See *electro-slag refining.*

SLAM Supersonic low-altitude missile.

slam acceleration Most rapid possible acceleration of engine, esp gas turbine, typified by violent forward movement of power lever to limit.

slamming Impact of front or rear wheels (depending on design geometry) of bogie on ground at vertical velocity greater than that of aircraft, caused by added velocity imparted by rotation of bogie beam.

Slammr Sideways-looking airborne multimission radar.

Slams Surface look-alike mine systems.

slant course line Intersection of course surface and plane of nominal ILS glidepath.

slant range LOS range between aircraft (aerial target) and fixed ground station; not same as range plotted on map, hence ** correction to radio navaid distances which is small until aircraft height is greater than 20% of **.

SLAP Saboted light-armour penetrator.

SLAR Side-looking airborne (or aircraft) radar.

slash mark Oblique stroke /.

slash rating For electric machines and other accessories, normally 150% of base load.

SLAT, Slat 1 Slow, low, airborne target.
2 Supersonic low-altitude target.

slat 1 Movable portion of leading edge of aerofoil, esp wing, which in cruising flight is recessed against main surface and forms part of profile; at high angle of attack either lifts away under its own aerodynamic load or is driven hydraulically to move forward and down and leave intervening slot.
2 Fixed leading-edge portion of aerofoil, either wing or tailplane (in latter case often inverted, lying along underside), forming slot ahead of main surface. Both (1, 2) postpone flow breakaway at high AOA and thus delay stall.

slave 1 See *slaving.*
2 See *slave station.*

slave aerial Mechanically scanned aerial slaved to another, eg SSR slaved to surveillance radar but off-mounted.

slaved gyro One whose spin axis is maintained in alignment with an external direction, eg magnetic N or local vertical.

slave landing gear Temporary landing gear used in factory to move incomplete aircraft.

slave station Radio station whose emissions are controlled in exact synchronization or phase with a master station at different geographical location.

slaving 1 To constrain a body to maintain an attitude in exact alignment with another.
2 To key a transmitter to radiate in exact phase or synchronization with a master.

SLBM Submarine-launched ballistic missile.

SLC 1 Sidelobe clutter.
2 Software life cycle.
3 Space launch complex (pronounced slick).

SLCM 1 Submarine- (or sea-) launched cruise missile.
2 Survivable low communications system.

Sleaford Tech Derogatory term for RAF College, Cranwell.

sleet Precipitation of rain/snow mix or partially melted snow. Two special US usages: frozen rain in form of clear drops of ice, and glaze ice covering surface objects; both highly ambiguous.

sleeve 1 Banner target.
2 Plastics cylinder used as colour-coded electrical cable marker.
3 Central portion of sleeve-valve cylinder.
4 Fabric tube for filling gas aerostat.

sleeve valve Any of various techniques for piston-engine valve gear using one or two concentric sleeves between piston and cylinder with suitably shaped ports in walls lining up intermittently with inlet/exhaust connections on cylinder; usual is Burt-McCollum single sleeve.

slender body One of such large slenderness ratio that squares and higher powers of disturbances can be ignored.

slender delta Aeroplane whose wing has ogival delta plan with very low aspect ratio such that at Mach numbers exceeding 2 entire wing lies within conical shockwave from nose.

slenderness ratio Length/diameter of fuselage or other slender body. Generally, synonymous with fineness ratio.

SLEP Service-life extension programme.

slew 1 To rotate in azimuth.
2 To offset centre of P-type or similar display laterally, eg to study air traffic or surface feature off edge.
3 To rotate gyro spin axis by applied torque at 90°.

slewed flight Yawed, eg with applied rudder while holding height and with wings level.

slewing Slew; also defined as changing scale on radar display (not recommended).

slew-wing aeroplane One whose wing is pivoted as one unit about mid-point, thus as one tip moves forward opposite tip moves aft. In some forms there is no fuselage and wing obliquity to airflow is determined solely by tip fin(s) and engine pod angles.

SLF Shuttle landing facility.

SLFCS Survivable LF communications system (SAC, USAF).

SLG Satellite landing ground, ie auxiliary field.

SLI Staatliche Luftfahrtinspektion (DDR).

Slice Internationally agreed subdivision of international funds, usually allocated in * groups or as 1-year * for infrastructure (NATO).

slice(d) Maximum-performance hard nose-down turn, over 90° bank.

slick Any streamlined free-fall store, especially GP bomb.

slick wing One with no provision for pylons or hardpoints.

slidewire Wire carrying transport trolley providing emergency escape from top of space-launch service tower.

sliding window Figurative (electronic) window in SSR which looks into each range bin in turn, feeds any traffic or other reply found there to plot extractor and usually also defruits.

SLIM 1 Surface-launched interceptor missile (concept).
2 Software life-cycle management.
3 Simplified logistics and improved maintenance.

slim jet See *single-aisle*, *narrow-body*.

slinger ring Channel or pipe around propeller hub (inside spinner if fitted) to which controlled supply of deicing fluid fed for centrifugal distribution along blades.

slinging point Clearly indicated location on airframe or major assembly around which sling of crane can be passed as loop for hoisting.

slip 1 Sliding towards inside of turn as result of excessive bank.
2 Loosely, any yawed flight causing indication towards centre of turn or lower wing on turn/* indicator, eg forward *, side-*; in particular, controlled flight of helicopter in direction not in line with fore/aft axis.
3 Measure of loss of propulsive power of propeller defined as difference between geometric and effective pitch.
4 Slurry composed of finely divided ceramic or glass suspended in liquid, eg for coating surfaces in precision casting techniques.
5 Difference between speed of induction motor under load and synchronized speed, expressed as percentage.
6 Crystalline defect characterized by (usually local) displacement of atoms in one plane by one atomic space.
7 Launch slipway for marine aircraft.
8 Shackle for bomb or other dropped store (becoming arch).
9 To change flight crews at one stopping place on airline route.

slip cover Fabric cover previously cut to shape and sewn, available from store tailored to aircraft type.

slip crew Airline flight or cabin crew who leave or join as operating crew at intermediate point in multi-sector flight. In some cases crew may continue on same aircraft but off-duty.

slip flow Flow in extremely rarefied fluid where mean free path is comparable with dimensions of body (see *free-molecular flow*, *Newtonian flow*).

slip function Basic propeller parameter, also called effective pitch ratio, = V/nD where V is airspeed of aircraft, n is rpm and D diameter.

slip gauge Extremely accurate wafers of steel of known thicknesses which can be stacked to build blocks accurate to within 0·25 μ (10^{-5} in).

slip-in See *slip (1)*.

slip joint One permitting axial sliding, eg in exhaust manifold to allow for expansion.

slippage mark White rectangle painted on wheel/tyre to show relative rotation.

slip pattern Planned arrangement for slipping crews detailed on crew roster.

slipper 1 Adjective describing drop tank or other

air-dropped or externally carried store shaped to fit underside and leading edge of wing.
2 Generalized term for precision part designed to slide over another, eg in air or other fluid bearings, or in con-rod big end which mates on periphery of master rod and held in place by rings, there being no ring-type big end but only a driving *.

slippery Having large momentum and little drag (colloq).

slipping torque Torque at which piston-engine starter clutch will slip.

slipping turn Turn with slip (1).

slipring Conducting ring rotating with rotor of electrical machine to transfer current without commutating.

slip speed Supercharger rpm required to maintain given pressure differential between inlet and delivery manifold when no air is being delivered.

slipstream 1 Airflow immediately surrounding aircraft; if behind propeller * is propwash. Velocities measured relative to aircraft.
2 To follow direction of streamlines, eg in case of freely hinged nozzle tailfeathers.
3 To follow in wake of another aircraft. Note: (2, 3) are verbs.

slip tank A tank, eg fuel, water, oil, which can be jettisoned; used in airships and a few aeroplanes but now overtaken by drop tank.

SL-ISA Sea level, international standard atmosphere.

slitting shear Hand or powered shears of lever type used for heavy or very wide sheet (not plate).

sliver fraction Volume of slivers remaining in case at web burnout divided by total propellant volume, symbol λ_v.

slivers Fragments of solid propellant which are left unconsumed after rocket-motor burnout.

SLMG Self-launching motor-glider.

SLOC Sea lines of communication.

slope 1 Glideslope angle to horizontal, usually 3° but steeper for STOL or noise-reduction approaches.
2 Rate of change of one variable with respect to another, tangent from origin to any point on curve plotting y against x or dy/dx.

slope angle Acute angle measured in vertical plane between flightpath and local horizontal.

slope-line system Approach light system giving vertical guidance by appearance of ground lights (UK 1946–50).

slope of lift curve Unit is increment of lift per radian change in AOA.

SLOS 1 Star line of sight.
2 Stabilized long-range observation system (or optical sight).

sloshing Gross oscillatory motion of liquid in tank sufficient to impose severe structural stress or affect vehicle trajectory; one cause of pogo effect.

sloshing baffles Transverse perforated bulkheads, usually part of tank structure, to curb sloshing; strictly anti-**.

Slot Sequential logic tester.

slot 1 Suitably profiled gap between main aerofoil, esp wing or tailplane, and slat or other leading-edge portion through which airflow is accelerated at high AOA to prevent breakaway; usually curves up and back to direct air over upper surface but on tailplane often inverted.
2 Gap between wing and hinged trailing-edge surface, eg flap or aileron, through which air flows attached across movable upper surface.
3 Particular allotted time for using facility (eg gunnery range), for space launch (also called window) or for controlled aircraft departure or arrival, esp at busy airport; hence to secure a *, to miss one's *.
4 Physical aperture for * aerial.
5 Particular band of aircraft weight or flight performance.
6 Figurative situation or position, esp a target situation, eg in the * = correctly set up for landing.
7 In carrier flying, to enter landing pattern by flying up ship's starboard side followed by break downwind.

slot aerial Aerial (antenna), eg for DME, in form of slot cut in metal skin, often backed by reflective cavity and aerodynamically faired by dielectric; normally 0·5 wavelength long and 0·05 wide, with polarization usually 90° to plane of slot.

slotted aerofoil One incorporating a fixed slot (1); if slot results from motion of a slat correct adjective is slatted.

slotted aileron Aileron separated from wing by slot (2).

slotted flap Flap, usually not translating but simply hinged, forming whole of local trailing edge and separated from wing by slot (2).

slow-blow fuse Cartridge designed to withstand brief overload (electrical).

slow-CAP Combat air patrol to protect slow-flying aircraft.

slow roll Precision flight manoeuvre in which aircraft, usually fixed-wing, is rolled through 360° by ailerons (using rudder as necessary) while keeping longitudinal axis sensibly constant on original heading; unlike barrel roll imparts −1 g in inverted attitude. Can be performed with longitudinal axis at any inclination in vertical plane. In US called aileron roll.

slow-running cut-out Pilot-operated valve which stops PE (4) by turning off supply of metered fuel (carburettor engine only).

slow-running jet Fine carburettor jet which alone supplies fuel to PE (4) mixture when throttle at idling position.

SLP 1 Space-limited payload.
2 Survivor-locator package (USAF).

SLR Side-looking radar.
SLS 1 Side-lobe suppression.
2 Sea-level static.
3 Self-launching sailplane.
SLSAR Side-looking synthetic-aperture radar.
SLS-TO Sea-level static, take-off.
SLU 1 Surface-launched unit.
2 Stabilized laser unit.
3 Switching logic unit.
sludge Viscous slimy deposit gradually formed in lubricating oil by oxidation, water contamination and other reactions.
sludge chamber Cavity, tube or other region in crankshaft web or crankpin, supercharger drive gear or other rotary component in which sludge is deliberately trapped by centrifugal force.
SLUFAE Surface-launched unit fuel/air explosive.
slug 1 Non-SI (UK only) unit of mass = g lbf = 14·5939 kg.
2 Pre-rivet forming feedstock for Drivmatic riveting machine.
3 Ferritic cylinder for varying coil permeability or inductance.
4 Metal or dielectric cylinder for waveguide impedance transforming.
5 Slab of alloy from which SFF warhead is formed.
slugging Malfunction in vapour-cycle ECS in which the compressor pumps liquid refrigerant.
slung load Payload carried well below helicopter on single cable.
slurry Suspension of finely divided solid particles in liquid, usually capable of being pumped; physical form of experimental fuels, particles often being metal.
slush Mixture of snow and water with SG 0·5 to 0·8; below these values is called wet snow, above is called standing water.
slush fund Money allocated for bribery, usually of persons able to influence potential customers.
SLV Space launch vehicle (USAF).
SM 1 Statute mile (often s.m. or mile).
2 Static margin.
3 Service module.
4 Sandwich moulding.
5 Strategic missile (former DoD designation prefix).
6 Standard missile (USN).
sm, s.m. 1 Statute mile.
2 Short emission.
SMA 1 Squadron maintenance area.
2 Stato Maggiore Aeronautica (I).
SMAC Scene-matching area-correlator.
SMAE Society of Model Aeronautical Engineers (UK).
small aircraft One of below 12,500 lb (5,670 kg) MTOW.
small arms All arms up to and including 0·6 in (15·24 mm) calibre (DoD). Despite "all arms", term normally means guns.
small business One with fewer than 500 employees.
small circle That described on spherical or spheroidal surface by intersection of plane not passing through body's centre.
small end End of connecting rod pin-jointed to piston.
small perturbation One for which 2nd and higher-order terms ignored.
small-scale integration Usually 10 or fewer gates or other functions per IC.
SMAP, Smap Systems-management analysis project (AFSC).
smart 1 Capable of being guided, by self-homing or external command, to achieve direct hit on point target.
2 Generalized term for clever, eg smart jammer listens for hostile emission and then jams on correct wavelength.
smartlet Smart bomblet.
Smash Southeast Asia multisensor armament system, helicopter.
SMAW Shielded metal-arc welding.
SMB Side marker board, for airport parking guidance.
SMC 1 Standard mean chord.
2 Surface movement control (ICAO).
SmCo Samarium cobalt, chief rare-earth magnetical material accepting 20 to 30 times normal current for short overload periods.
SMCS 1 Spoiler mode control system.
2 Structural mode control system.
SMD 1 Shop modification drawing.
2 System management directive.
3 Surface-mounted device (electronics).
SMDC Shielded mild detonating cord.
SME Small/medium enterprise.
smear Degraded radio reception due to another transmission on same frequency or degraded TV picture due to ghost image closely following primary image.
smear camera See *streak camera.*
smearer Subcircuit to eliminate pulse-amplification overshoot.
smear metal Metal melted by high-speed machining or welding and deposited on workpiece.
smectic phase Liquid-crystal phase having layered structure with constant preferred direction; flow is abnormal and X-ray diffraction pattern obtained from one direction only.
SMES Strategic Missile Evaluation Squadron.
SMET Simulated mission endurance testing.
SMFA Service du Matériel de la Formation Aéronautique (F).
SMG 1 Sync/message/guard time-slot.
2 Spinning-mass gyro.
SMI Structural merit index.
SMILS Sonobuoy missile-impact location system.
Smith diagram Standard plot for solving electrical transmission-line problems.

SMM 1 Space manufacturing module.
2 Solar maximum mission.
3 Service Météorologique Métropolitain (F).
SMMR Scanning multi-frequency microwave radiometer.
SMO 1 Supplementary met office.
2 Synchronized modulated oscillator.
3 Shelter management office.
smog Fog contaminated by liquid and/or solid industrial pollutants, particularly smoke.
SMOH Since major overhaul; suffixes LE, RE = left/right engine.
smoke apparatus Aircraft installation for leaving either a smoke screen or, today more often, smoke trail of desired colour for display purposes.
smoke bomb Air-dropped pyrotechnic, able to float, for indicating wind velocity.
smoke box Container of slow-burning fuel for producing wind-indicating smoke trail.
smoke float See *smoke bomb.*
smoke pot Remotely triggered device fired to indicate a hit on ground target.
smoke tunnel Not precisely defined; wind tunnel in which either general recirculating smoke or discrete streams give visible flow indication.
SMP Self-maintenance period (aircraft carrier).
SMR 1 Surface-movement radar (ICAO).
2 Svenska Mekanisters Riksförening (Sweden).
SMRS Stores management and release system.
SMS 1 Strategic missile squadron.
2 Stores-management set, or system.
3 Suspended manoeuvring system.
4 Supply and movements squadron (RAF).
SMT 1 Shadow-mask tube.
2 Système modulair thermique, IR common module (F).
SMTC Space and Missiles Test Center (Vandenberg AFB, USAF).
SMTI Selective moving-target indicator.
SMTO Space & Missile Test Organization (AFSC).
SMW Strategic missile wing.
SN 1 Snow (ICAO).
2 Since new (often S/N).
3 Shipping notice.
S/N 1 Stress against number of alternating load cycles to failure; S is normally ratio of alternating load to ultimate strength, so that for S=1 N=1, while for S=0·8 N may be 10^4.
2 Serial number.
3 Signal/noise ratio.
Sn Tin.
sn Sthène.
SNA Sindicato Nacional dos Aeronautas (Brazil).
SNABV Syndicat National des Agences et Bureaux de Voyages (F).
SNAEC, Snaec Special notice to aircraft and engine contractors (UK, MoD).
snag 1 Fault condition or impediment to progress (UK, colloq).
2 Dogtooth or other abrupt discontinuity in leading edge.
snake drill Hand drill with tool bit driven by flexible connection.
snake mode Control mode in which pursuing aircraft flies preprogrammed weaving path to allow time to accomplish ident functions (DoD).
snaking Natural oscillation in yaw at approximately constant amplitude.
SNAP, Snap Systems for nuclear auxiliary power; major programme for space electric power generation using RTG and similar methods.
Snapac Sindacarto Nazionale Autonomo Personale Aviazione Civile (I).
snap-action 1 Positive full-range movement of bistable device, esp mechanical, eg spring-loaded two-pole switch or valve.
2 Various meanings in electronics, esp abrupt jump in output of mag amplifier with large positive feedback.
snap-down Ability to see, engage and destroy target at much lower level, esp one very close to surface; hence * missile. Crucial factor is ability of radar or other sensor to see target against ground clutter.
snap gauge C-frame go/no-go gauge for shaft measures, usually with one anvil adjustable.
snap report Preliminary report by aircrew of observations, prior to compilation of mission report (ASCC etc). Term not to be used (DoD).
snap ring Sprung fastener which locks into peripheral groove on either inside or outside diameter.
snap roll See *flick roll.*
snap-shoot Traditionally, quickly aimed shot; in modern air combat, shooting with fixed gun without need for prior tracking using correctly interpreted HUDsight symbology, usually providing a tracer line.
snap start Prelaunch AAM condition in which weapon is pre-tuned to guidance radar and then returned to passive mode but ready for instant launch.
snap-up Rapid maximum-performance pull-up to engage targets at higher altitude.
snap-up missile AAM capable of engaging and destroying target aircraft at much greater height than launch platform.
SNAW School of Naval Air Warfare (UK).
SNC Standard nav computer; small leg-strapped box giving continuous moving-map readout showing aircraft position.
SNCTA Syndicat National des Contrôleurs du Trafic Aérien (F).
SNCTAA Syndicat National des Cadres et Techniciens de l'Aéronautique et de l'Astronautique (F).
SNDV Strategic nuclear delivery vehicles.
SNEA Sindicato Nacional das Empresas Aeroviarias (Brazilian Air Transport Association).

Snell law Law of index of refraction.
SNF Short-range nuclear force(s).
SNG Synthetic (or synthesized, or substitute) natural gas.
SNI Signal/noise index (Omega).
snifter(s) Small spin motors (often one motor with tangential nozzles) for small (eg anti-tank) missiles.
SNII Aeroplane scientific test institute of GVF (USSR).
Snipag Syndicat National des Industriels et Professionnels de l'Aviation Générale (F).
snips Hand shears for cutting sheet metal.
Snirfag Syndicat National des Industriels Réparateurs et Fournisseurs de l'Aviation Générale (F).
SNLE Sous-marin nucléaire lanceur engins (missile-firing submarine, F).
SNM Special nuclear material.
SNOE Smart noise operation equipment (ECM).
Snomac Syndicat National des Officiers Mécaniciens de l'Aviation Civile (F).
Snorac Syndicat National des Officiers Radios de l'Aviation Civile (F).
Snort Supersonic naval ordnance rocket track.
snort Submarine schnorkel pipe, esp tip seen on radar above ocean.
Snorun Snow on runway.
snow 1 Precipitation in form of feathery ice crystals (BSI); better definition is: small (under 1 mm) grains, granular *; long (2+ mm) grains, ice needles; large agglomerations in form of flakes, *. Dry * is SG 0·2 to 0·35; wet * is 0·35 to 0·5.
2 Speckled interference on electronic display.
3 Air-intercept code: sweep jamming (ie display looks like *).
snow gauge Combination of rain gauge and vertical measuring stick to determine snow moisture content.
snow lights Specially designed runway-edge lights which stand above level of any snow yet snap off if struck by aircraft.
snow pellets Small white opaque pellets of water/ice, softer than hail.
snow static Severe R/F interference caused by snow (suggest arch).
Snowtam Special-series Notam announcing presence or removal of hazardous conditions due to snow, ice, slush or standing water in association with these on movement area.
SNPL Syndicat National des Pilotes de Ligne (F).
SNPNAC Syndicat National du Personnel Navigant de l'Aéronautique Civile (F).
SNPO Space Nuclear-Propulsion Office (AEC, NASA).
SNSH Snow showers.
SNTA Syndicat National des Transporteurs Aériens (F).
snubber Device which greatly increases stiffness of elastic system whenever deflection or travel exceeds given limiting value, eg rubber block to arrest travel of shock suspension and special features (often hydraulic) to arrest travel of actuator at full stroke.
SNVS Stabilized night-vision system.
SOA 1 Spectrometric oil analysis.
2 State of the art.
3 Separate operating agency.
SOAP Spectrometric oil-analysis programme.
SOAR Shuttle Orbiter applications and requirements.
soar To maintain flight in sailplane by seeking thermals or other upcurrents.
SOB Souls on board, from traditional marine usage giving number of people aboard.
SOC 1 Shut-off cock.
2 Struck off charge, ie consigned to scrap.
3 Satellite (or systems) operations complex.
4 Sector operations centre (RAF).
socked-in Airfield shut by bad visibility (colloq).
sodium Sodium light (colloq).
sodium light Deep yellow lights in some cases used in approach lighting.
sodium line-reversal Line-reversal pyrometer using sodium line at $5{\cdot}896 \times 10^{-7}$ m.
SODP Start-of-deceleration point.
SOF 1 Service Officiel Français.
2 Special Operations Force (USAF).
3 Stand-off flare (IRCM).
4 Strategic offensive forces.
5 Supervisor of flying.
Sofag Special-ops force assistance group (USAF).
Sofar Sound fixing and ranging, technique for fixing position at sea by time-difference measures of sound from explosion (eg of depth charge, usually at considerable depth) or impact of spacecraft on surface. Hence * bomb, special sound-producing bomb.
SOFI Sprayed-on foam insulation.
S of S Secretary of State, = minister (UK).
SOFT Site operational functional test.
soft Not hardened against NW explosion.
soft failure 1 Usual interpretation is cessation of function without any incorrect function (eg no hardover signal).
2 In EDP, short term transient failure followed by return to normal; believed caused by alpha-particles.
soft flutter Flutter that is possibly severe but non-divergent, and confined within apparently safe amplitude limits.
soft hail See *snow pellets*.
soft-in-plane Semi-rigid helicopter main rotor of subcritical type, fundamental lag frequency being less than N_1 (rotor rpm) at normal operating speed.
soft iron Iron containing little carbon, as distinct from steel; loses nearly all magnetism when external field removed.
soft landing Gentle landing as distinct from hard (free fall); term applies to arrival on surfaces

other than Earth.

soft obstacle 1 Conceptual obstacle, eg 35 ft screen.
2 One physically present, eg ILS localizer, but whose engineering design minimizes damage to colliding aircraft.

soft radiation Unable to penetrate more than 100 mm of lead.

soft ride Ride, esp in lo-flying aircraft at high-subsonic speed, judged comfortable and in no way rough enough to impair crew functions. Normally defined as fewer than 2 0·5 g bumps per minute (see *LLDF*). Occasionally selected mode in TFR system.

soft tooling Tooling whose dimensions can be adjusted, normally within small limits (eg could not accommodate different 707 fin sizes, which required new tooling).

soft undocking One that does not influence subsequent trajectory, eg by not using separating thrusters.

soft valve Thermionic valve into which some air has leaked.

software 1 All programs and component parts of programs used in EDP (1), eg routines, assemblers, compilers and narrators. Divided into two parts. Basic *, usually provided by equipment manufacturer, is machine-oriented and is essential to permit or extend use of particular hardware; examples are diagnostic programs, compilers, I/O conversion routines and programs for file or data-management. Application * is normally user-oriented and often compiled by user or a subcontracted * house, to enable machine to handle specific tasks; may include GP packages, eg NC tool, payroll or airline booking, or locally created programs for highly specific tasks, eg exploring flight characteristics of unbuilt aircraft.
2 Ambiguously, also used for parachute or drogue packs, esp those installed in aircraft. This usage is potentially misleading.

SOH Since overhaul.

SOI Space object identification (Norad).

SOIS Silicon on insulating substrate.

SOJ Stand-off jammer.

SOLAP Shop-order location and reporting.

Solar Shared on-line automated reservation system.

solar apex Point on celestial sphere towards which Sun is moving.

solar array Large assembly of solar cells, on rigid frame, folding, in roll-up sheet or other geometric form.

solar atmospheric tides Cyclic variations in atmospheric pressure ascribed to Sun's gravitation, with primary 12-h component (about ± 1·5 mb at Equator, 0·5 in mid-latitudes) and much smaller 6-h and 8-h effects.

solar battery See *solar cell.*

solar cell Photoelectric (photovoltaic) device converting sunlight direct into electricity, usually by liberating electrons and holes in silicon p–n junction.

solar chamber Test chamber in which is simulated solar radiation outside atmosphere.

solar constant Rate at which solar radiation is received outside atmosphere on unit area normal to solar radiation at Earth's mean distance from Sun, about 1·38769 kW/m^2.

solar cycle Approx 11-year cycle in sunspot frequency.

solar day 1 Time between two successive solar transits of same meridian, ie time for Earth to rotate once on its axis with respect to Sun (mean or apparent).
2 Time for Sun to rotate once on its axis with respect to fixed stars.

solar flare See *flare (2).*

Solar Happ Solar high-altitude powered platform.

solari board Usual type of large electromechanical display at airports indicating flight arrivals and departures.

solar noise Solar radiation at RF frequencies.

solar paddle Solar-cell array on fixed frame resembling paddle.

solar panel Any fixed planar solar array.

solar-particle alert network Global observation system to warn astronauts of solar flares.

solar propulsion Loose term sometimes used for rocket systems based on electric power, which can be derived from solar cells, but best restricted to solar sailing using a sail (2).

solar sail See *sail (2).*

solar simulator Device for simulating solar radiation outside Earth's atmosphere.

solar wind Plasma radiating from Sun, assumed equally in all directions, which in vicinity of Earth has T about 200,000°K, V about 400 km/s and density of 3·5 particles/cm^3; grossly distorts terrestrial magnetic field, causing upstream shock front; another effect is to blow comet tails downstream from Sun.

solderless splice Joint made by crimping or machine-wrapping.

SOLE Start-of-life efficiency of thermoelectric module or other progressively degraded power-conversion device.

solenoid 1 Range of simple electromagnetic devices in which current in a coil (usually of cylindrical form) moves iron core, eg to operate a switch.
2 Tube formed in space by intersection of two surfaces at which a particular quantity (eg pressure, temp) is everywhere equal.

solid angle Portion of space viewed from given point and bounded by cone whose vertex is at that point, measured by area of sphere of unit radius centred at same point cut by bounding cone. SI unit is steradian.

solid conductor One containing single wire.

solid fuel Preferably reserved for fuel not used as

propellant but as energy source, eg for EPU. For rocket, see *solid propellant*.

solidity 1 Ratio of total area (not projected area but integral of chord lengths across length of blade) of propeller or rotor to disc area. Basic measure of proportion of disc occupied by blades.
2 At standard radius, ratio of sum of blade chords to circumference, which is not same as (1) since blades of different rotors or propellers are not all same plan shape.

solid motor Rocket filled with solid propellant.

solid nose Term to distinguish aircraft nose with metal skin or radome from others of same aircraft type which are glazed (colloq).

solid propellant Rocket propellant containing all ingredients for propulsive jet in solid form, either in cast, extruded or otherwise prepared grain or in granular, powder, multiple-rod or other form. Some definitions questionably exclude non-monolithic forms.

solids In ag-aviation, non-liquid chemicals, eg powders, dusts and granules.

solid-state devices Electronic devices using properties of solids, esp semiconductors.

solid-state oxygen Alkali-metal chlorates which can readily be made to yield free oxygen.

solid surface RF reflector, esp for very large aerial, whose reflective surface is not wire mesh but aluminium sheet.

solid target Not banner or drogue but towed aerodyne.

solid wire Single tinned or galvanized steel wire.

SOLL, Soll Special operations, lo-level.

solo According to some definitions, pilot flying unaccompanied by instructor, but this could admit passengers; invariably means pilot is only human occupant of aircraft.

solstice Either of two points on ecliptic furthest from celestial equator, direction of Sun's centre at max declination; N hemisphere is summer *, about 22 June; S hemisphere is winter *, about 23 December.

solstitial colure Celestial great circle through poles and solstices.

solution heat treatment First stage of heat treatment of certain light alloys in which salt bath (often $NaNO_3/KNO_3$) is used for accurate heating followed by room-temperature cooling or quenching.

solvent extraction Various processes in which solvents, often hydrocarbons, eg propane, are used to separate lube-oil products from pipe-still distillates. Other meanings in processing of coal.

SOM Stand-off missile.

somatic Affecting exposed individual only, as distinct from offspring.

Sommerfield matting Mass-produced airfield pavement in form of 75 ft (22·86 m) rolls of 13 SWG wire mesh reinforced at 8 in (203 mm) intervals by steel rod with hooked end linking to adjacent strip (UK, WW2).

SON Statement of operational need.

sonar Use of sound as method of communication under water, detection of surface or submerged targets and measurement of range, and in some cases bearing and relative speed. Analogous to radar, and similarly may be active (emitting high-energy sound waves of tailored form, eg from * transducer) and working with reflections from all submerged objects, or passive, in which receivers listen for sounds emitted from targets. From sound navigation and ranging.

sonar capsule Device giving enhanced echoes to sonar to assist location of marine object, eg floating RV or space payload.

sonar transducer Translates electrical energy into high-intensity sound, normally used in multiple to form a sonar stave (typically radiating 500–1,000 W), which in turn is used in multiple to form 360° or directional array.

sonde Airborne telemetry system, esp to transmit met or other atmospheric data.

sone Primitive unit of perceived noise equal to that from simple 1 kHz tone 40 dB above listener's threshold. Subjective judgement of any sound enables it to be expressed in sones (0·001 * = millisone); useless for noise investigation (see *noise*).

sonic 1 Pertaining to local speed of sound.
2 Approximately at local speed of sound.

sonic bang Noise heard as shockwave(s) from supersonic object pass hearer's ears; small object at close range generates sharp crack, normal aircraft at close range one or more loud bangs (resembling close thunder) and distant SST dull boom(s) resembling distant thunder (see *boom signature*). Crack of a whip and natural thunder are both examples.

sonic barrier See *sound barrier*.

sonic boom See *boom*, *sonic bang*.

sonic drilling See *ultrasonic machining*.

sonic erosion See *ultrasonic machining*.

sonic line Curved surface above or below wing or other body which has accelerated flow beyond Mach 1, at which M = 1, enclosing region of supersonic flow terminated at rear by shockwave.

sonics 1 Aggregate of installed sonars, sonobuoys and displays in platform, eg in aircraft.
2 Technology of applying sound to functions other than those related to hearing.

sonic soldering See *ultrasonic bonding*.

sonic speed Local speed of sound, symbol a.

sonic venturi Venturi in which sonic speed is reached at throat, thereby automatically limiting maximum flow (eg in bleed systems).

sonobuoy Discrete sonar devices immersed or dropped into water; can be active (emitting) or passive, directional or non-directional, and except when dunked by helicopter normally provide readout by radio, usually upon command.

sonodunking Action of dunking permanently attached sonobuoy.

SOO Standard operations (or operational) orders.

SOON, Soon Solar-observatory optical network (USAF).

SOP Standard operating procedure.

SOPC Shuttle operations and planning complex.

Sopemea Société pour le Perfectionnement des Matériels et Equipements Aérospatiaux (F).

SOR 1 Struck off records.
2 Specific operational requirement.

sorb To acquire gas by sorption; hence sorbent material.

Soreas Syndicat des Fabricants d'Organes et d'Equipements Aéronautiques et Spatiaux (F).

sorption Taking up of gas by absorption, adsorption, chemisorption or any combination of these.

SORT 1 Structures for orbiting radio telescopes.
2 Simulated optical range tester.

sortie An operational flight by one aircraft.

sortie capacity Maximum number of sorties mounted by unit or other airpower source (eg one airfield) in stated period, usually 24 h.

sortie generation Ability of combat unit to put its aircraft in air, especially around clock.

sortie number Reference identifying all images secured by all sensors on one air reconnaissance sortie.

sortie plot Map overlay representing area(s) covered by imagery during one sortie.

sortie rate Number of combat missions actually performed by unit (eg squadron or polk) in 24 h period.

sortie reference See *sortie number*.

SOS 1 International distress signal.
2 Silicon on sapphire.
3 Special operations squadron (USAF).
4 Sidewall overhead stowage in passenger airliner.
5 Stabilized optical sight.
6 Squadron Officer School (AU [2]).

Sosus Sound surveillance system.

SOT Stator outlet temperature.

Sotas Stand-off target acquisition system.

SOTD Stabilized optical tracking device.

sound attenuation Reduction in sound intensity, esp through deliberate conversion to other energy forms, eg heat, in absorbent or other layers of material.

sound barrier Conceptual barrier to manned flight at supersonic speed when this was extremely difficult, ie before about 1952.

sound energy Measure of either total emitted energy (for brief sound, eg explosion) or sustained rate of energy transfer for prolonged sound, in latter case measured in watts.

sounding 1 Any penetration of natural environment for observation or measurement.
2 Complete set of measures taken in and of upper atmosphere for met or other purpose. Hence * balloon, unmanned free balloon carrying upper-atmosphere instruments; * rocket, stabilized but usually unguided rocket carrying upper-atmosphere instruments.

sound intensity Average sound power passing at given point through unit area normal to propagation, expressed either in W/cm^2 or as sound level.

sound level Ratio of sound power to a zero reference, expressed in dB (see *noise*).

sound power Sound energy (rate for sustained sound) in watts.

sound pressure Total instantaneous pressure at point at given time minus static pressure; unit is N/m^2.

sound pressure level, SPL 20 log SPL/ref pressure (see *noise*).

sound probe Instrument responding to sound, eg sound pressure, without significantly altering sound field.

sound suppressor See *suppressor*.

sound wave Disturbance conveying sound in form of longitudinal alternate compressions/rarefactions through any medium. Frequency spread may be much greater than human aural range, and extreme-energy case is blast wave, which initially (like shockwave) travels faster than sound.

source 1 Contractor for entire article, eg aircraft or missile (see *second* *).
2 Origin of noise.
3 Origin of fluid in fluid flow.
4 Solid-state electrode connection corresponding to cathode.

source noise Generated noise, that emitted by source in all directions as distinct from that received by observer.

souris Inlet centrebody shock-cone (F, literally "mouse").

SOV 1 Shut-off valve.
2 Simulated operational vehicle.

SOV-AB New persistent Toxic-B lethal gas and dispenser system (USSR).

SOW 1 Stand-off weapon.
2 Statement of work.
3 Special operations wing.

SP 1 Stabilized platform.
2 Speed brake.
3 Scheduled passenger.
4 Staging post.

S/P Speed/power measurement point, ie 1 g, level flight, constant V.

SPA 1 Solar-powered aircraft.
2 Seaplane Pilots Association (US).

SPAAG Self-propelled anti-aircraft gun.

space Various precise definitions, but loosely volume in which celestial bodies move and esp local portion of solar system outside Earth's atmosphere.

space age Conceptual period in which human beings first learned to operate in space, beginning 1957 (first artificial satellite) or 1961 (first manned space flight).

space/air vehicle See *aerospace vehicle.*

space biology See *bioastronautics.*

spaceborne Travelling through space; suggest unnecessary word.

space capsule Environmentally controlled container in which device or living organism flies in space; suggest not used for human occupation.

space charge Negative charge carried by cathode electrons which unless continuously accelerated away bar further emission.

space-charge region See *depletion layer.*

SpaceCom Space Command (USAF).

spacecraft Self-contained space vehicle, manned or unmanned.

spaced armour Fitted in layers with sufficient gaps to defeat HEAT or hollow-charge weapons.

spaced-diversity Radio communications technique which avoids fading by using three or more receiver aerials spaced 10 or more wavelengths apart, all feeding separate amplifying channels.

space defence All measures designed to destroy attacking enemy vehicles, including missiles, while in space, or to nullify or reduce effectiveness of such attack (DoD).

space equivalent Region in atmosphere where one particular parameter is similar to that in space.

space-erectable Capable of being assembled in space, eg radiation shield.

space fabrication Manufacturing or building operations in space.

space-fixed reference 3-D cartesian co-ordinate system related to fixed stars.

spaceframe 3-D framed structure assembled from simple tubes or girders with pinned or fixed joints, usually built up from succession of triangulated assemblies for rigidity. Note: may have nothing to do with space.

space gyro Gyro having complete freedom about all axes, as distinct from rate *, tied * or Earth *.

spacelab Laboratory for operations in space; with capital S, ESA/NASA programme.

space medicine Branch of aerospace medicine concerned with health of human beings before, during and after spaceflight.

space probe See *probe (3).*

spacer See *shim.*

space simulator Simulator wherein is reproduced one or more parameters of space environment; arguably impossible to simulate all.

space station Permanent structure, probably manned, established in space; probably in Earth orbit, and with routine crew replacement and import of materials.

spacesuit Pressurized suit for EVA and other operations in space or on lunar surface.

Spacetrack Global system or radar, optical and radiometric sensors linked to computation/analysis centre at Norad for detection, tracking and cataloguing of all man-made objects in Earth orbit. USAF portion of Spadats (DoD).

space tug Propulsion vehicle for attachment to space materials, capsules, payloads and laboratories delivered to local area by Shuttle for onward exact positioning.

space wave Combined direct wave and ground-reflected wave from transmitter to receiver, as distinct from surface wave or ionosphere-refracted wave.

space zones Loose subdivision of local space into translunar (between or near Earth and Moon); interplanetary; interstellar.

SPAD Signal processing and display (sonar).

Spadats Space detection and tracking system; reports orbital parameters of all satellites and debris to central control facility (DoD).

SPADCCS Space command and control system.

SPADE, Spade Single programmable access demand exchange; for small comsat-users.

spade Term for several detail features of aircraft, notably fixed or retractable blades projecting into propulsive jet in attempt to enlarge and break up periphery, promote mixing and reduce noise.

Spadoc Space Defense Operations Center (USAF).

SPAe Service de la Production Aéronautique (F).

SPAF Svenska Privat och Affärs Flygföreningen (Swedish AOPA).

spaghetti 1 Complex masses of electrical, hydraulic or other pipes or cables (colloq).
2 Insulating or colour-identification tubing slipped over wires (colloq).

spalling Separation of pieces of armour from inner face of sheet after warhead impact.

SPAN, Span 1 Solar-particle (or proton) alert network (NOAA).
2 Stored-program alphanumerics.
3 Spacecraft analysis (NASA).

span 1 Distance between extremities of wingtips. Term not normally applied to rotorcraft and often a nominal dimension which may or may not take into account tip tanks, ECM pods, winglets and similar extras. When folding wings folded term is width. With variable sweep or slew wing, angular position must be quoted. Symbol b.
2 Operative radial distance from root to tip of rotating aerofoil (eg of helicopter, gas-turbine compressor blade) discounting any inner portion not of aerofoil section. Not defined whether should include tip appendage, eg shroud or tip-drive propulsion. Note: effectively radius, not diameter.
3 Elapsed time within which NW should detonate.
4 Axial distance between centres of bearings supporting a shaft, or between supports of a beam or truss.

spanloader Aeroplane (conceivably, glider) carrying payload distributed across most or all of span (1), normally in ISO containers fitting within wing profile.

span loading Weight of aeroplane or glider divided by square of span (1); W/b^2.

spanwise In a lateral (transverse) direction, esp along wing towards tip.

spanwise lift distribution Plot in front elevation of actual wing lift for elemental chordwise sections of wing from tip to tip, normally (ie without active ailerons or DLC) having semi-elliptic form falling to zero at tips.

SPAR 1 Semi-permanent airfield runway.
2 Solid-state phased-array radar.

spar 1 Major structural member of slender form projecting out from one end (which may be pinned or fixed).
2 Specif, main spanwise structural member(s) of wing or rotorcraft rotor. Wing may have one to many discrete *, or two may be made into single strong box-* (often integral tank) to which secondary leading and trailing structures added. D-* is box structure formed by thick leading edge and a * forming upright of D.

spare Article of repair or replacement nature.

spareable Item capable of being supplied, esp from stock, as spare.

spark discharge Electrical discharge, usually brief, resulting in very large electron flow linking points of high potential difference along narrow and brilliantly luminous path.

spark erosion See *spark machining*.

spark-ignition engine PE (4), eg of Otto type, in which hot electrical spark is used to ignite mixture before each power stroke; thus, not diesel.

sparking plug Term (spark plug in N America) reserved for plugs for spark-ignition engine and those few jet engines where ordinary commercial plug is used; term for gas turbines generally is igniter plug.

spark machining Precision machining of extremely hard or otherwise difficult material by minute stream of HF sparks struck between anode tool (often of particular shape) and cathode workpiece; usually action is purely thermal.

spark photography 1 Simplest and oldest method of high-speed photography, in which shutter is left open and scene illuminated by single point-source brilliant spark at exact time.
2 Various techniques, eg in tunnels, in which hot spark used to create local airstream of contrasting refractive index.

Sparta Special anti-missile research tests, Australia (US programme).

SPAS Shuttle pallet satellite.

Spasm Self-propelled air-to-surface missile.

Spasur Space-surveillance system with mission of detecting and establishing orbital parameters of every man-made object in Earth orbit, using fan of CW across Conus; US Navy portion of Spadats.

spatial Relating to space, not in sense of this dictionary but 3-D volume of any (poss very small) size; thus, concerned with geometric position. Normal dictionary entries need revision.

spatial disorientation Colloquially, not knowing which way is up; eg after losing control of aircraft in cloud.

spatial resolution Ability of sensor to distinguish between two very close distant objects; thus an angular measure (see *resolution*).

spatiography Mapping ("geography") of space.

spats Aerodynamic fairings over fixed landing wheels; purists do not allow term to encompass trousers.

SPC Synthetic particulate chaff.

SPCD Space Communications Division (USAF).

SPD 1 Spectral power distribution.
2 Speed.
3 System programme director.

SPE 1 Solar-particle event.
2 Solid-polymer electrolyte.

spec Specification.

special air mission One conducted by 89 Military Airlift Wing, whether or not President is aboard (USAF).

special-assignment airlift Airlift which for any reason cannot be accommodated by channel airlift (DoD).

special cargo Item requiring unusual handling, eg detonators, precision instruments.

special flight One set up to move a specific load.

special material Nuclear, esp fissile, material not naturally occurring, eg Pu-239, U-235 or enriched uranium.

special pilot ratings Instrument rating and instructor rating (US).

special reconnaissance Flight made covertly, ie without hostile detection.

special rules zone Protected airspace surrounding minor airfield which does not justify a control zone; extends from surface to published (usually low) FL.

special summary drawing Prepared for each production aircraft in GA and certain other civil categories listing all equipment fits, avionic fits, furnishing fabrics/colours, seat types and similar customer choices.

special technical instruction Commands urgent non-recurrent action to remedy serious defect (UK military).

special VFR Particular weather minima below those for normal VFR but which permit VFR flight; hence ** operations = flight within control zone under special VFR clearance.

specific air range See *specific range*.

specification 1 Numerical and descriptive statement of capabilities required of new hardware item, eg future type of aircraft.
2 Concise list of basic numerical measure describing existing hardware item.

specific consumption See *specific fuel consumption.*

specific energy Energy per unit mass, whether released by chemical combustion or degradation of KE, units (SI) J kg^{-1}.

specific excess power Propulsion power available over and above that needed to propel aircraft in level flight at given reference speed, and thus available for climb or manoeuvres at sustained high speed. With streamlined aeroplane (fighter) drag is small in relation to thrust at most reference speeds so SEP is broadly governed by thrust/weight ratio T/W.

specific fuel consumption Symbol c′, rate of consumption of fuel for unit power or thrust, and thus basic measure of efficiency of prime mover; term confined to air-breathing engines (equivalent for rocket is specific impulse). SI unit for jet engines (turbojet with/without afterburner, turbofan, ramjet or pulsejet) is mg/Ns (milligrammes per Newton-second); traditional imperial measure is lb/h/lb thrust, measure being for SL static ISA condition unless otherwise specified. For shaft engines (turboshaft, turboprop, piston) SI unit is μg/J (microgrammes per joule); traditional Imperial unit is lb/h/hp, hp being qualified for turboprop as shaft or equivalent.

specific gravity, SG Density of material expressed as decimal fraction (less than or greater than unity) of density of water at 4°C.

specific heat Quantity of heat required to raise unit mass of material by unit temperature, usually from 0° to 1°C; unit is J/(kg·K); traditional measure is calories per gramme. For gases thermodynamic process must be stated; C_p (** at constant pressure) and C_v (** at constant volume) are not same.

specific humidity Ratio of mass of water vapour to total mass of moist gas (dimensionless).

specific impulse I_{sp}, basic performance parameter of rocket, = thrust divided by rate of consumption of propellants, = total impulse divided by total mass of propellants (see also *motor* **, = total impulse divided by total loaded mass of solid motor), in general = effective jet velocity divided by g; unit = s (seconds) derived from force × seconds divided by mass (strictly, in SI Newtons force cannot be divided by kilogrammes mass).

specific power Not defined; often used for thrust/weight ratio of prime mover, esp including electrical battery or fuel cell. For air-breathing engines see *power/* or *thrust/weight ratio.*

specific propellant consumption Reciprocal of I_{sp}, mass flow of propellants or rate of burning to generate unit thrust in rocket.

specific range Air distance flown for unit consumption of fuel; traditional unit is nam/lb (NAMP), while SI would be air-km/kg.

specific search Reconnaissance of limited number of points for specific information.

specific speed Basic performance parameter of hydraulic turbine; trad unit is rpm at which 1 hp is generated with head of 1 ft. SI unit needed for modern hydraulic motors and rocket turbopumps.

specific stiffness Stiffness (usually Young's modulus) divided by density, E/ρ.

specific strength Ratio of ultimate (tensile/compressive/shear/bending) strength to density.

specific tasking Planning phase in which commanders designate actual units to fill force list of operation plan.

specific thrust Propulsive thrust divided by engine mass (not normally applicable to rockets).

specific volume Volume per unit mass, reciprocal of density.

specific weight 1 Symbol $\bar{p}$, engine mass divided by net thrust (air-breathing jet engines).
2 Symbol $\bar{\omega}$, engine mass divided by net thrust (stub) power (PE [4]). Strictly, mass should include cowl and propeller.

specified approach funnel Funnel extending about 0·5° above and below PAR glidepath and ±2° of PAR centreline within which aircraft is not advised by PAR controller monitoring ILS approach.

specimen performance Performance, esp of civil transport, worked out and presented as guide to correct procedures and methods of compliance.

specklegram Photograph taken by laser speckle interferometry.

spectacles 1 Main control wheel when made with single left/right curved handlebars, suggest synonymous with ram's horn; yoke is generalized term for all configurations.
2 Vertical figure-eight manoeuvre.

spectrograph Spectroscope with camera or other recorder.

spectroheliograph Takes pictures of Sun in monochromatic light; spectrohelioscope is for direct viewing.

spectrometric analysis Usual method is to analyse spectrum of light as test substance is burned in electric arc.

spectrophotometer Photometer which measures variation of radiant intensity with wavelength.

spectropyrheliometer Measures variation of intensity of direct solar radiation with wavelength.

spectroradiometer See *spectrophotometer.*

spectroscope Instrument for dispersing light into spectrum.

spectrum 1 Visual display, EDP (1) printout, photo record or other presentation of variation of radiation intensity (sometimes other parameters, eg sound pressure level) with wavelength/frequency or other variables.
2 Continuous range of wavelength/frequencies within which radiation forms common grouping, eg visible *, IR *.

3 Stylized colours of visible *.
4 Various specialized terms in maths, mechanics etc.

specular Offering a smooth reflecting surface; rule-of-thumb demarcation from diffuse scattering surface is roughness factor. In low-level attack with radar/laser, reflection from calm sea is * up to grazing angles near 20°; steeper attack gives diffuse.

SPED Supersonic planetary entry decelerator.

speech inversion Simple form of scrambling in which frequencies relative to a reference value are inverted.

speed Scalar quantity, as distinct from velocity.

speed brake See *airbrake*.

speed bump See *shock body*.

speed capsule See *shock body*.

speed control computer Accepts flight data from aircraft sensors and provides flight-management outputs for throttle servoactuators and, probably, flight director and autopilot.

speed course Accurately surveyed course for attempts on speed records; previously also used for true G/S measures (now arch).

speed lock Autopilot sub-mode in which TAS (in some systems G/S) is held constant.

speed probe Basic sensor for rpm of shaft, with magnetic pole piece held close to teeth of existing gear or special toothed disc to generate emf whose frequency is transmitted. Has virtually replaced tachometer except in GA aircraft.

speed range Aircraft max level speed minus min speed, eg V_{mc}.

speed reference system Chief meaning is subsystem on advanced transports (Airbus) which provides flight director visual guidance on how far pilot should haul back in windshear without triggering stick shaker.

speed rotor Main input sensor of anti-skid brake, whose angular velocity is held against sudden reductions.

speed stability Condition such that aircraft tends to return to preset speed following any excursion, thrust remaining constant; condition not obtained below V_{md}.

Speedtape Commercially available thin aluminium sheet, resembling stout foil, with adhesive backing revealed by peeling off skin.

speed trend Additional protection against windshear, shows what speed/AOA will be 10s hence with no action by pilot.

spelk Unusable bundles of short lengths of surplus reinforcing fibres.

spent fuel Fissile fuel whose allotted life has been consumed, though material still fissile.

SPER, Sper 1 Syndicat de Industries de Matériel Professionnel Electronique et Radio-Electrique (F).
2 Strategic-planning executive review.

SPERT, Spert Scheduled (or simplified) programme evaluation and review technique.

SPET Solid-propellant electric thruster.

SPF Svensk Pilotförening (Sweden).

SPFDB See *SFDB*.

SPGG Solid-propellant gas generator.

sphere of influence Volume around body in space within which small particle is attracted to body; in case of Earth not true sphere.

spherical angle Angle between two great circles.

spherical data system Long-range nav system, eg for maritime patrol aircraft, related to spherical Earth surface.

spherical triangle Formed by arcs of three great circles.

spherics See *sferics*.

spherodizing Hot soaking of irons and steels close to (usually just below) critical temperature followed by slow cooling.

spherometer Instrument with three legs and central micrometer leg for measuring radius of convex or concave spherical surfaces.

sphygmomanometer Instrument for measuring blood pressure.

SPI 1 Surface position indicator.
2 Short pulse insertion (SSR).
3 Spike-position indicator.
4 Scatter-plate interferometer.
5 Special position identification (pulse).

spicules Long, bright filaments briefly extended from chromosphere.

spider 1 Structural heart of propeller or helicopter rotor in form of hub integral with radial members which bear all stresses from attached blades.
2 Multi-finger plate securing structural members grouped at a common joint, each finger being aligned with and secured to one of members, all being in same plane.

spider beam Spaceframe-type rocket interstage structure.

spigot See *sprag*.

spike 1 Conical inlet centrebody of supersonic airbreathing engine, usually designed to translate.
2 Short-duration transient (in signal, current, radar display or any other oscillating variable) in which amplitude makes large excursion beyond normal.
3 Centrebody of * nozzle.

Spikebuoy Tactical warfare sensor dropped from air to land on spike (eg in jungle), thereafter transmitting on command ground tremors which can be attributed to human beings or vehicles.

spike inlet 2-D airbreathing inlet in form of body of revolution with central spike (1).

spike nozzle Rocket nozzle in which gas escapes through ring around centrebody in form of concave-profile cone.

spill To cause spilling.

spillage 1 Amount by which mass flow into airbreathing inlet is less than datum flow (in UK

called intake flow).
2 Flow of air from below wing to above at tip.
spillage drag Difference between drag at given engine airflow and drag at datum flow.
spill burner Gas-turbine burner in which fuel is supplied at constant high pressure, giving good swirl and atomization, excess over requirement being "spilt" back through second pipe for reuse.
spill door Auxiliary door, usually spring-loaded to open outwards, through which excess engine airflow (spillage) escapes with min drag, eg in high-speed cruise or on letdown.
spilling Escape of air at one part of parachute canopy periphery, either through instability or for directional control of trajectory.
spillover Airflow deflected to pass outside air-breathing inlet in supersonic flight with detached shock.
SPILS, Spils Spin-prevention and incidence-(AOA is meant) limiting system.
spin 1 Sustained spiral descent of fixed-wing aerodyne with AOA beyond stalling angle; in most cases a stable autorotation (see *flat* *, *inverted* *, *upward* *).
2 To shape by spinning.
spin axis Axis of rotation of gyro wheel.
spin chute Anti-spin parachute.
spin dimpling Dimpling by coldworking tool (eg 60° cone) around hole for rivet or other fastener under pressure, without cutting.
spindle Structural frame in rigid airship to which mooring cone is attached.
spindling Generalized term for machining of wooden parts, esp to effect particular desired uniform cross-section to spar, longeron or other structural member. * machine basically resembles spar mill.
spine Non-structural fairing along dorsal centreline of aircraft or along outside of ballistic vehicle (occasionally in other locations but always parallel to longitudinal axis) covering pipes, controls or other services. In some cases merely drag-reducing fairing linking canopy to fin.
spine hood Hinged or removal cover over equipment in spine.
spin motor Rocket(s) imparting spin (rotation about longitudinal axis) to missile or other vehicle.
spinner Streamlined fairing over propeller hub; not used for similar fairing over helicopter rotor hub.
spinning 1 Sheet-metal shaping by forcing against spinning die of desired profile (suitable only for bodies of revolution).
2 To engage in spin (1).
spinning nose dive Flight manoeuvre in which aeroplane or glider is rolled with ailerons while in steep dive (US usage).
spinning test 1 To explore spin characteristics of aircraft.
2 Basic strength test of propeller or helicopter rotor, or (often at very high speed in low-drag environment) other rotating assemblies, eg turbine or fan rotor.
spinning tunnel See *spin tunnel.*
spin-off Predicted or unexpected advances in one technology caused by transfer of technical solutions from another. Also called fallout.
spin parachute See *anti-spin parachute.*
spin-recovery parachute See *anti-spin parachute.*
spin rocket See *spin motor.*
spin-stabilized Given gyroscopic directional pointing stability by high-speed rotation about longitudinal axis.
spin table Large disc on which objects, including human beings, can be rapidly rotated about vertical axis for various test and research purposes; ** in turn may be mounted on arm of centrifuge to give additional sustained unidirectional lateral acceleration.
Spintcom Special-intelligence communications.
spin tunnel Wind tunnel in which flow through working section is vertically upwards, thus free model supported by airflow can be examined for spinning characteristics.
spiral Flight manoeuvre in which at least 360° change in heading is effected while in glide or shallow dive (chiefly US usage).
spiral aerial (antenna) Aerial in form of single conductor wound as spiral on conical dielectric support; common as passive RWR receiver.
spiral angle Angle between pitch cone generator of bevel gear and tangent to tooth trace; positive for right-hand gear.
spiral bevel gear Crown gear has pitch curves inclined to pitch element and usually circular arcs.
spiral dive Extremely dangerous flight manoeuvre in which aircraft, invariably fixed-wing aerodyne, is unwittingly in spiral descent with neither turn nor slip indicating and in absence of external cues few indications other than horizon instrument if fitted.
spiral glide Sustained gliding turn.
spiral instability Faulty aeroplane flight characteristic in which there is an inherent tendency to depart from straight and level flight into oscillating sideslip and bank, latter always being too great for tendency to turn. (Long-established UK definition unrelated to Dutch roll, which is high-subsonic phenomenon; suggest arch.)
spiral scan See *scan types.*
spiral stability Desired aeroplane flight characteristic in which, in co-ordinated turn, it automatically resumes straight and level flight on release of flight controls. Note that no slip is present to assist recovery, and that large fin (needed for other reasons) exerts adverse influence.

spiroid gear Patented gear in which conical worm engages with many teeth simultaneously of face-type gear, possibly with exceptionally large reduction ratio.

spitting Air intercept code: "I am about to lay sonobuoys and may be out of radio contact (because very low) for few minutes" (DoD).

SPL 1 Sound pressure level.
2 Standard parts list(ing).
3 Supplementary flightplan message.
4 Sun-pumped laser.
5 Signature and Propagation Laboratory (USA).

splash 1 Code word sent to observer 5 s before estimated impact of weapon(s).
2 Target destruction verified by visual or radar means.
3 Generalized term for action of destroying aerial target.

splash-detection radar Pinpoints impact of vehicle with ocean to facilitate positive scoring and vehicle recovery (eg in Nike X ABM tests).

splashdown End of space mission in which spacecraft, capsule or other recoverable object impacts ocean surface; defined as either a time or a location.

splashed Air-intercept code: enemy aircraft shot down (followed by number and type) (DoD).

splash lubrication Use of small lips or vanes on con rods or crankpins to splash oil inside crankcase; rare in aviation.

splatter 1 Adjacent-channel interference in pulsed transmissions, measured as amount of spectrum energy that can appear in adjacent channel; varies greatly with different pulse waveforms.
2 Cloud of canopy fragments after MDC detonation.

splice Structural joint made by plate overlapping both members, plus doubler.

spline 1 Axial groove in shaft for meshing with driven member; hence splined shaft has entire periphery formed into splines, invariably of rectangular section.
2 Flexible non-structural strip (various materials) bent to required curvature in construction of fairing.

split-altitude profile Flight profile has two main flight levels.

split-axle gear Landing gear on simple low-performance aircraft in which there is no axle or other linking transverse member. Also called divided *.

split basing Division of tactical-aircraft unit's resources between two operating bases.

split cameras Two or more cameras fixed so that imagery of one overlaps that of neighbour(s) by selected amount.

split charter Commercial flight flown on charter to two companies, each providing part of payload.

split-compressor engine Gas turbine in which compression is performed by two separate rotating assemblies running at different speeds; can be axial + centrifugal, while term two-spool suggests axial + axial.

split courses See *multiple courses.*

split-distance Two friendly fighters fly apart for mutual interception practice, usually head-on.

split flap Flap formed from only underside of aerofoil, depressed with plain hinge leaving upper surface unaltered; gives high drag but little extra lift.

split gear Landing gear trucks (B-52) attempt to steer left and right simultaneously.

split landing gear See *split-axle gear.*

split line Small flash-projecting ridge round surface of die forging.

split mission Several meanings, including (1) profile part hi and part lo, (2) task includes recon and attack, (3) task includes two surface targets, and (4) transport flight carries loads for two destinations.

split needles Various flight-instrument indications for aeroplane (eg dual engine-speed indicator with one engine at flight idle) or helicopter.

split pair See *split vertical photography.*

split patch Patch for reinforcing end of surface (eg fabric air inlet) at junction with airship envelope.

split-plane manoeuvring Air-combat manoeuvring for mutual support following a defensive split.

split-S Flight manoeuvre comprising half flick (snap) roll followed by second half of loop, resulting in loss of height and 180° change in heading.

split-surface control Powered flight-control surface subdivided into two or three portions, each of same area and each driven by its own independent power unit (thus, VC10 has four elevators and three rudders).

splitter 1 Fixed or laterally movable surface dividing fluid flow in duct, eg to feed two engines.
2 Machine which divides or apportions signals, shaft power or other services among selected recipients.

splitter box Divides one rotary input among two outputs in variable proportions, eg to control either or both of two roll-control or DLC spoilers.

splitter panels Sound-absorbing panels, usually radial struts or concentric rings, to reduce noise from jet-engine inlet.

split vertical photography Simultaneously triggered reconnaissance cameras, each tilted same angle to left or right of centreline, with small centreline overlap.

split-work blade Gas-turbine fan blade comprising inner (compressor supercharging) portion separated from outer (fan) portion by part-span shroud.

SPM 1 System planning manual.
2 Solar-proton (or particle) monitor.

spm Shots per minute.
SPO 1 System program office.
2 Safety petty officer (carrier catapult).
SPOH Since part (or partial) overhaul.
spoil To reduce, not necessarily to zero (eg lift or thrust).
spoiler 1 Hinged or otherwise movable surface on upper rear surface of wing which when open reduces lift, and usually also increases drag. Most are essentially flat plates hinged at or beyond leading edge and power-driven to open upwards, either symmetrically to incline flight path downwards or differentially to command or augment roll. Supersonic combat aircraft often have no other roll control, while others use * only at low speeds or high speeds. * operation on commercial transports is invariably linked with position of speed brakes, response depending on whether differential or not. DLC * is primary flight control, esp during landing. Many * used as lift-dumpers after touchdown.
2 Movable deflector used to kill most if not all residual thrust from engine or from turbofan core; much simpler and lighter than reverser.
3 Comb, flap or other device extended on command to break up local airstream, eg ahead of open weapons bay.
spoiler initiation angle Rotary deflection of pilot's roll control (spectacles, handwheel, etc) at which spoiler(s) start to open (from wings-level flight with speedbrake fully closed).
spoiler-mode control system Governs spoilers as DLC in flight and auto dumper and speedbrake on touchdown.
spoking Regular or erratic flashing of rotating timebase on PPI or other radial display.
spongy Descriptive but vague term for flight controls where response appears unduly delayed or uncertain.
sponson 1 Projection from flying-boat hull in form of short thick wing to provide stability on water in place of outer-wing floats.
2 Projection from helicopter fuselage in form of short thick wing to provide attachment for main landing gear (and in some cases retraction stowage space).
3 Short, thin wing-like structure projecting from aeroplane or helicopter fuselage to carry weapons, external tanks, guns and possibly other devices.
spoof To copy hostile IFF reply code.
spoofer Air-intercept code: contact is employing electronic or tactical deception measures (DoD).
spool 1 One complete axial compressor rotor, in case of multi-shaft engine forming LP, IP or HP portion of complete compressor.
2 Attachment anchor, usually one of L/R pair, for catapult bridle of carrier aircraft.
3 To open throttles (colloq).
spool down To allow engine rpm to decay to zero, eg after closing HP cocks; normally turbofan or turbojet.
spool up To accelerate engine rpm, esp to TO power or at least to much higher level than previously; normally turbofan or turbojet.
sporadic E Irregular radio-reception and disturbance caused by abnormal variation in E-layer.
SPOT 1 Spot wind (ICAO).
2 Speed, position, track.
spot 1 To form up aircraft in close ranks on carrier deck ready for free or catapult take-offs.
2 Designated place on airfield where landing is to be made.
3 Bright region where electrons strike fluorescent tube face in CRT and many other displays or image converters.
4 To determine, by observation, deviations of ordnance from target for purpose of supplying necessary information for adjustment of fire (DoD).
5 To search from ground for hostile aircraft in own airspace (UK civillian usage, WW2; hence spotter, spotting).
spot annealing Annealing local area of hard steel, eg to drill and tap fixing hole.
spot elevation See *spot height* (US usage).
spot facing Local surface machining round hole or other point, to improve surface finish, adjust dimensions or provide square-on surface for bolt head.
spot fuel Uplifted and paid for on the spot, as distinct from part of ongoing contract.
spot height Height of point, esp mountain peak or other high point, marked on map or chart.
spot hover Helicopter training manoeuvre in which machine is hovered at low level over point, turning through four successive headings 90° apart.
spot jamming Jamming of specific frequency or channel.
spot landing Aeroplane landing made from specified position and height AGL on to spot (2); form of accuracy landing.
spot net Com net used for spot (4) information.
spot size Diameter of spot (3).
spotter Person assigned to task of watching for and identifying hostile aircraft; hence raid *, official responsible for immediate warning of imminent attack to high-priority establishment whose personnel would remain at work throughout air raids (UK civilian usage, WW2).
spotting Act of arranging aircraft on flight deck (see *spot* [1]).
spotting factor Ease with which particular aircraft type can be spot (1) positioned on deck; not quantified but takes into account overall folded dimensions and turn radius and possibly stability and laden weight.
spot weld Local, usually circular, weld quickly made by electrical resistance jaws working on sheet; tool can be point or roll electrodes.

spot wind Wind measured at one geographical location.

spot wobble Technique for imposing small SHM waveform on each line of TV or similar raster display to blur separate lines.

SPR 1 Solid-propellant rocket.
2 A commercial rain-repellant for transparencies.
3 Secretarial program review (US).

SPRA Special-purpose reconnaissance aircraft; one designed to cross hostile frontier covertly.

sprag 1 Pivoting and lockable projection in cargo-floor guiderails to prevent movement of pallets or platforms. Occasionally (not DoD) called spigot.
2 Pawl or spigot-type mechanical lock for a rotary member, as in next two entries.

sprag brake Positive mechanical brake for locking wheels of helicopter parked in confined area on rolling deck.

sprag clutch Positive mechanical lock for failed engine and drive of helicopter, while permitting continued overrunning or autorotation of rotor system.

spray To dispense liquid from aircraft for agricultural, defoliant or ECM purposes.

spray bar Bar for spraying, esp for ag purposes, arranged spanwise below trailing edge and usually freely swinging to knock upward on impact with obstruction.

spray dam Strip projecting along forebody chine of marine aircraft, as far as possible following streamline in cruising flight, to deflect water spray downward.

spray dome Mound of water thrown into atmosphere when shockwave from underwater NW reaches surface.

Spraymat Patented (Napier, now Lucas) electrothermal anticing and de-icing mats featuring sprayed-on metallic layers.

spray strip See *spray dam.*

spreader 1 Ag aircraft rigged for solids.
2 Mechanical (usually centrifugal) dispenser of ag solids.

spreader bar 1 Horizontal member(s) separating and joining floats of twin-float seaplane.
2 Horizontal axle-like member joining left/right landing gears of early or light aircraft, other than true live axle.

spread spectrum Vast and growing technology forming complete division of electronics, esp military and avionics, fundamental of which is use of PN, FH, TH or any combination of these to modulate signal whose bandwidth is much wider than that of plain message; latter is conventional (eg biphase digital or PDM analog) and is merged digitally with ** modulation to generate emitted signal. Advantages are very great anti-jam capability, military security, multiple access, low detectability, Selcal capability and auto transmitter ident, multipath tolerance, and inherent precision-nav capability.

springback Angular distance through which metal bent to new shape springs back after bending force is removed; allowed for in making tooling or in hand operations.

spring feel Simplest form of artificial feel, in which force is applied to pilot's flight control (eg stick or yoke in pitch) by linear spring, thus within limits exactly proportional to deflection and unvarying with dynamic pressure.

spring sheet-holder Small lock in form of cylinder and spring-loaded rod with locking end inserted by pliers through holes in sheets to hold location during riveting.

spring tab Servo tab whose deflection relative to surface is resisted by spring, usually torsion bar, which is often preloaded so that at gentle inputs pilot moves surface and tab unaided; at higher inputs spring is overcome and tab deflects to assist.

Sprite Signal processing in the element.

SPS 1 Secondary power system.
2 Samples per second.
3 Blown flap (USSR).
4 Self-protection subsystem.
5 Simplified processing station.
6 Service (module) propulsion system.

SPSS Statistical package for social sciences (CAA).

SPT-B Selectable-performance target, ballistic.

Spur Space-power unit reactor.

Sputnik Russian word for "fellow traveller", name of first artificial satellite, launched 4 October 1957; later colloq for any satellite or even any spacecraft.

sputtering Ejection of metal atoms from cold cathode by evaporation or ion bombardment, either as nuisance, to form fine coating on substrate or to form colloidal metal solution.

SPW 1 Self-protection weapon (tactical attack aircraft AAM, not necessarily fired in forwards direction).
2 Secondary power.

SQ 1 Squall (ICAO).
2 Software quality.

SQA Software quality assurance; /CM adds "/configuration management".

SQM 1 Software quality metrics, quantified measures of SQ.
2 Software quality management.

squadron Any of many types of military or naval unit, including common admin unit of combat aircraft; according to many services "consisting of two or more flights". Foreign-language equivalents include staffel (G), escadron (F), escuadron (S), polk (R).

squall Strong but intermittent wind; gust whose effect lasts minutes rather than seconds and extends a kilometre/mile or more horizontally.

squarco Radar beam squinting plus area coverage.

square 1 See *signal area.*
2 Large square marked on remote part of airfield, to be accurately flown by pupil helicopter pilot, especially in presence of wind.
square course Airfield circuit (US usage); does not mean literally square.
square/cube law Basic geometric law: areas of similar-shaped solid bodies are proportional to squares of linear dimensions, and volumes (ie for equal densities, masses) to cubes. Thus if two aeroplanes are of same shape but one has twice linear dimensions, it will have four times wing area and eight times weight, hence W/S is doubled.
square engine PE (4) whose bore equals stroke.
square parachute Canopy is approximately square when laid out flat.
square search Various standard air/surface search patterns in form of overlapping rectangles (usually squares or near-squares) so that after period aircraft has examined large strip or rectangle with minimal duplication.
square stall Stall with wings level throughout.
square thread Thread with vertical faces (unlike Acme) used for transmitting power as linear thrust in either direction.
square wave EM or other periodic wave which alternates between steady positive and steady negative values in time extremely brief by comparison with steady periods.
square-wing biplane Upper and lower spans equal; according to some authorities, also without stagger.
squaring shears Hand or power shears for cutting sheet positioned on marked-out platform.
squash head Warhead for use against hard targets, esp those with single thick metallic armour layer; basic principle is transmission of intense shockwave through armour, causing transverse acceleration high enough to spall pieces off inside face.
squat(ting) Downward movement of marine-aircraft c.g. (eg due to trough) in essentially level attitude while running on water.
squat switch Bistable switch triggered by sustained (usually 2·5 s) compression of main or nose landing-gear struts on touchdown, to operate lift dumpers, reversers and/or other devices. Term also applicable when input is rotation of MLG bogie beam.
squawk Generalized word for airborne transponder or IFF operation and keying; when used alone usually a ground-radar ATC command to switch to normal or to directed mode.
squawk alt Switch to active Mode C with auto altitude-reporting.
squawk flash Operate IFF I/P switch.
squawk ident Engage ident feature; civil counterpart of flash.
squawking Air/ground code: showing IFF/transponder in mode/code indicated.
squawk low Switch to low sensitivity.
squawk Mayday Switch to emergency position; for civil transponder Mode A, Code 7700; for mil IFF Mode 3, Code 7700 plus emergency feature.
squawk mike Operate IFF MIC switch and key transmitter as directed.
squawk normal Switch to normal sensitivity.
squawk (number) Operate in Mode A/3 on designated code.
squawk (number) and ident Operate on specified code in Mode A/3 and engage ident (mil IFF I/P).
squawk standby Switch to standby position.
squeeze-film bearing Has small annular space between outer track and housing filled with pressure-feed lube oil which cushions dynamic radial loads, thus reducing engine vibration and possible fatigue.
squeeze riveting Rivet closure by single sustained force instead of blows.
squelch Subcircuit in communications receiver which holds down volume until a signal is received, or which greatly reduces volume on operator command.
squib Any small pyrotechnic used as source of hot gas, eg to fire igniter of rocket or fuel in some (eg missile/RPV) gas turbines.
squib valve Ambiguous, has been used to mean (1) solenoid valve for controlling thrusters (Apollo RCS) or (2) squib-actuated valve.
squid 1 Superconducting quantum interference device; pair of Josephson junctions.
2 Semi-stable part-opened regime of parachute canopy, normally encountered only at extreme airspeeds.
squidding Operation of parachute in squid position.
squint Small angular error between actual and theoretical direction of main-lobe axis or direction of max radiation of radar or other directional emitter (see * *angle*).
squint angle 1 Maximum angle away from missile axis at which homing head (IR, radar, EO, etc) can acquire and lock on to normal emitting point target at significant distance.
2 Angle of squint.
squirrel cage 1 Air-combat dogfight (colloq).
2 Induction motor whose rotor comprises axial bars joined to rings at each end, pulled round by rotating field.
squirt Jet aircraft (colloq, c 1942–48).
SR 1 Short-range.
2 Search rate.
3 Sunlight-readable.
4 Special rules.
5 Single rotation.
6 Sortie rate.
7 Strategic reconnaissance, role designator (US), originally transposition of RS (recon/strike).
8 Sunrise.

9 Service report.
10 Solid rocket.
sr Steradian.
SRA 1 Special-rules area.
2 Shop-replaceable assemblies.
3 Surveillance radar approach (CAA).
4 Spin reference axis.
5 Specialized repair activity (DoD).
SRAM Short-range attack missile.
SRARM Short-range anti-radiation missile.
SRB Solid-rocket booster.
SRBM Short-range ballistic missile.
SRC 1 Science Research Council.
2 Secondary radar code.
3 Sample return container (spaceflight).
SRCC Structures Research Consultative Committee (SBAC).
SRD 1 Short-range diversion.
2 Service-revealed difficulty.
SRDE 1 Signals Research & Development Establishment (Christchurch, UK, now closed).
2 Search-radar data extractor.
SRDS Systems Research and Development Service (FAA).
SRE Surveillance radar element; portion of GCA.
S_{ref} Reference area.
SREM Software requirements engineering methodology.
SRFCS Self-repairing flight-control system.
SRFW Schweizerische Rettungsflugwacht (= GASS) (Switz).
SRM 1 Solid-rocket motor.
2 Short-range missiles (HUD selection).
SRO 1 Station routine orders.
2 Senior ranking officer.
SROB Short-range omnidirectional beacon.
SRP 1 Steep RP (GGS selection, arch).
2 Software rapid prototyping.
SRR 1 Search/rescue area (ICAO).
2 System requirements review.
3 Software requirements review.
4 Short-range recovery.
SRS 1 Sonobuoy reference system.
2 Speed reference system.
SRSK Short-range station-keeping.
SRT System readiness test(s).
SRU Shop-replaceable unit.
SRW Strategic reconnaissance wing (USAF).
SRWBR Short-range wideband radio.
SRZ Special-rules zone.
SS 1 Sunset.
2 Source-substantiation.
3 System(s) simulation.
4 Spread-spectrum.
S_s Torsional stress.
SSA 1 The Soaring Society of America.
2 Self-Soar Association (US).
3 Static stability augmentation.
SSAC 1 Source-selection advisory council.
2 Solid-state air-cooled.
SSADP System station annunciator display panel.
SSALS Simplified short approach-light system.
SSALSR SSALS with RAIL.
SS-Anars Space sextant, autonomous navigation and attitude reference system.
SSAP Survival stabilator actuator package.
SSB 1 Single sideband.
2 Space Science Board (US, NAS).
3 Split system breakers.
4 Ballistic-missile-firing conventional submarine.
SSBN US Navy code for ballistic-missile-firing nuclear-propelled submarine.
SSBS Sol/sol balistique stratégique; French MRBM.
SSC Sidestick controller.
SSD Systems and Simulation Division (USAF).
SSEB Source-selection evaluation board.
SSF Station services flight (RAF).
SSFM Steerable sensor-fuzed munition.
SSGN US Navy code for winged guided (cruise) missile nuclear submarine.
SSI 1 Small-scale integration.
2 Significant structural item(s), or structurally significant item.
3 Sandwich speckle interferometry.
SSIA Service de Surveillance Industrielle de l'Armement (F).
SSID Supplementary structural inspection document.
SSKP Single-shot kill probability.
SSL Speed select lever.
SSM 1 Surface-to-surface missile.
2 Seat statute-mile; non-SI traffic unit.
3 Special-shape market (advert-shape aerostats).
4 Ambiguously, sea-skimmer missile.
SSMA Spread-spectrum multiple access.
SSME Space Shuttle main engine.
SSMO Summary of synoptic met observations.
SSP 1 Sensor select panel.
2 Source selection panel (NATO).
SSPA Solid-state phased array.
SSPI Sighting system passive/IR.
SSPS Satellite solar power system (or station).
SSR 1 Secondary surveillance radar.
2 Solid-state relay.
SSR video Video (raw radar) signal processed in SSR computer to exclude unwanted information and leave graphical display of targets, data and other displayed information in correct display positions.
SSS 1 Space surveillance system (USN).
2 Small scientific satellite.
3 Strategic satellite system; also survivable strategic satellite.
4 Stick-shaker speed.
SST 1 Supersonic transport.
2 Static-strength test(ing).
3 Static storage tank.

SSTDMA Satellite system TDMA.
SSTO Single stage to orbit.
S-stoff RFNA (G).
SSU 1 Signal summing unit.
2 Semiconductor storage unit.
3 Sensor surveying unit.
SSUS Spinning solid upper stage.
SSW Swept square wave (ECM).
ST 1 Standard time.
2 Stairway (passenger, not powered).
3 Sharp-transition (VASI).
S_T 1 Area of horizontal tail.
2 Distance between centres of contact areas of tandem wheels.
St 1 Stratus.
2 Stokes.
3 Stanton number.
S_t Unit tensile stress.
STA 1 Service des Transports Aériens (F).
2 Station (also Sta), eg fuselage STA 307·8.
3 Supersonic transport aircraft (duplicates SST).
4 Static-test article.
5 Straight-in approach (ICAO).
6 Shuttle training aircraft.
7 Structural test airframe (or article).
8 Section Technique de l'Armée (F).
stabbing Assembly process in which finished gas-turbine stator blade (IGV) is fired at high speed through unbroken ring to form tight-gripping joint.
stabbing band Narrow projecting band around rotating part, eg turbine disc, from which metal can be removed for dynamic balancing (now generally superseded).
STABE Second-time-around beacon echo.
stabilator Slab horizontal tail used as single primary control (stabilizer/elevator). Normally pitch axis only, but occasionally used to mean taileron (US usage).
stability Generally, quality of resisting disturbance from existing condition and tendency to restore or return to that condition when disturbance is removed. For aircraft, meaning is confined to basic flight control and defined as tendency to resume original (normally straight/level but not necessarily) attitude after upset (rotation about any axis); qualified according to axis and whether stick-fixed or stick-free. For atmosphere, temperature distribution such that particle tends to stay at original level. For structure, ability to develop internal forces resisting those externally applied. For materials other than structural, usually ability to withstand harsh environment (eg high temperature) without even gradual physical or chemical change.
stability augmentation Various species of auxiliary subsystem added to primary flight-control system (usually of advanced aeroplane or spacecraft) to achieved desired vehicle characteristics by selection of variable gains in feedback loops from surfaces. In some forms surfaces are commanded, eg yaw damper. Modern fighters with relaxed longitudinal stability would be dangerously unstable without **. Usually has limited authority and does not move pilot's controls.
stability derivatives See *derivatives*.
stability factor Ratio of change in transistor collector current to change in I_{co} (DC collector current for zero emitted current).
stability limits One meaning (suggest arch) is extreme angles of incidence to which taxiing seaplane can be trimmed without porpoising.
stability margin See *static margin*.
stabilization Positive action to maintain stability, esp of spacecraft or payload, when term invariably refers to attitude. Passive * is any method requiring no sensing, logic or power, eg gravity-gradient, spin and solar-wind/aerodynamic pressure. Semi-passive requires stored momentum, eg gravity-gradient plus CMG. Semi-active introduces limited thrust/torqueing, eg on one axis. Active features sensing, logic and control about all axes. Hybrid are systems with more than three degrees of freedom, eg to control despun/gimballed/independent secondary devices such as aerials or telescopes.
stabilized gyro Usually means aligned with a desired direction, eg Earth centre, true N or magnetic meridian.
stabilized platform Invariably, platform maintained always horizontal at any place near Earth, ie perpendicular to local vertical.
stabilizer 1 Tailplane or slab horizontal tail (US).
2 Loosely, any fixed tail surface, but fin normally prefaced by "vertical" (US).
3 Low-density core (eg foamed-in-place plastics, balsa, honeycomb) filling interior of secondary structure, control surface, flap, door or similar structure.
4 Additives to retard chemical reactions.
5 Gyro subsystem to stabilize pivoted or gimballed device, eg radar aerial.
stabilizing altitude Altitude at which actual rate of climb is zero.
stabilizing floats Small seaplane-type floats mounted well outboard under wing of flying boat to provide roll stability when afloat.
stabilizing gears Small landing gears carried well outboard (eg near or at tips of wing) of landplane with centreline main gears to provide lateral stability on ground, esp when turning. Also called outriggers.
stabilizing parachute Used to stabilize fall of otherwise unstable paradropped load.
Stabimatic Simple modular autopilot for GA, buildable from wing-leveller using vacuum aileron input to fully coupled 3-axis system capable of capturing desired FL.
stable aerofoil Complete wing whose CP travel is very small.

stable air Air mass in which actual lapse rate is less than adiabatic lapse rate (dry or saturated, depending on humidity), in extreme cases becoming negative (ie inversion).

stable-base film Reconnaissance or scientific film of extremely high dimensional stability.

stable oscillation 1 Oscillation whose amplitude is constant.
2 Oscillation whose amplitude decreases (BS).

stable platform See *stabilized platform*.

Stabo Anti-runway munition (G).

STAC Supersonic Transport Aircraft Committee (UK, 1956–62).

stack 1 Superimposed series of holding patterns, each at assigned FL.
2 To assign to hold in *.
3 Exhaust pipe, of any length or configuration, from PE (4) (US).
4 To assemble multi-stage launch vehicle (colloq), and vehicle thus assembled.

Staco Standing Committee for the Study of Principles of Standardization (ISO).

Stadan Space tracking and data network.

stadiametric aiming Optical aiming using lead angle calculated from apparent size of target and aspect, using various methods including subjective judgement (suggest arch).

stadiametric ranging Estimating target range from knowledge of its true size.

stadiametric warning Based on range-closure derived from apparent size of other body.

STAé Service Technique Aéronautique (F).

Staff, STAFF Smart target-activated fire-and-forget.

Staffel Squadron (G).

Staff-Pak Four interlinked laboratory modules designed for installation in transport (C–130) to provide electrically noise-free environment.

staff pilot Experienced military pilot assigned to special duties, ie not with operational or training unit.

Stag Simultaneous telemetry and graphics.

stag Stagnation.

stage 1 One complete element of propulsion, jettisoned (staged) when propellants are consumed (normally applied to rocket). In a multi-* vehicle each * fires in sequence following separation of predecessor to reduce mass remaining.
2 One complete element of multi-* process, normally compression or expansion, through which fluid is passed; eg compression or expansion by passage through single long diffuser, venturi or other tapering duct is not * but use of several in succession causes each to become one *.
3 One complete element of fluid-flow compressing or expanding (eg power-extracting) device; eg one planar assembly of compressor rotor blades and associated ring of stator blades.
4 Sector (4) or, military, portion of air route between two staging units; sometimes, for flight planning, one point on route.
5 Various meanings in electronics, EDP (1) and other disciplines.

stage cost Direct operating cost of flying one (mean, or one specific) stage (4).

staged crew Prepositioned at staging unit to take over incoming flight.

stage flight One flight forming part of longer multi-stage journey.

stage fuel Fuel burned in flying one stage (4); hence ** carpet, plot of variables for flight-planning purposes.

stage length Air-route distance between two staging points; in commercial use normally synonymous with sector distance.

stage time 1 Planned or actual time of staging (1).
2 Sector time.

Stagg Small turbine advanced gas-generator.

stagger 1 Distance measured parallel to aircraft longitudinal axis between biplane lower-wing leading edge and vertical projection on to lower-wing chord line of upper-wing leading edge at same spanwise station (UK). Negative when upper plane is aft of lower.
2 Acute angle measured in vertical plane parallel to aircraft longitudinal axis between leading edges of lower and upper planes at same spanwise location (US). Negative when upper plane is aft of lower.
3 PRF variation by various means involving interleaving trains separated by offset interval; alternative EW technique to PRF jitter.

stagger angle Acute angle in plane parallel to aircraft longitudinal axis between line joining points equidistant from centreline on upper and lower LE (see *stagger* [1]) and local vertical.

stagger tuning Increasing pass-bandwidth of RF receiver by tuning different output stages (one meaning of stage [5]) slightly above or below central frequency.

stagger-wing Biplane of any make with negative stagger.

stagger wire Diagonal wire joining lower and upper wings of biplane and lying approx in plane parallel to axis of symmetry. (US term; UK = incidence wire.)

staging 1 Separation of one stage (1) from next.
2 Time at which * (1) is scheduled or actually occurs.
3 Flying by separate stages (4), with or without changes of crew.

staging area Geographical area between mounting area of exercise and objective, esp for airborne or amphibious operation.

staging base Landing and take-off area with minimum servicing, supply and shelter provided for temporary occupancy of military aircraft during course of movement from one location to another (DoD).

staging point, unit Place or organization linking two stages (4).

stagnation point Point on surface of body in viscous fluid flow (one facing upstream and one down) where fluid is at rest with respect to body, flow in boundary layer on each side of ** being in opposite directions.

stagnation pressure Pressure at stagnation point, normally same as total head, total pressure or pitot pressure, = sum of local atmospheric plus dynamic pressures.

stagnation region Region close to upstream stagnation point.

stagnation stall Several related afflictions of afterburning turbofans normally occurring on afterburner light-up at high altitude at modest airspeed, in most cases with rapid pressure pulses (say, seven per second) in fan duct causing oscillating stall of fan and then core engine.

stagnation streamline That which in any representation of 2D flow passes through front and rear stagnation points on immersed body.

stagnation temperature That at stagnation point, when all relative kinetic energy has been converted isentropically to heat.

stake Anvil-type bench tool for sheet.

stake out To picket aircraft.

staking Swageing terminal on to electrical conductor.

stale track Shown on radar display at last known position, even though it has not appeared on subsequent updates.

stall 1 Gross change in fluid flow around aerofoil, usually occurring suddenly at any 2-D section aligned with flow, at AOA just beyond limit for attached flow (at which lift coefficient is max); characterized by complete separation of boundary layer from upper surface and large reduction in lift. Traditional wings normally * at AOA near 16°–18°, which can be attained at any airspeed depending on applied vertical acceleration. AOA for * is increased by slat, and some highly swept (variable-sweep at max sweep) wings exhibit no * even at AOA beyond 60°.
2 Sudden breakdown in fluid flow previously attached to solid surface, caused by changed angle of either surface or flow, violent pressure pulse in flow (esp travelling upstream) or other severe disturbance, eg flutter.
3 Point at which opposing linear force or torque overcomes that of driving member (eg PFCU, hydraulic motor, airbrake actuator), causing a commanded movement to be arrested.

stalled pressure Delivery pressure at which delivery from variable-stroke fluid pump, centrifugal compressor or certain other pumps falls to zero; also called reacted pressure.

stall fence Fence whose purpose is primarily to improve behaviour at stall.

stalling flutter Flutter in one or more degrees of freedom near angle of stall (1).

stalling speed Any speed at which stalling AOA is reached, esp that at 1 g when ** is at lowest value (when depends on aircraft weight, aircraft configuration and air density, among other variables). Usually assumptions include SL ISA, gear/flaps down, power off.

stall line Boundary between acceptable operating conditions for gas turbine and stall zone for any given altitude as plotted on compressor map.

stall margin Difference, normally expressed as available spread of rpm, between gas-turbine operating line (at any altitude and for transient slam accelerations, etc) and stall line.

stall out To stall as result of attempting too steep a climb or for any other reason, esp when chasing opponent, thus leaving manoeuvre incomplete or failing to get into firing position.

stall protection system Aeroplane flight-control subsystem sensitive to AOA (sometimes sensed at points on either side of centreline to cater for rapid-roll AOAs) which at given value triggers positive action to prevent stall; obvious example is stick-pusher, but stick-shaker sometimes considered for inclusion.

stall quality Pilot's subjective opinion of behaviour of aircraft (normally fixed-wing aerodyne) at stall, assessed for all types of stall (eg accelerated) and configurations (eg dirty).

stall strip Strip or other projection added to skin of aeroplane, often in spanwise direction along lower leading edge, to serve as stall-promoter and ensure stall (1) occurs first at that point.

stall tolerance Generally non-quantified quality of gas-turbine fan and/or compressor to accept distorted airflow or other disturbance (eg gun gas, ingested jet gas, birds, hail or pressure pulses moving upstream) without stalling.

stall turn Flight manoeuvre in which aircraft (aeroplane or glider) is pulled up into very steep climb, usually with engine cut well back, until on point of stall full rudder is applied to cause rapid rotation in yaw, with wings rotating in near-vertical plane; ends in dive and pullout on to desired heading (generally on to reciprocal). In US hammerhead stall (also see *wingover*).

stall warning Anything giving pilot warning of impending stall, eg natural buffet, inbuilt ** system sensing AOA and giving visible or aural warning, or stick-shaker with or without knocker.

stall warning and identification system SWIS, system commanded by AOA vane (12) whose signals are analysed for AOA and rate; because of natural lag or hysteresis in system, trigger is fired progressively earlier as rate is increased by building in a phase advance giving protection at all rates.

stall zone Region beyond stall line of gas turbine at any altitude where attempted pressure ratio is too great for rpm and airflow.

stalo Stabilized local oscillator.

Staloc Self-tracking automatic lock-on circuit.

stamo Stabilized master oscillator.

Stamp 1 Small tactical aerial mobility platform (USMC).
2 Single-tube auto multipoint.

STAN Sum total and nose gear.

Stanags Standardization agreements, esp of terminology (NATO).

stand-alone Generalized term meaning equipment, eg radar, is not integrated directly into existing system of radars or other sensors, computers and com network. Increasingly being used for fixed-base weapon systems and even vehicle-mounted (eg airborne) equipments.

standard air munitions package Conventional air/ground munitions required for 30-day support of one aircraft of specific type, air-transportable in three equal increments.

standard atmosphere Model atmosphere defined in terms of pressure, density and temperature for all heights, assuming perfect gas, devoid of any form of water or suspended matter; approximates to real atmosphere and taken as reference for aircraft performance and all other quantitative measures. First NACA 1925 (see *model atmosphere*), later refined 12 times, current 1980 ICAO Doc 7488. Physical constants: P_0 10132·5 N/m^2; T_0 288·15°K; M_0 2·89644 × 10^{-4} kg/mole; ρ_0 1·2250 kg/m^3; R* 8·31432 J k-mol^{-1} K^{-1}; temperature gradient from –5,000 m (5 km below SL) to altitude (11,000 m) at which T is –56·5°C is –0·0065°C per standard geopotential metre; from 11 to 20 km temperature gradient is zero; from 20 to 32 km temperature gradient is +0·0010°C per standard geopotential metre.

standard bird Two (occasionally three) specifications for birds used as inert (sometimes frozen) bodies fired into gas turbine in ingestion testing.

standard body Not wide-body, ie width in region of 3 m. Loosely synonymous with narrow-body, slimjet, single-aisle.

Standard Class Sailplane competition class limiting span to 15 m and prohibiting high-lift flaps and certain other features.

standard conditions Standard temperature and pressure.

standard data message NATO message format for digital com between national or international units or facilities; an example of a Stanag result.

standard day of supply Total amount of supplies needed for average day as defined by NATO Standing Group rates.

standard design memo, SDM International standardized proforma for hard-copy communication of information affecting hardware design, often as computer printout.

standard deviation Quantification of dispersion of data points about mean value: square root of average of all squares of variances (amount by which each point differs from mean); symbol σ.

standard empty weight No longer definable, most precise equivalent is APS.

standard gravity See *gravity*.

standard industry fare level Hypothetical revenue rate per mile invented by CAB as guide to IRS in taxing GA aircraft on non-company business (US).

standard instrument departure Preplanned, coded ATC IFR departure routing, preprinted for pilot use in textual form often (at major traffic points) supplemented by graphics; abb SID.

standardization Objective of achieving interoperability through use of either uniform or at least compatible hardware; significant that definition of word by DoD, NATO, SEATO, CENTO and IADB is in each case different, while definition "standardization agreement" differs for NATO, SEATO and CENTO.

standardized product One conforming to specifications resulting from same technical requirements (NATO, CENTO, IADB).

standard load One preplanned as to dimensions, weight and balance, and designated by a number or other classification (NATO).

standard mean chord Gross wing area divided by span; position defined by co-ordinates of quarter-chord point and an inclination found by integrating (three methods). Also called geometric mean chord, symbol $\bar{c}$. Numerically equal to chord of rectangular wing of same span and gross area.

standard NPL tunnel Closed-jet, no return flow.

standard of build Precise description of which of various options were followed in construction and equipment of aircraft, esp of prototype or development aircraft where *** changes between one aircraft and next.

standard option Choice of build standard, engine, avionics, finish colours, furnishing or other variables offered to all customers (eg improved stopping on Advanced 727).

standard parallel Parallel on map or chart along which scale is as stated.

standard pitch Geometric pitch at two-thirds (sometimes other fraction or percentage) of propeller radius, also called nominal; symbol P_s.

standard pressure 1 Standard SL atmospheric pressure of 10132·5 N/m^2.
2 In met, usually 1000 mb = 10^5 N/m^2.

standard propagation Assumes smooth spherical Earth of uniform dielectric constant and conductivity under standard atmospheric refraction decreasing uniformly with height.

standard radio atmosphere One having excess modified refractive index (also see *standard refraction*).

standard-rate turn Usually heading change 3°/s or 2 min for 360°.

standard refraction Idealized radio refraction decreasing uniformly with height at 39 × 10^{-6} units per km; included in groundwave calculations by enlarging Earth radius to 8·5 × 10^6 m.

standard structure Not normally used in aerospace; elsewhere often structure whose dimensions are everywhere mid-way between tolerance limits.

standard temperature Value upon which a temperature scale is based, in physics normally 273·15°K (0°C), but for practical (eg gas-turbine rating) purposes normally that at zero height in standard atmosphere, 288·15°K.

standard terminal arrival route Preplanned, coded ATC IFR arrival routing, preprinted for pilot use in textual form often (at major traffic points) supplemented by graphics; abb Star.

standard time, ST Universally adopted time for all countries, based on zone time but modified to suit country's longitude span and if necessary zoned such that difference between ST and GMT is always divisible exactly by 0·5 h.

standard turn See *standard-rate turn*.

standard weight 1 Term formerly used in FAA certification as being certificated gross weight.
2 Also used to mean assumed mass of such loads as adult passenger, parachute, and unit volumes (eg litre or US gal) of fluids.

standby, stand-by 1 Generalized term for being available at short notice, in some cases (eg redundant flight-control channel) instantaneous and in others (interceptor on *) at minutes.
2 R/T code: "I must pause for a few seconds"; if followed by "out" means pause may be much longer but channel must be kept clear for resumption, the meaning becoming "other stations please do not transmit on this frequency".
3 Able to board flight if seat available.

standby item One duplicating another in function and used following failure of that normally operative.

standby mode One of several basic operating modes for equipment, eg radar, normally characterized by receivers shut down but transmitters warmed and ready for immediate power.

standby redundancy System design such that redundant duplicative channels do not normally operate (as in parallel) but are switched in following failure of those normally operative.

stand-down Particular aircraft, though serviceable, remains for long period on ground, for whatever reason.

standing water Defined as mean depth exceeding 12·7 mm (0·5 in).

standing wave 1 Oscillatory motion in vertical plane of air downwind of steep hill or mountain face in which troughs and peaks (latter usually marked by cloud at various levels) remain roughly stationary; at lowest level cloud often rotor, high levels usually lenticular. Strong ** often associated with jetstream.
2 Stationary wave formed in transmission line by superposition of travelling wave reflected back from point of impedance-change of forward wave.
3 Stationary wave or wave-pattern formed in vibrating body, eg turbine disc, by reflection.

standing wire That length of cable consumed in making splice.

stand-off 1 Distance from target surface to reference point on hollow-charge warhead (usually apex of cone).
2 To remain outside airfield circuit or pattern, normally following command or positive decision to do so, eg following landing-gear failure or obstruction on runway.
3 To have to park too far from terminal to use airbridges.
4 To remain outside effective range of enemy defences, esp when making an attack with * missile.

stand-off ability Capability of forcing, eg by outgunning, similar enemy vehicles to remain beyond their own firing range.

stand-off armour Armour fixed on outside of existing armour with sufficient spacing to protect against hollow-charge piercing weapons. Unlike spaced armour, normally only two layers in all.

stand-off bomb ASM (1) launched beyond enemy defence perimeter. Odd UK term usually synonymous with * missile.

stand-off flare 1 IRCM payload dispensed at sufficient distance to protect against enemy heat-homing missiles.
2 Rarely, illuminating flare dropped on parachute beside rather than over surface target, usually in helicopter ASV attack.

stand-off missile One which may be launched at a distance (from target) sufficient to allow attacking personnel to evade defensive fire from target area (USAF).

stand-off steps Passenger/crew stairways kept at terminal equipped with loading bridges at gates to cater for overflow traffic that has to stand-off (3).

stand up To become operational at commissioning ceremony of new squadron (US).

Staneval Standards evaluation.

stang fairing Fairing on sides of jet engine installation, especially pod, covering reverser bucket hinges and rams.

Stano Surveillance, target acquisition and night observation.

Stansit Statistics on non-scheduled air transport (ECAC).

Stanton number Non-dimensional number defining heat transfer through surface; $St = -q/\rho VC_p\, dT$ where q is total quantity of heat, ρ is density of fluid (eg air), V is relative velocity, C_p is specific heat at constant pressure and dT is recovery temperature minus wall temperature.

STAPL Ship-tethered aerial platform (Kaman).

STAR, Star 1 Standard terminal arrival route.
2 Ship tactical airborne RPV.
3 Surface-to-air recovery.
4 Satellite de télécommunications, d'applications et de recherche (F).

Star 1 Starboard.
2 Satellite for telecom, applications and research.
star 1 Star-shaped empty space along centre of solid rocket propellant grain.
2 Formation formed all on same level by wrist-linked team of free-fall parachutists.
3 Helicopter main-rotor control radial arms, located under hub; one fixed, one rotating.
Staran Association processor in ARTS–II (FAR).
starboard Naval-derived term for right, right-hand or towards right.
star-centred Cast or extruded with star (1).
stardust Air-impingement haze in acrylic finish.
staring focal-plane array Important class of sensors which operate like human eye, with focal plane of input optics covered with dense micro-mosaic of 2-D receptors which "look" continuously, direction of target being determined from knowledge of which detector(s) see it.
starplates Upper and lower spiders each formed from one piece of composite material or metal (titanium) forging and forming structural basis of modern non-articulated helicopter rotor hub. Not to be confused with star (3).
Stars 1 Silent tactical-attack/reconnaissance system.
2 Plural STAR, Star (1).
3 Standard terminal and arrival reporting system (CAA).
4 Small transportable aerostat relocatable system.
5 Surveillance and target-attack radar system.
START Strategic arms reduction talks.
Start 1 Spacecraft technology and advanced re-entry test.
2 Solid-state angular rate transducer.
starter Device for cranking any prime mover of rotary type during starting; 14 basic species.
starter exhaust Overboard discharge of exhaust from starter of cartridge, monofuel, fuel/air, bipropellant or air-bleed types; potentially dangerous.
starter gearbox Box containing gear train through which starter cranks engine; may be reduction or step-up gears.
starter/generator Single electrical machine, usually DC, serving both as electrical starter and electrical generator (see *CSDS*).
starter magneto Various forms of magneto designed to provide powerful spark during start, eg hand-cranked, impulse starter and LT boosting energy transfer.
starting chamber 1 Combustion chamber in multi-chamber (multi-can) gas turbine in which igniter is fitted, flame thereafter being carried round by inter-chamber pipes.
2 Liquid-rocket precombustion chamber.
starting coil Auxiliary induction coil used as HT booster when starting PE (4); alternatively used as energy-transfer LT booster.
starting pressure Minimum rocket combustion pressure at which nozzle exit plane is shock-free.
starting transients Temporary variations in pressures, flows, velocities and temperatures during complete start sequence of rocket.
starting vortex Transverse vortex left by lifting wing at start of motion providing essential link behind subsequent vortices from wingtips (part-theoretical concept, since at start wing may not be lifting and ** has zero strength, but necessary because trailing tip vortices cannot have free ends).
star tracker Optical or opto-electronic sensor which automatically locks on to preselected celestial body or bodies to provide input to astro or astro-inertial nav system.
start-up costs Those incurred in launching new type of aircraft. Precise definition lacking; general opinion is that it covers all costs up to certification, excluding related engine and systems.
star washer Hard steel washer with multiple twisted radial projections which bite into superimposed nut and prevent it becoming loose.
state 1 Readiness condition of combat aircraft, from cockpit-alert to unserviceable (different national subdivisions).
2 In connection with runway, condition of surface or traffic occupancy.
3 For rotorcraft main rotor, three subdivisions: propeller *, vortex-ring * and windmill-brake *.
4 Amount of main propulsion fuel remaining, esp when running low.
State chicken Air-intercept code: fuel state requires recovery, tanker or diversion (DoD).
State lamb Air-intercept code: "I do not have enough fuel for intercept plus reserve required for carrier recovery".
state of the art Level to which technology and science have at any designated cutoff time been developed in any given industry or group of industries (USAF).
State tiger Air-intercept code: "I have fuel for completion of mission".
static 1 At rest, or at rest relative to solid surface.
2 Structural test with application of single increasing load.
3 Radio and other com interference, esp that due to discharge of * electricity.
static balance 1 Aircraft condition in which there is no resulting moment about any axis.
2 Control-surface condition in which in absence of any applied torque surface is freely balanced about hinge axis, either because c.g. lies on that axis or because mass balance has been added.
3 Propeller condition in which when supported on rod on knife-edge it rests in any position.
static cable Longitudinal cable supported at each end along interior of transport to which parachute strops of troops and airdropped loads are

attached. Also called strop line, anchor cable.

static ceiling Altitude at which airship in ISA (in some definitions, without forward speed) neither gains nor loses altitude after all ballast has been dumped.

static characteristic Basic plot of thermionic valve, grid volts against anode current.

static conversion Conversion of energy from one form to another without use of moving parts; common example is solid-state AC/DC converter.

static discharger Device for harmless distribution of static electricity, esp static wick.

static electricity Electric charge built up on non-conductive surface or insulated body by deposition of electrons or positive charges, eg by friction between air and aircraft or between fuel and hose, ultimately reaching very large potential difference.

static firing Firing test of rocket motor, engine or vehicle while attached to test stand.

static flux Magnetic field through magneto frame and inductor with latter stationary.

static friction Force required to initiate relative movement between surfaces in contact.

static gearing ratio Ratio of angular deflection of vehicle (esp missile or RPV) control surface to rotation of vehicle axis which caused surface to deflect.

static ground line Connection of aircraft earthing (grounding) system to earth.

static head Pitot head measuring ambient static pressure.

static instability Unlikely design fault in which aircraft, once disturbed from straight/level flight, suffers increasing upset in absence of aerodynamic inputs, eg due to high c.g.

static lift Difference in mass between gas contained in aerostat at rest and air displaced by whole aerostat.

static line Links parachute with static cable so that as parachute leaves aircraft it is opened automatically. Hence ** jump.

static load Applied force is unidirectional, either held constant or increasing in programmed way from zero to maximum.

static margin Basic measure of aeroplane static stability (primarily in pitch), normally defined as distance of c.g. ahead of neutral point expressed as %MAC. Measured stick-fixed or stick-free; former is %MAC proportional to stick displacement with percentage change of speed from trimmed value, called positive when direction backward for lower speed; stick-free is %MAC proportional to rate of change of applied stick force with percentage change of speed from trimmed speed, positive when pull needed by lower speed.

static marking Flaws on imagery caused by static-electricity discharge.

static moment Product of area and its distance (measured from centroid) from reference axis.

Staticon Photoconductive type of TV camera tube.

static pin Wire fitting, usually multi-pin, pulled from parachute by static line.

static port See *static vent.*

static pressure That exerted by fluid on surface of body moving with it; alternatively on a surface exactly parallel to local free-stream velocity vector and which does not affect local flow.

static pressure gradient Possible progressive increase in static pressure along constant-section duct, eg tunnel working section, resulting from progressive thickening of boundary layer.

static pressure head See *static head.*

static propeller thrust That generated by propeller restrained against movement.

static radar One whose scanning is electronic, not mechanical.

static radial Normal radial engine, as contrasted with rotary.

static rail Rigid form of static cable.

static sag Angular or vertical-linear distance through which a structure sags at rest compared with some other reference condition, esp wingtip at rest compared with 1 g flight.

static soaring Soaring (US term, now arch, contrasting normal soaring with so-called dynamic soaring).

static-source selector Pilot switch connecting static system with chosen sensor.

static stability 1 Measure of tendency of aeroplane to return to or diverge from initial trimmed condition following upset.
2 Study of applied moments imposed on aeroplane following upset of AOA or speed from initial trimmed condition.
3 Quality of having positive static margin.
4 Stability attained by action of weight, esp by having low c.g. (common US definition entirely unrelated to normal).

static-stability compensator Limited-authority subsystem which ensures aeroplane has stick-free static stability by applying small nose-down trim force immediately before stalling AOA is reached, esp to avoid self-stall; in UK called static-stability augmenter.

static system Barometric-ASI plumbing incorporating pitot/static head or static vent and connections to ASI, VSI and possibly other devices; usually includes alternative sources.

static temperature T_s, temperature measured by thermometer moving with fluid and thus measuring true temperature due to random motion of fluid molecules.

static test For aeroplane or other structure, test under static loads; hence constructor may build ** specimen as well as fatigue specimen. For some vehicles, usually means test while fixed in one place.

static 3-D radar Large surface radar with fixed

aerial scanned electronically.

static thrust Thrust of jet engine restrained against forward motion.

static tube That transmitting pressure from static head.

static turn indicator Rare instrument driven by pressure difference from two modified static heads mounted transversely at wingtips.

static vent Carefully designed opening in plate aligned with skin of aircraft which under most flight conditions senses true static pressure.

static wick Device for discharging static electricity from fine (very small radius) tips of thousands of conductive wires, braid or graphite particles built into flexible wick projecting behind trailing edge.

statimeter Instrument for measuring static thrust.

station 1 Dimension measured from * origin locating all planes along fuselage normal to longitudinal axis, or along wing normal to transverse axis. Every structural part, bolt hole, equipment item, door frame and every other dimensional reference is in form of * number (in US in inches, elsewhere in mm). Corresponding term for vertical distances is WL, waterlevel.
2 Fixed base for military, naval, air or research operations.
3 Airport (esp staff, facilities and costs); airline usage, usually adjective only.
4 Planned or actual position of geostationary satellite.
5 Position relative to other vehicles of one vehicle in group (esp of aircraft and ships).
6 Location of radio transmitter.

stationary front Front without significant motion over Earth's surface.

stationary orbit See *geostationary*.

stationary shockwave Shockwave at rest relative to solid surface.

station box Human interface with intercom and/or remote radio transmitter/receiver, comprising jack socket, tuning and other controls; in large aircraft links up to 12 audio (crew) inputs to intercom and com systems.

station capacity Maximum number of aircraft that can be handled simultaneously by radio (esp DME) station.

station costs Airline costs attributed to staff and facilities at airports.

station ident Ident feature keyed into ILS localizer transmission at regular intervals, either three- (military, two-) letter code or voice.

stationkeeping Ability of vehicle, esp satellite, to maintain prescribed orbit or station (most geosynchronous satellites follow small figure-eight shapes).

station-keeping equipment Small 360° radar scanning horizontally and used to assist several aircraft to hold exact formation up to several miles apart (eg for precisely sequenced airdrops).

station passage Flight directly overhead a station (6), esp a point navaid, eg VOR, when "to" becomes "from" and bearings become reciprocals.

station set Selected items of mission-support equipment prepositioned at designated locations for support of war or emergency ops (USAF).

station time Time at which crews, pax and cargo are to be on board military transport ready for take-off (DoD).

station zero Origin of aircraft measurement system along each axis; for longitudinal axis may be at or in front of nose.

statistical accelerometer Instrument recording number of times particular acceleration is exceeded; apparently synonymous with recording accelerometer.

stator Fixed part of rotary machine, eg electrical generator or motor, hydraulic pump, gas turbine compressor or turbine, RC engine casing or brake discs keyed to landing-gear axle.

stator blade Fixed (except for variable incidence) blade attached to axial compressor stator casing.

stator casing That enclosing axial compressor.

statoscope Instrument (now arch) for detecting small excursions from preset pressure altitude.

status Assessment of target, esp one seen only on radar, as friend, foe or unknown.

statute mile Non-SI unit of length = 1609·344 m.

Stavka Supreme-command staff (USSR).

stay Lateral strut or bracing member, esp for landing gear.

STBA Service Technique des Bases Aériennes (F).

STBY Standby.

STC 1 Sensitivity time constant (rarely, control).
2 Self-test capability.
3 Supplemental type certificate (US).
4 Satellite test centre.

STCR System test and checkout report.

STD System technology demonstration.

STD bus Traditional US-developed 8-bit bus for computers and related EDP.

STDMA Space/time-division multiple access.

STDN Space-tracking data network.

STE Sun-tracking error.

Steady Air-intercept code: "I am on prescribed heading", or "Straighten out on present heading".

steady flow See *time-invariant flow*.

steady initial-climb speed See V_4.

steady-state condition One that is time-invariant, as applied to signal or flutter amplitude, physical or chemical properties or any other variable.

stealth Technology for making tangible objects, initially aircraft, as invisible and undetectable as possible. It covers all EM wavelengths as well as sound, and is increasingly essential for survival in defended airspace. Also adjective.

steam bombing Visual manual attack on target of opportunity using free-fall bombs, especially by

advanced automated aircraft.
steam gauge Traditional dial instrument in a modern cockpit (colloq).
steam(ing) fog Forms when supersaturated freezing air with inversion moves over warm water; also called Arctic smoke or sea smoke.
STE bus New international standard 8-bit computer bus originally developed for Eurocards, which has rendered STD obsolete.
STEC Solar/thermal energy conversion.
steel drag chute Reverser (colloq).
steep approach That adopted by helicopter pilot descending into obstructed, eg urban, heliport, begun at 90 m/200 ft above selected landing spot and made straight-in at close to 50° from downwind.
steepest-descent method Basic method of optimization in which all contour lines (each representing a plotted variable) are crossed perpendicularly.
steep gliding turn Steep turn performed in glide, if continued resulting in tight spiral (more common in US).
steep turn Various definitions with bank angles: over 50°; 45°–70°; over 60°.
Stefan-Boltzmann constant That in Stefan-Boltzmann law, $\sigma = 5{\cdot}66961 \times 10^{-8}$ W/m^2 K^4.
Stefan-Boltzmann law Basic law of thermal radiation: total radiation from black body is proportional to 4th power of absolute temperature, $E = \sigma T^4$.
stellar guidance See *astronavigation*.
stellar/inertial guidance Inertial navigation intermittently updated and refined by astro.
Stellite Large family of hard alloys of Co (30–80%), Cr (10–40%), W (0·25–14%) and Mo (0·1–5%), and in one case with 30% Ni and 5% Fe. Some cannot be machined; common use is PE (4) valve heads and seats.
St Elmo's fire See *corona discharge*.
Stem 1 Shaped-tube electrolytic machining.
2 System (or spaceflight) trainer and exercise module.
Step Standard equipment package.
step 1 Segment of climb from one FL to next, each normally begun either on ATC clearance or upon arriving at suitable gross weight from burning fuel.
2 Sharp or angled discontinuity in planing bottom (float or hull) to improve planing characteristics and ease take-off.
3 Stage (1), latter being preferable.
step pad Secondary structure built externally on top of fuselage, esp of helicopter, for ground crew or other persons.
stepped climb Climb in series of steps (1), separated by slow (drift-up) climb or level flight.
stepped formation One in which successive aircraft or elements are at higher or lower level.
stepped solvents Solvents in liquid (eg aircraft finish) which evaporate at very different rates.
steradian SI unit of solid angle, abb sr: that solid angle which, having as its vertex centre of sphere, cuts off area equal to that of square whose sides are equal to radius of sphere.
stereogram Stereoscopic set of graphics or imagery arranged for viewing.
stereographic See *polar stereographic*.
stereographic coverage Air reconnaissance cover by overlapping imagery to provide 3-D picture; 53% overlap is minimum and 60% normal.
stereoscopic cover NATO term for stereographic cover.
stereoscopic pair Two images for stereographic (stereoscopic) viewing of same scene.
sterilization-proof Solid rocket propellants (fuels and, esp, binders) are often degraded by sterilization (4) heating, and * binders have had to be developed.
sterilize 1 To mark off portion of runway as unusable, for any reason.
2 To blank off portion of instrument panel, eg for equipment not yet available.
3 To prohibit unauthorized access, esp to entire airside of civil airport for security reasons.
4 Normal meaning for spacecraft either departing for or arriving from other planet.
5 To deactivate device, eg mine, eg after preset period.
6 To render unusable or unavailable; eg F-111 MLG * underside of fuselage for weapon carriage.
stern attack Air-intercept attack which terminates with crossing angle of 45° or less (DoD).
stern-droop Structural sag of rear end of airship.
sternheaviness Tailheaviness of airship.
sternpost Single vertical member marking rear termination of fuselage, hull or float (BS adds "not to be confused with rudder post", but often same member).
sternpost angle Acute angle between horizontal (usually same as longitudinal axis or underside of keel ahead of step) and line joining top of step on centreline and afterbody terminator, bottom of sternpost or rearstep (heel) of flying-boat hull.
stern wave Formed at low speeds as swell ahead of stern of taxiing marine aircraft.
Stevenson screen Standard Met Office slatted box for ground instruments.
Stevi Sperry turbine-engine vibration indicator.
STF Self-test facility.
StF Stratiform cloud.
STFV Sensor total field of view.
StG Stukageschwader, dive-bomber wing (G, WW2).
sthène Non-SI unit of force formerly standard in French legal system, that giving 1 tonne acceleration of 1 m/s^2; sn = 1,000 N = 1 kN.
STI Standard (or special) technical instruction.
stick 1 Control column (colloq).
2 Any primary pilot input in pitch, in figurative sense, eg *-free static stability.

3 Succession of missiles (ordnance items would be better) fired or released separately at predetermined intervals from single aircraft (ASCC).
4 Number of parachutists who jump from one aperture of aircraft during one run over DZ.
5 A pilot, especially fighter or aerobatic (colloq).

stick and string Generalized term for construction of pre-WW1 aeroplanes (colloq).

stick-canceller Electronic box in FCS whose output is a stick movement, eg in ASW hover.

stick-fixed Elevator or tailplane is trimmed to hold level flight and thereafter held in this position.

stick-fixed static stability See *static stability*.

stick force per g Pilot's applied force in direct fore/aft movement divided by vertical acceleration resulting.

stick-free Elevator or tailplane is trimmed to hold level flight and thereafter left free (not applicable with irreversible flight controls).

sticking Tendency of planing bottom of marine aircraft to adhere to water on take-off.

stick knocker Stall-warning device added to stick shaker to give loud knocking aural warning.

stick movement per g Linear fore/aft movement measured at top of stick divided by resulting vertical acceleration.

stick pusher Positive stall-prevention system which when triggered (by sensed AOA or AOA factored by a rate-of-increase term) forces stick forward, commanding aircraft to rotate from climb to shallow dive.

stick shaker Stall-warning system which when triggered by AOA passing preset value (occasionally factored to allow for high rate of increase) applies large oscillating force which shakes stick (2) (normally a large yoke) rapidly through small angle in fore/aft plane.

stick-shaker speed SSS or S-cubed, really a misnomer as system is triggered by AOA, not a particular speed.

stick spacing Linear distance on ground between ordnance items dropped in a stick (3).

stick time Logged pilot time, esp as PIC.

stick travel Total range of travel of stick in either fore/aft or side-to-side direction, normally measured not as angle but as linear distance.

stick travel per g See *stick movement per g*.

stiction See *static friction*.

stiffener Normally a strip or beam attached to sheet to resist load normal to surface. An integral * is formed in skin itself, usually as pressed channel parallel to relative wind.

stiffening bead Integral stiffener (see above).

stiffness Ability of system to resist a prescribed deviation. In structural member within elastic limit, ratio of steady applied force to resulting displacement or ratio of applied torque to resulting angular deflection. For many dynamic parts such as servo-actuators there is static * and dynamic *. Static * is stiffness characteristic exhibited in steady-state condition, normally trying to approach infinity (discounting compliance within actuator and deflection of surrounding structure) up to stall thrust. Dynamic * is an apparent value dependent on ability of flow power of servo valve to hold output against oscillating load, up to critical frequency (varying with magnitude of applied load) at which servo contribution becomes negligible to give a degraded "infinite-frequency *".

stiffness criterion Relationship between stiffness and other properties of structure which when satisfied ensures prevention of flutter or other type of instability or loss of control (BS).

stiff nut Nut provided with means for gripping male thread to provide sustained torque resisting rotation once tightened.

stiff pavement One able to accept any input aircraft bending moment; stiffness measure is tonnes vertical load to produce 1 cm deflection under point of application.

stiff wing Fixed-wing, ie not helicopter (colloq).

stilb Non-SI unit of luminance, sb = 10^4 cd/m^2.

Stiletto Becoming generalized term for anti-radar D/F and passive-ranging systems.

stiletto criterion Basic design case for passenger floors: intensity of * heel loading of 100 kg/cm^2, or over 100 times limit for distributed heavy cargo.

stiletto weld One whose cross-section resembles exclamation mark without stop.

still-air range Not a normal performance or flight manual figure; in general a vague estimate of distance under ideal still-air cruise conditions aircraft could fly without air refuelling, ignoring ATC constraints, reserves or any other factor. Unlike ferry range, usually assumes some (occassionally max) payload.

stilling chamber Large volume in which fluid flow eddies and gross turbulence are brought to rest; generally synonymous with settling chamber.

sting 1 Long cantilever tube projecting upstream in tunnel to which model is attached with minimum interference from mount.
2 Long cantilever tube projecting directly ahead of nose of aircraft to carry instrumentation with minimum interference from mount or following aircraft; also called probe, instrumentation boom.

sting hook Normal form of arrester hook in form of single strong tube (term introduced WW2 to distinguish from A-frame).

Stings Stellar-inertial guidance system.

sting switch Activated by springy rod projecting below RPV (or other aircraft), triggered on landing.

Stirling cycle Heat-engine cycle in which heat added at CV followed by isothermal expansion with heat addition, heat then rejected at CV followed by isothermal compression with heat rejection; very efficient, esp with regenerator, but mechanical problems (Philips use patented Rollsock to seal reciprocating parts). So far used in aerospace mainly for space power generation.

STLO Science and technology (or scientific and technical) liaison office.

STM 1 Short-term memory.
2 Supersonic tactical missile.

STN Station.

STNA Service Technique de la Navigation Aérienne (F).

Stn No Stanton number.

STNR Stationary.

STO 1 Short take-off.
2 System test objective.
3 Station telecom officer.
4 Council (Soviet) for labour and defence (USSR).

stochastic Implying presence of unknown or random variable; thus, * process is ordered set of observations, each a sample from a probability distribution.

stock 1 Raw material pre-formed to standard dimensions as sheet, strip, tube etc.
2 Material surplus to a part's finished dimensions, to be removed during manufacture.

stockpile/target sequence, STS Order of events in removing NW from storage and assembling, testing, transporting and delivering to target (DoD).

stocks Shaped supports on which flying-boat hull is built, but in no sense tooling.

stock template, ST One developed by trial and error, esp for parts undergoing severe deformation.

Stoddard Common naphtha-like hydrocarbon solvent.

STOH Since top overhaul.

stoichiometric Provided in exact proportions required for complete chemical combination, esp of fuel/air mixtures.

stokes Non-SI unit of kinematic viscosity, 1 St = 10^{-4} m^2/s.

STOL, Stol Short take-off and landing.

Stolport Airport, esp urban (metropolitan), configured and designated for Stol operations.

Stol runway Runway, normally 900 m (2,000 ft), specifically designated and marked for Stol operations (FAA). Letters STOL at threshold and a TDAP.

stoneguards Various mesh screens or deflectors; undefineable.

stooping Atmospheric refractive phenomenon in which image (mirage) of distant object is vertically foreshortened.

stop(s) Mechanical limiter(s) to permissible travel of flying-control or other mechanism.

stop alt squawk Turn off altitude-reporting switch and continue Mode C framing pulses.

stop countersink Fitted with collar to limit penetration.

stop drill 1 See *stop hole.*
2 Drill with collar to limit penetration.

stop hole Hole drilled in end of fatigue crack to provide larger radius and halt further spread.

stop nut Nut which stops in place without further action (such as wiring or bending up tabs).

stopover 1 Stop by pax at intermediate airport authorized by ticket for prescribed period.
2 One-night stay away from base by slip crew (civil or military).

stopped-rotor aircraft See *stowed-rotor aircraft.*

stopping Loosely, sealer for cracks, esp in pressurized riveted joint.

stop squawk Switch off transponder.

stopway 1 Defined rectangular area on ground at end of runway in direction of take-off (ie beyond upwind end, symmetrical about extended CL) prepared as a suitable area in which aircraft can be stopped in case of abandoned TO (ICAO).
2 Area beyond TO runway, no less wide and centred upon extended CL, able to support aircraft during aborted TO without causing structural damage to it and thus designated for use (FAA).

stop weld, stop-weld Material, usually in form of paint, which has so high a melting point that it is not fused by any conventional welding process. Inserted as applied layer along line followed in seam-welding, results in treated area remaining unwelded.

storage Generalized term for any kind of memory in EDP (1), display technology, EW and similar disciplines. Not normally used in aircraft for fuel or other consumables, ordnance and ammunition or cargo/baggage.

storage CRT Various families able to write information into a storage surface by adding/subtracting from an initial potential; most configurations have symmetrical layout with write gun at one end and read gun at other.

storage system In electronics and displays, basically comprising direct-view storage tubes and those in which storage is entirely separate; invariably concerned with vector inputs, converted into analog (beam-deflection) signals and varying luminance.

storage tube See *storage CRT.*

store 1 Basic element of EDP (1) storage; normally a bistable device accommodating one bit.
2 Generalized adjective for storage, hence * address, etc.
3 To place information in memory for future reading.
4 Generalized term for any mission-related payload carried by combat aircraft in form of discrete streamlined device either carried externally or released from internal bay; anything that occupies a pylon or ERU, whether intended for air-dropping or not.
5 Plural, domestic supplies and consumables for large aircraft on long flight (suggest usage arch).

stores inventory display Cockpit readout, often illuminated on command, showing locations and identities of all stores (4) remaining on board.

stores pylon Pylon for carrying, and if necessary releasing, a store or stores (4).

storm cell Central region of most intense turbulence, indicated in cockpit radar display by black hole or red colour.

storm-warning radar See *airborne weather radar*.

Stovie Pilot of Stovl aircraft (colloq).

Stovl, STOVL Short take-off, vertical landing; common operating mode of Harrier family.

Stow System for take-off weight and c.g., also called IWBS, with display readout driven by sensors on all landing gears.

stowed-rotor aircraft Rotorcraft, normally helicopter in vertical mode, whose lifting rotor(s) can be stopped and retracted in wingborne cruising flight.

stow position Travelling or inoperative az/el attitude of large radar or other steerable aerial, normally pointing to zenith.

STP 1 Standard temperature and pressure.
2 Space test program (DoD).
3 System-test procedure.
4 Systems technology program (DoD).
5 Short-term planning.

STPAé Service Technique des Programmes Aéronautiques (F).

STPD Standard temperature and pressure, dry.

STPE Service Technique des Poudres et Explosifs (F).

STR 1 Service trials report.
2 Sidetone ranging modulation; EW (spread-spectrum) acquisition assistance technique.
3 Sustained turn rate.
4 Software trouble report.
5 Systems technology radar.
6 Standard test rack.

STRA Simultaneous turnround actions.

Stradographe Runway friction measurer with braked or toed-in wheels and EDP (1) for nine variables.

strafe To rake with fire, eg from automatic guns; spelt as shown (not straff) and rhyming with chafe, not chaff. US usage is in favour of "straff", and "straffing" is becoming predominant.

straighteners Vanes, cascades or, in tunnel, transverse flat-plate honeycomb, to remove swirl or turbulence from flow.

straight-flow Gas turbine of normal, ie not reverse-flow, layout.

straight-in approach In IFR, an instrument approach wherein final approach is begun without a prior procedure turn. In VFR, entry of traffic pattern by interception of extended runway CL without executing any other portion of traffic pattern (FAA).

straight leading edge Having no taper; likewise for trailing edge.

straight-pass attack One using on-board weapon-aiming system to hit point surface target without search or visual acquisition, thus in straight run at highest speed at lo level.

straight roller bearing Not tapered, thus no axial load expected.

straight-run Hydrocarbon distillate, eg from original crude, representing all products separating out between specified upper/lower temperature limits; not normally used as aviation fuel.

straight stall One performed with minimal yaw, using rudder if necessary to hold heading.

straight-through duct Inlet duct to trijet centre engine in which flow from inlet to nozzle is essentially straight, as distinct from S-duct.

straight wing Of traditional planform, specif not swept.

strain Deformation under stress expressed as % original dimension; length, area or volume.

strain energy Elastic energy recovered from body by removing stress.

straingauge Device for transducing strain into electrical signal, usually by extremely accurate measurement of change of resistance of conductor.

strain hardening Increase in hardness and reduction in ductility caused by strain, esp by cold-working (eg rolling); only way to harden some wrought light alloys. Introduces a strain exponent (expressed as n) into stress/strain equations.

strain rate Strain per unit time, normally under uniform stress and often synonymous with creep.

strain viewer Instrument giving pictorial strain pattern using polarized light.

strake 1 Long but shallow surface normal to skin and aligned with local airflow; extremely low-aspect-ratio fin.
2 One row from stem to stern of single plates cladding marine aircraft.

stranded conductor Electric cable containing numerous conductive wires twisted together within single insulating sheath.

Stranger Air-intercept code: unidentified target. Normally ground/air message followed by bearing, distance, altitude in that order.

strangle General term meaning please switch off a particular emitter (military usage).

Strangle Parrot Ground/air code: switch off IFF.

strapdown Generalized adjective for device mounted so that its attitude changes with that of aircraft or spacecraft; specif, one not gimballed about three axes.

strapdown INS Simplified INS using strapdown platform.

strapdown platform Platform for INS on which sensing gyros and accelerometers are fastened without relative motion; usually there are three mutually perpendicular gyro/accelerometer units, and in some cases an element of redundancy from a fourth mounted at 45° to all three others.

strapping 1 Interconnection of resonant chambers of cavity magnetron or related oscillator to give one stable preferred mode.
2 Calibration of storage tank so that measurement of contents depth can be related to actual volume.
3 Metal or other straps, wire or other ties around palletized, igloo or other loose cargo.

strategic Concerned with broad politico-military objectives and enemy's warmaking potential.

strategic aeromedical evacuation Airlift of patients out of theatre of operations to main support area.

strategic airlift In support of all arms between area commands or between home state and overseas area.

strategic air transport According to DoD, one in accord with strategic plan; according to NATO, movement between theatres by scheduled service, special flight, air logistic support or medevac.

strategic air warfare Air operations designed to effect progressive destruction and disintegration of enemy's warmaking capacity (NATO; DoD is much longer definition which adds nothing).

strategic attack Aerospace attack on selected vital targets of enemy nation to destroy warmaking capacity or will to fight.

strategic bomber Delivery-system aeroplane for strategic attack.

Strategic Defense Initiative Plan by Reagan administration to enhance US ability to conduct warfare in space, eg by using direct-impact missiles or super-power lasers to incapacitate satellites.

strategic plan Plan for overall conduct of war.

strategic psywar Actions designed to undermine enemy's will to fight.

strategic transport Aircraft for transport between theatres.

strategic warning Notification that enemy-initiated hostilities may be imminent. Hence ** lead-time, time elapsing between ** receipt and beginning of hostilities. May include two action periods, ** predecision time and ** postdecision time in which national commander takes a decision to respond positively to **.

stratified-charge engine PE (4) in which mixture strength is varied in controlled manner during induction stroke, with min turbulence, if possible to leave layered charge in cylinder with highest density near source of ignition. Now especially important in RC (1) engines.

stratiform cloud Sheets in stable thin layers.

stratocumulus, Sc Layer of connected cloudlets at low-cloud level often arranged in aligned rows.

stratopause Atmospheric layer at top of stratosphere where inversion ceases at 270·65°K.

stratosphere Atmospheric region between tropopause and stratopause within which temperature remains essentially constant and then, at upper levels, rises with altitude.

stratus Uniform layer of low (usually grey) cloud, well clear of surface.

stray Naturally occurring EM signals, eg static.

streak 1 Horizontal smear, usually white, following moving image on TV or other raster display.
2 Long flame in airbreathing engine normally denoting abnormal combustion at one point.

streak camera Family of cameras for ultrafast photography in which changing scene is viewed through slit perpendicular to main image variation and optics sweep image along fixed arc of film. Basic type has rotating mirror, normally projecting through array of biconvex lenses.

streaking 1 Unwanted manifestation of streak (1).
2 Unwanted manifestation of streak (2) in gas turbine, usually resulting in reduced life or damage to NGVs.

stream 1 To release parachute retarding horizontal motion, eg braking parachute.
2 To dispense chaff as solid, at random intervals or as bursts (DoD).
3 Jet stream (colloq).
4 Shower of meteoroids with similar orbits and timing.

streamering 1 Visible brush discharge.
2 Unreefed parachute canopy opens but fails to deploy fully; hence streamered.

stream function Basic parameter of 2-D non-divergent fluid flow with value (symbol ψ) constant along each streamline related to velocities along each axis by $u = -d\,\psi/dy$, $v = +d\,\psi/dx$.

stream landing Landings by group of aircraft in quick succession.

streamline Line marking path of particle of fluid in homogenous flow, esp in streamline flow; line whose tangent is everywhere parallel to instantaneous velocity at that point.

streamlined 3-D body shaped such that fluid drag is a minimum.

streamline flow Fluid flow that is laminar and time-invariant, and in which each streamline is devoid of a closed curve or sudden change in direction.

streamline position That in which a hinged or pivoted body, eg control surface or pylon on a variable-sweep wing, is aligned with relative wind.

streamline wire One whose section is streamlined, though seldom optimum (usually two intersecting circular arcs).

stream take-off Take-off by group of aircraft in quick succession with departure in trail formation.

stream thrust Total of pressure force and time rate of momentum flow across any cross-section in fluid flow, $F = PA + \rho AV^2$.

stream tube In a laminar fluid flow, volume of flow enclosed by streamlines passing through upstream and downstream closed loops (not necessarily circular) placed normal to flow. At any point velocity is inversely proportional to ** cross-section area.

streamwise tip Wingtip of high-subsonic or transonic aeroplane in which leading edge is curved progressively back parallel to local airflow to eliminate outward sweep of isobars; one form is *Küchemann tip*.

street Regular procession of straightline vortices shed alternately from above and below body, eg naturally oscillating cylinder or wire, each vortex following same path as next-but-one predecessor. Normally if D is body diam, two half-streets are separated by 1·2 D and spacing between vortices in each half-street is 4·3 D. Also called Kármán * or von Kármán *.

streetcar STCR (colloq).

strength 1 Physical * is ability to withstand stress without rupture; normally subdivided into compressive, shear and tensile.
2 In radio and related fields, signal amplitude in W or dB.
3 Dielectric * is max potential gradient in V/mm.

strength test See *static test*.

strength/weight ratio For material, ultimate tensile strength/density; for a structural member, breaking stress/weight.

stress 1 Condition within elastic material caused by applied load, temperature gradient or any other force-producing mechanism, measured as force divided by area. Unit * is force per unit area normal to direction of force. It is this force that resists externally applied loads.
2 Generalized term for psychological, physiological or mental load on organism, esp human, which reduces proficiency.
3 Measure of resistance of viscous fluid to shear between adjacent layers (see *viscosity*, *Newton's laws*).

stress analysis Determination of all loads borne by all elements of structure in all flight conditions, external reaction points of application and direction, and allowable and actual stresses in each member.

stress-bearing Required to resist applied load(s).

stress concentration Localized region of increased stress caused by sudden changes in section, poor design and manufacturing imperfections, eg tooling marks (see *stress raiser*).

stress concentration factor Peak actual local stress divided by stress for member calculated by any standard method without presence of stress raisers.

stress corrosion 1 Metal cracking due to residual stress from manufacturing processes or concentrated stresses caused by flight loads and/or poor design.
2 Exfoliation corrosion.

stress cycle Complete cycle of variation of stress with time, repeated more or less identically (very numerous for PE [4] conrod, less for wing LCF).

stress distribution Variation of stress across cross-section of member.

stressed skin Form of semi-monocoque construction in which skin, nearly always metal, bears significant proportion of flight loads, and makes principal contribution to stiffness.

stress-free stock Selection of stock material for primary structure in which presence of residual internal stress results in rejection or return for further treatment; important in heavy plate for wing skins, etc.

stressing Stress analysis of structural members, usually while altering their design to attain optimized structure.

Stresskin Patented metal (esp stainless) sandwich panels requiring no supporting structure over surface.

stress raiser Local abrupt change in section resulting in stress concentration; severity varies inversely with radius, so that a single scratch (eg from emery particle) can over a period initiate a fatigue crack that will eventually prove catastrophic.

stress ratio Max to min ratio in one stress cycle.

stress-relief annealing Heating to beyond critical temp and slow cooling to relieve internal stress, eg after cold working or welding.

stress/strain curve Plot of strain resulting from all stresses from zero to yield point and on to rupture. Normally linear over most of plot to yield point (limit of proportionality where elastic deformation gives way to plastic).

stress wave Sonic pulse propagated through various devices, eg magnetostrictive-tablet display; also called strain wave.

stress wrinkle Visible wrinkling of skin caused by applied load, esp in secondary structure, eg sagging of rear fuselage of B–52.

stretch 1 Increase in capacity of transport by adding plugs to fuselage, normally both in front of and behind wing.
2 To apply tensile stress exceeding elastic limit.

stretchability Potential for stretch (1).

stretching 1 Process of introducing a stretch (1).
2 See *stretch-wrap forming*.

stretching press See *stretch press*.

stretch-levelling Stretching sheet or plate just beyond elastic limit (typically 1·5% elongation) to remove all irregularities; also called stress-levelling.

stretchout Agreement, initiated by contractor or customer, to reduce rate of production without altering quantity to be built.

stretch point Fuselage station at which plug is to be added for stretch (1). Possibly allowed for in original design for stretch planned far in future.

stretch press Any of several families of press,

mainly hydraulic (eg Hufford, Sheridan), in which sheet is pulled beyond elastic limit over a 3-D die or tool of correct profile; invariably both ends are pulled equally over die at centre.

stretch-wrap forming Use of stretch press of various types (eg Hufford) in which as rams operate their axes simultaneously rotate to wrap workpiece around tool.

streuwaffen Cluster bomblets or scatter weapons (G).

strew To lay down sonobuoys in prescribed sonar pattern.

strike 1 An attack designed to inflict damage on, seize or destroy an objective (DoD, NATO).
2 As (1) is suggestive of target within reach of surface forces, better definition is a tactical close-support or interdiction attack on surface target, with conventional or nuclear weapons.
3 Significant impact(s) with foreign objects, esp with bird(s) while on TO run or airborne.

strike aircraft Aircraft, normally aeroplane, for carrying out strike (2); definition which restricts term to naval carrier-based aircraft is arch, though such aircraft are not excluded.

strike-camera Camera, usually optical wavelengths but can include radar and/or IR, operated automatically or on command during process of carrying out strike to record fall of ordnance and give preliminary indication of likely results.

strike-CAP Fighter role in which strike task is predominant; offensive ordnance jettisoned only under direct attack.

strike control and reconnaissance Mission flown for primary purpose of acquiring and reporting air-interdiction targets and controlling air strikes against such targets (USAF).

strike down To fold aircraft on deck and transfer to hangar(s) aboard carrier; term comes from traditional verb to secure items so that they cannot move relative to deck.

strike force Composed of units capable of conducting strikes (1), attack or assault ops (DoD); not necessarily aviation.

strike photography Imagery secured during air strike.

striker Various meanings in connection with firing ammunition and ordnance devices by impact on percussion cap, in some case syn with firing pin, and in others a hammer for hitting firing pin or intermediate between pin and percussion device. Hence * pin.

striking voltage Critical potential difference across gas-discharge tube and certain other devices at which discharge occurs and current flows.

string 1 Generalized term for assembly of devices in essentially linear sequence through which signal or object passes.
2 Literally piece of string or wool used as crude indication of relative wind, esp of yaw (probably obs).

stringer Longitudinal member (ie in fuselage more or less aligned with longitudinal axis and in wing and tail surfaces more or less perpendicular to this axis) which gives airframe its shape and provides basis for skin. In fuselage they link frames and in aerofoils they link ribs. Most existing definitions are arch, describing * as light auxiliary or fill-in member; modern transport fuselage has no other longitudinals apart from possible underfloor keel on centreline. Integral skin removes need for * except in some structures where integral stiffeners are used plus * at 90° different orientation.

strip 1 To dismantle, also called teardown.
2 See *strip stock*.

strip and digit Instrument based on roller-blind tape (usually with variable indices) plus alphanumeric window.

Stripline Microwave transmission line formed from two close strip conductors face-to-face or single strip close to conductive surface; also called Microstrip and other registered names.

strip map Various forms (eg folded paper or film for projection) of topographical map in strip form, either covering entire flightplan track or sector(s) between WPs or other intermediate points.

strip plot Portion of map or overlay upon which is delineated coverage of air-reconnaissance imagery without indicating outlines of individual prints.

strip stock Standard forms of metallic raw material in long, narrow strips, often coiled; suggest term confined to flat section only.

strobe 1 High-intensity flashing anti-collision beacon, esp on aircraft.
2 Less often, rotating anti-collision beacon, esp on aircraft.
3 See *strobe marker*.
4 To select particular portion of waveform, timebase or other time subdivision in cyclic phenomenon.
5 Stroboscope.

strobe marker Small bright spot, short gap or other discontinuity on radar timebase or other portion of cyclically scanned display to indicate portion receiving special attention.

strobe pulse See *strobe marker*.

strobe timebase Small section of timebase containing target blip, Loran signal or other object of interest which is extracted by strobe marker and expanded to fill original timebase width. In some cases process can be repeated, giving expanded **.

Strobokerr Camera for ultrafast photography in which incoming scene passes through Kerr cell, which divides it into pulses of light so brief that clear non-smeared images are projected on to film revolving at high speed inside drum.

stroboscope Instrument for apparently bringing rotating or oscillating objects to rest by intermittent phased illumination, eg for inspection of deflection under high-speed operation or to determine rpm or Hz.

stroke 1 Linear distance moved by piston of PE (4) from TDC to BDC or vice versa.
2 One complete translation of piston from TDC to BDC or vice versa, performing particular operation; thus inlet or induction *, compression *, power * and exhaust *.
3 Linear or angular distance travelled by electron or other beam forming timebase or any other electronic display.
4 Linear or angular distance travelled by output of actuator.
5 One major "flash" making up flash of lightning, normally repeated every 38–45 ms until discharge complete.
6 Basic element of *-generated writing.

stroke font Total repertoire of alphanumeric characters possible with particular stroke-generated writing system.

stroke-generated writing Major branch of alphanumerics for electronic displays in which each character is assembled from one or more straight-line strokes. Simplest format is 7-segment, also called DHW (double-hung window), all segments of which are used in figure 8 (this is common in LED and LCD displays for min cost, eg watches); another common system is 14 or 16-bit starburst, and best results are yielded by multi-stroke systems using five lengths and up to 40 orientations.

stroking Correct cycling of pump plungers over full stroke, eg in variable-displacement pump with swash drive from one end.

strongback Structural member added over large surface to provide rigidity in bending and torsion, eg across E-3A rotodome.

strongpoint See *hardpoint*.

strop 1 Flexible loop connecting deck accelerator (catapult) to aircraft not fitted with nose towbar.
2 Length of webbing connecting static line of airdrop load to anchor cable (DoD).
3 Also see *suspension* *.

Strouhal number Constant for particular bodies in fluid flows giving rise to street of vortices and from which frequency of shedding can be derived; S = fD/V where f is frequency, D is diameter of cylinder or wire and V is velocity; for regular cylinder S is 0·2 for R between 10^3 and 2×10^5.

structural damping Total damping of assembled structure.

structural design Total task of designing structure.

structural element Subdivisions of structure such as a spar, rib or frame which may themselves be assembled from smaller structural members.

structural factor Seldom used ratio of structure mass to sum of structural plus propellant mass for rocket vehicle; mass ratio more common.

structural failure Breakage under load.

structural limit Greatest weight (mass) at which aircraft can show compliance with structural requirements, usually MRW.

structural machine screw HT-steel machine screw (term often synonymous with bolt) with unthreaded shank portion below head.

structural member Portion of structure, esp any part important in bearing loads; can be single piece of material or assembly, and demarcation with structural element blurred.

structural merit index Ratio E/ρ, modulus divided by density.

structural-mode control system Flight-control subsystem whose purpose is to reduce stresses in structure during flight through gusts and other turbulence, esp in dense air, or to reduce vertical accelerations experienced by crew compartment. Requires vertical-acceleration sensor and separate horizontal (occasionally inclined or vertical) surfaces to apply required countering forces to airframe, usually symmetric about c.g. to avoid applying pitch moment. Often similar in principle to active controls.

structural placard Structurally significant airspeeds, eg max gear-down or flap-extension IAS, displayed in cockpit.

structural section Cross-section, not necessarily unvarying, of structural materials or finished parts of high fineness ratio, produced mainly by extrusion or rolling between shaped dies but often machined to incorporate tapers or discontinuities.

strut 1 Externally mounted structural member intended to bear compressive loads, and usually of streamline section.
2 Ambiguously, stub wing or pylon * attaching engine pod to fuselage or wing (loads mainly tensile and bending).
3 Loosely, any major bracing member or portion of truss even if load is tensile (suggest incorrect usage), thus a lift * on braced high-wing monoplane is misnomer.

STS Space transportation system.

STST Strategic transportable satellite terminal.

STTA Service Technique des Télécommunications de l'Air (F).

STTE Special-to-type test equipment.

STTEA Service Technique des Télécommunications et des Equipements de l'Aéronautique (F).

STU Service trials unit (UK).

stub 1 Portion of transmission line up to ¼-wavelength long connected in shunt with dipole feeder to match impedances.
2 Short straight exhaust stack carrying gas from one PE (4) cylinder direct to atmosphere just clear of cowl or any downstream structure.

stub antenna/aerial Stub (1), usually for DME, IFF and beacon transponders.

stub exhaust See *stub (2)*.

stub float See *sponson*.
stub plane Short length of wing projecting from fuselage or hull to which main wings are attached (BS).
stub power Net power available for propulsion.
stub thrust Net thrust available for propulsion.
stub wing Short aerofoil projecting approx horizontally from fuselage serving either as structural bracing member (eg linking lift struts and main gears) or as beam supporting external stores racks or other loads.
stub-wing stabilizer See *sponson*.
stud 1 Headless bolt threaded at both ends but not in mid-portion used for attaching items to threaded hole in casting or forging.
2 Projecting pin used as fulcrum, pivot, locating dowel or for other purpose.
student pilot Person authorized and licensed to receive flying instruction from rated flight instructor.
study Formal investigation, with or without research or any kind of testing, of possible solutions to a stated future requirement; hence * phase, * program, * contract.
stuffed Large component, eg wing panel or section of fuselage, manufactured complete with internal systems and equipment.
S-turn Flight manoeuvre in form of S in horizontal plane, often carried out across road, railway, fence or similar axis.
STV 1 Structural (or systems) test vehicle.
2 Satellite transfer vehicle.
STVA Self-tuning vibration analyzer.
STW Special-to-weapon equipment, eg in WAS.
STWA Short(er) trailing-wire antenna.
styrene Liquid hydrocarbon ($C_6H_5CH{:}CH_2$) polymerized into polystyrene for use as basis of important resins, low-density foams (expanded polystyrene) and many other products.
SuAwacs Soviet Union Awacs.
sub Smaller contractor, partner chosen in *teaming down*; could be thought of as short for subordinate or subcontractor.
subassembly Assembly (eg structural, electronic or for some other system) forming part of a larger item.
subcarrier Subsidiary modulated carrier which in turn modulates primary RF.
subchannel Intermittent fraction of telemetry channel conveying one repeatedly sensed measurement.
sub-cloud car Car suspended below above-cloud aerostat giving view of ground.
subcommutation Commutation of additional telemetry channels, output of which is fed to primary commutator; called synchronous if commutation frequency is multiple of that of primary (which it usually is).
subcontract Agreement other than prime contract to perform work on same programme. Normally between subcontractor and prime contractor for work assisting latter to complete task. Does not necessarily confer design authority or responsibility.
subcritical mass Mass of fissile material inadequate in magnitude or configuration to sustain chain reaction.
subcritical rotor Rotor, esp main lifting rotor of helicopter, whose fundamental flapping, lagplane or other resonant frequency is less than normal operating frequency, latter being frequency with which one blade passes given angular position.
subcritical wing Wing designed not to exceed M_{crit}.
subgrade Soil underlying airport pavement.
subgrade code ICAO coding for strength of subgrade: A, 150 MN m^{-3}; B, 80; C, 40; D, 20.
subgravity Condition in which apparent vertical acceleration is between zero and + 1 g.
subharmonic SHM or other sinusoidal waveform whose frequency is exact multiple of related fundamental waveform.
sub-idle Operating speed or control selection below idling, occasionally provided with main engines for ground operation.
subjective Dependent upon personal opinion; in such fields as aircraft noise and visible smoke unquantified * measures play major role, usually as result of seeking views of statistically significant population.
sub-kiloton weapon NW whose yield is below 1 kT.
sublimation Direct transition, in either direction, between solid and vapour state.
sublimator Solid material designed to reject waste heat by sublimation to space environment, usually a porous plate and in some cases not a true * but medium for evaporating liquid heat-transfer medium.
submarine missile Missile launched by submarine, eg against distant city target.
submarine rocket Submarine-launched rocket (Subroc) weapon for use against other submarines.
submarine striking force Force of submarines having guided or ballistic missile capabilities and formed to launch offensive nuclear strike (DoD).
submodulator AF power amplifier preceding main RF modulator.
subprogram See *subroutine*.
subroutine Set of instructions necessary to enable EDP (1) computer to carry out well defined mathematical or logical operation, forming part of complete program; unit of routine for specific sub-task, usually written in relative or symbolic coding even if full routine is not, and in closed * entered by jump path from main routine and reverting to main routine at completion (one sequence of instructions can be * and at same time a main routine with respect to its own *).
subsatellite Satellite of Moon.

subsatellite point Point on Earth's surface directly below Earth satellite, ie on local vertical through satellite, at any time.

subsidence 1 Extensive sinking of air mass, eg in polar high, in which air forced to descend is warmed by compression, increasing stability.
2 Disturbance to aircraft flight which dies away without oscillation (UK usage only and unusual).

subsonic Slower than speed of sound in surrounding medium.

subsonic flow Flow whose velocity is less than speed of sound within it; in contracting duct accelerates and rarefies slightly and in expanding duct decelerates and compresses slightly.

substantiation Formal demonstration of compliance, eg with design fatigue-life requirement.

substrate In various kinds of planar technology, structural layer upon which operative layers and/or devices are formed; in most microelectronics has low resistance, while in solid-state circuitry and solar cells * is normally an insulator.

substratosphere Imprecise, generally taken to mean upper layer of troposphere.

subsystem There is no clear demarcation between system and * though it is simple to give examples of dynamic organized groupings of devices that can be seen to be one or other. It could be argued, eg, that because it is part of a flight-control system a stability-augmentation system is a subsystem. Again, a BITE can be considered an integral part of a system or as a *.

suck-in Traditional verbal command in hand-starting PE (4), meaning "Do not energize ignition (cylinders are being filled with mixture)".

suck-out Deliberate or undesired characteristic of some vertically translating airbrakes, esp on sailplanes, of being pulled open by local depression under certain flight conditions.

suction Withdrawal of fluid through a region of local depression.

suction-cup gun Spraygun using ejector effect to withdraw medium.

suction face That side of propeller blade, normally facing forward, formed from upper surface of its aerofoil.

suction flap Not normal term; in some cases means blown flap.

suction gauge Instrument measuring pressure below atmospheric, usually by aneroid or bourdon tube.

suction gyro Gyro whose rotor is driven by atmospheric air jets trying to fill evacuated case.

suction stroke Induction or inlet stroke.

suction wing 1 One whose boundary layer is continually sucked away by powered suction system.
2 Wing of deep section with upper rear discontinuity from which large airflow is removed by suction to maintain attached flow (rare after 1949).

sudden ionospheric disturbance Abnormal behaviour of ionosphere following passage of radiation travelling at speed of light from source of solar flare, affecting Earth's sunlit face. Gradual return over following hour or more.

sudden pull-up Basic stressing case in which severe positive g is applied by violent nose-up elevator command (US term).

sudden stoppage Inflight stoppage of PE (4) apparently within about one turn of propeller, or of turbine within about one second, indicative of severe internal damage.

suite Aggregate of all equipments, not necessarily integrated or forming a common system, of similar general type carried in vehicle. Especially favoured in connection with ECM, for which as many as 14 separate electronic equipments may be carried (and in theory operated simultaneously) though most are linked only through common power supplies and possibly cockpit displays. For GA avionics * is supplanting "fit".

sulfated Sulphated (UK spelling is used in following entries).

sulphate Salt or ester of sulphuric acid.

sulphated Condition of lead/acid battery after prolonged discharge; lead sulphate plates cannot be restored by charging.

sulphidation corrosion Accelerated metallic corrosion due to sea salts in atmosphere.

sulphide Compound of sulphur with element or radical.

SUM Surface-to-underwater missile.

sumerian cobalt One of many cobalt-steel materials used for permanent magnets.

summer solstice Point on ecliptic, or time thereof, at which Sun reaches max N declination; about 21 June.

summing gear Differential gears which add or subtract motions of two members.

summing unit Device whose output is sum of inputs; can be mechanical, fluid flow (including pneumatic logic) or electronic.

sump Low region of fluid (liquid) system where liquid tends to collect by gravity. In PE (4) bottom of crankcase, which in wet-* engine also serves as oil tank. In fuel tank lowest point with tank in normal attitude, where water collects and may be drained.

sump jar Container in vent line from battery box in which alkaline chemicals neutralize battery-charging gas.

Sums, SUMS Shallow underwater missile system.

Sun Apparent diameter (optical) $13{\cdot}92 \times 10^8$ m, mean density $1{\cdot}41$ g/cm^3, mass c $1{\cdot}99 \times 10^{30}$ kg, surface gravity $274\ ms^{-2}$ ($26{\cdot}9$ g), rotational period c 25·4 days at Equator (33 at 75° N/S lat), radiating surface temperature 5,800 K, chromosphere up to c 10^6 K, emits various corpuscular and EM radiation at total energy c $3{\cdot}93 \times 10^{26}$ Js^{-1}; moving with solar system through interstellar medium in local arm of galaxy at c 20 km s^{-1}. See *solar wind.*

Sun compass Compass based on az/el of Sun, formerly used near magnetic poles, where magnetic compass unreliable.

Sundstrand drive Infinitely variable hydraulic CSD.

sun gear Central gearwheel in planetary reduction gear.

sunk costs Costs incurred in development which are paid off in another programme.

sunlight-readable Readable in illumination (illuminance) of 10^4 lx; basic requirement of military cockpit warning panels and displays.

Sup Aéro See *ENSAé.*

superadiabatic lapse rate Greater than DALR, such that potential temperature decreases with height.

superaerodynamics Aerodynamics involving such high relative velocities and such low densities that body has passed before air molecules can collide with others and exchange energy; also called free-molecule flow and Newtonian aerodynamics, and akin to MHD flow.

superalloy Any alloy designed for extremely severe conditions, esp at very high temperatures.

superboom Boom from SST or other aircraft which, because of reflection and/or refraction in atmosphere, is heard up to 250 km from flightpath.

supercharged core engine Turbofan in which fan is regarded as "supercharger" for core; no need for this concept.

supercharged harness Pressurized harness to reduce ignition arcing.

supercharged turboprop Powerful turboprop derated and matched with lower-capacity gearbox and propeller, thus giving same power at all airfield heights and temperatures; essentially synonymous with flat-rated.

supercharger Compressor driven by crankshaft step-up gears or by exhaust turbine which increases density of air or mixture supplied to cylinders of PE (4), either to boost power or, more often, to assist in maintaining power at high altitudes. Virtually all are single-stage centrifugal, in some cases with choice of gear ratios and formerly (WW2) with two consecutive stages and intercooler. Term also (suggest formerly) used for cabin blower.

supercirculation Increase of wing lift by increasing circulation by positive power-consuming means; secondary gains include postponement of stall, reduction of drag (both by improving flow and by enabling wing and other aerofoils to be smaller) and as means towards realizing laminar flow. Commonest form is blown flap, with more ambitious schemes discharging supersonic bleed air along upper part of leading edge or other places; used facing to rear to accelerate boundary layer and as by-product impart thrust.

supercompression PE (4) has such high compression ratio that it must not be operated at full throttle below given height (today arch).

superconductivity Near-zero electrical resistance exhibited by some metals, esp particular mixtures, as 0°K is approached. Extremely powerful currents and magnetic fields are possible. One branch of cryogenics.

supercooled Vapour and finely dispersed water droplets can exist as vapour and liquid at below 0°C, freezing immediately on contact with solid object.

supercritical wing Wing designed to cruise at above M_{crit}, characterized by bluff leading edge, flattish top, bulged underside and downcurved trailing edge; by reducing peak suction max acceleration and shock formation are delayed and wing can be deeper, have less sweep, house more fuel and weigh less than conventional wing for same cruise M.

supercruise See *supersonic cruise.*

supercruiser Aircraft designed to cruise at supersonic speed, usually Mach 1·5 to 2.

superheat 1 Temperature difference between aerostat gas or hot air and surrounding atmosphere; called positive if gas is warmer than atmosphere.
2 Heat energy added to gas or vapour after evaporation has been completed.

superheated vapour Vapour heated above its boiling point for given pressure.

superheterodyne Radio receiver in which received signal is mixed (heterodyned) with local oscillatory frequency to give intermediate frequency which is then amplified with various advantages.

superhigh frequency See *frequency, radio.*

superior planets Those further out than Earth, ie Mars to Pluto.

superplasticity Property of flowing like hot glass at elevated temperatures under modest applied pressures with no tendency to necking or fracture; possessed by many alloys, eg Prestal at 250°–260°C.

superposition 1 Principle in stress analysis that aggregate of all strains caused by a load system may be considered to be sum of all individual strains experienced by each member taken in isolation.
2 Identical principle for algebraic sum of currents or voltages in linear network.

superpressure Pressure difference between gas in aerostat at any point and surrounding atmosphere at same height; called positive if gas pressure greater (as it usually is).

superpressure balloon Unvented envelope strong enough not to burst in long-duration voyage at constant pressure height.

superrefraction Warm air over cold sea, extends radio/radar ranges.

super search mode Radar scans entire HUD field of view.

supersonic 1 Faster than speed of sound in surrounding medium.
2 R/T callsign, suffix when actually cruising at * speed.

supersonic combustion ramjet Ramjet whose combustion system is (very unusually) designed to function at supersonic speed; abb SCRJ or scramjet.

supersonic compressor Axial compressor in which fluid velocity is supersonic relative to whole length of rotor blades, stator blades or both, with oblique shocks giving greatest possible pressure rise per stage. (Some axial and centrifugal compressors not classed as * do in fact have local flow over Mach 1 at periphery at max rpm.)

supersonic diffuser Contracting duct (see *supersonic flow*).

supersonic flow Flow which relative to immersed body or surrounding walls is supersonic. In contracting duct decelerates and compresses; in expanding duct accelerates and rarefies.

supersonic inlet Air inlet designed for supersonic flow both past and through it for at least part of flight; ideally has centrebody or side wedge to create attached oblique shock and various forms of variable geometry and auxiliary doors.

supersonic jet 1 Propulsive jet from rocket, ramjet or afterburner whose velocity relative to source is supersonic.
2 Supersonic aeroplane (colloq).

supersonic nozzle Propulsive or wind-tunnel nozzle through which relative flow velocity is supersonic. Ideally of con/di form with variable profile and area.

supersonic propeller Propeller whose blades are designed to operate with supersonic relative velocity over major portion of surface.

supersonic tunnel Wind tunnel capable of supersonic speed in working section (either brief or sustained).

supersonic turbine Turbine of any kind designed to operate with flow velocity relative to rotor blades supersonic (rare).

superstall Progressive stall attainable by certain aeroplanes, eg T-tail with rear-mounted engines, in which (partly because at low speeds drag increases faster than lift when pilot pulls nose up) decay of speed and increasing AOA leads to stable condition in which aeroplane descends in approx constant attitude (not far removed from level flight) but with decaying speed and AOA increasing continuously so that after long period it approaches 90°. Root cause is combination of nose-up pitching moment plus immersion of horizontal tail in wing wake, destroying effectiveness. If not recoverable, called locked-in stall.

superstandard propagation Propagation with superrefraction.

superstructure 1 Secondary structure built above main fuselage or other part of aircraft (rare; eg not used for AWACS radar aerial).
2 Secondary fairing structure to streamline box-like truss.

supplemental carrier US air carrier operating under supplemental certificate, normally authorizing services of various kinds other than scheduled.

supplemental type certificate, STC Authorizes alteration to aircraft, engine or other item operating under approved type certificate (US).

supplementary aerodrome One designated for use by aircraft unable to reach its regular or alternate aerodrome (BS, suggest arch).

supply balloon Flexible container for storing gas at low pressure ready for aerostats; normally too heavy to fly even if free. Hence supply main, supply tube, links * with aerostat needing supply of gas.

support 1 All services and material needed or provided to assist operator after delivery (see * *items*).
2 Action taken to assist friendly unit in battle.
3 Part of force or unit held back at start of action as reserve.
4 Underpinning of new programme by R&D effort to ensure answers are available to technical problems.

supporting aircraft All active aircraft other than unit aircraft (DoD).

supporting surfaces Those aerofoil surfaces whose chief function is to provide lift for aerodyne; can be fixed or rotating.

support items For support (1) typically publications, training, simulator and instructional rigs, auxiliary ground equipment, spare parts, testing, warranty provisions and field modification kits.

support zone Designated surface area for air-landing or other operations in direct support of battle (no longer used DoD, NATO).

suppressant Active ingredient for suppressing an action, normally fire, esp one automatically released by sensitive pressure sensors in fuel tanks or similar regions; passive suppressing methods, eg reticulated foam, are not *.

suppressed 1 Installed so that item does not project beyond skin of aircraft; thus * aerial is synonymous with flush aerial.
2 Emitter, especially engine(s), designed or installed to minimize emissions; for civil aircraft noise predominates, for military aircraft IR radiation.

suppressive Intended to suppress hostile defensive fire by offensive action, eg direct attack with weapons and offensive ECM; hence * attack, * support, * weapons (eg ARMs).

suppressor 1 Jet-engine nozzle either configured for minimum noise or shielded by additional surrounding duct.
2 Various additions to electrical or electronic devices or circuits to reduce unwanted leakages, emissions or other phenomena, eg extra grids in thermionic valves (tubes) to stop secondary

emission and large series resistors in HT circuits to eliminate sparking.

SUPPS Regional supplementary procedures.

Supral Superplastic aluminium alloy marketed for SPFDB applications (BACo).

suprathermal ion detector One of ALSEP experiments left on Moon; measures energetic ions impacting surface to determine solar-wind energies.

SUPT Specialized undergraduate pilot training.

SURE, Sure Sensor update and refurbishment effort (USN).

surface 1 Aerofoil, esp large, eg wing (not small rotating, eg compressor blade).
2 Exterior of aircraft, eg * friction drag.
3 2-D layer corresponding to particular pressure altitude.
4 2-D layer in any plane corresponding to particular electronic radiation pattern or time difference.
5 Hinged or extendable area for flight control, lift augmentation or drag augmentation.
6 Generalized term for Earth's *, hence * target = one on land or water.

surface actuator Device which physically moves a surface, eg control surface; need not embody any form of control function.

surface boundary layer Atmosphere in contact with Earth's surface, extending up to base of Ekman layer (anemometer level).

surface burning Combustion of fuel for propulsion on outside of aeroplane, proposed for variable-geometry aircraft for Mach numbers of about 5, using variable body profile for ramjet effect..

surface cooler See *surface radiator.*

surface corrosion Galvanic (non-mechanical) attack on surface, eg by salt spray, often under paint film.

surface effect Effects on air-supported vehicle of close horizontal surface beneath (synonymous with helicopter ground effect), hence ** vehicle (= air-cushion vehicle), ** ship (US usage in latter cases).

surface-friction drag Drag due to all forces tangential to surface, notably shearing of boundary layer; added to form drag makes profile drag; added to pressure drag makes total drag.

surface gauge Gauge (US, gage) in form of precision stand moved about on surface plate carrying adjustable scriber for transfer of exact height measures, eg in marking out or in checking finished workpieces.

surface loading Mean normal force per unit area carried by a particular aerofoil under specified aerodynamic conditions (BS); in case of wing term is wing loading; suggest few cases where ** needed.

surface movement One vehicle (eg aircraft, or even bicycle) in motion on airfield movement area. Hence ** indicator, usually a PPI radar.

surface of discontinuity Sloping demarcation between warm and cold air masses.

surface plate Steel table with extremely flat and smooth surface.

surface power unit Surface actuator embodying control functions, eg control valves, feedback inputs, summing units and possibly redundancy provisions.

surface sampler Device for scooping up specimen of planetary or other surface for analysis or other study (eg to investigate for presence of life).

surface tape Pinked-edge strips of fabric doped over all seams, rib-stitching and edges of fabric covering. Also called finishing tape.

surface target One on land or sea.

surface-to-air missile Missile launched from surface (6) against target above surface (DoD). Hence ** envelope, that airspace within kill capabilities of particular SAM system; ** installation, a ** site with system installed; ** site, prepared plot of ground designated for but not occupied by SAM system.

surface-to-surface missile Surface (6) -launched missile designed to operate against target on surface (6), including those underground.

surface wave 1 Radio wave travelling round surface (6); most effective propagation mode of LW/LF.
2 Acoustic wave in surface-wave device.

surface-wave device New family of electronic devices based on surface acoustic waves sent across piezoelectric slab or other substrate; originally (1970s) used for delay lines and now for complex signal processing.

surface wind Generalized term for wind measured at surface (6); in US gradually switching from 20 ft (6·5 m) anemometer level to ICAO 10 m (32·8 ft) level.

surface zero See *ground zero.*

surfacing Improving low-drag quality of vehicle surface (in general task called stopping and *), eg by perfecting flatness and smoothness.

surfactant Surface active agent, material (usually liquid or particulate) which alters surface tension and/or performs other tasks at boundaries between dissimilar materials, eg detergent.

surge 1 Gross breakdown of airflow through compressor, normally of axial type, resulting from local stall and usually characterized by muffled bang and sudden increase in turbine temperature; hence * line, * point. Often used synonymously with stall.
2 Various abnormally large currents, signal amplitudes or voltages in electrics or radio, eg on first switching on or caused by lightning or static discharge.
3 Planned large increase in flying rate of military unit, eg to explore ultimate potential of personnel and hardware over short or longer term under crisis conditions.
4 Unplanned transient increase in flow of fuel in

aerial refuelling, sometimes causing a disconnect.
5 General change in atmospheric pressure at surface apparently superimposed on predicted diurnal or cyclonic change.

surge box Term used for various kinds of device in aircraft fuel system to reduce pressure/flow excursions caused by fuel momentum either in tanks (sloshing) or in pipelines, esp during high-rate refuelling.

surge diverter Protective semiconductor device having negative resistance/temperature coefficient to earth voltage surges.

surge line Boundary between gas-turbine operating region and region where surge of compressor is certain; locus of all surge points. Generally same as stall line.

surge point Any combination of airflow and pressure ratio for gas turbine at which surge occurs.

surging 1 Occurrence of surge (1).
2 Fault in wind tunnel characterized by erratic or low-frequency pulsations in velocity, flow and pressure.
3 See *sloshing*.

surveillance Systematic observation of aerospace, surface or subsurface objects by any kind(s) of sensor.

surveillance approach Instrument approach conducted in accordance with directions issued by ground controller referring to a surveillance radar display (DoD).

surveillance radar 1 Primary radar scanning in azimuth, often through 360°, supplying P-type display (PPI). Not normally giving elevation or height of aerial targets.
2 Specif, primary radar whose purpose is to determine az/el position, track and (with SSR) identity of all aerial targets, and to provide radar separation, navigational assistance, storm warning and vectoring for final approach (but not normally to handle complete radar approach).

surveillance radar element, SRE Portion of GCA system which vectors incoming traffic until established on ILS and handover to PAR.

surveillance system Any means of surveillance not contained wholly in one vehicle or site, eg RPV, electronic com and guidance, digital sensor data-link, and control/receive ground station.

survey 1 To examine damaged vehicle, eg crashed aircraft, often on behalf of insurers or underwriters, to establish damage, possibility of salvage (eg to fly out from crash site) and best course of action.
2 Normal meaning in photogrammetry and mapping.
3 Examination by surveyor (see below).

surveyor Technically qualified and designated official empowered to collaborate with aerospace design staff on behalf of national certification authority and examine subsequent hardware and design software to establish compliance with airworthiness requirements; usually concerned with particular design aspect, eg fluid systems, or with particular class of aircraft.

survivability Capability of a system to withstand a man-made hostile environment without suffering an abortive impairment of its ability to perform its designated mission (USAF). Refers specif to various effects of NW attack, eg degradation of volatile memory.

survival capsule Detachable crew compartment, normally of military aircraft, capable of separation in emergency and soft landing on land or water, thereafter serving as shelter.

survival kit Man-portable package containing items to help sustain life remote from other human beings.

survival radio Self-contained, portable, shockproof, floating radio emitting homing (and possibly voice) signals on 121·5 and 243 MHz.

Survsat Survivable sat-com system.

SUS Signal, underwater sound.

susceptance Reciprocal of reactance.

susceptibility Degree to which any hardware is open to attack as result of inherent weakness (DoD).

suspension 1 Linkage between aerostat and load, hence * band (fabric band linking envelope to * lines), * bar connecting suspension lines to basket ropes in balloon, * line, main connections between envelope and basket or suspended car, and * winch connecting kite balloon to surface (6).
2 System of particles dispersed through fluid, including atmosphere.

suspension strop Webbing or wire rope connecting helicopter and cargo sling.

sustained flight Time-invariant flight, ie steady lift and airspeed.

sustainer Propulsion, either rocket or airbreathing, that provides power for sustained flight following short high-thrust acceleration period under power of boost motor(s). Not normally applied to any stage of large or small multi-stage vehicle; must be long-duration propulsion system handling entire mission after separation or burnout of booster (in ramrockets and many related systems may in fact use booster case for combustion).

Sutherland law Gives temperature variation for viscosity of air.

Sutton harness Traditional (WW1) personal seat harness with two lapstraps, two shoulder straps and central pinned clip passing through all four.

SV 1 Satellite vehicle.
2 State vector.
3 Static vent.
4 Simulation validation.
5 Shop visit.

SVCBL Serviceable (ICAO).

SVD System verification diagram (software).

SVF Schweizerische Vereinigung für Flugwissenschaftten (Swiss).
SVFR Special VFR.
SVI Smoke volatility index.
SVLR Schweizerische Vereinigung für Luft- und Raumrecht (Swiss).
SVMS Space-vehicle motion simulator.
SVR 1 Slant visual range; attempt to give pilot on final approach idea of when he will acquire approach lighting, reported as either nominal contact height at which 150 m (500 ft) segment with one cross-bar will become visible, or, at top of shallow layer of fog, as expected minimum visible length of lighting.
2 Shop visit rate.
SVS Secure voice switch.
SV-stoff Mixed acids, eg 95% RFNA, 5% H_2SO_4 (G).
SVUOM State Research Institute for Protection of Materials (Czech).
SVWT Schweizerische Vereinigung für Weltraumtechnik (Switz).
SW 1 Single-wedge-type aerofoil for purely supersonic vehicles, eg missiles.
2 Short-wave.
3 Surface wave.
4 Strategic wing (USAF).
S/W Software.
swageing Joining by cold-squeezing one member around another, eg electrical or control-cable terminal or end-fitting on to end of cable.
swallowed shock Position of shockwave across airbreathing engine or other inlet inside duct, when mass flow is ρVA but internal pressure is low (normally equated with zero thrust). Can be feature of plain pitot inlet or any other type. Resumption of correct engine operation restores shock to normal position at inlet.
swallowing capacity Ability of inlet to handle large airflow, esp over wide range of air densities and Mach numbers.
Swaps Standing-wave acoustic parametric source.
swashplate Disc rigidly or pivotally mounted on shaft as drive mechanism for plungers or rams arranged parallel to shaft; when disc is normal to shaft, plunger stroke is zero, increasing to max at max * obliquity. In hydraulic motor * is driven, not driving, member.
swathe Width of area treated at one pass of ag-aircraft; hence * width. Normally no deliberate * overlap.
sway braces 1 See *crutches*.
2 Additional struts, not normally part of aircraft, required to brace particular large or winged store.
SWC 1 Special Weapons Center (Kirtland AFB, NM; USAF).
2 Sky-wave correction (Loran).
3 Solar wind composition.
SWD Surface-wave device.
SWE 1 Stress wave emission.
2 Software engineering.
sweat cooling See *transpiration cooling*.
sweating, sweated joint Joining two tinned members without additional solder or brazing metal.
sweep 1 Sweepback.
2 Total angular movement of aerial, eg surveillance radar, oscillating in azimuth (sector scan).
3 Total movement, normally expressed in linear measure, of time-base spot scanning across CRT or other display.
4 One complete cycle of VG wing.
5 Angular deviation of locus of centroids of propeller blade sections from radial line tangential thereto at propeller axis projected on plane of rotation (BS).
6 Offensive tactical mission against surface targets, normally targets of opportunity (WW2).
7 To range over continuous (usually large) band of frequencies.
8 To employ technical means to uncover covert surveillance devices (DoD).
sweepback Visibly obvious backwards inclination of aerofoil from root to tip so that leading edge meets relative wind obliquely.
sweepback angle Angle between normal to longitudinal OX axis (axis of symmetry in most aircraft) and reference line on aerofoil, normally 0·25 (one-quarter) chord line or, less often, leading edge; both normal line and reference line lie in same plane, which is usually that containing centroids of aerofoil sections from root to tip (thus for canted verticals, ** measured in plane of each surface).
sweepforward Visibly obvious forwards inclination of aerofoil from root to tip so that leading edge meets relative wind obliquely; hence * angle, or forward-sweep angle.
sweep jamming To emit narrow band of jamming swept (7) back and forth over wide operating band of frequencies.
sweep oscillator Signal generator whose frequency is varied periodically by fixed amount at constant amplitude above and below central fixed frequency; also called Wobbulator (UK), sweep generator (US) or scanning generator.
sweep-tip blade Helicopter rotor (main or tail) blade whose locus of centroids is radial from root to near tip and then sharply inclines back.
swept Incorporating sweepback (never used of forward sweep).
SW/FR Slow write, fast read.
SWG 1 Standard wire gauge (UK); standard range of sheet thicknesses.
2 Square waffle grid (space structures).
swing 1 Involuntary and often uncontrollable divergent excursion from desired track of tailwheel-type aeroplane running on ground.
2 To turn propeller by hand to start PE (4); if not engaged in starting engine, or with turboprop, term is to hand-turn or pull-through.

3 To calibrate compass deviation by recording its value at regular intervals, usually 15°, during 360° rotation of aircraft on compass base.
4 Distortion of radio range; also called night effect.
5 Sudden yaw of aeroplane consequent upon loss of power of engine mounted away from centreline.

swing-by Close pass of planet or other celestial body by spacecraft on Grand Tour.

swing force Aircraft or complete combat unit can fly air/air and air/ground in same mission.

swinging base Compass base; also called deviation clock.

swinging compass Magnetic compass used as standard for calibrating that in aircraft.

swing-piston engine Various topological families of PE (4) in which two, three or more pistons oscillate around toroidal cylinder alternately compressing mixture between them, being driven by firing strokes or, in some, acting as pumps. Most do not have mechanical drive but supply gas, eg to drive turbine.

swing-wing aircraft Aeroplane with variable sweep (1); also called VG aircraft (colloq).

SWIP Super weight-improvement program (US).

Swipe Simulated weapon impact predicting equipment.

SWIR Short-wave IR.

swirl 1 Gross rotation of flow about axis approx aligned with flow direction, eg in propeller slipstream or induced by large drive fan in low-speed tunnel (removed by straighteners).
2 Rotation of air in whirling-arm room or other non-evacuated chamber containing high-speed rotating object.

swirl vanes Array, usually radial about an axis, of curved or oblique vanes for imparting swirl to fluid flow, eg immediately upstream of or surrounding primary gas-turbine fuel burner.

SWIS Stall-warning and identification system.

swishtail See *fishtail*.

switches off Traditional verbal command in hand-starting PE (4) to ensure that ignition is inoperative at start (actually switches normally closed, short-circuiting HT).

switches on Seldom used; normal call is "contact".

switching system Automatic switching in large network, eg military com, airline reservations or nationwide computer link; normally electromechanical pre-1960 and electronic later, allowing for on-line, real-time messages, data transfer, storage or display.

swivelling engine Entire engine, or liquid-rocket thrust chamber, is gimbal-mounted or pivoted so that thrust axes can rotate relative to vehicle.

Swizz SWIS (colloq).

SWL Strategic-weapon launcher.

SWR Standing wave ratio.

SWS Standard warning system (CAA).

SWTDL Surface-wave tapped delay line.

SWTL Surface-wave transmission line.

SWY Stopway.

SX Sheet explosive.

S_x Simplex.

SXT, Sxt Sextant.

SXTF Satellite X-ray test facility.

SYCAF Système de Couplage Automatique sur Faisceau (ILS coupler, F).

Sycep Syndicat des Industries de Composants Electroniques Passifs (F).

Sygnog System go/no-go.

sylphon Stack of aneroid capsules; sometimes called * tube.

symbology Symbols conveying meanings to human beings, the technology of their design and production and their incorporation in systems and displays; most important are alphanumerics, in various national languages, followed by more than 9,000 standard conventional symbols so far available for various technologies. About 50 different forms and variations have been agreed for HUDs, Hudsights and other weapon-aiming systems, eg simple cross or cross/ring reticles, range rings that unwind as range closes, and various aiming lines, wing bars and arrow or triangular markers.

symmetric aerofoil Wing profile whose mean line is straight.

symmetric double-wedge Wing profile in form of sharp-edged parallelogram, used mainly for supersonic missiles whose subsonic qualities are unimportant.

symmetric flight Both left/right wings equally loaded.

symmetric flutter Left/right symmetry in amplitude and direction.

symmetric immersion Both flying-boat tip floats in water equally (rare).

symmetric instrumentation Installation of experimental sensing equipment (such as pressure transducers) over the entire surface of aircraft such that for each sensing head on left half there is an exactly corresponding unit on right.

symmetric pull-out Pull-out from dive with wings level.

symmetric stall Stall with wings level, longitudinal axis rotating within plane of symmetry.

symmetry check Measurements to corresponding L/R points from centreline.

Syname Syndicat National de la Mesure Electrique et Electronique (F).

synchro Generalized term for bipolar AC synchronous systems in which a master unit or sensor commands identical response (eg angular position) by one or more instruments or other receivers. An alternative to voltage signalling by potentiometer and digital signalling by encoder.

synchronization 1 Commanding all aircraft engines to rotate at same speed.
2 Commanding automatic guns to fire at cyclic

rate forming exact fraction or multiple of blade-passing frequency of propeller; not same as interrupter.
3 Process of adjusting timing (epoch), frequency and phase of spread-spectrum receiver's PN correlation to match those of received signal.
4 Process of preadjusting outputs of two or more control (eg FCS) channels to reduce dead-zone if operated together or switchover transient if operated separately.

synchronous corridor Equatorial belt within which synchronous satellite must remain (normally describing small vertical figure-eights).

synchronous orbit See *geostationary orbit*.

synchronous satellite One whose rotation is synchronized with that of Earth; also called geostationary.

synchronous sighting See *tachymetric aiming*.

synchro pair Two aircraft which perform synchronized manoeuvres to entertain crowd while rest of team reposition.

synchrophasing Commanding all propellers of multi-engine aircraft to rotate in step, all blades instantaneously at same angular positions.

synchropter Helicopter lifted by two or more rotors whose blades intermesh (suggest colloq).

synchroscope Instrument for giving visual indication of synchronization, or lack of it, between two or more frequencies or speeds.

syncom Synchronous communications (satellite).

sync pulse/signal Sync is generalized term for synchronization between TV camera and receiver, or between any raster-scan sensor and display or output, hence * is integral part of transmitted waveform to maintain lock on synchronization.

syncrude Synthetic crude petroleum; starting point for various synthesized petroleum-type hydrocarbons.

synergic ascent Following synergic curve.

synergic curve Trajectory for departing spacecraft for min energy requirement, ie lowest propellant consumption for given position and velocity; on Earth initially vertical to leave denser atmosphere quickly and then curving to take advantage of Earth's rotation.

synergism Favourable interrelationship between variables such that overall benefit of a change is greater than sum of individual gains; eg scaling down aircraft size has synergistic effects on structure weight, drag, engine size, fuel consumption and fuel mass for given range.

Synjet Non-petroleum-based fuel produced to current (or broadened) Jet A specification.

synodic period Interval of time between identical positions of celestial body in solar system measured with respect to Sun.

synodic satellite Hypothetical Earth satellite located on Earth/Moon axis at 0·84 lunar distance from Earth.

synoptic chart Standardized map of weather, showing isobars, fronts and weather symbols, and covering large area for one particular time.

synoptic meteorology Collection of met information covering large area at one time (as near as possible to present), esp with view to forecasting.

syntactic foam Composite material consisting predominantly of premanufactured hollow microspheres embedded in resin; for radomes usually 30–140 μ spheres with 1·5 μ wall, in epoxy or polyimide matrix.

synthetic lubricants Post-1948 families of turbine oils originally based on esters of sebacic acid, esp dioctyl sebacate; later with complex thickeners added.

synthetic resins Too numerous to outline, but mainly polymers or copolymers and often thermosetting; used as bases for many materials (eg plastics and paints) and as adhesives, including nearly all those for aerospace bonding and for fibre-reinforced composites.

synthetic rubber Vast family of rubber-like materials originally (1917) based on isoprene and today nearly all based on copolymers of butadiene; includes many solid propellants.

synthetic training All training that simulates, eg with simple Link, mimic boards, system rigs, air-combat simulators and, esp, flight simulators. Also generally held to include actual flight training when something, eg absence of external vision, is simulated.

system 1 Generalized term for any dynamically functioning organization of man-made devices.
2 Portion of vehicle, eg aircraft, missile, etc, forming integral network of related and inter-controlled devices to accomplish set of specif related functions.
3 Composite of equipment, skills and techniques capable of performing and/or supporting operational role (USAF).
4 Often used to mean (1) plus supporting equipment, documents, training devices and all other products and services, as in weapon *.
5 Incorrectly used to mean mere assemblage of mechanical parts, eg engine LP *.

system concept Integrated approach to design, procurement or operation of system in sense (4).

system discharge indicator, SDI Yellow disc or blow-out plug in aircraft skin to indicate fire-extinguishing system discharged for reason other than fire or overheat warning.

Système International In full, SI d'Unités, system of unified units of measurement adopted by all principal industrialized countries since 1960 and gradually being implemented; seven base units, metre (m), kilogram(me) (kg), second (s), ampere (A), kelvin (K), candela (cd) and mole (mol).

system(s) engineering Not briefly definable, but extremely broad discipline akin to operations research whose main objective is to apply broad overview of entire system (4) in order to advise customer and/or management on objectives and possibilities and refine and integrate all subsystems before start of hardware design.

system source selection Selection by government of industrial source, known as system prime.

system turnover Formal acceptance by customer of responsibility for system (4).

Systo Systems Command program officer (USAF).

syzygy Point on orbit, esp that of Moon, at which body is in conjunction or opposition.

SZH Schweizerische Zentrale für Handelsförderung (Switz).

T

T 1 Temperature.
2 Tesla (nT is more common).
3 Prefix, tera = 10^{12}.
4 Suffix, true heading.
5 Time of day, or in countdown.
6 Period.
7 Kinetic energy.
8 Thrust (F is preferred).
9 Torque.
10 Transmitting.
11 Tower (ATC service).
12 Code: aircraft has transponder with 64-code.
13 Tensile stress in fibres in composite material.
14 Short tons (NW yields only; ton not normally abb).
15 Aircraft designation prefix/suffix: trainer.
16 US piston-engine prefix: turbocharged.
17 US military engine prefix: turboprop.
18 Navaid terminal-category.
19 Tee, see *basic T*.
20 Threshold lighting.
21 Electrochem transport number (also t).

t 1 Elapsed time in seconds.
2 Tonnes.
3 Thickness, esp max of aerofoil.
4 Triode.
5 Turns.
6 One revolution (F), as in t/min = rpm.
7 Often, non-abs temperature.
8 Threads, eg in tpi (per inch).
9 Trend landing forecast.

T^2 Total time.

T^3 Tailplane trimming tank.

T-rail Standard longitudinal floor rail tailored to receive pax seats, cargo tie-down rings or stretcher racking.

T-section Shaped like T or inverted T in cross-section.

T-stoff HTP plus a little oxyquinoline (G).

T-tail Aeroplane tail with horizontal surface mounted on top of fin.

T-time Reference time in countdown, often that for start of engine ignition.

TA 1 Trunnion angle (INS).
2 Telescoped ammunition.
3 Twin-aisle (transport).
4 Target alert.
5 Terrain avoidance.
6 Tuition assistance.
7 Transition altitude (CAA).

T_a Actual temperature.

t_a Action time.

TAA Transport Association of America.

TAAM 1 Terminal-area altitude monitoring.
2 Tactical (or Tomahawk) airfield attack missile.

tab 1 Hinged rear portion of flight-control surface used for trimming (trim *), to reduce hinge moment and increase control power (servo *, balance *, spring * etc) or to increase hinge moment and effectiveness of powered surface (anti-balance *, flap *).
2 Rarely, auxiliary aerofoil hinged to trailing edge of flight-control surface, usually of servo type.
3 Small hinged spoilers on inside of propelling nozzle or rocket expansion nozzle for thrust vectoring or noise reduction.
4 Abb tabulator, alphanumeric display.

tabbed flap Flap, usually Fowler, whose trailing edge is hinged (without slot) and at max landing setting is pulled down by linkage to much greater angle than main surface.

tablock Flat washer with rectangular tab bent up to lock nut (sometimes capital T).

TABMS Tactical air battle management system.

Tabs Telephone automated briefing system.

tabulated altitude Altitude of celestial body read from a table.

TABV Theatre airbase vulnerability.

Tab-Vee Alternative form of TABV, common name for NATO hardened shelter for aircraft.

TAC 1 Tactical Air Command (USAF).
2 Trim augmentation computer.
3 Turbo-alternator compressor (Brayton cycle).
4 Thermosetting asbestos composite.
5 Terminal-area chart.

Tac Usually tactical.

Tacamo Take charge and move out (USN airborne strategic communications system, mainly NCA to SSBNs).

Tacan Tactical air navigation; UHF R-theta-type navaid giving bearing/distance of aircraft from an interrogated ground station.

Tacbe Lightweight tactical beacon equipment; 2-channel ground/air voice (Burndept).

TACC Tactical Air Command (or Control) Center (USAF).

Taccar Time-average clutter coherent airborne radar.

TACCO, Tacco Tactical co-ordinator; flight-crew member of ASW aircraft responsible for overall mission management during search or attack operations.

TACCS Tactical air command and control specialist.

Tacdew Tactical combat direction and EW.

TA/CE Technical analysis and cost estimate.

Taceval Tactical evaluation, hardware development, test and training effort on production air weapon systems (USAF).

Tacfax Tactical digital facsimile.

TACG Tactical air control group (USAF).

tachogenerator, tachometer generator Tachometer of electric synchronous-motor type with shaft-driven generator feeding AC to one or more synchronized displays.

tachometer Instrument for indicating speed of rotating shaft.

tachometer cable Rotary drive of mechanical tachometer, two spiral layers of steel wire.

tachometric Tachymetric.

tachymetric aiming Aiming of gunfire, bomb or other weapon by continuously maintaining sightline on target, thus determining speed relative to surface target and in some cases track through target.

Tacjam Tactical jamming of UHF/VHF com.

tack coat Very thin coat of finish, usually dope, which precedes the full-density wet coat.

tack rag Soft lint-free rag slightly damp with thinner.

tack weld Small dab of weld metal making local link to hold parts in correct location (but capable of easy rupture if in error) while main weld is made.

TACMS Tactical missile system (DoD).

Tacom Tactical-area com.

Taconis oscillation Pulsating (c 1 Hz) resonance in cryogenic refrigerant.

Tacor Threat-assessment and control receiver (ECM).

Tacos Tactical airborne countermeasures or strike.

TACP Theatre air control party.

TacR Tactical reconnaissance.

TACS 1 Tactical air control system.
2 Less often, tactical air control squadron.
3 Thruster, attitude-control system.

Tacsatcom Tactical satellite communications (DoD).

TACSI, Tacsi TACS (1) improvements.

TACT Transonic aircraft technology (NASA).

Tactec Totally advanced com technology (RCA).

tactical Generalized term meaning concerned with warfare against directly opposing forces, usually involving air, land and sea forces together, and in limited theatre of operations.

tactical aeromedical evacuation From combat zone to outside it, or between points in combat zone.

tactical air command center Theatre HQ of USMC air operations.

tactical air control centre Principal centre, shore or ship-based, from which all tactical air is controlled.

tactical air co-ordinator Directs, from aircraft, air close support of surface forces.

tactical aircraft shelter Normally protects against conventional attack but may be extended to offer protection against NW blast and radiation and CBW.

tactical air officer (afloat) Responsible under amphibious task-force commander for all supporting air operations until control is passed ashore.

tactical bomb line See *bomb line*.

Tactifs Tactical integrated flight system.

tactile faceplate Electronic display screen sensitive to fingertip touch for reprogramming, selecting from menu, changing scale or operating mode, or adjusting any variable.

Tacts Tactical aircrew combat training system (ACMI).

TAD 1 Turbo-alternator drive.
2 Target assembly date.

Tadds Target alert data display sets; part of FAAR.

Tadil 1 Tactical digital intelligence link (C adds "command", J adds "joint" [service]).
2 Tactical aircraft digital information link.

Tadjet Transport, airdrop, jettison.

tadpole Track of moving target on radar display presented with comet-like tail to show direction of travel. Most air-defence radars can select tadpoles on or off.

tadpole profile Aerodynamic profile with conventional nose followed by single-surface construction downstream (eg fin of A–4 followed by single-skin rudder).

TADS, Tads 1 Tactical air defense sight (US).
2 Target acquisition (or airborne) and designation (or data) system (or sight) (USA).

TAEM Terminal-area energy management.

TAF 1 Tactical air force.
2 Terminal area forecast (Int met figure-code).

TAFI Turn-around fault isolation.

TAFIIS TAF (1) integrated information system.

TAFS Airfield met forecast.

Tafseg Tactical air force systems engineering group.

TAG 1 Telegraphist air gunner (Royal Navy, WW2).
2 Thrust-alleviated gyroscope.
3 Tactical airlift group.
4 Telescoped-ammunition gun.

Tags Technology for automated generation of systems.

TAH Transfer and hold.

TAI 1 Total active inventory.
2 Thermal anti-icing.

tail 1 Rear part of aircraft, where applicable.
2 Assembly of aerofoils whose main purpose is stability and control, normally located at rear of aerodyne or airship.
3 Trailing luminous area behind blip of moving target.
4 Normal verb meaning in air-intercept shadowing from astern.

tail boom Tubular cantilever(s) carrying tail (2)

attached either above short fuselage nacelle or as L/R pair to wings.

tail bumper Projecting or reinforced structure under tail designed to withstand impacts and scraping on runway.

tail chute See *tail parachute*.

tailcone Conical fairing of rear of body, esp downstream of turbine disc in jetpipe.

tail drag Restraining mass free to slide on ground to which moored airship stern attached.

taildragger Aircraft with tailwheel or tailskid (colloq).

tailed delta Aircraft with delta wing and horizontal tail.

tail-end charlie Rear gunner at extreme rear of fuselage (colloq). Also, last aircraft in line-ahead formation.

taileron Single-piece horizontal tail surface, one of two forming tailplane whose left/right halves can operate in unison (as tailplane commanding pitch) or differentially (as ailerons commanding roll). Term preferable to ailavator or rolling tailplane. Elevon differs in that it is hinged to wing. US term stabilator is ambiguous and can mean * or slab tailplane.

tailfeathers Free-floating flaps forming periphery of supersonic airbreathing propulsive nozzle, usually as outer boundary of large secondary nozzle; take up slipstreaming angular positions aligned with streamlines.

tail fin Fixed stabilizing fin at rear, "tail" normally being redundant.

tail float Float supporting tail of float seaplane (now arch).

tail group Complete tail (2), considered as design task or as element of total aircraft mass.

tail guy Secures tail of moored airship, often to tail drag.

tailheaviness Condition in which aircraft rotates nose-up unless prevented.

tailless aircraft Normally applied to aeroplanes and gliders only and usually meaning that there is no separate horizontal stabilizing or control surface, though there may be a vertical tail (2). In extreme case there is no tail surface, and (esp if fuselage vestigial or absent) this is more often called flying-wing aircraft.

tail load Vertical up or down thrust acting on tailplane.

tail logo Bold logo of operator displayed on tail; hence ** light, also valuable as anti-collision beacon.

tail number See *serial (2)*.

tailored fuel Synthesized to meet specific operational specification.

tail parachute Parachute attached to tail, normally for anti-spin or anti-superstall purpose. Not used as braking parachute.

tailpipe Exhaust pipe of turboprop or turboshaft; according to some, PE (4) exhaust pipe downstream of collector or manifold.

tailplane 1 More or less horizontal aerofoil at tail of aerodyne (invariably fixed-wing) providing stability in pitch; fixed or adjustable only for trim, and carrying elevators (US = stabilizer).
2 Aerofoil pivoted at tail about horizontal axis and driven directly by pilot of fixed-wing aerodyne or rotorcraft as primary flight control in pitch in translational flight; forms complete surface without separate elevators (US = stabilizer).

tail rotor Helicopter anti-torque rotor, rotating at tail about more or less horizontal axis. Not used for rear tandem rotor.

tail setting angle Angle between chord lines of wing and tailplane (1).

tailsitter VTOL aerodyne whose fuselage is approx perpendicular at take-off, in hovering mode and at landing.

tailskid Projection supporting tail of aerodyne on ground, esp one whose c.g. is well aft of main landing gear.

tailskid shoe Replaceable pad on end of tailskid which slides on ground.

tailslide Transient flight condition of fixed-wing aerodyne in which relative wind is from astern, eg in stall from near-vertical climbing attitude.

tailspin Spin (arch).

tail surface Any aerofoil forming part of tail (2).

tail undercarriage Rearmost unit of tailwheel-type landing gear (rare, suggest arch).

tail unit Complete tail (2) of horizontal, vertical and/or canted surfaces, often including ventral fins or strakes. Also called empennage.

tail view Tail-on view showing object from directly astern; not normal aspect for layout drawing.

tail warning radar Aft-facing radar, usually of active type, intended to detect other aircraft (and possibly SAMs) intercepting from behind.

tailwheel 1 Rear wheel of * type landing gear, supporting tail on ground.
2 Auxiliary wheel under tail of aircraft with nosewheel-type landing gear (eg Albemarle); fitted in place of tail bumper.

tailwheel landing gear Landing gear comprising left/right main units ahead of c.g. and tailwheel at rear.

tailwind Wind blowing approx from astern of aircraft and thus increasing groundspeed.

TAINS, Tains Tercom and inertial navigation system.

TAIR Terminal-area instrumentation radar.

TAIRCW Tactical air control wing.

TAIS Tactical air intelligence systems.

Take 5 Traffic crossing airway must maintain prescribed separation of 5 nm horizontally and 5,000 ft vertically from any GAT track in airway.

take-off 1 Procedure in which aerodyne becomes airborne; not normally used for launch of glider (except on aerotow) or high-acceleration launch of missile or RPV, and never for any ballistic vertical-liftoff vehicle.

2 Moment or place at which aerodyne leaves ground or water.
3 Net flightpath from brakes-release to screen height.
4 Power * for extraction of shaft power.

take-off boost Boost pressure permitted for take-off, usually 2 min limit.

take-off distance, TOD Field length measured from brake-release to reference zero (at screen); can be longer than runway and extreme limit TOD_a = entire runway + stopway + clearway = $TOR_a \times 1{\cdot}5$. For multi-engine aeroplanes usually factored acording to number of operative engines, thus TOD_4 or TOD_3. TOD_r = TOD required for particular aircraft and WAT, not normally to exceed 0·87 TOD_a.

take-off limit No general meaning.

take-off mass Not normal term; for rocket or space launcher usually liftoff mass or launch mass.

take-off noise Measured on extended runway centreline 3·5 nm (strictly 6,485·5 m, but taken as 6·5 km) from brakes-release. A second reference point, not used for certification, is at side of runway opposite supposed start of run 1 nm from centreline.

take-off power Power authorized for PE (4) or turboprop for take-off, usually 2½-min rating for turbine engines. In case of turboshaft, a lower rating than 2½-min contingency.

take-off rating 1 Boost/manifold pressure/rpm figures authorized for PE (4) at take-off.
2 Thrust published for turbojet or turbofan at take-off, normally achieved by engine control system rather than set directly by pilot, and subject to ATR or FTO techniques.

take-off run 1 Loosely, distance travelled over land or water in aeroplane or aerotow-glider take-off to point of becoming airborne.
2 TOR, field length measured from brake-release to end of ground run plus one-third of airborne distance to screen height. TOR_a = TOR available = length of runway; TOR_4, TOR_3 are factored for engine-out cases, and TOR_r = TOR required.

take-off safety speed V_2, lowest speed at which aeroplane complies with required handling criteria for climb-out following engine failure at take-off.

take-off weight 1 See *MTOW*.
2 Actual weight at take-off (2) on particular occasion.

TAKEOVER In HUD or as caption: autopilot has disconnected.

T-AKX Ro/Ro ship commandeered in emergency for RDF.

talbot MKS unit of luminous energy; 1 */s = 1 lm.

TALC Tactical airlift center (USAF).

TALCM Tactical air-launched cruise missile.

talkdown Landing, esp in bad visibility, using GCA.

talk-through Facility whereby two mobile radio stations communicate via a base station.

tall aircraft One calling for LEW technique or experience.

tall-aircraft VASI See *T-VASI*.

Tally Ho Air-intercept code: target visually sighted. Normally followed by Heads Up or Pounce (DoD).

Talon(s) Tactical airborne Loran (system).

TALT Tactical arms limitation talks.

TAM Technical acknowledgement message (ICAO).

T&E Test and evaluation.

tandem actuator Has two pistons or jacks on same axis, with linear output.

tandem bicycle gear Two main landing gears on centreline.

tandem boost Rocket boost motor(s) mounted directly behind main vehicle, staging rearwards at burnout.

tandem-fan engine Gas turbine with single core driving front and rear fans on common shaft projecting ahead of engine; fans can have shared inlet for conventional flight or valved separate inlets and exits for V/STOL.

tandem main gears Two or more similar main gears in tandem on left and on right.

tandem rotors Helicopter lifted by two (usually identical but handed) rotors, designated front and rear.

tandem seating One behind other, in combat aircraft usually with rear seat at higher level.

tandem vehicle One assembled from portions, eg stages, assembled in tandem and staged axially, in contrast to strap-on, lateral or other configurations such as Space Shuttle.

tandem-wheel gear Two or more similar wheels in tandem on one leg, ie not a bogie.

tandem-wing aircraft One lifted by two wings in tandem, neither bearing more than 80% of total weight.

Tanegashima Principal Japanese satellite launch facility.

tangential ellipse See *Hohmann*.

tangential landing Running landing by rotorcraft or VTOL.

tangent modulus Slope of tangent to stress/strain curve at any point.

tangent of camber In aerofoil profile, line drawn tangent to mean camber line at intersection with leading edge, in modern wings occasionally negative (sloping up to front).

Tango Standard ground position marker in shape of T (RAF).

tank Container of all fuel, liquid, propellant, lube oil, hydraulic fluid, anticing fluid or toilet chemical, and often gun ammunition; not used for containers of breathing Lox, air-conditioning refrigerants, potable water or suppressant/extinguishant.

tankage Aggregate capacity of all tanks for particular fluid.

tank circuit Tuned RF circuit with capacitor and inductor in parallel.

tanker Aircraft equipped for inflight refuelling of others.

tank farm Cluster of storage tanks, usually for fuel, at airport.

tank sealer Various thermoplastic liquids, resistant to hydrocarbon fuels, sloshed inside integral tankage to seal all interior surface; today superior methods of wet assembly and multiple coatings are used.

Tans Tactical air navigation system; airborne computer storing many waypoints and fed by other inputs (eg Doppler, magnetic heading).

TAOC Tactical air operations centre (RAF/Army).

tap 1 To bleed; hence tapping, pipe for bleed air.
2 To cut threads in drilled hole; also tool for doing this.
3 Electrical power wire connected to main conductor at point along latter.
4 Engine throttle or power lever (colloq).

TAPA 3-D antenna pattern analyser (USAF ECM).

tape 1 Main meaning in aerospace is as medium for software, usually magnetic or punched-paper.
2 One form of CF prepreg, used for layups or moulding but seldom for filament winding.
3 Pinked-edge fabric strip used for surface finishing (surface *).

tape control Automatic control, eg of machine tool, by tape (1).

tape instrument Cockpit instrument whose presentation is based on linear tape driven over end spools, usually in conjunction with fixed and/or movable index pointers or bars. Usually vertical, as in VSFI.

tape mission Reconnaissance of Elint type in which digital (eg signal) or digitized pictorial information is stored on 7-track magnetic tape from which whole mission profile can be assigned to exact ground track, with each hostile emitter or other target assigned to precise location and timing.

taper For given wing section profile * equal in plan and thickness, usually defined as straight or compound; in some aerofoils * not equal in plan/ thickness so section profile changes.

tapered sheet Thickness varies (usually at uniform rate) along one axis.

taper ratio Normally defined as ratio of tip chord C_t to either root chord or equivalent centreline chord C_c.

taper reamer Used to smooth and true previously tapped hole.

taper tap Hand-turned tap (2) to initiate thread cutting.

Tapley meter Damped pendulum in heavy stable case whose limit of swing feeds record of instantaneous or max vehicle deceleration; not suitable for runway friction measures.

Taps Tercom aircraft-positioning system.

taps Throttles (colloq).

TAR 1 Terminal-area surveillance radar (ICAO).
2 Thrust auto reduce (SST).
3 Threat-avoidance receiver; passive ECM.

Taran 1 Tactical radar and navigation.
2 Test and repair as necessary.

TARC 1 Transport Aircraft Requirements Committee (UK 1956–62).
2 Tactical air reconnaissance center (USAF).

Tarcap Target combat aircraft practice (practise).

tare General adjective meaning unladen; uncommon in aerospace.

tare effect Forces and moments on tunnel model caused by support-structure interference.

TAREWS, Tarews Tactical air reconnaissance and electronic-warfare support; RPV (USAF).

target 1 Objective of air-combat mission, either in air or on surface.
2 Objective of intelligence or Elint activity.
3 Any true echo (blip) seen on radar, and object causing it.
4 Objective of any missile.
5 To insert position co-ordinates of fixed surface * into guidance software of ballistic or cruise missile; also called targeting.
6 Unpiloted (towed or RPV) aerodyne serving as target for friendly fire.

target acquisition Detection, identification and location in sufficient detail for effective employment of weapons.

target alert EFIS warning of future turbulence.

target allocation In air-defence weapon assignment, process of assigning particular target or airspace to particular interceptor or SAM unit (NESN).

target approach point Nav checkpoint, usually prominent land feature similar to initial point, over which final turn in to DZ or LZ is made.

target CAP Target combat air patrol; patrol of fighters over enemy target area to destroy hostile aircraft and cover friendly surface forces.

target crossing speed Relative lateral velocity or sightline spin (angular rate) of aerial target seen from interceptor.

target date Date on which particular planned event should take place.

target designation Marking or otherwise pointing out a target, or setting it into HUD or fire-control system.

target designation control Throttle thumbswitch for slewing sight (or HUD) brackets to contain a surface target.

target director post Positions friendly aircraft, in all weathers, over predetermined geographical positions, eg targets.

target discrimination See *discrimination.*
target dossier File of assembled intelligence information on target, normally including multisensor readouts and Elint.
target drone Pilotless target aircraft, today often an RPV.
target-following radar One locked on to target.
target indicator, TI Visible pyrotechnic, electronic homing beacon or other device airdropped on surface target.
targeting 1 To target (5).
2 Distribution of targets assigned to weapons, esp ICBMs and SLBMs.
target marker Visible pyrotechnic dropped on surface target, or aircraft dropping same.
target of opportunity 1 Target visible to a sensor or observer and within range of weapons and against which fire has not been scheduled or requested (DoD).
2 Target which appears during combat and which can be reached by weapons and against which fire has not been scheduled (NATO).
3 NW target detected after operation begins that should be attacked as soon as possible within time limits for co-ordination and warning friendly forces.
target pattern Flightpath of aircraft (meaning is normally in plan view) during attack phase (DoD).
target price That hoped to be achieved, eg in incentive-type contract.
target reverser Jet-engine (TJ or TF) thrust reverser comprising two deflectors (also called clamshells or buckets) which swing down to meet downstream of nozzle.
target tape Basic software for programming missile guidance of inertial and certain other species.
target tug Manned aircraft or RPV towing target (6) for live air/air or surface/air firing.
tarmac Colloq UK (esp non-aviation people) for paved apron; US = hardstand.
TARN Telegraphic auto routing network (NATO, Litton).
TARPS Tactical-aircraft reconnaissance-pod system (USN).
Tarsp Tactical air radar signal processor.
TAS 1 True airspeed.
2 Training aggressor squadron.
3 Target-acquisition system.
TASC Touch-activated screen control.
TASD Trajectory and signature data.
Tases Tactical airborne signal exploitation system.
Tasi True airspeed indicator (pronounced "tarzi").
task 1 Specific assignment to one air vehicle, or any other military force, normally involving operational or simulated mission or particular training exercise or programme.
2 Specific assignment to competitors in sailplane championships, eg speed round triangle, distance or declared goal, not disclosed prior to day.
tasked Required to fulfil certain tasks (1), either of variable operational or routine nature. Not available for other missions.
tasking Process of assigning tasks to available units or individual aircraft or crews to fulfil all mission requirements.
TASM 1 Total available seat-miles.
2 Tactical anti-ship missile.
Tasmo Tactical air support for maritime operations.
TASST Tentative airworthiness standards for SST (FAA).
Tasval Tac aircraft survivability against armour; post-Jaws programme by joint USA/USAF task force.
TAT 1 Total (or true) air temperature.
2 Tactical armament turret (helicopters).
TATF Terminal Automation Test Facility (FAA).
TATI Trim and tailplane incidence (indicator) (pronounced "tatty").
TATS Tactical aircraft training system.
TAV Transatmospheric vehicle.
TAW Thrust-augmented wing.
TAWC Tactical Air Warfare Center.
TAWDS Target-acquisition/weapon-delivery system, with Pave Mover.
tax Taxiway lights (ICAO).
taxi To move aircraft on surface (land or water) under own power.
taxi-holding position Designated point at which all vehicles may be required to hold to provide adequate clearance for arrivals/departures on runway.
taxiing Participle/gerund from taxi; note spelling.
taxilane Path on large apron or other paved area to be followed by nose gear, marked by continuous white line.
taxitrack Assigned taxiing route at land airfield, not necessarily paved. Most or all may be perimeter track.
taxiway Assigned taxiing route at land airfield, paved.
Taylor diagram Plot of dry and saturated adiabatic curves on axes of pressure and volume (reciprocal of density) showing loss of pitot pressure in moist air.
Taylor/Maccoll Original more exact solution for pressure over unyawed circular cone in supersonic flow (1932).
Taylor/Maclaurin Mathematical expansion of $f(x)$ for values near $x = 0$.
Taylor series Power series of $f(x)$ in ascending powers of $x - a$ where $f(x)$ and derivatives are continuous near $x = a$.
TB 1 Turn/bank.
2 Timebase.
3 Torpedo-bomber.

4 Terminal block.
5 Heavy bomber (USSR).

t_b Burn time.

TBA Total (helicopter rotor) blade area.

T-bar system See *T-VASI.*

TBC 1 Toss-bombing computer.
2 Tailored-bloom chaff.
3 Tactical bombing competition.

TBD To be determined.

TBE Timebase error.

TBH Truck-bed height.

TBI Turn/bank indicator = turn/slip.

TbIG Terbium iron garnet.

TBO Time between overhauls.

TBPA Torso back protective armour.

TBR Torpedo-bomber reconnaissance.

TBRP Timebase recurrence period.

TBT Turbine-bearing temperature (also used to mean turbine temperature).

tBX Air temperature (USSR).

TC 1 Toilet cart.
2 Time constant.
3 Twin co-ordinator.
4 Thermocouple.
5 Test-complete (verb).

T_c 1 Tropical continental air mass.
2 PN code bit length, also called chip width.
3 Adiabatic flame temperature of rocket.

t/c Thickness/chord ratio of aerofoil.

TCA 1 Terminal control area.
2 Télécommande automatique; IR/optical + wire guidance for missile. Operator merely keeps sight on target.
3 Time of closest approach.
4 Track crossing angle.
5 Temperature control amplifier.

TCAS Traffic alert and collision-avoidance system.

TCC 1 Thermal-control coating.
2 Thrust control computer.
3 Telecommunications center(s).

TCCF Tactical combat control facility (USAF).

TCH Threshold crossing height.

TCI Tape-controlled inspection.

TCJ Tactical communications jamming.

TCM Trim-control module.

TCMA Time-co-ordinated multiple access.

TCML Target co-ordinate map locator.

TCO 1 Total cost of ownership.
2 Tape-controlled oscillator.

TCP 1 Transfer-of-control point.
2 Transmission control program (Autodin).
3 Tri-cresyl phosphate.

TCR Terrain-closure rate.

TCS 1 Tilt-control switch (tilt-wing V/STOL).
2 TV camera set (F-14).
3 Tracking and communications subsystem (ACMI).
4 Telemetry and command system (satcom).
5 Turbulence control structure.
6 Tactical control squadron.

Tc/s Teracycles per second = THz.

TCSS Terminal coms switching system.

TCT 1 Transverse-current tube.
2 Tactical computer terminal.
3 Target-centred tracker.

TCTO Time-compliance technical order.

TCU 1 Towering cumulus (alternatively TCu).
2 Tracking control unit.
3 Thermal cueing unit.

TCV Terminal-configured vehicle.

TCX Transfer-of-control cancellation message.

TD 1 Touchdown.
2 Transposition docking.
3 Tunnel diode.
4 Time delay.
5 Time duplex.
6 Test directive.

T_d, Td Dewpoint temperature.

t_d Ignition delay time (rocket).

TDA Tunnel-diode amplifier.

TD&E 1 Transposition, docking and LM ejection.
2 Tactics development and evaluation.

TDAP Touchdown aim point.

TDAS Test-data acquisition system.

TDC 1 Top dead centre.
2 Through-deck cruiser (for STOVLs).
3 Target designator (or designation) control.

TDCS Traffic data collection system.

TDF Tactical digital facsimile.

TDG 1 Triggered discharge gauge.
2 Two-displacement gyro.

TDI 1 Triple display indicator.
2 Tapped delay input.

TDL Tapped delay line.

TDM Time-division multiplex.

TDMA Time-division multiple access.

TDO Tornado.

TDoA Time difference of arrival.

TDR Technical despatch reliability.

TDRS Tracking and data-relay satellite; TDRSS adds system (NASA).

TDS 1 Tactical data system.
2 Time/distance/speed scale.
3 Thermal diffuse scattering.
4 Training depot station (RFC).
5 Tactical drone squadron.

TDV Technology development vehicle.

TDY Temporary duty.

TDZ Touchdown zone.

TDZL Touchdown-zone lights (FAA).

TDZ marking White axial stripe on each side of runway.

TE, t.e. 1 Trailing edge.
2 Tactical evaluation.

TEA 1 Tri-ethyl alcohol, hypergolic igniter.
2 Transferred-electron amplifier.

teaming agreement 1 Inter-company agreement for marketing purposes, not involving licensing or co-production.
2 Now coming to mean any inter-company

agreement to assist penetration of markets.

teaming down Getting into partnership with smaller companies.

teaming up Getting into partnership with one or more giant companies, one of which may be eventual prime.

teampack Packaged for carrying across rough terrain by team of 2 to 8 personnel.

Teams Tactical evaluation and monitoring system (Northrop).

tearaway connector Umbilical pull-off coupling.

teardown Dismantling into component parts.

teardrop Standard procedure flying pattern similar to racetrack but with one end large-radius and the other small.

teardrop canopy Of smoothly streamlined shape, usually moulded from one transparent sheet.

tearoff cap Lightly sewn fabric parachute cover torn off pack by static line.

tease Faulty operation of circuit-breaker in which snap-action is absent; hence *-free.

TEB Tri-ethyl borane.

TEC 1 Trans-Earth coast.
2 Thermal (or thermoelectric) energy converter.

tech mod Technology modernization.

Technamation Technical animation, methods for training and educational displays giving illusion of motion, eg flow through pipes, rotation of shafts, etc.

technical delay Delay ascribed to fault in hardware, lasting longer than 5 (sometimes 15) min.

technical despatch reliability Percentage of scheduled flights which are unaffected by any prior technical fault, but ignoring delays due to other causes.

technical electrics All services other than commercial electrics.

technically closed Problem has been solved.

technical stop Stop by commercial transport at airport for reasons other than traffic or refuelling; not shown in timetable.

technical survey Inspection for monitoring (bugging) systems (DoD).

Techroll Patented (CSD) configuration for vectoring nozzle of solid-propellant rocket motor in which nozzle drive forces are reduced by fluid-filled constant-volume surround sealed by flex diaphragm.

TECR Technical reason (ICAO).

Tecstat Nazionale Associazione Tecnici di Stato (I).

TED 1 Transferred-electron device.
2 Tactical (or threat) evaluation display.

TEDS Tactical expendable drone system, for ECM saturation jamming.

TEE Tubular extendible element, produced by unrolling steel tape.

tee Air/ground wind-direction indicator in shape of large T in white, either placed on ground and occasionally rotated or pivoted to base (and in a few cases moved by weathercocking). Cross-piece is at downwind end of upright.

tee connector T-shaped plumbing connector.

tee gearbox One rotary shaft geared to another at 90° at a point other than one end.

tee junction T-shaped connection of two microwave waveguides.

teetering rotor Helicopter main rotor with two blades freely pivoted as one unit about horizontal axis transverse to line joining blade tips.

TEF Total environment facility, for processing reconnaissance data.

Teflon Family name (du Pont) for large family of fluorocarbon-resin rubbers and plastics.

tehp Thrust equivalent horsepower, normally same as ehp.

TEI Trans-Earth insertion (or injection).

tektites Small glassy bodies unrelated to surrounding Earth and believed of extraterrestrial origin.

TEL 1 Tetraethyl lead.
2 Telebrief.
3 Telephonic (ICAO).
4 Transporter/erector/launcher.

Telar TEL (4) and radar, on one vehicle.

TELATS Tactical electronic locating and targeting system (USAF).

telebrief Direct telephone com between ground personnel, eg air controller or ground crew, and military aircrew seated in aircraft.

telecommunications Transmission, emission or reception of signs, signals, writing, images or sounds by wire, radio, visual or other EM system; abb telecom.

teleconference Conference between participants linked by telecom system.

telegraph Telecom using succession of identical electrical pulses.

telemetry Transmission of real-time data by radio link, eg from missile to ground station; today invariably digital and important in RPVs and unmanned reconnaissance systems. Data can be pressure, velocity, surface angular position or any other instrument output, or any form of reconnaissance output. Telemeter is verb; use as noun arch. Noun is * system or telemetering system.

telephone Transmission of sounds, signals or images by wire or other discrete-path link, eg microwave beam or optical link using free coherent beam or fibres.

telephotography Photography of distant objects on Earth.

telephotometer Visibility meter.

teleprinter Telegraphy with keyed input and printed written output.

teleprocessing EDP (1) by computer fed by telecom system.

teleran Television radar air navigation; use of ground radar to feed airborne TV display.

telescope 1 In astronomy, instrument for collecting EM radiation (esp light, radio, IR and X-ray)

from extraterrestrial sources.
2 To reduce overall dimensions by folding or, esp, linear retraction, eg helicopter rotor.
3 To reduce propeller diameter by cropping tips.

telescoped ammunition Projectile is carried largely within case, reducing length and increasing propellant energy per unit overall volume.

telescope gauge Precision rod sliding in tube and locked to measured dimension, eg hole diameter, subsequently measured by micrometer.

telescoping To telescope (2, 3).

telescramble To render telecom, usually telephone, conversation unintelligible by scrambling.

teletype US term for teleprinter; hence *-writer; often capital T (regd name).

television Transmission and reception of real-time imagery by electronic means. Link usually by radio but may be any other telecom form, and imagery usually keyed to sound channel, entire received signal also being recordable by receiver; abb TV.

television command See *television guidance*.

television guidance Command guidance by radio link sending steering commands from operator watching TV picture taken by camera in nose of vehicle.

telint Telemetry intelligence.

telling See *track telling*.

TELS Turbine-engine loads simulator.

TEM 1 Transmission electron microscope (or microscopy).
2 Technical error message (ICAO).
3 Thermally expanded metal.

Temp Test & evaluation master plan (AFSC).

temper Degree of hardness introduced to metal by heat treatment, cold-working or other process.

temperature 1 Property of material systems, commonly called intensity of heat, determining whether they are in thermodynamic equilibrium. Normally a measure of translation kinetic energy of atoms or molecules. SI unit is K, not necessarily written °K. Specified as reported *, a local actual value, or as forecast * or declared *, read from statistical tables.

temperature accountability All factors in aircraft design, operation and certification determined by reduction in propulsive thrust and wing lift caused by increase in atmospheric temperature.

temperature coefficient Rates of change of variable per unit change in temperature.

temperature coefficient of pressure For a given ratio of solid-rocket-motor surface to throat area, $\pi_K = \delta I_n P_c / \delta T$ where I_n is initial specific impulse, P_c is chamber pressure and T is temperature.

temperature correction Correction applied to bring instrument reading to STP conditions.

temperature gradient Rate of change of temperature with unit distance through material in direction normal to isotherm surfaces, esp rate of change of temperature in atmosphere with unit increase in height.

temperature inversion See *inversion*.

temperature recovery factor Usually, equilibrium temperature of solid surface in high-speed flow, varying according to turbulence of boundary layer; usually T_w and given by several formulae.

temperature stress Stress caused by temperature, esp changes in temperature between different parts of body.

temperature traverse Series of temperature (usually stagnation/total temperature) measures taken either over area or along straight line perpendicular to fluid flow, eg at exit from combustion chamber.

template Simple pattern, usually planar, either cut to shape of a part or with shape and dimensions marked on surface, used as guide in repeated marking out of desired shape. In UK occasionally written templet.

Temple-Yarwood Formula for pressure coefficient at low Mach as function of critical Mach M_c: $C_{po} = 1 - 0{\cdot}522\,(1 + 0{\cdot}2\,M_c^2)^3 / M_c^2\,(1 - 0{\cdot}05\,M_c^2)^2$.

Tempo Technical military planning operation.

temporary flight restriction Order prohibiting unauthorized aircraft from airspace above major accident or natural disaster.

TEMS Turbine-engine monitoring system.

Tencap Tactical exploitation of national capabilities (US).

tendency Variation with respect to time, esp change in atmospheric pressure in 3 h prior to an observation.

Tenley Secure voice system for Tri-Tac, for NSA (US).

Tensabarrier Seat-belt-type barrier to control people movements at airport (BAA); quickly closed or opened.

tensile strength Tensile force per unit cross-section required to cause rupture.

tensile stress That produced by two external forces acting in direct opposition tending to increase distance between their points of application.

tensiometer Measures actual tensile stress in flexible cable, such as flight-control circuit; can be used for flexible bracing wires.

tensioner Self-contained mechanism inserted into cable carrying tensile load, eg in manual flight-control system, which maintains desired (usually constant) tension throughout; often in form of spring-loaded quadrants.

tension field Surface within which tensile force acts, with direction parallel to forces. Hence * beam, eg wing spar, within which * acts diagonally in vertical plane, tending to pull upper and lower booms together.

tension regulator See *tensioner*.

TEO Transferred-electron oscillator.

TEOSS, Teoss Tactical emitter operational support system; EW locator (USAF).

TEP Tactical electronic plot.

tephigram Graphical plot of atmospheric temperature and entropy on grid of intersecting isothermals and isentropic lines against vertical axis of height (decreasing pressure levels); also written Tϕ gram, for temperature and entropy. Pronounced tee-fie-gram.

Tepigen Television picture generation (or generator).

Tepop Tracking-error propagation and orbit prediction.

TER 1 Triple ejector rack.
2 Total-energy requirements.

tera Prefix = $\times 10^{12}$.

Tercom Navigation technique in which vehicle guidance memory compares profiles of terrain below, sensed as unique sequences of digital height measures, with those already stored; hence, each match with terrain increases accuracy of refinement of basic (eg INS) guidance, whereas most systems degrade with time. From terrain çomparison or terrain contour-matching.

térébenthine Refined turpentine used as rocket fuel (F).

Terec Tactical electronic reconnaissance (Litton).

TERLS Thumba Equatorial Rocket Launching Station (UN facility in India).

terminal 1 Building, either discrete or dispersed, at airport which links airside with landside and through which all passenger traffic passes, or through which all cargo traffic passes. Very seldom is there one * for pax and cargo, and at many major hub airports each airline or group of airlines has its own *.
2 Downtown (city-centre) building at which passengers may check in for flights and from which they may be conveyed by public transport with baggage already checked to pass straight through a * (1).
3 Normal general meaning of being connecting point through which flow passes into or out of a system in electric circuits, air traffic, data processing and many other disciplines.
4 Final portion of flight of missile between midcourse and target.

terminal airport That at which flight terminates. Also correctly used as point at which particular item of traffic leaves flight.

terminal alternate Alternate named in flightplan as second-choice terminal (3) if for any reason normal destination unattainable.

terminal area See *terminal control area.*

terminal ballistics Behaviour of projectiles at impact with, and penetration of, target.

terminal building See *terminal (1).*

terminal clearance capacity Max amount of cargo or personnel that can be moved through terminal (3) daily (DoD).

terminal communications Com services or facilities within terminal control area other than those used for approach or ground movement.

terminal control area Airspace control area, or portion thereof, normally at confluence of airways or air traffic service routes in vicinity of one or more airports. Extends from surface or from higher FL to specified FL and within it all aircraft are subject to specified rules and requirements. Often a TMA.

terminal count Final portion of countdown ending in liftoff.

terminal guidance 1 That governing trajectory from end of midcourse to impact with or detonation beside target.
2 Electronic, mechanical, visual or other assistance given aircraft pilot to facilitate arrival at, operation within or over, or departure from, air landing or airdrop facility (DoD).

terminal manoeuvring area, TMA Controlled airspace region surrounding busiest airports (usually large city with many airfields); normally permanent IFR with other traffic by dispensation.

terminal nosedive Vertical dive at full power (arch).

terminal phase 1 Portion of trajectory of ballistic missile between re-entry and target. Also other meanings in particular space missions, eg LM/CM redocking in lunar orbit.
2 Final part of missile trajectory after missile's own seeker has detected and locked on to target.

terminal radar service area Primarily an electronic environment, not extending below a floor at medium FL, providing radar vectoring and sequencing of all IFR and VFR aircraft landing at primary airport, separation of all aircraft in TRSA service area, and advisories on all unidentified aircraft on a workload-permitting basis.

terminal velocity 1 Highest speed of which aeroplane (rarely, other aircraft) is capable, reached at end of infinitely long vertical dive at full power through uniform atmosphere (suggest arch).
2 Ultimate speed reached by inert body in free fall through particular prescribed atmosphere.

terminal VOR VOR located at or near airport at which particular flight terminates and specified as navaid used in final approach clearance.

terminating bar lights Red lights between final-approach lights and wing-bar lights.

terminator 1 Solid-propellant rocket subsystem comprising signal input, squib or detonator and blow-out port(s) for causing immediate thrust termination.
2 Boundary between sunlit and dark sides of planetary body, eg Earth, Moon.

Terms Terminal management system.

Tern, TERN Terminal and en route nav.

ternary Device capable of three states, normally called 0, 1, x.

terne plate US term for tinned, or lead-coated, mild steel sheet (not plate).

Terp Turbine-engine reliability programme.

terprom Terrain profile matching, usually similar in principle to tercom.

terps Terminal instrument procedures (US).

terrain-avoidance system System, usually radar-based, providing pilot or other crew member with situation display of ground or obstacles ahead which project above either horizontal plane parallel to aircraft or plane containing aircraft pitch/roll axes so that pilot can manoeuvre aircraft laterally to avoid obstruction.

terrain-clearance system System, usually radar-based, providing pilot or autopilot with climb/dive signals such that aircraft maintains preset flight level while clearing peaks within selected height in vertical plane through flight vector. Unlike terrain-following, after each protruding peak aircraft levels out at prescribed FL.

terrain comparison See *Tercom.*

terrain database Comprises computer-stored 2-D grid of ground spot heights plus land culture information.

terrain-following system System, usually radar-based, which provides pilot or autopilot with climb/dive signals such that aircraft will maintain as closely as possible a selected lo height above ground contour in vertical plane through flight vector. In effect system projects radar ski-toe locus which slides over terrain ahead to give minimum safe clearance.

terrain masking Obscuration of aerial and other targets by hills or buildings, esp as seen at acute grazing angles by overland downlook radar.

terrain orientation Holding topographical map so that aircraft heading is at top of sheet or folded sheet.

terrain profile Outline of profile of ground surface, usually with vertical scale × 5 (sometimes × 10) published on approach chart or other documents to assist pilots.

terrain-profile recorder Airborne instrument, recording sensitive radar or laser altimeter, giving hard-copy readout for mapping and surveying.

terrestrial radiation Earth IR radiation; also called eradiation.

terrestrial reference guidance Any method providing steering intelligence from characteristics (usually stored as quantified digital measures) of surface being overflown, thereby achieving flight along predetermined path. One example is Tercom.

terrestrial refraction Refraction observed in light from source within Earth atmosphere; thus caused only by inhomogeneities of atmosphere itself.

terrestrial scintillation Generalized term for scintillation effects observed in light from sources within Earth atmosphere; also called atmospheric boil, optical haze and shimmer.

Tersi Series of EW jamming and aerial-pattern simulators.

tertiary airflow That passing through tertiary holes.

tertiary holes Apertures in gas-turbine flame tube or combustor downstream of secondary holes admitting air purely for dilution and cooling purposes to achieve desired uniform gas temperature across chamber exit plane.

TES Test and evaluation squadron.

tesla SI unit of magnetic flux density; 1T = 1 $Wb/m^2 = 10^4G$.

tesla coil Induction coil without iron core normally giving HF output.

testbed Mounting, either on ground or in form of aircraft, upon which item can be mounted or installed for test purposes. When an aircraft may be, but not necessarily, prefixed by "flight" or "flying". In US normally called test stand.

test cell Usually horizontal test stand, eg for rocket motor, surrounded except on operative side by protective shelter giving protection from weather and limited protection externally from explosion inside.

test chamber Environmentally controlled sealed chamber in which test can take place; eg can simulate stratosphere or hard vacuum with space solar radiation.

test clip Spring-steel clip for quick electrical connections to terminals.

test club Club propeller making no pretence at aerofoil shape but merely having stubby projecting arms in correct balance.

test diamond Region in supersonic tunnel working section within which model is placed and within which flow conditions are essentially constant at any one time.

test firing Firing of rocket, of any type, while mounted on testbed (test stand).

test flight Flight by aircraft, winged spacecraft or cruise missile for purpose of evaluating or measuring performance, handling or system operation.

test pattern Geometric pattern used in testing electronic displays.

test rig See *rig (3).*

test section 1 Tunnel working section.
2 Special glove aerofoil carried on flying testbed.

test set Packaged equipment, either versatile or for testing specific system, of electronic, hydraulic, pneumatic, microwave/RF or any other character, which can readily be brought to aircraft or have device brought to it.

test vehicle Air vehicle, normally unmanned, built to test major element of its design, construction or systems and thus prove new concept.

TET 1 Turbine-entry temperature.
2 Tolerable exposure time, esp with reference to aircraft in high-speed flight in gusts.

TETA Triethylene-triamine.

Tetra Turbine-engine transient response analyser (EDP code).

tetraethyl lead Liquid added to some petrols (gasolines) to improve resistance to detonation (anti-knock value or fuel grade); base material is $Pb(C_2H_5)_4$. Resulting fuel is called leaded.

tetrode Thermionic valve (tube) containing cathode, plate and two other electrodes.

Tetwog Turbine-engine testing working group.

Tewa, TEWA Threat-evaluation and weapon assignment.

TEWS, Tews Tactical electronic-warfare suite (or system).

Textolite Obsolete "plastic"; hot pressed canvas/resin laminates.

textured visuals Visuals whose CGI tones are not plain grey shades but have texture corresponding to real life, giving enhanced spatial cues.

TF 1 Trip fuel.
2 US military engine designation prefix: turbofan.
3 US military aircraft designation prefix: fighter trainer; dual version of established fighter.
4 Turbine fuel is available.
5 Toroidal field.
6 Technology forecasting.
7 Torpedo (armed) fighter (UK, pre-1953).
8 Terrain-following.
9 Thin film.

Tf Flexible temperature (flexible TO rating).

TFC 1 Total final consumption.
2 Tactical fusion centre (AAFCE).

tfc Traffic (FAA).

TFE 1 Terrain-following E-scope.
2 Thermionic fuel element.
3 Tetrafluoroethylene, major additive to magnesium IRCM.

TFEC Tactical fighter electronic combat.

TFEL Thin-film electroluminescent display, a CRT alternative.

TFG 1 Thrust-floated gyro.
2 Tactical fighter group.

TFM Tactical flight management system.

TFMS Tactical frequency-management system.

TFPA Torso front protective armour.

TFR 1 Terrain-following radar.
2 Total fuel remaining.

TFSB Tungsten-filament seven-bar display.

TFSF Time to first system failure.

TFT 1 Thin-film transistor.
2 Trim for take-off.

TFTS Tactical fighter training squadron.

TFW Tactical fighter wing.

TFWC Tactical Fighter Weapons Center.

TG Transportgeschwader (G).

TGB Transfer gearbox.

TGC Travel-group charter (US term, basically = UK ABC).

TGG Third-generation gyro (Northrop).

TGL Touch-and-go landing (ICAO).

TGP Twin-gyro platform.

TGS 1 Triglycine sulphate; pyroelectric IR detector material.
2 Taxiing guidance system (ICAO).

TGSM Terminally guided sub-munition.

TGT 1 Turbine gas temperature.
2 Target.

Tgt Opp Target of opportunity.

TGW Terminally guided weapon.

T_H Total temperature.

THAD Terminal-homing accuracy demonstrator.

THAR Tyre height above runway.

THAWS, Thaws Tactical homing and warning system (RCA).

theatre Geographical area of military operations in which commander of unified or specified command has complete responsibility; today used as adjective, often synonymous with tactical.

theodolite Optical sight or telescope whose az/el can be accurately read off angular scales.

theoretical gravity That at Earth's surface if Earth's mass was reshaped as perfect sphere.

theoretical thrust coefficient A thrust/time value for solid-propellant rockets computed from large equation involving an effective value and assumed conditions for various areas and pressures. Symbol C°_{I}.

therapeutic adaptor Coupled to continuous-flow oxygen mask, approx triples flow rate; used for pax with respiratory or heart problem.

therapeutic oxygen Administered primarily to treat ailment, eg pulmonary or cardiac faults.

thermal 1 Local column of rising air in atmosphere, usually caused by surface heat source.
2 To use (1) as energy input for soaring flight.

thermal anticing Anticing by heating affected surface.

thermal barrier Notional barrier to further increase in some variable, eg flight speed in atmosphere or turbine entry temperature in engines, caused by inability of materials to withstand increased temperatures. Continually being eroded by new refractory materials.

thermal battery Electrical cell stored inactive and activated chemically for one-shot high-power output.

thermal blooming See *blooming*.

thermal coefficient of expansion Increase of (1) length per unit length, or (2) area per unit area, or (3) volume per unit volume, caused by rise in temperature of 1°C (often defined as from 0° to 1°C, or from 15° to 16°C).

thermal conductivity Time rate of flow of heat through unit area normal to temperature gradient per unit T° difference. Symbol λ or k, rate given by Fourier's law. SI unit is $Wm^{-1} K^{-1}$ or $Ns^{-1} K^{-1}$.

thermal cueing unit Adjunct to FLIR-based attack system which puts marker boxes round all likely surface targets, picking them according to their high temperature, and which automatically feeds target co-ordinates to the attack system if any of these boxes is touched by the pilot on the HDD touch display.

thermal cycling Oscillating between low and high temperatures.

thermal de-icing De-icing by heating affected surface.

thermal diffusivity Measure of transfer of heat by diffusion analogous to viscous motion; symbol $\alpha = \lambda/\rho C_p$.

thermal efficiency Basic efficiency parameter of heat engine, defined as percentage ratio of work done in given time to mechanical equivalent of heat energy burned in fuel supplied in same period.

thermal emission EM radiation solely due to body's temperature (which if hot enough contains strong visible radiation).

thermal excitation Acquisition of excess energy by atoms or molecules as result of collisions.

thermal expansion Increase in dimensions caused by increase in temperature.

thermal exposure Calories/cm^2 received by normal surface in course of complete NW detonation (DoD).

thermal fatigue Mechanical fatigue caused by stresses repeatedly imposed by thermal cycling.

thermal gradient See *temperature gradient*.

thermal gradiometer Airborne instrument for detecting thermals by thermocouples on wingtips which, in presence of temperature difference, sends electrical signal to cockpit indicator.

thermal heating Tautological; kinetic heating is meant.

thermal imagery Produced by measuring and electronically recording thermal radiation from objects (NATO). Normally IR wavelengths only are implied. Hence thermal imaging, to produce pictorial displays or printouts showing variation of temperature over field of view.

thermal instability Any combination of temperature gradient, thermal conductivity and viscosity resulting in convective currents, eg wind in atmosphere.

thermal lift 1 Lift due to thermal (1).
2 Lift imparted to air mass because of greater density of cold surrounding air, not quite synonymous with (1).

thermal load Imprecise term usually meaning temperature gradient or temperature stress.

thermally expanded metal Fabrication of parts from aluminium alloy sheets rolled together with intervening patterns of "ink"; the latter prevents the sheets bonding and, on subsequent heating, expands to force the unbonded parts to fit a mould.

thermal neutron Neutron slowed, eg in moderator, to thermal equilibrium with surroundings at about 2,200 m/s (so-called slow neutron).

thermal noise RF noise caused by thermal agitation in dissipative body; also called Johnson noise.

thermal protection Protection against kinetic heating during atmospheric entry (re-entry) of spacecraft structure, RV or other body, esp one intended for repeated space missions.

thermal pulse Total IR emission from NW detonation, or plot of IR flux against time during complete burst and fireball climb.

thermal radiation 1 See *thermal emission*.
2 Total heat and light radiation produced by NW detonation (DoD).

thermal relief valve Safety valve in fluid system to guard against excessive pressure caused by overheating.

thermal runaway 1 Fault condition with element of danger affecting Ni/Cd batteries characterized by particular cells losing resistance (possibly because of high temperature) and thus taking increased current, lowering resistance still further in chain-reactive process.
2 Similar divergent overheating in current-carrying transistor.

thermal shock Severe mechanical stress resulting from sudden extreme temperature gradient.

thermal soaring See *soaring*.

thermal stress See *temperature stress*.

thermal switch Switch activated by temperature difference or particular temperature.

thermal thicket Flight conditions in which kinetic heating (or other thermal problems) is a factor to be considered but does not yet impose a thermal barrier (colloq).

thermal wind Notional vector difference between winds at different heights caused by horizontal variation of atmospheric temp and hence pressure at all upper heights (note: not pressure surfaces).

thermal X-rays EM radiation, mainly in soft (low-energy) X-ray region, emitted by extremely hot NW debris.

thermel Any device based on Seebeck, eg thermocouple, thermopile.

thermie Non-SI unit of work (mechanical energy); 1 th = 4·1855 MJ.

thermionic Involving electrons emitted from hot bodies.

thermionic converter Electric generator powered by hot emitter and cold collector.

thermionic rectifier Depends on unidirectional electron flow from cathode to anode.

thermionic tube See *thermionic valve*.

thermionic valve Evacuated capsule, usually glass, containing heated cathode emitting electrons attracted to anode, usually via one or more intervening control electrodes usually called grids. In US called vacuum tube.

thermistor Protective resistor based on semiconductor having high negative temperature coefficient of resistance.

thermite Mixture of finely divided magnesium

and iron oxide used as heat source in welding and as incendiary filling; originally spelled with capital T.

thermochemistry Branch of chemistry concerned with thermally induced reactions and relationship between chemical changes and heat.

thermochromic tube CRT with phosphor replaced by heat-sensitive layer.

thermocline Sharp submarine temperature gradient.

thermocouple Instrument based on Seebeck effect which measures temperature difference between pair of dissimilar-metal junctions; much used for high-temperature measures using refractory metals, and in common copper/constantan junction at room temperature, eg for met observation.

thermodynamics Science based upon heat flow and temperature changes, esp those in moving fluids.

thermodynamic cycle Operating cycle of any heat engine. In some, eg virtually all PE (4), one parcel of fluid at a time goes through complete ** in same enclosed (usually variable-size) volume; in others, eg gas turbines, continuous flow of fluid goes through ** by passing from one part of device to another, each component handling only one part of **.

thermodynamic efficiency See *thermal efficiency*.

thermodynamic energy equations Exact expressions of variation of pressure, volume and temperature in reversible processes in perfect gas.

thermodynamic equilibrium Time-invariant state in which all processes are balanced by reverse process and entropy production vanishes.

thermoelectric cooling Local cooling using Peltier and cooling "hot" junction; "cold" junction then falls to desired level at –20 to –30°C.

thermoelectric generator Electric generator based on thermocouples using Seebeck, Thompson, Kelvin or Peltier effects; common spacecraft systems use nuclear reactor or radioisotope to heat junction often based on Ge/Si alloy.

thermogram 1 Single-line output of traditional thermograph.
2 Pictorial output of thermographic camera.

thermograph Recording thermometer using pen/chart or light-spot trace on film. Output is a thermogram.

thermographic camera IR camera, usually of IRLS type.

thermography Translation of temperature changes in a scanned scene into visual picture, today important in military and civil aerial reconnaissance, industrial process control, medicine and many other fields. Either black/white (black = cold, white = hot) or colour.

thermohydrometer Hydrometer with thermometer, giving two chart outputs.

thermometer Instrument for measuring temperature.

thermometer screen Louvred box screening thermometer from direct sunlight; usually contains other met instruments and in US called instrument shelter.

thermonuclear Processes in which extremely high temperatures are used to initiate fusion of light nuclei.

thermonuclear weapon Device in which fission provides high temperature for fusion of nuclei of hydrogen isotopes (deuterium, tritium); not normally abb TNW, though this has been used.

thermopile Thermoelectric generator comprising stack of thermocouples.

thermoplastic recording Patented (GE) process for recording sound or video signals via electron beam direct on thermoplastic layer heated by microscopic currents induced in underlying conductive layer.

thermoplastics Large class of synthetic polymers which may be repeatedly softened and remoulded by heating.

thermosetting plastics Synthetic polymers that are chemically changed irreversibly by heat, generally setting hard.

thermosphere Outermost region of atmosphere from top of mesosphere outwards into space, characterized by more or less steadily increasing temperature with distance from Earth.

thermostat Device for maintaining a desired temperature by taking action at preset limits of low and high temperature.

thermotropic model Atmosphere used in forecasting one temperature and one pressure surface.

theta Greek letter θ, used for many parameters, including pitch angle (thus, $\dot{\theta}$ = pitch rate) and azimuth (hence Rθ).

thickened fuel Aircraft fuel designed to resist fine dispersion and instead to break down in crash into globules with near-zero surrounding vapour; generally synonymous with gelled fuel.

thick-film Very diverse technology of electronics involving processing, high-current devices, current-generation (inc solar cells) and many other topics, mainly using insulating substrates but often with semiconductor layer.

thickness Of wing, maximum straight-line distance from external skin of upper surface to external skin of lower surface measured in plane of aerofoil profile and perpendicular to chord line.

thickness/chord Ratio of thickness to chord of wing, both measured in plane of aerofoil profile at same station.

thickness distance Distance aft of leading edge of max thickness of supersonic rhomboidal or double-wedge wing, expressed as % chord.

thickness gauge See *feeler gauge*.

thickness lines Join points on chart where vertical distance between pressure surfaces is everywhere same.

thickness ratio Wing t/c ratio.

thimble 1 Pear-shaped eye around which end of control cable is spliced.
2 Ratched turning knob of hand micrometer.
3 Pimple-like radome, especially on nose.

thin-case bomb Conventional bomb for blast effect against soft target. Also called light-case (UK, WW2).

thindown Progressive energy loss by primary cosmic rays in ionizing surrounding medium.

thin-film circuit Electrical or electronic circuit formed by depositing thin film on (usually insulating) substrate; normal manufacturing methods are vacuum deposition and cathode sputtering. Films may be conductive, semiconductor or insulating.

thin-film lubrication Imperfect, with occasional metal/metal contact.

thin-film transistor IGFET constructed by evaporating on to insulating substrate metal electrodes, semiconductor layer(s), insulating upper layer and metallic gate; abb TFT.

think tank Centralized group of people normally working for government or large corporation engaged in futures, forecasting, ultra-new technologies and other disciplines calling for visionary judgement.

thinner(s) Solvents for paint, dope and other liquids to reduce viscosity.

thin route Airline route, usually intercontinental, offering only modest traffic.

third-angle projection Convention in engineering drawing in which front view, side elevation and plan each show face nearest to it in adjacent view; traditional US arrangement becoming standard in European aerospace.

third-level carrier Generalized term for "third tier" of scheduled airline operations, also called feeder or commuter and often of radial nature serving single city hub. No clear demarcation separating from second-level (local-service or regional).

thixotropic Becoming liquid when vibrated or stirred, setting on standing for a period.

Thornel Tradename for carbon and graphite fibres.

thou Thousandth of an inch, 25·4 μ.

THP 1 Thrust horsepower, often thp.
2 Through-hole plated.
3 Turbo-hydraulic pump.

THR Threshold.

thread chaser Tool for removing contamination, eg paint or dirt, from thread.

thread gauge Hand gauge with many specimen threads, one of which is matched with part.

threading the needle Process of accurately flying through a small gate in airspace, eg in setting a speed record (colloq).

thread insert Steel helix screwed into soft (eg aluminium) hole.

threat Hostile anti-aircraft defences, especially air-defence radars, SAM systems, AAA and fighters.

threat evaluation Process of detecting, analysing and classifying hostile offensive systems, either in warning of attack or during penetration of hostile territory when systems are surface-to-air.

threat library Numerical characteristics of hostile threats, especially EM emitters, stored in friendly computer (eg of RWR receiver).

threat simulation Simulation of hostile offensive systems, eg by add-ons to RPV target to include emissions, dispensed payloads and jamming.

three-axis autopilot Has authority in pitch, roll and yaw.

3-bar VASI Comprises VASI plus additional pair of upwind (210 m, 700 ft) wing bars symmetrically disposed about centreline each having at least two light units, for use by LEW aircraft.

three-body problem Mechanics of motion of small body in gravity of two others.

three-control aeroplane Conventional, with separate pilot input for each rotational axis.

3-D cam Cam whose profile varies across its width and which moves axially as well as rotationally.

3-D flow Fluid flow which cannot be represented fully in 2-D, eg flow over a real wing.

3-D radar Radar enabling position of target to be determined in 3-D space, either by Cartesian methods or, more often, by az/el plus slant range.

3-D tool Jig or fixture used to define exact shape of finished assembly, eg complex hydraulic piping or wiring loom.

three greens Landing gear is down and locked (colloq).

three-moment equation For solving bending moments and other loads at ends of two adjoining spans of continuous beam.

3-P Planning, production, progress.

three-phase current Alternating electrical current made up of three phases, each with vector separation of 120°; carried by triple wire.

three-phase equilibrium See *triple point*.

3ϕ Three-phase current.

three-pointer altimeter Dial instrument with short needle for thousands (ft or m), mid-length for hundreds and longest for tens.

three-point landing Correctly judged landing by tailwheel-type aeroplane in which main and tail wheels touch ground simultaneously with wing stalled.

three-point mooring Mooring for aerostat in which three lines are run (often from single point, eg nose of airship) to three ground anchors, usually at apices of equilateral triangle.

3-pole switch Opens and closes three conductors or circuits.

three-shaft Gas turbine has LP, IP and HP shaft systems.

3-view drawing GA drawing, normally showing

elevation (left side), front and plan.

3-way switch Routes input along either of two outputs.

3 wire Target of most carrier arrested landings, No 3 wire; hence ** landing.

3-wire circuit Neutral wire between two outer wires, latter having potential difference from neutral equal to half that between them.

threshold 1 Beginning of usable portion of runway, ie downwind end.
2 In automatic control systems, point at which response is first noticed, usually defined in terms of input displacement (see * level).
3 Flight condition when fixed-wing aerodyne is on point of stall.
4 Point at which sound just becomes audible (* of audibility or of hearing), normally 2×10^{-5} N/m^2.
5 EAS giving lowest comfortable cruising, possibly higher than that for min fuel.

threshold contrast Smallest contrast in luminance visible under given conditions.

threshold curve Plot of sound frequency against noise level in dB (or other noise measure) just audible against quiet background, eg anechoic chamber.

threshold displacement Linear distance between end of full-strength runway pavement and displaced threshold, with latter shown on airfield charts as white bar across runway crossed by narrow black line, and expressed as minus quantity in certain navaid figures, eg Vorloc II = −380 ft.

threshold dose Min quantity of radiation producing detectable biological effect.

threshold illuminance Min value of illuminance eye can detect under given dark adaptation and target size; also called flux-density threshold.

threshold level Threshold (2), esp in rate gyro developing electrical output as function of rate of turn; that angular rate after rotational acceleration from rest at which there is first indication of output, or change in output; normal unit is °/s $\times 10^{-6}$.

threshold lights If fitted, bidirectional units, showing green towards approach and red towards runway, in continuous row across threshold (rare at displaced threshold).

threshold limit value Average concentration of toxic gas normal person can withstand 8 h per day, 5 days per week.

threshold marking For simple runway, runway number in white, visible to pilot on approach; if displaced threshold, preceded by white transverse bar touched by four arrowheads pointing upwind and preceded by series of centreline arrows. For instrument runway, four bold white axial stripes in rectangular group on each side preceding runway number.

thresholds Limits on programme monetary changes imposed by US Defense Secretary.

threshold sampling time Time since overhaul at which engines are removed and inspected in preparation for extension in TBO; * may be less or more than new TBO.

threshold speed V_T, V_{AT} and $V_{T\ max}$.

throat Point of smallest cross-section in duct, esp that in con/di nozzle, supersonic tunnel upstream of working section and rocket engine or motor thrust chamber and nozzle.

throatable Jet or fluid flow controllable by changing shape or area of throat (unusual except in tunnels).

throat control In gas turbines, system controlling flow through nozzle guide vanes upstream of turbine.

throatless chamber Rocket thrust chamber without throat yet still achieving supersonic expansion, eg multi-chamber toroidal type.

throatless shear(s) Power shear for cutting large sheet or plate which may be rotated during cut to leave curved edge.

throat microphone Microphone held against skin of throat; better for deep or guttural voices or languages.

throttle 1 Input control, usually hand lever rotating through arc, for main vehicle propulsion.
2 System responsible under pilot for varying engine power.
3 Valve in carburettor or fuel control which governs admission to engine of either air, fuel or (piston engine only) mixture.
4 To reduce power of engine, also called to * back.
5 To constrict fluid flow path and thus reduce mass flow.

throttle back To reduce power.

throttle friction Pilot-operated device which greatly increases resistance of throttle lever(s) to movement, effectively locking them in set position; also called friction lock.

throttle icing Ice accretion in carburettor near or on partially closed throttle (3).

throttle lock See *throttle friction.*

throttle tension Locking resistance value of friction lock.

throttling capability Range of thrust expressed as percentage to 100, over which liquid rocket (occasionally other type of engine or propulsion) is designed to operate.

through-deck ship Generally, one with flight deck unobstructed by any full-width superstructure, even though not necessarily extending to bow.

throw 1 Part of crankshaft to which conrod attached, comprising webs and crankpin.
2 Loose measurement of distance to which ECS fresh-air inlet projects, in absence of bulk cabin air movement.

throw weight Total mass of payload carried by ballistic missile, in case of ICBM including warheads, RVs, decoys and other penaids, post-boost propulsion and terminal guidance systems.

THRP Port throttle (caption).

THRS Starboard throttle (caption).

thrust Force, esp that imparting propulsion. SI unit is newton (N), conveniently multiplied in DaN and (most common) kN.

thrust angle Acute angle between axis of nozzle of canted solid motor and centreline axis of vehicle, measured in plane passing through both axes if possible.

thrust augmentation Usually means afterburning, but also applied to water injection, and to PE (4) ejector-exhaust schemes.

thrust axis Axis along which resultant thrust acts, normally aligned with propelling nozzle or axis of rotation of propeller; not normally used with helicopter.

thrust bearing Bearing, usually tapered roller, needle or ball, that resists axial shaft load due to propeller thrust.

thrust buildup Sequence of programmed events in large rocket engine between ignition and liftoff.

thrust chamber Complete thrust-producing portion of liquid rocket engine comprising combustion chamber and nozzle, often mounted on gimbals; not applicable to other types of engine.

thrust coefficient 1 For propeller, basic performance calculation method based on Drzwiecki method of plotting grading curve of thrust against blade radius, yielding value K_t, constant for each value of advance ratio J; ** then equals K_tJ^2 and has symbol C_T. This is also measured thrust divided by ρn^2D^4 where ρ is density, n rpm and D diameter.
2 For rocket motor, measured ** is thrust: time integral over action-time interval divided by product of average throat area and integral of chamber pressure: time over action-time interval, symbol C_f.

thrust component In propeller theory (Drzwiecki), force on one element parallel to axis of rotation, T_c; convenient to plot T_c as ordinate against blade radius, area under curve being measure of total thrust, $T = N\frac{1}{2}\rho V^2 \int_0^r T_c\, dr$ where N is number of blades, $\frac{1}{2}\rho V^2$ dynamic head and T_c integrated between axis of rotation (or, in practice, spinner diameter) and tip radius r.

thrust cutoff See *cutoff*.

thrust decay Gradual falloff in thrust of solid motor, usually a slow fall from peak to cutoff or burnout followed by rapid ** over 2 to 8 s and to zero after perhaps 10–12 s.

thrust equivalent horsepower See *thrust horsepower*.

thruster Small propulsor, normally any of many kinds of rocket, used for spacecraft attitude control or fine adjustment of velocity.

thrust face Side of propeller blade corresponding to underside of aerofoil.

thrust/frontal area Jet-engine thrust divided by engine's nominal or published frontal area; fair criterion in early days of jet propulsion but today meaningless. Important only in highly supersonic aircraft, in which area of propelling nozzle exceeds that of engine.

thrust horsepower Seldom-used measure attempting to determine power imparted to aircraft. For propeller aircraft normally engine bhp or shp multiplied by prop efficiency (in case of turboprop plus a variable component due to exhaust thrust). For jet engines, basically thrust actually imparted to aircraft multiplied by TAS, keeping units compatible.

thrust lever Jet-engine throttle, or power lever.

thrust line Thrust axis, either of one propulsor or of whole aircraft.

thrust loading Total mass (weight) of jet-propelled vehicle divided by aggregate thrust, usually calculated for SLS-TO condition.

thrust meter Instrument for measuring thrust, more commonly of jet engine.

thrust power Appears always to be synonymous with thrust horsepower.

thrust rating computer Central element in auto power management system (ATS).

thrust reverser See *reverser*.

thrust section Portion of vehicle, esp slender rocket, containing propulsion.

thrust specific fuel consumption See *specific fuel consumption*.

thrust spoiler Pilot-controlled spoiler which when actuated diverts jet from jet engine (esp from turbofan core) to reduce thrust close to zero. Lighter and simpler than a reverser and merely eliminates possibly embarrassing idling thrust.

thrust structure In large ballistic vehicle propelled by multiple rocket chambers, structure which transmits thrust from all chambers and diffuses it into airframe. Normally large tubular truss structure at rear but can include side structures for laterally attached motors, eg SRBs.

thrust terminator Any quick-acting device for terminating thrust of solid rocket motor, including blow-off ports, nozzle ejection and inert-liquid injection into case.

thrust-vector control Control of vehicle trajectory by rotating thrust line, esp that of rocket; may involve gimballed chamber, rotation of chamber about skewed axis, inert-liquid injection at nozzle-skirt periphery, jet tabs, spoilers, refractory vanes and other methods; abb TVC.

thrust/weight ratio Basic measure of combat aeroplane performance: thrust (normally SLS-TO) divided by total mass of aircraft.

thrust wire Diagonal bracing wire transmitting airship thrust to envelope.

THT Transient heat transfer.

THUM, Thum Met readings of temperature and humidity, hence * flight.

thumbstick Pilot input controller, eg for RPV or

anti-tank missile, in form of miniature stick operated by thumb, typically attached to pistol grip and with * pivots between vertical thumb and operator.

Thump Met readings of temperature, humidity and pressure.

thunderstorm effect Error, possibly approaching 180°, of ADF in vicinity of thunderstorm; needle may point to nearby Cb or flick over, giving false indication of station passage.

thyratron Gas-discharge triode used as relay, switch or sawtooth generator.

thyristor Multilayer semiconductor device also called Si-controlled rectifier; bistable, in one state high-impedance in both directions, in other high-impedance in one direction only.

THz Terahertz.

TI 1 Target indicator.
2 Thermal infra-red.
3 Training instructor.

TIA Type inspection authorization; allows FAA to fly new aircraft.

ti-aluminites Alloys of titanium and aluminium.

TIAS Target identification and acquisition system (ARMs).

TIB Technical Intelligence Bureau (former UK government department, still a title in many contries).

TIC 1 Technical information centre.
2 Tantalum integrated circuit.
3 Target-insertion controller.

tic Visual marking pulse on telemetry readout indicating time intervals, often every 0·5 s (see *time* *).

tick Audible marking pulse serving as regular (often infrequent, eg each 10 s or 60 s) time signal.

TICM Thermal-imaging common module(s).

Ticonal Magnetic alloy of Ni/Co plus a little Al/Cu.

TID Tactical information display.

tiddleywinks effect Tendency of nose gear to project stones and other loose objects laterally.

TIDP Telemetry and image data processing.

tie Structural member normally loaded in tension.

tie bar Filament-wound tension member connecting helicopter main-rotor blade to hub; fatigue-proof because of large number of load-bearing members. Also called dog-bone.

tied gyro Gyro whose rigidity is related to Earth rather than space; eg that in traditional horizon has axis tied by gravity aligned with local vertical.

tied on Air-intercept code: "Aircraft indicated is in formation with me".

tiedown 1 Picketing arrangement for aircraft left in open (US).
2 Cargo lashing.

tiedown diagram Drawing illustrating method of securing particular type or item of cargo in particular vehicle (DoD).

tiedown point Permanent attachment point for cargo provided on or in vehicle (DoD); hence * pattern.

tiedown test Rocket engine static test.

tie rod General term for tie of rod-like form, esp with threaded ends.

Ties Tactical information exchange system.

TIF, Tif Take-off inhibit function, temporarily suppresses all non-essential cockpit warnings.

TIFS Total in-flight simulator.

TIG 1 Tungsten inert-gas welding.
2 Time of ignition.

tightening Tendency of aeroplane or glider trimmed for level flight to increase rate of a commanded turn or dive pull-out, demanding a push force on stick or yoke to hold constant g.

TIGO Prefix, US piston engine, turbocharged, direct-injection, geared, opposed cylinders.

til Until.

tile 1 Thin-film or thick-film substrate; also occasionally used for substrate of solar cell.
2 Discrete unit of surface thermal-protection system for RV or large spacecraft, eg Space Shuttle, inspectable and replaceable.

TILS Tactical ILS.

tilt 1 Angular deviation of locus of centroids of sections of helicopter main-rotor blade from plane of rotation (BSI). Measured as forward or backward though actually up/down.
2 Angular movement or offset of camera axis about aircraft longitudinal axis (NATO).

tilt angle Angle between axis of air camera and aircraft vertical (OZ) axis; normally angle at perspective centre between photograph perpendicular and plumb line (NATO).

tilting-duct VTOL VTOL aeroplane which in hovering mode is lifted by ducted propellers or fans rotated through approx 90° for translational flight.

tilting-engine/jet/propeller/wing Same definition as above but for different pivoted component. Tilting-jet means entire engine is pivoted.

tilting fuselage Unusual class of VTOL aeroplanes in which fuselage can be pivoted near mid-length, in some cases complete with attached wing, in order for jet thrust to act vertically. Also called nutcracker aircraft.

tilting head 1 Rotorcraft head pivoted about lateral axis relative to supporting structure.
2 Machine-tool cutter and drive pivoted about horizontal axis.

tilting-nozzle VTOL Not used; term is jet-deflection, vectored-thrust, lift/cruise, vectoring or vectored-jet.

tilt-rotor VTOL aeroplane lifted in hovering mode by one or (usually) more rotors which are rotated through approx 90° for translational flight.

tilt-wing VTOL aeroplane whose wing, carrying complete propulsion system, is pivoted upwards through approx 90° in vertical mode, thrust then exceeding total weight.

time Normally measured by subatomic frequency reference, eg crystal clock, but defined according to position of celestial reference point; depending on which point chosen * called solar (Sun), lunar (Moon) or sidereal (vernal equinox), solar being subdivided into mean or apparent according to which Sun. Practical time designated GMT or according to designated longitude zone. SI unit is s, 3,600 to h, 86,400 to week.

timebase 1 Straight line traced by spot on CRT or other display of cartesian and several other types providing timescale for measurement, eg of target range.
2 Straight line, regularly incorporating time tic, on data readout.

time between overhauls, TBO Period recommended by manufacturer and beyond which all warranties become invalid and operation may be in violation of certification.

time box Small box, usually rectangular or square, which moves along cockpit display future track, according to flight plan, at selected groundspeed.

time-change item One whose operation is limited to number of operating hours, number of operating cycles or (rarely) passage of time, and which must be periodically replaced on this basis.

time circle Basic symbology of many HUDs and other attack systems in which bright circle starts at 60 s and unwinds anticlockwise to 180° at 30 s and to vanish at 0 s.

time constant Usually, time taken from start of input signal for instrument to indicate specified % final reading; for exponential response, eg thermometer, time to reach 63·2% final reading; also called relaxation time, lag coefficient. Same meaning in charge/discharge of electrical C/R circuit or current in L/R circuit.

time dilation Apparent slowing-down of time as observer's speed reaches significant fraction of that of light; also called clock paradox or twin paradox.

time/distance/speed scale Simple written scale, either purchased (in which case of sliderule type) or prepared before flight, with which unknown distance or speed can be immediately read if other two factors are known.

time-division multiplex Dividing several continuous measures, eg in telemetry system, or several input signals, to form single continuous interlaced pulse train sent over single channel.

time hack Time at which a future event is scheduled, eg at which a particular squadron is to start engines (colloq, chiefly military).

time in service For maintenance time records, aircraft log and similar purposes, elapsed time from aircraft leaving surface until touching it again on landing (FAA).

time lag Any delay between stimulus and response, or cause and effect, esp that between start of signal and full indication by instrument.

time of flight Elapsed time from weapon launch, release or departure from gun muzzle to instant it strikes target or detonates.

time-of-flight spectrometer Instrument sorting particles, esp neutrons, according to time to travel known distance.

time of origin Local time message is released for transmission.

time of useful consciousness See *time reserve*.

time on target 1 Time, either planned or actual, at which aircraft attacks or photographs target.
2 Time at which NW detonation is planned at specified GZ (DoD).

time over target Time at which aircraft arrive(s) over designated point for purpose of conducting an air mission on a target (USAF).

time pulse distributor EDP (1) circuit that generates timing pulses during machine cycle, gated by command generator to carry out commanded operations.

time reserve Time between sudden total loss of oxygen supply and time when human can no longer be relied upon to function normally or rationally.

Time-Rite Patented indicator of piston position for timing (1).

time series Sequence of time-variant measures, either continuous (eg barograph trace) or discrete (eg hourly met pressure readings).

time sharing Use of one EDP (1) processor or computer, usually large and beyond means or requirements of each customer, by a number of customers or users whose programs are run in short bursts in time-division multiplexed form switched according to cyclic formula agreed between users (in simplest form, a round robin).

time signal 1 Broadcast signal used as very accurate time reference.
2 Time reference mark along border of reconnaissance imagery or other film.

time/size plot Diagram whose ordinate is a measure of aircraft size, eg MTOW or pax seats, and abscissa is time in years.

time/speed scale Scale for given groundspeed used in conjunction with plotting chart or topographical map.

timeswitch Electrical switch activated by time of day or elapsed time from a start point.

time tic Time reference mark along telemetry readout; usually small inverted V every second along straight timebase.

time tick Regular time signal of one or more audible brief sounds.

time to go In air intercept, time to fly to offset point from any other initial position; after offset point, time to fly to intercept point (DoD).

time zone Regions of local standard time, esp over sea areas, where they are exactly divided by 15° widths of longitude.

timing 1 Angular positions of PE (4) crankshaft at which valves first rise from seats or touch them again, and at which spark occurs; also called valve *, ignition *.
2 In US, assessment of human pilot's ability to co-ordinate flight controls on correct time basis for smooth manoeuvres; not often regarded as a topic elsewhere unless demonstrably faulty.
timing consideration Measure of time missile (or, possibly, other weapon such as aeroplane) is exposed on ground between withdrawal from hardened shelter and launch (probably arch).
timing disc Disc, engraved marking or other feature on PE (4) to assist establishing exact crankshaft angular positions for timing purposes.
timing parallax Film distance between time signal (2) and corresponding frame of imagery.
timing pulse Pulse used as time reference in telemetry, radar and SSR and other electronic systems.
Timos Total-implant MOS device or circuit.
TIMS Technology integration of missile subsystems.
tin 1 Soft white metal, SG 7·31, MPt 231·85°C, symbol Sn.
2 To coat surface of mild steel sheet with tin to prevent corrosion.
3 To coat metal surface with solder before making joint.
Tina Thermal-imaging navigation aid.
TIO US piston engine designation: turbocharged, direct injection, opposed.
TIP 1 Message code: until past specified waypoint or other point (ICAO).
2 Tracking and impact prediction.
3 Technical information panel (Agard).
4 Test integration plan.
5 Tailored instruction program (US).
tip 1 Extremity of aerofoil.
2 Angle of rotation of reconnaissance camera about aircraft transverse axis; also called pitch.
3 Wing-tip fuel tank (DoD) (colloq).
tip aileron Aileron forming most or all of tip of wing.
tip cargo Special cargo, eg radioactive isotopes, carried in small compartment in wingtip of some transports.
tip chord Chord at tip of aerofoil, esp wing, normally measured parallel to plane of symmetry of wing (for variable-sweep, at min sweep angle) between points where straight leading/trailing edges meet curvature at tip. Where both edges have pronounced sweep at tip, or where they are joined by line not parallel to plane of symmetry (eg Lightning, Tornado) other definitions apply, often unique to type.
tip cropping Cutting off at Mach angle.
tip dragger Spoiler above wingtip used asymmetrically to cause yaw.
tip drive Rotation of main rotor(s) of rotorcraft by thrust applied at or near tips.
tip droop Downward folding of wingtips through large angle, usually 60°–80°, to move forward aerodynamic centre of wing at supersonic speed and decrease trim drag; in some aircraft (XB–70) also generates compression lift.
tip float See *stabilizing float.*
TIPI Tactical information processing and interpretation system (USAF).
tip loss Inefficiency of tip of aerofoil in lifting mode caused by spanwise deflection of isobars and relative wind, in some transonic cases approaching 90° and making tip mere dead weight.
tip loss factor Correcting factor in calculating lift of rotorcraft lifting rotor to allow for tip loss.
tip-path plane Plane containing path of tips of helicopter or other rotorcraft main lifting rotor, tilted in direction of travel or horizontal acceleration.
tipping See *propeller tipping.*
tip pod Streamlined container carried centred on or below tip of aerofoil.
tip radius Usually synonymous with radius.
tip rake See *rake.*
TIPS 1 Total integrated pneumatic system (C–5).
2 Telemetry integrated processing system (AFSC).
tipsail See *winglet.*
tip speed Tangential speed of rotating tip due solely to its rotation and ignoring superimposed vehicle airspeed; ie angular velocity times radius.
tip stall Stall of tip of aerofoil, esp wing, while remainder of surface remains unstalled; common condition caused mainly by higher lift coefficient at tip unless stall strip applied inboard.
tip tank Fuel tank formed as streamlined body, jettisonable or otherwise, carried centred on or below wingtip.
tip trailing vortex See *vortex.*
tip vortex See *vortex.*
TIR 1 Total indicator reading.
2 Traffic information radar.
3 Target-illuminating radar.
4 Thermal infra-red.
tire US spelling; "tyre" is used in this dictionary.
tiredness General deterioration of airframe caused by long and intensive use, primarily manifest in repeated cyclic loading and successive severe gusts but also including superficial damage caused by impact of steps, ground vehicles, stones etc; no significant crack need be present but many structural parts will not be original and many boltholes will be oversized and re-reamed for bolts of increased diameter.
Tirp Terminal instrument radar procedure.
TIRSS Theatre intelligence, reconnaissance and surveillance study (USAF).
TIS, tis 1 Tracking information (or instrumentation) subsystem.
2 Thermal-imaging system.
3 Tactical intelligence squadron.

4 Traffic information service(s) (CAA).

Tiseo Target-identification system (or sensor), electro-optical.

TIT Turbine inlet temperature.

tit Any control button (UK, colloq, WW2).

titanium White metal, SG (bulk solid) 4·5, MPt 1,800°C, passivated by oxide/nitride coating forming in atmosphere. Used as high-purity 99·9%, commercially pure 99% and in many alloys, esp those with Al and V.

Tite Tews intermediate test equipment.

title block Standardized rectangular format on drawing, usually lower right corner, listing title, part numbers, mod states, names of draughtsmen/tracers etc, dates and other information.

titles Name of owner or operator painted on commercial or GA aircraft, to be read from a distance.

TIVO US piston-engine designation: turbocharged, direct injection, vertical crankshaft (for helicopter), opposed.

TJ Turbojet.

TJAG The Judge Advocate-General.

TJRJ Turbojet/ramjet or turboramjet.

TJS Tactical jamming system.

Tk Track.

TKE, TkE Track angle error.

TKF Tactical combat aircraft (G).

TKM Tonne-kilometres.

TKOF, tkof Take-off.

TKP Tonne-km performed; basic measure of airline traffic.

TKS Chemical de-icing pastes, now used mainly for protection on ground; from Tecalemit/Kilfrost/Sheepbridge-Stokes (today Kilfrost subsidiary).

TL 1 Thermoluminescence.
2 Transition level.

TLA 1 Towed linear-array sonar.
2 Throttle-lever angle.

TLBR Tactical laser beam recorder.

TLC Trans-lunar coast.

TLD Technical-log defect.

TLG Tail landing gear.

TLI Trans-lunar insertion.

TLM Telemetry.

TLO Terminal learning objective.

TLS 1 Tactical landing system.
2 Translunar shuttle.

TLSI Technical-log special inspection.

TM 1 Training memoranda.
2 Tactical missile.
3 Trade mark.
4 Ton-mile (seldom abb).
5 Transcendental meditation, relevant to aerospace.
6 Transverse magnetic EM propagation mode.
7 Telemetry.

Tm Tropical maritime.

TMA 1 Terminal manoeuvring (or control) area, ie terminal airspace.
2 Trimethylamine.

TMC Thrust-management computer.

TMD Tactical munitions dispenser.

TME Total mission energy, normally in non-SI kWh.

TMEL Trimethyl-ethyl lead.

TMF True-mass flowmeter.

TMG 1 Track made good.
2 Thermal/meteoroid garment.

T/MGS Transportable/mobile ground station.

TMIS Technicians maintenance information system.

TML Tetramethyl lead.

TMM Tantalum manganese-oxide metal device.

TMN True Mach number.

TMO Traffic management office (AFSC).

TMP 1 Transverse-magnetized plasma.
2 Twin machine-gun pod.

tmpr, tmprly Temporarily.

TMRC Technical-manual reference card.

TMS 1 Thrust-management system.
2 Test and monitoring station.

TMSA 1 Trainer-mission simulator aircraft.
2 Technical Marketing Society of America.

TMXO Tactical miniature crystal oscillator.

TN 1 Nuclear, thermonuclear (weapon prefix, USSR).
2 Technology need.

TNA Truth in Negotiations Act (US Congress).

TNAV, T-nav So-called four-dimensional navigation system commanding three spatial dimensions and time.

TNE Tungsten nuclear engine.

TNF Theatre nuclear forces (S^3 or S-cubed adds "survivability, security and safety").

TNI Total noise index.

TNR Transfer of control message, non-radar.

TNS Technical news-sheet.

TNT Advanced supercritical wing (G).

TNT equivalent See *yield*.

TNW 1 Theatre nuclear weapon.
2 Tactical nuclear warfare.

TO, T-O 1 Take-off.
2 Technical order.

TOA 1 Total obligational authority, sum that may be obligated in coming FY for contracts possibly running for many years hence.
2 Time of arrival, hence TOA/DME.
3 Usually plural, transportation operating agencies (MAC, MSC and MTMC, US).

TO&E Table of organization and equipment.

toboggan In-flight refuelling technique in which shallow dive is maintained to match speeds of fast tanker (if necessary with spoilers or airbrakes) and slow receiver.

TOC 1 Top of climb.
2 Total operating cost (often t.o.c.).
3 Travel order card.
4 Tactical operations center (US).

TOCG Take-off c.g. position.

TOCS, Tocs Terminal operations control system.

TOD 1 Top of descent.
2 Take-off distance.

TOD_a, TODA Take-off distance available.

TO dist Take-off distance.

TOD_r, TODR Take-off distance required.

TOE Ton (usually tonne or short ton) of oil equivalent; measure of energy.

toe 1 Figurative forward extremity of ski shape whose contact with ground is commanded by TFR.
2 Any lateral extremity at foot of graphical plot.

toe brakes See *wheelbrakes.*

toed in Left/right (eg engines) have axes which in horizontal plane are inclined to meet aircraft centreline ahead of nose. Hence, toed out: axes meet centreline to rear, as in case of engines whose axes are perpendicular to tapered leading edge (eg Ju 52/3 m).

toe plates Hinged tapered plates along outer edges of cargo-aircraft vehicle ramp.

TOF Take-off fuel; quantity aboard at take-off.

TO-FLX Derated (flexible) take-off.

TOFP Take-off flightpath.

to/from Indication of whether certain radio navaids are moving towards or away from ground station, either by caption window in instrument or by various switches or procedures; also called sense indication.

TOGW Take-off gross weight, either published MTOW or that at one particular take-off.

TOJ Track on jam(ming).

TOLA Take-off and landing analysis.

Told card Take-off and landing data, kept handy in cockpit.

tolerance 1 Maximum departure permitted between dimension of an actual part and its nominal value; usually part may be either over or undersize (eg 653 ± 0·1 mm) but occasionally * is unilateral (eg 653–0·1 mm).
2 Maximum error permissible in calibration of instrument or other device.
3 Maximum quantity of harmful radiation which may be received by particular person with negligible results, also called * dose.
4 Ability of individual to withstand cumulative doses of drug.

toluene Flammable liquid used as solvent and thinner; also called methyl benzene ($C_6H_5CH_3$) or toluol.

tomodromic Heading to intersect a particular line, eg trajectory of another aircraft.

ton 1 Standard SI-related unit is tonne (t), = 1,000 kg = 1 Mg = 0·984207 long ton = 2,204·6236 lb. In Americas 2,000 lb, commonly called short * (not abb), = 907·18474 kg. In UK and Commonwealth 2,240 lb, commonly called long * (not abb), = 1,016·0469088 kg. In aerospace much confusion exists because of these three values, especially 12% difference between short and long * in aircraft payloads, airline traffic (usually short *), airfield pavements (mainly metric) and many other areas.
2 In air-conditioning and refrigeration, rate of removal of heat sufficient to freeze 1 short * of ice each 24 h = 3,140·05 W (if long * is basis, 3,516·85 W).

tonal balance Can refer to audio frequencies (balance across pitches of sounds as heard by listener) or to white/grey/black or colour tones in radar or other electronic display.

tone 1 Sound of one pitch containing no harmonics, usually synonymous with mono-*.
2 Specifically, in AAM launch, aural note which changes to singing or growling after IR lock-on.

tone localizer Localizer whose L/R indications are received as contrasting tones heard on each side of glideslope centreline.

tonf Ton force, non-SI unit of force, = 9,964·02 N.

Tonka See *R-stoff.*

TOO Target of opportunity.

tool Though obviously normal meaning applies in aerospace, an added meaning is extension to include any device or construction facilitating manufacture or assembly, even when it plays no part in shaping workpiece. Examples include assembly structures of kind in most cases preferably called jigs (more explicit), as well as temporary fixtures, struts and props, inflatable bags, rubber press-* and dies of all kinds, and devices for holding or locating during tests or other operations.

tool bit Small cutting tool, usually from square steel bar with super-hard added tip, fixed in place on machine tool; not used for drills and millers.

tool design Design of tools for particular programme, esp design of all required jigs, fixtures, templates, gauges and special-purpose tools, eg for checking dimensions and alignment of large parts.

tooling See *tool.*

toolmaker Skilled person, usually previously machinist, responsible for making many special-purpose in-plant tools (both jigs/fixtures and cutting tools) and in particular for setting up machines for semi-skilled operatives and minders, today often versed in NC.

toolroom Originally room where cutting tool bits were kept, today clean (often in strict sense) environment for super-accurate measures, gauges and manufacturing operations calling for abnormal standards of accuracy.

tool steel High-carbon steels retaining extreme hardness at elevated temperatures (note: bits are now usually carbides, cermets or other materials).

TOP 1 Total obscuring power; basic measure of chaff or aerosol, in US expressed in non-SI units sq ft/lb (cross-section of sky per unit mass dispensed), for 80% opaqueness to hostile radar or other sensor.
2 Tube à ondes progressives = TWT (F).

3 Take-off power.

top chord Main transverse (end-to-end) upper member of truss.

top cover Defending friendly fighters watching over bomber or attack aircraft from higher level, esp while over hostile territory.

top dead centre Instantaneous position of PE (4) or reciprocating-pump crankpin in which centre-line of crankshaft, crankpin and cylinder are all in line with piston at extreme top of stroke; hence also corresponding position of piston.

top dressing Application of ag-chemical to land or growing crops from above; normally method of applying fertilizers rather than insecticides, for which technique may be to coat undersides of leaves also.

top-hat Family of standard structural sections based on five straight surfaces, each at 90° to neighbour(s); resemble top hat in shape.

top loading Increasing apparent (effective) height of radiating aerial by adding metal plate, mesh or radial wires at extremity.

topocentric Referred to observer's position; measures, usually linear distance or az/el, based on observer's position as origin.

topographic Representing physical features of Earth's surface, both natural and man-made; hence * display, * map. DoD definition of * map: one which presents vertical position of features in measurable form as well as horizontal. Normally, essential feature is use of contour lines, as well as normal positional information.

top overhaul Overhaul of PE (4) cylinders (valve grinding, ring replacement, decarb etc) without opening crankcase.

topping Operating cycle of liquid-propellant turbopump for rocket engine in which cryogenic fuel is heated, producing high-pressure gas used to drive turbine(s); this gas then passes at lower pressure to combustion chamber (different nozzles from main flow), where it burns. Hence * cycle, * engine.

topping off Replacement of cryogenic propellant lost by boiloff.

topping up Replenishment of gas-filled aerostat, eg after a flight.

topple Real or apparent wander in vertical plane of gyro-axis (see *toppled*).

toppled Gyro whose gimbals have for any reason ceased to maintain its correct axis in space, so that further rotation of mounting results in violent direct precession. Traditional gyro instruments can be * by aerobatics or any rotation of aircraft axes beyond defined limits, instrument then being useless as attitude reference until gyro has settled again into normal operation. New term is needed for either topple or toppled.

TOPS Thermoelectric outer-planet spacecraft.

Top Secret High grade of defence classification for material whose unauthorized disclosure might result in severance of diplomatic relations, war, or collapse of defence planning.

Topsep Targeting optimization for solar-electric propulsion.

topside On carrier flight deck, esp movement thereto by elevator; eg coming * as aircraft or other item appears level with deck (USN).

top-temperature control Any subsystem limiting a temperature to a specified safe limit, esp that for TET, TGT or equivalents.

TOR, T-OR Take-off run.

tor Torr.

TOR$_a$, TORA Take-off run available, usually = TOR.

torch igniter Combined igniter plug and fuel atomizer emitting jet of flame from burning fuel. Very rare in gas turbines but occasionally used in afterburners and a minority of liquid and solid rockets.

torching 1 Faulty operation of gas turbine, esp jet engine, in which unburned fuel travels past turbine and results in flames travelling down jetpipe, often expelled from nozzle.

2 Faulty operation of PE (4) in which unburned fuel travels through exhaust valve and burns in exhaust pipe, often causing visible flame beyond nozzle.

3 Degassing in ultrahigh-vacuum technology by applying gas flame to walls.

toric Having a surface described by a segment of a conic section.

toric combiner Optical lens assembly used to combine a generated-information display with an image of real world.

tornado Localized violent whirlwind east of Rockies in US with such low pressure in core as to explode structures in its path, usually pendant under a Cb. Also used for Gulf of Guinea thunder squalls advancing westwards in line.

toroidal Shaped like doughnut.

toroidal vanes Rings of curved section guiding air to eye of centrifugal compressor.

torpedo director Traditional optical sight for aerial torpedo attack; user sets target size/speed and receives azimuth guidance.

torque For all practical purposes, synonymous with turning moment or couple; a rigorous definition is effectiveness of a force in setting a body into rotation, according to which trying to loosen a tight nut unsuccessfully or rotate free end of rod fixed at other is not application of * (though in second case it is torsion). Often invertedly defined as resistance to a twisting action. For propellers see * *component*.

torque box Box-like structure, eg wing torsion box, designed to resist applied torque.

torque brake Variable brake on rotating shaft, eg slat drive, triggered at particular point of system travel.

torque coefficient Product of prop torque divided by $\rho N^2 D^5$.

torque component Q_c, tangential force acting in

plane of propeller rotation on any elementary chordwise lamina; thus total propeller torque Q = N$\frac{1}{2}\rho V^2 \times$ integral of Q_c from axis to tip with respect to radius.

torque dynamometer Measures shaft power by measuring N (rpm) and torque.

torque effect Reaction on vehicle of torque applied to propeller or rotor (** for rotodome usually ignored); in helicopter countered by tail rotor.

torque horsepower Shaft horsepower, often same as brake horsepower.

torque link Pivoted links preventing relative rotation between cylinder and piston of oleo shock-strut; limiting factor with bogie main gears on allowable steering angle of nose gear. Also called scissors or nutcracker.

torquemeter Device, either instrument or component part of engine, for measuring torque; in turboprop or some PE (4), usually oil-pressure system sensing axial load on reduction-gear planetary helical gears or, less often, tangential reactive load around annulus gear.

torquer Device imparting torque to an axis of freedom of a gyro, usually in response to signal input.

torque-set screw Can be repeatedly unscrewed without losing original torque needed to release; used to latch long-MTBM panels.

torque stand Test stand for engines, esp aircraft piston engines.

torque tube Tubular member designed to withstand torque, either one applied inevitably and to be resisted or one to be transmitted as part of drive system (eg in primary flight controls).

torque wrench Hand tool with dial or other direct readout of torque imparted.

torquing Input to gyro from torquer for slaving, capturing, slewing, cageing etc.

torr Non-SI unit of high-vacuum pressure, = 133·322 N/m^2 = 0·0193368 lbf/in^2. Originally (and still very nearly) 1 mm Hg.

torsion Deflection, usually within elastic range, caused by twisting, ie applied torque (note: * is result, not an applied stress).

torsional instability Characteristic of structural member such that, when loaded in compression or bending, it will twist before reaching ultimate compressive stress.

torsional load One imparting turning moment or torque.

torsional stress Stress resulting from applied torque; for torque tube $S_s = Tc/J$ where T is torque, c is radius and J polar moment of inertia.

torsion balance Instrument containing light horizontal rod suspended by fine fibre for measuring weak forces, eg gravitation, radiation.

torsion-bar tab See *spring tab*.

torsion box Main structural basis of wing, comprising front and rear spars joined by strong upper and lower skins; also called wing box, inter-spar box.

torso harness Normal seat harness of military pilot restraining torso over full length.

TOS Transfer orbit stage (eg Shuttle to geo-synchronous).

toss bombing Method of attack on surface target with free-fall bomb, esp NW, in which aircraft flies toward target, pulls up in vertical plane and releases bomb at angle that compensates for effect of gravity drop; similar to loft-bombing and unrestricted as to altitude but normally entered from lo. Two main varieties: forward **, in which bomb is released at angle short of 90° (usually about 70°), after which aircraft continues with Immelmann-type manoeuvre; and over-the-shoulder **, in which aircraft overflies target and releases at angle beyond 90°.

TOT 1 Time on (or over) target.
2 Turbine outlet temperature.

total blade-width ratio Ratio of propeller diameter to product of number of blades and max blade chord; also called total prop-width ratio.

total conductivity Sum of electrical conductivities of all free ions, positive and negative, in given volume of atmosphere.

total curvature Change in direction of ray between object and observer.

total drag Component of total aerodynamic force on unducted body parallel to free-stream direction, = induced plus profile = pressure plus surface friction.

total-energy equation Expression for sum of pressure, kinetic and potential energies of given volume of atmosphere as result of combining mechanical energy equation with thermodynamic.

total head See *total pressure*.

total impulse Basic measure of quantity of energy imparted to vehicle by rocket, = integral of thrust versus time over total operating time, abb I_T, expressed in Ns, kNs or (US) lbf-s.

totalizer Indicator showing quantity of variable (fuel, ammunition etc) that has passed sensing point (see *detotalizing*).

totalled Damaged beyond repair.

total lift Component along lift axis of resultant force on aircraft.

total noise rating See *noise*.

total obligational authority Money for 5-year defence programme or any portion for a given FY (DoD).

total operating time Time between ignition of solid-propellant rocket motor and time when thrust decays to zero; this is usually at least 15% longer than burn time and 10% longer than action time.

total-package procurement Award of one very large prime contract for entire operative system from conceptual stage through R&D to engineering design, test and production.

total pressure Pressure that would be reached in

fluid moving past body if its relative velocity were to be brought to zero adiabatically and isentropically. For low speeds taken as $p + \frac{1}{2}\rho V^2$ and at high Mach numbers as $p(1 + [\gamma-1] M^2/2)^{\gamma/\gamma-1}$. Also called total head or stagnation pressure, and usually same as pitot pressure, impact pressure.

total-pressure head See *pitot head.*

total propeller width ratio See *total blade-width ratio.*

total refraction Curvature of radiation out of layer or medium.

total system Entire system supplied to virgin site, including accommodation, civil engineering, power supply, refuse disposal etc, as turnkey contract.

total temperature Temperature of particle of fluid at stagnation point or otherwise brought to rest adiabatically and isentropically. If T_s is static temperature, $T_H = T_s (1 + \frac{1}{2} [\gamma-1] M^2)^{-1}$.

TOTE, Tote Tracker optical thermally enhanced.

tote board Display board presenting written information in tabular form, esp in ATC (1) flight-progress board or cockpit alphanumeric tab (4).

touch-and-go Practice landing in which aeroplane is permitted to touch runway briefly; in many cases flaps are moved to take-off setting while weight is on wheels.

touchdown Moment of contact of aircraft with surface on landing or of soft-landing spacecraft with designated destination surface.

touchdown aim-point Area of runway on which pilot intends to land. This is usually in touchdown zone, but on STOL runway a ** marker is provided in form of 90 m (200 ft) broad axial white strip on each runway edge projecting inwards from white edge strips.

touchdown ROD Touchdown rate of descent; value shown by ** indicator, usually sensitive VSI based on radio altimeter or laser altimeter.

touchdown RVR Touchdown runway visual range; RVR at time and place of landing.

touchdown zone That portion of runway selected by most pilots as touchdown zone; on precision instrument runway marked by three close axial white bars 90 m (200 ft) long on each side between centreline and edge, beginning 150 m (500 ft) beyond threshold.

touch screen technology Ability of advanced displays to interface with humans by direct fingertip touch of part of display of interest, notably by touching particular line or word in alphanumeric readout.

touchwire Human input to electronic display in form of matrix of fine wires, any of which, when touched, switches enlarged local region of display to fill entire area (or, in alphanumerics, switches in amplified readout of that particular item).

toughness Ability of structural material to absorb mechanical energy in plastic deformation without fracture.

tourist Originally (1949) special high-density airline accommodation usually synonymous with coach; today standard type of seating, denoted by symbol Y.

TOW 1 Take-off weight, usually meaning MTOW.

2 Tube-launched optically-tracked wire-guided (missile).

tow 1 Standard manufactured form of reinforcing fibre, eg carbon, graphite, as long unwoven staple.

2 Aero * for one flight of sailplane; hence on *, * release.

towbar Connects tug and nose gear for towing or pushing away from gate.

tow dart Dart-type aerial target towed by RPV or target drone or, in some cases and on 900 m (2,000 ft) line, by manned aircraft.

towed body Remote sensing unit of helicopter MAD or airborne magnetometer.

towed glider Glider on aero tow.

towel rack Rail-like aerial (antenna) for HF com or Loran (colloq).

tower Airport or airfield control tower, esp service or facility based therein; hence * airport, * frequency, * controlled.

towering Opposite of stooping, refraction phenomenon in atmosphere in which visual image of distant object appears extended vertically.

tower shaft Radial shaft transmitting drive from engine spool to accessory gearbox or other unit.

tower-snag recovery Recovery of RPV or other winged vehicle by flying it to hook a line suspended between arms on tower built for this purpose; in most systems line imparts decelerative drag which stalls vehicle within distance significantly less than height of tower.

towhook Pilot-operated coupling/release for glider tow cable.

towing basin See *towing tank.*

towing eye Eye (structural ring) attached to nose gear or other part of aircraft for towing on ground.

towing sleeve Towed sleeve (drogue) target.

towing tank Long, narrow water tank for hydrodynamic tests, also called seaplane basin/tank, along with models of hull/float forms, skis, ACVs and other objects are towed. Seldom used for wavemaking.

Townend ring Pioneer ring-type cowling for radial engines, with chord seldom greater than external diameter of cylinders and no pretension at true aerofoil shape, though usually with tube around inner side of leading edge.

Townsend avalanche Cascade multiplication of ions in gas-filled counting-tube technology.

Townsend coefficients In DC gas-discharge, First ** = η = number of electron/ion pairs per volt; Second ** = γ = number of secondary electrons emitted from cathode per impacting positive ion.

Townsend discharge DC discharge between two electrodes immersed in gas and requiring cathode electron emission.

Townsend ionization coefficients Average number of ionizing collisions electron makes in drifting unit distance along applied field.

Toxic-B Most common lethal air-dropped or dispensed war agent (USSR).

TP 1 Turbulence plot.
2 Thermoplastics.
3 Test pilot.
4 Target practice.

TPA 1 Target-practice ammunition.
2 Traffic-pattern altitude.
3 Taildragger Pilots' Association (US).

TPAR Tactical penetration-aid rocket, carrying expendable ECM payloads.

TPC Total programme cost(s).

TPDR Transponder (more often TXP or XPDR).

TPF Terminal phase, final.

TPG 1 TV picture generator, Tepigen.
2 Technology planning guide.

TPI 1 Terminal phase initiation.
2 Turns (or threads) per inch (often t.p.i.).

TPM Terrain profile matching.

TPN Technical procedure notice; (L) adds "electronic".

TPP 1 Tip-path plane.
2 Total-package procurement (C adds "concept").
3 Technology program plan (AFSC).

TPR 1 Terrain profile recorder.
2 Thermoplastic rubber.

TPS 1 Thermal protection system.
2 Test program set.
3 Technical problem-solving.

TPT, TP/T Target practice, tracer.

TPU 1 Terminal position update (RV).
2 Tactical processing unit.

TPWG Test planning working group.

TR, Tr 1 Track.
2 Thrust reverser.
3 Tracking rate.
4 Tactical reconnaissance (US, role prefix).
5 Torpedo reconnaissance (UK, defunct).
6 Technical report.

T/R 1 Transmitter/receiver, com radio.
2 Transformer/rectifier (or TR).

TRA 1 Track angle.
2 Radar transfer-of-control message.
3 Thrust-reverser aft (SST).

TRAAMS Time-referenced angle-of-arrival measurement system.

TRAC 1 Telescoping-rotor aircraft.
2 Trials recording and analysis console, giving immediate video-tape of fire-control system performance.

Traca Total radar aperture-control antenna.

Tracals Traffic-control approach and landing system (USAF).

trace 1 Line on CRT and many other displays made by electron beam, successive sweeps being linked by retraces.
2 Line of data on any linear graphic printout visible to eye (thus, not applicable to magnetic tape).
3 EDP (1) diagnostic technique which analyses each instruction and writes it on an output device as each is executed.

tracer 1 Ammunition whose projectiles leave bright visible trails.
2 Substance added (usually in very small proportion) to main flow in order latter may be followed accurately through process, living organism etc; * may be physical, chemical or, often, radioactive.

tracer display True historic display on Hudwac or similar sight system featuring tracer line and other symbology for snapshooting, normally with real target scene through combiner glass.

tracer line Bright line on sight system showing locus of points where a projectile would now be had it been fired during preceding few seconds, ie where projectiles from continuous burst would now be. A range marker, usually a ring, is superimposed at actual target range. Pilot must then place this ring over target in order to hit it, or arrange for target to pass through ring.

track 1 Path of aircraft over Earth's surface from take-off to touchdown.
2 At any time in flight, angle between a reference datum, and actual flightpath of aircraft over Earth's surface, measured clockwise from 000° round to 360°. Magnetic * is referred to magnetic N; true * is referred to true N and is * normally used in plotting; required * is that desired; * made good is that found by inspection to have been achieved; great-circle and rhumb-line * are those which are thus represented on chart.
3 To observe or plot a * (1), eg by radar or on plotting board.
4 Series of related contacts on plotting board.
5 To display or record successive positions of moving object.
6 To lock on to source of radiation and obtain guidance therefrom.
7 Path traced by tips of propeller, rotor or similar rotating radial-arm assembly.
8 Distance measured as straight line between centre of contact area of left mainwheel, or geometric centre of left main-gear bogie, and corresponding centre on right; in case of aircraft with centreline gears and outriggers, measured between outriggers; if landing gears are skids, measured between lines of contact; if main gears are skis, measured between centrelines; if gears are inflatable pontoons, measured between centres of ground contact area; not normally applied to marine aircraft, but would presumably be distance between CLs of two floats.

9 As plural, rails along which travel area-increasing flaps or certain translating leading-edge slats, carrying these surfaces out approx in direction normal to leading or trailing edge.
10 To keep device aimed at moving target.
11 Conductive path on printed-circuit board.

track angle See *track (1)*.

track ball Basic human interface with electronic displays, either for inputting data or calling up portion of display for any reason; comprises ball recessed into console rolled by operator's palm in any direction to generate either stream of digital pulses or analog voltage about two co-ordinate axes to achieve desired place, eg particular aircraft on display.

track beacon See *NDB*.

track clearance Clearance to fly stated track (1) as far as particular fix.

track correlation Correlating track information using all available data, for identification purposes.

track crossing angle 1 Angular difference between tracks of interceptor and target at time of intercept (DoD).
2 Generally, angle between two flight paths measured from tail of reference aircraft.

track handover Process of transferring responsibility for production of air-defence track from one track-production area to next (NESN).

tracking 1 Air intercept code: "By my evaluation, target is steering true course indicated" (DoD).
2 Precise and continuous position-finding of targets by radar, optical or other means (NESN). Hence synonymous with track (5).
3 Measure of correct rotation of separate blades of helicopter main rotor in that each should follow exactly behind its predecessor, ie all tips should lie in common tip-path plane.
4 Procedure to ensure * (3) by holding paper or fabric against painted blade tips and adjusting hub settings until a single spot results.
5 Correct holding of frequency relationships between all receiver circuits tuned from same shaft to maintain constant intermediate frequency in superhet or constant difference frequencies.
6 Keeping device, eg fighter aircraft, aimed at target; hence synonymous with track (10).

tracking station Fixed station for tracking (2) objects in air or space.

track initiator Person responsible for taking decision on appearance of unknown blip on air-defence or other surveillance radar that it represents a target whose track is to be determined and assigned an identity.

track intervals Convenient time/distance divisions between checkpoints when navigating visually.

track lock lever Hand lever locking flight-crew seat in desired position.

track made good See *track (1)*.

track marker Symbology on display indicating track, eg straight black or bright line, with or without arrowhead, cross, ring or square, depending on type of display; absent when display is auto track-oriented.

track oriented Aligned with current track at 12 o'clock position, eg hand-held map, projected-map display, etc.

track production Function of air-surveillance organization in which active and passive radar inputs are correlated into coherent position reports together with historic positions, identities, heights, strengths and direction of flight (NESN).

track-production area Area in which tracks are produced by one radar station.

track repetition Time between exact overflights of spot on Earth by satellite.

track separation 1 Lateral distance between aircraft tracks imposed by ATC.
2 Distance (often at Equator) or angular longitude difference between successive passes of Earth satellite.

track symbology Symbols used to display and identify tracks on radar or data-readout console or other electronic display.

track telling Process of communicating air-defence, surveillance and tactical-data information between command and control systems and facilities: back tell, transfer from higher to lower echelon of command; cross tell, between facilities at same level; forward tell, to higher level; lateral tell, across front at same level; overlap tell, to adjacent facility concerning tracks detected in latter's area; and relateral tell, via third party.

trackway Standard prefabricated military track for land vehicles, quickly laid for recovery of force-landed aircraft or across infilled bomb crater on airfield, esp to speed reopening of bombed runway.

track-while-scan Radar/ECM scan produced by two unidirectional sector scans simultaneously scanning in two planes, usually one vertically and one horizontally, allowing target common to both to be accurately tracked in az/el.

Tracon Terminal radar approach control (FAA).

Tracs Terminal radar and control systems (Canada DND).

tractor Adjective meaning pulling, hence * aeroplane is pulled by propellers, not pushed; * propeller is in front of engine and driven by shaft in tension.

tradeoff Generalized term for fair exchange between inter-related variables; thus in aircraft design there are numerous and continuing examples of * between wing area, thrust, fuel consumption, gust response, structure weight and many other parameters; in aircraft flight management pilot can * (used as verb) speed for height, etc.

Tradoc Training & Doctrine Command (USA).

traffic 1 Quantity of vehicles, eg aircraft, in operation; measured as number in flight in region at one time, number under positive control, or general number in vicinity, or as number in given period. For control purposes includes * on movement area.
2 Number of landings and take-offs at airport in given period, eg one calendar year.
3 One aircraft in flight as reported to or noticed by another in vicinity.
4 Output of commercial or military air transport operator, measured in such units as number of pax or mass of cargo carried multiplied by mean distance each is transported, eg passenger-miles or tonne-km (standard units compatible with SI are needed).
5 Number or frequency of messages on telecom system.

traffic information, radar Information issued to alert aircraft to radar target observed on ground radar display which may be in such proximity to its position or intended route as to warrant its attention.

traffic lights Any red/amber/green presentation, especially that by a radar altimeter referenced to a preselected low (minimum safe) height setting.

traffic pattern See *circuit*; * usually used for tracks/profiles of arrivals and departures of non-GA traffic, ie military or commercial.

TrAG Training air group (RN, WW2).

trail 1 Relative motion of dropped store, eg free-fall bomb, behind aircraft flying at constant V, broken down into * distance, cross-*, range component of *, * angle, cross-* angle, and range component of cross-*.
2 Distance between centre of tyre contact area with ground and intersection with ground of free castoring axis; not relevant to power-steered aircraft.
3 To shadow another aircraft or hostile ship(s).
4 Tendency of freely hinged (ie, not irreversible) control surface to align itself with relative wind. Normally negative, surface "floating" in line with wind, but an overbalanced surface has positive * and unless restrained will be blown to limit of its deflection.

trail angle 1 Angle between vertical and line joining bomb impact to aircraft at time of impact.
2 Several angles in landing gear, including acute angle between bogie beam and aircraft horizontal plane (negative when front axle lower than rear) and acute angle between local aircraft vertical and axis of main-gear oleo strut (often not relevant because of gear geometry).

trail distance Horizontal distance between point of bomb impact and point vertically below aircraft at time of bomb impact.

trailer 1 Aircraft following and keeping under surveillance a designated airborne contact (DoD).
2 General US term for towed road vehicle for human occupancy.

trail formation In direct line-ahead, each aircraft or element being in front of those following.

trailing aerial Aerial in form of long wire, usually with weight or drogue on end, capable of being wound in or out from underside of aircraft.

trailing area Area of flight control or other pivoted surface on aircraft downwind of hinge axis.

trailing edge 1 Rear edge of aerofoil or streamlined strut.
2 Outline of pulse as amplitude falls from peak to zero or min positive value.

trailing finger Extra electrode in PE (4) magneto distributor which transmits large current from booster magneto to cylinder next in firing order when engine is started.

trailing sweep Sweep (5) when deviation is towards trailing edge.

trailing vortex See *vortex (1)*.

trail length Length of cable deployed connecting aircraft to braking parachute, anti-spin parachute or other drag device.

trail rope 1 Trailed by balloon over ground to check groundspeed and assist in regulating height.
2 Carried in airship for ground handling.

train Bombs dropped in short intervals or sequence (DoD).

train bombing Two or more bombs released at predetermined interval from one aircraft as result of single actuation of release mechanism (USAF).

trainer Aircraft for training flight personnel, esp pilots.

training aids Items whose primary purpose is to assist instruction and growth of operator skill/familiarity, such as publications, tapes, films, mimic boards, systems rigs, procedure trainers and simulators.

training package Self-contained arrangement to train personnel (eg of an air force or of purchaser of GA aircraft) for fixed fee or fixed outlay per month; may include design of trainer, construction of facilities as turnkey contract or part of larger programme.

trajectory Flightpath in 3-D of any object, eg aeroplane or electron or other particle, with exception of orbits and other closed paths. Can be ballistic, acted on only by atmospheric drag and gravity, or controlled by various external forces.

trajectory band Webbing strip round top of aerostat envelope to reduce distortion.

trajectory scorer Instrument carried by aerial target which continuously defines position of intercepting missile in sphere whose centre coincides with origin or target's co-ordinate axes; readout is time-history record of missile range and angular position commensurate with scoring requirements.

trajectory shift Distance or angular measure of deviation of missile from ballistic trajectory under influence of a thrust mechanism (ASCC).

TRAM, Tram Target-recognition attack multisensor; DRS adds "detection and ranging set".

tram Trammel bar, or as verb to use same (colloq).

tramline pointer(s) Twin parallel lines or bars between which is to be aligned instrument needle.

tramlines Guidance lines on flight deck for V/STOL aircraft proceeding under own power.

trammel bar Hand gauge in form of straight bar, set-square, triangle or other shape provided with precision locating feet; used in checking dimensions, angles and alignments of large structures; abb tram or tram bar.

tramming Use of trammel bar.

transatmospheric vehicle Also called aerospace plane, spacecraft capable of atmospheric flight with full propulsion, lift and control, and recovery at base similar to that of aeroplane. Launch may be either by vertical rocket or horizontal take-off.

transattack period From initiation of NW attack to its termination (DoD).

transborder Crossing frontier, eg airborne pollution, fallout, virus etc.

transceiver Radio transmitter and receiver sharing common case and subcircuits, precluding simultaneous transmission and reception.

transcowl Translating (fore/aft-moving) structure of fan reverser.

transducer Device for translating energy from one form to another, eg mechanical strain to electrical signal (straingauge), temperature to electrical signal (thermocouple), or electrical signal into sound (earphone or loudspeaker).

transducer gain Ratio, usually expressed in dB, of power delivered to transducer load (output) to available power input.

transductor Any magnetic device, eg saturable reactor or magnetic amplifier, in which nonlinear characteristic controls circuit.

transerter Device for sampling unknown surface material, eg on planet other than Earth; hence * auger, tube containing spiral auger which carries sample material to various instruments and experiments.

transfer ellipse See *transfer orbit*.

transfer loader Wheeled or tracked vehicle with platform positioned at any convenient height and with horizontal adjustment, used in transfer of cargo or casualties between modes of transport.

transfer function Mathematical treatment of ratio of output response to input signal, usually a Laplace transform, expressible as plot of frequency and in closed-loop systems controlling sensitivity of output to system error.

transfer of control Action whereby responsibility for provision of separation of an aircraft is transferred from one controller to another (see *handover*, *handoff*).

transfer orbit Elliptical orbit linking two other orbits, eg one round Moon and one round Earth; for min energy invariably tangential to both linked orbits (see *Hohmann*).

transfer punch Centre punch for transferring positions of template holes to sheet beneath.

transferred position line PL redrawn to slightly later time, parallel to original but displaced by calculated ground distance.

transformation Methods (Laplace, Fourier) of simplifying solution of differential equations; hence Laplace or Fourier transform or inverse transform.

transformer Device for transferring energy from one electrical circuit to another, usually with change of voltage, by magnetic induction.

transient 1 Temporary surge or excursion of variable, eg on first switching on.

2 Short-duration electrical impulse having steep leading edge and repeated irregularly.

3 Awaiting orders, or staging through en route to another destination.

transient distortion Inability of equipment to reproduce very brief signals.

transient performance Air-combat performance sustainable for a few seconds only (eg by trading height for speed or vice versa, or allowing speed/energy to decay).

transient response Response to sudden changes in demand, eg hydraulic system or liquid rocket engine, where this factor is significant.

transient trimmer Short-duration input to longitudinal trim system to counter known disturbing moment, eg when extension of rocket pack under F-86D caused pitch-attitude changes affecting aiming.

transistor Electric/electronic device for amplification or control consisting of semiconductor material to which are attached metal electrodes. Name comes from transfer resistor, and in simplest form one electrode is emitter, connected to p-type material, separated from other p-type electrode (called collector) by layer of n-type.

transistor amplifier Amplifier employing one or more transistors arranged in any of several configurations, eg common-emitter, common-collector or common-base.

transit 1 Passage of celestial body across meridian.

2 Passage of one aircraft through controlled airspace.

3 Instrument used to determine (1).

4 Apparent passage of celestial body across face of another.

5 Condition in which three points are aligned, eg observer and two objects on Earth's surface,

prefaced by "in" (ie said to be in *).

transit bearing Measuring time at which two surface features have same (measured) bearing from aircraft in flight.

transit mode Configuration of mobile system, eg SAM missile, radars and support facilities, for moving on ground to new location with radars folded, missiles packed, launchers at 0° elevation and doors closed, etc.

transition 1 One meaning in aerospace is change from jet-supported VTOL flight to wing-borne translational flight and vice versa.
2 Another is sudden switch from blind instrument approach to visual on first sighting runway lights.

transitional surface Specified surface sloping up and out from edge of approach surface and from line originating at end of inner edge of each approach area, drawn parallel to runway centreline in direction of landing (ICAO).

transition altitude Altitude in vicinity of airfield at or below which aircraft control is referred to true altitude (see *transition level*).

transition distance Ground distance covered in transition (1 or 2).

transition down Change in helicopter flight level to dunk sonobuoy in sea in ASW operation.

transition envelope That portion of flight envelope in which trimmed controllable flight is possible in powered flight regime, bounded by airspeed, height, ROC, power, conversion angle, AOA, control margins, etc (USAF).

transition flight Flight at TAS below power-off stall speed, where lift is derived from both wing and powerplant.

transition layer Airspace between transition level and transition altitude (NESN).

transition level Lowest flight level available for use above transition altitude (DoD).

transition manoeuvre Aeroplane manoeuvre linking two glidepaths or approach trajectories; unusual except at airports where approach has to be on instrument runway and actual landing on a parallel runway.

transition point Point on 2-D aerofoil or other surface at which boundary layer changes from laminar to turbulent, extremely sensitive to surface roughness, temperature difference, steadiness of upstream flow and other factors, and difficult to locate accurately in model testing.

transition strip Area of airfield adjacent to runway or taxiway suitably paved to allow aircraft to taxi across it in all weathers.

transition temperature Many meanings in which particular temperature-dependent change takes place, but esp temp range in which metal ductility or fracture mode changes rapidly.

transition up Change in helicopter flight level to pull sonobuoy out of water.

transition zone 1 Narrow atmospheric region along front where characteristics change rapidly, values lying between those of dissimilar air masses on either side.
2 Short section of glidepath within which average pilot makes transition from IFR to visual.

transitron Pentode oscillator with negative resistance and near-constant sum of anode/screen current.

transit time 1 Elapsed time between instant of filing message with AFTN station for transmission and instant it is made available to addressee.
2 Elapsed time between electrodes in valve or other device for any electron.

translating Moving in straight line relative to surroundings.

translating centrebody Supersonic-inlet centrebody able to move linearly into or out of inlet under control of automatic control system.

translating nozzle Jet engine nozzle which in reverse mode moves to rear, further from engine, opening gap in jetpipe for gas deflected by reverser clamshells.

translation Motion in more or less straight line, from A to B, with no rotation about any axis.

translational flight Flight at sensible airspeed, such that wing generates lift; loosely from A to B, moving under power from one place to another.

translational lift Additional lift gained by helicopter in translational flight resulting from induced airflow through main rotor(s) gained from forward airspeed.

translation bearing Mechanical bearing permitting sliding motion, eg air-cushion pad or oil-film bearing.

translation rocket Separation or staging rocket.

translatory resistance derivatives Those expressing moments and forces caused by small changes in translational velocity.

Transloc Transportable (ground station) Loran-C.

translucent Permitting EM radiation, esp light, to pass through but diffused in direction.

translucent rime See *glazed ice*.

translunar Different definitions: most authorities agree word means extending from Earth to just beyond Moon, but a minority claim it excludes all space inside lunar orbit.

transmission 1 Process by which EM radiation or any other radiated flux is propagated, esp through tangible medium transparent to such radiation.
2 Process of sending signal via telecom network.
3 Signal or message thus sent.
4 Mechanism transmitting mechanical energy or power, eg between helicopter engine(s) and rotors.

transmission anomaly Deviation from inverse-square-law propagation, esp in underwater sound; symbol A, = H − 20 log R where H is transmission loss in dB and R is horizontal range in m.

transmission coefficient Radiant energy which

remains after passage through a layer, medium interface or other intervening material to that incident upon it; expressed as fraction or percentage; symbol τ, = $e^{-\sigma}$.

transmission factor Ratio of dose inside shielding material or hard shelter to that received outside in NW attack.

transmission grating Diffraction grating ruled on transparent substrate.

transmission level Ratio, usually expressed in dB, of signal power at any point in network to power at reference point.

transmission limit Particular frequency or wavelength above or below which almost all power is absorbed or diffused by medium.

transmission line Conductor conveying electrical power or signal, esp wire, coaxial cable or waveguide carrying information signals.

transmission loss Decrease in power, usually expressed in dB, between energy sent and that received; in radio propagation through space, ratio expressed in dB of power received by aerial to that sent out by identical transmitting aerial; in underwater sound, symbol H, = S – L where S is sound level (dB) and L is SPL in dB above reference rms value (traditionally 1 dyne/cm^2).

transmission modes Possible configurations of electric and magnetic field patterns in waveguide; TM (transverse magnetic) has magnetic vector perpendicular to direction of travel; TE (transverse electric) has electrical vector perpendicular to direction of travel; TEM has both vectors perpendicular to direction of travel. There are an infinite number of modes, but each cannot function below a particular cutoff frequency, and two parallel wires send nothing but DC.

transmissivity Ratio of radiation transmitted through medium, or through unit distance of it, to that incident upon it, usually expressed as %. In some usages transmission coefficient is synonymous.

transmissometer Invariably, synonymous with visibility meter, telephotometer; instrument for measuring atmospheric extinction coefficient and determination of visual range.

transmittance 1 Ratio of EM radiation transmitted through medium to that incident upon it, $T = I/I_0$ (essentially synonymous with transmissivity).
2 Ratio of luminance of surface at which light leaves medium to illuminance of incident surface (provided units are compatible).

transmitter Equipment for converting code, sound or video signals into modulated RF signal and amplifying and broadcasting latter.

transmitter chain Klystron with TWT driver.

transmitter/receiver See *transceiver*.

transmutation Conversion of atoms into different element(s) by nuclear radiation.

transom Traditional name for near-vertical transverse bulkhead at stern of boat; occasionally present in seaplane float.

transonic General term for fluid flow in which relative velocity around immersed body or surrounding duct is subsonic in some places (seldom below M 0·8) and supersonic in others (seldom above 1·2). Thus * range depends on shape of body, being very narrow and close to M 1·0 for slender body with pointed nose of small semi-angle and very thin wings.

transonic blading Rotating blading whose surrounding fluid has subsonic relative velocity at root and supersonic at tip.

transonic transport Aircraft designed to cruise at about M 1·15 without producing sonic bang at ground level.

transonic tunnel Wind tunnel whose working section can operate at Mach numbers close to that of sound, say 0·8 to 1·2.

transosonde Balloon, normally for met purposes, designed to maintain constant pressure level.

transparency 1 Portions of airframe optically transparent, eg windows, canopies, moulded noses, etc; also called glazing.
2 Imagery fixed on transparent base for viewing by transmitted light, often synonymous with diapositive.

transparent plasma Plasma through which EM radiation, esp that used by com system, can pass; generally plasma is transparent to frequencies higher than its own.

transpiration Flow of fluid through passages very long in relation to diameter (eg interstices of porous solid) yet large enough for flow to depend chiefly on pressure difference and fluid viscosity.

transpiration cooling Cooling of hot solid by fluid, eg air, passed under pressure through its porous wall.

transponder Transmitter/responder; radio device which when triggered by correct received signal sends out precoded reply on same (rarely different) wavelength; received signal usually called interrogation, and reply usually coded pulse train. Used in many systems, eg telemetry, DME, SSR, IFF; abb xponder, XPDR, TPDR.

transponder india Code for ICAO SSR (DoD).

transponder sierra Code for IFF Mk X/SIF (DoD).

transponder tango Code for IFF Mk X basic (DoD).

transport Aircraft designed for carrying ten or more passengers or equivalent cargo and having MTOW greater than 12,500 lb (5,670 kg). Note: this was originally US usage, where * normal meaning is called transportation, which also means ticket.

transport equation Complicated integral/differential equation by Boltzmann for distribution function in fluid, eg gas at low pressure, subject to flow and intermolecular collision.

transporter Land vehicle, usually large and for

off-road use, for carrying large missile or other mobile system or system element.

transporter/erector Transporter for ballistic missile or large radar which also erects its load into firing or operating position. Hence **/launcher, which also fires missile.

transporter tower Airship mooring mast on mobile base.

transport joint Joint between major portions of structure, eg between centre section and outer panel, dismantled for transport (US transportation) in another vehicle.

transport wander See *apparent wander.*

trans-sonic See *transonic.*

transuranium elements Also called transuranic, those of atomic number higher than 92 (uranium); not occurring in nature but produced by nuclear reactions.

transverse Though this means athwartships or sideways it is often related not to basic vehicle axes but to major axis of local part; thus in wing it is common to consider spar as longitudinal and ribs as *.

transverse axis OY, pitch axis, parallel to line through wingtips.

transverse bulkhead See *bulkhead.*

transverse electric See *transmission modes.*

transverse-flow effect In helicopter translational flight air passing through main rotor is initially at higher level and that passing through rear of rotor disc is accelerated as it passes across top of rotor; ** is differential lift caused by this difference in relative wind, causing blades in rear part of disc to flap upward. Note: in this usage transverse means longitudinal.

transverse load One acting more or less normal to major axis of member, thus tending to bend it; thus weight of fuselage forms ** on wing.

transverse magnetic See *transmission modes.*

transverse member Structural member running across from side to side; in wing and other aerofoils often interpreted as in chordwise direction.

transverse Mercator Map projection in which meridian is used as false equator; map is that produced by light source at Earth centre projecting on to cylinder wrapped round Earth touching along selected meridian, if necessary passing across pole. Parallels near pole almost circular but become ellipses of increasing elongation until Equator is straight line; great circles are straight lines parallel to selected meridian, otherwise complex curves.

transverse pitch Perpendicular distance between two rows of rivets.

transverse wave Displacement direction of each particle is parallel to wave front and normal to direction of propagation; includes EM waves and water-surface waves.

TRAP Terminal radiation airborne measurements program; note, not Tramp.

trap 1 Radio receiver subcircuit which absorbs unwanted signals.
2 In ultrahigh-vacuum technology, device which prevents vapour pressure of mercury or oil in diffusion pump from reaching evacuated region.
3 Rare filter in solid-propellant rocket to prevent escape through nozzle of unburned propellant.
4 Hollows in PE (4) rotating components in which oil sludge collects by centrifugal action.
5 See *flame trap.*
6 Verb, to make arrested landing on carrier.
7 Landing made as in (6).

Trapatt Trapped-plasma avalanche-triggered transit device.

trapeze bar 1 Transverse bar linking balloon basket suspension to riggings attached to envelope, permitting relative pitch but not yaw or roll; also called suspension bar.
2 Transverse bar on underside of airship or large aerodyne for attachment (including inflight release and recovery) of aeroplane or other aircraft.

trapeze beam General name for beam pivoted to two swinging parallel arms; also called bifilar suspension.

trapezium distortion Distortion of basic rectangular image on CRT or other display, eg caused by unbalanced deflection voltages.

trapezoidal modulation Involves changing waveform from sinusoidal to near-trapezoidal, with 95% modulation straight-top.

trapezoidal section Supersonic wing section looking like extremely flat shallow triangle with flat top, and bottom joined by wedge leading/trailing edges both on upper surface.

trapezoidal wing Usually means one whose plan is *, not section; in most cases leading and trailing edges are both at 90° to flightpath, joined by tips with straight rake which is usually negative and at Mach angle.

trapping Process by which particles are caught in radiation belts.

trap weight Maximum weight permitted for arrested carrier landing.

travelling-wave aerial One in which sinusoidal waves travel from feeder to terminated end.

travelling-wave amplifier Microwave amplifier depending on interaction between slow-wave field (eg travelling along wire helix) and electron beam directed along axis.

travelling-wave magnetron Usual type of modern multi-cavity magnetron in which TWT-type amplification is used at high power.

travelling-wave tube Various species of microwave amplifiers in which interaction takes place between electron beam and RF field travelling in same or opposite direction or in other arrangement; abb TWT.

travel pod Streamlined external pod carried on normal pylon or hardpoint for pilot or crew personal baggage.

traverse 1 To rotate in azimuth.
2 To take set of readings along line, either discrete readings (eg pitot pressure) at selected points or forming continuous plot, eg from point well below trailing edge of wing to point well above (in this case called pitot *).
3 Surveying method based on accurate distance/angle measures between fixed points.
4 Aircraft track made up of large number of straight legs linked by turns usually of 90° or 180°; used in search, patrol or EW duties (US term).

traverse flying Flying along traverse (4).

TR box Transmit/receive switch in system (eg radar) using one aerial for emitting and reception; prevents emitted signal from being passed to receiver.

TRC 1 True track.
2 Thrust rating computer.

TRCS Techniques for determining near-field radar cross-section of space vehicles, eg RVs.

TRD 1 Western rendering of turbo-reaktor dvigatel, Russian for turbojet.
2 Torsional resonance damper.

TRE 1 Has been used in classified ads to mean transmitter/receiver equipment, transponder equipment and even tailored radio equipment.
2 Telecommunications Research Establishment, Malvern, now RSRE.
3 Type-rating examiner.
4 Target-rich environment.

tread 1 Track (8) (US usage).
2 Normal meaning for * of tyre.

TREE, Tree Transient radiation effects on electronics.

TREF Transient Radiation Effects Laboratory (USAF).

trefoil Cluster of three parachutes.

trenched 1 Of rotating shaft, provided with multiple grooves forming labyrinth seal.
2 Of passenger-transport aisle, lower than rest of floor on which seats are mounted.

TRF 1 Tuned radio frequency.
2 Threat radar frequency (SU adds "spectrum utilization").

TRIA 1 Tungsten-reinforced iron alloy.
2 Tracking range instrumented aircraft (USAF).

Triac 1 Test Resources Improvement Advisory Council (AFSC).
2 Without initial capital, semiconductor gate (switch) similar to silicon rectifier but triggered by either positive or negative pulse.

Triad US deterrent concept based on simultaneous demonstration of hard land missiles, SLBMs and recallable bombers.

triage Urgent investigation of casualties of NW attack to determine which need and will respond to medical treatment.

triangle of velocities Basic triangle in DR navigation with sides representing heading (course) and TAS, track and G/S, and wind velocity.

triangular parachute One whose canopy is approx triangular when laid out flat.

triangular pattern Regular repeated flight pattern flown by aircraft with radio failure, equilateral triangle with sides 1 min (jet) or 2 min (others); flown left or right-handed depending on whether transmitter and receiver failed or only transmitter.

triangulation Mensuration technique, eg in sheet metalwork, in which whole area is divided into equal adjoining triangles.

triangulation balloon Small balloon used as sighting mark in triangulation survey.

triangulation station Point on land whose position is determined by triangulation; also called trig point.

triboelectrification Electricity produced by frictional processes.

tribology Study of solid surfaces sliding over one another, with or without interposed fluid.

tribometer Instrument for measuring sliding friction, usually small-scale between smooth surfaces (ie not of runway).

tri-camera photography Simultaneous exposures by fan of three overlapping reconnaissance cameras.

trichlorethylene Common solvent and cleaner, $CH_2\ HCl_4$.

trichromatic Three-colour, eg TV or electronic display.

tricycle Though nearly all aeroplane landing gears support at three points (ignoring different number of wheels at each point) this adjective means use of nosewheel instead of tailwheel. Becoming redundant, adj being needed only to distinguish tailwheel-type aircraft.

Tridop CW Doppler trajectory measurement using three fixed receivers.

Trigatron Pulse modulator having DC-charged hemispherical electrode discharged by pulse from hemispherical trigger in gas-filled envelope.

trigger 1 Pulse used in electronic circuits to start or stop an operation.
2 To initiate action using pulse of EM energy, eg leading edge of first pulse in SSR pulse train begins * of transponder.

triggering transformer Extra-high-voltage transformer connected in series with high-energy igniter to ionize gap and allow triggering spark to occur.

trijet Aeroplane propelled by three jet engines.

trike 1 Microlight using trike unit.
2 Trike unit.

trike unit Three-wheel chassis, usually with engine and seat(s), forming basic body and landing gear of microlight or ULA.

Trim 1 Trail/roads interdiction multisensor.
2 Time-related instruction management.

trim 1 Basic measure of any residual moments about aircraft c.g. in hands-off flight.
2 Condition in which sum of all such moments is zero.
3 Condition in which aircraft is in static balance in pitch (BSI).
4 To adjust trimmers or other devices to obtain desired hands-off aircraft attitude (according to BSI, in pitch only).
5 Angle between longitudinal axis (OX) and local horizontal, esp of airship, marine aircraft or seaplane float on water.
6 To make fine adjustment to value of any variable, eg velocity at cutoff of ICBM or capacitance deposited on circuit.
7 To make fine adjustments to flap, LG door, external access panel or other part of aircraft surface so that when closed there are no surface discontinuities.

trim angle Trim (5), positive when bow higher than stern.

trim cord Short length of cord doped above or below trailing edge of control surface of simple aircraft to adjust trim.

trim curve Plot of elevator or tailplane angle (η_t) against airspeed or Mach for each c.g. position, altitude etc.

trim die Die which trims to final dimensions.

trim drag Sustained increment of drag caused by need to deflect pivoted surface continuously in order to achieve required trim, eg to deflect elevons or tailplane up to counteract rearward shift in CP as aeroplane accelerates to supersonic cruise speed. Normally eliminated in SST by pumping fuel aft to shift c.g.

trimetrogon Reconnaissance camera installation of three cameras in which one is vertical and others take high obliques at 90° to line of flight at inclination of 60° from vertical.

trim for take-off Automatic trimming of flight-control system to correct settings for take-off; abb TFT.

trimmed In correct trim; also called trimmed-out.

trimmer Trimming system about any axis (as plural, about all axes); normally trim tab(s).

trimming moment Moment about reference point, usually c.g., exerted by trimming system or by seaplane float or hull when held in water at particular fixed trim angle.

trimming strip Strip of metal (occasionally length of cord or wire) attached to trailing edge of control surface and adjustable on ground to achieve desired trim.

trimming system Flight-control subsystem through which pilot inputs bias controls about all axes to obtain desired trim; in most cases ** operates via hinged trim tabs but in supersonic fighters inputs are separate irreversible actuators driving into surface power units and in simple lightplanes often a spring-loading device in cockpit.

trimming tab See *trim tab*.

trimming tanks Fuel tanks (occasionally for other liquids) located as far as possible from c.g. between which fuel can be pumped to achieve desired trim in pitch, eg in SST at subsonic or supersonic speed.

trim size Finished size of map or chart sheet.

trim tab Small hinged portion of trailing edge of primary flight-control surface whose setting relative to surface is set by pilot via screwthread, powered trimmer actuator or other system preventing subsequent rotation under air loads and whose effect is to hold main surface in desired neutral position for trimmed flight.

trim tank See *trimming tanks*.

trim template Template used for marking finished shape of part already formed (eg in press) but not trimmed.

triode Thermionic valve containing cathode, anode and control grid.

trip 1 One complete flown sector; hence * time, * fuel.
2 To activate an electrical circuit-protection system, or action thus triggered, eg overvoltage *.

triplane Aeroplane with three main lifting surfaces, in practice superimposed but not required by definition.

triple Triple seat unit.

triple-A Anti-aircraft gunfire or defences (colloq).

triple ejector rack Rack for carrying and forcibly ejecting three dropped stores.

triple modular redundancy Majority-vote redundancy by three parallel systems.

triple-output GPU Normally means electrical supply offers three voltages, eg 28 V DC, 115 and 415 V AC.

triple point 1 Unique thermodynamic-equilibrium value of temperature/pressure at which substance can exist as solid, liquid and vapour.
2 In any diametral vertical section through NW air burst, intersection of incident, reflected and fused (Mach) shock fronts; height above surface (called Mach stem) increases with distance from GZ (DoD).

triple release Release of three free-fall bombs at short intervals, with middle one intended to be on target (suggest arch).

triple seat Passenger seat unit for three persons side-by-side.

triple-tandem Actuator of tandem type with three actuation sections.

triplex Among many meanings, use of three parallel systems with minimal or zero interconnection to provide majority-vote redundancy; such triple redundancy provides for continued operation after single failure (SFO/FS operation). In * detection/correction system one channel may be a model.

triplexer Dual duplexer allowing one aerial to

feed two radar receivers simultaneously, both isolated during transmission of each pulse.

trip line Cable or thin rope attached to far end of sea-anchor drogue which when pulled spills out contents to allow drogue to be wound back on board.

tripod Vectored-thrust turbofan with one hot and two cold nozzles.

Tri-Tac Tri-service tactical communications (DoD).

TRL 1 Tyre rolling limit; upper right boundary of aeroplane TO performance carpet.
2 Thrust-reduction limit in flexible take-off.

TRM Technical requirements manual.

Tr (M) Magnetic track.

TRO Technical resources operation.

trochoid Path traced by point on circle which rolls along straight line. Commonest RC engine pistons are of related profiles.

troland Unit of retinal illuminance; that produced by viewing surface whose luminance is 1 cd/m^2 through artificial pupil of 1 mm^2 area centred on natural pupil. E (in *) = LA (in nits × mm^2).

troll(ing) Flying of random pattern by EW/Elint /ECM aircraft to try to trigger and detect hostile emissions.

trombone Adjustable U-shaped length of wave-guide or co-axial line used for phasing.

trombone lever Cockpit lever with linear trombone-like movement, eg F-111 wing sweep.

tromboning See *path stretching*.

tropical air mass Warm air originating at low latitudes, esp in subtropical high-pressure system.

tropical conditions Various standardized conditions normally including ambient temperature at least 30°C (usually higher) and 100% RH, ie hot and moist.

tropical continental T_c, air mass characterized by high temperature and low humidity; usually unstable and associated with clear sky.

tropical maritime T_m, air mass characterized by high temperature and humidity.

Tropical Maximum Atmosphere One of many artificially averaged atmospheres.

tropical revolving storm Largest and most violent form of thermal depression, generally less than 800 km (500 miles) diameter but with pressure falling in centre to about 960 mb, originating in ITCZ in 5° to 15° N or S; called cyclone (Bay of Bengal, Arabian Sea), hurricane (S Indian Ocean and W Indies), typhoon (China Sea) and willy willy (W Australia).

tropical trials Trials under tropical conditions, for most equipments simulated exactly to specification but for aircraft by flying to actual hot/high airfield; for aircraft low atmospheric density more important than high humidity specified for, say, avionics.

tropical year Period of revolution of Earth round Sun with respect to vernal equinox (with respect to stars, 359°9′ 59·7″); 365 d 5 h 48 min 45·80 s in 1980, increasing 0·0053 s per year.

tropopause Boundary between troposphere and stratosphere characterized by abrupt change in lapse rate (except at high latitudes in winter when change almost undetectable); height fixed in ISA at 11 km (36,089·3 ft), at pressure about 230 mb and relative density about 30%. In practice much higher over Equator (about 15–20 km) than over poles (about 8–10 km).

troposphere Lowest portion of atmosphere, extending from surface to stratosphere; characterized by lapse rate, humidity, vertical air movements and weather. Subdivided into surface boundary layer, Ekman layer and free atmosphere.

tropospheric scatter OTH radio propagation by reflection or scattering from irregularly ionized regions of troposphere; using forward scatter at about 25–60 MHz, ranges of about 1,400 km (870 miles) are possible.

tropospheric wave Radio wave propagated by reflection from place of rapid change in dielectric constant (high ionization gradient).

troubleshooting Process of investigating and detecting cause of hardware malfunction (US usage; UK = fault diagnosis).

trough 1 Long but narrow region of low atmospheric pressure, opposite of ridge, with curving isobars at apex and absence of recognizable front.
2 Point in space where gravitational fields (eg Earth/Moon) cancel out (colloq); neutral point but has implication of vagueness in location.

trouser Fairing for fixed landing gear in form of continuous streamline section round leg and upper part of wheel, usually tapering in chord and thickness.

Trowal Trough of warm air aloft (suggest arch).

TRP 1 Threat-recognition processor (USN).
2 Tuition refund program(me).
3 Thrust-rating panel.
4 Terminal rendezvous point (usually TRV).

TRR 1 Total removal rate.
2 Tyre (tire) rolling radius.

TrReq Track required.

TRRR, TR³ Trilateration range and range-rate.

TRS 1 Tropical revolving storm.
2 Tactical reconnaissance system (or sensor, or squadron).
3 Teleoperator retrieval system.

TRSA Terminal radar service area.

TRSB Time-reference scanning beam MLS.

TRT Terec remote terminal.

Tr (T) True track.

TRTG Tactical radar threat generator.

TRTT Tactical-record traffic teletypewriter.

TRU Transformer/rectifier unit, eg for converting raw AC into low-voltage DC.

truck Bogie of main landing gear (US usage);

same word also used for four MLGs of B–52, which are twin-wheel, one axle.

truck-bed height Commonly accepted average height of payload bed of road truck (UK, lorry), usually taken as 41 in (1,030 mm).

TRUD, trud Time remaining until dive; count of time from launch of upper-atmosphere or other vertical ballistic rocket vehicle and time at which it reaches apogee and begins dive; hence * count.

true airspeed See *airspeed.*

true altitude 1 Actual height above SL; calibrated altitude corrected for air temperature.
2 Observed altitude of body.

true bearing Angle between meridian plane (referred to true N) at observer and vertical plane through observer and observed point.

true course Angle between aircraft longitudinal axis OX and plane of local meridian (referred to true N).

true heading See *true course.*

true-historic display One showing what would have happened, eg line showing locus of impact points of where bullets would have hit had they been fired.

true meridian Great circle through geographical poles.

true north Direction towards N pole of meridian through observer.

true position Position of celestial body or spacecraft computed from orbital parameters of Earth and body without allowance for light time.

true power Actual (I^2R) power of AC electrical circuit.

true prime vertical Vertical circle through true E and W points of horizon.

true Sun Sun as it appears to Earth observer, as distinct from mean, dynamic mean.

true track Angle between true N and aircraft path over ground.

true vertical Local vertical, line passing through observer and centre of Earth.

Tru-Loc Flight-control cable end-fitting swaged on by machine with tensile rating equal to breaking strength of cable.

truncation error EDP (1) error resulting from use of only finite number of terms of infinite series, or from other simplifying techniques, eg calculus of finite differences.

trunk 1 Large-section lightweight air-conditioning duct, or duct between gasbag valve and gas hood.
2 Major route, with or without surrounding box conduit, for large number of controls, services, cables, wire looms and other lines.
3 * route or operator.

trunk route 1 Most important type of commercial route, eg between largest cities in country or countries, offering highest traffic. Hence * operator, domestic *, * traffic.
2 Established air route along which strategic moves of military forces can take place (NATO).
4 Compartment for baggage or general storage in GA aircraft (US usage derived from cars).

truss Rigid load-bearing planar structure of spaceframe type, usually comprising essentially horizontal upper and lower chords linked by various vertical and diagonal members.

truth table EDP (1) technique where coded signals are allocated to particular addresses, comprising series of adjacent address codes and output codes.

TRV Terminal rendezvous point (army helicopters).

TRVR Touchdown runway visual range.

try-again missile Conceptual AAM of 1950s which, finding initial interception was outside design manoeuvre limits, made programmed turn for second attempt.

tryptique See *carnet.*

TS Thunderstorm (ICAO).

Ts Static temperature.

TsAGI Common Western rendition of Central Aero and Hydrodynamics (research) Institute (USSR).

TSC 1 Transportation Systems Center (US DoT).
2 Triple store carrier.

TSCM Technical surveillance countermeasure(s).

TSD Tactical situation display.

TSF 1 Technical supply flight (RAF).
2 Radio (F, arch).

TSFC Thrust specific fuel consumption; SFC of turboprop based on EHP.

TSFE Thermally stimulated field emission.

TSGR Thunderstorm plus hail (ICAO).

TSI Turn and slip indicator.

TsIAM Common Western rendition of Central (research) Institute for Aviation Motors (USSR).

TSIO US piston-engine code: turbosupercharged, direct injection, opposed; now called TIO.

TSIR Total system-integration responsibility (AFSC).

TsKB Common Western rendition of Central Constructor (design) Bureau (USSR).

TSMO Time since major overhaul.

TSMP Terminal-segment master plan.

TSMTR Transmitter (alternative to XMTR).

TSN Time since new.

TSO 1 Technical service order (FAA).
2 Time since overhaul.
3 Technical standard(s) order (CAA).

TSOR Tentative specific operational requirement.

TSP Turret stabilized platform.

TSPI Time/space/position information.

TSPR Total system-performance responsibility (AFSC).

TSR 1 Torpedo spotter reconnaissance.
2 Tactical strike/reconnaissance.

TSS 1 SST (F).
2 Tunable solid-state (laser).

TSSA Thunderstorm plus duststorm or sandstorm (ICAO).
TSSC Technical supply subcommittee (AEA).
TSSTS Tactical sigint system training simulator.
TST 1 Transonic transport (usually M 1·1–1·2).
2 Threshold sampling time.
TSUS Tariff schedules of the US.
TSV Time since shop visit.
TSW Tactical supply wing (RAF).
TT 1 Total time.
2 Dry-bulb temperature.
3 Teletypewriter (ICAO).
4 Turnround time.
5 Target towing (role prefix, UK, defunct).
6 All-weather (F).
TTAE, TTAf&E Total time, airframe and engine.
TT&C Telemetry, tracking and (command or control).
TTB Tanker/transport/bomber (aircrew grading).
TTBT Threshold test-ban treaty.
TTC 1 Tracking, telemetry and command.
2 Technical Training Command (RAF) or Center (US ATC).
3 Top-temperature control.
TTCP The technical co-operation programme (UK/US).
TTCR Time/temperature cycle recording.
TTF 1 Target-towing flight.
2 Tanker task force.
3 Threat-training facility (US).
TTG Time to go.
TTL Transistor/transistor logic; hence TTLIC adds integrated circuit.
TTP Time to protection, measured from firing chaff, aerosol, flare or other dispensed payload to time aircraft can be judged protected.
TTR Target-tracking radar.
TTS 1 Time to station; time in min to fly to tuned DME station. Often a selectable mode on DME panel instrument.
2 Technology Transfer Society (US).
TTSMOH Total time since major overhaul.
TTSN Total time since new.
TTT 1 Tactical technical requirement(s), basis of each military specification (USSR).
2 Tailplane trimming tank.
TTU Triplex transducer unit feeding height, TAS and Mach to CSAS.
TTW Time of tension to war.
TTWS Terminal threat warning system.
TTY Telephone/teletypewriter.
tub Tub section has been used to describe lower rear half of main fuselage structure of pod/boom helicopters.
Tuballoy Depleted uranium.
tube 1 Vacuum tube (UK = thermionic valve).
2 Large modern jetliner (colloq).
tube oil Liquid primer for hot-coating interior of inaccessible workpieces.
tube yawmeter Array of four (or five if one in centre) pitot tubes each inclined at about 45° to flow and radially spaced at 90° so that dP across each opposite pair gives flow inclination in that plane; provides complete picture of local flow direction, little affected by yaw up to 15°.
tubo-annular Gas-turbine combustor of annular type within which are separate tubular flame tubes. Latter are sometimes linked into common turbine NGV ring.
tubular combustor Gas-turbine combustion chamber whose upstream part is of essentially circular cross-section with central fuel burner. Engine may have one or more (early turbojets had as many as 16 because annular technology unknown).
tubular rivet Rivet whose shank is a tube, usually inserted on central mandrel which on withdrawal closes upset head.
tuck, tuck-under Strong tendency to dive, especially at high Mach number.
tufting Covering aircraft, model or portion thereof with tufts.
tufts Short pieces of wool, thread or other very light, flexible and easily visible material which give qualitative picture of local airflow direction and (from steadiness or oscillatory motion or turbulence) vorticity or turbulence. Mounted on aircraft or model skin or at varying distances from it.
tug 1 Self-contained propulsion system for long-term use in space tasked with taking payloads, satellites, spacecraft and portions of station structure, propellants and other supplies (eg delivered by Shuttle Orbiter) and placing them in desired locations or trajectories. Intention is to attach * to all types of payload, in various ways giving thrust without rotation, and to replenish propellants in space.
2 Aeroplane for towing glider, target or other unpowered object. Hence * pilot, * queue, * landing area.
tulip valve PE (4) exhaust valve shaped in side elevation like tulip.
tumble 1 To rotate about lateral axis, ie end over end; rare in aircraft but not uncommon in spacecraft.
2 To rotate metal parts in drum, often with powder abrasive, to remove flash or burrs and obtain polished surface.
3 Of gyro wheel, to precess to limit after toppling.
tumblehome Distance measured parallel to transverse axis from vertical line through widest extremity of seaplane or flying-boat hull to any point on skin above.
tumblehome line Abrupt change of curvature of mould line, buttock line or water-level contour at end of contour.
tumble limit Angular displacement in pitch or roll at which traditional gyro instrument is on gimbal stops.

tumbler Drum in which metal parts are tumbled (2).

tumbler switch Snap-action electrical switch with short operating lever.

tumbling To tumble (1, 2, 3); to be in that condition.

tunable beam approach Pre-ILS landing system (BABS, SBA) in which pilot tuned receiver to particular airfield (arch).

tuned bandpass transformer Typically, primary winding forms anode load of one stage and secondary drives grid of succeeding stage.

tuned circuit Oscillatory circuit with capacitance and inductance selected or tuned to resonate at frequency of applied signal.

tuned grid circuit Parallel resonant circuit linking grid and cathode with max response at resonant frequency; similar resonant response for tuned anode circuit.

tuner Subcircuit or device in RF receiver which selects desired frequency and rejects all others.

tungsten Silvery-grey metal, symbol W, SG 19·3, MPt 3,380°C; one of densest, hardest, strongest and most refractory metals known.

tuning Fine adjustment over continuous analog range of values to obtain that desired, eg RF frequency/wavelength or optimum operating condition of engine or other device.

tunnel 1 See *wind tunnel.*
2 Axial fairing along body, eg to cover pipes, cables and other lines routed outside skin.

tunnel diode Semiconductor device having single p-n junction across which electrons flow by quantum tunnelling; when biased to centre of a negative-resistance mode can operate as amplifier, oscillator or switch.

tunnel shock That occurring immediately downstream of supersonic working section in wind tunnel when Mach number falls below unity; intense if no second throat.

tunnel vision Inability to perceive anything outside an extremely small angular range of FOV, as if one were in a tunnel; caused by disease or high g.

TUP Technology utilization program (NASA, technology transfer).

turbidity Any condition of atmosphere which reduces its transparency to radiation, esp optical; term normally applied to cloud-free sky where * is caused by suspended matter, scintillation and other effects.

turbine 1 Gas-turbine shaft power, ie turboshaft or turboprop; eg helicopter has * power.
2 Prime mover whose power is obtained by action of working fluid (water, steam, cold or hot gas etc) reacting on blades or shaped passages which rotate a shaft. One or more are source of power in all gas-* engines and in turbochargers, turbopumps and many other rotary machines.

turbine-airscrew unit See *turboprop.*

turbine bearing Bearing supporting turbine shaft.

turbine blade Radial aerofoil mounted in edge of turbine disc whose tangential force rotates turbine rotor. Each turbine stage has many blades, occasionally fabricated in groups of two or more.

turbine disc Central member upon which turbine rotor blades are mounted; in many multi-stage turbines there are no flat discs, first and last stages being conical and intervening stages being rings gripped between them.

turbine entry temperature See *turbine gas temperature.*

turbine gas temperature Measured gas temperature at specified plane at inlet to 1st-stage stators, between 1st-stage stators and rotor or at some inter-stage location; best parameter for engine control and monitoring, though EGT more common.

turbine rotor Complete turbine rotating assembly with 1–6 stages; stators are excluded.

turbine shroud Shroud around periphery of turbine rotor stage.

turbine stage Single turbine disc or ring with inserted blades; for completeness should be associated with preceding stator (IGV) stage.

turbine stator Ring of fixed blades, also called inlet guide vanes, upstream of each turbine rotor stage, on to whose blades ** directs gas with optimum distribution of V and pressure for maximum turbine work and efficiency (eg changes radial pressure/V, previously uniform, so that with increasing radius pressure increases while V falls).

turbine vane US usage for turbine blade.

turbine wheel US usage for either one complete stage or complete turbine, possibly of several stages.

Turbinlite Airborne searchlight for night interception experiments (RAF, WW2).

turbo 1 Turbocharger.
2 Generalized prefix meaning driven by or associated with gas turbine, eg *-supercharger, *-ramjet.

turboblower Air blower driven by exhaust-gas turbine to sweep burnt mixture from cylinders of two-stroke diesel. (Today also often used, esp with diesels, to mean turbocharger.)

turbocharger PE (4) supercharger driven by exhaust-gas turbine.

turbofan Most important form of propulsion for all except slow aeroplanes (say, below 600 km/h, 375 mph); comprises gas-turbine core engine, essentially a simple turbojet, plus extra turbine stages (usually on separate LP shaft) driving large-diameter fan ducting very large propulsive airflow round core engine and generating most of thrust. For given fuel consumption generates much more TO thrust than turbojet, with many times less noise, but performance falls off more rapidly with forward speed. A few * engines have aft fan whose blades form outward extensions of those of driving gas turbine.

turbofan-prop Turbofan driving propfan mounted ahead of inlet and acting as fan booster stage to give jet as well as shaft power.

turbojet Simplest form of gas turbine, comprising compressor, combustion chamber and turbine, latter extracting only just enough energy from gas flow to drive compressor. Most of energy remains in gas, which is expanded to atmosphere at high velocity through constricting propelling nozzle. In supersonic aircraft often fitted with afterburner.

turboprop Gas turbine similar to turbofan but with extra turbine power geared down to drive propeller. Difference between two forms of engine is of degree only; fan of turbofan is invariably shrouded, running inside profiled case, while propeller (but not propulsor) normally operates unshrouded and unducted. In general * has much higher bypass ratio than turbofan and is tailored to slower aircraft.

turbopump Pump driven by turbine turned by gas from rocket propellants.

turboramjet Combination of turbojet and ramjet as integrated propulsion for supersonic aircraft. In theory afterburning turbojet or turbofan can be classed as *, but in practice true * is very large ramjet within or upstream of which is turbojet for starting from rest and acceleration to ramjet lightup speed. In some forms large valves or duct-diversion doors are needed to change internal flows between turbojet and ramjet modes.

turborocket Various combinations of gas turbine and rocket in one engine.

turboshaft Gas turbine for delivering shaft power, eg to power helicopter, ACV or other non-flying vehicle. Essentially a turbofan or turboprop with fan or propeller removed. Often can deliver power at either end, and usually has at least one stage of speed-reducing gearbox.

turbostarter Main-engine starter driven by turbine turned by gas from cartridge, IPN or other fuel.

turbosupercharger See *turbocharger*.

turbulator See *vortex generator*.

turbulence Time-variant random motion of fluid in which velocity of any particle, or at any point, is characterized by wild and unpredictable fluctuations which are extremely effective in conveying heat, momentum and material from one part of fluid to others. Called isentropic if rms velocity is same in all directions.

turbulence cloud Cloud formed because of atmospheric turbulence, usually distinctive layer above condensation level about 30 mb (say, 300 m, 1,000 ft) thick in otherwise stable air.

turbulence control structure Gigantic "golfball" with porous walls attached to inlet of engine on outdoor test to eliminate effect of wind.

turbulence number R (4) at which C_d of smooth sphere becomes 0·3.

turbulence plot Term has been used for plotting wake turbulence behind aircraft, building or other bodies, and also for recording geographical locations of severe atmospheric turbulence, including CAT, over a long period.

turbulence screen Screen across wind tunnel to reduce turbulence, usually rectilinear array of crossing sharp-edged strips.

turbulent boundary layer One that is no longer laminar, characterized by gross random lateral motions, and Reynolds stresses much larger than viscous; all boundary layers become turbulent at R = 250,000+ unless surface unusually smooth, though apart from rise in skin-friction drag there should be no other significant effect and no separation.

turbulent bursts Microscopic eruptions which occur constantly over aircraft (or other) surface, beginning at surface; responsible for most of skin-friction drag and nearly half total aerodynamic drag.

turbulent flow Flow having turbulence superimposed on main movement, measured as velocity increments about all three axes expressed as fraction or % of mean flow velocity.

turkey 1 Badly designed aircraft, especially aeroplane, with sluggish or dangerous handling.
2 Aircraft with performance so poor as to be useless.

turn Angular change of track; thus 30° * does not mean 30° bank.

turn and slip Flight instrument indicating rate of turn by needle and slip/skid by ball or bubble in lateral tube; in related turn and bank slip/skid is indicated by second needle at 6 o'clock position.

turnaround Elapsed time between aircraft parking at stopping point and moving off to continue flight or carry out fresh mission.

turnaround cycle Comprises loading time at home base, time to/from destination, unloading/loading time at destination, unloading time at home, planned maintenance and, where applicable, time awaiting facilities (DoD).

turnaround time See *turnaround*.

turnbuckle Double-ended threaded barrel for adjusting wire/cable tension.

turn co-ordinator Rate gyro which senses rotation about both roll and yaw axes (US).

turn errors See *turning errors*.

turn indicator Flight instrument indicating rate of turn about aircraft vertical axis, almost always combined with slip/skid.

turning 1 General term for operation on a lathe in which workpiece is rotated.
2 Manoeuvre by which parachutist rotates to face direction of drift.
3 Rotating engine by means other than own power; also called cranking.

turning errors Those due to instrument deficiencies, eg due to acceleration; see *acceleration error*.

turn-in point Point in space at which aircraft starts to turn from approach direction to line of attack (DoD, NATO, becoming arch).

turnoff Point at which aeroplane leaves runway to taxi to parking place, normally also junction of runway and paved taxiway; high-speed * is configured with gentle turn radius to avoid overstressing landing gear laterally.

turnoff lights Flush lights at 15 m (50 ft) intervals defining curved path from runway centreline to taxiway centreline.

turnover structure Strong structure intended to protect aircraft occupants in event of overturn on ground, eg crash arch, crash pylon.

turnover voltage Reverse V of point-contact semiconductor device (corresponding to reverse breakdown V of junction device) beyond which control over reverse current is lost.

turn radius Half lateral distance required to change heading 180°.

turnstile aerial Crossed dipoles, equal quadrature-phase signals.

turpentine One of terpenes, BPt 155°C; solvent, thinner and (in France) rocket fuel.

turret Enclosure for airborne gun(s) able to rotate in azimuth and with provision for gun rotation in elevation; manned or remotely controlled, but in latter case merged indefinably into barbette. Later forms (c1955) exactly resembled sting of wasp and did not fit traditional configuration. Today found mainly on helicopters, always with remote control.

turret lathe Equipped with rotating toolholder for six (occasionally more or less) tools, indexed to work in sequence; large capstan with power operation and turret on main bed-slides.

Tuslog The US Logistics Group (USAFE).

TV 1 Television.
2 Terminal velocity.
3 Theatre of war (USSR).

TVAT Television air trainer; carried on external pylon.

TVBS Television broadcast satellite.

TVC Thrust-vector control.

TV command Guidance by human operator watching TV picture from camera in nose of controlled vehicle.

TVD 1 Turboprop (USSR).
2 Theatre(s) of military operation (USSR).

TVDU Television display unit.

TV homing Automatic homing guidance by comparing TV picture with sequence stored in missile memory.

TVI TV interference.

TVLA Tuned vertical line array (sonobuoy).

TVM Track-via-missile.

TVOR Terminal VOR.

TVR 1 Track-via-missile radar guidance.
2 Trajectory/velocity radar.

TVSU Television sight unit.

T/W Thrust/weight ratio.

T_w Torque applied at wing root.

TW/AA Tactical warning and attack assessment (Norad).

TWC Tungsten-wire composite.

TWD Touchwire display.

TWEB Transcribed weather broadcast.

20-minute rating A-h rating of battery indicating current required to discharge from max to zero in 20 min.

TWI Tail-warning indicator.

twilight Designated civil *, nautical * or astronomical * at Sun zenith angles of 96°, 102°, 108°, respectively.

twilight band Strip, about one aircraft span in width, where radio range A/N just detected against steady note (obs).

twilight effect Faulty indications of radio navaids of pre-1950 (occasionally, later) during twilight ascribed to ionospheric distortions.

twilight zone Bi-signal radio range zone where only A or N heard (obs).

twin Aircraft, usually aeroplane, powered by two similar engines; arguably, one powered by two different engines, eg jet plus prop.

twin-aisle aircraft Usually synonymous with widebody, transport whose passenger seating is divided by two axial aisles.

twin cable Plastic extrusion containing two side-by-side untwisted conductors.

twin-contact tyre See *twin-tread.*

twin-float seaplane One supported on water by two similar floats side-by-side.

twin-gyro platform Platform housing two gimballed gyros providing precision heading and attitude reference for military aircraft; additional input, eg Doppler, needed for position readout.

twinkle roll Has been used to mean flick roll, but normally max-rate slow roll performed by two or more formating aircraft simultaneously, each about own axis.

twin paradox See *time dilation.*

twin-row PE (4) of radial type with two rows each occupying one plane and each driving on one crankpin.

twin tail Conventional tail with twin vertical surfaces, often at extremities of tailplane (1935–50).

twin-tread tyre Contacting ground around two circular treads separated by deep groove; intended to reduce shimmy, esp of tailwheels.

twist 1 Variation in angle of incidence along aerofoil, always present in rotating blades. In wing, normally subdivided into aerodynamic (defined as variation in no-lift direction along span) and geometric (variation in angle between chord and fixed datum along span).
2 Roll about longitudinal axis, as in * and steer.

twist and steer Control method for fixed-wing aerodynes in which it is necessary first to roll to (normally large) bank angle and then pull round on to desired heading. Only possible method on aircraft (eg missiles) with no control surfaces other than main wings, which pivot independently for twist and together for steer. Not applicable to conventional aeroplanes, which make simultaneous harmonized movements about all axes.

twitch factor Unquantified factor degrading flight-crew (esp pilot) performance in terms of accuracy or incidence of errors, resulting from fear or high workload. In simplest case ** magnifies task of adding 2+2.

twizzle Various manoeuvres adopted by combat pilots, esp of large aircraft, to throw fighter off aim or effect escape; always involved steep climb and/or dive, combined with limited rolling manoeuvres (WW2).

2½-axis machining Continuous-path machining (usually with NC) on x, y axes, intermittent on z.

two-axis autopilot Simple autopilot having authority in roll and yaw only.

two-axis homing head Scans in az and el.

two-bay biplane Wing assembly comprises two bays on each side of fuselage, each comprising rectangular cellule of inner and outer interplane struts with points of attachment linked by diagonal wires. Loosely, biplane with interplane struts at two different spanwise locations on each side of fuselage (discounting any struts near fuselage).

two-colour pyrometer High temperature is measured at two wavelengths.

two-control aircraft Aeroplane or glider with flight-control system in two axes only, invariably ailerons and elevator; thus no rudder or pedals.

two-crew Misnomer; what is meant is that flight crew comprises two persons; * operation is often matter between airline employers and ALPA, and largest transports frequently have flight crew of at least three.

two-cycle See *two-stroke*.

2-D flow Flow which can be described completely in one plane, eg around wing of infinite length.

2-D inlet Two-dimensional inlets are in theory ideal for operation over wide range of Mach numbers but because of end and corner effects and mechanical complexity are in practice inferior to axi-symmetric.

2-D nozzle Jet nozzle of rectangular cross-section and constant longitudinal profile which can vector thrust in vertical plane. Studied for future US fighter.

2-D radar Radar giving target position in two dimensions, eg az/el, or range/bearing; only a few can pinpoint body's location in 3-D space.

2-D wing Usually one of infinite span, whose tip effects can thus be ignored; for some purposes constant-section wing joined at each end to tunnel wall is approximation.

2 Gins Two-gimbal INS (colloq).

two-level bridge (jetway) Passenger bridge whose landside end can be raised or lowered to mate with arrival or departure floor level.

two-man rule Philosophy under which no individual is allowed unaccompanied access to NW or certain designated components or associated system interfaces.

two-meal service Two meals served on one sector.

2-minute turn Rate 1 turn; also called standard turn or procedure turn.

two-moment equation Relates simple-beam loading to shear and BM.

2 on 1 injector Rocket liquid injector spraying two streams of liquid A on to one stream of liquid B, the three meeting at various angles depending on design.

2-place Two-seater.

2-point landing See *wheeler*.

2-point suspension Bifilar, hung from two points at same level on two filaments or cables.

2-pole switch Opens or closes both sides of same circuit or two separate circuits simultaneously.

two-position propeller Has only two settings, fine and coarse.

two-pulse rocket Motor, usually solid propellant, comprising two stages (eg, boost and sustain) fired in series.

two-rate oleo Normally, first part of travel is at low rate (ie, large deflection per unit load); after loading to given limit, often close to static position at gross weight, high-rate law applies. Common on naval helicopters.

two-row radial See *twin-row*.

two-segment approach Two angles, usually 6°/3°.

two-shaft engine Gas turbine with mechanically independent LP and HP sections each running at its own speed.

two-speed supercharger Driven by step-up gearbox which, usually automatically at given pressure height, changes ratio to increase impeller speed at high altitudes.

two-spool Gas turbine of two-shaft type, originally confined to engines having two axial compressors (LP and HP) in series. Also called split-compressor engine.

two-stage amber Synthetic day/night pilot training aid in which pilot wears blue goggles and flies trainer with amber transparencies (or vice versa) which effectively eliminates external cues while allowing view of instruments.

two-stage igniter Generally, one that first ignites a local flame, burning main fuel, which in turn is used to ignite main combustion (gas turbine or rocket but now rare).

two-stage supercharger One having two impellers in series, either both driven by crankshaft or with first (LP impeller) driven by exhaust turbo.

2-star red Standard distress pyrotechnic, fired without pistol.

two-step supercharge Throttle is gated below rated altitude, thereafter being free to move to max position; used with or without two drive ratios.

2-stick Dual-control aircraft (colloq).

2-stroke PE (4) cycle in which every upstroke expels exhaust and compresses mixture and every downstroke provides power. Often abb 2T. Not normally certificatable for unrestricted aviation use.
2-view drawing Orthographic projection drawing comprising two views, usually left-side elevation and plan.
TWR 1 Tail-warning radar.
2 Tower (or twr).
twr Tower; one definition = aerodrome control (arch).
T/W ratio Thrust/weight ratio.
TWS 1 Tail-warning set (or system).
2 Track while scan.
3 Threat-warning system.
TWSC Thin-wall steel case.
TWSS TOW weapon subsystem.
TWT Travelling-wave tube.
TWTA TWT amplifier.
TWU Tactical weapons unit (RAF).
TWX Theatre war exercise.
TWY, twy Taxiway.
TWYL Taxiway link.
TX, Tx Transmitter, transmit, transmission.
TXP Transponder (also TPDR, XPDR etc).
TYD, TY$ Then-year dollars, ie based on a selected historic monetary value.
TYP Type of aircraft (ICAO).
type Particular type of aircraft, ignoring marks or subdivisions.
type certificate Legal document, in USA issued by FAA, allowing manufacturer to offer item (eg aircraft, engine) for sale.
type certificate data sheets Official specifications to which each unit (aircraft, engine, propeller) commercially offered for sale must conform (FAA).
type conversion Clearance of qualified pilot to fly additional type of aircraft.
type record Various books or document dossiers, some of which have legal status, but basically complete record of all decisions made regarding particular type of hardware, esp including all modifications, service difficulties, airworthiness directives and alterations made locally, eg on owner's initiative.
type test Basic government test clearing engine or other machine for production and acceptance by government customers (UK).
TYPH Typhoon.
typhoon Tropical revolving storm.
typical In structures and structural components, arithmetical mean structure calculated by measuring sections of all members actually produced.
Ty-Rap Patented nylon strap for tying bundles (looms) of electrical wiring.
tyre sizes Usual sequence of three numerical values is overall diameter, section width (undeflected) and (if present) inner diameter or bead size.
Tyuratam Soviet test centre for ICBMs, various other missiles and FOBS, and launch centre for Cosmos military satellites; location 45·8°N, 63·4°E.
TZD True zenith distance.

U

U 1 Overall heat-transfer coefficient.
2 US military-aircraft designation prefix, utility (and 1946–52 unpiloted, USN).
3 US missile designation prefix, first letter, underwater-launched.
4 US missile designation prefix, second (middle) letter, underwater attack, ie ASW.
5 Linear acceleration (arch, today symbol is A or a).
6 Internal or intrinsic energy quantity.
7 IFR flightplan code, transponder with 4096-code capability.
8 Occasionally used for fluid velocities.
9 Umrüst, factory mod (G, WW2).
10 Unmanned.

u 1 Force in structural member due to unit load.
2 Tangential velocity, or linear velocity of point in rotating structure, eg aeroplane in roll.

U-alpha Free-stream fluid velocity.

U-code See *U (7).*

U-index Monthly mean of differences between consecutive daily mean values of horizontal component of Earth's magnetic field.

u-index Value of U-index divided by sin of magnetic co-latitude multiplied by cos of angle between magnetic meridian and horizontal component.

U-tail Aeroplane tail with twin verticals attached to fuselage (can be inclined outwards but not applicable to butterfly tail).

UA 1 Uncontrolled airspace.
2 Unit of account; standard accounting unit (EEC).

Uα See *U-alpha.*

UAAA Ultralight Aircraft Association of Australia.

UAB Until advised by (ICAO).

UAC Upper-area control centre, or (CAA) upper airspace control.

UAI Union Astronomique Internationale (Int).

UANC Upper-airspace nav chart.

UAR 1 Upper air route.
2 Unattended radar.

UARS Unattended radar station.

UAS 1 University air squadron(s).
2 Upper airspace.

UATI Union des Associations Techniques Internationales (Int).

UATP Universal air travel plan (Int, US-based).

UAV Unmanned air vehicle.

UBE Ultra-bypass engine.

Ubee, ubie Alternatives to UBE.

U/c Undercarriage, ie landing gear.

UCCEGA Union des Chambres de Commerce et Établissements Gestionnaires d'Aéroports (F).

Uchinoura Rocket launch site, Kagoshima prefecture (J).

UCI Unit contruction index; measure of flotation (ability of landplane to use soft airfields) based on MTOW, tyre pressure and gear geometry.

UCIMU Unione Costruttori Italiani Macchine Utensili.

UCL Up command link; opposite of DDL.

UCNI Unified com/nav/ident.

UDF 1 UHF D/F.
2 Unducted fan.
3 Unit development folder (software).

UDMH Unsymmetrical dimethyl hydrazine; hypergolic rocket fuel of hydrazine family.

UDMU Universal decoder memory unit (GPATS).

UDS 1 Unidirectional solidification.
2 Unsupported (or unstocked) dispersed (or dispersal) site.

UDT Unidirectional transducer.

UE, u/e 1 Under-excitation.
2 (UE only) Unit equipment (or establishment), list of aircraft or other items serving with combat units.

UEO WEU (F).

UER Unscheduled engine removal.

UFA Until further advised.

UFC 1 Unified fuel control.
2 Up-front control.

UFDR Universal flight-data recorder.

UFN Until further notice.

UFO Unidentified flying object.

UFS Ultimate factor of safety.

UFTAS, Uftas Uniform flight-test analysis system (AFFTC).

UGAI Unione Giornalisti Aerospaziali Italiani.

Ugine-Séjournet Method of extruding steel using glass as lubricant.

UGM US prefix, underwater-launched surface-attack missile.

UGS Upgraded silo.

UHC Unburned hydrocarbons.

UHD Ultra-high density.

UHF, uhf Ultra-high frequency, usually 300–3,000 MHz (see *frequency*).

UHPT Undergraduate helicopter pilot training.

UHS Ultra-high-speed (logic).

UHV Ultra-high vacuum.

UIC Upper information centre (ICAO).

UIF Unfavourable information file.

UIP Upgrading instructor pilots.

UIR Upper-airspace FIR.

UIS Upper information system.

UIT Union Internationale des Télécommunications.
UJT Unijunction transistor.
UK Training centre (USSR).
UKACC UK Air Cargo Club.
UKADR UK air-defence region.
UKF Unbemanntes Kampfflugzeug = RPV (G).
UKISC UK Industrial Space Committee.
UKMF UK Mobile Force.
UKOOA UK Offshore Operators' Association.
UKRAOC UK Regional Air Operations Centre.
UKWMO UK Warning and Monitoring Organization.
UL Ultralight.
ULA 1 Uncommitted logic array.
2 Ultra-light aircraft.
ULAA Ultra-Light Aircraft Association (Aust).
Ulana Unified local-area network architecture.
ULC Unit load container.
ULD Unit load device, eg LD-3 container, 88 × 125 pallet etc.
ULEA Ultra-long-endurance aircraft.
ULL Ullage (NASA).
ULLA Ultra-low-level airdrop.
ullage Volume above liquid in a tank; occasionally misused to mean last dregs of liquid itself remaining in tank.
ullage engine Rocket motor fired in ullage manoeuvre.
ullage manoeuvre Applying axial thrust to space-launcher stage or other liquid propellant tank(s) that are nearly empty in order to collect remaining propellants around delivery pipe connection.
ullage motor See *ullage engine.*
ullage space See *ullage.*
ULM Microlight (F).
ULMS Undersea-launch (or long-range) missile system.
ULSA Ultra-low-sidelobe antenna.
ultimate factor of safety Number by which limit (or proof) load is multiplied to obtain ultimate load; purely arbitrary factor of safety, usually varying from 1·5 to 2, representing best humanly attainable compromise between economic structure weight and aircraft that will not break.
ultimate load Greatest load that any structural member is required to carry without breaking; that at which it may legally be on verge of breaking. Usually, limit load × UFS.
ultimate strength Strength required to bear ultimate loads, or product of greatest load considered possible in service multiplied by UFS. Gust requirements, in which structure oscillates, can be superimposed on max static load to demand even greater design strength in "flappable" parts of structure, esp wing.
ultimate stress That in a member or piece of material at moment of fracture; in theory that in structural member loaded to ultimate strength.
ultra-bypass engine One name for unducted prop-fan, with BPR from 20 to 50 (Boeing).
ultra-high-density seating Seating configuration for maximum number of passengers (invariably greater than original certification limit) with minimal pitch and no galley.
ultra-high frequency See *frequency.*
ultra-high magnetic field Generally, greater than 10 T (100 kG).
ultra-high-speed photography Rate higher than 10^4 images/s.
ultra-high vacuum Previously below 10^{-13} torr (not apparently yet defined in SI, but this is 13·33 × 10^{-11} N/m^2); density 3·2 × 10^3 mol/cc, mean free path (air 25°C) 3·8 × 10^{10}.
ultralight aircraft 1 Most common definition: not exceeding 1,000 lb (454 kg) MTOW.
2 In US, term for microlight, with severe upper limit of 254 lb (115 kg) empty.
ultra-low airdrop Below 15 m (50 ft) AGL.
ultramicroscope Instrument for observing extremely small particles by intense illumination causing diffraction rings on dark background.
ultrasonic Mechanical vibrations, eg sound waves, of frequency too high to be audible to humans (opp = infrasonic). Generally frequencies above 15 kHz, usually generated by electro-acoustic transducer and propagated through solids, liquids and gases.
ultrasonic bonding Techniques of joining materials with ultrasonic energy, in some cases to cause local melting (ultrasonic welding) and in others to cause either enhanced local diffusion or merely heating to accelerate curing or setting of adhesive.
ultrasonic inspection NDT method in which cracks are revealed by discontinuity in propagation of ultrasonic waves through metal.
ultrasonic machining Techniques for shaping solids, esp those too hard or for any other reason not machinable by conventional means. Most common method uses shaped tool oscillated vertically above work at frequency about 20 kHz, with space between tool/work fed with hard powder abrasive.
ultrasonic rolling Transducer (typically 20 kHz) placed inside one (rarely both) of rolls in rolling mill with objective of reducing rolling energy needed, reducing temperature of metal being rolled and rolling to thinner gauges.
ultrasonic welding Techniques in which ultrasonic vibrations are used to join two metal (rarely, either or both is non-metal) surfaces brought into contact. Surfaces usually well mating, atomically clean and pressed together under pressure. In some techniques ultrasonic rearrangement of atoms is sufficient to cause melting at interface, while in others there is diffusion between two faces.
ultrasound Sound of ultrasonic frequency.
ultraviolet Band of EM radiation of frequency just too high (ie wavelength too short) to be visible; links visible violet at about 3·8 to 4 ×

10^{-7} m to limiting wavelength put at from 1 to $1{\cdot}36 \times 10^{-8}$ m (next major part of spectrum is X-ray at appreciably shorter wavelength).

ultraviolet detector Material, eg single-crystal silicon carbide, not triggered by sunlight or other incident radiation but instantly reponsive to UV.

ultraviolet imagery That produced by sensing UV reflected from target (DoD).

ULV Upper limit of video (HUD).

UMA 1 Unmanned aircraft, drone or RPV.
2 Uffici Meteorologici Aeronautica (I).

umbilical 1 Connection between vehicle, esp space launcher, missile (esp large ballistic) or RPV, and ground along which pass all necessary supplies and signals up to moment of liftoff or launch, eg computer data-link, telemetry, guidance programming, topping-up liquid propellants, electrical/hydraulic/pneumatic power and instrumentation. Generally grouped into one (at most two) large multiple conduit with pull-off connector. Occasionally plural.
2 Jetway, loading bridge (colloq).

umbilical cord See *umbilical (1)*.

umbilical tower Tower carrying umbilical lines and connector to an upper stage of a ballistic vehicle; often called umbilical mast.

umkehr Anomaly of relative zenith intensities of scattered sunlight near UV as Sun nears horizon; often cap U.

UMLS Universal MLS.

UMT Universal military training.

unaccelerated flight Flight without imposed acceleration; but there are several conflicting definitions and term needs redefining to avoid ambiguity: (1) flight along straight-line trajectory at constant V, thus no imposed acceleration; (2) rectilinear flight (eg straight and level) with no imposed acceleration normal to flightpath other than Earth gravity; (3) weightless flight with not even gravitational acceleration, thus curvilinear around Earth. Definition (2) allows acceleration along line of flight, thus speed can vary.

unaccompanied baggage Not carried on same aircraft with pax or crew to which it belongs; DoD definition adds "not carried free on ticket used for personal travel".

unanimity rule Basic IATA principle that fare from A to B is same as from B to A and same for all carriers offering identical service; can be relaxed.

unapproved Not tested to established requirements, thus flown on special dispensation, eg VW engines with single ignition.

unaugmented In case of turbojet or turbofan, not equipped with afterburner, or not using afterburner.

unavbl Unavailable.

Unavia Italian national organization for study and development of aircraft technology.

unbalanced cell Cell of Ni/Cd battery which has discharged more than others; first step to thermal runaway.

unbalanced field Any TO in which accelerate/stop distance is not same as normal TO to 35 ft.

unbalanced turn One with slip or skid.

unblown 1 STOL aircraft with USB, IBF, EBF or other powered blowing system in flight mode with blowing inoperative.
2 Of PE (4), not equipped with supercharger.

UNC United Nations command; followed by various force initials.

uncertain Category of aircraft whose safety is not known, normally applied when 30 min elapsed since arrival message or ETA and not answering radio call; hence uncertainty phase.

Uncertificated Aircraft category; airworthiness not established.

UNCL Unified numerical-control language.

Unclassified 1 Security category for official matter which requires no safeguards but may be controlled for other reasons.
2 Performance category for aircraft, usually civil transports, in service prior to 1951 and thus not built to CAR.4(b), BCARs or SR.422A/B.

uncontrolled airspace Airspace where no ATC service is provided.

uncontrolled mosaic Made up of uncorrected images matched from print to print without ground control or other orientation, giving mosaic on which distances and bearings cannot be accurately measured.

unco-operative Generalized adjective for vehicle of essentially passive nature, devoid of helpful emissions and not responding when electronically challenged; applied to aircraft, usually means not equipped with transponder.

uncoupled Vibration mode wholly independent of others at same time.

Unctad UN Conference on Trade and Development.

unctld Uncontrolled airspace (FAA).

undamped Vibration of free nature dependent only on internal mass and elastic and inertia forces.

underbreathing See *hypoventilation*.

undercarriage Landing gear (UK usage; original BSI definition included main wheels, skids or floats and support and explicitly excluded tail wheel or skid). Hence * door, indicator, same as landing gear door, indicator. Note: floats are usually called alighting gear.

underdeck spray Pad deluge directed up at underside of launch pad (in some cases deliberately including underside of base of launched vehicle).

undergraduate Aircrew (esp pilot) pupil who has not yet qualified, ie won his/her wings.

underlay Lowest stratum of layered defence.

undershoot 1 Faulty approach by aerodyne, usually fixed-wing, which if continued to ground results in landing short of desired area, eg before threshold; normally corrected by go-around

(overshoot).
2 To fail to capture desired flight condition, eg altitude, IAS, by falling short when approaching value from below, normally through small lack of aircraft energy.

under the hood Instrument flight training in which pupil is prevented from seeing outside aircraft by folding opaque hood.

under the radar Flight levels as close as possible (see *lo*) to ground in attempt to thwart hostile attempts to obtain positive track by defensive radars; becoming pious hope.

underwater missile Launched below water surface.

underwedge bleed Secondary airflow extracted from location on underside of variable wedge above supersonic airbreathing inlet, usually from point of max wedge depth at throat.

undevelopable 3-D curvature, which cannot be drawn accurately on flat sheet and can be made only by some type of sheet forming.

undk Undock(ing).

undock 1 To separate two vehicles in space previously joined and with intercommunication but not necessarily sharing common atmosphere.
2 To remove airship from hangar.

unducted fan Engine in which gas-turbine core drives fan blades external to cowled engine. Blades can be mounted on ring of turbine blades in gas flow, or driven via gearbox (tractor or pusher) from separate turbine.

unduplicated Generalized term meaning that in assessing airline or other route network each sector is counted once only, in one direction only, despite fact it is used in both directions and may form part of from 2 to 48 distinct routes; hence * route mileage, * network.

unfactored Not multiplied by a factor, eg factor of safety; hence * load = limit load.

unfavourable unbalanced field One whose clearway allows TO at increased weight at which TOD to 35 ft exceeds accelerate/stop distance.

unfeather To restore propeller from feathered state to normal operative state with engine transmitting power and blades in positive pitch.

UNGA Unione Nazionale Giovanile Aeronautica (I).

UNICE, Unice Union of the industrial federations of the EEC countries.

Unicom Private radio com service on five frequencies based at airports, heliports etc, with or without tower in addition; used for various advisories other than ATC purposes (US).

unidirectional aerial Single well defined direction of max gain.

unidirectional composite All fibres are parallel, usually aligned with direction of applied load.

unidirectional current Flowing in one direction only, eg signal pulses or DC with superimposed AC.

unidirectional solidification Techniques for obtaining strongly preferred direction of crystalline grains on solidification of alloy from melt such that each crystal forms long string or column aligned with max applied load, transverse intercrystalline joints being rare. In most applications, eg turbine rotor blades, a step on route to single-crystal material.

Unido UN Industrial Development Organization.

unified fuel control Control system for supersonic airbreathing engine governing engine acceleration, pilot input response, Mach, nozzle and Plap.

unified system Generalized term for TTC (tracking, telemetry and command) system handling all digital, video and voice signals in integrated system for large or complex air vehicle or space payload. Also applied to simpler satellites, missiles and RPVs, eg unified-pulse systems in which each command pulse train comprises vehicle address, message ident and message, synchronized with radar ranging pulses; thus one ** handles all vehicles.

Unified thread Standard 60° screwthread (US, UK, Can).

uniform acceleration Time-invariant, giving straight line on V/T plot.

uniformly distributed load One imposing constant force on each unit area of horizontal floor.

uniformly varying load Magnitude varies directly with distance along straight-line axis from reference point.

uniform photo-interpretation report Third-phase machine-formatted intelligence report of particular objective containing detailed information extracted from photo sensor imagery (USAF).

uniform velocity Time-invariant speed and direction.

Unihedd Universal head-down display; standardized cockpit display of information from many sources and sensors with high brightness.

unijunction transistor Bar of doped semiconductor with p-n junction near centre, normally forward-biased, giving sharp peak of emitter V for small emitter current; used in pulse generators and sweep circuits. Also called double-base diode.

unilateral instrumentation In unyawed flight each half-aircraft behaves as mirror image of other and, therefore, full information can be obtained by fitting pressure-sensing heads, transducers and other experimental equipment on one half only; ** can be used during yawed flight.

unipole Hypothetical aerial radiating equally in all directions.

Unishear Fixed or portable high-speed shear for thin sheet.

unit 1 Fundamental subdivision of any numerical measure. Nearly all industrial countries have in theory adopted SI system. Base * of seven fundamental quantities are: length, metre (m);

mass, kilogram(me) (kg); time, second (s); electrical current, ampere (A); thermodynamic temperature, kelvin (K); luminous intensity, candela (cd); and amount of substance, mole (mol).
2 Military element whose structure is prescribed by competent authority; normally applied only to small organizations at field level, eg squadron, or to element of airborne force, eg two or three aircraft.
3 One complete item from production, eg production *, * sales.

unit aircraft Those provided to an aircraft unit for performance of a flying mission (DoD).

unit area That equal to square of unit length on each side.

unitary Having a single warhead, as distinct from dispensed or clustered munitions.

unit deformation Deformation divided by original length or other undistorted measure; ie deformation per unit of original length.

unitized load Any collection of loads packaged in container, on pallet or in any other way securely joined so that they can be treated as one item.

unit of issue Measure in which commodity (esp military) is issued, eg by number, dozens, metres, feet, US gal etc.

units of measurement See *unit (1)*. In most practical flying SI is still unattained, distances being in nautical miles, pressures in lb/sq in, atmos, bar and various other units (hardly ever Pa), acceleration in g and volumes in litres, US gal or Imp gal instead of m^3 (but fuel mass generally in kg).

unit stress Usually load divided by cross-section area normal to applied force direction.

universal gas constant See *gas constant*.

universal motor Electrical motor operative on AC or DC.

universal polar stereographic Grid for regions from 80° latitude to poles.

universal receiver Operative on AC or DC, with various protective devices.

universal shunt Resistor in parallel with ammeter to increase range of currents measured as FSD is approached.

universal tester Usually, hand meter for measuring AC/DC current, AC/DC volts and resistances.

universal time Defined by rotation of Earth, and thus not absolutely uniform; ** 1, corrected for polar motion; ** 2, corrected for seasonal variation in rotation.

universal timing disc Disc graduated 000°/360° with associated pendulous pointer, attached to PE (4) crankshaft.

universal transmission function Attempts to describe mathematically IR propagation in atmosphere.

universal transverse Mercator Grid co-ordinate sytem from 84°N to 80°S.

unlgtd Unlighted (FAA).

unlimited ceiling Traditionally, less than 50% cloud, base 9,750+ ft AGL (FAA).

unload To reduce g (normal acceleration), usually to restore lost speed.

unlocked Automatic-gun action in which at moment of firing breech is not locked (eg to barrel or case) but is of sufficient mass (inertia) for pressure to have fallen to safe level before breech opens significantly; usually synonymous with blowback action.

unmask Point at which vehicle becomes visible to defending surveillance systems.

unmkd Unmarked (eg obstruction) (FAA).

UNOSC UN Outer Space Committee.

unprepared airfield Usual meaning is without permanent paved runway.

unrefuelled range Range on fuel carried at take-off; usually applied to aircraft with provision for inflight refuelling.

unreinforced ablator Without honeycomb filling or backing.

unrotated projectile Cover name for unguided air and ground-launched rockets (UK, WW2).

unscheduled maintenance Those unpredictable maintenance requirements not previously planned or programmed but which require prompt attention and must be added to, integrated with or substituted for previously scheduled workloads (USAF).

unsealed strip Runway with no waterproof coating; normally means compacted earth, gravel or other substrate with or without top layer of rolled material or prefabricated mat of mesh or steel planking.

unstable aerofoil Generally means one with extensive CP travel.

unstable air Air in which temperature decreases with height at rate greater than DALR or SALR; thus a parcel given small vertical movement (up or down) will continue with increasing speed.

unstart Explosively violent breakdown of correct (ie started) airflow through supersonic inlet to airbreathing engine, notably with expulsion of shockwave(s) and temporary reversal of flow. All highly supersonic (M 3+) engines have such large contraction ratio and constricted throat that any yaw, spillage from neighbouring engine, gunfire or other disturbance can cause *.

unstick Point at which fixed-wing aerodyne leaves surface of land or water; hence * run = ground or water run; * speed = that at which aircraft becomes airborne, usually about 25% of way from V_R to V_2 (symbol V_{us}).

unstick-speed ratio V/V_{us}, ratio of aircraft speed to unstick speed either as % or as fraction. Usually plotted as abscissa on take-off performance graph, esp of marine aircraft.

unstressed Not bearing significant external load.

unsupported site Possible operating site for V/STOL aircraft but devoid of prestocked supplies, eg POL, ammunition etc.

unsymmetrical Generalized chemical description

of molecular structure where left is not mirror image of right; eg in 1,2 dimethylhydrazine each N has an H and a CH_3 attached (symmetrical), but 1,1 dimethylhydrazine has left N joined to two H atoms and right N joined to two CH_3 methyl molecules and is unsymmetrical (UDMH).

unsymmetrical flight Condition in which aircraft (aeroplane or glider) is not balanced about longitudinal axis, due to roll, roll/yaw or other rotary manoeuvre causing gross alteration to normally symmetric wing lift. Hence unsymmetric load, that in unsymmetric flight.

unsymmetric thrust Thrust with one failed engine away from centreline; normally called asymmetric.

UNT Undergraduate navigator training (S adds "system") (USAF).

UNTSO UN Truce Supervisory Organization.

unusable fuel All fuel which cannot be fed to engines, eg trapped in tank sumps, pockets beyond ribs in integral tanks, in gullies between spanwise stiffeners in machined panels and in supply system. See Addenda.

unwarned exposed Friendly troops are in open when NW detonates on near enemy.

UP 1 Unrotated projectile (1938–1944 name for British rockets).
2 Unguided projectile.

UPC Unit production cost.

UPCF Union des Pilotes Civils de France.

up-chaff Normally, between chaff cloud or stream and target; hence * interception, with good radar view of target.

update To refresh memory, radar picture or other electronic device with later information; hence * rate, rate (possibly as often as kHz) at which a system or input is scanned for new or changed values.

updraught carburettor One fed by duct conveying air upwards from below; in US updraft carbureter.

Uped UV pre-initiation electrical discharge (laser).

up-front control Single small panel in fighter cockpit giving complete control of all CNI functions.

up gear US voice command or check for raising landing gear.

UPI Undercarriage position indicator (UK).

Upkeep Water-skipping dambusting bomb (UK, 1943).

upkeep Generalized US term for all tasks aimed at preventing deterioration of hardware, eg GA aircraft; less used for commercial and military.

uplift 1 Total disposable load of cargo aircraft, or cargo taken on board.
2 Fuel taken on board, esp away from home base.
3 Fuel taken aboard from air-refuelling tanker aircraft.

uplink Telemetry, command, data or other electronic link between Earth and spacecraft.

upload Load acting vertically upwards, or vertical component of loads, eg airloads due to lift.

uplock Mechanical lock securing device, eg landing gear, in up or housed position.

upper air Portion of atmosphere above lower troposphere, normally (eg for synoptic purposes) that above pressure height 850 mb.

upper-air observation Observation of upper air, above effective range of surface measures; also called sounding.

upper airspace Normally all FLs above 250 (7,620 m/25,000 ft).

upper atmosphere Not strict term, normally interpreted as above tropopause.

upper branch That half of meridian or celestial meridian passing through observer or observer's zenith.

upper culmination See *upper transit*.

upper limb That half of limb of celestial body having greater altitude.

upper stage Second or subsquent stage in multistage rocket, in most vehicles not fired in atmosphere.

upper-surface aileron Split surface forming part of upper surface of wing only, used for roll or, in some cases, as spoiler (rare).

upper-surface blowing Discharge of main propulsive jets (flattened laterally to cover more span) across top of wing; in high-lift mode deflected down by Coanda-effect attachment to upper surface of large trailing-edge flaps to give augmented lift.

upper wing Top wing of biplane; hence ** pylon, pylon attached above top plane of biplane.

uprated Cleared to deliver more power, usually but not necessarily after incorporation of improvements; eg * gas turbine may operate at higher TGT or incorporate more efficient blades with reduced tip clearance; hence uprating, process of authorizing greater output from existing or improved machine.

UPS 1 Uninterruptable power supplies.
2 UV photoelectron spectroscopy.

upset 1 Sudden externally imposed or undesired disturbance to flightpath, eg by violent gust.
2 Metalworking process akin to forging in which rod or other slender workpiece is heated and placed under axial compression to reduce length and increase diameter, usually at particular place and to attain particular longitudinal profile.

upstage In a direction towards nose of multistage ballistic vehicle; * and downstage can refer to positions of items within same stage or to items in different stages.

upstairs 1 Aloft, ie flying (colloq).
2 High altitude, especially as distinct from low.

UPT Undergraduate pilot training (S adds "system").

UPU Universal Postal Union (Int).
upward Charlie Upward roll (UK colloq).
upward ident Upward identification light; white light visible from hemisphere above aircraft and usually provided with manual keying from cockpit, eg to send aircraft callsign or other message.
upward roll Roll while in steep climb, usually zoom after fast low pass.
upwash 1 Upward movement of air around outside of trailing vortex behind wing giving positive lift, usually curving over to become downwash on inner side of vortex.
2 Upward movement of air ahead of leading edge of subsonic wing giving positive lift.
URA Unrestricted article.
URIPS, Urips Undersea radioisotope power supply.
URR Ultra-reliable radar.
URSI Union Radio Scientifique International (Int, pronounced as word).
URV Unmanned research vehicle.
US 1 Unserviceable (often U/S).
2 Under instruction.
3 Ultrasonic (sometimes U/S).
USA 1 United States Army.
2 Useful screen area (HDD).
USAAAVS USA Agency for Aviation Safety.
USAAC USA Air Corps (1926–41).
USAAF USA Air Force (1941–47).
USAARL USA Aeromedical Research Laboratory.
USAAS USA Air Service (1918–26).
USAAVSCOM USA Aviation Systems Command.
usable lift 1 Thermal worth a sojourn.
2 See *useful lift*.
USABMD USA Ballistic Missile Defense Agency.
USAC Urban Systems Inter-Agency Advisory Committee (US).
USAEC USA Electronics Command.
USAETL USA Engineering Topographical Laboratory.
USAF US Air Force.
USAFA USAF Academy.
USAFE USAF Europe.
USAFSS USAF Security Service.
usage Usually means hours flown.
USAID US Agency for International Development.
USAMICOM USA Missile Command.
US&RAeC United Service and Royal Aero Club (UK, formerly).
USB 1 Upper-surface blowing.
2 Upper sideband.
USCG US Coast Guard.
USDAO US defense attaché offices.
USDOC, Usdoc US Department of Commerce.
USDR&E Under-SecDef for Research & Engineering.
use Workpiece metallurgically flawless and suitable for forging.
useful lift Difference between fixed weight of aerostat and gross weight, available for fuel, oil, consumables and payload.
useful load For any aircraft, difference between empty weight and laden weight; for certificated aircraft, difference between OEW and MTOW (or MRW if separately authorized). Note: OEW includes many items previously loosely included in **.
USG US gallon, non-SI unit of volume = $3{\cdot}785411784 \times 10^{-3}$ m^3 = 0·83267 Imp gal.
USGPM US gal/min.
USGS US Geological Survey.
USHGA US Hang Gliding Association.
USIA US Information Agency.
USIAS Union Syndicale des Industries Aéronautiques et Spatiales, now Gifas (F).
USMC US Marine Corps.
USN US Navy.
USNI US Naval Institute.
USNO US Naval Observatory.
USNTPS USN Test Pilot School, Patuxent River.
USPA US Parachute Association.
USPS US Postal Service.
USR Special work control (USSR).
USSP Universal sensor signal processor.
USSTAF US Strategic Air Forces.
UST Upper-surface transition, location of shockwave.
US ton Short ton (see *ton*).
USV Uberschall verkehrsflugzeug = SST (G).
US/VTOL Ultra-short or VTOL.
USW Ultrasonic welding.
USWB US Weather Bureau.
UT 1 Under training (often u/t).
2 Universal time.
UTA 1 Unit training assembly.
2 Upper control area (CAA).
UTC Co-ordinated universal time.
UTIAS University of Toronto Institute for Aerospace Studies.
utilidor Thermally insulated, heated conduit (above or below ground level) conveying water, steam, sewage and fire pipes between buildings (USAF).
utility Generalized term for aircraft intended for, or assigned to, variety of missions mainly of transport nature but with limited payload; usually denotes odd-job status but occasionally equipped with mission interfaces, eg for photography, target towing or even surface attack in limited war. FAA = limited aerobatic category. * glider = secondary or training glider.
utility finish Provides fabric with tautness and fill, but lacks gloss.
utility system Usually loose term meaning that system serves routine domestic functions not crucial to safety of vehicle; occasionally means special system for specific purpose, and sometimes even a standby emergency system. Never

involves flight control, navigation or weapon release.

utilization 1 Proportion of total time an equipment is available for use or is actually used.
2 Number of hours per year something, eg aircraft, is used.

utilization rate 1 For civil transports, utilization (2).
2 For combat aircraft, flying rate, usually qualified according to conditions, eg normal ** based on (say) 40-h week; emergency ** based on max attainable on 6-day week; wartime ** based on 7-day week with wartime crew, maintenance and safety criteria.

UTM Universal transverse Mercator.

UTP Unit test pilot.

UUM Unification of Units of Measurement Panel (ICAO).

UUPI Ultrasonic undercarriage position indicator.

UUT Unit under test.

UV 1 Ultraviolet.
2 Under-voltage.
3 Unmanned vehicle.

UVAS Unmanned vehicle for aerial surveillance.

UVDS UV flame detector system.

Uverom Ultraviolet erasable read-only memory.

UVL UV laser.

UVS UV spectrometer, or spectrometry.

UVV Ultimate vertical velocity; landing-gear design case.

UVVS Administration of the air force (USSR).

UW Unconventional warfare.

UWB Ultra-wideband.

UWR Ungrooved wet runway.

UXB Unexploded bomb.

V

V 1 Volts.
2 Velocity, including TAS, EAS or ASIR.
3 US piston-engine designation prefix, vee-type (obs).
4 US piston-engine designation prefix, vertical orientation, ie crankshaft vertical (currently in use).
5 JETDS code, visible light(s).
6 Volume, or volume-fraction.
7 Total shear stress.
8 US military-aircraft designation prefix, V/STOL.
9 US military-aircraft designation modifying prefix, staff/VIP.
10 Potential energy.
11 Vanadium.
12 Prefix, Victor airway.

v 1 Specific volume of gas.
2 Component of RMS velocity; phase velocity of EM wave.
3 Thermionic valve.
4 Linear velocity of point due to rotation of body in pitch.
5 Relative velocity between two moving bodies or points.
6 Propwash velocity relative to undisturbed air, ie "true" V_p.

$\bar{V}$ Horizontal tail volume coefficient ($l_t S_t / \bar{c} S$).

V′ Radial inflow velocity, eg to eye of centrifugal compressor.

V_1 Decision speed; ASIR defining decision point on take-off at which, should one engine fail, pilot can elect to abandon take-off or continue. Calculated by WAT and runway friction index for each take-off, never less than V_{MCG}. Can also be regarded as engine-failure recognition speed, made up of V_{EF} plus increment due to pilot thinking time.

V_2 Take-off safety speed; lowest ASIR at which aeroplane complies with those handling criteria associated with climb following one engine failure, normally obtained by factoring V_{MCA}, V_{MSL} and pre-stall buffet speed. Aeroplane should reach V_2 at screen after engine failure at V_1 and climb out to 120 m height without speed falling below V_2.

V_3 Normal screen speed with all engines operating, at which aeroplane is assumed to pass through screen height in normal take-off; usually about V_2 + 10 kt.

V_4 Steady initial climb speed for first-segment noise-abatement climb with all engines operating.

V-aerial Two rod conductors balance-fed at apex with geometry giving desired directional propagation.

V-band Original radar frequency band 46–56 GHz (obs).

V-belt Drive belt of tapering cross-section, often coming to narrow inner edge like V, mating with pulleys having inclined inner peripheral faces.

V-block Hardwood block with large V-notch used in hand sheet-metalwork.

V-engine See *vee engine*.

V-tail See *butterfly tail.*

VA 1 Volts × amperes; basic measure of AC or reactive electrical power.
2 Visual aids (ICAO panel).
3 Code: fixed-wing attack (USN).
4 Voice-activated.

V_A Design manoeuvring speed; on basic manoeuvring envelope speed at intersection of positive stall curve (assumed in cruise configuration) with n_1 (limiting positive manoeuvring load factor). Highest EAS at which limit load factor can be pulled.

V_a Aquaplaning speed; speed (usually ASIR) at which wheels lose effective contact with runway covered with standing water.

VA 1 Air army (USSR).
2 Veterans Administration (US).

VAB Vehicle (originally Vertical) Assembly Building.

V_{AB} Resultant velocity of circulation of two particles A and B.

VABI, Vabi Variable-area bypass injector.

VAC Vintage Aircraft Club (UK).

VACA The Vintage Aircraft Club of Australia.

vacancy Unoccupied lattice site in crystal.

Vacbi Video and computer-based instruction.

vacuum gauge Instrument for measuring low fluid pressure, eg Pirani, Knudsen, ionization, McLeod.

vacuum orbit Orbit of satellite round incomparably more massive body in complete absence of any atmosphere.

vacuum pump Device for establishing unidirectional flow of gas molecules and thus of evacuating fixed enclosure; following gross evacuation by mechanical pump alternatives include vapour/ diffusion pump, cryopump etc. Also, in aircraft, simple device for applying depression to air-driven flight instruments, eg venturi.

vacuum specific impulse Specific impulse in vacuum operation.

vacuum thrust Thrust of rocket in vacuum, typically about 25% greater than at sea level (actual thrust rises by approx product of atmospheric pressure and rocket nozzle area).

vacuum tube Electronic tube whose internal pressure is so low that residual gas or vapour atoms or molecules have no significant effect on operation; also called thermionic valve in most common form.

vacuum tunnel Wind tunnel operated at much less than sea-level pressure.

VAD 1 Velocity/azimuth display.
2 VHPIC applications demonstration (programme).

VADS Vulcan air-defense system (USA).

VAES Voice-activated electronic system.

VAFA Vintage Aircraft and Flying Association (UK).

VAFB Vandenberg AFB (USAF).

VAI Voice-interactive avionics (not VIA).

validation Generalized word for activity intended to re-validate licence or authority, esp training, examinations and emergency-procedure practice of graduate pilot, instructor or ATC (1) officer. ATC * involves simulated crises worse than any normally encountered.

validation phase Period when major-programme characteristics are refined by study and test to validate alternatives and decide whether to proceed to FSD.

Value Validated aircraft logistics utilization evaluation.

value engineering Complete engineering discipline devoted to seeking ways to achieve desired hardware performance, quality and reliability at minimal total cost, eg by elimination of unnecessary items, changes in material and manufacturing method, and simplification of design.

valve 1 Device for controlling fluid flow, eg into and out of PE (4) cylinder (inlet *, exhaust *) or into/out of aerostat, esp airship.
2 Numerous other fluid-flow control devices in hydraulics, oxygen, propellants, hot-gas, bleed air, carburetion and other systems.
3 Vacuum tube (UK usage).
4 To release air or gas from aerostat into atmosphere.

valve clearance Gap between end of stem and rocker arm.

valve duration Time or angular crankshaft movement during which a valve remains open.

valve face Mating edge of valve (1) bedded by grinding into seat.

valve gear Mechanism driving valves of PE (4).

valve hood Umbrella-like cowl protecting main airship gas valve against rain or icing.

valve lag Angular motion of crankshaft between either maker's specified valve-closing position or TDC and point at which valve actually closes.

valve lead Angular motion of crankshaft between closure of valve and either maker's specified closing position or TDC.

valve lift Total linear motion of poppet valve, ie cam stroke × mechanical advantage of valve gear.

valve line Cord operating aerostat gas valve.

valve petticoat See *petticoat.*

valve ports Inlet and exhaust passages forming part of cylinder head.

valve rigging Linkage inside aerostat (airship) envelope by means of which automatic valve is opened.

valve seat Angled ring, usually made of hard erosion-resistant material, forming poppet-valve mating face in cylinder head.

valve timing Exact plot of crankshaft angular positions at which PE (4) valves open and close.

VAM 1 Variable aerofoil machanism; usually infinitely reprofilable leading edge.
2 Visual anamorphic movie; external-scene (OTW) add-on to simulator.
3 Visual approach monitor (wide-body HUDs).

Vamom Visite d'aptitude à la mise en oeuvre et à la maintenance (Armée de l'Air, F).

Vamp Variable anamorphic motion picture; see *VAM (2).*

VAMS Vector airspeed measuring system.

van Generalized US term for air-conditioned towed vehicles for major support operations, eg strip, check and reassemble major avionic systems in field. Not called trailer.

vanadium Hard silver-white metal, symbol V, SG 5·96, MPt 1,730°C; important alloying element, esp in steels.

Van Allen belt(s) Inner and outer zones of high-intensity particulate radiation trapped in Earth's magnetic field around Equator (inner mainly protons, outer mainly electrons) at radii from Earth's centre from 8,700 to 26,000 km.

Van de Graaff Registered name of high-voltage generator using rotating belt to convey electrical charges.

Vandenberg Originally Point Cooke, Calif, today main West Coast rocket test base, head of Pacific Missile Range, address Lompoc (USAF).

Van der Waals equation Best known equation describing behaviour of real (as distinct from perfect) gas: $(p + a/b^2)(v - b) = RT$, where p is pressure, v volume, R universal gas constant, T °K and a, b constants.

Van der Waals forces Interatomic or intermolecular attractive forces between interacting varying dipole moments; varies as seventh power of radius.

V&F Vinyl and fabric.

V&V Verification and validation (software).

Van Dyke Generalized theory of aerodynamics at Mach 5+ (hypersonics), where air can no longer be assumed perfect gas owing to molecular vibration, dissociation, electronic excitation and ionization.

vane 1 Generalized term for thin (flat or, usually, curved but not necessarily aerofoil section) aerodynamic surface either fixed in order to turn air or gas flow, or freely pivoted and thus align-

ing itself with fluid direction.
2 Stator blade (compressor or turbine, US usage).
3 Gas guide surfaces at nozzle to turbine (US and UK); abb IGV.
4 Strips, usually of circular-arc section, in cascade at corners of wind tunnel.
5 Radial strips around fuel burner of gas turbine imparting rotary vortex motion, often called swirl *.
6 Curved surfaces in cascade at angle bends of airflow in many gas turbines; again called swirl * though rotation is not desired.
7 Curved forward extensions to radial arms of centrifugal compressor or supercharger impeller, called rotating guide *, RGV.
8 Alternative (unusual) term for fence.
9 Swinging retractable leading-edge flaps normally housed in fixed glove of swing-wing aircraft and extended in high-lift mode, called glove *.
10 Common term for slat fixed to leading edge of flap and for various other auxiliary aerofoils carried on flaps, ailerons, droops and leading-edge slats.
11 Normal meaning for weathercocking surface.
12 Particular application of (1) is sensor in SWIS.

vane pump Large family of fluid pumps in which flat surfaces oscillate and rotate inside chamber eccentrically arranged around drive shaft.

vane rate Angular velocity of vane in SWIS indicating rate at which aircraft is approaching stall.

Vanvis Visual and near-visual intercept system; visual refers to EM frequency used, not to pilot acquiring target visually.

Van Zelm Catcher installation for catapult bridles thrown from aircraft carriers.

VAP 1 Vortex avoidance procedure.
2 Visual-Aids Panel (ICAO).
3 Spray dispensing system for HCN or persistent Toxic-B (USSR).

VAPI Visual approach-path indicator; also known as vertical speed and approach-path indicator.

Vapo Velocity at apogee.

vaporizing combustor Gas-turbine combustion system in which fuel is vaporized prior to passage through burner and ignition (as in common blowlamp); hence vaporizer, in which fuel is vaporized, and vaporizing burner.

vapour Substance in gaseous state but below critical temperature; thus can be converted by pressure alone to liquid or solid (US = vapor).

vapour cycle Closed-circuit refrigeration, eg for air-conditioning, in which heat is extracted by refrigerant alternately evaporated and condensed.

vapour degreasing Immersion in hot solvent vapour, eg trichlorethylene.

vapour lock Complete breakdown of supply of liquid, eg fuel from tank, because of blockage by bubble of vapour at high point in pipe.

vapour-phase inhibitor Nitrite-based chemical, often locked in paper or other solid, which protects metal parts against corrosion by preventing formation of vapour, esp of water; usually white powder which volatilizes and recrystallizes on metal surface.

vapour pressure Pressure exerted by molecules of vapour on walls of container; with mixture sum of partial pressures.

vapour tension Max attainable vapour pressure exerted by plane liquid surface with vapour above, varies with temperature.

vapour trail See *condensation trail.*

vapour-type thermometer Needle is driven by Bourdon tube sensing pressure from vapour capsule whose temperature is that indicated.

V_{APP} Approach speed.

Vaptar Variable-parameter terrain-avoidance radar.

VAQ Code: airborne early-warning squadron (USN).

VAR 1 VHF aural range, ie radio range.
2 Visual/aural range; radio range in which two airways are located by A/N signals and two by visual means, eg panel instrument (obs).
3 Volt/ampere(s) reactive; unit of wattless (reactive) electrical power.
4 Vacuum-arc remelting.

varactor Device employing p-n junction whose capacitance is varied by reverse voltage; important in parametric amplifiers.

variable-area nozzle Propelling nozzle whose cross-section area can be altered (usually, together with profile) to match changed Mach number and afterburner operation (airbreather) or atmospheric pressure from SL to vacuum (rocket).

variable-area wing Rare arrangements have included variable span and even retractable lower wing of biplane into upper, but modern flap systems (eg Fowler) give significant change.

variable camber 1 Apart from experimental aircraft, most important methods are hinged leading and trailing edges, eg F-16, which can each be pivoted slightly up, centred or pivoted fully down independently to suit desired flight condition. Tail surfaces, eg tailplane, rudder, are today often divided spanwise to change not just surface angle but also camber, for greater power.
2 Camber is varied by elastic deformation of surface, eg 747 Krügers.

variable-camber flap 1 See *anti-balance tab.*
2 Flap, usually Krüger, whose profile changes on extension.

variable-cycle engine Jet engine in which path of working fluid can be altered by shutters/valves/doors, eg to convert from turbojet or turbofan to ramjet, hybrid rocket or other form in cruising flight.

variable-datum boost control Auto boost control for supercharged PE (4) in which governed boost pressure increases as pilot's throttle lever is opened.

variable-delivery pump Fluid pump whose output can be varied independently of drive speed, usually by variable-angle swashplate driving stroked plungers (at 90° stroke is zero).

variable-density tunnel Wind tunnel whose pressure can be varied over wide range (usually not below atmospheric) while in operation; normally pressurized to achieve desired R.

variable-discharge turbine Gas turbine, eg driven by PE (4) exhaust, whose thoughput (mass flow) can be controlled by valve (often called waste gate) to match turbine power to altitude and other variables.

variable-displacement pump Fluid pump whose output (mass flow) can be varied over wide range, often to zero, for any given input drive speed; often synonymous with variable-delivery, but ** explicitly implies reciprocating plungers or pistons whose stroke is variable.

variable-floor-level bridge Passenger bridge whose airside end can be raised and lowered to match aircraft sill height.

variable-geometry aircraft Aircraft whose shape can be varied in gross manner, ie more fundamentally than by retractable landing gear or flaps; term has come to mean variable wing sweep, so should not now be used in any other context, unless circumstances change or clear explanation of meaning is furnished.

variable-geometry engine Invariably refers to air-breathing engine for supersonic propulsion in which for reasonable efficiency it is essential to have not only fully variable inlet and nozzle but also variation in flow path in engine itself, eg to divert flow around HP compressor in supersonic mode or to convert engine into ramjet. Needs explanation when used.

variable-geometry inlet See *variable inlet*.

variable-incidence Pivotally mounted so that angle of incidence can be altered. * guide vane: stator blade or turbine inlet guide vane whose incidence is altered for best compromise between flow incident on leading edge and flow angle leaving trailing edge, invariably auto scheduled by engine control system. * tail: tailplane whose incidence is varied either for trimming, with elevators as primary flight-control surfaces, or as primary flight control. * wing: wing pivoted on transverse axis so that over full (large) range of flight AOA fuselage can remain more or less level, eg to improve pilot view on approach or permit short landing gear.

variable inlet Variable-geometry airbreathing engine inlet whose area, lip/wedge/centrebody axial position and duct profile can all be adjusted to match required flight shock position and mass flow. Mere downstream auxiliary inlets or spill doors do not qualify.

variable-inlet guide vane Gas-turbine IGV whose incidence varies according to engine operating regime; very rare upstream of turbine but common upstream of first and often subsequent stages of axial compressor, to match airflow mass flow and whirl to rotor blade conditions and avoid stall; controlled by auto system always sensitive to rotor speed and inlet air temperature and occasionally to other variables.

variable overhead Varies with number of particular item manufactured.

variable-pitch Synonymous with variable-incidence. Normally confined to propellers, where incidence called pitch. Usually means pitch can be varied on ground, or by pilot in flight, often only as choice of either coarse or fine pitch, without auto control such as constant-speed. Fine distinction between * and adjustable-pitch, latter explicitly meaning ground-adjustable only.

variable-ratio Two main applications: in shaft-drive gearbox * drive is usually synonymous with constant-speed drive, ie ratio varied to hold output speed constant despite varying input; * bypass engine or turbofan (rare) has bleeds, two-position shutters or doors to change ratio of airflow between bypass duct and core.

variable-stator Usually means gas turbine with not just one but several rows of variable-incidence stator blades (vanes) in axial compressor(s).

variable-stroke Though many reciprocating engines and machines patented with *, invariably means axial oscillating plunger liquid pump driven by variable-angle swashplate (see *variable-delivery*).

variable-sweep Aerofoil is pivoted so that sweep angle can be varied. Mainly applicable to main wings, where left/right wings are made separate from rest of structure and attached by large-diam fatigue-free pivots so that both surfaces can be scheduled (either auto or by pilot command) by actuator over wide range of sweep angles, symmetrical about axis of symmetry. Does not mean slew-wing. (Helicopter blade rotation about drag hinge is strictly * but is not called such). Also called variable-geometry (VG) and (colloq) swing-wing.

variance Mathematical average of square of deviations from mean value.

variance rate Difference between standard and actual wages (US hourly-paid workers).

variant Different version of same basic aircraft type.

variation 1 Horizontal angle between local magnetic and geographical meridians, expressed as E or W to indicate direction of magnetic pole from true. Also called declination (see *grid* *).
2 Detailed schedule of change orders or special

furnishings or equipment specific to one customer.
3 Small periodic change in astronomical latitude of Earth locations due to wandering of poles.

varicam Pioneer variable-camber aerofoil.

vario VSI (F, colloq).

variocoupler RF transformer with fixed and moving windings.

variometer 1 Aneroid-type VSI (traditional term of gliding fraternity; * used to seek thermals).
2 Variable inductive coupler with fixed and moving coils or rotors for comparing magnetic fields, esp Earth's field.

varistor Two-electrode semiconductor resistor characterized by R varying inversely with applied V, in either direction of current.

Varite Resistor characterized by negative temperature coefficient of R.

varnish Solutions of resins, eg common gum or wood rosin, in drying oil, eg linseed.

Varsol Naphtha-like petroleum solvent.

VAS 1 Visual augmentation system.
2 Voice-activated system.

VASI, Vasi Visual approach slope indicator; systems of visible lights arranged in pattern at landing end of runway providing descent (and limited lateral guidance) information. Basic * comprises downwind and upwind bars each of three lights on each side of runway (in US, total of 2, 4 or 12), each light projecting red lower segment and white upper. When correctly on glidepath, as close as possible to ILS glidepath, pilot sees pink. For long or LEW aircraft extra light bars are added in 3-bar *, 3-bar Avasi, T-Vasi and AT-Vasi.

Vasis Vasi system (UK style).

Vast 1 Versatile avionics shop test(er); packaged lab for installation on Navy airfields or carriers; occasionally "a" held to mean automatic.
2 Vibration analysis systems technique (GE).

VAT 1 Value-added tax; applicable to light and sport aircraft (UK).
2 Vernier axial thruster.

V_{AT} Target threshold speed; scheduled speed for arrival at threshold at screen height of 10 m after steady stable approach at angle of descent not less than 3°. Usually about 1·3 V_s. (see V_{Tmax}, V_{Tmin}). Subdivided: V_{AT_0}, all engines operating; V_{AT_1}, one engine failed; V_{AT_2}, two engines failed, etc.

Vat-B Short-form weather report comprising word "weather" and four numbers (DoD, from vis, amount [cloud], top [cloud] and base [cloud]).

VATLS Visual airborne target-locator system; laser ranging device.

Vatol Vertical-attitude take-off and landing, ie "tail-standing".

VATS, Vats Video-augmented tracking system.

VAWT Vertical-axis wind turbine.

V_B Max speed at which specified gust (eg + 66/−66 ft/s) can be withstood without airframe damage; hence speed at left lower and upper corners of basic gust envelope. Usually more than half V_c and less than half V_D. Plays role in establishing V_{RA}.

VBC Velocity blast contour.

V_{BE} Speed for best flight endurance.

VBI Verband Beratender Ingenieure (G).

V_{BR} Speed for best range.

V_{BROC} ASIR for best rate of climb of helicopter, usually at or near SL.

VBS Visual bootstrapping subsystem.

VBV Variable-bypass valve.

VC 1 US military aircraft prefix, staff/VIP transport.
2 Variable-camber.
3 Vanadium carbide.

V_c 1 Design cruising speed, usually one of speeds used in establishing structural strength.
2 Relative closing speed between two aerial targets.

v_c 1 Rate of climb (note: not IAS or other horizontal speed in climb).
2 Circular velocity, ie speed of satellite in linear measure.

VCAA Vintage and Classical Aeroplane Association.

VCASS Visually coupled airborne systems simulator.

VCC 1 Vehicle control centre.
2 Colour CRT display (F).

VCE Variable-cycle engine.

VCID Voice-controlled interactive device.

VCM Visual countermeasures (OCM is usual term).

VCO 1 Voltage-controlled oscillator.
2 Variable clock oscillator.

VCOS Vice-Chief of Staff.

VCR Video cassette recorder.

VCS 1 Vertical cross-section.
2 Vehicle control system.

VCSS Vinyl-coated stainless steel.

VCTS Variable cockpit training system.

VCXI Visual course cross-pointer indicator.

VD 1 Video detector.
2 Video disc.

V_D Design diving speed; highest speed at which aircraft is permitted to fly, forming vertical right-hand boundary to both basic manoeuvring envelope and basic gust envelope. One of speeds used in establishing structural strength.

v_d Rate of descent (note: not airspeed during descent).

v.d. Vapour density.

VDC 1 Variable-datum control.
2 Volts, direct current.

VDDL Verband Deutscher Drachenfluglehrer eV (G).

VDE Verein Deutscher Elektrotechniker (G, technical society).

VDF 1 VHF D/F; most common ground D/F.
2 Verein Deutscher Flugzeugführer (G, PIC society).
3 Verband Deutscher Flugleiter (G).

V_{DF} Maximum demonstrated flight diving speed; highest IAS at which aircraft is ever flown (normally only during certification); associated with poor strength margins (inadequate for severe gust) and possibly poor handling, yet attainable at full power even in climb on many transports at low/medium levels.

VDFT Verband Deutscher Flugsicherungs-Techniker eV (G).

VDI 1 Vertical displacement indicator.
2 Variable-depth (sonar) output indicator.
3 Verband Deutscher Ingenieure (G, technical society).

VDL Verband Deutscher Luftfahrt-techniker (G, aerospace engineers' society).

VDLU Verband Deutscher Luftfahrt-Unternehmen (G).

VDMA 1 Variable-destination multiple access.
2 Verein Deutscher Maschinenbau-Anstalten (G).

VDP Validation demonstration phase.

VDR, v.d.r. Voltage-dependent resistor.

VDRT Visual display research tool.

VDS 1 Variable-depth sonar.
2 Video-disc simulation.

VDT 1 Variable-discharge turbine (driven by PE [4] exhaust).
2 Vapour-deposited tungsten.
3 Variable-deflection thruster.

VDU Visual (or video) display unit.

VE Value engineering.

V_E Generalized symbol for a limiting speed at which something (eg flap, landing gear) may be selected and extended; suffixes, eg V_{E1}, V_{EN}, used for specific items and additional terms are used (more are needed) for limiting speeds for subsequent retraction. Main gears (usually not nose) are generally cleared to V_{MO} when locked down, but V_E always applies for selecting down or up.

Vebal Vertical ballistic (downward-firing anti-armour).

VECP Value-engineering change proposal.

vector 1 Adj, having both magnitude and direction; thus velocity is * quantity, speed is not.
2 Line representing quantity's magnitude, direction and point of application, eg force on structure, or aircraft heading /TAS.
3 A particular aircraft heading or track, eg that needed to arrive at destination.
4 To issue headings to aircraft to provide nav guidance (DoD definition adds "by radar").
5 On HUD, aircraft direction of travel, usually synonymous with track.
6 To control trajectory by altering thrust axis of propulsion.
7 In translations from Italian, rocket launch vehicle.
8 Prefix to three-digit heading passed to interceptor engaged in interception (for recovery corresponding word is steer).

vector computer Device for solving vector triangles, eg CSC.

vectored attack Surface attack in which weapon carrier is vectored (4) to weapon-delivery point by unit which holds contact on target (DoD, NATO).

vectored thrust Propulsive thrust whose axis can be rotated to control vehicle trajectory; term normally applied to swivelling-nozzle jet engine of aeroplane, corresponding term for space and military rockets being usually TVC.

vector flight control Control of trajectory by vectored thrust.

vectoring in forward flight See *viff*.

vector quantity One that has magnitude and direction.

vector sight Traditional type of bomb sight incorporating mechanical representation of vectors of relevant vector triangle.

vector steering Control of trajectory by vectored thrust.

vector triangle Closed figure formed from three vectors, eg (1) heading/TAS, track/GS and W/V, or (2) lift, drag and resultant force on lifting wing.

vee depression Vee-shaped low extending between two highs, usually with squall.

veeder counter Stepping digital counter, eg odometer; today often LED or LCD.

vee engine PE (4) whose cylinders are arranged in two inclined in-line rows (banks) in V form seen from either end; hence vee-12 (often called V-12) ** with six cylinders in each bank.

vee formation Aircraft formation in shape of horizontal V proceeding apex-first, with odd number of aircraft.

veering Change of wind direction clockwise seen from above; now applies in either hemisphere.

vee tail See *butterfly*.

V_{EF} Speed at which critical engine failure occurs in accelerate/stop take-off; defines V_1 as * plus speed gained with critical engine inoperative during time pilot takes to recognize situation and respond.

vegetable Mine laid by aircraft at sea (RAF, WW2), hence gardening.

vegetable oil Several, esp castor, used for engine lubrication pre-1935.

vehicle Self-propelled, boosted or towed conveyance for transporting a burden on land, sea or through air or space. This is DoD/NATO wording. Only possible word in most generalized contexts, and also only word covering aircraft, spacecraft, missiles, RPVs etc. Air * identifies flying portion of weapon or reconnaissance system that has extensive non-flying portions.

vehicle axes Axes, usually cartesian, related to vehicle rather than to Earth or space.

vehicle mass ratio Ratio of final mass of vehicle, usually M_f, after cutoff or burnout of propulsion to initial mass, usually M_0. Normally applied to rocket vehicles.

veil cloud Loose term meaning either Cs or cloud forming thin veil on mountain.

vela sensor Usually measures velocity and angle of attack.

velocimeter 1 Generalized term for velocity meter.
2 CW-reflection Doppler system for measurment of radial velocity, ie speed of approach or recession relative to observer.

velocity 1 Measure of motion: speed (linear or angular) in specified direction.
2 Loosely (though common in fluid flow), speed. SI unit is m/s.

velocity blast contour Plot of jet wake velocities immediately behind large jet as it proceeds from gate to take-off point. Usual measure is 6 ft (1·83 m) above ground at distance 50 ft (15·24 m) behind tail.

velocity factor Ratio of speed of RF wave along conductor to its speed in free space, usually 0·6 to almost unity; symbol k. In typical co-axial cable value is about 0·66, often expressed as a percentage.

velocity gate Basic ability of CW, Doppler and certain other tracking radars to sense and lock on to particular radial velocity characteristic of target. Hence ** pull-off, basic ECM technique to pull radar off target signal and thus give infinite JSRR.

velocity gradient Rate of change of fluid speed per unit distance traversed perpendicular to streamlines, eg in boundary layer.

velocity head Not accepted term; usually means pitot pressure but has even been used to mean kinetic energy of unit mass of fluid.

velocity jump Angle between launch line and line of departure (ASCC).

velocity microphone Electrical output = f(V) where V is mean speed of particles on which sound waves impact.

velocity modulation Various techniques of modulating electron beams, eg by HF transverse field which impresses sinusoidal velocity contour causing corresponding variation in intensity of scanning spot.

velocity of advance Airspeed past propeller blades ignoring speed due to rotation of blades; essentially synonymous with slipstream speed. Always greater than aircraft airspeed, provided engine is operating.

velocity of light In vacuum $2{\cdot}9979250 \times 10^8$ m/s, symbol c.

velocity of propagation Speed at which EM wave travels along conductor, eg co-axial cable, waveguide; usually = kc, where k is velocity factor.

velocity potential Integral of flow velocity parallel to surface, = ϕ, = Ux + Vy (rectilinear) = $\frac{\Gamma\theta}{2\pi}$ (vortex).

velocity profile Plot of velocity of viscous fluid in transverse perpendicular to flow direction; thus for laminar flow through small tube ** = parabolic curve.

velocity signature Record of Doppler track of aerial or other moving target.

velocity transducer Generates electrical output proportional to imparted V.

velocity vector Main reference on primary flight display showing desired flight path on which aircraft is held by FCS.

VEN Variable-exhaust nozzle.

vendor Supplier to a programme, almost always manufacturer of finished parts, devices and equipment though * list often includes suppliers of raw material and even services.

vendor audit Survey of vendor profiles, eg before entering into discussion or negotiation.

vendor profile Detailed standardized description of vendor companies for benefit of large-system prime contractors.

veneer Thin sheet wood; when applied to plastics usually means sheet with simulated woodgrain.

vent 1 Opening to atmosphere, eg from fuel tank, to spill excess air or vapour.
2 Opening in centre of parachute for stabilization.
3 Precision aperture in aircraft skin sensing true local static pressure; called static *.
4 Spanwise aperture on USB flap either to entrain air from below or to reduce Coanda attachment to upper surface.

vent cap Vent (2) patch.

vent hem Reinforced hem around vent (2).

ventilated shock Shockwave at UST point which has been stabilized and weakened by a double slot system.

ventilated suit Partial-pressure or pressure suit provided with ventilation system, eg by ventilation garment.

ventilation garment Light inner suit, forming part of pressure suit, through which dry air is pumped to control body surface temperature and evaporation.

venting cycle Regular or automatic venting of Ni/Cd battery to avoid cell imbalance or thermal runaway.

vent patch Piece of fabric covering vent (2) sewn to hem.

ventral 1 On belly side of body; hence with horizontal fuselage, on underside, or, occasionally, firing or directed through underside.
2 Underfin, vertical or inclined, fixed or hinged.

ventral container/pallet/pod/radar/tank Criterion for adj "ventral", as distinct from fuselage, centreline etc, is that item either forms underside of fuselage or is flush against it, eg Beech 99 baggage container, RF-111 reconnaissance pallet, Harrier gun pod, RA-5C SLAR and Attacker drop tank.

ventral fin Fixed or movable fin on underside of body, usually but not necessarily at tail.

ventral inlet Inlet on underside of body far enough from nose not to be called chin inlet.

venturi Duct for fluid flow which contracts to min cross-section at throat and then expands (usually to same area as inlet); pressure at throat falls to min value which can be used to drive vacuum instruments or as measure of airspeed; flow throughout always subsonic.

venturi meter Instrument for measuring fluid mass flow through calibrated horizontal venturi;

$$\text{flow } Q = A_2 \sqrt{\frac{p_1 - p_2}{\rho/2[1-(A_2/A_1)^2]}}$$

where p_1, A_1 are pressure and cross-sectional area at start of venturi, p_2,A_2 are values at throat and ρ is density. Assumes flow laminar and compressibility ignored.

venturi pitot Combination of pitot tube and venturi, one giving pressure above atmospheric, other below.

venturi tube See *venturi*.

VER Vertical ejector rack; VER-2 is twin-store *.

Verdan Versatile digital analyser; pioneer airborne computer used in Reins (Autonetics); colloq translated as "Very Effective Replacement for Dumb-Ass Navigators".

verge ring 000°–360° ring rotated to set traditional magnetic compass.

verification 1 To ensure meaning and phraseology of transmitted message conveys exact intention of originator.
2 Any action, including surveillance, inspection, identification, detection, to confirm compliance with arms-control agreement.
3 Confirmation of flightplan by inserting waypoints and destination in nav system and causing system and displays to drive through to destination in accelerated timescale.
4 Re-examination of firing data (DoD).

vernal equinox Intersection of ecliptic and celestial equator occupied by Sun about March 21 as it passes through zero declination, changing from S declination to N.

vernier Aid to fine adjustment or refined measurement, eg linear scale placed adjacent to main scale, with which, by aligning two corresponding divisions, main scale can be read with enhanced accuracy.

vernier engine/motor/rocket Small rocket which remains operating or is started after shutdown of main propulsion to provide accurate final adjustment of vehicle velocity; in most cases has thrust-vector control to refine exact direction of trajectory.

vertex 1 Highest point of trajectory.
2 Point on great circle nearest a pole.
3 Node or branch point in network.

vertical 1 Vertical or slightly inclined fin on supersonic fighter-type aircraft, either at nose or tail, usually under body at nose but usually above at tail, often serving as primary flight-control surface. Twin canted * form U-tail.
2 See *vertical air photograph*.

vertical air photograph Taken with optical axis of camera perpendicular to Earth's surface (note: definition, agreed by DoD, NATO, does not say with axis vertical).

vertical and/or short take-off and landing, V/STOL Usually means that aircraft is fixed-wing aerodyne with jet lift giving VTOL capability at less than max weight but that in most missions STO is preferred; landing at light weight may be vertical, thus STOVL. Some VTOLs do not have STO capability and thus cannot be called V/STOL.

vertical axis 1 Vehicle-related ** is that axis perpendicular to both longitudinal and transverse horizontal axes; in aircraft called OZ, usually lying in plane of symmetry.
2 Local vertical.

vertical bank Aircraft is rolled through 90°, so that OZ axis is horizontal and OY axis is coincident with local vertical.

vertical camera Optical axis is perpendicular to Earth's surface; according to other authorities, axis is aligned with local vertical, no matter what local terrain may be. Need for clarification.

vertical circle Great circle of celestial sphere passing through zenith and nadir.

vertical development Depth of cloud from base to top.

vertical ejector rack Carries two superimposed external stores; much lower clean drag than TER.

vertical engine 1 PE (4) or gas turbine whose main rotating member rotates about vertical axis, usually for helicopter drive or jet lift.
2 PE (4) whose cylinders are vertical above or below crankshaft (arch).

vertical envelopment Tactical manoeuvre in which troops, air-dropped or air-landed, attack flanks and rear of hostile surface force, effectively cutting off latter.

vertical fin Fin; traditional fin is fixed, and word "vertical" alone usually means powered fin serving as flight control.

vertical force Vertical component of Earth's magnetic field.

vertical gust Gust; but in V-g recorder V signifies vehicle velocity.

vertical gyro Two-degrees-of-freedom gyro torqued on gimbal mounts to hold spin axis vertical, thus giving output signals proportional to rotation about two orthogonal axes, usually pitch/roll.

vertical interval Difference in height between two locations, eg between two targets or between observer and target.

vertical launch Launch of vehicle on initially vertical trajectory where such trajectory is not inevitable; eg non-ballistic vehicle or one launched from launcher of variable elevation.
vertical layout Traditional layout of project or design office in which an administrative or seniority hierarchy takes precedence over (1) project or programme, and (2) type of work.
vertical lift Lift force along local vertical generated by aerodyne wing, rotor or engine.
vertical navigation, VNav Guidance of flight trajectory in vertical plane, eg to minimize pilot workload in letdowns, holding patterns and during climb or descent to ATC cleared FLs along particular routes or on early stages of approach; provided by modern transport nav systems, esp those of energy-management type.
vertical overspill Vertical beam from weather radar reflected from surface below to give height ring on display; not present in mapping mode.
vertical pincer Any DA engagement, one low and one high (relative to target).
vertical pressure gradient Change of atmospheric pressure per unit change in height (traditionally per 1,000 ft in UK/US).
vertical probable error Product of range probable error and slope of fall.
vertical reference Earth-related vertical axis, ie local vertical normally approximated by vertical gyro.
vertical replenishment Use of VTOL (helicopter) or V/STOL aircraft for transfer of stores and/or ammunition from ship to ship or to shore (NESN).
vertical reverse Aerobatic manoeuvre related to half flick (snap) roll; begun from tight turn by pulling hard back and applying full top rudder to flick inverted, thereafter completing second half of loop to recover level flight; often called vertical reversement (US term).
vertical riser Flat-rising VTOL, ie jet-lift aircraft taking off with fuselage horizontal.
vertical rolling scissors Defensive descending manoeuvre in vertical plane in attempt to make enemy overshoot and fly into attacker's future flight path.
vertical separation Specified difference in FL between air traffic on conflicting courses; normally published for (1) tracks 000–179 and (2) tracks 180–359, and for FLs 0–180, 180–290 and 290+.
vertical speed Rate of change of height.
vertical speed indicator Panel instrument indicating vertical speed, ie rate of climb/descent; invariably one pointer zeroed at 9 o'clock.
vertical spin tunnel See *spinning tunnel.*
vertical stabilizer Fin (US).
vertical stiffeners Angle or other sections riveted or bonded at intervals along spar web or fuselage keel to resist buckling in vertical plane.
vertical strip Single flightline of overlapping vertical reconnaissance images, eg of beach or road.
vertical tail Traditionally, fin(s) and rudder(s); hence ** area, aggregate area in side elevation of fin(s) and rudder(s), together with dorsal fin and any ventral fin(s) but exclusive of fillets, fairings or bullets.
vertical take-off and landing Aerodyne has capability of rising from surface without airspeed, hovering and returning to soft landing again without airspeed, generating lift greater than its weight by rotors, ducted fans, jets, deflected propulsion or other internally energized means.
vertical tape instrument Display has roller blind translating vertically, against which are read fixed and/or moving index markers; usually engine instruments are grouped in multi-engine aircraft so that in correct operation all readouts are at same levels.
vertical translation Motion of aeroplane in vertical plane, esp under direct lift force, without change of pitch attitude; can be achieved, eg, by Harrier viff or by F-16 symmetric wing-flaperon deflection with scheduled flaperon/tailplane interconnect gain and with pitch hold engaged.
vertical tunnel See *spinning tunnel.*
vertical turn Turn with approx 90° bank.
vertical virage Turn with approx 90° bank (arch).
vertical wind tunnel See *spinning tunnel.*
vertigo Subjective sensations caused by faults in inner-ear semicircular canals: subjective * = external world is moving past sufferer; objective * = external world is rotating.
vertiplane VTOL aircraft having fixed wing with flaps powerful enough to lift aircraft at zero forward speed by deflecting propwash; FAI category E-4 but no records and (it would appear) no actual flying examples.
vertrep See *vertical replenishment.*
Very Patented signal pistol, standard Allied aviation from WW1; hence * light, * pistol etc.
very high Above FL 500 (DoD).
very high frequency See *frequency.*
very high frequency omni-range See *VOR.*
very high speed photography Image rate 500 to 10^4/s.
very-large-scale integration Commonly accepted as over 10^5 devices per chip.
very low frequency See *frequency.*
V_{esc} Escape velocity = $\sqrt{2K/R}$ where K is a constant (universal gravitational constant × primary-body mass) and R is distance from centre of primary body.
VF 1 Voice frequency.
2 Unit prefix, fighter squadron (USN).
V_F Design flap limiting speed; replaced by V_{FE}.
V_f 1 Surface wind.
2 Volume fraction of fibre or whisker reinforced composites; expressed as % volume occupied by fibre.
3 Fuel flow (CAA).
V_{FC} Maximum speed for flight stability (FC = full control); usually synonymous with V_{MO}, little

used outside US and suggest passing from use.

V_{FE} Maximum flaps-extended speed; usually an ASIR and in most flight manuals precise meaning is explained, eg whether limit is for landing setting or any lesser setting. Note: this is invariably a limit for an established flap setting; it does not allow for changed settings.

VFLB Variable-floor-level bridge.

VFO Variable-frequency oscillator.

VFR Visual flight rules.

VfR Verein für Raumschiffahrt, society for space travel (G).

V_{FTO} V_1 for flexible take-off, factored for full power in event of one engine failure at reduced rating.

VG 1 Variable-geometry.
2 Vertical gyro.

V_g Geostrophic wind.

V-g Aircraft speed and normal acceleration; hence * diagram, graphic plot of these parameters; * recorder, instrument recording speed and applied loading in vertical (very rarely in other) plane.

Vgh Aircraft speed, vertical acceleration and height; hence * recorder, continuous recording of these parameters on wire or tape (arch).

VGI Vertical gyro instrument.

VGK Supreme military command (USSR).

VGPO Velocity-gate pull-off; basic ECM technique for making hostile tracking radar break lock.

V_{grad} Gradient wind.

VGS 1 Velocity-gate steal; usually synonymous with VGPO.
2 Volunteer gliding school.

VH Designation, very heavy aircraft or unit (SAC, obs).

V_H Max speed in level flight with max continuous power; little used outside US. Definition should add that H = high altitude; this power at med/low FLs would usually exceed V_{DF}.

V/H Velocity/height ratio in taking reconnaissance imagery and in sensor design (M0·9/200 ft gives * = 5).

VHF Very high frequency.

VHF Omni-range See *VOR*.

VHFRT VHF R/T.

VHPIC Very-high-performance (or power) integrated circuit.

VHRR Very-high-resolution radiometer.

VHSIC Very-high-speed integrated circuit; Si or SOS, then GaAs, two to three orders of magnitude faster than MSI/LSI.

VI 1 Visual ident mode.
2 Viscosity index.

V_i 1 IAS (not ASIR).
2 EAS.

VIAM All-union (research) institute for aviation materials (USSR).

vibrating voltage regulator Controller of DC machines which uses vibrating points to sense voltage and adjust resistance controlling field current.

vibration indicator Instrument or sensor for either recording or indicating mechanical vibration either remote from crew or of too high a frequency to be obvious, esp emanating from turbine engines.

vibration isolater See *isolator*.

vibration meter Instrument for recording vibration frequency; very rarely either amplitude or acceleration in addition.

vibration mode Most physical objects, from wings to quartz crystals, can vibrate at a fundamental mode having lowest frequency, or (depending on dimensions, mounting, impressed forces, coupling and other factors) at any of numerous modes having different visual pattern and higher frequency. Fundamental mode usually longitudinal over length or thickness, flexural over width or thickness, and shear over thickness or face.

vibrator 1 Mechanical source of high-power sinusoidal vibration for test purposes; also called vibration generator.
2 Rapid-action switch for alternately reversing polarity of transformer primary fed from DC to give raw AC output; also called vibrator converter.

vibratory torque control Mechanical coupling for rotary output, eg from PE (4), which at low rpm locks drive into hydraulically stiff configuration but at higher rpm unlocks and allows drive to be taken through slender quill shaft.

vibrograph Seismic instrument giving record of vibration displacement/time.

vicious cycling Standard maintenance procedure for Ni/Cd battery in which charge is violently drained and replaced; also called deep cycling.

Vicon Visual confirmation (of voice instruction or clearance, especially to take off).

Victor airway Airway linking VORs, thus virtually all airways in US and many other countries; identified by prefix V.

VID 1 Visual identification.
2 Virtual (confusingly, also visual) image display.

video Generalized adj or noun for electronic transmission of visual information.

video compression Retuning of video from primary radar so that each radial trace is briefly stored and then written more quickly; this allows time for cursive writing of synthetic data.

video detector Diode which demodulates video signal.

video display Electronic display which, whether or not it presents alphanumerics, symbology and other information, presents pictures.

video extractor System for analysing all signals, selecting all that form part of useful image (eg in TV or radar) and excluding all others; in SSR usually synonymous with plot extractor.

video link Telecom system conveying pictorial

information.

video map(ping) Superimposition on radar display of fixed information or picture, usually derived from fine-grain photo plate scanned in synchronization with rotation of radar aerial or from computer memory, eg in GCA to show exact relationship of aircraft to surface obstructions.

video signal Telecom signal conveying video information.

vidicon Most common form of video (TV) picture (camera) tube, in which light pattern is stored on photoconductive surface; this is then scanned by electron beam, which deposits electrons to neutralize charge and thus generate output signal.

Vidissector Modern form of pioneer Farnsworth camera tube; used in space TV surveillance (ITT).

Les Vieilles Racines French association of aerospace pioneers and professionals.

Les Vieilles Tiges French association of pioneer pilots.

Vierendeel Girder (truss) comprising upper/lower chords and verticals, without diagonals or shear web, designed for flexure.

VIEWS, Views Vibration indicator early-warning system.

viff Vectoring in forward flight; pilot control of trajectory by direct control of propulsive thrust axis of jet-lift V/STOL aeroplane, selecting downwards for lift (normal acceleration) and forward of vertical for deceleration, thus performing combat manoeuvres unmatchable by any conventional aircraft. Hence verb to viff, viffing etc.

VIGV Variable-incidence (or inlet) guide vane.

VIM Vacuum-induction melting.

V_{IMD} Speed (IAS) for minimum drag.

V_{IMP} Speed (IAS) for minimum power, not necessarily same as above.

$V_{i\text{-}mp}$ Speed corresponding with lowest power at which both height and speed can be maintained, ie min speed for continuous cruise (arch).

V (Int)^2 Vehicle integrated intelligence.

vinyl ester Low-viscosity solventless liquid resin used as alternative to epoxy in wet lay-up of FRC materials.

VIP 1 Value improvement report.
2 Common meaning, very important person.
3 Video integrated presentation.

virage Tight turn (pre-WW1).

virga Streaks of water or ice particles falling from cloud but evaporating before reaching surface.

virgin fibre Continuous tow, long staple.

VIRSS Visual and IR sensor systems.

virtual gravity Terrestrial acceleration acting on parcel of atmosphere, reduced by centrifugal force due to parcel's relative motion; symbol g* = approx 99·99% g.

virtual height Apparent height of ionized atmospheric layer calculated from time for radio pulse to complete vertical round trip.

virtual image One visible in mirror but not projectable on surface.

virtual image display Small CRT giving binocular colour high-resolution image of surface target, which appears to be 935 mm (36·9 in) behind face of magnifying lens.

virtual inertia That part of inertia forces acting on oscillating body due to surrounding fluid (eg air) and proportional to fluid density.

virtual level Energy level of subatomic nuclear system for which excitation energy exceeds lowest nuclear-particle dissociated energy.

virtual mass Actual mass plus apparent mass.

virtual piston Pumping effect caused by collapse of launch tube by explosive lens.

virtual stress See *Reynolds stress.*

virtual temperature Temperature parcel of air would have had had it been entirely free of water vapour, symbol $T_v = (1 + 0{\cdot}61\ q)\ T$ where T is measured temperature and q is specific humidity.

vis Invariably means visibility, but ambiguous.

viscoelasticity Behaviour of material which has hereditary or prior stress-history memory and exhibits viscous and delayed elastic response to stress superimposed on normal instantaneous elastic strain.

viscosimeter Instrument for measuring viscosity; Saybolt and Engler * are simple calibrated containers with narrow orifice, result being obtained by timing run-off; accurate (absolute) * include Stokes (falling speed of small sphere), rotating-cylinder (outer cylinder drives inner via fluid interface whose drive torque is measured), capillary tube (Poiseuille), and oscillating disc (parallel and close to plane surface).

viscosity 1 Loosely, internal friction in fluid; property which enables fluid to generate tangential force and offer dissipative resistance to flow, defined as ratio of shear stress to strain; in air almost unaffected by pressure but increases with temperature. Symbol μ; for air $1{\cdot}78593 \times 10^{-5}$ Ns/m^2 (in traditional units 3·73 slug/ft-s).
2 Kinematic * is μ/ρ where ρ is density; varies with pressure as well as temperature, units are m^2/s; also called dynamic *. Symbol ν.

viscosity coefficient Synonymous with viscosity.

viscosity index Usually synonymous with viscosity, or its variation with temperature.

viscosity manometer Instrument for measuring very low fluid pressure by torque exerted on disc suspended on quartz fibre very close to spinning disc; examples are Dushman and Langmuir gauges.

viscosity valve Liquid-system control valve controlled by viscosity of medium, eg in bypassing lube-oil cooler.

viscous aquaplaning Occurs when runway merely damp, with water film not penetrated by tyres; important on smooth surfaces, esp coated with deposited tyre rubber; persists to low speeds.

viscous damping Energy dissipation in vibrating system in which motion is opposed by force proportional to relative velocity.

viscous flow Flow in which viscosity is important; can be laminar or turbulent but criterion is that smallest cross-section of flow must be very large in relation to mean free path. At very low R inertia becomes unimportant and flow governed by Stokes equations.

viscous fluid One in which viscosity is significant.

viscous force Force per unit mass or volume due to tangential shear in fluid.

viscous stress Fluid shear stress, symbol $\tau = \mu du/dy$ (Newton's law) where u is velocity distant y from surface or other reference layer.

visibility Distance at which large dark object can just be seen against horizon sky in daylight (DoD, NATO) (see *RVR*, *RVV*).

visibility meter Instrument for measurement of visibility, visual range, eg telephotometer, transmissometer, nephelometer etc.

visibility prevailing Distance at which known fixed objects can be seen round at least half horizon.

visible horizon Circle around observer where Earth and sky appear to meet (USAF). Also called natural horizon.

visible light EM radiation to which human eye responds, typically with wavelength from 400 to 700 nm, 0·4 to 0·7 μ.

visible line Full line on drawing representing a line visible in assembled subject item.

visible radiation See *visible light.*

visible spectrum See *visible light.*

visiónics Collective term for optical and electronic devices, operating at many EM wavelengths, which enhance human vision, especially at night and in bad weather or smoke or other adverse conditions.

vision-in-turn window Eyebrow window in roof of flight deck or cockpit.

visol Butyl ether +15% aniline; rocket fuel with Salbei.

visor 1 Pivoting or translating fairing for forward-facing cockpit windows of supersonic or hypersonic aircraft, in latter case with thermal-protection coating.
2 Hinged screen, transparent clear or tinted, protecting eyes against solar radiation and/or micrometeorites in space exploration.

VISSR Visible IR spin/scan radiometry (or radiometer).

Vista 1 Very intelligent surveillance and target acquisition.
2 Variable-stability inflight simulator test aircraft.

visual See *visual contact.*

visual acuity Human ability to see clearly, as tested by letter card.

visual approach Landing approach conducted by visual reference to surface, esp by aircraft on IFR flightplan having received authorization.

visual approach path indicator Simple Vasi for private pilots with red/amber/green sectors (Lockheed-Georgia).

visual approach slope indicator See *Vasi.*

visual/aural range See *VAR.*

visual contact To catch sight of, or a sudden good view of, target or Earth's surface.

visual cue Visual contact of object outside vehicle, esp Earth, sufficient to give pilot orientation and position information; eg first glimpse of ground through cloud during letdown in remote place far from electronic aids.

visual envelope Plot of flight-deck window areas showing range of P1 vision.

visual flight rules Those rules as prescribed by national authority for visual flight, with corresponding relaxed requirements for flight instruments, radio etc; in US typically demands flight visibility not less than 3 st miles (1 outside controlled airspace) with specified distances from any clouds.

visual head-up image Imagery created outside simulator, focused at infinity; alternative or addition to terrain model or CGI.

visual identification Control (ATC) mode in which aircraft follows a radar target and is automatically positioned to allow visual identification (NESN); ie interceptor pilot looks at other aircraft and, if possible, identifies type and registration or other identity.

visualization Making fluid flow visible, eg by injection of carbon tetrachloride or other liquid, smoke, tufting, Schlieren, spark photo, hot wire, china clay on surface, liquid films etc.

visual met conditions Generally UK counterpart to VFR, with similar requirements.

visual/optical countermeasures Those directed against human eyesight, ranging from reduction of night adaptation to blinding.

visual photometry Luminous intensity judged by eye.

visual range Distance under day conditions at which apparent contrast between specified target and background becomes equal to contrast threshold of observer (there is also night *) (see *RVR*, *RVV*).

visual reference Earth's surface, esp that clearly identified and thus giving geographical position as well as attitude and orientation guidance, used as reference in controlling flight trajectory, if necessary down to touchdown.

visual report Identical to hot report but based solely on aircrew observations; answers specific questions concerning target for which sortie was flown (ASCC). (ASCC adds "not to be confused with inflight report", while DoD states "not to be used; use inflight report").

visuals Lifelike scenes viewed through windows of simulator, almost always wholly synthetic and computer-generated (but in old simulators achieved by optics moved over a large model).

visual separation Basic method of avoiding collisions in TMAs or on ground by seeing and avoiding; normally involves conflicts between arrivals and departures and can be accomplished by pilot action or tower instruction.

VIT Vision in a turn.

vital actions Rigorously learned sequences instilled into all pilots, either specific to type or, more often, as general good airmanship; necessary eg before entering aircraft, upon entering cockpit, on starting engine, before taxiing, before TO etc. Generally remembered by mnemonics, eg Bumpf (check brakes, u/c, mixture, pitch, flaps) or HTMPFFG (hood/harness, trim/throttle friction, mixture, pitch, fuel cocks, flaps, gills/gyros).

vital area Designated area or installation to be defended by air-defence units (DoD).

vitrifying Transformation of ceramic from crystalline phase into amorphous or glassy state.

V_J Velocity of propulsive jet, normally measured relative to vehicle.

VKIFD Von Kármán Institute for Fluid Dynamics (Int).

VL 1 Vertical landing.
2 Tunnel velocity × characteristic model length.

V_L Relative velocity of flow on underside of aerofoil.

VLAC Vertical-Lift Aircraft Council (US Aerospace Industries Association).

VLAD Vertical line-array Difar.

VLCHV Very low cost harrassment vehicle.

V_{LE} Maximum speed with landing gear locked down (E = extended); flight manual specifies whether maingear only or all units. Always significantly higher than V_E values.

VLEA Very-long-endurance aircraft, powered by solar cells charging fuel cell(s) for continuous flight of months or years.

VLED Visible LED.

VLF 1 Very low frequency.
2 Vectored-lift fighter.

V_{LG} No longer used; was vague max-landing-gear speed, now replaced by V_{E1}, V_{LE}.

V_{LO} US term, maximum speed for landing-gear operation; synonymous with UK V_E.

V_{LOF} Lift-off speed, at which aeroplane becomes airborne.

VLR 1 Very long range (arch).
2 Telecom code for long-range search/rescue aircraft.
3 Velocity/length Reynolds number.

VLS 1 Visible light sensor.
2 Vertical launch system.

VLSI Very-large-scale integration.

VM, v.m. Velocity modulation.

V_M Speed at which precipitation (esp slush) drag is maximum; always well below aquaplaning speed.

V_m Volume fraction of composite material occupied by matrix.

VMA Code, fixed-wing attack squadron (USMC).

VMC Visual met conditions.

V_{MC} Minimum-control speed; more precisely specified as following three entries:

V_{MCA} Minimum speed at which aeroplane can be controlled in air; defined as limiting speed above which possible to climb away with not more than 5° bank and with yaw arrested after suffering failure of critical engine in TO configuration, with engine windmilling and c.g. at aft limit. There are usually suggested limits of required rudder-pedal force and on absolute value of *.

V_{MCG} Minimum speed at which aeroplane can be controlled on ground; defined as that above which pilot can maintain directional control after failure of critical engine without applying more than 70 kg pedal force, without going off runway and if possible while holding centreline, with 7+ kt crosswind and wet surface.

V_{MCL} Minimum speed at which aeroplane can be controlled in landing configuration, while applying max possible variations of power on remaining engines after failure of critical engine.

V_{MCP} Speed, usually EAS, at maximum continuous power in level flight; V_H is more commonly used.

VMDI Vector miss-distance indicator.

VMF 1 Code, fixed-wing fighter squadron (USMC).
2 Navy (USSR).

VMFA Code, fixed-wing fighter/attack squadron (USMC).

V_{MG} Vertical main landing gear strut load at MRW and with full aft c.g.

VMM Veille Météorologique Mondiale = World Weather Watch.

VMO Variable metering orifice.

V_{MO} Maximum permitted operating speed under any condition (less than V_{DF} but latter is wholly exceptional limit not intended to be reached except during certification flying).

VMS Vehicle monitoring system (robotics).

V_{MS} Minimum EAS observed during normal symmetric stall, usually less than V_S.

V_{MU} Minimum demonstrated unstick speed, at which with all engines operating, and without regard to safety, noise-abatement or any other factor, aeroplane will leave ground and hold positive climb.

VN Speed multiplied by (or plotted against) normal acceleration; more commonly V-n.

V_n Component of wind acting perpendicular to heading; also called normal wind, normal component or (loosely) crosswind component.

V-n Flight speed (usually EAS) multiplied by or plotted against normal acceleration; hence * diagram, of which two forms: basic manoeuvring envelope and basic gust envelope.

V_{NA} Noise-abatement climb speed, usually synonymous with V_4 (1st segment) but also commonly used for 4th segment Fuss.

VNAV, V-nav Vertical navigation; generalized topic of control of flight trajectory in vertical plane; now becoming automatic in transport-aircraft energy-managing flight systems.

V_{NE} Never-exceed speed; an exceptional permitted maximum beyond V_{MO} of which captain may avail himself in unusual circumstances. Implication is that * must be reported and explained.

V_{NO} Max permitted normal-operating speed, generally replaced by V_{MO}; in smooth air can be exceeded "with caution".

VNRT Very near real time.

VO 1 Visual optics.
2 US piston engine code, vertical crankshaft, opposed.

VOA Velocity of arrival.

VOC 1 Validation of concept.
2 Visual optical countermeasures.

Vocoder Voice coder; device responding to spoken input (usually previously stored) to generate synthetic speech output.

VOCS Voice-operated carrier suppressor.

VOD 1 Vertical on-board delivery; usually synonymous with vertrep.
2 Vertical obstruction data.

Voder Device with keyboard input controlling generation of electronic sounds, esp synthetic speech output.

VÖFVL Verband Österreichischer Flugverkehrsleiter (Austria).

Vogad Voice-operated gain-adjusting device; auto volume compressor or expander.

voice frequency Normally taken as 25 Hz to 3 kHz for telecom, much greater range for hi-fi.

voice-grade channel Covers about 300–3,000 Hz, for speech, analog, digital or facsimile.

voice keying System enabling telecom to use common R/T transmit and receive sites and similar frequencies but with voice-operated carrier suppressor and delay network to switch outgoing signal to transmitter and incoming to receiver. Remote stations normally switch automatically to receive mode except when user is speaking.

voiceless homing Any electronic homing system not using speech; traditionally meant radio range (arch).

voice-operated relay See *voice keying*.

voice rotating beacon Short-range radio navaid transmitting stored-speech headings (usually QDMs) which differ from 000° round to 359°; a form of talking VOR, also called talking beacon, abb VRB (arch).

void Undesired gap in welded joint.

void fraction Percentage of total frontal area of jet engine through which airflow passes.

Voigt effect Double refraction (associated with Zeeman) of light passing through vapour perpendicular to strong magnetic field.

VOIR Venus-orbiting imaging radar.

Voiska-PVO Troops of air defence of homeland (USSR).

VOL Vertical on-board landing.

volatile 1 EDP (1) memory which dissipates stored information when electrical power is switched off; thus, next morning or after weekend all bits must be restored before computer operation. Also means electrical transients cause corruption (though this may be only temporary). Hence * data, * memory, volatility.
2 Having high vapour pressure, and thus low boiling or subliming temperature at SL pressure; hence volatility.

vol à voile(s) Gliding, soaring (F).

vol d'abeille Beeline, straight-line distance (F).

Volmet Routine ground-to-air broadcast of met information (ICAO).

vol piqué Dive (F).

vol plané Planing (inclined) flight, ie glide by powered aeroplane (arch except in F).

volt SI unit of EMF, = W/A.

voltage amplifier Delivers small current to high impedance to obtain voltage gain.

voltage-dependent resistor Ohms = f(V); resistance varies directly with applied voltage.

voltage drop PD across any impedance carrying current.

voltage-fed aerial (antenna) Fed from one end, where signal potential is maximum.

voltage gain Ratio of output/input voltages.

voltage standing-wave ratio Ratio max/min V along waveguide or coaxial.

volt-ampere SI unit of AC power; true power VA multiplied by power factor cos θ.

voltmeter Instrument for measuring potential difference, ie V.

volume SI unit is m^3 (conversion factors for non-SI measures, $ft^3 \times 0{\cdot}02831684$, from UK gal $\times$ $0{\cdot}004546087$ and US gal $\times$ $0{\cdot}003785411$).

volume fraction 1 Proportion, usually %, of reinforced composite (FRP) occupied by reinforcing fibres.
2 Generally, proportion of whole volume occupied by particular substance.
3 For aerostats see *air* *, *gas* *.

volumetric efficiency Volume of combustible mixture (in diesel, air) actually drawn into cylinder of PE (4) on each operating cycle divided by capacity (swept volume) of cylinder, usually expressed as %. Symbol η_v or e_v.

volumetric loading Also called * density, total volume of solid rocket motor propellant divided by total volume of unloaded case, usually expressed as %. Symbol λ.

volume unit Measure of audio volume to be outputted by electrical current, expressed in dB equal to ratio of magnitude of electrical waves to magnitude of reference volume, usually 1 mW; abb VU.

volute Spiral or planar helix; thus, spiral casing of

centrifugal compressor or supercharger impeller (becoming arch).

Vom Volts/ohms/milliamps tester.

Von Brand Standard method of measuring jet smoke by passing measured gas volume through filter and then recording intensity of calibrated light reflected by filter pad. Gives quantified measure of particulate matter trapped by chosen filter. Hence * scale for visible smoke.

Von Kármán street See *street* (also called Kármán street).

VOP Variation of price.

VOR VHF omnidirectional radio range, most common of all radio navaids; fixed beacon emits fixed circular horizontal radiation pattern on which is superimposed rotating directional pattern at 30 Hz giving output whose phase modulation is unique for each bearing from beacon. Thus airborne station can read from panel instrument bearing of aircraft from station, called inbound or outbound radial. Each fixed station identified by three-letter keyed intermittent transmission (sometimes voice).

VOR/DME VOR steering guidance with DME distance information.

Vorgen Voltage generator.

VOR/ILS Linkage of VOR signals to aircraft ILS so that left/right steering guidance is given by ILS panel instrument. Latter often called * deviation indicator, or Vorloc.

Vorloc, VOR/Loc Panel instrument giving steering guidance from received VOR signals; can be complete ILS indicator or (esp in light aircraft) simple VOR receiver with localizer needle only.

Vortac Combination of VOR and Tacan (occasionally written VOR/TAC) offering from one fixed station VOR az, Tacan az and Tacan (DME) distance information; ident codes prove VOR and Tacan signals are both from same fixed station. Normally * is end-product of trying to integrate civil (VOR/DME) with military (Tacan) navaids, latter being UHF and therefore inherently incompatible.

vortex Fluid in rotational motion (possessing vorticity), eg streamed behind wingtip or across leading edge of slender delta.

vortex breakdown Sudden separation of large vortex from leading edge of slender delta (naturally followed by its decay) at particular AOA (higher than stalling AOA for most wings); essentially represents stall of slender-delta wing.

vortex burst See *vortex breakdown*.

vortex dissipator Bleed-air jet(s) blown down below and ahead of jet-engine inlet to prevent ingestion of material from unpaved airfields or contaminated runway.

vortex drag Drag caused by vortex formation; not normally a recognised part of aircraft drag.

vortex filament Line along which intense (theoretically infinite, at R = 0) vorticity is concentrated; either closed loop or extended to infinity.

vortex flow Fluid flow combining rotation with translational motion.

vortex generator Small blade perpendicular to skin of aircraft or other body set at angle to airflow to cause vortex which stirs boundary layer, usually to increase relative speed of boundary layer and keep it attached to surface; also called turbulator.

vortex hazard Danger to aircraft, esp light aircraft, from powerful vortices trailed behind wingtips of large aircraft; also called wake hazard, wake turbulence.

vortex lift Lift generated by slender delta or similar wing having sharp, acutely swept leading edge (subsonic relative velocity normal to leading edge): large and powerful vortex is shed evenly on left/right wings, adding major nonlinear increment to lift; also postpones stall to lower speed and extreme AOA, but with high drag.

vortex line Line whose direction at every point coincides with rotation vector; all of whose tangents are parallel with local direction of vorticity. Must be closed curve or extend to infinity, or to edge of fluid or to a point on an infinitely intense vortex sheet.

vortex ring Vortex forming closed ring (eg smoke ring); collar vortex.

vortex-ring state Operating state of rotorcraft (esp helicopter) main rotor in which direction of flow through rotor is in opposite sense to relative vertical flow outside rotor disc and opposite to rotor thrust. Occurs in autorotative landing.

vortex separation Filtration of different types of particle from fluid by different centrifugal force in vortex motion.

vortex sheet Theoretical infinitely thin layer of fluid characterized by infinite vorticity; in practice layer of finite thickness formed by large number of small vortices, eg as trailed behind lifting wing (where much of vorticity is quickly rolled up into two large tip vortices).

vortex street See *street*.

vortex strength Circulation round any body or other closed system, symbol Γ, constant at all points on a vortex filament.

vortex trail Visible (white) trail from wingtip, propeller tip etc, caused by intense vortex.

vortex tube Device devoid of moving parts in which pressure difference induces fluid flow through tangential slots into tube; violent vortex divides flow into surrounding warm flow and cold (about 40°C cooler) core.

vorticity Vector measure of local rotation in fluid; in uniformly rotating fluid proportional to angular velocity (in UK, exactly defined as twice angular velocity). Symbol $q = \nabla V$ where ∇ is del (mathematical operator) and V is vector velocity (= curl V in US; often called rot V, from rotation, in Europe).

vorticity component Circulation around elementary surface normal to direction of vorticity divided by area of surface; more strictly, limit of circulation as area of element approaches zero.

vortillon Name coined by McDonnell Douglas to describe DC-9 fence around underside of wing leading edge controlling boundary-layer direction.

VOT, Vot VOR test signal; ground facility for testing accuracy of VOR receivers.

voter Binary logic element or device which compares signal condition in two or more channels and changes state whenever a predetermined signal mismatch occurs, usually to exclude a minority "outvoted" signal.

voter threshold Difference between signals at which voter is switched or triggered; normally difference between one selected signal and mid-value signal from all others in parallel system.

voting system System in which outputs of several parallel channels are sensed and compared by voter so that any single malfunctioning channel may be excluded.

Votol Vertical-only take-off and landing.

VOWS, Vows Valuation of weight saved; measure of financial reward (usually in increased annual earning power) from cutting each unit of mass (kg or lb) from empty weight; eg VOWS for Concorde in 1974 currency was £50/lb.

Vox Voice keying (NASA).

VP 1 Variable-pitch.
2 Code: fixed-wing patrol squadron (USN).
3 Vector processor.
4 Video processor.

V_p 1 Propellant volume; that volume occupied by solid propellant in a rocket motor.
2 Propwash velocity, V + v.

v.p. Vapour pressure.

VPC Vertical-path computer.

$V_{\%}$ Best angle-of-climb speed (UK usage, US = V_X).

$V_{\%SE}$ Best angle-of-climb speed, single-engine (UK).

V_{peri} Perigee velocity, maximum speed of satellite

$$= \sqrt{K\frac{2}{r_0} - \frac{1}{a}}$$

where K is const, r_0 is distance to centre of primary and a is semi-axis of elliptical orbit.

VPI 1 Vapour-phase inhibitor.
2 Vertical-position indicator.

VPN 1 Vickers pyramid number.
2 Variable primary nozzle.

VPTAR See *Vaptar*.

VPTS Voice-processing training system.

VR 1 Resultant velocity.
2 Vernier radial (thruster).
3 Visual reconnaissance.

V_R Rotation speed, at which PIC starts to pull back on yoke to rotate aeroplane in pitch; normally determined by one of following: not less than 1·05 V_{MCA}; not less than 1·1 V_{MS}; not less than 1·05 or 1·1 V_{MU}; and it must allow 1·1 V_{MCA} or 1·2 V_{MS} to be achieved at screen after one engine failure.

V_r Radial velocity, eg component of velocity along sightline to target (rate of change of range) or fluid speed along radial direction in centrifugal compressor measured relative to compressor (ie eliminating tangential component).

V_{RA} Rough-air speed; max recommended EAS for flight in turbulence.

VRB Voice rotating beacon.

VRC Vendor reject crib.

V_{REF} 1 Loosely, any reference speed.
2 In jet transports, a typical approach speed at about 1·35 V_{s0} (chiefly US).

VRF Code: ferry squadron (USN).

vrille Spin (arch).

VROC Vertical rate of climb (helicopter).

VRP Visual reporting post.

VRU Vertical reference unit.

VS 1 USN squadron code: fixed-wing ASW.
2 Vestigial sideband.

V_s 1 Stalling speed; IAS at which aeroplane exhibits characteristics or behaviour accepted as defining stall (in US, FAA adds "or min steady flight speed at which airplane is controllable", which is not same thing and is separately defined in UK as V_{MCA}).
2 Velocity of slip (propeller).
3 Slipstream velocity, relative to aircraft.

V_{s0} Stalling speed with flaps at landing setting.

V_{s1} Stalling speed in a specified configuration other than clean.

V_{s1g} Stalling speed under 1 g vertical (normal) acceleration; obtained from V_s by correcting for any imposed normal acceleration that may have been present during an actual measured stall; a "pure" V_s not normally entering into performance calculations.

VSA By visual reference to ground (ICAO code).

VSAM Vestigial-sideband amplitude modulation.

vsby Visibility.

VSC Video scan converter.

VSCF Variable-speed constant-frequency.

VSD Vertical situation display.

VSER Vertical speed and energy rate.

VSFI Vertical-scale flight instrument.

VSI 1 Vertical speed indicator; output is rate of climb or descent.
2 Vertical soft-iron; component of Earth's field.
3 Vapour space inhibitor = VPI.

VSS 1 Video signal simulator.
2 Variable-stability system.
3 Vehicle systems simulator.

VSSC Vikram Sarabhai Space Centre (India).

V_{SSE} Minimum speed, selected by manufacturer, for intentionally shutting down one engine in flight for pilot training; in UK and some other countries this is prohibited below 3,000 ft (907 m) AGL.

VST Variable-stability trainer; aeroplane with

avionics and flight-control surfaces added to enable it precisely to duplicate flight characteristics of other types.

V-Star Variable search and track air-defence radar.

V/STOL Vertical or short take-off and landing.

VSTT Variable-speed training target.

VSV Variable-stator valve.

VSW 1 Vertical speed and windshear; hence VSWI = * indicator.
2 Variable-sweep wing.

VSWR Voltage standing-wave ratio.

VT 1 Vernier thruster.
2 Voltage transient.
3 Vectored thrust.
4 Internal prison, eg for design teams (USSR, 1929–42).

V_t True airspeed, in aerodynamics.

VTA 1 Military transport aviation (USSR).
2 Vibration tuning amplifier.

VTAS Visual target acquisition system.

VTC 1 Vectored (or vectoring) thrust control.
2 Vernier thruster control.
3 Variable time-constant.
4 Vibratory torque control (Teledyne Continental).

VTM Voltage-tunable magnetron.

V_{Tmax} Maximum threshold speed, above which risk of overrunning is judged unacceptable; usually V_{AT} + 15 kt.

V_{Tmin} Minimum threshold speed, below which risk of stall (esp in windshear) is judged unacceptable; usually V_{AT} – 5 kt.

VTO 1 Vertical take-off.
2 Varactor-tuned oscillator.

VTO grid Vertical take-off grid designed to reduce erosion and reingestion problems in operating jet-lift aircraft from unprepared surfaces.

VTOL Vertical take-off and landing; VTOVL adds redundant second "vertical".

VTR Video tape recorder.

V_U 1 Relative airspeed across upper (positively cambered or lifting) surface of aerofoil.
2 Tangential velocity, eg of flow leaving centrifugal compressor.

Vulkollan Hard erosion-resistant thermosetting plastics material (trade name, F).

V_{US} Unstick speed of marine aircraft.

VV 1 Visibility code (synoptic chart).
2 Valve voltmeter.
3 Validation and verification (sometimes V&V).
4 Velocity vector.

VVA Zhukovskii Air Force Academy (USSR).

VVIA VVA engineering academy (USSR).

VVS Air force (USSR).

VWF Verband der Wissenschaftler an Forschungsinstituten eV (G).

VWRS Vibrating-wire rate sensor.

VWS Air-ventilated wet suit (aircrew anti-exposure suit).

V_X Speed for best angle of climb, segment not specified (US usage).

VXO Variable crystal oscillator.

V_Y Speed for best rate of climb (US usage).

V_{YSE} Best rate-of-climb speed, single-engine.

V_{ZRC} Zero rate-of-climb speed, at which with one engine inoperative drag reduces climb gradient to zero.

W

W 1 Watts.
2 Generalized symbol for weight, usually synonymous with mass.
3 Generalized symbol for force or applied load.
4 Generalized symbol for energy (W = work).
5 Titanium.
6 US aircraft modified mission prefix, weather reconnaissance.
7 USN aircraft mission prefix 1952–62, electronic search.
8 US aircraft modified mission suffix, special research (= AEW).
9 JETDS code, armament; or automatic flight or remotely piloted.
10 Weather.
11 West, western longitudes.
12 Weapon.
13 Wave(s).
14 IFR flightplan code: no transponder but approved R-nav.
15 Wing (military unit).

w 1 Generalized symbol for special fluid velocities, eg vertical gust, wing downwash, propeller slipstream etc.
2 Warm (air mass).
3 Load per unit distance or per unit area.
4 Specific loading.
5 Linear velocity due to yaw.
6 Suffix, wing; thus W_w = wing weight.
7 Rate term for weight or mass, eg per unit time.

W-code W (14): approved R-nav but no transponder.

W-engine PE (4) with three linear banks of cylinders about 50°–60° apart; also called broad-arrow.

W-wing Shaped like W in planform with sweepback inboard and forward sweep outboard.

WA Work authorization.

W_A Equipped airframe weight.

W_a Air mass flow, eg passing through engine per second.

WAAF Women's Auxiliary Air Force (UK, WW2).

WAAM Wide-area anti-armour munition.

WAAS Wide-area active surveillance (radar).

WAASA Women's Aviation Association of South Africa.

WAC chart World aeronautical chart.

WACA World Airlines Clubs Association (Int).

WACRA, Wacra World Airline Customer Relations Association (Int).

WAD Workload assessment device.

WAF Women in the (US) Air Force.

wafer Complete (near-circular) slice of single-crystal (usually epitaxial) semiconductor material on which numerous electronic devices are constructed, subsequently separated by scribing and cleavage to make chips.

WAFS Women's Auxiliary Ferrying Squadron (US).

Wagner bar Pioneer spoiler-type flight control with bang/bang solenoid operation for radio command guidance of missiles (from 1937).

Wagner beam Idealized pure tension-field beam assumed to have zero compressive strength and thus to react loads as diagonal tensions; generally, a beam designed to buckle in operation.

wagon wheel See *wheel (6).*

waist 1 Amidships portion of fuselage or hull; thus * gunner, firing laterally (often from rear midships area).
2 Middle portion of gas-turbine engine, esp where diameter here is less than elsewhere; thus * gearbox for accessories.

waisting 1 Local reduction in diameter caused by plastic flow under tension at point of failure.
2 Local reduction of cross-sectional area of body, eg to conform to Area Rule.

wake 1 Fluid downstream of body where total head has been changed by body's presence; usually to some degree turbulent.
2 Wingtip vortices left in atmosphere behind aircraft whose wing is generating lift, in case of large/heavy aircraft very powerful and persistent.

wake-interaction noise Generated by impingement of wake from moving blade on object downstream, eg stator; in gas turbine hundreds of such interactions can generate noises at blade-passing frequency of each stage of blading, plus harmonics.

wake turbulence Turbulence due to wakes (2).

walkaround oxygen Bottle carried by aircrew when disconnected from system.

walkback Return of deck arrester wire (pendant) to "cocked" or ready position.

walking 1 To advance engine throttles (power levers) in asymmetric steps, left/right/left etc.
2 To keep straight on take-off with violent left/right rudder.

walking beam Pivoted beam transmitting force or power, eg to retract landing gear; secondary meaning arising by chance from repeated usage is landing-gear retraction and bracing beam whose upper end is not pivoted direct to airframe but to a sliding or translating member, eg long jackscrew.

walking in/out Movement of aerostat, esp airship, in or out of hangar by large groundcrew walking while holding tethering lines.

walkingstick 1 Laser used to indicate ground clearance in descent through cloud with ceiling close to terrain (not colloq; name derives from stick of blind person).
2 Gas-turbine vaporizing burner in which fuel is heated in tube with 180° bend looking like top of *.

walk-on service Airline service in which anyone can walk directly on board, buying ticket from cabin crew.

walkway Catwalk or other narrow structure provided in airship (rarely, large aircraft of other type) to provide human access to other part; hence * girder.

wall 1 Sides of tyre, between wheel and tread.
2 Internal boundaries of wind tunnel; not only vertical sides.

wall constraint Distortion of flow round model in tunnel with closed working section arising from fact that walls are seldom aligned with streamlines.

wall energy Energy per unit area of boundary between oppositely oriented magnetic domains.

Wallops Flight Center NASA complex on US east coast used for nearly all small rocket launches.

wallowing Uncommanded motions about all three axes.

Walter Emergency locator beacon emitting battery-powered pulses.

WAM Window addressable memory.

Wams, WAMS Weapon-aiming mode selector.

wander See *apparent precession.*

wandering Slow and apparently steady uncommanded change in heading; may from time to time reverse itself.

War Air Service Program American (DoD) plan for civil airline routes, equipment and services following withdrawal of CRAF aircraft.

war consumables All essential expendables directly related to hardware of a weapon/support system or combat/support activity (USAF).

WARC-ST World Administrative Radio Conference for Space Telecommunications (ITU).

war game Simulation, by whatever means, of warfare using rules intended to depict real life.

war gas Chemical agent designed for use against human body directly; eg not used to contaminate water supply. Liquid, solid or vapour.

warhead Portion of munition containing HE, nuclear, thermonuclear, CBI or inert materials intended to inflict damage. DoD recognizes additional term, * section, as assembled * including appropriate skin sections and related components; thus * alone need not include fuzing, arming, safety and other subsystems.

warhud Wide-angle raster HUD.

Warmaps Wartime Manpower Planning System (US).

war materiel All purchasable items required to support US and Allied forces after M-day.

warm front Locus of points along Earth's surface at which advancing warm air leaves surface and rises over cold air.

warm gas thruster Propulsive jet composed of HP stored gas heated (eg by main rocket combustion) before expulsion through separate nozzle; various configurations but common for vehicle roll control.

warm sector Portion of depression, esp recently formed, occupied by warm air.

warm-up Generalized term for process, or necessary elapsed time, in which device is operated solely for purpose of bringing it to steady-state operating condition, with steady running speed, temperature, pressure or other variables. Examples: PE (4), gyro.

warm-up time Published time for device, eg gyro, to reach specified performance from moment of energization.

warned exposed Friendly forces are lying prone with all skin covered and wearing at least two-layer summer uniform.

warned protected Friendly forces are in armoured vehicles or crouched in holes with improvized overhead shielding.

warning area Airspace over international waters in which, because of military exercise, non-participating aircraft may be at risk.

warning indicator Device intended to give visual or aural warning of hazard, eg fault condition or hostile attack.

warning net Designated telecom system for disseminating information on enemy activity to all commands.

warning order Preliminary notice of friendly action to follow.

warning panel Area of display, eg on flight deck, containing numerous designated warning indicators or captions, with or without attention-getting master *.

warning receiver Passive EM receiver with primary function of warning user his unit/vehicle/location is being illuminated by an EM signal of interest.

warning red Attack by hostile aircraft/missiles imminent or in progress.

warning streamer Brightly coloured fabric strip, flexible in light breeze, drawing attention to protective cover or other item which must be removed from vehicle before flight or launch.

warp Threads in fabric parallel to selvage, continuous full length of material.

warping Twisting wings asymmetrically to obtain lateral stability and control; usually imposed by diagonal downward pull on rear of wing near tip to twist (wash-in) and increase camber.

warp-sheet Standard raw-material form of reinforcing fibre, esp carbon/graphite, in which broad sheet is made up entirely of parallel fibres;

usually used cut to shape in multiple laminate structure.

warpwheel Pre-1914 lateral control wheel.

WARR Wissenschaftliche Arbeitsgemeinschaft für Raketentechnik und Raumfahrt (G).

war-readiness spares kit Prepared kit of spares and repair parts to sustain planned wartime or contingency operations of weapon system for specified period (USAF).

Warren Structure in form of frame truss comprising upper/lower chords joined by symmetric diagonal members only; hence * bracing, * girder, * struts, * truss. * biplane in front view has only diagonal interplane struts, forming continuous zig-zag along wing.

war reserve Inactive stocks, subdivided into nuclear and other, of all forms of supplies (US = materiel) drawn upon in event of war.

wartime rate Max attainable flying rate based on seven-day week and with wartime maintenance/ safety criteria.

warting Pitting of metal surface, esp in gas turbine, caused by combined actions of carbon and atmospheric salts plus thermal cycling.

WAS Weapon-aiming system.

WASG 1 Warranty and service guarantee.
2 World Airline Suppliers' Guide.

wash Wake (colloq).

wash-in, washin Inbuilt wing twist resulting in angle of incidence increasing towards tips.

washing machine Trainer used by commanding officer or CFI and thus that in which failed pupil makes last flight (US colloq, arch).

wash-out, washout 1 Inbuilt wing twist resulting in angle of incidence reducing towards tips.
2 To fail course of flight (pilot) instruction.
3 Failed pupil pilot.
4 Removal of particulate matter from atmosphere by rain.

wash primer Self-etching primer to prepare surface of Al or Mg for subsequent priming or painting.

WASP, Wasp 1 War Air Service Program (DoD).
2 Wide-area special projectile (USAF).
3 Women's Airforce Service Pilots (US, WW2).

Waspaloy Trademark (Pratt & Whitney) of nickel alloys for gas-turbine rotor blading and similar purposes, typically with about 19% Cr, 14% Co and also Mo, Ti, Al etc.

wastage Those pupils who fail a course of instruction, esp aircrew; hence * rate, % failing.

waste gate Controllable nozzle box for exhaust gas turbine of turbocharged PE (4); hence ** valve, which when open allows gas to bypass turbine but which gradually closes with height until at rated full-throttle or other selected height valve is closed completely.

wastes 1 Human body wastes, esp fecal.
2 Surplus radioactive equipment and materials.

WAT Weight/altitude/temperature; factors independent of runway which govern each take-off and determine whether aeroplane can meet specified positive climb criteria after engine failure at V_1; pronounced "watt", but invariably written all in capitals.

WATA World Association of Travel Agents (Int).

WAT curve Graphical plot of WAT limitation for particular aircraft type; hence WAT limit, limiting value(s) of WAT at which performance is minimum for compliance with requirements.

water bag Polythene bag carrying water ballast.

water ballast Standard ballast carried by competitive sailplane.

water barrier 1 Runway overrun barrier using water as retarding material.
2 Notional barrier to prolonged spaceflight caused by fact that plants used for fresh food/oxygen continuously convert more material to water than they return in consumable form.

water bias See *sea bias*.

water bomber Aircraft designed for surface (eg forest) firefighting by dropping large masses of water; usually marine aircraft with means for quick on-water replenishment using ram inlet on planing bottom.

water cart Dispenser of water to aircraft on apron, with supplies of either or both demineralized water for engines or potable water for passengers.

water-collecting sump Low point in any system where water could collect, esp fuel tank and tray under vapour-cycle air-conditioning coils, from which water can be extracted.

water-displacing fluid Commercial liquids (eg LPS-3) which preferentially attach themselves to metal surface in place of local droplets of moisture, thus arresting corrosion.

water equivalent depth Measure of depth of precipitation contamination on runway; WED = actual depth × density, thus 20 mm slush with SG 0·5 gives WED 10 mm. For water, *** = actual depth.

waterfall Basic model of software life cycle.

water flaps Surfaces hinged about near-vertical axis near afterbody keel of marine aircraft (esp jet) which when under water are used differentially for steering and together for braking (usually also used as airbrakes).

water gauge Pressure expressed as height of column of water; 1 in H_2O = 249·089 N/m^2.

water injection Injection of demineralized water, either pure or with 30–67% alcohol or (more commonly) 44–60% methanol, into cylinders of PE (4) to cool charge and eliminate detonation at max BMEP or into compressor or combustor of gas turbine to cool air and increase mass flow and thus power.

water jacket Container for cooling water around cylinder (arch).

water level Generalized term in lofting and aerospace construction generally (except vehicles whose major axis is vertical, eg most space launchers) to denote measures in vertical planes; thus ** 193 = 193 mm above aircraft reference datum for measures in vertical plane, which may be longitudinal axis OX or some other essentially horizontal reference; abb WL. Note: in US unit is often still inches.

waterline 1 Intersection of body exterior profile and a horizontal plane; often used as synonymous with water level, thus WL 0 is lowest point of body and all subsequent slicing planes are parallel to prime longitudinal axis or other horizontal reference. Thus * view, * plot (all waterlines drawn on common axis of symmetry).
2 Any horizontal reference other than local Earth surface used in aircraft attitude instrument or HUD.

water loop Inadvertent turn by marine aircraft on water, eg after dipping wingtip float at high speed (full ** rare because usually aircraft rolls in opposite direction through centripetal force).

water recovery Recovery of usable water from propulsion exhaust, esp aboard airship for use as ballast.

water resistance Drag caused by water to aircraft moving through it, made up of skin friction and wavemaking.

water rudder Small surface usually hinged on centreline of marine aircraft to sternpost or rear-step heel; used for directional control on water.

waterspout Visible water-filled tornado over sea.

water suit Anti-g suit in which interlining is filled with water which automatically provides approx required hydrostatic pressures under large normal accelerations.

water tunnel Similar to wind tunnel but using water as working fluid for large R at low V.

water twister Rotary liquid-turbine device which absorbs energy in MAG arrested landings.

WATOG, Watog World Airlines Technical Operations Glossary.

watt SI unit of power (not only electrical power), W = J/s. Conversion factors: hp 745·7 exactly, CV 735·499, Btu/min 17·5725, in each case to convert from * to unit stated.

watt-hour SI unit of energy = 3,600 J.

wattless power Reactive power VAR; also called wattless component.

wattmeter Instrument for measuring electrical power

wave 1 Disturbance propagated in medium such that at any point displacement = f(time) and at any time displacement of point = f(position); any time-varying quantity that is also a f(position). This definition falls down for light and other EM radiation, which appear not to need a "medium" for propagation (f = function of).
2 Formation of assault vehicles (land, sea or air) timed to hit hostile territory at about same time.

wave angle Angle between upstream free-stream direction and an oblique shock, symbol α.

waveband Particular portion of EM spectrum in telecom frequency region assigned by national authority for specific purpose.

wave crest Peak of waveform.

wave disturbance Discontinuity or distortion along a met front.

wave drag Additional increment of aerodynamic drag caused by shockwave formation, made up of distribution of volume along length (longitudinal axis) and drag due to lift; symbol for coefficient of ** C_{DW}.

waveform Shape of a repetitive (eg sinusoidal) wave when plotted as amplitude against time-base or when displayed on CRT.

waveform generator Converts DC or raw AC into any desired waveform output, with any frequency, amplitude (both time-varying if required) or other characteristic, eg for testing all airborne electronic systems.

wave front 1 Leading edge of shockwave group, or of blast wave from explosion.
2 In repetitive wave a surface formed by points which all have same phase at a given time.

waveguide Conductor for EM radiation in reverse sense to normal conductor in that radiation travels through insulator (usually atmosphere) surrounded by metal walls, usually rectangular cross-section, along which waves propagate by multiple internal reflection. Rarer form is dielectric cylinder along whose outer surface EM radiation propagates.

waveguide modes See *propagation modes.*

waveguide mode suppressor Filter matching particular waveguide cross-section designed to suppress undesirable propagation modes.

wavelength Distance between successive wave crests; symbol $\lambda = v/f$ where v is velocity of EM radiation (usually close to speed of light) and f is frequency.

wavelet Small shockwave, usually present in large numbers in boundary layers and around surface of supersonic body. Sometimes called Mach wave or Mach *.

wave lift Lift on lee side of ridge or mountains.

wavemaking resistance Drag of taxiing marine aircraft caused by gross displacement of water in waves; reaches max at about 20–30% of unstick speed.

wave motion Oscillatory motion of particle(s) caused by passage of wave(s), usually involving little or no net translation, ie particle resumes near-original position after wave has passed. Direction of ** varies with transverse waves (eg EM radiation), longitudinal waves (sound) or other forms, eg surface (water/air interface) waves.

wave number Reciprocal of wavelength $1/\lambda$ or (alternatively) $2\pi/\lambda$.

wave-off Overshoot commanded by carrier deck-landing control officer.

wave period Elapsed time between successive

crests, 1/f, where f = frequency.

wave soaring Using wave lift.

wave trough Point of minimum, or max-negative, amplitude, usually half-way between crests.

waviness Surface irregularities with spacing greater than for roughness; height is mean difference between peaks and valleys and spacing is distance between peaks.

waybill Document listing description of each item of cargo, consignor, consignee, route, destination, flight number, date and other information.

waypoint 1 Predetermined and accurately known geographical position forming start or end of route segment.
2 In US, as (1) but with addition "whose position is defined relative to a Vortac station" (FAA).
3 In military operations, a point or series of points in space to which an aircraft may be vectored.

WB Weather Bureau (US, NOAA).

Wb Weber.

WBF Wing-borne flight (jet V/STOL).

WBO Wien-bridge oscillator.

WBPT Wet-bulb potential temperature.

WBS 1 Work breakdown structure.
2 Weight and balance system.

WBSV Wide-band secure voice.

WBTM Weather Bureau technical memoranda.

WBVTR Wideband video tape recorder.

W/C Wavechange.

WCG Water-cooled garment.

WCNS Weapon control and navigation system.

W/comp Wind component.

WCQL Worst-cycle quality level.

WCR Weight/capacity ratio.

WCS 1 Weapon control system.
2 Waveguide com system.
3 Writable control store (EDP).

WCTB Wing carry-through box (swing-wing aircraft).

WCTL Worst-condition time-lag.

WD Wind direction.

WDAU Weapon-dispenser arming unit.

WDD Western Development Division (USAF, 1954–62, now Samso).

WDEL Weapons Development and Engineering Laboratories (US).

WDF Water-displacing fluid.

WDM Wavelength division multiplexing.

WDNS Weapon-delivery and navigation system.

WDS Wavelength dispersive spectrometer.

WDU Wireless Development Unit (UK, WW2).

W_E Mass of propulsion system (from weight of engine[s]).

WEAA Western European Airports Association (Int).

WEAAC West European Airport Authorities Conference (Int).

WEAAP West European Association for Aviation Psychology (Int).

weakest maintained Weakest fuel mixture at which under specified conditions max power can be maintained; also called WMMP (suggest arch).

weak link Point at which structure, esp hold-back tie (eg on aircraft about to be catapult-launched), is designed to break when normal operating load is applied. Catapult launch ** resists full thrust of aircraft engines but breaks when catapult thrust is added. Occasionally a safety feature fracturing only on overload.

weak mixture Fuel/air ratio for PE (4) below stoichiometric; economical for cruising but engine runs hot. Hence ** rating, max power permitted for specified conditions cruising with **; ** knock rating, fuel performance-number grade under economical-cruise conditions.

weak tie Structural weak link designed to fail in normal operation (eg holdback on cat take-off).

weapon-aiming system That governing launch trajectory of unguided weapon.

weapon bay Internal compartment for carriage of weapons, esp of varied types, eg AAMs, ASMs, NWs, free-fall bombs, guns, sensors, cruise missiles etc. If for one type of weapon preferable to be more explicit. Derived terms include ** door, ** fuel tank, ** hosereel pack.

weapon control system Avionics and possibly other subsystems (eg optics) built into launching aircraft to manage weapons before release and release them at correct points along desired trajectories. Should not be used to mean radio command or other form of guidance system of missile.

weapon debris Residue of NW after explosion; not usually well defined but generally means all solids (assumed recondensed from vapour) originally forming casing, fuzing and other parts, plus unexpended Pu, U-235 or other fissile material.

weapon delivery Total action required to locate target, establish release conditions and maintain guidance to target if required (ASCC).

weaponeering Process of determining quantity of specific weapon necesssary for required degree of damage to particular (surface) target. Takes into account defences, errors, reliabilities etc.

weapon line See *bomb line.*

weapons assignment Process by which weapons are assigned to individual air weapons controllers for an assigned mission (DoD).

weapons of mass destruction For arms-control purposes, strategic NW, C, B, R devices with potential of killing large numbers of people, but exclusive of delivery systems.

weapons recommendation sheet Defines intention of attack and recommends nature of weapons, tonnage, fuzing, spacing, desired mean points of impact, intervals of reattack and expected damage.

weapons state of readiness In DoD usage, lists of

numbers of air-defence weapons and reaction times: 2 min, 5 min, 15 min, 30 min, 1 h, 3 h, and released from readiness.

weapon system 1 A weapon and those components required for its operation (DoD, NATO). This could simply be part of manned aircraft, eg radar, HUD, WCS.
2 Composite of equipment, skills and techniques that form instrument of combat which usually, but not necessarily, has aerospace vehicle as its major operational element (USAF). As originally conceived in 1951, includes all type-specific GSE, training aids, publications and every other item necessary for sustained deployment.

weapon-systems physical security Concerned to protect aerospace operational resources against physical damage.

weapon/target line Sightline (straight line) from weapon to target.

weather Short-term variations in atmosphere, esp lower atmosphere.

weather advisory Expression of hazardous weather likely to affect air traffic, not predicted when area forecast was made.

weather beam Emitted by radar operating in weather mode, conical pencil of approx 5° total angle projecting horizontally ahead (thus filling whole troposphere about 100 km ahead).

weather categories 1 Traditional **, eg US Cat C (contact), N (instrument) and X (closed), common today in many countries.
2 Precise measures of DH/RVR as they affect arrivals: Cat 1, DH 60 m/200 ft or better, RVR 800 m/2,600 ft or more; Cat 2, 60–30 m/200–100 ft, 800–400 m/2,600–1,300 ft; Cat 3a, 0 along runway, 200 m/700 ft in final descent phase; Cat 3b, 0, 50 m/150 ft; Cat 3c, 0, 0 (visual taxiing impossible).

weather central Organization collecting, processing and outputting all local weather information.

weathercocking Tendency of aerodynamic vehicle to align longitudinal axis with relative wind; note that this affects pitch as well as yaw.

weathercock stability Basic directional stability of air vehicle or re-entering spacecraft; in CCV (eg modern fighter) this is degraded to ultimate degree and replaced by synthetic * applied by avionics linking sensors to flight controls.

weather forecast Prediction of weather within area, at point or along route for specified period.

weather map Shows weather prevailing, or predicted to prevail.

weather minima Worst weather under which flight operations may be conducted, subdivided into VFR and IFR; usually defined in terms of ceiling, visibility and specific hazards to flight.

weather radar Airborne radar (less often, surface radar) whose purpose is indication of weather along planned track; traditional output is picture of heavy precipitation, but modern * can indicate severe turbulence (in meaningful colours) even if precipitation absent.

weather reconnaissance Flight undertaken to take measurements (traditionally = thum = temp + humidity) at specified flight levels up to near aircraft ceiling; today rare but also includes all forms of weather research.

weather report 1 Broadcast * by national weather service, eg each hour.
2 An actual, transmitted by airborne flight crew.

weather satellite See *met satellite*.

weathervane US term for weathercock (eg on building), and for weathercocking tendency of aircraft on ground to face into wind. Hence * effect of vertical tail, which progressively gives directional stability to VTOL aircraft as forward speed increases.

weave To make continuous and smooth changes of direction and height while over a period following a desired track; weaving assigned to proportion of fighters escorting slower aircraft so that continuous watch could be kept astern and in other difficult areas. Hence, weaver.

WEB Web effective burn (time).

web 1 Principal vertical member of a beam, spar or other primary structure running length of wing or fuselage, providing strength necessary to resist shear and keep upper and lower booms (chords) correct distance apart. Occasionally expanded to * member, * plate.
2 In solid-propellant rocket, distance through which propellant burning surface will advance from initial surface until * burnout as defined by two-tangent method; usually measured as linear distance perpendicular to initial surface, symbol τ_W.
3 Any material form resembling sheet, either as discrete pieces or continuous * unrolled from drum or coil; esp sheet form of prepreg, supplied in standard widths.

web average burning-surface area Total volume of solid rocket propellant, excluding slivers, divided by web; thus has dimensions of an area, symbol A_s.

webbing Strong close-woven fabric strip produced to specified UTS, used eg for securing loose (bulk) cargo.

weber, Wb SI unit of magnetic flux; that flux which, linking circuit of one turn, produces EMF of 1 V as it is reduced to zero uniformly in 1 s. To convert from maxwell, multiply by 10^{-8}.

web fraction Web (2) divided by internal radius of motor case or chamber; symbol f, expressed as %.

WEC World Energy Conference.

WECMC Wing electronic-combat managers' course.

WECPNL Weighted Equivalent Continuous Perceived Noise Level (see *noise*).

WED Water equivalent depth.

wedge 1 Air mass having wedge shape in plan, esp such a mass of high pressure extending bet-

ween two lows.
2 Sharp-edged essentially 2-D * forming one wall of 2-D inlet for supersonic airbreathing engine, extending ahead of inlet so that its shock may be focused on inlet lip and normally extending rearwards as a variable ramp. Hence * inlet.

WEDS Weapons effects display system.

weeds, in the At lowest possible level, on TFR or manually.

Wefax Weather facsimile format; one selectable mode of data transmission between weather satellite and ground printout.

Weft Wings/engine(s)/fuselage/tail; most basic of mnemonics used in early aircraft-recognition instruction.

Wehnelt Type of cathode whose emissivity is enhanced by coating of radioactive-metal oxides.

Weick Coefficient for calculating propeller characteristic, also called speed/power coefficient, $C_s = V \sqrt[5]{\rho/PN^2}$, where V is velocity of advance, ρ is density, P is power and N is rpm.

weighing points Locations, published in engineering documents and stencilled on aircraft, where jacks are applied in weighing process.

weigh-off Free ballooning of airship before casting off to refine trim.

weight Force exerted on a mass by Earth's gravity; thus a figure unique to a particular location which by international agreement is any at which g (free-fall acceleration) is 9·80665 m/s. For mdern aeroplane some measures include: empty, complete aircraft plus systems measured in accord with specification (eg in US military usage, MIL-STD-3374); CCDR, empty minus items listed in STD-25140A such as engine(s), starter(s), electrics and avionics where removable direct from racking, wheels/brakes/tyres/tubes; standard empty, standard aircraft plus unusable fuel, full oil and full operating fluids (thus excluding potable water); basic empty, standard empty plus optional equipment; structure, bare airframe without systems and equipment other than wing/tail movables and flight-control power units, flap actuation, landing gear and actuation, and equipped engine installation(s) minus engines; useful load, those items that when added to * empty (for transport, OEW) will add up to gross weight for design mission (transport, MTOW); operating weight (military), empty plus useful load minus expendable fuel (internal/external), ammunition and stores; operating empty (OEW), equipped empty + all consumables (fuel, lube, filled galleys and bonded stocks, toiletries etc) + removable furnishings, reading and entertainment materials, cutlery, flight and cabin crews and their baggage, ship's papers; zero-fuel (ZFW), (military) operating plus ammunition/missiles/stores, (transport) MTOW minus usable fuel; gross (military), MTOW (civil), allowable at moment of take-off; ramp (MRW), (civil) allowable at moment of starting engines; basic flight design (military), TO with full internal fuel and useful load for primary mission; min flying gross (military), empty + min crew, 5% usable/unusable fuel (zero fuel for flutter) and lube consistent with fuel; max design, military equivalent of MRW allowing for full internal/external fuel (in some cases extended to higher figure still after air refuelling); max landing (MLW, civil), figure specified for each type between ZFW and MTOW; landplane landing design gross, basic flight design gross plus empty external tanks and pylons minus 60% internal fuel; max landing design gross, max design minus dropped tanks, fuel expended in one go-around (overshoot) or 3 min (whichever is less) and any items routinely dropped immediately after take-off; bogey, also called target bogey, established 4% below specification * (in practice * tends to rise, and bogey usually pious hope); specification (military), that number written into original agreed specification; job-package target/bogey, series of targets for each * group parts-breakdown; current, also called current status, that representing best available information, obtained by adding to previously reported status all subsequent revisions.

weight and balance sheet Document carried with transport (military/civil) recording distribution of weight and c.g. at TO and (military) landing.

weight breakdown Subdivision of aircraft weight (usually a design gross) into broad headings: structure (itself divided into wing group, tail group, fuselage and landing gear), powerplant, equipment services, and disposable load (latter divided into fuel/consumable items and payload).

weight coefficients Dimensionless ratios BF/TOW, BF/ZFW, RSV/LW, RSV/ZFW, TOW/LW and TOW/ZFW.

weight flow See *mass flow*.

weight gradient Required change in weight, eg MTOW, for unit change in temperature, usually expressed in kg/°C, in such corrections as QNH variation, air-conditioning and anti-icing.

weight in running order Traditional measure of PE (4) weight in which radiator/coolant/pipes/controls were added, plus oil within engine, but excluding tanks/fuel/oil/reserve coolant/exhaust tailpipes/instruments.

weightless Condition in which no observer within system can detect any gravitational acceleration; can be produced either in free fall near a massive attracting body, eg Earth satellite, or remote from any attractive body.

weight-limited Payload that can be carried is limited by restriction on aircraft MTOW or ZFW and not by available space.

weight per horsepower Usually incorrectly called power/weight ratio, basic measure of piston or other shaft-output engine; dry weight divided by max power (latter can be $2\frac{1}{2}$-min contingency).

weight per unit thrust Dry weight (mass) of jet engine divided by a specified measure of thrust (for turbojets/turbofans usually SLS/TO); in SI it is not possible to divide a mass by a force, but a meaningful ratio is still obtainable provided units of both are compatible and specified.

weld bead Metal deposited along welded joint.

weldbonding Combination of resistance spot-welding and adhesive bonding with properties superior to either alone.

weld continuity Specified as tack, intermittent, continuous.

welded patch Thin sheet steel patch welded over local damage in steel tubular airframe.

welded wing Pair in unvarying side-by-side formation about 500 ft apart.

welded steel blade Propeller blade assembled by edge-welding two shaped sheets of steel to form aerofoil; also called hollow-steel.

weld fusion zone Width of bead.

welding Joining metal parts by local melting, with or without addition of filler metal to increase strength of joint, using gas torch, electric arc, electrical resistance, friction, explosive, ultrasonic vibration and other methods, often with local atmosphere of inert gas. Techniques generally called diffusion bonding are closely allied but often require no heat or added metal and rely upon natural bonding of two clean surfaces in intimate contact.

welding flux Material, eg provided as coating on welding rod, which melts and flows over joint, excluding oxygen.

welding jig Fixture for holding parts to be welded in exact relative positions while joints are made.

welding machine Invariably an electric machine welding workpieces by spot, roll or seam methods.

welding rod Consumable rod of correct metal to act as joint filler which also conveys current to form arc struck against workpiece; diameter selected according to current and usually with flux coating.

well Generalized code word (including air intercept) = serviceable.

WEM Warning electronic module.

WEMA Western Electronic Manufacturers Association (US).

WEP 1 Weapon effect planning.
2 War emergency power.

wet 1 To come in contact with surface of body; hence wetted area.
2 With water injection.

wet adiabatic See *saturated adiabatic*, *SALR*.

wet assembly Important technique for modern aerospace structures in which all primary components are not only given successive surface treatments but are put together and joined while their surfaces are still wet; eg each component would be anodized, then coated with primer and finally with Thiokol sealer, rivets or bolts also being coated with Thiokol except in case of interference-fit bolts (Taper-loks or radius-nose Hi-locks) which fill holes completely.

wet boost Boost pressure permissible for PE (4) with water injection in operation.

wet builder Manufacturer employing wet-assembly techniques.

wet-bulb potential Temperature air parcel would have if adiabatically cooled to 100% RH and then adiabatically brought to 1,000 mb level.

wet-bulb thermometer Has sensitive element surrounded by muslin kept moist by supply of water, hence reading gives indirect measure of relative humidity.

wet emplacement Rocket test emplacement or vehicle launch pad whose flame deflector and nearby parts are cooled by deluge of water.

wet filter Particles are retained by a liquid film on element surface.

wet H-bomb Thermonuclear device whose fusion material is liquid, cryogenic or otherwise.

wet lease Hire of commercial transport from another carrier complete with crew (at least flight crew, but often not cabin crew) and in effect forming continuation of previous operation, with major servicing performed by owner, but with hirer's logo and insignia temporarily applied.

wet pad Wet emplacement for launches.

wet point Plumbed hardpoint.

wet rating Power or thrust with water or water/methanol (rarely, water/alcohol) injection.

wet-run anti-icing Surfaces kept continuously above that at which droplets freeze.

wet sensor ASW sensor dropped or dunked into ocean, eg sonobuoy.

wet start Faulty start of gas turbine in which unburned or burning fuel is ejected from tailpipe.

wet suit Standard anti-exposure suit for working in sea; eg in winch rescue.

wet sump PE (4) sump which serves as container for entire supply of lube oil.

wet take-off Take-off with water injection.

wetted area Total area of surface of body over which fluid flow passes and on which boundary layer forms. In case of aircraft usually simplified to visible external skin, ignoring inner surfaces of air inlets, ducts, jetpipes and air-conditioning system.

wetted fibre Fibre, eg carbon/graphite/boron/glass, coated with same resin as will be used to form matrix of finished part.

wetted surface See *wetted area*.

wetting agent Surface-active agent which, usually by destroying surface tension, causes liquid to spread quickly over entire surface of solid or be absorbed thereby.

wet weight Weight of devices plus liquids normally present when operating.

wet wing Wing whose structure forms integral fuel tank. Note: not merely a wing in which fuel tanks are housed.

wet workshop Space workshop launched as operative rocket stage whose propellants must be consumed or removed prior to equipping as workshop in space.

WEU 1 Assembly of Western European Union.
2 Warning electronic unit.

WEWO Wing electronic-warfare officer (RAF).

WF US mission prefix, weather-reconnaissance fighter.

W_F Total weight of fuel.

W_f Mass flow rate of fuel.

WFC Wallops Flight Center (NASA).

WFOV Wide field of view.

WFU Withdrawn from use.

WG 1 Working group.
2 Water gauge.

WGD Windshield (windscreen) guidance display, transport-aircraft HUD for Cat IIIB operations.

WGN White gaussian noise.

Wh Watt-hour.

Wheatstone bridge Circuit with known and variable resistances with which unknown resistance can be accurately measured.

wheel 1 Early aircraft commonly had such an input to lateral control system (cf ships, road vehicles); terminology is needed for all forms of stick/control column/handwheel/yoke/spectacles. No modern captain would say "take the *".
2 Complete turbine rotor, esp single-stage or small size.
3 High-speed gyrostabilizing device (see *reaction* *, *internal* *, *gyro* *).
4 One normal usage is applicable to change of heading by formation of aircraft.
5 To introduce 3-D curvature to sheet (see *wheeling*).
6 Defensive formation of several aircraft circling in horizontal plane.

wheelbarrowing Violent oscillation in pitch on ground, either because of undulating surface or harsh brake application and rebounds from nose gear (not applicable to tailwheel aircraft).

wheelbase Distance in side elevation between wheel centres of nose and main landing gears; where there are two sets of MLGs (eg 747, C-5A) measure is to point mid-way between mean points of contact of front and rear MLGs.

wheelbrakes Brakes acting on landing wheels.

wheelcase Compartment containing train of drive gears to accessories which are usually mounted thereon, the whole tailored to fit against or around prime mover, eg turbofan. There is only a shade of difference between this and accessory gearbox.

wheel door(s) Covers landing wheels when retracted.

wheeler Tail-high landing by tailwheel aeroplane, subsequently sinking into three-point position.

wheeling Sheet-metal forming (on * machine) by locally squeezing it between upper and lower rollers; workpiece hand-positioned to achieve desired 3-D curvatures.

wheel landing See *wheeler*.

wheel load Vertical force exerted by each landing wheel on ground; hence ** capacity, but this is imprecise in comparison with various soil-mechanics and civil-engineering measures, eg CBR.

wheel mode Satellite or portion thereof rotates, often fairly slowly (5–30 rpm), for attitude stabilization.

wheel satellite Made in shape of wheel, normally rotating either for attitude stabilization or to impart artificial gravity to occupants.

wheel well Compartment in which one unit of landing gear is housed when retracted.

which transponder R/T code: please state type of transponder fitted, IFF, SSR or ATCRBS (DoD).

whifferdil Nose up 30°–60°, bank 90°, change heading 180° ending in dive.

whiffletree 1 Also rendered whippletree, originally horse-traction linkage, which equalized and distributed pull of many horses to one point; now used to distribute pull of one large hydraulic jack or other load applicator over an area of airframe via array of beams pulled at intermediate point (usually mid-point) and transmitting pull from both ends to other such beams.
2 Also used loosely for any pivoted beam or bellcrank (US).

whip One meaning is upward jump of carrier pendant or runway arrester wire after being depressed by passage across it of landing wheel.

whip aerial Flexible aerial (antenna), either quarter-wave or of arbitrary length and usually vertically polarized, projecting at about 90° from skin.

whip stall US term which appears ill-defined; existing definitions agree only on fact that * is complete, violent and involves large positive change in pitch attitude; some suggest it is entered with flick manoeuvre, others make this impossible by suggesting possible tail-slide at outset. One authority gives tail-slide as alternative meaning.

whirl Rotational flow of fluid, esp when in translational motion in duct, in which case streamlines follow spiral paths.

whirling arm Large family of instruments or installations in which object (usually under test) is whirled on end of balanced beam rotated about vertical axis; originally used for aerodynamic tests, eg of aerofoils, but now more often used to apply sustained high-g acceleration to human beings and devices, sometimes in evacuated chamber.

whirling mode Vibratory mode of shaft in which elastic bending is suffered, giving severe out-of-straightness either at end or along length, pronounced only at certain critical rpm (critical whirling speeds).

whirl-mode flutter Aeroelastic flutter, esp of wing, in which input energy is derived, at least initially, from whirling of engine or propeller shaft (catastrophic on one type of turboprop transport).

whirlwind Small local tornado in dry air, without cloud or rain.

whirlybird Helicopter (rarely, other rotorcraft) (colloq).

whisker 1 Small single crystal, usually lenticular, whose strength is very close to max theoretically attainable.
2 Sharpened contact pressed against semiconductor in point-contact transistor or solid-state diode; hence * resistance.

whistle Annoying high-pitched note, usually slightly varying in pitch, caused by interference carrier in superhet reception.

whistler RF signal, usually in form of falling note, generated by lightning; heard on Earth and Jupiter and in former case bounces to and fro along magnetic-field lines between N/S hemispheres.

Whitcomb body Streamlined body added to aircraft (eg wing trailing edge) to improve Area Rule volume distribution; many other names, eg speed bump, Küchemann carrot.

white, white hot TV/video mode giving positive (normal) picture, which in IR display renders hot as white and cold as dark.

white body Hypothetical surface which does not absorb EM radiation at any wavelength, ie absorptivity always zero.

white level Max permissible video signal, 100% + modulation or 0% –.

white noise 1 Strictly, noise having constant energy per unit bandwidth (Hz).
2 Commonly used to mean spectrum of generally uniform level on constant-% bandwidth basis (so-called broadband noise) without discrete-frequency components.

whiteout 1 Loss of orientation with respect to horizon caused by overcast sky and sunlight reflecting off snow.
2 Zero visibility caused by what one authority calls ping-pong-ball snow.

white room 1 Super-clean room in which air is continuously filtered to eliminate micron-size and larger particles and special rules almost eliminate introduction of contaminants by human beings or objects (eg indiarubber, pencils, handkerchiefs, etc, prohibited).
2 Anechoic chamber.

White Sands Chief USA missile range, large area of New Mexico; abb WSMR.

white-tailed 1 Not bearing markings of a commercial air carrier.
2 Available for wet or dry lease with peelable logo and name.
3 Completed but unsold.

Whitney punch Hand-operated tool for punching holes of selected sizes in metal sheet.

Whitworth Traditional UK screwthread with radiused crest and root and 55° angle.

whizzer TV zoom lens.

whizzkid Civilian analyst or adviser in DoD.

whizzo WSO (colloq).

WHO World Health Organization (UN agency, HQ Switz).

whole-aircraft charter Operator charters complete aircraft for one flight or for a period, in contrast to split charter.

whole-body counter Nucleonic instrument for identifying and measuring body burden (whole-body received radiation) of human beings and other living organisms.

whole-range distance Horizontal distance between point vertically below release point and whole-range point.

whole-range point Point on surface vertically below aircraft at moment of impact of bomb released by it, assuming constant aircraft velocity (ASCC).

WI Welding Institute (BWRA).

WIA Wounded in action.

WIC 1 Warning information correlation.
2 Women, infants and children.

wick 1 See *static wick*.
2 Throttle(s); eg to turn up * = increase propulsion power (colloq). Invariably used for turbine engine(s), esp jet(s).

Wide Wide-angle infinity display equipment.

wideband amplifier One offering uniform response over many decades of frequency.

wideband dipole Large ratio diameter/length.

wideband ratio Ratio of occupied frequency bandwidth to intelligence bandwidth.

wide body See *wide-body aircraft*.

wide-body aircraft Commercial transport with internal cabin width sufficient for normal passenger seating to be divided into three axial groups by two aisles; in practice this means not less than 4·72 m (15 ft 6 in) (B.767, narrowest **).

wide-cut Generalized term for aviation turbine fuels assembled from wider range of hydrocarbon fractions than kerosine-type fuels; more accurate term is wide boiling-point range, for whereas little or no kerosine-type boils below 174°C, * begins to boil at 52–53° and fractions continue to boil off at up to about 220°. SG well below 0·76, compared with 0·79–0·8 for kerosine-type fuels. Widely held to be dangerous because of high volatility (arguable) but operational fuel of nearly all air forces and one or two airlines.

wide-deck Lycoming term for cylinder having wide base held by hexagonal nuts on large-radius circle; no separate hold-down rings or plates.

Widia German range of sintered tungsten carbides with 3–13% Co.
width of sheaf Lateral interval between centres of flak bursts or impacts (DoD).
Wiedemann-Franz law States ratio of electrical to thermal conductivity for all metals is proportional to °K (for most metals observed value is slightly higher than ** figure).
Wien bridge Stabilized oscillator whose frequency is determined by circuits incorporating resistances, capacitances and two triodes.
Wien law States wavelength of peak radiation from hot-body source is inversely proportional to °K.
WII light Wing ice inspection light.
Wilco R/T code: "I will comply with your instruction".
Wildhaber-Novikov Best known form of conformal gears with mating profiles formed by convex/concave circular arcs whose centres of curvature are on or near pitch circles.
Wild Weasel Though original name of specific programme, now generalized term for dedicated EW platform based on airframe of combat (eg attack or fighter) aircraft.
Williot Diagram which graphically portrays deflections of all joints (panel points) of loaded planar truss; result contains small inaccuracies which are removed by Mohr correction diagram, result being called Williot-Mohr.
will not fire Code sent to spotter or other requesting agency affirming that target will not be engaged by surface fire.
Wimet British range of tungsten carbides.
WIN WWMCCS intercomputer network.
winch launch Launch of glider by winch, usually locally built on road vehicle and driven by latter's engine.
winchman Member of aircrew of rescue helicopter in charge of winching payloads, eg rescuees, and who may hand winch to colleague and descend to organize pick-up of incapacitated rescuee from below.
winch suspension Rigging joining kite balloon and flying cable.
wind across Horizontal component of wind at 90° to catapult or centreline of (axial or angled) deck.
windage Loss of rpm of rotating device caused by air drag.
windage jump Vertical jump (up or down) of bullet trajectory caused by crosswind, eg firing laterally from bomber or gunship.
wind angle Angle between wind direction and true course (heading), measured 000–180°L or R of course.
wind axes Three rectilinear axes (u, v, w) with origin within aircraft, usually at c.g., and directions each representing component of relative wind in longitudinal, lateral and vertical planes. Traditional names: lift axis, + upward; drag axis, + to rear; crosswind axis, + to left.
wind cone See *windsock*.
wind correction angle Difference between course and track.
wind diagram US for triangle of velocities.
wind direction That from which wind is blowing, expressed as number from 000° to 359°.
wind down Horizontal component of wind along axis of catapult or centreline of (axial or angled) deck.
Windee Wind-tunnel data encoding and evaluation (EDP [1]).
wind factor Net effect of wind on aircraft progress, expressed as ± knots (USAF).
wind-gauge sight Drift meter; instrument which (BSI definition), by determining track on two or more courses, enables air, wind and ground speeds to be represented by vectors (arch).
wind gradient Rate of change of wind with unit increase in height AGL; usually factor of interest is component of wind along runway, thus direct headwind at 150 m/500 ft veering to crosswind at threshold is regarded as change equal to full wind speed, even though actual wind speed does not alter.
windmill 1 Of inoperative engine, to be driven by propeller (piston or turboprop) or by ram airflow through it (turbojet or turbofan); hence windmilling, windmilling drag etc.
2 Small propeller-like * used on older aircraft to drive electrical generator and occasionally other machines; in principle like modern RAT but permanently operating.
3 Of free-turbine propeller, to spin idly in wind when aircraft parked (normally prevented by brake).
windmill-brake state Operating condition of helicopter rotor in which thrust, flow through disc and flow outside disc are all in same direction.
window 1 Transparent area in skin for aircraft optics or IR (in latter case not apparently "transparent").
2 WW2 code for frequency-cut metal reflective slivers, wire, foil etc that later became better known by US name chaff (ECM).
3 Launch opportunity, defined as unique and possibly brief time period in which spacecraft can be launched from its particular site and accomplish its mission; usually recurring after a matter of days to months.
4 Small band of wavelengths in EM spectrum to which Earth atmosphere is transparent; there are many such, though together they account for only small part of total spectrum, rest being blocked. Like (3) this meaning rests on * being a small transparent gap in a dark continuum.
5 Verb, to enlarge local part of drawing or graphic display, usually to show greater detail.
wind rose Polar plot for fixed station showing frequency of winds and strengths over given (long) period from 000°–359°.

wind rotor Multi-blade rotor for forced air-cooling of landing wheel brake of which it forms part.

winds aloft US term for upper winds.

windscreen Windows through which pilot(s) look ahead, called windshield in US. Originally on open-cockpit aeroplanes complete assembly of frame and windows ahead of pilot's head. On modern flight deck less obvious and generally replaced by such term as flight-deck windows/transparencies/fenestration,which includes side and roof (eyebrow) windows.

windscreen wiper Term confined to mechanical devices with oscillating blades; rotary-disc and air-blast (eg for rain-shedding) excluded.

windshear Exceptionally large local wind gradient. Originally defined as "change of wind velocity with distance along an axis at right angles to wind direction, specified vertical or horizontal" (BSI), which is same as wind gradient. Today recognized as extremely dangerous phenomenon because encountered chiefly at low altitude (in squall or local frontal systems) in approach configuration at speed where * makes sudden and potentially disastrous difference to airspeed and thus lift. In practice pilot must take into account air movement in vertical plane (see *downburst*) because sudden encounter with downward gust is more serious than mere fall-off in headwind. Often accompanied by severe turbulence and precipitation which can make traditional ASI under-read.

windshear indicator Modern electronic displays will show * situation well, but with traditional instruments an extra dial is avoided by adding an energy-rate pointer to VSI; this striped needle is driven by combined vertical-speed rate and rate of change of airspeed to show rate of change of aircraft energy. Pilot can readily work throttles to keep this needle coincident with VSI needle.

windshield See *windscreen*.

wind shift Sudden change in wind direction.

windsock Traditional fabric sleeve hung from mast to give approx local wind strength/direction; also called wind sleeve, wind cone.

wind star Plot for determining wind by drawing drifts measured on two headings (US, arch).

wind tee White T-shaped indicator displayed in signals area to show pilots wind direction; also called wind T.

wind tetrahedron US counterpart of wind T (which is also used in US): large pyramid shape indicating wind direction, rotated on pedestal.

wind tunnel Any of family of devices in which fluid is pumped through duct to flow past object under test. Duct can be closed circuit or open at both ends. Working section, containing body under test, can be closed or open (called open-jet). Fluid can be air at any temperature/pressure or various other gases or vapours. Operation can be continuous, intermittent or as brief as a millisec. Particular species is spinning tunnel.

wind-tunnel balance Apparatus for measuring forces and moments on object tested in tunnel; originally included actual mounting for object but today usually electrical force-transducers built into sting.

wind-up turn Turn by winged aerodyne that becomes ever-tighter.

wind vane Small pivoted blade which aligns itself with local airflow; usually drives via rotary viscous damper to one or more precision potentiometers to form a flow-angle transmitter for AOA or yaw.

wing 1 Main supporting surface of fixed-* aerodyne; despite term "rotating-* aircraft" not used in that context, usual word for rotating supporting aerofoils being blade. Normally taken to cover all parts within main aerofoil envelope, including all movable surfaces. Many terms, eg * jig, * structure, are judged self-explanatory.

2 Numerous types of military unit, eg in RAF basic combat administration organization comprising perhaps three squadrons, often all sharing same base, but in US usually a larger unit comprising one or more groups with support organizations, in Navy a self-contained unit for deployment of organic air power at sea (one * per carrier) or land, and in Marines those aviation elements for support of one division.

3 Extreme left or right of battlefront or aerial formation.

4 To be positioned on *, to be alongside in formation (see *wingman*).

5 To fly (colloq).

wing area 1 Area of surface encompassed by planview outline drawn along leading and trailing edges, including all movable surfaces in cruise configuration, and including areas of fuselage, nacelles or other bodies enclosed by lines joining intersections of leading or trailing edges with such bodies. Thus, pods hung below wing are excluded; wingtip tanks are normally excluded; winglets are included as they appear in plan view; in case where root meets fuselage or nacelle at sweep approaching 90° (eg with forebody strakes) most authorities include these portions and fuselage up to intersection of leading edge with fuselage. VG "swing wing" is measured at min sweep. Foreplane or lifting canard is counted separately.

2 Net ** excludes projected areas of fuselage, nacelles etc. Note: US measures of gross * often omit all areas outside basic wing trapezium, such as exist if taper is greater at root.

wing arrangement 1 Basic configuration of aeroplane.

2 Height at which monoplane wing is mounted, eg low, mid etc; also called *wing position*.

wing axis Locus of all aerodynamic centres.

wing bar Row of approach lights perpendicular to

runway and starting beyond runway edge (Vasi).

wing bending moment Bending moment. Hence *** relief, reduction afforded by masses (fuel, engine pods) distributed across span instead of being located in or on fuselage.

wing bending torsion mode Aeroelastic deflection, sustained or flutter, of swept wing in which bending introduces twist.

wing box Primary structure of modern stressed-skin wing in which all loads are taken by cantilever beam comprising upper and lower (usually machined) skins joined to front and rear spars, plus small number of ribs (occasionally additional spar[s] within *). This strong structure, usually by far heaviest single piece of airframe, is usually sealed to form integral tank.

wing car Airship car suspended to L or R of centreline.

wing cell See *cell, cellule.*

wing chord See *chord.*

wing drag When lifting, induced plus profile drags.

wing drop Sudden loss of lift on one wing, eg near stalling AOA, causing rapid roll not recoverable by aileron.

wing fence See *fence.*

wing fillet See *fillet*; avoided if possible, but often introduced because of need to accommodate main landing gears and in some cases combined with air-conditioning ducts on underside.

wing flaps See *flap.*

wing flutter See *flutter.*

wing guns Guns mounted within or attached to wing.

wing heavy Tending to roll in one direction.

winglet 1 Upturned wingtip or added auxiliary aerofoil(s) above and/or below tip to increase efficiency of wing in cruise, usually by reducing tip vortex and thus recovering energy lost therein and improving circulation and lift of outer portion of wing.
2 Miniature wing mounted horizontally on fuselage (not at nose or tail), on interplane struts or elsewhere (eg nacelles), often not so much for lift as to carry external load or connect bracing struts or main gears to fuselage.

wing leveller Simple single-axis autopilot with authority only in roll; often with heading lock and VOR/ADF coupler.

wing loading Gross weight or MTOW divided by wing area (1).

wingman Second in element of two combat aircraft, esp interceptors; term loosely applied to pilot or aircraft. Flies off wingtip of element leader except when required to perform manoeuvres, eg day/visual intercept of unidentified aircraft.

wingover US flight manoeuvre most briefly defined as climbing turn followed by diving turn; at apogee aircraft (usually trainer) is almost stalled, and rotation continues in pitch and roll so that recovery takes place at lower level by diving out on reciprocal. Almost = stall turn.

wing overhang See *overhang.*

wing panel See *panel.*

wing pivot That on which VG wing ("swing wing") is attached.

wing plan Shape of wing outline seen from above (see *plan*).

wing position Height at which wing (1) is mounted relative to fuselage, esp as seen from front, eg low, shoulder, parasol etc.

wing profile See *profile.*

wing radiator Cooling radiator mounted on wing, esp inside wing, fed by leading-edge inlet.

wing reactions Those forces applied to fuselage by wing.

wing rib See *rib.*

wing rock See *rock.*

wing root Junction of wing with fuselage (not with nacelle or any other body). Some authorities include junction of wing with opposite wing, eg on centreline (upper wing of biplane). Hence ** chord, ** fillet, ** thickness.

wing section Appears to be synonymous with aerofoil section, wing profile; only ambiguity caused by erroneously using term to mean portion of wing, eg centre section, outer panel.

wing setting See *angle of incidence.*

wing skid Protective skid on underside near tip (rare since 1916).

wing skin Usually refers to large stressed skin forming upper or lower surface of wing box; largest single piece of material in aircraft.

wing slot Slot built into wing; extends through outer wing from lower to upper surface with profile generally similar to that left by open slat.

wing spar Principal spanwise member of wing, in traditional or light aircraft usually isolated but in modern stressed-skin wing forming one face of wing box which itself behaves as a spar. Always extends full available depth, unlike stringers. In slender delta built perpendicular to longitudinal axis, thus in such wings there are many spars which terminate not at tip but at points along leading edge.

wing spread See *span.*

wing strut Primary structural links joining wings of biplane (interplane struts) or bracing monoplane diagonally to fuselage (such are actually ties in flight but struts on ground).

wing sweep See *sweep.*

wing tanks 1 Tanks, normally for fuel, either accommodated in wing or (integral) formed by sealing wing box.
2 Wing-mounted drop tanks where aircraft also has drop tanks elsewhere; thus ** may be dropped first to permit sweep to be increased.

wingtip Outer extremity of wing; either extreme tip or general area.

wingtip aileron Aileron forming entire tip of wing, either with chordwise inner end or extending inboard along trailing edge.

wingtip fence Winglet of very low aspect ratio.

wingtip flare Pyrotechnic attached to wingtip and ignited by pilot to assist night landing (obs technique).

wingtip float Stabilizing float of flying boat, usually inboard from wingtip.

wingtip handler Person walking beside sailplane being towed across airfield holding one tip so that wings are level.

wingtip rake See *rake*.

wingtip tank External fuel tank, jettisonable or not, carried on wingtip.

wingtip vortices See *vortex*; always present off tips of conventional wings at lifting AOA, and when intense (eg tight turn) pressure at centre falls so low that moisture condenses to leave white visible trail.

wing truss Wing plus all bracing struts and wires transmitting loads to or from fuselage.

wing walk Area marked on upper surface where maintenance engineers may walk in soft shoes.

wing warping Lateral control by warping; called primary on lower wing of biplane, secondary on upper. Relies on wing torsional flexure (see also *warping*).

wing yawmeter Yawmeter in form of miniature wing, aligned vertically or (for AOA) horizontally, with sensing holes at 0·15 or less chord.

WINS Workshop in negotiating skills.

winterization Process of equipping aircraft for flight in Arctic-type environment; obviously includes full anti-icing and de-icing of airframe, engines, propellers and flight-deck windows and extends to landing gears, systems, fluid specifications and many items of GSE.

winter solstice Point on ecliptic occupied by Sun at max southerly declination, around 22 December.

WIP Work in progress (check if field is open).

wipe-off switch Attached externally under belly or engine pod in most vulnerable place in belly landing; triggers safety system, eg tank inerting or fire extinguishers.

WIPO World Intellectual Property Organization; handles copyright (eg EDP data); Berne Convention.

wire braid Woven covering over ignition cables to prevent escape of emissions causing radio interference.

wire bundle Large group of electrical wires individually tagged, clipped together and then attached as a unit to structure, usually in wireway.

wired program One employing wired storage; also called fixed program.

wire-drawing 1 Manufacture of wire of required diameter by drawing through circular die.
2 Part-throttling fluid flow by passage through constriction (colloq).

wired storage EDP (1) storage which was originally literally wired in and could be erased only by physically removing it; today any indestructible storage (usually for ROM).

wire edge Sharp burr along edge of sheet freshly cut by shear.

wire gauge 1 Measure of diameter of wire or thickness of sheet; in UK by SWG measure based on Imperial (0·001–0·5 in).
2 Hand gauge for measurement of sheet thickness or wire diameter.

wire group Several wires routed to common destination and tied by clips.

wire guidance Command guidance of missile or other vehicle by electrical signals (bang/bang or analog) conveyed along fine wires unrolled behind vehicle and used to position surfaces governing trajectory.

wireless telegraphy Transmission of Morse by radio.

wire link Telemetry in which signals are conveyed by wire instead of by radio; also called hard-wire telemetry.

wire locking Tying a group of nuts with safety wire to prevent rotation.

wiresonde Met balloon at low altitude transmitting data over fine wire(s).

wire-strike protection system Measures taken to reduce lethality of impact between low-flying small helicopter and electric (eg national grid supply) cable.

wireway Large conduit forming secondary structure along which numerous electrical circuits (occasionally also fluid lines) are routed; in civil aircraft under floor or behind trim.

wire-wound Constructed by wrapping ultra-high-strength (eg tungsten) wire on mandrel and bonding by plasma spray or other method. Usually final part, eg rocket case or nozzle, has several layers aligned in different directions; also called wire-wrapped.

wire-wrapped connection Connection between single electrical signal wire and terminal made not with solder but by fast wrapping by machine designed for purpose.

wiring All internal wires in airship, divided into structural *, preserving cross-section and other dimensions, and gasbag *, distributing lift and preventing chafing.

WIS WWMCCS information system.

witness Hole, groove, recess or slot cut in reference plate by each tool used in complete NC program for producing a large machined workpiece; if each * is dimensionally correct this proves software and cutters and reference plate is stored as master for that program.

WJAC Women's Junior Air Corps (UK).

wksp Workshop.

WL 1 Water level, eg WL 365·8.
2 Weight-limited.

WL reference Horizontal axis used to define water levels, usually synonymous with static ground line (usually not synonymous longitudi-

nal axis).
WLDP Warning-light display panel.
WLFL Wet landing field length.
WM 1 Water/methanol.
2 Weak mixture.
WM50 Water 50%, methanol 50%; today MW is preferred usage.
WMD Weapons of mass destruction.
WMMP Weak-mixture max power.
WMO World Meteorological Organization (UN agency, HQ Switz).
WMS World Magnetic Survey (ICSU).
W-N Wildhaber-Novikov.
WO 1 Work order.
2 Winch operator.
W/O Written off.
W_O Gross weight.
wobble plate See *swash plate*.
wobble pump Cockpit hand pump, eg for building up fuel pressure before starting PE (4).
wobbulator FM signal generator, varied at constant amplitude above/below central frequency.
WOC Wing operations center (US).
WOD Wind over deck (carrier).
WOFF Weight of fuel flow; usually means per unit time.
wooden bomb Hypothetical concept of weapon or device which has 100% reliability, infinite shelf life and requires no special storage, surveillance or handling (DoD).
wooden round Missile that fails to work when needed (colloq).
Wood's metal Low-melting alloy (70°C), typically 50% Bi, 25% Pb, 12·5% Sn, 12·5% Cd.
woolpack Cumulus.
Woomera Location in South Australia of Weapons Research Establishment, head of missile range extending northwest to Indian Ocean.
WOP Wet (integral-tank) outer panel.
Wop Radio ("wireless") operator (colloq, arch); hence *AG adds air gunner.
Word Wind-oriented rocket deployment (Stencel seat).
word Basic group of ordered characters or digits handled in EDP (1) as one unit.
word rate Frequency derived from elapsed time between end of one word and start of next.
work Transfer of energy, defined as force multiplied by distance through which output moves.
work function 1 Thermodynamic: Helmholtz free energy A = U – TS where U is internal energy, T °K and S entropy.
2 Electronic: energy (usually measured in eV) supplied to electron at Fermi level in metal to remove it to infinite distance; governs thermionic emission.
work hardening See *strain hardening*.
working fluid Fluid used as medium for transfer of energy within system or (in wind tunnel) for study of flow around body. Can be gas, vapour or liquid.
working load That borne by structure or member in normal operation; in aircraft varies greatly with turbulence, manoeuvres and with aeroelastic forces and max is usually taken. Hence working stress experienced.
working pressure/temperature Value typical in normal operation.
working section Where model or other body is placed in wind tunnel; may be of open-jet form, without bounding walls, in traditional low-speed tunnel.
work lights Powerful airborne floodlights, directed forward or laterally, to enable agricultural aircraft to work in the dark.
work package 1 Quantified series of operations required to make one part of airframe; eg one wing panel may be divided into 15 **, each of short time period and exactly costable.
2 Basic unit of manufacturing effort into which entire product (eg aeroplane) is divided for allocation of effort in agreed manner between programme partners.
world fare Averaged value(s) of IATA tariffs.
World Geographic Reference System Better known as Georef, uniform position-fixing and designation system for control of aircraft, targeting of ICBMs and many other functions (USAF).
World Weather Watch Scheme to share and disseminate weather-satellite information internationally; also called VMM (WMO).
Wortmann Family of wing profiles designed by Prof F. X. Wortmann of Stuttgart; tailored to R appropriate to sailplanes (ie small chord), with outstandingly low drag which, esp for flapped sections, extends over wide range of C_L.
WOT Wide-open throttle.
WP Warsaw Pact.
W_P Payload.
W/P Waypoint; also WP or WPT.
WPA Wheelchair Pilots Association (USA).
WPFC World Precision-Flying Championships.
WPM, wpm Words per minute.
Wpn, wpn Weapon.
WPR Wideband programmable receiver.
WPU Weapon programming unit.
WR Work request.
WRA Weapon replaceable assembly.
WRAF Women's Royal Air Force (UK).
wrap-around Wrap-round.
wrapped connection Not soldered but made by wire wrapping.
wrapping cord See *serving cord*.
wrap-round boosts Boost motors, usually four spaced at 90°, arranged around sides of body of missile or test vehicle and at burnout separated laterally.
wrap-round engine Turboramjet comprising turbojet core surrounded by separate ramjet duct; differs from afterburning turbofan in that there is no fan, all burners are in roughly same axial

plane, and in RJ mode core is shut off.

wrap-round fins Rocket or missile fins which, before launch, are recessed around the curved body; they open after launch on hinges parallel to the longitudinal axis.

wrap-round windscreen Manufactured as single blown or vacuum-formed moulding forming entire front of cabin, extending from nose or engine cowl to behind front seats.

wrap-round wing skin Single sheet wrapped around leading edge and forming upper/lower wing surfaces back to rear spar.

WRCS Weapon-release computer set (USAF).

WRD War-reserve drop-tank (large quantities in store).

WRE Weapons Research Establishment (Woomera, Aust).

Wrebus WRE break-up system; radio-commanded method of explosively disintegrating errant vehicle.

wring (wringing it) out To demonstrate one's piloting skill by flamboyant demonstration intended to push aircraft (usually training aeroplane) to limits; has overtones of poor airmanship.

wrinkle Visible buckle in skin.

wrist pin Attaches radial-engine con-rod (articulated rod) to master rod; also called knuckle pin. Hence ** end of articulated rod.

write In EDP (1), to enter in memory, to record input information.

write off To damage an aircraft so severely it is not judged worth repairing.

write-off 1 An aircraft so severely damaged that repair is uneconomic.
2 Unobligated balance of funds removed from account involved.

written off Struck off charge, for reasons of obsolescence, unrepairable damage, total cannibalization or destruction in crash; does not include sale or MIA.

WRM War-readiness (or reserve) materiel (USAF).

wrought Generally, shaped by plastic deformation, eg by forging or other hot or cold working as distinct from casting, sintering or machining.

WRP Wing reference plane.

WRS Word-recognition system.

WRSK War-readiness spares (or supply) kit (USAF).

WS 1 Weapon system.
2 Weather service (FAA).
3 Water service (cart).

W/S 1 Warning/status (display).
2 Windscreen.

W_s Structure weight.

WS^3 See *WSSS*.

WSC Weapon-system controller.

WSD Wind speed and direction (W/V is preferred).

WSDL Weapon-system data-link.

WSEG Weapon-systems evaluation group (DoD).

WSF Weapon storage flight (RAF).

WSFO Weather Service Forecast Office (US).

WSI 1 Windshear indicator.
2 Water-separometer index (fuel).

WSIM Water-separometer (or separation) index, modified.

WSL Weapon-system level (programme lead responsibility).

WSMC Western Space & Missile Center (Vandenberg, USAF).

WSMIS Weapon-system management information system.

WSMR White Sands Missile Range (USA).

WSMS Windshear monitor system.

WSO Weapon-system operator (or weapon systems officer).

WSPO Weapon-system project (or program[me]) office.

WSPS 1 Weapon-system physics section.
2 Wire-strike protection system.

WSRN Western Satellite Research Network (US).

WSSD Weapon-system support development.

WSSP Weapon-system support program(me).

WSSS Weapon storage and security system.

WST Weapon-system trainer.

WT 1 Waste ticket (US manufacturing).
2 Water twister.

W/T Wireless telegraphy.

wt Weight.

WTA Water, turbine aircraft; demineralized water for injection with or without methanol.

WTI Weapon and tactics instructor.

W_u Useful load.

Würzburg Standard German aerial-target tracking radar, 1940–45.

WV, W/V Wind velocity.

WVR Within visual range.

WW Present weather (ICAO).

WWABNCP Worldwide airborne national command post(s) (US).

WWACPS Worldwide airborne command-post system.

WWMCCS Worldwide military command and control system.

WWO Wing weapon officer (RAF).

WWP World Weather Programme.

WWS Wild Weasel Squadron (USAF).

WWSSN Worldwide Standardized Seismograph Network.

$W_{(w+u)}$ Weight of wing plus undercarriage (landing gear).

WWV NBS exact-time service (US).

WWW World Weather Watch (WMO).

W(WW) Past weather (ICAO).

WX 1 Weather.
2 Airborne weather radar.

WX/C Weather controller.

X

X 1 Longitudinal axis (more strictly OX); all measurements parallel to this direction, esp force.
2 Generalized term for reactance.
3 US DoD aircraft designation mission prefix: experimental; and also modified mission prefix: experimental.
4 JETDS code: (1) ident, recognition; (2) facsimile, TV.
5 Length of beam or other structural member.
6 IFR flight plan code: transponder, no code.

x 1 Generalized term for unknown quantity.
2 Horizontal axis (abscissa) of cartesian co-ordinates or graphical figure.
3 Any value measured parallel to (2) or any co-ordinate point measured along that axis.
4 Reactance (X is more common).
5 Longitudinal displacement; distances measured along OX axis.
6 Mole fraction.

$\bar{x}$ Position of c.g. as co-ordinate along OX axis.

X-aerial Crossed rods, two longer (dipoles) and two shorter (director).

X-allocation First 126 paired (1–63, 64–126) DME interrogation frequencies.

X-axis X (1) or x (2).

X-band Former common-usage radar frequency band centred on wavelength of 3 cm, later amended to 2·73–5·77 cm (about 10·9–5·2 GHz) (obs).

X-channel DME or Tacan channel associated or paired with another radio service on same frequency.

X-cut crystal Cut parallel to Z-axis, perpendicular to X-axis.

X-engine PE(4) with four banks of cylinders arranged in form of X seen from end.

X-plates Vertically parallel deflection plates in CRT whose potential difference deflects beam horizontally and creates timebase.

X-ray Extremely short-wavelength EM radiation, with frequency higher than any other except gamma and nuclear radiations; typical wavelengths 10–1,000 pm.

X-ray analysis Based on diffraction of X-rays by crystalline solids.

X-ray astronomy Study of X-rays arriving at Earth from space (not possible at surface of Earth because of attenuation by atmosphere).

X-unit Non-SI unit of length = 10^{-11} cm = 0·1 pm = 100 fm.

X-wing Aircraft able to operate as helicopter or, with special four-blade rotor stopped with blades diagonal to airstream, as aeroplane.

Xbar Crossbar (ICAO).

X_c Capacitive reactance.

X/C Percentage of aerofoil chord.

XCTR Exciter.

X-Cty Cross-country.

xducer Transducer.

Xe Xenon.

xenon Rare inert gas used in some gas-discharge tubes and as pumping source for lasers, symbol Xe.

xfer Transfer.

xfmr Transformer.

XG Centre of gravity (c.g. is preferred).

X_L Inductive reactance.

x_L Moment of total lift of aerodyne about c.g.

XM Extra marker.

xmt Transmit.

xmtr Transmitter.

XO Executive officer (USN).

xponder Transponder; also rendered TPDR etc.

XPR Transponder.

XPS X-ray photoelectron spectroscopy.

XR Code: Office of CoS Plans and Programs (AFSC).

XRD X-ray diffractometer (diffractometry).

XRED X-ray energy dispersion analysis.

XRF/NAA X-ray fluorescence neutron-activation analysis.

XS Atmospherics (ICAO).

xsmn Transmission (both radio and helicopter dynamic parts).

XSS Expendable sonobuoy (or sonar or sonic) sensor(s).

xtal Crystal.

XTK Across-track distance error.

X_{w+u} Distance along OX axis of c.g. of wing and undercarriage (landing gear) from c.g. of complete aircraft.

xylene Family of toxic aromatic and inflammable hydrocarbons in some ways rersembling benzene, with general formula C_8H_{10}, widely used as solvent; also called xylol.

XXX CW transmission for uncertainty or alert (Int).

Y

Y 1 Yaw, yaw angle.
2 Lateral axis (strictly OY) or any measure or component along that axis, esp lateral force.
3 Generalized term for admittance.
4 IATA symbol, tourist class.
5 US DoD aircraft designation prefix: prototype/service-test quantity.
6 Yttrium.
7 Year (eg in FY, fiscal *).
8 Adiabatic factor.
9 Bessel function.

y 1 Vertical axis (ordinate) of cartesian co-ordinates or graphical figure.
2 Any value measured parallel to (1) or any co-ordinate point measured along that axis.
3 Admittance (Y is more common).

$\bar{y}$ Co-ordinate position of c.g. along OY axis (normally at or near origin).

Y-allocation Second group of 126 frequency-paired Tacan/civil DME channels.

Y-alloys British aluminium alloys originally developed by HDA for pistons, typically with 4% Cu, 2% Ni, 1·5% Mg and other elements, which because of their retention of strength to 250–300°C were chosen as ruling materials for Concorde with names RR.58 (UK) and AU2GN (F).

Y-axis See *Y (2), y (1)*.

Y-channel Second group of civil DME/Tacan frequencies paired with other radio services.

Y-connection One end of all three coils of 3-phase electrical machine is connected to common point while other ends constitute 3-phase line; alternative to delta.

Y-cut crystal Cut parallel to Z-axis, perpendicular to Y-axis; thus parallel to one face of hexagon.

Y-plates Horizontally parallel deflector plates whose potential difference positions electron beam vertically in CRT.

Y-scale Scale along line of principal vertical in oblique reconnaissance imagery, or along any other line which on ground area shown would be parallel to this.

Y-section Structural section resembling Y, often with flanged or beaded edges.

Y-valve Lube-oil drain valve from dry sump.

Y-winding See *Y-connection*.

YAF Yesterday's Air Force, Calif (US).

YAG Yttrium aluminium garnet (laser).

Yagi aerial Directional aerial comprising dipole, reflector and one or more (usually linear array) directors spaced at 0·15–0·25 wavelength.

yard Traditional Imperial unit of length = 0·9144 m exactly.

yaw Rotation of aircraft about vertical (OZ) axis; positive = clockwise seen from above.

yaw damper Automatic subsystem in aeroplane (usually only jet) FCS which senses onset of yaw and immediately applies correct rudder to eliminate it. Early types were called parallel because they operated whole circuit including pedals; modern series ** has no effect on FCS further forward than fin (though sensing gyro may be near c.g.) and its activity is unnoticed by pilot. Most aircraft are flyable throughout virtually whole flight envelope with ** inoperative.

yawed wing Wing proceeding obliquely to relative wind; slewed wing.

yaw guy Cable along ground under mooring airship for attachment of yaw-guy wires.

yaw-guy wires Ropes or cables dropped from bow of airship before mooring for securing to yaw guys; stop nose from swinging.

yawhead Yaw sensor, eg angled pitots, on pivoted vane (but not gyro).

yawing moment Moment tending to rotate aircraft about vertical OZ axis, symbol usually $N = qS\bar{c}C_n$ where q is constant, S is wing area, $\bar{c}$ is mean chord and C_n ** coefficient; measured positive if clockwise seen from above.

yaw lines Yaw-guy wires (US).

yawmeter Instrument or sensor for detecting yaw; can be simple device, eg two or more pitot tubes at different inclinations whose pressure difference is sensed, or electronic gyro-fed subsystem.

yaw pointing Additional flight-control mode for modern fighters in which yaw can be controlled without changing flight trajectory by varying sideslip angle while holding zero lateral acceleration. Usually achieved by deflecting vertical canard while linked to rudder(s) and roll-control surfaces. Gives much quicker and better gun-aiming.

yaw string Crude yawmeter comprising string, wool or other filament allowed to align with relative wind, eg ahead of windscreen.

YBC Years between calibrations.

YCZ Yellow caution zone.

Y/D Yaw damper.

yd, yds Yards.

yellow arc Range on dial instrument indicating caution, or higher than normal.

yellow caution zone Region (of runway or glide-slope angle) marked by yellow or amber lights.

yellow gear/stuff Aircraft GSE vehicles on airfield or carrier (service carts, tugs, dollies, handling and store-loading equipment).

yellow sector Area on left of ILS centreline (on right if using back course).

yield 1 Explosive power of NW, measured in TNT equivalent weights and usually given as: very low, under 1 kt; low, 1–10 kt; medium, 10–50 kt; high, 50–500 kt; very high, over 500 kt (0·5 Mt).

2 Revenue per traffic unit, eg per tonne-km, pax-mile, etc.

yield factor of safety Specified factor used in some airworthiness requirements to prevent permanent deformation of structures.

yield load Limit load × yield factor of safety.

yield point Unit stress at which deformation continues (to breakage) without further increase in applied load; slightly higher than elastic limit and not normally approached in practice.

yield strength Unit stress corresponding to a specified permanent elongation, for light alloys usually taken as 0·2%.

yield stress Ambiguously, stress at yield point, not at yield strength; greatest that material can reach.

YIG Yttrium indium garnet.

Yoder rolling Manufacture of complex sections by sequential precision rolling operations tailored to each section.

yoke 1 Control column of large aircraft in which roll input is by separate handwheel/spectacles pivoted at top of column.

2 Main magnetic structure of electrical machine supporting poles and conveying flux round linkage on each side of armature.

3 Frame on which are wound CRT deflection coils, or case of high-permeability metal surrounding such coils.

4 Interconnecting cross-member or tie.

5 Forked mounting, eg passing both sides of nosewheel.

YOS Years of service.

YOT "You over there," man in right-hand seat of F-111, A-6 etc (colloq).

Young-Helmholtz Original theory of colour vision, based on receptors for red/green/blue.

Youngman flap Patented (Fairey) trailing-edge flap carried on struts below trailing edge and in addition to normal deflection also having a negative (usually −30°) setting for use as dive brake.

Young's modulus Basic measure of material strength under tension, ratio of normal stress (within limit of proportionality) to strain, ie ratio of tensile load per unit cross-section area to elongation per unit length, within elastic limit, symbol E. SI units, kN/m^2 or MN/m^2.

yo-yo Family of air-combat manoeuvres in horizontal and vertical planes intended to reduce angle-off or hold nose/tail separation and thus prevent overshoot of defender's turn. Hi-speed * trades speed for height, lo-speed * opposite.

YP Yield point.

YPA Young Pilots' Association (UK).

yr Year.

YS Yield strength (or stress).

YTD Year to date.

yttrium Transition metal important as constituent of garnets used as lasing materials; symbol Y.

Yukawa potential Describes meson field about a nucleon.

Z

Z 1 Vertical (normal) axis (OZ), or any parameter measured along that axis or parallel to it, eg normal force (also called N by some authorities) or structural depth.
2 Impedance.
3 Zenith.
4 Section modulus (also called S).
5 Time suffix: GMT.
6 US military aircraft designation prefix: obsolete (pre-1962).
7 US military aircraft modified mission designation prefix: planning (ie still a project).
8 USN aircraft modified mission suffix: administrative version (pre-1962).
9 Fluid pressure-difference (dP) due to difference in level (in Earth gravity).
10 Viscosity relative to absolute viscosity of water.
11 Atomic number.

z 1 Normal axis of cartesian figure (ie perpendicular to plane of paper).
2 Distance measured along normal axis (also Z).
3 Impedance (Z more common).
4 Zone.

$\bar{z}$ Co-ordinate position of c.g. measured along OZ normal axis.

Z-axis See *Z (1)*.

Z-battery Fired unguided AA rockets (UK, WW2).

Z-beacon See *Z-marker*.

Z-correction Correction for coriolis applied by moving celestial fix or PL at 90° to track (ASCC).

Z-marker Traditionally fan or cone marker filling radio-range cone of silence, typically with audible 3 MHz tone (arch).

Z-meter Impedance meter.

Z-scale Scale used in calculating height of object in oblique reconnaissance image.

Z-section Structural section in general form resembling Z, often with flanged or beaded edges and all angles 90°.

Z-stoff Aqueous solution of calcium (rarely sodium) permanganate.

ZA Zone aerial (antenna).

ZAB Incendiary bomb or warhead (USSR).

Zahn cup Container with calibrated hole for measurement of liquid viscosity.

Zamak Family of Zn-based die-casting and stamping alloys, some of them important for rubber-press dies.

ZAP Box or dispenser of ZABs (USSR).

zap 1 To smash or render unserviceable (colloq).
2 To adorn visiting friendly combat aircraft from competitor unit with unauthorized insignia or paint scheme.
3 To disable a (usually aerial) target in one pass.

Zapp flap Early form of trailing-edge flap resembling split type but with leading edge translated aft along guides and with pivoted arms holding surface at about mid-chord; thus small hinge moment or operating force.

ZAR 1 Zero-airspeed radius, at which retreating blade of helicopter is at rest with respect to local airflow.
2 Zeus acquisition radar.

ZAW Zentralausschuss der Werbewirtschaft (G).

ZBTA Zinc-base trial alloy; cheap material for NC reference plates.

ZD Zero defects.

z_D Moment arm of total drag about c.g., ie vertical distance from drag resultant to c.g.

Zeeman Effect in which line spectra, eg of incandescent gas, are each split up into two or more components by passage through magnetic field.

Zell Zero-length launching, or to make such a launch; hence zelling.

Zelmal Zero-length launch and mat landing.

zener current That flowing in insulator in electrical field sufficiently intense to excite electrons direct from valence to conduction.

zener diode Si junction diode having specific peak inverse voltage, at which point it breaks down and suddenly allows flow.

zener voltage 1 Field strength needed to initiate zener current, about 10^6V/mm.
2 That associated with reverse-VA of semiconductor, more or less constant over wide range of current.

zenith Point on celestial sphere directly overhead. Strictly this is observer's *; astronomical * is where plumb-line would intersect celestial sphere, and geographical * is where line perpendicular to smooth Earth would do so. These are not all synonymous.

zenith attraction Effect of planetary gravity on free-falling body (originally, Earth on meteorite) of increasing velocity and moving radiant toward zenith.

zenith distance Angular distance from celestial body to zenith.

zenographic Of positions on surface of Jupiter, referred to planet's equator and specified reference meridian.

zero defects Conceptual (sometimes attainable) goal of perfection, esp in manufacturing industry (chiefly US).

zero-delivery pressure In delivery line of variable-displacement pump at normal operating speed with zero stroke.

zero-draft forging One without draft, ie to finish dimensions.

Zerodur Glass ceramic with very low coefficient of expansion; used for LINS.

zero-force separation Interstage separation in which all mechanical links have already been broken, including all electrical connectors, etc.

zero-fuel weight MTOW minus total usable-fuel weight; usually limiting case for wing bending moment because wings are empty and fuselage is full.

zero-g See *zero gravity.*

zero gravity Free-fall, weightlessness; or in deep space remote from massive bodies.

zero gross gradient Altitude at which gross climb gradient (all engines operating, corrected for anti-icing) is zero.

Zerol Zero-spiral angle bevel gear.

zero lash Mechanism adjusted until there is no play or backlash, eg by using hydraulic valve lifters in PE (4).

zero-length Launching of vehicle from aircraft or surface launcher in such a way that it is free as soon as it begins to move; in case of aircraft rockets, hung on clips instead of being accelerated along tubes or rails; for surface launch, thrust away by rocket boost motors whose vertical component is greater than weight. Sometimes abb to zero-launch; usually abb Zell, * launch(ing).

zero-lifed Restored to new condition after having seen service; applies esp to airframe or engine and follows meticulous inspection and rectification such that flight-time count may legally be restarted.

zero lift Angle of attack at which aerofoil generates neither positive nor negative lift, ie no force normal to airstream; usually an obvious negative AOA for traditional cambered wing at below M_{crit} but for supersonic wing can be positive angle.

zero-lift line Drawn through trailing edge of aerofoil parallel to relative wind when lift is zero.

zero meridian Prime Meridian, 0° longitude.

zero point Location of centre of NW at moment of detonation; usually above or below ground zero.

zero-power transfer switch Automatically switches load when waveform is passing through zero.

zero-range ring Circular arc forming range origin of most weather radars, and many other polar-type displays (most unusual to have origin as point).

Zero Reader Pioneer (Sperry) flight instrument with cross-pointers driven by various selectable sensors to control trajectory (vertical and lateral displacement), pilot steering aircraft so that both needles cross at origin of display in method first used on ILS meter. In most respects similar to ILS meter that can accept inputs from en route navaids and attitude sensors.

zero stage Extra axial stage added on front of existing multi-stage axial compressor in new uprated version of gas-turbine engine; can be overhung ahead of front bearing and may or may not be preceded by inlet guide vanes. Hence, more rarely, zero-**, for a second additional upstream stage.

zero-thrust pitch Distance propeller advances in one revolution when operating at normal speed but moved through still air by external force so that it generates no thrust; also called exponential mean pitch.

zero-torque pitch Distance propeller advances in one revolution when moved through still air by external force at such a speed that its drive torque is zero; ie when windmilling in frictionless bearings. Symbol P_a.

zero/zero landing or option Totally blind helicopter landing enveloped in snow.

zero/zero seat Ejection seat qualified for operation at zero height, zero airspeed; ie pilot can safely eject from parked aircraft.

zero-zero stage See *zero stage.*

Zerstörer Destroyer, large or long-range fighter (G, WW2).

ZF Zero force (separation)

z_F Moment arm about c.g. of net propulsive force; usually acts in purely vertical plane, ie in pitch.

ZFCG Zero-fuel c.g.

ZFT Zero flight time (of hardware, or pilot conversion entirely on simulator).

ZFW Zero-fuel weight.

ZG 1 Zero gravity (also Z-g).
2 Zerstörergeschwader (G, WW2).

ZGG Zero gross gradient.

ZHR Zenithal hourly rate (measure of meteor shower).

ZI Zone of the Interior (US).

zinc Blue-white metal, SG 6·92, MPt 907°C, cheap and important in alloys for casting (esp die-casting) and dies, form blocks and other press tooling. Symbol Zn.

zinc/air Electrical battery in which KOH (potassium hydroxide) electrolyte is pumped through cell with Zn cathode (converted to oxide) and anode of porous Ni through which is pumped air (oxygen). Much higher energy density than lead/acid.

zinc chromate $ZnCrO_4$, yellow pigment used as basis for yellow-green * primer; mixed with alkyd resins to give strongly adhering anti-corrosive treatment almost universal in metal aircraft construction whose chromate ions are released by moisture.

zinger Snag (US colloq, especially in air combat).

zip fuel Exotic or high-energy fuel for airbreathing engines, esp ethyl diborane and other liquids based on boron compounds.

Zipper Target CAP (combat air patrol) at dawn or dusk (DoD).

zipstring Something simple and cheap.

Zirconium Zr, white, ductile metal, SG 6·44, MPt 1,857°C; important as alloying element and in nuclear applications.

ZLA Zero-lift angle.

ZLBH Zero-lifed bare hull.

ZLDI Zentralstelle für Luft- und Raumfahrtdokumentation und Info (G).

ZM Z-marker, VHF station location.

Zn Zinc.

ZNKJRK All-Japan Air Transport and Service Association.

zodiac Band of celestial sphere centred on ecliptic extending 8° on either side and containing Sun, Moon and all planets used for nav purposes (except, sometimes, Venus).

zodiacal counterglow See *gegenschein*.

zodiacal light Faint cone of light seen (esp in tropics) pointing towards ecliptic after sunset or before sunrise.

ZOK Factory for experimental construction (USSR).

ZOMP Weapon(s) of mass destruction (USSR).

zone 1 Administrative region of airspace, esp controlled airspace.
2 Portion of drawing (see *zoning*).
3 Quadrant of radio range, portions of (early-type) Decca coverage and other navaid subdivisions.
4 Sector of Earth sharing common time, bounded by two standard meridians; there are 24.
5 Circular areas centred on NW explosion: I, within MSD (min safe dist), within which all friendly forces evacuated; II, all personnel max protection; III, min protection.
6 Portion of aircraft/spacecraft with separately controllable ECS.

zone marker See *Z-marker*.

zone numbers Those locating an item on zoned drawing.

zone of intersection Portion of civil airway overlapping or lying within any other airway (US chiefly, eg CAR 60.104).

zone of protection Within cone of 45° total apex angle whose apex is top of lightning conductor (eg on airport building or tower).

zone signals Radio-range quadrant signals (see *zone* [*3*]).

zone time 1 Civil time of meridian passing through centre of a time zone.
2 Time kept in sea areas in 15° zone of longitude or multiple of 15° from prime meridian (ASCC).

zoning Dividing large engineering drawing into numbered/lettered grid so that items can more quickly be located by assigning each a grid reference.

zoom 1 Abnormally steep climb trading speed for height; applies chiefly to majority of aeroplanes whose T/W ratio is much less than 1 even at near SL; normally a manoeuvre in low-level display flying.
2 Optimized steep climb by high-performance jet (which at SL might have T/W greater than 1 and thus could make sustained climb at 90°) at high altitude, normally starting at max level Mach, trading speed for height in order to reach exceptional height far above sustained level ceiling.
3 To enlarge or reduce image of object in TV, video etc, using lens of continuously variable focal length (* lens).
4 In air-combat, unloaded low-alpha climb to gain max height for min dissipation of energy.

zoom boundary Limits of flight envelope (in which ordinate is altitude) attained by zooming (see *zoom* [*2*]).

zoom ceiling That attained with a zoom (2).

zoom climb Zoom (1); second word redundant.

ZP Zone of protection.

ZPI Zone position indicator radar.

ZPU Hostile AAA using close-range automatic weapons (US colloq, from a Soviet 14·5 mm AA gun designation).

Zr Zirconium.

ZROC Zero rate of climb; hence * speed, that (on back of drag curve) at which slender delta can just hold height at full power near SL.

ZTDL Zero-thrust descent and landing.

ZTV Zone trim valve, governs each zone (6) in controlling ECS.

zulu Phonetic word for Z, thus * time = GMT, * flightplan = one wherein all times are GMT.

ZVEI Zentralverband der Elektrotechnischen Industrie (G).

Zyglo NDT process in which part is coated with penetrant dye, cleaned, coated with powder developer (which extracts dye from any cracks) and examined in UV when dye fluoresces.

Greek symbols

The following are some of the aerospace usages of letters of the Greek alphabet; in a few cases the capital forms are also used.

α (alpha) 1 Angle of attack.
2 Generalized symbol for acceleration, angular as well as linear.
3 Absorption factor.
4 Propeller or compressor axial inflow factor.
5 Decay coefficient (radio).
6 Coefficient of linear thermal expansion.
7 Degree of dissociation.

β (beta) 1 Angle of sideslip.
2 Angle between fluid and tangentially moving blade; pitch angle.
3 Particular operating regime of propeller when pilot selects pitch directly (available on ground only).
4 Angle of incidence of tailplane relative to main wing.
5 Angle of yaw (airships).
6 Coefficient of cubic thermal expansion (γ preferred).
7 Luminance factor.

γ (gamma) 1 Ratio of specific heats of gas (at constant pressure and constant volume).
2 Surface tension.
3 Dihedral angle.
4 Gliding angle.
5 Electrical conductivity.
6 γ_c, angle of climb.
7 γ_d, angle of dive, descent.
8 Magnetometer field strength, in MAD operations usually = 10^{-5} oersted.
9 Coefficient of cubic thermal expansion.
10 Shear strain (also ϕ).

Γ (capital gamma) 1 Circulation, vortex strength, rotation in fluid flow.
2 Gamma function (mathematics).

δ (delta) 1 Small increment.
2 Control-surface angle (but see other letters for angles of individual surfaces).
3 Unit elongation or structural deflection.

Δ (capital delta) 1 Generalized prefix, difference, differential.
2 Deflection of panel point of truss.
3 Increment, not necessarily small.
4 Load on water, hydrostatic lift.
5 ∇, operator del (mathematics).

ϵ (epsilon) 1 Angle of downwash.
2 Permittivity.
3 Eccentricity or (suffix) eccentric load.
4 Emissivity.
5 Dielectric constant.
6 Strain, direct (also e).

ζ (zeta) Rudder angle.

η (eta) 1 Generalized symbol for efficiency, with suffixes indicating which, eg η_m mechanical, η_r radiation.
2 Elevator angle.
3 Ratio of wing lifts (biplane).
4 Electrolytic polarization.

θ (theta) 1 Generalized symbol for bearing, azimuth direction, angular distance, polar co-ordinate angle etc.
2 Angle of pitch (all meanings, eg of fuselage attitude, propeller or rotor blade etc).
3 Angle of twist or torsional strain.
4 Phase displacement (also ϕ).
5 Temperature.
6 Semi-vertex angle of Mach cone.
7 Spherical co-ordinate measures.

θ/R Bearing and range.

Θ (capital theta) Absolute temperature.

ι (iota) No common meanings.

κ (kappa) 1 Permittivity (dielectric constant) (also ϵ).
2 Compressibility.
3 Conductivity.

λ (lambda) 1 Wavelength.
2 Area ratio of wing S/b^2.
3 Damping coefficient.
4 Scale ratio (model: full size).
5 Thermal conductivity.
6 Solid rocket volumetric loading.

Λ (capital lambda) 1 Angle of sweepback, measured at 0·25 chord; if leading or trailing edge with suffixes LE, TE.
2 Permeance.
3 Ionic/molar conductance.

μ (mu) 1 Generalized prefix, micro ($\times 10^{-6}$).
2 Without any other symbol, micrometre or micron.
3 Coefficient of friction.
4 Permeability.
5 Joule-Thompson coefficient.
6 Viscosity.
7 Aircraft velocity (non-dimensional).
8 Ratio of spans (biplane).
9 Amplification factor (radio).
10 μ_1, relative density of aircraft (aerodyne).
11 μ', specific fuel consumption of jet engine

(normally air-breathing). Not to be confused with fact that sfc of shaft-drive engines is measured in μg/J (microgrammes per joule).

ν (nu) 1 Kinematic viscosity.
2 Wave velocity.
3 Reluctance, reluctivity.
4 ν_c, molecular diffusivity.
5 Frequency in s^{-1}.
6 Poisson's ratio.
7 $\bar{\nu}$, wave number (also ν).

ξ (xi) Aileron angle.

o (omicron) Too like o to be used separately.

π (pi) 1 Ratio of circumference of circle to diameter, 3·141592653.
2 Solar parallax.

Π (capital pi) Generalized symbol for a product (math).

ρ (rho) 1 Generalized symbol for density; ρ_0 air density at SL.
2 Radius of curvature.
3 Resistivity.
4 Reflection factor, reflectance.
5 Electric charge density (MHD, C/m^3).

σ (sigma) 1 Relative density, eg to standard atmosphere.
2 Poisson's ratio (ν preferred).
3 Prandtl interference factor (biplane).
4 Stefan-Boltzmann constant.
5 Electrical conductivity.
6 Velocity error, eg in INS readout.
7 Standard deviation.
8 Normal stress (also f).

Σ (capital sigma) Generalized symbol for a summation.

τ (tau) 1 Temperature (also θ).
2 Transmission factor.
3 Shear stress in fluid (also q).
4 Period (growth or decay).

υ (upsilon) No aerospace meanings, loosely used interchangeably with v.

ϕ (phi) 1 Generalized function symbol.
2 Angle of bank or roll.
3 Effective helix angle.
4 Latitude, terrestrial, celestial etc.
5 Shear strain (also γ).
6 Polar co-ordinate angle.
7 Radiant or magnetic flux.
8 Entropy.
9 Velocity potential.
10 Phase displacement (also θ).
11 Cos ϕ, electrical power factor.

Φ (capital phi) Flux, magnetic or luminous.

χ (chi) 1 Mass.
2 Magnetic susceptibility.

ψ (psi) 1 Stream function.
2 Angle of yaw.
3 Azimuth angle.
4 Electrostatic, dielectric or luminous flux.
5 Helmholtz free energy.
6 Sweep angle of fin (vertical stabilizer), often with suffix number giving % chord of point of measurement.

Ψ (capital psi) Electric flux.

ω (omega) 1 Frequency.
2 Angular velocity.
3 $\bar{\omega}$, specific weight (PE [4]).

Ω (capital omega) 1 Ohms resistance.
2 Resultant velocity.
3 Solid angle (also ω).

NATO reporting names

Since the Second World War there have been three successive series of "type numbers" or "reporting names" assigned to Soviet aircraft by the NATO Air Standardization Co-ordinating Committee. When first assigned each new name is classified; this extends even to the suffix letters which identify important modifications of each basic design. Thus as this is written the reporting name "Fitter-K" is classified, even though it has leaked out (like all its predecessors) and is known to refer to an Su-17/20 with a ram-air inlet in the fin. The following list covers current types only. Bomber and reconnaissance names begin with B, transports with C, fighters with F, helicopters with H and other types with M. A single-syllable name (except for helicopters) denotes a propeller aircraft.

Backfire Tu-22M/Tu-26
Badger Tu-16 (Tu-88 and Chinese H-6)
Beagle Chinese H-5 (licence-built Il-28)
Bear Tu-20 (Tu-95, Tu-142)
Blackjack New swing-wing strategic bomber
Blinder Tu-22 (Tu-105)
Brewer Most Yak-28 tactical versions
Camber Il-86
Candid Il-76
Careless Tu-154, Tu-164
Cash An-28
Clank An-30
Classic Il-62
Cline An-32 (IAF Sutlej)
Clobber Yak-42
Coaler An-72, An-74
Cock An-22
Codling Yak-40
Coke An-24 (Chinese Y-7)
Colt An-2 (Chinese Y-5)
Condor An-124
Coot Il-18 and variants
Crate Il-14
Crusty Tu-134
Cub An-12 and variants
Curl An-26
Farmer MiG-19 (and Chinese J-6, including trainer)
Fencer Su-24
Fiddler Tu-28P/128 (Tu-102)
Firebar Yak-28P
Fishbed MiG-21 (single-seaters and Chinese J-7)
Fishpot Su-11 (single-seater)
Fitter Su-7, Su-17, Su-20 and Su-22 (including some two-seaters)
Flagon Su-15 (including two-seaters)
Flanker Su-27
Flogger MiG-23 and MiG-27 (including trainers)
Forger Yak-38
Foxbat MiG-25 (including reconnaissance and trainer versions)
Foxhound MiG-31
Frogfoot Su-25
Fulcrum MiG-29
Halo Mi-26
Harke Mi-10
Havoc Mi-28 attack helicopter.
Haze Mi-14
Helix Ka-27, Ka-32
Hind Mi-24
Hip Mi-8, Mi-17
Hokum Kamov (Ka-29?) air-combat helicopter.
Hoodlum Ka-26
Hook Mi-6
Hoplite Mi-2 (excluding PZL variants)
Hormone Ka-25
Hound Mi-4 (and Chinese Z-5)
Maestro Yak-28U
Mail Be-12/M-12
Mainstay Il-76 Awacs version
Mascot Chinese HJ-5 (licence-built Il-28U)
Max Yak-18 (and most variants except Yak-50/52/55 and Chinese CJ-6)
May Il-38
Midget MiG-15UTI
Mongol MiG-21U variants
Moose Yak-11
Moss Tu-126
Moujik Su-7U (but not later trainer versions)

Chinese aircraft
Most indigenous Chinese designs have not been assigned NATO reporting names. Two exceptions which have been published are:
Fantan Nanchang Q-5
Finback Shenyang J-8

US military aircraft designations

Since 1962 all US military aircraft have been designated according to a common system which assigns a letter for the basic mission, followed after a hyphen by a number for the aircraft basic type. A simple example is B-1, signifying bomber type 1. Modifying letters are then added to give information on permanent changes to the basic mission, and occasionally a status prefix is added to show that the vehicle is "not standard because of its test, modification, experimental, or prototype design". Between the basic mission letter and the hyphen a further letter can be added to denote the following "vehicle types": rotary-wing, V/STOL, glider, lighter-than-air. To the right of the number is added a series number, running consecutively from A (for the first production version) onwards, omitting I and O; the series letter is changed for each "major modification that alters significantly the relationship of the aerospace vehicle to its non-expendable system components or changes its logistics support". Finally, in the fullest form of each designation, a block number is added to identify identical aircraft forming one production "block"; these numbers are usually multiples of 5, intermediate numbers then being assigned to identify later field modifications.

Status prefix

G Permanently grounded
J Special test (temporary)
N Special test (permanent)
X Experimental
Y Prototype
Z Planning

Modified mission

A Attack
C Transport
D Director
E Special electronic installation
F Fighter
H Search and rescue
K Tanker
L Cold weather
M Multi-mission
O Observation
P Patrol
Q Drone
R Reconnaissance
S Anti-submarine
T Trainer
U Utility
W Weather

Basic mission

A Attack
B Bomber
C Transport
E Special electronic installation
F Fighter
O Observation
P Patrol
R Reconnaissance
S Anti-submarine
T Trainer
U Utility
X Research

Vehicle type

G Glider
H Helicopter
V VTOL/STOL
Z Lighter-than-air vehicle

As an example of how the system works, if there were a special test version of the trainer variant of the US Marine Corps Harrier it would be the NTAV-8A (ignoring any block number): **N**, special test; **T**, trainer; **A**, attack; **V**, V/STOL; **8**, eighth V/STOL type; **A**, first production model.

US missile and RPV designations

US unmanned air vehicles have their own designation system. This works in the same way as that for military aircraft, though it should be noted that manned aircraft converted as remotely piloted drones or targets retain their original designation, with the drone prefix (Q) added. An exception is the Sperry (Convair) PQM-102, which is in accord with neither system.

Status prefix

C Captive
D Dummy
J Special test (temporary)
M Maintenance
N Special test (permanent)
X Experimental
Y Prototype
Z Planning

Launch environment

A Air
B Multiple
C Coffin
F Individual
G Runway
H Silo-stored
L Silo-launched
M Mobile
P Soft pad
R Ship
U Underwater attack

Mission

D Decoy
E Special electronic installation
G Surface attack
I Aerial intercept
Q Drone
T Training
U Underwater attack
W Weather

Type

M Guided missile/drone
N Probe
R Rocket

As an example, the original experimental Lockheed Aquila RPVs were designated XMQM-105: **X**, experimental; **M**, mobile launcher; **Q**, drone; **M**, guided (missile or drone); 105th guided type.

US engine designations

Each US engine manufacturer has its own entirely individual designation system. Department of Defense designations for jet engines are governed by a common scheme, though the Navy still uses a strictly numerical sequence of Mk (Mark) numbers for its solid rocket motors. The following is the DoD scheme for jet and turbine engines:

Status prefix

J Special test
X Experimental
Y Prototype

Engine category

F Turbofan (current)
J Turbojet
LR Liquid rocket
RJ Ramjet
SR Solid rocket
T Turboprop/turboshaft
TF Turbofan (formerly)

Manufacturer code

A Allison Division, General Motors
AJ Aerojet-General
GA Garrett
GE General Electric
L Avco Lycoming, Stratford
LD Avco Lycoming, Williamsport
MA Marquardt
NA Rocketdyne (North American)
P Pratt & Whitney (formerly)
PW Pratt & Whitney (current)
RR Rolls-Royce
T Teledyne CAE
TC Morton Thiokol
WR Williams Research

Designations are completed by a suffix model number. In 1945 these began at 1 for AF numbers, using odd numbers only, and at 2 for the Navy, using even numbers. Thus the prototype C-130 had YT56-A-1 turboprops, while the first F-8 Crusader had a J57-P-12 turbojet. Today AF numbers start at 100 or 200 and Navy numbers at 400. Thus the F-15C has the F100-PW-220 turbofan.

Piston engines are designated by a letter giving the geometrical configuration of the cylinders, followed by a number giving the cubic capacity (displacement, or swept volume) in cubic inches rounded off to the nearest multiple of 5. Prefix letters can then be added (if necessary in multiple) giving further information. Suffix letters (previously numbers) indicate successive models of the same basic design.

Status prefix

X Experimental
Y Prototype

Prefix letter

A Aerobatic
G Geared
H Helicopter
I Direct fuel injection
T Turbo-supercharged
V Vertical mounting

Configuration letter

L Inline (upright or inverted)
O Horizontally opposed
R Radial
RC Rotating combustion (Wankel type)
V Vee

Thus Avco Lycoming's TIGO-541-E is the fifth model in a family of opposed engines of 541·5 cu in capacity (in a new series distinguished by the number 541 from the original series rounded off to 540) with turbocharger, direct injection and geared drive.

Joint Electronics Type Designation System

US military electronic equipment is designated with the following series of letters and numbers, reading left to right:

1 Prefix AN, indicating that the equipment is a formally designated military system.

2 Three-letter equipment indicator code. The first letter indicates the platform from which the equipment operates, the second its type and the third its function. Platform letters are: **A** Piloted aircraft, **B** Underwater mobile (submarine), **D** Pilotless carrier, **F** Fixed ground, **G** General ground use, **K** Amphibious, **M** Mobile (ground), **P** Portable, **S** Water (surface), **T** Ground-transportable, **U** General utility, **V** Vehicular (ground), **W** Water (surface and underwater combination), **Z** Piloted/pilotless airborne vehicle combination.

Type indicators are: **A** Invisible light/heat radiation, **C** Carrier, **D** Radiac, **G** Telegraph or teletype, **I** Interphone/public address, **J** Electro-mechanical or inertial wire-covered, **K** Telemetry, **L** Countermeasures, **M** Meteorological, **N** Sound in air, **P** Radar, **Q** Sonar/underwater sound, **R** Radio, **S** Special or combination of types, **T** Telephone (wire), **V** Visual and visible light, **W** Armament, **X** Facsimile or television, **Y** Data-processing.

Function indicators are: **A** Attachment, **B** Bombing, **C** Communications, **D** Direction-finding, reconnaissance and/or surveillance, **E** Ejection and/or release, **G** Fire control or searchlight direction, **H** Recording and/or reproducing, **K** Computing, **M** Maintenance and/or test assembly, **N** Navigation aid, **Q** Special or combination of purposes, **R** Receiving/passive detection, **S** Detection and/or range and bearing search, **T** Transmitting, **W** Automatic flight/remote control, **X** Identification and recognition, **Y** Surveillance and control.

3 Number indicating place in the chronological sequence of all such systems to have entered service.
4 Designation modifying suffix giving additional information: **A, B, C etc** Successive major variants, **(V)** Available in various configurations, **(V)1, 2, 3 etc** Indicates the variant used in a particular installation, **(X)** Under development, **( )** Not yet formally designated.
As an example, AN/ALR-62(V)4 indicates the 62nd type of piloted-aircraft countermeasures receiver/passive-detection system; this variant, the fourth, is specific to the EF-111A electronic-warfare aircraft.

FAI categories

For the purpose of homologating records the FAI groups all aircraft into the following categories:

A (Balloons)
AA Gas balloons, **AM** Mixed (hot air plus gas), **AX** Hot air. The following suffix numbers indicate envelope volume: **1** up to 250 m^3, **2** 250–400, **3** 400–600, **4** 600–900, **5** 900–1,200, **6** 1,200–1,600, **7** 1,600–2,200, **8** 2,200–3,000, **9** 3,000–4,000, **10** 4,000–6,000, **11** 6,000–9,000, **12** 9,000–12,000, **13** 12,000–16,000, **14** 16,000–22,000, **15** 22,000+.

B (Dirigibles)
BX Hot-air airships. Suffixes: **3** 900–1,600 m^3, **4** 1,600–3,000, **5** 3,000–6,000, **6** 6,000–12,000, **7** 12,000–25,000, **8** 25,000–50,000, **9** 50,000–100,000, **10** 100,000+.

C (Aeroplanes)
Group I Piston-engined. **C-1-a/o** Landplanes. Up to 300 kg empty weight, **C-1-a** 300–500, **b** 500–1,000, **c** 1,000–1,750, **d** 1,750–3,000, **e** 3,000–6,000, **f** 6,000–8,000.
C-2 Seaplanes. **a** up to 600 kg, **b** 600–1,200, **c** 1,200–2,100, **d** 2,100–3,400.
C-3 Amphibians. Weight classes as C-2.
Group II Turboprops. As Group I plus **C-1-g** 8,000–12,000 kg, **h** 12,000–16,000, **i** 16,000–20,000, **j** 20,000–25,000.
Group III Turbojets and turbofans. As above plus **C-1-k** 25,000–35,000 kg, **l** 35,000–45,000, **m** 45,000–55,000.
Group IV Rocket aircraft.
D (Gliders): D-1 Single-seat, **D-2** Multi-seat.
DM (Motor-gliders): DM-1 Single-seat, **DM-2** multi-seat.
E (Rotorcraft)
E-1 Land rotorcraft. Weight categories as for Group I landplanes.
E-2 Convertiplanes, **E-3** Autogyros, **E-4** Vertiplanes.
F (Model aircraft): Various subsections.
G (Parachuting): Various subsections.
H (Jet-lift aircraft): This section is growing.
I (man-powered aircraft)
K (spacecraft): Various types of mission.
N (STOL aircraft): The FAI is having difficulty in drawing demarcation lines for this category.
O (hang-gliders): Various subsections.
P (aerospace craft): Category new in 1985.
R (microlights): Category new in 1985, various subsections.

Civil aircraft registrations

AP	Pakistan
A2	Botswana
A3	Tonga
A5	Bhutan
A6	United Arab Emirates
A7	Qatar
A9C	Bahrain
A40	Oman
B	China
C, CF	Canada
CC	Chile
CCCP [SSSR]	Union of Soviet Socialist Republics
CN	Morocco
CP	Bolivia
(CR-C)	Cape Verde Republic (now D4)
CR, CS	Portugal
CU	Cuba
CX	Uruguay
C2	Nauru
C5	The Gambia
C6	Bahamas
C9	Mozambique
D	Germany, Federal Republic of
DDR (previously) DM)	Germany, Democratic Republic of
DQ	Fiji
D2	Angola
D4	Cape Verde
D6	Comoro, Republic of
EC	Spain
EI, EJ	Ireland, Republic of
EL	Liberia
EP	Iran
ET	Ethiopia
F	France, French overseas departments and territories
G	United Kingdom
HA	Hungary
HB + national emblem	Liechtenstein
HB + national emblem	Switzerland
HC	Ecuador
HH	Haiti
HI	Dominican Republic
HK	Colombia
HL	Korea, Republic of
(HMAY), (6HMAY)	Mongolia
HP	Panama
HR	Honduras
HS	Thailand
HZ	Saudi Arabia
H4	Solomon Islands
I	Italy
JA	Japan
JY	Jordan
J2	Djibouti
J3	Grenada
J5	Guinea-Bissau
J6	St Lucia
J7	Dominica (not Dominican Republic)
LN	Norway
LQ, LV	Argentina
LX	Luxembourg
LZ	Bulgaria
N	United States of America and outlying territories
OB	Peru
OD	Lebanon
OE	Austria
OH	Finland
OK	Czechoslovakia
OO	Belgium
OY	Denmark
P	Korea, Democratic People's Republic of
PH	Netherlands, Kingdom of the
PJ	Netherlands Antilles
PK	Indonesia and West Irian
PP, PT	Brazil
PZ	Suriname
P2	Papua New Guinea
RDPL	Lao, People's Democratic Republic of
RP	Philippines
SE	Sweden
SP	Poland
SSSR	Union of Soviet Socialist Republics (has appearance of CCCP)
ST	Sudan
SU	Egypt
SX	Greece
S2	Bangladesh
S7	Seychelles
S9	São Tôme and Principe
TC	Turkey
TF	Iceland
TG	Guatemala
TI	Costa Rica
TJ	Cameroun
TL	Central African Republic
TN	Congo, People's Republic of the
TR	Gabon
TS	Tunisia
TT	Chad
TU	Ivory Coast
TY	Benin
TZ	Mali
T3	Kiribati
VH	Australia
VP, VQ, VR	United Kingdom colonies and protectorates

VP-F	Falkland Islands
(VP-H)	Belize (formerly)
VP-LAA to VP-LJZ	Antigua/Barbuda
VP-LKA TO VP-LLZ	St Kitts/Nevis/Anguilla
VP-LMA TO VP-LUZ	Montserrat
VP-LVA to VP-LZZ	Virgin Islands
VP-V	St Vincent and Grenadines
(VP-W, VP-Y)	Zimbabwe (formerly)
VQ-G	Grenada (formerly)
VQ-H	St Helena
VQ-L	St Lucia (formerly)
VQ-T	Turks and Caicos Islands
VR-B	Bermuda
VR-C	Cayman Islands
VR-G	Gibraltar
VR-H	Hong Kong
VR-U	Brunei (formerly)
VT	India
V3	Belize
V8	Brunei Darussalam
XA, XB, XC	Mexico
XT	Burkina Faso (formerly Upper Volta)
XU	Democratic Kampuchea (Cambodia)
XV	Vietnam
XY, XZ	Burma
YA	Afghanistan
YI	Iraq
YJ	Vanuatu
YK	Syrian Arab Republic
YN	Nicaragua
YR	Romania
YS	El Salvador
YU	Yugoslavia
YV	Venezuela
Z	Zimbabwe
ZA	Albania
ZK, ZL, ZM	New Zealand
ZP	Paraguay
ZS, ZT, ZU	South Africa
(ZS)	Transkei
(ZS)	Namibia (South-West Africa)
(ZS)	Boputhatswana
3A	Monaco

3B	Mauritius
3C	Equatorial Guinea
3D	Swaziland
3X	Guinea, Republic of
4R	Sri Lanka
4W	Yemen, Arab Republic of the
4X	Israel
5A	Libyan Arab Jamahiriya
5B	Cyprus
5H	Tanzania, United Republic of
5N	Nigeria
5R	Malagasy Republic (Madagascar)
5T	Mauritania
5U	Niger
5V	Togo
5W	Western Samoa
5X	Uganda
5Y	Kenya
6O	Somalia
6V, 6W	Senegal
6Y	Jamaica
7O	Yemen, Democratic People's Republic of the
7P	Lesotho
7Q-Y	Malawi
7T	Algeria
8P	Barbados
8Q	Maldives
8R	Guyana
9G	Ghana
9H	Malta
9J	Zambia
9K	Kuwait
9L	Sierra Leone
9M	Malaysia
9N	Nepal
9Q	Zaïre
9U	Burundi
9V	Singapore
9XR	Rwanda
9Y	Trinidad and Tobago

Registration marks shown in brackets denote either those countries which have become independent during the last few years but have not yet notified ICAO of a new nationality mark, or the use of unofficial marks. Into the latter category come (HMAY) and (6HMAY).

Addenda

A (18) Accepted (nav display).
AB/LD Airbrakes/lift dumpers.
ABMD Advanced ballistic missile defence; I adds "initiative", P "program" and S "system" (US).
AD (12) Accidental damage.
Adex, ADEX Air defence exercise.
Adisp Automated data interchange systems panel (ICAO)
AFIRMS Air Force integrated readiness measurement system (US).
AFS (4) Automatic flight system, ie AFCS.
aft-loaded wing Supercritical wing.
AGM (2) Missile range instrumentation ship (US code).
airbridge Becoming preferred term for passenger loading bridge.
AIRMIS, Airmis Airline management information system, EDP for smaller airlines.
Alithalite Minimum-density general-purpose Al-Li alloys (Alcoa).
Al-Li Aluminium-lithium alloys.
annunciator (3) In a CDU, the alpha and numeric keys providing part of the operator interface.
ANT Antrieb (powerplant) neuer technologie (G).
AR5 RAF NBC hood and respirator.
AS (5) Air start (service vehicle).
ASO (2) Acoustic sensor operator.
ASOC Air Support Operations Centre (RAF).
ASP (6) Adaptive signal processing.
ASRS (2) or **AS/RS** Automatic storage/retrieval system.
Ataco Air tactical control officer.
ATM (3) Air task message, request for a particular combat mission to be flown (RAF).
ATO (2) Airborne tactical officer.
ATS (8) Aviation training ship (RN).
AVS Air vehicle specification.
BAZ (2) Back azimuth unit, MLS addition used for departures and overshoots.
BCIU Bus control and interface unit.
BL (2) Bulk loader.
bomb-burst Standard manoeuvre by formation aerobatic teams in which entire team commences vertical dive, usually from top of loop, in tight formation; on command, trailing smoke, members roll toward different azimuth directions and pull out of dive, disappearing at low level "in all directions".
Braduskill "Slow kill" technique proposed in SDI for gradual closure on hostile satellites.
broad goods Carbon-fibre and other fibre-reinforced sheet as delivered in the bale.
bulk loader Self-drive belt conveyor vehicle for loading bulk cargo.
bund Fuel/ammo dump sufficient for one day's operations from a STOVL hide, replenished daily from Logspark (RAF).
BWFT Ministry of research and technology (G).
CAIG Cost/Analysis Improvement Group (DoD).
CATCS Central Air Traffic Control School (RAF Shawbury).
cavalry charge Distinctive trumpet-like aural warning, usually signifying autopilot disconnect.
CCH Close-combat helicopter.
CFDS Centralized fault display system.
CFES Continuous-flow electrophoresis system.
Chase Coronal helium-abundance Spacelab experiment.
chicken bolts Temporary fasteners used in metal airframe assembly.
Cidin, CIDIN Common ICAO data-interchange network.
CNS (3) Common nacelle system, able to accept different types of engine.
COD (2) Component operating data.
commitment Announced decision to purchase an aircraft type, usually commercial transport.
continuous half rolls Display/competition manoeuvre in which numerous half rolls are made, marking being on accuracy of intermediate wings-level positions, which are held very briefly.
CORE Control of requirements expression, discipline which defines software design (BAe).
CSH Combat support helicopter, basically transport role.
Cute Common-user terminal equipment.
CXR Helicopter configuration, co-axial rotors.
DARC, Darc Direct-access radar channel.
Datacs Digital autonomous terminal access communications system (Boeing).
Datar Detection and tactical alert of radar (helo RWR).
DBS (3) Database storage.
DDP Declaration of design and performance, formal statement by manufacturer accompanying every functioning item or modification.
DDRMI DME/VOR radio-magnetic indicator, usually duplicated and partnered by two ADF/RMIs.
delayed-flap approach Function added to AFCS (TCC) to permit a decelerated approach following late selection of flap to reduce fuel burn or noise.
Demoval Demonstration and validation.
DERD Display of extracted radar data.
DFR (2) Digital flight recorder.
DHV Deutscher Hubschrauber Verband, helicopter association (G).
DME-P DME, precision.
DMV Deutscher Modellflieger Verband eV.
DNC Also can mean direction numerical control.
dolly (4) Each truck in airport baggage train.
domsat Domestic (usually communications) satellite.
DRV Deutscher Reisebüro-Verband eV, association of travel agents (G).

DS (4) Dynamic simulator.
DVO (2) Direct voice output, synthesized speech confirming correct acceptance of DVI, and possibly giving audio warnings.
Dynarohr Registered name of sound-suppressing lining of engine fan ducts and cases.
Ebsicon Standard NATO word for all SIT image tubes.
EBU (2) Engine build-up.
ECB (2) Electronic control box (APU).
ED (5) Environmental damage.
effective pitch ratio Basic propeller characteristic V/nd where V is airspeed, n propeller speed and d propeller diameter (units being compatible).
electronic flight-control unit Computer controlling surfaces used as spoilers and airbrakes, with or without roll-control function.
ELP Electroluminescent panel.
ERAPDS Enhanced recognized air picture dissemination system.
ERMA, Erma Extended-red multi-alkali (Gen II image intensifiers).
ESB Elevating sliding bridge, simpler than apron-drive bridge.
escape slide Becoming the favoured term for an escape chute. A dual-lane slide conveys two separated streams of pax simultaneously.
ESCS (2) Electrical system controller subsystem.
ESMO, Esmo ESM operator.
exceedence Single event, recordable on all HUM systems, in which engine or other device suffers an excursion in operating regime beyond allowable limits.
Fang Federation of Heathrow Anti-Noise Groups.
FCU (3) Flight control unit.
FE (2) Flight engineer.
FEM (2) Finite element mesh.
Fibredux Family of CFRP and hybrid prepreg resins (Ciba-Geigy).
FIDS, Fids Fault identification and detection system.
flick roll Essentially a horizontal spin, made by slowing to spin-entry speed with engine throttled back and then applying hard back stick and full rudder. Result should be a controlled, very rapid 360° roll.
flight augmentation computer Main AFCS component providing yaw damping, pitch trim and flight-envelope monitoring and protection.
flight control unit Primary flight-deck interface with AFCS providing autopilot modes, SPD/MACH, HDG SEL, ALT SEL and similar functions.
flight envelope monitoring AFCS function providing computation of V_{MIN}, V_{MAX} and angle-of-attack limits.
FMGC Flight management and guidance control (system).
FNCP Flight navigation control panel.
FOD (2) Fibre-optic databus.
FOT (2) Fibre-optic twister for making inverted image upright.
four greens Gear down and full flaps.
FS (5) Front spar.
fudge factor Multiplying factor to allow for unknowns (colloq).
gamma (3) SI unit of very small changes in magnetic flux density; 1 * (not abbr) = 1 nT.
Gasil General aviation safety information leaflet (CAA).
GC (3) Goggles-compatible; C adds "cockpit". Goggles = NVG.
GeCu Germanium/copper (IR detector).
generator line contactor Main circuit-breaker between generator and AC bus.
GEO Geostationary Earth orbit.
GRBL Green raster brightness level.
group technology General term for philosophy that links CAD with CAM to give CIM, based on recognizing similarities between discrete parts.
GS (6) Galley service vehicle.
GTO Geostationary transfer orbit.
HACS Helicopter armoured crashworthy seat.
HALE High altitude, long endurance.
Halon Family of halogen-based fluids, stored as liquid under pressure and used as fire extinguishants and for inerting space above fuel in tanks.
HEDI, Hedi High endoatmospheric defense interceptor (SDI).
HLWE Helicopter laser warning equipment.
Honcho Big chief, hence fighter ace or high-ranking officer (US, colloq).
Horus Hypersonic orbital return upper stage.
HPS (3) Hardened personnel shelter.
HRTS High-resolution telescope and spectrograph.
hybrid display One in which alphanumerics and symbology are cursively written on top of a raster background.
IJMS Alternatively "S" stands for "structure".
ILA (2) Internationales Luftfahrt-Archiv (G).
IMS (7) Integrated mission system.
IOR Immediate operational requirement.
IPE (2) Improved-performance engine.
ITC (2) Investment tax credit.
JATE Joint Air Transport Establishment (UK).
Jeep Escort carrier or CVE (WW2).
JHSU Joint Helicopter Support Unit (UK).
KCK Composite sandwich, Kevlar/carbon/Kevlar.
KPS Knowledge processing system.
lane (3) Passenger guideway down escape slide.
LCFD Low-cycle fatigue damage.
LD (4) Load device, prefix of standard family of cargo containers and pallets, each of particular dimensions and with certificated permissible load.
le_v Tip chord of vertical tail.
Lima Laser ionization mass analyser.
li_v Root chord of vertical tail.

LLH Light liaison helicopter.
LMT (2) Locally manufactured tools, made to exact specification of an OEM.
LOC inertia smoothing Extra AFCS function, usually customer option, added in FCCs to alleviate effect of ILS localizer noise. Typically Setting 1 smoothes approach and provides survival after LOC failure below 100 ft/30 m; Setting 2 reduces minima on Cat II or III ILS.
LRC (3) Light reflective capacitor.
LS (3) Lavatory service vehicle.
l_v Moment arm of vertical tail (usually from 25% MAC).
LWTR Licence with type rating.
Magnamite Family of CFRP composites (Hercules Inc).
maldrop Faulty delivery of helo slung load caused by equipment malfunction.
MDOF Multiple degrees of freedom.
MEP (5) Member of European Parliament.
MFFC Mixed fighter force concept (Phantoms/Hawks, RAF).
MGIR Motor-glider/instrument rating.
MIRTS Modularized infra-red transmitting set.
MLS (2) Multi-level security.
MPED Multi-purpose electronic display.
MPP (2) Maintenance programme proposal.
mppa Million passengers per annum.
MPR (2) Military power reserve.
MSF (2) Multi-sensor fusion.
MTR (4) Helicopter configuration, main and tail rotors.
MTTS Multi-task training system.
Nadin National airspace data interchange network (FAA).
Naegis, NAEGIS NATO airborne early warning ground environment integration segment.
NCC Nickel-coated carbon.
NGC Nylon/graphite composite.
Nicral Nickel/chromium/aluminium (plasma spray).
nodalization Equipment of helicopter with anti-vibration couplings between rotor head and fuselage.
Noda-Matic Patented (Bell) system of damping out major blade-passing frequency vibrations by hanging fuselage from steel rods linked with flexible glassfibre straps carrying resonant masses.
NST (2) NATO staff target.
OAFS Open apron, free-standing (ie, no passenger loading bridges).
OMM (2) Oxygen-mask microphone.
OSC Overhead stowage compartment(s) for carry-on baggage.
OTAA Office of Trade Adjustment Assistance (US).
OTAEF Operational Test & Evaluation Force (but see OTE, OT&E).
overlay Upper (exoatmospheric) layer of layered defense system (SDI).
pack (3) In particular, ECS air-conditioning comprising bootstrap ACM (1) plus air/air heat-exchanger.
PBF Personnel (aircrew) briefing facility.
PDU (3) Pneumatic drive unit.
penny-farthing Helicopter of normal configuration (colloq).
PFR (2) Passenger flow rate.
piccolo tube Tube perforated by (usually linear) row of holes from which hot de-icing air is blown, usually to impinge on interior of wing, slat or cowl LE.
picture manoeuvre Manoeuvre made by large aerobatic team involving wide separation of aircraft to fill large part of display area, eg bomb-burst.
PID (2) Passive identification device.
pilot pushing Unlawfully urging aircrew to work excessive hours.
PLB (2) Passenger loading bridge.
Plod Passenger on-board delivery (literally, landed on deck).
PM (6) Pressurized module.
POC (2) Proof of concept.
popouts Fast-inflating buoyancy bags (helicopter).
power performance index Computerized index of helicopter engine power corrected for atmospheric pressure/temperature, altitude, power-turbine temperature and drive torque.
power transfer unit Interconnection between otherwise totally separated hydraulic systems, eg Green and Yellow, which enables power to be transferred from one to the other.
P/PATM Passengers per air-transport movement.
PPI (2) Power performance index.
press-ups Short vertical climbs and descents (jet-VTOL).
PS (7) Payload specialist (spaceflight).
PVC (2) Permanent virtual circuit.
PVD (3) Peripheral vision display.
PVU Portable ventilator unit.
PW (3) Potable water.
Queen Mary Articulated road vehicle for transporting dismantled or crashed aircraft (RAF).
R (28) Reject, rejected (nav display).
RAP (8) Rack and pinion.
RAPT Radar procedures trainer.
RCC (5) Repetitive chime clacker.
RCTC Rear-crew trainer cabin.
Remro Remote radar operator.
repetitive chime clacker This aural warning is generally used to indicate a fault in aircraft configuration.
RHL Rudder hinge line.
RIC Also translated as Recon Intelligence Centre.
riser (3) Main duct conveying ECS fresh air to top of cabin or flight deck.
RISLS Receiver interrogation sidelobe suppression.

RMP (4) Radio management panel.
roll off Uncommanded roll when at high AOA (eg climbing at low speed) due to contamination by ice or other matter.
RR (3) Rain-repellant liquid.
RS (4) Rear spar.
RTO (5) Runway turnoff lights.
run-back ice Very dangerous accretion of ice which formed on LE, was melted by de-icing system and refroze further back.
runway turnoff lights Fixed wide-beam white lights on each side of the forward fuselage giving lateral illumination.
SD (5) System display.
SDOF Single degree of freedom.
SDOM Six degrees of motion (simulator).
Selcal Selective calling, system enabling ground controller to call a single aircraft without the crew having to listen out on that frequency or any other aircraft being bothered. Selcal code for each aircraft is four-letter code using letters A through M. Airborne decoder usually accepts inputs from VHF or HF. See *Adsel*, *DABS*.
service door Door in outer skin covering a maintenance or control panel, eg for cargo loading.
SFEA Survival and Flight Equipment Association (UK).
SG (7) Symbol generator.
shelter marshal Officer controlling all who enter or leave HAS, PBF or HPS.
SHS Since hot-section inspection.
SJR Helicopter configuration, single jet (i.e. tip drive) rotor.
slave actuator Actuator, usually ballscrew, forming one of a number transmitting motion originating at a remote power source. Distributed along wing to move flaps, and around engine to move reverser blocker doors.
S/LD Spoiler(s)/lift dumper(s).
slide raft Escape slide which can be detached and used subsequently as fully equipped life raft.
slot (7) Further to existing entry as verb, also window in sky about 300 ft to right (stbd) of ship's bows and 600 ft above sea.
SLST Sea-level static thrust.
snapshot "Photograph" of all input parameters recorded at particular points in time by HUM to log steady-state conditions for future long-term analysis.
splitter gearbox 1 Splitter box.
2 Gearbox dividing input along two channels, eg from reverser PDU (3) to left/right half-rings of slave actuators.
SRHIT Small radar-homing interceptor technology (SDI).
SSUP Space station users panel (Int).
STAJ Short-term anti-jam.
Syrca Système de restitution de combat aérien (F).
SYSTRID, Systrid Powerful 3-D CAD technique (Battelle).
TAOR Tactical area of operations (UK).
tapelayer Computer-controlled tool for laying up prepreg tape in manufacture of composites; automatically positions, starts, stops and dumps material rejected during prior editing.
TASC Touch-activated simulator control.
TBC (4) Thermal barrier coating.
TCC (4) Turbine case cooling.
2DOF Two degrees of freedom.
TFH Thick-film hybrid.
TFSUSS Task force on scientific users of space station.
THL Tailplane hinge line.
threat cloud Total collection of warheads, chaff and other penetration aids in ICBM attack.
thrust control computer AFCS computer providing control of engine N_1 and thrust, computation of engine limit parameters, and autothrottle.
thrust rating panel AFCS display of limiting and target values of engine parameters and selectors for operating mode (climb, cruise, MCT or TO/GA) or FTO temperature.
TLLF Tactical low-level flight.
TLSS Tactical life-support system (USAF flight suit).
TMMS TOW mast-mounted sight.
TO/GA Take-off/go-around.
torque roll Performable only by aircraft with fast-responding (eg piston) engine on centreline giving very large torque in relation to aircraft weight; approach is made at flight idle at minimum safe flying speed, whereupon throttle is banged wide open to cause rapid roll in opposite direction as aircraft accelerates, pilot recovering to wings-level with aileron.
TPTA Tailplane trim actuator.
triform Structural member, usually an extruded section, providing attachment faces along three planes passing through same line.
trim air Hot bleed air added downstream of ECS pack(s) to achieve desired cabin air temperature.
TSR (3) Helicopter configuration, twin side-by-side rotors.
TTR (2) Helicopter configuration, twin tandem rotors.
unusable fuel Replacing definition on p.511, fuel that cannot be used in flight with wings level and at cruise AOA (or nose 3° up). **Trapped fuel** is that fuel remaining in worst case on ground using booster pumps for defuelling and switching off immediately associated LP warning lights illuminate. **Unpumpable fuel** is that fuel deliberately designed to be unpumpable (through levels of sumps and booster pump inlets) to avoid ingestion of water (certification requirement).
USMS Pronounced uz'ms, utility systems management system, software programmed into six microprocessors which control a complete fighter (BAe).

VDDL Verband Deutscher Drachenfluglehrer eV (G).
War Executive Officer in RAF squadron responsible for detailed planning of all combat missions.
Warlord War Executive (colloq).
WD (2) Warning display.
windshield guidance display Optical projector in glareshield giving HUD-type ground roll guidance after blind touchdown.
Wisp Waves in space plasma.
WSCP Warning system control panel.
WSSG Warning-system signal generator.
X-licences Range of licences for ground engineers, with endorsements for many disciplines or equipments (UK).
XRT X-ray telescope.
YC Tourist class.